The Handbook of Rationality

The Handbook of Rationality

Edited by Markus Knauff and Wolfgang Spohn

The MIT Press
Cambridge, Massachusetts
London, England

The MIT Press would like to thank the anonymous peer reviewers who provided comments on drafts of this book. The generous work of academic experts is essential for establishing the authority and quality of our publications. We acknowledge with gratitude the contributions of these otherwise uncredited readers.

This book was set in Stone Serif and Stone Sans by Westchester Publishing Services. Printed and bound in the United States of America.

Library of Congress Cataloging-in-Publication Data

Names: Knauff, Markus, editor. | Spohn, Wolfgang, editor.
Title: The handbook of rationality / edited by Markus Knauff and Wolfgang Spohn.
Description: Cambridge : The MIT Press, 2021. | Includes bibliographical references and index.
Identifiers: LCCN 2020048455 | ISBN 9780262045070 (hardcover)
Subjects: LCSH: Reasoning (Psychology) | Reason. | Cognitive psychology. | Logic. | Philosophy of mind.
Classification: LCC BF442 .H36 2021 | DDC 153.4/3—dc23
LC record available at https://lccn.loc.gov/2020048455

10 9 8 7 6 5 4 3 2 1

Contents

Preface

This book is for students and scholars who want to think about the big questions: Are we rational? What at all does it mean to be rational? Why do we sometimes deviate from the norms of rationality? What happens in our brain when we reason and decide? What is a good or bad decision? How do we come to sensible beliefs and knowledge? These are questions that people have wondered about since the beginning of humankind. And these are questions about which, since ancient times, thinkers have had much to say. The aim of *The Handbook of Rationality* is to show how (analytic) philosophers and (cognitive) psychologists think about these big questions. In the past decades, both disciplines have made tremendous advances in better understanding the very nature of human rationality. Yet, the interaction between the fields is developed only in rudimentary forms. The consequence is that no publications exist that provide students and scholars with an easily accessible integrated overview about the state of the art in the psychology and philosophy of rationality. *The Handbook of Rationality* seeks to fill this gap and to be a novel resource for students and scholars in psychology and philosophy, but also in neighboring areas, such as economics, neuroscience, artificial intelligence, linguistics, law, sociology, anthropology, or education.

This handbook has not become so voluminous because we wanted it to be. A little less work with the publication of this volume would have been all right. But a less complete treatment would not have done justice to the importance of the subject. Of course we were fully aware of the complexity and breadth of the topic of rationality when we started to think about the chapters and authors that we wanted to collect in this handbook. However, during our work, we repeatedly encountered new topics and issues that we had not considered at the beginning because they only arose from our interdisciplinary cooperation. If we had merely concentrated on our respective home disciplines, we would have missed many of these interesting questions lying in the no man's land between the disciplines.

What has greatly facilitated our interdisciplinary collaboration and made it so productive is the fact that we already collaborated for many years within the Priority Program SPP 1516, *New Frameworks of Rationality*, which has been funded by the German Research Foundation (Deutsche Forschungsgemeinschaft [DFG]) from 2011 to 2018. The program consisted of 15 research projects from psychology, philosophy, and artificial intelligence and often combined methods from these disciplines within highly interdisciplinary projects. The program was headed by Markus Knauff and coinitiated by Wolfgang Spohn, the cognitive psychologists Ralph Hertwig and Michael Waldmann, and the philosopher Gerhard Schurz. Later, the computer scientist Gabriele Kern-Isberner joined this governing board. In this priority program, we gathered many times, worked together very intensively, worked on many joint publications, and spent many, many hours of intensive discussion (www.spp1516.de). All this resulted in enormous shared benefit and progress in mutual understanding.

At some point, we thought we should share our progress with a broader scientific public, and so the idea for this handbook was born in the summer of 2015. We are very grateful to the DFG for the generous funding of the interdisciplinary research program. We are aware that Germany is one of the few countries that still invest so much in basic research without expecting results that are directly economically exploitable. This is a privilege we greatly appreciate. Many scientists from this program have also contributed to this handbook, although we have taken great care to include many other experts from the worldwide community of rationality research. The handbook is a truly international enterprise. We are very pleased that most of the world's leading experts were very enthusiastic and immediately agreed to contribute a chapter to this handbook. The different disciplines use quite different formats for footnotes, indexing, references, and other stylistic matters. To establish uniformity throughout the handbook, we have chosen to use the rules from the *Publication Manual of the American Psychological Association* (6th edition, 2010).

Our list of acknowledgments is long. First, we thank all our authors. In order to cover our field comprehensively, we had to engage the best authors across the disciplines. We expected that many authors would deliver too late, that we would have to wait a very long time for many contributions, that we would have to send out dozens of reminders, and so on. But we were too pessimistic. Most of our authors were very reliable, delivered their contributions fairly punctually, and were also very cooperative and responsive. We are very sad that one of our initial authors, Sieghard Beller, passed away before he could deliver his chapter and that another author, Arthur Merin, died shortly after finishing his chapter. We are most grateful to all of the authors for the exemplary cooperation.

We are also indebted to all our reviewers. In general, each chapter was reviewed by two reviewers, one from psychology and one from philosophy. In most cases, this was possible, and sometimes we even received three or four reviews from colleagues with different disciplinary backgrounds. For each chapter, at least one review came from an author of another chapter and the other from an external reviewer. We are very thankful to the following external reviewers:

Robert Audi	Simon Handley
Anthony Blair	Sven Ove Hansson
Thomas Blanchard	Tim Henning
Luc Bovens	Risto Hilpinen
Matteo Colombo	Franz Huber
Vincenzo Crupi	Carsten Koenneker
David Danks	Jakob Koscholke
Lorraine Daston	Annette Lessmöllmann
Eddie Dekel	Tania Lombrozo
Emmanuelle-Anna	Hugo Mercier
Dietz Saldanha	Philippe Mongin
Igor Douven	Martin Monti
Frank Fischer	Josef Perner
Christian Freksa	Martin Peterson
Peter Gärdenfors	Eric Raidl
Valeria Giardino	Christian Ruff
Itzhak Gilboa	Giovanni Sartor
Anthony Gillies	Walter Schaeken
Herbert Gintis	Mark Schroeder
Geoffrey Goodwin	Armin Schulz
Werner Güth	Keith Stenning
Jaap Hage	Bob Sternberg
Jakub Szymanik	Peter Wakker
Raimo Tuomela	Paul Weirich
Gottfried Vosgerau	Gregory Wheeler
Thomas Voss	Kevin Zollman

We are very sad that Christian Freksa died in an accident before the book came out.

We mentioned that we could build on a long previous cooperation within the Priority Program *New Frameworks of Rationality*. Here we want to thank all members of the program for the lively and productive cooperation. It was most instructive.

We are deeply grateful to Christopher von Bülow, who checked all the manuscripts and brought them into the final uniform shape. He did so with unsurpassable care, at times at which we editors were already exhausted. It is also his merit that this project came to a good end.

We also thank Lupita Estefania Gazzo Castañeda, Andreas Kemmerling, and Christoph Klauer for their comments on an earlier version of the introductory chapter. Knauff wrote parts of the long introductory chapter at the University of California, Santa Barbara (UCSB), and thus wants to thank Daniel Montello for his hospitality and the perfect environment for focused thinking and writing. Knauff also wants to thank Estefania and his whole team for the many interesting and lively discussions. Spohn is doubly grateful to the DFG for support: in 2016–2018 through the research unit FOR 1614 at the University of Konstanz and in 2019–2020 through the excellence cluster EXC 2064/1 (project number 390727645) at the University of Tübingen.

We are also most grateful to Philip Laughlin and the team from MIT Press. The interaction with all of them was always pleasant and professional. This handbook was a huge enterprise and thus, in the final phase, an enormous work for the publisher, which was performed in an excellent way.

Normally, the editors form a unit that cannot thank itself. Still, we feel the urge to do this. We can only thank each other for the untiring willingness to work together over such a long period of time on such a large project. It was a great scientific experience that neither of us wants to miss, even though we spent so many days and nights discussing and Skyping, as well as exchanging (about 3,000) emails. Our spouses, Erika and Ulli, have faithfully accompanied our work with their understanding, charity, and support. We dedicate our work to them!

Markus Knauff and Wolfgang Spohn

Gießen and Konstanz, January 2021

Overview of the Handbook

Markus Knauff and Wolfgang Spohn

This handbook seeks to cover its topic, rationality, in an interdisciplinary way, more than any other existing handbook on this topic. Surely it cannot fulfill this intention completely. The topic is just too large. But it intends to reach broad representativity at least in its two main disciplines, philosophy and psychology. In addition, it presents some insights from disciplines such as artificial intelligence, behavioral and microeconomics, and neuroscience, although these areas are only partially covered here. The current program is already very ambitious.

The handbook is not organized along disciplinary lines of psychology and philosophy. This would be in clear contradiction to our interdisciplinary endeavor. Rather, it is organized along the common themes that occur in both fields, albeit often in different forms and under different names. In our introductory chapter, we propose various distinctions for classifying the overwhelming research on rationality. Certainly, the most important one is that between theoretical and practical rationality. This distinction is mirrored in the headings of part II and part III, the two main parts of this handbook. These parts are preceded by part I, which addresses some fundamental or propaedeutic matters concerning the history, some paradigms, and some key issues of research on rationality. The handbook concludes with part IV, which presents further facets of rationality that are relevant for both theoretical and practical rationality.

Each of the four parts consists of several sections, which in turn each contain several chapters, so that we ended up with 15 sections and 65 chapters. The order of chapters within a section does not have a particular significance. We just had to find one that creates connections between the different disciplinary backgrounds of the authors. The assignment of chapters to sections was more difficult. The field is so variegated and interlinked that no classification can be perfect. Still, we hope that we have organized the handbook in a way that makes it easy to read.

In our **introductory chapter**, we systematize the field by means of four binary distinctions: theoretical versus practical rationality, normative versus descriptive theories of rationality, individual versus social rationality, and outcome- versus process-oriented accounts of rationality. Since the distinctions can be combined with each other, we end up with a system of 16 compartments. Our chapter says something about most of the 16 compartments and uses this system not only to give guidance to the readers of this handbook but also to suggest a new classification system for researchers in human rationality. We recommend all readers to study this introductory chapter. This will certainly make it easier to orient oneself in, and to contextualize, the rich material collected in this handbook.

Part I: Origins and Key Issues of Rationality

The chapters in **section 1** deal with the history of rationality research and the cortical and evolutionary foundations of rational thinking. In chapter 1.1, *Thomas Sturm* describes the history of philosophical conceptions of rationality, which were just as much psychological conceptions. Since this history is so enormously rich, he just focuses on the emergence of the distinction between a descriptive and a normative perspective, which is so important for the entire handbook. This philosophical chapter is then complemented by chapter 1.2, in which *Jonathan Evans* summarizes the younger history of the psychology of reasoning from his personal point of view. Then, in chapter 1.3, *Gerhard Schurz* looks even farther back in time by speculating about the evolution of rationality. In chapter 1.4, *Vinod Goel* deals with the material basis of rationality, the brain. He presents findings from cognitive neuroscience and brain imaging on which cortical networks are involved in different kinds of rational thinking.

The chapters in **section 2** are concerned with some substantial philosophical and psychological topics of rationality. The section begins with chapter 2.1, in which

John Broome deals with the relation between rationality and reasoning. Both keywords are extensively used in this handbook, but they do not seem to denote exactly the same matter. The terms are also used quite differently in philosophy and psychology. In chapter 2.2, *Ralph Wedgwood* discusses the distinction between theoretical and practical rationality. This is also the top-level distinction in this handbook. The next chapters present various cognitive approaches to rationality. In chapter 2.3, *Philip Johnson-Laird* gives an overview of the theory of mental models, which is a far-reaching framework for explaining accurate and fallacious human reasoning. Another cognitive framework is the heuristics and biases approach, which is presented and critically discussed in chapter 2.4 by *Klaus Fiedler, Johannes Prager*, and *Linda McCaughey*. An equally influential framework is established by dual-process theories of reasoning. This framework and the related empirical evidence are presented by *Karl Christoph Klauer* in chapter 2.5. We could have placed the Bayesian reasoning theories here, too, as these theories also have a quite universal ambition. However, the relevant chapter by Chater and Oaksford fits even better in the section that is particularly dedicated to probabilistic reasoning. The last chapter in section 2 is a big jump toward artificial intelligence. In their chapter 2.6, *Johan van Benthem, Fenrong Liu*, and *Sonja Smets* outline the logico-computational perspective on rationality. They show how this approach can be efficiently used to solve rationality problems in computers. This chapter also creates a bridge to section 3 in part II, which is about logical and deductive reasoning.

Part II: Theoretical Rationality

Sections 3–7 are concerned with the key issues of theoretical (or "epistemic" or "doxastic") rationality and reasoning. The first term originates from philosophy and is, in fact, not as common in psychology, in which cognitive theories of reasoning are a very active research field. However, the philosophical term is much broader and thus serves as the heading for the following sections. Not surprisingly, this part of the handbook starts with the two dominating reasoning paradigms: deductive logic as already conceived in ancient philosophy and probabilistic thinking as developed since the middle of the 17th century. These two paradigms are represented in sections 3 and 4, respectively.

The chapters in **section 3** are concerned with deductive logic and reasoning. Of course, the principles of propositional and first-order logic have long been part of the basic knowledge of our disciplines. Yet we wanted

to explain at least in one place of this handbook what a logical proof is and how it is related to rational belief. This is done in chapter 3.1 by *Florian Steinberger*, who added some novel ideas to this classical topic. Propositional and first-order logic are also important in cognitive research on human reasoning. In chapter 3.2, *David O'Brien* presents the natural-logic account and some evidence in support of this account. In chapter 3.3, *Sangeet Khemlani* explains the cognitive foundations of syllogistic reasoning, which is a fragment of first-order logic. His approach is based on mental models as described in chapter 2.3.

Then, **section 4** is concerned with approaches from probability theory. In chapter 4.1, *Alan Hájek* and *Julia Staffel* explain the normative foundations of Bayesianism. Chapter 4.2 by *Stephan Hartmann* complements this by explaining the usefulness of the theory of so-called Bayes nets. A further complement is given in chapter 4.3 by *Arthur Merin*, which unfolds the probabilistic core of all considerations of relevance. This is important for accounts of reasoning, since premises or arguments are usually assumed to be relevant for their conclusions. In chapter 4.4, *Niki Pfeifer* presents his probability logic. In chapter 4.5, *Nick Chater* and *Mike Oaksford* defend their psychological perspective on rationality as conceived within Bayesianism. This, in turn, is complemented by chapter 4.6 by *Klaus Oberauer* and *Danielle Pessach*, who explain what, for (some) psychologists, conditionals have to do with probability. Chapter 4.7 by *Didier Dubois* and *Henri Prade* also proceeds from the assumption that beliefs, or epistemic states in general, come in degrees. This is why the chapter is placed in this section. Yet, in their account, rational degrees of belief do not behave like probabilities but in certain other ways. We think it is important to acknowledge that such alternatives exist.

Logic and probability theory are, of course, not the only paradigms of theoretical rationality and reasoning. In fact, many other frameworks exist, and many of them are not restricted to quantitative conceptions of degrees of belief, as in Bayesianism. These alternative accounts are represented in the remaining sections of part II.

The chapters in **section 5** are concerned with qualitative representations of belief and the related accounts of reasoning. In chapter 5.1, *Hans van Ditmarsch* presents the basic theory of doxastic and epistemic logic. Then, belief revision theory is concerned with the rational change of epistemic states. This normative account of belief revision is described in chapter 5.2 by *Hans Rott*. The dynamic account of belief revision theory is completed by ranking theory. This theory is laid out in chapter 5.3 by *Gabriele Kern-Isberner, Niels Skovgaard-Olsen,* and *Wolfgang Spohn*. The dynamics of belief is also important in cognitive theories of defeasible reasoning. In chapter 5.4, *Lupita Estefania Gazzo Castañeda* and *Markus Knauff* describe the empirical results on human belief revision and defeasible reasoning. They also explain how qualitative and quantitative theories try to account for the empirical findings. Defeasible reasoning, finally, is closely related to argumentation, where (rational) arguments drive (rational) epistemic change. The psychological perspective on argumentation is presented in chapter 5.5 by *Ulrike Hahn* and *Peter Collins*. Chapter 5.6 by *John Woods* deals with the same matter from a philosophical perspective.

There is also a close relation between epistemic change, on the one hand, and conditional and counterfactual reasoning, on the other, although the relation is not easy to specify. The broad range of approaches to reasoning with conditionals and counterfactuals is the topic of **section 6**. It begins with chapter 6.1 by *William Starr*, who reviews the different attempts to capture the logic of conditional and counterfactual constructions. In psychology, so-called supposition theory has become prominent. This attempt to explain human conditional reasoning is represented in chapter 6.2 by *David Over* and *Nicole Cruz*. In chapter 6.3, *Ruth Byrne* and *Orlando Espino* explain how human reasoners deal with counterfactual inferences. A special, but psychologically particularly relevant, case of conditional reasoning, so-called utility conditionals, is treated by *Jean-François Bonnefon* in chapter 6.4.

From counterfactual reasoning, it is just a small step to causal reasoning and its subform, diagnostic reasoning. The importance of causal reasoning among our many reasoning activities cannot be overemphasized. This is a huge topic on its own. Thus, we devote the complete **section 7** to it. In chapter 7.1, *Judea Pearl* presents his influential account of causal (and counterfactual) inference in probabilistic or statistical terms. How this kind of reasoning actually works in humans is discussed by *Michael Waldmann* in chapter 7.2. The special case of diagnostic reasoning is dealt with by *Björn Meder* and *Ralf Mayrhofer* in chapter 7.3. This is what this handbook offers concerning theoretical or epistemic rationality and reasoning.

Part III: Practical Rationality

Sections 8–12 deal with the core topics of practical rationality and decision making. Again, the first term is familiar in philosophy, and the second is mostly used in psychology. The terms are not equivalent, but they are close enough to be treated under one heading. We think that in this way, we can best highlight the interdisciplinary links and show how both research areas are connected.

This part of the handbook should of course start with accounts of individual decision making, which is the topic of **section 8**. In chapter 8.1, *Till Grüne-Yanoff* explains the main ideas of preference and utility theory. Next, chapter 8.2 by *Martin Peterson* presents standard decision theory and the normative arguments in its favor. However, this account has been criticized from the psychological as well as from the economic side. Indeed, prospect theory arose from the observation that people often deviate from expected utility maximization. This theory is explained by *Andreas Glöckner* in chapter 8.3. However, there are still further ways how beliefs and desires can be combined to determine a rational decision. Some of them are presented in chapter 8.4 by *Brian Hill*. Another critical view on economic decision theory comes from the theory of bounded rationality, which is influential in psychology and economics. This approach is reviewed in chapter 8.5 by *Ralph Hertwig* and *Anastasia Kozyreva*. The section ends with a chapter 8.6 by *Valerie Thompson, Shira Elqayam*, and *Rakefet Ackerman*, which is concerned with the connection between rationality and metacognition. The authors report empirical results that show that metacognitive control and monitoring processes are important for promoting our rational performance. Thinking about our own thinking makes us more rational. This chapter is quite general, and it might have fit under other headlines, but it also connects well to the previous chapters on practical rationality.

Then, **section 9** deals with game theory, the other standard theory of practical rationality. We treat here only so-called noncooperative game theory, which is, strictly speaking, still about individual practical rationality but in a social context. Chapter 9.1 by *Max Albert* and *Hartmut Kliemt* presents the classical theory, as it has been developed since the 1940s. In the 1980s, epistemic game theory emerged, which promises a more rigorous rationalization of the norms of game theory. This theory is described in chapter 9.2 by *Andrés Perea*. A different perspective is offered by evolutionary game theory, which is outlined by *J. McKenzie Alexander* in chapter 9.3. Again, these theories are criticized from the empirical point of view. This led to the rise of behavioral and psychoeconomics, which are sometimes (mistakenly) seen as fields of psychology but are still driven by the interests of economists. The many different ideas in this area are presented in chapter 9.4 by *Sanjit Dhami* and *Ali al-Nowaihi*.

In our introductory chapter, we explain why most research on rationality focuses on individual rationality.

In fact, **section 10** is the only section that really deals with social rationality in the proper sense. Therefore, it is not strictly limited to practical matters. In chapter 10.1, *Franz Dietrich* and *Kai Spiekermann* deal with social epistemology, which discusses the normative standards for group belief formation. Recently, related topics such as we-intentionality and collective rationality have developed into a larger philosophical field. These topics are presented in chapter 10.2 by *Hans Bernhard Schmid*. Communication, particularly linguistic communication, is a very special social and rational activity, which is the topic of chapter 10.3 by *Georg Meggle*. Rational choice theory has also become a strong paradigm in the social sciences. This line of research is explained by *Werner Raub* in chapter 10.4. A look into the history of philosophy, into political philosophy, and so on reveals that the standard theories of practical rationality go far beyond the scope of instrumental decision making. At least chapter 10.5 by *Julian Nida-Rümelin, Rebecca Gutwald,* and *Niina Zuber* discusses such extensions and under the label of "structural rationality." Finally, in chapter 10.6, *Leda Cosmides* and *John Tooby* present their adaptationist account of rationality. Although their evolutionary account seeks to explain all forms of human rationality, their chapter mainly focuses on practical rationality.

Practical reasoning need not proceed in quantitative terms of probabilities, utilities, or the like. Deontic and legal reasoning, for instance, can rely on qualitative representations and processes. Such qualitative accounts are the topic of **section 11**. In chapter 11.1, *John Horty* and *Olivier Roy* present the current state of the art in deontic logic (i.e., the logic of obligations and permissions). Then, in chapter 11.2, *Shira Elqayam* deals with the psychological theories of deontic reasoning. Legal reasoning is sufficiently different from deontic logic to deserve a separate treatment. This is given in chapter 11.3 by *Eric Hilgendorf* and in chapter 11.4 by *Henry Prakken*. A surplus of these chapters is that they represent different legal cultures, which heavily determine the field: chapter 11.3 is embedded in German law, while chapter 11.4 is based on the Anglo-Saxon legal system.

Whenever we are engaged in practical reasoning and decision making, moral issues come into play. In fact, they often are more important than issues of rationality. However, morality is a different topic, which we do not tackle in this handbook. We only wanted to address a few connections to rationality. These connections are the topic of **section 12**, which consists of two philosophical chapters and a psychological one. Chapter 12.1 by *Christoph Fehige* and *Ulla Wessels* describes the relation between rationality and morality in philosophy, while chapter 12.2 by *Michael Smith* deals with moral reasons, also from a philosophical point of view. But moral judgments are also a big topic in psychology. In chapter 12.3, *Alexander Wiegmann* and *Hanno Sauer* describe the psychological view on moral judgments and their relation to rationality. This concludes our chapters on practical rationality.

Part IV: Facets of Rationality

Sections 13–15 deal with further facets of rationality. Their topics are so distinct that they could not be subsumed under the headings of theoretical and practical rationality. In fact, most of the questions arise in *both* areas of human reasoning. For example, **section 13** is concerned with visual and spatial thinking. Such cognitive processes are an important complement to the many chapters in which reasoning is conceptualized, more or less explicitly, in linguistic or propositional terms. An important question in this context is the connection between logical and diagrammatic reasoning. In chapter 13.1, *Mateja Jamnik* takes the perspective of a computer scientist and argues that diagrams can help to reason logically. Another question is how humans and artificial intelligence systems reason about space and time. In chapter 13.2, *Marco Ragni* describes some differences and explains why people sometimes commit errors in spatial and temporal reasoning. In chapter 13.3, *Markus Knauff* explores whether visualization supports or hampers human reasoning. The empirical results show that it can indeed sometimes impede reasoning.

Another distinct topic is scientific rationality, which has theoretical as well as practical aspects, and which is, of course, of essential concern to us as scientists. The three chapters in **section 14** deal with different aspects of scientific rationality. One idea is that the sciences exercise a particularly sophisticated form of epistemic rationality, which may have stronger claims on objectivity. This idea is pursued in chapter 14.1 by *Line Edslev Andersen* and *Hanne Andersen*. Another topic is the value-freedom of science. This is the topic of chapter 14.2 by *Anke Bueter*. Finally, it is also important how scientific results are communicated to and perceived by the public. In chapter 14.3, *Rainer Bromme* and *Lukas Gierth* discuss the public understanding of science and its relation to scientific rationality.

The final **section 15** consists of still more chapters of psychological and even political interest. Chapter 15.1 by *Henry Markovits* explains how children learn to reason rationally. Chapter 15.2 by *Keith Stanovich, Maggie Toplak,* and *Richard West* presents the psychological

findings concerning the connection between intelligence and rationality—it is not quite as close as it may
seem. Chapter 15.3 by *Stephanie de Oliveira* and *Richard Nisbett*, finally, reports how training can improve
people's rational thinking and that such training does
not even need to be intensive or time-consuming.

This is also an optimistic message at the end of this
handbook, which offers a big journey through the state of
the art in rationality research. We can only recommend
again to start this journey with our introductory chapter.
It gives a systematic structure to the field and should
help our readers to make cross-connections between the
philosophical and psychological methods, theories, and
ideas represented in this handbook. Research in these
(and some other) areas has been conducted separately
for too long, an anachronism that this handbook aims
to overcome.

Psychological and Philosophical Frameworks of Rationality—A Systematic Introduction

Markus Knauff and Wolfgang Spohn

Summary

Rationality is a vast issue. It has occupied the most brilliant thinkers since the beginning of human culture. It mirrors our capacity for self-reflection, our desire to recognize the potentials and limitations of our mental abilities, and our determination to understand how our mind works. Such questions can be frightening, as they seem so complex that we may never find conclusive answers. But we can try. In this introductory chapter to *The Handbook of Rationality*, we unfold the concept of rationality and the relations between theoretical and practical rationality, normative and descriptive theories of rationality, individual and social rationality, and an output-oriented and a process-oriented perspective on rationality. We also describe some cognitive preconditions and domains of rationality and the different intellectual traditions in psychology and philosophy—where they intersect, fall apart, and converge. We hope that this overview helps readers to orient themselves in the widely ramified research on human rationality.

1. Why Study Rationality?

Humans are rational animals. Since Aristotle, this has been considered the essence of *Homo sapiens*. Of course, much about this view can be criticized: one objection might be that irrationality seems to prevail. The human species is about to make our planet uninhabitable. The globe is full of political, economic, cultural, and religious conflicts. Often, people seem to be driven not by rationality but rather by power struggles, anxiety, greed, prejudice, and so on. All of this can scarcely be called rational. However, it is hard to argue about the relative amounts of rationality and irrationality. Rationality attributions are delicate, as this handbook will richly display. We well know that a collective of rational individuals may end up in an irrational collective disaster (Janis, 1972; Moorhead, Ference, & Neck, 1991; Reason, 1987). And the extent of the irrationality often depends on the perspective of the assessor. In fact, to call humans rational is to say that they are *gifted* with rationality, but not that they *exercise* this gift all the time. In Chomsky's terms, rationality is a *competence*, and the *performance* may well be defective.

A second objection to the definition of humans as rational animals might be that rationality is not unique to humans. Today we know that many other animals have mental capacities that were traditionally assumed to distinguish *Homo sapiens*. Of course, it is important to study continuities and differences between potentially rational creatures, and there is no doubt that human rationality has evolved from more fundamental neural mechanisms of adaptation and cognition. However, our mental powers of representation, of differentiation, of inner and outer experience, are much larger than those of any other animal. We find full-blown rationality only in us, although it also exists in basic forms in other species.

A third objection to the definition of humans as rational animals might be that various other characteristics distinguish our species from other animals as well: language, religion, pedagogics, arts, aesthetics, humor—the list is endless. What is so special about rationality? Take language, for instance: one might say that language is not only for communication but also for so many other social and individual purposes. Hence, compared to rationality, it is the far more comprehensive phenomenon. However, the many uses of language depend on the serious and sincere exchange of information and expression of beliefs and desires (see D. Lewis, 1969, and chapter 10.3 by Meggle, this handbook). Without the formation of such beliefs and desires—an essentially rational process—language would not be what it is.

Still, some of these objections may be justified. However, if we really were to doubt that rationality is at the heart of our existence as human beings, we would have difficulties explaining how people think, reason, judge, make decisions, solve problems, and act. Only on this rationality assumption can we explain how we interact, communicate, and form societies. And only on this assumption can we explain how humans developed science, culture, and modern technology. All these abilities and achievements rely on our mental powers of representation and processing, which are implemented in the neural structure of our brains. Thus, it is legitimate to question how rational humans are, but shifting priority away from rationality would be a mistake.

Today we are facing many vital challenges, from fighting starvation and curtailing war to coping with climate change, pandemics, worldwide diseases, and the drawbacks of new technologies and globalization. Indeed, for the first time in history, it seems that these challenges may concern humankind globally and substantially. One may well argue about what will help us most in meeting these challenges: empathy, justice, modesty, peacefulness, honesty, solidarity, the right values, religion? Surely rationality alone won't do. Yet there is no way to bypass rationality. Theories of justice refer to theories of rationality; we must act rationally to reach our moral, ethical, and humanist goals; and even religion cannot ignore rationality. So, whichever human features we promote, whatever measures we propose in order to meet our challenges, they should be well reasoned and withstand our critical reflection.

These first thoughts already reveal the two fundamental aims of this handbook. One aim is to bring together mainly psychological and mainly philosophical accounts of rationality and to display what both disciplines can learn from each other. The other aim is to substantiate the relation between the *normative* and the *descriptive* perspective on human rationality,[1] between what human thinking *ought* to be like and what it actually *is* like.

One might say that philosophy has been attempting to grasp rationality for 2,500 years, while psychology has been doing so only since the end of the 19th century. But this does not adequately represent the intellectual history. Large parts of this history, beginning already with the Presocratics (see Lorenz, 2009), were concerned with psychological issues, even if called philosophy, and were indeed labeled "philosophical psychology." Yet, for a long time, the methodology of this field was not clear, and hence the normative and the descriptive perspective were often not clearly defined and separated (see chapter 1.1 by Sturm, this handbook). This changed at the end of the 19th century, when psychology became an independent academic discipline, and one of its first topics was indeed to empirically investigate the specific conditions under which cognition and action conform to or deviate from what were at that time regarded as the norms of rationality (Störring, 1908; Wundt, 1896/2010).

Interestingly, while this happened in scientific psychology, logic developed in the opposite direction and became determinedly antipsychologistic under the influence of Frege (1884). We will return to this below.

Nowadays, the roles of psychology and philosophy in the academic landscape are clarified. Psychology is now the empirical science of the mind, with its own methodology, while philosophy sees more clearly that the normative issues belong to its proper domain. Perhaps because each discipline first had to find its role, the discourse between the two disciplines was initially poor or appeared quite unproductive. However, this has recently changed to a high degree, as will be amply displayed in this handbook.[2]

Clearly, both perspectives, the descriptive and the normative one, are each important and valuable in themselves. Their relation, however, is utterly contested. Some scholars argue that psychology should ignore the normative work in philosophy and simply go ahead describing how people think and reason (Elqayam & Evans, 2011). Other scholars refer to the fundamental distinction between competence and performance and argue that it is in principle impossible to empirically confirm that humans are irrational (Cohen, 1981). If either of these camps is right, there is no tension between the normative and the descriptive perspective on rationality—they are either independent or basically coincide. However, both views are minority positions in the community of rationality researchers. The premise of this handbook is instead that a comprehensive account of human rationality requires both empirically evaluated descriptive theories and elaborated normative theories as a positive or a negative point of comparison.

The aim of this handbook is to give a comprehensive overview of the state of the art in philosophy and psychology, and of the areas where the two disciplines are connected, where they share interests and concepts, and where they have diverging interests and research agendas. The handbook has many contributions from both disciplines and also includes some approaches from economics, sociology, artificial intelligence, and cognitive neuroscience, even if it cannot do so exhaustively.

This introductory chapter provides some structure to our huge topic. It does not follow the table of contents, which we already explained in the overview of the handbook. Rather, the arrangement of the chapter reflects some basic distinctions, conceptual frameworks, and research lines in the philosophical and psychological exploration of human rationality. It is organized as follows:

In section 2 of this chapter, we start with a preliminary contour of the subject matter we are talking about when we use the term "rationality." Then, in section 3, we describe the essential distinction between *theoretical* or *epistemic* and *practical* rationality. Psychologists are used to making a related distinction between *reasoning* and *decision making*. Next, in section 4, we will discuss the important tension between a *normative* and a *descriptive* perspective on rationality. As we just stated, this is one of the most crucial issues in our psychological-philosophical enterprise. In section 5, we introduce the distinction between *individual* and *collective* or *social* rationality, a topic that overlaps with topics from economics and other social sciences. Section 6 is concerned with the distinction between what we call *output-oriented* and *process-oriented* theories of rationality. While the former is the dominant approach in philosophy, the latter lies at the heart of cognitive psychology, which seeks to understand the processes that lead from the input given to the cognitive system to the output generated by the system. The four distinctions in sections 3 to 6 can be combined with each other and thus lead to a *systematic framework* that, for the first time, combines philosophical and psychological theories and empirical results on human rationality. We think that this framework not only guides the readers through this handbook but also provides a new perspective on research on human rationality. Then, in section 7, we will be concerned with some of the *preconditions* of human rationality, that is, the neural underpinnings of thinking and reasoning, as well as the relation between rationality and intelligence, memory, and other cognitive functions. In section 8, we discuss whether special forms of rationality exist, for example, in science, communication, or artificial intelligence (AI). We also discuss some connections of rationality to concepts such as emotion, morality, and culture. In the (final) section 9 of this chapter, we conclude with some open questions and problems that interdisciplinary rationality research still has to tackle.

2. Contours of the Concept of Rationality

Rationality constitutes a rich conceptual field. Broadly construed, it is about our higher cognitive faculties, about acquiring concepts, forming beliefs, gaining knowledge, inferring and reasoning, thinking, judging, decision making, planning, deliberating, calculating, and satisfying wishes, needs, and desires. In recent terms, we may also say that it is about our intentional and propositional attitudes. Being endowed with self-reflection, humans have surely wondered about these issues all along and so started thinking about how the human mind might work.

The field is also most confusing, tentatively tamed by many incongruous terms. If we look only at the five languages in which essential parts of early philosophical psychology were conducted—Greek, Latin, French, English, and German—we find many relevant terms, often accompanied by full doctrines. These terms are neither easily translatable, nor do they have a stable meaning. Think, for instance, of the distinction between *understanding* and *reason*, which acquired ever more significance in 17th- and 18th-century philosophy, culminating in Immanuel Kant, for whom that distinction between "Verstand" and "Vernunft" took a very specific form closely intertwined with his entire philosophical edifice. In the hermeneutic tradition, then, understanding was rather something opposed to explanation (von Wright, 1971). Nowadays, these distinctions are treated as spurious by most authors, with the exception, perhaps, of orthodox Kantians.

Or consider the English term "reason," which has three different meanings, for which German, for example, uses three different words. First, "reason" without a determiner (*Vernunft*) is just a general colloquial term for our higher cognitive faculties (which derives from medieval German *vernehmen* which meant *erfassen*, i.e., "to grasp"). Second, "a reason" (*Grund*) is what we have or give or accept in order to explain or justify whatever may be explained or justified. And finally, "to reason" (*räsonieren, schließen*) is something we do when we argue, make inferences, or arrive at conclusions. The connection presumably is that when we reason or give reasons, we *use* our reason. However, when we give reasons, we are not necessarily reasoning, and reasoning need not proceed from reasons (but only from premises the status of which may be left open). That's the English muddle, which will occupy us later. Other languages have their own difficulties.

The etymological origin of these terms is the Greek *raetos*, which, among other things, means "rational," in the sense of rational numbers, and thus displays another aspect of our conceptual field. The term ratio owed its tremendous career also to the fact that Cicero (106–43 BC) established it as the standard translation of the Greek *logos*. The term *rationalitas*, however, is first found in Tertullian (ca. 150–220 AD), for whom rationality was one of the essential attributes of the soul. Since possessing rationality simply amounts to being endowed with reason, the term *ratio*/"reason" was the widely used one, while *rationalitas*/"rationality" played a minor role in history.[3]

The term "rationality" became fashionable only at the end of the 19th century under the influence of economics and social science. In philosophy, this was perhaps due to the fact that the normative dimension was seen more clearly in the 20th century and was more closely associated with the new term "rationality" than with old terms like "reason" and "understanding." In psychology, the term became important in controversies about the role of empirical psychological laws for logic and epistemology (Wundt, 1910). More details on the historical developments in philosophy and psychology are represented in chapter 1.1 by Sturm and chapter 1.2 by Evans (both in this handbook).[4] In the following, we just describe some milestones in the recent history of our topic that were particularly influential in shaping the current state of the art and still guide ongoing debates in philosophy and psychology.

2.1   Milestones in the Study of Rationality

Let us start our milestones with *George Boole* and *Gottlob Frege*, two eminent philosophers and logicians. Boole (1854/1951) developed a new logic, now called "propositional calculus," which forms the basic part of classical logic. The title of his book, *An Investigation of the Laws of Thought*, indicates that Boole was concerned with human thinking, not with the abstract truth of propositions. He wrote, "The design of the following treatise is to investigate the fundamental laws of those operations of the mind by which reasoning is performed" (Boole, 1854/1951, p. 1).

Frege later argued in exactly the opposite direction and became one of the most influential thinkers in the history of logic and philosophy. One reason for the importance of Frege's work is that his revolution of logic was not only a revolution of large areas of philosophy. It was also the seed of what later emerged as analytic philosophy. Another reason is that his famous *antipsychologism* was essential for the separation of philosophy and psychology, a gap that we want to bridge in this handbook. In his *Grundlagen der Arithmetik*, Frege suggested, as one of three fundamental principles,[5] that one must "separate sharply the psychological from the logical, the subjective from the objective" (Frege, 1884, p. x). For Frege, grasping a thought such as Newton's law of gravitation "is a process which takes place at the very confines of the mental and which for that reason cannot be completely understood from a purely psychological standpoint. For in grasping the law something comes into view whose nature is no longer mental in the proper sense, namely the thought" (Frege, 1979, p. 145). Here, one must observe Frege's peculiar use of "thought." For him, thoughts are objective entities belonging to the abstract, nonmental "third realm" of "senses" (i.e., meanings). Thus, logic is about the laws of truth, which resemble laws of nature. These laws entail laws of thought. "Rules for asserting, thinking, judging, inferring follow from the laws of truth"

(Frege, 1918, p. 289). Here, "rules" translates *Vorschriften*; they are prescriptions. So, Frege's main claim was that the laws of thought are normative laws, not empirical laws of nature.

In the same year in which Frege published his *Begriffsschrift*, namely in 1879, *Wilhelm Wundt* founded the first German institute for experimental psychology at the University of Leipzig. Wundt, who called Frege's antipsychological approach *logicism*, argued that logic is a result of psychological functions and thus a branch of psychology. Although already the philosopher David Hume took psychology to be the backbone of philosophy, Wundt was probably the first psychologist who systematically applied this view to the study of logic and rationality (Wundt, 1910).

Wundt still retained the connection of psychology to philosophy. However, his work was also a milestone with regard to the academic independence of psychology. One of his many scholars, *Gustav W. Störring*, may be said to have started empirical research on cognition and rationality when he published a 127-page article (Störring, 1908) in which he described the first systematic experiments on human reasoning. In his studies, volunteers had to solve a battery of deductive reasoning problems, while their verbal responses, response times, eye movements, gestures, and even the expansion and contraction of their chest were measured. This work was part of Störring's attempt to develop psychology into a more experimental-physiological direction. He was occupied with the laws of thought in a descriptive, not a normative, sense.

Probably one of the most influential thinkers about the nature of human rationality was the developmental psychologist *Jean Piaget*. In his quite extensive publications, Piaget placed great importance on the development of rationality in children, culminating in his pioneering book *The Growth of Logical Thinking* (Inhelder & Piaget, 1955/1958). His approach was often referred to as "genetic epistemology," as he explained how people acquire knowledge and reasoning abilities by the biological functions of accommodation and assimilation to the environment. According to Piaget, people develop cognitively from birth throughout their lives in four different stages: the sensorimotor (birth to age 2), preoperational (2–7 years), concrete operational (7–11 years), and formal operational (11 years and onward) stages. For Piaget, rational reasoning is practically identical with logical reasoning and develops primarily in the formal operational phase, where children learn to think about abstract entities and concepts and develop logical abilities for reasoning, planning, and problem solving. For Piaget, these rational abilities rely on abstract formal rules of inference stored in long-term memory, which he thought to be almost equivalent to the rules of formal logic. Later, this approach was developed further in rule-based theories of reasoning, one of which is described in chapter 3.2 by O'Brien (this handbook). This approach, however, has been attacked by the adherents of several other theories, notably the *theory of mental models*, which has been developed by Philip Johnson-Laird from the 1980s onward and is another milestone in the history of the psychology of reasoning (Johnson-Laird, 1983, 2006, 2010; Johnson-Laird, Khemlani, & Goodwin, 2015). This theory is described in detail in chapter 2.3 by Johnson-Laird (this handbook; see also chapter 3.3 by Khemlani and chapter 6.3 by Byrne & Espino, both in this handbook).

Another major episode in the discussion of rationality was the *Popper–Kuhn controversy*. With Frege's new logic as a firm base, *Karl Popper* (1934/1989) developed his falsificationist hypothetico-deductive scientific methodology, which purported to state how science has to rationally proceed. Popper had a dispute about this with his positivistic colleagues in the Vienna Circle (e.g., Rudolf Carnap). And there was further controversy, notably in the social sciences, where the Frankfurt school (Habermas, Adorno, etc.) accused Popper of being a positivist. However, the dispute widened dramatically when *Thomas S. Kuhn* (1962) meticulously showed that scientists actually do not follow Popper's methodology. One may conclude, as Popper did, that scientists proceed irrationally. But this is implausible. Isn't rationality the hallmark of the sciences? Thus, Kuhn's observations were taken rather as an empirical argument against Popper's normative methodology—a most remarkable fact—and as a challenge to find a more adequate normative methodology that does not accuse scientists of being irrational. *Imre Lakatos's* (1978) "methodology of scientific research programmes" was such an attempt. This gave rise to what is nowadays called "science and technology studies," an amalgam of history, sociology, and philosophy of science. However, the controversy also motivated more abstract investigations into the logic of theory change and the nature of defeasible reasoning, which surface in many places in this handbook.

Popper's logic-inspired falsificationism was a direct reference point for another milestone of cognitive rationality research: Wason's selection task, which is sometimes called "the Drosophila of the psychology of reasoning" (Beller, 1999). The task is named after *Peter Wason*, a cognitive psychologist who wanted to explore whether people indeed seek to falsify logical rules as required by Popper (Wason, 1966, 1968). The task is as follows:

You are shown a set of four cards placed on a table, each of which has a number on one side and a letter on the other

side. The visible faces of the cards show A, D, 4, and 7. The cards obey the rule: if there is a vowel on the one side of the card, there is an even number on the other side of that card. Which card(s) must you turn over in order to test the truth of this rule?

According to the norms of propositional logic, participants should check the validity of the given rule by turning over the cards with the A (modus ponens) and the 7 (modus tollens) on the front. Typically, fewer than 10 out of 100 participants can solve this problem correctly (J. St. B. T. Evans, Newstead, & Byrne, 1993). Why is that so? Psychologists have given many different answers to this question, some of which are discussed in chapter 1.3 by Schurz, chapter 2.3 by Johnson-Laird, chapter 4.5 by Chater and Oaksford, and chapter 15.1 by Markovits (all in this handbook).

Such tasks were enormously important for the psychology of reasoning. They were designed to study *pure* inference processes without the contamination of true or false prior beliefs. The participants were instructed to draw logically valid conclusions from the premises, which they had to consider as facts that could not be questioned. Since the content of the premises did not matter, the only possible error seemed to be that the conclusion was inferred from the given premises in an unreasonable way. Although this paradigm has many advantages, it also led researchers to pay little attention to the contents of the beliefs represented in the premises. Participants are simply forced to believe that the premises are true; otherwise their responses are often excluded from the analysis.

Later, the pendulum swung more toward the study of beliefs and away from processes of reasoning. This development proceeded roughly in two steps. The first step consisted in research paradigms invented to study interactions between logicality and believability. In such experiments, a conclusion that is logically valid or invalid in relation to the premises could be true or false in relation to the participants' prior beliefs. This introduces the *belief bias*, which arises if a participant in an experiment infers a logically invalid conclusion from the premises because it agrees with her prior beliefs or, conversely, if she rejects a conclusion because it is implausible, even though it is logically valid (J. St. B. T. Evans, 1989). Such *content-effects* became very prominent in cognitive psychology and have been identified in almost all areas of human reasoning (J. St. B. T. Evans, 1993).

The second step in developing more belief-related reasoning research was even more fundamental. Under the label *new paradigm psychology of reasoning* (Oaksford & Chater, 2007, 2020), researchers started around the year 2010 to focus on the subjective degrees of belief that people might have in the contents of the premises and on how those affect the acceptability of the conclusion. Here, a different connection between psychology and Kuhn's theory of progress in science shows up. Proponents of this approach see their new paradigm as "revolutionary" in the Kuhnian sense,[6] as it replaces the norms of two-valued classical logic by those of Bayesian probability theory (Oaksford & Chater, 2007). While the previous research regarded the influence of beliefs as a bias, the new paradigm puts such subjective degrees of belief center stage. The positions of the leading figures in the field are represented in chapter 4.5 by Chater and Oaksford and chapter 6.2 by Over and Cruz (both in this handbook).

These developments in psychology go back to another milestone on the normative side of rationality: the development of probability theory by Reverend *Thomas Bayes* and *Frank P. Ramsey*. Thomas Bayes (1764/1970) is often credited with having established an important theorem in his groundbreaking work (but see Stigler, 1983), a theorem showing how to infer new posterior probabilities from old prior probabilities and the evidence. Today, this framework is called *Bayesianism* and is very popular in philosophy and psychology. An important later step was the work by Ramsey, who—in his 1926 essay "Truth and Probability" (in Ramsey, 1978)—strongly emphasized the subjective interpretation of probability. Unfortunately, Ramsey's writings are few, since he died in 1930 at the age of 26. However, the widespread popularity of Bayesianism in epistemology, philosophy of science, statistics, and cognitive psychology owes much to Ramsey, who made degrees of belief central to epistemology. Ramsey also insisted that degrees of belief must be conceived as subjective probabilities and even proposed, for the first time, a method for measuring those subjective degrees of belief. In all fairness, though, *Bruno de Finetti* (1937) must be mentioned as another pioneer in subjective probability theory. (See also Gillies, 2000; Krüger, Daston, & Heidelberger, 1987; Krüger, Gigerenzer, & Morgan, 1987.)

Nowadays, most psychological research in the "new paradigm" is within this Bayesian framework (see the chapters in section 4 of this handbook).[7] Slowly, however, the insight gains strength that subjective degrees of belief do not need to be modeled as probabilities. Models can also build on other systems for representing degrees of belief and their revision. These options will be richly dealt with in this handbook from both a normative and

a descriptive point of view (see chapter 4.7 by Dubois & Prade and the chapters in section 5 of this handbook).

Ramsey was also important for another paradigm of rationality research, namely decision theory, which builds on probability theory and assumes decisions and actions to be guided by utilities. In his paper "Truth and Probability," he proposed for the first time a method for simultaneously measuring subjective probabilities *and* utilities, and thus laid the foundations of standard decision theory as it is taught today. This measurement method also offered novel ways of justifying decision theory, which later gained ever more importance (see chapter 8.2 by Peterson, this handbook). And this was only possible by interpreting probability in a subjectivist way. Nowadays, standard decision theory is the most widely accepted account of practical rationality and decision making. It serves as the basis of economics and of much of the social sciences, at least as far as it is pursued within the *rational choice paradigm* (see sections 8 and 9 and chapter 10.4 by Raub, all in this handbook).

Another milestone in rationality research was the work of *Herbert A. Simon* (Simon, 1957, 1959), who was as strongly influenced by standard decision theory as Wason was inspired by Popper's falsificationism. Simon criticized standard decision theory for its idealizations and claimed that human decision making is limited in various ways: by incomplete knowledge, by imperfect memory, by restricted capacities of representation and computation, and so on. Thus, he developed the concept of *bounded rationality*, which has by now ramified into various research fields. Simon was also the first to suggest that human behavior is determined by *heuristics* rather than by a decision calculus. He proposed the *satisficing heuristic*, according to which the search for options is stopped as soon as an option is found that reaches a preset achievement level such that no further optimization is needed. Simon's work first radiated into economics but was soon recognized in cognitive psychology, where it initiated much experimental work (Gigerenzer & Selten, 2002). As a pioneer of cognitive-oriented artificial intelligence (together with Allen Newell, he invented the general problem solver [GPS]), Simon was also a founding father of cognitive science (Newell & Simon, 1972), where rationality research looms large as well.

The idea of heuristics was extended in the work of *Amos Tversky* and *Daniel Kahneman* (1974), two psychologists who were also interested in economics. Their experiments showed that human judgment does not conform to the rules of the probability calculus but is rather guided by various heuristics, cognitive rules of thumb that are often helpful because they save cognitive resources, but may also lead to systematic deviations from the norms of mathematical probability and decision theory. This approach was further developed in the program of *ecological rationality* by *Gerd Gigerenzer* and coworkers, although in this theoretical framework, heuristics have a less negative connotation. In fact, these researchers argue that heuristics are highly adaptive and rational (Gigerenzer, Hertwig, & Pachur, 2011). Kahneman and Tversky (1979) also developed *prospect theory* and presented many experimental findings showing that this psychological theory is empirically more adequate than standard decision theory from economics. Today, prospect theory and the *heuristics and biases program* can be found in most textbooks of psychology and behavioral economics (see chapter 8.3 by Glöckner and chapter 8.5 by Hertwig & Kozyreva, both in this handbook).

It should be noted, though, that most of the reported theories somehow oscillate between the normative and the descriptive view on rationality. Piaget was perhaps most explicit in using logic as both a descriptive and a normative framework for human rational reasoning. In contrast, standard decision theory was normatively motivated, and the hope was that it would not be too strongly empirically idealized. Kahneman and Tversky decidedly took the empirical point of view but also uncovered the many empirical inadequacies of normative decision theory. Simon's work was more ambiguous in this respect, as it attempted to define what rationality normatively requires, given the constraints of the cognitive system and the environment. Ecological rationality mirrors a similar view on human rationality. We shall return to these issues in section 4 of this introductory chapter and repeatedly in this handbook. Of course, these few milestones cannot stand in for a comprehensive history of our topic, not even for the past 150 years. But they should prepare the readers to better follow the further structure of this chapter.

2.2 Basic Concepts of Rationality Assessment

Let us turn to some basic concepts of rationality assessments. We already mentioned that the history of the terminology of rationality research does not provide a stable starting point. Neither is it advisable to start amid the current discussion—there is too much theory-ladenness. So, where should we start? Perhaps with ordinary language, which, albeit imperfect, is the best initial starting point we have: it preserves the insights of our ancestors, if also their confusions. Hence, let us briefly look at how we talk about rationality in our everyday life. Which are the things we call reasonable, rational, and so on? And what is it about those things that makes us call them rational?

First, there are many things that are not subject to our rationality assessment, for instance, the weather. We may call them *arational*. They are outside of our consideration. Many things, though, are assessable as *rational*—and these kinds of things may also be assessed as *irrational*. For instance, people are upset about the irrationality of certain traffic regulations and road constructions, and the presentation of products in a supermarket may be very reasonable, at least from the manager's point of view. Still, two categories stand out as primary objects of our rationality assessments: *beliefs* or *epistemic attitudes* in general[8] and *actions*. This is why we carefully distinguish theoretical rationality—which is concerned with beliefs—and practical rationality—which is concerned with actions (see section 3 of this introductory chapter).

Other things may then be called rational or irrational in a derived sense, insofar as they are caused by, or causally connected to, those primary objects, such as the road constructions or the presentation of products in the grocery store. Of course, we also assess persons and other animals as rational or irrational, insofar as they act and believe in a rational or irrational way. Whether we can evaluate emotions as rational or irrational as well is a matter of much controversy (see, e.g., de Sousa, 2011; D. Evans & Cruse, 2004; Helm, 2001). We will return to this question only briefly in section 8 of this introductory chapter. Our overall focus is on what we just called the primary objects. We will explain below that intentions and desires should be included as well.

We should emphasize, though, that it is *actions* and not behavior in general that are judged as rational. Psychologists typically say that action is behavior controlled by the mind. Philosophers typically say that action is intentional behavior (i.e., caused by an intention). Many psychologists study unintentional behavior, which is not, or hardly, under cognitive control. Sneezing, or salivation in the presence of food, is an arational, unintentional behavior. Sure, there is also intentional sneezing, which may or may not serve its purpose and may hence be assessed as rational—but this is fake sneezing. This restriction of rationality assessments to actions is important, but also delicate. Actions may be characterized as intentional and controlled even though the intention need not be consciously conceived and the control need not be actually exercised. So, actions may very well be purely habitual. Certainly, boundaries are vague. Moreover, the intentionality of a behavior is only a necessary, not a sufficient, condition for its being an action.[9]

Let us look a bit more carefully at our rationality assessments. A noteworthy point is that our judgments can take a *relative* and a *categorical* form.[10] In certain situations, your belief that it will rain this afternoon may seem obviously unreasonable. This would be a categorical judgment. However, given that your only information is some outdated weather forecast, your belief may be reasonable after all, namely, relative to this information. Therefore, carrying an umbrella for your shopping tour may be categorically called unreasonable, but relative to your unreasonable belief, it is perhaps not so unreasonable after all.

Similarly, in psychological experiments, volunteers are typically instructed to assume that the premises of a reasoning problem are true. In the experimental setting, the premises are simply given and serve as input to be taken for granted by the participants. Thus, such experiments often assume a *closed world*, in which only the information from the premises is relevant for the conclusion. Relative to the closed world of the experiment, it would be reasonable to reach the conclusion. However, the assumption may be wrong. If the participant reaches a different conclusion, this need not be unreasonable—it may be reasonable relative to the participant's prior knowledge, which is not contained in the closed world but still taken to be relevant by the person. Psychologists then tend to speak of the *belief bias* (J. St. B. T. Evans, 1993), but in fact, such responses just demonstrate that human reasoning is often defeasible and, thus, does not conform to the monotonicity principle of classical logic, according to which no valid inference can be turned invalid by simply adding premises. Many experimental findings and different theories of defeasible reasoning are described in chapter 5.4 by Gazzo Castañeda and Knauff (this handbook).

Relative judgments of rationality seem more basic, while categorical judgments seem derived: they are something like judgments *all things considered* or relative to the total evidence. We will yet look at how they might be derived. But let us first discuss what the relative judgments are relative *to*. In the case of beliefs, the answer seems straightforward: a belief is assessed as rational relative to the evidence the agent has, relative to her reasons or the information she has, or more generally relative to her other beliefs or the epistemic state the belief is embedded in. This raises the question whether a belief can also be rational relative to other than epistemic matters, desires perhaps, or emotions.[11] This is a difficult issue. It seems that a certain desire or emotion may be a *cause* of a belief but never a reason for it. We would disapprove of such a belief as wishful thinking. But what might be the psychological mechanisms behind such phenomena? We shall return to this point in subsection 3.3 of this introductory chapter and when we discuss the distinction between the normative and the descriptive perspective and that

between output-oriented and process-oriented investigations of rationality (section 6 of this chapter).

If we accept that only beliefs can be reasons for a belief,[12] it seems clear how we arrive at categorical judgments about beliefs: a belief is categorically reasonable if it is reasonable relative to other epistemic items that are in turn categorically reasonable. So, two things may go wrong with a belief: it may be based already on unreasonable premises, or it may be inferred from reasonable premises in an unreasonable way.[13] Again, both can happen in psychological experiments: participants can either deviate from a logical conclusion because they do not accept the premises or proceed from other premises, or they take the premises for granted but commit errors in the reasoning process itself.

So much for how we assess the rationality of beliefs. But what about actions, the other main class amenable to rationality assessments? Somehow our actions have to serve our goals according to our own lights. That is, the rationality of our actions is assessed *relative* to our beliefs and desires. This is the familiar and highly important notion of *instrumental rationality*, which is at home everywhere, in philosophy, psychology, and economics. "Beliefs and desires" must not be taken literally here. In fact, the term "beliefs" stands for an entire epistemic or cognitive complex, while "desires" stands for the aggregate of volitional, optative, conative, or buletic attitudes ("pro-attitude" is a more neutral but less familiar term). In ordinary language, we speak of drives, wishes, wants, interests, inclinations, goals or aims, norms and values, and the like. Psychologists distinguish between relatively stable behavioral dispositions, such as the needs for achievement, affiliation, or power, and short-term driving forces in a particular situation (Heckhausen & Heckhausen, 2018). However, let us here stick to "desire" as a generic term for this plenitude of concepts and classifications for the driving forces for actions.

Is there also a categorical rationality assessment of actions? Clearly, an action is categorically irrational if it derives from the given beliefs and desires in an irrational way. However, it may be categorically irrational even if it derives rationally from the given beliefs and desires. This would be the case if the beliefs or desires are themselves irrational. We have already seen how beliefs may be judged to be categorically irrational. So, this judgment immediately enters the practical assessment of actions. But are desires themselves also subject to a rationality assessment? Instrumental rationality refuses to judge desires in this way. It simply takes them as given. Recall the famous dictum of Hume (1739–1740/1975c, p. 415) that reason is and ought only to be the slave of the passions. This leaves no room for a rationality assessment of the passions themselves. If so, the rationality of actions would ground only on the rationality of beliefs.

However, the case is a bit more complicated. We need to distinguish between *intrinsic* desires, the objects of which are desired in themselves, and *extrinsic* desires, the fulfillment of which only serves some other (extrinsic or intrinsic) desires. A similar distinction in moral philosophy is that between entities good as such, or as an end, and entities good as means. Relatedly, psychologists distinguish between extrinsic and intrinsic motivations. An action is extrinsically motivated when it serves to avoid something unpleasant or to get a return, a good grade, money, or the like. By contrast, it is intrinsically motivated when it is rewarding or enjoyable for its own sake, such as solving a puzzle or editing a book.

With this distinction at hand, we can extend rationality assessments also to desires: extrinsic motivations or desires may be irrational given certain intrinsic motivations or desires. If I want to become a millionaire (in order to satisfy my intrinsic desire for a carefree life) and *therefore* want to become a philosopher, then something is wrong with me. Either I have weird beliefs about the profitability of philosophy, or my leaning toward philosophy is simply irrational. That much we can say even from the point of view of instrumental rationality.

This points to the real issue. Can we assess as rational also *intrinsic* desires, motivations, or values? Many analytic philosophers still deny this and accept Hume's position. They would only grant that actions and (intrinsic) desires can be judged as moral or immoral. However, the majority disagrees. A crucial point is that it becomes difficult here to separate considerations of rationality and of morality (see chapter 12.1 by Fehige & Wessels, this handbook). The difficulty already commences with Kant, for whom the *categorical imperative* is the first a priori principle of practical reason. As such, it is a principle of rationality. But at the same time, it is the highest moral principle (see also chapter 1.1 by Sturm, this handbook). Max Weber (1921–1922) famously developed a concept of value rationality as opposed to instrumental rationality. These threads of Kant and Weber were taken up by many, notably by Jürgen Habermas (1981), who sees practical rationality as going beyond instrumental rationality, for instance, in communication, which is characterized by treating one's fellows as ends in themselves (in chapter 10.3 in this handbook, Meggle argues that communicative rationality is just instrumental rationality). Habermas (1973) criticized Weber's concept of instrumental rationality as a subordination to a fundamentally opportunistic lifestyle, while proper practical rationality

also means becoming aware of and changing one's own role in society. Quite different attempts to make sense of rational (and not already moral) assessments of intrinsic desires may be found in Nozick (1993), H. S. Richardson (1994), and Kusser and Spohn (1992). However, this is an obscure topic in philosophy and not treated in cognitive psychology. Therefore, we will not further pursue it in this handbook. In any case, it is important to be clear about the ways in which our rationality assessments are relative and can be categorical.[14]

Having thus elucidated the objects of rationality judgments and their relational character, we have still not addressed the most essential question: What precisely makes one object—belief, desire, action—rational relative to other objects? What is the content of this relation? In short, what *is* rationality? In a way, the entire handbook is about this relation. Ordinary language is of little help here. We might say that the things relative to which some belief or desire or action is rational are the *reasons* for that belief or desire or action. But what is this reason relation? We certainly have a good intuitive understanding of it. We might say that some kind of *reasoning* or *inference* leads from the former to the latter. But what are the mechanisms governing such inferences? Questions like these are omnipresent in this handbook.

David Marr (1982) has famously introduced the distinction between the computational, the algorithmic, and the implementational level of explaining cognitive phenomena, a distinction governing cognitive science ever since. On the *computational* level, the goal of a computation is set: what is its function, and what is to be achieved by it? The *algorithmic* level inquires *how* this computational goal is reached: which processes or algorithms are used to reach the goal defined on the computational level? The *implementational* level then addresses how these processes are physically realized. Searle (1992) argued that mental processes can be explained only on the basis of the biological, not just the functional, properties of the brain, a position he calls "biological naturalism." By contrast, functionalist philosophers of mind from Putnam (1967/1975) and Fodor (1975, 2000) onward have claimed that the implementational level is irrelevant for understanding the functioning of the mind: it is largely irrelevant whether the processes on the algorithmic level are realized in a biological system such as the human brain or by the silicon chips of a computer. And there are even modern dualists, like Chalmers (1996) and others, who argue that some mental characteristics cannot be reduced even to functional characteristics. We briefly discuss this topic in section 7 of this introductory chapter about the preconditions of rationality. A more detailed description of the cortical basis of human rationality is given in chapter 1.4 by Goel (this handbook). However, in the main, this handbook sticks to the computational and the algorithmic level of description, because we think that these are the most relevant levels if one really wants to understand what makes people rational—and sometimes irrational.

However, Marr's terminology is a bit confusing. His conceptual distinctions are immensely important, but the terms are misleading. For instance, a computation consists just in running an algorithm, and the algorithmic level also determines the kind of representations on which the algorithm works. Therefore, we here use the terminology already indicated at the end of section 1 of this introductory chapter, where we called Marr's computational level the *output-oriented* and the algorithmic level the *process-oriented* level of explanation.[15] This terminology is less biased toward computer science and also signals a slight shift of meaning. In particular, we want to make room for the idea that the computational goals are set by normative and not by empirical considerations. We will return to this issue in section 6 of this introductory chapter. But before that, we now come to the top-level distinction in most conceptions of rationality, in both psychology and philosophy.

3. Theoretical and Practical Rationality

Let us start with an example. Many current public and scientific debates revolve around the increased mortality of honeybees. This Sunday, there is a demonstration against the use of the herbicide glyphosate, which is feared to be harmful to animals, including bees. Should you participate? If you want to make a rational decision, your thinking should proceed in two steps: in the first step, you should weigh the arguments that speak for or against the assumption that glyphosate kills bees, evaluate the inferences made by the supporters and opponents of this claim, consider empirical evidence from scientific research, and so on. Let us assume that, based on these considerations, you conclude that glyphosate indeed can cause increased mortality among honeybees. Now, in a second step, you have to decide whether or not you will go to the demonstration. On the one hand, you would prefer a relaxed Sunday on your sofa, it is likely to rain, and you fear that there may be some violence at the demonstration. On the other hand, you think, based on your previous conclusion, that glyphosate should be banned and that it is important to fight for nature, and many of your friends will be there too. Both considerations are important for your decision to attend the demonstration.

Of course, in daily life, these kinds of thoughts are not so clearly separated, your decision might be influenced by many other internal and external factors, and so on. Yet, our example makes clear that we must carefully distinguish two aspects of rationality. The first part of the example refers to *theoretical* rationality, which is about the rational justification of beliefs, inferences, and explanations, or of our epistemic states in general. This issue lies at the center of epistemology and constitutes one of the main topics in the philosophy and psychology of rationality. Thus, the Wason selection task and the large amount of studies on human conditional, syllogistic, probabilistic, counterfactual, causal, or relational reasoning are concerned with theoretical or epistemic rationality. The so-called Linda problem, which is almost as famous as the Wason selection task (see chapter 1.2 by Evans, chapter 4.3 by Merin, and chapter 8.5 by Hertwig & Kozyreva, in this handbook), is also concerned with the theoretical rationality, or irrationality, of human reasoners.

The second part of the example above is concerned with *practical* rationality, which is about assessing actions or pro-attitudes in general. Here, psychologists explore how people choose between different alternatives that have different values for them. Thus, it is concerned with what people decide in a particular situation where different options are considered.

John Searle, who is just as important for philosophy as for the cognitive sciences, expressed the distinction in terms of "direction of fit": theoretical rationality is about how to *represent* the world, how to make our mind correspond to it, whereas practical rationality is about how to *shape* the world, how to make it correspond to our mind (Searle, 1983). The distinction can also be made with respect to the kinds of reasons that can be adduced for the attitudes in question. Beliefs, or epistemic attitudes in general, can only be justified with reference to further epistemic elements—knowledge, beliefs, evidence, perceptions, or testimony—while the reasons for actions and intentions lie not only in those epistemic elements but also in our motives and desires or pro-attitudes. So, to account for theoretical rationality, we need to talk only about epistemic matters, whereas accounting for practical rationality also requires talking about motivations, aims, values, and so on.

Some psychologists might have problems with this sharp separation of theoretical and practical rationality. Their main argument is that in daily life, the functions of reasoning and decision making are intertwined. Both kinds of rationality, theoretical and practical, are strongly interlinked, because reasoning typically serves good decision making. This is also substantiated in the account of *utility conditionals* in chapter 6.4 by Bonnefon (this handbook). Recently, the proponents of the "new paradigm" in psychology have argued in this direction, as they consider subjective psychological value, or utility, and social pragmatics as important for reasoning. Accordingly, this approach aims to integrate the psychology of reasoning with the study of decision making (Elqayam & Over, 2013).

It is certainly true that we can find reasoning on both sides. There is even an influential philosophical position to the reverse effect, namely, that we can find decision making on both sides: we decide not only what to do but also what to believe. This doctrine, *doxastic voluntarism*, goes back to Descartes (1641, Fourth Meditation). It tries to understand belief formation as deciding which beliefs best fulfill our epistemic aims, where truth is the central epistemic aim, but may be accompanied by other aims. The result is what is called *epistemic decision theory* (Konek & Levinstein, 2019; Levi, 1967). We can even say that participants in reasoning experiments decide between conclusions having different values—a reasoner may value the valid conclusion more than the invalid one. In these psychological and philosophical lines of thought, there is no clear distinction between theoretical and practical rationality.

Here we disagree. In our view, practical rationality *presupposes* theoretical rationality. We think that we can study theoretical rationality independently from practical rationality but not the other way around. The reason is that to account for theoretical rationality, we need to talk only about epistemic matters, whereas accounting for practical rationality also requires talking about motivations, aims, values, and so on. The activity of reasoning is necessary on both sides. However, theoretical reasoning proceeds from factual or empirical premises to factual or empirical conclusions. Practical reasoning takes motives or values or pro-attitudes as additional premises and arrives at practical conclusions (intentions or actions). Of course, these are different reasoning tasks requiring different cognitive processes. In the following two subsections, 3.1 and 3.2, we unfold the distinction in more detail. In subsection 3.3 of this introductory chapter, we will return to the issue of what may be special about theoretical rationality.

3.1 Theoretical Rationality

Theoretical rationality is often also called "epistemic rationality," even though, as mentioned in subsection 2.2 of this introductory chapter, "doxastic rationality" would be the more appropriate term. There, the fundamental issue concerns the rational form of epistemic attitudes, states,

or processes. Concerning this shape, the first question is which entities those attitudes are about, that is, what it is that we believe, take to be plausible, and so on. Usually, these are taken to be linguistic entities. And usually, these entities are taken not to be sentences or utterances themselves—otherwise the Frenchman and the Spaniard would not be able to share beliefs—but rather the *contents* or meanings of sentences and utterances. Philosophers and cognitive psychologists call the latter "propositions," while there is some disagreement about a suitable characterization of propositions as objects or contents of epistemic attitudes.[16] This handbook is silent on those issues—deliberately, because treating them fairly would mean engaging deeply the most fundamental questions of the cognitive sciences and the philosophy of mind and language, which are not our topic. The consequence of our decision not to enter into issues of mental contents is that this handbook remains quite vague about the objects or contents of our beliefs and, when it comes to practical rationality, about the contents of our motives and desires.

The next important point pertains to the many possible forms of epistemic states. Basically, these forms may be conceived in a *qualitative*, a *comparative*, and a *quantitative* way. In many situations in everyday life (and in the psychological lab), people just believe or disbelieve something, for example, that glyphosate is harmful to bees. For many decades, psychologists have mainly explored inferences based on such qualitative beliefs. Usually, participants had just two response options: belief or disbelief, entailed by the premises or not, logically valid or invalid, yes or no, and so forth. Today we have a relatively clear empirical understanding of how people deal with such inferences, although there is less agreement regarding the underlying cognitive processes (J. St. B. T. Evans et al., 1993; Johnson-Laird & Byrne, 1991). Thus, an epistemic state (of a certain person at a certain time) may be qualitatively characterized simply as a set of beliefs, a set of contents that are accepted (or "endorsed" or "maintained") by the person at that time. If this set may be an arbitrary set, philosophers call it a *belief base*; if it is to satisfy the fundamental and strong but contested rationality postulates of consistency and deductive closure, it is called a *belief set*. Philosophers also discuss weaker rationality requirements. This conception is unfolded in several chapters of this handbook (see in particular chapter 5.1 by van Ditmarsch and chapter 5.2 by Rott).

However, this characterization neglects that beliefs usually come in degrees. Our beliefs are often more or less uncertain. For instance, you think that glyphosate probably kills bees, but you are not certain. We have a very rich vocabulary for describing this *quantitative* level; we speak of probability, plausibility, believability, uncertainty, credibility, likeliness, and so on. In rationality research, the dominant quantitative account by far is *Bayesianism*, the doctrine that degrees of uncertainty or belief are subjective probabilities. This holds for psychology just as much as for philosophy (Jeffrey, 1992; Oaksford & Chater, 2007). However, one should observe that the label "Bayesianism" is only legitimate when these degrees conform to the axioms of mathematical probability theory.

Probabilistic thinking has evolved for about 350 years (cf. Hacking, 1975).[17] It has pervaded most scientific disciplines (Krüger, Daston, & Heidelberger, 1987; Krüger, Gigerenzer, & Morgan, 1987; for a philosophical introduction, see Gillies, 2000), and so it comes as little surprise that Bayesianism has become a very strong paradigm in psychology and philosophy as well. For this reason, we devote the entire section 4 of the handbook to it. From the Bayesian perspective, the qualitative characterization of belief states looks hopelessly vague, if not useless. Bayesianism has thus developed imperialistic tendencies, claiming that it is the *only* basic characterization of epistemic attitudes. If so, this would bring laudable unification to the field. However, this imperialism has come under pressure from two sides.

One reason is that one cannot simply discard the qualitative level, as advocated by the "radical probabilism" of Jeffrey (1992). The qualitative notion of belief seems too deeply entrenched in everyday discourse. So, the only option is to somehow reduce the one to the other level. However, there is no good reductive account. How weak may a belief be and still count as a belief? Can we accept the so-called Lockean thesis that something is believed if and only if the degree of belief in it is above a certain vague and perhaps contextually given threshold, just as we might say that a man is tall within the entire population if his size is above a certain threshold, say 6'3", and tall among basketball players if he is above 7 feet? In fact, the relation between belief and probabilities is currently the matter of many controversies.[18]

The second reason why it is not cogent to represent degrees of belief as probabilities is that the past 50 years have seen a plethora of alternative proposals (some of which have earlier precursors). An important driving force was artificial intelligence, which needed algorithms for uncertain reasoning, found probability theory infeasible, and hence looked for alternatives. In the meantime, Markov and Bayes net theorizing have much improved the computational manageability of

probabilities (Pearl, 1988). However, alternatives are available (Halpern, 2003). The need was also raised from the psychological and the economic side, which found subjective probabilities descriptively wanting. Even the theory of mental models, which for a long time was dominated by binary classical logic, has more recently developed approaches for how to deal with degrees of belief without using Bayesian probabilities (Hinterecker, Knauff, & Johnson-Laird, 2016; Johnson-Laird et al., 2015; Johnson-Laird & Ragni, 2019). This increases justificational pressure on the normative side—why this rather than that mathematical format?—as well as on the descriptive side—why model human uncertainty in this rather than in that way? Discussions of such alternatives may be found in chapter 4.7 by Dubois and Prade; chapter 5.3 by Kern-Isberner, Skovgaard-Olsen, and Spohn; chapter 8.3 by Glöckner; and chapter 8.4 by Hill (all in this handbook).

Between the qualitative and the quantitative, there is, moreover, the *comparative* level, where one proposition is taken as more plausible or credible or certain than another. For instance, you may take it to be more credible that glyphosate harms bees than that sugar does. One may even take the comparative level as basic, since it also allows one to state the threshold idea of qualitative belief and certainly underlies any quantitative measure of uncertainty (like one may base judgments of tallness as well as measurements of height on the relation "taller than"). Indeed, one may wonder which formal properties of those plausibility comparisons lead to which numerical measure. Again, this opens a large space of normative dispute as well as issues of empirical adequacy. In Knauff (2013), a comparative level has been developed for beliefs about spatial relations, which also plays a role in chapter 13.2 by Ragni and chapter 13.3 by Knauff (both in this handbook). However, this comparative level is not further developed in this handbook.

So far, our remarks about how to conceive of epistemic states were made only from a *static* or *synchronic* perspective. They have dealt with how epistemic states are at a given moment. This is basic—and only preparatory for taking a *dynamic* or *diachronic* perspective. It is important, we think, to distinguish two different aspects of the epistemic dynamics, *internally* and *externally* driven dynamics. Both are subject to considerations of empirical and normative adequacy, and they lead to different kinds of dynamic issues, as we will describe now.

There is, first, an *internally driven dynamics*, which is, roughly, brought about by *thinking*. It runs on the process-oriented (or "algorithmic") level. As indicated at the end of subsection 2.2 of this introductory chapter, this is a crucial issue for cognitive scientists but also for logicians. It is one thing to analyze the formal structure of what a reasoner receives as input—typically the set of premises and their formal structure—and then to look at the output the reasoner generates—typically a conclusion that is evaluated as (more or less) justified according to some normative standard. However, the history of psychology shows that approaches occupied only with input–output associations have limited explanatory and predictive power. It is another and much more demanding thing to understand how the input to the human cognitive system is *processed* and why this leads to the observed output, for instance, a particular conclusion or belief. What mental machinery, involving which cognitive operations, lies behind these internally driven dynamics of human rationality? Here, the main questions are: On the basis of which mental processes are people *competent* to reason rationally? Why are these processes sometimes *error-prone*? And how do reasoning processes *interact* with other cognitive and noncognitive psychological mechanisms? This is a more dynamic description of the main questions that have occupied the psychology of reasoning since its very beginning (Wilkins, 1928; Woodworth & Sells, 1935). The most influential theories from the psychology of reasoning are sketched in sections 2 and 4 of this handbook, which compare normative and descriptive theories of rationality in reasoning.

The logical side also deals with a multitude of issues. There are not only actual but also many potential reasoning processes, which may or may not be followed and which may or may not be correct (as is presupposed by speaking of errors). For instance, logic is full of different (but equivalent) sound calculi or proof procedures, whether or not they are actually used. In a qualitative picture, reasoning is, above all, logical inference. Piaget would have said the same about psychology. It is amply clear, though, that logical inference must not be restricted to deductive or monotonic inference. The logical zoo has become quite diversified, and there is a growing field of nonmonotonic logic, default logic, and defeasible reasoning. This field is partially represented in sections 5 and 6 of this handbook. Each logic comes with its own sound procedures or derivation rules and possibly with a semantic justification. And each of them may or may not be adequate for describing how the human mind works, and may or may not be approximated by mental models or heuristics. In a quantitative picture, reasoning is first and foremost probabilistic reasoning.

Again, though, one must emphasize that all the other quantitative formats come with their own accounts of reasoning. Again, this opens a rich field of normative and empirical assessment.

However, this does not exhaust the dynamic perspective. There is, second, an *externally driven dynamics*. This is about how epistemic states change under the influence of external input: experience, perception, learning, or information in general. Again, one may say that this is about drawing inferences from this input. Cognitive psychologists have developed several descriptive theories of how people account for new information, how they detect and resolve inconsistencies with prior beliefs, how they take into account the order in which new information comes in, how this affects the person's entire set of beliefs, how they consider the trustworthiness and reliability of the source of new information, and how people deal with the fact that new information is inconsistent with their prior beliefs (Elio & Pelletier, 1997; Johnson-Laird, Girotto, & Legrenzi, 2004; Politzer & Carles, 2001; Revlin, Cate, & Rouss, 2001). Such theories are either motivated by approaches from artificial intelligence and philosophy, for example, principles of minimal change or epistemic entrenchment (Alchourrón, Gärdenfors, & Makinson, 1985; Gärdenfors, 1984, 1992), or they have genuine roots in psychology, for instance, when people consider the trustworthiness of information sources or try to explain conflicts between prior beliefs and new facts (Khemlani & Johnson-Laird, 2011; Wolf, Rieger, & Knauff, 2012).

While these process-oriented theories seek to reconstruct how the cognitive system deals with new information, we can also take an output-oriented perspective (Marr's computational perspective) on the externally driven dynamics of belief. This is not about how the input is processed, but about how the goals of computation change through the input—they have their dynamics as well. Put plainly, it is about *what* to believe upon receiving some data and not about *how to arrive* at it. According to Spohn (2012, chapter 1), this problem of revision is tantamount to the venerable problem of induction. Above, we mentioned the limitations of looking only at the input–output associations. However, this must not distract from the fact that these associations have their own dynamics, which is not treated by focusing on the internally driven dynamics but rather needs to be studied on its own. The traditional probabilistic account of this issue consists in learning by conditionalization or, equivalently, by Bayes' theorem; it has by now been considerably refined. However, each epistemic format comes with its own learning theory. Philosophers tend to emphasize the externally driven dynamics; it is addressed in chapter 4.1 by Hájek and Staffel, chapter 4.2 by Hartmann, chapter 5.2 by Rott, and chapter 5.3 by Kern-Isberner, Skovgaard-Olsen, and Spohn (all in this handbook).

Of course, the opposition between internally and externally driven dynamics is not as clear-cut as it may seem. For instance, suppositional reasoning also belongs to the internally driven dynamics, because a person does not require external input in order to work out what she *would* believe given a certain supposition. Then, however, there seems to be a close connection between internally and externally driven dynamics. Isn't what we accept under a given supposition precisely that which we would accept after getting the information (through whatever channels) that the supposition actually holds? This is a topic that has been intensely discussed under the heading "supposing vs. updating" (see Spohn, 2012, chapter 9; Zhao et al., 2012), but we cannot go into the details here. We just want to emphasize once more that all theories of the dynamic aspects of rationality can be assessed in a normative and a descriptive dimension. We return to this in section 6 of this introductory chapter.

3.2 Practical Rationality

We have quite extensively discussed theoretical rationality. Many points, however, apply not to theoretical rationality specifically but similarly to practical rationality, that is, to the rationality of actions or intentions and motivational attitudes in general. Recall that practical rationality presupposes and thus includes theoretical rationality. For this reason, we can carry over most of the conceptual distinctions we have already made, and we can be much briefer even though the topic is wider. The topic is wider because we now deal with a more comprehensive conceptual field. We now deal not only with epistemic attitudes but also with decision making and motivational matters, which we have already unfolded in subsection 2.2 of this introductory chapter.

A first issue of practical rationality is whether it is legitimate to lump together drives, wishes, motivations, wants, goals, aims, norms and values, etc., as the topics of practical rationality. In a theoretical spirit, many philosophers tend to do so by simply speaking of desires, and many economists do so by measuring all of them on a single utility scale. Basically, they treat drives and motivations in the same manner as other-regarding preferences and moral obligations. However, other philosophers, economists, and psychologists emphasize the differences between the various kinds of such pro-attitudes. We do not

want to take a stance here, because in the end, it is a theoretical issue whether or not the various kinds of desires work according to different theories. Most psychologists would probably argue for such a separation. Perhaps the "imperialism" of utility theory here is as problematic as that of Bayesianism in theoretical rationality. Still, here we ignore these potential differences, as our conceptual points about desires presumably apply across the board.

So, let us focus on the same questions as we already did on the side of epistemic rationality. The first question is again which entities those desires are about, that is, what it is that we want, desire, etc. On the epistemic side, we already mentioned some problems regarding propositions as the contents of beliefs. Now, in practical rationality, the issue is even more obscure with respect to desires and motivational attitudes. If we treat the contents of desires as propositions, this might not fully capture our ordinary talk. We may also say that the entities we desire are objects or goods or actions, and so on. There is perhaps a unification: "I want this bike" is the same as "I desire that I possess this bike," and "I intend to do *a*" is the same as "I intend that I perform action *a*" (Jeffrey, 1965, chapter 4). One may think that objects, goods, and actions are clearer contents of desires than propositions. But this is not true. For instance, I love Wonder Woman, but she does not actually exist. So, what is the object of my desire here? Another example: I may want to meet Dr. Jekyll but not Mr. Hyde. This is not possible, because both are the same person. There are no two different objects here that I could desire to meet or not to meet. Philosophers then tend to say that desires are about intentional objects or about objects under a description. And they prefer to consider the contents of desires to be propositions because propositions can contain objects in such a modified understanding.[19] We are aware that this may sound weird to people who are not usually involved in such theoretical considerations. Hence, we better avoid delving into those problems and simply stick to propositions, for ease of exposition.

The next distinction that we made within the realm of theoretical rationality was the one between qualitative, quantitative, and comparative shapes of mental states. We have the same distinction on the side of practical rationality. In a qualitative framework, propositions can simply be or not be desired, wanted, mandatory, or the like. Recall, though, that we need to account for the interaction between desires and beliefs or epistemic attitudes in general, because we must respect the basic distinction between intrinsic and extrinsic desires, which we introduced in subsection 2.2 of this introductory chapter. Extrinsic desires involve such interaction, since they are directed at propositions somehow believed to be conducive to intrinsically desired propositions. Within a purely qualitative framework, however, it is very difficult to account for this interaction (for a proposal and further references, see Spohn, 2020).

So, just as on the epistemic side, where we talked about degrees of belief, we should allow desires, too, to come in varying strength. If we do so, we again have two options. We can either express the strengths of desires quantitatively, or we can express them merely on a comparative level. In fact, the comparative level is more prominent on the practical than on the epistemic side. Thus, we enter the large field of preference theory. The relation between the comparative and the qualitative level has been intensely studied in the *theory of revealed preferences* (Samuelson, 1938), which originally served as a way of operationalizing preferences by means of choice behavior. The comparative and the quantitative level are also closely tied together, namely, by the famous von Neumann–Morgenstern utility theory (von Neumann & Morgenstern, 1944, chapter I.3), which showed how we can measure utility functions on the basis of preferences in a sufficiently unique way. Before this theory, numerical strengths of desires seemed to be elusive things of doubtful scientific status, but afterward, talk of numerical utilities seemed legitimate (for all this, see chapter 8.1 by Grüne-Yanoff, this handbook).

Thus, talk of utility functions measuring motivational strengths by real numbers is by now well established and even dominant, at least in economics and philosophy. The great advantage of utility functions is that they combine so smoothly with probability measures, the prevalent numerical representation on the epistemic side. This combination culminates in the principle of maximizing expected utility, the fundamental rationality principle of standard decision theory (see chapter 8.2 by Peterson, this handbook). No wonder there is a tremendous debate about the normative foundations of this quantitative framework and its empirical adequacy. In the past decades, three cognitive and social scientists—Herbert Simon, Daniel Kahneman, and Richard Thaler—were awarded the Nobel Prize in Economics for showing how and explaining why people deviate from these norms of standard decision theory. This stimulated a number of alternative accounts, which are represented in chapter 8.3 by Glöckner, chapter 8.4 by Hill, and chapter 9.4 by Dhami and al-Nowaihi (all in this handbook). Today, this field is quite variegated and heterogeneous. It is sometimes called "qualitative decision theory," where "qualitative" does not have the meaning from above, but only signals that the field moves

to a less fine-grained way of description than offered by probabilities and utilities, although it does not belong to the qualitative level as we use the term here.

Let us now come to the next distinction we made on the side of epistemic rationality. Is there also an *internally and an externally driven dynamics* of desires, wishes, motivations, and so on? Yes, and in principle we can make the same distinctions, although the algorithms of practical reasoning certainly differ from those of theoretical reasoning. On the side of the internally driven dynamics, it is tempting to interpret decision theory as an algorithmic theory about practical reasoning processes at the process-oriented level. Decision theory provides general rules for calculating expected utilities and thus for determining which options maximize expected utility.[20] However, thus interpreted, decision theory seems descriptively inadequate—people apparently do not follow these rules and do not do any of these calculations. Hence, we may prefer to locate decision theory on what we call the output-oriented level. In the attempt to be more adequate on the process-oriented level, cognitive psychologists have developed various accounts of so-called fast and frugal heuristics and bounded rationality (see chapter 8.5 by Hertwig & Kozyreva, this handbook). However, the goal of developing psychologically plausible algorithmic theories of practical rationality is still daunting, probably even more demanding than in the area of epistemic rationality.

What about the externally driven dynamics? In practical rationality, this becomes relevant as soon as we conceive decision theory not as a process-oriented but as an output-oriented theory that only states which intention should result from the given utilities and probabilities. Here again, epistemic rationality comes into play. Desires and intentions are also affected by the externally driven dynamics of epistemic states. We learn—and thereby change our extrinsic desires and indeed our intentions. This is a very important point, which many people forget when talking about desires and practical rationality: we often wait for, or indeed seek, information, and depending on what we learn, we intend to do either this or that. For instance, once I have learned the weather forecast for this afternoon, I may change my desire to go to the demonstration in a T-shirt. There are sophisticated theories elaborating on *strategic rationality*, that is, on plans or strategies optimally responding to various possible pieces of information (see Raiffa, 1968).

So far, we have only discussed the dynamics of extrinsic desires or intentions. Is there also an externally driven dynamics of intrinsic motivational states? Sure. Psychologists have extensively studied this, although not under the heading of rationality. In motivation research, our actions are seen as either driven by relatively stable behavioral dispositions or triggered by stimuli from the environment, for instance, when you see appetizing food or a physically attractive person (Heckhausen & Heckhausen, 2018). However, even the relatively stable dispositions are not immutable. Our pro-attitudes change and evolve all the time, due to all kinds of influences apart from information. Their dynamics is also driven by education in general and moral education in particular, by the social environment, by fashion and advertising, by saturation, boredom, and curiosity, simply by maturing and aging, etc. So, this is usually considered an issue of descriptive theories of rationality.

However, the dynamics of desires is not only a factual matter to be studied empirically. It also raises normative issues. And many of them are of a moral kind. We want people to acquire the *right* values and try to prevent them from adopting wrong ones, whatever they are. These topics fall outside the scope of this handbook, since they are about morality, not rationality.

Another field that is not represented in this handbook, although it has to do with externally driven dynamics of desires, is *dynamic decision theory* (McClennen, 1990). It deals with what economists call "endogenous preference change" (Bowles, 1998; Loewenstein & Elster, 1992), that is, precisely with our present issue. The question there is not which changes would be rational but how to behave rationally in view of preferences changing due to whatever influences.[21] Thus, the change itself need not be assessed as rational, but when deciding, we may take such changes into account in a rational way. However, such considerations seem to require optimizing according to two different points of view, the old preferences and the new preferences. From the perspective of standard decision theory, which is about optimizing within only one point of view, this is a difficult, if not unsolvable, question. So, this field has thus far remained esoteric and will be neglected here.

3.3   Truth and Rationality

At the beginning of this section, we argued that it is important to distinguish between theoretical and practical rationality—a distinction that roughly matches the distinction between reasoning and decision making in psychology. But then we saw that these two types of rationality have many things in common and that theoretical rationality is a precondition of, and thus part of, practical rationality. So, why still sharply distinguish the two types of rationality?

The answer seems simple: because epistemic rationality is about the rational justification of beliefs, whereas practical rationality is about the rationality of means to an end. But this is too simple. A crucial point is that epistemic rationality is not merely about the justification of beliefs. Things are a bit more complicated because we need to distinguish between epistemic and nonepistemic justifications of beliefs. A standard example, slightly modified, comes from BonJour (1985, p. 6): a mother ought to, and does, believe in the innocence of her son even in the face of overwhelming evidence that he committed a severe crime. One may take it to be morally required that at least the mother backs her son, and given this moral requirement, she may be rationally justified in having that belief. However, given the evidence, the mother would be epistemically irrational to believe in her son's innocence. This shows that the assessment of beliefs as rationally justified may involve more criteria than merely epistemic justification. The justification of beliefs may also take nonepistemic and specifically moral forms. This is not per se a reason to call such beliefs irrational. They are, however, epistemically irrational. Theoretical rationality is restricted to the epistemic part of justification.

But then the question arises whether nonepistemic justifications of beliefs are still acceptable. We agree with BonJour (1985) and many other philosophers that the rational formation of beliefs and doxastic states is subject only to the norms of epistemic rationality. This also means endorsing the normative principle that any external considerations demanding epistemic irrationality must be rejected. It was a long historic fight to establish this principle, starting with the progress from myth (*mythos*) to reason (*logos*) in Ancient Greek philosophy, and continuing with the rise of scientific methods, with Cartesian methodical doubt (which called into question any dogmatic truth), with Kant's characterization of Enlightenment: "Have courage to use your own mind!" or with the claim of Peirce (1877) that the scientific method is the only appropriate one for fixing our beliefs. Certainly, the fight is not over.

The task, then, is to more substantially describe this narrow sense of epistemic rationality. Philosophers discuss various specific aims of our epistemic activities, and, as mentioned at the beginning of this section, some proceed to formulate a specific epistemic decision theory. Among these aims may be simplicity, systematicity, relevance, explanatory power, and perhaps even aesthetic values. For sure, though, the primary aim is *truth*. This first maxim was stated by William James (1896/1956, section VII) in the shortest possible way: "Believe truth! Shun error!" Some philosophers prefer to identify *knowledge* instead of merely truth as the aim of belief (see Chan, 2013; Williamson, 2000, chapter 11). However, in subsection 2.2 of this introductory chapter, we already decided to leave the difference between knowledge and true belief undiscussed.

Of course, belief formation does no more than *aim* at truth. Rational belief formation in no way guarantees truth. This is part of our skeptical heritage. We may accidentally hit the truth in irrational ways. And we may arrive at false beliefs in a perfectly rational way, not only individually but also collectively. In principle, the possibility of misleading evidence and false theorizing can never be excluded. This is why philosophers say that the reasons we have for our beliefs are no more than truth-*conducive*. The point of the normative principle stated above is only that there is no alternative way to approach the truth.

It is a difficult issue whether all other epistemic goals can be reduced to the aim of truth or knowledge. Willard Van Orman Quine once famously claimed that "normative epistemology is a branch of engineering," namely, "the technology of truth-seeking." "The normative here, as elsewhere in engineering, becomes descriptive when the terminal parameter is expressed" (Quine, 1986, pp. 663–664). We doubt this (see Spohn, 2012, chapter 1) but need not expand on the issue now (see also subsection 4.3 of this introductory chapter). Moreover, it is quite unclear how to transfer the aim of truth to the other epistemic formats mentioned above. For instance, it makes no sense to call subjective probabilities true or false. A key notion discussed instead is that probabilities may be justified by their greater or lesser *accuracy* (see chapter 4.1 by Hájek & Staffel, this handbook). In fact, this point speaks in favor of the indispensability of the notion of belief, because beliefs can plainly be called true or false. However, these problems cannot distract from the fact or the norm that our epistemic activities essentially aim at truth.

But then the question is: what is truth? Aristotle's correspondence theory is the traditional paradigm. It has been amended by many modern versions: Tarski's semantic theory, deflationary and disquotational truth theories, the redundancy theory, and so on (see Kirkham, 1992; Künne, 2003). All these theories are epistemologically not enlightening and do not intend to be so. Because of this deficit, a host of further truth theories have been proposed: Peirce's and James's pragmatic theory, Habermas's consensus theory, idealistic and constructivist truth theories in various forms, the coherence theory of truth, again in various forms, and so forth.[22] Already the labels indicate that we are moving into a very controversial terrain that we cannot discuss here.

The main difficulty is that psychologists and philosophers tend to have quite different opinions and theories about the notion of truth. On the one hand, the opinions of most psychologists are shaped by countless experimental findings showing how deeply people's understanding of the world is mediated by their experience and constructed by their brains. This may suggest that the idea that we perceive the "real world" is just an illusion, a concern that can be traced back to the work of Piaget, the learning theory of Bruner (1957), and the social constructivism of Vygotsky (1978).[23] The skepticism goes even further back to the groundbreaking work of Sir Frederic Bartlett, who already emphasized the reconstructive character of the human mind and the social factors that play an important role in what we think is true of the world (Bartlett, 1932/1995).

On the other hand, philosophers could say that such findings about the constructive achievements of the brain inform us about how people come to take something to be true, but not about truth itself; they may find the empirical views of psychologists only indirectly relevant to the multitude of philosophical truth theories mentioned above. However, it is important for philosophers that the notion of truth is distinct from, and does not collapse into, the notion of belief (of *taking* something to be true). This leads to an objective or at least intersubjective notion of truth, which most psychologists might accept as an ideal but often question on empirical grounds.

Maybe psychologists and philosophers should try to collaborate more on the topics of truth and objectivity. This is also important for the distinction between practical and epistemic rationality. If beliefs aim at truth, as said above, and do so in some not entirely subjective sense, then epistemic rationality serves a distinguished, not entirely subjective, aim. By contrast, it is very doubtful that there are corresponding aims on the side of practical rationality. We rationally pursue our practical goals, which may be many. And we all should behave morally. However, whether there is an (objective) moral truth is the highly contested issue of moral realism, which is still undecided (see, e.g., Sayre-McCord, 1988; Schroeder, 2018). And even if there is something like moral truth, it is highly contested in turn whether the pursuit of this moral truth is a demand of rationality (see also chapter 12.1 by Fehige & Wessels, this handbook). By contrast, the pursuit of truth might be the key issue of epistemic rationality. This is what sets apart the two domains and makes epistemic rationality special.

4.   Normative and Descriptive Theories of Rationality

Whenever we talk about theoretical and practical rationality, about beliefs, knowledge, desires, pro-attitudes, etc., this has a *normative* and a *descriptive* dimension. The distinction is at the heart of our philosophico-psychological enterprise and was already implicitly used in the previous sections. The core of the distinction can be made explicit in three questions: We *ought* to be rational, but what exactly does this require from us? We think we *are* rational (at least to some extent), but how do we *actually* reason? And what is the relation between these two questions?

The first question is more at home in philosophy. It is concerned with the *normative* standards of rationality. The second question has been at the center of psychological research since the very beginning of the discipline. It is concerned with the development of *descriptive* theories of human rationality. The third question might be the most controversial. It is motivated by the seemingly huge gap between normative and descriptive approaches to human rationality: people seem to commit many errors and suffer from biases in their thinking, reasoning, and decision making, when measured against standard normative systems such as classical logic, probability theory, or decision theory. Our aim in this section is to illustrate the normative–descriptive distinction by explaining some central issues and results on both sides. We also want to discuss the relation between the two perspectives, where they diverge, where they converge, and how we conceive of this relation, to obtain a comprehensive theoretical and empirical understanding of human rationality.

We should start by clarifying what we mean when we speak of "ought" or "norms." These terms are systematically ambiguous. We may speak about empirical or about genuine normativity. *Empirical* normativity concerns the norms we empirically find in a certain community. For instance, when we come to Great Britain, we learn that one ought to drive on the left; this is an empirical fact about Great Britain. Another example is religion. We can empirically study which religious norms are accepted in certain communities and thus mostly followed. In a *genuinely* normative perspective, by contrast, the empirical norms can never settle whether we should really drive on the left or whether we ought to follow certain religious norms. We cannot find out merely by empirical investigation what ought to be the case and what we genuinely ought to do or believe. It is a matter of normative deliberation and of taking or accepting a normative stance. In this sense, it is basically a fallacy to derive *ought* from *is* (and also *is* from *ought*).

This is an important distinction, which is often blurred. Within legal philosophy, it has been strongly emphasized by Hart (1961, pp. 54ff.), who distinguished normativity viewed from an *external* and from an *internal* perspective. From the external perspective, it is just an empirical question what is normative in a given group ("In the UK, they recognize the law to drive on the left . . ."). In this perspective, there is no opposition between the descriptive and the normative. An outside observer can recognize an empirical norm to hold just by noticing that it is mostly followed and that deviations are sanctioned by law or by the fellow people. In the internal perspective, by contrast, we ourselves accept norms as standards governing how we ought to act. Only in this internal perspective do we have a real distinction between normative and descriptive theories of rationality. It is this real distinction that we want to discuss in this section. Here, normativity is always understood in the genuine or internal sense.

4.1   Normative Theories

Genuine normativity is a huge topic for philosophers. But, we argue, it is also highly relevant for psychology: empirical research needs some normative reference points as benchmarks for human thinking and action. This is sometimes denied by cognitive scientists (Elqayam & Evans, 2011; Elqayam & Over, 2016), a position that we will discuss in subsection 4.3 of this introductory chapter. However, psychological research is full of examples where empirical findings are compared to normative standards: memories are compared with actual events, responses to visual stimuli are classified as "hits" or "misses," spoken sentences are evaluated as syntactically correct or incorrect, and even emotional states are judged as either "normal" or pathological, to name just a few examples. In rationality research, normative standards usually come from logic, probability theory, and decision theory. So, let us start by discussing these theories, although we shall see that the normative discussion is much broader.

Let us first look at the norms of *theoretical* rationality. Already classical logic is a complicated case. If we follow Frege, logic is not a normative theory at all. It states the abstract laws of truth, which are what they are independently of any thinker. Logical truth is like mathematical truth. $2+3 = 5$—this is an atemporal mathematical truth. Logic becomes normative only when we add that the laws of truth are at the same time the laws of correct thinking, that our reasoning ought to follow the deductive rules of logic (see also chapter 3.1 by Steinberger, this handbook).

Indeed, this seems to go without saying. Didn't we say that truth is the central aim of rational belief? Now, the crucial feature of logical deduction is truth preservation: it inevitably carries us from true premises—if they *are* true—to true conclusions. This is the standard argument for the epistemic value of classical logic. But it does not necessarily speak in favor of classical logic. It could also speak for other systems of logic, such as intuitionistic, relevance, or paraconsistent logic.

Moreover, the fact that logic is truth-preserving does not imply that only logical deduction is rationally legitimate. Recall James's maxim quoted above: "Believe truth! Shun error!" Of course, logic is perfect for the second aim, shunning error: if we do not start from an error, logic does not introduce one. However, concerning the first aim, believing truth, logic does very poorly. It does not tell us what to believe, apart from the logical consequences of the given information. In other words, what is not implicit in the premises never becomes believed; logic is not ampliative. Obviously, we need more than logic. We shall return to this point.

Another point is that classical logic cannot handle degrees of belief. Yet we need a normative theory that tells us how we ought to deal with uncertainty. And there, probabilities are the natural choice. Again, though, probability theory is just a mathematical theory to be used for many and variegated purposes and under different interpretations. One of them is to interpret probabilities as subjective degrees of belief, and then it is obviously a normative requirement that these degrees should behave like mathematical probabilities. Indeed, philosophers have devised a plethora of justifications for considering degrees of belief as probabilities, the most important of which are represented in chapter 4.1 by Hájek and Staffel (this handbook). One type of justification, accuracy arguments, even attempts to justify probabilities in terms of truth approximation (=accuracy), that is, with reference to truth as the epistemic goal (Joyce, 2009; Pettigrew, 2016). These justifications extend to learning rules, that is, rules for how to revise one's degrees of belief after receiving new information or evidence. Bayes' theorem, which is tantamount to the rule of conditionalization, is the main rule. There are, however, various other learning rules studied in the literature (see again chapter 4.1 by Hájek & Staffel, this handbook).

Note that this is a purely normative discussion. If there should be other normatively defensible conceptions of degrees of belief, the entire argument cannot be cogent. Indeed, some alternative conceptions of degrees of belief will be mentioned below. It is therefore unfortunate that psychologists widely take the normative requirements of subjective probability theory for granted and thus focus only on empirical evidence for, or counterevidence against, this assumption.

It should also be noted that the requirement to obey the mathematical axioms of probability is still very weak. It is, for instance, compatible with the strange anti-inductive inference to ever *less* expect the next swan to be white, the *more* white swans one has seen. So, epistemic rationality requires much more than these axioms, but it is highly contested what this might be (see, e.g., the attempts of Carnap, 1971, 1980, at an inductive logic, or D. Lewis, 1980; see also chapter 4.1 by Hájek & Staffel, this handbook).

Let us now turn to the norms of *practical* rationality. For a long time, this field was dominated by decision theory. This is, however, as problematic as the use of classical logic and Bayesian probability for epistemic rationality. The standard decision theory has been codified by Savage (1954). Its central principle of maximizing expected utility, often called "Bayes' principle" (not to be confused with Bayes' theorem), is primarily a normative principle. Again, some psychologists seem to see it as the only normative option and then raise their empirical objections. Economists and philosophers, by contrast, have devoted much effort to justifying the standard theory. The main argument is: If your preferences have certain normatively commendable features, then your utilities can be measured on an interval scale such that your most preferred option is one that maximizes expected utility. Savage (1954) was the first to extend this kind of argument to probabilities and utilities simultaneously (see also chapter 8.2 by Peterson, this handbook). Indeed, probability and utility theory support each other, and neither is easily replaced by some alternative in the presence of the other.

There are many empirical criticisms of standard decision theory, as we will see below. However, there is also considerable normative criticism, as illustrated, for example, by Ellsberg's paradox (Ellsberg, 1961). There is an urn before you with 90 balls of various colors. More precisely, it contains 30 red balls and 60 other balls. The 60 other balls are either black or white in an unknown ratio. You have to guess the color of the ball drawn next. If you are right, you get a reward; otherwise, you get nothing. Now, you first have a choice between guessing that the ball is red and guessing that it is white. Most subjects prefer to bet on red—the known risk regarding red seems preferable to the uncertainty regarding white. Second, you are offered a choice between guessing that the next ball is black-or-red and guessing that it is black-or-white. Here, most people prefer to bet on black-or-white, perhaps because this time, black-or-white has a known risk while black-or-red has not. But this seems irrational—and is so according to standard decision theory. The first preference seems to take red as more likely than white, while the second preference seems to take red as less likely, which is inconsistent.

The point of the example is not to show that people are somehow irrational. Rather, it intends to demonstrate that the allegedly inconsistent preferences are at least plausible and that there is something wrong with standard decision theory. Thus, we also find attempts to *rationalize* (not merely explain) these preferences, for instance, by introducing a nonadditive kind of probability with a corresponding alternative way of calculating expected utilities. Generally, such normative criticisms tend to generate alternative theories that can cope with the objections, and we could enter here a variegated landscape of normative disputes in decision theory (see chapter 8.4 by Hill, this handbook).

So much about the difficulties with the norms most often used in psychological research. We should, however, emphasize that the normative discussion about rationality is much richer and has produced many more proposals than we can report here. In fact, the entire handbook is supposed to give such an overview, but even this is incomplete. We have briefly mentioned an alternative to standard decision theory, and we have already announced that probability theory is by far not the only account of degrees of belief. Alternatives are plausibility theory, the Dempster–Shafer theory of belief functions, possibility theory, ranking theory, imprecise probabilities, and so on (see the overview in Halpern, 2003). Some of these accounts are represented in this handbook (see chapter 4.7 by Dubois & Prade and chapter 5.3 by Kern-Isberner, Skovgaard-Olsen, & Spohn). Moreover, these alternative approaches often come with their own decision theory (see Halpern, 2003, chapter 5; Spohn, 2017, 2020). We mentioned that, with the exception of chapter 8.1 by Grüne-Yanoff, treating preference theory, the comparative level is neglected in this handbook. However, this is an area of rich normative theorizing as well, concerning both practical and theoretical rationality. Certainly, cognitive research would benefit from more seriously considering such approaches, rather than limiting itself to the traditional reference points of logic, probability theory, and decision theory.

The gap between the normative and the descriptive perspective is perhaps most apparent in the realm of logic. Above, we mentioned the strengths and the weaknesses of classical logic. It is obvious that most of our reasoning is ampliative. This is not something to be criticized

but to be accounted for. The need to go beyond classical logic was perhaps most strongly felt concerning the conditional "if–then." The long-known paradoxes of material implication[24] clearly showed that material implication as provided by propositional logic does *not* represent the conditional.[25] This point had dramatic effects. One may even say that the philosophy of logical empiricism failed due to the inability of classical logic to cope with the conditional.[26] The situation changed only with the discovery of conditional logic by Stalnaker (1968),[27] which in turn was based on developments in modal logic about 10 years earlier. Since then, the field has exploded.

Another strong influence came from AI in the 1970s, which also saw the need to go beyond classical logic in order to build ampliative inference into the computer. To that purpose, AI researchers developed various kinds of nonmonotonic, default, and circumscription logics. These developments are not represented in this handbook (but see Gabbay, Hogger, & Robinson, 1994). However, the field soon merged with philosophical attempts, and today a quite confusing multitude of nonmonotonic reasoning systems is available. This is partially a play with formal possibilities, but it is also a field of serious normative dispute. There are stronger and weaker systems, and it is a normative issue whether the axioms and rules of these systems are acceptable. This field is incompletely represented in sections 5 to 7 of this handbook (but see also Koons, 2017). Again, we feel that this is a domain where the distance between psychologists and philosophers should be reduced. A recent attempt in this direction can be found in Ragni, Eichhorn, Bock, Kern-Isberner, and Tse (2017). The work of Stenning and van Lambalgen (2008) is another attempt that uses logic programming to model human nonmonotonic reasoning.

Is there a common methodology behind all these ways of normative theorizing? Indeed there is, although as far as we know, it has received little systematic discussion. What one finds are somewhat vague approaches in which the methodology of normative theorizing is assimilated to the methodology of empirical theorizing. The idea of such approaches is not that normative theorizing would be in search of something like normative truth. This would be a problematic idea. Rather, it is in search of normative agreement, of the better and maybe decisive argument. Normative theories of rationality are just *hypotheses* about what is the best way to come to rationally justified beliefs or actions.

This normative theorizing is much the same as cognitive or empirical theorizing in general. Typically, cognitive psychologists use data from controlled experiments and develop theories that agree with these data.

Normative theories also build on a kind of basic data, although not on empirical or experimental data, but rather on basic normative reference points, such as normative intuitions or primitive normative assessments. For instance, imagine you know that, so far, I have seen very many white and very few black swans, and now you see me betting that the next swan will be black. Spontaneously and indignantly, you comment, "That's silly!" This is a primitive assessment of epistemic rationality, not yet backed by a theory, but very likely to be respected by any systematization of such assessments. Or imagine you give me a voucher for a movie ticket and I throw it away. You are upset and shout, "You are crazy!" This expresses an intuition of practical irrationality. Or you might reproach me: "You can't say that we should host the refugees and at the same time vote for the nationalists!" Obviously, some of these primitive normative assessments are very strong, while others are weaker and fleeting: what at first looks unreasonable perhaps turns out after closer inspection not to be so.

These primitive normative intuitions are not data in the sense typically used in science, but they are similar. Experimental data are the reference points for empirical theories: the data should be predicted and explained by the theory. Similarly, normative theories must agree with and justify those normative reference points or primitive assessments. So, from there on, normative theorizing proceeds just like cognitive theorizing. We try to find rules of moderate generality, we think about first principles, and we systematize our normative assertions. Whenever there is an incoherence in our structure—for example, when a rule or principle has counterintuitive consequences—we try to solve the problem, and so forth. Perhaps the situation is more fluid than in empirical theorizing, because our primitive normative assessments are not fixed and may sometimes change under the influence of convincing theories. In the end, however, we are satisfied when we have brought everything into a good and stable *reflective equilibrium*.[28] Yet this equilibrium is not a matter of taste; it is a matter of careful and ever more embracing argument, even though the result cannot be called "the truth." We will return to this issue in subsection 4.3 of this introductory chapter.

4.2 Descriptive Theories

Let us now turn to *descriptive* theories of rationality. Developing empirical theories about the cognitive processes underlying human rationality is arguably one of the most challenging endeavors of the cognitive sciences. Research from the past decades shows that people often draw unjustified conclusions from given premises, that is, they

do not adhere to the principles of theoretical rationality. Similarly, most people and organizations accept more losses for a potential high gain than the rules of rational choice theory seem to allow; national lotteries and equity trading are good examples. Hence, people deviate from the alleged norms of practical rationality. So, does the design of the human mind really provide the necessary cognitive abilities to solve the complex problems we are facing, as individuals and as society? What are the limitations of our rational capacities? Many chapters of this handbook seek to give answers to such questions; some are empirically robust and well founded, and others are more tentative and still under debate.

Let us first look at the descriptive side of *theoretical rationality*. Almost all research in this area is concerned with reasoning, that is, people's ability to draw rationally justified conclusions from given premises. As we already said, the three main cognitive issues are competence, errors, and the interaction of reasoning with prior knowledge and beliefs.

Different descriptive theories explain human reasoning *competence*. Some accounts assume that people reason epistemically by syntactic, language-based mental derivations. On the output-oriented level, such accounts explore whether or not the output of the reasoning process is logically justified by the input the reasoner received. On the process-oriented level, the key idea consists in a repertoire of inference rules represented in long-term memory. These rules are derived from general knowledge and refer to sentential connectives such as "if–then" and quantifiers like "all" and "some." Reasoning is a process of transferring the inference rules into working memory and applying them to the given premises, which are also represented in a language-like format. The result is a language-based conclusion (e.g., Rips, 1994). Obviously, such accounts, in order to model reasoning processes, use logic as both a normative standard and a descriptive framework. The core idea is that human reasoning proceeds in analogy to the proofs of formal logic, although some logical connectives might be understood in a way diverging from propositional logic. The natural-logic account in chapter 3.2 by O'Brien (this handbook) is an instance of this theoretical framework. Many versions of this account were inspired by the developments in AI since the 1960s, where most AI systems were so-called production systems and inspired by the idea of a general problem solver (Newell, 1973; Newell & Simon, 1972).

Other descriptive theories assume that human reasoning competence consists in processing subjective probabilities (e.g., Oaksford & Chater, 2007, 2020). We already mentioned this account in our list of milestones in subsection 2.1 of this introductory chapter. Such theories primarily focus on the output-oriented level. They take the premises as input to the cognitive system and then compare the output to the alleged normative standards of Bayesian probability calculus. Although most proponents of this account say that they are not interested in explaining mental operations on the process-oriented level, it is clear that such accounts must rely on some kind of rules for the computation of subjective probabilities. An important rule in this context is Bayes' theorem (Oaksford & Chater, 2001, 2007). However, since proponents of this account just formulate (output-oriented) computational-level theories, they do not explain how the cognitive system actually computes conclusions according to this theorem.

Yet other theories adopt the position that people use mental simulations to draw epistemically rational inferences. On the output-oriented level, people use the input to create models of what would be the case if the input is true and produce the output as a result of a mental simulation performed to find new information not explicitly contained in the input. These mental models capture possibilities of how the world is, or could be, under certain conditions (e.g., Johnson-Laird, 2001, 2010; Johnson-Laird & Byrne, 1991). The key assumption on the process-oriented level is that reasoning does not rely on syntactic operations, as in rule-based approaches, but on the construction and manipulation of mental models. A mental model represents a possible "state of affairs" described in the premises of an inference problem. It only represents what is true according to the premises but not what is false. The common assumption of most mental-model accounts (Johnson-Laird, 1983; Johnson-Laird & Byrne, 1991) is that reasoning is a cognitive process in which spatially organized or iconic models of the premises are constructed and then alternative models are sequentially generated and inspected. A conclusion is true if it holds in all models that agree with the premises (Johnson-Laird, 1999, 2001; Johnson-Laird & Byrne, 1991; Johnson-Laird & Khemlani, 2013). This theory is supported by many experimental results and is also implemented in several computer programs (Bara, Bucciarelli, & Lombardo, 2001; Khemlani & Johnson-Laird, 2013; Krumnack, Bucher, Nejasmic, Nebel, & Knauff, 2011; Ragni & Knauff, 2013). Chapter 2.3 by Johnson-Laird, chapter 3.3 by Khemlani, chapter 6.3 by Byrne and Espino, chapter 13.2 by Ragni, and chapter 13.3 by Knauff (all in this handbook) are related to this theory.

Rule-based theories and the model theory also explain why human reasoning is sometimes error-prone—the second question that a descriptive theory of reasoning should account for. In one respect, both theories agree: both assume that, in theory, humans have the competence to think and act rationally according to certain normative standards, but that, in practice, this competence is limited by many internal and external conditions. Apart from this commonality, however, there are large differences between the different theories.

According to rule-based theories, errors can creep in due to the variety of processes that are necessary for reasoning. Where exactly they may creep in is determined by the core assumptions of the theory. Before a mental representation of the premises can be stored in working memory, the premises must first be encoded by processes of understanding. Here, interpretation errors may occur because the use of logical expressions in natural language differs from that in logic. Once the premises are available in working memory, abstract reasoning schemata must be applied to the premises to derive valid inferences. Here, the wrong rules might be selected or the coordination of different rules might fail. And invalid conclusions may be produced by still other processing errors (Braine, 1990; Braine & O'Brien, 1998; Rips, 1994).

In the model theory, reasoning errors are mainly caused by the limited capacities of working memory: reasoning becomes more difficult when multiple models need to be generated from the premises, and it takes additional time to discover inconsistencies between tokens in a model. Errors can occur when potential alternative models are not generated or inconsistencies are overlooked. A crucial assumption is that inferences derived from initially formed explicit models are simpler than inferences that can be performed only by elaborating other implicit models. The distinction between implicit and explicit models is described in chapter 2.3 by Johnson-Laird (this handbook). In the past years, the *preferred models theory* has explained why people usually prefer to construct just a single easy, simple model of the given information, why they ignore alternative interpretations that are also logically valid, and why this leads to inferences that deviate from the norms of classical logic (Knauff, 2013; Ragni & Knauff, 2013).

We should also mention that other theories claim that humans are fundamentally irrational. Some of these theories are about theoretical rationality, but they are particularly prominent in research on practical rationality, notably in areas overlapping with behavioral economics. The most prominent accounts are the different versions of dual-process theories, according to which people reason by means of two different systems or processes. Kahneman (2011) argues that one system, System 1, processes the incoming information fast, largely unconsciously, and relatively effortless. It relies on heuristics that often lead to good solutions but sometimes also to violations of rational norms. The other system, System 2, processes the incoming information slowly, consciously, and relatively effortfully. This more "rational" system might rely on logical rules, mental models, probabilities, and so on. Because people tend to use System 1 more often, they are considered largely irrational. Although the account is intuitively appealing—which might be one reason for its success inside and outside psychology and even in the public—it has many empirical and theoretical problems (Gigerenzer, 2010; Osman, 2004). Today, most reasoning researchers prefer to speak of two "processes" rather than "systems" and also admit that the simplified functional distinction in earlier theories cannot be upheld (J. St. B. T. Evans, 2018). In chapter 2.5 of this handbook, Klauer gives an up-to-date review of dual-process theories of human reasoning.

How do reasoning processes interact with prior beliefs, background knowledge, or attitudes? We have already said a bit on this topic in subsection 2.2 of this introductory chapter, but we want to emphasize again that no descriptive reasoning theory denies that in daily life, all of these strongly interact. Proponents of the Bayesian approach often allege that other theories do not pay attention to knowledge and prior beliefs. Yet, this is not really true. In fact, a more adequate description is that present reasoning theories differ in the role they assign to prior beliefs. Broadly construed, one position is that reasoning should be studied and described in its pure form. In such accounts, prior knowledge is a moderating factor that can influence reasoning by supporting or hindering it. This is where the term "bias" comes from. The other position is that people do not do much reasoning anyway but mostly use their beliefs to deal with inference problems. "Reasoning" is basically the retrieval of beliefs from memory and some less relevant cognitive procedures to process this knowledge. Most serious theories lie in between the two extremes, and there are several approaches that can deal with the effects of prior beliefs and the resulting nonmonotonicity of human reasoning within logical theoretical frameworks (see chapter 5.4 by Gazzo Castañeda & Knauff, this handbook).

Let us also briefly look at the descriptive side of *practical rationality*. Again, we can be shorter here, although the issue is broader, because we now deal not only with epistemic matters but also with how people actually judge and decide based on their desires, attitudes, values, and so forth. That is, we discuss the topics of judgment and decision making under the heading of practical rationality, even though philosophers would carefully distinguish between decisions and judgments. Here, we just stay away from vocabulary and briefly look at the psychology of judgment and decision making. Nor do we look at what some psychologists and neuroscientists call "decisions" too, for example, to lift one's arm or not, or to choose between two equally meaningless stimuli on the left and right side of the visual field (Heekeren, Marrett, & Ungerleider, 2008). Such "perceptual decisions" are largely *arational* and thus not considered here.

In principle, we could ask the same questions for practical rationality as we did for theoretical rationality: how can we explain people's competence, their performance, and the interaction of the underlying cognitive processes? The situation is a bit more complicated, however. One reason is that prior knowledge and beliefs are now complemented, and maybe even overruled, by values, attitudes, and so on. Another reason is that the discussions that we already mentioned on the normative side carry over to descriptive theories of human judgment and decision making as well. In fact, there is hardly any consensus concerning the best norms by which the judgments and decisions of people should be assessed. Accordingly, it is quite hard to say what a "good" judgment or decision is and what we should call an "error." So, many deep questions touching upon moral, ethical, cultural, economic, political, and ideological issues are involved in such appraisals. Hence, we better stay away from such evaluations in the next few paragraphs.

Broadly construed, a judgment occurs when somebody assigns a value to an object on a particular dimension. The objects of judgment can be things, situations, actions, people, abstract stimuli, and so on. In a judgment, the attitudes, values, preferences, etc. of a person become visible. So, philosophers would rather call it a "value judgment."

The most important theoretical framework for human judgment is the *lens model*, which was developed around the 1950s by Egon Brunswik (1956) and promoted by Kenneth R. Hammond (1996) in research on social judgments. The basic idea of this quite metaphorical model is that people cannot perceive the environment directly and objectively but instead use cues to make inferences, judgments, and choices. Some of these cues are more obvious in the environment while others must be inferred from further cues. People must select the cues they consider relevant, assign relative weights to them, and finally use these cues as aids in the judgment process. Since the real properties of the object are not directly accessible to the perception of the judging person, the cues' usefulness depends on how well they represent the real properties of the judged object. This is why Brunswik is considered one inventor of the concept of *ecological validity*, which indicates how accurately the cues represent the actual properties of the object within the given environment. The higher the correlation is, the more the cue helps to come to an appropriate judgment.

The lens model is also popular in descriptive theories of human decision making. When individuals decide, they must choose between at least two different alternatives, for example, going to the anti-glyphosate demonstration or having a relaxed Sunday on the sofa. Whereas judgments stand for evaluations and preferences, decisions indicate an intention to pursue a particular course of action (Hardman, 2009, p. 3). The common feature is that both processes are based on cues that reflect more or less well the actual characteristics of the judged object or decision alternatives.

Descriptive theories of decision making seek to explain how people identify different choice options, how they gather and weigh information and cues, how they evaluate the feasibility and desirability of the alternatives, etc., in order to make choices that lead to the best outcome, including all costs and benefits. It is obvious that this is the place where all the normative questions reappear, which we ignore in this subsection. What we can say, however, is that an important turnaround happened when Kahneman and Tversky (1979) developed their prospect theory (see chapter 8.3 by Glöckner, this handbook), in which they argued that the formal structure of probability and utility functions is empirically incorrect and hence proposed more adequate alternatives. Since then, several descriptive theories have been developed that also account for decisions that they may or may not want to call "errors" because they deviate from the classical economic norms. Good examples are the approaches of bounded rationality (see chapter 8.5 by Hertwig & Kozyreva, this handbook), which criticize the standard decision theory as computationally overdemanding. And since classical economists are prone to load utility functions with egoistic or self-concerned connotations, it is easy to denounce the thus-laden theory as highly ideological, which in a way questions its normative status (see subsection 5.1 of this introductory chapter).

4.3 On the Relation between Descriptive and Normative Theories of Rationality

How are normative and descriptive theories of rationality related to each other? For Piaget, the answer was clear: typically developed adolescents and ultimately adults should attain the ability to reason according to the rules of formal logic. In this view, human reasoning research actually is, and should be, driven by the

norms of logic and by a comparison of mental reasoning with these logical norms (Inhelder & Piaget, 1958). Certainly, this is a strong idealization, which not many people would support today. We cannot go into all the details of this complicated methodological issue.[29] What we can do, however, is to briefly present our opinion on the matter, which, we feel, does not accord with any of the opinions we find in the literature.

First, the normative and the descriptive perspectives are both legitimate and important. This goes without saying for the descriptive perspective. And it holds for the normative perspective precisely in the genuine, internal sense explained at the beginning of this section. We are bound to act, and thus we are bound to take a normative stance—even the decision to let things go is a normative stance. So, the normative perspective is absolutely unavoidable. And it extends to rationality, if rationality is normative. This raises the issue of how the normative and the descriptive perspective on rationality are related.

One possibility is that the two perspectives are simply independent. Apparently, they are at least logically independent. Anyone inferring an *ought* from an *is* commits a naturalistic fallacy—this philosophical lesson from Hume and Moore seems to stand.[30] And surely, reversely inferring an *is* from an *ought* is no less of a fallacy. These assertions are cleared up from the logical point of view in Schurz (1997), although there are quite a few logical subtleties. As a consequence, Elqayam and Evans (2011) argue that much of cognitive rationality research commits these fallacies, to its detriment. The authors do not deny the legitimacy of the normative point of view, but they suggest that empirical research better makes itself completely independent from it.

Another possibility is that "rationality" simply means different things in the two perspectives: there may be rationality₁ and rationality₂ (J. St. B. T. Evans & Over, 1996). Then, however, our joint handbook would merely rest on a big equivocation. Neither possibility gives us the positive relation we are looking for.

A natural way of building a bridge between the normative and the descriptive side, particularly in our scientifically minded times, is to try to *naturalize* rationality. "Naturalized epistemology" is a slogan introduced by Quine (1969), and if the quote from Quine (1986) in subsection 3.3 of this introductory chapter were true, epistemic rationality would indeed reduce to "technology," to a branch of empirical science. In the same vein, Schurz and Hertwig (2019) explain rationality in cognition to be instrumental for the predefined aim of cognitive success, which is measured in terms of ecological validity and applicability.

Cohen (1981), although not avowedly in the naturalistic camp, in effect argues in a similar way. As mentioned in subsection 4.1 of this chapter, he distinguishes a narrow and a wide reflective equilibrium in normative theory construction. The narrow equilibrium takes normative intuitions or primitive normative assessments as fixed reference points and attempts to systematize only them. By contrast, the wide equilibrium accounts for more comprehensive considerations even of an expert nature, e.g., logical theorizing or only theoretically explicable principles of minimal change. Then normative intuitions may be negotiable. At the end of subsection 4.1 of this chapter, we, too, used the metaphor of a reflective equilibrium. There, we definitely intended it in the wide sense. Cohen, by contrast, ties rationality theory firmly to what he calls the narrow equilibrium. Thereby, however, normative theory construction amounts to empirical theory construction over the fixed normative reference points: it determines our cognitive competence. The cognitive performance may deviate—a little. But the competence can never turn out to be irrational. This is Cohen's way of empirically controlling normativity. Again, we do not find genuine normativity in Cohen's picture.

In our view, these strategies of naturalistic reduction do not properly respect the autonomy of the normative discourse, which is so amply displayed in this handbook. Let us explain this with respect to the three naturalistic strategies just mentioned.

First, if rationality judgments were just a matter of a priori normative intuition, as suggested by Schurz and Hertwig (2019), their criticism of the normative perspective would be justified. But they are not; they are a matter of normative argument. They are also right when they complain that the metaphor of reflective equilibrium entails circular justification. But that's perhaps the nature of justification, and we simply need further criteria to distinguish good from bad circles. In their alternative account of rationality as instrumental for cognitive success, they simply take the aim of cognitive success for granted, and thus rationality research reduces to the exploration of this instrumental relation. We, by contrast, want to maintain the autonomy of normative discourse by interpreting their account as a normative one. It may well be normatively convincing to strive for cognitive success in their precise sense, weighing ecological validity and applicability, and to include this success in our normative reflective equilibrium. But this requires further normative argument.

Second, the quote from Quine (1986) in subsection 3.3 of this introductory chapter underrates this autonomy in

a similar way. Beliefs aim at truth. Granted, but perhaps this slogan does not assign to epistemology the task of providing a "technology" for reaching a predetermined aim. Perhaps the aim itself is explicable only within the reflective equilibrium for epistemic rationality (see Spohn, 2016).

Third, Cohen, too, seems to underrate the autonomy when he firmly ties rationality to normative intuitions. But these intuitions, even though we have no better starting point, are often dim or confused, and the task is not just to quasi-empirically systematize but to normatively develop and explicate them. Thereby, they become an indispensable ingredient of the wide normative reflective equilibrium.

Our rejection of naturalistic reduction and our emphasis on the autonomy of the normative discourse can be seen to reflect Kant's doctrine of the autonomy of pure practical reason (Kant, 1788/1908): it has the positive freedom of self-legislation, and it has the capacity and the mission to fill that freedom. However, this emphasis at the same time deepens the gulf separating it from the empirical realm. As stated, logic cannot bridge the gap between the normative and the descriptive. What else can?

The crucial point, we think, is that there are plenty of *defeasible* (not deductively valid) inference relations between the normative and descriptive perspectives on rationality. In practice, we find these inferences everywhere, but we do not find this point clearly addressed in the methodological literature. In moral philosophy, another chapter of genuinely normative theorizing, people talk of prima facie reasons. For instance, if something is (descriptively) pleasing, then, prima facie, it is (normatively) good. This assertion is not analytically true—it is an open question whether something pleasing really is good. But the assertion is plausible—and acceptable when there are no opposing reasons.

The general reason for these defeasible inference relations is that humans are receptive to norms. We tend to follow the norms, not only the norms somehow externally given but also the genuine norms internally endorsed. This receptivity is sometimes strong, sometimes weak. There are uncountable confounding factors. Still, the receptivity exists. The point is often expressed[31] by stating that the normative theory simultaneously serves as an empirically idealized theory. This entails, quite generally, the following defeasible connection: whenever something that is under human control *ought* to be the case, there is some plausibility that it actually is the case. And reversely, whenever something under human control is the case, then there is some plausibility that it *ought* to be the case. This may sound far too strong, and you will be able to immediately cite hundreds of counterexamples. Granted, but the abstract point could only be put so starkly. Of course, it must be applied with all due caution and reservation. But this does not make the point go away.

In rationality research, this point is ubiquitous. Philosophers explicate reasoning in the hope that people then obey their explications. And psychologists refer to normative standards in order to specify precisely this defeasible relation. Now, whenever there seems to be a discrepancy between norm and empirical fact, an exception to our defeasible rules, there are several ways to go. There is, first, the standard way of sticking to the norm and finding additional explanations why observed facts deviate from the norm. The Wason selection task, for example, leaves the norms of deductive logic unshaken and postulates certain biases disturbing the performance. In this case, the criticism concerns the people: they are somehow irrational. There is, second, the possibility of modifying the normative reference point in order to reduce or possibly eliminate the discrepancy. This was the Bayesian strategy: perhaps people's inferences should not be tested against logic but against probability theory. Then the criticism turns against the scientist for choosing an inappropriate normative reference point.

So far, the discrepancies had no consequences for the normative discussion. Neither logic nor probability theory are demoted in their normative status by the Wason task. It is just a matter of finding out empirically which normative ideal we should take people to adhere to and how much deviation from the normative ideal we should admit. The default assumption is perhaps: the less deviation, the better. But this is not so clear—the psychology of reasoning might still have to find its own descriptive reflective equilibrium.

Then, however, there is the third case, where the discrepancy is taken as a normative criticism. Take the Popper–Kuhn controversy (subsection 2.1 of this introductory chapter): Kuhn described the behavior of scientists. Popper took the first way and criticized this behavior as irrational. But this response seemed implausible, since the scientists did not accept the charge of irrationality. So, it rather seems that something is wrong with Popper's normative theory. Or take the Ellsberg paradox (subsection 4.1 of this chapter): people turn out to be stubborn. Even when we explain to them why their preferences are allegedly irrational, they stick to them. So, are they stubbornly irrational? Perhaps something is wrong with standard decision theory if it does not distinguish between known risk and a numerically identical subjective probability. This is an empirical input, which needs to be respected in the normative discussion. Here we have indeed a defeasible inference from *is* to *ought*.

It is unnecessary to give further instances of the various ways of this defeasible inference. If we are aware of its possibility, we easily see it everywhere. Defeasibility is not a one-way bridge. Defeasible inference carries us both ways, from the normative to the descriptive, and inversely. Empirical theorizing has to find its own reflective equilibrium (most descriptions of scientific methodology are much more detailed, though). Normative theorizing has to find its own reflective equilibrium as well, as sketched above and exemplified in the handbook. However, the two equilibria are correlated. One cannot optimize the one in disregard, or at the cost, of the other.[32] This is very vague but already gives an idea of our main point. It is methodologically important to observe the logic of descriptive–normative reasoning and to keep this *double equilibrium* in mind.

Elqayam and Evans (2011) fear to thereby get entangled in the arbitration problem, the problem of deciding between competing normative accounts: "Psychologists can . . . and do get involved in arguments about which norm is right—perhaps an odd activity for empirical scientists" (p. 239). The arbitration problem is indeed severe and will perhaps never be conclusively decided. Still, we think this activity is not odd at all but unavoidable.[33] It is an activity, or discourse, that philosophers and cognitive psychologists should jointly engage in. It is necessary in order to make progress in our understanding of human rationality. In fact, this handbook is one large plea for this joint enterprise.

5.   Individual and Social Rationality

There is another basic classification of theories of rationality, namely, into those about individual and those about social or collective rationality. This classification cuts across the distinctions we have already introduced. We find a lot of empirical research on both individual and social rationality, as well as on many normative issues. And just like individual rationality, social rationality has an epistemic as well as a practical dimension.

At first glance, the distinction is clear: individual rationality is about the epistemic or practical reasoning processes of a single person and their outcomes, while social rationality is about group processes, their results, and their rational organization, concerning the purely epistemic formation of a joint opinion as well as the agreement on a joint strategy or the settlement of practical conflicts. We have seen many issues and instances of individual rationality in the previous sections. In fact, those sections were severely biased toward individual rationality, while neglecting the social dimension of rationality.

There are reasons for this bias. The main reason is that philosophy and psychology both have an individualistic bias. Our Humean heritage gives a nice little example of this bias. On the one hand, one should think that communication provides the foundation for all social epistemology. On the other hand, Hume (1748/1975b, section X, "Of Miracles") classified testimony (through listening and reading) as a mode of perception among others, to be compared with perceptions in other modes, and the processing of any kind of perception is a matter of individual rationality. If so, social epistemology seems reduced to individual epistemology. Not that we would maintain this nowadays, but the bias persists. Another reason is that the notion of social rationality is much less clear than that of individual rationality, as we will indicate below. So, the field of social rationality is still more tentative.

For these reasons, this handbook has an individualistic bias as well. However, the social dimension of rationality is so important that we also wanted it to be represented in this handbook, even if only in a more exemplary way (see sections 9, 10, 12, and 14 in this handbook). In the following, we want to discuss at least briefly some central issues of social rationality.

5.1   Individual Rationality in a Social Context

Game theory is the classical paradigm of rational choice theory in a social context.[34] It has developed into *the* foundational economic theory and pervades economic theorizing. It is clearly about practical rationality, and it is clearly about how to behave rationally in social contexts, that is, when other agents are involved. However, is it also about collective rationality?

There is no clear answer. In order to understand this, we must look at the basic distinction between cooperative and noncooperative game theory. Cooperative game theory is not necessarily about cooperation. But it doesn't turn cooperation into a problem. It assumes that communication among all participants or players is free and available and that agreements can be reached, which may or may not be enforceable. This may, but need not, result in cooperation. Perhaps you only compromise with your opponents, or you find partners in order to defeat your opponents, and so on. Large parts of cooperative game theory are indeed about suitable coalition-forming. Thus, it certainly addresses important forms of collective rationality.

Similarly, noncooperative game theory is not necessarily about noncooperative agents. Clearly, players are allowed to cooperate also in noncooperative games. However, cooperation requires action, e.g., communicating, bargaining, entering into agreements, etc., and from the point of view of the noncooperative theory, each kind of action must be explicitly represented as a move in the game. This is why game theorists consider the noncooperative part as basic and the cooperative one as derived. And it is for this reason that only noncooperative game theory is represented in this handbook.

Now, it is important to see that noncooperative game theory is not about collective rationality. It is rather about individual rationality in a social context.[35] How can you pursue your interests in an environment of other agents who also pursue their interests? This is not simply a matter of instrumental rationality, of manipulating our fellows like instruments. It means recognizing that our fellows try to reach their aims as well. It is, one might say in Kantian terms, a matter of treating our fellows as ends and not as means.

In this perspective, the notion of a Nash equilibrium takes center stage. The actions or strategies of the players are in equilibrium if, given the choices of the other(s), no one can improve by changing her choice. Only then does everybody individually optimize. Also, only such equilibria can be publicly known and shared—no public advice can deviate from such an equilibrium without giving to at least one player a reason to reject the advice.

It may be asked whether such individual rationality in a social setting is really something special. Classical game theory assumes so up to the present day and considers standard decision theory as the limiting case of "one-person games" or "games against nature." Bayesian game theory, as developed by Harsanyi (1967, 1968a, 1968b), and its successor, epistemic game theory (Bernheim, 1984; Pearce, 1984; Spohn, 1982), try to reverse the direction of derivation and to conceive of equilibrium behavior as ordinary expected utility maximization under certain special conditions (see chapter 9.2 by Perea, this handbook, for how far this strategy carries). Quite a different interpretation of noncooperative game theory is found in *evolutionary game theory* (see chapter 9.3 by Alexander, this handbook), which is particularly germane for evolutionary explanations of various cognitive features (see chapter 1.3 by Schurz and chapter 10.6 by Cosmides & Tooby, both in this handbook).

The fact that individual rationality in a social setting does not per se amount to social or collective rationality becomes obvious in the *prisoners' dilemma*, which is perhaps the most famous paradigm of game theory:[36] two prisoners in two separate cells may or may not confess an alleged common crime. In other words, they can either defect (from his partner in crime, *D* = confess) or cooperate (with his partner in crime, *C* = not confess). If a prisoner is the only one to confess, he is set free as a key witness (this has the highest utility, 3—the exact figures are not relevant), while the other one is imprisoned for a long time (this has the lowest utility, 0). If both confess, they get imprisoned for a slightly shorter time: at least they have admitted their guilt (utility 1). And if both don't confess, they can be convicted only of a lesser crime, entailing a small punishment (utility 2). The story has thousands of variations and occurs to us every day. The only important matter is the structure of the situation, the distribution of utilities, as given, for example, in table 1. There, the utilities of the row chooser are given in the lower left corner of each field and those of the column chooser in the upper right corner. Obviously, the only Nash equilibrium is (*D, D*). However, the justification of *D* is even stronger: each player fares better by defecting, not only given that the other defects, but *whatever* the other does: either he gets 3 instead of 2 or 1 instead of 0. In other words, cooperating is strictly *dominated* by defecting. Hence, there is almost unanimous agreement that defecting is the only rational thing to do, at least if the game is played only once; iterated playing is a more complicated issue. From the point of view of individual rationality, this seems undisputable. From the point of view of collective rationality, however, this is silly. Both would fare better by cooperating: they would get 2 instead of the 1 from the equilibrium.

The prisoners' dilemma is often characterized as a conflict between individual and collective rationality. But what is collectively rational? In the prisoners' dilemma, it seems obvious that joint defection is collectively irrational because it is strictly Pareto-dominated, as economists say: both could fare better by jointly cooperating. However, not being Pareto-dominated (i.e., being "efficient") provides at best a necessary condition for collective rationality, indeed a very weak one. Note that the *C/D* combinations are also efficient. In fact, it is quite unclear what collective rationality might mean beyond the criterion of efficiency, and in some contexts, the latter is even doubtful as a necessary condition (e.g.,

in the liberal paradox invented by Sen, 1970, chapter 6). However, this only shows that social rationality is a much more difficult and obscure notion than individual rationality. We will return to this later.

What is the normative status of game theory? It is quite unclear. The vacillating interpretation of rationality as individual or social adds to the uncertainty. However, the uncertainty persists even if we focus on individual rationality. Noncooperative game theory clearly has strong normative foundations, but there are also many situations, like the iterated prisoners' dilemma and the ultimatum game (Güth, Schmittberger, & Schwarze, 1982), where the recommendations of game theory are ignored by agents and indeed doubtful even from a normative point of view. Therefore, many economists have become quite guarded about game theory as a normative theory and prefer to characterize it as an idealized model that can claim validity only under restricted conditions. How far this validity extends, though, has become quite contentious in turn.

Most of the descriptive work in the field uses experimental paradigms inspired by noncooperative game theory. Dozens of experiments use different versions of the prisoners' dilemma, the ultimatum game, and many other games, to study the roots of rational cooperation and conflicts between individuals and social groups (Axelrod, 1997).

We should emphasize, however, that such experiments were criticized for many reasons (Bowles, 1998; Gomberg, 1989; Ostrom, 1998, 2010). The main criticism is that they derive from the tradition of classical economic thinking and the self-adjusting mechanisms of free markets. Indeed, it may be argued that an indoctrination with rational choice theory biases people toward acting more selfishly than they would otherwise. Therefore, one may suspect that the theory and its related paradigms create the type of selfish people they axiomatically assume (Frey & Meier, 2002). The main cause for this criticism is that economic theory tends to load the notion of utility with egoism and to assume that people decide like selfish maximizers, promoting only their individual benefit. However, many empirical results in the social and behavioral sciences, particularly in behavioral economics, show that this draws a too one-sided picture of human rationality, downplaying the fact that collaboration lies at the heart of the human species and our society (Tomasello, 2009a, 2009b).

5.2 Social Rationality

Let us now turn to proper social rationality. This is about the attempts of social groups to form joint beliefs and joint preferences and to perform joint actions in a

Table 1
Prisoner's dilemma

		Player B	
		C	*D*
Player A	*C*	2, 2	0, 3
	D	3, 0	1, 1

rational way. When the group cooperates to arrive at rationally justified beliefs, conclusions, or explanations, we speak of *epistemic* social rationality. When the group cooperates to form joint decisions, we talk about *practical* social rationality. A group cooperates if each group member coordinates its activities with that of other group members, and all group members work toward a goal that benefits the entire group (Witte & Davis, 1996).

There are many situations in our society where groups of people work together to achieve shared epistemic goals. For instance, a group of jurors wants to reach the verdict "guilty" if the defendant is guilty and "not guilty" if he is innocent. The members of an ethics committee, in psychology or medicine, reason together to anticipate possible risks resulting from experimental treatments or medical therapies. Scientists working together on a joint publication draw inferences from empirical data and want to come to justified conclusions and explanations. Politicians and parliaments seek to find the right means to reduce climate change or to fight pandemics. It is obvious that we urgently need a better understanding of this form of social rationality, since it lies at the heart of democratic societies, which assign central legislative, executive, and judicial decisions to *groups* of people, ranging from boards, cabinets, commissions, and parliaments to the entire electorate.

The easiest way of responding to normative questions here might be to use the same standards as for the epistemic rationality of individuals. The problem with this transfer, however, is that individual rationality is usually defined with respect to the formation, combination, and revision of individuals' beliefs, but it is unclear whether it makes sense to attribute beliefs to collectives (e.g., List & Pettit, 2011; Theiner, Allen, & Goldstone, 2010). Some accounts have been suggested, but none of them are as prevalent as the logical and Bayesian frameworks of individual epistemic rationality (see also chapter 10.1 by Dietrich & Spiekermann, this handbook).

So, let us look at the descriptive side of epistemic social reasoning, where interacting groups evaluate the "correctness," "truth," and "believability" of epistemic inferences and beliefs. Only a small number of studies investigate this topic. One stream of research started from work by Patrick R. Laughlin and coworkers, who showed that the mere announcement of a purely "intellectual task" with an "objectively correct solution" triggers other social combination processes than the announcement that different opinions should be merged to come to a joint decision (Laughlin & Adamopoulos, 1980; Laughlin & Ellis, 1986). The authors use the concept of "demonstrability" to explain which beliefs are

shared by a cooperating group: a "demonstrably" correct solution is the one for which the group members can demonstrate that it is correct. They also found a correlation between demonstrability and the number of group members required for a collective solution (Laughlin & Ellis, 1986). Similarly, a study by Moshman and Geil (1998) on deductive reasoning in groups indicates that groups are better at avoiding typical reasoning biases. They asked 143 college undergraduates—32 individuals and 20 groups of 5 or 6 interacting peers—to solve the Wason selection task and found a clear group advantage. After intensive deliberation, groups turned out to be less prone to the confirmation bias than individuals. There are a few more studies on the epistemic rationality of groups, but overall, this field has started to flourish only recently (e.g., Claidière, Trouche, & Mercier, 2017; Jern, Chang, & Kemp, 2014).

A second branch of research is in the tradition of Gricean and neo-Gricean theories of communication (Grice, 1975, 1989; Levinson, 2000), the "relevance theory" (Sperber et al., 2010; Sperber & Wilson, 1995), or the "argumentative theory of reasoning" (Mercier & Sperber, 2017). All these approaches to communicative rationality could be said to view human conversation as a form of social epistemic rationality. Some of these approaches are discussed in chapter 5.5 by Hahn and Collins and chapter 5.6 by Woods (both in this handbook).

Research on *practical* social rationality is concerned with many topics that are not important when groups perform epistemic tasks. This research has in fact a much longer tradition; there is a huge amount of literature from philosophy, psychology, economics, political science, etc. Nevertheless, the area is still quite diffuse, and the empirical and the normative side seem less well integrated than in the case of individual rationality. One reason is perhaps that in the field of practical social rationality, normative benchmarks stand out much less clearly than in that of individual rationality. For this reason, we have treated the topic in this handbook rather in an illustrative than in a fully representative way.

On the normative side, about the earliest attempts to get a theoretical grip on the topic come from democratic voting and social choice theory, which are not represented in the handbook (see List, 2013). They stand in the utilitarian tradition and try to determine (procedures for finding) the common good or a social preference on the basis of individual preferences. Group decision making has the similar problem of integrating diverging aims into one common aim and of finding a joint strategy for proceeding. Such group decision processes often presuppose the sharing of information and the integration of diverging opinions into a common one. This is a central issue in social epistemology; see chapter 10.1 by Dietrich and Spiekermann (this handbook).

Clearly, these fields are full of normative issues. One of the earliest and most baffling ones was Arrow's impossibility theorem, showing that there is no way of aggregating arbitrary individual preferences into a social preference, provided this aggregation is to satisfy certain apparently very plausible conditions, one of which is non-dictatorship (i.e., that the social preference is not identical with the preference of a given individual or "dictator"; List, 2013, section 3). It is much less clear, though, whether such normative issues are issues of rationality. In the case of group opinion and group decision making, the answer seems to be positive: they seem quite analogous to individual decision making. However, already the case of preference aggregation seems different. Is the non-dictatorship condition (if it can be interpreted as the absence of a dictator in the ordinary sense) a demand of rationality? Hardly. It rather appears to be a moral requirement. In other words, in these social matters, normative issues tend to take on a moral character. And the extent to which morality is a question of rationality is very much open. If we could reduce morality to rationality, moral maxims would be just as well justified as postulates of rationality. Presumably, though, morality should keep its autonomy. This unclear situation is another reason why this handbook does not fully engage in normative social issues. At least section 12 of this handbook is devoted to the relation between rationality and morality.

Cognitive psychologists have also investigated practical rationality and decision making in groups. Most of this research is motivated by the hope that social rationality is in principle superior to that of individuals. However, the empirical findings suggest otherwise: while several experimental studies showed that sometimes groups do indeed perform better than individuals, other studies did not find any group advantage, or even reported that individuals outperform groups.

The superiority of groups can have several reasons. One prominent finding is that groups often outperform individuals because groups tend to weight the input from more competent members more strongly. This often leads to a "truth wins" principle in a social combination process in which the existence of a single knowledgeable group member is necessary and sufficient for a correct group response (Laughlin & Ellis, 1986). However, Schultze, Mojzisch, and Schulz-Hardt (2012) have provided an alternative explanation, namely, that group members actually become more accurate individually during group interaction. Many other comparisons between groups and individuals are reported and often suggest different explanations for group advantages (Liang, Moreland, & Argote, 1995). The "wisdom of crowds" literature (Surowiecki, 2004) shows that collectives can decide rationally, or optimally, even when the individuals in the collective are irrational (Lyon & Pacuit, 2013). Yet, this only works when the decisions of the group members are independent from each other. Good overviews about the advantages of groups in judgment and decision making can be found in Esser (1998) and Witte and Davis (1996).

Inferiority of groups can also have several reasons. For instance, classical studies have demonstrated that *conformity* can lead to suboptimal group performance (Asch, 1951, 1956). Janis (1972, 1982) showed that members of groups often avoid raising controversial issues due to the desire to preserve harmony. This "groupthink" can lead to irrational decisions in groups of potentially rational agents—and has been considered the main reason for the fact that in 1986, critical information was neglected when planning the launch of the space shuttle *Challenger*, which ultimately cost seven astronauts their lives (Moorhead et al., 1991). Research in the hidden-profile paradigm—in which some information is shared among group members, whereas other pieces of information are unshared—demonstrates the potential disadvantages of group collaboration (Brodbeck, Kerschreiter, Mojzisch, & Schulz-Hardt, 2007; Jönsson, Hahn, & Olsson, 2015; Kerr, MacCoun, & Kramer, 1996; Kerr & Tindale, 2004; Mojzisch & Schulz-Hardt, 2006; Stasser & Titus, 2003). Research in social-choice and game theory demonstrates that groups of fully rational agents can end up making decisions that are collectively irrational (Bacharach, 2006; McAdams, 2009). Today, we know many more biases that appear in groups and prevent them from making good decisions. For instance, we could hardly call it rational if the order in which group members speak influences the joint decision. In group meetings, however, those who speak first often have a higher impact on the outcome of the group decision. This often leads to suboptimal output of the group (Bazerman, 2012; Hartmann & Rafiee Rad, 2020) and may polarize the group members (Zhu, 2013). Again, this can make it hard for groups to reason and decide optimally or rationally.

In sum, the study of social rationality, both from the normative and from the descriptive perspective, is a highly complex issue that we are just beginning to understand. It is also related to many other research fields in social psychology, economics, political science, and even anthropology and legal theory. This does not make the topic simpler. It is obvious, however, that such questions are immensely important for our society, and it is good that recently more psychologists and philosophers have become interested in these topics.

6. Rationality in a Process-Oriented and an Output-Oriented Perspective

We now want to return to the important distinction between output-oriented and process-oriented approaches to human rationality. At first glance, the two perspectives seem quite closely attached to the two disciplines at the center of this handbook: output-oriented approaches dominate in philosophy, and process-oriented approaches are crucial for the cognitive psychology of human rationality. Although these connections to the two disciplines are not entirely wrong and can largely be seen in the historical developments of both disciplines, we will later show that the distinction also cuts across the disciplines: some psychologists are also satisfied with output-oriented approaches, and process-oriented approaches are, to some extent, also pertinent in philosophical approaches to rationality.

Our distinction between theories that focus either on the output or on the process of cognition is of course not new. Marr's (1982) well-known distinction between the *computational* and the *algorithmic* level of description is certainly the most influential version of this distinction, at least in cognitive science. In subsection 2.2 of this introductory chapter, we already explained why we nevertheless decided to use another terminology here: Marr's terminology often leads to misunderstandings, it is too closely attached to the computer metaphor of human cognition, and our terminology leaves more space for the vital normative–descriptive distinction.

Of course, Marr's approach had precursors. Already in the early days of modern logic, there was, on the one hand, the so-called syntactic approach, according to which logic just consists in the manipulation of symbols in conformity to certain rules, for example, in writing strings of symbols (i.e., formulae) that have the shape of axioms and then transforming them in ways defined by inference rules; no understanding of the formulae is involved. On the other hand, there was the so-called semantic approach, according to which formulae receive truth conditions (i.e., meanings), and inference is defined by necessary truth preservation, which leaves open how the inference is syntactically, that is, algorithmically, realized. The two approaches indeed diverge, as became clear through the epoch-making incompleteness results of Kurt Gödel (1931): logical or mathematical

provability does not exhaust logical or mathematical truth! This in turn led to investigation of what is computable, i.e., algorithmically accessible, in the first place. Three very different theories—the theories of Turing machines, recursive functions, and λ-definability—were proved to be equivalent, which corroborated the *Church–Turing thesis* that each of these theories indeed captures our intuitive notion of computability (see Kleene, 1967, chapter V; Kripke, 2013). This was the birth of all our modern conceptions of algorithms.

Another, related terminological distinction is the separation of the *symbol level* and the *knowledge level* in the classical work by Newell and Simon (Newell, 1982; Newell & Simon, 1972). On Marr's computational level, we ask, "What is the goal of the computation?" We ask "why" a cognitive process is performed by the cognitive agent. This is similar to Newell's *knowledge level*, which is an abstract level of description for knowledge, intentions, and reasons. On Marr's algorithmic level, the question is: "How are the goals and strategies at the computational level algorithmically realized?" which is similar to Newell's *symbol level*. On this level, we ask "how" the cognitive system works: which properties the representations have, how the relationship between input and output is mediated, and which transformations are made to get from the former to the latter.

The distinction is also reflected in more recent philosophical writings, where we find the distinction between *reason-based* and *mechanistic* models of a given phenomenon. Bechtel and Abrahamsen (2005) state that process-oriented explanations in the life sciences often consist of models of the mechanisms taken to be responsible for a given phenomenon. This differs from the output-oriented, nomological explanations commonly presented in philosophy. Bechtel and Abrahamsen identify several differences between these approaches. First, whereas in philosophy, the focus on language is quite typical, scientists who develop mechanistic explanation are not limited to linguistic representations. Second, the fact that mechanisms involve assumptions about component parts and operations provides direction to both the discovery and testing of mechanistic explanations (Bechtel & Abrahamsen, 2005, p. 421). Finally, mechanistic approaches are developed for specific cognitive phenomena and are initially not phrased in terms of universally quantified statements. Later generalization then involves the investigation of both the similarities of a new phenomenon with those already studied and the differences between them. In the past decades, Marr's approach and the related suggestions to distinguish between different levels of explanation have received many variations, extensions, and criticisms (Peebles & Cooper, 2015). A good overview is given in the collection of articles in Colombo and Knauff (2020), which brings together about a dozen authors from philosophy, psychology, biology, AI, robotics, anthropology, and other fields.

We have already mentioned many examples of output- and process-oriented approaches to rationality. In principle, when cognitive psychologists explore rational reasoning, be it of a logical, conditional, causal, probabilistic nature or whatever, they aim at understanding the underlying cognitive representations and processes. The difficulty of this attempt, however, arises from the fact that human reasoning belongs to the area of complex cognition, a particularly challenging subfield of cognitive psychology. One crucial feature of complex cognition is that it is embedded in a multitude of cognitive processes interacting with one another and with other, noncognitive processes. For instance, reasoning involves a large amount of strongly interlinked processes of language processing, perception, working memory, long-term memory, etc. (Knauff & Wolf, 2010). We return to some of these subsystems in the next section on the preconditions of rationality. In the past, a small number of models tried to account for this plurality of processes, but this was often at the expense of the theoretical rigor of these approaches (Dörner, 1999). Most approaches nowadays focus instead on subcomponents of reasoning and decision-making processes, with the goal of complementing them stepwise with additional processes. Most cognitive approaches in this handbook belong to this class of theories. Another process-oriented approach is the theory of *computational rationality*, which combines models from AI, cognitive science, and neuroscience to reconceive processes of reasoning and action under uncertainty through the lens of computation (Gershman, Horvitz, & Tenenbaum, 2015). We had a chapter on this approach planned, but, unfortunately, it did not work out.

We have also mentioned many examples of philosophers' attempts to state and justify normative standards for these reasoning processes. Typically, those accounts took an output-oriented perspective: They were about establishing a criterion for logically valid deductive inference, about probabilistic norms for degrees of belief, about establishing maximizing expected utility as the rational way to act, and so on. And they were about all the possible alternatives and amendments to these standards. Such approaches also make clear that the two perspectives apply to practical rationality just as well as to theoretical rationality. Of course, there are reasoning processes on both sides. However, the process-oriented

perspective may be more difficult to implement on the practical side.

We mentioned already in subsection 3.1 of this introductory chapter how the distinction between the process- and the output-oriented perspective relates to the distinction between internally and externally driven dynamics of epistemic states: the internal dynamics is about how our thinking does or should proceed, and it is thus bound to take a process-oriented perspective. By contrast, the external dynamics is about how we do or should respond to external input; in determining this response, we primarily take the output-oriented perspective, although we may and should investigate by which internal dynamics the response is mediated.

We would like to point out, though, that there is a deeper question underlying our considerations, namely, the question of how to conceive of an epistemic state (or a mental state in general) at all. This question is not pursued in the handbook and will only be touched on here.

Philosophers are used to distinguishing two conceptions of epistemic (or mental) states: as something occurrent or as a disposition. For them, a state is *occurrent* if it presently "goes through our mind" or "is before our eyes," that is, if we are presently aware of it. Thus, philosophers tend to conceive occurrent epistemic states as conscious. For instance, when asked for your opinion about glyphosate, or any other question, you form an occurrent thought or belief, which did perhaps not occur to you a moment before, and you are aware of this belief. We treat an epistemic state as *dispositional*, by contrast, if we take it to consist in a behavioral disposition that may manifest in our verbal and other behavior, to be elicited by questions or tests, but without further input that could change the disposition. The disposition lies, so to speak, in the background of the occurrent beliefs, and we are typically not aware of it.

To illustrate this distinction: in subsection 3.1 of this introductory chapter, we distinguished between belief bases and belief sets. A belief *base* can be understood in the *occurrent* sense, as just a few beliefs of which we are currently thinking and which we may then algorithmically evolve. By contrast, a belief *set* is deductively closed, that is, it contains all logical consequences of the belief base. Thus, we can never have it wholly present in our conscious mind. Belief sets can be understood only in the *dispositional* sense. Similarly, if the Bayesian conceives an epistemic state as a probability measure, he can do so only in the dispositional sense. At any given moment, at most tiny parts of such a measure can be grasped.

The distinction between occurrent and dispositional belief has been developed independently of cognitive psychology.[37] It is tempting, however, to identify occurrent and dispositional beliefs with representations in working memory and in long-term memory, respectively. Occurrent beliefs might be those that are momentarily active in working memory, often conceptualized as a "global workspace" (Baars, 2002; Dehaene, Changeux, & Naccache, 2011). Only the representations of beliefs in working memory are consciously accessible and available for cognitive processing. Process-oriented approaches from cognitive psychology are concerned with the format, representation, update, and revision of such temporarily active, occurrent beliefs in working memory. Dispositional beliefs, in contrast, may be seen as contents of long-term memory to which we do not have immediate access. Here, our epistemic states are represented as a kind of accumulation of myriads of occasions in which the relevant topics have been presented to our mind. So, consistency checks, for instance, can be performed only for the occurrent beliefs in working memory but not for all beliefs represented in long-term memory, even though this might be normatively desirable.

Some of these ideas about the relationship between the philosophical concepts of occurrent and dispositional beliefs and the psychological conceptions of memory can also be found in Goldman (1978). Here we may leave open whether this cognitive account perfectly matches the philosophical intentions. In any case, the lesson for us is: When our study of rationality pursues an output-oriented or a process-oriented perspective and explores an internally or an externally driven dynamics, we implicitly presuppose this or that conception of epistemic states, and we must be clear beforehand which conception we apply.

Many of our previous thoughts may suggest that philosophers are only interested in the output-oriented perspective and psychologists only in the process-oriented one. Yet, this is not entirely true. Many *micro-theories* from the early stages of psychological reasoning research took an output-oriented perspective (for an overview, see J. St. B. T. Evans et al., 1993). Even today, most psychological studies about choice are still about the resulting choices and not about the underlying thought processes. Similarly, we have mentioned that the "new paradigm," which is closely associated with Bayesianism, is also just output oriented—and our criticism was precisely that it does not attempt to move onward to the process-oriented level. So, clearly, we also find output-oriented approaches in psychology.

Conversely, philosophers are also interested in the process-oriented perspective, and this is where approaches from philosophy overlap with research in AI, for which the process-oriented perspective on algorithms is even the defining characteristic. This overlap is due to the common ancestry in logic. Indeed, in logic, a precise definition of logical validity (output oriented) was preceded by an account of valid proof procedures (process oriented). Many surprisingly different proof procedures have been invented for logics with all kinds of purposes (see chapter 3.1 by Steinberger, this handbook). For instance, axiomatic calculi are very convenient for metamathematical purposes, while they are hardly manageable for actual proofs. No empirical claim is associated with such algorithms. Their point is, rather, to establish the feasibility of the standards assumed, for example, those of logical validity in the case of logic. This feasibility is not guaranteed at all, as is displayed by the incompleteness results in logic (see Shoenfield, 1967, chapter 6), and recursion theory tells us that the computable functions form only a small subclass of all possible functions (Shoenfield, 1967, chapter 7).

The interest in algorithms goes far beyond deductive logic. AI researchers try to find efficient algorithms, and philosophers belabor the wide field of philosophical logics, where we find epistemic, conditional, counterfactual, and nonmonotonic logics, among others (see chapter 5.1 by van Ditmarsch, chapter 5.2 by Rott, chapter 6.1 by Starr, and chapter 7.1 by Pearl, all in this handbook). There is also probability logic (see chapter 4.4 by Pfeifer, this handbook). And the so-called roll-back analysis of decision trees (Raiffa, 1968) is an algorithm for finding a strategy with maximal expected utility. It works perfectly, but precisely its unrealistic nature has motivated the search for more realistic accounts in the field of bounded rationality (Rubinstein, 1998).

This interest in algorithms is not an empirical one, and it is not backed up by any empirical research on thought processes. Indeed, today few AI researchers share the concerns of cognitive science. They are rather interested in the space of possibilities for specifying processes and finding algorithms that arrive at the intended results and besides have other kinds of desirable properties. Similarly, philosophers often discuss normatively plausible axioms before having a semantic benchmark, and if they possess it, they are interested in its computational feasibility—algorithmic inaccessibility might, but need not, be regarded as a blemish on the benchmark. In any case, we may conclude that the process-oriented and the output-oriented perspective are not exclusively assignable to psychology or, respectively, philosophy.

This brings us to the main issue of this section, the relation of the output- and process-oriented perspectives to the normative–descriptive distinction, which is so central to our enterprise. Certainly, one may suspect strong relationships between the two dichotomies. One obvious connection is that the conclusions of our reasoning processes should satisfy certain normative benchmarks. Such normative evaluations always take place within the output-oriented perspective. Since the normative evaluations lie mainly in the scope of philosophy, this would explain why philosophy is mostly engaged with this perspective. Conversely, the study of our actual reasoning processes can, it seems, only be an empirical, that is, descriptive, matter.

By and large, this is correct. However, we should note that the normative perspective also extends to the process-oriented level. Of course, we normatively evaluate also algorithms in themselves and not only the results they deliver. Computer scientists do this all the time. However, the relevant normative considerations in AI differ from those in philosophy. Typically, the former are about computational complexity classes, that is, about the time and storage space needed by an algorithm to solve a problem. For computer scientists, complexity considerations are based on the behavior of an algorithm in worst-case scenarios (Papadimitriou, 1994). The nightmare for computer scientists are NP-hard problems, which cannot in general be solved efficiently. Although such results from complexity theory take a process-oriented perspective and are obviously relevant for software development, it is questionable whether computational complexity is indeed relevant from a cognitive point of view. One criticism is that it might be better to look at the average performance of algorithms (Sedgewick & Flajolet, 2013). Another criticism might be that people usually do not have to deal with an infinite number of choice alternatives, which typically is the basis of complexity theory. An example is the traveling salesman problem (finding the shortest possible route that visits a given set of cities), which is an NP-hard problem but is manageable when just a small number of cities must be visited. Moreover, people often do not aim for optimality but rather are satisfied with a good solution. Altogether, complexity theory provides potentially interesting process-oriented normative insights for rationality research, but it is questionable how relevant such findings are for descriptive theories of rationality. Some psychologists emphatically deny their relevance (Gigerenzer et al., 2011), while others consider complexity theory highly relevant for the cognitive sciences (Otworowska, Blokpoel, Sweers, Wareham, & van Rooij, 2018).

A different example of a normative process-oriented perspective is given by the attempts in decision theory to consider decision costs. The various ways of reaching a decision require efforts of varying costs, which should enter into the expected utilities of the available options. Yet this idea is apparently threatened by circularity. The costs of a decision procedure depend on the decision problem at hand, and to determine them, one would need to already have completed the decision procedure for the given problem. For this reason, the problem seems quite intractable (Bossaerts & Murawski, 2017; Gottinger, 1982). A radical response to it consists in satisficing: by fixing in advance an appropriate satisfaction level, we avoid the costs of further optimization.

The literature on bounded rationality is quite ambiguous with regard to the normative–descriptive distinction. Treatments from the economic side can also be read as normative recommendations for how to get along with all kinds of restricted resources (Rubinstein, 1998), while psychologists usually explore how people proceed under actual restrictions concerning time, working memory, etc., without evaluating the thinking processes themselves (see also chapter 8.5 by Hertwig & Kozyreva, this handbook). Such evaluations come up only when we ask how to improve our thinking or our rational capacities in general (see chapter 15.2 by Stanovich, Toplak, & West, this handbook).

So, what is the connection between normative and descriptive theories, on the one hand, and output- and process-oriented accounts of rationality, on the other? The lesson from this section is that that the two distinctions are independent and that all four areas generated by them are important and indeed pursued in both psychology and philosophy. The two disciplines have different preferences; this is nothing to complain about. However, it would be particularly helpful if the field were to make progress also on the normative side of process-oriented theories and explain what good thinking ought to be like, not only regarding the output but also regarding the thought processes themselves. Such a development would be helpful not only for education and the development of curricula in schools and universities: it would indeed help all of us if we better understood *how* we should think in order to develop and update rational beliefs and come to good decisions and actions.

7. Preconditions of Rationality

This section is concerned with a topic that also could have stood at the very beginning of this introductory chapter: What are the preconditions of human rationality that distinguish us from other animals? What enables people to think and act (more or less) rationally? Of course, the answer to these questions is: our larger brains! Certainly, the increase in brain size is the most striking feature of human evolution. Probably no other organ in mammalian evolution has evolved as quickly as the human brain: the brain volume of the prehominid *Australopithecus africanus* was about 400 to 500 cm^3, which is about the brain size of today's chimpanzees; the brain size of the first tool-making *Homo habilis* was between 500 and 750 cm^3; and that of *Homo erectus*, who lived about 1.5 to 0.3 million years ago, reached a volume between 880 and 1,100 cm^3. The now living humans have a very wide range of brain volumes. According to Beals et al. (1984), the average worldwide brain volume is 1,350 cm^3, although the existing differences are substantial—the brain volume of a healthy adult can lie between approximately 900 and 2,000 cm^3.

Obviously, brain volume alone is not decisive for cognitive ability. In absolute terms, the sperm whale has the largest brain, and even the brains of horses and elephants are larger than those of humans. In relative terms—when we relate brain size to body mass—the shrew mouse has the largest brain, although it does not have the cognitive abilities of humans. Among humans, males have about 10% larger brains than females—but on average, there are no differences in cognitive abilities between men and women as assessed by traditional intelligence tests. Nowadays, most researchers say that about 10% of cognitive ability can be explained by brain size (Pietschnig, Penke, Wicherts, Zeiler, & Voracek, 2015).

In fact, the connection between brain size and cognitive abilities is much more complex. Some studies show that the correlation is higher when specific areas of the brain are considered. For instance, cognitive abilities are more strongly correlated with the size and thickness of the prefrontal cortical areas (Menary et al., 2013), they are correlated with the connectivity of neurons in the brain (Cole, Yarkoni, Repovš, Anticevic, & Braver, 2012; Genç et al., 2018), and there is also evidence that the brains of people with lower cognitive abilities consume more energy in particular brain areas than the brains of people with higher cognitive abilities (Neubauer & Fink, 2009). However, even hard-core neuroscientists acknowledge that we need much more research to better understand the connection between cognitive abilities and the anatomy, structure, and functioning of the brain (Luders, Narr, Thompson, & Toga, 2009). This may be one reason why, in trying to understand rationality and other cognitive capacities of humans, many cognitive psychologists for a long time refrained from studying the brain.

Another reason for the long-lasting skepticism of cognitive psychologists toward brain research is even more fundamental. It lies in the assumption of *multiple realizability*, which was the core argument for the rise of *functionalism* in cognitive psychology and philosophy of mind. As a rule of thumb, we can say that the more complex the cognitive ability was that the researchers were interested in, the less they believed that these abilities could be understood by studying how the brain works. Since rationality ranges very high on this scale of cognitive complexity, this position was particularly prominent in rationality research. Right around the millennium, this changed, and today many researchers use functional brain imaging and other neuroscientific methods to study the cognitive and cortical foundations of human rationality. In the following sections, we discuss which of these systems appear to be so fundamental that they can actually be regarded as essential preconditions of rationality. We also argue that the development of *Homo sapiens* into a highly social being was an essential precondition for the evolution of rationality.

7.1 Cortical Preconditions of Rationality

The first neuroscientific studies on rationality focused primarily on epistemic rationality. Meanwhile, however, there are also many findings on the neural preconditions of practical rationality. The early experiments with functional brain imaging, which we can assign to the field of epistemic rationality, have primarily investigated logical thinking. The pioneering work was carried out by Vinod Goel (see chapter 1.4 by Goel, this handbook) and our own group (e.g., Knauff, Mulack, Kassubek, Salih, & Greenlee, 2002). It would not be helpful to present all the complex neural activation patterns that have since been reported in many neuroscientific articles. These results can be found in excellent meta-analyses of the several dozen published studies on logical reasoning with conditionals, relations, and quantifiers, which used quite different methods for presenting premises and asking for conclusions (Goel, 2007; Monti, Osherson, Martinez, & Parsons, 2007). Instead, we believe, it is much more useful to bring some structure to the field by describing the *core networks* associated with logical thinking.

We think there are three: the first core network covers large areas of the brain that have often been associated with the processing and production of language. These include areas in the left frontal and prefrontal cortex (PFC), also comprising Broca's area, a region involved in semantic tasks and language production (Goel, Gold, Kapur, & Houle, 1998; Prado et al., 2015). Although this system is highly sensitive to the format of the presentation and the like (hence the activations vary enormously in the studies), the fact that many imaging studies found activations in these brain regions is usually interpreted as indicating that logical thought processes are based on linguistic processes and representations. It is obvious that this is consistent with one of the central theoretical assumptions of rationality research, to which we return later.

The second core network involved in logical reasoning covers areas in the parietal cortex, the inferior temporal cortex, and the occipital cortex, which are involved not only in visual perception but also in visual mental imagery (Fangmeier, Knauff, Ruff, & Sloutsky, 2006; Knauff, 2009; Knauff, Fangmeier, Ruff, & Johnson-Laird, 2003; Knauff et al., 2002; Ruff, Knauff, Fangmeier, & Spreer, 2003). These findings are often interpreted as indicating that in logical thinking, not only linguistic processes are involved but also processes in which people vividly imagine what would be the case if the premises of a given logical problem were true. These mental simulations—or models, or images—can be of very different precision. Some may be almost like representations caused by visual perception, while others are less vivid but still much more concrete than purely linguistic representations. While linguistic representations agree with the intuitions of many rationality researchers, visual representations fit well with the intuitions of laypeople, who often report experiencing their own thinking as "seeing before the inner eye" (Knauff, 2009, 2013). We will also return to this topic in a moment.

The third core network includes areas of the brain that are associated with executive control processes and conflict resolution. These cover areas in the right lateral/dorsolateral PFC, detecting conflicts between the logical structure of a task and its content. These areas are activated, for example, when a conclusion is logically valid but implausible, because it does not agree with our beliefs. Typically, this is associated with content effects at the cognitive level and the so-called *belief bias*. Since these experiments were carried out mainly by Goel and his collaborators, they take up a lot of space in his chapter (chapter 1.4 in this handbook). In addition, these findings are related to other results showing that in conflicts between form and content, the normatively correct answer is available in the brain—even if it cannot subsequently prevail against the intuitively more plausible response. De Neys interprets this as indicating that people have logical intuitions that are deeply rooted in our brains (De Neys, 2012).

A methodological problem with imaging techniques is that they only establish correlations (but no causal relations) between cortical and cognitive processes. There is a

certain irony in the fact that, despite these relatively weak statistical accounts, the results of brain-scanning experiments are so overrated by many people, scientists and laypeople alike (McCabe & Castel, 2008; Munro & Munro, 2014). One way to more directly study the causal connections between cortical and cognitive functions is to examine patients suffering from damages of certain brain regions due to strokes, tumors, accidents, etc. In fact, a variety of studies have yielded interesting results that, fortunately, match well with the results from imaging studies. For instance, a strong argument for the necessity of certain brain regions for a specific cognitive capacity is given when a specific kind of brain damage leads to impairment in tasks of type A but does not in tasks of type B, while another kind of brain damage leads to impairments in B but not in A. Such *double dissociations* have also shown that damage to certain brain regions leads to impairment in linguistic or imagery-based reasoning, while other regions are necessary for abstract or concrete reasoning processes that may or may not agree with our beliefs (Goel, Shuren, Sheesley, & Grafman, 2004). Brief overviews of patient studies on rational reasoning and its impairments can be found in Knauff (2009) and Reverberi, Shallice, D'Agostini, Skrap, and Bonatti (2009).

Patient studies, however, also have problems because they are often based on a small number of participants, sometimes just on a single case. In addition, the brain damage is often not clearly delimited or associated with other damages. Recently, new methods have become available in which specific brain regions can be selectively blocked for a short period of time (by applying a strong magnetic field). This corresponds to a temporary lesion—even if this description is greatly simplified and not really correct. In any case, with this method, it is possible to investigate which cognitive tasks can and cannot be performed by an individual whose information processing in the particular brain region is temporarily disrupted. Such studies have also shown that the application of the disrupting magnetic field to areas in the parietal and visual cortex can affect logical thinking, which indicates that mental imagery indeed plays a causal role in logical thinking (Hamburger et al., 2018; Ragni, Franzmeier, Maier, & Knauff, 2016).

Recently, some Bayesians have criticized the fact that the background knowledge and prior beliefs of participants have not been sufficiently taken into account in previous experiments. Moreover, they argued that the instructions were always biased toward the standards of classical logic and did not allow the participants to sufficiently consider their various beliefs when solving a problem (Oaksford, 2015). However, this camp of rationality research has not yet carried out its own brain-imaging experiments to support these arguments. To take up their criticism, Gazzo Castañeda, Sklarek, Dal Mas, and Knauff (2021) conducted a combined cognitive and brain-scanning experiment in which conditionals with high and low conditional probabilities as well as abstract conditionals were embedded in valid and invalid inferences. Some participants had to evaluate the probability of the conclusion, while others received deductive instructions. Both groups of participants could freely choose between a dichotomous and a graded response. During brain scanning, the participants having received deductive instructions showed elevated cortical activity in regions typically associated with conflict detection and inhibition of prior knowledge. In contrast, when given probabilistic instructions, additional activity was found in areas correlated with the influence of prior knowledge. Moreover, the results revealed that many participants, even those having received probabilistic instructions, preferred dichotomous responses. Content only affected inferences under probabilistic, but not deductive, instructions. These findings suggest that people's consideration of prior knowledge and their preference for graded responses are not universal. In fact, people can flexibly activate or suppress background knowledge when they reason probabilistically or deductively. This calls into question how meaningful and empirically justified the disputes between the different camps in the psychology of epistemic rationality are.

There are also many neuroscientific studies on practical rationality. It has been known for decades that damage to the brain's frontal lobe impairs people's ability to think and make choices and decisions (Stuss & Levine, 2002), and these findings were supported by functional brain-imaging studies of the intact brain (Koechlin, Ody, & Kouneiher, 2003). Many of these studies were concerned with the neural correlates of dealing with moral dilemmas, the prisoners' dilemma, and several other paradigms and tasks from social psychology and behavioral economics (Sanfey, Loewenstein, McClure, & Cohen, 2006; Sanfey, Rilling, Aronson, Nystrom, & Cohen, 2003). It is not surprising that the pattern of neural activation depends strongly on the given paradigms and tasks. Here, again, we just want to give a summary of the cortical networks whose functioning seems to be a precondition of decision making and practical rationality (although neuroscientists probably wouldn't use the latter term).

Neuroscientific studies have usually distinguished between two components of decision-making processes: the first core network is related to cognitive control

functions, including task switching, response inhibition, error detection, response conflict, and working memory (Miyake et al., 2000). Several studies have found correlated activity in the dorsolateral PFC and the anterior cingulate cortex, as well as other parts of the PFC (Botvinick, Nystrom, Fissell, Carter, & Cohen, 1999; MacDonald, Cohen, Stenger, & Carter, 2000; Stuss, 2011). The second core network is connected to assigning values to choices. These functions have been mainly associated with ventral and medial parts of the PFC (Gläscher et al., 2012; Grabenhorst & Rolls, 2011).

What is interesting about these studies are not the exact cortical localizations but rather the finding that the cortical networks for "cognitive control" and "value-based processes" seem to be quite distinct, even though they interact in many ways. This is interesting, because it has implications for our understanding of practical rationality: the involved processes are not a muddle; rather, they can be clearly separated and have differential effects on the outcome of the decision process. This finding has also been supported by brain-imaging studies with patients who had lesions to one or the other network (Gläscher et al., 2012). That these damages in fact resulted in an impairment of either the cognitive or the value-based processes supports the assumption that these networks are not only correlated but both necessary for rational decision making and practical rationality.

7.2 Cognitive Preconditions of Rationality

Having a brain is obviously a necessary precondition for rationality. But which cognitive preconditions must a "system" meet to enable rationality? Certainly, the system must be *intelligent*, at least to a certain degree. However, recent research has demonstrated that the concept of intelligence should not be identified with the concept of rationality, for many reasons. In particular, "intelligence" is an individual-difference concept for distinguishing between cognitive abilities of individuals, whereas "rationality" is typically not used in this way. There is the essential distinction between epistemic and practical rationality but no corresponding distinction in most intelligence tests. And the concept of intelligence frequently served as a door-opener for racism, eugenics, and the justification of social injustice, whereas rationality is not associated with such ideological matters.

In the present context, however, the most striking problem with equating intelligence and rationality arises from the many experimental findings signifying that even very intelligent people are prone to irrationality. The present handbook is full of such findings. There are even studies in which classical reasoning problems were combined with IQ tests and other measures of cognitive ability, and they often found that performance on such tasks correlates relatively weakly with cognitive abilities and often only under very specific task instructions. For example, people with higher cognitive abilities performed better only if there was an advanced warning that biased processing must be avoided (Stanovich & West, 2008). Based on this evidence, Stanovich and coworkers argued that rationality is a very different concept than intelligence. A certain level of intelligence is a necessary but not sufficient precondition of rationality, which also relies on the ability for reflective thought (Stanovich, 2009) to control and monitor one's own thinking and to correct it from faulty beliefs and reasoning biases. Chapter 15.2 by Stanovich, Toplak, and West (this handbook) further explains why rationality is substantially different from intelligence.

Psychologists have extensively studied control and monitoring processes under the label *metacognition* (Koriat, 1993, 2018). This research largely relies on the investigation of metamemory, which has long recognized the difference between the processes responsible for retrieving information from memory and the processes responsible for monitoring that information (Dunlosky & Bjork, 2008; Koriat, 1993). According to Koriat, monitoring refers to the "subjective assessment of one's own cognitive processes and knowledge" (Koriat, Ma'ayan, & Nussinson, 2006, p. 38). This assessment can rely on implicit cues, for example, the ease with which a memory comes to mind (Koriat & Ma'ayan, 2005), or it can be based on explicit cues, for example, on the person's assessment of her own abilities (Prowse Turner & Thompson, 2009). For example, a basic metacognitive control function is confidence judgment (Koriat, 2008): following an initial response, a second response about the correctness of the previous response is generated (Fleming, Dolan, & Frith, 2012). This is often accompanied by a metacognitive experience, called the *feeling of rightness* (FOR), which can signal when additional cognitive work is needed (Thompson & Johnson, 2014; Thompson, Prowse Turner, & Pennycook, 2011; Thompson, Therriault, & Newman, 2016). Another metacognitive experience is *fluency*, the ease with which information is processed (Unkelbach & Greifeneder, 2013).

While most of the past metacognitive research was concerned with the regulation and monitoring of retrieval processes from memory, there has recently been a growing interest in the metacognitive processes that accompany more complex cognitive tasks, such as problem solving and rational reasoning (Ackerman & Thompson, 2017).

For instance, Thompson and coworkers have shown that metacognitive control processes support individuals in solving conditional reasoning problems (Thompson et al., 2011). The difference between classical reasoning experiments and these studies was that participants were asked to generate an initial response, then a FOR rating, and then a final conclusion. Thompson et al. (2011) found that individuals use FOR as a metacognitive control function, which can signal when additional analysis and cognitive effort is needed, which finally leads to better reasoning performance in individuals. More on the essential connection between rationality and metacognition can be found in chapter 8.6 by Thompson, Elqayam, and Ackerman (this handbook).

Psychological research on metacognition mainly proceeds from what we have called the process-oriented perspective. However, the issues may also be tackled from an output-oriented perspective. This has been done by philosophers, although they do not use the label "metacognition" but rather speak about "higher-order attitudes," which are concerned with beliefs about one's own beliefs or also about one's own desires, or with desires concerning one's own desires, etc. Higher-order attitudes are hence concerned with all kinds of self-referential propositional attitudes. The origin of this topic lay in the fact that constructions like "It is necessary that *p*," "It is permitted that *p*," "*a* wants that *p*," "*a* thinks that *p*," etc. can be iterated, just as we can form arbitrarily long conjunctions. For instance, I can say, "I think that I think" or "I believe that I want," etc. So, the question arose what the logic of these constructions is, a question that is extensively treated in modal and intensional logic (Montague, 1974; Zalta, 1988). It became clear, however, that this question is not just a matter of logic, of the linguistic rules governing these constructions, but also a matter of rationality requirements. For instance, it is disputed whether we are always aware of our beliefs, i.e., whether when I believe that *p*, I also *believe* that I believe that *p*. And conversely, the question is whether a second-order belief is rationally infallible, i.e., when I *believe* that I believe that *p*, then I do in fact believe that *p* (Kemmerling, 2017, chapters 13–20). The topic is not restricted to epistemic matters. For instance, Frankfurt (1971) has famously proposed that free will is essentially connected to our ability to form effective second-order desires and volitions.

Higher-order attitudes have also a dynamic dimension. So-called auto-epistemology deals with the connections between one's present first-order beliefs and one's (present) second-order beliefs about one's future (or past) beliefs. For instance, if you believe that tomorrow you will believe that it will rain the day after tomorrow, shouldn't you already *now* believe that it will rain the day after tomorrow? One is inclined to say, "Yes, you should" (Binkley, 1968). The issue has interesting consequences, particularly in probabilistic terms (van Fraassen, 1984). Finally, interpersonal versions of the topic loom large in everyday life: we permanently think about the attitudes and beliefs of our fellows, about what they think about us, and how common knowledge emerges in social or public processes. This is particularly relevant in social discourse and communication (see chapter 10.3 by Meggle, this handbook). It is also related to the *false-belief task*, to which we return in the next section. Although the task is mainly used to study the development of "theory of mind" in children, it has also been used to investigate the development of rationality in children, for instance, when they learn to reason counterfactually (Rafetseder, Schwitalla, & Perner, 2013).

Although the false-belief task is a good example of the overlap between philosophy and psychology in the study of epistemic rationality, there are still many ideas in philosophy that have not yet been taken up by psychologists. For instance, the philosophical work on one's present beliefs and one's beliefs about one's future beliefs would also be very interesting from a cognitive point of view. Overall, research on metacognition and second-order attitudes and beliefs would have many implications, for example, concerning the rationality of nonhuman animals: other animals seem to have little or no reflective metacognitive capacities. So, does rationality require such capacities? This raises not only empirical but also conceptual questions.

Let us turn to the relation between rationality and *memory*, one of the grand concepts and research topics of cognitive psychology. We can hardly doubt that rationality would be impossible without an ability to learn, store, represent, and remember knowledge, beliefs, attitudes, values, and all kinds of mental states. An interesting observation, though, is that theories of rationality are often not connected with accounts of the structure, functions, and constraints of human memory. This applies particularly to normative and output-oriented theories of rationality, notably in philosophy, but not only there. These theories presuppose a highly idealized model of the agents' memory. A good example for such an idealization is the rational requirement of the consistency of one's belief system. Thus, the evaluation of one belief requires the evaluation of others to maintain consistency. While this is a normative demand, the belief systems of actual human beings are often not adjusted in that way. This becomes apparent when we consider

that knowledge and beliefs must be maintained over certain periods of time, must be recalled when required, and must be consciously manipulated to draw inferences and to make decisions. Beliefs can also change over time, and, importantly, they can be forgotten or may sometimes not be retrievable from memory.

All these processes take place in two subsystems of human memory: working memory and long-term memory. *Working memory* is conceived as a memory system that allows us to maintain information for certain relatively short periods. Even though older theories of short-term memory specified particular time frames in seconds, the really critical issue is how much information needs to be maintained and whether new information is coming in while you are still occupied with maintaining the previous information. In theory, information in working memory can be rehearsed for a very long period if no new information is coming in. The limiting factor is thus not time but capacity. An early quantification of the capacity limit was suggested by Miller (1956), who claimed that the information-processing capacity of working memory is around seven elements. Later research, however, showed that this number depends on many other factors; for instance, working memory span is lower for long than for short words (Baddeley, 1986). The "elements" can be words or visuospatial information or any other kind of meaningful verbal or perceptual units. They can come bottom-up, from the perception of external stimuli, or top-down, from thought processes. Forgetting in working memory can have different causes. One cause is that elements in working memory decay over time, unless decay is prevented by rehearsal. Another cause can be that new elements replace older ones, and different elements interfere with each other or might compete during the retrieval process. Since all these can lead to forgetting, elements in working memory can have two fates. If they are not rehearsed, they are irretrievably lost. Or they can, via consolidation, become elements of long-term memory.

Long-term memory, in contrast, is the storage system for all we have learned through our lifetime. The elements in long-term memory might not currently be used but are needed to enable rational thinking and actions. Elements in working memory can become elements of long-term memory when we make meaningful associations to what is already stored in long-term memory. One subsystem of long-term memory—called *semantic memory* by psychologists since Tulving (1972)—is a repository for the long-lasting storage of world knowledge, ideas, concepts, beliefs, attitudes, and everything that we have accumulated throughout our lives that helps us to make sense of the world. For example, the knowledge that glyphosate is a pesticide is stored in semantic memory and also the belief that it may kill bees. All such memory traces are strongly interconnected in a network-like structure, which follows several complex organizational principles (for an overview, see J. R. Anderson, 2000). Another subsystem of long-term memory—which psychologists call *episodic memory*—is related to biographical information and events, which are represented including the temporal and spatial context. Surely, the memory traces of such personal experiences are of great importance for the formation and revision of beliefs, but we cannot discuss this topic here.

The structure and functioning of working and long-term memory are often the limiting factors of human rationality. They can explain the large gap between how people actually reason and how they *should* reason according to certain normative theories. For example, when a person is confronted with the set of premises of a reasoning problem, the amount of information can simply exceed the capacity of working memory. Hence, the individual might forget some of the relevant propositions and thus come to incorrect conclusions. Or the information from an earlier premise might interfere with new information, and so forth. There are many other reasons for human irrationality that are related to the structure and limitations of human memory. People might also use strategies to deal with these limitations, for instance, by using heuristics, which we have already mentioned several times. Mental model theory, too, is inspired by the limitations of working memory, since deviations from the norms of rational reasoning may occur because the number of mental models to be considered exceeds the capacity of working memory (Klauer, 1997; Knauff, Strube, Jola, Rauh, & Schlieder, 2004; Süß, Oberauer, Wittmann, Wilhelm, & Schulze, 2002; Vandierendonck & De Vooght, 1997). Other rationality theories are more strongly related to long-term memory, for instance, by assuming that we solve problems by retrieving domain-specific knowledge that we have acquired in the past (Cheng & Holyoak, 1985; Oaksford & Chater, 2012, 2020).

It is essential to understand that working memory is the window to our beliefs, knowledge, attitudes, opinions, and so on. It brings together the outer and inner world. Only what is momentarily active in working memory can be consciously experienced, reflected upon, and checked for consistency or correspondence to the actual world. It is the "global workspace," corresponding to all the momentarily active, subjectively experienced mental states of a person (Baars, 2002; Dehaene et al., 2011). In section 6, we already argued that occurrent beliefs may be seen as elements in working memory, whereas dispositional beliefs are (sets of) elements stored in long-term memory. This connection between occurrent beliefs and working memory, on the one hand, and dispositional beliefs and long-term memory, on the other, fits well with the fact that we are aware of what is currently in working memory but do not have direct conscious access to long-term memory, just as we are not permanently aware of the countless beliefs that we have accumulated throughout life. We can, of course, become aware of a small subset of beliefs when they are transferred into working memory in the service of ongoing tasks. In principle, such a task can also consist in evaluating and readjusting beliefs in the way normative accounts propose. In working memory, though, only a small number of beliefs can be considered at the same time, which obviously provides strong constraints on consistency checks as suggested in many theories of belief systems and belief revision in philosophy and AI research (Alchourrón et al., 1985; Gärdenfors, 1992). This is an idea that deserves more empirical investigation. In any case, there can be no doubt that having a functioning memory is a central, if not *the* central, precondition that a system must meet to enable rationality.

You may object that *language* is at least equally important for rationality. This is certainly true. Indeed, many philosophers and psychologists share the view that language and thought, and thus rationality, are closely related. Already Plato lets Theaetetus consent to the claim, "Well, then, thought and speech are the same; only the former, which is a silent inner conversation of the soul with itself, has been given the special name of thought" (*Sophist* 263e, quoted from Plato, 1921). Similarly, Aristotle (1984, book III) saw thinking as an ability that only those living beings can have who can "consult with themselves". This is a widespread position in contemporary philosophy and psychology that we cannot spell out in detail here. However, the opposite position is that language, thinking, and rationality might not be as closely linked as it seems. Pinker (1994), for instance, says that the idea that thought and language are the same is a "conventional absurdity" (p. 49). With this position, of course, we are drifting into a huge field of research, where philosophers and psychologists have suggested many different theories of human thought that are less closely linked to language and more to mental models, spatial representations and processes, mental simulations, bodily experiences, and so forth (de Vega, Glenberg, & Graesser, 2008).

Why is the role of language for human rational think-ing so controversial? One reason is that many people, unnecessarily, associate the topic with their opinions about the influence of culture, cultural relativity, and the universality of cognitive principles. This leads directly to opinions about the relation between nature and nurture that have always been more ideological than scientific. The other reason, although perhaps not independent of the first point, is that empirical evidence is equivocal. Indeed, the famous Sapir–Whorf hypothesis, that the structure of a language determines its speakers' way of thinking and conceptualizing the world, has led to an enormous amount of empirical research. It will not come as a surprise that many linguists interpret the results as support for linguistic relativism, whereas cognitive psychologists tend to think more in terms of universal principles that all people have in common, independent from the language they may use. The universalistic posi-tion can be found in Li and Gleitman (2002), and a rela-tivistic reply can be found in Levinson, Kita, Haun, and Rasch (2002).

So, is language a precondition of rationality? Cur-rently, the best answer is: yes and no. In the previous section, we reported findings from cognitive neurosci-ence indicating that one cortical system involved in rational thinking comprises several language-related brain areas. And there are tons of findings from cogni-tive psychology that point in the same direction. On the other hand, this does not mean that language is nec-essary for all kinds of rational thinking. Can't children think rationally before they learn to speak? Don't non-human animals also exhibit forms of rationality? Maybe the equation "thinking = speaking" is too strong, but it also would be mistaken to deny that language plays an important role in human rationality.

Visual imagination is another capacity of the human mind that many cognitive scientists consider an impor-tant precondition of human rationality. And if we accept this, it would again challenge the position that thinking is determined by language. Sometimes, as we just said, people indeed experience their thinking as inner speech, but they also very often report seeing "mental pictures" during thinking, "seeing before their inner eye," for instance, what would be the case if the premises of a reasoning problem were true. They often say that they can scan such visual mental images to find new infor-mation that is not explicitly given. This idea is as old as the idea of thinking in language. Aristotle, for instance, used the term *phantasma* for ideas closely akin to percep-tual experiences. Today, most psychologists accept that visual mental images are a kind of mental representation

that closely resembles perceptual experience. Empirical evidence shows that people can scan visual images to obtain information that has never been intentionally stored, and it shows that the experience of visual mental imagery is accompanied by activity in the primary and secondary visual cortex, which supports the idea of a strong overlap between visual imagination and visual perception (Kosslyn, 1994, 2005).

Today, most cognitive scientists accept that visual mental images are a special kind of mental representa-tion in human memory. However, as with language, there is still disagreement about the actual importance of visual imagery for human rationality. On the normative side, skepticism prevails. Among philosophers and logi-cians, the idea is prominent that rational thoughts must be based on some kind of formal language. Although it is generally accepted that pictures and diagrams are suit-able for explaining logical relationships, they are rarely accepted as an independent method of proof. This way of thinking has a very long tradition, and there are rela-tively few exceptions to this view (see chapter 13.1 by Jamnik, this handbook).

On the descriptive side, the evidence is inconclusive. Most cognitive experiments were based on the following hypothesis: if visual images are important for rational thinking, then tasks with an easily imaginable content should be easier to solve than tasks with a content that is difficult to imagine. One can already see that this question is primarily concerned with content effects, which, by definition, should not really play a role in logic. However, the empirical findings suggest other-wise. Some experiments have shown that logical think-ing is indeed easier with easily imaginable materials. Other experiments did not find any connection at all, or the results even pointed to the opposite direction. These contradictory findings have only been resolved in recent years, when it was shown that the different results were caused by the fact that very different types of imaginal thinking were investigated (Knauff, 2013; Knauff et al., 2003; Knauff & Johnson-Laird, 2002). Summarizing the more recent results, it appears that the assumptions on the normative side are not entirely wrong. Indeed, some studies could show that visual mental images can even be a nuisance to logical thinking. This may, however, not be true for all kinds of rational thinking. But in gen-eral, the results allow two conclusions. On the one hand, mental representations, in which people mentally simu-late what is or could be the case, may be an important precondition for rational thinking. This is, for instance, the core assumption of mental model theory. Mental simulations help us to think about alternatives to reality, to think counterfactually, to think about the possible consequences of our own actions, and also to understand the mental states of others. On the other hand, this does not say anything specific about the representational format of such mental simulations or models. It is indeed premature to assume that these kinds of representations are equivalent to pictorial mental images (see chapter 13.3 by Knauff, this handbook). Rather, several experimental findings and computational models show that they can be more abstract and only account for information that is relevant for the given task (Knauff, 2013). These findings underscore what cognitive psychologists repeat—with good reasons—like a mantra: we may sometime have conscious access to the outcome of our thoughts, but it is almost impossible to experience the cognitive processes that have led to this outcome. This is why we need both output-oriented and process-oriented research to understand the true nature of human rationality. Introspection cannot be relied upon; it can and does mislead us and makes us believe that our thinking works in the way we experience it.

7.3   Social and Evolutionary Preconditions of Rationality

Our previous thoughts on the preconditions of rationality have been limited to the individual. For instance, visual mental images may be some of the most private things we all have, which may be why they are so interesting and prominent in the psychological literature. In subsection 5.2 of this introductory chapter, however, we have already stressed the social aspects of rationality. We would now like to follow up on these thoughts and argue that not only the biological and the related cognitive development of our brain was important for the evolution of rationality. Equally important was the development of *Homo sapiens* into a social being that cooperates, thinks about the beliefs and actions of other people, and tries to convince them of its own beliefs and plans.

One of the earliest theories in this direction was developed by Leda Cosmides and John Tooby (Barkow, Cosmides, & Tooby, 1992; Cosmides, 1989). In their evolutionary theory of rationality, they argue that human reasoning abilities have developed because in bartering and trading, compliance with contracts must be checked and cheaters must be sanctioned. For instance, to hunt a mammoth, many men must work together. To accomplish this, other men must guard the village. That is why a contract is made: if the hunters share their prey, the others take care of their families in their absence. This is a rule that must be followed by both sides to ensure

the common good of the entire village. Violations of the rule must be punished. According to Cosmides and Tooby, that is why, in the course of evolution, a system for detecting cheaters has developed in our brain that helps us to draw logically valid conclusions. The idea of a "cheater-detection module" goes back to the evolutionary biologist Robert Trivers and his theory of *reciprocal altruism* (Trivers, 1971). In the past decades, Cosmides and Tooby have developed their account further into an adaptationist program, according to which human cognitive architecture was shaped by natural selection to solve exact and nonintuitive information-processing problems. The researchers describe this approach in chapter 10.6 (in this handbook).

Other social theories are based on the close connection between thinking, argumentation, and communication. For instance, logic helps us in developing a sound argumentation; it brings structure to our arguments and makes them comprehensible for the interlocutor. We generally find illogical and incoherent arguments unconvincing. Already the father of logic, Aristotle (1985), in his *Rhetoric*, dealt with how to persuade other people of one's opinion with good arguments and correct conclusions in meetings and discussions (see also chapter 5.6 by Woods, this handbook). Recently, Hugo Mercier and Dan Sperber (2017) have developed an evolutionary theory of human reasoning according to which the human species developed its tremendous reasoning competence because logically sound arguments are the best means to persuade others.

Mercier and Sperber (2017) also provide a socioevolutionary account of why reasoning is so often unreliable. Reason, they argue, is not directed toward solitary use, toward achieving rational beliefs and decisions by our own efforts. Rather, reason helps us to justify our beliefs and actions to others, to persuade them through argumentation, and to evaluate the justifications and arguments that others direct at us. Although we agree that social cooperation may be an important evolutionary foundation of rationality, there are also many objections. The most general reservation applies to all evolutionary theories of cognition: they are often "just-so stories," that is, unverifiable narrative explanations for human behavior (Gould, 1978; R. C. Richardson, 2007).

Mercier and Sperber also try to explain why reason did not evolve in other animals. Their answer is: because no other animal has language. This position is related to theories that place the close interaction between social cognition and language at the center of the evolution of cognitive abilities (Heyes, 2012). In essence, these theories state that tasks such as hunting a mammoth require a high degree of cooperation, which is only possible in combination with language. Of course, other animals, such as groups of lions or societies of ants and bees, also collaborate, in a more instinctive way, but they do not communicate their goals and plans, which is only possible by means of language.

Although we do not have anything like "cognitive fossils," whereas other disciplines can draw on petrified remains of plants or animals from earlier geological ages, many empirical findings emphasize the importance of sociality and cooperation for human theoretical and practical rationality. In a nutshell, these studies show that many mistakes people make, for example, in conditional reasoning, no longer occur if the task is placed in a social context. A famous study was conducted by Cheng and Holyoak (1985), who presented students with a social version of the Wason task. Imagine you are a customs officer at the airport. You should check the following rule: "If there is 'entry' on one side of the form, then 'cholera' is in the list of diseases on the other side. This is to make sure that entering passengers are protected against the disease." When the participants were presented with the corresponding cards for the cases P (entry), $\neg P$ (no entry), Q (cholera on the list), and $\neg Q$ (no cholera on the list), they performed significantly better than in the original Wason task and also better than in tasks embedded in a nonsocial context. Other studies showed that improved performance in the Wason task involving social agreements only occurs when the subjects put themselves in the role of a "rule supervisor" but not when they are only "observers" (Gigerenzer & Hug, 1992). In subsection 5.2 of this introductory chapter, we already reported several studies showing that groups of people are better at solving epistemic reasoning tasks than the individuals that constitute the group.

In the same section, we also reported some results on the social aspects of practical rationality. Game theory, as we already described, is by far the most far-reaching theoretical and empirical framework in this context. A special version of it is *evolutionary game theory* (Weibull, 1997; see also chapter 9.3 by Alexander, this handbook), where successful cooperative action has been explored with different kinds of games, e.g., signaling games (D. Lewis, 1969), the so-called stag hunt (Skyrms, 2004), and the already mentioned prisoners' dilemma (Axelrod, 1997). These studies showed that, depending on the game, successful cooperative action can, but need not, emerge in groups and need not always be evolutionarily stable. Tit for Tat, for instance, can lead to the escalation of conflict. An alternative is the so-called GRIT ("graduated and reciprocated initiatives in tension reduction") strategy, which can lead

to stepwise de-escalation. According to this method, it is rational for the party willing to mediate to approach the opponent by first publicly declaring his willingness to reconcile and then making as many concessions as possible, as long as these do not cause any major damage. If the opposing party reciprocates these steps toward reconciliation, chances are good that the conflict can be resolved in a rational way (Osgood, 1962).

Even Charles Darwin was aware of the importance of cooperation in human evolution. In his groundbreaking book *On the Origin of Species*, he wrote that man's low physical strength and speed and his lack of natural weapons, etc., are more than compensated by man's social cooperation, which has enabled him to help and receive help from others (Darwin, 1859). Apparently, this is miles away from the idea of Thomas Hobbes and his famous dictum "Homo homini lupus"—"Man is a wolf to man." It also far away from many, sometimes intentional, misinterpretations of Darwin's theory as a scientific justification for egoism and selfishness. Today, for many researchers the development of social cooperation has been the main cause for the development of the cognitive abilities of humans, which go so far beyond those of all other species. The cognitive anthropologist Michael Tomasello found that other animals and even great apes do not, or at least to a much lesser extent, have the capacity to understand their conspecifics as intentional beings (Call & Tomasello, 2008). This research is closely related to studies on the *false-belief task* that explores the development of "theory of mind" in human infants. For example, in the unexpected-transfer task (Dennett, 1978; Perner & Wimmer, 1985), a child sees that Mary puts a candy into a red box. Then she goes out of the room and the child sees that another child moves the candy from the red into a blue box. Now the observing child is asked, where will Mary look for the Candy? The correct answer, of course, is that Mary will look in the red box. Such theory-of-mind tasks have received great attention in psychology and philosophy, as they require an awareness in the child that another individual does not necessarily possess the same beliefs or knowledge that they themselves possess. They are also important when we want to understand the evolutionary roots of human rationality. From our view, it is hard to imagine that rationality could have evolved without the ability to infer the mental states of others and oneself. Both go hand in hand, and presumably reflections on the mental states of others even came first, before humans ever started to think about their own mental states. In any case, both seem to be essential preconditions for human rational thinking. More details on the evolutionary foundations of human rationality can be found in chapter 1.3 by Schurz (this handbook).

8. Frontiers of Rationality

In the previous sections, we have provided a systematic overview of the psychological and philosophical frameworks for rationality. Most of these frameworks attempt to characterize rationality as a general concept that can be applied to many domains and fields of human thinking, reasoning, judgment, and decision making. This generality is important, because we do not want to have too many small, highly specific empirical results about, and conceptions of, rationality for every single domain or area of discourse. But, one might object, does this generality sufficiently recognize that there are many fields and applications of accounts of rationality—in everyday life, in public discourse, and in the sciences and humanities— that might require more specific concepts of rationality? In the following, we want to mention at least the more important of these domains and discuss whether they indeed require special concepts of rationality or can be subsumed under the more general theories presented in this handbook. That is the first aim of this final section of this introductory chapter. And then there are so many other topics in philosophy and psychology that overlap with rationality. We could not cover all these neighboring fields, but we should at least briefly explain how these areas are related to the empirical and theoretical conceptions of rationality presented in this handbook. That is the second aim of this section.

8.1   Are There Special Forms of Rationality?

When we ask this question, we first think of counterexamples, for example, logic, which is a general discipline par excellence—the canons of logic apply everywhere. Similarly, game theory is intended to apply to all agents in social situations. It has thus invaded large areas of the social sciences (see chapter 10.4 by Raub, this handbook) and is not restricted to particular domains, for example, to games in the ordinary sense or financial bargaining, even though it has to say special things about particular situations. And so on. While these theories are very general and largely domain independent, it has been argued that there are also domain-specific forms of rationality not to be subsumed under the existing general accounts. We want to briefly address three of these areas: Do we need a particular concept of scientific rationality? Is there a special communicative rationality designed for social interaction? Do we need a particular account of rationality for artificial intelligence?

Let us begin with the sciences (both natural and otherwise), which consider themselves to be the heartland of rationality. Does this mean that there is a special *scientific rationality*? It is clear where the self-confidence of scientists comes from: from the rise of the scientific method through figures like Francis Bacon, Galileo Galilei, and René Descartes and the subsequent Age of Enlightenment and the accompanying successes in the explanation, prediction, and manipulation of natural phenomena. This was the beginning of the evolution of our modern rational practices and our conceptions of epistemic rationality. The main goal of science is to acquire new knowledge and justified beliefs by means of empirical methods and systematic theorizing. This epistemic work lies at the heart of science, and scientists seem to pursue their epistemic work in a particularly rational way. At least, this is the common ideology.

But scientists also have problems of a practical nature. For instance, the choice of research issues is a practical decision. Should I try to develop vaccinations against a pandemic virus or rather implants for cosmetic surgery? Who should benefit from my research, the whole society or a privileged group? These are clearly value-laden decisions, for the budgeting institution as well as for the individual researcher. The legal claim of the freedom of science is a practical decision as well, based on general freedom rights and assumptions about how to best motivate and organize research. Many scientists think that such practical problems are not scientific problems, which could be solved by scientific methods, and hence do not lie in their specific competence. Scientific rationality does not refer to such problems.

But then, the so-called value-freedom of science is often cited as a hallmark of scientific rationality. This attitude was famously promoted by Weber (1917/1973) but has also been criticized as being itself a normative and value-laden position. Habermas, as one of the main critical voices, has emphasized the normative and philosophical presuppositions of empirical research (Habermas, 1968). The *positivism controversy* centered on such issues (Adorno et al., 1969). There is also a lot of contemporary discussion about the postulate of the value-freedom of science, which is represented in chapter 14.2 by Bueter (this handbook).

Thus, it seems that it cannot be denied that scientific practice is suffused with norms and values. Can we still defend the idea of the value-freedom of science as an ingredient of scientific rationality? The only way to do so is to limit the idea to the area of epistemic rationality. Only here can value-freedom be a norm for science, and then it means that no value judgment or normative conclusion can be inferred from scientific methods alone. Doing so would be to commit the naturalistic fallacy, that is, to violate Hume's principle that no *ought* can be inferred from an *is* (see subsection 3.3 of this introductory chapter and chapter 1.1 by Sturm, this handbook). This also implies that scientific knowledge can support normative conclusions only in the presence of normative premises. These premises, however, must be made explicit and transparent to the public. From what we have written in subsection 4.3 of this chapter, it is clear that we want to modify this requirement. There, we concluded that the empirical scientist must also engage in normative theorizing, because she must deal with all the defeasible bridges between descriptive and normative considerations.

So, if science has a special claim on rationality, this can only refer to the epistemic side. Scientists are supposed to be specialists in epistemic rationality. Cognitive psychologists, for instance, have highly sophisticated methodologies of hypothesis testing, mostly of a statistical and an experimental nature. This is what cognitive psychology has in common with the other sciences, although each field has its own challenges and its own responses. Scientific methodologies are sophisticated normative canons of epistemic rationality, and they exceed everyday epistemic rationality by far. This is a competence science has built up over centuries. Of course, this is not to say that scientists always follow these canons. Phenomena like "p-hacking" (performing statistical tests until they yield a significant result) or publication bias (that mostly statistically significant results are published in journals) indicate that science also suffers from certain structural deficits in epistemic rationality.[38]

However, such deficiencies should not distract from the special normative competence of the sciences with regard to epistemic rationality. Still, this is not yet to say that there are special principles of scientific rationality (chapter 14.1 by Andersen & Andersen in this handbook offers some relevant considerations). Ockham's razor (Sober, 2015), classical statistics, and inference to the best explanation (Lipton, 1991) may be claimed to constitute specifically scientific methods. The general trend in philosophy of science, though, is to subsume such scientific methods under general principles of epistemic rationality. For instance, the Popper–Kuhn controversy mentioned above seemed to suggest otherwise. According to Kuhn, scientists follow a special practice, which Lakatos (1978) and others have attempted to rationalize. This rationalization seemed to be tailored for the sciences. However, it later led to investigations into the logic of theory change in general, and this in turn resulted in belief revision theory as displayed in chapter 5.2 by Rott (this handbook)

and related accounts. Thus, what seemed special to science has been subsumed under general epistemic rationality. Another example is provided by the dispute between classical and Bayesian statistics, where classical statistics defends special principles of statistical inference, while Bayesian statistics tries to get along with general principles of Bayesian learning. Thus, it is at least open whether the sciences can or must appeal to special principles of epistemic rationality. This question is also important when we consider the connection between scientific rationality and the public understanding of science, which is discussed by Bromme and Gierth in chapter 14.3 of this handbook.

A different topic is *communicative rationality*. In subsection 7.2 of this introductory chapter, we briefly discussed the extent to which rationality presupposes linguistic faculties. A complementary idea is that social linguistic behavior, that is, communication, embodies a particular form of rationality. A prominent example is the approach developed in the *opus magnum* of Habermas (1981), who has a democratic notion of practical rationality that transcends that of instrumental rationality. For him, rationality is rather a form of public justification. Thus, communicative rationality is manifested in a discourse free of domination, in which all arguments are considered without prejudice. Rationality and validity claims in general have to prove themselves in such a discourse. Ultimately, communicative rationality is guided by a very Kantian idea: namely, that in rational discourse, we should treat our fellow humans not as means to our ends but as ends in themselves. If so, communicative rationality would be entailed by one version of Kant's categorical imperative.

Not surprisingly, Habermas enthusiastically received the program of *inferentialist semantics* consistently expounded by Robert Brandom (1994, 2000). There, a linguistic community is represented as a community of reason-givers and reason-takers, and it is explained how this exchange of reasons and inferential relations is constitutive of linguistic meanings. Yet, Brandom's notion of a reason is based on the ordinary notion of an epistemic or practical reason, as it is widely discussed in this handbook (e.g., in chapter 2.1 by Broome and chapter 5.3 by Kern-Isberner, Skovgaard-Olsen, & Spohn, both in this handbook). So Brandom does not support the assumption of a particular communicative rationality. In fact, he only insists that linguistic practice is an essentially rational practice.

A different account of communicative rationality is provided by H. P. Grice's (1975, 1989) theory of *conversational implicatures*. This theory conceives of communication as a rational activity governed by the so-called cooperative principle. Based on this principle, Grice shows why communication works the way it does. However, nothing more than the rationality of cooperation is assumed in his approach, and again no deeper theory of rationality is involved.

Indeed, in analytic philosophy, there is a general tendency to deny a special form of communicative rationality. This tendency is clearly expressed in chapter 10.3 by Meggle (this handbook). It mainly attempts to develop the *intentionalistic account* of meaning of Grice (1957), which, in contrast to Habermas, does not assume richer forms of rationality transcending instrumental rationality. David Lewis (1969) then builds on Grice's approach and amends it by his account of conventions in general and meaning-constitutive linguistic conventions in particular. This account appeals to game-theoretic rationality, which Meggle conceives as a form of instrumental rationality and not as a more embracive kind of practical rationality, as Habermas sees communicative rationality. However, we can interpret the game-theoretic approach also in a different way and argue that according to game theory, individuals treat each other as persons with their own aims and interests and thus respect the Kantian formula. If so, the analysis of conventions and linguistic communication provided by this approach conforms to Habermas's ideal as well.

These are just three examples that illustrate that the forms of rationality that play a role in communication are already well covered in this handbook. They also explain why we did not further pursue the prospect of a special theory of communicative rationality. This also conforms to our general attitude of not dealing with philosophical and psychological issues of concept formation and linguistic meaning in this handbook.

A very different issue is the relation between rationality and *artificial intelligence* (AI). In section 7 of this introductory chapter, we already learned that rationality should not be identified with intelligence. Hence, we will not discuss here whether AI systems are intelligent. But we should discuss whether such systems are rational in the general sense or whether we need a special account of rationality to evaluate AI systems. To do this, it is important to distinguish two different approaches to artificial intelligence: *symbolic* AI, in which knowledge and reasoning procedures are explicitly represented by symbols, constraints, and predicates, and AI based on *machine learning*, which uses statistical methods to identify patterns and regularities in massive data sets to make predictions and decisions. Interestingly, the two approaches have opposite strengths when it comes to epistemic and practical rationality. In symbolic AI systems, all information is explicitly represented in the "knowledge base" and is more or less readable for

humans. Importantly, it is required to keep the knowledge base consistent, and where inconsistencies appear, rational updating and revision strategies are used to reestablish consistency. Consistency is one of the general requirements of epistemic rationality and a research area where AI, philosophy, and cognitive psychology are intimately linked. In fact, many theories of rational belief revision were developed at the intersection of these disciplines (Gärdenfors, 1992) and are already represented in the general concept of epistemic rationality.

The epistemic rationality of symbolic AI systems is not limited to the way knowledge is represented but concerns also the way inferences are performed. Typically, symbolic AI systems use the general principles of classical logic to draw inferences and evaluate conclusions. However, this approach has many difficulties and thus has been complemented by systems of nonmonotonic logic, which mirror different, sometimes weaker, rationality requirements. Still, the goal of symbolic AI is to build systems that resemble or even exceed human epistemic rationality.

The problem, though, is that even if these systems reach a relatively high level of epistemic rationality, they often fail to show satisfactory performance when it comes to decision making. For instance, so-called expert systems, which boomed in the 1980s, were not very good at making decisions in medicine, law, or other domains. To understand the limits of these systems, we did not need a special conception of AI rationality.

The limited practical rationality of symbolic AI systems is probably one reason why, in areas where decisions are concerned, they have been largely ousted by AI systems based on machine learning. These methods are not new; they were already used in the 1980s, but with limited success (Rumelhart, McClelland, & the PDP Research Group, 1986). Today, AI systems based on machine learning perform much better, mainly due to the enormously increased computing power of computers and the huge amounts of data from the Internet and social media. The AI systems are trained with these giant data sets and learn to recognize statistical patterns, which are then applied to new data. The training makes it possible to make decisions, for example, which costumer receives a loan, which stocks are bought or sold, for which patient a certain medical therapy is still worthwhile or not, how new employees are hired, and so on. Many of these systems do not perform as well as Silicon Valley wants us to believe. Still, we can evaluate their outputs with our general understanding of what is a good decision.

The problem, though, is that nobody can understand how such systems reach their decisions. We can only evaluate the output of the system but not the processes that led to its decision. Moreover, such systems have very limited epistemic rationality. They rely on massive data sets but not on knowledge in the sense of true and justified beliefs. They cannot explain how they acquire a belief, which we usually see as a requirement for rationality in humans. An alternative might be *explainable artificial intelligence*, which combines machine learning and symbolic AI to make its knowledge and decisions more transparent. Such systems might have the potential to exhibit more theoretical and practical rationality—in the general sense of these concepts.

8.2 Overlaps with Rationality

There is hardly any situation in which our beliefs and actions are driven, or should be evaluated, solely by the norms of rationality. We are of course aware that rationality issues strongly overlap with many other concepts and facets of human mental life that we have not yet covered in this introductory chapter. So, let us at least briefly address some relations to other topics in the vicinity of rationality.

One such topic is the relation between rationality and *emotions*. It surfaces a little in some chapters (chapter 8.1 by Grüne-Yanoff; chapter 8.6 by Thompson, Elqayam, & Ackerman; and chapter 9.4 by Dhami & al-Nowaihi), but it is nowhere systematically treated in this handbook. This may be surprising for psychologists, as the general relationship between emotion and cognition is an issue with enormous theoretical and practical relevance (Damasio, 1994). The situation looks quite different, however, if we look at the more specific relationship between rationality and emotions. On the one hand, neuropsychological studies with patients show that people who are incapable of experiencing emotions but have retained their cognitive abilities can be seriously impaired in certain decision-making tasks (Dimitrov, Phipps, Zahn, & Grafman, 1999). On the other hand, psychological laypeople and experts tend to assume that a "cool head" thinks better than a "hot head" (Blanchette, 2013). Only recently has this topic been systematically investigated in the field of logical thinking. Several questions have played a role here: one is whether the current affective state of a person influences her ability to think logically, independently of the content of the reasoning task. Another question is whether the emotional content of a reasoning problem itself influences logical reasoning performance. And, of course, we can ask how the emotional content of the problem and the person's affective state interact with each other.

The findings are still equivocal. Some empirical studies show that people who witnessed the terror attack in London on July 7, 2005, at close range and who thus had strong emotions performed better than other people on logical reasoning problems with contents related to terror (Blanchette, Richards, Melnyk, & Lavda, 2007). But laboratory studies show that both positive and negative emotional states of the reasoner can have a devastating effect on reasoning performance (Jung, Wranke, Hamburger, & Knauff, 2014). And interactions between the emotional value of a content and the emotional state of the reasoner have seldom been reported (Blanchette, Caparos, & Trémolière, 2018). We cannot discuss all these interesting findings, although it is likely that inconsistent results might also have to do with the fact that for ethical reasons, strong emotions cannot be induced in laboratory research. Moreover, the many different kinds of long-term and short-term affects, sentiments, feelings, emotions, or moods may interact with rational reasoning in many different ways. They can, for instance, put additional load on the cognitive system, but can also be adaptive and help to avoid hazardous situations (De Jong, Mayer, & Van Den Hout, 1997). Already in the 1960s, Schachter and Singer (1962) argued that the experience of emotion consists of two components: physical arousal and its cognitive interpretation. In other words, when an emotion is felt, the person is in a particular physiological state and must figure out what caused this arousal. Since our bodily states are highly ambiguous, emotion itself requires a cognitive act of interpretation and thus of reasoning (Chater, 2018).

This leads directly to the main question for philosophers: can emotions, too, be normatively assessed with regard to their rationality? In subsection 2.2 of this introductory chapter, we put this question aside and focused on actions and beliefs as the sole objects of rationality assessments. However, one might also have the view that emotions do not only have causes, do not just befall us, but also have reasons and are amenable to reasons. We all are familiar with the fact that one is not automatically exculpated by saying, "Sorry, but this is how I feel." People can have more or less "adequate" or, conversely, exaggerated or even pathogenic feelings and emotional responses. They can have persistent emotions whose cause has long dissolved. There can be an absence (or suppression) of feelings where one would expect them to show, and so on. However, even if we grant that emotions are responsive to reasons, one may doubt that assessing the adequacy of emotions is already a judgment about their rationality. The topic is richly discussed (see, e.g., de Sousa, 2011; D. Evans & Cruse, 2004; Helm, 2001).

Again, we have preferred not to cover this topic in the handbook.

A connected topic is the relation between rationality and *morality*. Here, emotions are involved in two ways. The immediate connection is pursued by *sentimentalist theories* of morality, which we find, for instance, in the work of Hume (1751/1975a) and Schopenhauer (1841). Clearly, we have particularly moral sentiments: positive ones like love, sympathy, and compassion and negative ones like remorse, shame, blame, and outrage. For sentimentalist theories, these moral feelings constitute the origin of morality and lie at the basis for justifying moral judgments.

However, there is also an indirect connection. Even if one does not appeal to particular moral sentiments, one may locate the basis of all our evaluations in how we feel. We ultimately strive for happiness, and we seek pleasure and try to avoid pain—this is the traditional terminology, which had a broader meaning than it does in current everyday language. In this picture, our ultimate values refer to certain positive or negative emotional or hedonic states, and then morality is about the general realization of these values. Ancient *hedonism* refers to pleasures and pains, and *eudaimonism* to happiness (though *eudaimonia* was not conceived as just a psychological state). We find it in Jeremy Bentham's (1789/1970) *hedonic utilitarianism*. And we find it in *contractarianism*, originating with Hobbes (1651/1994), where the social contract enforcing moral behavior guarantees the individual pursuit of happiness and the avoidance of mutual harm. These moral conceptions were driven by the idea that morality can be explained by the rational pursuit of these ultimate values. This would be a strong justification: morality would just be instrumental to those values.

Kant (1788/1908), of course, is famous for being opposed to all of this. For him, the first principle of morality, the *categorical imperative*, is an a priori principle of pure reason and thus derives from rationality alone, without appealing to pains, pleasures, or other feelings. This obviously presupposes a stronger notion of rationality, transcending Hume's instrumental rationality (see also chapter 1.1 by Sturm, this handbook). Again, these issues are much too large to be treated properly in this handbook, but they are reflected in chapter 12.1 by Fehige and Wessels and chapter 12.2 by Smith (both in this handbook).

While philosophers reflect on the nature of moral sentiments and their relation to rationality, cognitive psychology and neuroscience conduct experiments to understand how moral values interact with other cognitive and noncognitive processes. One active research area in this context is *law and legal reasoning*. Do people use moral values or legal norms when thinking about crimes and court decisions? Empirical studies show that legal experts such as judges and lawyers can follow the rational norms of the legal system, whereas laypeople often base their opinions about crimes and how they should be punished on their intuitive feeling of rightness and morality. In particular, if they judge a crime to be exceptionally immoral and evil, they tend to demand very severe punishments. And they do this even if, from a legal point of view, there are mitigating or even exculpatory conditions, just to satisfy their need to punish the transgressor (Gazzo Castañeda & Knauff, 2016). Legal logic and logical models of legal argumentation are the topics of two quite different chapters in this handbook: chapter 11.3 by Hilgendorf and chapter 11.4 by Prakken.

Another line of experimental research actually goes back to a thought experiment by Karl Engisch, a German jurist and philosopher of law who considered in his study (1930, p. 288) what is nowadays called the "trolley problem": imagine you are standing on a footbridge and see a runaway trolley moving toward five people lying on the main track. You are standing next to a lever that controls a switch. If you pull the lever, the trolley will be redirected onto a side track and the five people on the main track will be saved. However, there is a single person lying on the side track. You have two options: do nothing and allow the trolley to kill the five people on the main track, or pull the lever and cause the trolley to kill the single person on the side track. Is one option morally required?

This thought experiment has been much discussed in philosophy to differentiate between deontological and consequentialist, in particular utilitarian, accounts of morality. Utilitarians would recommend pulling the lever because it results in fewer fatalities, while deontologists would object that human life is an incommensurable good, so that counting fatalities is not necessarily an argument. Instead, they would argue that it is better to *refrain* from pulling the lever, and thus to let a greater number die, than to actively *do* something resulting in the death of (fewer) persons.

Psychologists, by contrast, investigate how people actually (say they would) decide. They also compare this scenario to a slightly different situation: imagine that a fat man is standing next to you, and instead of turning a lever, you could push him on the track, thereby again saving the group of five but killing the fat man. How do people decide now?

Most of the empirical results show that persons decide differently in the two situations (Awad, Dsouza, Shariff, Rahwan, & Bonnefon, 2020). Is that rational? A PhD student of philosophy, Joshua Greene, was the first who used this dilemma in a functional brain-imaging study. Greene, Sommerville, Nystrom, Darley, and Cohen (2001) showed that the "personal" dilemma (pushing the man off the footbridge) activated brain regions associated with emotions, while the "impersonal" dilemma (flipping a lever) engages regions associated with controlled, rational reasoning. These findings (and their interpretations) are problematic for many reasons (Waldmann, Nagel, & Wiegmann, 2012) but are related to many different versions of dual-process accounts of moral decision making, which are still popular in cognitive research (see chapter 2.5 by Klauer, this handbook). Again, the topic of morality is too comprehensive to receive treatment in this handbook, but chapter 12.3 by Wiegmann and Sauer (in this handbook) comments on it from the psychological side.

By the way, it may be unsurprising that moral research today gets much attention from military and engineering fields, for instance, in the development of autonomous weapons or vehicles, for which situations can occur where one or another potentially fatal collision is unavoidable. In such cases, the car's software has to "decide" what to crash into and which casualties to hazard. In 2018, the *moral machine project* began to collect information on such decisions on an Internet platform, where people can choose between two different destructive outcomes (Awad et al., 2018). This looks like a dangerous blurring of the border between empirical decisions and normative theories, which we try so carefully to reflect in this handbook. Since Hume, we know that we better avoid inferences from *is* to *ought*.

We may widen the perspective even further, from the relation between morality and rationality to the general *cultural, social, and political embedding* of rationality. Since Plato, political philosophy has been deeply involved in rationality issues. In the *Republic*, he described how society should be rationally organized: ruled by a leading class that is rational, intelligent, and self-controlled (Plato, 2013). Hobbes, in the *Leviathan* (1651/1994), argued that citizens are only protected from mutual violence, as it occurs in the state of nature, if power is delegated to a reasonable sovereign. Later, Jean-Jacques Rousseau, in The Social Contract (1762), challenged this view by arguing that all members of a society have the same rights and duties. Although these issues are superimposed by more important topics of political philosophy, such as liberty, equality, and justice, they also have much to do with rationality, yet could not be covered in this handbook. The lack becomes particularly apparent when we consider John Rawls's (1971) theory of *justice as fairness*, in which the so-called veil of ignorance creates a situation in which rational decision making arguably

justifies his principles of justice. But then those principles are at issue, not the theory of rational decision making. Hence, we preferred to abstain from extending the handbook in this direction. It is reflected, to some extent, in chapter 10.5 by Nida-Rümelin, Gutwald, and Zuber (this handbook).

We have also abstained from engaging in intercultural considerations. A review of the sparse literature that is available in languages we understand suggests that there is a wide range of alternatives to the conception of rationality in the European and Northern American intellectual tradition. For instance, a core distinction in this handbook is that between normative and descriptive views on rationality. The distinction is so deeply entrenched in our scientific thinking that we can hardly imagine doing without it. But is this necessary? Perhaps not. For instance, the distinction does not seem to be clearly present in traditional Chinese thought. Instead of this dichotomy, we find a general tendency in traditional China to think of things as interactions of *yin* and *yang* and the more fluid interactions of the "five phases," water, wood, fire, metal, and earth (Marchal & Wenzel, 2017). Comparative ethnologists have found various other examples of fundamentally different ways of thinking: people in Melanesia and Polynesia use numerical systems that significantly differ from the norms of Western mathematics (Beller & Bender, 2008). Islam provides alternatives to the economic rationality of rational choice theory (Tafer, Boussahmine, & Bouanini, 2016), and studies looking at the brains of people playing a fairness game found very different neural activities in Buddhist meditators and Western participants (Kirk, Downar, & Montague, 2011). Such observations challenge our arrogance that rationality everywhere can be judged by a single set of criteria. It is not unlikely that we find diverging conceptions of rationality in different cultures and different socioeconomic systems (Lloyd, 2017). Relatedly, Henrich, Heine, and Norenzayan (2010) have argued that psychologists and other behavioral scientists often make strong claims based on experiments with samples drawn entirely from *Western, Educated, Industrialized, Rich, and Democratic* (WEIRD) societies. However, we confess that the present handbook is committed to the Western intellectual tradition in psychology and philosophy.

This is the right place to finally explain why so many areas of so-called continental philosophy are not represented in this handbook. We do not want to suggest that they have nothing substantial to say about rationality. In fact, reflections on reasons and rationality are ubiquitous also in these parts of philosophy. For instance, consistency is central to most concepts of rationality, but in *dialectic thinking*, contradictory hypotheses lead to a rational synthesis. Dialectic thinking is important in the work of Georg Friedrich Wilhelm Hegel but not represented in this handbook. Friedrich Nietzsche is famous for his *perspectivism* (see, e.g., R. L. Anderson, 1998; Danto, 1965, chapter 3). His criticism was that scholars are often not aware of their own perspective and thus do not control its influence on their work. Maybe his demand for perspectival deconstruction should be part of epistemic rationality, but it is neither treated in this handbook nor exercised by its authors. Or think of the profound criticism of Western *technical rationality* in Martin Heidegger (1954). It may be misguided, but it springs from deep philosophical sources and is still disconcerting. Michel Foucault has tremendously shaped our intellectual landscape. The power to delimit reason is key to his early analysis of madness (Foucault, 1972), and his later analysis of government shows the historical conditions of the possibility of different types of governmental rationalities (Foucault, 2008). And when Habermas is discussed under the heading of *communicative rationality* in chapter 10.3 by Meggle (this handbook), this is restricted to the narrow perspective of philosophy of language and neglects Habermas's much more general political intentions concerning the theory of democracy.

So why is this handbook so obviously biased against continental philosophy and toward analytic philosophy, and within the latter toward the more formal ways of philosophizing? Before we explain this bias, we should mention that already the alleged opposition between "analytic" and "continental" is infelicitous, as "analytic" vaguely designates certain ways of philosophizing, which, however, still diverge dramatically, and "continental" refers to geographical locations and origins, which again have produced quite different strands and styles of philosophizing.

But even if we follow this traditional terminology, there are several reasons why this handbook is biased toward analytic philosophy. One reason is that the handbook brings together cognitive psychology and philosophy, and that analytic philosophy lies much closer to the intersection of these disciplines than other parts of philosophy. This is why analytic philosophers and cognitive psychologists collaborate so successfully under the roof of cognitive science. For the same reason, we also have a certain bias toward the more formal side of analytic philosophy, which provides more specific and detailed accounts of the norms of rationality and has more natural connections to cognitive theories of the human mind. Another, even more fundamental reason for our bias is that continental philosophy often focuses on the wider cultural and political context of rationality, as our examples above indicate. However, as just said, this handbook is silent on this wider context. Although such perspectives on rationality are interesting and important, they do not fit well with the predominantly individualistic orientation of psychological research and are difficult to investigate with the methods of experimental psychology. Analytic philosophers have similar reservations. For them, such political and cultural associations are often sweeping and hard to seize—none of these accounts could be subject to sophisticated normative discourse, as we find it in analytic philosophy. So, we had to draw a line here, and therefore the contributions of continental philosophy are not represented in this handbook.

9. Conclusion and Open Issues

The aim of this chapter was to offer the reader a guide through the complex field of rationality research. We hope that we could make clear why we consider the distinction between theoretical and practical rationality so important. We also emphasized why the relation between normative and descriptive theories of rationality is so essential for our philosophico-psychological enterprise and why research on social rationality should go beyond competitive scenarios in which the actors consider only their individual benefits. We moreover explained why it is important that rationality research should focus not only on the rational assessment of the *results* of a thought process but also on the cognitive processes that led from the input to the output. Furthermore, we introduced the concept of a *double equilibrium*, which requires empirical research to listen to normative theorizing and, conversely, reminds us that normative considerations are always defeasible and must withstand critical, empirical scrutiny. In the final sections, we have discussed some cognitive prerequisites of rationality and argued against the view that many specialized conceptions of rationality are needed for the many areas of everyday life, public discourse, and the sciences and humanities. Our position was that rationality assessments in most of these areas can rely on general concepts of rationality.

Of course, there are many open questions and tasks that rationality research still has to tackle. One of these tasks is to overcome the fixation on logic and probability theory, which is particularly prevalent in the psychology of reasoning. We have shown that philosophy has for decades offered a richer set of options. Psychology has just started to pay more attention to such accounts, and we hope that it will continue to go in that direction.

While this is an issue of epistemic rationality, we see a related challenge in practical rationality. Here, we think that it is important to further develop alternatives to standard decision theory. Again, we believe that psychology and philosophy should go beyond these limitations of present research. There are developments in this direction, but currently it seems that philosophy and psychology are pursuing different lines. More integration would be highly desirable.

A related observation is that most of the current work under the heading of social rationality is in fact about individual rationality in social situations. We think this is a weakness of current research that should be overcome. In so many situations of everyday life, people seek to think and act rationally in concert with others, but we do not know much about the underlying cognitive processes. This is a research field where social psychologists, cognitive psychologists, social scientists, and philosophers should collaborate more intensively. We also do not know enough about the normative standards that we should use to assess collective forms of rationality, which is another area where we think that rationality research must make progress. This is important for many fundamental questions of rationality, but it is obvious that this topic is also socially and politically highly relevant.

Another characteristic of current research is that in the various subfields, the process-oriented perspective is pursued sometimes more and sometimes less intensely. Surely this is partly due to the varying difficulties of doing so. Some theories from philosophy may be really hard to bring into process-oriented terms. Still, we think that this is desirable and would strengthen rationality research where it is yet weak.

We also think that philosophers and psychologists should try to reach more consensus on the nature of concepts and meaning. Currently, there seems to be much dissent on this topic between the two disciplines, which affects many areas of cognitive science. For the same reason, we did not discuss the extremely complex question of what we mean when we talk about truth. Truth may be the ultimate goal of rationality, but there are so many different conceptions of truth in philosophy and psychology that it is hard to see how a more unified view could be reached in the near future. Fortunately, this is not primarily a disagreement about rationality. However, in the long run, more collaboration on the topics of meaning and truth between the different disciplines would certainly also be beneficial for our understanding of rationality.

All these issues for future research need much more interdisciplinary interaction. We think that this handbook can show that by such interdisciplinary collaboration, research into rationality has already made great progress. Today we have much more knowledge about human rationality than each of our disciplines could have achieved on their own. However, there is still room for further progress. For instance, it is essential that all parties in rationality research be clear about the complex relation between descriptive and normative considerations. Unclarity about this relation still seems to hamper research. And we think that other disciplines should be more involved in our interdisciplinary endeavor. Linguists, for instance, have substantial things to say about rationality in language, pedagogues and education researchers should systematically study how students can be trained in rational thinking, and anthropologists and ethnologists can help us better understand how rationality is rooted in different cultures and human lifestyles.

We began this introductory chapter with the Aristotelian definition of humans as rational animals, and we mentioned a few objections to this view. So, are we really as rational and smart as we think? Or is this more a self-delusion that persists even if empirical findings put our optimism into question? The intention of this chapter was not to answer this question. That would be presumptuous. What we wanted to do, rather, was to help our readers to develop their own opinion on this question. This opinion, we hope, might benefit from the interdisciplinary and multiperspectival character of this handbook. In the end, we may agree with how Christof Koch (2016, p. 25) summarized the lessons from Darwin's theory of evolution: we are unique, but so is every other species, each in its own way.

Notes

1. There are many labels for this and similar distinctions: "is–ought," "ontic–deontic," "descriptive–prescriptive," "empirical–normative." We have decided to use the pair "descriptive–normative" throughout and thus follow the usage of Elqayam and Evans (2011).

2. The Priority Programme "New Frameworks of Rationality" was an important step toward improved collaboration between psychologists, philosophers, and researchers in other disciplines in the field of rationality (see the Preface in this handbook).

3. For all this, see the entries "Ratio," "Rationalität," and "Vernunft, Verstand" in Ritter, Gründer, and Gabriel (1971–2007).

4. Section 2 of chapter 2.1 by Broome (this handbook) also contains detailed remarks about the etymology of the English word "reason."

5. The other two are his famous context principle and his rule to strictly distinguish object and concept.

6. Knauff and Gazzo Castañeda (2021) criticize this terminology and show that the new trend toward subjective degrees of belief cannot be regarded as a "paradigm shift" in the Kuhnian sense.

7. Note, though, that the label "Bayesianism" can take on quite different meanings in psychology (Jones & Love, 2011).

8. A philosopher would say that we should talk of "doxastic" attitudes. Doxastic attitudes are belief attitudes in all shades and grades: opinions, doubts, subjective probabilities, and so on, while epistemic attitudes, strictly speaking, concern only knowledge. Still, "epistemic attitude" is the more familiar term. We shall therefore use it always in the wide sense of "doxastic attitude." Of course, the philosopher would add that this simply conforms to the sloppy talk about knowledge found throughout the sciences, which hardly distinguishes knowledge from true or certain belief. And he would refer to the huge philosophical discussion of knowledge in the past 50 years, which is largely neglected by those sciences (see, e.g., Bernecker & Pritchard, 2011, particularly parts II–IV and VII; or Williamson, 2000), as the main advocate of the opinion that the study of knowledge even precedes the study of belief). Let this remark suffice to indicate that the concept of knowledge is a philosophical snakepit from which we keep a healthy distance despite our talk of epistemic attitudes.

9. After the heyday of so-called logical behaviorists like Ryle, who subsumed action explanations under the special kind of (noncausal) dispositional explanation, the view that actions are caused by intentions in an ordinary way was initiated by Hempel (1961–1962) and Davidson (1963) and is still the dominant view in philosophy (see, e.g., Bratman, 2006).

10. The following considerations are more extensively explained in Spohn (2002).

11. Many philosophers say that beliefs are rational relative to the *facts*, or that only facts can be reasons for beliefs. But then these facts must be perceived and hence believed, and so this relation is again mediated by beliefs.

12. This may be taken as a modern version of Berkeley's dictum that only ideas can be similar to ideas (cf. Berkeley, 1710/1949, section 8).

13. The situation resembles the Agrippa trilemma, often taken to be the basic problem of epistemology (see, e.g., BonJour, 1985), according to which a belief is justified only if it is justified by reasons that are in turn justified. The trilemma has a skeptical force: how, then, could a belief ever be justified? Many versions of foundationalism and coherentism, justification internalism, reliabilism, and contextualism respond to this challenge. We need not discuss this here. Certainly, though, the situation vis-à-vis reasonableness is similar.

14. Spohn (2002) expands on these workings of rationality assessments.

15. Elqayam and Evans (2011, p. 238) similarly speak of the "product" and the "process level."

16. In philosophy, the debate about the nature of those contents, as well as about the nature of the concepts of which they consist, is endless, most sophisticated, and without conclusion (see, e.g., Fodor, 1998; García-Carpintero & Macià, 2006).

17. In fact, the formal structure of probabilities was not firmly fixed from the beginning; see Shafer (1978).

18. A plea for *epistemic dualism* is found in Spohn (2012, chapter 10); one of several attempts for unification is found in Leitgeb (2017). For an attempt at connecting the issue with psychological research, see also Weisberg (2020).

19. The conception of intentional objects goes back to Brentano (1874). The book that alerted philosophers most to the problems surrounding intentional objects, propositions, and the like was Quine (1960). The subsequent discussion is endless, without a clear conclusion. For a recent treatment, see Recanati (2016). See also the plea for intentional objects in Spohn (2009, chapter 16).

20. The basic algorithm is the so-called roll-back analysis as explained in Raiffa (1968).

21. Often, though, the issue is not individual rationality but rather policy: how to design institutions when these are assumed to have an influence on people's preferences?

22. There is no textbook summarizing all these alternative accounts. One may consult the comprehensive handbook by Glanzberg (2018). The richest, though critical, discussion of antirealistic truth theories in analytic philosophy is found in Devitt (1991).

23. Such thoughts are also vigorously discussed in philosophy, based on the brain-in-a-vat scenario of Putnam (1981, chapter 1).

24. The term was introduced by C. I. Lewis (1912).

25. Grice (1975) tried to save this representation with the help of his theory of implicatures; see also Bennett (2003, chapter 2).

26. It was a central problem for logical empiricism to explain how we can understand not directly observable dispositions on the basis of observational language and extensional logic. (Philosophers have a more general understanding of dispositions than psychologists; see also section 6 of this introductory chapter.) This problem turned out to be a Kuhnian anomaly for logical empiricism and eventually led to its breakdown. The problem was precisely how to understand conditionals like "If that piece of sugar were put into water, it would dissolve" (= "That piece of sugar is soluble") in terms of classical logic and material implication. Compare, for example,

Hempel's paradigmatic movement from Hempel (1958) to Hempel (1973).

27. So, this was at the same time when the Wason selection task was designed. A different attempt to cope with these paradoxes is relevance logic, developed by A. R. Anderson and Belnap (1975). It is, however, not represented in this handbook.

28. Cohen (1981) gives a similar description, although he distinguishes between a narrow and a wide reflective equilibrium.

29. It is not only a methodological issue. Philosophers tend to carry it deeply into metaphysics and semantics, usually within metaethics and with respect to morality and moral predicates and properties. However, with respect to normativity, morality and rationality are in the same boat. Thus, these philosophical discussions, as pursued, for example, in Ridge (2019), carry over to rationality.

30. Hume (1975c, p. 469) famously said, referring to the transition from *is* to *ought*, that it "seems altogether inconceivable, how this new relation can be a deduction from others, which are entirely different from it." The label "naturalistic fallacy" was introduced by Moore (1903, §10) for the related fallacy of inferring the goodness of something solely from its descriptive properties.

31. Paradigmatically by Hempel (1961–1962).

32. To our knowledge, this kind of double reflective equilibrium was first described in Spohn (1993, p. 188). See also Spohn (2002, section 4).

33. As is more fully argued in Spohn (2011).

34. Rich textbooks are, for example, Myerson (1991) and Maschler, Solan, and Zamir (2013). See also section 9.1 by Albert and Kliemt (this handbook).

35. The distinction between social rationality and individual rationality we are invoking here is more salient in the practical context. That is why only this context is considered here. However, it pertains just as well to the epistemic context: see the distinction between epistemology *in* groups versus epistemology *of* groups in chapter 10.1 by Dietrich and Spiekermann (this handbook).

36. There are, however, many other paradigmatic and highly interesting social situations modeled by game theory. For brief explanations, see chapter 9.1 by Albert and Kliemt (this handbook).

37. It has been important to logical behaviorists like Ryle (1949). For a general discussion of the philosophical intricacies of dispositions, see, for example, Mumford (1998).

38. One could argue that the so-called replication crisis (that many published findings could not be replicated in subsequent studies) is a drastic example of the limits of epistemic rationality in science (Open Science Collaboration, 2015). However, many researchers have questioned the popular interpretation of the relatively low rates of replicability. For example, a "regression to the mean" can explain why even if the initial results were correct, they need not be replicable in subsequent studies (Fiedler & Prager, 2018). The publication from 2015 also showed that replicability varies strongly over different subdisciplines and that replication rates in cognitive psychology are better than in other areas.

References

Ackerman, R., & Thompson, V. A. (2017). Meta-reasoning: Monitoring and control of thinking and reasoning. *Trends in Cognitive Sciences, 21*(8), 607–617.

Adorno, T. W., Dahrendorf, R., Pilot, H., Albert, H., Habermas, J., & Popper, K. R. (1969). *Der Positivismusstreit in der deutschen Soziologie* [The positivism dispute in German sociology]. Neuwied, Germany: Luchterhand.

Alchourrón, C. E., Gärdenfors, P., & Makinson, D. (1985). On the logic of theory change: Partial meet contraction and revision functions. *Journal of Symbolic Logic, 50*(2), 510–530.

Anderson, A. R., & Belnap, N. D., jr. (1975). *Entailment: The logic of relevance and necessity.* Princeton, NJ: Princeton University Press.

Anderson, J. R. (2000). *Cognitive psychology and its implications* (5th ed.). New York, NY: Worth.

Anderson, R. L. (1998). Truth and objectivity in perspectivism. *Synthese, 115*, 1–32.

Aristotle. (1984). *De anima* [On the soul]. In J. Barnes (Ed.), *Complete works of Aristotle: Vol. 1. The revised Oxford translation* (pp. 641–692). Princeton, NJ: Princeton University Press.

Aristotle. (1985). *Ars rhetorica* [Rhetoric]. In J. Barnes (Ed.), *Complete works of Aristotle: Vol. 2. The revised Oxford translation* (pp. 2152–2269). Princeton, NJ: Princeton University Press.

Asch, S. E. (1951). Effects of group pressure upon the modification and distortion of judgments. In L. W. Porter, H. L. Angle, & R. W. Allen (Eds.), *Organizational influence processes* (2nd ed., pp. 295–303). Armonk, NY: Sharpe. (Original work published 1951)

Asch, S. E. (1956). Studies of independence and conformity: I. A minority of one against a unanimous majority. *Psychological Monographs: General and Applied, 70*(9), 1–70.

Awad, E., Dsouza, S., Kim, R., Schulz, J., Henrich, J., Shariff, A., . . . Rahwan, I. (2018). The Moral Machine experiment. *Nature, 563*(7729), 59–64.

Awad, E., Dsouza, S., Shariff, A., Rahwan, I., & Bonnefon, J.-F. (2020). Universals and variations in moral decisions made in 42 countries by 70,000 participants. *Proceedings of the National Academy of Sciences, 117*(5), 2332–2337.

Axelrod, R. (1997). *The complexity of cooperation: Agent-based models of competition and collaboration.* Princeton, NJ: Princeton University Press.

Baars, B. J. (2002). The conscious access hypothesis: Origins and recent evidence. *Trends in Cognitive Sciences, 6*(1), 47–52.

Bacharach, M. (2006). *Beyond individual choice: Teams and frames in game theory.* Princeton, NJ: Princeton University Press.

Baddeley, A. D. (1986). *Working memory.* Oxford, England: Oxford University Press.

Bara, B. G., Bucciarelli, M., & Lombardo, V. (2001). Model theory of deduction: A unified computational approach. *Cognitive Science, 25*(6), 839–901.

Barkow, J. H., Cosmides, L., & Tooby, J. (Eds.). (1992). *The adapted mind: Evolutionary psychology and the generation of culture.* New York, NY: Oxford University Press.

Bartlett, F. (1995). *Remembering: A study in experimental and social psychology.* Cambridge, England: Cambridge University Press. (Original work published 1932)

Bayes, T. (1970). An essay towards solving a problem in the doctrine of chances. In E. S. Pearson & M. G. Kendall (Eds.), *Studies in the history of statistics and probability.* London, England: Griffin. (Original work published 1764)

Bazerman, M. H., & Moore, D. A. (2013). *Judgment in managerial decision making* (8th ed.). Chichester, England: Wiley. (Original work published 2002)

Beals, K. L., Smith, C. L., Dodd, S. M., Angel, J. L., Armstrong, E., Blumenberg, B., . . . Trinkaus, E. (1984). Brain size, cranial morphology, climate, and time machines. *Current Anthropology, 25*(3), 301–330.

Bechtel, W., & Abrahamsen, A. (2005). Explanation: A mechanist alternative. *Studies in History and Philosophy of Biological and Biomedical Sciences, 36,* 421–441.

Beller, S. (1999). Wenn Wissen logisches Denken erleichtert bzw. zu verhindern scheint: Inhaltseffekte in Wasons Wahlaufgabe [When knowledge facilitates or apparently prevents logical reasoning: Content effects in the Wason selection task]. In H. Gruber, W. Mack, & A. Ziegler (Eds.), *Wissen und Denken: Beiträge aus Problemlösepsychologie und Wissenspsychologie* (pp. 35–52). Wiesbaden, Germany: Deutscher Universitätsverlag.

Beller, S., & Bender, A. (2008). The limits of counting: Numerical cognition between evolution and culture. *Science, 319*(5860), 213–215.

Bennett, J. (2003). *A philosophical guide to conditionals.* Oxford, England: Oxford University Press.

Bentham, J. (1970). *An introduction to the principles of morals and legislation* (H. Burns & H. L. A. Hart, Eds.). Oxford, England: Oxford University Press. (Original work published 1789)

Berkeley, G. (1949). *Treatise concerning the principles of human knowledge.* In A. A. Luce & T. E. Jessop (Eds.), *The works of George Berkeley, Bishop of Cloyne* (Vol. 2). London, England: Thomas Nelson & Sons. (Original work published 1710)

Bernecker, S., & Pritchard, D. (Eds.). (2011). *The Routledge companion to epistemology.* London, England: Routledge.

Bernheim, B. D. (1984). Rationalizable strategic behavior. *Econometrica, 52,* 1007–1028.

Binkley, R. (1968). The surprise examination in modal logic. *Journal of Philosophy, 65,* 127–136.

Blanchette, I. (Ed.). (2013). *Emotion and reasoning.* Hove, England: Routledge.

Blanchette, I., Caparos, S., & Trémolière, B. (2018). Emotion and reasoning. In *The Routledge international handbook of thinking and reasoning* (pp. 57–70). New York, NY: Routledge/Taylor & Francis.

Blanchette, I., Richards, A., Melnyk, L., & Lavda, A. (2007). Reasoning about emotional contents following shocking terrorist attacks: A tale of three cities. *Journal of Experimental Psychology: Applied, 13*(1), 47–56.

BonJour, L. (1985). *The structure of empirical knowledge.* Cambridge, MA: Harvard University Press.

Boole, G. (1951). *An investigation of the laws of thought, on which are founded the mathematical theories of logic and probabilities.* New York, NY: Dover. (Original work published 1854)

Bossaerts, P., & Murawski, C. (2017). Computational complexity and human decision-making. *Trends in Cognitive Science, 12,* 917–929.

Botvinick, M., Nystrom, L. E., Fissell, K., Carter, C. S., & Cohen, J. D. (1999). Conflict monitoring versus selection-for-action in anterior cingulate cortex. *Nature, 402*(6758), 179–181.

Bowles, S. (1998). Endogenous preferences: The cultural consequences of markets and other economic institutions. *Journal of Economic Literature, 36*(1), 75–111.

Braine, M. D. S. (1990). The "natural logic" approach to reasoning. In W. F. Overton (Ed.), *Reasoning, necessity, and logic: Developmental perspectives* (The Jean Piaget Symposium Series 16, pp. 133–157). Hillsdale, NJ: Erlbaum.

Braine, M. D. S., & O'Brien, D. P. (Eds.). (1998). *Mental logic.* Mahwah, NJ: Erlbaum.

Brandom, R. B. (1994). *Making it explicit: Reasoning, representing, and discursive commitment.* Cambridge, MA: Harvard University Press.

Brandom, R. B. (2000). *Articulating reasons: An introduction to inferentialism.* Cambridge, MA: Harvard University Press.

Bratman, M. E. (2006). *Structures of agency: Essays.* Oxford, England: Oxford University Press.

Brentano, F. (1874). *Psychologie vom empirischen Standpunkte* [Psychology from an empirical point of view]. Leipzig, Germany: Duncker & Humblot.

Brodbeck, F. C., Kerschreiter, R., Mojzisch, A., & Schulz-Hardt, S. (2007). Group decision making under conditions of distributed knowledge: The information asymmetries model. *Academy of Management Review, 32*(2), 459–479.

Bruner, J. S. (1957). Going beyond the information given. In E. H. Gruber, K. R. Hammond, & R. Jessor (Eds.), *Contemporary approaches to cognition: A symposium held at the University of Colorado* (pp. 41–69). Cambridge, MA: Harvard University Press.

Brunswik, E. (1956). *Perception and the representative design of psychological experiments* (2nd ed.). Berkeley: University of California Press.

Call, J., & Tomasello, M. (2008). Does the chimpanzee have a theory of mind? 30 years later. *Trends in Cognitive Sciences, 12*(5), 187–192.

Carnap, R. (1971). A basic system of inductive logic: Part I. In R. Carnap & R. C. Jeffrey (Eds.), *Studies in inductive logic and probability* (Vol. I, pp. 33–165). Berkeley: University of California Press.

Carnap, R. (1980). A basic system of inductive logic: Part II. In R. C. Jeffrey (Ed.), *Studies in inductive logic and probability* (Vol. II, pp. 7–155). Berkeley: University of California Press.

Chalmers, D. J. (1996). *The conscious mind: In search of a fundamental theory*. Oxford, England: Oxford University Press.

Chan, T. (Ed.). (2013). *The aim of belief*. Oxford, England: Oxford University Press.

Chater, N. (2018). *The mind is flat: The illusion of mental depth and the improvised mind*. New Haven, CT: Yale University Press.

Cheng, P. W., & Holyoak, K. J. (1985). Pragmatic reasoning schemas. *Cognitive Psychology, 17*(4), 391–416.

Claidière, N., Trouche, E., & Mercier, H. (2017). Argumentation and the diffusion of counter-intuitive beliefs. *Journal of Experimental Psychology: General, 146*(7), 1052–1066.

Cohen, L. J. (1981). Can human irrationality be experimentally demonstrated? *Behavioral and Brain Sciences, 4*(3), 317–331.

Cole, M. W., Yarkoni, T., Repovš, G., Anticevic, A., & Braver, T. S. (2012). Global connectivity of prefrontal cortex predicts cognitive control and intelligence. *Journal of Neuroscience, 32*(26), 8988–8999.

Colombo, M., & Knauff, M. (2020). Editors' review and introduction: Levels of explanation: From molecules to culture. *Topics in Cognitive Science, 12*, 1224–1240.

Cosmides, L. (1989). The logic of social exchange: Has natural selection shaped how humans reason? Studies with the Wason selection task. *Cognition, 31*(3), 187–276.

Damasio, A. R. (1994). *Descartes' error: Emotion, reason and the human brain*. New York, NY: Grosset/Putnam.

Danto, A. C. (1965). *Nietzsche as philosopher*. New York, NY: Columbia University Press.

Darwin, C. (1859). *On the origin of species*. London, England: Murray.

Davidson, D. (1963). Actions, reasons, and causes. *Journal of Philosophy, 60*, 685–700.

de Finetti, B. (1937). La prévision: Ses lois logiques, ses sources subjectives. [Foresight: Its logical laws, its subjective sources]. *Annales de l'Institut Henri Poincaré, 7*, 1–68.

Dehaene, S., Changeux, J.-P., & Naccache, L. (2011). The global neuronal workspace model of conscious access: From neuronal architectures to clinical applications. In S. Dehaene & Y. Christen (Eds.), *Characterizing consciousness: From cognition to the clinic?* (pp. 55–84). Berlin, Germany: Springer.

De Jong, P. J., Mayer, B., & Van Den Hout, M. (1997). Conditional reasoning and phobic fear: Evidence for a fear-confirming reasoning pattern. *Behaviour Research and Therapy, 35*(6), 507–516.

De Neys, W. (2012). Bias and conflict: A case for logical intuitions. *Perspectives on Psychological Science, 7*(1), 28–38.

Dennett, D. C. (1978). *Brainstorms: Philosophical essays on mind and psychology*. Cambridge, MA: MIT Press.

Descartes, R. (1641). *Meditationes de prima philosophia* [Meditations on first philosophy]. Paris, France: Michel Soly.

de Sousa, R. (2011). *Emotional truth*. Oxford, England: Oxford University Press.

de Vega, M., Glenberg, A. M., & Graesser, A. C. (2008). *Symbols and embodiment: Debates on meaning and cognition*. Oxford, England: Oxford University Press.

Devitt, M. (1991). *Realism and truth* (2nd ed.) Oxford, England: Blackwell.

Dimitrov, M., Phipps, M., Zahn, T. P., & Grafman, J. (1999). A thoroughly modern Gage. *Neurocase, 5*(4), 345–354.

Dörner, D. (1999). *Bauplan für eine Seele* [Construction plan for a soul]. Reinbek bei Hamburg, Germany: Rowohlt.

Dunlosky, J., & Bjork, R. A. (2008). The integrated nature of metamemory and memory. In J. Dunlosky & R. A. Bjork (Eds.), *Handbook of metamemory and memory* (pp. 11–28). New York, NY: Psychology Press.

Elio, R., & Pelletier, F. J. (1997). Belief change as propositional update. *Cognitive Science, 21*(4), 419–460.

Ellsberg, D. (1961). Risk, ambiguity, and the Savage axioms. *Quarterly Journal of Economics, 75*, 643–669.

Elqayam, S., & Evans, J. St. B. T. (2011). Subtracting "ought" from "is": Descriptivism versus normativism in the study of human thinking. *Behavioral and Brain Sciences, 34*(5), 233–248.

Elqayam, S., & Over, D. E. (2013). New paradigm psychology of reasoning: An introduction to the special issue edited by Elqayam, Bonnefon, and Over. *Thinking & Reasoning, 19*(3–4), 249–265.

Elqayam, S., & Over, D. E. (2016). From is to ought: The place of normative models in the study of human thought. *Frontiers in Psychology, 7*, 628.

Engisch, K. (1930). *Untersuchungen über Vorsatz und Fahrlässigkeit im Strafrecht* [Studies on intent and negligence in criminal law]. Berlin, Germany: Liebermann.

Esser, J. K. (1998). Alive and well after 25 years: A review of groupthink research. *Organizational Behavior and Human Decision Processes, 73*(2–3), 116–141.

Evans, D., & Cruse, P. (Eds.). (2004). *Emotion, evolution, and rationality*. Oxford, England: Oxford University Press.

Evans, J. St. B. T. (1989). *Bias in human reasoning: Causes and consequences*. Hove, England: Erlbaum.

Evans, J. St. B. T. (1993). Bias and rationality. In K. I. Manktelow & D. E. Over (Eds.), *Rationality: Psychological and philosophical perspectives* (pp. 6–30). London, England: Taylor & Francis/Routledge.

Evans, J. St. B. T. (2018). Dual-process theories. In *The Routledge international handbook of thinking and reasoning* (pp. 151–166). New York, NY: Routledge/Taylor & Francis.

Evans, J. St. B. T., Newstead, S. E., & Byrne, R. M. J. (1993). *Human reasoning: The psychology of deduction*. Hove, England: Erlbaum.

Evans, J. St. B. T., & Over, D. E. (1996). *Rationality and reasoning*. Hove, England: Psychology Press.

Fangmeier, T., Knauff, M., Ruff, C. C., & Sloutsky, V. (2006). fMRI evidence for a three-stage model of deductive reasoning. *Journal of Cognitive Neuroscience, 18*(3), 320–334.

Fiedler, K., & Prager, J. (2018). The regression trap and other pitfalls of replication science—Illustrated by the report of the Open Science Collaboration. *Basic and Applied Social Psychology, 40*, 1–10.

Fleming, S. M., Dolan, R. J., & Frith, C. D. (2012). Metacognition: Computation, biology and function introduction. *Philosophical Transactions of the Royal Society of London B: Biological Sciences, 367*(1594), 1280–1286.

Fodor, J. A. (1975). *The language of thought*. Cambridge, MA: Harvard University Press.

Fodor, J. A. (1998). *Concepts: Where cognitive science went wrong*. Oxford, England: Clarendon Press.

Fodor, J. A. (2000). *The mind doesn't work that way*. Cambridge, MA: MIT Press.

Foucault, M. (1972). *L'histoire de la folie à l'âge classique* [History of madness]. Paris, France: Gallimard. (First published as *Folie et déraison*, Paris, France: Plon, 1961)

Foucault, M. (2008). *Le gouvernement de soi et des autres: Cours au Collège de France 1982–1983* [The government of self and others]. Paris, France: du Seull/Gallimard.

Frankfurt, H. G. (1971). Freedom of the will and the concept of a person. *Journal of Philosophy, 68*, 5–20.

Frege, G. (1884). *Die Grundlagen der Arithmetik: Eine logisch mathematische Untersuchung über den Begriff der Zahl* [The foundations of arithmetic: A logico-mathematical enquiry into the concept of number]. Breslau, Germany: Marcus.

Frege, G. (1918). Der Gedanke. Eine logische Untersuchung [The thought]. *Beiträge zur Philosophie des deutschen Idealismus, 1*, 58–77.

Frege, G. (1979). *Posthumous writings* (H. Hermes, F. Kambartel, & F. Kaulbach, Eds.; P. Long & R. White, Trans.). Oxford, England: Blackwell.

Frey, B. S., & Meier, S. (2002). *Two concerns about rational choice: Indoctrination and imperialism* (Zurich IEER Working Paper No. 104). Zurich, Switzerland: University of Zurich. Available at SSRN: https://ssrn.com/abstract=301867 or http://dx.doi.org/10.2139/ssrn.301867

Gabbay, D. M., Hogger, C. J., & Robinson, J. A. (Eds.). (1994). *Handbook of logic in artificial intelligence and logic programming: Vol. 3. Nonmonotonic reasoning and uncertain reasoning*. Oxford, England: Oxford University Press.

García-Carpintero, M., & Macià, J. (2006). *Two-dimensional semantics*. Oxford, England: Clarendon Press.

Gärdenfors, P. (1984). Epistemic importance and minimal changes of belief. *Australasian Journal of Philosophy, 62*(2), 136–157.

Gärdenfors, P. (Ed.). (1992). *Belief revision*. Cambridge, England: Cambridge University Press.

Gazzo Castañeda, L. E., & Knauff, M. (2016). Defeasible reasoning with legal conditionals. *Memory & Cognition, 44*(3), 499–517.

Gazzo Castañeda, L. E., Sklarek, B., Dal Mas, D., & Knauff, M. (2021). Probabilistic and deductive reasoning in the human brain. Manuscript submitted for publication.

Genç, E., Fraenz, C., Schlüter, C., Friedrich, P., Hossiep, R., Voelkle, M. C., . . . Jung, R. E. (2018). Diffusion markers of dendritic density and arborization in gray matter predict differences in intelligence. *Nature Communications, 9*(1), 1905.

Gershman, S. J., Horvitz, E. J., & Tenenbaum, J. B. (2015). Computational rationality: A converging paradigm for intelligence in brains, minds, and machines. *Science, 349*(6245), 273–278.

Gigerenzer, G. (2010). Personal reflections on theory and psychology. *Theory & Psychology, 20*(6), 733–743.

Gigerenzer, G., Hertwig, R., & Pachur, T. (Eds.). (2011). *Heuristics: The foundations of adaptive behavior*. New York, NY: Oxford University Press.

Gigerenzer, G., & Hug, K. (1992). Domain-specific reasoning: Social contracts, cheating, and perspective change. *Cognition, 43*(2), 127–171.

Gigerenzer, G., & Selten, R. (Eds.). (2002). *Bounded rationality: The adaptive toolbox*. Cambridge, MA: MIT Press.

Gillies, D. (2000). *Philosophical theories of probability*. London, England: Routledge.

Glanzberg, M. (Ed.). (2018). *The Oxford handbook of truth*. Oxford, England: Oxford University Press.

Gläscher, J., Adolphs, R., Damasio, H., Bechara, A., Rudrauf, D., Calamia, M., . . . Tranel, D. (2012). Lesion mapping of cognitive control and value-based decision making in the prefrontal cortex. *Proceedings of the National Academy of Sciences, 109*(36), 14681–14686.

Gödel, K. (1931). Über formal unentscheidbare Sätze der ‚Principia mathematica' und verwandter Systeme I [On formally undecidable propositions of Principia Mathematica and related systems]. *Monatshefte für Mathematik und Physik, 38*, 173–198.

Goel, V. (2007). Anatomy of deductive reasoning. *Trends in Cognitive Sciences, 11*(10), 435–441.

Goel, V., Gold, B., Kapur, S., & Houle, S. (1998). Neuroanatomical correlates of human reasoning. *Journal of Cognitive Neuroscience, 10*(3), 293–302.

Goel, V., Shuren, J., Sheesley, L., & Grafman, J. (2004). Asymmetrical involvement of frontal lobes in social reasoning. *Brain, 127*(4), 783–790.

Goldman, A. I. (1978). Epistemology and the psychology of belief. *The Monist, 61*(4), 525–535.

Gomberg, P. (1989). Marxism and rationality. *American Philosophical Quarterly, 26*(1), 53–62.

Gottinger, H. W. (1982). Computational costs and bounded rationality. In W. Stegmüller, W. Balzer, & W. Spohn (Eds.), *Philosophy of economics* (pp. 223–238). Berlin, Germany: Springer.

Gould, S. J. (1978). Sociobiology: The art of storytelling. *New Scientist, 80*(1129), 530–533.

Grabenhorst, F., & Rolls, E. T. (2011). Value, pleasure and choice in the ventral prefrontal cortex. *Trends in Cognitive Sciences, 15*(2), 56–67.

Greene, J. D., Sommerville, R. B., Nystrom, L. E., Darley, J. M., & Cohen, J. D. (2001). An fMRI investigation of emotional engagement in moral judgment. *Science, 293*(5537), 2105–2108.

Grice, H. P. (1957). Meaning. *Philosophical Review, 66*, 377–388.

Grice, H. P. (1975). Logic and conversation. In P. Cole & J. L. Morgan (Eds.), *Syntax and semantics: Vol. 3. Speech acts* (pp. 41–58). New York, NY: Academic Press.

Grice, H. P. (1989). *Studies in the way of words*. Cambridge, MA: Harvard University Press.

Güth, W., Schmittberger, R., & Schwarze, B. (1982). An experimental analysis of ultimatum bargaining. *Journal of Economic Behavior & Organization, 3*, 367–388.

Habermas, J. (1968). *Erkenntnis und Interesse* [Knowledge and human interests]. Frankfurt/Main, Germany: Suhrkamp.

Habermas, J. (1973). *Legitimationsprobleme im Spätkapitalismus* [Legitimation problems in late capitalism]. Frankfurt/Main, Germany: Suhrkamp.

Habermas, J. (1981). *Theorie des kommunikativen Handelns* [The theory of communicative action] (Vols. 1–2). Frankfurt/Main, Germany: Suhrkamp.

Hacking, I. (1975). *The emergence of probability*. Cambridge, England: Cambridge University Press.

Halpern, J. Y. (2003). *Reasoning about uncertainty*. Cambridge, MA: MIT Press.

Hamburger, K., Ragni, M., Karimpur, H., Franzmeier, I., Wedell, F., & Knauff, M. (2018). TMS applied to V1 can facilitate reasoning. *Experimental Brain Research, 236*(8), 2277–2286.

Hammond, K. R. (1996). *Human judgment and social policy: Irreducible uncertainty, inevitable error, unavoidable injustice*. New York, NY: Oxford University Press.

Hardman, D. (2009). *Judgment and decision making: Psychological perspectives*. Hoboken, NJ: Wiley.

Harsanyi, J. C. (1967). Games with incomplete information played by "Bayesian" players, Part I. The basic model. *Management Science, 14*, 159–182.

Harsanyi, J. C. (1968a). Games with incomplete information played by "Bayesian" players, Part II. Bayesian equilibrium points. *Management Science, 14*, 320–334.

Harsanyi, J. C. (1968b). Games with incomplete information played by "Bayesian" players, Part III. The basic probability distribution of the game. *Management Science, 14*, 486–502.

Hart, H. L. A. (1961). *The concept of law*. Oxford, England: Oxford University Press.

Hartmann, S., & Rafiee Rad, S. (2020). Anchoring in deliberations. *Erkenntnis, 85*, 1041–1069.

Heckhausen, J., & Heckhausen, H. (2018). *Motivation and action*. Berlin, Germany: Springer.

Heekeren, H. R., Marrett, S., & Ungerleider, L. G. (2008). The neural systems that mediate human perceptual decision making. *Nature Reviews Neuroscience, 9*, 467–479.

Heidegger, M. (1954). Die Frage nach der Technik [The question concerning technology]. In *Vorträge und Aufsätze* (pp. 9–40). Pfullingen, Germany: Neske.

Helm, B. W. (2001). *Emotional reason: Deliberation, motivation, and the nature of value*. Cambridge, England: Cambridge University Press.

Hempel, C. G. (1961–1962). Rational action. *Proceedings and Addresses of the American Philosophical Association, 35,* 5–23.

Hempel, C. G. (1958). The theoretician's dilemma: A study in the logic of theory construction. In H. Feigl, M. Scriven, & G. Maxwell (Eds.), *Minnesota studies in the philosophy of science* (Vol. II, pp. 37–98). Minneapolis: University of Minnesota Press.

Hempel, C. G. (1973). The meaning of theoretical terms: A critique of the standard empiricist construal. In P. Suppes, L. Henkin, A. Joja, & G. C. Moisil (Eds.), *Logic, methodology and philosophy of science IV* (pp. 367–378). Amsterdam, Netherlands: North-Holland.

Henrich, J., Heine, S. J., & Norenzayan, A. (2010). The weirdest people in the world? *Behavioral and Brain Sciences, 33*(2–3), 61–83.

Heyes, C. (2012). New thinking: The evolution of human cognition. *Philosophical Transactions of the Royal Society of London B: Biological Sciences, 367*(1599), 2091–2096.

Hinterecker, T., Knauff, M., & Johnson-Laird, P. N. (2016). Modality, probability, and mental models. *Journal of Experimental Psychology: Learning, Memory, and Cognition, 42*(10), 1606–1620.

Hobbes, T. (1994). *Leviathan*. In E. Curley (Ed.), *Leviathan, with selected variants from the Latin edition of 1668*. Indianapolis, IN: Hackett. (Original work published 1651)

Hume, D. (1975a). *An enquiry concerning the principles of morals* (L. A. Selby-Bigge, Ed., 3rd ed., revised by P. H. Nidditch). Oxford, England: Clarendon Press. (Original work published 1751)

Hume, D. (1975b). *An enquiry into human understanding* (L. A. Selby-Bigge, Ed., 3rd ed., revised by P. H. Nidditch). Oxford, England: Clarendon Press. (Original work published 1748)

Hume, D. (1975c). *A treatise of human nature* (L. A. Selby-Bigge, Ed., 2nd ed., revised by P. H. Nidditch). Oxford, England: Clarendon Press. (Original work published 1739–1740)

Inhelder, B., & Piaget, J. (1958). *The growth of logical thinking from childhood to adolescence: An essay on the construction of formal operational structures*. New York, NY: Basic Books. (Original work published 1955)

James, W. (1956). The will to believe. In W. James, *The will to believe and other essays in popular philosophy* (pp. 1–31). New York, NY: Dover. (Original work published 1896)

Janis, I. L. (1972). *Victims of groupthink: A psychological study of foreign-policy decisions and fiascoes*. Oxford, England: Houghton Mifflin.

Janis, I. L. (1982). *Groupthink: Psychological studies of policy decisions and fiascoes* (2nd ed.). Boston, MA: Houghton Mifflin Harcourt.

Jeffrey, R. C. (1965). *The logic of decision*. Chicago, IL: University of Chicago Press.

Jeffrey, R. C. (1992). *Probability and the art of judgment*. Cambridge, England: Cambridge University Press.

Jern, A., Chang, K.-M. K., & Kemp, C. (2014). Belief polarization is not always irrational. *Psychological Review, 121*(2), 206–224.

Johnson-Laird, P. N. (1983). *Mental models: Towards a cognitive science of language, inference, and consciousness*. Cambridge, MA: Harvard University Press.

Johnson-Laird, P. N. (1999). Deductive reasoning. *Annual Review of Psychology, 50*(1), 109–135.

Johnson-Laird, P. N. (2001). Mental models and deduction. *Trends in Cognitive Sciences, 5*(10), 434–442.

Johnson-Laird, P. N. (2006). *How we reason*. New York, NY: Oxford University Press.

Johnson-Laird, P. N. (2010). Mental models and human reasoning. *Proceedings of the National Academy of Sciences, 107*(43), 18243–18250.

Johnson-Laird, P. N., & Byrne, R. M. J. (1991). *Deduction*. Hove, England: Erlbaum.

Johnson-Laird, P. N., Girotto, V., & Legrenzi, P. (2004). Reasoning from inconsistency to consistency. *Psychological Review, 111*(3), 640–661.

Johnson-Laird, P. N., & Khemlani, S. (2013). Toward a unified theory of reasoning. *Psychology of Learning and Motivation, 59,* 1–42.

Johnson-Laird, P. N., Khemlani, S., & Goodwin, G. P. (2015). Logic, probability, and human reasoning. *Trends in Cognitive Sciences, 19*(4), 201–214.

Johnson-Laird, P. N., & Ragni, M. (2019). Possibilities as the foundation of reasoning. *Cognition, 193,* 103950.

Jones, M., & Love, B. C. (2011). Bayesian fundamentalism or enlightenment? On the explanatory status and theoretical contributions of Bayesian models of cognition. *Behavioral and Brain Sciences, 34,* 169–231.

Jönsson, M. L., Hahn, U., & Olsson, E. J. (2015). The kind of group you want to belong to: Effects of group structure on group accuracy. *Cognition, 142,* 191–204.

Joyce, J. M. (2009). Accuracy and coherence: Prospects for an alethic epistemology of partial belief. In F. Huber & C.

Schmidt-Petri (Eds.), *Degrees of belief* (pp. 263–297). Dordrecht, Netherlands: Springer.

Jung, N., Wranke, C., Hamburger, K., & Knauff, M. (2014). How emotions affect logical reasoning: Evidence from experiments with mood-manipulated participants, spider phobics, and people with exam anxiety. *Frontiers in Psychology*, 5, 1–12.

Kahneman, D. (2011). *Thinking, fast and slow*. New York, NY: Farrar, Straus and Giroux.

Kahneman, D., & Tversky, A. (1979). Prospect theory: An analysis of decision under risk. *Econometrica*, 47(2), 263–291.

Kant, I. (1908). *Kritik der praktischen Vernunft* [Critique of practical reason]. In *Kant's gesammelte Schriften* (Vol. 5, pp. 1–163). Berlin, Germany: Reimer. (Original work published 1788)

Kemmerling, A. (2017). *Glauben: Essay über einen Begriff* [Belief: An essay on a concept]. Frankfurt/Main, Germany: Klostermann.

Kerr, N. L., MacCoun, R. J., & Kramer, G. P. (1996). Bias in judgment: Comparing individuals and groups. *Psychological Review*, 103(4), 687–719.

Kerr, N. L., & Tindale, R. S. (2004). Group performance and decision making. *Annual Review of Psychology*, 55(1), 623–655.

Khemlani, S., & Johnson-Laird, P. N. (2011). The need to explain. *Quarterly Journal of Experimental Psychology*, 64(11), 2276–2288.

Khemlani, S., & Johnson-Laird, P. N. (2013). The processes of inference. *Argument & Computation*, 4, 4–20.

Kirk, U., Downar, J., & Montague, P. R. (2011). Interoception drives increased rational decision-making in meditators playing the ultimatum game. *Frontiers in Neuroscience*, 5, 49. doi:10.3389/fnins.2011.00049

Kirkham, R. L. (1992). *Theories of truth*. Cambridge, MA: MIT Press.

Klauer, K. C. (1997). Working memory involvement in propositional and spatial reasoning. *Thinking & Reasoning*, 3(1), 9–47.

Kleene, S. C. (1967). *Mathematical logic*. New York, NY: Wiley.

Knauff, M. (1999). The cognitive adequacy of Allen's interval calculus for qualitative spatial representation and reasoning. *Spatial Cognition and Computation*, 1(3), 261–290.

Knauff, M. (2009). Reasoning. In M. D. Binder, N. Hirokawa, & U. Windhorst (Eds.), *Encyclopedia of neuroscience* (pp. 3377–3382). Berlin, Germany: Springer.

Knauff, M. (2013). *Space to reason: A spatial theory of human thought*. Cambridge, MA: MIT Press.

Knauff, M., & Gazzo Castañeda, L. E. (2021). When nomenclature matters: Is the "new paradigm" really a new paradigm for the psychology of reasoning? *Thinking & Reasoning*.

Knauff, M., Fangmeier, T., Ruff, C. C., & Johnson-Laird, P. N. (2003). Reasoning, models, and images: Behavioral measures and cortical activity. *Journal of Cognitive Neuroscience*, 15(4), 559–573.

Knauff, M., & Johnson-Laird, P. N. (2002). Visual imagery can impede reasoning. *Memory & Cognition*, 30(3), 363–371.

Knauff, M., Mulack, T., Kassubek, J., Salih, H. R., & Greenlee, M. W. (2002). Spatial imagery in deductive reasoning: A functional MRI study. *Cognitive Brain Research*, 13(2), 203–212.

Knauff, M., Strube, G., Jola, C., Rauh, R., & Schlieder, C. (2004). The psychological validity of qualitative spatial reasoning in one dimension. *Spatial Cognition & Computation*, 4(2), 167–188.

Knauff, M., & Wolf, A. G. (2010). Complex cognition: The science of human reasoning, problem-solving, and decision-making. *Cognitive Processing*, 11(2), 99–102.

Koch, C. (2016). Does brain size matter? *Scientific American*, 27(1), 22–25.

Koechlin, E., Ody, C., & Kouneiher, F. (2003). The architecture of cognitive control in the human prefrontal cortex. *Science*, 302(5648), 1181–1185.

Konek, J., & Levinstein, B. A. (2019). The foundations of epistemic decision theory. *Mind*, 128, 69–107.

Koons, R. (2017). Defeasible reasoning. In E. N. Zalta (Ed.), *The Stanford encyclopedia of philosophy*. Retrieved from https://plato.stanford.edu/archives/win2017/entries/reasoning-defeasible/

Koriat, A. (1993). How do we know that we know? The accessibility model of the feeling of knowing. *Psychological Review*, 100(4), 609–639.

Koriat, A. (2008). Subjective confidence in one's answers: The consensuality principle. *Journal of Experimental Psychology: Learning, Memory, and Cognition*, 34(4), 945–959.

Koriat, A. (2018). When reality is out of focus: Can people tell whether their beliefs and judgments are correct or wrong? *Journal of Experimental Psychology: General*, 147(5), 613–631.

Koriat, A., & Ma'ayan, H. (2005). The effects of encoding fluency and retrieval fluency on judgments of learning. *Journal of Memory and Language*, 52(4), 478–492.

Koriat, A., Ma'ayan, H., & Nussinson, R. (2006). The intricate relationships between monitoring and control in metacognition: Lessons for the cause-and-effect relation between subjective experience and behavior. *Journal of Experimental Psychology: General*, 135(1), 36–69.

Kosslyn, S. M. (1994). *Image and brain: The resolution of the imagery debate*. Cambridge, MA: MIT Press.

Kosslyn, S. M. (2005). Mental images and the brain. *Cognitive Neuropsychology*, 22(3–4), 333–347.

Kripke, S. A. (2013). The Church–Turing "Thesis" as a special corollary of Gödel's completeness theorem. In B. J. Copeland,

C. Posy, & O. Shagrir (Eds.), *Computability: Turing, Gödel, Church, and beyond* (pp. 77–104). Cambridge, MA: MIT Press.

Krüger, L., Daston, L. J., & Heidelberger, M. (Eds.). (1987). *The probabilistic revolution: Vol. 1. Ideas in history*. Cambridge, MA: MIT Press.

Krüger, L., Gigerenzer, G., & Morgan, M. S. (Eds.). (1987). *The probabilistic revolution: Vol. 2. Ideas in the sciences*. Cambridge, MA: MIT Press.

Krumnack, A., Bucher, L., Nejasmic, J., Nebel, B., & Knauff, M. (2011). A model for relational reasoning as verbal reasoning. *Cognitive Systems Research, 12*(3–4), 377–392.

Kuhn, T. S. (1962). *The structure of scientific revolutions*. Chicago, IL: University of Chicago Press.

Künne, W. (2003). *Conceptions of truth*. Oxford, England: Oxford University Press.

Kusser, A., & Spohn, W. (1992). The utility of pleasure is a pain for decision theory. *Journal of Philosophy, 89*, 10–29.

Lakatos, I. (1978). *The methodology of scientific research programmes* (Philosophical Papers, Vol. I; J. Worrall & G. Currie, Eds.). Cambridge, England: Cambridge University Press.

Laughlin, P. R., & Adamopoulos, J. (1980). Social combination processes and individual learning for six-person cooperative groups on an intellective task. *Journal of Personality and Social Psychology, 38*(6), 941–947.

Laughlin, P. R., & Ellis, A. L. (1986). Demonstrability and social combination processes on mathematical intellective tasks. *Journal of Experimental Social Psychology, 22*(3), 177–189.

Leitgeb, H. (2017). *The stability of belief: How rational belief coheres with probability*. Oxford, England: Oxford University Press.

Levi, I. (1967). *Gambling with truth: An essay on induction and the aims of science*. New York, NY: Knopf.

Levinson, S. C. (2000). *Presumptive meanings: The theory of generalized conversational implicature*. Cambridge, MA: MIT Press.

Levinson, S. C., Kita, S., Haun, D. B. M., & Rasch, B. H. (2002). Returning the tables: Language affects spatial reasoning. *Cognition, 84*(2), 155–188.

Lewis, C. I. (1912). Implication and the algebra of logic. *Mind, 21*, 522–531.

Lewis, D. (1969). *Convention: A philosophical study*. Cambridge, MA: Harvard University Press.

Lewis, D. (1980). A subjectivist's guide to objective chance. In R. C. Jeffrey (Ed.), *Studies in inductive logic and probability* (Vol. II, pp. 263–293). Berkeley: University of California Press.

Li, P., & Gleitman, L. (2002). Turning the tables: Language and spatial reasoning. *Cognition, 83*(3), 265–294.

Liang, D. W., Moreland, R., & Argote, L. (1995). Group versus individual training and group performance: The mediating factor of transactive memory. *Personality and Social Psychology Bulletin, 21*(4), 384–393.

Lipton, P. (1991). *Inference to the best explanation*. London, England: Routledge.

List, C. (2013). Social choice theory. In E. N. Zalta (Ed.), *The Stanford encyclopedia of philosophy*. Retrieved from https://plato.stanford.edu/archives/win2013/entries/social-choice/

List, C., & Pettit, P. (2011). *Group agency: The possibility, design, and status of corporate agents*. Oxford, England: Oxford University Press.

Lloyd, G. E. R. (2017). *The ambivalences of rationality: Ancient and modern cross-cultural explorations*. Cambridge, MA: Cambridge University Press.

Loewenstein, G., & Elster, J. (Eds.). (1992). *Choice over time*. New York, NY: Russell Sage Foundation.

Lorenz, H. (2009). Ancient theories of soul. In E. N. Zalta (Ed.), *The Stanford encyclopedia of philosophy*. Retrieved from https://plato.stanford.edu/archives/sum2009/entries/ancient-soul/

Luders, E., Narr, K. L., Thompson, P. M., & Toga, A. W. (2009). Neuroanatomical correlates of intelligence. *Intelligence, 37*(2), 156–163.

Lyon, A., & Pacuit, E. (2013). The wisdom of crowds: Methods of human judgement aggregation. In P. Michelucci (Ed.), *Handbook of human computation* (pp. 599–614). New York, NY: Springer.

MacDonald, A. W., Cohen, J. D., Stenger, V. A., & Carter, C. S. (2000). Dissociating the role of the dorsolateral prefrontal and anterior cingulate cortex in cognitive control. *Science, 288*(5472), 1835–1838.

Manktelow, K. (2012). *Thinking and reasoning: An introduction to the psychology of reason, judgment and decision making*. Hove, England: Psychology Press.

Marchal, K., & Wenzel, C. H. (2017). Chinese perspectives on free will. In K. Timpe, M. Griffith, & N. Levy (Eds.), *The Routledge companion to free will* (pp. 374–388). New York, NY: Routledge.

Marr, D. (1982). *Vision: A computational investigation into the human representation and processing of visual information*. San Francisco, CA: W. H. Freeman.

Maschler, M., Solan, E., & Zamir, S. (2013). *Game theory*. Cambridge, England: Cambridge University Press.

McAdams, R. H. (2009). Beyond the prisoners' dilemma: Coordination, game theory, and law. *Southern California Law Review, 82*(209), 209–258.

McCabe, D. P., & Castel, A. D. (2008). Seeing is believing: The effect of brain images on judgments of scientific reasoning. *Cognition, 107*(1), 343–352.

McClennen, E. F. (1990). *Rationality and dynamic choice*. Cambridge, England: Cambridge University Press.

Menary, K., Collins, P. F., Porter, J. N., Muetzel, R., Olson, E. A., Kumar, V., . . . Luciana, M. (2013). Associations between cortical thickness and general intelligence in children, adolescents and young adults. *Intelligence, 41*(5), 597–606.

Mercier, H., & Sperber, D. (2017). *The enigma of reason*. Cambridge, MA: Harvard University Press.

Miller, G. A. (1956). The magical number seven, plus or minus two: Some limits on our capacity for processing information. *Psychological Review, 63*(2), 81–97.

Miyake, A., Friedman, N. P., Emerson, M. J., Witzki, A. H., Howerter, A., & Wager, T. D. (2000). The unity and diversity of executive functions and their contributions to complex "frontal lobe" tasks: A latent variable analysis. *Cognitive Psychology, 41*(1), 49–100.

Mojzisch, A., & Schulz-Hardt, S. (2006). Information sampling in group decision making: Sampling biases and their consequences. In K. Fiedler & P. Juslin (Eds.), *Information sampling and adaptive cognition* (pp. 299–326). Cambridge, England: Cambridge University Press.

Montague, R. (1974). *Formal philosophy: Selected papers of Richard Montague* (R. H. Thomason, Ed.). New Haven, CT: Yale University Press.

Monti, M. M., Osherson, D. N., Martinez, M. J., & Parsons, L. M. (2007). Functional neuroanatomy of deductive inference: A language-independent distributed network. *Neuroimage, 37*(3), 1005–1016.

Moore, G. E. (1903). *Principia ethica*. Cambridge, England: Cambridge University Press.

Moorhead, G., Ference, R., & Neck, C. P. (1991). Group decision fiascoes continue: Space shuttle Challenger and a revised groupthink framework. *Human Relations, 44*(6), 539–550.

Moshman, D., & Geil, M. (1998). Collaborative reasoning: Evidence for collective rationality. *Thinking & Reasoning, 4*(3), 231–248.

Mumford, S. (1998). *Dispositions*. Oxford, England: Oxford University Press.

Munro, G. D., & Munro, C. A. (2014). "Soft" versus "hard" psychological science: Biased evaluations of scientific evidence that threatens or supports a strongly held political identity. *Basic and Applied Social Psychology, 36*(6), 533–543.

Myerson, R. B. (1991). *Game theory: Analysis of conflict*. Cambridge, MA: Harvard University Press.

Neubauer, A. C., & Fink, A. (2009). Intelligence and neural efficiency. *Neuroscience & Biobehavioral Reviews, 33*(7), 1004–1023.

Newell, A. (1973). Production systems: Models of control structures. In W. G. Chase (Ed.), *Visual information processing* (pp. 463–526). New York, NY: Academic Press.

Newell, A. (1982). The knowledge level. *Artificial Intelligence, 18*, 87–127.

Newell, A., & Simon, H. A. (1972). *Human problem solving*. Oxford, England: Prentice-Hall.

Nozick, R. (1993). *The nature of rationality*. Princeton, NJ: Princeton University Press.

Oaksford, M. (2015). Imaging deductive reasoning and the new paradigm. *Frontiers in Human Neuroscience, 9*, 101. doi:10.3389/fnhum.2015.00101

Oaksford, M., & Chater, N. (2001). The probabilistic approach to human reasoning. *Trends in Cognitive Sciences, 5*(8), 349–357.

Oaksford, M., & Chater, N. (2007). *Bayesian rationality: The probabilistic approach to human reasoning*. Oxford, England: Oxford University Press.

Oaksford, M., & Chater, N. (2012). Dual processes, probabilities, and cognitive architecture. *Mind & Society, 11*(1), 15–26.

Oaksford, M., & Chater, N. (2020). New paradigms in the psychology of reasoning. *Annual Review of Psychology, 71*, 305–330.

Open Science Collaboration. (2015). Estimating the reproducibility of psychological science. *Science, 349*(6251), aac4716.

Osgood, C. E. (1962). *An alternative to war or surrender*. Urbana: University of Illinois Press.

Osman, M. (2004). An evaluation of dual-process theories of reasoning. *Psychonomic Bulletin & Review, 11*(6), 988–1010.

Ostrom, E. (1998). A behavioral approach to the rational choice theory of collective action: Presidential address, American Political Science Association, 1997. *American Political Science Review, 92*(1), 1–22.

Ostrom, E. (2010). Beyond markets and states: Polycentric governance of complex economic systems. *American Economic Review, 100*(3), 641–672.

Otworowska, M., Blokpoel, M., Sweers, M., Wareham, T., & van Rooij, I. (2018). Demons of ecological rationality. *Cognitive Science, 42*(3), 1057–1066.

Papadimitriou, C. H. (1994). *Computational complexity*. Reading, MA: Addison-Wesley.

Pearce, D. G. (1984). Rationalizable strategic behavior and the problem of perfection. *Econometrica, 52*, 1029–1050.

Pearl, J. (1988). *Probabilistic reasoning in intelligent systems: Networks of plausible inference*. San Mateo, CA: Morgan Kaufmann.

Peebles, D., & Cooper, R. P. (2015). Thirty years after Marr's *Vision*: Levels of analysis in cognitive science. *Topics in Cognitive Science, 7*(2), 187–190.

Peirce, C. S. (1877). The fixation of belief. *Popular Science Monthly, 12*, 1–15.

Perner, J., & Wimmer, H. (1985). "John *thinks* that Mary *thinks* that . . .": Attribution of second-order beliefs by 5- to 10-year-old children. *Journal of Experimental Child Psychology, 39*(3), 437–471.

Pettigrew, R. (2016). *Accuracy and the laws of credence.* Oxford, England: Oxford University Press.

Pietschnig, J., Penke, L., Wicherts, J. M., Zeiler, M., & Voracek, M. (2015). Meta-analysis of associations between human brain volume and intelligence differences: How strong are they and what do they mean? *Neuroscience & Biobehavioral Reviews, 57*, 411–432.

Pinker, S. (1994). *The language instinct.* London, England: Allen Lane.

Plato. (1921). *Sophist* (H. N. Fowler, Trans.). In *Plato in 12 volumes* (Vol. VII, Loeb Classical Library). Cambridge, MA: Harvard University Press.

Plato. (2013). *Republic* (Vols. I–II, C. Emlyn-Jones & W. Preddy, Eds. & Trans.). Cambridge, MA: Harvard University Press.

Politzer, G., & Carles, L. (2001). Belief revision and uncertain reasoning. *Thinking & Reasoning, 7*(3), 217–234.

Popper, K. R. (1945). *The open society and its enemies* (Vols. 1–2). London, England: Routledge & Kegan Paul.

Popper, K. R. (1989). *Logik der Forschung* [The logic of scientific discovery] (9th ed.). Tübingen, Germany: Mohr. (Original work published 1934)

Prado, J., Spotorno, N., Koun, E., Hewitt, E., Van der Henst, J.-B., Sperber, D., & Noveck, I. A. (2015). Neural interaction between logical reasoning and pragmatic processing in narrative discourse. *Journal of Cognitive Neuroscience, 27*(4), 692–704.

Prowse Turner, J. A., & Thompson, V. A. (2009). The role of training, alternative models, and logical necessity in determining confidence in syllogistic reasoning. *Thinking & Reasoning, 15*(1), 69–100.

Putnam, H. (1960). Minds and machines. In S. Hook (Ed.), *Dimensions of mind: A symposium* (pp. 138–164). New York, NY: Collier Books.

Putnam, H. (1975). The nature of mental states. In *Mind, language and reality* (Philosophical Papers, Vol. 2, pp. 429–440). Cambridge, England: Cambridge University Press. (Original work published 1967)

Putnam, H. (1981). *Reason, truth and history* (Philosophical Papers, Vol. 1). Cambridge, England: Cambridge University Press.

Quine, W. V. O. (1960). *Word and object.* Cambridge, MA: MIT Press.

Quine, W. V. O. (1969). Epistemology naturalized. In *Ontological relativity and other essays* (pp. 69–90). New York, NY: Columbia University Press.

Rafetseder, E., Schwitalla, M., & Perner, J. (2013). Counterfactual reasoning: From childhood to adulthood. *Journal of Experimental Child Psychology, 114*(3), 389–404.

Ragni, M., Eichhorn, C., Bock, T., Kern-Isberner, G., & Tse, A. P. P. (2017). Formal nonmonotonic theories and properties of human defeasible reasoning. *Minds and Machines, 27*(1), 79–117.

Ragni, M., Franzmeier, I., Maier, S., & Knauff, M. (2016). Uncertain relational reasoning in the parietal cortex. *Brain and Cognition, 104*, 72–81.

Ragni, M., & Knauff, M. (2013). A theory and a computational model of spatial reasoning with preferred mental models. *Psychological Review, 120*(3), 561–588.

Raiffa, H. (1968). *Decision analysis: Introductory lectures on choice under uncertainty.* New York, NY: Random House.

Ramsey, F. P. (1978). *Foundations: Essays in philosophy, logic, mathematics and economics* (D. H. Mellor, Ed.). London, England: Routledge & Kegan Paul.

Rawls, J. (1971). *A theory of justice.* Cambridge, MA: Belknap Press.

Reason, J. T. (1987). The Chernobyl errors. *Bulletin of the British Psychological Society, 40*, A46.

Recanati, F. (2016). *Mental files in flux.* Oxford, England: Oxford University Press.

Reverberi, C., Shallice, T., D'Agostini, S., Skrap, M., & Bonatti, L. L. (2009). Cortical bases of elementary deductive reasoning: Inference, memory, and metadeduction. *Neuropsychologia, 47*(4), 1107–1116.

Revlin, R., Cate, C. L., & Rouss, T. S. (2001). Reasoning counterfactually: Combining and rending. *Memory & Cognition, 29*(8), 1196–1208.

Richardson, H. S. (1994). *Practical reasoning about final ends.* Cambridge, England: Cambridge University Press.

Richardson, R. C. (2007). *Evolutionary psychology as maladapted psychology.* Cambridge, MA: MIT Press.

Ridge, M. (2019). Moral non-naturalism. In E. N. Zalta (Ed.), *The Stanford encyclopedia of philosophy.* Retrieved from https://plato.stanford.edu/archives/fall2019/entries/moral-non-naturalism/

Rips, L. J. (1994). *The psychology of proof: Deductive reasoning in human thinking.* Cambridge, MA: MIT Press.

Ritter, J., Gründer, K., & Gabriel, G. (Eds.). (1971–2007). *Historisches Wörterbuch der Philosophie* [Historical dictionary of philosophy] (Vols. 1–13). Basel, Switzerland: Schwabe.

Rousseau, J.-J. (1762). *Du contrat social; ou Principes du droit politique* [On the social contract; or, Principles of political rights]. Amsterdam, Netherlands: Rey.

Rubinstein, A. (1998). *Modeling bounded rationality*. Cambridge, MA: MIT Press.

Ruff, C. C., Knauff, M., Fangmeier, T., & Spreer, J. (2003). Reasoning and working memory: Common and distinct neuronal processes. *Neuropsychologia*, *41*(9), 1241–1253.

Rumelhart, D. E., McClelland, J. L., & the PDP Research Group. (1986). *Parallel distributed processing: Explorations in the microstructure of cognition* (Vols. 1–2). Cambridge, MA: MIT Press.

Ryle, G. (1949). *The concept of mind*. London, England: Hutchinson.

Samuelson, P. A. (1938). A note on the pure theory of consumer's behaviour. *Economica*, *5*, 61–71.

Sanfey, A. G., Loewenstein, G., McClure, S. M., & Cohen, J. D. (2006). Neuroeconomics: Cross-currents in research on decision-making. *Trends in Cognitive Sciences*, *10*(3), 108–116.

Sanfey, A. G., Rilling, J. K., Aronson, J. A., Nystrom, L. E., & Cohen, J. D. (2003). The neural basis of economic decision-making in the ultimatum game. *Science*, *300*(5626), 1755–1758.

Savage, L. J. (1954). *The foundations of statistics*. New York, NY: Wiley.

Sayre-McCord, G. (Ed.). (1988). *Essays on moral realism*. Ithaca, NY: Cornell University Press.

Schachter, S., & Singer, J. (1962). Cognitive, social, and physiological determinants of emotional state. *Psychological Review*, *69*(5), 379–399.

Schopenhauer, A. (1841). *Die beiden Grundprobleme der Ethik* [The two fundamental problems of ethics]. Frankfurt/Main, Germany: Hermannsche Buchhandlung.

Schroeder, M. (2018). The moral truth. In M. Glanzberg (Ed.), *The Oxford handbook of truth* (pp. 579–601). Oxford, England: Oxford University Press.

Schultze, T., Mojzisch, A., & Schulz-Hardt, S. (2012). Why groups perform better than individuals at quantitative judgment tasks: Group-to-individual transfer as an alternative to differential weighting. *Organizational Behavior and Human Decision Processes*, *118*(1), 24–36.

Schurz, G. (1997). *The is–ought problem: A study in philosophical logic*. Dordrecht, Netherlands: Kluwer.

Schurz, G., & Hertwig, R. (2019). Cognitive success: A consequentialist account of rationality in cognition. *Topics in Cognitive Science*, *11*, 1–30.

Searle, J. R. (1983). *Intentionality: An essay in the philosophy of mind*. Cambridge, England: Cambridge University Press.

Searle, J. R. (1992). *The rediscovery of the mind*. Cambridge, MA: MIT Press.

Sedgewick, R., & Flajolet, P. (2013). *An introduction to the analysis of algorithms* (2nd ed.). Boston, MA: Addison-Wesley.

Sen, A. K. (1970). *Collective choice and social welfare*. San Francisco, CA: Holden-Day.

Shafer, G. (1978). Non-additive probabilities in the work of Bernoulli and Lambert. *Archive for History of Exact Sciences*, *19*, 309–370.

Shoenfield, J. R. (1967). *Mathematical logic*. Reading, MA: Addison-Wesley.

Simon, H. A. (1947). *Administrative behavior*. New York, NY: Macmillan.

Simon, H. A. (1957). *Models of man, social and rational: Mathematical essays on rational human behavior in a social setting*. New York, NY: Wiley.

Simon, H. A. (1959). Theories of decision making in economics and behavioral science. *American Economic Review*, *49*(3), 253–283.

Skyrms, B. (2004). *The stag hunt and the evolution of social structure*. Cambridge, England: Cambridge University Press.

Sober, E. (2015). *Ockham's razor: A user's manual*. Cambridge, England: Cambridge University Press.

Sperber, D., Clément, F., Heintz, C., Mascaro, O., Mercier, H., Origgi, G., & Wilson, D. (2010). Epistemic vigilance. *Mind & Language*, *25*(4), 359–393.

Sperber, D., & Wilson, D. (1995). *Relevance: Communication and cognition* (2nd ed.). Oxford, England: Blackwell.

Spohn, W. (1982). How to make sense of game theory. In W. Stegmüller, W. Balzer, & W. Spohn (Eds.), *Philosophy of economics* (pp. 239–270). Berlin, Germany: Springer.

Spohn, W. (1993). Wie kann die Theorie der Rationalität normativ und empirisch zugleich sein? [How can the theory of rationality be normative and empirical at the same time?] In L. Eckensberger & U. Gähde (Eds.), *Ethik und Empirie: Zum Zusammenspiel von begrifflicher Analyse und erfahrungswissenschaftlicher Forschung in der Ethik* (pp. 151–196). Frankfurt/Main, Germany: Suhrkamp.

Spohn, W. (2002). The many facets of the theory of rationality. *Croatian Journal of Philosophy*, *2*, 247–262.

Spohn, W. (2009). *Causation, coherence, and concepts: A collection of essays*. Dordrecht, Netherlands: Springer.

Spohn, W. (2011). Normativity is the key to the difference between the human and the natural sciences. In D. Dieks, W. J.

Gonzalez, S. Hartmann, T. Uebel, & M. Weber (Eds.), *Explanation, prediction, and confirmation* (pp. 241–251). Dordrecht, Netherlands: Springer.

Spohn, W. (2012). *The laws of belief: Ranking theory and its philosophical applications*. Oxford, England: Oxford University Press.

Spohn, W. (2016). Truth and rationality. *Tomsk State University Journal of Philosophy, Sociology, and Political Science, 36*, 7–19.

Spohn, W. (2017). Knightian uncertainty meets ranking theory. *Homo Oeconomicus, 34*, 293–311.

Spohn, W. (2020). Defeasible normative reasoning. *Synthese, 197*, 1391–1428.

Stalnaker, R. C. (1968). A theory of conditionals. In N. Rescher (Ed.), *Studies in logical theory* (pp. 98–112). Oxford, England: Blackwell.

Stanovich, K. E. (2009). *What intelligence tests miss: The psychology of rational thought*. New Haven, CT: Yale University Press.

Stanovich, K. E., & West, R. F. (2008). On the relative independence of thinking biases and cognitive ability. *Journal of Personality and Social Psychology, 94*(4), 672–695.

Stasser, G., & Titus, W. (2003). Hidden profiles: A brief history. *Psychological Inquiry, 14*(3–4), 304–313.

Stenning, K., & van Lambalgen, M. (2008). *Human reasoning and cognitive science*. Cambridge, MA: MIT Press.

Stigler, S. M. (1983). Who discovered Bayes's theorem? *The American Statistician, 37*, 290–296.

Störring, G. (1908). Experimentelle Untersuchungen über einfache Schlussprozesse [Experimental studies on basic inference processes]. *Archiv für die gesamte Psychologie, 11*, 1–127.

Stuss, D. T. (2011). Functions of the frontal lobes: Relation to executive functions. *Journal of the International Neuropsychological Society, 17*(5), 759–765.

Stuss, D. T., & Levine, B. (2002). Adult clinical neuropsychology: Lessons from studies of the frontal lobes. *Annual Review of Psychology, 53*, 401–433.

Surowiecki, J. (2004). *The wisdom of crowds: Why the many are smarter than the few and how collective wisdom shapes business, economies, societies and nations*. New York, NY: Doubleday.

Süß, H.-M., Oberauer, K., Wittmann, W. W., Wilhelm, O., & Schulze, R. (2002). Working-memory capacity explains reasoning ability—and a little bit more. *Intelligence, 30*(3), 261–288.

Tafer, Z., Boussahmine, A., & Bouanini, S. (2016). Behavioral economic, rationality and Islamic ethics. *Journal of Business and Economics, 7*(5), 871–888.

Theiner, G., Allen, C., & Goldstone, R. L. (2010). Recognizing group cognition. *Cognitive Systems Research, 11*(4), 378–395.

Thompson, V. A., & Johnson, S. C. (2014). Conflict, metacognition, and analytic thinking. *Thinking & Reasoning, 20*(2), 215–244.

Thompson, V. A., Prowse Turner, J. A., & Pennycook, G. (2011). Intuition, reason, and metacognition. *Cognitive Psychology, 63*(3), 107–140.

Thompson, V. A., Therriault, N. H., & Newman, I. R. (2016). Meta-reasoning: Monitoring and control of reasoning, decision making, and problem solving. In L. Macchi, M. Bagassi, & R. Viale (Eds.), *Cognitive unconscious and human rationality* (pp. 275–299). Cambridge, MA: MIT Press.

Tomasello, M. (2009a). *The cultural origins of human cognition*. Cambridge, MA: Harvard University Press.

Tomasello, M. (2009b). *Why we cooperate*. Cambridge, MA: MIT Press.

Trivers, R. L. (1971). The evolution of reciprocal altruism. *Quarterly Review of Biology, 46*(1), 35–57.

Tulving, E. (1972). Episodic and semantic memory. In E. Tulving & W. Donaldson (Eds.), *Organization of memory* (pp. 381–403). New York, NY: Academic Press.

Tversky, A., & Kahneman, D. (1974). Judgment under uncertainty: Heuristics and biases. *Science, 185*(4157), 1124–1131.

Unkelbach, C., & Greifeneder, R. (2013). A general model of fluency effects in judgment and decision making. In C. Unkelbach & R. Greifeneder (Eds.), *The experience of thinking: How the fluency of mental processes influences cognition and behavior* (pp. 11–32). Hove, England: Psychology Press.

Vandierendonck, A., & De Vooght, G. (1997). Working memory constraints on linear reasoning with spatial and temporal contents. *Quarterly Journal of Experimental Psychology: Section A, 50*(4), 803–820.

van Fraassen, B. C. (1984). Belief and the will. *Journal of Philosophy, 81*, 235–256.

von Neumann, J., & Morgenstern, O. (1944). *Theory of games and economic behavior*. Princeton, NJ: Princeton University Press.

von Wright, G. H. (1971). *Explanation and understanding*. Ithaca, NY: Cornell University Press.

Vygotsky, L. S. (1978). *Mind in society: Development of higher psychological processes*. Cambridge, MA: Harvard University Press.

Waldmann, M. R., Nagel, J., & Wiegmann, A. (2012). Moral judgment. In K. J. Holyoak & R. G. Morrison (Eds.), *The Oxford handbook of thinking and reasoning* (pp. 364–389). New York, NY: Oxford University Press.

Wason, P. C. (1966). Reasoning. In B. M. Foss (Ed.), *New horizons in psychology*. Harmondsworth, England: Penguin.

Wason, P. C. (1968). Reasoning about a rule. *Quarterly Journal of Experimental Psychology, 20*(3), 273–281.

Weber, M. (1973). Der Sinn der „Wertfreiheit" der soziologischen und ökonomischen Wissenschaften [The meaning of "value freedom" in the sociological and economic sciences]. In *Gesammelte Aufsätze zur Wissenschaftslehre* (4th ed., pp. 146–214). Tübingen, Germany: Mohr. (Original work published 1917)

Weber, M. (1921–1922). *Wirtschaft und Gesellschaft* [Economy and society] (Vols. I–II). Tübingen, Germany: Mohr (Siebeck).

Weibull, J. W. (1997). *Evolutionary game theory* (Vol. 1). Cambridge, MA: MIT Press.

Weisberg, J. (2020). Belief in psyontology. *Philosophers' Imprint, 20*(11), 1–27.

Wilkins, M. C. (1928). *The effect of changed material on ability to do formal syllogistic reasoning* (Archives of Psychology, No. 102). New York, NY: Woodworth.

Williamson, T. (2000). *Knowledge and its limits*. Oxford, England: Oxford University Press.

Witte, E. H., & Davis, J. H. (Eds.). (1996). *Understanding group behavior* (Vols. 1–2). Mahwah, NJ: Erlbaum.

Wolf, A. G., Rieger, S., & Knauff, M. (2012). The effects of source trustworthiness and inference type on human belief revision. *Thinking & Reasoning, 18*(4), 417–440.

Woodworth, R. S., & Sells, S. B. (1935). An atmosphere effect in formal syllogistic reasoning. *Journal of Experimental Psychology, 18*(4), 451–460.

Wundt, W. M. (1910). Psychologismus und Logizismus [Psychologism and logicism]. In *Kleine Schriften* (Vol. 1, pp. 511–634). Leipzig, Germany: Engelmann.

Wundt, W. M. (2010). *Grundriss der Psychologie* [Outlines of psychology]. Leipzig, Germany: Engelmann. (Original work published 1896)

Zalta, E. N. (1988). *Intensional logic and the metaphysics of intentionality*. Cambridge, MA: MIT Press.

Zhao, J., Crupi, V., Tentori, K., Fitelson, B., & Osherson, D. (2012). Updating: Learning versus supposing. *Cognition, 124*, 373–378.

Zhu, D. H. (2013). Group polarization on corporate boards: Theory and evidence on board decisions about acquisition premiums. *Strategic Management Journal, 34*(7), 800–822.

Part I   Origins and Key Issues of Rationality

Section 1   Origins of Rationality

1.1 Theories of Rationality and the Descriptive–Normative Divide: A Historical Approach

Thomas Sturm

Summary

This chapter considers the history of the descriptive–normative divide, viewed as fundamental to, but also challenging, current research on rationality. I show how, from antiquity through early modernity, this distinction never became a topic of discussion. That changed, however, during the Enlightenment, through a mixture of new metaphysical and scientific ideas, and, importantly, due to philosophical struggles over the nature, potential, and limits of rationality. The distinction first became explicit in moral philosophy through Hume's "is–ought distinction." In Kant, we find more sophisticated uses of it in both his theoretical and practical philosophy. Criticisms of those views are also discussed, with an eye toward examining how they are related to competing concepts of rationality. Finally, I show how this distinction became emphasized but also criticized in the psychologism debate and still shapes discussions in philosophy and psychology today.

1. Three Historical Stages of the Distinction

"Reason"[1] or "rationality" has been called an "accordion term" (Burian, 1977): its meaning is highly ambiguous (Black, 1986), "fragmented" (Evans, 1991; Stich, 1990), or hard to give a unified account of (Audi, 2001). A historical review proves this to be true. Plato's *logos* differs substantially from Descartes's *ratio* or *intellectus*, both differ from Hume's notion of reason as well as from Kant's complex concept of *Vernunft*, and so on. The Greek, Latin, French, German, and English terms *logos* and *nous*, *ratio* and *intellectus*, *raison* and *entendement*, *Vernunft* versus *Verstand*, or "reason" and "understanding" (or "intellect") are related but different and at times even at odds (Rapp et al., 2001). Again, the disciplines dealing with rationality—philosophy, statistics, and the cognitive and social sciences—have different agendas and therefore use different concepts. It is common to say, for instance, that philosophy deals with rationality only from a normative point of view, whereas psychology is merely descriptive. Historical inquiry can reveal how this view came into existence but also why it is historically as well as philosophically problematic.

In the following analysis, I focus on a number of exemplary authors writing up until the early 20th century. In doing so, I concentrate on the *concept* of rationality (and *theories* in which it figures) in relation to the descriptive–normative distinction instead of mere terminology or etymology. Also, the discussion will be *problem centered*: When, where, and why did the descriptive–normative distinction emerge? What assumptions and arguments make it possible? How is it related to accounts of rationality? What problems has it faced, and what debates has it led to?

One can distinguish between three main phases here. I start, in section 2, by showing that from antiquity through early modernity, despite important contributions to theories of rationality, the descriptive–normative distinction existed at best implicitly and thus never became a topic of discussion. Next, section 3 argues that it was only through a mixture of new metaphysical and scientific developments, and importantly a struggle over the nature of rationality, that the descriptive–normative distinction became prominent, refined, and problematized. This started in moral philosophy, through David Hume's famous is–ought divide. This, however, is fraught with problems. In Immanuel Kant, we find more sophisticated uses of the is–ought distinction, partly fed by the beginnings of empirical research into thinking and decision making. Finally, section 4 is devoted to how the distinction became sharpened but also contested in the late 19th century, within the psychologism debate, paving the way for current attitudes toward it.

2. Antiquity: An Implicit Distinction

Ancient Greek philosophy has been described as the arduous attempt of moving from mythical to rational accounts of the world and our place in it (Buxton, 2002;

Guthrie, 1952). Indeed, the writings of ancient philosophers are replete with discussions that influenced subsequent debates concerning rationality (Frede & Striker, 1996). Considering the most influential of those authors, Plato and Aristotle, will provide a basis for later ideas. It will also show the way ancient thinkers argued over how far rationality can control other faculties but not over whether facts and norms of reasoning are related or not.

2.1   Plato: Logos, the Theory of Forms, and the Good

In Plato's dialogues, several expressions, such as *logos*, *nous*, or *dianoia*, are used when discussing aspects of rationality, in varying and sometimes confusing ways.[2] However, some basic points are uncontroversial. Plato's teacher Socrates emerges as a defender of rationality against the senses, passions, and mere obedience to authority. This notion of rationality does not, therefore, merely exclude religious or mythical accounts, although it does that, too. When speaking of the *logos*, *nous*, and so on, Socrates and Plato refer to faculties of *thinking*: something over and above perceptions, passions, or feelings; something we *do* to reflect on or take a critical stance toward such mental states or dispositions. This idea has roots in common ways of talking and acting we encounter throughout the ancient world, as seen already in Homer's work. With the advent of philosophy, the relevant notions became technical terms and part of theories with substantive and controversial assumptions (Frede, 1996b).

For Plato, rationality is not merely a faculty of the mind. He thinks that we can exercise rationality because we take part in a cosmological order that is itself governed by a *logos*. That order guarantees that someone who reasons well will be right. But how is such taking part in the order of the cosmos possible? Plato's answer is, by means of "forms" or "ideas" (*ideai* or *eide*): abstract entities that make objects what they are.[3] While not all dialogues contain this same doctrine of "Platonism" or of Platonic forms (Patzig, 1970/1996b; Ryle, 1966), some (*Phaidon*, the *Republic*, or *Theaitetos*) defend the influential claim that perception (*aisthesis*) is insufficient to guarantee knowledge (*episteme*) of true reality, as it only provides access to particular passing appearances (e.g., to sensations of color, tastes, or smells). Forms, in contrast, can be grasped by the *logos*, now understood as a mental faculty, and they capture invariant and general features of reality. Among other things, forms enable us to gain mathematical knowledge—for Plato, as for many others, a paradigm of nonempirical knowledge.

Plato's account of reason can be further explained by reference to his doctrine of three powers of the soul: first, a rational part (*to logistokon*, serving to find the true and the good and designed to rule the other powers of the soul); second, a power of will or courage (*to thymoeides*); third and finally, an appetitive or desiring part (*to epithumetikon*, directed toward procuring nourishment and sex) (*Republic* IV.435b–445b). When the rational power rules the soul, the person flourishes, enjoying a healthy soul and true happiness. In the early dialogue *Protagoras*, Socrates even argues that having rational insight into the good and the right *guarantees* that one will act accordingly and that therefore weakness of the will (*akrasia*) is impossible (although in later texts, this position shifts). There is, then, no sharp division between theoretical and practical rationality; insight into the true and insight into the good are intimately entwined with one another.

2.2   Aristotle: Logic and Science

Aristotle famously criticizes Plato's doctrine of forms, assigning perception or experience a stronger role in his epistemology and also separating theoretical from practical rationality more clearly (*akrasia*, for instance, is a definite possibility for him). More important here, he understands *nous* as a power of intuitive insight into basic truths of logic, mathematics, and reality, whereas *logos* is a capacity for judgment and for combining judgments in arguments or syllogisms. Without these insights gained via *nous*, our reasoning would never come to an end and have no foundation.

One of Aristotle's most lasting contributions to Western philosophy is his development of (syllogistic) logic (Lear, 1980; Patzig, 1968). He delivers not only a systematic account of deduction—the study of correctly inferring conclusions from premises on the basis of purely formal features—but also already a modal logic—possibilities, necessities, and degrees. Furthermore, in his *On Sophistical Refutations*, Aristotle develops a taxonomy of "disputed" (*eristikos*) or sophistic arguments that do not really justify the conclusions they claim to. For example, the fallacy of "affirming the consequent" contains a genuine violation of deductive logic. As Aristotle explains, using an example, "since after rain the ground is wet in consequence, we suppose that if the ground is wet, it has been raining; whereas that does not necessarily follow" (*De Sophisticis Elenchis*, chapter 5). He is fully aware that people easily produce flawed arguments, but he does not enter into any deeper empirical explanations of why this is so or how facts about human reasoning could be related to its normative rules.

Aristotle's views about *nous* and *logos*, together with his theory of syllogism, furthermore shape his influential account of science (*episteme*). Deductive validity is not enough for science; what is also needed is reasoning that amounts to *demonstration* (*apodeixis*; see *Analytica Posteriora*). This requires a rigorous regime of reasoning. For instance, one needs to structure knowledge in such a way that arguments are reordered to begin from initial premises that are "better known by nature" (*gnorimoteron phusei*; e.g., *APo* 71b33–72a25; cf. *EN* 1095b2–4) or that are necessarily true and known to be so. This does not mean that all initial premises must be known intuitively, or through *nous*, although ultimately such a state of intellectual insight is preferred by Aristotle (*APo* 99b20–100b17; Barnes, 1994). Moreover, scientific demonstrations not only report facts but also *explain* them and, again, explain them in a way that organizes the plurality of explanations so that we start from those that are best known. Science is thereby given an axiomatic structure. The ideal of scientific rationality does not mean for Aristotle that all good reasoning must proceed from necessary principles; in many areas, we do not (yet) have such starting points. We must start with widely shared opinions (*endoxa*), which, if not shared by everyone, must at least be shared by the majority or by experts (again, at least in their majority) (*Top.* 100b21–23). From *endoxa*, we then reason our way toward beliefs about controversial matters by means of dialectical arguments (which must also respect the doctrines of syllogistic logic).

One problem for Aristotle's views might arise from his claim that observation is a genuine source of knowledge—a point vivid in *De partibus animalium* or *Historia animalium*, in which he claims that systematized observations prepare even knowledge of axioms. However, while he affirms this, he also emphasizes that true knowledge or science essentially depends on his strong notion of rationality (Frede, 1996a). Another issue concerns whether we can say that Aristotle understands his ideal of scientific rationality as a model for the representation of scientific knowledge and its justification, instead of how to proceed in inquiry (Barnes, 1975). Not all followers of Aristotle agree here, and it might be anachronistic to read him this way. In any case, his ideal had a lasting influence, shaping the views of Descartes, Leibniz, the *Logique de Port-Royal* (Arnauld & Nicole, 1662/1992), Thomas Hobbes, and much later authors like Bernhard Bolzano (cf. De Jong & Betti, 2010). This is so even though it was increasingly challenged, especially through recognition of the fallibility of theories that had stood for millennia.

2.3 Why Was the Descriptive–Normative Distinction Not Debated?

Ancient authors thus developed determinate, as well as quite divergent, views on what rationality is in theory and practice. In addition, authors adopted normative viewpoints, partly in response to the possibility of error. Plato, Aristotle, and others are clear that failures of rationality do occur, even frequently (e.g., in Plato's *Republic* 438d–439e). In his *Nicomachean Ethics*, Aristotle talks of *right* reason when he declares that to "act according to the *orthos logos* is a commonplace, and should be assumed" (*EN* 1103b31–2). Still, while flaws of human reasoning and their sources (such as passions or sensory illusions) were noted, this did not lead to a discussion of how reasoning proceeds as a matter of fact compared to how it *ought* to proceed. Why was there no *explicit* discussion of the distinction and relation between descriptive and normative perspectives on rationality?

No literature addresses this question, but we can imagine possible explanations. One might point to the widely held assumption that human rationality participates in a cosmological order governed by the *logos*—understood now as a principle of the world's rational structure. Another explanation relates to influential assumptions of Aristotelian metaphysics: the nature of all beings is determined by their form; forms are also always *final* causes. In the case of living beings, the form thus determines what the animal is supposed to become and do under normal circumstances. If we observe dogs for long enough, we discover that they bark, as they should, or else they are not truly dogs. If we observe humans for long enough, we realize that, *qua* rational animals, they will usually reason as they ought to, despite occasional mistakes, or else their rational nature is in doubt. Thus, ancient philosophy possessed bridges between "is" and "ought." Normativity seemed to be rooted in reality or nature: a foundation safe enough for thinkers from antiquity to the Middle Ages. Even though views on rationality or reason evolved,[4] novel assumptions had to emerge for the descriptive–normative divide to become fully explicit and a topic worthy of serious debate.

3. Early Modernity: Toward an Explicit Distinction and Debate

Fast forward to early modernity. Few developments had a greater impact on the history of philosophy and science than the so-called Scientific Revolution (despite being a contested concept) (Henry, 2008; Wootton, 2015). After murmurings in the late Middle Ages, its course spanned the work of Copernicus, Kepler, Galileo, Newton, and numerous others. It overthrew Ptolemaic cosmology, Aristotelian physics, Galenic medicine, and other long-accepted belief systems. It changed metaphysical assumptions, and it invited renewed debates about what counts as legitimate knowledge and what methods can rationally support it.

Three points matter most for the subsequent history of the is–ought distinction. First, developments in philosophy and science undermined the idea that our rationality somehow participates in a cosmological order guided by the *logos* or by God. Reason or rationality increasingly became understood as merely a faculty of individual, typically human, minds. In addition, the "new science" undermined the Aristotelian view of final causes as a metaphysical bridge between "is" and "ought." Scientific explanations were only to be given in terms of efficient causality. Second, the Scientific Revolution expanded into new domains, such as anatomy, chemistry, economics, or psychology. We will see examples of this in specific contexts. Third, and most important, rational criticism was applied to ever broader domains. The Scientific Revolution prepared the way for the Enlightenment of the 18th century: the movement that, in Kant's expression, encouraged human beings to "think for themselves" by using the "freedom to make public use of one's reason in all matters," in order to overcome their "self-incurred minority" (Kant, 1900ff., volume VIII, pp. 35, 37). Ancient philosophy already contained the germ of such a critical function of reason, but it had not spread widely. Now, Enlightenment ideas became known all over Europe and beyond. In particular, criticism became applied to reason itself. The is–ought distinction first became fully recognized within a lively dispute over reason's nature, potential, and limits, in both theory and practice; then it was refined, revised, or criticized. This required several arduous steps in philosophical reflection and debate; consider René Descartes, David Hume, and Immanuel Kant.

3.1 Descartes: Rationalism without a Theory of Rationality

Descartes, the founder of modern rationalism (Hatfield, 2008/2018; Williams, 1978), offers a radical reaction to the Scientific Revolution. He is highly optimistic about reason's powers and opens the *Discourse on Method* (1637), if somewhat jokingly, describing "good sense" (*bon sens*) or "reason" (*raison*) thus:

> Good sense is of all things in the world the most equally distributed, for each thinks himself so abundantly provided with it, that even those most difficult to please in all

other matters do not commonly desire more of it than they already possess. It is unlikely that this is an error on their part; it seems rather to be evidence that the power of forming a good judgment and of distinguishing the true from the false, which is properly what we call Good sense or Reason, is by nature equal in all men. Hence too it will show that the diversity of our opinions does not come from some men being more rational than others, but solely from the fact that our thoughts pass through diverse channels and the same objects are not considered by all. (Descartes, 1964ff., volume VI, pp. 1–2)[5]

This equally distributed faculty of reason is the main tool for Descartes's most influential project: *Meditationes de prima philosophiae* (1641). Here, he proposes that every serious thinker, once in a lifetime, can and should think through the possibility that all beliefs about reality are completely false. If beliefs about, say, the positions of the Earth and the Sun, the motions of wandering and fixed stars, or the whole Ptolemaic system are based on but a gigantic perceptual illusion, why think that any other beliefs handed to us by tradition are better off? Therefore, Descartes argues, one ought to start from scratch if one wants to "establish any firm and permanent structure in the sciences." He views it as a demand of "reason . . . to withhold my assent from matters which are not entirely certain and indubitable" (Descartes, 1964ff., volume VII, p. 18). Accordingly, his first arguments in the *Meditations* involve a radical thought experiment: there might be a demon who deceives us about everything we believe—particularly our empirical beliefs about reality. This skepticism needs to be countered before one can take controlled steps that erect a new foundation of knowledge.

In this, Descartes builds upon his earlier views about what rules one should follow in the exercise of one's own reason. One ought to proceed in a clear order by first breaking down all problems into their most simple versions, for which one can find simple and clear answers. His view on such answers employs the Aristotelian distinction between *nous* and *logos* in a Latinized form: *intuitio* and *demonstratio* are two main sources of rational knowledge; the former immediate, the latter derivative. In the early *Regulae ad directionem ingenii*, Descartes writes that "intuition is the undoubting conception of a clear and attentive mind, and springs from the light of reason (*rationis luce*) alone" (Descartes, 1964ff., volume X, p. 368). In his later work, "clear and distinct perception" adopts the role of *intuitio*, but the core idea remains the same: we can know some things with absolute certainty because they represent simple, directly evident starting points for all inquiry. Ideally, all other knowledge can be derived from them via chains of highly controlled steps of reasoning (Descartes, 1964ff., volume VI, p. 19).

The famous *Cogito* and *Sum* offered in *Meditation* II (cf. Perler, 1998, pp. 139–149; Williams, 1978, chapter 3) are instances of such basic certainties: our conviction in them "is so strong that we have no reason to doubt" (Descartes, 1964ff., volume VII, pp. 144–145). Descartes adds to these certainties the claim that we are equally certain about the *contents* of our mental states. Even when the demon might deceive me that there is a light before me, I can still be certain that *I believe that I see such a light*. The same with all contents of our beliefs: we might be mistaken that they are true, but we cannot be mistaken that we hold them when we do. There are, then, three basic Cartesian certainties: I can be absolutely certain that I exist, that I think, and what I think. No more is achieved at that stage. When Descartes wonders, "What am I?" looking for further information about his thinking self, and considers the traditional idea of the human being as *animal rationale*, he declares,

> Shall I say a rational animal? No; for then I should have to inquire what an animal is, and what is rational; and thus from a single question I should insensibly fall into an infinitude of others more difficult. (Descartes, 1964ff., volume VII, p. 25)

The rationalist Descartes does not begin from a *theory* of rationality. One consequence of this is that he is not clear about the status of basic certainties: they surely cannot be a result of voluntary decisions (Newman, 2007). Are they, then, something one is *psychologically* compelled to accept? Am I compelled to believe that *I think* whenever I consider the possibility of an evil demon deceiving me, or am I *justified* in being certain that *I think* whenever I entertain radical doubt? Descartes does not consider this vital question. However, the former option would seem too weak, as he needs real certainty. But what could guarantee *that*? He points out that overcoming radical doubt requires that one has entertained all possible reasons against a proposition. But in order to be more than the consideration of doubts a thinker happens to entertain, a systematic account of rationality would be required. As we have seen, Descartes believes that this would be too problematic a starting point. Instead, in Meditation III, he famously sets out that these certainties ultimately depend on the existence of a benevolent God, and he therefore presents a proof of God's existence and perfection.

Thus, paradoxically, Descartes's rationalism is not founded on a theory of rationality but on philosophical theology. This is a burden that scholars agree his arguments cannot bear (e.g., Perler, 1998, pp. 187–202; Williams, 1978, chapter 5). While Descartes defends the priority and authority of reason over the senses with arguments that must be taken seriously (Frankfurt, 1970; Loeb, 1990), and while he allows for a role of the senses in science, once reason's authority is granted, he offers no satisfactory rational basis for our beliefs. He is not even clear about the status of our most deeply held convictions. He simply assumes that there is a "natural light" (*lumen naturale*) of reason that we can rely on: that all humans are endowed equally with the power of reasoning and that we can and should "apply it well." His approach lacks a systematic and critical account of the nature, foundations, and limits of reason.

3.2 Hume: Practical Reason and the Is–Ought Distinction

We can take a step forward by turning to Hume. His system of empiricism is meant to achieve two things at once: to undermine rationalist metaphysics and to promote the idea of a thoroughly empirical "science of human nature"—hence his celebrated title, *A Treatise of Human Nature: Being an Attempt to Introduce the Experimental Method of Reasoning Into Moral Subjects* (1739–1740/1978; cf. Garrett, 2015; Stroud, 1977). Hume feels obliged to consider the nature and limits of reason in two main respects: What can reason deliver concerning *knowledge*? How can it guide our *actions*? His answer to the second question presupposes that to the first but not the other way around. Both answers include a severe restriction of the powers of reason, and they ultimately lead him toward his is–ought distinction.

Concerning its relation to knowledge, Hume claims that reason is "the discovery of truth and falsehood." More specifically, reason can discover knowledge of either "matters of fact" or "relations of ideas" (Hume, 1739–1740/1978, pp. 448, 458): empirical knowledge (which is contingent and uncertain) or a priori knowledge (the contradictory opposite of which is impossible). The latter can—using the *intuitio–demonstratio* distinction—be either intuitively or demonstratively certain. Hume also assumes that all we can know empirically must be based on sensory "impressions," and thus, all our claims about the future or past, and about what causes impressions, have at best the status of probable belief. Most important, he rejects the idea that reason can deliver *metaphysical* knowledge (e.g., of God, or of a difference between mind and body). To demonstrate this, he develops skeptical arguments concerning the possibility of such allegedly rational knowledge in three areas: causal or inductive inference, the existence of material bodies, and personal identity. Nonetheless, Hume thinks that human nature means we must hold beliefs on these topics. This is not a rational "must" or "ought"; it is simply something we unavoidably do in our normal lives. Thus, although there is no demonstrative proof of the "principle of induction," we cannot but form inductive beliefs based on habits and experiences (Winters, 1979, pp. 26–29).

As to action, two points are paramount. On the one hand, against the tradition going back to Socrates, Hume rejects the idea that we have rational control over our passions or desires. He famously declares, *"Reason is, and ought only to be the slave of the passions*, and can never pretend to any other office than to serve and obey them" (Hume, 1739–1740/1978, p. 415) and offers colorful examples:

> 'Tis not contrary to reason to prefer the destruction of the whole world to the scratching of my finger. 'Tis not contrary to reason for me to chuse my total ruin, to prevent the least uneasiness of an Indian or person wholly unknown to me. 'Tis as little contrary to reason to prefer even my own acknowledg'd lesser good to my greater, and have a more ardent affection for the former than for the latter. (Hume, 1739–1740/1978, p. 416)

Some important clarifications are needed here. First, although reason is entirely inactive concerning *passions*, it can still help to direct our *actions*. Based on our best beliefs concerning cause–effect relations, we can determine rational ways to realize our passions (Hume, 1739–1740/1978, p. 459). This is called *instrumentalism* about practical reason (e.g., Audi, 2001, p. 5; Korsgaard, 1996, p. 312), but this instrumentalism involves reason as only a passive, cognitive faculty. It therefore cannot provide a foundation of morality. It is precisely at this stage of his thought that Hume introduces his famous is–ought distinction:

> In every system of morality, which I have hitherto met with . . . the author proceeds for some time in the ordinary way of reasoning, and establishes the being of a God, or makes observations concerning human affairs; when of a sudden I am surprised to find, that instead of the usual copulations of propositions, *is*, and *is not*, I meet with no proposition that is not connected with an *ought*, or an *ought not*. This change is imperceptible; but is, however, of the last consequence. For as this *ought*, or *ought not*, expresses some new relation or affirmation, 'tis necessary that it should be observed and explained; and at the same time that a reason should be given, for what seems altogether inconceivable, how this new relation can be a deduction from others, which are entirely different from it. But as authors do not commonly use this precaution, I shall presume to recommend it to the readers; and am persuaded, that this small attention would subvert all the vulgar systems of morality, and let us see, that the distinction of vice and virtue is not founded merely on the relations of objects, nor is perceived by reason. (Hume, 1739–1740/1978, p. 469)

Hume thinks that the is–ought distinction—"Hume's Guillotine" (Black, 1964)—subverts theories of ethical rationalism such as those of Samuel Clarke (1675–1729) or (perhaps) John Balguy (1686–1748),[6] which start from factual assumptions about human conduct or—more subtly—about how actions are related to one another in an imagined rational order (created by God). For instance, giving a gift is "in agreement with" gratitude, or committing a crime is "in agreement with" punishment (Balguy, 1729, pp. 7–8). Balguy thought that reason affords, through comparing ideas of actions, (partly self-evident) knowledge of what actions agree or disagree with one another and *hence* what our obligations are. Hume claims that such an inference constitutes a fallacy and that, once this is recognized, rationalistic moral theories are left hanging in the air. For Hume, reason is a purely cognitive faculty: all it provides us with is knowledge of either empirical "matters of fact" or analytical "relations of ideas." Neither option provides knowledge of how humans *ought* to act. The real basis of "ought" statements must therefore be sought outside reason: in the passions.

Hume's views have attracted many followers. Bertrand Russell accepted instrumentalism about reason—albeit not in terms of "passions" but of "ends":

> 'Reason' has a perfectly clear and precise meaning. It signifies the choice of the right means to an end that you wish to achieve. It has nothing whatever to do with the choice of ends. (Russell, 1954, p. viii)

More recently, Robert Nozick has noted that instrumentalism is "powerful and natural": indeed, the "default theory" of practical rationality (Nozick, 1993, p. 133).

In its explicitness, Hume's is–ought distinction constitutes a historical novelty. It provoked critical reactions, for instance, from Adam Ferguson (1769) or Thomas Reid (1788/1827) and other, more positive reactions, such as those of 20th-century ethical emotivists (e.g., Stevenson, 1944). Hume paved the way for discussions on the relation between facts and norms, descriptions and prescriptions, and similar issues. However, neither his instrumentalism nor the is–ought distinction are unproblematic. I now present two problems concerning each of these two ideas (cf. Hepfer, 1997, pp. 115–126; Setiya, 2004; Stroud, 1977, pp. 154–170).

First, even though the is–ought distinction appears to express a clear, logically valid point, there may be counterexamples to it. Consider the following:

1. Beer drinking is common in Belgium.

Therefore, beer drinking is common in Belgium or all banks ought to be socialized.

Such an argument seems to refute the universal validity of Hume's Guillotine. Debates about such examples continue (cf. Hudson, 1969; Pigden, 2010; Schurz, 1997, 2010; Singer, 2015; Wolf, 2015), but Hume's view may be an overgeneralization. However, this misses a core point of his argument. Not only do such counterexamples use tricks (such as extension by disjunction), but it is also doubtful that Hume wanted to establish a completely general is–ought distinction. He has a distinct target: rationalistic ethical theories that derive moral norms from facts about behavior or from alleged necessary relations between ideas of actions. Hume gives an account of what reason can and cannot do, and on the basis of that account, he views the inference to "oughts" as unwarranted.

The second problem concerns Hume's assumption that any "ought" is derived from passions. Even if one grants that reason cannot determine or control passions, one may doubt that these can justify any "ought" statements. To say that passions have strong causal force and that we feel pushed by them to carry out certain actions does not imply that we therefore *ought* to perform those actions. Conversely, even if there are instrumental reasons to perform an action (e.g., we ought to avoid unhealthy food if we want to stay healthy), it does not follow that we will be motivated to do so. If Hume throws the is–ought distinction at rationalists, they can throw it back at him just as well.

A related point comes from Thomas Reid. He attacks Hume by claiming that ethical theories do not start from factual premises but from normative ones: from basic, "self-evident" insights into our duties from which normative conclusions can be deduced (Reid, 1788/1827, p. 661). If one accepts this strategy, the fallacy is circumvented. However, this requires a convincing account of the normative basis of reason. Otherwise, one might end in skepticism concerning the normativity of practical reason *in general*—not only noninstrumental or moral but instrumental reason as well. There is suggestion that Hume is forced to adopt such skepticism (Hampton, 1995; Korsgaard, 1996, pp. 312–314; Setiya, 2004).

Third, one might claim that reason functions not only to steer actions toward the realization of goals but also to systematize our plurality of goals. An assessment and hierarchical ordering of ends serves the long-term or enlightened self-interest of agents and helps to exclude certain forms of irrationality. For these reasons, one should incorporate this requirement (as current theories of rational choice do). Hume, however, as we have seen, maintains that it is not "contrary to reason for me to chuse my total ruin" and "as little contrary to reason to prefer even my own acknowledg'd lesser good to my

greater" (Hume, 1739–1740/1978, p. 416). This follows from his assumptions that reason *only* functions cognitively and that passions never represent objects and so cannot be true or false (Hume, 1739–1740/1978, pp. 415, 458). This might be questioned as well: even though passions cannot be true or false, they have an "intentional" character—they concern what states of affairs we intend to bring into existence. Such intentions, at least many of them, might be evaluable by reason too (Stroud, 1977, pp. 162ff.).

Finally, is instrumentalism about reason a *descriptive* or a *normative* claim? The problematic answer is: it is both. In line with his "science of human nature," Hume aims to discover an empirically correct conception of the relation between reason and passion (Hampton, 1995; Winters, 1979). One might push this further by saying that his goal is not to *justify* "oughts" but merely to *explain* them. That is, Hume would only be interested in telling us why people speak in normative terms in their moral conduct, without endorsing the validity of such talk. However, his writings are replete with normativity. He introduces his instrumentalism as a claim that reason "*is, and ought to be*" the slave of the passions. Given Hume's Guillotine, the phrase "*is, and ought to be*" can hardly be a slip of the tongue. Furthermore, his instrumentalism is directed against rationalistic ethical theories that have, of course, normative pretensions: "Every rational creature, it is said, is *obliged to regulate* his actions by reason; and if any other motive or principle challenge the direction of his conduct, he *ought to oppose* it, till it be entirely subdued, or at least be brought into conformity with that superior principle" (Hume, 1739–1740/1978, p. 413, italics added). If Hume's instrumentalism regarding practical reason is merely a descriptive thesis, then ethical rationalists can reject it as irrelevant—and perhaps false, by claiming that we can (sometimes) control our passions, even if it is difficult. The arguments that Hume develops against such a view must express more than just empirical truths about reason and passion. Indeed, his considerations seem more fundamental. Hume's Guillotine appeals to a basic logical distinction. Similarly, his assumption that passions are never representations of objects, and so cannot be true or false, does not look like an empirical claim either. Hume wants to be a scientist of human nature. While this leads him to grapple with the problem of how descriptive and normative claims concerning morality are related, he adopts no clear and convincing stance on it. Perhaps this is because he introduces the is–ought distinction in order to attack versions of ethical rationalism. Hume lays the ground for the distinction between the descriptive and the normative, but he does not consider its extent or think through its implications. Most important, he has no convincing account of the sources of normative rationality.

3.3 Kant: Applying the Is–Ought Distinction in Philosophy and Science

The "all-crushing Kant" (Moses Mendelssohn) delivers an even more devastating criticism of rationalistic metaphysics, as represented especially by Leibniz, Christian Wolff (1679–1754), and their followers. Kant radically reconceptualizes reason, restricting its theoretical functions and expanding its role in practical philosophy (see Guyer, 2006). In doing so, he understands and uses the is–ought distinction in more sophisticated ways than Hume. Although the debate over the foundations of morality remains important (Guyer, 2016), Kant's use of the distinction also goes beyond that domain. He repeatedly claims that while certain disciplines or sciences deal with facts of human thought, desire, and action, others are concerned with irreducibly normative questions concerning such mental activities, and he thoroughly addresses the sources of normativity.

In his opus magnum, the *Critique of Pure Reason* (1781, 2nd rev. ed. 1787),[7] Kant argues that reason cannot arrive at knowledge claims about reality independently of experience. Attempts to prove metaphysical claims (about an immaterial soul, the existence of God, etc.) lead to "transcendent illusions." Pure reason, through its infinite quest for explanations and justifications, leads us to fall prey to these illusions (Kant, 1781/²1787/1998, A vii). We cannot avoid them, just as we cannot avoid perceptual illusions (Kant, 1781/²1787/1998, A295/B351–352, A297/B354), even when we grasp their illusory nature.[8] Therefore, metaphysics is an inescapable consequence of human reason (Willaschek, 2018). However, we can learn to see through illusions of reason by scrutinizing the reasoning patterns that lead to them. It is this project that leads Kant to think that reason is not only there to criticize existing systems of belief and conduct but that in order to do so properly, it must critically examine itself.

Such self-examination, to pick out the central constructive doctrine of the *Critique*, leads Kant to determine the boundaries of all possible knowledge claims by means of a system of synthetic a priori principles: principles (such as the principle of causality) that are necessary for making "objective" knowledge claims (i.e., claims that are determinately true or false about real objects). Synthetic a priori principles are *not* a product of reason but of cooperation between "pure understanding" (*reiner Verstand*) with its concepts, the so-called categories, and "pure sensibility" (*reine Anschauung*) with its a priori forms of space and time. Kant describes understanding, alongside the faculties of judgment (*Urteilskraft*) and reason, as the "higher" cognitive faculties (and sometimes speaks of reason in confusing ways where he means understanding[9]). Traditional rationalist metaphysics violates the limits of possible knowledge determined by synthetic a priori principles.

Having refuted traditional rationalist metaphysics, Kant thinks that he needs to show that reason has wholly different tasks. This can be connected to the is–ought distinction. While he agrees with Hume that the distinction between "is" (*Sein*) and "ought" (*Sollen*) is deep, fundamental, and in principle easy to recognize, Kant connects it with quite different accounts of reason. First and foremost, he uses the is–ought distinction to separate theoretical from practical philosophy: the former concerns only nature, the realm of things as they are in fact, whereas the latter concerns only how things ought to be (Kant, 1781/²1787/1998, A319/B375, A633/B661). This is a distinction between the main *domains* of philosophy, and Kant also states that reason is a "lawgiver" for each of them: laws of nature for theoretical and laws of morality for practical philosophy (A840/B868). Second, and cutting across this basic divide, Kant applies the is–ought distinction at the level of our making *judgments*, or even of pursuing systematic or scientific investigations within these two domains. Thus, in theoretical philosophy, although its claims concern *Sein*, we can nonetheless make normative claims too and actually have to: theoretical disciplines such as logic provide norms for good reasoning about facts. Here, Kant applies the is–ought distinction via warning that one cannot derive the laws of logic from psychological laws of actual human thinking. It is a frequent Kantian claim that when dealing with recognizably different tasks, one ought to treat them in different disciplines: we "do not enlarge but disfigure sciences, if we allow them to trespass upon one another's territory" (Kant, 1781/²1787/1998, B viii).

Kant's sophisticated use of the is–ought distinction has advantages over Hume's. Given Hume's project of an empirical "science of human nature," perhaps he wishes only to explain why people make demands or state obligations, without considering the normative validity of such talk. But Hume's instrumentalist account of practical reason certainly looks normative, and as argued above, he makes many other normative claims too. In contrast, Kant recognizes that we make normative judgments and that such discourse can then be studied either descriptively (and empirically) or normatively (ethically). This provides greater clarity.

Now, in moral philosophy, Kant's specific concept of "pure practical reason" plays a strong, constitutive role. While Hume uses the is–ought distinction to fight ethical rationalism, Kant applies it to defend his own rational ethics. For Hume, practical reason is at best applied theoretical reason; for Kant, it governs a territory where it sets up its own, irreducibly normative, moral laws of conduct. Reason, he argues in *Groundwork to the Metaphysics of Morals* (1785), helps to identify and justify the basic principle of morality: the "categorical imperative."[10] We understand moral rules as *universal* and *necessary*. First, if a moral norm is valid, then it is valid for all agents in similar situations: making an exception for oneself is immoral. The fact that we often observe people deviating from norms does not invalidate them. Second, a moral rule does not describe, but *demands*, certain courses of conduct; this "ought" character constitutes its necessity. Kant combines these claims with different disciplines studying rational action. Thus, "anthropology" studies human actions empirically; ethics or the "metaphysics of morals," in contrast, considers how we ought to act (Kant, 1900ff., volume IV, pp. 389, 411–412; volume VI, pp. 216–217). Note that when Kant speaks this way, he means only moral normativity. As should be emphasized, he also makes room for "merely" instrumental normativity, in his terminology, "hypothetical imperatives"; also, despite differences in detail, he agrees with Hume that such imperatives can be grounded in empirical means–ends knowledge (Kant, 1900ff., volume IV, pp. 413–419).

How does Kant, again in contrast to Hume, also come to use the is–ought distinction within *theoretical* philosophy? As indicated above, the Scientific Revolution led to the idea that not only the physical but also the mental world might be studied scientifically. Kant is quite familiar with ideas of a distinct empirical science of human thinking, feeling, desiring, and acting (Sturm, 2009, chapter 4). Such studies were usually called "empirical psychology": a pet project of Wolff and his disciple Alexander Baumgarten (1714–1762), but also of critics such as Johann Nikolaus Tetens (1736–1807). They refined folk-psychological or folk-philosophical theories of the architecture of mental faculties. At the same time, some psychologists developed the first experimental and quantitative studies of mental processes. Thus, Gottlob Friedrich Hagen (1710–1774) measures the "perfection" of the power of reason by the number of extended syllogisms one can understand and repeat and by how many demonstrations one can produce (Hagen, 1733, §§12–14). Kant emphasizes that such empirical psychology contains factual, contingent statements and explanations (Kant,

1900ff., volume XXV, pp. 243, 473). Because of this, it would be misguided to use psychology in logic:

> Logic is a science of reason . . . a science *a priori* of the necessary laws of thought . . . hence a science of the correct use of the understanding and of reason in general, not subjectively, however, i.e., not according to empirical (psychological) principles for how the understanding does think, but objectively, i.e., according to principles *a priori* for how it ought to think. (Kant, 1900ff., volume IX, p. 16)

This richer background distinguishes Kant from Hume, who is unclear about how far his "science of human nature" is descriptive, normative, or both. Still, whether the Kantian view of logic is convincing became much disputed in the next century (cf. section 4).

Alongside ethics and logic, a further area where Kant brings the is–ought distinction into play is in his philosophy of science. "Reason," now in the sense of an alleged power of metaphysical knowledge, produces its own concepts and principles (Kant, 1781/²1787/1998, A289–290/B355–356). As noted, even when Kant claims that we should not view such "ideas of pure reason" as representing possible knowledge of real objects, he thinks that we cannot avoid having them. Do these ideas have any other function, then? The critical examination of reason leads Kant to redefine one of reason's functions: ideas of God, the soul, or the world as a whole play a positive role in our theoretical knowledge by becoming used not "constitutively" (i.e., referring to objects) but "regulatively" in scientific research (Kant, 1781/²1787/1998, A311/B368, A647/B675): they *prescribe* to us how to systematically connect pluralities of empirical judgments into systematized bodies of theories. Thus, judgments about stellar positions, motions, velocities, and the structure of the planetary system become unified in the theories of Copernicus, Kepler, Newton, and in Kant's own cosmology (containing his famous nebular hypothesis), all guided by the notion of the world as a systematic whole governed everywhere by the same laws. The normative aspect here is that we ought to aim at such systematized bodies of knowledge. Moreover, there are certain normative rules, such as Occam's razor ("Do not multiply entities beyond necessity") or continuity of forms (e.g., "Make division of things in chemical, biological, or psychological taxonomies as complete as possible"), for doing so (Kant, 1900ff., volume V, p. 182). We might never arrive at a truly finished theory in any domain; in Kant's view, science is forever an open-ended enterprise. However, that we ought to pursue that aim is reasonable: it supports scientific progress (cf. Neiman, 1994, chapter 2; Wartenberg, 1979).

What to think of Kant's claims concerning the is–ought distinction? I will briefly consider here two objections concerning his ethics. First, he claims that reason is not a purely passive, cognitive capacity but "has causality, or that at least we can represent something of the sort in it" and that this "is clear from the imperatives that we propose . . . in everything practical" (Kant, 1781/²1787/1998, A547/B575). Hume would reject this claim about a "causality" or activity of practical reason, and we might wonder on whom the burden of proof falls here. Kant's qualification "at least we can represent something of the sort in it" shows he is perhaps not certain about the causality either. In his ethical works, he develops deeper accounts of "pure practical reason"— although it is disputed whether he succeeds with them.

Second, and related to this, his notion of pure practical reason seems too idealistic to provide norms that are realistic for human beings. Can we really follow moral norms? Is Hume not right that reason is—and ought to be—dominated by passions? This popular objection might explain why Kant's notion of pure practical reason did not become incorporated into modern theories of rational choice or economic decisions, which usually tried to be ethically neutral (but see Bjorndahl, London, & Zollman, 2017; Churchman, 1970).

Partly, this criticism rests on a limited understanding of Kant's practical thought. He does not maintain that empirical study of human history or society treats thought or action as arational processes. His "pragmatic anthropology" aims to study the human being as an animal endowed with reason (Kant, 1900ff., volume VII, p. 119; Sturm, 2009, chapter 8). In this, Kant accepts that human beings can be *imperfectly* rational: the possession of the faculty of reason is one thing; its improvement is another. In this vein, he writes that the human being is not an *animal rationale* but an *animal rationabile* (Kant, 1900ff., volume VII, p. 321). Anthropology can teach us how to improve our rationality—not the basic competence that normal human beings all possess but our performance in following its norms (or developing better ones)—in "pragmatic" ways, that is, in ways that in the long run might even support the realization of morality. This requires a historical and social process. Still, Kant insists that one cannot derive categorical moral norms from such pragmatic anthropological knowledge.

To sum up: while Descartes neither has an account of rationality that could ground his normative pretensions nor states the is–ought distinction, his attempt to establish metaphysical truths provokes Hume as well as Kant to develop accounts of reason that enable them to attack such metaphysics. Against this background, Hume uses his strictly cognitive notion of reason to argue that ethical rationalism fails, among other things, because reason cannot establish any moral "oughts." Kant, who accepts the is–ought distinction, develops its uses further by applying it both to distinguish the two realms of theoretical and practical reason and to distinguish between descriptive and normative investigations within each domain. The background motivation for this is his sustained attempt to understand reason and all its potential as well as its limits. While his account can be considered an advance on Hume's, it is not without problems either.

4.   The Psychologism Debate

Two major developments determined the subsequent evolution of the debate over the is–ought divide, primarily within contexts of theoretical philosophy. To understand why they matter for us, we must keep in mind that rationality has traditionally been connected to the theory of logical inference.[11] In Kant, this is the most basic and uncontroversial notion of reason, and his view of logic is unapologetically normative and antipsychologistic. Even in the 20th century, much cognitive psychology concerning reasoning starts from experiments involving logical tasks (see Sturm, 2012; chapter 1.2 by Evans, this handbook).

The first development consists in the rise of neo-Kantianism in the late 19th century, after a phase of dominance of German idealism (Heis, 2018). Philosophers such as Eduard Zeller (1814–1908) or Hermann Cohen (1842–1918) demand that the metaphysics of Hegel and his followers be replaced by a logic married to *Erkenntnistheorie* ("epistemology": a term invented in the early 19th century; Vaihinger, 1876). As Zeller declared in an influential speech, idealists view logic as "not only knowledge of the forms of thought but knowledge of reality as well" (Zeller, 1877, p. 480).

The second development concerns the further advance of scientific psychology. In the 19th century, this led to the formation of the first academic professorships, institutions, and societies of psychology (Gundlach, 2005). Much of the work began with perception, but it also became wedded to previous empiricist conceptions of logic and reasoning. As Zeller (1877) declares, there was "an older" notion of logic "concerned with the activity of thought as such" (pp. 479–480). Indeed, many early modern thinkers saw logic as normative but *nonetheless* dependent on actual mental processes and their laws (Hatfield, 1997). John Stuart Mill, for whom logic is the "science of reasoning" and studies "the operations of the human understanding in the pursuit of truth"

(Mill, 1843/1973–1974, p. 6), declares that knowledge of laws of thinking is based on "the ordinary methods of experimental inquiry" (Mill, 1843/1973–1974, p. 853) and maintains that "the general laws of association prevail among these more intricate states of mind" (Mill, 1843/1973–1974, p. 856). A similar view can be found in Wilhelm Wundt (1832–1920), the influential organizer of psychology in the late 19th century.

Against such conceptions, Zeller emphasizes the strictly formal character of logic and demands it to be separated from both metaphysics and psychology and connected with epistemology instead. Thus, he declares logic to be "the scientific methodology" that is presupposed in any "investigation of the real" (Zeller, 1877, p. 481), pointing to Kant for this (Zeller, 1877, p. 490). The rise of neo-Kantianism and academic psychology thus leads to a clash over the status of logic (and, thereafter, of epistemology and philosophy of science). The psychologism debate involves numerous contenders; only some major positions and arguments can be considered here (Carl, 1994, chapter 1; Kusch, 1995, 2007/2015).

For instance, Theodor Lipps (1817–1881) criticizes the Mill–Wundt view that logic, while based on empirical laws of thinking, is a normative discipline. Lipps describes logic "as the physics, not the ethics, of thinking" (Lipps, 1880, p. 529). In addition, he argues that we must distinguish between prescriptivity and normativity: logical rules are not the result of a voluntary decision or an "authoritative will" and so should not be called prescriptive. Instead, laws of logic are like "norms of nature," and therefore logic is "the physics of thinking or it is nothing at all" (Lipps, 1880, p. 531). Wundt responds that Lipps presents a false dichotomy, since logic can be based on descriptive laws insofar as "our thinking is not influenced by anything which interferes with its being governed by these laws" (Wundt, 1882, p. 345).

One of the most influential antipsychologism positions—often characterized as decisive—stems from the philosopher–mathematician Gottlob Frege (1848–1925). Frege is viewed as the founder of modern logic and of analytic philosophy. His main project was to show that mathematics or, more specifically, arithmetic could be reduced to logical laws. Since the logics as Frege found them were insufficient to support such a reduction, he created a novel, axiomatized predicate calculus with innovative tools for complex quantification.

Frege's antipsychologism concerning logic (cf. Carl, 1994, chapter 1) is based on the claim that the *validity* of logical laws is independent of how people *actually* think and reason (Frege, 1956). As he explains, "We can inquire, on the one hand, how we have gradually arrived at a given proposition and, on the other, how we can finally provide it with the most secure foundation" (Frege, 1879/1972, Preface). This may seem like a merely gradual distinction, but it is not. To begin, Frege points out that the "first question may have to be answered differently for different persons; the second is more definite, and the answer to it is connected with the inner nature of the proposition considered" (Frege, 1879/1972, Preface). With the latter point, what Frege refers to is what logic should be all about. Logical rules determine whether inferences from premises to conclusions are truth preserving. Whether the premises of an argument are true often cannot be decided by logic alone, but logic tells us which inference patterns guarantee that if the premises are true, then the conclusion will also be. The validity of logical laws has to do, among other things, with the meaning of the logical connectives[12] such as "$\rightarrow$" ("if–then," interpreted as the material conditional) and "$\neg$" (negation). Thus, *modus ponens* ($p \rightarrow q$; p; therefore, q) and *modus tollens* ($p \rightarrow q$; $\neg q$; therefore, $\neg p$) are deductively valid under any interpretation of the propositional variables p and q. That human reasoning sometimes violates and sometimes conforms to such valid patterns is, from Frege's point of view, irrelevant for their logical validity.

As seen in Mill and Wundt, psychologism can be combined with the view that logic is normative—something Kant would have denied. Conversely, proponents of antipsychologism do not necessarily think of logic as normative or as a kind of "ethics of thinking." Frege writes more cautiously that logic is "different from psychology, related to ethics" (Frege, 1979, p. 1): he avoids calling logic a *kind* of ethics. He does not view the laws of logic as inherently normative but descriptive. Descriptive of *what*? Logical rules purport to be universal and timelessly true. Frege (1956, p. 302) speaks of a "third realm" (*drittes Reich*) beyond the material world and our subjective mental representations. This is much disputed (Carl, 1994, chapter 8), but what matters is the innocent point accepted since Aristotle: given that we can make mistakes in our reasoning, we ought to correct them, and this can involve logical laws. So, these laws have a prescriptive function, but in themselves, Frege views them as strictly descriptive.

Frege's antipsychologism, his distinction between an explanation of the genesis of our beliefs and a possible justification of their grounds, revived Kantian views. While his and others' (especially Edmund Husserl's) arguments became widely accepted, the status of logic remains an open question. Logic after Frege has become enriched not only by deontic and modal logics but also by more specialized or applied theories such as relevance logic or defeasible reasoning. These address special problems of human reasoning where classical logic often fails (see, e.g., sections 3, 5, 6, and 7, this handbook).

5.   Conclusion: Toward the 20th Century

While Frege focused on logic, his views became influential beyond this domain, especially through the epistemology and philosophy of science of the Vienna Circle. Not all reasoning, after all, is deductive. Many nondeductive inferences, for instance, about scientific hypotheses, use probabilities. Logical Empiricists such as Hans Reichenbach (1938) or Rudolf Carnap (1950) tried to do for probability theory or "inductive logic" what Frege had achieved for deductive logic: design an axiomatic system with clear tasks, limits, and structure. They spoke of epistemology and philosophy of science as "logic of science." However, this requires that all philosophical claims about knowledge or science are formal or logical truths, which is highly implausible. Still, even critics of the Vienna Circle, like Karl Popper, viewed the distinction between "is" and "ought" as basic for (not as identical to!) both that between "discovery" and "justification" of beliefs (Reichenbach, 1938, chapter 1) and that between tasks of psychology and epistemology. They thereby tried to carve out a territory for philosophy of science and its theories of confirmation, falsification, scientific explanation, and so on (Peckhaus, 2006). With the advent of naturalistic and historicist approaches in philosophy of science, the is–ought, the descriptive–normative, and the discovery–justification distinctions have become questioned again (e.g., Giere, 1989; Hoyningen-Huene, 1987; Kitcher, 1992; Schickore & Steinle, 2006).

I have argued that the divide between the descriptive and the normative has an interesting and instructive history. Meanings, together with debates over them, often depended on competing assumptions about what rationality is, as well as on agendas and assumptions in metaphysics, ethics, logic, and the sciences. Bringing this history to light should enable us to understand better and to reconsider our current situation. The divide often poses challenges for current approaches to the study of rationality, particularly in interdisciplinary contexts. In any case, the demand to reflect critically on the normative authority of rationality and its relation to empirical studies of human reasoning is alive and well.

Acknowledgments

For comments and criticisms, I am grateful to Lorraine Daston, Karl Hepfer, Markus Knauff, Dominik Perler, Wolfgang Spohn, Jens Timmermann, two anonymous

referees, and Christopher Evans. This work was supported by the Ministry of Science and Higher Education of the Russian Federation grant no. 075-15-2019-1929, project *Kantian Rationality and Its Impact in Contemporary Science, Technology, and Social Institutions*, Immanuel Kant Baltic Federal University (IKBFU), Kaliningrad; and—for language correction costs—by the Spanish Ministry for Economics, Industry, and Competition (MINECO), research project *Naturalism and the Sciences of Rationality: An Integrated Philosophy and History* (FFI2016-79923-P).

Notes

1. I do not here consider the view that reason and rationality should be distinguished, in a way that approximately mirrors the distinction between instrumental, means–ends reasoning and (moral) reasoning concerning ends themselves (e.g., Elster, 2009; Rawls, 1980; Sibley, 1953). For criticisms of such distinctions, see Patzig (1994/1996a) and Sturm (2019). In collaboration with others (Erickson et al., 2013), I have argued that the term "rationality" gained popularity during the 20th century, tending to push "reason" into the background. "Rationality" often became understood as a mechanical or "mindless" application of formal and optimizing rules, exemplified by logic and rational choice theory, whereas "reason" referred more to a mindful or reflective use of the faculty of thinking. The terminology has not always followed this trend, however. Here, I will speak primarily of "rationality" and resort to "reason" and related terms when this is more appropriate for the source texts.

2. For example, *logos* has many meanings, such as "speech," "explanation," "ratio," "computation," "reckoning," "argument," and "law," to mention but a few from the entry in Liddell–Scott's Greek–English Lexicon. Moss (2014) argues that *logos* in Plato's and Aristotle's works should often be translated as "reason," related to explanatory accounts.

3. The masterful analogies of the sun, the divided line, and the cave (*Republic* VI. 507b–511e; VII. 514a–520a) present the theory of forms in vivid ways.

4. Cf. Frede and Striker (1996) and Rapp et al. (2001). In an independent philosophical tradition, namely Chinese, we find related philosophical problems (e.g., about norms for our practical lives and an awareness that these can be violated), but again, there is a lack of philosophical reflection on the is–ought gap even beyond the Middle Ages (Marchal & Wenzel, 2017). How far the emergence of a discussion of the is–ought gap is an exclusively Western development deserves further investigation.

5. While the references use the standard Adam–Tannery edition, translations are from the edition by Haldane and Ross (Descartes, 1934ff.), with minor improvements.

6. It is not certain that Hume knew Balguy's views. However, they provide an illuminating background, since Balguy debated the basis of moral obligations with an anonymous critic. The

critic's objections were published together with Balguy's replies (Balguy, 1729). For instance, the critic argued that if we do not already have the idea of obligation or "ought," then the inspection of ideas such as "Gratitude, Ingratitude, and Bounty" by reason, independently of sentiments, "could never so much as afford us a general Idea of Obligation in itself; or inform us what is meant by that Term; much less could we be able to deduce the particular Obligation to Gratitude from these Ideas" (Balguy, 1729, p. 8 [Art. IV of the criticism]; cf. pp. 9–16). This is an anticipation of Hume's is–ought distinction and similar to other aspects of his ethical theory.

7. Citations from Kant's work refer to the Academy edition (Kant, 1900ff.) and the standard edition of the first *Critique* but using the English translations (Kant, 1992ff.).

8. A similar analogy is drawn in current heuristics-and-biases psychology but with more contested examples of cognitive illusions (see Margalit, 1986; Sturm, 2012).

9. Kant distinguishes several meanings of *Vernunft* (cf. the relevant entries in Willaschek, 2015). Only in the broadest sense of *Vernunft*, in which it is the "highest faculty of the mind" and opposed to experience (Kant, 1781/1787, A835/B863), is it true that understanding is part of reason.

10. In fact, each special moral duty is a categorical imperative. When Kantians speak of *the* categorical imperative, they refer to the principle underlying *all* moral norms.

11. Today, many argue that logic not only *prescribes* how to reason well but is even *constitutive* of rationality (as an interpretation of, e.g., Kant's position, see Hanna, 2006). One may also claim, more weakly, that only *some* logical rules are constitutive of rationality (Cherniak, 2001); rationality is constituted not only by logic but also by other formal systems, especially probabilistic and decision-theoretic rules (Stein, 1996; for historical background, see Erickson et al., 2013); not only logic but also the semantics and pragmatics of language matter for good reasoning (Grice, 2001; this is also accepted by the bounded rationality program of Gigerenzer, Todd, & ABC Research Group, 1999); or logic is not constitutive of rationality at all (Mercier & Sperber, 2017).

12. I do not use Frege's logical notation but the more familiar Hilbert–Ackermann notation.

References

Aristotle. (1984). *The complete works of Aristotle* (J. Barnes, Ed.). Princeton, NJ: Princeton University Press.

Arnauld, A., & Nicole, P. (1992). *La logique, ou l'art de penser* [Logic, or the art of thinking]. Paris, France: Gallimard. (Original work published 1662)

Audi, R. (2001). *The architecture of reason.* Oxford, England: Oxford University Press.

Balguy, J. (1729). *The second part of the foundation of moral goodness*. London, England: Pemberton.

Barnes, J. (1975). Aristotle's theory of demonstration. In J. Barnes, M. Schofield, & R. Sorabji (Eds.), *Articles on Aristotle: Vol. I. Science* (pp. 65–87). London, England: Duckworth.

Barnes, J. (1994). Commentary. In Aristotle, *Posterior analytics* (J. Barnes, Trans. & Comm., pp. 81–271). Oxford, England: Clarendon Press.

Bjorndahl, A., London, A. J., & Zollman, K. J. S. (2017). Kantian decision making under uncertainty: Dignity, price, and consistency. *Philosophers' Imprint, 17*(7). Retrieved from http://hdl.handle.net/2027/spo.3521354.0017.007.

Black, M. (1964). The gap between "is" and "should." *Philosophical Review, 73*, 165–181.

Black, M. (1986). Ambiguities of rationality. In N. Garver & P. Hare (Eds.), *Naturalism and rationality* (pp. 25–40). Buffalo, NY: Prometheus.

Burian, R. M. (1977). More than a marriage of convenience: On the inextricability of the history and philosophy of science. *Philosophy of Science, 44*, 1–42.

Buxton, R. (2002). *From myth to reason? Studies in the development of Greek thought.* Oxford, England: Clarendon Press.

Carl, W. (1994). *Frege's theory of sense and reference.* Cambridge, England: Cambridge University Press.

Carnap, R. (1950). *Logical foundations of probability.* Chicago, IL: University of Chicago Press.

Cherniak, C. (1986). *Minimal rationality.* Cambridge, MA: MIT Press.

Churchman, C. W. (1970). Kant—a decision theorist? *Theory and Decision, 1*, 107–116.

De Jong, W., & Betti, A. (2010). The classical model of science: A millennia-old model of scientific rationality. *Synthese, 174*, 185–203.

Descartes, R. (1934ff.). *The philosophical works of Descartes* (E. S. Haldane & G. R. T. Ross, Trans., 2 vols.). Cambridge, England: Cambridge University Press.

Descartes, R. (1964ff.). *Œuvres de Descartes* [Descartes' works] (C. Adam & P. Tannery, Eds., 10 vols.). Paris, France: Vrin.

Elster, J. (2009). *Reason and rationality.* Princeton, NJ: Princeton University Press.

Erickson, P., Klein, J., Daston, L., Lemov, R., Sturm, T., & Gordin, M. D. (2013). *How reason almost lost its mind.* Chicago, IL: University of Chicago Press.

Evans, J. St. B. T. (1991). Theories of human reasoning: The fragmented state of the art. *Theory & Psychology, 1*, 83–105.

Ferguson, A. (1769). *Institutes of moral philosophy.* Edinburgh, Scotland: Kincaid & Bell.

Frankfurt, H. (1970). *Demons, dreamers, and madmen.* Indianapolis, IN: Bobbs-Merrill.

Frede, M. (1996a). Aristotle's rationalism. In M. Frede & G. Striker (Eds.), *Rationality in Greek thought* (pp. 157–173). Oxford, England: Clarendon Press.

Frede, M. (1996b). Introduction. In M. Frede & G. Striker (Eds.), *Rationality in Greek thought* (pp. 1–28). Oxford, England: Clarendon Press.

Frede, M., & Striker, G. (Eds.). (1996). *Rationality in Greek thought*. Oxford, England: Clarendon Press.

Frege, G. (1956). The thought. *Mind, 65*, 289–311. (Original work published 1918–1919)

Frege, G. (1972). *Conceptual notation* (T. W. Bynum, Ed.). Oxford, England: Oxford University Press. (Original work published 1879)

Frege, G. (1979). *Posthumous writings* (P. Long, R. White, & R. Hargraves, Trans.). Oxford, England: Oxford University Press.

Garrett, D. (2015). *Hume*. London, England: Routledge.

Giere, R. N. (1989). Scientific rationality as instrumental rationality. *Studies in History and Philosophy of Science, 20*, 377–384.

Gigerenzer, G., Todd, P. M., & ABC Research Group. (1999). *Simple heuristics that make us smart*. New York, NY: Oxford University Press.

Grice, H. P. (2001). *Aspects of reason*. Oxford, England: Oxford University Press.

Gundlach, H. (2005). Reine Psychologie, Angewandte Psychologie und die Institutionalisierung der Psychologie [Pure psychology, applied psychology, and the institutionalization of psychology]. *Zeitschrift für Psychologie, 212*, 183–199.

Guthrie, W. C. K. (1952, December). *Myth and reason*. Oration delivered at the London School of Economics and Political Science, London, England.

Guyer, P. (2006). *Kant*. Oxford, England: Routledge.

Guyer, P. (2016). Is and ought: From Hume to Kant, and now. In P. Guyer, *Virtues of freedom* (pp. 21–35). Oxford, England: Oxford University Press.

Hagen, G. F. (1733). *Dissertatio mathematica de mensurandis viribus propriis atque alienis* [Mathematical dissertation on the measurement of the powers of one's own and others]. Gießen, Germany: Litteris Mullerianis.

Hampton, J. (1995). Does Hume have an instrumental conception of practical reason? *Hume Studies, 21*, 57–74.

Hanna, R. (2006). *Rationality and logic*. Cambridge, MA: MIT Press.

Hatfield, G. (1997). The workings of the intellect: Mind and psychology. In P. Easton (Ed.), *Logic and the workings of the mind* (pp. 21–45). Atascadero, CA: Ridgeview.

Hatfield, G. (2018). René Descartes. In E. N. Zalta (Ed.), *The Stanford encyclopedia of philosophy*. Retrieved from https://plato.stanford.edu/archives/sum2018/entries/descartes/

Heis, J. (2018). Neo-Kantianism. In E. N. Zalta (Ed.), *The Stanford encyclopedia of philosophy*. Retrieved from https://plato.stanford.edu/archives/sum2018/entries/neo-kantianism/

Henry, J. (2008). *The scientific revolution and the origins of modern science*. New York, NY: Palgrave.

Hepfer, K. (1997). *Motivation und Bewertung: Eine Studie zur praktischen Philosophie Humes und Kants* [Motivation and evaluation: A study of Hume's and Kant's practical philosophy]. Göttingen, Germany: Vandenhoeck & Ruprecht.

Hoyningen-Huene, P. (1987). Context of discovery and context of justification. *Studies in the History and Philosophy of Science, 18*, 501–515.

Hudson, W. D. (1969). *The is–ought question*. London, England: Palgrave Macmillan.

Hume, D. (1978). *A treatise of human nature* (L. A. Selby-Bigge, Ed., 2nd ed.). Oxford, England: Oxford University Press. (Original work published 1739–1740)

Kant, I. (1900ff.). *Gesammelte Schriften* [Collected writings] (Academy edition). Berlin, Germany: De Gruyter.

Kant, I. (1992ff.). *Cambridge edition of the works of Immanuel Kant* (P. Guyer & A. Wood, Eds.). Cambridge, England: Cambridge University Press.

Kant, I. (1998). *Kritik der reinen Vernunft* [Critique of pure reason] (J. Timmermann, Ed.). Hamburg, Germany: Meiner. (Original work published 1781/^{2}1787)

Kitcher, P. (1992). The naturalists return. *Philosophical Review, 101*, 53–114.

Korsgaard, C. (1996). *Creating the kingdom of ends*. Cambridge, England: Cambridge University Press.

Kusch, M. (1995). *Psychologism*. London, England: Routledge.

Kusch, M. (2015). Psychologism. In E. N. Zalta (Ed.), *The Stanford encyclopedia of philosophy*. Retrieved from https://plato.stanford.edu/archives/win2015/entries/psychologism/

Lear, J. (1980). *Aristotle and logical theory*. Cambridge, England: Cambridge University Press.

Lipps, T. (1880). Die Aufgabe der Erkenntnistheorie und die Wundt'sche Logik [The task of epistemology and Wundt's logic]. *Philosophische Monatshefte, 16*, 529–539.

Loeb, L. E. (1990). The priority of reason in Descartes. *Philosophical Review, 99*, 3–43.

Marchal, K., & Wenzel, C. H. (2017). Chinese perspectives on free will. In K. Timpe, M. Griffith, & N. Levy (Eds.), *The Routledge companion to free will* (pp. 374–388). London, England: Routledge.

Margalit, A. (1986). The past of an illusion. In E. Ullmann-Margalit (Ed.), *The kaleidoscope of science* (pp. 89–94). Dordrecht, Netherlands: Springer.

Mercier, H., & Sperber, D. (2017). *The enigma of reason.* Cambridge, MA: Harvard University Press.

Mill, J. S. (1973–1974). *A system of logic, ratiocinative and inductive.* In J. M. Robson (Ed.), *The collected works of John Stuart Mill* (Vols. 7–8). London, England: Routledge & University of Toronto Press. (Original work published 1843)

Moss, J. (2014). Right reason in Plato and Aristotle: On the meaning of *logos. Phronesis, 59,* 181–230.

Neiman, S. (1994). *The unity of reason.* New York, NY: Oxford University Press.

Newman, L. (2007). Descartes on the will in judgment. In J. Broughton & J. Carriero (Eds.), *A companion to Descartes* (pp. 334–352). Oxford, England: Wiley-Blackwell.

Nozick, R. (1993). *The nature of rationality.* Princeton, NJ: Princeton University Press.

Patzig, G. (1968). *Aristotle's theory of the syllogism* (J. Barnes, Trans.). Dordrecht, Netherlands: Reidel.

Patzig, G. (1996a). Aspekte der Rationalität [Aspects of rationality]. In G. Patzig, *Gesammelte Schriften* [Collected writings] (Vol. 4, pp. 99–116). Göttingen, Germany: Wallstein. (Original work published 1994)

Patzig, G. (1996b). Platons Ideenlehre, kritisch betrachtet [Plato's doctrine of ideas, critically viewed]. In G. Patzig, *Gesammelte Schriften* [Collected writings] (Vol. 3, pp. 1–24). Göttingen, Germany: Wallstein. (Original work published 1970)

Peckhaus, V. (2006). Psychologism and the distinction between discovery and justification. In J. Schickore & F. Steinle (Eds.), *Revisiting discovery and justification* (pp. 99–116). Berlin, Germany: Springer.

Perler, D. (1998). *Descartes.* Munich, Germany: Piper.

Pigden, C. R. (Ed.). (2010). *Hume on 'is' and 'ought'.* Hampshire, England: Palgrave Macmillan.

Plato. (1997). *Complete works* (J. M. Cooper, Ed.). Indianapolis, IN: Hackett.

Rapp, C., Horn, C., Enders, M., Speer, A., Hasse, D., Largier, N., . . . Bremer, M. (2001). Vernunft; Verstand [Reason; understanding]. In G. Gabriel (Ed.), *Historisches Wörterbuch der Philosophie* [Historical dictionary of philosophy] (Vol. 11, pp. 748–866). Basel, Switzerland: Schwabe.

Rawls, J. (1980). Kantian constructivism in moral theory. *Journal of Philosophy, 77,* 515–572.

Reichenbach, H. (1938). *Experience and prediction.* Chicago, IL: University of Chicago Press.

Reid, T. (1827). *Essay on the active powers of the human mind* (D. Stewart, Ed.). London, England: Thomas Tegg. (Original work published 1788)

Russell, B. (1954). *Human society in ethics and politics.* London, England: Allen & Unwin.

Ryle, G. (1966). *Plato's progress.* Cambridge, England: Cambridge University Press.

Schickore, J., & Steinle, F. (2006). *Revisiting discovery and justification.* New York, NY: Springer.

Schurz, G. (1997). *The is–ought problem.* Dordrecht, Netherlands: Kluwer.

Schurz, G. (2010). Non-trivial versions of Hume's is–ought thesis. In C. Pigden (Ed.), *Hume on 'is' and 'ought'* (pp. 198–216). Hampshire, England: Palgrave Macmillan.

Setiya, K. (2004). Hume on practical reason. *Philosophical Perspectives, 18,* 365–389.

Sibley, W. M. (1953). The rational versus the reasonable. *Philosophical Review, 62,* 554–560.

Singer, D. J. (2015). Mind the is–ought gap. *Journal of Philosophy, 112,* 193–210.

Stein, E. (1996). *Without good reason.* Oxford, England: Clarendon Press.

Stevenson, C. L. (1944). *Ethics and language.* New Haven, CT: Yale University Press.

Stich, S. P. (1990). *The fragmentation of reason.* Cambridge, MA: MIT Press.

Stroud, B. (1977). *Hume.* London, England: Routledge & Kegan Paul.

Sturm, T. (2009). *Kant und die Wissenschaften vom Menschen* [Kant and the sciences of man]. Paderborn, Germany: Mentis.

Sturm, T. (2012). The "rationality wars" in psychology: Where they are and where they could go. *Inquiry, 55,* 66–81.

Sturm, T. (2019). Rationalität versus Vernunft? Über eine Unterscheidung bei John Rawls (und anderen) [Rationality versus reason? On a distinction in John Rawls (and others)]. In D. Sölch (Ed.), *Philosophische Sprache zwischen Tradition und Innovation* [Philosophical language between tradition and innovation] (pp. 211–233). Frankfurt/Main, Germany: Lang.

Vaihinger, H. (1876). Über den Ursprung des Wortes „Erkenntnistheorie" [On the origin of the word "epistemology"]. *Philosophische Monatshefte, 12,* 84–90.

Wartenberg, T. E. (1979). Order through reason. *Kant-Studien, 70,* 409–424.

Willaschek, M. (2015). *Kant-Lexikon* [Kant lexicon]. Berlin, Germany: De Gruyter.

Willaschek, M. (2018). *Kant on the sources of metaphysics.* Cambridge, England: Cambridge University Press.

Williams, B. (1978). *Descartes.* London, England: Penguin.

Winters, B. (1979). Hume on reason. *Hume Studies, 5,* 20–35.

Wolf, A. (2015). Giving up Hume's guillotine. *Australasian Journal of Philosophy, 93,* 109–125.

Wolff, C. (1962ff.). *Gesammelte Werke* [Collected works] (J. École, Ed.). Hildesheim, Germany: Olms.

Wootton, D. (2015). *The invention of science: A new history of the Scientific Revolution.* New York, NY: Penguin.

Wundt, W. (1882). Logische Streitfragen [Logical contestations]. *Vierteljahrsschrift für wissenschaftliche Philosophie, 6,* 340–355.

Zeller, E. (1877). Über Bedeutung und Aufgabe der Erkenntnistheorie [On the significance and task of epistemology]. In E. Zeller, *Vorträge und Abhandlungen* [Lectures and treatises] (pp. 479–526). Leipzig, Germany: Fues's Verlag.

1.2   The Rationality Debate in the Psychology of Reasoning: A Historical Review

Jonathan St. B. T. Evans

Summary

For most of its history, the psychology of reasoning assessed people's ability to evaluate logical arguments without formal training, on the assumption that logic provided the standard for rational reasoning. Numerous experiments reported from the 1960s onward showed that logical errors, cognitive biases, and content-dependent reasoning were common in university student populations. This led to great debate about human rationality, in which philosophers were involved. This chapter traces the history of that debate, in which arguments were made (a) that standard logic is not necessarily the correct norm for human reasoning, (b) that apparent errors may reflect misinterpretations of the task or instructions, and (c) that the experiments were artificial and unrepresentative. Ultimately, most psychologists were unwilling to label participants as irrational, and toward the end of the 20th century, a number of them formed a "new paradigm" psychology of reasoning, in which use of prior beliefs and expressions of uncertainty in reasoning is regarded as rational.

1.   Reasoning and Rationality

Traditionally, the psychology of reasoning has consisted of giving people logic problems to solve and comparing their deductions with the dictates of formal logic. The first such experiments were published early in the 20th century (Wilkins, 1928; Woodworth & Sells, 1935) and immediately demonstrated some basic findings that were to occupy a large field of study in the later part of the century (see Evans, Newstead, & Byrne, 1993). First, people make many logical errors. In particular, they frequently endorse fallacies, that is, arguments whose conclusions could be true, given the premises, but are not necessary and valid inferences. People are also systematically biased by both the linguistic form of the premises and their content or meaning. That is to say, they respond to factors unrelated to the logic of the problems. Given a philosophical heritage that decreed that logic was the basis for rational human thought (Henle, 1962), the rationality debate in the psychology of reasoning was always likely to happen. It eventually exploded into life in the 1980s and has never really subsided. It did, however, cause many researchers in the field radically to rethink their approach, eventually leading to a "new paradigm" in 21st-century psychology of reasoning. How and why the field got from its origins to this new paradigm is a story I shall tell in this chapter.

Although the concept of rationality has become much more nuanced in recent publications, a basic distinction that is commonly made is between instrumental and epistemic rationality. *Instrumental* rationality is served when organisms achieve their goals, and it is evident that both humans and other animals need to be rational in this sense if they are going to survive and reproduce. *Epistemic* rationality is the acquisition and maintenance of true beliefs about the world. Epistemic rationality is clearly subservient to instrumental rationality, but in higher organisms and especially humans, the maintenance of accurate knowledge about the world around us is required for rational actions that will achieve our goals. For example, animals must know how and where to obtain food and water. The notion of epistemic rationality is much more developed for us human beings with our unique possession of language, explicit belief systems, and capacity for reasoning. Deductive reasoning is a key concept. If the fundamental beliefs we hold are true, then so too will be any others that we deduce from them in a logically valid manner—hence, the traditional emphasis on logic as the basis of rational thought.

The modern psychology of reasoning dates from the 1960s and especially the pioneering studies of the British psychologist Peter Wason (see Wason & Johnson-Laird, 1972). The belief that logic provided the basis for rational thinking was dominant at that time and not simply because psychologists were influenced by philosophical tradition (Henle, 1962). It was also the heyday of Jean Piaget, whose theoretical work was enormously influential. Piaget had proposed that children's thinking and reasoning develop through a series of stages culminating in "formal operations" by adulthood (Inhelder & Piaget, 1958). In other words, Piaget proposed not only that logic *should* be the basis for adult human thought but that this actually was so. By contrast, Wason accepted logic as the normative framework but argued strongly that people were illogical and therefore irrational. His ideas presaged the rationality debate to come.

2. Origins of the Rationality Debate: The Case for Irrational Reasoning

2.1 Bias and Error in Deductive Reasoning

As already mentioned, the field of deductive reasoning has early origins. Prior to Wason, most studies used syllogistic reasoning tasks, based on the simple system of logic developed by Aristotle. Even the earliest studies showed evidence of cognitive bias. For example, Woodworth and Sells (1935) showed an "atmosphere effect" in abstract syllogistic reasoning, the first in a number of accounts of why people often accept invalid conclusions due to syntactic properties of the premises or their arrangement (see Evans, Newstead, et al., 1993, chapter 7). Consider the example

(1) Some *A* are *B*

Some *C* are *B*

Therefore, some *A* are *C*

Due to the atmosphere effect, it was argued, people are more likely to accept this conclusion as valid than, say, "All *A* are *C*," which has a different quantifier. Both conclusions are, of course, logically invalid. (Consider, for example, "Some men are golfers; some women are golfers; therefore, some men are women.")

Mental model theorists later described a "figural bias," which operates through the order of terms (Johnson-Laird & Steedman, 1978). Consider the case

(2) Some *A* are *B*

Some *B* are *C*

Therefore, some *A* are *C*

The terms *A*, *B*, and *C* here run in the same order in premises and conclusions, leading to more inferences of this type than if the order is as shown in (1). Perhaps the most sophisticated model of bias in abstract syllogistic reasoning is that presented by Chater and Oaksford (1999). They showed that people tend to choose the form of the least informative premise as that of the conclusion—the "min-heuristic." Hence, the validity judgments of participants,

normally university students, are often influenced by factors other than logical reasoning.

From the very beginning, it was also discovered that when content-rich syllogisms are used, the prior beliefs that people hold influence their decisions. In particular, as first shown by Wilkins (1928), there is a tendency for people to endorse the validity of syllogisms because they believe the conclusion to be true, now generally known as the *belief bias effect*. This was followed by various studies of questionable methodology, reviewed by Evans, Barston, and Pollard (1983). These authors, however, established the reality of the belief bias phenomenon with improved methodology leading to the modern study of the effect, with many studies published (for recent examples, see Dube, Rotello, & Heit, 2010; Klauer, Musch, & Naumer, 2000; Trippas, Handley, & Verde, 2013). Consider the following two syllogisms used by Evans et al. (1983):

(3) No addictive things are inexpensive

Some cigarettes are inexpensive

Therefore, some addictive things are not cigarettes

(4) No millionaires are hard workers

Some rich people are hard workers

Therefore, some millionaires are not rich people

The two syllogisms have the same form, and neither is valid. However, Evans et al. found that 71% of participants declared (3) a valid syllogism, compared with only 10% for (4). The difference, of course, is that the conclusion to (3) is believable and that to (4) unbelievable. The importance of this basic finding for the present purposes is quite simple: despite being told clearly to assume the premises and decide whether the conclusion follows logically, participants are quite unable to ignore their (logically irrelevant) actual beliefs. By the time I wrote my first book reviewing the psychology of deductive reasoning (Evans, 1982), much evidence had been accumulated in the psychological experiments for illogicality, systematic biases, and a strong influence of prior belief on reasoning, and many similar findings have continued to be reported since.

One influential approach has been the proposal that people reason by use of mental logics comprising inference rules, which are instantiated for particular contexts (Braine & O'Brien, 1991, 1998; Rips, 1983, 1994). However, this approach attracted strong criticisms (Johnson-Laird, 1983; Oaksford & Chater, 1991), and Philip Johnson-Laird and colleagues went on to propose a highly influential alternative view that people reason by imagining and eliminating possibilities. This led to a major research program (see Johnson-Laird, 2006; Johnson-Laird & Byrne,

The Rationality Debate in the Psychology of Reasoning

1991) with a very large number of published studies. However, mental model theorists have mostly stuck quite closely to propositional logic as a normative account (e.g., Byrne & Johnson-Laird, 2009; Johnson-Laird & Byrne, 2002), in contrast with researchers in the new paradigm, discussed later. Mental model theorists have, however, investigated cognitive biases and demonstrated cognitive illusions (Johnson-Laird & Savary, 1999). Their approach to rationality is that people are rational in principle but fallible in practice, mostly due to limitations in working memory capacity. This complies with the idea of bounded rationality proposed by Herbert Simon (1982). Johnson-Laird (2006) has further clarified two fundamental bases for human rationality within the mental models framework: first, our ability to search for and identify counterexamples, which demonstrate invalid arguments, and, second, our ability to understand and reason about what is true or false. He maintains that rationality does not depend upon any form of internalized logical rules, providing a radical alternative account of epistemic rationality and deductive competence.

2.2 Peter Wason and the 2 4 6 and Selection Tasks

Peter Wason was a British psychologist who published most of his work in a 25-year period starting in the late 1950s. He is particularly famous for the invention of two fiendish reasoning tasks that appeared simple but led most of his participants into logical errors. Of equal significance for this historical review are the comments that he wrote about these tasks, now rarely read and noted by modern students of the field. His first invention was the "2 4 6" task (Wason, 1960), which continues to be studied to the present day (for a full history, see Evans, 2016b). For a highly cited paper, it is actually quite short, only a few pages long; it contains a study that can scarcely be described as an experiment as it has only one condition with no controls or comparisons. As with the first presentation of his even more famous four-card selection task (Wason, 1966), the reader is simply invited to compare the poor performance of his participants to their prior expectation that the task *ought* to be simple and straightforward. While the selection task enjoyed a "heyday" between 1980 and 2000, in which it was the focus of a number of major theoretical papers, citations of Wason's work on this continue to increase to the current day (Evans, 2016a)— remarkable when one considers that Wason's last major article was published more than 30 years ago.

Wason considered his tasks to be "deceptively simple," by which he presumably meant that they look easy while being anything but. The 2 4 6 task is so called because participants are told that the experimenter has a rule in mind that classifies triples of whole numbers, an example of which is 2 4 6. They were to discover the rule by offering triples of their own and being told whether or not they conformed to the experimenter's rule. The actual rule was any ascending sequence, much more general than the deliberately biasing example of 2 4 6 suggests. Two unusual features of the task were that participants were required to write down the reasons for their choice of triples and also, if they announced the wrong rule, to continue with the task until they gave up or a time limit was reached. What happened was that people formed more specific hypotheses, such as "ascending with equal intervals," and got repeated positive feedback from testing triples that conformed to it (e.g., 1 3 5, 10 30 50, etc.). Few people solved this problem without announcing wrong rules and some not at all. Wason also published example protocols that showed how people became convinced their hypotheses were correct and often reformulated them in other words, so that they announced the same wrong rule more than once.

The 2 4 6 task does not fit the standard psychological paradigm for studying deductive reasoning and perhaps has elements of induction (rule discovery) or abduction (reasoning to best explanation), although it also requires hypothetico-deductive reasoning. The selection task also falls outside of the standard paradigm, although it is generally studied by deduction researchers. In the original task (Wason, 1966), participants were told that a set of cards each had a letter on one side and a number on the other. They were also told that a rule applied to these four cards and might be true or false:

If a card has a vowel on one side then it has an even number on the other side.

They were then shown four cards lying on a table, whose facing sides displayed a vowel, a consonant, an even number, and an odd number, for example,

E P 4 7

Their task was to choose those cards, and only those cards, that needed to be turned over in order to decide whether the rule was true or false. Wason maintained (and most subsequent researchers of the task agreed) that the correct choices were the vowel and the odd number, E and 7 in the example given. This is because, logically, the conditional statement can only be false if a card is found with a vowel on one side that does not have an even number on the other. Very few people gave this response—later estimates of solution rates are usually around 10%. Most people chose only the vowel or the vowel and the even number (e.g., E and 4).

Wason believed that both tasks showed that partici- pants had a verification bias, contrary to the then fash- ionable view of Karl Popper (1959) that the rational approach to hypothesis testing was falsification. Verifi- cation or confirmation bias has not withstood the scru- tiny of later researchers as an account of either task, but the details are not relevant for current purposes. Of more concern is Wason's assertion that failure on these tasks demonstrated that people were fundamentally irratio- nal. The case was made in a number of places, including an influential book of the time (Wason & Johnson- Laird, 1972). People were illogical and therefore did not develop formal operational thought as Inhelder and Piaget (1958) had proposed; they verified, and failed to falsify, their hypotheses and hence violated Popper's pre- scriptions for scientific thinking. Wason drew the stron- gest possible conclusion from his studies of the selection task:

> The selection task reflects [a tendency towards irrationality in argument] to the extent that subjects get it wrong. . . . It could be argued that irrationality rather than rationality is the norm. People succumb all too readily to logical failures. (Wason, 1983, p. 59)

Wason was being controversial and knew it. He was well aware, for example, of a philosophical traditional that made logical reasoning the centerpiece of rational thinking, as he included the famous paper of Henle (1962) in one of his edited collections. Henle asserted that apparent failures of logical reasoning were due to misinterpretations, additions of unstated premises, or "failures to accept the logical task." Wason railed against this tradition as well as the overbearing influence of Piaget in the psychology of that period (Wason, 1977). He believed—with clear justification at the time—that he had discovered that people were fundamentally irra- tional in their reasoning.

The selection task went on to have a long and distin- guished history of study (Evans, 2016a), especially since it was discovered quite early on that versions using real- istic content could be much easier to solve. This led ini- tially to a belief in a general thematic facilitation effect, but the story proved a good deal more complicated as later research unfolded (see Evans, Newstead, et al., 1993; Evans & Over, 2004). Not all realistic versions facilitate, and most versions that do switch from indicative to deontic logic (Manktelow & Over, 1991). That is, the rules tested are typically permissions and obligations such as "If you drink alcohol in a bar you must be over 18 years of age," and the task set is to check whether or not the rule is obeyed rather than whether it is true or false. I will

return to this when discussing evolutionary approaches later in the chapter.

2.3 The Influence of Kahneman and Tversky

In the early 1970s, when belief in irrationality was at its peak in the psychology of reasoning, the consensus view in a different tradition—the study of judgment and deci- sion making (JDM)—was still that people were essentially rational. This changed dramatically when Amos Tversky and Daniel Kahneman started to publish their famous papers, which launched the heuristics and biases tradi- tion (e.g., Tversky & Kahneman, 1974). Many reasoning researchers also paid attention to the JDM literature, so this work also had a clear influence on the rationality debate in the psychology of reasoning. The combined impact of studies in both fields helped to stimulate the historically important critique of Cohen (1981).

Prior to Kahneman and Tversky's work, the field of judgment and decision making had enjoyed a belief in rationality similar to that promoted by Henle and Piaget. While it was conceded that people had calibra- tion issues, they were generally considered to be good intuitive statisticians (Peterson & Beach, 1967). Eco- nomic decision theory had exerted a big influence on the study of decision making thanks to Ward Edwards (e.g., 1961) with attempts to propose subjective ver- sions of both probability and utility to fit the model of the "rational man" of decision theory. Tversky and Kahneman blew this cozy consensus away with a series of papers demonstrating systematic biases in statistical judgment and, later, decision making (Kahneman & Tversky, 1979). Kahneman would later receive the Nobel Prize for Economics for this work, Tversky being by then deceased and hence ineligible. I have no space to review this work in detail here but note that the fundamental idea was that people relied on simple heuristics such as representativeness (Kahneman & Tversky, 1972) and availability (Tversky & Kahneman, 1973) when making probability judgments. While sometimes helpful, these heuristics frequently led to systematic biases, as they demonstrated in a series of ingenious experiments. This set in train a tradition known as "heuristics and biases," which was studied mostly by other researchers as time went on, with collections of papers later published in influential books (Gilovich, Griffin, & Kahneman, 2002; Kahneman, Slovic, & Tversky, 1982).

Unlike Wason, Kahneman and Tversky generally avoided direct claims that they were showing people to be irrational, but their many admirers and followers were less restrained. There was great interest in this work in business schools, which increased when they added *prospect theory* (Kahneman & Tversky, 1979; Tversky & Kahneman, 1992) as an alternative to the traditional teaching of economic decision theory. By the end of the 1970s, the idea that people were fundamentally irrational in their reasoning, judgment, and decision making was taking a strong hold and was no longer confined to the journals of academic psychology. However, strong reaction from the defenders of human rationality was soon to arrive.

3. The Critique of Cohen (1981): Three Arguments for Rationality

Jonathan Cohen (1981) was by no means the only author to react against the emerging view of irrational thinking in the psychology of reasoning and judgment, but his paper was the most influential both at the time and for many years to follow. Cohen was a philosopher and hence employed a priori rather than empirical arguments. He posed the question, *"Can human irrationality be experimentally demonstrated?"* in his title and provided the answer in his text: an emphatic "no." This position and ones like it have later been dubbed "Panglossian" (see, e.g., Stanovich, 2011). However, Cohen made several strong arguments, which I later described (Evans, 1993) as falling into three broad categories:

1. The normative system problem
2. The interpretation problem
3. The external validity problem

Since these three basic arguments have also arisen in much more recent debate about rationality, I will take a little time to explain them. The *normative system problem* is that when participants are judged to be right or wrong, this is done by reference to a standard normative theory, which itself may be subject to argument and alternatives. For example, Wason adopted Popper's falsificationist philosophy of science as his normative system when he described verification bias on the 2 4 6 and selection tasks. Popper argued that lawful scientific statements were strict universals and could hence never be logically verified, but they could be falsified. Popper's work was recent and "hot" in the early 1960s when Wason was writing but has been severely challenged since, with some contemporary philosophers and psychologists preferring a Bayesian philosophy of science, in which belief in scientific theories is gradually acquired or lost as evidence accumulates (Howson & Urbach, 2006).

Although most psychologists still accept Wason's statement of the correct solution to the selection task, it has been famously challenged (Oaksford & Chater, 1994). The

"alternative norms" problem later became a major issue for debate in the field (e.g., Stanovich, 1999): you do not need to be a Panglossian to recognize the ambiguity of norms. Much more recently, Elqayam and Evans (2011) discussed the normative system problem at length in order to argue their case that normative rationality has no useful part to play in cognitive psychology. They argued that ambiguous and alternative norms frequently arise in the literatures on reasoning and decision making. Indeed, it was the questioning of standard binary logic and the material conditional that led to the new paradigm psychology of reasoning discussed at the end of this chapter. So the problem here is that if people are to be judged to be reasoning incorrectly and therefore to be irrational, do we actually have a clear and agreed standard for what is right and wrong in the first place?

The second type of objection made by Cohen and others (e.g., Henle, 1962; Smedslund, 1970, 1990) is that participants may not *interpret* the problem in the way that the experimenter intends and assumes. If, for example, they add or omit premises in deductive reasoning, then the conclusion that follows from their interpretation cannot be compared to the one prescribed by the experimenter. In the statistical reasoning literature, a good example is provided by the *conjunction fallacy* originally reported by Tversky and Kahneman (1983) and much investigated thereafter: people make an apparent logical error by judging that the probability of $A \& B$ is greater than the probability of A. In the famous Linda problem, Linda is described in a way that provokes the stereotype of a feminist. Participants then appear to judge as more likely that she is a bank teller *and* a feminist than just a bank teller, which is logically impossible. However, the context might suggest that the "bank teller" option is really a bank teller who is *not* a feminist. With this interpretation, there is no error. A number of authors have proposed interpretational explanations of the conjunction fallacy (see Moro, 2011).

Cohen also suggested that psychological experiments can be artificial and unrepresentative, dismissing the Wason selection task as a "cognitive illusion." I call this the *external validity problem*. Actually, the 2 4 6 task is a better example, as Wason's claim that it showed a verification bias was challenged almost immediately (Wetherick, 1962). Wetherick's objections were essentially correct, if largely ignored at the time. The paper was not even cited by Wason and Johnson-Laird (1972), whose well-read book popularized Wason's original view of the task. However, from the 1980s onward, other authors made similar arguments, which are now widely accepted (see Evans, 2016b). The general argument is that while

Wason showed a preference for positive hypothesis test-ing, this only has the effect of repeated confirmation due to the fact that participants have been induced to adopt a hypothesis much more specific than the actual rule. Moreover, it has been argued that positive test-ing of hypotheses is generally an adaptive and effective strategy in science (Klayman & Ha, 1987) and that the adoption of a verification or falsification attitude in sci-entific hypothesis testing makes no practical difference to the effectiveness of a test (Poletiek, 2001).

In the case of the 2 4 6 task, it does seem that the origi-nal experiment suggested a misleading interpretation and that it was indeed unrepresentative of real life. However, of Cohen's three arguments, the external validity prob-lem is the weakest. It is implausible to suggest that psy-chologists consistently contrive to demonstrate errors and biases that only occur as laboratory tricks. As a bias researcher myself, I can attest that there is no conspiracy among psychologists to denigrate human rationality. In the heuristics and biases tradition, in particular, most of the cognitive biases discovered originally on artificial laboratory tasks have been shown to apply in real-world contexts and with expert groups (Gilovich et al., 2002; Kahneman et al., 1982). That is not to say, of course, that there may not be a number of laboratory demonstrations of biases that are, nevertheless, misleading.

4.   The Great Rationality Debate

Following Cohen's critique, psychologists started to stake out various positions on the topic of rationality in reasoning and decision making, most of which were well established by the end of the 20th century. This has been labeled the "Great Rationality Debate" by Keith Stano-vich (2011), who in an earlier book (Stanovich, 1999) had already categorized them as falling into three main camps, which he termed "Panglossian," "Meliorist," and "Apologist." *Panglossians*, like Cohen, essentially argue that irrationality is not possible and there can be no gap between descriptive and normative accounts of human behavior (see Stanovich, 1999, p. 5). *Meliorists* claim that, while often convergent, such accounts can also diverge. People can fail to attain normative rationality due to lack of education, cognitive capacity, or rational attitude, but much of this divergence can be addressed by education and training. Meliorists include Jonathan Baron (1985) and Stanovich himself. Finally, *Apolo-gists* attempt to explain why performance often fails to match normative standards and give much weight to computational limitations, in the "bounded rationality" tradition of Simon (1982).

Not all authors seem to fall into Stanovich's catego-ries, including myself, as I have consistently questioned the value of normative theory itself in the study of rea-soning since my earliest papers on the topic (Evans, 1972). Authors such as myself may be termed "descrip-tivists" (Elqayam & Evans, 2011). In a collaboration with the philosopher David Over, I suggested that normative rationality (conforming to a normative theory) may often diverge from instrumental rationality (acting in such a way as to achieve one's goals)—see Evans and Over (1996). (This view was strongly contested by Stanovich, e.g., 1999, who regards normative theory as a descrip-tion of instrumental rationality.) Evans and Over (1996, chapter 2) strongly critiqued decision theory as lacking the boundaries and constraints necessary for compari-son with actual human choices. In a later book (Evans & Over, 2004), we challenged the binary logic that has traditionally been used to provide norms for deductive reasoning. We also picked out the case of belief bias as being a possible artifact of experimentation (Cohen's external validity problem), since in the real world, it is rational to reason from *all relevant belief* (Evans, Over, & Manktelow, 1993). However, when I later teamed up with Shira Elqayam to present a critique of normative rationality in *Behavioral and Brain Sciences* (Elqayam & Evans, 2011), it became apparent from the commentar-ies it elicited that many authors disagree with us. The majority view appears to be that normative theory does have an important role to play in the descriptive study of reasoning and decision making, and it continues to play a central role in the Great Rationality Debate.

4.1   Evolutionary and Ecological Approaches

While the philosopher Cohen started the rationalist backlash against research on cognitive biases, the cause of humanity was taken up later by psychological writers. Many assumed that human beings would be adapted to their environment, which may occur by evolution, learn-ing, or a combination of both. Gerd Gigerenzer emerged in the early 1990s and launched a series of attacks on the work of Tversky and Kahneman as providing an inac-curate picture of irrationality. This eventually led to an exchange in the pages of *Psychological Review* (Gigeren-zer, 1996a; Kahneman & Tversky, 1996). One of Gigeren-zer's arguments was that people are able to reason with frequencies rather than probabilities and that appar-ent cognitive biases could be made to disappear when frequencies were used in the word problems instead (Gigerenzer, 1991; Gigerenzer & Hoffrage, 1995). This was supported by the work of the evolutionary psychol-ogists Cosmides and Tooby (1996; see also chapter 10.6

by Cosmides & Tooby, this handbook) but resulted in a flurry of papers from skeptical authors disputing their explanation, based on Darwinian algorithms, of *why* frequency formats make the problems easier. Critics suggested that these formats cued simple mental models that facilitated reasoning for reasons that have nothing to do with evolution (see Barbey & Sloman, 2007, for an extended review and discussion of this debate).

Leda Cosmides and John Tooby set out a more general argument for *massive modularity* (Cosmides & Tooby, 1992; Tooby & Cosmides, 1992), according to which the mind comprises specialized self-contained modules. Cognitive modularity was originally proposed by Fodor (1983), who argued that many such modules are contained in the mind, for example, to process language or to explain much of the working of the visual system (although many authors, including Cosmides and Tooby, use the term without subscribing to all of his detailed definitions). Importantly, Fodor argued that there must also be a general-purpose reasoning system, something that the massive modularity hypothesis denied. Cosmides and Tooby argued that evolution could only produce specialized and not general reasoning systems, a highly controversial position. Fodor (2000, 2001) was among their many critics, and a more general coverage of the debate can be found in Over (2003). In the psychology of reasoning, these arguments were first applied to the Wason selection task. As mentioned earlier, the task is much easier when people are asked to check whether permissions or obligations have been complied with. Cosmides (1989) argued that these versions of the task comprised *social contracts*, for which we have evolved a cheater-detection module. In later work, this was extended to include a hazard-avoidance module (Fiddick, Cosmides, & Tooby, 2000), which applied to other contexts in which the task becomes easy. Their work on this topic was also supported by Gigerenzer and Hug (1992).

Again, all these claims proved highly controversial and were subject to critiques of methodology and the postulation of rival theoretical accounts that did not depend upon evolutionary mechanisms. The relevance to the rationality debate is the argument that people have an inherent rationality in their reasoning provided by their genes but that it only operates in contexts of evolutionary relevance. One problem with this approach, pointed out by Stanovich (2004), is that evolutionary and individual rationality can diverge. For example, the craving for fat and sugar that made great evolutionary sense to our ancestors has become a cause of obesity and health problems in the modern world. Modern humans can also frustrate the goals of the genes, for example, by enjoying sex with contraception. Stanovich (2004) claimed that human beings are the one species that the genes lost control of, as we can think for ourselves and pursue our own goals.

Gigerenzer and colleagues developed a major research program based on "fast and frugal heuristics," which are taken from an evolutionary toolbox similar to the modular mind of Cosmides and Tooby (Gigerenzer, 1996b; Gigerenzer & Todd, 1999). In contrast with the heuristics of Kahneman and Tversky, these are portrayed as making us smart without the need for much thought. In fact, Gigerenzer claims that ignorance often trumps knowledge and that we are better off relying on gut feelings than engaging in reasoning (Gigerenzer, 2007), ironically employing masterful reasoning to make his case! Again, we see the denial of any useful role for a general-purpose reasoning system, which puts this approach in direct conflict with the "dual-process theories" of reasoning discussed later. In fact, Gigerenzer is a strong critic of dual-process theory (Gigerenzer, 2011; Kruglanski & Gigerenzer, 2011). Evolutionary arguments have also been made to explain our reasoning facility without attributing to it the importance that is usually attached to it. For example, it has been proposed that reasoning evolved for purposes of argumentation rather than problem solving (Mercier & Sperber, 2011), although a more complex story has emerged in later development of this account (Mercier & Sperber, 2017).

Evolutionary rationality is closely related to the idea of *ecological rationality*, which comes from a slightly different tradition. *Rational analysis* is a tradition in cognitive psychology started by John Anderson (1990, 1991). It has been applied to the psychology of reasoning and decision making in a major program led by Mike Oaksford and Nick Chater (Oaksford & Chater, 1998, 2007), one of the roots of the new paradigm psychology of reasoning discussed at the end of this chapter. It is grounded in ecological rationality and related to other psychological work in this tradition (Gibson, 1979; Hammond, 1966). The approach takes it as axiomatic that behavior is adapted to the environment and that therefore an analysis of the environment itself is required in order to see what is the optimal way of operating within it (a computational-level account, in the terminology of Marr, 1982). Normative theory, while important, is subservient to this ecological rationality. As Oaksford and Chater (2007) put it, "According to this viewpoint, formal rational principles relate to explaining everyday rationality because they specify the optimal ways in which the goals of the system can be attained in a particular environment" (p. 32). They also make it perfectly clear that if a *standard* normative theory, such as binary or propositional logic, fails to achieve that task, then it must be discarded and replaced. Much of their research, starting with their alternative account of the selection task (Oaksford & Chater, 1994), has consisted of providing alternative normative accounts that render the actual observed behavior of their participants normatively correct, in the spirit of Cohen's first argument for rationality. Their position evolved to one of arguing that human rationality is based upon Bayesian decision theory rather than logic.

4.2 Dual Processes, General Intelligence, and Individual Differences

Dual-process theories have become highly popular in cognitive and social psychology but have diverse origins and many different implementations (Evans, 2008; Evans & Frankish, 2009). They are also highly controversial and have been the subject of many criticisms (see Evans & Stanovich, 2013). What all have in common is the assertion that there are two fundamentally different kinds of human thought, one fast and intuitive and the other slow and reflective. These are sometimes referred to as "System 1" and "System 2" (Kahneman, 2011; Stanovich, 1999), although both Stanovich and I now prefer the more neutral reference to "Type 1" and "Type 2" processing in our current writing. Modular cognitive processes would be automated and therefore Type 1, but such processing can also be based upon associative or implicit learning (Reber, 1993) or on the automation of knowledge that was once explicit and processed in a Type 2 manner. Type 2 processing is now generally thought to involve a central general-purpose working memory system of the kind described by Baddeley and his coworkers (Baddeley, 2007) and is related to the concept of general intelligence. Hence, dual-process theory strongly asserts the existence of a general-purpose reasoning system, in stark contrast with the evolutionary approaches discussed above.

Traditionally, dual-process theory has largely explained cognitive biases in terms of Type 1 processing and correct reasoning or decision making in terms of Type 2 processing (Evans, 1989; Kahneman, 2011; Stanovich, 1999). However, this is oversimplified as intuitions can be sound and reasoning flawed. Type 2 processing, for example, can go awry because of insufficient cognitive capacity, lack of relevant education or "mindware," or a failure to adopt a rational attitude in which intuitions are carefully checked (Evans, 2007; Evans & Stanovich, 2013; Stanovich, 2011). Nevertheless, much has been made of

the fact that general intelligence correlates not only with working memory capacity but also with normatively correct answers on a wide range of laboratory reasoning and decision tasks (Stanovich, 1999, 2011; Stanovich, West, & Toplak, 2016). However, the same research program has always shown that another individual difference factor has a significant, if weaker, influence on correct performance. This is *rational thinking disposition*, assessed by a variety of psychometric scales. In essence, it appears that some people are inclined to make quick decisions based on intuitions and others to check them out by more careful reasoning. Since most laboratory tasks are designed to minimize the value of Type 1 processing (previous experience is unhelpful) and to maximize the role of Type 2 processing (problems are novel and load on working memory), a rational thinking disposition aids solution.

Stanovich has discussed rationality at length in his several books on this topic. In the first of these (Stanovich, 1999), he rejected the argument that rationality is revealed by alternative normative accounts, as in the work of Oaksford and Chater. He points out that on most of the tasks studied, general intelligence correlates with the *standard* normative solution. For example, even on the very difficult abstract form of the Wason selection task, the few people who do manage to solve it have higher IQs than those who do not (Stanovich & West, 1998). In this and later books, he develops the idea that rationality should not, however, be equated with IQ, due to the strong influence of rational thinking dispositions: there should therefore be a Rationality Quotient, or "RQ," that takes both into account (Stanovich, 2009). In a recent major project, Stanovich and his collaborators explored the psychometric properties of a large range of laboratory tasks in order to provide the foundation for such an RQ test (Stanovich et al., 2016), called the "CART." The CART still loads heavily on general intelligence, however, and currently lacks both the item analysis and validation studies that would be required to turn it into a marketable psychometric instrument.

Dual-process theory is currently undergoing something of a crisis due to a series of results that show that correct answers on many of the tasks may be given quickly and intuitively and do not benefit from slower reflection (see a number of papers in the collection edited by De Neys, 2017). Recent findings also suggest that people of higher IQ have better intuitions, or Type 1 processes, than those of lower intelligence. This is leading some authors, previously sympathetic to the theory, to question the traditional dual-process account in which better normative performance on tasks and by individuals is achieved by Type 2 processing. I personally think that some of this pessimism is premature as much of the research is based on relatively simple tasks in which "logical intuitions" are less surprising than they would be on tasks that load heavily on working memory.

5.   The Rejection of Binary Logic and the New Paradigm Psychology of Reasoning

The rationality debate in the psychology of reasoning has produced a tangible result: a new paradigm. The traditional method, dubbed the "deduction paradigm" by Evans (2002), provides no instruction in logic and rejects participants who have such training. The idea is to see if people are inherently logical in their reasoning. Generally, they are given the premises of an argument, which they are told to assume are true, and asked if a particular conclusion necessarily follows. Their decisions are then compared with those of standard logics such as the propositional calculus. By the end of the 20th century, this paradigm was creaking under the weight of a mass of findings showing that people (a) made many logical errors, (b) had systematic cognitive biases, and (c) frequently reasoned on the basis of actual beliefs rather than assumed premises. A Kuhnian crisis was looming (Evans, 2002). There were also powerful voices, such as those of Oaksford and Chater, arguing that the normative model of standard logic was simply inapplicable to everyday reasoning.

With the pressure of the rationality debate upon them, reasoning researchers were basically faced with two choices: (1) defend the deduction paradigm and conclude that people were massively irrational or (2) abandon the paradigm, especially the use of standard logic as a normative system. Many chose the latter and moved toward a new paradigm, although the mental model camp put up a lot of resistance, as mentioned earlier. A key development came in the study of conditionals when psychologists became increasingly aware that many philosophers reject the material conditional of propositional logic as a description of the ordinary conditional of natural language (Edgington, 1995). In this system, "If p then q" means the same thing as "Either not-p or q," but this leads to absurd and paradoxical inferences. Such a conditional is always true when either p is false or q is true (or both). So, for example, "If Hillary Clinton is president of the USA, then the UK is at war with France" must be true.

Also taking the lead from philosophers, psychologists became interested in the Ramsey test, which is the idea that we evaluate a conditional by imagining p to be true and then evaluating q in a thought experiment. This was quickly linked with the idea that the probability of the conditional statement should be equal to the conditional probability, that is, $P(\text{if } p \text{ then } q) = P(q \mid p)$, known as "the Equation." Psychological evidence was gathered in the early 2000s that many people complied with the Equation and not with the material conditional when judging the likelihood of conditional statements (Evans, Handley, & Over, 2003; Oberauer & Wilhelm, 2003; Over, Hadjichristidis, Evans, Handley, & Sloman, 2007). David Over and I presented a suppositional theory of conditionals (Evans & Over, 2004) developing the Ramsey test into a full-blown psychological account of conditional thought. This work, together with the case that Oaksford and Chater had been pressing for a decade, led to what is now clearly recognized as a new paradigm in the psychology of reasoning (Elqayam & Over, 2012; Oaksford & Chater, 2010; Over, 2009).

Researchers within the new paradigm agree on the rejection both of the traditional methods and of the normativity of binary logic. Participants are more likely to be asked to reason with uncertain premises or to draw uncertain conclusions. For some researchers, pragmatics is key, as ordinary reasoning is now recognized as belief based. For others, the new paradigm involves an attempt to replace binary logic with alternative normative systems, such as Bayesian decision theory, probability logic, or multivalued logic. Researchers may, however, subscribe to Bayesianism in a strict or a soft sense (Elqayam & Evans, 2013), the latter falling short of a normative theory. The new paradigm is a looser concept than the old paradigm as there is as yet no real general agreement as to its priorities, methods, and theoretical framing (Evans, 2012). It is quite clear, though, that the rationality debate played a critical role in moving most of the field away from the traditional study of logicality in human reasoning.

6.   Final Thoughts

Throughout cognitive psychology generally, there is no rationality debate. People are not accused of irrationality for having poor memories or succumbing to visual illusions, for example. Even within the psychology of thinking, cognitive failure does not necessarily lead to imputations of irrationality; for example, we do not expect that people should be able to compute the square roots of large numbers in their heads. The issue is pretty much restricted to the study of reasoning and decision making—so what makes these topics different? The only answer that I can see is the strong influence of philosophy. The issue of rationality and the efficacy of different normative systems are primarily philosophical issues. However, psychologists have become highly engaged with them, to the extent that they have tried to approach rationality as an *empirical* question. The only way that this can be done is to observe human reasoning and decision making and compare it with a normative standard. But as we have seen, such a path is rife with pitfalls. All normative theories are themselves contestable, for example. And should the outcome be that people appear irrational, strong arguments can be made on evolutionary and ecological grounds that this conclusion must be false. But if the conclusion is prejudged, why bother asking the question at all?

Personally, I think these fields of study should be approached like the rest of cognitive psychology: we should describe the processes of reasoning and decision making and simply try to understand the cognitive and neural mechanisms that underlie them. However, this is not the popular view and certainly does not reflect the history of the psychology of reasoning in which the rationality debate has played such a prominent role. If the Panglossians are right, then they must agree with me on how the topic should be studied, as the issue of human rationality can never be decided empirically. In reality, however, most psychologists have focused on right and wrong answers throughout the history of the psychology of reasoning and continue to do so today. When people were shown to be intransigent in their lack of respect for logic, standard logic was abandoned in favor of a new paradigm. People are now seen to concern themselves with degrees of belief rather than truth and to reason with uncertainty. But normativity has not gone away in this new paradigm: psychologists continue to hunt for new formal descriptions of rational reasoning that might account for the way in which people actually think.

One final thought is on the nature of evolution. We are what we are and do what we do because we evolved to do so. We became the dominant species on Earth—at least by some measures—as a result and certainly the only one that can with great power and flexibility design its environment to suit itself, rather than evolving to fit the environment. Our scientific and cultural achievements are extraordinary. But the 21st-century world is also striven with war, disease, and starvation, and there is every likelihood that the failure to control climate change or to manage the horrendous weapons of mass destruction we have designed will bring the whole project to a premature end. To me this reflects the nature of evolution: it does not optimize, and it does not accord with some philosophical ideal of rationality. It is driven mostly by what worked in the past and provides no guarantees for the future. The ultimate test of the rationality of human beings lies not in the psychologist's laboratory but in the highly uncertain and dangerous real world that we inhabit.

Acknowledgments

I thank Shira Elqayam and Thomas Sturm for detailed critical reading of an earlier draft of this chapter.

References

Anderson, J. R. (1990). *The adaptive character of thought*. Hillsdale, NJ: Erlbaum.

Anderson, J. R. (1991). Is human cognition adaptive? *Behavioral and Brain Sciences, 14*, 471–517.

Baddeley, A. (2007). *Working memory, thought, and action*. Oxford, England: Oxford University Press.

Barbey, A. K., & Sloman, S. A. (2007). Base-rate respect: From ecological validity to dual processes. *Behavioral and Brain Sciences, 30*, 241–297.

Baron, J. (1985). *Rationality and intelligence*. Cambridge, England: Cambridge University Press.

Braine, M. D. S., & O'Brien, D. P. (1991). A theory of if: A lexical entry, reasoning program, and pragmatic principles. *Psychological Review, 98*, 182–203.

Braine, M. D. S., & O'Brien, D. P. (1998). The theory of mental-propositional logic: Description and illustration. In M. D. S. Braine & D. P. O'Brien (Eds.), *Mental logic* (pp. 79–89). Mahwah, NJ: Erlbaum.

Byrne, R. M. J., & Johnson-Laird, P. N. (2009). 'If' and the problems of conditional reasoning. *Trends in Cognitive Sciences, 13*, 282–287.

Chater, N., & Oaksford, M. (1999). The probability heuristics model of syllogistic reasoning. *Cognitive Psychology, 38*, 191–258.

Cohen, L. J. (1981). Can human irrationality be experimentally demonstrated? *Behavioral and Brain Sciences, 4*, 317–370.

Cosmides, L. (1989). The logic of social exchange: Has natural selection shaped how humans reason? *Cognition, 31*, 187–276.

Cosmides, L., & Tooby, J. (1992). Cognitive adapations for social exchange. In J. H. Barkow, L. Cosmides, & J. Tooby (Eds.), *The adapted mind: Evolutionary psychology and the generation of culture* (pp. 163–228). New York, NY: Oxford University Press.

Cosmides, L., & Tooby, J. (1996). Are humans good intuitive statisticians after all? Rethinking some conclusions from the literature on judgment under uncertainty. *Cognition, 58*, 1–73.

De Neys, W. (Ed.). (2017). *Dual process theory 2.0*. London, England: Routledge.

Dube, C., Rotello, C. M., & Heit, E. (2010). Assessing the belief bias effect with ROCs: It's a response bias effect. *Psychological Review, 117*, 831–863.

Edgington, D. (1995). On conditionals. *Mind, 104*, 235–329.

Edwards, W. (1961). Behavioral decision theory. *Annual Review of Psychology, 67*, 441–452.

Elqayam, S., & Evans, J. St. B. T. (2011). Subtracting "ought" from "is": Descriptivism versus normativism in the study of human thinking. *Behavioral and Brain Sciences, 34*, 233–290.

Elqayam, S., & Evans, J. St. B. T. (2013). Rationality in the new paradigm: Strict versus soft Bayesian approaches. *Thinking & Reasoning, 19*(3–4), 453–470.

Elqayam, S., & Over, D. E. (2012). Probabilities, beliefs, and dual processing: The paradigm shift in the psychology of reasoning. *Mind & Society, 11*, 27–40.

Evans, J. St. B. T. (1972). On the problems of interpreting reasoning data: Logical and psychological approaches. *Cognition, 1*, 373–384.

Evans, J. St. B. T. (1982). *The psychology of deductive reasoning.* London, England: Routledge.

Evans, J. St. B. T. (1989). *Bias in human reasoning: Causes and consequences.* Brighton, England: Erlbaum.

Evans, J. St. B. T. (1993). Bias and rationality. In K. I. Manktelow & D. E. Over (Eds.), *Rationality: Psychological and philosophical perspectives* (pp. 6–30). London, England: Routledge.

Evans, J. St. B. T. (2002). Logic and human reasoning: An assessment of the deduction paradigm. *Psychological Bulletin, 128*, 978–996.

Evans, J. St. B. T. (2007). *Hypothetical thinking: Dual processes in reasoning and judgement.* Hove, England: Psychology Press.

Evans, J. St. B. T. (2008). Dual-processing accounts of reasoning, judgment, and social cognition. *Annual Review of Psychology, 59*, 255–278.

Evans, J. St. B. T. (2012). Questions and challenges for the new psychology of reasoning. *Thinking & Reasoning, 18*, 5–31.

Evans, J. St. B. T. (2016a). A brief history of the Wason selection task. In N. Galbraith (Ed.), *The thinking mind: The use of thinking in everyday life* (pp. 1–14). Hove, England: Psychology Press.

Evans, J. St. B. T. (2016b). Reasoning, biases and dual processes: The lasting impact of Wason (1960). *Quarterly Journal of Experimental Psychology, 69*, 2076–2092.

Evans, J. St. B. T., Barston, J. L., & Pollard, P. (1983). On the conflict between logic and belief in syllogistic reasoning. *Memory & Cognition, 11*, 295–306.

Evans, J. St. B. T., & Frankish, K. (Eds.). (2009). *In two minds: Dual processes and beyond.* Oxford, England: Oxford University Press.

Evans, J. St. B. T., Handley, S. J., & Over, D. E. (2003). Conditionals and conditional probability. *Journal of Experimental Psychology: Learning, Memory, and Cognition, 29*, 321–335.

Evans, J. St. B. T., Newstead, S. E., & Byrne, R. M. J. (1993). *Human reasoning: The psychology of deduction.* Hove, England: Erlbaum.

Evans, J. St. B. T., & Over, D. E. (1996). *Rationality and reasoning.* Hove, England: Psychology Press.

Evans, J. St. B. T., & Over, D. E. (2004). *If.* Oxford, England: Oxford University Press.

Evans, J. St. B. T., Over, D. E., & Manktelow, K. I. (1993). Reasoning, decision making and rationality. *Cognition, 49*, 165–187.

Evans, J. St. B. T., & Stanovich, K. E. (2013). Dual-process theories of higher cognition: Advancing the debate. *Perspectives on Psychological Science, 8*, 223–241.

Fiddick, L., Cosmides, L., & Tooby, J. (2000). No interpretation without representation: The role of domain-specific representations and inferences in the Wason selection task. *Cognition, 77*, 1–79.

Fodor, J. (1983). *The modularity of mind: An essay on faculty psychology.* Scranton, PA: Crowell.

Fodor, J. (2000). Why we are so good at catching cheaters. *Cognition, 75*, 29–32.

Fodor, J. (2001). *The mind doesn't work that way: The scope and limits of computational psychology.* Cambridge, MA: MIT Press.

Gibson, J. J. (1979). *The ecological approach to visual perception.* Boston, MA: Houghton Mifflin.

Gigerenzer, G. (1991). How to make cognitive illusions disappear: Beyond "heuristics and biases." *European Review of Social Psychology, 2*(1), 83–115.

Gigerenzer, G. (1996a). On narrow norms and vague heuristics: A reply to Kahneman and Tversky (1996). *Psychological Review, 103*, 592–596.

Gigerenzer, G. (1996b). Reasoning the fast and frugal way: Models of bounded rationality. *Psychological Review, 103*, 650–669.

Gigerenzer, G. (2007). *Gut feelings: The intelligence of the unconscious.* London, England: Penguin.

Gigerenzer, G. (2011). Personal reflections on theory and psychology. *Theory & Psychology, 20*(6), 733–743.

Gigerenzer, G., & Hoffrage, U. (1995). How to improve Bayesian reasoning without instruction: Frequency formats. *Psychological Review, 102*, 684–704.

Gigerenzer, G., & Hug, K. (1992). Domain-specific reasoning: Social contracts, cheating, and perspective change. *Cognition, 43*, 127–171.

Gigerenzer, G., & Todd, P. M. (1999). *Simple heuristics that make us smart.* New York, NY: Oxford University Press.

Gilovich, T., Griffin, D., & Kahneman, D. (2002). *Heuristics and biases: The psychology of intuitive judgement.* Cambridge, England: Cambridge University Press.

Hammond, K. R. (Ed.). (1966). *The psychology of Egon Brunswik.* New York, NY: Holt, Reinhart & Winston.

Henle, M. (1962). On the relation between logic and thinking. *Psychological Review, 69,* 366–378.

Howson, C., & Urbach, P. (2006). *Scientific reasoning: The Bayesian approach* (3rd ed.). Chicago, IL: Open Court.

Inhelder, B., & Piaget, J. (1958). *The growth of logical thinking.* New York, NY: Basic Books.

Johnson-Laird, P. N. (1983). *Mental models: Towards a cognitive science of language, inference, and consciousness.* Cambridge, England: Cambridge University Press.

Johnson-Laird, P. N. (2006). *How we reason.* Oxford, England: Oxford University Press.

Johnson-Laird, P. N., & Byrne, R. M. J. (1991). *Deduction.* Hove, England: Erlbaum.

Johnson-Laird, P. N., & Byrne, R. M. J. (2002). Conditionals: A theory of meaning, pragmatics, and inference. *Psychological Review, 109,* 646–678.

Johnson-Laird, P. N., & Savary, F. (1999). Illusory inferences: A novel class of erroneous deductions. *Cognition, 71,* 191–299.

Johnson-Laird, P. N., & Steedman, M. (1978). The psychology of syllogisms. *Cognitive Psychology, 10,* 64–99.

Kahneman, D. (2011). *Thinking, fast and slow.* New York, NY: Farrar, Straus and Giroux.

Kahneman, D., Slovic, P., & Tversky, A. (Eds.). (1982). *Judgment under uncertainty: Heuristics and biases.* Cambridge, England: Cambridge University Press.

Kahneman, D., & Tversky, A. (1972). Subjective probability: A judgment of representativeness. *Cognitive Psychology, 3,* 430–454.

Kahneman, D., & Tversky, A. (1979). Prospect theory: An analysis of decision under risk. *Econometrica, 47,* 263–291.

Kahneman, D., & Tversky, A. (1996). On the reality of cognitive illusions: A reply to Gigerenzer's critique. *Psychological Review, 103,* 582–591.

Klauer, K. C., Musch, J., & Naumer, B. (2000). On belief bias in syllogistic reasoning. *Psychological Review, 107,* 852–884.

Klayman, J., & Ha, Y.-W. (1987). Confirmation, disconfirmation, and information in hypothesis testing. *Psychological Review, 94,* 211–228.

Kruglanski, A. W., & Gigerenzer, G. (2011). Intuitive and deliberate judgments are based on common principles. *Psychological Review, 118*(1), 97–109.

Manktelow, K. I., & Over, D. E. (1991). Social roles and utilities in reasoning with deontic conditionals. *Cognition, 39,* 85–105.

Marr, D. (1982). *Vision: A computational investigation into the human representation and processing of visual information.* San Francisco, CA: Freeman.

Mercier, H., & Sperber, D. (2011). Why do humans reason? Arguments for an argumentative theory. *Behavioral and Brain Sciences, 34,* 57–111.

Mercier, H., & Sperber, D. (2017). *The enigma of reason: A new theory of human understanding.* Cambridge, MA: Harvard University Press.

Moro, R. (2011). On the nature of the conjunction fallacy. *Synthese, 171,* 1–24.

Oaksford, M., & Chater, N. (1991). Against logicist cognitive science. *Mind & Language, 6,* 1–38.

Oaksford, M., & Chater, N. (1994). A rational analysis of the selection task as optimal data selection. *Psychological Review, 101,* 608–631.

Oaksford, M., & Chater, N. (1998). *Rationality in an uncertain world.* Hove, England: Psychology Press.

Oaksford, M., & Chater, N. (2007). *Bayesian rationality: The probabilistic approach to human reasoning.* Oxford, England: Oxford University Press.

Oaksford, M., & Chater, N. (2010). Cognition and conditionals: An introduction. In M. Oaksford & N. Chater (Eds.), *Cognition and conditionals: Probability and logic in human thinking* (pp. 3–36). Oxford, England: Oxford University Press.

Oberauer, K., & Wilhelm, O. (2003). The meaning(s) of conditionals: Conditional probabilities, mental models, and personal utilities. *Journal of Experimental Psychology: Learning, Memory, and Cognition, 29,* 680–693.

Over, D. E. (Ed.). (2003). *Evolution and the psychology of thinking: The debate.* Hove, England: Psychology Press.

Over, D. E. (2009). New paradigm psychology of reasoning. *Thinking & Reasoning, 15,* 431–438.

Over, D. E., Hadjichristidis, C., Evans, J. St. B. T., Handley, S. J., & Sloman, S. A. (2007). The probability of causal conditionals. *Cognitive Psychology, 54,* 62–97.

Peterson, C. R., & Beach, L. R. (1967). Man as an intuitive statistician. *Psychological Bulletin, 68,* 29–46.

Poletiek, F. H. (2001). *Hypothesis-testing behaviour.* Hove, England: Psychology Press.

Popper, K. R. (1959). *The logic of scientific discovery.* London, England: Hutchinson.

Reber, A. S. (1993). *Implicit learning and tacit knowledge.* Oxford, England: Oxford University Press.

Rips, L. J. (1983). Cognitive processes in propositional reasoning. *Psychological Review, 90*, 38–71.

Rips, L. J. (1994). *The psychology of proof: Deductive reasoning in human thinking*. Cambridge, MA: MIT Press.

Simon, H. A. (1982). *Models of bounded rationality*. Cambridge, MA: MIT Press.

Smedslund, J. (1970). Circular relation between understanding and logic. *Scandinavian Journal of Psychology, 11*, 217–219.

Smedslund, J. (1990). A critique of Tversky and Kahneman's distinction between fallacy and misunderstanding. *Scandinavian Journal of Psychology, 31*, 110–120.

Stanovich, K. E. (1999). *Who is rational? Studies of individual differences in reasoning*. Mahwah, NJ: Erlbaum.

Stanovich, K. E. (2004). *The robot's rebellion: Finding meaning in the age of Darwin*. Chicago, IL: University of Chicago Press.

Stanovich, K. E. (2009). *What intelligence tests miss: The psychology of rational thought*. New Haven, CT: Yale University Press.

Stanovich, K. E. (2011). *Rationality and the reflective mind*. Oxford, England: Oxford University Press.

Stanovich, K. E., & West, R. F. (1998). Cognitive ability and variation in selection task performance. *Thinking & Reasoning, 4*, 193–230.

Stanovich, K. E., West, R. F., & Toplak, M. E. (2016). *The rationality quotient: Toward a test of rational thinking*. Cambridge, MA: MIT Press.

Tooby, J., & Cosmides, L. (1992). The psychological foundations of culture. In J. H. Barkow, L. Cosmides, & J. Tooby (Eds.), *The adapted mind: Evolutionary psychology and the generation of culture* (pp. 19–136). New York, NY: Oxford University Press.

Trippas, D., Handley, S. J., & Verde, M. F. (2013). The SDT model of belief bias: Complexity, time, and cognitive ability mediate the effects of believability. *Journal of Experimental Psychology: Learning, Memory, and Cognition, 39*(5), 1393–1402.

Tversky, A., & Kahneman, D. (1973). Availability: A heuristic for judging frequency and probability. *Cognitive Psychology, 5*, 207–232.

Tversky, A., & Kahneman, D. (1974). Judgment under uncertainty: Heuristics and biases. *Science, 185*(4157), 1124–1131.

Tversky, A., & Kahneman, D. (1983). Extensional versus intuitive reasoning: The conjunction fallacy in probability judgment. *Psychological Review, 90*, 293–315.

Tversky, A., & Kahneman, D. (1992). Advances in prospect theory: Cumulative representation of uncertainty. *Journal of Risk and Uncertainty, 5*, 297–323.

Wason, P. C. (1960). On the failure to eliminate hypotheses in a conceptual task. *Quarterly Journal of Experimental Psychology, 12*(3), 129–140.

Wason, P. C. (1966). Reasoning. In B. M. Foss (Ed.), *New horizons in psychology* (Vol. I, pp. 106–137). Harmondsworth, England: Penguin.

Wason, P. C. (1977). The theory of formal operations: A critique. In B. A. Geber (Ed.), *Piaget and knowing: Studies in genetic epistemology* (pp. 119–135). Abingdon, England: Routledge.

Wason, P. C. (1983). Realism and rationality in the selection task. In J. St. B. T. Evans (Ed.), *Thinking and reasoning: Psychological approaches* (pp. 44–75). London, England: Routledge & Kegan Paul.

Wason, P. C., & Johnson-Laird, P. N. (1972). *Psychology of reasoning: Structure and content*. Cambridge, MA: Harvard University Press.

Wetherick, N. E. (1962). Eliminative and enumerative behaviour in a conceptual task. *Quarterly Journal of Experimental Psychology, 14*, 246–249.

Wilkins, M. C. (1928). *The effect of changed material on ability to do formal syllogistic reasoning* (Archives of Psychology No. 102). New York, NY: Woodworth.

Woodworth, R. S., & Sells, S. B. (1935). An atmosphere effect in syllogistic reasoning. *Journal of Experimental Psychology, 18*, 451–460.

1.3 Evolution of Rationality

Gerhard Schurz

Summary

Rationality is a normative concept that has two dimensions: theoretical and practical rationality. We suggest evaluating theoretical rationality in terms of its cognitive success and practical rationality in terms of its promoting well-being and social cooperation. Based on this understanding, the evolution of rationality is investigated from both a philosophical and a psychological point of view. Human cognition has evolved in three dimensions: biological, cultural, and individual. Many researchers believe that human cognition originates from a sociolinguistic coevolution. A second characteristic is the coevolution of human technology and humans' ability to reason causally. Based on evolutionary considerations, it is argued that human cognition is based on a division of labor between specialized cognitive modules and general-purpose nonspecific mechanisms of learning and reasoning.

1. Factual and Normative Aspects of Rationality

Is it rational to increase human power by means of science and technology? Is it rational to believe in God because this increases one's happiness? The answer to these questions cannot solely be based on empirical facts and logical reasoning, because the notion of rationality used in these questions has a *normative* component. Since David Hume (1739/2004), we know that there is no logically valid inference from facts to norms.[1] The notion of rationality that one assumes relies at least partially on a value-loaded decision.

Thus, before we start this chapter, we have to explicate our assumed concept of rationality. In philosophy, one usually distinguishes between *theoretical* rationality (i.e., the rationality of our picture of the world) and *practical* rationality (i.e., the rationality of our norms of action). The aim of theoretical rationality is to acquire predictively and causally or explanatorily relevant information about the world. More generally formulated, the goal of theoretical rationality is *cognitive success*. Schurz and Hertwig (2019) define the cognitive success of a method as the product of its validity (percentage of correct inferences among all inferences rendered) times its applicability (percentage of all problems to which the method is applicable among all problems of a given type). When speaking in what follows about the evolution of theoretical rationality, we will focus on the dimension of cognitive success. There are other, *noncognitive* aspects of human belief systems that are the target of evolutionary selection. One such aspect played a significant role in the evolution of religion and has been called the *generalized placebo effect*: the mere *belief* in a benevolent God (or system of Gods) equipped with supernatural powers, who protects one and rewards godly behavior (either earthly or afterworldly), has strongly positive effects on one's psychological well-being, entirely *independently* of the cognitive implausibility of this belief.[2] Thus, although we focus our chapter on cognitive aspects, it should be clear that also certain noncognitive aspects play a role in the evolution of human belief systems.

The goal of practical rationality is usually seen in guiding our actions in the interest of human beings. What this means depends on one's accepted value judgments. As a minimal core meaning, we assume that acting in a practically rational way implies acting (a) in the egoistic interest of the acting person and (b) promoting cooperation between the interaction partners (possibly extended to nonhuman beings). Our discussion will focus on these two aspects of the evolution of practical rationality.

The remainder of this article is structured as follows: after laying down the foundations of generalized evolution theory, including biological, cultural, and individual evolution (section 2), we discuss the interrelation between the evolutions of social cognition, language, and mental representation (section 3); the evolution of causal reasoning (section 4); the relation between modularity and universality in human cognition (section 5); and the evolution of induction and deduction (section 6), and we conclude with a synopsis of the evolutionary architecture of human cognition (section 7).

2. Cognitive Evolution in Three Dimensions

Since our notion of rationality coincides with that of successful cognition, including social cognition, the focus of our chapter will be on the evolution of cognition. Human cognition is the product of *evolution in three dimensions*. The first and second dimension are:

(1) *Biological evolution* based on the biological inheritance of genetic information, including some recently explored epigenetic effects[3]

(2) *Cultural evolution*, where "culture" is intended in the broad sense of the generation-wise tradition of *acquired* information

According to a more recent research paradigm, also the evolution of culture can be understood as a generalized Darwinian process, which is not based on the inheritance of genes through (sexual) reproduction but rather on the inheritance of cultural traits based on *social learning*.[4] The latter notion means learning from other individuals, as opposed to individual learning from one's own trial-and-error experience. Knowledge acquired through social learning (i.e., not inherited genetically) is passed on from generation to generation and is successively improved through ongoing iterations of variation–selection (or trial-and-error) cycles.[5] This is most impressively exemplified in the evolution of technology from the tools of the Stone Age to contemporary automated industry and electronic intelligence (cf. Basalla, 1988), but it applies equally to other domains in the evolution of culture, such as science or law.

Biological- as well as cultural-level evolution is based on the three Darwinian modules of *reproduction, variation*, and *selection* (Dennett, 1995; Schurz, 2011; Sober, 1993). In contrast, the third dimension of evolution is only partially Darwinian:

(3) *Individual learning*, including inductive association learning and behavioral trial-and-error learning

The latter kind of learning, also called "operant conditioning," consists of Darwinian processes of variation and selection, insofar as spontaneously exhibited patterns of behavior are increased or decreased in their frequency, depending on whether the feedback from the environment was positive or negative for the individual. The only difference to a "full" evolutionary process is that individual trial-and-error learning is confined to the life span of an individual and cannot transcend it (except

when it becomes part of cultural evolution). Therefore, Schurz (2011, section 11.1.2) characterizes individual trial-and-error learning by means of Campbell's notion of "retention" instead of reproduction (Campbell, 1960).

In association learning, two stimuli frequently observed to be co-occurring are associated with each other in one's mind and projected as expectation in regard to future observations. This is nothing but the inference of *induction* as described by David Hume. A subcase of association learning is *classical conditioning*, as exemplified by Pavlov's dog, which, due to a repeated ringing of a bell shortly prior to the giving of food, soon expects food already at hearing the bell ringing and starts to salivate. Classical conditioning is a biologically widespread mechanism demonstrated almost everywhere in the animal world, even in worms (Bitterman, 2000; Delius, Jitsumori, & Siemann, 2000). Moreover, classical conditioning has a known neurological basis, the so-called *Hebb rule* (Rojas, 1996, pp. 21, 258–259). However, classical conditioning does not follow the Darwinian modules of spontaneous variation and selection but is a form of directly environment-driven (or "Lamarckian") learning, since association hypotheses are directly "imprinted" upon the individual by the observed environment via inductive generalization.

In conclusion, and in anticipation of the following, human cognition turns out to be based on the interaction of a large diversity of cognitive processes. Some of them are genetically determined, others are culturally acquired, and still others are individually learned. Even if a cognitive trait has a genetic basis, it does *not* usually determine a person's mind (thinking this would be so constitutes a frequent misunderstanding of genetic accounts). Because of the enormous plasticity of the human mind (cf. Churchland, 1979), there is presumably not even a single innate cognitive mechanism that cannot be overruled or corrected by other cognitive processes and in particular by culturally acquired abilities. For example, humans experience the Mueller–Lyer illusion as illustrated in figure 1.3.1: they see the left line as longer than the right line, although their length is equal. But the human mind is plastic enough to recognize this error, to correct it by reflective reasoning, and to recognize its cause, which lies in mechanisms of three-dimensional vision: the arrows and inverted arrows produce the three-dimensional effect that the right stroke appears to be *closer* to the observer than the left stroke and is thus perceived as being shorter (Rock, 1984, pp. 163, 167).

As another example, consider Immanuel Kant's famous thesis that perceived space is necessarily Euclidean and has three dimensions, because this is dictated by the innate (a priori) structure of human visual cognition. However,

Figure 1.3.1
Mueller–Lyer illusion.

mathematically educated humans can overcome even this innate "barrier" and conceptualize non-Euclidean or more-than-three-dimensional spaces. Further examples will be given below.

In the next four sections, we start with the most basic components of human cognition and proceed to increasingly higher forms.

3. Social Cognition and the Evolution of Language and Mental Representation

The most striking cognitive difference between humans and all other animals is the human faculty of language. Rudimentary abilities of symbolic signaling systems are already present in nonhuman animals, but while great apes can be trained to learn at most a few hundred words, humans can, in principle, produce a potentially infinite number of grammatical sentences from several hundreds of thousands of words and abstract syntax. According to a well-known theory advanced by philosophers (Gehlen, 1956/1977; Topitsch, 1979), anthropologists (Boyer, 1994; Wenegrat, 1990), and biologists (Tomasello, 1999), the hominids' language faculties have developed (presumably over 1–2 millions of years) in close interrelation with their *social* cognition, their abilities to communicate and to organize joint intention and cooperative action. One also speaks of a *sociolinguistic coevolution* (Heyes, 2012).

Early hominid groups could even hunt animals as large as mammoths; a high amount of social coordination and division of labor is required for tasks of this sort. Although many aspects of human language appear to be cognitively universal, the initial selection pressure for the evolution of language was cultural. According to Thompson, Kirby, and Smith (2016), a strong genetic inheritance of grammar as postulated by Chomsky (1968) is not needed to explain the apparent universality of generative grammar in humans. By means of computer simulations and mathematical proof, Thompson et al. (2016) demonstrate that a *weak genetic bias* in favor of a certain trait is sufficient to explain its universality in human cultures, since the genetic bias is amplified by learning processes. Jablonka and Lamb (2005) and Jablonka, Ginsburg, and Dor (2012) explain language evolution as a *gene–cultural coevolution*: various aspects of language initially invented culturally were later genetically assimilated (i.e., became partially

"inborn") because they created genetic selection pressures for cognitive abilities specialized for learning language; this mechanism is also known as the "Baldwin effect" (cf. Dennett, 1995, pp. 77ff.).

Communication—both about one's intentions and about the environment—is an obvious presupposition of successful cooperation. The coordination of individual actions that is required for successful cooperative behavior is systematically analyzed in *evolutionary game theory*. Weibull (1995, pp. 28–31) classifies symmetric two-person games into three kinds: (1) Hawk–Dove games (possessing a stable mixed equilibrium); (2) coordination games (possessing several stable pure equilibria), including the famous stag-hunt game as a subcase; and (3) prisoner's dilemma games (possessing no nontrivial equilibria). For the stag-hunt game, it has been shown (Skyrms, 2004) that under many but not all circumstances, successful cooperative behavior can emerge but is *not* evolutionary stable: if too many individuals prefer hunting small prey ("hare") instead of large prey ("stag"), because the success of the former does not depend on one's partners' cooperative behavior, then the evolution of cooperation breaks down.

Even worse are the evolutionary chances of cooperation in the prisoner's dilemma game, in which a cooperative action of *helping the other* is only optimal if it occurs reciprocally. Here cooperative actions are doomed to get exploited by "egoists" (i.e., noncooperators who receive help but don't return help). Thus, in the prisoner's dilemma game, cooperative action has to be stabilized by additional mechanisms of reciprocation.[6] Axelrod (1984/2006) proposed the strategy *tit-for-tat* (TFT), which initially acts cooperatively and then retaliates (i.e., punishes defection with defection and rewards cooperation with cooperation). Axelrod suggested that TFT is universally optimal and thus evolutionary stable, but later studies showed that there is no generally stable cooperation strategy in the iterated prisoner's dilemma, since for every refined cooperation strategy, one can invent a refined deception strategy that exploits it (Lorberbaum, 1994). However, there are strategies by means of which the evolution of cooperation can at least be imperfectly stabilized, such as (1) group selection with ongoing regrouping (Sober & Wilson, 1998), (2) communication and correlated pairing (Skyrms, 2010), (3) institutionalized sanctioning mechanisms (Gürerk, Irlenbusch, & Rockenbach, 2006), (4) cognitive mechanisms of cheater detection (Cosmides & Tooby, 1992), and (5) religious stabilization of altruism (Wilson, 2002).

Historically, the strategy of reciprocation has first been rationally reflected in the earliest human juridical document, the Babylonian *Codex Hammurabi* (1800 B.C.). But the evolution of cooperation begins much earlier. Many hunter–gatherer tribes were quite egalitarian, especially in places where food supply was instable and food sharing important. In their comparatively small groups, egoistic behavior of group members was much more easily controllable than in large agricultural societies (Sober & Wilson, 1998, p. 178; Wilson, 2002, p. 21). Thus, in all known hunter–gatherer societies, mechanisms of social reputation ("This guy is a good one") have developed and are still effective in peer groups of contemporary societies. In contrast, in the larger communities of premodern states formed after the agrarian revolution around 10,000 years B.C., reputation mechanisms alone were not enough to prevent cooperation-undermining exploitation behavior, whence in all known agrarian proto-states, governmental institutions that sanctioned rule breakers and collected taxes (among other activities) have emerged.

As explained, social cooperation requires the evolution of language and communication. But more is required for cooperation: the ability to represent and understand the minds, intentions, and beliefs of the other persons with whom one cooperates. Already at a very young age, human children develop the competence to understand other persons. At about nine months, babies acquire the ability of *joint attention* (i.e., they recognize the other's gaze direction and can direct their own gaze to the object spotted by the other). Tomasello (1999, pp. 61–70) speaks of the "nine-month revolution." A few months later, human start to *point* toward objects of their interest, and at the same time, the learning of language begins. Later, children acquire the ability of recognizing not only the others' intentions but also their *beliefs* and of distinguishing these from their own beliefs. This capability of children has been called their ability of *mindreading* (Nichols & Stich, 2003), or their possession of a *theory of mind* (Leslie, 1987). It was first demonstrated in an experiment called the *false-belief task* (Wimmer & Perner, 1983). The result of this experiment was based on verbal instructions and supported the conclusion that the ability to understand (false) beliefs of others emerges in children at the age of about four years. However, more recent experiments using a violation-of-expectation method demonstrated that already 15-month-old infants seem to understand when a person has a false belief (Onishi & Baillargeon, 2005).

There is a controversy about the nature of the cognitive mechanisms underlying human "mindreading" (cf. Nichols & Stich, 2003). While the theory-of-mind approach assumes a specialized cognitive module responsible for this faculty, the *simulation approach* argues for the view that the same cognitive mechanism by which people represent their own intentions and beliefs is used for simulations of the (different) intentions and beliefs of other persons. An argument in favor of the simulation view is evolutionary simplicity: one cognitive system is evolutionary less costly and thus more probable than two separate systems (Schulz, 2011).[7]

At about six years, children acquire the *conventional moral stage*, according to the findings of Piaget (1932) and Kohlberg (1984). This stage goes beyond the mere understanding of the nature of reciprocity, which children already possess at the age of three or four (Cosmides & Tooby, 2015, p. 653): it consists in the interpersonal generalization of reciprocal rules (if I have to follow the rule, you have to follow it, too), which establishes a basic sense of normative justice together with the ability to detect rule-breakers.

On a deeper level, the sociocognitive coevolution was not confined to communication and mindreading but went hand in hand with the evolution of higher-level systems of *mental representation* that intermediated between stimulus and response. A recent analysis of their evolutionary advantage has been given by Schulz (2018), thereby integrating and complementing related accounts of Millikan (2002), Papineau (2003), and Sterelny (2012). Schulz argues that connecting behavioral responses not directly with perceived stimuli, but rather with mental representations of object-configurations (which are in a separate step related to perceived stimuli), is cognitively much more efficient, because behaviorally relevant object-configurations (e.g., a bear in a cave) are not always indicated by the same perceptual cue but by different cues, dependent on the given circumstances. Note that in a similar way, the cognitive efficiency of the common-cause abduction of unobservable entities has been demonstrated in the philosophy of science (Schurz, 2016).

4. Evolution of Causal Reasoning

Besides the sociolinguistic, there was a second coevolution, the *sociotechnical* coevolution (Heyes, 2012; Jablonka et al., 2012). Essential to it was the ability of hominids not only to predict but also to *manipulate* the objects in their environment, from fire usage, stone tools, weaving techniques, and advanced weapons to the invention of agriculture. Crucial for these developments was humans' capability to reason about the *causes* of (pleasant or unpleasant) events. Causal reasoning goes beyond mere inductive correlational reasoning: it reflects the causal structure behind correlations. For example, if

A is observed to be correlated with *B*, then according to widely accepted causal principles, (1) *A* may be the cause of *B*, or (2) *B* may be the cause of *A*, or (3) *A* and *B* may be effects of a common cause (which follows from the so-called *causal Markov* condition; cf. Pearl, 2009; Schurz & Gebharter, 2016; Spirtes, Glymour, & Scheines, 2000). Only in case (1) is it possible to change *B* by manipulating *A*. Thus, causal knowledge is crucial for humans' ability to manipulate and control their environment.

The earlier conjecture that causal reasoning is an exclusively human ability turned out to be untenable. Chimpanzees possess many systematic reasoning abilities, ranging from object classification and spatial movement to intelligent tool usage (Tomasello, 1999, pp. 16ff.). Also, other higher mammals have rudimentary causal reasoning abilities (Dunbar, 2000). However, humans' capabilities of causal modeling are by far superior compared to their nonhuman relatives. Herrmann, Call, Hernández-Lloreda, Hare, and Tomasello (2007) found that in simple causal tasks, great apes are about equivalent to three-year-old toddlers.[8] However, the big steps in causal and analytic reasoning in child development are achieved at ages between five and seven years (Brainerd, 1978). The ability to mentally slip into the role or "mind" of someone else is the basis of pretend play in childhood (Leslie, 1987) and a crucial precondition for the child's ability of *counterfactual* reasoning (Buchsbaum, Bridgers, Weisberg, & Gopnik, 2012; Rafetseder, Schwitalla, & Perner, 2013), which, according to philosophical analysis, constitutes an essential component of causal reasoning (Lewis, 1973; Woodward, 2003).

According to the anthropological theory mentioned at the beginning of section 3, humans' *causal* reasoning evolved in close relation to their *intentional* reasoning (i.e., their interpretation of the intentions of living beings as the causes of their behaviors). This hypothesis fits well with the cultural history of ideas. From the Stone Age to premodern civilizations, the major causal explanation mechanisms of *Homo sapiens* were twofold: some natural processes are causally explained by mechanical contact forces (e.g., throwing a stone or carrying a load), but natural processes such as the weather or the movement of celestial bodies, which emerge apparently unforced without an obvious mechanical cause, were explained by their *animation* according to the intentional model (Gehlen, 1956/1977). More specifically, these natural processes were explained in terms of the same social-intentional model that explained the behavior of fellow men—a process that led to the polytheistic conception of natural Gods (Topitsch, 1979; Wenegrat, 1990).

As rationality progressed in the cultural history of ideas, such explanations were rejected because of their lack of empirical support and replaced by explanations in terms of physical causes. However, the epistemic justification of the assumption of causes governing the behavior of nonliving physical objects turned out to be a highly difficult problem in the history of philosophy. The idealist philosopher Leibniz attempted to reduce causal to generalized intentional notions. In contrast, the empiricist philosopher David Hume argued that observational evidence can give us only knowledge of correlations, not of causation. In contemporary philosophy, there are promising suggestions for solving Hume's challenge (Schurz & Gebharter, 2016; Spirtes et al., 2000), but the controversy is still going on.

The fact that human reasoning operates with two distinct models of causality, physical versus intentional causality, is confirmed by several psychological studies. Already at the age of six months, babies distinguish between nonliving and living objects, to which they spontaneously apply these different paradigms of causality: while nonliving objects can be put in motion only by the transmission of a force by physical contact, living objects are also able to move spontaneously without any external force, solely due to their intentions or internal causes. This finding was established by means of experiments in which babies were presented two sequences of events (cf. Leslie, 1982; Spelke, Phillips, & Woodward, 1995): in the "normal" sequence, a nonliving object (e.g., a ball) moves toward a second one, hits it, and pushes it away. In the "abnormal" sequence, the first object likewise moves toward the second one but already comes to a halt a distance away from it, and the second object moves away without touch "as if by magic." While the babies hardly paid attention to the first event, their attention (measured by the duration of looking) lingers very long on the second event. In a second series, the same experiment was conducted with people instead of objects: a boy walks toward a girl; in the first case, he touches her and the girl walks away, and in the second case, he remains standing in front of the girl and the girl walks away after that. In this case, no significant difference was found in the duration of the babies' gaze: both sequences are equally "normal" for the babies.

Thus, already babies distinguish spontaneously between contact causality for nonliving and intentional causality for living objects. In other words, intentional causality and contact causality are two distinct causal models. This also fits with the human history of ideas, insofar as contact causality for inanimate objects not only is well anchored in intuitive human physics but has also dominated the history of philosophy from Aristotle to Descartes. In contrast, objects that moved without apparent contact causes (e.g., heavenly bodies) were consistently explained using the intentional model of causality as explained above.

5. Modularity and Universality in Human Cognition

A further debate concerns the generality versus specificity of human mechanisms of cognition. According to the school of evolutionary psychology (Barkow, Cosmides, & Tooby, 1992), the human mind consists of various cognitive "modules." These are cognitive programs or processes that were evolutionarily selected for specific tasks (Carruthers & Chamberlain, 2000); they are genetically anchored and little mutable.[9] Examples of such area- and purpose-specific modules have already been mentioned and include (1) the module of shared attention (developed in the child at 9 months), (2) the modules of intentionality and contact causality (at 6 months), (3) the modules of language acquisition (from 18 months on), (4) mental spatial models (at the sensorimotorical level with 0–2 years; Brainerd, 1978), (5) the theory-of-mind module (at 4–5 years), and (6) the development of conventional moral values (at 6 years), which are associated with (7) the module of cheater detection (see below). Related is the research program of *adaptive rationality* according to which humans' cognitive mechanisms are adapted to the structure of local environments, being tailored to the specific tasks for which they provide highly efficient solutions (Gigerenzer, Todd, & ABC Research Group, 1999; Todd, Gigerenzer, & ABC Research Group, 2012). A second claim of proponents of this paradigm is that simple heuristics are frequently more successful than computationally costly general reasoning mechanisms, following the slogan "Less can be more."

The *modularist* view has also been called the "Swiss-Army-knife model" (Tooby & Cosmides, 1992) or the "adaptive toolbox model" (Gigerenzer & Selten, 2001). The opposite view is defended by *generalist* models of cognition. Until the 1960s, classical logic was the dominant generalistic normative standard of theories of rational reasoning (cf. Inhelder & Piaget, 1958). When psychologists discovered empirically that, in many domains, human reasoning did not accord with the principles of logic (e.g., Wason, 1966), these findings were interpreted as signs of human irrationality (cf. Evans, 2002). It was suggested that psychologists should adopt an alternative general normative system, such as Bayesian probability and decision theory (e.g., Oaksford & Chater, 1991). However, human reasoning has been observed to deviate from the norms of probability, too

(Barbey & Sloman, 2007; Kahneman 2011; Kahneman & Tversky, 1972). These findings led many psychologists toward a general skepticism in regard to the "normative" status of generalist reasoning accounts.

Contemporary defenders of generalistic reasoning accounts are more modest. They agree that the early models of reasoning in terms of logic or probability theory were too simplistic or unrealistic from a cognitive viewpoint. Moreover, they do not deny that human cognition hosts a multitude of specialized cognitive abilities or modules. However, it is implausible that the human brain contains a module for each specific purpose or area of application; 200,000 years of evolution are insufficient to allow so many modules (Tomasello, 1999, pp. 204–205). Thus, contemporary generalists argue that besides specialized modules, cognitive evolution has also brought about a set of purpose- and area-unspecific cognition mechanisms. This view has been defended, among others, by Over (2003, p. 122), Almor (2003, p. 104), and, more recently, in a volume on the "new thinking on the evolution of cognition" (Heyes & Frith, 2012). The authors in this book argue that human cognition is largely purpose-general; it does not resemble a Swiss army knife but rather a human *hand*, whose fingers can serve an extremely large variety of purposes (Heyes, 2012, p. 2092).

Recently, the existence of generalistic "improvisational intelligence" has also been acknowledged by some evolutionary psychologists (Barrett, Cosmides, & Tooby, 2007). For semantic clarification, note that the notion of a "cognitive module" in the mainstream literature is always related to a *specific type* of adaptation problem, purpose, and environmental condition. Thus, one should not also call general cognitive abilities "modules," as this would stretch this notion over all bounds and make it empty. Moreover, it has been argued that purpose-specificity and domain-specificity are two distinct *dimensions* of specificity/generality (Duchaine, Cosmides, & Tooby, 2001). However, it seems that the two—although clearly nonidentical—are definitely not independent: the number and diversity of *purposes* to which a cognitive ability is applicable and the number and diversity of *conditions* under which it is applicable are strongly correlated.

6. Induction and Deduction

The most important example of a *general* mechanism of learning and cognition is *induction*, which in the form of conditioning or association learning is common in virtually all animals (recall section 2). In the view of earlier behavioral psychologists such as Thorndike (1911), the merit of learning by conditioning consists precisely in the fact that this is *not* an area-specific but a universally applicable learning process.

General cognitive mechanisms are particularly important for sustainable evolutionary success under environmental conditions that are changing in unforeseeable ways—a situation that occurred frequently in hominid evolution and even more frequently in human cultural evolution. This argument has been put forward by Schurz and Thorn (2016) as an important complementation of the adaptive rationality account (see also Thorn & Schurz, 2019). To be cognitively successful under changing environments, one needs strategies for learning which cognitive methods perform best in which environment, or temporal phase of the environment (a point acknowledged within the adaptive rationality program; cf. Todd et al., 2012, p. 15). Obviously, meta-cognitive selection strategies have to be sufficiently general; otherwise, they could not serve the purpose of selecting the optimal method in a wide range of environments.

In view of the cognitive universality of inductive reasoning, it is remarkable that the epistemic justification of induction turned out to be one of the most difficult problems in philosophy. According to David Hume's famous challenge (1748/2006, chapters 4, 6), the reliability of inductive reasoning from past observations to future (unobserved) cases can in no way be demonstrated by rational argument, neither strictly nor even probabilistically. For this reasoning rests on the assumption that nature is uniform, that is, that the future resembles the past, but that this is so cannot be demonstrated by observation or by logic or probability calculus. Nor can it be demonstrated by induction, by pointing to the fact that so far, the method of induction was successful, because the "inductive justification of induction" is circular and thus epistemically worthless (for details, see Schurz, 2019, chapters 2–4). An alternative account of justifying induction, going back to Reichenbach (1935, §80), is based on the idea that induction is the *best that we can do* in order to achieve successful predictions. Schurz (2019) defends this idea in terms of the *optimality of meta-induction*. Given a type of problem (a prediction task or decision problem), the meta-inductivist observes the success records of all methods that are *accessible* to her (i.e., methods whose success rates can be observed). Using these records, the meta-inductivist constructs an optimal combination of these methods, in the form of a weighted average that is periodically updated in a success-dependent way. Based on discoveries in computational learning theory (Cesa-Bianchi & Lugosi, 2006),

Schurz (2008, 2019) demonstrates that meta-induction achieves an optimal success rate in all possible worlds among all competing methods (of prediction or decision) that are accessible to the given agent. Schurz (2008, 2019) proposes meta-induction as a new solution to Hume's problem of the justification of induction mentioned above. Since meta-induction is primarily a strategy of social learning—that is, learning from the performance of others (although it can also be applied to the results of individual learning)—its optimality constitutes an essential foundation for the social propagation of information in cultural evolution (Schurz, 2012, 2019, section 10.2).

Induction and meta-induction are presumably the evolutionarily most fundamental cognitive mechanisms (insofar as all forms of learning from feedback are based on induction). They are certainly more fundamental than causal reasoning (see below) and even more fundamental than logical reasoning and deduction. On the other hand, from a philosophical viewpoint, logical reasoning is more fundamental than induction, since induction presupposes some amount of logic and not vice versa.

Psychologists have found out that human reasoning's fit with the rules of classical propositional logic is weak: while some rules (e.g., modus ponens) are already mastered by children at the age of six (Brainerd, 1978), other logical rules such as modus tollens are not even understood by the majority of adults. The latter fact is exemplified in Wason's famous card selection task (Evans, 1982, chapter 9; Wason, 1966). In this experiment (which is also discussed in chapter 2.3 by Johnson-Laird and chapter 4.5 by Chater & Oaksford, both in this handbook), test persons (TPs) are given four cards from a pack of cards and told to test the following rule: *If there is an A on the front side, then there is a 1 on the back side.* They see four cards lying on a table, two face-up, showing an A and a B, respectively, and two face-down, showing a 1 and a 2. They were asked, Which of these four cards do you have to turn over in order to check whether the rule actually applies to these cards? According to the laws of classical propositional logic, one should turn over the first (A) card and the fourth (2) card: turning over the first card corresponds to the logically valid modus ponens (MP) inference (from "If A, then 1" and "A" infer "1"). Turning the fourth card over corresponds to the equally valid modus tollens (MT) inference (from "If A, then 1" and "not-1" follows "not-A"). In contrast, turning over the second and third cards corresponds respectively to the two invalid inferences NA ("negating the antecedent") and AC ("affirming the consequent"). The empirical performance results, however, are MP 100%,

NA 5%, AC 10%, and MT 5%. Thus, normal adults only master the logical rule of MP but fail to master the rule of MT and apply equally or even more often the invalid rules NA and AC.

To defenders of the logico-generalist paradigm of cognition, these results came like a shock. Equally disturbing was a further result first brought to light by Griggs and Cox (1982), which demonstrated that people apply the rule of MT perfectly in the social context of detecting rule breakers. Here, the TPs had to test the following consumption regulation in a youth club: *Those who drink alcohol have to be at least 16 years old.* TPs were confronted with four adolescents; of two of them (Berta, Klaus), they only knew the beverage but not the age (14 y., 18 y.), and of the other two (Lisa, Martin), they only knew the age but not the beverage (cola, beer). They were asked, *Whom do you have to check to determine whether he/she has broken the rule?* with the result that now all TPs checked Berta and Martin (i.e., they mastered this instantiation of MP and MT perfectly).

As an explanation of this finding, Cosmides and Tooby (1992) proposed the thesis that humans have a *specialized* cognitive module for the detection of social rule breakers (*cheater detection*), which emerged through selection in the course of social evolution. More recent experiments (Cosmides, Barrett, & Tooby, 2010) show that these special inference abilities cannot be explained by assuming a more specific but still general reasoning ability, such as reasoning with *deontic rules* (as argued by Cheng & Holyoak, 1985) or with *utilities* (as proposed by Manktelow & Over, 1991); rather, they are restricted to situations in which the action (whose permission is governed by the rule) is beneficial to the actor, so that the rule violation is indeed an act of cheating.

The fact that humans possess *content-specific* reasoning modules does not exclude that their conditional reasoning is also governed by some purpose-general rules. In this line, psychologists and philosophers have developed the following evolutionary explanation for the deviations of human conditional (if–then) reasoning from the norms of deductive logic: the if–then relations of our natural environment are almost never *strictly* true but rather admit of exceptions (i.e., they express relations of *high conditional probability*). The hypothesis of ordinary "If *A* then *B*" statements as high-conditional-probability statements ("*B* is highly probable given *A*") was proposed by Adams (1975), evolutionarily substantiated by Schurz (2001), and has been experimentally confirmed (Evans, Handley, & Over, 2003; more on this in chapter 4.6 by Oberauer & Pessach, this handbook). It turns out that humans' intuitive if–then reasoning fits well with the rules of probability logic (Pfeifer & Kleiter,

2010; Schurz & Thorn, 2012). The MT rule, however, is *not* probabilistically valid. This may explain why subjects rarely apply the rule MT in Wason's experiment. In contrast, in the second experiment, the more specific cognitive module of cheater detection becomes activated. In this way, both the specific module of cheater detection and the general reasoning mechanisms for uncertain conditionals have a plausible evolutionary explanation, *without* coming into conflict with each other.

The interaction of specialized cognitive modules and general cognition mechanisms can also be excellently studied in the abovementioned example of the cognitive model of contact causality. Although already toddlers have the intuition that inanimate objects can be put in motion only through contact or pushing, people have known *magnets* for centuries. After a short time of adaptation to the seemingly "magical" actions at a distance of magnets, children just as adults become used to this without problems. Apparently, the innate but, in the case of magnets, inapplicable model of contact forces is *overwritten* and thus corrected by inductive learning. We learn from this consideration that humans' innate model of causality does not determine human reasoning by "a priori necessity" (as believed by the philosopher Immanuel Kant) but just acts as one innate disposition among many others and can be *corrected* by other cognitive mechanisms.

Through examples like these, it becomes apparent how efficiently special and general cognitive mechanisms can work together. Over (2003, pp. 124–125) and Schurz and Thorn (2016) therefore propose a *dualistic* theory of cognition, according to which human cognition consists of area-specific modules *and* general cognitive processes. A further example splendidly illustrating the division of labor between general and purpose-specific reasoning processes is the selection of optimal methods from the "adaptive toolbox" by the general strategy of meta-induction, as explained above.

7. The Evolutionary Architecture of Human Cognition

We have so far explained the cooperation of specialized (modular) and general mechanisms of human cognition. Most modular cognitive processes, but also many general processes such as inductive inference, work for the most part *unconsciously*. What, then, is the actual role of *consciousness* in human cognition? There is no consensus on this question. An extreme view is *epiphenomenalism*, which claims that consciousness only plays the role of a subsequently summarizing *reporter* on our unconscious mental processes but not the role of a causal trigger (Block, Flanagan, & Güzeldere, 1996,

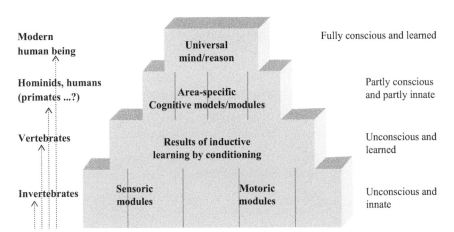

	Fully conscious and learned
Modern human being	
Universal mind/reason	
Hominids, humans (primates ...?)	Partly conscious and partly innate
Area-specific Cognitive models/modules	
Vertebrates	Unconscious and learned
Results of inductive learning by conditioning	
Invertebrates	Unconscious and innate
Sensoric modules Motoric modules	

Figure 1.3.2

Evolution-based structure of human cognition.

chapter 19). This view appears to me to be exaggerated and even untenable. It is admittedly true that the conscious human mind reflects only a fraction of the unconscious processes of human cognition and often does play the role of a summarizing reporter. Moreover, conscious human reasoning processes are slower by decimal powers than the unconscious and modular cognitive processes. When we make practical decisions, for example, about whether we should quickly run across an intersection or not, then we gauge the involved probabilities and utilities intuitively within a fraction of a second, while if we calculated such an estimate in a precise decision-theoretic manner, it would take us hours.

And still, this much-slower conscious, logical mind has proved in the course of cultural evolution that—if given enough time to pursue its activities in a secure space—it is capable of much greater cognitive achievements than all cognitive heuristics put together. Its merits lie in its ability to comprehend its domain of application in a logically consistent and empirically controlled way, without any simplifications and distortions. Through the institutional establishment of an area of research and education sealed off from the pressure of practical demands, the conscious mind was able to develop our scientific-technological civilization. Conscious reasoning, from Euclid and Leonardo to Minkowski, was able to systematize the mechanisms of Euclidean geometry and perspectival projection so completely that all deceptions of our perception modules were resolvable. From Aristotle to Boole and Gödel, conscious reasoning was able to capture the laws of logical and probabilistic inference so completely that thereby all intuitive errors of reasoning could be recognized. Through abstract-mathematical thinking, human reasoning could penetrate areas exceeding everything that natural imagination and heuristics could ever give us:

with Einstein (and others), it was able to intrude into the laws of the unimaginably large, with Bohr (and others) into the laws of the unimaginably small, with Darwin (and others) to transcend the border between the nonliving and the living, and with computer technology to transcend the border between nature and mind. The overwhelming success of conscious scientific reasoning extended humans' range of possible action and gave space to new technologies from synthetic chemistry and nuclear energy to genetic engineering and robotics.

In conclusion, we can say that even if humans' general reasoning capacities become efficient through their interaction with the specialized cognitive modules, it is still these general reasoning capacities that made scientific-technological progress possible and distinguish humans most strongly from their nonhuman ancestors. This leads to modeling human cognition as a layered structure that is illustrated by the architectonics displayed in figure 1.3.2.

Notes

1. For the proof of this thesis of Hume by means of modern logic, see Schurz (1997).

2. Religious beliefs in this generalized characterization are found in more or less all known cultures (including China), although there are important differences, for example, concerning polytheism versus monotheism or omnipotence versus power-restricted Gods or "ghosts" and so on (cf. Boyer, 1994; Schurz, 2011, section 17.5; Wenegrat, 1990). Another approach to the generalized placebo effect is the psychology of "positive thinking" (Taylor, 1989).

3. Cf. Jablonka and Lamb (2005). These authors even postulate four dimensions of evolution.

4. Pioneers of the modern theory of cultural evolution (to mention just a few) are Cavalli-Sforza and Feldman (1973), Dawkins (1976/1987), Boyd and Richerson (1985), and Donald

Campbell (1960); more recent developments are Mesoudi (2011), Schurz (2011), Dennett (2018), and Henrich (2016).

5. Dawkins (1976/1987) suggested the name "meme" for the cultural counterpart to genes, but this notion is controversial (cf. Aunger, 2002).

6. The notions of "egoistic" versus "cooperative" action are understood here in the evolutionary sense, not necessarily implying a corresponding psychological mechanism (cf. Sober & Wilson, 1998, p. 201).

7. Tomasello (1999) has argued that the ability to conceive their conspecifics as *intentional* beings discriminates humans from nonhuman great apes. Later this thesis turned out to be untenably strong: also great apes possess these abilities but to a much lesser extent (Call & Tomasello, 2008). Based on the observation of eye movements, Krupenye, Kano, Hirata, Call, and Tomasello (2016) found evidence that chimps can even anticipate actions based on false beliefs.

8. In contrast, human toddlers are better than chimpanzees in tasks of social cognition.

9. Note that a genetically determined mechanism need not be "single track" but can be "multitrack," leading to different behaviors in different environments (like a hardwired computer program can react differently to different inputs). In such a case, the *phenotypic manifestation* of the hardwired mechanism is mutable but not the mechanism itself.

References

Adams, E. W. (1975). *The logic of conditionals*. Dordrecht, Netherlands: Reidel.

Almor, A. (2003). Specialized behaviour without specialized modules. In D. E. Over (Ed.), *Evolution and the psychology of thinking* (pp. 101–120). Hove, England: Psychology Press.

Aunger, R. (Ed.). (2002). *Darwinizing culture: The status of memetics as a science*. Oxford, England: Oxford University Press.

Axelrod, R. (2006). *The evolution of cooperation* (Rev. ed.). New York, NY: Basic Books. (Original work published 1984)

Barbey, A. K., & Sloman, S. A. (2007). Base rate respect: From ecological rationality to dual processes. *Behavioral and Brain Sciences*, *30*, 241–297.

Barkow, J. H., Cosmides, L., & Tooby, J. (Eds.). (1992). *The adapted mind: Evolutionary psychology and the generation of culture*. New York, NY: Oxford University Press.

Barrett, H. C., Cosmides, L., & Tooby, J. (2007). The hominid entry into the cognitive niche. In S. W. Gangestad & J. A. Simpson (Eds.), *The evolution of mind: Fundamental questions and controversies* (pp. 241–248). New York, NY: Guilford Press.

Basalla, G. (1988). *The evolution of technology*. Cambridge, England: Cambridge University Press.

Bitterman, M. E. (2000). Cognitive evolution. In C. M. Heyes & L. Huber (Eds.), *The evolution of cognition* (pp. 61–80). Cambridge, MA: MIT Press.

Block, N., Flanagan, O., & Güzeldere, G. (Eds.). (1996). *The nature of consciousness*. Cambridge, MA: MIT Press.

Boyd, R., & Richerson, P. J. (1985). *Culture and the evolutionary process*. Chicago, IL: University of Chicago Press.

Boyer, P. (1994). *The naturalness of religious ideas*. Berkeley: University of California Press.

Brainerd, C. J. (1978). *Piaget's theory of intelligence*. Englewood Cliffs, NJ: Prentice-Hall.

Buchsbaum, D., Bridgers, S., Weisberg, D. S., & Gopnik, A. (2012). The power of possibility: Causal learning, counterfactual reasoning, and pretend play. *Philosophical Transactions of the Royal Society B*, *367*(1599), 2202–2212.

Call, J., & Tomasello, M. (2008). Does the chimpanzee have a theory of mind? 30 years later. *Trends in Cognitive Sciences*, *12*(5), 187–192.

Campbell, D. T. (1960). Blind variation and selective retention in creative thought as in other knowledge processes. *Psychological Review*, *67*, 380–400.

Carruthers, P., & Chamberlain, A. (Eds.). (2000). *Evolution and the human mind*. Cambridge, England: Cambridge University Press.

Cavalli-Sforza, L. L., & Feldman, M. W. (1973). Models for cultural inheritance I: Group mean and within group variation. *Theoretical Population Biology*, *4*, 42–55.

Cesa-Bianchi, N., & Lugosi, G. (2006). *Prediction, learning, and games*. Cambridge, England: Cambridge University Press.

Cheng, P. W., & Holyoak, K. J. (1985). Pragmatic reasoning schemas. *Cognitive Psychology*, *17*, 391–416.

Chomsky, N. (1968). *Language and mind*. New York, NY: Harcourt Brace & World.

Churchland, P. M. (1979). *Scientific realism and the plasticity of the mind*. Cambridge, England: Cambridge University Press.

Cosmides, L., Barrett, H. C., & Tooby, J. (2010). Adaptive specializations, social exchange, and the evolution of human intelligence. *PNAS*, *107*(Suppl. 2), 9007–9014.

Cosmides, L., & Tooby, J. (1992). Cognitive adaptations for social exchange. In J. H. Barkow, L. Cosmides, & J. Tooby (Eds.), *The adapted mind: Evolutionary psychology and the generation of culture* (pp. 163–228). New York, NY: Oxford University Press.

Cosmides, L., & Tooby, J. (2015). Adaptations for reasoning about social exchange. In D. M. Buss (Ed.), *The handbook of evolutionary psychology: Vol. 2. Integrations* (2nd ed., pp. 625–668). Hoboken, NJ: Wiley.

Dawkins, R. (1987). *The selfish gene* (2nd, expanded ed.). Oxford, England: Oxford University Press. (Original work published 1976)

Delius, J., Jitsumori, M., & Siemann, M. (2000). Stimulus equivalencies through discrimination research. In C. M. Heyes & L. Huber (Eds.), *The evolution of cognition* (pp. 103–122). Cambridge, MA: MIT Press.

Dennett, D. C. (1995). *Darwin's dangerous idea: Evolution and the meanings of life*. New York, NY: Simon & Schuster.

Dennett, D. C. (2018). *From bacteria to Bach and back: The evolution of minds*. London, England: Penguin Books.

Duchaine, B., Cosmides, L., & Tooby, J. (2001). Evolutionary psychology and the brain. *Current Opinion in Neurobiology, 11*(2), 225–230.

Dunbar, R. M. (2000). Causal reasoning and mental rehearsal in primates. In C. M. Heyes & L. Huber (Eds.), *The evolution of cognition* (pp. 205–220). Cambridge, MA: MIT Press.

Evans, J. St. B. T. (1982). *The psychology of deductive reasoning*. London, England: Routledge & Kegan Paul.

Evans, J. St. B. T. (2002). Logic and human reasoning. *Psychological Bulletin, 128*, 978–996.

Evans, J. St. B. T., Handley, S. J., & Over, D. E. (2003). Conditionals and conditional probability. *Journal of Experimental Psychology: Learning, Memory, and Cognition, 29*(2), 321–335.

Gehlen, A. (1977). *Urmensch und Spätkultur* [Prehistoric man and late culture] (4th corr. ed.). Frankfurt/Main, Germany: Athenaion. (Original work published 1956)

Gigerenzer, G., & Selten, R. (Eds.). (2001). *Bounded rationality: The adaptive toolbox*. Cambridge, MA: MIT Press.

Gigerenzer, G., Todd, P. M., & ABC Research Group (1999). *Simple heuristics that make us smart*. Oxford, England: Oxford University Press.

Griggs, R. A., & Cox, J. R. (1982). The elusive thematic-materials effect in Wason's selection task. *British Journal of Psychology, 73*, 407–420.

Gürerk, Ö., Irlenbusch, B., & Rockenbach, B. (2006). The competitive advantage of sanctioning institutions. *Science, 312*, 108–111.

Henrich, J. (2016). *The secret of our success: How culture is driving human evolution, domesticating our species, and making us smarter*. Princeton, NJ: Princeton University Press.

Herrmann, E., Call, J., Hernández-Lloreda, M. V., Hare, B., & Tomasello, M. (2007). Humans have evolved specialized skills of social cognition: The cultural intelligence hypothesis. *Science, 317*, 1360–1366.

Heyes, C. M. (2012). Introduction. *Philosophical Transactions of the Royal Society B, 367*(1599), 2091–2096.

Heyes, C. M., & Frith, U. (Eds.). (2012). New thinking: The evolution of human cognition. *Philosophical Transactions of the Royal Society B, 367*(1599).

Hume, D. (2004). *A treatise of human nature: Book III. Of morals*. Mineola, NY: Dover. (Original work published 1739)

Hume, D. (2006). *An inquiry concerning human understanding* (S. Butler, Ed.). Fairford, England: Echo Library. (Original work published 1748)

Inhelder, B., & Piaget, J. (1958). *The growth of logical thinking: From childhood to adolescence*. New York, NY: Basic Books.

Jablonka, E., Ginsburg, S., & Dor, D. (2012). The co-evolution of language and emotions. *Philosophical Transactions of the Royal Society B, 367*(1599), 2152–2159.

Jablonka, E., & Lamb, M. J. (2005). *Evolution in four dimensions: Genetic, epigenetic, behavioral, and symbolic variation in the history of life*. Cambridge, MA: MIT Press.

Kahneman, D. (2011). *Thinking, fast and slow*. London, England: Macmillan.

Kahneman, D., & Tversky, A. (1972). Subjective probability: A judgment of representativeness. *Cognitive Psychology, 3*(3), 430–454.

Kohlberg, L. (1984). *The psychology of moral development*. New York, NY: Harper & Row.

Krupenye, C., Kano, F., Hirata, S., Call, J., & Tomasello, M. (2016). Great apes anticipate that other individuals will act according to false beliefs. *Science, 354*(6308), 110–114.

Leslie, A. M. (1982). The perception of causality in infants. *Perception, 11*, 173–186.

Leslie, A. M. (1987). Pretense and representation: The origins of "theory of mind." *Psychological Review, 94*(4), 412–426.

Lewis, D. (1973). Causation. *Journal of Philosophy, 70*, 556–567.

Lorberbaum, J. (1994). No strategy is evolutionarily stable in the repeated prisoner's dilemma. *Journal of Theoretical Biology, 168*(2), 117–130.

Manktelow, K., & Over, D. (1991). Social roles and utilities in reasoning with deontic conditionals. *Cognition, 39*, 85–105.

Mesoudi, A. (2011). *Cultural evolution: How Darwinian theory can explain human culture and synthesize the social sciences*. Chicago, IL: University of Chicago Press.

Millikan, R. G. (2002). *Varieties of meaning: The 2002 Jean Nicod Lectures*. Cambridge, MA: MIT Press.

Nichols, S., & Stich, S. (2003). *Mindreading: An integrated account of pretense, self-awareness, and understanding other minds*. Oxford, England: Clarendon Press.

Oaksford, M., & Chater, N. (1991). Against logicist cognitive science. *Mind & Language, 6*, 1–38.

Onishi, K. H., & Baillargeon, R. (2005). Do 15-month-old infants understand false beliefs? *Science, 308*(5719), 355–358.

Over, D. E. (2003). From massive modularity to metarepresentation: The evolution of higher cognition. In D. E. Over (Ed.), *Evolution and the psychology of thinking* (pp. 121–144). Hove, England: Psychology Press.

Papineau, D. (2003). *The roots of reason: Philosophical essays on rationality, evolution, and probability*. Oxford, England: Clarendon Press.

Pearl, J. (2009). *Causality: Models, reasoning, and inference* (2nd ed.). Cambridge, England: Cambridge University Press. (Original work published 2000)

Pfeifer, N., & Kleiter, G. D. (2010). The conditional in mental probability logic. In M. Oaksford & N. Chater (Eds.), *Cognition and conditionals: Probability and logic in human thinking* (pp. 153–173). Oxford, England: Oxford University Press.

Piaget, J. (1932). *The moral judgement of the child*. London, England: Kegan Paul, Trench, Trubner.

Rafetseder, E., Schwitalla, M., & Perner, J. (2013). Counterfactual reasoning: From childhood to adulthood. *Journal of Experimental Child Psychology, 114*(3), 389–404.

Reichenbach, H. (1938). *Experience and prediction: An analysis of the foundations and the structure of knowledge*. Chicago, IL: University of Chicago Press.

Rock, I. (1984). *Perception*. New York, NY:: Scientific American Books.

Rojas, R. (1996). *Neural networks: A systematic introduction*. Berlin, Germany: Springer.

Schulz, A. W. (2011). Simulation, simplicity, and selection: An evolutionary perspective on high-level mindreading. *Philosophical Studies, 152*(2), 271–285.

Schulz, A. W. (2018). *Efficient cognition: The evolution of representational decision making*. Cambridge, MA: MIT Press.

Schurz, G. (1997). *The is–ought problem: An investigation in philosophical logic* (Studia Logica Library). Dordrecht, Netherlands: Kluwer.

Schurz, G. (2001). What is 'normal'? An evolution-theoretic foundation of normic laws and their relation to statistical normality. *Philosophy of Science, 68*(4), 476–497.

Schurz, G. (2008). The meta-inductivist's winning strategy in the prediction game: A new approach to Hume's problem. *Philosophy of Science, 75*(3), 278–305.

Schurz, G. (2011). *Evolution in Natur und Kultur: Eine Einführung in die verallgemeinerte Evolutionstheorie* [Evolution in nature and culture: An introduction to the generalized theory of evolution]. Heidelberg, Germany: Spektrum Akademischer Verlag.

Schurz, G. (2012). Meta-induction in epistemic networks and the social spread of knowledge. *Episteme, 9*(2), 151–170.

Schurz, G. (2016). Common cause abduction: The formation of theoretical concepts and models in science. *Logic Journal of the IGPL, 24*(4), 494–509.

Schurz, G. (2019). *Hume's problem solved: The optimality of meta-induction*. Cambridge, MA: MIT Press.

Schurz, G., & Gebharter, A. (2016). Causality as a theoretical concept. *Synthese, 193*(4), 1071–1103.

Schurz, G., & Hertwig, R. (2019). Cognitive success: A consequentialist account of rationality in cognition. *Topics in Cognitive Science, 11*(1), 7–36.

Schurz, G., & Thorn, P. D. (2012). Reward versus risk in uncertain inference: Theorems and simulations. *Review of Symbolic Logic, 5*(4), 574–612.

Schurz, G., & Thorn, P. D. (2016). The revenge of ecological rationality: Strategy-selection by meta-induction within changing environments. *Minds and Machines, 26*(1–2), 31–59.

Skyrms, B. (2004). *The stag hunt and the evolution of social structure*. Cambridge, England: Cambridge University Press.

Skyrms, B. (2010). *Signals: Evolution, learning, & information*. Oxford, England: Oxford University Press.

Sober, E. (1993). *Philosophy of biology*. Boulder, CO: Westview Press.

Sober, E., & Wilson, D. S. (1998). *Unto others: The evolution and psychology of unselfish behavior*. Cambridge, MA: Harvard University Press.

Spelke, E. S., Phillips, A., & Woodward, A. L. (1995). Infants' knowledge of object motion and human action. In D. Sperber, D. Premack, & A. J. Premack (Eds.), *Causal cognition: A multidisciplinary debate* (pp. 44–78). Oxford, England: Clarendon Press.

Spirtes, P., Glymour, C., & Scheines, R. (2000). *Causation, prediction, and search* (Lecture Notes in Statistics, Vol. 81). Cambridge, MA: MIT Press.

Sterelny, K. (2012). *The evolved apprentice: How evolution made humans unique*. Cambridge, MA: MIT Press.

Taylor, S. E. (1989). *Positive illusions: Creative self-deception and the healthy mind*. New York, NY: Basic Books.

Thompson, B., Kirby, S., & Smith, K. (2016). Culture shapes the evolution of cognition. *PNAS, 113*(16), 4530–4535.

Thorn, P. D., & Schurz, G. (2019). Meta-inductive prediction based on attractivity weighting: Mathematical and empirical performance evaluation. *Journal of Mathematical Psychology, 89*, 13–30.

Thorndike, E. L. (1911). *Animal intelligence: Experimental studies*. New York, NY: Macmillan.

Todd, P. M., Gigerenzer, G., & ABC Research Group (Eds.). (2012). *Ecological rationality: Intelligence in the world*. Oxford, England: Oxford University Press.

Tomasello, M. (1999). *The cultural origins of human cognition*. Cambridge, MA: Harvard University Press.

Tooby, J., & Cosmides, L. (1992). The psychological foundations of culture. In J. H. Barkow, L. Cosmides, & J. Tooby (Eds.), *The adapted mind: Evolutionary psychology and the generation of culture* (pp. 19–136). New York, NY: Oxford University Press.

Topitsch, E. (1979). *Erkenntnis und Illusion* [Knowledge and illusion]. Hamburg, Germany: Hoffman & Campe.

Wason, P. C. (1966). Reasoning. In B. M. Foss (Ed.), *New horizons in psychology* (Vol. 1, pp. 106–136). London, England: Penguin.

Weibull, J. W. (1995). *Evolutionary game theory*. Cambridge, MA: MIT Press.

Wenegrat, B. (1990). *The divine archetype: The sociobiology and psychology of religion*. Lexington, MA: Lexington Books.

Wilson, D. S. (2002). *Darwin's cathedral: Evolution, religion, and the nature of society*. Chicago, IL: University of Chicago Press.

Wimmer, H., & Perner, J. (1983). Beliefs about beliefs: Representation and constraining function of wrong beliefs in young children's understanding of deception. *Cognition, 13*(1), 103–128.

Woodward, J. (2003). *Making things happen: A theory of causal explanation*. Oxford, England: Oxford University Press.

1.4   Rationality and the Brain

Vinod Goel

Summary

We are widely considered the "rational animal." Coherence relations are essential for rationality. They allow for inferences that can identify actions consistent (or inconsistent) with achieving our goals, in the context of our beliefs. Consistent inferences can be further differentiated into those that are certain, plausible, and indeterminate. Twenty-plus years of research on the neural basis of reasoning reveals no single unitary mechanism of inference. Rather, our system of reasoning can be understood in terms of multiple systems for generating inferences and a common system for detecting and/ or resolving conflict or inconsistency. The systems for inference generation vary as a function of conceptual and logical relations. Within logical relations, they further vary as a function of argument form, presence or absence of belief-laden content, argument presentation modality, and determinacy of the conclusion. In this chapter, we organize and present the research on the neuroscience of reasoning along these lines.

1.   The Reasoning Animal

Within the Western intellectual tradition, humans are widely regarded as the reasoning or rational animal. This is to say that our behaviors are explained by postulating beliefs and desires, as well as a principle of "coherence" that guides our pursuit of the latter in the context of the former. Rationality is instrumental: it is a means to an end. A rational choice is a deliberate choice or action (selected from a large/unbounded set) that moves an organism closer to its goals in a manner consistent with its knowledge and beliefs.[1] For example, if I am thirsty and desire to drink water, and if I believe that there is a glass of water within reach on the right-hand side of my desk, and I reach out to the right-hand side, grasp it, bring it to my lips, and drink, my action is coherent or reasonable (in the context of my beliefs and desires). However, if, given the same desires and beliefs, I reach out to the left-hand side of my desk, that action would be incoherent or unreasonable because it is inconsistent with my belief that the glass is on the right-hand side and would not fulfill my desire to drink from it.[2]

I am using the term "coherence" as a basic, primitive, intuitive notion, meaning roughly "making sense." It is a relationship between propositions. Consider the following example: "If George is taller than Michael, then Michael is shorter than George." You will recognize this statement as self-evident and true. But suppose that you refuse to accept its truth and ask me to prove it. What do I do? How can I possibly prove this to you? I can't. It is like being asked to prove a postulate from Euclidean geometry. They are self-evident. Either you understand them or you don't. But we all do understand Euclidean postulates and simple logical relations as self-evident. These basic, intuitive notions are enhanced and elaborated into sophisticated systems of reasoning, involving formal logic and probability theory, that we can learn and use, with varying degrees of success.

Coherence relations can be broken down into semantic, logical, and conceptual relations. Semantic relations hold by virtue of the meaning of open-class words in the language. For example, a widow is a woman whose husband has died. Logical relations hold by virtue of the "closed-class" words in a language, such as "and," "or," "if–then," "all," "some," and "none," and prepositional phrases, such as "greater than," "inside of," and so on. Each is associated with a fixed pattern of inference. Technically, such inferences need not involve any knowledge of the world, only the language (but see below). Conceptual relations, on the other hand, involve evaluation of propositions in light of our understanding of the world, including co-occurrence experiences and causal knowledge. For example, I may conclude that all dogs have tails, because all dogs that I have seen have had tails (co-occurrence), or I may conclude that the seasons are caused by tilting of the Earth on its axis, by having a (causal) model of the Earth's orbit around the sun. We will confine our discussion to logical and conceptual relations.

Broadly speaking, the above types of inferences are our basic mechanisms for determining rational actions. They constitute a system for generating new beliefs from observations and/or existing beliefs and for maintaining consistency among our beliefs (i.e., our mental representations of the world). They allow us to generate possible actions and identify those that are consistent or inconsistent with achieving goals. Inconsistent actions can be ruled out. Consistent actions can be further broken down into those that are certain, plausible (but not certain), and indeterminate. Any creature whose actions are a function of representations of the world—in particular, representations that have propositional content—will need some system of coherence maintenance to perform these dual functions of generating inferences (to guide actions) from perceptual input and existing beliefs, as well as maintaining the consistency of beliefs.

We are creatures whose behavior is a function of our beliefs about the world rather than of the world itself. Beliefs are psychological attitudes toward mental representations that have propositional content. If I have the attitude of "belief" to the proposition "There is a tiger under my desk," then I am asserting that a certain state of affairs is true of the world (namely, that there is a tiger under my desk). The source of this knowledge can be direct perception or inference based upon perception and/or other beliefs.

Irrespective of source, to be useful and to facilitate survival and propagation, beliefs need to meet certain constraints: in the case of direct perception, they need to be veridical. In the case of inference, they need to be consistent. The issue of veridicality is largely self-evident. For instance, in the above tiger example, if my tiger beliefs are veridical with respect to the actual state of affairs, my engagement in tiger-avoiding behavior will be appropriate and conducive to my survival. If there is a mismatch between my beliefs and the facts in the world, my actions will be inappropriate. If there is no tiger under my desk but I believe there to be one, I will run away unnecessarily. If there *is* a tiger under my desk but I do not believe that there is one, I will be eaten. Beliefs that are not veridical are typically not useful (and may be harmful). Veridicality is a relationship between a representation and the world. Much perceptual and cognitive neural machinery is devoted to getting this relationship largely right, most of the time.

Apart from perception, inference constitutes the other important source of knowledge for humans. Inferences are drawn from perceptual input and/or existing belief networks. For example, suppose I observe 12 white swans on Lake Simcoe. I may be tempted to conclude, "All swans are white." This constitutes new knowledge, based on a conceptual inference from the observation of 12 swans. I then need to maintain this belief for it to guide my future behavior. There exist sophisticated long-term memory systems for this purpose. At some future date, I see a black swan at the zoo. This observation is inconsistent with my previous belief "All swans are white." If the inconsistency is detected, my belief "All swans are white" will need to be revised to "Most swans are white." In the absence of this conflict detection and revision, I would entertain the beliefs "All swans are white" and "All swans are not white" (because at least one swan is black). They cannot both be true of the world.

The importance of the consistency of beliefs is not always appreciated but is as critical as the veridicality of beliefs. For example, if I hold the belief "Tigers are extremely dangerous" and also the belief "Tigers are not extremely dangerous," what is it that I believe? More importantly, when confronted with a tiger, do I run away or do I ignore it and do nothing? Two different actions are mandated; one will lead to survival, the other to death. It is therefore not surprising that there should be considerable neural machinery devoted to maintaining consistency of beliefs.

Therefore, a creature whose actions are a function of mental representations (with propositional content) needs to ensure that these representations meet the following requirements:

1. Veridicality
2. Maintenance
3. Inference
4. Inconsistency detection
5. Updating/belief revision

The veridicality requirement falls outside the scope of inference because, as noted above, veridicality is a relationship between propositions and the world. It is largely delegated to the perceptual systems. The maintenance requirement ensures retention of beliefs and falls within the purview of memory systems. The system of inference is very much about generating new beliefs based on coherence relations between propositions. The inconsistency detection system ensures the detection of inconsistency among beliefs (which would be harmful to the organism). The belief revision requirement draws upon the inference and inconsistency detection apparatus to maintain consistency among beliefs.

The goal of this chapter is to summarize what we know about brain systems involved in inference generation and consistency maintenance. In presenting this summary, I will not use cognitive theories of reasoning as organizing principles. Many articles and chapters organize data along one of the standard theories, be it mental models, mental logic, dual mechanisms, or a probabilistic account (Elqayam & Over, 2012; Evans, 2003; Henle, 1962; Johnson-Laird, 1994). These theories were developed to account for behavioral data. They may provide some insights for organizing the neuroscience data. However, making a priori wholesale commitments to one theory or another is probably counterproductive. My approach is to look at the neuroscience of reasoning data agnostically and to see what type of story might be embedded in it.

It is for this reason that I have tried to step back and ask, "What is the evolutionary problem that our system of inference developed to solve?" The answer that I proposed above is inference generation and inconsistency detection in service of "truth preservation." The neuroscience data on reasoning suggest that the proximal mechanisms for solving these problems are multiple systems for generating inferences and a separate common system for detecting and/or resolving inconsistency. The inference systems include a left prefrontal cortex (PFC) "interpreter" system that draws upon linguistic relations, a visuospatial system located in the parietal cortex that is engaged in linear comparisons, a visual-spatial system in the right PFC for set-inclusion determinations, and a system for dealing with indeterminate inferences, located in the right ventral lateral PFC. The generation systems do not guarantee consistency. A separate mechanism, located in the right lateral PFC, is provided for this purpose. It seems to be concerned with both the consistency between new information provided by the perceptual system and existing beliefs, as well as the consistency among propositional representations (internal or external). It is the latter issue that is relevant here.[3]

2. Neurological Systems for Generating Conceptual and Logical Inferences

The neuropsychological work on identifying systems for inference (or hypothesis generation) perhaps began with Sperry and Gazzaniga's studies on split-brain patients and Gazzaniga's conclusions regarding the dominance of the left hemisphere in generating inferences. In one classic experiment (Gazzaniga, 1989), split-brain patients were presented with a picture of a chicken claw projected to the right visual field (left hemisphere) and a picture of a snowy winter scene projected to the left visual field (right hemisphere). The patient then had to select (one with each hand), from an array of other pictures, the two most closely related to the projected pictures. The patient selected a shovel with the left hand (because the right hemisphere, controlling that hand, had processed a snowy winter scene) and a chicken with the right hand (because the left hemisphere, controlling that hand, had processed a chicken claw). Upon being asked to explain the choice of the shovel with the left hand (guided by the right hemisphere), the patient's left hemisphere (dominant for language) had no access to the information about the snowy scene processed by the right hemisphere. However, instead of responding "I don't know," the patient fabricated a plausible story, based upon background knowledge, and responded that the shovel was required to clean the chicken coop. Findings such as these led Gazzaniga (1998) to conclude that the left hemisphere was critical in drawing inferences. In fact, it couldn't help itself. It seems compelled to complete patterns and impose order on an uncertain world.

These initial findings have been fine-tuned over the years, and there is now considerable data to support the role of the left PFC in both semantic inference and simple logical inference. Unsurprisingly, however, the emerging picture is much more complex. The left PFC is not the only inference generation system in the brain. It is certainly involved in conceptual inferences. But its role in formal logical inferences seems much more constrained. Formal logical inferences engage the bilateral PFC and the parietal cortex. The relative engagement of these two regions is a function of logical form and even presentation modality. Furthermore, within the same logical form, content effects and argument determinacy modulate the specific neural systems engaged.

2.1 Conceptual Inference

Conceptual relations involve evaluation of propositions in light of our understanding of the world, including co-occurrence experiences and causal knowledge. Consider the following example:

(A) Eve is 42 years old. She is a serious and orderly woman. She loves a glass of good wine and playing chess. She tries to watch the news on foreign TV stations every day.

From this information, participants are much more likely to draw the inference "Eve is a librarian" than the inference "Eve likes to watch football." Notice that neither of these statements appear in the given information, nor follow logically from the given information. Nevertheless, the former inference is considered more plausible than the latter, given what we have been told about Eve and what we know about the world in general (Tversky & Kahneman, 1974).

Such content-based inferences are examples of inductive inferences. They draw upon our beliefs and knowledge about the world and tend to preferentially activate the left PFC. Goel, Gold, Kapur, and Houle (1997) and Goel and Dolan (2004) carried out inductive-inference studies with arguments such as in (B) and, with respect to the PFC, reported activation in the left dorsolateral PFC, Brodmann Areas (BA) 9, 8, and 45 (figure 1.4.1a).

(B) Snakes are cold-blooded;

Alligators are cold-blooded;

∴ All reptiles are cold-blooded.

In a recent follow-up to these imaging studies, Goel, Marling, Raymont, Krueger, and Grafman (2019) had neurological patients with penetrating focal lesions engage in simple inductive inferences involving conclusions of variable believability, such as in the following arguments:

(C) Rexdale is a German shepherd;

Rexdale lives in Düsseldorf;

∴ All German shepherds live in Düsseldorf.

(D) Lipstick is moist and glossy;

Fish scales are moist and glossy;

∴ Lipstick is made from fish scales.

(E) Snakes are reptiles;

Snakes are cold-blooded;

∴ All reptiles are cold-blooded.

The conclusion of argument (C) is highly unbelievable because we know that German shepherds can be found in many cities. The conclusion of argument (E) (although technically false) is highly believable, given what we have been taught about reptiles. Most of us do not have strong beliefs about the manufacture of lipstick and thus rate the believability of the conclusion in argument (D) as less certain. The authors report that patients with unilateral focal lesions to left BA (9 and 10) have less intense beliefs about moderately believable (argument (D)) and highly believable (argument (E)) conclusions and are less likely to accept these arguments as plausible.

Inductive reasoning has also been examined through the use of analogical mapping tasks. Wharton et al. (2000) had participants examine pictures of colored geometric shapes and determine whether the shapes were analogous (analogy condition) or identical (literal condition)

to a source picture of shapes. They reported enhanced brain activation in the medial frontal cortex (BA 8), the left PFC (BA 6, 10, 44, 45, 46, and 47), the anterior insula, and the left inferior parietal cortex (BA 40) when subjects made analogical-match judgments.

Other studies have imaged brain activation associated with judgment of analogous word pairs as in example (F) (Green, Fugelsang, Kraemer, Shamosh, & Dunbar, 2006) and verbal analogies, as in example (G) (Luo et al., 2003).

(F) Planet : sun   versus   electron : nucleus

(G) Soldier is to army as drummer is to band.

Green and colleagues (2006) found enhanced activation of parietal-frontal regions, most notably the left superior frontal gyrus (BA 9 and 10) for word pair stimuli (example (F)). Examining analogous concepts (example (G)), Luo and colleagues (2003) reported activation in the bilateral PFC (BA 45, 47, and 11), the left temporal lobe (BA 22), and the hippocampus.

Overall, studies evaluating language-based inductive arguments generally indicate activation in large areas of the brain, including the left frontal and parietal lobes. These regions overlap with the cortical regions involved in deductive reasoning with familiar material (discussed below). However, the evaluation of inductive arguments seems to be distinguished from the evaluation of deductive arguments (such as arguments (H), (N), (O), (P) below) by greater involvement of the left middle frontal gyrus (BA 9) (Goel et al., 1997; Goel & Dolan, 2004).

2.2 Logical Inference

Logical relationships between propositions are set up by closed-class (logical) terms of language rather than open-class (content) terms. For example, categorical syllogisms deal with quantification and negation, involving reasoning with the words "all," "some," and "none," as in arguments (H) and (I) below. Transitive arguments involve prepositional phrases such as "on top of," "shorter than," "more expensive than," "inside of," and "outside of" that can be used to build hierarchical relations, as in arguments (J) and (K) below. Finally, studies involving sentential connectives are generally confined to conditionals and disjunctions, as in arguments (L) and (M) below (Eimontaite et al., 2018; Noveck, Goel, & Smith, 2004). Logical arguments are designed to be valid by virtue of their structure rather than content. In terms of formal logical inference, it doesn't matter whether the arguments are about the color of broccoli or Julius Caesar crossing the Rubicon: they will be valid or invalid by virtue of their logical structure. Interestingly, this does matter to the brain. We return to this issue below.

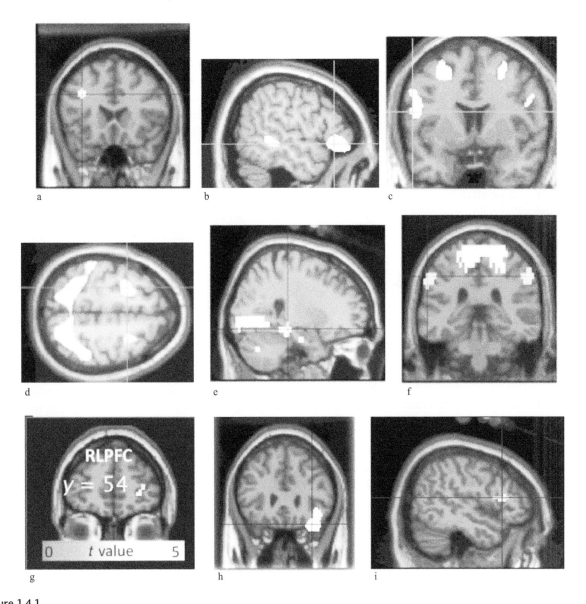

Figure 1.4.1

Systems for inference generation and conflict detection. (a–h) *Systems for inference generation*: (a) A left dorsolateral prefrontal cortex (PFC) system is involved in generating inductive inferences (reproduced with permission from Goel & Dolan, 2004); (b) a left lateral frontal-temporal linguistic system is activated during syllogistic reasoning involving content that we have beliefs about (reproduced with permission from Goel, Buchel, Frith, & Dolan, 2000); (c, d) bilateral frontal and parietal spatial systems are involved in formal syllogistic reasoning lacking meaningful semantic content (reproduced with permission from Goel et al., 2000); (e) linguistic transitive arguments with conclusions that we have beliefs about engage the left parahippocampal gyrus and the bilateral hippocampus (reproduced with permission from Goel, Makale, & Grafman, 2004); (f) linguistic transitive arguments involving conclusions that we have *no* beliefs about engage spatial systems in the bilateral parietal cortex (reproduced with permission from Goel, Makale, et al., 2004); (g) transitive arguments with pictorial stimuli engage the right rostral lateral PFC (reproduced with permission from Wendelken & Bunge, 2010); (h) indeterminate arguments with content that we have no beliefs about engage the right ventral lateral PFC (reproduced with permission from Goel, Stollstorff, Nakic, Knutson, & Grafman, 2009). (i) *A common system for conflict detection*: a common right lateral/dorsolateral PFC (BA 44/45) system seems to be engaged in detecting and/or resolving conflict or inconsistency (reproduced with permission from Goel & Dolan, 2003).

(H) All broccoli are vegetables;

     All vegetables are green;

     ∴ All broccoli are green.

(I) All *A* are *B*;

     All *B* are *C*;

     ∴ All *A* are *C*.

(J) London is north of Paris;

     Paris is north of Rome;

     ∴ London is north of Rome.

(K) *A* is north of *B*;

     B is north of *C*;

     ∴ *A* is north of *C*.

(L) If it rains on Saturday, then Linda will not come to the barbecue;

     It is raining on Saturday;

     ∴ Linda will not come to the barbecue.

(M) Tom went to the movies with either Linda or Mary;

     Tom did not go to the movies with Mary;

     ∴ Tom went to the movies with Linda.

Types of logical argument forms Interestingly, different logical forms seem to call upon different neural machinery. A qualitative review by Goel (2007) and a quantitative meta-analysis by Prado, Chadha, and Booth (2011) identified different brain systems engaged by categorical syllogisms, transitive arguments, and conditional arguments.

For categorical syllogisms, as in examples (H) and (I), Prado et al. (2011) report activation in the left PFC (BA 9 and 44), the left precentral gyrus (BA 4), the right caudate, and the left putamen. For arguments involving transitive relations, such as (J) and (K), they report activation in the bilateral parietal lobes (BA 7 and 40) and the bilateral dorsal PFC (BA 6). For conditional arguments, as in example (L), they report activation in a left hemisphere system involving the dorsal PFC (BA 6) and the angular gyrus (BA 39). The results regarding categorical syllogisms and transitive inference are consistent with the qualitative summary provided by Goel (2007). The results for conditional arguments may be less robust due to the insufficient number of studies. As an example, a recent study by Baggio et al. (2016) reports activation in the left PFC (BA 44 and 47) for conditional reasoning.

A number of individual neuroimaging and patient studies highlight the differential brain response to syllogistic reasoning and transitive reasoning. Activation in the left lateral and dorsal PFC (BA 6, 44, and 45) is widely reported for syllogistic reasoning tasks in neuroimaging studies (Goel, 2007; Goel et al., 2000; Goel & Dolan, 2003; Reverberi et al., 2012). One of the few lesion studies of deductive reasoning also reports that patients with left lateral and superior medial frontal lesions performed poorly on elementary deductive reasoning problems (Reverberi, Shallice, D'Agostini, Skrap, & Bonatti, 2009). Linguistically presented logical arguments involving transitive relations (examples (J) and (K)) activate the parietal cortex to a greater extent than the PFC (Goel, 2007; Knauff, Fangmeier, Ruff, & Johnson-Laird, 2003; Modroño et al., 2018; Prado et al., 2011). In a patient study directly comparing reasoning in transitive arguments with categorical syllogisms, it was reported that patients with lesions to the parietal cortex were impaired in the former task but not the latter (Waechter, Goel, Raymont, Kruger, & Grafman, 2013).

The meta-analysis study of logical form tells us something important: there is no single mechanism for logical inference. Examination of individual studies paints an even more nuanced picture. Generation systems vary as a function of content effects, type of spatial relations involved, argument determinacy, and even modality of argument presentation (linguistic versus pictorial).

Content and logical form As indicated above, deductive arguments are valid or invalid by virtue of their logical structure. Logically, the semantic content of the premises does not make a difference to the validity of arguments. However, psychologically and neurologically, the semantic content of deductive arguments makes a significant difference. Given that we do not typically reason about *A*s and *B*s but about whether climate change causes hurricanes or whether one should buy a new or a used car, the issue of prior beliefs and semantic content of propositions becomes a central one.

Psychologically, the *content effect* is the finding that, despite deductive reasoning being a function of logical form, argument content modulates response. It is one of the oldest findings in the cognitive psychology literature. Wilkins (1928) noted that valid logical arguments with *believable* conclusions are much more likely to be rated as valid than valid arguments with *unbelievable* conclusions.

A robust consequence of the content effect is the *belief bias effect*. In reasoning with content one has beliefs about, one will encounter either a congruency or an incongruency between the logical response and conclusion believability. *Congruent* arguments are either valid with believable conclusions (argument (H)) or invalid with unbelievable conclusions (argument (N)). *Incongruent* arguments are either valid with unbelievable conclusions (argument (O)) or invalid with believable conclusions (argument (P)).

(N) No apples are fruit;

    All fruit contain calories;

    ∴ No apples contain calories.

(O) All apples are fruit;

    All fruit are poisonous;

    ∴ All apples are poisonous.

(P) No apples are fruit;

    All fruit contain calories;

    ∴ All apples contain calories.

Cognitive neuroscientists have examined content effects by comparing inferences involving meaningful contents (argument (H)) with inferences involving non-meaningful contents (argument (I)), as well as inferences involving different types of contents (arguments (H) and (J)). These studies allow us to determine not only the effect of content on neural processing but also the effects of different *types* of contents. The neuroimaging data indicate that the main effect of comparing categorical syllogism arguments with meaningful semantic content (such as in (H)) with arguments without semantic content (such as in (I)) results in activation of a left frontal-temporal system, even after controlling for content (figure 1.4.1b) (Goel et al., 2000; Goel & Dolan, 2003). Comparing arguments without semantic content (such as (I)) with logically equivalent arguments *with* semantic content (such as (H)) activates bilateral PFC (figure 1.4.1c,d), along with bilateral parietal and occipital regions. This result persists even after controlling for the presence or absence of content (Goel et al., 2000; Goel & Dolan, 2003). When these comparisons are carried out with transitive arguments involving geographical knowledge, such as argument (J), and equivalent transitive arguments without semantic content, such as argument (K), the results show involvement of the left parahippocampal gyrus and the bilateral hippocampus in the meaningful-content condition and bilateral parietal involvement in the no-content condition, even after controlling for content (Goel, Makale, & Grafman, 2004) (figure 1.4.1e,f).

The involvement of the left frontal-temporal system in reasoning about familiar, meaningful content has also been demonstrated in neurological patients. In one study, Vartanian, Goel, Tierney, Huey, and Grafman (2009) administered three-term transitive inference arguments to patients with frontotemporal dementia. The parietal lobes of these patients were largely spared. As predicted by the imaging studies, they performed normally on arguments that they could have no beliefs about (such as (K)) but were selectively impaired in arguments involving meaningful content (such as (J)).

In another patient study, Goel, Shuren, Sheesley, and Grafman (2004) administered the Wason card selection task to patients with focal lesions to either the left or the right PFC and found that all patients performed as well as normal controls on the arbitrary version of the task, but unlike the normal controls, patients failed to benefit from the presentation of familiar content in the meaningful version of the task. In fact, consistent with the neuroimaging data, the latter result was driven by the exceptionally poor performance of patients with lesions to the left PFC. Patients with lesions to the right PFC performed as well as normal controls.

These results have been interpreted in the literature as the recruitment of different neural systems for contentful and noncontentful reasoning (Evans, 2003; Goel, 2007). However, repetitive transcranial magnetic stimulation (rTMS) studies suggest an even finer-grained distinction: Tsujii, Masuda, Akiyama, and Watanabe (2010) and Tsujii, Sakatani, Masuda, Akiyama, and Watanabe (2011) show that rTMS disruption of the left PFC specifically reduces reasoning accuracy only on a subset of contentful reasoning trials, the congruent trials. If this is the case, it suggests that the left PFC's role in logical inference may be limited to belief bias and conceptual connections and simple logical connectives. Arguments involving complex logical relations need to draw upon additional cognitive resources.

Formal logical inference   Logical arguments such as (I) and (K), which lack any meaningful semantic content that participants could have beliefs about, must be evaluated with formal machinery (i.e., based purely on structure). Arguments of the forms (H), (N), (O), and (P) result in engagement of the left PFC, while logically identical arguments lacking familiar content, as in argument (I), engage the bilateral lateral PFC along with the bilateral parietal and occipital lobes (Goel et al., 2000; Goel & Dolan, 2003) (figure 1.4.1c,d).

One interpretation of these results would be that, while the left PFC may be necessary and sufficient to deal with logical inference involving familiar material that participants have beliefs about (at least in congruent trials), the right PFC is part of the system required to deal with logical inference in purely formal situations. Disruption of left PFC functioning would impair the content-sensitive inference system, resulting in poor performance on congruent trials. The rTMS data reported above support this prediction (Tsujii et al., 2010; Tsujii et al., 2011).

Stimuli presentation modality One seeming inconsistency in the neuropsychology-of-reasoning literature has to do with the neural basis of transitive inference. Above, we have reported that transitive inference involves largely the bilateral parietal cortex (Goel, 2007; Prado et al., 2011). However, at least two studies (Fangmeier, Knauff, Ruff, & Sloutsky, 2006; Wendelken & Bunge, 2010) have focused on right BA 10 (medial anterior prefrontal and right rostrolateral PFC, respectively) as critical to "relational integration" (figure 1.4.1g). Interestingly, these two studies differ from other studies of transitive reasoning in the neuroimaging literature by virtue of using nonlinguistic/pictorial stimuli. Waechter et al. (2013) propose that the modality difference across the studies (linguistic versus pictorial) may account for the differences in results. In particular, when linguistic stimuli are used, greater effort and resources are required to map the stimuli onto spatial mental models as a prerequisite to solution (Johnson-Laird, 1983; Mani & Johnson-Laird, 1982). This requires the parietal cortex (Cohen et al., 1996; Goel & Dolan, 2001; Knauff et al., 2003). In the case of pictorial stimuli, the spatial relations are actually exemplified or embodied in the stimuli; thus, this mapping has already been done in the task presentation, rendering the involvement of the parietal cortex less critical and perhaps shifting processing to the PFC.

3. Indeterminacy Tolerance

The purpose of inference is to generate new information (or at least make information that was previously implicit, explicit). For example, given the information in (Q), I can make a determinate inference about the relative population sizes of Toronto and Guelph—namely, that Toronto has a greater population than Guelph—without explicitly being told so. I can also be certain that the population of Guelph is not greater than the population of Toronto, because that would contradict the given information.

(Q) The population of Toronto is greater than the population of Hamilton. The population of Hamilton is greater than the population of Guelph.

(R) The population of Toronto is greater than the population of Hamilton. The population of Toronto is greater than the population of Guelph.

But what happens in the case of example (R)? Here we are explicitly told about the relative population sizes of Toronto and Hamilton, and of Toronto and Guelph. What inference can we draw about the relative population sizes of Guelph and Hamilton? In the absence of any additional information, relying solely on the premises, nothing follows. In this case, any conclusions we derive about the relative population sizes of Guelph and Hamilton will be invalid, not because of inconsistency with the premises but because of indeterminacy given the premises: there is no fact of the matter regarding the relative population sizes of Guelph and Hamilton, given the information provided in the premises.

Goel et al. (2007) show that neurological patients with focal lesions to the left PFC have generalized impaired reasoning, including arguments with complete information (i.e., determinate, as in argument (Q)), while patients with right PFC lesions are selectively impaired only in arguments with incomplete information (i.e., indeterminate, as in argument (R)). This patient study demonstrates a double dissociation across left and right PFC along the dimension of determinacy. Neuroimaging studies involving similar transitive arguments reveal similar results (figure 1.4.1h) (Brzezicka et al., 2011; Goel et al., 2009). A study examining deductive reasoning with indeterminate syllogistic arguments also reported activation in the right PFC instead of the left PFC (Parsons & Osherson, 2001). (A transcranial magnetic stimulation study involving spatial relational reasoning also showed involvement of the right superior parietal cortex in dealing with uncertainty, although it did not test for the involvement of the right PFC; Ragni, Franzmeier, Maier, & Knauff, 2016.)

These data seem to suggest that we have developed special brain systems for dealing with indeterminate inferences. In cases where we can, we will fill in the missing information, using the left hemisphere interpreter, but in cases where this is not possible, a right ventrallateral PFC system is engaged to tolerate or accommodate the indeterminacy.

4. Detection of Conflict or Inconsistency

Conflict or inconsistency can arise among one's existing beliefs, between existing beliefs and incoming new information, and from inferences drawn from existing beliefs or external propositions. Detecting and/or resolving these conflicts or inconsistencies seems to be a generalized function of the right lateral and dorsolateral PFC (figure 1.4.1i).[4]

In the context of logical reasoning, we are specifically referring to an inconsistency between a response cued by our beliefs about the world and a response cued by the logical structure of the argument. Within incongruent trials (as in arguments (O) and (P)), the prepotent response is the incorrect response associated with the believability of the conclusion. Incorrect responses in such trials indicate that subjects failed to detect and/ or overcome the conflict between their beliefs and the logical inference and/or to inhibit the prepotent response associated with the belief bias. These belief-biased responses activate the ventral-medial PFC (BA 11 and 32), highlighting its role in nonlogical, belief-based responses (Goel & Dolan, 2003). The correct response indicates that subjects detected the conflict between their beliefs and the logical inference, inhibited the prepotent response associated with the belief bias, and engaged a formal reasoning mechanism. The detection of this conflict requires engagement of the right lateral and the dorsal-lateral PFC (BA 45 and 46) (see figure 1.4.1i), while generating the logical response calls upon the visuospatial machinery in the parietal cortex (De Neys, Vartanian, & Goel, 2008; Goel et al., 2000; Goel & Dolan, 2003; Prado & Noveck, 2007; Stollstorff, Vartanian, & Goel, 2012). Knauff and colleagues (Hamburger et al., 2018; Knauff, 2013) make a related point in the context of visual imagery impeding spatial reasoning.

These functional magnetic resonance imaging results have been replicated by rTMS studies demonstrating that stimulation of the right PFC specifically impairs performance on incongruent reasoning trials such as in arguments (O) and (P) (Tsujii et al., 2010; Tsujii et al., 2011). The rTMS data also show that disruption of the left PFC results not only in decreased performance in congruent trials but also in *improved* performance in *in*congruent trials. That is, when the left PFC is impaired, participants are less likely to go with the believability of the conclusion and will recruit other cortical regions to formally evaluate the argument.

One early demonstration of this conflict detection system with lesion data was carried out by Caramazza, Gordon, Zurif, and DeLuca (1976) using simple two-term reasoning problems such as the following: "Mike is taller than George. Who is taller? Who is shorter?" They reported that left hemisphere patients were impaired in all forms of the problem but—consistent with imaging data (Goel et al., 2000; Goel & Dolan, 2003; Stollstorff et al., 2012)—right hemisphere patients were only impaired when the form of the question was inconsistent with the premise (i.e., "Who is shorter?").

A final example of the role of the right PFC in conflict detection is provided by Reverberi, Lavaroni, Gigli, Skrap, and Bonatti (2005). They carried out a revised version of the Brixton task involving rule-induction and rule-conflict conditions. They reported that while patients with lesions to the left PFC showed an impairment in

rule induction, patients with lesions to the right PFC were impaired specifically in the rule-conflict condition.

This conflict detection role of the right lateral and the dorsal PFC is a generalized phenomenon that has been documented in a wide range of paradigms in the cognitive neuroscience literature (Fink et al., 1999; Picton, Stuss, Shallice, Alexander, & Gillingham, 2006; Vallesi, Mussoni et al., 2007; Vallesi, Shallice, & Walsh, 2007). Marinsek, Turner, Gazzaniga, and Miller (2014) have actually suggested that conflict or inconsistency detection is the main role of the right PFC in cognitive functioning.

5. Summary and Conclusion

Humans are creatures whose behavior is a function of their beliefs about the world rather than the world itself. Not only do we have beliefs, but our beliefs also have propositional content. Beliefs with propositional content allow us to generate new knowledge by drawing inferences that take us beyond direct perception and differentiate between what is necessarily the case, what might be the case, and what absolutely *cannot* be the case. One can even time-travel with such mental representations and entertain past and future possibilities, including counterfactuals.

For such a system to be useful, the inferences must be coherent, and this coherence must be maintained over the whole system of beliefs. There must also be a system for detecting inconsistencies in inferences and among beliefs. Twenty-plus years of neuroscience studies of logical reasoning have revealed that there is no unitary reasoning module in the brain for undertaking this. Rather, our ability to reason seems to be underwritten by two separate classes of mechanisms: (1) mechanisms for hypothesis generation and inference and (2) a mechanism for detecting conflict or inconsistency.

Inference generation calls upon several different systems, including (1) a left PFC interpreter system sensitive to semantic, conceptual, and simple logical relations; (2) multiple visuospatial systems; and (3) a system for tolerating indeterminacy. The first of these systems deals largely with semantic and conceptual relations and simple syntactic inferences. The second deals with more formal processing of logical arguments. The neurological basis of this system seems to vary as a function of logical form and argument presentation modality. For example, set-inclusion relationships call upon the right and/or bilateral PFC, while linear comparisons call upon the parietal cortex. Additionally, linear comparisons presented pictorially activate the right rostral PFC, while the same linear comparisons presented linguistically activate bilateral parietal cortex. Finally, a system in the right ventral-lateral PFC seems to play a critical role in allowing for indeterminate inferences.

These systems of inference generation do not, in and of themselves, guarantee consistency. We seem to have a separate system in the right lateral/dorsolateral PFC for detecting conflict or inconsistency between external and internal representations and among internal representations. Together, these various systems account for the ability to draw correct inferences and maintain the consistency of the overall belief network.

The details of this overall account will undoubtedly change as additional studies are carried out and new data are generated and added to our knowledge base. However, after 20 years of research, the broader picture that is emerging may be reasonably secure: specifically, that there is no single system of logical reasoning in the brain. Our ability to engage in rational thought is underwritten by several different types of inference systems and a common system for detecting inconsistencies.

Notes

1. Additionally, while not germane to our purposes here, it is important to note that there is a "gap" between input and output conditions in rational actions. The antecedent condition is never causally sufficient for the consequent condition (Cassirer, 1944).

2. Sometimes a distinction is made between "instrumental" rationality and "epistemic" rationality (Stanovich, 1999). The latter is an evaluation of the fit between an individual's beliefs and the facts in the world. However, this is simply instrumental rationality applied to belief evaluation and revision. It is not clear that this distinction needs to be dealt with separately, at least for our purposes.

3. Most neuroscience discussions of rationality usually begin and end with delusions. While delusions are an important topic, I will not touch upon them here. I think we will increase our chances of understanding delusions if we can first understand the neural basis of normal rationality.

4. Experimental data have thus far not clearly distinguished between the right PFC's role in detecting and resolving the conflict.

References

Baggio, G., Cherubini, P., Pischedda, D., Blumenthal, A., Haynes, J.-D., & Reverberi, C. (2016). Multiple neural representations of elementary logical connectives. *NeuroImage, 135*, 300–310.

Brzezicka, A., Sędek, G., Marchewka, A., Gola, M., Jednoróg, K., Królicki, L., & Wróbel, A. (2011). A role for the right prefrontal and bilateral parietal cortex in four-term transitive reasoning: An fMRI study with abstract linear syllogism tasks. *Acta Neurobiologiae Experimentalis, 71*, 479–495.

Caramazza, A., Gordon, J., Zurif, E. B., & DeLuca, D. (1976). Right-hemispheric damage and verbal problem solving behavior. *Brain and Language, 3*, 41–46.

Cassirer, E. (1944). *An essay on man: An introduction to a philosophy of human culture*. New Haven, CT: Yale University Press.

Cohen, M. S., Kosslyn, S. M., Breiter, H. C., DiGirolamo, G. J., Thompson, W. L., Anderson, A. K., . . . Belliveau, J. W. (1996). Changes in cortical activity during mental rotation: A mapping study using functional MRI. *Brain, 119*, 89–100.

De Neys, W., Vartanian, O., & Goel, V. (2008). Smarter than we think: When our brains detect that we are biased. *Psychological Science, 19*, 483–489.

Eimontaite, I., Goel, V., Raymont, V., Krueger, F., Schindler, I., & Grafman, J. (2018). Differential roles of polar orbital prefrontal cortex and parietal lobes in logical reasoning with neutral and negative emotional content. *Neuropsychologia, 119*, 320–329.

Elqayam, S., & Over, D. (2012). Probabilities, beliefs, and dual processing: The paradigm shift in the psychology of reasoning. *Mind & Society, 11*, 27–40.

Evans, J. St. B. T. (2003). In two minds: Dual-process accounts of reasoning. *Trends in Cognitive Science, 7*, 454–459.

Fangmeier, T., Knauff, M., Ruff, C. C., & Sloutsky, V. (2006). fMRI evidence for a three-stage model of deductive reasoning. *Journal of Cognitive Neuroscience, 18*, 320–334.

Fink, G. R., Marshall, J. C., Halligan, P. W., Frith, C. D., Driver, J., Frackowiak, R. S. J., & Dolan, R. J. (1999). The neural consequences of conflict between intention and the senses. *Brain, 122*, 497–512.

Gazzaniga, M. S. (1989). Organization of the human brain. *Science, 245*, 947–952.

Gazzaniga, M. S. (1998). *The mind's past*. Berkeley: University of California Press.

Goel, V. (2007). Anatomy of deductive reasoning. *Trends in Cognitive Science, 11*, 435–441.

Goel, V., Buchel, C., Frith, C., & Dolan, R. J. (2000). Dissociation of mechanisms underlying syllogistic reasoning. *NeuroImage, 12*, 504–514.

Goel, V., & Dolan, R. J. (2001). Functional neuroanatomy of three-term relational reasoning. *Neuropsychologia, 39*, 901–909.

Goel, V., & Dolan, R. J. (2003). Explaining modulation of reasoning by belief. *Cognition, 87*, B11–B22.

Goel, V., & Dolan, R. J. (2004). Differential involvement of left prefrontal cortex in inductive and deductive reasoning. *Cognition, 93*, B109–B121.

Goel, V., Gold, B., Kapur, S., & Houle, S. (1997). The seats of reason? A localization study of deductive and inductive reasoning. *NeuroReport, 8*, 1305–1310.

Goel, V., Makale, M., & Grafman, J. (2004). The hippocampal system mediates logical reasoning about familiar spatial environments. *Journal of Cognitive Neuroscience, 16*, 654–664.

Goel, V., Marling, M., Raymont, V., Krueger, F., & Grafman, J. (2019). Patients with lesions to left prefrontal cortex (BA 9 and BA 10) have less entrenched beliefs and are more sceptical reasoners. *Journal of Cognitive Neuroscience, 31*, 1674–1688.

Goel, V., Shuren, J., Sheesley, L., & Grafman, J. (2004). Asymmetrical involvement of frontal lobes in social reasoning. *Brain, 127*, 783–790.

Goel, V., Stollstorff, M., Nakic, M., Knutson, K., & Grafman, J. (2009). A role for right ventrolateral prefrontal cortex in reasoning about indeterminate relations. *Neuropsychologia, 47*, 2790–2797.

Goel, V., Tierney, M., Sheesley, L., Bartolo, A., Vartanian, O., & Grafman, J. (2007). Hemispheric specialization in human prefrontal cortex for resolving certain and uncertain inferences. *Cerebral Cortex, 17*, 2245–2250.

Green, A. E., Fugelsang, J. A., Kraemer, D. J. M., Shamosh, N. A., & Dunbar, K. N. (2006). Frontopolar cortex mediates abstract integration in analogy. *Brain Research, 109*, 125–137.

Hamburger, K., Ragni, M., Karimpur, H., Franzmeier, I., Wedell, F., & Knauff, M. (2018). TMS applied to V1 can facilitate reasoning. *Experimental Brain Research, 236*, 2277–2286.

Henle, M. (1962). On the relation between logic and thinking. *Psychological Review, 69*, 366–378.

Johnson-Laird, P. N. (1983). *Mental models: Towards a cognitive science of language, inference, and consciousness*. Cambridge, MA: Harvard University Press.

Johnson-Laird, P. N. (1994). Mental models, deductive reasoning, and the brain. In M. S. Gazzaniga (Ed.), *The cognitive neurosciences* (pp. 999–1008). Cambridge, MA: MIT Press.

Knauff, M. (2013). *Space to reason: A spatial theory of human thought*. Cambridge, MA: MIT Press.

Knauff, M., Fangmeier, T., Ruff, C. C., & Johnson-Laird, P. N. (2003). Reasoning, models, and images: Behavioral measures and cortical activity. *Journal of Cognitive Neuroscience, 15*, 559–573.

Luo, Q., Perry, C., Peng, D., Jin, Z., Xu, D., Ding, G., & Xu, S. (2003). The neural substrate of analogical reasoning: An fMRI study. *Cognitive Brain Research, 17*, 527–534.

Mani, K., & Johnson-Laird, P. N. (1982). The mental representation of spatial descriptions. *Memory & Cognition, 10*, 181–187.

Marinsek, N., Turner, B. O., Gazzaniga, M., & Miller, M. B. (2014). Divergent hemispheric reasoning strategies: Reducing uncertainty versus resolving inconsistency. *Frontiers in Human Neuroscience, 8*, 839.

Modroño, C., Navarrete, G., Nicolle, A., González-Mora, J. L., Smith, K. W., Marling, M., & Goel, V. (2018). Developmental grey matter changes in superior parietal cortex accompany improved transitive reasoning. *Thinking & Reasoning, 25*, 151–170.

Noveck, I. A., Goel, V., & Smith, K. W. (2004). The neural basis of conditional reasoning with arbitrary content. *Cortex, 40*, 613–622.

Parsons, L. M., & Osherson, D. (2001). New evidence for distinct right and left brain systems for deductive versus probabilistic reasoning. *Cerebral Cortex, 11*, 954–965.

Picton, T. W., Stuss, D. T., Shallice, T., Alexander, M. P., & Gillingham, S. (2006). Keeping time: Effects of focal frontal lesions. *Neuropsychologia, 44*, 1195–1209.

Prado, J., Chadha, A., & Booth, J. R. (2011). The brain network for deductive reasoning: A quantitative meta-analysis of 28 neuroimaging studies. *Journal of Cognitive Neuroscience, 23*, 3483–3497.

Prado, J., & Noveck, I. A. (2007). Overcoming perceptual features in logical reasoning: A parametric functional magnetic resonance imaging study. *Journal of Cognitive Neuroscience, 19*, 642–657.

Ragni, M., Franzmeier, I., Maier, S., & Knauff, M. (2016). Uncertain relational reasoning in the parietal cortex. *Brain and Cognition, 104*, 72–81.

Reverberi, C., Bonatti, L. L., Frackowiak, R. S. J., Paulesu, E., Cherubini, P., & Macaluso, E. (2012). Large scale brain activations predict reasoning profiles. *NeuroImage, 59*, 1752–1764.

Reverberi, C., Lavaroni, A., Gigli, G. L., Skrap, M., & Shallice, T. (2005). Specific impairments of rule induction in different frontal lobe subgroups. *Neuropsychologia, 43*(3), 460–472.

Reverberi, C., Shallice, T., D'Agostini, S., Skrap, M., & Bonatti, L. L. (2009). Cortical bases of elementary deductive reasoning: Inference, memory, and metadeduction. *Neuropsychologia, 47*, 1107–1116.

Stanovich, K. E. (1999). *Who is rational? Studies of individual differences in reasoning*. New York, NY: Psychology Press.

Stollstorff, M., Vartanian, O., & Goel, V. (2012). Levels of conflict in reasoning modulate right lateral prefrontal cortex. *Brain Research, 1428*, 24–32.

Tsujii, T., Masuda, S., Akiyama, T., & Watanabe, S. (2010). The role of inferior frontal cortex in belief-bias reasoning: An rTMS study. *Neuropsychologia, 48*, 2005–2008.

Tsujii, T., Sakatani, K., Masuda, S., Akiyama, T., & Watanabe, S. (2011). Evaluating the roles of the inferior frontal gyrus and superior parietal lobule in deductive reasoning: An rTMS study. *NeuroImage, 58*, 640–646.

Tversky, A., & Kahneman, D. (1974). Judgment under uncertainty: Heuristics and biases. *Science, 185*, 1124–1131.

Vallesi, A., Mussoni, A., Mondani, M., Budai, R., Skrap, M., & Shallice, T. (2007). The neural basis of temporal preparation: Insights from brain tumor patients. *Neuropsychologia, 45*, 2755–2763.

Vallesi, A., Shallice, T., & Walsh, V. (2007). Role of the prefrontal cortex in the foreperiod effect: TMS evidence for dual mechanisms in temporal preparation. *Cerebral Cortex, 17*, 466–474.

Vartanian, O., Goel, V., Tierney, M., Huey, E. D., & Grafman, J. (2009). Frontotemporal dementia selectively impairs transitive reasoning about familiar spatial environments. *Neuropsychology, 23*, 619–626.

Waechter, R. L., Goel, V., Raymont, V., Kruger, F., & Grafman, J. (2013). Transitive inference reasoning is impaired by focal lesions in parietal cortex rather than rostrolateral prefrontal cortex. *Neuropsychologia, 51*, 464–471.

Wendelken, C., & Bunge, S. A. (2010). Transitive inference: Distinct contributions of rostrolateral prefrontal cortex and the hippocampus. *Journal of Cognitive Neuroscience, 22*, 837–847.

Wharton, C. M., Grafman, J., Flitman, S. S., Hansen, E. K., Brauner, J., Marks, A., & Honda, M. (2000). Toward neuroanatomical models of analogy: A positron emission tomography study of analogical mapping. *Cognitive Psychology, 40*, 173–197.

Wilkins, M. C. (1928). *The effect of changed material on the ability to do formal syllogistic reasoning* (Archives of Psychology No. 102). New York, NY: Woodworth.

Section 2   Key Issues of Rationality

2.1 Reasons and Rationality

John Broome

Summary

I explore the relationship between rationality and reasons, particularly the reductive idea that rationality can be defined in terms of reasons. I start with an analysis of the meaning of "rationality" in order to clarify the issue. Then I assess the view that rationality consists in responding correctly to reasons. To this I oppose a "quick objection," describe the defenses the view has against this objection, and argue that these defenses are unappealing. Next, I assess various related views, including the view that rationality consists in responding correctly to beliefs about reasons and argue against each of them. Eventually, I identify the kernel of truth that lies within them, which is that rationality requires you to intend to F if you believe you ought to F. I call this principle "enkrasia." It is only one requirement of rationality among many, so it licenses no reduction of rationality.

1. Normativity and Reasons

Knauff and Spohn explain in their Introduction that a recurring theme in this handbook is the relation between positive and normative approaches to the study of rationality. This chapter investigates some fundamental aspects of the normative approach.

The word "normative" has various meanings, and at least two are current in philosophy. One meaning is "involving correctness." Rules and requirements are by definition normative in this sense. Any rule or requirement sets up a standard of correctness, so that complying with it is correct according to the rule or requirement. Rationality requires things of us, or—to put it differently—it prescribes things to us. For instance, it prescribes that we intend means to ends that we intend and that we do not have contradictory beliefs. So rationality is inevitably normative in this sense of "normative." Much of the study of rationality is concerned with its normativity in this sense, investigating just what rationality requires of us.

Many things besides rationality are normative in this sense. Fashion is an example. Fashion these days prescribes that men do not wear bellbottom trousers. But this does not automatically imply that a man ought not to wear bellbottom trousers or has any reason not to. It is a real question whether we have any reason to dress as fashion requires. This a question about the normativity of fashion in a different sense. In this sense, "normative" means "involving ought or reasons."

We may ask the same question about rationality: when rationality requires something of us—such as to intend means to ends we intend—does that imply we ought to do it or have any reason to do it? Although rationality is inevitably normative in the first sense, there are real questions about its normativity in the second sense. These and related questions are the topic of this chapter. In this chapter, "normative" has the second sense.

Reasons are a paradigmatic feature of normativity in this sense. They became an important object of study for philosophers only in the middle of the 20th century. A significant achievement of the philosophy of normativity since that time has been to make a sharp distinction between motivating reasons and normative reasons (e.g., Smith, 1994). (To be precise, the distinction is between the property of being a motivating reason and the property of being a normative reason. Many particular things have both properties.) Motivating reasons explain, or help to explain, why a person does something. Normative reasons explain, or help to explain, why a person ought to do something (Broome, 2013, chapter 4). It is normative reasons that figure in this chapter, since the chapter is about the relation between rationality and normativity.

In the past few decades, reasons have come to dominate the philosophy of normativity. As a result of what is often called "the reasons-first movement,"[1] many philosophers now think that rationality can be given a reductive definition in terms of reasons. If that were true, it would mean that the study of rationality is nothing more than the study of reasons. One aim of this chapter is to explore this reductive idea.

2. The Meaning of "Rationality"

It is natural to associate reasons with rationality. The words "reason" and "rationality" have a common origin in the Latin word "ratio." But this simple etymological association covers up a tangle of meanings that connect the two words. I need to start with some disentanglement.

The word "reason" entered English from French along with the Norman invasion of England in 1066. Its first recorded occurrence in English is in a book called the *Ancrene Riwle*, whose earliest manuscript dates from about 1225 (Day, 1952). "Reason" appears there in various different senses, all of which survive today. Sometimes it means simply "explanation," as it still does in such sentences as "The reason for the long delay was incompetence." Often it refers to a motivating reason, which is a special sort of explanation of why a person does something.

Just once in the *Ancrene Riwle*, "reason" refers to a normative reason. This is in the sentence (translated into modern English):

> The third reason for fleeing the world is the gaining of heaven. (p. 73, folio 43)

Just previously in the text, the author says he will describe "eight reasons why one ought to flee the world," which is to say, eight explanations of why one ought to flee the world. Then he starts to enumerate them, and when he comes to the third, he describes it as a reason for fleeing the world. So "a reason for fleeing the world" refers to an explanation of why one ought to flee the world. This is a normative reason.[2]

In all those senses, "reason" is a count noun. It also appears once in the *Ancrene Riwle* as a non–count noun naming a property that people possess. We still call this property the "faculty of reason." Since it is a mental faculty, let us call this "reason in the mental sense." The original text needs some exegesis:

> Wummon is the reisun—thet is, wittes skile—hwen hit unstrengeth. (p. 121, folio 73)

The author has just recounted a parable from the Bible. He is saying that the woman in the parable represents the faculty of reason (spelled "reisun"). Because the word "reason" had only recently acquired the mental sense, he glosses it using an older English term for the faculty of reason: he says (in modern spelling), "that is, wit's skill." Since this earliest mention of the faculty of reason is obscure, here is a clearer one from Shakespeare (*A Midsummer Night's Dream*, Act 2, Scene 2):

> The will of man is by his reason sway'd.

The adjective "rational" is first recorded by the *Oxford English Dictionary* (*OED*) in 1398. From its beginning, it was cognate to "reason" in the mental sense and in that sense only. It had the meaning "having the faculty of reason." It had this meaning and no other for about two hundred years. The *OED* shows that for all that period, it was applied as a predicate only to people, creatures, souls, minds, and suchlike: all things that could possess the faculty of reason. This meaning of "rational" persists today.

The noun "rationality" appeared in 1627 as the name of the property that is ascribed by this adjective. Since this property is just reason in the mental sense, "rationality" and "reason" in this sense were originally synonyms.

However, the meaning of "rationality" has by now broadened. "Reason" in the mental sense refers only to a faculty. "Rationality" today refers to the same faculty and also to a state of mind—roughly, a state of mind that could have arisen from the exercise of the faculty of reason, which is to say, a coherent state of mind. The term "structural rationality" is often used today for the rationality of states.[3] These days, we would not count a person as fully rational if she had the faculty of rationality but not structural rationality. For instance, a person is not fully rational if she does not intend means to her ends, even if she has the ability to ensure that she does intend means to her ends. Ability is not enough; we expect it to be exercised.

Nevertheless, even in this broadened sense, rationality retains one central feature: it is a property of a person and specifically a property of her mind. Moreover, it depends on the other properties of the person's mind: as Wedgwood (2002) puts it, rationality supervenes on—depends only on—the mind. If a person might be in either of two possible situations, but her mental properties apart from rationality would be the same in either, she would be equally as rational in one as in the other. So, even though the meaning of "rationality" has broadened beyond the mental sense of "reason," it still refers to a mental property.

However, from 1598 onward, the *OED* records "rational" used as a predicate of things that do not have minds. These days, we apply "rational" to acts, beliefs, city plans, and many other things without minds. These uses of "rational" for nonmental things are derived from the original, mental sense applied to people. A city plan is rational if it could have been designed by rational people. A person's act is rational if, were she to do it, she would be no less rational than if she were not to do it. And so on. Nevertheless, in its core meaning, "rationality" still refers to a property of people, and it is a mental property. It supervenes on the mind.

Some philosophers assign a different meaning to "rational." For example, Kolodny and Brunero (2013/2018) say,

> "What would it be rational for an agent to do or intend?" could mean:
>
> 1. By doing or intending what would the agent make her responses (i.e., her attitudes and actions) cohere with one another? . . .
> 2. What does the agent have reason, or ought she, to do or intend?

The first of these meanings is roughly structural rationality. The second is sometimes called "substantive rationality,"[4] but it is not a normal meaning of "rational" at all. To see this, think of a case where you ought to turn left, but you firmly believe on the basis of strong but misleading evidence that you ought to turn right. In the substantive sense, it would be rational for you to turn left. But no nonphilosopher would say it is rational. No one would call it rational to do the opposite of what you firmly believe you ought to do.[5]

The substantive meaning of "rational" could be etymologically justified. "Rational" is a cognate word to "reason," and it could in principle be cognate to "reason" in the normative sense. Actually, however, "rational" has never had this normative meaning in common English. It has always been cognate to "reason" in the mental sense only. I use it only with this common meaning.

3.   Rationality as Responding Correctly to Reasons

Although "rationality" does not have a normative *meaning*, it is a popular view among philosophers and others that, as a *substantive* matter, rationality is nevertheless intimately connected with normativity. A strong version of this view is the claim that rationality consists in responding correctly to reasons.[6] This is a reductive claim: it claims that rationality is reducible to reasons in this way.

It is subject to something I call the "quick objection."[7] The property of rationality supervenes on the mind, whereas the property of responding correctly to reasons does not. These therefore cannot be the same property. This section examines the quick objection and defenses against it.

Take a particular person called "you." You are rational to some degree, and this degree supervenes on your mind. I insisted in section 2 that this is part of the meaning of "rational."

What about responding correctly to reasons? First, what is this property, more exactly? You have many reasons. Each is a reason for some particular thing: a reason for you to do something, or to believe something, or not to intend some particular end without also intending a means to it, or for something else. I use a schematic letter to represent this generality: a reason of yours is a reason for you to F. Responding correctly to reasons cannot be simply Fing whenever you have a reason to F. Often you have a reason to F and also a reason not to F. You cannot both F and not F, so if responding correctly to reasons required you to F whenever you have a reason to F, you often could not respond correctly to reasons. That cannot be so.

Instead, we must recognize that your reasons in some way combine together. They may weigh against each other, some may override others, some may cancel others, and so on. Your reasons together require various things of you. They require you to F, to G, and so on. Another way of putting this is that you ought to F, to G, and so on.

In this section, I assume that, to respond correctly to reasons, you must comply with reasons, by which I mean you must F whenever your reasons together require you to F.[8] An alternative interpretation is that you must intend to F whenever your reasons together require you to F. I consider that interpretation in section 5. Your reasons together could not both require you to F and require you not to F, so the previous problem does not arise for either interpretation.

Responding correctly to reasons may imply not just complying with reasons but also doing so because your reasons require you to. In this section, I assume that complying with reasons is at least a part of responding correctly to reasons. Consequently, the property of responding correctly to reasons cannot supervene on your mind unless the property of complying with reasons does. I shall argue that complying with reasons supervenes on your mind only if some unappealing philosophical theories are true.

For you to comply with reasons is for the following universal conditional proposition to be true: that, for any F, you F if reasons require you to F. This conditional supervenes on your mind if both sides of it do—that is, if, for any F, first, whether or not reasons require you to F supervenes on your mind and, second, if reasons require you to F, whether or not you F supervenes on your mind. As I shall put it: first, what your reasons require supervenes on your mind and, second, your performance supervenes on your mind. It is conceivable that the property of complying with reasons could supervene on your mind even if one of these conditions was not satisfied, but I cannot see how this could actually happen. So the quick objection divides into two objections. The first is that what your reasons require does not supervene on your mind. The second is that your performance does not supervene on your mind. Either is enough to refute the claim that rationality consists in responding correctly to reasons. I shall develop these two objections in turn.

3.1 First Objection

Does what your reasons require of you—in other words, what you ought—supervene on your mind? The claim that it does is a sort of subjectivism about ought. Various subjectivist theories support it. For example, one is the theory that you ought to F if and only if Fing has the greatest expected value for you out of all the alternatives, where expected values are given by your own credences and your own judgments of value.

Many philosophers find subjectivism about ought an unappealing theory. It conflicts with common sense, if nothing else. Common sense tells us that external facts can influence what you ought to believe or do. For example, the fact that lowering clouds are gathering is a reason to expect rain, and the fact that your child is badly hurt is a reason to take her to the hospital.

Kiesewetter (2017, chapter 7) offers a means of easing this discomfort with subjectivism. He agrees with common sense that reasons are often features of the external world and argues that this can be made consistent with subjectivism about ought.

His argument is this. A feature of the external world is a reason for you only if it available to you, by which he means it is part of your body of evidence. Indeed, he assumes that what you ought is determined by your total body of evidence together with features of your mind such as your likes and dislikes. He now applies a strong dose of externalism about the mind, taking his lead from Williamson (2000). According to Williamson, your evidence is what you know, and your knowledge is a mental state of yours. Given this, your body of evidence is a feature of your mind. So what you ought is entirely determined by features of your mind, even though reasons are features of the external world.

I doubt this will ease many philosophers' discomfort.[9] It conflicts equally with common sense. In effect, it expands the notion of the mind to include whatever facts in the world constitute reasons. Subjectivism about ought remains an unappealing theory.

3.2 Second Objection

Your reasons often require you to act in the external world. For example, your reasons may require you to insure your house. Even reasons that are features of your own mind may require this. For example, perhaps you want to avoid risk of financial ruin and believe that insuring your house is necessary for that purpose, and perhaps this desire and belief constitute a reason for you to insure your house.

So let us assume your reasons require you to insure your house. Suppose you take the usual steps to do so: you complete an application form, glance through the contract, pay the premium, and so on. Compare two cases. In the first, by these steps, you successfully insure your house. In the other, a clause in the contract, which you do not read, says your house is insured only if it is roofed with metal, tiles, or slate. Your house is roofed with cedar shingles, so you do not successfully insure it. But suppose you never claim on insurance, and your failure never comes to light. Then your mind has all the same properties in both cases. Nevertheless, in one you do as your reasons require and in the other you do not. So your performance does not supervene on your mind.

This second objection could be overcome if we could accept a sort of subjectivism about performance. We could say that reasons cannot require you to do something unless the criterion for whether or not you do it is internal to your mind. We could deny in the example that your reasons require you to insure your house. We could say instead that they require you to act in a way that appears to you to be insuring your house or, alternatively, that they require you to intend to insure your house.

This sort of subjectivism has been defended,[10] but it, too, is unappealing. The relevant reason in this case is a reason of self-interest: it is in your interest to insure your house. It is not in your interest to do something that appears to you to be insuring your house or to intend to insure your house, except insofar as either leads you to actually insuring it.

3.3 Conclusion

The claim that rationality consists in responding correctly to reasons can be defended against the quick objection only by showing that complying with reasons supervenes on your mind. There are two objections to this claim, which can be overcome only on the basis of unappealing philosophical theories. The quick objection is vindicated to this extent.

In any case, blocking the quick objection is far from sufficient to establish that rationality consists in responding correctly to reasons. There are other, independent objections. One is that it is often moral reasons that require you to do some act. Suppose you respond correctly to these reasons by doing this act because your moral reasons require you to. If responding correctly to reasons constituted rationality, this would exhibit your rationality. But actually, it exhibits your morality rather than your rationality (see chapter 12.1 by Fehige & Wessels, this handbook).

The same would be true even if your responding to reasons supervened on your mind. Suppose that your moral reasons require you not to have racist beliefs or not to have evil intentions, for example. Again, responding correctly to these reasons exhibits your morality and not your rationality.

The claim that rationality consists in responding correctly to reasons remains dubious.

4. Rationality as Entailed by Responding Correctly to Reasons

A weaker claim is that rationality is *entailed* by responding correctly to reasons.[11] This, too, may be intended as a reductive claim that rationality is nothing more than a part of responding correctly to reasons.

4.1 Structural Reasons

On one version of this view, everyone has reasons—call them "structural reasons"—to have her mind in good coherent order. You have a reason not to have contradictory beliefs, a reason to intend means to ends that you intend, and so on. The view is that rationality consists in responding correctly to your structural reasons. Responding correctly to structural reasons would supervene on your mind, so this view is immune to the quick objection.

But it misunderstands responding correctly to reasons. To respond correctly to reasons, you must F when your reasons together require you to F, not when you have a single reason to F. Even if you have a structural reason not to have contradictory beliefs, you might have another reason to have contradictory beliefs. For example, an evil demon might announce it will destroy the world unless you have some contradictory beliefs. In a case like this, your reasons together may require you to have contradictory beliefs, so that responding correctly to reasons would imply having contradictory beliefs. Nevertheless, if you do have contradictory beliefs, you will not be fully rational. This shows that rationality is not a part of responding correctly to reasons, even if structural reasons indeed exist.

This argument illustrates a fundamental difficulty that stands in the way of reducing rationality to reasons. Rationality imposes *strict* requirements on us, and if we violate them, we are necessarily not fully rational. But what reasons require of us is generally defeasible; it can be overridden by further reasons. So reasons are not well suited to account for rationality.

4.2 Myth Theory

Another version of the view that rationality is entailed by responding correctly to reasons is known as "myth theory."[12] It is the view that rationality in the mental sense is a myth, or at least that structural rationality is a myth. Structural rationality is the property a person has when she has consistent beliefs and intentions, intends means to ends she intends, and so on. Myth theorists do not deny that this property exists. But they think it is an uninteresting property, because if a person responds correctly to reasons, she will possess it automatically as a consequence. They think that, if your mind is properly aligned with the world—so you believe what your reasons require you to believe, you intend whatever your reasons require you to intend, and so on—a necessary consequence is that your mind will be properly aligned internally. You will have consistent beliefs and intentions, intend means to ends you intend, and so on; you will be structurally rational.

Kolodny (2007) expresses his version of myth theory by denying that "there are rational requirements of formal coherence as such." Rationality definitely has requirements in one sense. Any necessary condition for something to possess a property may be called a requirement of the property. For example, a necessary condition for being bald is not having much hair, so we may say that baldness requires you not to have much hair. In this sense, rationality definitely requires you not to have contradictory beliefs. But Kolodny is using "requires" in a different sense. This is the sense that appears in my expression "your reasons require you to F." To say rationality requires you to F is to say that rationality prescribes Fing to you. Kolodny denies that rationality issues prescriptions.

I have two replies to myth theory. One is to deny it. I deny that if your mind is properly aligned with the world, it will necessarily be properly aligned internally. An example is where your reasons permit you to do something and also permit you not to do it. Cases like this are common: your reasons for going to Paris may neither outweigh nor be outweighed by your reasons for not going to Paris. Then, even if your mind is properly aligned with the world, you may intend to go to Paris, and also you may intend not to go to Paris. Furthermore, the world may give you no reason for not having both intentions; having both might even be helpful because it leads you to prepare for both eventualities. So, even if your mind is properly aligned with the world, you may have both intentions. But then your mind is not properly aligned internally: you are not fully rational if you have contradictory intentions. You respond correctly to reasons, but you are not fully rational. In reaction to examples like this, Kolodny (2007) urges us to abandon the idea that you are necessarily not fully rational if you have contradictory intentions. That seems to me a desperate expedient.

The second reply is to point out that often you cannot respond correctly to reasons except by engaging your rationality. For instance, if you are to intend means to an end you intend, you may need to work out by theoretical reasoning what is a means to your end, and you may then need to do some instrumental reasoning in order to come to intend the means. Reasoning is a rule-governed process that takes you from some existing premise-attitudes of yours, such as existing beliefs and intentions, to a new conclusion-attitude. Correct reasoning is reasoning that follows correct rules. What rules are correct is determined by principles of rationality that connect the conclusion-attitude to the premise-attitudes.[13] These principles are independent of what your reasons require of you. They have to be independent, because reasoning proceeds in exactly the same way whether or not your reasons require you to have the premise-attitudes or the conclusion-attitude. You can reason equally well from false beliefs and bad intentions as from true beliefs and good intentions.

So even if it were true that responding correctly to reasons entails rationality, it would not follow that rationality can be reduced to responding correctly to reasons. Responding correctly to reasons itself depends on rationality.

5. Rationality as Responding Correctly to Beliefs about Reasons

A different reductive claim is that rationality consists in responding correctly to beliefs about reasons. This is subject to various interpretations. According to one proposed by Parfit (2011, chapter 5), responding correctly to beliefs about reasons implies Fing whenever you are required to F by the reasons you believe there to be. But this is ruled out by the first objection in section 3, because what is required by the reasons you believe there to be does not supervene on your mind. The way these reasons combine together to determine what is required may depend on something external to you.

On a second interpretation, supported by Kolodny (2008b), responding correctly to beliefs about reasons implies Fing whenever you believe your reasons require

you to F. Because your belief is a mental state, it supervenes on your mind, so this interpretation is immune to the first objection. We can make it immune to the second objection by confining the response to mental states: we may say that rationality consists in having mental states that are correct responses to beliefs about reasons.

A third interpretation takes responding correctly to beliefs about reasons to imply intending to F—rather than actually Fing—when you believe your reasons require you to F. Once again, this is immune to the first objection, because your belief is a mental state. It is also immune to the second objection, because intending to F is a mental state and so supervenes on your mind.

Neither the second nor the third interpretation is vulnerable to the quick objection. Nevertheless, both are mistaken. The problem with them is that they do not cover all of rationality. There are many necessary conditions for rationality that are not implied by this claim. For example, you are necessarily not fully rational if you have contradictory beliefs or intentions, even if you yourself believe there is nothing wrong with having contradictory beliefs or intentions. These conditions of rationality impose "strict liability," as I put it (Broome, 2013, p. 75).

Still, the third interpretation of the claim does contain a truth. Rationality does not *consist in* responding correctly to beliefs about reasons, but it does *require* responding correctly to beliefs about reasons. That is:

Rationality requires of you that you intend to F if you
    believe your reasons require you to F.

This is one among many requirements of rationality. I call it *enkrasia*. It is only a rough formulation of *enkrasia*; an accurate formulation is more complicated and appears in Broome (2013, pp. 170–171). The state of believing your reasons require you to F while not intending to F is known as *akrasia*. Akrasia has traditionally been taken to be irrational (Aristotle, *Nicomachean Ethics*, Book 7; Davidson, 1969), and *enkrasia* asserts that it is irrational.

6. Conclusion

Enkrasia is an important connection between reasons and rationality. It is a kernel of truth that is hidden inside the grander reductive views I have argued against: the view that rationality consists in responding correctly to reasons and the view that it consists in responding correctly to beliefs about reasons. Those reductive views are false.

Acknowledgments

My thanks to Jonathan Dancy, Terry Irwin, Benjamin Kiesewetter, Franziska Poprawe, Wlodek Rabinowicz, Ben Sacks, Kurt Sylvan, and Michael Waldmann for comments and advice. Research for this chapter was supported by ARC Discovery Grants DP140102468 and DP180100355.

Notes

1. Leading works in this movement are Nagel (1970), Parfit (2011), and Scanlon (1998).

2. See the definitions in Broome (2013, sections 4.2 and 4.3).

3. For example, by Scanlon (2007) and Wallace (2003/2018). In chapter 10.5 by Nida-Rümelin, Gutwald, and Zuber (this handbook), the term is used differently.

4. I believe this term originates with Max Weber; see Kalberg (1980).

5. I assume you have no attitude that favors turning left. Arpaly (2003) argues that sometimes it is genuinely rational to do something you believe you ought not to do. Her prime example is Huck Finn, who believes he ought not to conceal the escaped slave Jim but does so. She claims his decision is rational because it coheres well with other attitudes of Huck's apart from his belief. My case is not like that.

6. The most thoroughgoing defence of this view is Kiesewetter's (2017, chapter 7). Other examples are in Gibbard (1990, p. 161) and Lord (2017). Lord's view is that rationality consists in doing what you ought to do, but it will quickly appear that this amounts to the same thing.

7. See Broome (2013, chapter 5), where parts of the following argument are developed in more detail. There is also a fuller development in Broome (2021).

8. This is Kiesewetter's (2017) and Lord's (2017) interpretation.

9. There is a full discussion of the argument in Broome (2021).

10. Kurt Sylvan pointed out to me that it is defended by Prichard (2002, pp. 95–97). Even with Jonathan Dancy's help, I have not been able to extract a credible argument from Prichard's text.

11. The following arguments are set out more fully in Broome (2013, section 5.4).

12. The leading proponents are Kolodny (2008a) and Raz (2005).

13. Specifically by what I call "basing permissions of rationality" (see Broome, 2013, sections 13.7 and 14.2).

References

Aristotle. (2014). *Nicomachean ethics* (C. D. C. Reeve, Trans.). Indianapolis, IN: Hackett.

Arpaly, N. (2003). *Unprincipled virtue: An inquiry into moral agency*. Oxford, England: Oxford University Press.

Broome, J. (2013). *Rationality through reasoning*. Chichester, England: Wiley-Blackwell.

Broome, J. (2021). Rationality versus normativity. In *Normativity, rationality and reasoning: Selected essays*. Oxford, England: Oxford University Press.

Davidson, D. (1969). How is weakness of the will possible? In J. Feinberg (Ed.), *Moral concepts* (pp. 93–113). Oxford, England: Oxford University Press.

Day, M. (Ed.). (1952). *The English text of the* Ancrene Riwle *edited from Cotton MS. Nero A. XIV by Mabel Day on the basis of a transcript by J. A. Herbert*. Oxford, England: Oxford University Press.

Gibbard, A. (1990). *Wise choices, apt feelings: A theory of normative judgment*. Oxford, England: Clarendon Press.

Kalberg, S. (1980). Max Weber's types of rationality: Cornerstones for the analysis of rationalization processes in history. *American Journal of Sociology, 85*, 1145–1179.

Kiesewetter, B. (2017). *The normativity of rationality*. Oxford, England: Oxford University Press.

Kolodny, N. (2007). How does coherence matter? *Proceedings of the Aristotelian Society, 107*, 229–263.

Kolodny, N. (2008a). The myth of practical consistency. *European Journal of Philosophy, 16*, 366–402.

Kolodny, N. (2008b). Why be disposed to be coherent? *Ethics, 118*, 437–463.

Kolodny, N., & Brunero, J. (2018). Instrumental rationality. In E. N. Zalta (Ed.), *Stanford encyclopedia of philosophy*. Retrieved from https://plato.stanford.edu/archives/win2018/entries/rationality-instrumental/

Lord, E. (2017). What you're rationally required to do and what you ought to do (are the same thing!). *Mind, 126*, 1109–1154.

Nagel, T. (1970). *The possibility of altruism*. Oxford, England: Clarendon Press.

Parfit, D. (2011). *On what matters* (Vol. 1). Oxford, England: Oxford University Press.

Prichard, H. A. (2002). Duty and ignorance of fact. In *Moral writings* (J. MacAdam, Ed., pp. 84–101). Oxford, England: Clarendon Press.

Raz, J. (2005). The myth of instrumental rationality. *Journal of Ethics & Social Philosophy, 1*.

Scanlon, T. M. (1998). *What we owe to each other*. Cambridge, MA: Harvard University Press.

Scanlon, T. M. (2007). Structural irrationality. In G. Brennan, R. Goodin, F. Jackson, & M. Smith (Eds.), *Common minds:*

Themes from the philosophy of Philip Pettit (pp. 84–103). Oxford, England: Clarendon Press.

Smith, M. (1994). *The moral problem*. Oxford, England: Blackwell.

Wallace, R. J. (2018). Practical reason. In E. N. Zalta (Ed.), *Stanford encyclopedia of philosophy*. Retrieved from https://plato.stanford.edu/archives/spr2018/entries/practical-reason/

Wedgwood, R. (2002). Internalism explained. *Philosophy and Phenomenological Research, 65*, 349–369.

Williamson, T. (2000). *Knowledge and its limits*. Oxford, England: Oxford University Press.

2.2 Practical and Theoretical Rationality

Ralph Wedgwood

Summary

Philosophers have long distinguished between practical and theoretical rationality. The first section of this chapter begins by discussing the ways in which this distinction was drawn by Aristotle and Kant; then it sketches what seems to be the general consensus today about how, at least roughly, the distinction should be drawn. The rest of this chapter explores what practical and theoretical rationality have in common: in the second section, several parallels between practical and theoretical rationality are outlined, and it is argued that these parallels make it plausible that a unifying account of rationality can be given. Finally, in the third section, a number of such unifying accounts of practical and theoretical rationality are surveyed. These include accounts that are inspired in various ways by Hume and by the results of formal decision theory, as well as views that appeal to reasons and to the distinctive value of correct or appropriate attitudes.

1.   The Distinction between Practical and Theoretical Rationality

1.1   Aristotle and Kant

The terms "practical" and "theoretical" derived from ancient Greek philosophy, particularly from the work of Aristotle, who in his ethical writings gave a theory of both *theōria* and *praxis*.

In the relevant senses,[1] both *theōria* and *praxis* are "activities" (*Nicomachean Ethics*, 1146b33) of "the parts of the soul that involve reason" (1095a10). In Aristotle's teleological system, the nature of both these activities is explained in terms of their essential *function*—that is, the role that these activities *ideally should* play in a human life. Specifically, when it fulfills its essential function, *theōria*—often translated as "contemplation" or "study"—is the manifestation of the virtue of *theoretical wisdom* (*sophia*, 1177a26), which in turn is explained as consisting of *scientific knowledge* (*epistēmē*), resting on *comprehension* (*nous*) of the first principles of science (1141a18). *Theōria*, then, involves actively exercising one's understanding of a body of scientific knowledge that one possesses (1177a26).

However, theoretical wisdom (*sophia*) is only *one* of the so-called intellectual virtues. The intellectual virtues also include technical skill (*technē*, 1140a1–25) and practical wisdom or prudence (*phronēsis*, 1140a26–b29). While theoretical wisdom is manifested in contemplation (*theōria*), practical wisdom is manifested in *action* (*praxis*). At least when the agent's capacities are all fulfilling their essential functions, a *praxis* is the execution of a "decision" (*prohairesis*, 1094a5). According to Aristotle, a "decision" is a special kind of desire to perform an action here and now, which arises from a wish (*boulēsis*, 1111b26) for an end that is viewed as good, through deliberation or rational calculation (*logismos*) about how best to achieve that end (1112b15). When we manifest the virtue of practical wisdom, we make a correct decision, on the basis of grasping the truth about the good and correctly perceiving our practical situation, and by executing such a decision, we perform a correct action (1140b21).

This is the general picture of Aristotle's distinction between theoretical and practical wisdom, as well as the activities, *theōria* and *praxis*, that under favorable circumstances are their manifestations. However, at one point, he seems to give a rather different general characterization of the difference between theoretical and practical reason (1139a3–15):

> Let us assume there are two parts [of the soul] that have reason: with one we study beings whose principles do not admit of being otherwise than they are, and with the other we study beings whose principles admit of being otherwise. . . . Let us call one of these the scientific part, and the other the rationally calculating part; for deliberating is the same as rationally calculating, and no one deliberates about what cannot be otherwise. Hence the rationally calculating part is one part of the part of the soul that has reason.[2]

Here Aristotle seems to say that theoretical reason is concerned with necessary truths, and practical reason—the sort of reason that consists in the capacity for "delib-erating"—is concerned with contingent truths. This characterization strikes contemporary philosophers as puzzling—since it leaves no room for theoretical reasoning about contingent facts that have no practical relevance (such as the contingent facts of ancient history, paleontology, and the like).

In a similar way, Kant also seems to seek an account of the difference between practical and theoretical reason in terms of the "cognitions" (*Erkenntnisse*) that they are concerned with. One short statement of this distinction appears in the presentation of Kant's *Logic* that was published toward the end of Kant's life by his colleague Gottlob Jäsche. In the primary sense, according to Kant (1900ff., volume 9, p. 86), a "practical cognition" is an "imperative"—where an imperative is defined as "a proposition that expresses a possible free action, whereby a certain end is to be made real." By contrast, theoretical cognitions are characterized as follows:

> Theoretical cognitions are ones that express not what ought to be but rather what is, hence they have as their object not an acting (*kein Handeln*) but rather a being (*ein Sein*).[3]

Kant (1900ff., volume 4, p. 413) gives a different account of "imperatives" in his *Groundwork of the Metaphysics of Morals*, where he says that an imperative is the "formula" of "a representation of an objective principle, so far as it is necessitating for a will." What this seems to mean is that an imperative is the expression of a cognition that represents a principle that the will of every rational being ought to follow—or, more precisely, the application of such a principle to an agent who does not necessarily follow or conform to this principle. As he says, "All imperatives are expressed through an *ought*. . . . They say that it would be good to do or refrain from something" (p. 413).

This approach, however, brought Kant into difficulties with his account of so-called *hypothetical* imperatives (p. 414). Every hypothetical imperative presupposes a certain *end* or *purpose* and merely indicates the *means* that are necessary for achieving that end or purpose. In other words, the imperative asserts that a certain action is *good for* the relevant end. Like all imperatives, each hypothetical imperative represents a principle that should in a sense guide the will of all rational beings—at least in the weak sense that no rational beings may will the end without also willing the necessary means.

However, as Kant later pointed out in the "First Introduction" to the *Critique of the Power of Judgment*, each of these hypothetical imperatives is in a way nothing more than a *theoretical* proposition about the causal connection between an action and the relevant end. His puzzling conclusion is that hypothetical imperatives—which he now prefers to call "technical" rather than "hypothetical"—are practical "in form" but not "in content" (Kant, 1900ff., volume 20, pp. 196, 200).

These difficulties seem to reveal that it is a mistake to account for the difference between practical and theoretical reason in terms of two kinds of propositions (or "cognitions"). Instead, the difference should be explained by distinguishing between two kinds of *attitudes* that can be taken toward propositions. This is the approach that has become the consensus among contemporary philosophers.

1.2 Belief versus Practical Attitudes

Virtually all contemporary philosophers seem to agree on the following way of distinguishing between theoretical and practical rationality:

(a) Theoretical rationality is the rationality of *beliefs* and other belief-like or belief-involving phenomena (such as credences, inferences, and the like).

(b) Practical rationality is the rationality of the *practical attitudes*—like intentions, decisions (or choices), and preferences—that are directly executed in action.

According to this approach, theoretical and practical rationality do not differ with respect to the *propositions* that they are concerned with. They differ with respect to the kinds of *attitude* that are in question—(a) belief in the case of theoretical rationality and (b) intention or decision or preference in the case of practical rationality.

In appealing to this contrast between (a) belief and (b) intention or decision, are contemporary philosophers departing from the philosophical tradition? There are reasons for thinking that they are not. As we have seen, Aristotle firmly associates theoretical wisdom with *knowledge*, and many contemporary philosophers think of knowledge as in a sense the *best case* of a certain central kind of belief—specifically, of the kind of belief that is sometimes called "full belief." On the practical side, Aristotle clearly associates practical wisdom with decision and action. Similarly, Kant (1900ff., volume 4, p. 412) states that "the will is nothing other than practical reason"—and what Kant calls a "determination of the will" seems essentially to involve an intention to act.

Still, on both the theoretical and the practical side, several different kinds of attitudes could be considered. On the theoretical side, there is not just full or outright belief but also various levels of confidence or credence that we can take toward the propositions that concern us. In addition, there may be various further belief-like attitudes that should be considered. Some philosophers—like Joyce (2010)—have suggested that there are both precise and imprecise credences; some—like Edgington (1995)—have argued that there are both conditional and unconditional credences; and others—like Friedman (2013)—have argued that suspension of judgment is a distinctive belief-like attitude as well. If all these different belief-like attitudes exist, they can all be assessed as rational or irrational. If any attitudes of these kinds are rational, they would count as theoretically rational.

In addition to these belief-like mental states, there are also belief-involving *events* of belief revision or theoretical reasoning. Such events involve *changing* one's beliefs in some way—either by coming to have a belief-like attitude toward a proposition toward which one previously had no attitude at all, by replacing an old belief-like attitude that one had in a proposition with a new attitude toward the very same proposition, or simply by reaffirming the old attitude that one had toward that proposition. If any of these mental events are rational, they would also count as theoretically rational.

On the practical side, intentions are mental *states*—while choices or decisions are mental *events* in which one forms or revises one's intentions about how to act. If an intention is executed or carried out, the execution of the intention is an action. It seems that all of these states and events—intentions, decisions (or choices), and actions—can be assessed as rational or irrational. If rational, they would count as practically rational. Some philosophers have argued that, among these states and events, some are more fundamental to practical rationality than others. For example, Harman (1986, p. 77) has argued that practical rationality fundamentally concerns processes in which agents changes their plans or intentions, and Broome (2013, p. 250) has argued that requirements of rationality fundamentally apply to sets of mental states, like beliefs and intentions, rather than applying directly to actions.

In formal accounts of practical rationality—such as those of Jeffrey (1983) and Joyce (1999)—the focus is typically not on intentions, decisions, or actions but on *preferences*. In these accounts, the fundamental instances of practical rationality are preferences and events in which our preferences change in response to changes in our beliefs. This approach is also akin to the contemporary consensus, since theorists who focus on preferences typically assume that there is an intimate connection between preferences, on the one hand, and choices (or decisions) and actions, on the other. Specifically, these theorists typically assume that no agent ever chooses (or decides on), or intentionally takes, any course of action A if there is an available alternative B that the agent prefers over A.

Admittedly, some contemporary philosophers would have reason to reject this way of distinguishing practical and theoretical rationality. Specifically, some philosophers—such as Setiya (2007, p. 49)—actually identify intentions with a special kind of belief. On this view, for you to intend to F is just for you to have a belief of a certain kind that you will F. What is special about intention, on this view, is that it is a belief that tends in a certain distinctive way to *cause* the truth of its propositional content (unlike most beliefs, which are caused by the truth of their content or else causally independent of the truth of their content). On this view, it could happen that when you have a belief of this sort, to the effect that you will F, this belief is *theoretically* rational (if when you have the belief, it is clear to you, given your evidence, that you will indeed F), even if it is not *practically* rational (if Fing is obviously a foolish or inadvisable course of action). For this reason, these theorists would have to find a different way to distinguish between practical and theoretical rationality.

Otherwise, however, virtually all contemporary theorists would distinguish between practical and theoretical rationality in the way that I have described. According to this contemporary consensus, then, theoretical rationality is the rationality of beliefs, while practical rationality is the rationality of intentions and choices and the like.

2. Unifying Practical and Theoretical Rationality?

Practical and theoretical rationality have some striking features in common. First, similar vocabulary is used on both the practical and the theoretical sides. The term "rational" itself is applicable, in what appears to be the very same sense, to both beliefs and choices. We can talk about the beliefs and choices that are "rationally required" of a thinker in a given situation and of the beliefs and choices that are "rationally permissible" for the thinker in the situation. We can talk about "reasonable" or "justified" beliefs and choices—where there is one way of using terms like "reasonable" and "justified" on which they express the very same concept as the term "rational." Finally, we can talk about "reasons" for beliefs and "reasons" for choices and actions; on both the practical and theoretical sides, we distinguish between there being "some reason" for a belief or an action, "sufficient reason" for the belief or action, and "compelling" or "overriding reason" for the belief or action. It seems unlikely that these terms—"justified," "reasonable," and "reason"—are used in different senses when applied to belief as opposed to choice or action.

Moreover, both theoretical and practical rationality come in *degrees*: while some beliefs are only slightly irrational, other beliefs are extremely irrational—and the same is obviously true of choices or decisions as well. On both the theoretical and the practical side, there is some plausibility in the following definition of a "rational requirement": meeting a certain condition *C* is rationally required of you at a time *t* if and only if, in all relevantly available worlds in which you are thinking as rationally as you can at *t*, you meet condition *C*.[4]

Finally, on both the practical and the theoretical side, we can draw the same distinction between your having beliefs (or intentions) that *it is rational* for you to have and your having those beliefs (or intentions) *rationally*. An irrational agent might, through sheer dumb luck, choose an option that it is rational for him to choose, but he would not thereby be choosing this option rationally. In epistemology, this distinction is typically drawn by means of the terminology of "propositional" and "doxastic" justification: an irrational thinker might, through sheer dumb luck, believe a proposition that there is *propositional* justification for her to believe, but this irrational thinker would not thereby be believing the proposition in a *doxastically* justified manner. Clearly, however, the same distinction can be drawn among choices and intentions as well.

With all these parallels between practical and theoretical rationality, it seems plausible that there should in principle be some way of *unifying* the two domains. What would such a unification of these two domains amount to? First, it would consist of a collection of general principles, capturing features of rationality in general, including both practical and theoretical rationality. Second, it would have to be true that the specific features of theoretical rationality can all be derived by combining these general principles about rationality with some specific principles about belief, while the specific features of practical rationality can all be derived by combining these general principles about rationality with some specific principles about intention, choice, and action. As we shall see in the following section, a number of different attempts have been made to provide a unifying account of this sort.

3.   Unifying Accounts of Rationality

3.1   Formal Theories of Rationality

Philosophers who have devised formal theories of rationality have typically offered unified accounts of both practical and theoretical rationality. One particularly common idea is that to be rational, an agent's beliefs

and preferences must be capable of being "*represented*" by a probability function and a utility function. There are some subtle differences between different versions of this idea, but according to the most common version, to say that a probability function and utility function "represent" the agent's beliefs and preferences is to say that the following conditions hold:

(a)   For any two relevant prospects *A* and *B*, the agent prefers *A* over *B* if and only if the utility function assigns a higher utility to *A* than to *B*, and the agent is indifferent between *A* and *B* if and only if the utility function assigns the same utility to *A* as to *B*.

(b)   For any two propositions *p* and *q*, the agent believes *p* with greater confidence than *q* if and only if the probability function assigns a higher probability to *p* than to *q*, and the agent believes *p* and *q* with the same degree of confidence if and only if the probability function assigns the same probability to *p* as to *q*.

(c)   For every prospect *A*, the utility that the utility function assigns to *A* is equal to *A*'s expected utility according to this probability function.

Suppose that every rational agent's beliefs and preferences can be represented by a utility function and probability function in this way. Then it follows that every rational agent's preferences give a *complete* ordering of the prospects over which they are defined: that is, for any two such prospects *A* and *B*, the agent prefers *A* over *B*, prefers *B* over *A*, or is indifferent between the two. Rational agents' preferences must also be *transitive*: if the agent prefers *A* over *B* and either prefers *B* over *C* or is indifferent between *B* and *C*, then the agent must also prefer *A* over *C*.

The centerpiece of many such formal accounts of rationality is a "representation theorem"—a formal proof that so long as the agent's beliefs and preferences meet certain conditions, they can be "represented" by a probability function and utility function in something like this sense. One of the most powerful of these theorems is due to Savage (1972). Savage only needs to assume that the agent's preferences meet a number of conditions—no special assumptions about the agent's beliefs are required—and on this basis he proves that these preferences can be represented by a *unique* probability function and a utility function that is unique up to an arbitrary choice of a unit and zero point.

How could this account of rationality unify practical and theoretical rationality? One approach might rely on Savage's theorem and attempt simply to *reduce* theoretical rationality to practical rationality. If it is true that every rational agent's preferences must satisfy the conditions that Savage's proof assumes, then a rational agent's beliefs can simply be *identified* with the probability function that is uniquely determined by these preferences. This might suggest a *pragmatist* view of belief—in effect, the view that a practically rational agent has no beliefs except for those that are already implicit in her preferences. If this pragmatist view of belief is correct, and if there is no more to be said about rational belief than that rational beliefs must be capable of being represented by a probability function, then this approach would be able to reduce theoretical rationality to practical rationality in this way.

However, this pragmatist view of belief—the view that the practically rational agent's beliefs are just those that are implicit in her preferences—is intensely controversial. Most more recent proponents of various forms of formal decision theory—such as Joyce (1999, p. 90)—would prefer to avoid being committed to this pragmatist view of belief.

If we reject the pragmatist view of belief, we would presumably accept that there are principles of rational belief that do *not* simply flow from the constraints on rational preferences. So, we would need an alternative way of unifying theoretical and practical rationality. One approach might stay close to the pragmatist tradition by arguing that the requirements of theoretical and practical rationality are both explained by the pragmatic defects of violating those requirements. For example, we might rely on the well-known "money pump" and "Dutch book" arguments, which seek to show that violating any requirements of either practical or theoretical rationality makes us willing to accept a collection of bets that will result in our losing money no matter what happens.[5]

These "Dutch book" and "money pump" arguments are fascinating, but their significance is disputed. Even if your beliefs do make you willing to accept a collection of bets that would result in your losing money whatever happens, you might never be offered such a collection of bets, and your beliefs might provide extremely useful guidance for the practical situations that you are actually in. So, it is controversial whether these arguments can really show what fundamentally unifies practical and theoretical rationality.

3.2 Reducing Theoretical Rationality to Instrumental Rationality?

Many contemporary philosophers who work on rationality seek to develop certain key insights of Hume (1739/1740, II.iii.3). The interpretation of Hume is controversial.[6] However, some contemporary philosophers take Hume's key lesson for us to be that the primary

form of rationality is what is often called "instrumental rationality"—the rationality of taking what (at least according to the evidence) seem to be optimal means to one's goals.

If instrumental rationality is primary in this way, then it appears that theoretical rationality must consist in having a set of beliefs that (according to the evidence) seems to be an optimal means to our distinctive *cognitive* or *epistemic* goals. Most versions of this approach assume that the relevant cognitive goal is that of believing the *truth*, and not believing what is *false*, about the questions that we have considered. It is then rationally permissible for a thinker to have a given set of beliefs if and only if these beliefs seem (at least according to the evidence) to be optimal means to this goal.[7]

This approach has also been criticized by a number of philosophers—notably, by Thomas Kelly (2003). Kelly's main concern is that it does not seem true that ordinary thinkers have very general cognitive aims or goals. It is not obviously true that it genuinely is one of my goals simply to believe true propositions as such. My cognitive goals tend to be more local and specific. I have the goal of having a true belief about the times of the flights from Denmark to California—but this is only because I need to make a choice about how to fly from Denmark to California sometime soon. I have no comparable interest in having true beliefs about the times of the flights between Cairo and Moscow, since I currently have no reason to think that I am likely to fly on that route at any point in the foreseeable future.

If Kelly (2003) is right that our cognitive goals are local and specific in this way, then these goals are *idiosyncratic*. Different people will have different cognitive goals. But the requirements of theoretical rationality seem to be in a way *intersubjective*. If you are compelled by an annoying interlocutor to consider a certain proposition *p*, and at the same time you are confronted with compelling evidence for *p*, then it seems that theoretical rationality will require you to believe *p*—even if it was not in any sense one of your goals to have a true belief about *p*. According to Kelly (2003, p. 623), it is hard to see how the instrumentalist conception of theoretical rationality can explain this fact.

For these reasons, then, while the project of reducing theoretical rationality to practical rationality is still worth investigating, few contemporary philosophers are optimistic about its chances of success.

3.3 Rationality and Reasons

Many contemporary philosophers seek to understand rationality in terms of *reasons*. Roughly, to think rationally, on this reasons-centered approach, is to respond appropriately to one's reasons. This approach holds out the hope of unifying practical and theoretical rationality. Practical rationality consists in making choices—and forming, revising, and maintaining one's intentions or plans about how to act—in a way that involves responding appropriately to reasons for such choices and intentions. Theoretical rationality consists in forming, revising, and maintaining one's beliefs in a way that involves responding appropriately to reasons for belief.

A number of variations on this basic theme are possible. Thus, for example, Joseph Raz does not directly identify rationality with responding appropriately to reasons. Instead, Raz (2011, p. 89) identifies *irrationality* with the "malfunctioning" of our "rational powers"—and Raz (2011, p. 85) identifies these "rational powers" with our capacities for recognizing and responding appropriately to reasons. Kolodny (2005, 2008) hopes to reduce rationality to responding correctly to the reasons that one believes there to be.

A different refinement of this approach postulates a distinction between "objective" and "subjective" (or "apparent") reasons. On one interpretation of this distinction, there is a certain set of propositions—call them the "reason-propositions" for the relevant agent at the time in question—such that the "objective reasons" for the agent at that time can be identified with the reason-propositions that are *true*, and the "subjective reasons" that the agent has at that time can be identified with the reason-propositions that the agent *believes* at that time. Then we could say that rationality consists of responding appropriately to one's subjective (or apparent) reasons.[8]

The attempt to give a unified account of rationality in terms of responding appropriately to reasons has been criticized in many ways. (For some powerful and interesting criticisms, see chapter 2.1 by John Broome, this handbook, and also Broome, 2013, chapters 5 and 6.) Perhaps the most fundamental objection, however, is the following: How exactly does this account of rationality as responding appropriately to ("subjective") reasons succeed in unifying practical and theoretical rationality? Is the notion of a "reason" itself a sufficiently unified notion?

According to Raz (2011, p. 36), *epistemic* reasons for believing a proposition are so "through being facts that are part of a case for (belief in) its truth." Even if there are also nonepistemic reasons for belief (for example, if an evil demon credibly threatens to punish me unless I believe something), the epistemic reasons are, in his terminology, "standard reasons" (Raz, 2011, p. 40). By contrast, Raz (1999) says that reasons for *actions* "are facts in virtue of which those actions are good in some respect

and to some degree" (p. 23). In other words, standard reasons for belief have to do with *truth*, while reasons for action have to do with *value*, and Raz (2011, pp. 42–43) denies that truth, or true belief, is a value.

For Raz (1999, pp. 22–23), standard reasons are unified by the role that they play in guiding our thought and by the principle that it is constitutive of being an intelligible thinker and agent in the first place that one has a general disposition to act in ways that one believes to be good, as well as to believe propositions that are supported by truth-related considerations. Raz's ideas on this topic are controversial: some philosophers, like Setiya (2007, pp. 21–67), argue strenuously against Raz's idea that intentional action is always undertaken under "the guise of the good." So, it is debated whether Raz successfully answers the unification challenge.

Schroeder (2007, p. 29) takes a different approach. He aims to combine the appeal to reasons with the idea that is suggested by Hume's (1739/1740, II.iii.3) focus on *desire*. Thus, he proposes that all reasons are grounded in desires (or in some other similar attitude). Every reason for you to F is grounded in some state of affairs S that you desire, in such a way that the reason consists of some fact that helps to explain why your Fing promotes, or raises the chances of, S (compared to the relevant alternative to Fing). Schroeder (2007, p. 113) expresses sympathy for the idea that a view of this kind gives a good account of reasons for belief as well as of reasons for action.

On the face of it, this is a surprising account of reasons for belief. But Schroeder ingeniously argues for several auxiliary theses that could explain why this view is more plausible than it first appears. First, a reason for believing p only needs to explain why believing p raises the chances of some desired state of affairs to some degree (compared to the relevant alternative). So, as Schroeder (2007, p. 113) argues, whatever states of affairs the agent may desire, if a proposition p is false, then not believing p will at least somewhat raise the chances of these desired states of affairs; so there is always some reason for not believing p if p is false. Second, he argues that the *weight* of a reason is not automatically determined either by the strength of the desire or by the degree to which the reason raises the chances of the desired state of affairs—instead, he proposes an entirely different account of the weight of reasons (Schroeder, 2007, chapter 7); so even if all reasons for belief are grounded in desires, it may still normally be the case that there is *most reason* for us not to believe p if p is false.

Schroeder's Hume-inspired account of reasons is still a work in progress. It remains to be seen exactly how an account developed along these lines can accommodate enough of our intuitive judgments about the rationality of belief, choice, and action. Still, if an account of this kind can be developed in detail, it would be an illuminating way of unifying practical and theoretical rationality.

3.4 A Values-Centered Approach

As we saw in section 2, rationality comes in *degrees*. Of the alternative possible beliefs and choices that you might have in response to your cognitive situation, some are more irrational than others. In this sense, rationality "ranks" some beliefs and choices as more rational than others.

This ranking of beliefs and choices will typically play a *reasoning-guiding* role: normally, your judgments of rationality will guide you toward having the beliefs or making the choices that you judge to be more rational and away from the beliefs and choices that you judge to be less rational. In this respect, rationality resembles a *value*, or a way of being good. (Specifically, it resembles the values that rank some alternatives as in the relevant way *better* than others and guide us toward *preferring* the alternatives that we judge to be better over those that we judge to be worse.) Thus, it seems plausible that rationality is itself a value. Rational thinking is thinking that is *good* in a certain way; irrational thinking is thinking that is *bad* in a corresponding way, and in general, the more irrational your thinking is, the *worse* your thinking is in this way.[9]

While we can evaluate both beliefs and choices as rational, there are also other ways of evaluating beliefs and choices. There is one particularly fundamental way of evaluating beliefs and choices that I shall pick out by using the terms "correct" and "incorrect," or "appropriate" and "inappropriate." (In everyday language, many other terms are also used: for example, we could talk about your "getting things right" in your belief about a certain question, or about your making "the wrong choice" among a set of options, and so on.)

As I am using the terms here, the key difference between rationality and appropriateness (or correctness) is this: whether or not a belief or choice is *rational* depends solely on its relations to the mental states and events that are present in the thinker's mind at the relevant times—whereas whether or not a belief or choice is *appropriate* (or correct) depends on its relations to the external world. Thus, whether a *belief* is correct or appropriate depends on whether the proposition that is believed really is true, and whether a *choice* is correct or appropriate depends on whether the chosen option is feasible and really a good thing to do.

Appropriateness also seems to come in degrees: some choices are more inappropriate—more badly wrong or incorrect—than others. I propose that something similar is true of beliefs. Specifically, I propose that what formal epistemologists like Joyce (1998) refer to as a belief's "degree of inaccuracy" can be interpreted as a measure of the degree to which a belief is incorrect or inappropriate. Thus, if the proposition p is true, then the greater the confidence with which you believe p, the more appropriate or correct your belief is, while if p is false, then the greater the confidence with which you believe it, the more inappropriate or incorrect your belief is. (So, on this interpretation of what it is to "get things right" or to "believe appropriately," it is possible for an irrational belief to get things right, through sheer dumb luck.)

In reasoning, we can form *expectations* of how correct different alternative beliefs or choices would be, and these expectations can guide us toward the beliefs and choices that we expect to be more correct and away from those that we expect to be less correct. In this way, it seems that the concept of correct or appropriate attitudes is—like the concept of rational attitudes—an alternative-ranking reasoning-guiding concept. This gives us reason to think that it is also a value concept.

It may be plausible that correctness is, to put it metaphorically, the "aim" of rationality. Suppose that for every agent and every point in time, there is a special probability function P such that the credences the agent has at that time are rational precisely in proportion to how closely they *approximate* that probability function P. Let us call this the "rational probability function" for the agent at that time. Then the key connection between rationality and correctness could be this: irrationality is, according to this rational probability function, *bad news* about correctness. The more irrational your beliefs or choices are, the greater the degree to which this probability function expects your beliefs and choices to fall short of the highest expected degree of correctness that is available to you in your situation.

Some philosophers will certainly object to this approach. As we have seen, Raz (2011, p. 40) denies that true belief is a value, and it may seem strange to say that correctness or appropriateness is a value. It must be conceded that it is a distinctive kind of value: the value of being admirable is exemplified by the *objects* of admiration, and when we judge that an object is admirable, that judgment guides us toward having certain attitudes toward the object in question. By contrast, the value of correctness or appropriateness is exemplified by various different *attitudes* (including attitudes of admiration).

A judgment about which attitudes are correct does not guide us toward having any (higher-order) attitude toward those attitudes; it directly guides us toward having the lower-order attitudes themselves. On reflection, it is clear that judgments about rationality guide us in the same way. Once it is conceded that both correctness and rationality are distinctive in this way, it seems more plausible that correctness and rationality are values, which are connected to each other in the probabilistic way that I have described.

The important point for our present purposes is that this approach to rationality clearly unifies both practical and theoretical rationality: it applies the same account— the idea of irrationality as bad news about correctness—to both practical and theoretical rationality. The difference between practical and theoretical rationality lies in the different standards of correctness that apply to belief and choice: the appropriate object of *belief* is what is *true*, and the appropriate object of *choice* is what is *feasible and good*. The specific features of rational *belief* result from applying this unified account of rationality to the standard of correctness for belief, while the specific features of rational *choice* result from applying the same account of rationality to the standard of correctness for choice. If this account can be successfully defended, then on both the practical and the theoretical side, rationality consists in thinking in a way that provides good news about its own correctness.

Notes

1. Aristotle seems sometimes to use *"praxis"* in a broader sense in which it includes all voluntary behavior (1111a26). But in this broader sense, action does not require "reason" (since even nonhuman animals and children are capable of *praxis* in this sense). So I shall focus on the narrower usage of the term here.

2. This translation is from Aristotle (1999).

3. These translations are from Kant (1992, 2000, 2012).

4. For a different conception of "rational requirements," see Broome (2013, chapter 7).

5. For an illuminating discussion of the "Dutch book" and related arguments, see Hájek (2005).

6. Millgram (1995) has powerfully argued that Hume himself was a *nihilist* about practical rationality—that is, he denied that terms like "rational" can strictly and literally be used to evaluate actions and volitions at all.

7. For a clear recent example of this approach, see Steglich-Petersen (2018).

8. One prominent philosopher who gives an account of rationality in terms of what one has sufficient "apparent reason" to do is Parfit (2011, pp. 34–35).

9. For a more detailed account of this values-centered approach, see Wedgwood (2017).

References

Aristotle. (1999). *Nicomachean ethics* (T. H. Irwin, Ed., 2nd ed.). Indianapolis, IN: Hackett.

Broome, J. (2013). *Rationality through reasoning.* Chichester, England: Wiley-Blackwell.

Edgington, D. (1995). On conditionals. *Mind, 104*(414), 235–329.

Friedman, J. (2013). Suspended judgment. *Philosophical Studies, 162*(2), 165–181.

Hájek, A. (2005). Scotching Dutch books? *Philosophical Perspectives, 19,* 140–151.

Harman, G. (1986). *Change in view.* Cambridge, MA: MIT Press.

Hume, D. (1739/1740). *Treatise of human nature.* In A. Merivale & P. Millican (Eds.), *Hume texts online.* Retrieved from https://davidhume.org/. References are to book, part, and section number.

Jeffrey, R. C. (1983). *The logic of decision* (2nd ed.). Chicago, IL: University of Chicago Press.

Joyce, J. M. (1998). A nonpragmatic vindication of probabilism. *Philosophy of Science, 65*(4), 575–603.

Joyce, J. M. (1999). *Foundations of causal decision theory.* Cambridge, England: Cambridge University Press.

Joyce, J. M. (2010). A defense of imprecise credences in inference and decision making. *Philosophical Perspectives, 24,* 281–323.

Kant, I. (1900ff.). *Gesammelte Schriften* [Collected writings]. Berlin, Germany: Reimer (later De Gruyter).

Kant, I. (1992). *Logic* (G. B. Jäsche, Ed.). In J. M. Young (Ed. & Trans.), *Lectures on logic* (pp. 521–642). Cambridge, England: Cambridge University Press.

Kant, I. (2000). First introduction to the *Critique of the power of judgment.* In P. Guyer (Ed.), *Critique of the power of judgment* (P. Guyer & E. Matthews, Trans., pp. 1–52). Cambridge, England: Cambridge University Press.

Kant, I. (2012). *Groundwork of the metaphysics of morals* (M. Gregor & J. Timmermann, Eds. & Trans., Rev. ed.). Cambridge, England: Cambridge University Press.

Kelly, T. (2003). Epistemic rationality as instrumental rationality: A critique. *Philosophy and Phenomenological Research, 66*(3), 612–640.

Kolodny, N. (2005). Why be rational? *Mind, 114*(455), 509–563.

Kolodny, N. (2008). Why be disposed to be coherent? *Ethics*, *118*(3), 437–463.

Millgram, E. (1995). Was Hume a Humean? *Hume Studies*, *21*(1), 75–94.

Parfit, D. (2011). *On what matters* (S. Scheffler, Ed., Vol. 1). Oxford, England: Oxford University Press.

Raz, J. (1999). *Engaging reason*. Oxford, England: Oxford University Press.

Raz, J. (2011). *From normativity to responsibility*. Oxford, England: Oxford University Press.

Savage, L. J. (1972). *The foundations of statistics* (2nd rev. ed.). New York, NY: Dover.

Schroeder, M. (2007). *Slaves of the passions*. Oxford, England: Oxford University Press.

Setiya, K. (2007). *Reasons without rationalism*. Princeton, NJ: Princeton University Press.

Steglich-Petersen, A. (2018). Epistemic instrumentalism, permissibility, and reasons for belief. In C. McHugh, J. Way, & D. Whiting (Eds.), *Normativity: epistemic and practical* (pp. 260–280). Oxford, England: Oxford University Press.

Wedgwood, R. (2017). *The value of rationality*. Oxford, England: Oxford University Press.

2.3 Mental Models, Reasoning, and Rationality

P. N. Johnson-Laird

Summary

Is human reasoning rational? This chapter argues that it is not based on logic or the probability calculus. Instead, rational inferences aim for accurate models of the world. An *intuitive* system uses mental models that ignore what is false in possibilities. So, it succumbs to systematic fallacies. But a *deliberative* system relies on explicit models that also represent what is false in possibilities, and so it can correct the fallacies. Neither system knows how to infer the probability of compound assertions, such as conjunctions. So, inferences often violate the probability calculus. In tests of inductive hypotheses, intuitions aim to establish examples of them; with feedback from tests, deliberations aim also to show no counterexamples exist. Reasoners can abduce algorithms and explanations that resolve inconsistencies between facts and inferences. The biggest risk in all reasoning is to overlook a possibility, but a simple procedure helps to prevent this sort of irrationality.

> Science can never grapple with the irrational.
>
> —Oscar Wilde, *An Ideal Husband*

1. Rationality and Why It Is Not Based on Logic

Biden will impose new tariffs on Russia or China, or both of them.

So, it is possible that he will impose new tariffs on Russia.

People make this inference, and it seems sensible. But, as you will soon see, it violates any logic dealing with possibilities. The inference raises the questions of what counts as rational in reasoning and to what extent naïve individuals—those untrained in logic or cognate disciplines—meet this target.

Certain beliefs, inferences, and actions are irrational. For instance, it is irrational to believe that the Earth is flat (Sperber, 1995), to infer that the bow doors of a car ferry are closed because no one has reported otherwise (Yardley, 2014), or, like the man who kissed his pet, to fail to think of an obvious possibility (his pet was a rattlesnake; see Northcutt, 2002). Yet, theorists disagree about what counts as rational thinking (see section 1 of this volume). It could be, for example, any mental process that is

- conducive to the survival of your genes and those of your relatives,
- predisposed to yield truth, and
- based on logic, the probability calculus, decision theory, or game theory.

The unqualified use of "logic" here and throughout this chapter refers to classical logic (see chapter 3.1 by Steinberger, this handbook).

It is hard to resolve the theorists' disagreement, because the only satisfactory way to do so should itself be rational. And, alas, that is what's at stake. This difficulty has sometimes led psychologists to describe inferences rather than to evaluate them (cf. chapter 1.2 by Evans, in this handbook). But if psychologists want to improve reasoning, they cannot ignore criteria for rationality. So, what are they?

Once upon a time, the answer was simple. It was: reasoning based on the laws of thought or, as Boole (1854) remarked, the laws of "right reasoning," which he formalized in a single algebra for both logic and probability. It should yield *valid* inferences, which are those whose conclusions are true in every case in which all their premises are true (Jeffrey, 1981, p. 1). It also requires that there is at least one case in which all the premises are true. Otherwise, contradictions imply any conclusion whatsoever, and naïve individuals do not make such inferences. The preceding definition of validity is independent of logic. It hinges on the meanings of assertions, as the following inference illustrates:

Necessarily, the plants will die.

Therefore, the plants will die.

It is valid in one logic of possibilities, but not in another, because the meaning of "necessarily" differs between the two logics (Girle, 2009).

Logic depends on formal rules that yield proofs of valid inferences, and psychologists once took for granted that logic underlies everyday reasoning (see chapter 3.2 by O'Brien, this handbook). But an immediate problem is that logic concerns the relations between sentences in a formal language, whereas inferences in daily life concern the propositions that utterances in natural language convey in context. To apply a formal rule of inference, such as *A or B or both; not A; therefore, B*, to propositions, you need to know their "logical forms." No algorithm yet exists that can carry out this task; it is extraordinarily difficult (see Keene, 1992). Consider, for example, the disjunctive premise in our opening example: *Biden will impose new tariffs on Russia or China, or both of them.* You understand that "or China" abbreviates the clause, "Biden will impose new tariffs on China." So, the logical form of the premise matches *A or B or both* in the formal rule above. This understanding depends on knowledge. And once you take that into account, there is no need for logical form. As we'll see, reasoning can be based on representations of the meaning and reference of assertions. Another difficulty is that sensible inferences in daily life often diverge from logic. The opening example calls for a logic dealing with possibilities, of which there are infinitely many (Girle, 2009). Yet, the inference is invalid in all of them (Johnson-Laird & Ragni, 2019). In logic, the disjunction is true given that at least one of its two clauses is true. But it isn't valid to infer just one clause from a disjunction of two. Imagine, say, that it is impossible that Biden will impose new tariffs on Russia but true that he will impose them on China. In this case, the disjunctive premise is true, but the conclusion is false, and so the inference is invalid in logic. We see later why people nonetheless make it and why some psychologists have given up logic in favor of the probability calculus as the basis of reasoning (cf. section 4 of this volume).

The present chapter advocates neither logic nor probability as the foundation of reasoning. Instead, its starting point is evolution. Living creatures evolved with a perceptual apparatus that constructs internal models of the world. The word "model" here signifies representations that are finite and iconic in that their structure mirrors the structure of the world. These models were adaptive for our evolutionary ancestors, and we have inherited the ability to construct them. And in almost any situation, we have the ability to conceive of a small exhaustive set of possibilities. Anything common to all of them is certain, unless they concern what is permissible—in which case, it is obligatory. Craik (1943) wrote that our minds construct small-scale models of the world to help us to make decisions but that reasoning depends on verbal

rules. The modern "model" theory assumes that reasoning relies instead on iconic models, which represent possibilities (Johnson-Laird, 1983). It postulates this working definition of right reasoning:

Rational inferences yield accurate models both of possibilities and of their implications.

The chapter overlooks strategic reasoning in games (see section 9 of this volume); as the late Reinhard Selten once remarked, "Game theory is rational theology" (Steingold & Johnson-Laird, 2002). Otherwise, it deals with all the main sorts of reasoning: deductive, probabilistic, inductive, and abductive.

2. Deduction and Possibilities

2.1 The Theory of Mental Models

The theory postulates a mechanism that builds models from the premises of inferences, which can be perceptions, descriptions, knowledge, or beliefs. So, a premise such as

There's a circle or a triangle, or both, in the picture

has a set of three models of the possibilities in the picture:

○

    Δ

○   Δ

Together, they have the force of an exhaustive conjunction of three possibilities that each hold in default of knowledge to the contrary. And so the first model yields the inference

Therefore it is possible that there is a circle in the picture.

The inference is valid by default, and individuals make it (Hinterecker, Knauff, & Johnson-Laird, 2016). They also estimate the conclusion as less probable than that of the premise, which is contrary to probabilistic approaches to reasoning (see below). Defaults are commonplace. You infer that a bird flies until you discover that it is an emu. The model theory extends the concept to the meaning of connectives, such as "or."

Logic stipulates that a conditional, *If A then B*, is true in every case except when *A* is true and *B* is false—in this case alone, the conditional is false. The general claim

If anything is a swan then it is white

is thus equivalent in logic to

If anything isn't white then it isn't a swan.

So, a green carnation is corroboratory evidence for the claim: it isn't white, and it isn't a swan. This well-known "paradox" of confirmation holds for logic (Hempel, 1945).

In contrast, the model theory treats the conditional about swans as true provided that examples are possible (swans that are white), but counterexamples are impossible (swans that are not white). Things that are not swans are possible whether the conditional is true or false (Johnson-Laird, Khemlani, & Goodwin, 2015). So, carnations are irrelevant. Likewise, scientists need examples of a hypothesis that they are trying to induce or test from observations (Nicod, 1950). Science, as many major practitioners have acknowledged, is a refinement of everyday thinking (see chapter 14.1 by Andersen & Andersen, this handbook). And a famous test illustrates scientific practice. The general theory of relativity predicts that if a ray of light passes close to a heavenly body, it should deviate toward the latter by a precise amount depending on its distance from the heavenly body's center (Einstein, 1920/2004, p. 111). To test this hypothesis, astronomers measured the apparent positions of "fixed" stars when light from them passed close to the sun—a situation satisfying the *if*-clause of the hypothesis—to determine whether examples of the hypothesis occurred and counterexamples did not. The results corroborated general relativity.

Reasoning based on models can deliver valid inferences, and validity is part of right reasoning. But it is not enough. The validity of the following inference is undeniable:

It's raining.

Therefore, it's raining and it's raining.

Yet, it is useless, except as an illustration of logic. In rational inferences, individuals should draw conclusions that are parsimonious (not just conjunctions of premises), that are novel (not just a repetition of a premise), and that maintain the semantic information in the premises, which guarantees validity (Johnson-Laird & Byrne, 1991). These are all emergent properties of inferences based on models. One notable consequence occurs with premises such as

None of the teachers is poor.

None of the parents is poor.

Most people respond, "Nothing follows." The premises establish no definite relation that meets the constraints above (see chapter 3.3 by Khemlani, this handbook). But the response is contrary to logic, which allows that infinitely many valid conclusions follow from any set of premises. They include some that people might accept if they were presented to them (e.g., "It is possible that some of the teachers are parents"). And many deductions are downright stupid. The model theory distinguishes those that reasoners draw for themselves within those that are valid.

2.2 Irrational Deductions

Do naïve individuals reason rationally? Even Boole (1854) recognized that people err, and so the nature of their mistakes matters. Haphazard errors don't threaten rationality. They fail to reflect underlying competence. But the model theory predicts systematic fallacies. It assumes that intuitive inferences rely on a simpler system of reasoning than deliberations: it postulates two systems of reasoning, which can lead to different conclusions. Wason was the first to formulate such a dual-system theory, but unlike other theories (see chapter 1.2 by Evans and chapter 2.5 by Klauer, both in this handbook), both its intuitive and deliberative systems were described in an algorithm (Johnson-Laird & Wason, 1970a).

The crux for rationality is whether reasoners rely on *mental* models or on *explicit* models. As earlier, the set of mental models for the disjunction "There's a circle or a triangle, or both" is:

```
○
        △
○       △
```

The first model represents the possibility that there is a circle but nothing about triangles. Mental models in the intuitive system represent only what is true in each possibility. In contrast, explicit models in the deliberative system use negation to represent clauses in the premises that are false in a possibility. The explicit models of the preceding disjunction are therefore:

```
○   ¬△
¬○   △
○    △
```

where "¬" denotes the mental symbol for negation, which is linked to its semantics.

The fallacies from mental models are ubiquitous (Khemlani & Johnson-Laird, 2017). Here is an example from two *exclusive* disjunctions, that is, those in which not both of their clauses can hold:

Either the circle is in the picture or else the triangle is.

Either the circle is in the picture or else the triangle isn't.

Could both of these assertions be true at the same time? Most people say, "Yes." They rely on mental models and grasp that the circle is possible according to each premise. They overlook that when one clause of an exclusive disjunction is true, the other clause is false. So, the explicit models of the two premises tell a different story. The first premise has the explicit models

```
○   ¬△
¬○   △
```

And the second premise has the explicit models

$$\begin{array}{cc} \circ & \Delta \\ \neg\circ & \neg\Delta \end{array}$$

The two sets have no possibility in common. And so the two disjunctions cannot both be true. Mental models lead to predictable fallacies, as in this example, but explicit models can correct them—in principle, they yield only deductions that are valid according to the model theory. They demand more capacity in working memory, and an inference can exceed that capacity—a factor that differs from one person to another. Deductions based on "or" and other connectives are computationally intractable—the required processing for many distinct premises can exceed the capacity of a finite brain, even one as big as the universe (see the Overview by Knauff & Spohn, this handbook).

The model theory applies to all the main domains of deductive reasoning (see chapter 3.3 by Khemlani on syllogisms, chapter 6.3 by Byrne & Espino on counterfactual reasoning, chapter 13.2 by Ragni on spatial and temporal relations, and chapter 13.3 by Knauff on visualization, all in this handbook). And its intuitive and deliberative processes have been implemented in a computer program for reasoning with connectives (mSentential, available at https://mentalmodels.princeton.edu/models/). It takes as input premises in natural language and evaluates the validity of a putative conclusion as necessary, possible, or impossible. From simple premises, it can construct its own conclusion in keeping with the constraints on rationality described earlier. It embodies the idea that individuals can make valid deductions, but in certain cases, they succumb to illusory inferences. They are a consequence of mental models, which represent only what is true in possibilities. This focus may reflect the perceptual origins of models.

3. Probability

3.1 Probabilistic Theories of Reasoning

The probability calculus is nothing more than common sense according to Laplace (1951/1820). This view is embodied in psychological theories that replace logic with probabilities as the foundation of everyday reasoning (see chapter 4.4 by Pfeifer, chapter 4.5 by Chater & Oaksford, and chapter 4.6 by Oberauer & Pessach, all in this handbook). On this "probabilist" account, the inference

If it rained then it was cold

It rained

Therefore, it was cold

depends neither on logical rules nor on mental models but on probabilities. These theories hark back to a treatment of probabilistic validity (p-validity) applying to conditionals (Adams, 1998). In its simplest definition, an inference is p-valid if its conclusion is no less probable than its premises in any consistent assignment of probabilities to its clauses.

Probabilists have criticized the model theory for diverging from orthodox logic (e.g., Oaksford, Over, & Cruz, 2019). But, by design, the model theory differs from logic. And probabilism has difficulties of its own, which illuminate the nature of rationality. One difficulty is that possibilities can underlie probabilities, but probabilities cannot underlie possibilities. Consider a museum attendant who gives this permission to a visitor:

If you have a ticket you can enter now or in fifteen minutes.

Knowing that she has a ticket, the visitor infers:

I can enter now.

The inference is sensible, and the model theory predicts it. Yet, it violates logic to infer one clause from a disjunction, and the inference is also p-invalid. But, as probabilists can point out, probabilities don't apply to permissions. However, the same sort of inference occurs with descriptions:

If the weather is bad, then it's possible that there's flooding or a tornado.

The weather is bad.

So, it's possible that there's flooding.

The model theory assigns probabilities to possibilities, and other methods exist to do so too (e.g., Lewis, 1981). And the preceding inference creates a dilemma for probabilism: either a probabilistic account does not apply to it, or else it violates p-validity.

How do naïve individuals infer probabilities? The answer is: with difficulty. And probabilist theories have so far failed to explain the process. Consider two assertions:

A: Trump is the Republican presidential candidate in 2024.

B: Warren is the Democratic presidential candidate in 2024.

Given such assertions, participants are happy to make estimates of the following triples of probabilities (Khemlani, Lotstein, & Johnson-Laird, 2015):

- P(A), P(B), and P(A and B).
- P(A), P(B), and P(A or B or both).
- P(A), P(B), and P(B given A).

Each set fixes the complete joint probability distribution (JPD) for A and B i.e., the probabilities of each of these four exhaustive cases: $A \& B$, $A \&$ not-B, not-$A \& B$, and not-$A \&$ not-B. Estimates of the triples above showed that the participants making them didn't know how to compute probabilities. Their estimates yielded robust violations of the principle that the percentage probabilities in the JPD should sum to 100%. Likewise, individuals do not infer numerical probabilities unless the premises or the task prompts them to do so (Goodwin, 2014), which is just as well, given their incompetence with numerical probabilities.

3.2   The Model Theory of Probabilities

The model theory provides a mechanism for computing probabilities. Given this problem:

- There is a box in which there is a yellow card, or a brown card, or both.
- What is the probability that in the box there is a yellow card and a brown card?

individuals tend to respond with estimates of around 33%. Such "extensional" inferences follow from the proportion of models of possibilities in which the outcome occurs, and the disjunction yields three possibilities, of which only one satisfies the case in question (Johnson-Laird, Legrenzi, Girotto, Legrenzi, & Caverni, 1999). Similar models with numerical tags allow individuals to infer answers to problems of the following sort (Girotto & Gonzalez, 2002), in which the arithmetic is simple and calls for separate estimates of the denominator and the numerator of a probability:

> The chances that Pat has the disease are 4 out of 10. If she has the disease, then the chances are 3 out of 4 that she has the symptom. If she does not have the disease, then the chances are 2 out of 6 that she has the symptom. Pat has the symptom. Pat has __ chances of having the symptom, and among them she has __ chances of having the disease.

As the example shows, frequencies from observations are not necessary for correct inferences. The participants can envisage the solution in a series of updates to models of Pat's chances. The initial model establishes that Pat has 4 chances of having the disease and 6 chances of not having the disease. Likewise, her chances of having the disease divide into 3 chances in which she has the symptom and 1 chance in which she does not. And her chances of not having the disease divide into 2 chances in which she has the symptom and 4 chances in which she does not. In sum, the models of Pat's chances are as follows:

disease	symptom:	3
disease	¬ symptom:	1
¬ disease	symptom:	2
¬ disease	¬ symptom:	4

So, Pat has 5 chances (out of 10) of having the symptom, and given the symptom, she has 3 chances out of 5 of having the disease. So, with simple arithmetic, naïve individuals can use this "subset principle" (rather than Bayes' theorem) to estimate a conditional probability.

Many events in daily life have no frequencies of occurrence (e.g., the event that Trump will be reelected). Yet people in such cases can infer numerical probabilities. The mystery is where the numbers come from. The model theory offers a solution: they come from the proportions of models, not of the event itself, but of its occurrence in models of relevant evidence (Khemlani et al., 2015). So, for Trump's reelection, relevant evidence is, for example, that most U.S. presidents running for a second term haven't been reelected. This yields a model in which the smaller proportion of possibilities supports reelection. The model theory postulates that the intuitive system represents this magnitude in a nonnumerical way. It relies on an icon akin to one that nonnumerate individuals use to represent magnitudes (see, e.g., Carey, 2009). The greater the length of the icon, the greater the magnitude that it represents, so a probability is represented as a pointer on a "scale" extending from an origin representing impossibility to an end point representing certainty:

|--- |

Subsequent evidence can shift the pointer one way or another, and its end point yields intuitive judgments, such as, "it's unlikely." The deliberative system can transform the icon into a numerical estimate: the probability is 40%. These two systems implemented in a computer program yield corroborated predictions about how individuals estimate the probabilities of assertions, including their propensity to make estimates that violate the probability calculus (Khemlani et al., 2015). The origins of these estimates, which go back to infancy, are in sampling from models of the world (e.g., Téglás et al., 2011).

Inferences of numerical probabilities should not flout the probability calculus. If they do, those who make them are vulnerable in principle to a "Dutch book"—a series of bets in which they are bound to lose money. Most of us, alas, often make irrational estimates of probabilities. Only rare experts are rational. A question that tests expertise in probabilities is about two events, A and B: what three sorts of estimate akin to the triples in section 3.1 fix the JPD in a way that always sums to 100%

whatever their numerical values? A correct answer is given below, but meanwhile readers should remember that the question stumps almost everyone.

Granted that most of us are not experts, our rationality can be assessed from our performance with simple extensional problems. But even in these tasks, we are vulnerable to illusory inferences:

There is a box in which there is at least a red marble, or else there is a green marble and there is a blue marble in the box, but not all three marbles.

What's the probability that there is a red marble and a blue marble in the box?

Most people say their joint occurrence is impossible, and so the probability is zero. Their estimate is predicted from the two mental models of the premise:

red marble

green marble   blue marble

But these models fail to take into account the cases in which it is true that there is a red marble but false that there is both a green and blue marble. The latter conjunction is false, for instance, in case there is only a blue marble. So, there can be both a red marble and a blue marble in the box, and the zero estimate is an illusion.

In sum, probabilities are tricky. Their calculus is not intrinsic to human mentality. That is why it was necessary to invent it. As a philosopher once remarked, "Anyone who played dice in Roman times armed with the probability calculus would soon have won the whole of Gaul" (Hacking, 1975). Ignorance can underlie irrationality.

4.   Induction and the Testing of Hypotheses

4.1   Constraints on Induction

Much of our reasoning is outside the scope of logic: it is inductive rather than deductive. And, as Hume (1748/1988, section IV) argued, induction relies on habit, not necessity. Theorists sometimes propose formal rules for induction, but they inherit the problem of logical form, and knowledge is crucial. If the first 10 people that you meet on a desert island are overweight men, knowledge guides you to induce that its inhabitants are more likely to be all obese than all men.

Knowledge can modulate the interpretation of connectives (e.g., Quelhas & Johnson-Laird, 2017), and so the boundary between deduction and induction is often unclear. The following inference is valid:

If it rained then the plants didn't die.

In fact, the plants did die.

So, it didn't rain.

Here's another example of what looks like the same sort of premises:

If it rained then it didn't pour.

In fact, it did pour.

Yet, you would be irrational to infer that it didn't rain. You know that in a meteorological context, *it pours* means that it rained hard. This knowledge modulates the interpretation of the conditional so that it refers only to a conjunction of two default possibilities: it rained but didn't pour, and it didn't rain. So the second premise contradicts the first premise. The example illustrates again the difficulty of pinning down logical form.

Beliefs underlie many inductions, which are therefore bound to be dangerous. Theorists have developed systems for updating beliefs (see section 5 of this volume). The trouble is that a set of beliefs can be consistent but irrational, because they do not correspond to reality. As a commentator on accidents at sea wrote, "Despite the increasingly sophisticated equipment, captains still inexplicably turn at the last minute and ram each other . . . they built perfectly reasonable mental models of the world, which work almost all the time, but occasionally turn out to be almost an inversion of what really exists" (Perrow, 1984, p. 230).

After a month, a new sort of light bulb fails in your living room. You draw an analogy: the bulbs are like some new flawed computer chips. You formulate an explanation: the bulbs' filaments contain an impurity. These processes are inductive, and they play a part in science and technology. For example, epidemiology was founded in part on John Snow's discovery of how cholera was communicated from one person to another. His discovery began with inductions: cholera is a single specific disease, its spread follows the trade routes—an infected mariner arriving in a port spreads the disease to those in contact with him. More puzzling was that it could leap over considerable distances. Snow inferred that "particles" of the disease in sewage contaminated drinking water. He was right: the particles were bacteria. The Wright brothers' invention of the airplane rested on numerous inductions. Maxim, the inventor of the eponymous gun, said, "Give us a motor, and we will very soon give you a successful flying machine." Wilbur Wright countered with an induction from the fate of glider pilots: if the engine fails and the pilot has no control of the aircraft, then it will crash and the pilot could be killed. The brothers' first goal was therefore to discover how to control a glider rather than to build a motor (for Snow's and the Wrights' inductions and deductions, see Johnson-Laird, 2006).

One constraint on inductions is similarity: similar causes have similar effects (Hume, 1748/1988, p. 80; Tversky & Kahneman, 1983). Another constraint is that physical contact implies causation—it is one cause of cholera's transmission. Yet, these constraints are those on which magical thinking and superstitions rely, and all cultures are susceptible: similarity yields homeopathic magic (take an antibiotic for a sore throat), and contact yields contagious magic (don't touch people with AIDS). There are too many possible hypotheses, and so any constraints are better than none. The only security in an induction comes from tests of its conclusion.

4.2 Testing Hypotheses

A test of an induction needs to determine whether its consequence is true or false. The most influential experimental paradigm for such studies is Wason's (1968) selection task. The experimenter chooses four cards from a pack in which, as the participants know, all the cards have a letter on one side and a number on the other side. They are laid out in front of the participant: *E F 2 3*. The participant's task is to select only those cards that, if turned over, would show whether a hypothesis about the four cards is true or false:

If there is an "E" on one side of a card then there is a "2" on the other side.

With an abstract hypothesis of this sort, participants tend to select the *E* and *2* cards or the *E* card alone. The selection of the *2* card is pointless, because nothing on its other side can refute the hypothesis. But what stunned psychologists and launched hundreds of studies was the participants' failure to select the *3* card. If it had an *E* on its other side, the hypothesis above would be false. Wason, his colleagues, and many others took this oversight to be irrational. Not everyone agreed. One view was that the probabilities of events related to everyday conditionals made it rational not to select the *3* card (see chapter 4.5 by Chater & Oaksford, this handbook). In contrast, the model theory's algorithm (Johnson-Laird & Wason, 1970a) predicts that the test of a general hypothesis needs to establish the possibility of examples and the impossibility of counterexamples (see section 2.1). So, selections of cards should be dependent on one another (e.g., the *2* card should tend to be selected with the *E* card, as an example of the hypothesis). It also predicts that the task and its contents should affect the likelihood that participants select potential counterexamples. A meta-analysis of over 200 experiments corroborated these predictions (Ragni, Kola, & Johnson-Laird, 2018).

In a repeated version of the selection task, participants had to make an efficient test of the generalization "All the triangles are white" about the shapes in two boxes (Johnson-Laird & Wason, 1970b). As the participants knew, one box contained 15 white shapes and the other box contained 15 black shapes. On each trial, they selected a shape from one of the two boxes to see whether or not it was a triangle. The participants started by selecting white shapes—potential examples of the hypothesis—but sooner or later switched to black shapes—potential counterexamples—and examined them all. The original selection task fooled them: they had just one opportunity to make a rational selection. In the repeated task, sooner or later they had the insight to select all potential counterexamples.

5. Abduction, Explanations, and Nonmonotonicity

5.1 Abductions That Resolve Inconsistencies

The author and a friend were sitting BC (before cell phones and corona virus) outside a café in Provence. We were waiting for two other friends, and we knew two things:

They had gone to pick up the car.
And if so then they would be back within 10 minutes.

So, we thought, they'll be back in 10 minutes. When they hadn't returned after 20 minutes, we realized that our conclusion was false. In logic, such a contradiction implies any conclusion whatsoever. Logic is "monotonic": a fact that contradicts an earlier valid inference does not call for its retraction. In daily life, however, it is rational to withdraw conclusions that facts refute, and so reasoning is "nonmonotonic" or "defeasible" (see chapter 5.4 by Gazzo Castañeda & Knauff, this handbook). William James (1907) wrote, "The new fact preserves the older stock of truths with a minimum of modification, stretching them just enough to make them admit the novelty" (p. 59). And this "minimalist" view has many defenders (see chapter 5.2 by Rott, this handbook). Yet, outside the café, we were concerned to figure out what had happened to our friends. We thought of various explanations: they had got lost (improbable), the police had stopped them (improbable), they had been in an accident (improbable), and they couldn't start the car (possible, because it had happened before). So, we reasoned, we had better sit tight and wait, because if we went in search of them and they returned to the café, they wouldn't know where we were. Sure enough, soon afterward, the car came spluttering into view, and we hopped in while its engine was running. It had needed a tow to start.

This anecdote exemplifies the model theory. Any inference in daily life is made in default of knowledge to the contrary. That is, conclusions can turn out to be false. In such cases, the primary task is to create an explanation that resolves the inconsistency between the fact and the premises. It can then guide the decision about what to do.

The creation of an explanation is known as "abduction." Its rational goal, as ever, is accuracy. The process is a sort of induction, because it goes beyond the premises and depends on knowledge. But it seeks more than an inductive generalization: it introduces new concepts in order to create an explanation. The process is triggered when a fact refutes an earlier inference. For example, the premises

If a cobra bit her then she'll die

A cobra bit her

elicit a mental model of a snake biting her and her death. But the fact

She did not die

is inconsistent with this model. And the conflict triggers an attempt to resolve the inconsistency. The whole non-monotonic process is implemented in the mSentential program. Its first step is to compute the facts: a cobra bit her but she did not die. It searches its knowledge for such cases and their causes. The program's knowledge base contains several relevant possibilities in which a deadly snake bites someone who subsequently does not die:

- the person takes an antidote,
- a tourniquet blocks the poison, and
- someone sucks out the poison.

Human reasoners tend to assess the probability of each explanation. The program chooses one at random and then uses the same procedure again to search for *its* cause. If successful, the result is a causal chain, for example,

A cobra bit her and she used a tourniquet and the tourniquet blocked the poison and she did not die.

It also uses the first cause in the chain to construct a counterfactual description of what would otherwise have occurred:

If she had not used a tourniquet then she would have died.

Of course, the program is itself only a partial model of human reasoning. When you create explanations, you work with real models of the world (not strings of words), and you assess the relative probability of different putative explanations.

The theory predicts that individuals should prefer causal explanations to minimal revisions. Indeed, they rate a causal chain as more probable than either the cause alone or the effect alone (Johnson-Laird, Girotto, & Legrenzi, 2004)—a violation of minimalism and a fallacy in which a conjunction is judged to be more probable than its constituents (see Tversky & Kahneman, 1983). When individuals are asked what follows from inconsistent premises, they tend to explain the inconsistency. And they judge such explanations as more probable than minimal revisions to the premises that restore consistency (Khemlani & Johnson-Laird, 2011). When they formulate an explanation first, a striking phenomenon occurs. They find it harder to detect the inconsistency—they seem to have explained it away (Khemlani & Johnson-Laird, 2012).

A consistent set of beliefs may not correspond to reality, but an inconsistent set of beliefs *cannot* correspond to reality. The ability to detect inconsistencies is therefore a hallmark of rationality. Reasoners can assess consistency by trying to find a model of all the information they have. Such a model shows that the information is consistent; otherwise, it is inconsistent. Once reasoners have detected an inconsistency, they can use their knowledge to abduce causal models to explain the origins of the inconsistency. It is rational to try to do so, but no guarantee exists that the explanation is correct. And, on rare occasions, people fail to find any explanation whatsoever for an inconsistency. Yet, they outperform any existing computer program in creating explanations. The crux is that they know more and are better at retrieving pertinent information.

5.2 Abductions of Algorithms

Individuals who know nothing of programming can create simple informal algorithms. Deductions alone won't work, and probabilities are no use, either. Algorithms are an explanation of how to do something, and so reasoners need to introduce new concepts and to organize them in a temporal sequence. In a word, they *abduce* algorithms. For example, they can do so to rearrange the order of cars in a train on a railway track (Khemlani, Mackiewicz, Bucciarelli, & Johnson-Laird, 2013). The track runs from left to right on a computer screen, and it has one siding, which is entered from, and exited to, the left track. As an illustration, you might like to think how you would use the siding to rearrange the cars ABCD on the left side of the track so that they arrive in reverse order on the right side of the track. The more moves (and the more cars in each move) that are called for to solve such a rearrangement, the harder the task is. A more substantial task, however, is to abduce an algorithm that solves such a rearrangement for a train with an arbitrary number of cars.

A computer program, mAbducer, creates such algorithms, which it can express in informal English. Here, for instance, is the algorithm it abduced to reverse the order of cars:

Move one less than the [number of] cars to the siding.

Move one car to the right track.

While there are more than zero cars on the siding,

    move one car to the left track,

    move one car to the right track.

The effect of the first move on ABCD on the left track is shown in the following diagram in which the square brackets demarcate the cars on the siding, and the dash shows that no cars are yet on the right side of the track.

Move one less than the cars to the siding: A[BCD] –

The second instruction is:

Move one car to the right track: – [BCD]A

Next, while there are more than zero cars on the siding, there is a loop of two instructions:

move one car to the left track: B[CD]A
move one car to the right track: – [CD]BA

The loop is repeated twice until the siding is empty, and the result is the cars in reverse order on the right track:

  – [ ]DCBA

As mAbducer illustrates, the model theory postulates that people who are not programmers abduce informal algorithms using a kinematic model that simulates a solution to an instance of the problem. They observe what happens in the simulation and transform their observations into an informal description of the required algorithm. In principle, individuals could determine the number of iterations of a loop of moves by solving two simultaneous linear equations. But, in fact, they more often observe the conditions in which the loop halts in their simulations (e.g., when the siding is empty in the algorithm above). The program implements both sorts of loop (in both a programming language and English).

Experiments showed that most individuals follow the observational procedure—they use a so-called while-loop of the sort illustrated above (Khemlani et al., 2013). Even 10-year-old children can abduce informal algorithms for real toy trains of a small number of cars.

In one experiment, they were not allowed to move the actual cars, and so they tended to make gestures corresponding to the moves their algorithms required. When they were prevented from gesturing, they were no longer so accurate in abducing algorithms (Bucciarelli, Mackiewicz, Khemlani, & Johnson-Laird, 2016). Gestures are an outward sign of inward simulation, and they reduce the load on working memory.

Different rearrangements vary in the difficulty of their abduction. For children and adults alike, it depends not on the number of cars that have to be moved but on the complexity of the algorithm. Various measures exist for this complexity, but a good proxy is the number of instructions in the algorithm. A rational algorithm should carry out the correct moves, and it should do so without unnecessary ones. It is tempting to say that it should be minimal (i.e., it should be of the shortest possible length in the given programming language). There is a problem, however: for all but the simplest algorithms, no way exists to prove that an algorithm is minimal (Chaitin, 1998).

6.   Conclusions

Rationality can seem remote, abstract, and complex. The message of this chapter, however, is that it matters. What counts as rational is a recursive question, because its correct answer depends on rational methods. It goes beyond reasoning to apply to beliefs, actions, and social interactions. But, for reasoning, the view defended here is:

Rational reasoning constructs accurate models of the world based on knowledge or beliefs; infers their consequences for what is possible, probable, or certain; and uses these models to formulate explanations.

In short, right reasoning aims for truth, parsimony, and explanatory power. But, as the chapter showed, humans are fallible. They can fail to think about what's false in making deductions and succumb to illusory deductions. They can make an incorrect inference of the likelihood of a compound assertion, because the probability calculus is not part of their mental equipment. One test of expertise is my earlier question: which three sorts of estimate concerning two events, A and B, fix the values of their joint probability distribution and, regardless of their numerical values, always yield a JPD that sums to 100%? One correct answer is: the probability of A, the conditional probability of B given A, and the conditional probability of B given *not-A*. My survey of psychologists showed that few of them could answer this question correctly.

Although humans are adept at induction, no constraints can guarantee the truth of its outcomes. The heuristics that humans use are often irrational. They yield truths and trumpery, science and superstitions. One rational criterion, however, is that hypotheses should be open to refutation—for Popper (1959), this criterion separates science from nonscience. But naïve individuals in the selection task often focus on corroboratory examples of a hypothesis. In its repeated version, however, they do realize the need to ensure that counterexamples do not exist.

When a fact refutes the conclusion of a valid inference, logic tells us nothing—it licenses any conclusion whatsoever from a contradiction. In contrast, individuals innocent of philosophical niceties follow neither logic nor the advice to make a minimal revision of their beliefs. Instead, they search for an explanation that resolves the inconsistency. These explanations are not always minimal. The truth, as Oscar Wilde remarked, is seldom simple (and never pure). One domain, however, does call for minimalism: the abduction of algorithms. Like so many ideals, alas, no way exists guaranteed to reach the shortest possible algorithm.

A common error in all sorts of reasoning is to overlook a possibility. Such failures are not inevitable, and the model theory suggests something that can be done to reduce them. You can learn to be more rational. You might imagine that you have to learn logic. It takes time—at least a semester's course for the rudiments—and it fails to generalize to novel sorts of inference (Cheng, Holyoak, Nisbett, & Oliver, 1986). In contrast, a good method for improving reasoning should be quick to learn, efficacious, and practical. One method meets these goals. Its single instruction is:

Try to construct all the possibilities consistent with the given information.

With pencil and paper, you can list them in separate columns, adding a column for each new possibility and crossing out a column if a premise refutes its possibility. It takes only a few minutes to learn this "model method," which Victoria Bell devised (see Johnson-Laird, 2006, p. 288). It speeds up reasoning and increases its accuracy (from 66% correct to 95% in one study). Once people have acquired the method, they can imagine the columns of possibilities and no longer need pencil and paper. It still works. And the method may be adaptable to other sorts of reasoning beyond deduction.

To answer the two questions in my opening paragraph: right reasoning should yield accurate models of the world, and human reasoning is rational in principle but often irrational in practice. Yet, humans realize their own shortcomings, and they are rational enough to develop tools that can help them. Contrary to Wilde's epigraph at the head of this chapter, psychology grapples with the irrational and even dispels it . . . sometimes.

References

Adams, E. W. (1998). *A primer of probability logic.* Stanford, CA: Center for the Study of Language and Information.

Boole, G. (1854). *An investigation of the laws of thought on which are founded the mathematical theories of logic and probabilities.* London, England: Macmillan.

Bucciarelli, M., Mackiewicz, R., Khemlani, S. S., & Johnson-Laird, P. N. (2016). Children's creation of algorithms: Simulations and gestures. *Journal of Cognitive Psychology, 28,* 297–318.

Carey, S. (2009). *The origin of concepts.* New York, NY: Oxford University Press.

Chaitin, G. J. (1998). *The limits of mathematics.* Singapore: Springer.

Cheng, P. W., Holyoak, K. J., Nisbett, R. E., & Oliver, L. M. (1986). Pragmatic versus syntactic approaches to training deductive reasoning. *Cognitive Psychology, 18,* 293–328.

Craik, K. J. W. (1943). *The nature of explanation.* Cambridge, England: Cambridge University Press.

Einstein, A. (2004). *Relativity.* New York, NY: Barnes & Noble. (Original work published 1920)

Girle, R. (2009). *Modal logics and philosophy* (2nd ed.). London, England: Routledge.

Girotto, V., & Gonzalez, M. (2002). Chances and frequencies in probabilistic reasoning. *Cognition, 84,* 353–359.

Goodwin, G. P. (2014). Is the basic conditional probabilistic? *Journal of Experimental Psychology: General, 143,* 1214–1241.

Hacking, I. (1975). *The emergence of probability.* Cambridge, England: Cambridge University Press.

Hempel, C. G. (1945). Studies in the logic of confirmation, Parts I and II. *Mind, 54,* 1–26, 97–121.

Hinterecker, T., Knauff, M., & Johnson-Laird, P. N. (2016). Modality, probability, and mental models. *Journal of Experimental Psychology: Learning, Memory, and Cognition, 42,* 1606–1620.

Hume, D. (1988). *An enquiry concerning human understanding* (A. Flew, Ed.). La Salle, IL: Open Court. (Original work published 1748)

James, W. (1907). *Pragmatism—a new name for some old ways of thinking.* New York, NY: Longmans.

Jeffrey, R. C. (1981). *Formal logic: Its scope and limits* (2nd ed.). New York, NY: McGraw-Hill.

Johnson-Laird, P. N. (1983). *Mental models*. Cambridge, MA: Harvard University Press.

Johnson-Laird, P. N. (2006). *How we reason*. New York, NY: Oxford University Press.

Johnson-Laird, P. N., & Byrne, R. M. J. (1991). *Deduction*. Hillsdale, NJ: Erlbaum.

Johnson-Laird, P. N., Girotto, V., & Legrenzi, P. (2004). Reasoning from inconsistency to consistency. *Psychological Review*, *111*, 640–661.

Johnson-Laird, P. N., Khemlani, S. S., & Goodwin, G. P. (2015). Logic, probability, and human reasoning. *Trends in Cognitive Sciences*, *19*, 201–214.

Johnson-Laird, P. N., Legrenzi, P., Girotto, V., Legrenzi, M. S., & Caverni, J.-P. (1999). Naive probability: A mental model theory of extensional reasoning. *Psychological Review*, *106*, 62–88.

Johnson-Laird, P. N., & Ragni, M. (2019). Possibilities as the foundation of reasoning. *Cognition*, *193*. doi:10.1016/j.cognition.2019.04.019

Johnson-Laird, P. N., & Wason, P. C. (1970a). A theoretical analysis of insight into a reasoning task. *Cognitive Psychology*, *1*, 134–148.

Johnson-Laird, P. N., & Wason, P. C. (1970b). Insight into a logical relation. *Quarterly Journal of Experimental Psychology*, *22*, 49–61.

Keene, G. B. (1992). *The foundations of rational argument*. Lewiston, NY: Edwin Mellen Press.

Khemlani, S. S., & Johnson-Laird, P. N. (2011). The need to explain. *Quarterly Journal of Experimental Psychology*, *64*, 276–288.

Khemlani, S. S., & Johnson-Laird, P. N. (2012). Hidden conflicts: Explanations make inconsistencies harder to detect. *Acta Psychologica*, *139*, 486–491.

Khemlani, S. S., & Johnson-Laird, P. N. (2017). Illusions in reasoning. *Minds and Machines*, *27*, 11–35.

Khemlani, S. S., Lotstein, M., & Johnson-Laird, P. N. (2015). Naive probability: Model-based estimates of unique events. *Cognitive Science*, *39*, 1216–1258.

Khemlani, S. S., Mackiewicz, R., Bucciarelli, M., & Johnson-Laird, P. N. (2013). Kinematic mental simulations in abduction and deduction. *Proceedings of the National Academy of Sciences*, *110*, 16766–16771.

Laplace, P. S. de. (1951). *Philosophical essay on probabilities*. New York, NY: Dover. (Original work published 1820)

Lewis, D. (1981). Ordering semantics and premise semantics for counterfactuals. *Journal of Philosophical Logic*, *10*, 217–234.

Nicod, J. (1950). *Foundations of geometry and induction*. New York, NY: The Humanities Press.

Northcutt, W. (2002). *The Darwin awards: Evolution in action*. New York, NY: Plume.

Oaksford, M., Over, D., & Cruz, N. (2019). Paradigms, possibilities, and probabilities: Comment on Hinterecker, Knauff, and Johnson-Laird (2016). *Journal of Experimental Psychology: Learning, Memory, and Cognition*, *45*, 288–297.

Perrow, C. (1984). *Normal accidents: Living with high-risk technologies*. New York, NY: Basic Books.

Popper, K. R. (1959). *The logic of scientific discovery*. New York, NY: Basic Books.

Quelhas, A. C., & Johnson-Laird, P. N. (2017). The modulation of disjunctive assertions. *Quarterly Journal of Experimental Psychology*, *70*, 703–717.

Ragni, M., Kola, I., & Johnson-Laird, P. N. (2018). On selecting evidence to test hypotheses. *Psychological Bulletin*, *144*, 779–796.

Sperber, D. (1995). Intuitive and reflective beliefs. *Mind & Language*, *12*, 67–83.

Steingold, E., & Johnson-Laird, P. N. (2002). Naive strategic thinking. In W. D. Gray & C. D. Schunn (Eds.), *Proceedings of the Twenty-Fourth Annual Conference of the Cognitive Science Society* (pp. 845–849). Fairfax, VA: Erlbaum.

Téglás, E., Vul, E., Girotto, V., Gonzalez, M., Tenenbaum, J. B., & Bonatti, L. L. (2011). Pure reasoning in 12-month-old infants as probabilistic inference. *Science*, *332*, 1054–1059.

Tversky, A., & Kahneman, D. (1983). Extension versus intuitive reasoning: The conjunction fallacy in probability judgment. *Psychological Review*, *90*, 292–315.

Wason, P. C. (1968). Reasoning about a rule. *Quarterly Journal of Experimental Psychology*, *20*, 273–281.

Yardley, I. (2014). *Ninety seconds at Zeebrugge: The Herald of Free Enterprise story*. Stroud, England: History Press.

2.4 Heuristics and Biases

Klaus Fiedler, Johannes Prager, and Linda McCaughey

Summary

Research on heuristics and biases has greatly influenced the current understanding of bounded rationality in psychology, behavioral economics, cultural anthropology, educational science, and neighboring disciplines of behavioral science (see chapter 1.1 by Sturm, chapter 1.2 by Evans, chapter 2.2. by Wedgwood, and chapter 8.4. by Hill, all in this handbook). The first major section of the present chapter is devoted to Kahneman and Tversky's seminal research program, providing a synopsis of the most prominent heuristics supposed to trigger violations of logical coherence rules: availability, representativeness, anchoring, and the simulation heuristic. A second major section deals with the contrast program advocated by Gigerenzer and colleagues, which highlights the adaptive value of fast and frugal heuristics. The last major section provides a critical discussion of the insights gained from half a century of fascinating research on heuristics and biases, and of the neglected issues—such as normative standards and underlying cognitive mechanisms—which constitute open questions to be resolved in future investigations of rational action and decision making.

1. The Notion of Heuristics and Biases in Rationality Research

Real life is replete with uncertainty. Even the most mundane and superficial actions and decisions involve assumptions about risks, probabilities, costs, and benefits, which are indeterminate and hard to evaluate objectively. Partner choice depends on social judgments of attractiveness, on trust, and on the likelihood of future scenarios. Consumers compare food items for healthiness and tastiness. Teachers' grading decisions entail inferences about students' competence and performance. Lying must be discriminated from veracious statements in everyday communication. Health-related behavior depends on the assessment of risks and dangers of nutrition, hobbies, sports, and transportation means. And predictions of future outcomes—in politics, sports, economic and ecological developments, and private affairs—are virtually never made with certainty.

In such an indeterminate world, there is often no absolute normative criterion for rational behavior. Scientifically approved answers to the question of how risks and costs can be minimized and how benefits and satisfaction can be maximized rely on rules of statistics and probability assessment, as a surrogate for a deterministic rule. Human judgments and decisions often do not rely on the mathematics of probability and statistics but rely instead on an alternative strategy, namely, on simplifying proxies or rules of thumb known as heuristics. Under specific conditions, "these heuristics are highly economical and usually effective, but they lead to systematic and predictable errors" (Tversky & Kahneman, 1974, p. 1131).

The purpose of the present chapter is to provide an overview of half a century of intensive research on heuristics and biases (Gigerenzer & Gaissmaier, 2011), which has had a strong impact on the study of rationality in behavioral science in general and on contemporary work in psychology, economics, biology, and philosophy in particular. Starting from a synopsis of Daniel Kahneman and Amos Tversky's groundbreaking demonstrations of the most prominent heuristics (Gilovich, Griffin, & Kahneman, 2002; Kahneman, Slovic, & Tversky, 1982; Tversky & Kahneman, 1974), we move on to a broader discussion of bounded rationality (Simon, 1982) and of the adaptive beauty of heuristic inferences considered from an alternative perspective by Gerd Gigerenzer and his colleagues (Gigerenzer & Goldstein, 1996; Gigerenzer, Todd, & ABC Research Group, 1999). Another major section will then be devoted to a critical assessment of what insights have been gained from this extremely prominent field of research and what neglected issues remain open to be answered in future investigations.

Let us illustrate the basic idea with an introductory discussion of one specific example, the availability heuristic

(Tversky & Kahneman, 1973), which is ideally suited to explain the manner in which heuristics allow for accurate judgments in many everyday situations, although in different situations the same heuristics may lead to serious biases and irrational decisions.

2.   The Availability Heuristic as an Illustrative Example

The availability heuristic (Tversky & Kahneman, 1973) affords a mental tool for estimating the frequency or probability of events. According to scientific methodology, normatively adequate estimations of such frequentist quantities should be inferred from proportions in representative samples. However, when human individuals estimate the risk (prevalence) of a disease or an accident, or the chances of winning a game in sports or of solving a problem, a representative sample is hardly ever available. Human memory is not a data repository; we do not keep representative data arrays in memory from which random samples could be drawn to inform unbiased frequency or probability estimates. Even when we do sometimes keep traces of ecological events in memory or in some external store, they cannot be expected to be representative of the current reality.

Nevertheless, we do have the capacity to estimate word frequencies or the likelihood of, say, different causes of death (Combs & Slovic, 1979) or of a certain team to become basketball champion, even though we do not keep in memory word counts, sports statistics, or epidemiological data about lethal events. We simply rely on the availability heuristic, which uses the availability of relevant memories as a proxy for frequency and probability estimation. The ease with which relevant memories (of word occurrences, sports teams' past success, lethal events) come to mind affords a useful heuristic of astounding validity. Most of the time, the likelihood of a target event is substantially correlated with the availability of relevant traces in memory. As memories of people dying from cancer are more available than memories of people dying from car accidents, which are more available than memories of suicide or lightning, heuristic estimates reflect the true ordering (cancer > car accidents > suicide > lightning), yielding a remarkably strong correlation between objective frequencies and the results generated by the availability heuristic.

It should be noted in passing that nonheuristic, purportedly normative methods also offer little more than approximate solutions; the norm distribution required to estimate a specific person's exact cancer, accident, or suicide risk is simply unknown. Available statistics that

pertain to the entire population, or to crude substrata thereof, may not apply to individual persons with their individual lifestyle, their genetic predispositions, and the countless resulting combinations of risk factors. So, for a fair comparison, heuristics should not be contrasted with "objectively true values" as in psychophysics experiments but with normative surrogate solutions that are themselves debatable and often misleading and that might also be called "heuristic." Yet, despite its relative usefulness, the availability heuristic can lead to distinct biases when available memories depend on distracting, extraneous factors. For instance, selective media reports may render certain infrequent causes of death (e.g., murder, lightning) more available than other, objectively more frequent, causes of death (e.g., suicide, household accidents). These misleading cases are the focus of most pertinent research.

3.   Synopsis of Prominent Heuristics, Their Domain, and Associated Biases

Let us now move from availability to a discussion of other prominent heuristics, their domain, and the kinds of biases that result from heuristic inference schemes.

3.1   Representativeness

The domain of the representativeness heuristic, one of the first heuristics specified by Kahneman and Tversky (1972), is categorization, or the diagnostic judgment of the likelihood that an event or person belongs to a category or group. A typical task may provide the following description of a target person: "very shy and withdrawn, invariably helpful, but with little interest in people, or in the world of reality." When participants are asked to judge whether the target person is more likely to be a farmer or a librarian (among other options), most people choose librarian, because the description is more representative of librarians than of farmers. They largely ignore the categories' base rates, that is, the fact that the proportion of farmers in the general population is much higher than the proportion of librarians. This pattern is obtained even when base rates are experimentally controlled (Tversky & Kahneman, 1974).

Similar to the availability heuristic, the representativeness heuristic enables accurate judgments, in this case of categorization, as long as the similarity of the target person description to the vocational stereotype constitutes a viable proxy of the actual profession. However, when likeness diverges from likelihood (Shweder, 1977)—that is, when the person sketch resembles a librarian, whereas farmer is much more likely—the heuristic will be misled by similarity, resulting in a distinct base-rate neglect (Bar-Hillel, 1984; Pennycook & Thompson, 2017).

According to the representativeness heuristic, samples are generally and uncritically assumed to be representative of an underlying entity or population. This assumption works in two directions: first, it is assumed that the properties of the underlying population can always be validly inferred from any given sample. Second, even small samples drawn from a given population are expected to have the same properties as the population at large, whereas deviating samples are supposed to be very unlikely. To illustrate the latter point, Kahneman and Tversky (1972) constructed the following task:

> All families of six children in a city were surveyed. In 72 families the exact order of births of boys [B] and girls [G] was G B G B B G. What is your estimate of the number of families surveyed in which the exact order of births was B G B B B B?

Both sequences are equally probable assuming invariance of the gender on successive births. However, they differ markedly from the well-known population distribution in terms of representativeness. Because the "law of small numbers" (Tversky & Kahneman, 1971) suggests that even in a short sequence, the ratio of boys and girls is close to 1:1, participants judged the first sequence to be more likely than the second sequence. In the same vein, seemingly regular chunks like B B B B or B G B G B G are poor representations of random events at odds of 1:1. Thus, series containing less regular chunks and seemingly less systematic content are judged to occur with a higher likelihood. The same reasoning mistake is at work when an explicit sampling task is replaced by a random process or "lottery": tossing a fair coin with the possible results of heads (H) and tails (T). Parallel to the above example, the series H T H T T H will be judged to occur more frequently than T H T T T T, which is again a mistake.

3.2 Anchoring and (Insufficient) Adjustment

Quantitative judgments—for instance, of total costs for a project, of the time required to complete a job, or of the appropriate punishment in a courtroom decision—constitute the domain of the anchoring heuristic (Epley & Gilovich, 2010; Strack, Bahník, & Mussweiler, 2016). Essential for this heuristic is the assumption that quantitative judgments typically start from an anchor, either a low or a high anchor, which is then adjusted in the light of relevant information. However, this adjustment process is typically insufficient or incomplete so that the resulting judgment remains biased toward the initial anchor. The anchoring heuristic thus produces

underestimation if the initial anchor is low and overestimation if the initial anchor is high. For example, the overall costs for a holiday trip tend to be underestimated if calculation starts from zero, or from the cheapest base price, to which predictable costs are added. In contrast, costs tend to be overestimated if an overly high anchor is used as a starting point and all nonapplicable costs are subtracted.

For a nice empirical illustration, the starting value of an auction often provides a good predictor of the final price (Ritov, 1996; but see Galinsky, Ku, & Mussweiler, 2009). An intriguing finding from many pertinent experiments says that exposure to fully irrelevant numerical stimuli—such as somebody's Social Security number—can cause an anchoring bias in judgments of completely unrelated quantities (Oppenheimer, LeBoeuf, & Brewer, 2008; Wilson, Houston, Etling, & Brekke, 1996). In a seminal demonstration by Tversky and Kahneman (1974), estimates of the number of African nations in the United Nations were biased toward random numbers generated by a wheel of fortune.

As in the case of the other heuristics, anchoring and (incomplete) adjustment will often result in rather accurate estimates. However, obviously, quantitative estimation can be biased to the extent that the starting anchors misrepresent the true value of the estimated quantity.

3.3 The Simulation Heuristic

The simulation heuristic is sensitive to expectancies resulting from mental simulation. What is easy to imagine or simulate mentally (e.g., one's favorite team winning or losing an important match) will determine not only one's expectations but also one's emotions of regret and disappointment elicited by unexpected outcomes. For instance, one is particularly disappointed and frustrated when one misses a train by a minute rather than by 30 minutes, because the counterfactual simulation of still reaching the train on time is so vivid in the former case. As a consequence, silver-medal winners in Olympic games can be less satisfied and more disappointed than bronze-medal winners, because counterfactual mental simulations may compare silver medals with a missed gold medal but bronze medals with a worse outcome of receiving no medal at all (Medvec, Madey, & Gilovich, 1995).

3.4 Availability

As introduced at the outset, the domain of the availability heuristic encompasses judgments of uncertain frequencies or probabilities. The ease with which memory content comes to mind affords a useful proxy for frequency estimation, although it can lead to distinct biases when ease of memory retrieval itself is biased, for instance, due to selective media coverage (Pachur, Hertwig, & Steinmann, 2012; Reber & Zupanek, 2002).

4. Lopsided Focus on Heuristics as Erroneous Inference Tools

The greatest part of the published literature, to be sure, is concerned with empirical demonstrations of biases and erroneous heuristic inferences; relatively little research has assessed how often heuristic judgments are valid and unbiased. For example, in one of the most frequently cited demonstrations of the availability heuristic, Tversky and Kahneman (1973) showed that the number of English words with the letter "k" in the first position was erroneously judged to be higher than the number of words with a "k" in the third position. This is apparently due to the fact that it is easier to retrieve words with a particular letter in the first than in the third position. However, as Sedlmeier, Hertwig, and Gigerenzer (1998) have shown, this finding does not generalize to all letters of the alphabet; it is peculiar to the few letters on which the original demonstration had concentrated.

Because of this lopsided research focus on the biasing potential of heuristics and the relative neglect of their ecological validity, accuracy, and adaptive value, heuristics are associated with negative connotations in the published literature. References to "heuristic inferences" often characterize them as sloppy, superficial, inaccurate, and irrational. The credo of many dual-process theories (Chaiken & Trope, 1999; Sloman, 1996; Smith & DeCoster, 2000) is that the remedy to cognitive biases and distortions is to switch from a heuristic to a systematic processing mode. Just as the alleged inferiority of heuristic processing is rarely demonstrated in a representative design, the alleged superiority of systematic processing modes is hardly ever proven.

Given the apparent analogy to the classic perceptual illusions of Gestalt psychologists, the negative image of biases or cognitive illusions resulting from distinct heuristics (such as availability) is hard to understand. Perceptual illusions (such as the contrast effect underlying the Ebbinghaus illusion[1]) had typically been understood as rare failures of adaptive and strikingly efficient processing. In the realm of cognitive illusions, though, most researchers inspired by Tversky and Kahneman's (1974) heuristics-and-biases approach were not interested in adaptive functions but almost exclusively in demonstrating cognitive limitations as causal origins of irrational behavior—even in the absence of cogent evidence.[2]

The aforementioned findings on availability effects in estimating causes of death (Combs & Slovic, 1979) provide testimony for this lopsided tendency. The same researchers who, for example, pointed out biased newspaper coverage as a plausible environmental cause of overestimated murder rates (relative to suicide rates) continued to treat these findings as evidence for an alleged cognitive bias in selective retrieval. Neither Combs and Slovic (1979) nor many other authors who drew on their findings ever made an attempt to rule out the possibility that (environmental) biases in newspaper reporting alone may account for the biased estimates even when retrieval processes are completely unbiased. Most researchers largely neglected the obvious fact that heuristics may serve a generally adaptive function, with only a few specific types of situations leading to the biases investigated so feverously by the research community.

5. Heuristics as Indispensable Modules of Adaptive Intelligence

We have already noted that for many real-life problems, there are hardly any purely logical, normative solutions. That is, so-called normative models of reality also rely on heuristics or proxies used to estimate the costs of a construction project, the severity of climate change, the utility of different ecological policies, or the number of participants required in an experiment. Indeed, heuristics are part and parcel of expert systems and of the most intelligent systems of adaptive cognition, as illustrated in the early work of Egon Brunswik. Long before Kahneman and Tversky unleashed this wave of research into the inadequacies of human judgment by cause of heuristics, Brunswik's (1955) *probabilistic functionalism* outlined a very different perspective on heuristics as indispensable modules of adaptive regulation.

According to Brunswik's lens model, the distal entities that are the focus of most judgments in real life—entities such as risk, danger, honesty, attraction, student ability, or interpersonal trust—are not amenable to direct perception through specialized sense organs. We somehow have to construe those distal entities from sets of proximal cues, which bear only weak or modest correlations to the distal entities. The inferential utilization of insufficient but useful and indispensable cues is nothing else but highly functional heuristic processing. For instance, to assess the distal variable of honesty in lie detection, we have to rely on such imperfect heuristic cues as speech hesitation, gaze aversion, pitch of the voice, or disguised smiling (Hartwig & Bond, 2011, 2014), all of which are rather low in diagnosticity. Although all

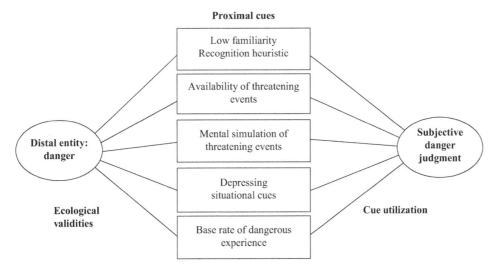

Figure 2.4.1
Brunswik's (1955) lens model applied to subjective judgments of danger, mediated by five proximal cues framed to resemble some of the most well-known heuristics (recognition, availability, simulation heuristic). Statistical relations of proximal cues to distal entities are called ecological validities. The weights given to cues in subjective judgments are referred to as cue utilization.

these cues provide only weak predictors of the distal criterion, honesty, they nevertheless allow for above-chance lie detection when an increasing number of cues of restricted validity are combined. Crucially, there is no alternative to relying on heuristic cues for lie detection, which is of utmost importance for adaptive behavior.

As we have learned from Brunswik (1955), we do not possess sense organs even for the literal perception of many obviously physical attributes, such as distance. Rather, in-depth perception organisms have to infer the perceived distance of an object using such proximal cues as disparity of the retinal images, movement parallax, surface texture, or blurredness of contours. All these cues afford only imperfect correlates of distance; in other words, they allow us to approximate the distal entity in question in a crude heuristic manner only. Moreover, the notion of *vicarious functioning* highlights the substitutability of all these cues, none of which is necessary or essential for depth perception. Still, they are functional because their ecological validity is better than chance and, as mentioned before, there is no alternative to relying on heuristic cues.

5.1 Using the Lens Model to Describe Heuristics' Adaptive Functions
By analogy, Hammond, Stewart, Brehmer, and Steinmann (1986) used the lens model in their social judgment theory to analyze the cues used as proxies for such distal entities as risk, attraction, trust, or required effort expenditure (see figure 2.4.1). Obviously, availability and other heuristics afford reasonable cues, the validity of

which is an open empirical question. However, as already mentioned, the validity of any alternative device supposed to be normatively appropriate is also an open empirical question. In any case, the lens model in figure 2.4.1, in analogy with social judgment theory (Hammond et al., 1986), suggests a framework within which the availability heuristic (along with other heuristic cues) serves an uncontested adaptive function—namely, to allow organisms to estimate danger under uncertainty, in the absence of immediate perceptual cues. Similar to Brunswik's in many aspects, an approach that focuses on the adaptive aspects of heuristics emerged after the heuristics-and-biases program had developed into one of the most prosperous research programs: the fast-and-frugal-heuristics program initiated by Gigerenzer et al. (1999).

6. Fast and Frugal Heuristics

Inspired by this Brunswikian perspective and previous work by Edwards (1965) and Peterson and Beach (1967) that had drawn a more optimistic picture of the fit between the human mind and the principles of logic and probability theory, Gigerenzer and colleagues returned to the notion that humans can be considered rational, giving an adaptive interpretation to the term. This alternative research program was deliberately opposed to the common research strategy of the Kahneman–Tversky program, in which cognitive performance was tested against the rules of classical logic and probability theory. Criticism of the standards against which decision making was assessed can be considered the starting point of the alternative program on "fast and frugal heuristics" (Gigerenzer, 1991, 1996; Todd & Gigerenzer, 2000).

Rejecting those normative standards as too narrow and unjustifiably content blind (Gigerenzer, 1996), this research group took a different approach to investigating heuristics. They viewed heuristics in the light of bounded rationality—a term coined by Simon (1982) for the view of rationality that takes into account human limitations and implores to relate them to the structure of the environments in which the decisions are made (Todd & Gigerenzer, 2000). Accordingly, they worked on the assumption that heuristics offer a reasonable way for humans to cope with the uncertainty of the world by making use of the structure of information environments. Different heuristics, metaphorically assorted in the "adaptive toolbox" (Gigerenzer et al., 1999), are suitable for different environments. Like a tradesperson who chooses the tool appropriate for the task at hand, we as decision makers are able to determine the heuristic suitable for the information structure of the decision and apply it. Gigerenzer and colleagues set out to discover such heuristic decision strategies, which environment structures they matched, and to what extent people actually use them. Instead of looking at and for heuristics as a way to explain fallacious reasoning and biases in judgment and decision making, they shifted the focus to the search for heuristics' benefits.

Arguably the two most important classes of heuristics advanced by this alternative approach, recognition and Take the Best, shall be illustrated in more detail in the following. Both of them represent single-cue heuristics in a Brunswikian sense. While the former always relies on the same primitive cue of recognition, the latter utilizes a more flexible, adaptive multicue toolbox, from which the best-suited cue is selected for the decision at hand.

6.1   The Recognition Heuristic

The recognition heuristic (Goldstein & Gigerenzer, 2002) is dependent on the decision maker's partial ignorance; indeed, it illustrates the class of ignorance-based decision-making strategies. Faced with a choice between two options, having to judge which has a higher value on some criterion, a decision maker applying this heuristic will simply choose the option that he or she recognizes (i.e., has seen or heard of before) as having a higher value than the unrecognized option. Naturally, the precondition for applying this heuristic is the recognition of merely one of the options—the partial ignorance mentioned above.[3] This heuristic will lead to accurate choices in environments where recognition is

a valid cue, positively correlated with the criterion the choice relates to. This may be the case in many domains, especially in the heavily studied paradigm of city population judgments, in which participants have to choose the city with the larger population out of two (Pachur, Todd, Gigerenzer, Schooler, & Goldstein, 2011). Cities that are recognized by many participants can indeed be expected to have larger population sizes. This heuristic affords an adaptive tool in situations characterized by partial ignorance, where only one option is known to them and more concrete knowledge is lacking (Pachur et al., 2011).

However, Oppenheimer (2003) pointed out that the recognition heuristic may not be as simple as initially portrayed. Participants did not apply the heuristic blindly in situations in which they had knowledge of the reason for the recognition, for example, if recognized towns were local or if they were well known due to unusual circumstances, as Chernobyl is for being the location of a nuclear disaster. Pachur et al. (2011) rejected Oppenheimer's criticism. If participants knew that the local towns were small, then they had what is called "criterion knowledge," objective knowledge about one of the options. If that criterion knowledge was conclusive, that is, "enable[d] the decision maker to deduce a solution" (Pachur et al., 2011, p. 7), then the recognition cues need not be utilized. Whether that is actually the case does not seem clear-cut. The fact that participants did not apply the heuristic when they recognized a city due to unusual circumstances connected to it was acknowledged and led to an extension of the recognition heuristic by another process—namely, judging whether the recognition cue is valid in this environment or not. This process comes after the recognition process and can explain the differential use in different environments, which would speak to the adaptive use of the heuristic.

Precisely this aspect of cue validity was also tested directly by examining the use of the recognition heuristic across different domains or environments with varying recognition validity (i.e., varying correlation between recognition and criterion). The higher the recognition validity, the more people relied on the recognition heuristic. Some studies even attest to a fascinating less-is-more effect with regard to knowledge. Those who know less, meaning that they recognize only a subset of options, may perform better under certain circumstances than those who recognize all options. This is because the latter cannot make use of the recognition heuristic but may not know enough about the options, beyond recognizing them, to compensate for it (Pachur et al., 2011).

As a cue, a special status is ascribed to recognition, because it is assumed to regulate the retrieval of all other cues. If an option is not recognized, one is unable to retrieve any other information about it. It can also be retrieved faster than any other cue. If the one cue that a decision is made upon is something other than recognition, one is in the realm of single-cue heuristics, the second class described here.

6.2 The Take the Best Heuristic

The second class of single-cue heuristics, also referred to as one-reason decision making, encompasses all heuristics that use only one cue, or reason, aside from recognition. The decision maker compares the options' values on that cue and chooses the option (among two or more options) with the highest value. These heuristics have in common the assumption that decision makers are capable of estimating the values of the cues they are comparing. Most assume that cue validity, the proportion of instances in which the cue correctly discriminated between options, can be estimated. Which of the cues is selected for the decision depends on the specific heuristic. Cue validity is the crucial dimension along which single-cue heuristics vary.

An entire class of so-called lexicographic strategies relies on the assumption that cues can be sorted by validity, from best to worst (Luce, 1956; Payne, Bettman, & Johnson, 1988; Tversky, 1969). Gigerenzer and Goldstein (1996) aptly termed the most prominent heuristic strategy from this class "Take the Best" (TTB). In TTB, the decision is based on the first cue in the ordering that discriminates between options, with the higher value on that cue determining the option to be chosen.

Gigerenzer and Goldstein (1996) also specified alternative ways of selecting the cue for the decision basis, the first of which is termed "minimalist algorithm" (p. 661) and selects cues at random until one discriminates, and the second of which is called "Take the Last algorithm" (p. 660) and selects the cue that was used for the decision last time. They compare some of those heuristics with complex models, considered normative models of decision making, like the weighted-additive decision strategy or Dawes' rule. Unsurprisingly, the more complex models could be fitted better to the data sets. However, more important, when the decision strategies were tested for how well they generalized to a new data set, TTB outperformed all other strategies in almost all cases. The reason for this is that their simplicity (or frugality) makes heuristics robust and avoids overfitting to one specific data set or environment.

7. Coherence and Correspondence

The issue of how the effectiveness of heuristics should be evaluated is broader than the contrast between fitting decision strategies to given data and generalizing to new data sets. Laws of probability and logic are traditionally the standard against which human decision making is evaluated. Rational, normative decision strategies are all coherent in that they do not violate any of those laws. However, heuristics do not have to adhere to normative rules to be able to lead to good decisions, illustrating the disconnect between the common evaluative standard and real-world requirements. A core aspect of the ABC Research Group's work is a standard for the evaluation of decision strategies that is in line with the view of bounded rationality. Instead of applying coherence criteria that examine whether any formal norms were violated (as was commonplace in research inspired by Kahneman and Tversky), bounded rationality requires that heuristics, and decision strategies in general, be evaluated by how well they correspond to the structure of the information environment in which they are used. They cannot be evaluated fully without at least some connection to real-world decision situations, because it is the heuristic's fit with the way the information is structured in the environment that needs to be assessed. The criteria used for this way of evaluating are called correspondence criteria, in contrast with the standard coherence criteria (Dunwoody, 2009), and include accuracy, frugality, and speed.

This leads to so-called *ecological rationality* (Todd, 2000), to be differentiated from "standard" rationality. Todd and Gigerenzer (2000) argue that this standard for evaluating heuristics is more sensible, because heuristics are not meant to be internally consistent but rather to provide reasonable inferences and adaptive guidance when time and resources are scarce. In addition, coherence criteria seem unsuitable because violations have not been found to be particularly costly (Arkes, Gigerenzer, & Hertwig, 2016). Once again, this highlights the immense advantage and absolute necessity of heuristic strategies: for many situations in the real world, no implementable standard normative strategy (optimization) exists, leaving heuristics as the only strategy.

8. Critical Appraisal of Scientific Insights Gained from Heuristics-and-Biases Approaches

There can be no doubt that the research program on heuristics and biases belongs to the most influential contributions to behavioral sciences. Current psychology would not be what it is today without the fascinating work reviewed so far in this chapter. The prototypical concept of the human mind in basic cognitive psychology was largely shaped by the twofold message that (a) human rationality is restricted, with experts and laypeople alike falling prey to a long list of biases and shortcomings (Nisbett & Ross, 1980), but that (b) apparent violations of manifold reasoning norms can nevertheless be adaptive in a complex and indeterminate world (Todd & Gigerenzer, 2003). Many domains of applied behavioral (and related) sciences have been influenced considerably by this research, including legal psychology (Gigerenzer & Engel, 2006), marketing and consumer research (Scholten, 1996), organizational psychology (West, Christodoulides, & Bonhomme, 2018), learning and education (Koriat & Levy-Sadot, 2001), and several major academic disciplines, such as economics, social and political sciences, and philosophy.

However, despite these signs of success, scientific fertility, and extraordinary impact, there is also room for critical appraisal. Although the empirical search for biases and shortcomings in diverse areas has flourished for almost five decades, some of the most essential theoretical underpinnings have been sorely neglected and never clarified. The debate about biases and alleged norm violations is still devoid of a clearly spelled-out normative fundament. The conceptual background of dual-process theories, in which most pertinent research is theoretically embedded, is weak and questionable. Most important, cogent empirical evidence from well-designed experiments concerning underlying mechanisms and delimiting conditions is conspicuously missing. Half a century after Tversky and Kahneman's first publication (1971), we know very little about what is going on in the black box of judges and decision makers.

8.1 Normative Benchmarks

A research program that relies heavily on empirical demonstrations of biases and transgressions of probability theorems is strongly contingent on normative benchmarks. To classify a judgment or decision as biased, one needs to have a norm or correctness criterion against which empirical deviations can be tested. Biases need to be systematic and of practically relevant size. Therefore, the tasks used for research on heuristics and biases provide participants with all information required to solve a probability calculation task or to choose a normatively optimal decision option (Edwards, Lindman, & Savage, 1963). Participants in the coin-tossing example knew

they were observing a fair coin. They could have known that the probability of any series must be equal to that of any other series of equal length since its probability is always ½ to the power of the series' length. The normative solution of this task can be considered stable and easy to implement.

When we take a closer look, however, we see that the coin-tossing example is essentially different from many everyday problems (Hahn & Warren, 2009). First, knowledge that a coin being tossed is fair, and thus the probabilities evenly distributed, may be a common feature in gambling, but it is a rare and highly improbable feature elsewhere. Assuming a trick coin landing heads with a probability of 0.8 and tails at 0.2, the task would be a little harder to solve normatively correctly. Moving away from gambling, we may end up wondering whether it is possible to calculate a normatively appropriate solution at all. We may never argue about how to calculate the odds of a fair coin-tossing series but may very well wonder how the true risk of, say, contracting the flu should be calculated. How can the best investment strategy on the stock market be determined? What is the normatively correct team to bet on in a basketball final? This demonstrates that there is a gap between experimental tasks involving betting and dice tossing, for which there exists an unequivocally correct solution, and real-life problems for which no normative solution exists, so that even experts and robot systems can only rely on approximations of normative solutions.

Unfortunately, this intricate problem of an essentially normative approach was not explicated from the beginning, although the entire research project focuses on deviations from norms. Kahneman and Tversky (1972) postulated a fundamental gap between probability theory and human responses. Despite their initial comparison of cognitive illusions with perceptual illusions (Tversky & Kahneman, 1983), they considered the representativeness heuristic to reflect an irrational illusion, the anchoring heuristic as a serious inability to ignore irrelevant numerical primes (Wilson et al., 1996), and sample-size neglect as a perfect example of irrational behavior.

The choice of normative criterion not only created a problem for Kahneman and Tversky's coherence-based approach to rationality but also represents an unresolved problem for the correspondence perspective taken by Gigerenzer et al. (1999) with its emphasis on the adaptive value of fast and frugal heuristics. Thus, when the Take the Best heuristic or the priority heuristic lead to 70% correct responses in pairwise choice tasks, the ultimate question is whether this can be interpreted as successful in terms of corresponding to the information environment or as a disappointing result. What is the normative benchmark for rationality, or bounded rationality, based on correspondence criteria: 65%, 70%, or maybe 80% correct predictions?

8.2 Dual-Process Theories

As a surrogate for the missing normative or psychophysical framework, many researchers in social cognition, but also in cognitive psychology, have located their work in the context of dual-process or dual-systems approaches (Chaiken & Trope, 1999; Sloman, 1996). These approaches postulate two systems, one that prefers fast and effortless thinking of the heuristic type (System 1) and one that can live up to systematic reasoning in line with rational procedures (System 2). Such a dual-process framework offers a seemingly plausible (but circular) explanation for virtually every empirical test of hypothetical heuristics. If participants exhibit the typical biases predicted by heuristic theories, they were apparently operating in System 1, presumably because time and effort were too costly compared to the reward for rational responding. But if they happen to provide normatively correct responses, then they must have entered System 2, presumably because they are personally involved or motivated to be accurate. Note that for such a conception to allow strict theory testing, a superordinate theory has to make clear-cut predictions about specific conditions under which either System 1 or System 2 is activated. For logical reasons alone, such a metatheory is almost impossible (Keren & Schul, 2009), because it is hard to see why System 1 and System 2 should not overlap, why there should be no more than two systems, and why a single system alone should not be able to account for all the evidence (Kruglanski & Thompson, 1999). In the absence of a metatheory that imposes strong constraints on the conditions that trigger different systems, it remains incomplete and of restricted scientific value. A promising suggestion for such a metatheory is delineated by Marewski and Schooler (2011).

8.3 Numeracy and Expertise

Especially the initial research on heuristics in the early 1970s did not take the perspective of dual-process theories' System 1: fast and loose guessing in the absence of motivation as well as mental and temporal resources. Numeracy and education level of participants have always been assumed to be sufficiently high. One of the first publications on the heuristics-and-biases project recruited members of illustrious psychological societies

as participants (Tversky & Kahneman, 1971). Further investigations recruited students in university courses, who were presumably well educated in stochastics. Nevertheless, cognitive biases were shown to be ubiquitous. Even experts fall prey to overconfidence (Klayman, Soll, González-Vallejo, & Barlas, 1999), exhibit marked anchoring biases (Englich, Mussweiler, & Strack, 2006), and fail to engage in proper logical reasoning (Wason, 1968). Risk assessment in highly consequential areas is replete with base-rate fallacies and failures to run appropriate cost–benefit analyses (Swets, Dawes, & Monahan, 2000). Even when it comes to taking advice from medical experts on existential health-related issues, doctors and patients do not live up to normative principles of risk assessment and rational action (Gigerenzer, Gaissmaier, Kurz-Milcke, Schwartz, & Woloshin, 2007; Wegwarth, Gaissmaier, & Gigerenzer, 2011).

8.4 The Conspicuous Lack of Insights on Underlying Cognitive Mechanisms

Perhaps the most striking critique of half a century of flourishing research on cognitive heuristics, however, refers to the paucity of solid experimental evidence on underlying mechanisms. Although Tversky and Kahneman (1974) introduced all of their heuristics as cognitive-process theories, very few experiments were conducted to isolate selective memory retrieval as the site of the availability heuristic, to specify the precise similarity function of the representativeness heuristic (Nilsson, Juslin, & Olsson, 2008), or to unravel distinct constraints on the (insufficient) adjustment process supposed to explain anchoring effects (Epley & Gilovich, 2001, 2006).

Although Gigerenzer and colleagues themselves started from this critique of missing experimental tests of Kahneman and Tversky's ideas, their own research suffered from the same paucity of unequivocal findings about cognitive mechanisms. There is hardly any cogent evidence on a miraculous algorithm that allows people to rank-order the validity of a longer list of candidate cues, as a precondition for Take the Best. Strict experimental tests of the recognition heuristic are rarely based on the deliberate manipulation of recognizable versus non-recognizable stimuli (e.g., in a paired comparison with selective repeated exposure). While Goldstein and Gigerenzer's (2002) famous demonstration that Germans outperform Americans when judging the population size of pairs of American cities was counted as evidence in favor of the recognition heuristic (because Germans know little about American cities except that they recognize some and do not recognize others), other research did not support this superiority of partial knowledge (Pachur & Biele, 2007; Pohl, 2006).

The cognitive algorithm supposed to underlie the priority heuristic (Brandstätter, Gigerenzer, & Hertwig, 2006) is specified in all detail: when choosing between two lotteries (each described by distinct outcomes *o* obtained with distinct probabilities *p*), the heuristic first tries to make a choice in favor of the lottery with the higher minimal *o*. Only when the lotteries' difference in minimal *o* is not sufficient, the heuristic tries to make a choice in favor of the lottery with the lower *p*(minimal *o*) of a minimal outcome. Only if the difference in *p*(minimal *o*) is not sufficient, the lexicographic algorithm switches to the highest outcome as a criterion, and only if this third cue still does not enable a clear-cut choice, the mechanism resorts to guessing. Although this is a precise algorithm, there is little experimental evidence available to support a cognitive process that follows exactly this sequence. The evidence marshalled in favor of the priority heuristic is, again, largely confined to reporting the percentage with which the heuristic chooses the lottery with the higher expected value. However, an accuracy rate of, say, 70% hardly reveals much about an underlying mechanism. When a deliberate attempt is made to test the heuristic experimentally, the evidence is incompatible with the heuristic: *p*(minimal *o*) exerts a strong impact regardless of whether the preceding minimal-*o* cue is high or low, and maximal *o* exerts a substantial influence regardless of whether both preceding cues, minimal *o* and *p*(minimal *o*), are high or low (Fiedler, 2010).

A continued debate revolves around the "adaptive toolbox" and its contents of many different heuristics. Alternative single-process models have been proposed as more parsimonious accounts and better fit some parameters, although it is very hard to test different approaches against each other (Glöckner, Hilbig, & Jekel, 2014; Söllner & Bröder, 2015). The lack of explanation regarding the way the decision maker chooses heuristics from the diversely filled toolbox in different environments is another important criticism (Cooper, 2000). Feeney (2000) points out that the choice among heuristics may lead to an infinite regress just like the one Gigerenzer and colleagues criticized about the Kahneman–Tversky research program. Todd and Gigerenzer (2000) admit that a metatheory that explains the choice of heuristics is sorely needed as a solution to this problem.

9. Concluding Remarks

By all standards, Kahneman and Tversky have certainly sparked a scientific revolution, and Gigerenzer and colleagues have arguably achieved their goal of initiating a slightly different revolution of the view of the mind. Other leading labs have also made strong contributions to research on rationality, including Fischhoff's (1975) work on the hindsight bias, Einhorn and Hogarth's (1981, 1986) seminal papers on friendly versus wicked decision environments, or recent work by Hertwig and colleagues (Hertwig, Herzog, Schooler, & Reimer, 2008; Hertwig, Hoffrage, & ABC Research Group, 2013; Hoffrage & Hertwig, 2012; see also chapter 8.5 by Hertwig & Kozyreva, this handbook) on the dialectic interplay of organisms and their environments. These pioneers have inspired a flood of research conducted in the spirit of their ideas. However, despite all this fruitful research and all the fascination it has elicited, the research program is far from being complete or exhaustive. We are only beginning to understand the interaction of environmental and cognitive sources of judgment biases and have only recently begun to replace the old-fashioned criterion of accuracy (relative to an elusive norm) by a new criterion of adaptive behavior, which is still conceptually underdeveloped.

To achieve this major developmental goal, it may be necessary, ironically, to rid ourselves of the old paradigmatic idea of a normatively correct solution for judgment and decision problems. The existence of such a normative criterion may turn out to be more the exception than the rule—quite like the dice-tossing problems and betting urns used in many traditional experiments are. As a consequence, there is no alternative to applying useful heuristics even when the smartest are motivated to find optimal solutions with the help of proper calculus (Bayes' theorem) and artificial intelligence (machine learning). Even under such auspicious conditions, there is hardly any normatively correct, conflict-free solution; there is always a need to find a heuristic solution that optimizes the adaptive value of intelligent action.

Acknowledgments

The work underlying this chapter was supported by a grant provided by the German Research Foundation (Deutsche Forschungsgemeinschaft) to the first author (FI 294/29-1).

Notes

1. The Ebbinghaus illusion produces a contrast effect such that a circle surrounded by four other circles appears larger (smaller) when the surrounding circles are smaller (larger).

2. Given that Kahneman and Tversky (1974) conceived cognitive illusions as analogous to perceptual illusions, this unequal treatment of the two types of illusions appears conspicuous.

3. Along the divide of compensatory and noncompensatory strategies, it is located on the noncompensatory side, meaning that it is not traded off against additional cues.

References

Arkes, H. R., Gigerenzer, G., & Hertwig, R. (2016). How bad is incoherence? *Decision*, *3*(1), 20–39.

Bar-Hillel, M. (1984). Representativeness and fallacies of probability judgment. *Acta Psychologica*, *55*(2), 91–107.

Brandstätter, E., Gigerenzer, G., & Hertwig, R. (2006). The priority heuristic: Making choices without trade-offs. *Psychological Review*, *113*(2), 409–432.

Brunswik, E. (1955). Representative design and probabilistic theory in a functional psychology. *Psychological Review*, *62*(3), 193–217.

Chaiken, S., & Trope, Y. (Eds.). (1999). *Dual-process theories in social psychology*. New York, NY: Guilford.

Combs, B., & Slovic, P. (1979). Newspaper coverage of causes of death. *Public Opinion Quarterly*, *56*, 837–843.

Cooper, R. (2000). Simple heuristics could make us smart, but which heuristics do we apply when? Commentary on Précis of *Simple heuristics that make us smart*. *Behavioral and Brain Sciences*, *23*(5), 746.

Dunwoody, P. T. (2009). Introduction to the special issue: Coherence and correspondence in judgment and decision making. *Judgment and Decision Making*, *4*(2), 113–115.

Edwards, W. (1965). Optimal strategies for seeking information: Models for statistics, choice reaction times, and human information processing. *Journal of Mathematical Psychology*, *2*(2), 312–329.

Edwards, W., Lindman, H., & Savage, L. J. (1963). Bayesian statistical inference for psychological research. *Psychological Review*, *70*(3), 193–242.

Einhorn, H. J., & Hogarth, R. M. (1981). Behavioral decision theory: Processes of judgment and choice. *Annual Review of Psychology*, *32*, 53–88.

Einhorn, H. J., & Hogarth, R. M. (1986). Judging probable cause. *Psychological Bulletin*, *99*(1), 3–19.

Englich, B., Mussweiler, T., & Strack, F. (2006). Playing dice with criminal sentences: The influence of irrelevant anchors on experts' judicial decision making. *Personality and Social Psychology Bulletin*, *32*(2), 188–200.

Epley, N., & Gilovich T. (2001). Putting adjustment back in the anchoring and adjustment heuristic: Differential processing of self-generated and experimenter-provided anchors. *Psychological Science*, *12*, 391–396.

Epley, N., & Gilovich, T. (2006). The anchoring-and-adjustment heuristic: Why the adjustments are insufficient. *Psychological Science*, *17*(4), 311–318.

Epley, N., & Gilovich, T. (2010). Anchoring unbound. *Journal of Consumer Psychology*, *20*(1), 20–24.

Feeney, A. (2000). Simple heuristics: From one infinite regress to another? Commentary on Précis of *Simple heuristics that make us smart*. *Behavioral and Brain Sciences*, *23*(5), 749.

Fiedler, K. (2010). How to study cognitive decision algorithms: The case of the priority heuristic. *Judgment and Decision Making*, *5*(1), 21–32.

Fischhoff, B. (1975). Hindsight is not equal to foresight: The effect of outcome knowledge on judgment under uncertainty. *Journal of Experimental Psychology: Human Perception and Performance*, *1*(3), 288–299.

Galinsky, A. D., Ku, G., & Mussweiler, T. (2009). To start low or to start high? The case of auctions versus negotiations. *Current Directions in Psychological Science*, *18*(6), 357–361.

Gigerenzer, G. (1991). How to make cognitive illusions disappear: Beyond "heuristics and biases." *European Review of Social Psychology*, *2*(1), 83–115.

Gigerenzer, G. (1996). On narrow norms and vague heuristics: A reply to Kahneman and Tversky. *Psychological Review*, *103*(3), 592–596.

Gigerenzer, G., & Engel, C. (Eds.). (2006). *Heuristics and the law*. Cambridge, MA: MIT Press.

Gigerenzer, G., & Gaissmaier, W. (2011). Heuristic decision making. *Annual Review of Psychology*, *62*, 451–482.

Gigerenzer, G., Gaissmaier, W., Kurz-Milcke, E., Schwartz, L. M., & Woloshin, S. (2007). Helping doctors and patients make sense of health statistics. *Psychological Science in the Public Interest*, *8*(2), 53–96.

Gigerenzer, G., & Goldstein, D. G. (1996). Reasoning the fast and frugal way. *Psychological Review*, *103*(4), 650–669.

Gigerenzer, G., Todd, P. M., & ABC Research Group (1999). *Simple heuristics that make us smart*. Oxford, England: Oxford University Press.

Gilovich, T., Griffin, D., & Kahneman, D. (Eds.). (2002). *Heuristics and biases: The psychology of intuitive judgment*. Cambridge, England: Cambridge University Press.

Glöckner, A., Hilbig, B. E., & Jekel, M. (2014). What is adaptive about adaptive decision making? A parallel constraint satisfaction account. *Cognition*, *133*(3), 641–666.

Goldstein, D. G., & Gigerenzer, G. (2002). Models of ecological rationality: The recognition heuristic. *Psychological Review*, *109*(1), 75–90.

Hahn, U., & Warren, P. A. (2009). Perceptions of randomness: Why three heads are better than four. *Psychological Review*, *116*(2), 454–461.

Hammond, K. R., Stewart, T. R., Brehmer, B., & Steinmann, D. O. (1986). Social judgment theory. In H. R. Arkes & K. R.

Hammond (Eds.), *Judgment and decision making: An interdisciplinary reader* (pp. 56–76). Cambridge, England: Cambridge University Press.

Hartwig, M., & Bond, C. F., Jr. (2011). Why do lie-catchers fail? A lens model meta-analysis of human lie judgments. *Psychological Bulletin, 137*(4), 643–659.

Hartwig, M., & Bond, C. F., Jr. (2014). Lie detection from multiple cues: A meta-analysis. *Applied Cognitive Psychology, 28*(5), 661–676.

Hertwig, R., Herzog, S. M., Schooler, L. J., & Reimer, T. (2008). Fluency heuristic: A model of how the mind exploits a by-product of information retrieval. *Journal of Experimental Psychology: Learning, Memory, and Cognition, 34*(5), 1191–1206.

Hertwig, R., Hoffrage, U., & ABC Research Group (Eds.). (2013). *Simple heuristics in a social world*. Oxford, England: Oxford University Press.

Hoffrage, U., & Hertwig, R. (2012). Simple heuristics in a complex social world. In J. I. Krueger (Ed.), *Social judgment and decision making* (pp. 135–150). New York, NY: Psychology Press.

Kahneman, D., Slovic, P., & Tversky, A. (Eds.). (1982). *Judgment under uncertainty: Heuristics and biases*. Cambridge, England: Cambridge University Press.

Kahneman, D., & Tversky, A. (1972). Subjective probability: A judgment of representativeness. *Cognitive Psychology, 3*(3), 430–454.

Keren, G., & Schul, Y. (2009). Two is not always better than one: A critical evaluation of two-system theories. *Perspectives on Psychological Science, 4*, 533–550.

Klayman, J., Soll, J. B., González-Vallejo, C., & Barlas, S. (1999). Overconfidence: It depends on how, what, and whom you ask. *Organizational Behavior and Human Decision Processes, 79*(3), 216–247.

Koriat, A., & Levy-Sadot, R. (2001). The combined contributions of the cue-familiarity and accessibility heuristics to feelings of knowing. *Journal of Experimental Psychology: Learning, Memory, and Cognition, 27*(1), 34–53.

Kruglanski, A. W., & Thompson, E. P. (1999). Persuasion by a single route: A view from the unimodel. *Psychological Inquiry, 10*(2), 83–109.

Luce, R. D. (1956). Semiorders and a theory of utility discrimination. *Econometrica, 24*(2), 178–191.

Marewski, J. N., & Schooler, L. J. (2011). Cognitive niches: An ecological model of strategy selection. *Psychological Review, 118*(3), 393–437.

Medvec, V. H., Madey, S. F., & Gilovich, T. (1995). When less is more: Counterfactual thinking and satisfaction among Olympic medalists. *Journal of Personality and Social Psychology, 69*(4), 603–610.

Nilsson, H., Juslin, P., & Olsson, H. (2008). Exemplars in the mist: The cognitive substrate of the representativeness heuristic. *Scandinavian Journal of Psychology, 49*(3), 201–212.

Nisbett, R. E., & Ross, L. (1980). *Human inference: Strategies and shortcomings of social judgment*. Englewood Cliffs, NJ: Prentice-Hall.

Oppenheimer, D. M. (2003). Not so fast! (and not so frugal!): Rethinking the recognition heuristic. *Cognition, 90*(1), B1–B9.

Oppenheimer, D. M., LeBoeuf, R. A., & Brewer, N. T. (2008). Anchors aweigh: A demonstration of cross-modality anchoring and magnitude priming. *Cognition, 106*(1), 13–26.

Pachur, T., & Biele, G. (2007). Forecasting from ignorance: The use and usefulness of recognition in lay predictions of sports events. *Acta Psychologica, 125*(1), 99–116.

Pachur, T., Hertwig, R., & Steinmann, F. (2012). How do people judge risks: Availability heuristic, affect heuristic, or both? *Journal of Experimental Psychology: Applied, 18*(3), 314–330.

Pachur, T., Todd, P. M., Gigerenzer, G., Schooler, L. J., & Goldstein, D. G. (2011). The recognition heuristic: A review of theory and tests. *Frontiers in Psychology, 2*, 1–14.

Payne, J. W., Bettman, J. R., & Johnson, E. J. (1988). Adaptive strategy selection in decision making. *Journal of Experimental Psychology: Learning, Memory, and Cognition, 14*(3), 534–552.

Pennycook, G., & Thompson, V. A. (2017). Base-rate neglect. In R. F. Pohl (Ed.), *Cognitive illusions: Intriguing phenomena in thinking, judgment and memory* (2nd ed., pp. 44–61). London, England: Routledge/Taylor & Francis.

Peterson, C. R., & Beach, L. R. (1967). Man as an intuitive statistician. *Psychological Bulletin, 68*(1), 29–46.

Pohl, R. F. (2006). Empirical tests of the recognition heuristic. *Journal of Behavioral Decision Making, 19*(3), 251–271.

Reber, R., & Zupanek, N. (2002). Effects of processing fluency on estimates of probability and frequency. In P. Sedlmeier & T. Betsch (Eds.), *Etc.: Frequency processing and cognition* (pp. 175–188). Oxford, England: Oxford University Press.

Ritov, I. (1996). Anchoring in simulated competitive market negotiation. *Organizational Behavior and Human Decision Processes, 67*(1), 16–25.

Scholten, M. (1996). Lost and found: The information-processing model of advertising effectiveness. *Journal of Business Research, 37*(2), 97–104.

Sedlmeier, P., Hertwig, R., & Gigerenzer, G. (1998). Are judgments of the positional frequencies of letters systematically biased due to availability? *Journal of Experimental Psychology: Learning, Memory, and Cognition, 24*(3), 754–770.

Shweder, R. A. (1977). Likeness and likelihood in everyday thought: Magical thinking in judgments about personality. *Current Anthropology, 18*(4), 637–658.

Simon, H. A. (1982). *Models of bounded rationality*. Cambridge, MA: MIT Press.

Sloman, S. A. (1996). The empirical case for two systems of reasoning. *Psychological Bulletin, 119*(1), 3–22.

Smith, E. R., & DeCoster, J. (2000). Dual-process models in social and cognitive psychology: Conceptual integration and links to underlying memory systems. *Personality and Social Psychology Review, 4*, 108–131.

Söllner, A., & Bröder, A. (2015). Toolbox or adjustable spanner? A critical comparison of two metaphors for adaptive decision making. *Journal of Experimental Psychology: Learning, Memory, and Cognition, 42*(2), 215–237.

Strack, F., Bahník, Š., & Mussweiler, T. (2016). Anchoring: Accessibility as a cause of judgmental assimilation. *Current Opinion in Psychology, 12*, 67–70.

Swets, J. A., Dawes, R. M., & Monahan, J. (2000). Psychological science can improve diagnostic decisions. *Psychological Science in the Public Interest, 1*(1), 1–26.

Todd, P. M. (2000). The ecological rationality of mechanisms evolved to make up minds. *American Behavioral Scientist, 43*(6), 940–956.

Todd, P. M., & Gigerenzer, G. (2000). Précis of *Simple heuristics that make us smart. Behavioral and Brain Sciences, 23*(5), 727–741.

Todd, P. M., & Gigerenzer, G. (2003). Bounding rationality to the world. *Journal of Economic Psychology, 24*(2), 143–165.

Tversky, A. (1969). Intransitivity of preferences. *Psychological Review, 76*(1), 31–48.

Tversky, A., & Kahneman, D. (1971). Belief in the law of small numbers. *Psychological Bulletin, 76*(2), 105–110.

Tversky, A., & Kahneman, D. (1973). Availability: A heuristic for judging frequency and probability. *Cognitive Psychology, 5*(2), 207–232.

Tversky, A., & Kahneman, D. (1974). Judgment under uncertainty: Heuristics and biases. *Science, 185*(4157), 1124–1131.

Tversky, A., & Kahneman, D. (1983). Extensional versus intuitive reasoning: The conjunction fallacy in probability judgment. *Psychological Review, 90*, 293–315.

Tversky, A., & Koehler, D. J. (1994). Support theory: A nonextensional representation of subjective probability. *Psychological Review, 101*(4), 547–567.

Wason, P. C. (1968). Reasoning about a rule. *Quarterly Journal of Experimental Psychology, 20*(3), 273–281.

Wegwarth, O., Gaissmaier, W., & Gigerenzer, G. (2011). Deceiving numbers: survival rates and their impact on doctors' risk communication. *Medical Decision Making, 31*(3), 386–394.

West, D. C., Christodoulides, G., & Bonhomme, J. (2018). How do heuristics influence creative decisions at advertising agencies? Factors that affect managerial decision making when choosing ideas to show the client. *Journal of Advertising Research, 58*(2), 189–201.

Wilson, T. D., Houston, C. E., Etling, K. M., & Brekke, N. (1996). A new look at anchoring effects: Basic anchoring and its antecedents. *Journal of Experimental Psychology: General, 125*(4), 387–402.

2.5 Dual-Process Theories of Deductive Reasoning

Karl Christoph Klauer

Summary

In dual-process models of thinking and reasoning, two kinds of processes are distinguished. Defining and correlated characteristics of these processes as well as their different functions are discussed along with the tasks and phenomena typically addressed by dual-process models. The relationships between the types of processes and normative, instrumental, and epistemic rationality are considered. A selective review of specific dual-process models illustrates the range of dual-process models that have been proposed. Evidence for and against dual-process models is summarized and evaluated.

1. Two Types of Thinking and Reasoning

Dual-process models of thinking and reasoning postulate that two kinds of processes are involved in reasoning. These two kinds of processes differ qualitatively and interact or compete in generating responses to reasoning problems. Dual-process models of thinking and reasoning are part of a larger family of such models in areas as diverse as social cognition, learning, and moral decision making. The guiding idea is that one kind of thinking, Type 1, delivers intuitive responses, gut feelings, or fast heuristic response proposals, whereas the other, Type 2, is responsible for slow but careful and systematic deliberations. Type 1 processes have been described as heuristic, context dependent, fast, and automatic; Type 2 processes as analytic, context independent, slow, and controlled.

In processing a reasoning problem, Type 1 processes are responsible for encoding the premises, for directing attention to relevant aspects in a bottom-up stimulus-driven manner, and for proposing an intuitive response. Furthermore, Type 1 processes have been argued to replace and add premises to reasoning problems through pragmatic implicatures and background knowledge. Finally, a Type 1 response can also function as a fallback response if Type 2 processes fail to propose a response.

Type 2 processes are involved in processing the task instructions. They assemble a task-set as an ad hoc constellation of processes coordinated to satisfy the task demands. Furthermore, Type 2 processes constantly monitor performance according to the implemented task-set and higher-level goals. They are responsible for strategic allocation of attentional resources and for processing that requires working-memory resources and mental simulation. They are activated when, for example, Type 1 processes do not suggest a response, suggest conflicting responses, or register surprise or a threat (e.g., Kahneman, 2011).

Within this framework, several dual-process models have been proposed that differ, among other issues, on how precisely Type 1 and Type 2 processes are characterized. Dual-process models have to explain as a minimum what distinguishes the two kinds of reasoning and how they interact in generating responses to reasoning problems.

Much debate has focused on which aspects distinguish Type 1 from Type 2 processes; table 2.5.1 provides a less than complete list of distinctions compiled in large part from a similar table in Evans and Stanovich (2013a). Furthermore, there is debate on how the two types of processes interact. One kind of interaction is that Type 1 processes shape the mental representations on which Type 2 processes operate (e.g., Evans, 2009; Thompson, 2013) by drawing attention to certain features of the problem at the expense of others and by adding premises from background knowledge. Type 1 processes also often generate a response proposal for the reasoning problem. In a default-interventionist model, this proposal serves as a default response that is acted upon unless Type 2 processes intervene (Evans, 2007). In contrast, in the parallel-competitive model, both kinds of processes operate in parallel, producing response proposals that compete for control of the eventual response (Evans, 2007).

Table 2.5.1
Some features and dimensions with respect to which Type 1 and Type 2 processes have been argued to differ

Type 1 process	Type 2 process
Fast	Slow
Automatic	Controlled
Nonconscious	Conscious
Unintentional	Intentional
Associative	Reflective
Parallel	Serial
Heuristic	Analytic
Experiential	Rational
Does not require working memory	Requires working memory

2. Stanovich's Tripartite Model

It is helpful to add a number of conceptual distinctions to this framework. In doing so, I will draw heavily on Stanovich's (2009) treatment. Stanovich (2009) defines the entirety of Type 1 processes as the set of autonomous systems in the brain that operate autonomously in response to specific triggering stimuli. These comprise many different and heterogeneous kinds of processes such as

- preattentive processes that supply content for focal attention and thereby shape the initial mental representation of the problem (Evans, 2009),
- domain-specific processing modules as discussed by evolutionary psychologists (e.g., Cosmides, 1989),
- domain-general processes of implicit learning and conditioning, and
- inference, decision-making, and classification rules (Mercier & Sperber, 2009) that are innate or have been practiced to automaticity (Anderson, 1983; Logan, 1985).

Type 2 processes, on the other hand, fall into two classes, termed the "algorithmic" and the "reflective" mind. The reflective level comprises control states that regulate processing according to higher-level goals. Processes at the reflective level initiate and guide processes at the algorithmic level. The algorithmic level implements processes and strategies such as those required in solving the tasks of a test of fluid intelligence, and it sustains hypothetical thought and mental simulations. The algorithmic processes can also inhibit and override the output of Type 1 processes. The reflective

processes, however, initiate such overriding as well as mental simulations.

Stanovich (2009) distinguishes between the algorithmic and the reflective level in part on theoretical grounds: they serve different functions in reasoning. But the distinction is also empirically motivated by individual-differences research on reasoning performance in different tasks (see Stanovich, 2009).

Finally, Stanovich's (2009) framework allows him to distinguish between different kinds of reasoning biases:

- First of all, Type 2 processes may be not engaged. This can come about where a biased Type 1 output is so compelling (for example, associated with high feelings of rightness; Thompson, 2009) that Type 2 processes are not invoked or where Type 2 processes are not available due to lack of capacity or effort.

- Type 2 processes may be engaged but fail to override a biased Type 1 response. Lack of effort is not the only possible cause of this outcome. Alternatively, the necessary Type 2 rules and algorithms may not have been learned and may therefore not be available, a situation that Stanovich (2009) called "mindware gaps."

- Relatedly, the available "mindware" may be faulty or itself biased, as in Stanovich's (2009) category of "biases due to contaminated mindware."

- Type 2 processes may operate on a mental representation that does not permit one to find the correct solution. For example, in Wason's (1966) selection task, shown in figure 2.5.1, preattentive Type 1 processes may draw attention to the cards whose visible sides show the elements explicitly mentioned in the rule ("A" and "3" in figure 2.5.1), which do not include one of the cards that needs to be selected ("1" in figure 2.5.1). Any reasoning focusing on only these cards would probably be seen as an instance of rationalization of an initial response (Evans, 1996; Evans & Wason, 1976). Stanovich (2009) terms this kind of Type 2 reasoning "serial associative cognition with a focal bias."

3. Type of Process and Rationality

3.1 Instrumental, Normative, and Epistemic Rationality

What is the relationship of the different processes to human rationality? Consider instrumental rationality, defined as choosing suitable means for one's ends. Type 1 processes, if defined as autonomous and triggered by specific cues, have no clear relationship to instrumental

The (reduced) base-rate task
(after Pennycook, Fugelsang, & Koehler, 2015)

General Instructions:
In a big research project a large number of studies were carried out where short personality descriptions of the participants were made. In every study there were participants from two population groups (e.g., carpenters and policemen). In each study one participant was drawn at random from the sample. You'll get to see a personality trait for this randomly chosen participant. You'll also get information about the composition of the population groups tested in the study in question. You'll be asked to indicate to which population group the participant most likely belongs.

One Trial:
This study contains: politicians and nannys.
Person 'A' is dishonest.
There are 999 nannys/5 politicians.
Is person 'A' more likely to be a) a politician or b) a nanny?

The belief-bias task
(after Handley, Newstead, & Trippas, 2011)

General Instructions:
You must assume each premise is true (even if in reality it is not true) and respond with the answer which logically follows from the statements presented.

For example: If you finish a drink, then the drink will be full.

Suppose that you finish your drink.
Will your drink be empty?

The logic-based answer to this problem is NO, because you must assume that if a drink is finished, then it will be full; therefore, the glass will not be empty.

One Trial:
If a child is crying, then he is happy.
Suppose a child is crying.
Is the child happy?

The Wason selection task
(after Klauer, Stahl, & Erdfelder, 2007)

Instructions:
Below you see a number of cards from a set of cards. Each card in the set has a capital letter on one side and a digit on the other. Naturally, only one side is visible in each case. For the set of cards, a rule has been stated. It is: "If there is an A on the letter side of the card, then there is a 3 on the number side." You must decide which card(s) displayed would have to be turned over in order to test the truth or falsity of the rule.

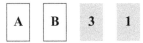

Figure 2.5.1
Three tasks frequently used in research on dual-process models.

rationality as they are not directly informed by one's goals. The same is true for normative rationality, defined as the adherence to given norms. There are, however, a couple of considerations that qualify these strong statements. Type 1 processes, it is now widely held, can also deliver normatively correct responses if they are well attuned to environmental regularities, and they can be sensitive to abstract structural features of a task. For example, the modus ponens rule describes the inference from a major premise of the form "If p then q" and a minor premise "p" to the conclusion "q." Modus ponens may be so well practiced that it can often be triggered habitually and fast in reading and comprehending the premises (e.g., Lea, 1995; Lea, O'Brien, Fisch, Noveck, & Braine, 1990; Rader & Sloutsky, 2002).

Relatedly, Type 1 processes are frequently argued to allow us to approach major goals that we hold (e.g., Evans, 2014), linking them to instrumental rationality. This is sometimes associated with dual-systems views

or two-minds theories in which Type 1 and Type 2 processes live in different systems, linked to different evolutionary stages (e.g., Evans, 2014; Mercier & Sperber, 2017; Stanovich, 2004).

For example, integrating one's beliefs about the likelihood of possible outcomes of acts and one's evaluations of these outcomes is seen as a hallmark of instrumental rationality in decision theory (Buchak, 2013). At the same time, relying on one's beliefs underlies the generation of belief bias (Wilkins, 1928) in deductive reasoning, which has sometimes been construed as reflecting the operation of a Type 1 process. Similarly, Gigerenzer and Todd (1999) have argued (a) that human beings have at their disposal a toolbox of fast and frugal heuristics, some of which would probably qualify as Type 1 processes for some authors, and (b) that these often mimic optimal strategies for achieving current goals by capitalizing on regularities in benign environments. And so, instrumental rationality alone is unlikely to be a good feature for cleanly distinguishing Type 1 from Type 2 processes. Similar points can be made regarding normative rationality.

Epistemic rationality regards the internal coherence and consistency of the beliefs that we hold and the support that we seek for them in terms of external evidence. In his two-minds theory, Evans (2014) argues that epistemic rationality is something pursued only by what he calls the "new" mind and Type 2 processes operating on explicit knowledge. In contrast, both the "old" mind, relying on Type 1 processes, and the "new" mind work in concert to achieve instrumental rationality. The next section explores the idea that justifications as an element of epistemic rationality might reliably distinguish between Type 1 and Type 2 processes.

3.2 Justification as a Qualitative Distinction between Type 1 and Type 2 Processes?

Consider an analogy to Plato's definition of knowledge as *justified* true belief. Type 1 processes can deliver correct responses (true belief), satisfying the requirements of instrumental rationality and sometimes normative rationality. But they do so for the wrong reasons (i.e., unjustified) inasmuch as they capitalize on environmental features that happen to covary with the normative or rational response most of the time. This can be revealed, for example, in experimental situations in which the auspicious environmental features are intentionally removed and in which responses then depart from the normative prescriptions or from those of instrumental rationality. Kahneman and Tversky's heuristics-and-biases program (see chapter 2.4 by Fiedler, Prager, & McCaughey, this handbook) has been characterized as having achieved just this (Kahneman, 2011). For example, one's beliefs will often provide an adequate guideline when judging the validity of the conclusion of a logical argument, but they may be misleading when the task is to disregard one's background knowledge and to judge the validity solely based on whether the proposition follows logically from the premises of the argument, taking these as true (see figure 2.5.1 for an example).

And thus, pursuing the above analogy a bit further, it may not be enough to focus on the correctness or reasonableness of the outcomes of the reasoning process. Instead, the outcome should also come about for the right reasons. One way to ensure this is to require that reasoners should be aware of reasons or justifications for the particular reasoning steps they engage in. This requirement to be able to justify one's responses as they are generated (rather than post hoc) captures nicely the distinction between Type 1 processes, characterized as intuitions being based

on gut feelings or reactions, and deliberate Type 2 processes, based on reflection. In a similar vein, Mercier and Sperber (2009) suggest distinguishing between intuitive and reflective inferences. Reflective inferences involve the representation of reasons for the inferred conclusions (see also Mercier & Sperber, 2017). Of course, conclusions that come with justifications need not be normatively or instrumentally correct, for reasons already discussed above, but this does not diminish their potential as a demarcation criterion.

4. Examples of Dual-Process Models and Related Models in the Reasoning Literature

4.1 Evans and Stanovich's Framework

Let us consider a number of dual-process models in the reasoning domain to illustrate these concepts further. Evans and Stanovich (2013a) have provided a framework for dual-process models in reasoning and decision making in which the two types of reasoning are distinguished in terms of two defining features: Type 2 processes load working memory, whereas Type 1 processes do not require working-memory resources. In addition, another hallmark of Type 2 processing is that it sustains hypothetical thought or cognitive decoupling (Stanovich, 2009), "the ability to distinguish supposition from belief and to aid rational choices by running thought experiments" (Evans & Stanovich, 2013a, p. 236). In particular, working memory and cognitive decoupling are involved when Type 1 processes and their outputs are to be inhibited.

The interaction between the two types of processing is described in terms of a default-interventionist model: Type 1 processes deliver a default intuition that can be overridden by Type 2 processes. Whether and when this occurs is a function of motivation, the availability of working-memory resources, and individual differences in thinking dispositions (Evans & Stanovich, 2013b). Type 2 processing can be algorithmic, proposing an alternative response to the default intuitive one or affirming it, and reflective, embodying goal-driven higher-level control responsible for the decision to override Type 1 processing. Evans and Stanovich (2013b) see this framework as a paradigm or metatheory that can inspire task-level theories.

4.2 A Single-Function Dual-Process Model

Oaksford and Chater (2010, 2012) propose what they called a "single-function dual-process model" targeted at reasoning with conditionals. Type 1 processes in this model are the processes of spreading activation in a constraint-satisfaction network identified with long-term memory. Type 2 processes operate on both long-term memory and working memory and are responsible for strategic queries of long-term memory, for representing the outputs of these queries in working memory, and for operations on these working-memory contents.

Upon encountering a conditional sentence of the form "If p then q," a part of long-term memory is activated, including the nodes for the antecedent (p) and the consequent (q) as well as other representations strongly linked with them. Given a minor premise, such as the antecedent in a modus ponens inference, reasoners will clamp on the node representing the minor premise and monitor the flow of activation in the activated portion of long-term memory until an equilibrium is reached. The activation of the node representing the consequent is then read off and the result of the Ramsey-style test (Bennett, 2003) stored in working memory. This representation in working memory is actually conceived of as similar to the mental-model representations in Johnson-Laird's mental model theory (see chapter 2.3 by Johnson-Laird, this handbook).

Strategic queries of long-term memory allow the constraint-satisfaction network to compute activation levels for hypothetical and counterfactual premises. Type 2 processes are also responsible for undoing changes in long-term memory due to a previous query, and they can have a decoupling function in screening the representations in working memory from the intrusion of contextualized long-term memory information. This can result in processing that mimics logical thought as a limiting case of the function primarily computed, namely conditional probabilities. Nevertheless, the two kinds of processes serve a single function: to assess probabilities.

Type 1 and Type 2 processes thus differ on several of the features listed in table 2.5.1, most conspicuously in terms of whether working memory is required. Furthermore, although Oaksford and Chater (2010) do not explicitly say so, processes of spreading activation in long-term memory would presumably be nonconscious and, once triggered by stimuli or a Type 2 query, run to completion automatically. In contrast, Type 2 processes would probably be seen as intentional and conscious attempts to probe long-term memory and to represent and integrate mental-model-like representations in working memory. The interaction of the two processes is twofold: Type 1 processes determine which parts of long-term memory are activated, shaping the problem representation. Furthermore, the interaction can take the default-interventionist form in that Oaksford and Chater (2010) assume that a default query of long-term

memory can occur stimulus driven and provide a default assessment of the relevant probabilities that can however be modified by Type 2 queries of long-term memory to the point where the default assessment is completely inhibited.

4.3 A Dual-Strategy Model

Verschueren, Schaeken, and d'Ydewalle (2005a, 2005b) have proposed a dual-strategy model of conditional reasoning that was further developed by Markovits and colleagues (e.g., Markovits, Brunet, Thompson, & Brisson, 2013; see also chapter 15.1 by Markovits, this handbook). The idea is that people can use a combination of statistical and counterexample forms of reasoning. Markovits et al. (2013) align the statistical strategy with the assessment of probabilities similar to the stimulus-driven long-term memory queries considered by Oaksford and Chater (2010, 2012). The statistical strategy is to accept conclusions that are assessed as probable (somewhat like in accepting believable conclusions in belief-bias problems, discussed in more detail below). It is associated with little cognitive cost, operates fast, and delivers a default response. Given time and cognitive capacity, the statistical strategy can be replaced by a counterexample strategy in which a conclusion is rejected if a counterexample can be found. Counterexamples can be implied by the given information, but as in Oaksford and Chater's (2010, 2012) single-function dual-process model, counterexamples can also be generated via a strategic search of relevant background knowledge. The counterexample strategy is slower, and it comes with a higher cognitive cost. One difference to Oaksford and Chater's approach is that the two kinds of processes compute different functions, the former being sensitive to probabilistic information, the latter being driven by the existence or absence of at least one counterexample. The approach is thus a dual-function dual-process one.

4.4 A Dual-Source Model

Consider finally the dual-source model of probabilistic conditional reasoning by Klauer, Beller, and Hütter (2010; see also Singmann, Klauer, & Beller, 2016). The model distinguishes between two processes in terms of whether they derive response proposals based on background knowledge or based on the form of the argument. The model is not committed to further possible distinctions between the two types of processes. Instead, it provides a methodology and experimental paradigm to disentangle the contributions of both kinds of processes, making them and their features amenable to empirical study.

5. Evidence in Favor of and Against Dual-Process Models

5.1 Working-Memory Load, Speeded Responses, and Patterns of Interference

Evidence in favor of dual-process models of reasoning has focused on tasks such as those shown in figure 2.5.1. In a classical belief-bias task, participants are to judge the validity of valid and invalid syllogisms with believable and unbelievable conclusions. Belief bias is the finding that participants are more likely to accept as logically valid syllogisms with believable conclusions than syllogisms with unbelievable conclusions. As already mentioned, the assessment of believability has been argued to be based on a Type 1 process that is fast and requires little working-memory involvement, whereas logic-based responses require slower and working-memory-dependent Type 2 processes. In line with these ideas, both time pressure (Evans & Curtis-Holmes, 2005) and working-memory load (De Neys, 2006) decreased participants' ability to discriminate between valid and invalid syllogisms, whereas the effect of believability was increased (Evans & Curtis-Holmes, 2005) or not affected (De Neys, 2006).

On the other hand, Handley, Newstead, and Trippas (2011) also asked a group of participants to judge the believability of the conclusion and found (a) that under some conditions, judgments of believability took more time than judgments of logical validity and (b) that judgments of believability were interfered with by the validity of the syllogism in question: conclusions of valid syllogisms were judged more believable than conclusions of invalid syllogisms. This suggests a more symmetric role for the belief-based and the logic-based process than the studies just cited.

Let us take a closer look at the actual problems employed. Evans and Curtis-Holmes (2005) and De Neys (2006) employed problems with conclusions that were either believable or unbelievable. For example, a believable conclusion was "Some healthy people are not astronauts"; a (relatively more) unbelievable conclusion was "Some astronauts are not healthy people." Judging the believability of these conclusions is probably driven by a spontaneous, stimulus-triggered memory query of the kind described by Oaksford and Chater's (2010) dual-process models. In contrast, Handley et al. (2011, experiments 1 to 3) employed the belief-bias problems exemplified in figure 2.5.1 in which the conclusion (e.g., "The child is happy") is neither believable or unbelievable. It only acquires a value on the believability dimension once the minor premise ("The child is crying") is integrated with relevant conditional background

knowledge (such as "If a child is crying, then he or she is not happy"). Note that assessing the believability of the conclusion thereby requires almost the same processing steps as assessing its validity, the only difference being that the minor premise needs to be integrated with the explicitly stated conditional rule ("If a child is crying, then he is happy") in checking validity, whereas it needs to be integrated with conditional background knowledge in assessing believability. In short, assessing validity and assessing believability appear to have been equated in terms of processing demands, questioning their alignment with different types of processing. Note also that this issue does not afflict Handley et al.'s experiments 4 and 5, which are, however, subject to a similar alternative account in terms of aftereffects of switching between assessing believability and validity from trial to trial (Handley et al., 2011, p. 40). Switching is known to inflate existing, and sometimes newly create, crosstalk between the two alternating tasks (Allport, Styles, & Hsieh, 1994).

Similar ideas have been further developed in Chun and Kruglanski's (2006; see also Kruglanski & Gigerenzer, 2011) work on the base-rate task. In base-rate problems, participants are to decide which of two professional categories a person is more likely a member of. Two kinds of information are provided: first, participants are told that the person in question is randomly drawn from a sample of persons with skewed base rates for the two categories. For example, there might be 800 lawyers and 200 engineers. The second source of information is a description of traits and/or habits of the person such as that the person is not married, is introverted, and likes to spend his free time reading science fiction and writing computer programs. The second source is thus a description that is stereotypical of members of one of the two categories, engineers in the present example. Relying on the stereotypical information is usually considered heuristic, processing the base-rate information can be analytical. A typical finding is that the base-rate information is neglected and only the stereotypical information used (Kahnemann & Tversky, 1973).

Chun and Kruglanski (2006) argue, however, that the decisive difference between the two kinds of information is not related to the different contents (base rates versus stereotypes) but to how easy it is to process them. In general, equating how easy it is to process the two kinds of information should make their impact on judgments in the base-rate task more symmetrical, as Chun and Kruglanski (2006) demonstrate across several studies.

We argued that this partly explains Handley et al.'s (2011) results: the task demands of some of their belief-bias problems equated processing ease for judging validity and judging believability. Indeed, when more difficult logical problems are used, the effects of validity on judgments of belief decrease, as shown by Trippas, Thompson, and Handley (2017), although these studies are also afflicted by the abovementioned switching confound that is known to inflate crosstalk between the two alternating tasks.

A similar issue may be in place for many recent studies using the base-rate task (e.g., Bago & De Neys, 2017; De Neys, 2012, 2014; De Neys & Glumicic, 2008; Newman, Gibb, & Thompson, 2017; Pennycook, Fugelsang, & Koehler, 2015; Pennycook & Thompson, 2012; Pennycook, Trippas, Handley, & Thompson, 2014). In the "reduced" version shown in figure 2.5.1, the issue is especially clear: base-rate information and stereotypical information are both presented in one sentence. Both can suggest responses via simple rules or heuristics—to pick the larger category for the base-rate information and to pick the matching category for the stereotypical information. Again, ease of processing appears to be closely comparable for the two heuristics, questioning their alignment with different kinds of processing.

For example, Pennycook et al. (2015) erect a three-stage dual-process model on complex patterns of interferences in response times and response rates in the reduced base-rate task. But as pointed out by Krajbich, Bartling, Hare, and Fehr (2015), these patterns of interference can also come about through simpler one-process models in which the strength of preference (operationalized in terms of relative choice frequency) for one of the responses (the stereotypical response versus the base-rate response) and response time are related so that the preferred response is generally given faster the stronger the preference for it. Pennycook, Fugelsang, Koehler, and Thompson (2016) acknowledge this argument and point to aspects of their data pattern that may still be difficult to account for by this alternative account.

A number of these studies also used a two-response paradigm (e.g., Bago & De Neys, 2017) in which participants give a first response as quickly as possible followed by a second response without time pressure. One major finding is that most participants stick to their first responses, so that changes from first to second response are rare. This has been argued to question a number of dual-process models: if the first response relies on Type 1 processes, whereas Type 2 processes are involved in the second response, changes should regularly occur when both kinds of processes suggest different responses. One clue to what is happening here may be that participants appear to be very consistent in their responses to the problems (e.g., De Neys, 2017a, p. 56): across problems,

they either rely on base-rate (or in other cases logical) information or on stereotypical information, suggesting that participants decide and perhaps justify to themselves early which of these two sources of information to rely on. They then stick to this deliberate decision rather than reevaluate it for each problem anew. If both sources of information are easy to process, both strategies are likely already available for the first response.

In fact, when problems are used in which logic and believability, or base-rates and stereotypical information, are arguably less strongly equated in processing difficulty, changes are more frequent (Thompson, Prowse Turner, & Pennycook, 2011). In addition, in Thompson et al.'s (2011) studies, feeling-of-rightness (FOR) judgments for the first responses significantly predicted how often changes occur and how long it takes to produce the second response after the first. In a default-interventionist dual-process framework, this is consistent with the idea that FOR determines the likelihood and intensity of the engagement of Type 2 processes.

5.2 Correlational Evidence

Another source of evidence is individual-differences research. Stanovich and West (2000; see also chapter 15.2 by Stanovich, Toplak, & West, this handbook) in particular have investigated correlates of task performance in many tasks in which the modal response deviates from the one considered normative. In many of these tasks, measures of intelligence correlated positively with normative correctness of the response and negatively with giving the modal response. Importantly, predictions can be made for where to expect such correlations (e.g., when Type 2 processing is substantially involved in a version of the task) and where not (e.g., when Type 1 processing usually generates the correct response without conflict with Type 2 processing)—predictions that have been confirmed in several instances (Evans & Stanovich, 2013a, 2013b).

6. Criticisms of Dual-Process Models

6.1 Distinguishing Type 1 and Type 2 Processes: The Dichotomous and the Gradual View

Criticisms of dual-source models have addressed many aspects of dual-process models (Keren, 2013; Keren & Schul, 2009; Kruglanski & Gigerenzer, 2011). A number of problems identified by Keren and Schul (2009) and Keren (2013) arise from the fact that dual-process models have historically characterized the two types of processes in terms of arrays of attributes (see table 2.5.1) that are not perfectly correlated, an approach explicitly eschewed in recent theoretical expositions of dual-process theory (e.g., Evans & Stanovich, 2013a).

This reminds one of earlier debates about the definition of automaticity (for a review, see Moors & De Houwer, 2006): some definitions understood the contrast between automatic and controlled processes as a dichotomy (Shiffrin & Schneider, 1977) with different clusters of features characterizing automatic and controlled processes, respectively, much like exemplified in table 2.5.1 for Type 1 and Type 2 processes. Other definitions considered a continuum of attentional requirements with automatic processes at one end and nonautomatic processes at the other (e.g., Anderson, 1983). The dichotomous view was challenged by studies finding lack of correlation among central features ascribed to automatic processes, such as that they are efficient, unintentional, uncontrollable, and unconscious (Bargh, 1992). Such challenges led to a decompositional view according to which different automaticity features are to be studied separately and independently, suggesting in the end the deconstruction of the concept of automaticity.

Gradual views, on the other hand, maintain the concept and align it with a transition from algorithmic computation to single-step memory retrieval (Logan, 1988) or algorithmic strengthening in skill acquisition (Anderson, 1983). Logan (1985) argued that the different automaticity features have different time courses of change with practice, and thus lack of co-occurrence among the features is no surprise. The gradual view would only be challenged if different features were found to develop in different directions as practice increases. Whereas the gradual view makes it difficult to draw a line distinguishing automatic from nonautomatic processes, another response was to raise one dichotomous feature to a definitional status and acknowledge that other features might, but need not, co-occur with it (Bargh, 1992).

Similar arguments and moves can be found in debates on dual-process models. For example, some, although not all, of the features in table 2.5.1 are inherently gradual and define processing continua rather than qualitative distinctions (Keren, 2013; Keren & Schul, 2009). Involvement of working-memory resources, a feature highlighted by Evans and Stanovich (2013a), clearly comes in degrees, and it is notoriously difficult to distinguish between small and zero involvement. In a similar vein, Kruglanski and Gigerenzer (2011; see also Chun & Kruglanski, 2006) acknowledge that different processes differ on many qualitative aspects such as the contents that they operate on but that graded differences in processing ease and in perceived ecological validity may be responsible for relatively greater impacts

on the ultimate response of some processes than others. Cognitive decoupling and mental simulations, another distinguishing feature highlighted by Evans and Stanovich (2013a), as well as the requirement to provide reasons (Mercier & Sperber, 2009), may be more difficult to construe as gradual distinctions, depending upon how exactly these features are defined and operationalized, and they may serve as defining features if a dichotomous view is preferred.

6.2 The Relative View

Alternatively, the conceptualization of dual-process models could also be based on a relative view as proposed by Moors and De Houwer (2006) for the distinction between automatic and controlled processes. This is a refinement of the gradual view in which a standard of comparison is introduced. In terms of dual-process models, a process would be called Type 1 relative to a standard of comparison appropriate to a given task context, whereas in another context, the same process might be classified as Type 2 relative to a standard of comparison that is prevalent in that context. For example, in the context of the Stroop task, a word-reading and a color-naming process are involved. When the task is to name the color in which a word is written, uninstructed, spontaneous reading of the word interferes with naming the word's color if the word itself denotes a different color. In contrast, if the task is to read the word, there is no analogous interference from the word's color in conflict trials (MacLeod, 1991). To account for this asymmetry, word reading has been considered relatively more automatic (Type 1) and color naming relatively more controlled (Type 2; Lindsay & Jacoby, 1994). In the tasks studied in the thinking-and-reasoning literature, both processes would, however, probably be considered Type 1 processes relative to the resource-depleting reasoning algorithms demanded by many reasoning tasks. Similarly, assessing believability in belief-bias tasks may normally tax cognitive resources so much less than assessing logical validity that the relative view would designate the former process Type 1 and the latter Type 2. If and when assessing believability is, however, made more difficult, and assessing logical validity easier, thereby equating both in terms of processing demands (as in Handley et al., 2011, experiments 1 to 3), the gradual view would deny such a designation in this context.

7. Conclusion

Keren (2013) argues that single-process accounts may be offered for a number of dual-process phenomena (see also Chun & Kruglanski, 2006; Krajbich et al., 2015; Kruglanski, 2013). Proponents of dual-process models need to defend their theories against this criticism, but the criticism is targeted at specific instances of dual-process models for specific tasks. At a more abstract level, dual-process models have proven to provide an immensely fruitful framework for integrating a diversity of related findings. They serve as a template for devising specific accounts of tasks of interest, providing a rich toolbox of concepts and process assumptions (Evans & Stanovich, 2013b). The framework has thereby generated an enormous amount of research producing new tasks and revealing new phenomena that might not have been uncovered without it, many of them summarized in the recent book by De Neys (2017b). There is little doubt that the core ideas and concepts of dual-process models will continue to be influential and to open up new and exciting avenues of research.

References

Allport, A., Styles, E. A., & Hsieh, S. (1994). Shifting intentional set: Exploring the dynamic control of tasks. In C. Umiltà & M. Moscovitch (Eds.), *Attention and performance XV: Conscious and nonconscious information processing* (pp. 421–452). Cambridge, MA: MIT Press.

Anderson, J. R. (1983). *The architecture of cognition.* Cambridge, MA: Harvard University Press.

Bago, B., & De Neys, W. (2017). Fast logic? Examining the time course assumption of dual process theory. *Cognition, 158,* 90–109.

Bargh, J. A. (1992). The ecology of automaticity: Toward establishing the conditions needed to produce automatic effects. *American Journal of Psychology, 105,* 181–199.

Bennett, J. (2003). *A philosophical guide to conditionals.* Oxford, England: Clarendon Press.

Buchak, L. (2013). *Risk and rationality.* Oxford, England: Oxford University Press.

Chun, W. Y., & Kruglanski, A. W. (2006). The role of task demands and processing resources in the use of base-rate and individuating information. *Journal of Personality and Social Psychology, 91,* 205–217.

Cosmides, L. (1989). The logic of social exchange: Has natural selection shaped how humans reason? Studies with the Wason selection task. *Cognition, 31,* 187–276.

De Neys, W. (2006). Dual processing in reasoning. *Psychological Science, 17,* 428–433.

De Neys, W. (2012). Bias and conflict: A case for logical intuitions. *Perspectives on Psychological Science, 7,* 28–38.

De Neys, W. (2014). Conflict detection, dual processes, and logical intuitions: Some clarifications. *Thinking & Reasoning, 20,* 169–187.

De Neys, W. (2017a). Bias, conflict, and fast logic: Towards a hybrid dual process future? In W. De Neys (Ed.), *Dual process theory 2.0* (pp. 47–65). Abingdon, England: Routledge.

De Neys, W. (Ed.). (2017b). *Dual process theory 2.0.* Abingdon, England: Routledge.

De Neys, W., & Glumicic, T. (2008). Conflict monitoring in dual process theories of thinking. *Cognition, 106,* 1248–1299.

Evans, J. St. B. T. (1996). Deciding before you think: Relevance and reasoning in the selection task. *British Journal of Psychology, 87,* 223–240.

Evans, J. St. B. T. (2007). On the resolution of conflict in dual process theories of reasoning. *Thinking & Reasoning, 13*(4), 321–339.

Evans, J. St. B. T. (2009). How many dual-process theories do we need? One, two, or many? In J. St. B. T. Evans & K. Frankish (Eds.), *In two minds: Dual processes and beyond* (pp. 33–54). New York, NY: Oxford University Press.

Evans, J. St. B. T. (2014). Two minds rationality. *Thinking & Reasoning, 20*(2), 129–146.

Evans, J. St. B. T., & Curtis-Holmes, J. (2005). Rapid responding increases belief bias: Evidence for the dual-process theory of reasoning. *Thinking & Reasoning, 11,* 382–389.

Evans, J. St. B. T., & Stanovich, K. E. (2013a). Dual-process theories of higher cognition: Advancing the debate. *Perspectives on Psychological Science, 8,* 223–241.

Evans, J. St. B. T., & Stanovich, K. E. (2013b). Theory and metatheory in the study of dual processing: Reply to comments. *Perspectives on Psychological Science, 8,* 263–271.

Evans, J. St. B. T., & Wason, P. C. (1976). Rationalization in a reasoning task. *British Journal of Psychology, 67,* 479–486.

Gigerenzer, G., & Todd, P. M. (1999). Fast and frugal heuristics: The adaptive toolbox. In G. Gigerenzer, P. M. Todd, & ABC Research Group (Eds.), *Simple heuristics that make us smart* (pp. 3–34). New York, NY: Oxford University Press.

Handley, S. J., Newstead, S. E., & Trippas, D. (2011). Logic, beliefs, and instruction: A test of the default interventionist account of belief bias. *Journal of Experimental Psychology: Learning, Memory, and Cognition, 37*(1), 28–43.

Kahneman, D. (2011). *Thinking, fast and slow.* New York, NY: Farrar, Straus and Giroux.

Kahneman, D., & Tversky, A. (1973). On the psychology of prediction. *Psychological Review, 80,* 237–251.

Keren, G. (2013). A tale of two systems: A scientific advance or a theoretical stone soup? Commentary on Evans & Stanovich (2013). *Perspectives on Psychological Science, 8,* 257–262.

Keren, G., & Schul, Y. (2009). Two is not always better than one: A critical evaluation of two-system theories. *Perspectives on Psychological Science, 4,* 533–550.

Klauer, K. C., Beller, S., & Hütter, M. (2010). Conditional reasoning in context: A dual-source model of probabilistic inference. *Journal of Experimental Psychology: Learning, Memory, and Cognition, 36,* 298–323.

Klauer, K. C., Stahl, C., & Erdfelder, E. (2007). The abstract selection task: New data and an almost comprehensive model. *Journal of Experimental Psychology: Learning, Memory, and Cognition, 33*(4), 680–703.

Krajbich, I., Bartling, B., Hare, T., & Fehr, E. (2015). Rethinking fast and slow based on a critique of reaction-time reverse inference. *Nature Communications, 6.*

Kruglanski, A. W. (2013). Only one? The default interventionist perspective as a unimodel—Commentary on Evans & Stanovich (2013). *Perspectives on Psychological Science, 8,* 242–247.

Kruglanski, A. W., & Gigerenzer, G. (2011). Intuitive and deliberate judgments are based on common principles. *Psychological Review, 118,* 97–109.

Lea, R. B. (1995). On-line evidence for elaborative logical inferences in text. *Journal of Experimental Psychology: Learning, Memory, and Cognition, 21*(6), 1469–1482.

Lea, R. B., O'Brien, D. P., Fisch, S. M., Noveck, I. A., & Braine, M. D. S. (1990). Predicting propositional logic inferences in text comprehension. *Journal of Memory and Language, 29,* 361–387.

Lindsay, D. S., & Jacoby, L. L. (1994). Stroop process dissociations: The relationship between facilitation and interference. *Journal of Experimental Psychology: Human Perception and Performance, 20,* 219–234.

Logan, G. D. (1985). Skill and automaticity: Relations, implications, and future directions. *Canadian Journal of Psychology/ Revue canadienne de psychologie, 39,* 367–386.

Logan, G. D. (1988). Toward an instance theory of automatization. *Psychological Review, 95,* 492–527.

MacLeod, C. M. (1991). Half a century of research on the Stroop effect: An integrative review. *Psychological Bulletin, 109,* 163–203.

Markovits, H., Brunet, M.-L., Thompson, V., & Brisson, J. (2013). Direct evidence for a dual process model of deductive inference. *Journal of Experimental Psychology: Learning, Memory, and Cognition, 39,* 1213–1222.

Mercier, H., & Sperber, D. (2009). Intuitive and reflective inferences. In J. St. B. T. Evans & K. Frankish (Eds.), *In two minds: Dual processes and beyond* (pp. 149–170). New York, NY: Oxford University Press.

Mercier, H., & Sperber, D. (2017). *The enigma of reason: A new theory of human understanding.* Cambridge, MA: Harvard University Press.

Moors, A., & De Houwer, J. (2006). Automaticity: A theoretical and conceptual analysis. *Psychological Bulletin, 132,* 297–326.

Newman, I. R., Gibb, M., & Thompson, V. A. (2017). Rule-based reasoning is fast and belief-based reasoning can be slow: Challenging current explanations of belief-bias and base-rate neglect. *Journal of Experimental Psychology: Learning, Memory, and Cognition, 43,* 1154–1170.

Oaksford, M., & Chater, N. (2010). Conditionals and constraint satisfaction: Reconciling mental models and the probabilistic approach? In M. Oaksford & N. Chater (Eds.), *Cognition and conditionals: Probability and logic in human thinking* (pp. 309–334). Oxford, England: Oxford University Press.

Oaksford, M., & Chater, N. (2012). Dual processes, probabilities, and cognitive architecture. *Mind & Society, 11,* 15–26.

Pennycook, G., Fugelsang, J. A., & Koehler, D. J. (2015). What makes us think? A three-stage dual-process model of analytic engagement. *Cognitive Psychology, 80,* 34–72.

Pennycook, G., Fugelsang, J. A., Koehler, D. J., & Thompson, V. A. (2016). Commentary: Rethinking fast and slow based on a critique of reaction-time reverse inference. *Frontiers in Psychology, 7,* Article 1174.

Pennycook, G., & Thompson, V. A. (2012). Reasoning with base rates is routine, relatively effortless, and context dependent. *Psychonomic Bulletin & Review, 19,* 528–534.

Pennycook, G., Trippas, D., Handley, S. J., & Thompson, V. A. (2014). Base rates: Both neglected and intuitive. *Journal of Experimental Psychology: Learning, Memory, and Cognition, 40,* 544–554.

Rader, A. W., & Sloutsky, V. M. (2002). Processing of logically valid and logically invalid conditional inferences in discourse comprehension. *Journal of Experimental Psychology: Learning, Memory, and Cognition, 28,* 59–68.

Shiffrin, R. M., & Schneider, W. (1977). Controlled and automatic human information processing: II. Perceptual learning, automatic attending and a general theory. *Psychological Review, 84,* 127–190.

Singmann, H., Klauer, K. C., & Beller, S. (2016). Probabilistic conditional reasoning: Disentangling form and content with the dual-source model. *Cognitive Psychology, 88,* 61–87.

Stanovich, K. E. (2004). *The robot's rebellion: Finding meaning in the age of Darwin.* Chicago, IL: University of Chicago Press.

Stanovich, K. E. (2009). Distinguishing the reflective, algorithmic, and autonomous minds: Is it time for a tri-process theory? In J. St. B. T. Evans & K. Frankish (Eds.), *In two minds: Dual processes and beyond* (pp. 55–88). New York, NY: Oxford University Press.

Stanovich, K. E., & West, R. F. (2000). Individual differences in reasoning: Implications for the rationality debate. *Behavioral and Brain Sciences, 23,* 645–726.

Thompson, V. A. (2009). Dual-process theories: A metacognitive perspective. In J. St. B. T. Evans & K. Frankish (Eds.), *In two minds: Dual processes and beyond* (pp. 171–196). New York, NY: Oxford University Press.

Thompson, V. A. (2013). Why it matters: The implications of autonomous processes for dual-process theories—Commentary on Evans & Stanovich (2013). *Perspectives on Psychological Science, 8*, 253–256.

Thompson, V. A., Prowse Turner, J. A., & Pennycook, G. (2011). Intuition, reason, and metacognition. *Cognitive Psychology, 63*(3), 107–140.

Trippas, D., Thompson, V. A., & Handley, S. J. (2017). When fast logic meets slow belief: Evidence for a parallel-processing model of belief bias. *Memory & Cognition, 45*(4), 539–552.

Verschueren, N., Schaeken, W., & d'Ydewalle, G. (2005a). A dual-process specification of causal conditional reasoning. *Thinking & Reasoning, 11*(3), 239–278.

Verschueren, N., Schaeken, W., & d'Ydewalle, G. (2005b). Everyday conditional reasoning: A working memory–dependent tradeoff between counterexample and likelihood use. *Memory & Cognition, 33*(1), 107–119.

Wason, P. C. (1966). Reasoning. In B. M. Foss (Ed.), *New horizons in psychology* (Vol. 1, pp. 135–151). Harmondsworth, England: Penguin.

Wilkins, M. C. (1928). *The effect of changed material on ability to do formal syllogistic reasoning* (Archives of Psychology No. 102). New York, NY: Woodworth.

2.6 Logico-Computational Aspects of Rationality

Johan van Benthem, Fenrong Liu, and Sonja Smets

Summary

Taking a broad historical line, we discuss major aspects of rationality that can be analyzed in a logical and computational perspective. Topics include classical notions from the foundations of computability, insights from the development of computer science and AI, and the richer picture of rationality emerging in current logical studies of agency. We also discuss two challenges: distributed computing and evolutionary games replace "high rationality" by "low rationality" in behavior; machine learning defies classical views of representation and inference. Both trigger new logical themes in the study of rationality.

1. Logic and Rationality

Rational behavior is a rich phenomenon, not captured in a single formula, but mapped out in this entire handbook. Let us take the commonsense view. We all believe, or try to believe, that our behavior is driven by reasons and reasoning and that we are susceptible to reason, changing our minds when confronted with new facts or considerations. This is how we see ourselves, how we justify our actions to others, and how academic organizations present themselves to a general public. Further important aspects of rationality, such as preferences and goals, will only be touched upon lightly in this chapter.

Since Antiquity, reasoning has been at the heart of logic. In fact, logic is often seen as rationality in its purest, perhaps its most intimidating, form. We will chart this match, without claiming logic is all there is to rationality.

Example: Valid and invalid consequence. Classical logic tells us things like this: an inference to $\neg B$ from $A \to B$ and $\neg A$ is invalid (B might hold for other reasons than A)—but the inference from $A \to B$ and $\neg B$ to $\neg A$ is valid and in fact the engine of refutation. Valid inferences, put together, form complex proofs that can yield surprising new insights. Over time, studying all this has

produced a rich semantic and syntactic discipline of logical systems. ⊣

Before we start, here is a distinction. Logical proof can be seen as a practical engine of rational behavior, and different logical systems can model rational reasoning practices. But there is also a theoretical foundational use of logic, as a study of the structure of rationality: its laws and its limitations. In this second sense, logical analysis can target any practice that rational agents engage in: not just reasoning but also observing, taking decisions, or debating. Both uses, practical and theoretical, will occur in what follows.

But here is a further step. Logical systems for analyzing reasoning are cultural artifacts that interact with human practices. In particular, they feed into computational devices. For instance, the valid inference from $A \to B$ and $\neg B$ to $\neg A$ is also a basic law of binary arithmetic, at work in one's computer. This association has proved fruitful in the foundations of mathematics, and practically, it has triggered the development of computers, information technology, and artificial intelligence, which are transforming our world. Some even fear that our logical tools have started overtaking us.

A full picture of reasoning requires that "us." Many themes in this chapter connect naturally to empirical cognitive psychology. This interface is beyond our scope, and we refer the reader to the empirical entries in this handbook. Instead, we proceed to the logico-computational perspective on rationality.

2. Mini-History of Reasoning and Computation

To understand the links between logic and rationality, a historical perspective is helpful (cf. Kneale & Kneale, 1962). Over time, many forms of reasoning were captured in logical systems by philosophers, mathematicians, and others—a process that is still continuing. Once discovered, these systems became intellectual tools that enhanced rational thinking. Then, in the early modern age, Llull and Leibniz realized that reasoning is close to computation. From there runs a straight road to the logic machines of Babbage and Lovelace and onward to modern computers and artificial intelligence (AI) systems. On the way, the notion of computation acquired sharper contours. A computing device need not be tied to one specific task; it can be programmable, the way a loom can weave different textiles depending on its book. Thus, computing means finding algorithms performing tasks, and since algorithms work on code, it also means finding data structures that represent information in appropriate ways.

All this is similar to the reality of logic itself. Textbooks say that logic is the study of inference or reasoning, but much more is involved. Reasoning presupposes a vehicle, often a language, representing the notions one reasons with. Thus, as noted in the perceptive essay (Beth 1971), logic has always been about a tandem of proof and definition or, if you wish, proof theory and model theory. Moreover, Beth added as a crucial third historical constant of logical thought the notion of algorithm, which combines the former two.

The history of computing is remarkable in that major principles were discovered before practical success, unlike in many other fields. Gödel (1931/1986) analyzed the limits of what logical proof systems can achieve, finding that systems whose expressive power suffices for encoding basic arithmetic are either inconsistent or incomplete: unable to prove all intuitive mathematical truths about their domain. Gödel's proof involved a deep analysis of computable ("recursive") functions, and subsequently, Turing (1936) defined machines that can compute all recursive functions. There is even a universal Turing machine that, given any program code and input, computes the effects of running that program on that input. In this setting, Church (1936) showed that standard reasoning systems such as first-order logic, although axiomatizable, are undecidable: no computing method can decide for arbitrary first-order consequence problems whether they are valid. Thus, a major trade-off came into view: increased expressivity of a language and complexity of its decision problem for validity are at odds.

This history highlights several points that seem crucial to understanding the nature and scope of rationality even today. The first is a practical issue of the modus operandi: if rationality has a computational engine, how should we understand its interplay of reasoning, information, and concept formation? The second point is theoretical: are there principled limitations to logical rationality—say, are some natural tasks beyond the scope of rational inquiry? Gödel's theorems keep generating discussion (Wang, 1996), and a common moral is that there is more

to rational thinking than what is captured in logical systems. Even so, whenever this "more" is explained systematically, the limitation theorems apply again. Finally, results on what proof systems and computing devices cannot do were in fact immensely helpful in the further development of systems of inference and computation that can do a lot. Likewise, the modern "challenges to rationality" discussed in this chapter may actually yield new insights into what rationality can achieve. Having said that, much current literature in AI or cognitive psychology is of the "can do" type: one seldom reads about exciting discoveries of deep new limitations.

The foundational era of Gödel and Turing showed what is provable or computable in principle. While this high abstraction level remains a valid perspective on rationality, subsequent history tells us many further things.

3.   Computer Science and Artificial Intelligence

3.1   Computer Science

The development of computing in the 20th century has generated major practical achievements but also an ever-growing insight into fine structure. There are different models for computation, from Turing machines to more recent formats for computing, and crucially, these models come in hierarchies. Some tasks are solved by simple finite automata; others require memory management to varying degrees. Likewise, there is a wide diversity of, poorer or richer, languages for specifying data structures and writing programs (Harel, 1987). Next, significantly, around 1980, computing architecture moved away from single Turing machines to distributed networks, the reality of computing today (Andrews, 2000). These developments have given rise to new fields such as automata theory (Chakraborty, Saxena, & Katti, 2011), complexity theory (Papadimitriou, 1994), and process algebra (Bergstra, Ponse, & Smolka, 2001), which chart the varieties of computation in different ways, many of them connected to logic. This process is still ongoing, and the foundations of computation remain under debate. For instance, there is no consensus on a definitive notion of algorithm, a more intensional notion than an extensional input–output record of Turing machines (Bonizzoni, Brattka, & Löwe, 2013; Haugeland, 1997). Computation today is rather a way of producing behavior.

All this comes with notions and insights that are relevant to rationality. The fine structure of computation gives a precise sense to the earlier double-edged modus operandi: reasoning engine and representational apparatus. And the variety in real computation suggests that the norm in rational performance is not one idealized super-device but "boundedness" of resources and powers. Matching this practical concern is a fundamental issue. In the complexity theory of space and time resources needed for computational tasks, we are really talking about the information in the world and how to process it. But this forces us to reflect on what is the information available to rational agents (Adriaans & van Benthem, 2008). And there is yet more to be learnt from the world of computing. If we think of the behavior of rational agents as performing many tasks at once, just as networks of computers do, then there is a fundamental issue of architecture. How do the different components of the overall system pass information and cooperate (Gabbay, 1999)?

3.2   Artificial Intelligence

Moving closer to humans, computer science blends seamlessly into AI. From the start, computers have been seen as a powerful model for human intelligence. In an interesting departure from the detailed internal analysis of computing by Turing machines, the famous Turing Test approaches intelligence in the tradition of measuring theoretical notions by external observable behavior (Turing, 1950). It proposes that a computer achieves intelligence if an observer using natural language cannot tell that computer apart from a human by asking questions and engaging in conversation. Over the years, computers started to pass variants of the Turing Test, or other types of intelligent behavior. Actually, none of these are usually considered conclusive, as the criteria are a moving target (van Harmelen, Lifschitz, & Porter, 2008). Passing the test is dismissed as not a display of "real intelligence," and then the demands are shifted a little further. But behavioral tests are crucial to judging human rationality as well. We seldom look inside people's heads to monitor their considerations but rather observe their words and actions.

A final intriguing feature of the Turing Test is its hybrid scenario where different types of agents, humans and machines, interact, presaging the reality of human–machine interactions in modern society. This scenario goes beyond classical emulation or competition concerns. How can societies of mixed agents, with different strengths and weaknesses, interact successfully (Wooldridge, 2002/2009)? The resulting diversity in agency is only beginning to be acknowledged more widely. Most paradigms in logic or philosophy assume that agents have similar abilities for reasoning, observing, and communicating, although their information and preferences may differ. Such uniformity assumptions underlie generic notions like "human"

or "rational actor," and they may even seem to embody moral imperatives, like treating everyone equally *qua* rights and duties. If one accepts diversity, however, notions and theories concerning rationality must be rethought.

The above trends in AI and computer science considerably extend the logical agenda for studying rationality. A major view of computation from the 1980s onward is that of behavior by agents (van Benthem, 2018), studied by merging ideas from computational and philosophical logic (Gabbay & Guenthner, 1983–; Gabbay, Hogger, & Robinson, 1993–1998). After explaining what is involved in such agency in section 4, concrete examples will be given in sections 5 and 6, showing logic at work in this modern setting. However, the logic-oriented agency approach has not gone unchallenged. In section 7, we will discuss the "high" versus "low rationality" competition in understanding social agency and in section 8 the rise of nonrepresentational machine-learning techniques. Both come with a greater emphasis on probability, the other main formal paradigm for studies of rationality. We will end with an assessment of the current landscape of logico-computational approaches to rationality.

4.   From Machines to Rich Rational Agents

4.1   A Conceptual Catalogue

Let us first think of what agents can do in general. Human agents have a much wider range of rational activities than just reasoning from given data, that is, elucidating what was already implicitly there. A rich dynamic information flow guides action. Agents constantly pick up new information from their environment by means of observation and communication, and they search their memory for information, too. Rationality is about picking up relevant information from whatever source is available, as much as reasoning, and that both in daily life and in science.

However, even rich processing of correct information is just one dimension. Information can be less or more reliable, and agents do not just accumulate knowledge but also form beliefs that can be shown wrong by new information. Thus, robustly rational agents are not those who are always correct but those who learn from errors and have a talent for correcting themselves (Kelly, 1996; Popper, 1963). Many facets of belief revision and learning are dealt with in chapter 5.2 by Rott; chapter 5.3 by Kern-Isberner, Skovgaard-Olsen, and Spohn; and chapter 5.4 by Gazzo Castañeda and Knauff (in this handbook). Logic is a major supplier of models here (van Benthem & Smets, 2015).

And rational agency does not stop here. Truly rational agents maintain a harmony between their information and beliefs with, on a par, their preferences, goals, or intentions. One can discuss which harmony is essential, whether maximizing expected utility (see chapter 8.2 by Peterson, this handbook) or some other option: logic cannot, and should not, decide. Real agents may differ widely in their distance from classical decision-theoretic views (cf. chapter 8.3 by Glöckner, this handbook), but the crucial point of rationality remains maintaining a workable balance between information, goals, and actions.

Summing up, a rational agent can gather information in a variety of ways, integrating observation, inference, and communication (van Benthem, 2011). In this process, the agent can form a rich variety of attitudes, from knowledge and belief to rejection or doubt. Also, the agent can function in environments where beliefs turn out wrong and learns from errors. And all this maintains a purpose, a balance between the agent's goals and its information or actions. And finally, even this is not yet a full picture: rational agents display their skills in social interaction, a topic that will return.

4.2 Multiagent Systems
Bits and pieces of this richer notion of rational agency have long been studied by philosophers and logicians: witness the *Handbook of Philosophical Logic* (Gabbay & Guenthner, 1983–). In the 1980s, a congenial picture of agency emerged in computer science and AI, namely in the study of multiagent systems (Fagin, Halpern, Moses, & Vardi, 1995; Shoham & Leyton-Brown, 2008; Wooldridge, 2010). This reflected a shift in thinking about computing, from machines to agents with a behavior analyzed in terms of features normally ascribed to humans. This extends to autonomous systems. Robots investigate their environment with sensors, decide, and act in performing their tasks (Cardon & Itmi, 2016). Again, tools from philosophical logic make sense (cf. Brafman, Latombe, Moses, & Shoham, 1997, on epistemic specifications for real robots, whose sensors have a margin of error). But conversely, notions from multiagent systems can be found in modern epistemology (Arló-Costa, Hendricks, & van Benthem, 2016). For instance, robots acting on evidence of varying quality have inspired new models for evidence-based belief (van Benthem & Pacuit, 2011).

4.3 Games
There is a natural confluence here with one more discipline. Agents that acquire information, choose actions, and pursue goals are like players in games. Indeed,

computer science has drawn closer to game theory (Nisan, Roughgarden, Tardos, & Vazirani, 2007), and logic, with its connections to games of argumentation (cf. chapter 5.5 by Hahn & Collins and chapter 5.6 by Woods, both in this handbook) and information seeking (Hintikka, 1973), is a natural partner. Indeed, epistemic game theory (see chapter 9.2 by Perea, this handbook) can be seen as a venture created by these contacts.

Even with all this, no canonical view has crystallized yet of what a rational agent is and does that would be similar in elegance and fertility to that of a computing machine, let alone of a "universal rational agent" comparable in its sweep to the universal Turing machine. In fact, the earlier recognition of diversity suggests a focus on different kinds of rational agents rather than uniqueness, changing standard assessments of performance. Is a rational agent someone who wields vast cognitive powers or someone doing their best with limited powers? Are rational agents those who perform well against other rational agents or those who cope with a large bandwidth of types of agents in their environment?

5. Logical Models of Rational Agency

In recent decades, many features of rational agents have been studied by logicians. Information update and knowledge change occur in temporal logics of agency (Belnap, Perloff, & Xu, 2001; Fagin et al., 1995; Parikh & Ramanujam, 2003). Another paradigm is dynamic epistemic logic (Baltag, Moss, & Solecki, 1998; van Ditmarsch, van der Hoek, & Kooi, 2007), which models processes whereby agents form and modify representations of the information at their disposal. Such updates are not inferences, but they can be described just as precisely in logical terms.

Example: Dynamics of information flow. In a simple two-party dialogue, agent 1, who is uncertain about the truth or falsity of *p*, asks "*p*?" A second agent then truthfully and publicly replies, "Yes." Analyzing the information flow in the dialogue, the first agent's question conveys that she doesn't know the answer but also that she thinks the second agent, a fully reliable source, does know. The second agent's answer conveys that 1's assumption about 2's knowledge was correct, and also, after 2's answer, *p* is common knowledge in the group of these two agents. ⊣

This mixture of knowledge of facts and knowledge about others is typical for communication. More complex scenarios, such as the famous Muddy Children puzzle (Fagin et al., 1995), illustrate how even truthful public announcements of ignorance following consecutive questions can gradually lead agents to knowledge.

A symbolic language capturing all of this has formulas [!φ]Kψ saying that the agent will know ψ after a public event carrying the information that φ is true. A key law in this logic of public announcements is the equivalence

$$[!\varphi]K\psi \leftrightarrow (\varphi \to K[!\varphi]\psi)$$

This relates new knowledge after the event φ happened to its "preencoding" before the event: the agent had conditional knowledge that the event !φ would result in the truth of ψ. Such logical laws interchange dynamic operators for events and epistemic operators for attitudes of agents, a crucial ingredient in understanding information update and knowledge change.

Logics of belief change under hard and soft information have been developed in the same style (cf. van Benthem & Smets, 2015), and other formal approaches, such as "AGM," occur in chapter 5.2 by Rott (this handbook). Learning fits well with belief revision: connections between dynamic logics of knowledge and belief, on the one hand, and formal learning theory, on the other, are found in Baltag, Gierasimczuk, and Smets (2011).

Example: Belief change. In the above two-party dialogue, now assume that agent 2 is *not* a fully reliable information source. Starting from the initial situation, in which agent 1 believes neither p nor ¬p, the answer of 2 to her question can trigger 1 to change her mind and, possibly, to adopt a wrong belief. Yet how exactly she changes her mind depends on the trust that 1 has in 2 as an information source about p. If 2's answer "Yes" is considered to be reliable but not infallible, the "belief upgrade" that it triggers can be more radical, inducing a strong belief in p, or more conservative, inducing a weak belief in p.   ⊣

Again, this process obeys logical laws. A formal language now has constructs [⇑φ] for effects of conservative upgrades and [⇑φ] for radical upgrades. This brings to light many principles of belief change. For instance, ¬K¬φ → [⇑φ]Bφ says, for factual statements φ, that the agent comes to believe that φ is the case, unless she already knew before the announcement that φ was false. In logical studies of learning, one studies iterations of such upgrades and analyzes how well they perform as a learning method.

The study of key features of rationality in this style continues. Further aspects of purposeful rational behavior brought into the scope of logic include the management of current "issues" that guide inquiry for tasks at hand (cf. the "inquisitive logics" of Ciardelli, Groenendijk, & Roelofsen, 2019).

Next, moving from informational tasks to agents' preferences and goals, which determine how they evaluate situations, Liu (2011) studies logics of preference change, which is connected to goal dynamics. Preference dynamics shows similarities with deontic logics, describing what is obligatory and permitted for agents in environments where new commands change the moral ordering of situations and actions (Yamada, 2008).

Finally, rational agents balance their information with preferences and goals. Logics combining all these features occur in influential frameworks for multiagent systems such as BDI (Rao & Georgeff, 1991), inspired by Bratman (1987), describing how agency is driven by a balance of beliefs, desires, and intentions. But perhaps the most active research area where all these features of rational agents meet is in the logical study of strategic behavior and equilibria in games (cf. chapter 9.1 by Albert & Kliemt, this handbook).

Example: Reasoning about extensive games. Consider a finite game with two players, A and E, and outcome-values written in the order A-value—E-value (see figure 2.6.1).

Intuitively, the outcome (99, 99) seems best for both players—but that is not what the standard game solution algorithm of backward induction yields. Looking from the bottom to the top, if E is to play, she will choose left, and so, if A believes that E will make this choice, she herself will play left in the first round and end the game, with outcome (1, 0).   ⊣

Analyzing why this game might play out in this non-Pareto-optimal manner involves many notions, including players' actions, beliefs, preferences, and plans. All of these have been studied by logicians, in different settings or as separate topics. For precise definitions of equilibria in games and a survey of logical game analysis, see van Benthem and Klein (2019).

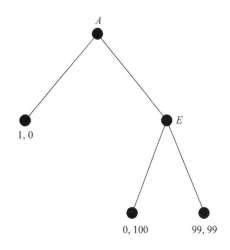

Figure 2.6.1

An extensive game.

All dynamic logics discussed here exemplify the earlier-mentioned tandem of algorithm and data in computation. The events that produce new information or new desires operate on well-chosen static models that support attitudes of knowledge, belief, preference, and the like.

Digression: Nonclassical logics. There are also approaches folding all of the above activities under varieties of inference, emphasizing departures from classical consequence to nonclassical, nonmonotonic logics (Horty, 2012) and resource-conscious substructural logics (Restall, 2000). For a comparison of the two methodologies, see van Benthem (2019).

Example: Nonmonotonic reasoning. Consequence in classical logic is monotonic; new premises do not invalidate earlier conclusions: if $\Gamma \vDash \varphi$ and $\Gamma \subseteq \Delta$, then $\Delta \vDash \varphi$. In contrast, default inferences are defeasible (cf. chapter 5.2 by Rott, this handbook): if I know that Tweety is a bird, I can conclude that Tweety can fly, yet with a further premise that Tweety is a penguin, it no longer follows that Tweety can fly. $\dashv$

The field of nonmonotonic logic studies properties of default reasoning. In contrast, dynamic logics of belief revision capture default phenomena on a classical base, locating the nonmonotonicity in belief change rather than in changing the inference rules: I believed that Tweety can fly, but after an event !*Penguin*(*Tweety*), I have lost that belief.

Discussion. The logical study of ever more aspects of agency aims at a not exclusively behavioral view of rationality by identifying key internal features and mechanisms. But combining logical and computational agendas does not make logical systems realistic software agents or human agents. Far more is needed for algorithms to work, and implementation requires further syntax. Recent studies mediate between semantic models and syntactic representations for computing agents (Halpern & Rêgo, 2013; Lorini, 2018).

Also, the development of ever richer models raises questions. Where is the boundary of agency, as more and more topics are taken on board, and what is "rational" about the activities so described? Are we describing what agents actually do, or are these logical systems normative? A common view holds that logics of agency describe idealized laws that may or may not be followed by actual agents. This tension may be just what is needed. We cannot say, for instance, that belief revision leads to "correction" of earlier beliefs unless we have a norm for what is correct in the given circumstances. These are big issues that we cannot settle here but that permeate much of this handbook.

6. Rationality in Interactive Social Settings

The modern study of agency reflects the fact that the paradigm of computing today are distributed systems, not single machines. Likewise, multiagent systems put interacting individual agents at center stage and, at a next level, view groups themselves as actors, up to crowds or societies. This shifts the location of rationality from single agents to the quality of their interactions. And it broadens the focus from individual desires and actions to include emergent properties of the social system.

The interactive perspective is not alien to logic. Ever since Antiquity, dialogue, argumentation, and debate have been paradigmatic scenarios, and the rich interface of logic and games has been noted already (Hodges & Väänänen, 2001/2019; van Benthem, 2014). A core topic in epistemic logic is the rational ability to reason about others, with iterated forms such as "Agent *i* knows that agent *j* knows that . . ." and analogues for belief and other attitudes (cf. chapter 5.5 by Hahn & Collins and chapter 5.6 by Woods, both in this handbook). This recursion to higher levels is widespread: we can even be afraid of fear, of fear of fear, and so on. The extent to which human agents truly display these abilities is studied in cognitive psychology under the heading of Theory of Mind (Isaac, Szymanik, & Verbrugge, 2014; Premack & Woodruff, 1978). Iterated knowledge is used in computer science in analyzing correctness and security of communication protocols (Fagin et al., 1995).

But there is much more to social interaction than epistemic reflection. Strategic action involves dependencies of one agent's behavior on that of others or, better, on her expectations about the behavior of others (Aumann, 1995). Here is a simple illustration, a computational task in an interactive setting.

Example: Sabotage game. A Traveler in a graph moves along edges to reach some specified goal region. This graph-reachability problem is solvable in polynomial time. But now, there is a malevolent Demon who cancels an edge after each move Traveler makes. After that, Traveler goes along some still existing edge, and so on. $\dashv$

This "sabotage game" models search tasks under adverse circumstances and other social-informational scenarios. The solution complexity of the sabotage game jumps from polynomial time to complete for polynomial space. Logic helps determine who has a winning strategy in a given sabotage game by defining the basic challenge–response pattern, and it helps reason about general properties of such games. This is one case where logic meets "gamifications" of agency scenarios, and logics can even be used to devise concrete new practical games, thus becoming a tool of design as much as of analysis (van Benthem, 2014).

With preference added, different notions of rationality have been investigated by logical means, such as those in game solution methods like backward induction, iterated removal of strictly dominated strategies (see chapter 9.1 by Albert & Kliemt and chapter 9.2 by Perea, both in this handbook), or iterated regret minimization (Halpern & Pass, 2009/2012). The structure of strategies in themselves is studied extensively at the interface of game theory, logic, and computer science (Brandenburger, 2014; van Benthem, Ghosh, & Verbrugge, 2015). Also, games influence computational logic: witness the "Boolean games" of Harrenstein, van der Hoek, Meyer, and Witteveen (2001), where players can manipulate truth values of propositions toward achieving their goals.

7. High and Low Rationality

At this point, a challenge arises to the preceding analyses of individual and social rationality. Classical game theory has agents that deliberate and design complex strategies, and rich epistemic and dynamic logics of agency reflect this. However, in evolutionary game theory (see chapter 9.3 by Alexander, this handbook), poor agents, perhaps hardwired biological types (Maynard Smith, 1982), do just as well.

Example: Evolutionary games. In a "Hawk–Dove" game, two individuals compete for a resource and can adopt either a Hawk or a Dove strategy. Hawks fight aggressively against other Hawks, in order to obtain the resource, until injury occurs and one retreats or, when facing a Dove, just take the resource, while a Dove retreats when facing a Hawk or shares the resource when facing a Dove.   ⊣

Game theory computes equilibria here, which are typically in mixed strategies. With repeated Hawk–Dove games, the appropriate notion is that of an "evolutionarily stable" strategy, and it can be shown that certain mixtures of Hawk–Dove populations are stable, when the value of obtaining the resource is greater than the cost associated with possible injury in a fight. One can think of these mixed strategies as complex behavior for individual reasoning agents but also as just percentages of a population consisting of two types of agents, each just doing what it does, perhaps for biological reasons.

In the terminology of Skyrms (2010), the realm of complex reasoning players is that of "high rationality" and the realm of hardwired simple agents that of "low rationality." Often the latter do as well as the former. For instance, a classical game-theoretic analysis might say that through some sophisticated Kantian or Rawlsian argument involving thinking about others, we all arrive at the conclusion that we should live by the principles of morality, with the exception perhaps of a few free riders. By contrast, a game-theoretic evolutionary stability argument may tell us that in the long run, a population with a certain proportion of simple law-abiders (the prey) and lawbreakers (the predators), who both cannot help being what they are, is stable. No reasoning need be involved at all; the morality is emergent system behavior. This influx from evolutionary game theory reinforces the message of distributed computing: a society of many simple agents can produce highly complex behavior.

Here is one more scenario of emergent long-term complex behavior.

Example: Limit behavior in social networks. The following network has an update rule that agents s (the nodes) adopt a belief p if p is held by all their neighbors (that is, all nodes with an arrow pointing to them from s). Applied iteratively, the following evolutions may occur with different initial situations (see figure 2.6.2).

Case 1: Initial $p = \{1\}$. The second stage has $p = \varnothing$ (no one believes p), and this remains the outcome ever after. Case 2: Initial $p = \{2\}$. The next successive stages are $\{3\}, \{4\}, \{2\}$; from this stage onward, the network activation states loop. Case 3: Initial $p = \{1, 2\}$. The next stage is $\{3\}$, and we get an oscillation as before in case 2. Case 4: Initial $p = \{1, 2, 3\}$. We get $\{1, 3, 4\}, \{2, 4\}, \{2, 3\}, \{1, 3, 4\}$, and an oscillation starts here.   ⊣

Thus, network update dynamics can stabilize in a single state (case 1) or oscillate in loops. Sometimes, successive models in a loop are isomorphic (cases 2 and 3), and sometimes the loop runs through different nonisomorphic network configurations (case 4). In infinite networks, an even further option is divergence toward ever different configurations. In all these scenarios, no logic seems involved in belief formation: behavior arises from agent types (the update rule) and the global structure of the network.

To put the challenge starkly, perhaps complex, logic-based rationality is not necessary to understand the behavior of human and artificial agents? But things are more complicated. In daily life, we think carefully in the "high" style about certain issues, but given our limited resources, we just follow, "low" style, our neighbors on perhaps the

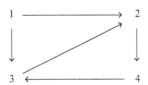

Figure 2.6.2
A social influence network.

majority of issues. This mixture calls for explanation, and current investigations are charting its details.

Combined high–low scenarios. Liu, Seligman, and Girard (2014) study agents in social networks that follow their neighbors' preferences, beliefs, or behavior via rules like following the majority or some other threshold, reflecting different agent types. The resulting diffusion process models the spread of fashions and new ideas. But agents still have epistemic states, and these can change dynamically as before, as described in an "epistemic friendship logic." In this setting, among other things, a logical characterization can be given of conditions for stabilization of agent's beliefs, thus predicting long-term system behavior. This framework combines ideas from sociology (Friedkin, 1998) with epistemic logic, adding an essential element to the earlier logical models of agency: the structure of the social network (see also chapter 10.1 by Dietrich & Spiekermann, this handbook). For a congenial study in another logical framework, see Xue (2017).

Other combined scenarios that have been studied include groups of individually rational agents who reason toward a common decision. Two things can happen: individual agents can enhance each other's reasoning power and bring about a higher level of group rationality, surpassing that of each individual agent. But groups may also get locked in irrational behavior. Whether the one will happen or the other is investigated in Baltag, Boddy, and Smets (2018), in terms of differences in interests and abilities between agents. One striking conclusion is that irrational group behavior is often not caused by irrational behavior of individual agents but by misalignments of their interests.

Other social phenomena that have been studied by logical techniques are informational cascades (Bikhchandani, Hirshleifer, & Welch, 1992), where a sequence of individual agents follows the decisions of their predecessors while ignoring their own private evidence. Baltag, Christoff, Hansen, and Smets (2013) ask whether individual rational agents, who use all their higher-order reasoning power, can stop a cascade from happening. The answer, surprisingly, is "no," and this fact can be proved by logical techniques that track information updates. However, the protocol regulating the agents' strategies matters: when agents have total communication and sharing of evidence, cascades can be stopped.

The preceding examples show how prima facie ideological differences between high and low rationality turn into a deeper study of how the two interface. This is logic at work at the ground level of agent activity. However, there is also a second, more methodological contact between the two sides: logic can analyze the structure of the dynamical systems theory underlying most low-rationality approaches and find patterns there, usually amounting to high-level qualitative descriptions of system behavior. Explorations in this direction include Kremer and Mints (2005) and Klein and Rendsvig (2017).

8. Machine Learning and Probability

In addition to the preceding tensions, the contemporary world of computing and AI offers a new challenge to logic-based views of rationality.

8.1 Machine Learning

Machine learning (Kelleher, Mac Namee, & D'Arcy, 2015) works well on large data sets, outperforming symbolic approaches, which tend to have problems of scalability. For instance, in supervised learning, a neural network is constructed, consisting of nodes with adjustable thresholds and links of adjustable strengths between nodes. Each setting for all of these produces an activity in the output layer given an input to the initial layer of the network. A cost function measures the distance of the current outputs to the desired ones on the training inputs. The network can then, by well-known techniques such as gradient descent (Russell & Norvig, 1994), adjust its weights and thresholds in the direction of lowering the cost function. In the end, stable optima in network activation are reached that work very well in new cases outside of the training set, in many computational and cognitive tasks. These networks, related to spin glass models in physics (Nishimori, 2001), use general statistical methods rather than specifically human agent features.

Neural networks in machine learning do not have any features immediately corresponding to classical logical models. There is no language and no representation, and the dynamic operations of the network do not reflect logical operations in any obvious manner. Also, very different stable states of the network resulting from training sessions can perform the same tasks, and invariants are hard to detect. Thus, whereas low-rationality methods raised the question whether logical analysis was necessary, machine-learning methods raise the question whether logical analysis is even possible.

It is far too early to adjudicate this debate. But here, too, there are some promising developments toward cooperation rather than antagonism. Integrating inference as used in neural networks and learning systems with symbolic reasoning is an active area of research (Baggio, van Lambalgen, & Hagoort, 2015; Balkenius & Gärdenfors,

2016; Leitgeb, 2004). "Explainable AI" seeks humanly intelligible qualitative patterns behind machine-learning systems, with topics such as causal reasoning (Halpern, 2016; Pearl, 2000; van Rooij & Schulz, 2019; see also chapter 7.1 by Pearl, this handbook), conditional logics as a way of classifying types of machine learning (Ibeling & Icard, 2018), or extracting logical inference modules from actual neural networks (Geiger, Icard, Lu, & Potts, 2021), while there is also a trend toward finding joint perspectives on learning in itself.

But also, recall a distinction made at the start of this chapter. If logic is only seen as a direct model for activities of reasoning or information update, other frameworks look like competitors. To some, the only question under debate is then whether logic can enhance such frameworks in terms of representation or computation. But in the more foundational sense of logic as an analysis of the structure of theories of computation and agency, even machine learning works on spaces with logical structure that can be described in logical terms (Leitgeb, 2017), and a meeting of the minds seems entirely feasible.

8.2 Probability

Continuing with methodological issues, here is one final contrast. A conspicuous feature of most studies of agency is the extensive use of probabilistic methods, a quantitative paradigm often seen as being at odds with qualitative logical analysis. Probability underlies many computational systems; it lies at the heart of game theory and dynamical systems theory, and in epistemology, probabilistic styles of analysis are at least as widespread as logic-based ones (see chapter 4.1 by Hájek & Staffel and chapter 4.7 by Dubois & Prade, both in this handbook).

The fruitful issue here is again one of combination. Qualitative and quantitative approaches naturally coexist, and the issue is just how. For instance, epistemic and doxastic logics, both static and dynamic, model uncertainty in terms of ranges of options (Adriaans & van Benthem, 2008), whereas Bayesian epistemology uses updates of probability functions (Bovens & Hartmann, 2003; Talbott, 2016). The compatibility of the two perspectives shows in combined systems (Halpern, 2003) that reason about both ontic and epistemic uncertainty, bringing together logic-based approaches with probabilistic conditioning. Other uses of probability concern action rather than information: witness the mixed strategies in game theory (for a logical perspective, see van Benthem & Klein, 2019). But there are also quite different interfaces of logic and probability, for instance, in the data-oriented parsing (DOP) architecture of Bod,

Scha, and Sima'an (2003) and Bod (2008), which combines classical rule-based models of language and reasoning with probabilistic pattern recognition in a memory of earlier performance. Finally, the foundations of probability were still close to logic in the work of Boole and de Finetti, and various strands of research link the two realms in new ways. Harrison-Trainor, Holliday, and Icard (2018) study low-complexity qualitative reasoning systems that admit of introducing probability measures, while Leitgeb (2017) derives qualitative notions of belief from richer probabilistic models.

There are many further philosophical and technical issues to be explored at this rich and growing set of interfaces that we cannot cover in this chapter; the reader is referred to Spohn (2012) and chapter 5.3 by Kern-Isberner, Skovgaard-Olsen, and Spohn (this handbook).

9. Conclusion

This chapter has presented broad perspectives from logic and computation on rational agency. These ranged from high-level foundational insights into information and proof to specific studies of various abilities of information- and goal-driven agents. A rational agent, in this light, is a reasoner, information-processor, concept-crafter, and purpose-seeker: fallible but talented. Is it also a human cognitive agent? On connecting logic and computation to cognitive reality, we refer to Stenning and van Lambalgen (2008).

The main thrust of a logical approach as we see it is theoretical, but the deep entanglement of logic and computation over the past century has added practical dimensions. Rationality as studied here can be programmed and put into intelligent systems, even though the path to feasibility is not easy or trivial. It is this very distance that allows logical theories to also be normative, providing an essential tension between the real and the ideal in the study of rational behavior, which keeps sparking further investigation.

We have not hidden the fact that the classical logico-computational paradigm faces challenges, coming from probability theory, dynamical systems theory, and machine learning. But we think this is all to the good, since these challenges suggest new interface topics of interest to all.

Finally, it should be clear that we have not claimed that logic is the only game in town. Neither is computation. The approach surveyed in this chapter does not hold the unique key to understanding the rich phenomenon of rationality, but it does offer one valid and illuminating perspective.

References

Adriaans, P., & van Benthem, J. (Eds.). (2008). *Handbook of the philosophy of science: Vol. 8. Philosophy of information.* Amsterdam, Netherlands: Elsevier.

Andrews, G. R. (2000). *Foundations of multithreaded, parallel, and distributed programming.* Reading, MA: Addison-Wesley.

Arló-Costa, H., Hendricks, V. F., & van Benthem, J. (Eds.). (2016). *Readings in formal epistemology: Sourcebook.* Cham, Switzerland: Springer.

Aumann, R. J. (1995). Backward induction and common knowledge of rationality. *Games and Economic Behavior, 8*(1), 6–9.

Baggio, G., van Lambalgen, M., & Hagoort, P. (2015). Logic as Marr's computational level: Four case studies. *Topics in Cognitive Science, 7*(2), 287–298.

Balkenius, C., & Gärdenfors, P. (2016). Spaces in the brain: From neurons to meanings. *Frontiers of Psychology, 7,* 1820.

Baltag, A., Boddy, R., & Smets, S. (2018). Group knowledge in interrogative epistemology. In H. van Ditmarsch & G. Sandu (Eds.), *Jaakko Hintikka on knowledge and game-theoretical semantics* (Outstanding Contributions to Logic, Vol. 12, pp. 131–164). Cham, Switzerland: Springer.

Baltag, A., Christoff, Z., Hansen, J. U., & Smets, S. (2013). Logical models of informational cascades. In J. van Benthem & F. Liu (Eds.), *Logic across the University: Foundations and Applications: Proceedings of the Tsinghua Logic Conference, Beijing, 2013* (Studies in Logic, Vol. 47, pp. 405–432). London, England: College Publications.

Baltag, A., Gierasimczuk, N., & Smets, S. (2011). Belief revision as a truth-tracking process. In K. Apt (Ed.), *Proceedings of the 13th Conference on Theoretical Aspects of Rationality and Knowledge, TARK XIII* (pp. 187–190). New York, NY: ACM.

Baltag, A., Moss, L. S., & Solecki, S. (1998). The logic of common knowledge, public announcements, and private suspicions. In I. Gilboa (Ed.), *Proceedings of the 7th Conference on Theoretical Aspects of Rationality and Knowledge (TARK 98)* (pp. 43–56). San Francisco, CA: Morgan Kaufmann.

Belnap, N., Perloff, M., & Xu, M. (2001). *Facing the future: Agents and choices in our indeterminist world.* Oxford, England: Oxford University Press.

Bergstra, J., Ponse, A., & Smolka, S. (2001). *Handbook of process algebra.* Amsterdam, Netherlands: Elsevier.

Beth, E. W. (1971). *Aspects of modern logic.* Dordrecht, Netherlands: Reidel.

Bikhchandani, S., Hirshleifer, D., & Welch, I. (1992). A theory of fads, fashion, custom, and cultural change as informational cascades. *Journal of Political Economy, 100*(5), 992–1026.

Bod, R. (2008). The data-oriented parsing approach: Theory and application. In J. Fulcher & L. C. Jain (Eds.), *Computational intelligence: A compendium* (Studies in Computational Intelligence, Vol. 115, pp. 307–348). Berlin, Germany: Springer.

Bod, R., Scha, R., & Sima'an, K. (2003). *Data-oriented parsing.* Stanford, CA: CSLI Publications.

Bonizzoni, P., Brattka, V., & Löwe, B. (Eds.). (2013). *The Nature of Computation: Logic, Algorithms, Applications: 9th Conference on Computability in Europe, CiE 2013, Milan, Italy, 2013, Proceedings* (LNCS 7921). Heidelberg, Germany: Springer.

Bovens, L., & Hartmann, S. (2003). *Bayesian epistemology.* Oxford, England: Clarendon Press.

Brafman, R. I., Latombe, J.-C., Moses, Y., & Shoham, Y. (1997). Applications of a logic of knowledge to motion planning under uncertainty. *Journal of the ACM, 44*(5), 633–668.

Brandenburger, A. (Ed.). (2014). *The language of game theory: Putting epistemics into the mathematics of games* (World Scientific Series in Economic Theory, Vol. 5). Singapore: World Scientific.

Bratman, M. E. (1987). *Intention, plans, and practical reason.* Cambridge, MA: Harvard University Press.

Cardon, A., & Itmi, M. (2016). *New autonomous systems* (Vol. 1). Hoboken, NJ: Wiley.

Chakraborty, P., Saxena, P. C., & Katti, C. P. (2011). Fifty years of automata simulation: A review. *ACM Inroads, 2*(4), 59–70.

Church, A. (1936). An unsolvable problem of elementary number theory. *American Journal of Mathematics, 58*(2), 345–363.

Ciardelli, I., Groenendijk, J., & Roelofsen, F. (2019). *Inquisitive semantics.* Oxford, England: Oxford University Press.

Fagin, R., Halpern, J. Y., Moses, Y., & Vardi, M. Y. (1995). *Reasoning about knowledge.* Cambridge, MA: MIT Press.

Friedkin, N. E. (1998). *A structural theory of social influence.* Cambridge, England: Cambridge University Press.

Gabbay, D. M. (1999). *Fibring logics.* Oxford, England: Clarendon Press.

Gabbay, D. M., & Guenthner, F. (Eds.). (1983–). *Handbook of philosophical logic.* Dordrecht, Netherlands: Springer.

Gabbay, D. M., Hogger, C. J., & Robinson, J. A. (Eds.). (1993–1998). *Handbook of logic in artificial intelligence and logic programming.* Oxford, England: Clarendon Press.

Geiger, A., Lu, H., Icard, T., & Potts, C. (2021). Causal abstractions of neural natural language inference models. CSLI, Stanford University.

Gödel, K. (1986). Über formal unentscheidbare Sätze der Principia Mathematica und verwandter Systeme I [On formally undecidable propositions of Principia Mathematica and related systems]. In S. Feferman, J. W. Dawson, Jr., S. C. Kleene, G. H.

Moore, R. M. Solovay, & J. van Heijenoort (Eds.), *Kurt Gödel: Collected works* (pp. 144–195). Oxford, England: Clarendon Press. (Original work published 1931)

Halpern, J. Y. (2003). *Reasoning about uncertainty*. Cambridge, MA: MIT Press.

Halpern, J. Y. (2016). *Actual causality*. Cambridge, MA: MIT Press.

Halpern, J. Y., & Pass, R. (2012). Iterated regret minimization: A new solution concept. *Games and Economic Behavior, 74*(1), 184–207. (Original work published 2009)

Halpern, J. Y., & Rêgo, L. C. (2013). Reasoning about knowledge of unawareness revisited. *Mathematical Social Sciences, 65*(2), 73–84.

Harel, D. (1987). *Algorithmics: The spirit of computing*. Reading, MA: Addison-Wesley.

Harrenstein, P., van der Hoek, W., Meyer, J.-J., & Witteveen, C. (2001). Boolean games. In J. van Benthem (Ed.), *Proceedings of the 8th Conference on Theoretical Aspects of Rationality and Knowledge* (pp. 287–298). San Francisco, CA: Morgan Kaufmann.

Harrison-Trainor, M., Holliday, W. H., & Icard, T. F., III. (2018). Inferring probability comparisons. *Mathematical Social Sciences, 91*, 62–70.

Haugeland, J. (Ed.). (1997). *Mind design II: Philosophy, psychology, artificial intelligence*. Cambridge, MA: MIT Press.

Hintikka, J. (1973). *Logic, language-games, and information: Kantian themes in the philosophy of logic*. Oxford, England: Clarendon Press.

Hodges, W., & Väänänen, J. (2019). *Logic and games*. In E. N. Zalta (Ed.), *The Stanford encyclopedia of philosophy*. Retrieved from https://plato.stanford.edu/archives/fall2019/entries/logic-games/

Horty, J. F. (2012). *Reasons as defaults*. Oxford, England: Oxford University Press.

Ibeling, D., & Icard, T. (2018). On the conditional logic of simulation models. In *Proceedings of the 27th International Joint Conference on Artificial Intelligence (IJCAI'18)* (pp. 1868–1874). Stockholm, Sweden: AAAI Press.

Isaac, A., Szymanik, J., & Verbrugge, R. (2014). Logic and complexity in cognitive science. In A. Baltag & S. Smets (Eds.), *Johan van Benthem on logic and information dynamics* (pp. 787–824). Cham, Switzerland: Springer.

Kelleher, J. D., Mac Namee, B., & D'Arcy, A. (2015). *Fundamentals of machine learning for predictive data analytics: Algorithms, worked examples, and case studies*. Cambridge, MA: MIT Press.

Kelly, K. T. (1996). *The logic of reliable inquiry*. New York, NY: Oxford University Press.

Klein, D., & Rendsvig, R. K. (2017). Convergence, continuity and recurrence in dynamic epistemic logic. In A. Baltag, J. Seligman, &

T. Yamada (Eds.), *Logic, Rationality, and Interaction: 6th International Workshop, LORI 2017, Sapporo, Japan, September 11–14, 2017, Proceedings* (pp. 108–122). Berlin, Germany: Springer.

Kneale, W., & Kneale, M. (1962). *The development of logic*. Oxford, England: Clarendon Press.

Kremer, P., & Mints, G. (2005). Dynamic topological logic. *Annals of Pure and Applied Logic, 131*(1–3), 133–158.

Leitgeb, H. (2004). *Inference on the low level: An investigation into deduction, nonmonotonic reasoning, and the philosophy of cognition*. Dordrecht, Netherlands: Kluwer Academic Publishers.

Leitgeb, H. (2017). *The stability of belief: How rational belief coheres with probability*. Oxford, England: Oxford University Press.

Liu, F. (2011). *Reasoning about preference dynamics* (Vol. 354 of Synthese Library). Dordrecht, Netherlands: Springer.

Liu, F., Seligman, J., & Girard, P. (2014). Logical dynamics of belief change in the community. *Synthese, 191*(11), 2403–2431.

Lorini, E. (2018). In praise of belief bases: Doing epistemic logic without possible worlds. In *Proceedings of the Thirty-Second AAAI Conference on Artificial Intelligence* (pp. 1915–1922).

Maynard Smith, J. (1982). *Evolution and the theory of games*. Cambridge, England: Cambridge University Press.

Nisan, N., Roughgarden, T., Tardos, E., & Vazirani, V. V. (Eds.). (2007). *Algorithmic game theory*. Cambridge, England: Cambridge University Press.

Nishimori, H. (2001). *Statistical physics of spin glasses and information processing: An introduction*. Oxford, England: Oxford University Press.

Papadimitriou, C. H. (1994). *Computational complexity*. Reading, MA: Addison-Wesley.

Parikh, R., & Ramanujam, R. (2003). A knowledge based semantics of messages. *Journal of Logic, Language and Information, 12*, 453–467.

Pearl, J. (2000). *Causality: Models, reasoning, and inference*. Cambridge, England: Cambridge University Press.

Popper, K. (1963). *Conjectures and refutations: The growth of scientific knowledge*. London, England: Routledge.

Premack, D., & Woodruff, G. (1978). Does the chimpanzee have a theory of mind? *Behavioral and Brain Sciences, 4*, 515–526.

Rao, A., & Georgeff, M. (1991). Modeling rational agents within a BDI-architecture. In J. Allen, R. Fikes, & E. Sandewall (Eds.), *Proceedings of the 2nd International Conference on Principles of Knowledge Representation and Reasoning* (pp. 473–484). San Mateo, CA: Morgan Kaufmann.

Restall, G. (2000). *An introduction to substructural logics*. London, England: Routledge.

Russell, S., & Norvig, P. (1994). *Artificial intelligence: A modern approach* (3rd ed.). Upper Saddle River, NJ: Prentice Hall.

Shoham, Y., & Leyton-Brown, K. (2008). *Multiagent systems: Algorithmic, game-theoretic, and logical foundations.* Cambridge, England: Cambridge University Press.

Skyrms, B. (2010). *Signals: Evolution, learning, & information.* Oxford, England: Oxford University Press.

Spohn, W. (2012). *The laws of belief: Ranking theory and its philosophical applications.* Oxford, England: Oxford University Press.

Stenning, K., & van Lambalgen, M. (2008). *Human reasoning and cognitive science.* Cambridge, MA: MIT Press.

Talbott, W. (2016). Bayesian epistemology. In E. N. Zalta (Ed.), *The Stanford encyclopedia of philosophy.* Retrieved from https://plato.stanford.edu/archives/win2016/entries/epistemology-bayesian/

Turing, A. M. (1936). On computable numbers, with an application to the Entscheidungsproblem. *Proceedings of the London Mathematical Society, 42,* 230–265.

Turing, A. M. (1950). Computing machinery and intelligence. *Mind, 236,* 433–460.

van Benthem, J. (2011). *Logical dynamics of information and interaction.* Cambridge, England: Cambridge University Press.

van Benthem, J. (2014). *Logic in games.* Cambridge, MA: MIT Press.

van Benthem, J. (2018). Computation as social agency: What, how and who. *Information and Computation, 261*(3), 519–535.

van Benthem, J. (2019). Implicit and explicit stances in logic. *Journal of Philosophical Logic, 48*(3), 571–601.

van Benthem, J., Ghosh, S., & Verbrugge, R. (Eds.). (2015). *Models of strategic reasoning: Logics, games, and communities* (Lecture Notes in Computer Science, Vol. 8972). Berlin, Germany: Springer.

van Benthem, J., & Klein, D. (2019). Logics for analyzing games. In E. N. Zalta (Ed.), *The Stanford encyclopedia of philosophy.* Retrieved from https://plato.stanford.edu/archives/spr2019/entries/logics-for-games/

van Benthem, J., & Pacuit, E. (2011). Dynamic logics of evidence-based beliefs. *Studia Logica, 99*(1–3), 61–92.

van Benthem, J., & Smets, S. (2015). Dynamic logics of belief change. In H. van Ditmarsch, J. Y. Halpern, W. van der Hoek, & B. Kooi (Eds.), *Handbook of logics for knowledge and belief* (pp. 313–393). London, England: College Publications.

van Ditmarsch, H., van der Hoek, W., & Kooi, B. (2007). *Dynamic epistemic logic.* Dordrecht, Netherlands: Springer.

van Harmelen, F., Lifschitz, V., & Porter, B. (Eds.). (2008). *Handbook of knowledge representation* (Foundations of Artificial Intelligence). Amsterdam, Netherlands: Elsevier.

van Rooij, R., & Schulz, K. (2019). Conditionals, causality and conditional probability. *Journal of Logic, Language and Information, 28*(1), 55–71.

Wang, H. (1996). *A logical journey: From Gödel to philosophy.* Cambridge, MA: MIT Press.

Wooldridge, M. (2009). *An introduction to multiagent systems* (2nd ed.). Chichester, England: Wiley. (Original work published 2002)

Wooldridge, M. (2010). *Reasoning about rational agents.* Cambridge, MA: MIT Press.

Xue, Y. (2017). *In search of homo sociologicus* (Unpublished doctoral dissertation). The Graduate Center, City University of New York.

Yamada, T. (2008). Logical dynamics of some speech acts that affect obligations and preferences. *Synthese, 165*(2), 295–315.

Part II   Theoretical Rationality

Section 3   Deductive Reasoning

3.1 Propositional and First-Order Logic

Florian Steinberger

Summary

This chapter addresses the question how (if at all) propositional logic (PL) and first-order logic (FOL) relate to epistemic rationality.[1] Rationality, it is often held, demands that our attitudes cohere in particular ways. Logic is often invoked as a source of such coherence requirements when it comes to belief: an ideally rational agent's beliefs are consistent and closed under logical consequence (i.e., the logical consequences of the agent's beliefs are also believed). However, this traditional picture has been challenged from various quarters. I begin by briefly reviewing the key concepts involved in PL and FOL. I then critically examine two distinct approaches to justifying logic-based requirements of rationality. The first lays down a set of desiderata codifying our intuitions and then seeks to formulate a principle articulating the link between logic and rational belief that satisfies them. The second starts by identifying our most fundamental epistemic aim and seeks to derive requirements of rationality based on their ability to promote this aim.

1. Logic and Rationality

Rational belief and correct belief are not the same thing. I may rationally believe a falsehood (for example, when the evidence available to me is misleading). If correctness for beliefs is truth, my belief may be correct though irrational (as when I engage in wishful thinking and, merely by a fluke, form a true belief). Part of what it means for beliefs to be rational is for them to cohere with the evidence. Another part is for them to cohere with each other.[2] Logic is often thought to be connected to the second sense of rationality. Logic is concerned with the relationships between the truth-values of different propositions. It tells us which propositions can and cannot be jointly true by dint of certain of their structural properties. For example, a chief concern of logic is logical consequence. If a proposition C is a logical consequence of a set of propositions, all of which I believe,

it is, in a specific sense, impossible for my beliefs to be true without C also being true. Conversely, if I know C to be false, I thereby know that my antecedent beliefs cannot all be true. Relatedly, if my beliefs are inconsistent (and I recognize them to be so), then, merely by virtue of the logical structure of my beliefs' contents, I am in a position to know that I must be mistaken with respect to at least one of them—my representation of the world is not merely false, but there is no logically possible world that makes all of my beliefs true. Logical coherence, then, is typically a necessary (though generally insufficient) condition for the truth of one's belief. And while I may not be rationally criticizable for having incorrect beliefs, many maintain that if my beliefs fail to be logically coherent, I am less than ideally rational. It is in this sense that logic is thought to offer up requirements of rationality.[3]

The relation between logic and epistemic rationality relies on our assumption that the entities with which logic is concerned are also (or are systematically related to) the contents of our rationally evaluable mental states. I use "proposition" to designate the type of entity that constitutes both the premises and the conclusions of logical arguments, on the one hand, and the contents of propositional attitudes, on the other. I assume there is a type of entity capable of performing both of these roles. Furthermore, as my talk of structure in the foregoing makes plain, I assume that propositions are (logically) structured entities. That said, what I go on to say is compatible with other, nonpropositional accounts of truthbearers.

The aim of this chapter is to provide a critical survey of the discussion over whether logic really does give rise to norms of epistemic rationality. More specifically, I ask whether classical propositional logic (PL) and its extension, first-order logic (FOL; also called "[first-order] predicate logic"), can be said to do so. The force of restricting ourselves to these logics comes to this. Logic, we said, is concerned with possible truth-value distributions across propositions based on their logically relevant structural features or *logical form*. The restriction to these particular logics is relevant in that different logics are sensitive to different structural features. The richer a logic's language, the more logical structure it is able to discern and hence the more rationally constraining it will be.

Another elementary but important observation is that PL and FOL are not themselves explicitly about rational belief formation and revision—their logical vocabulary contains no symbols that are naturally interpreted as designating doxastic states. In this they contrast with more sophisticated formalisms such as belief revision theory (Alchourrón, Gärdenfors, & Makinson, 1985), ranking theory (Spohn, 2012), and various Bayesian approaches whose express aim is to model the dynamics of belief management. The existence of these theories does not, however, render the present question nugatory. On the contrary, the fact that such theories incorporate implicit assumptions about the relationship between logic and rational belief renders the exploration of these foundational questions all the more urgent.

The chapter is structured as follows. The first two sections provide brief reviews of PL and FOL. Section 4 offers an overview of recent attempts at articulating the relationship between PL and FOL and rational belief as well as criticisms that have been leveled at such attempts. Section 5, finally, explores the relation of logical requirements of rationality to recent work in formal epistemology.

2.   Propositional Logic

For our purposes, we can conceive of a logic as a formal language along with a semantic (model-theoretic) and a syntactic (proof-theoretic) relation of logical consequence. When it comes to formal languages, the symbols making up the alphabet are usually divided into three separate categories:

1. Descriptive symbols
2. Logical symbols
3. Auxiliary symbols

Descriptive symbols, intuitively, are the expressions of the language that have variable semantic values and so may be used to express truth-evaluable claims about the world. PL treats propositions as atoms, that is, PL is blind to the internal structural features of propositions. For example, it is insensitive to the distinction between propositions expressed by sentences involving only a unary predicate and singular term ("Rachel is rational") and sentences composed of binary predicates and two singular terms ("Rachel is taller than Steve").

Consequently, the only descriptive symbols contained in PL's language are formulas that serve to represent propositions (as opposed to expressions serving to express subpropositional content). PL distinguishes between *atomic* propositions, whose logical form cannot be further analyzed (using the resources of PL) and which are represented by atomic formulas, and *complex* ones, represented by complex formulas. Extending the chemistry metaphor, complex formulas are molecular compound formulas held together by the bonds that are logical connectives. The logical connectives are the following unary or binary operators (their approximate English cognates are given in brackets): "¬" ("not"), "∧" ("and"), "∨" ("or"), "→" ("if . . . , then . . ."), and "↔" ("if and only if"). The only auxiliary symbols are parentheses, which are needed to avoid ambiguity among our expressions (e.g., to distinguish $(A \wedge (B \vee C))$ and $((A \wedge B) \vee C)$). In summary, the language contains

1. propositional variables: $\mathscr{P} = \{p_1, p_2, p_3, \ldots\}$ (in practice, we typically use "p," "q," "r," . . .);
2. logical connectives (or logical constants): ¬, ∧, ∨, →, ↔; and
3. auxiliary symbols: (,).

This alphabet constitutes the language of PL, $\mathscr{L}_{PL}$. Our next order of business is to specify a grammar for our language. Call a grammatically correct expression in our formal language a formula. We now define the set of formulas of PL, $\mathscr{F}_{PL}$, by means of the following inductive definition:

1. If $A \in \mathscr{P}$, then $A \in \mathscr{F}_{PL}$.
2. If $A \in \mathscr{F}_{PL}$, then so is ¬A.
3. If $A \in \mathscr{F}_{PL}$ and $B \in \mathscr{F}_{PL}$, then so are $(A \wedge B)$, $(A \vee B)$, $(A \rightarrow B)$, and $(A \leftrightarrow B)$.
4. Only such strings of symbols of our alphabet as can be constructed in accordance with rules 1 to 3 are formulas.

It is a direct consequence that every formula has a set of immediate subformulas (ISFs):

- If $A \in \mathscr{P}$, then ISF(A) = ∅.
- If $A = \neg B$, then ISF(A) = $\{B\}$.
- If $A = (B * C)$ (where $* \in \{\wedge, \vee, \rightarrow, \leftrightarrow\}$), then ISF($A$) = $\{B, C\}$.

2.1   Semantics of PL

The semantics of PL mirrors its syntax in the following sense. The truth-value of a complex formula is determined by the truth-values of its immediate subformulas and the logical connective governing it. Logical connectives are unary (negation) or binary truth functions from the truth-values of the subformula(s) to the truth-value of the complex formula to be evaluated. The atomic formulas, by contrast, are assigned their truth-values directly, by way of an interpretation. An interpretation in PL is a function $\Im$ whose domain is the set $\mathscr{P}$ of propositional variables and whose codomain is the set of $\{t, f\}$ of truth-values:

$$\Im: \mathscr{P} \to \{t, f\}.$$

Our grammar guarantees that every complex formula can be broken down uniquely into subformulas of decreasing complexity until the process bottoms out in the propositional variables figuring within it. Consequently, the truth-value of a complex formula too will ultimately depend on the truth-values assigned to the propositional variables it contains and on the particular mode of composition of the formula. Our aim now is to use this insight to extend each interpretation function into a corresponding valuation function that assigns to each formula a truth-value. More precisely, every interpretation $\Im$ in PL determines a unique corresponding valuation $\mathfrak{V}_\Im$ (as can easily be proved).

A valuation in PL (relative to an interpretation $\Im$), then, is a function $\mathfrak{V}_\Im: \mathscr{F}_{PL} \to \{t, f\}$ that satisfies the following semantic rules:

1. $\mathfrak{V}_\Im(p_i) = t$ iff $\Im(p_i) = t$,
2. $\mathfrak{V}_\Im(\neg A) = t$ iff $\mathfrak{V}_\Im(A) = f$,
3. $\mathfrak{V}_\Im((A \wedge B)) = t$ iff $\mathfrak{V}_\Im(A) = t$ and $\mathfrak{V}_\Im(B) = t$,
4. $\mathfrak{V}_\Im((A \vee B)) = t$ iff $\mathfrak{V}_\Im(A) = t$ or $\mathfrak{V}_\Im(B) = t$,
5. $\mathfrak{V}_\Im((A \to B)) = t$ iff $\mathfrak{V}_\Im(A) = f$ or $\mathfrak{V}_\Im(B) = t$,
6. $\mathfrak{V}_\Im((A \leftrightarrow B)) = t$ iff $\mathfrak{V}_\Im(A) = \mathfrak{V}_\Im(B)$.

Clauses 1 to 6 are also called "semantic rules." Rule 1 says that propositional variables are to be evaluated via $\mathfrak{V}_\Im$ just as the underlying interpretation $\Im$ dictates. In rules 2 through 6, complex decomposable formulas are evaluated in conformity with the meanings of the logical connectives.

With this definition in place, we can now define the central concepts of consequence, validity, and consistency:

1. For all formulas $A_1, \ldots, A_n$ and B of $\mathscr{F}_{PL}$: B is a *logical consequence* in PL of $A_1, \ldots, A_n$ (in symbols: $A_1, \ldots, A_n \vDash_{PL} B$) iff for all interpretations $\Im$ with $\mathfrak{V}_\Im(A_1) = t, \ldots, \mathfrak{V}_\Im(A_n) = t$, it holds that $\mathfrak{V}_\Im(B) = t$.
2. An argument form $A_1, \ldots, A_n \therefore B$ of PL is *valid* in PL iff $A_1, \ldots, A_n \vDash_{PL} B$.
3. A set of formulas $\{A_1, \ldots, A_n\}$ is *consistent* in PL iff there exists an interpretation $\Im$ such that $\mathfrak{V}_\Im(A_1) = t, \ldots, \mathfrak{V}_\Im(A_n) = t$.

It is not hard to see that the concepts of logical consequence and consistency are intimately related:

- A set of formulas $\{A_1, \ldots, A_n, B\}$ is inconsistent (i.e., not consistent) in PL iff $A_1, \ldots, A_n \vDash_{PL} \neg B$.

To see this, observe that $A_1, \ldots, A_n \vDash_{PL} \neg B$ iff there exists no interpretation $\Im$ such that $\mathfrak{V}_\Im(A_i) = t$ for $i = 1, \ldots, n$ and $\mathfrak{V}_\Im(\neg B) = f$. But $\mathfrak{V}_\Im(\neg B) = f$ iff $\mathfrak{V}_\Im(B) = t$. Hence, $A_1, \ldots, A_n \vDash_{PL} \neg B$ iff there exists no interpretation $\Im$ such that $\mathfrak{V}_\Im(A_i) = t$ for $i = 1, \ldots, n$ and $\mathfrak{V}_\Im(B) = t$, but that is tantamount to saying that $\{A_1, \ldots, A_n, B\}$ is inconsistent.

2.2 Natural Deduction for PL

Let us now turn to the syntactic or proof-theoretic analogues of these notions. The proof system introduced here is a variant of natural deduction. It originates with Gerhard Gentzen (1934/1969) and was designed as an alternative to existing Frege–Hilbert axiomatic systems, with the aim of representing deductive reasoning more closely. "$\Gamma \vdash A$" is to be read as "The formula A is derivable in our system from the set of formulas Γ." For the most part, each connective is characterized in terms of a pair of rules. The introduction rules ($\ldots$-I) state the conditions under which a proposition with the connective in question figuring as its main connective can be deduced; the elimination rules ($\ldots$-E) state the deductive consequences of having derived such a proposition. The system, which we refer to as ND_{PL}, is constituted by the following deductive rules:

$$\wedge\text{-I} \frac{\Gamma_1 \vdash A \quad \Gamma_2 \vdash B}{\Gamma_1 \cup \Gamma_2 \vdash A \wedge B} \qquad \wedge\text{-E} \frac{\Gamma \vdash A_1 \wedge A_2}{\Gamma \vdash A_i}$$

$$\vee\text{-I} \frac{\Gamma \vdash A_i}{\Gamma \vdash A_1 \vee A_2} \qquad \vee\text{-E} \frac{\Gamma_1 \vdash A \vee B \quad \Gamma_2, A \vdash C \quad \Gamma_3, B \vdash C}{\Gamma_1 \cup \Gamma_2 \cup \Gamma_3 \vdash C}$$

$$\to\text{-I} \frac{\Gamma, A \vdash B}{\Gamma \vdash A \to B} \qquad \to\text{-E} \frac{\Gamma_1 \vdash A \quad \Gamma_2 \vdash A \to B}{\Gamma_1 \cup \Gamma_2 \vdash B}$$

$$\neg\text{-I} \frac{\Gamma, A \vdash \bot}{\Gamma \vdash \neg A} \qquad \neg\text{-I} \frac{\Gamma_1 \vdash A \quad \Gamma_2 \vdash \neg A}{\Gamma_1 \cup \Gamma_2 \vdash \bot}$$

$$\text{ECQ} \frac{\Gamma \vdash \bot}{\Gamma \vdash A} \qquad \text{DNE} \frac{\Gamma \vdash \neg\neg A}{\Gamma \vdash A}$$

where, in ($\wedge$-E) and ($\vee$-I), $i = 1$ or $i = 2$.

"$\bot$" denotes a contradictory proposition. "ECQ" abbreviates "ex contradictione [sequitur] quodlibet" (i.e., "from a contradiction anything follows"). "DNE" abbreviates "double negation elimination." We can then say that

- for all formulas $A_1, \ldots, A_n$ and B of $\mathscr{F}_{PL}$: B is *derivable from* $A_1, \ldots, A_n$ in ND_{PL} (in symbols: $A_1, \ldots, A_n \vdash_{ND_{PL}} B$) just in case there is a derivation employing only ND_{PL} rules and relying solely on premises $A_1, \ldots, A_n$.

ND_{PL} can be proved to be *sound*:

Soundness: If $A_1, \ldots, A_n \vdash_{\mathrm{ND_{PL}}} B$, then $A_1, \ldots, A_n \vDash_{\mathrm{PL}} B$, and *complete* with respect to PL:

Completeness: If $A_1, \ldots, A_n \vDash_{\mathrm{PL}} B$, then $A_1, \ldots, A_n \vdash_{\mathrm{ND_{PL}}} B$.

3. First-Order Logic

The language of FOL is expressively richer than that of PL, which manifests itself in two principal respects: (1) FOL is sensitive to logically relevant subsentential structure, and (2) FOL is able to represent quantified sentences such as "All donkeys have soft noses" or "Some people annoy everybody." Since PL lacked the expressive resources to represent such sentences, it had to treat the propositions so expressed as unanalyzable. Consequently, PL is blind to the validity even of simple syllogisms such as

All donkeys have soft noses. Camillo is a donkey. Therefore, Camillo has a soft nose.

FOL allows us to capture not only arguments such as these but also propositions involving multiple generality, where several quantifiers interact, as in our previous example "Some people annoy everybody."

Let us turn to the vocabulary of the language of FOL, $\mathscr{L}_{\mathrm{FOL}}$:[4]

1. Descriptive symbols:
   (a) Countably many individual constants: $a_1, a_2, a_3, \ldots$
   (b) Countably many predicate symbols: $P_1^n, P_2^n, P_3^n \ldots$ (for all arities $n \geq 1$)
2. Logical symbols:
   (a) Connectives: $\neg, \wedge, \vee, \rightarrow, \leftrightarrow$
   (b) Quantifiers: $\exists$ ("there is"/"some"), $\forall$ ("all"/"every")
   (c) Countably many individual variables: $x_1, x_2, x_3, \ldots$
3. Auxiliary symbols:
   (a) ( )
   (b) ,

Different applications of FOL will necessitate different choices of descriptive symbols. Each choice determines a specific language of first-order logic. What is common to all such languages are the logical symbols. The propositions expressed by "Camillo is a donkey" and "All donkeys have soft noses" can be represented by the formulas "$D(c)$," where "c" is a constant—comparable to a proper name—that denotes Camillo, "D" is the predicate "is a donkey," and "S" is the predicate "has a soft nose." The latter sentence can then be represented by "$\forall x\,(D(x) \rightarrow S(x))$," where "$\forall x \ldots$" is to be read as "for

all $x, \ldots$"; roughly, the sentence can be paraphrased as "For all things x, if x is a donkey, then x has a soft nose."

In the following, I use lowercase letters "t," "c," "v" as meta-variables for singular terms, constants, and variables, respectively. Note that for all t, t is a singular term in $\mathscr{L}_{\mathrm{FOL}}$ iff t is an individual constant or an individual variable. We are now in a position to define the set of formulas of FOL, $\mathscr{F}_{\mathrm{FOL}}$:

If P^n is an n-ary predicate and $t_1, \ldots, t_n$ are singular terms, then $P^n(t_1, \ldots, t_n) \in \mathscr{F}_{\mathrm{FOL}}$. These are the atomic formulas of $\mathscr{L}_{\mathrm{FOL}}$.

If $A \in \mathscr{F}_{\mathrm{FOL}}$, then so is $\neg A$.

If $A \in \mathscr{F}_{\mathrm{FOL}}$ and $B \in \mathscr{F}_{\mathrm{FOL}}$, then so are $(A \wedge B)$, $(A \vee B)$, $(A \rightarrow B)$, and $(A \leftrightarrow B)$.

If $A \in \mathscr{F}_{\mathrm{FOL}}$ and v an individual variable, then $\exists v A \in \mathscr{F}_{\mathrm{FOL}}$.

If $A \in \mathscr{F}_{\mathrm{FOL}}$ and v an individual variable, then $\forall v A \in \mathscr{F}_{\mathrm{FOL}}$.

Only strings of symbols constructed in conformity with the clauses above qualify as well-formed formulas (wffs).

We can again define the notion of an immediate subformula in FOL:

- If A is atomic, then $\mathrm{ISF}(A) = \varnothing$.
- If $A = \neg B$, then $\mathrm{ISF}(A) = \{B\}$.
- If $A = (B \cdot C)$ (where $\cdot \in \{\wedge, \vee, \rightarrow, \leftrightarrow\}$), then $\mathrm{ISF}(A) = \{B, C\}$.
- If $A = \exists v B$, then $\mathrm{ISF}(A) = \{B\}$.
- If $A = \forall v B$, then $\mathrm{ISF}(A) = \{B\}$.

3.1 Semantics for FOL

The semantics for FOL takes a slightly different form due to the fact that we need an apparatus to assign semantic values to subsentential expressions. Moreover, we must specify a domain of objects for our quantifiers to range over.

- An *interpretation*, $\mathfrak{I}$, in FOL is a pair $\langle D, \varphi \rangle$ where
  1. D is a nonempty set of objects,
  2. φ is an interpretation function such that
  3. for all individual constants c, $\varphi(c) \in D$,
  4. for all n-ary predicates P^n, $\varphi(P^n) \subseteq D^n$, where D^n is the n-fold Cartesian product of D, that is, the set of n-tuples of elements of D.

Consider the atomic formula $P(x)$. Suppose we put $D = \mathbb{N}$ (where $\mathbb{N}$ is the set of natural numbers) and interpret P by letting $\varphi(P)$ be the set of prime numbers. The formula $P(x)$ cannot be said to be true or false, because we do not know what x refers to. We could prefix the formula with a quantificational expression as in $\exists x\,P(x)$, in which case the variable is said to be "bound" (as opposed to "free") and the formula does obtain a truth-value: it is true because the set of prime numbers is nonempty. Alternatively, we could replace x by a constant c, which we might interpret by setting $\varphi(c) = 28$. The result, $P(c)$, again has a truth-value (it is false because 28 is no prime number). Individual variables thus function somewhat like demonstratives (e.g., "that"): without the benefit of appropriate contextual information, we cannot determine their referent. But here is the challenge. As in the case of PL, we want to explain the truth-value of an FOL formula as a function of the truth-values of its immediate subformulas. But to account for the truth-value of, say, $\exists x\, P(x)$, we must then first account for "the" truth-value of $P(x)$. To handle such cases, we need the notion of a variable assignment. A variable assignment σ relative to a domain D is a function that assigns to each variable an arbitrary referent in D. More precisely:

- A variable assignment σ under an interpretation $\mathfrak{I} = \langle D, \varphi \rangle$ is a function that assigns to every individual variable v an element d of D.

In our example, the question as to the truth-value of $\exists x\, P(x)$ (relative to the above interpretation) turns on the question whether there exists a σ such that $\sigma(x)$ is a prime number. That is, we effectively account for quantification by appeal to quantification over variable assignments in the metalanguage.

For convenience sake, let us merge our variable assignments with our interpretation function. That is, given an interpretation $\mathfrak{I} = \langle D, \varphi \rangle$, the interpretation function φ and any variable assignment σ (under $\mathfrak{I}$) can be folded into a single function, φ_σ, which now covers all singular terms—individual constants and individual variables:

- Let $\mathfrak{I} = \langle D, \varphi \rangle$ be an interpretation in FOL and σ a variable assignment under $\mathfrak{I}$; then φ_σ is defined as follows:

  1. for all individual constants c: $\varphi_\sigma(c) = \varphi(c)$, and
  2. for all individual variables v: $\varphi_\sigma(v) = \sigma(v)$.

We have now outlined the machinery needed for determining whether a formula—any formula, including those with free variables—is true or false relative to a variable assignment. Just as in the case of propositional logic, what we are after is a function that assigns to every formula a truth-value on the basis of an interpretation and a variable assignment under that interpretation. We will again use φ_σ to designate that function. So, $\varphi_\sigma(A)$ designates the truth-value of the formula A relative to the interpretation $\mathfrak{I} = \langle D, \varphi \rangle$ and given the variable assignment σ. It is important to be clear about the double role the function φ_σ is playing here. In the last paragraph, φ_σ

played the role of a function that took singular terms—individual constants and individual variables—as inputs. Now we are charging it with the further task of taking formulas as inputs and outputting truth-values (relative to the interpretation and the variable assignment in question). In practice, the context will make it abundantly clear whether the function φ_σ is being applied to a singular term or to a formula.

With this we are in a position to define valuation functions for first-order logic:

- Let $\mathfrak{I} = \langle D, \varphi \rangle$ be an interpretation in first-order logic and σ a variable assignment under $\mathfrak{I}$. A *valuation* in first-order logic (relative to $\mathfrak{I}$ and σ) is a function φ_σ defined for all singular terms (as we explained in the previous section), which moreover assigns to every formula A of $\mathscr{L}_{\text{FOL}}$ of a given language of first-order logic the value t or f so as to satisfy the following semantic rules:

  1. $\varphi_\sigma(P^n(t_1, \ldots, t_n)) = $ t iff $\langle \varphi_\sigma(t_1), \ldots, \varphi_\sigma(t_n) \rangle \in \varphi(P^n)$,[5]
  2. $\varphi_\sigma(\neg A) = $ t iff $\varphi_\sigma(A) = $ f,
  3. $\varphi_\sigma((A \wedge B)) = $ t iff $\varphi_\sigma(A) = \varphi_\sigma(B) = $ t,
  4. $\varphi_\sigma((A \vee B)) = $ t iff $\varphi_\sigma(A) = $ t or $\varphi_\sigma(B) = $ t,
  5. $\varphi_\sigma((A \rightarrow B)) = $ t iff $\varphi_\sigma(A) = $ f or $\varphi_\sigma(B) = $ t,
  6. $\varphi_\sigma((A \leftrightarrow B)) = $ t iff $\varphi_\sigma(A) = \varphi_\sigma(B)$,
  7. $\varphi_\sigma(\forall v A) = $ t iff for all variable assignments σ' under $\mathfrak{I}$, if σ' is a v-variant of σ, then $\varphi_{\sigma'}(A) = $ t,
  8. $\varphi_\sigma(\exists v A) = $ t iff there is a variable assignment σ' under $\mathfrak{I}$ such that σ' is a v-variant of σ and $\varphi_{\sigma'}(A) = $ t.

If $\mathfrak{I} = \langle D, \varphi \rangle$ and $\varphi_\sigma(A) = $ t, we say that A is *true* relative to the interpretation $\mathfrak{I}$ and the variable assignment σ under $\mathfrak{I}$ (similarly for falsity).

Of course, to complete our definition, we must still define the notion of a v-variant:

- Let σ, σ' be variable assignments under an interpretation $\mathfrak{I}$. Then σ' is a *v-variant* of σ iff for all individual variables $v' \neq v$ it is the case that $\sigma'(v') = \sigma(v')$.

Hence, if σ' is a v-variant of σ, then σ' agrees with σ on all individual variables except possibly with respect to v, where the two may (but need not) diverge. It follows that every variable assignment is a v-variant of itself (with respect to any individual variable v).

We are now in a position to define our key logical notions for FOL:

- For all formulas $A_1, \ldots, A_n, B$: B is a *logical consequence* of $A_1, \ldots, A_n$ ($A_1, \ldots, A_n \vDash_{\text{FOL}} B$) iff for all interpretations $\mathfrak{I} = \langle D, \varphi \rangle$ and all variable assignments σ under $\mathfrak{I}$, if $\varphi_\sigma(A_1) = $ t, $\ldots$, $\varphi_\sigma(A_n) = $ t, then $\varphi_\sigma(B) = $ t.

- A set of formulas $\{A_1, \ldots, A_n\}$ is *consistent* in FOL iff there exists an interpretation $\mathfrak{I} = \langle D, \varphi \rangle$ and a variable assignment σ under $\mathfrak{I}$ such that $\varphi_\sigma(A_1) = \mathrm{t}, \ldots, \varphi_\sigma(A_n) = \mathrm{t}$.

3.2 Natural Deduction for FOL

Our system of natural deduction $\mathrm{ND_{PL}}$ can be extended into a system for FOL, $\mathrm{ND_{FOL}}$, by adding the following two pairs of rules for the quantifiers.

In the application of $\exists$-E, one must ensure that a does not occur in $\exists x\, A(x)$, in Γ_2, or in B.

$$\exists\text{-I}\ \frac{\Gamma \vdash A(t)}{\Gamma \vdash \exists x\, A(x)} \qquad \exists\text{-E}\ \frac{\Gamma_1 \vdash \exists x A(x) \quad \Gamma_2, A(a) \vdash B}{\Gamma_1 \cup \Gamma_2 \vdash B}$$

$$\forall\text{-I}\ \frac{\Gamma \vdash A(a)}{\Gamma \vdash \forall x\, A(x)} \qquad \forall\text{-E}\ \frac{\Gamma \vdash \forall x A(x)}{\Gamma \vdash A(t)}$$

In $\forall$-E, every free occurrence of x in A is replaced by t. In the application of $\forall$-I, one must ensure that a does not occur in Γ, and x must be uniformly substituted for a and must not be bound by any other quantifier in A. If a were allowed to occur among the hypotheses Γ upon which the conclusion depends, the following would be possible:

$$\forall\text{-I}\ \frac{A(t) \vdash A(t)}{A(t) \vdash \forall x\, A(x)}$$

Similar examples can readily be given for the remaining restrictions. As in the case of PL, FOL can be shown to be sound and complete with respect to $\mathrm{ND_{FOL}}$.[6]

4. Logic and Rationality

Having thus reminded ourselves of the key concepts and features of PL and FOL, let us return to the supposed connection between logic and rationality we sketched in the introduction.[7] How exactly do the laws of logic express constraints on rational belief?

We do sometimes talk as if logic were itself a theory of rational belief. For instance, the rules of our natural deduction systems are typically referred to as "inference rules" and are read accordingly (e.g., "From A and B, infer $A \wedge B$"). We already noted above that such talk cannot be taken literally. According to Gilbert Harman (1984, 1986), our talk of "inference rules" and the underlying identification of logic and rational belief formation is simply a category mistake. Logic is one thing; a theory of epistemic rationality is quite another. Logic has an abstract subject matter; it concerns itself with certain properties and relations of (sets of) propositions. A theory of rationality, by contrast, aims to give a normative account of how we should form and

revise our beliefs. Once we recognize the fundamental difference between these two theoretical enterprises, according to Harman, we realize that an explanatory gap separates them. Harman is doubtful that there is a substantive and informative story to be told as to how that gap might be closed.

But perhaps this is too quick. Even if deductive logic and a theory of rationality do not come to the same thing, there may still be an interesting normative connection between the two. Following John MacFarlane (2004), let us call a general principle that seeks to articulate such a relation between logic and norms governing belief, a *bridge principle*. Schematically, a bridge principle can be represented as follows:

(*) If $A_1, \ldots, A_n \vDash C$, then $N(\alpha(A_1), \ldots, \alpha(A_n), \beta(C))$.

Such principles take the form of a material conditional the antecedent of which expresses an instance of logical consequence and the consequent of which states a requirement of rationality governing the doxastic attitudes whose contents are so related: "N" stands for the governing norm, and "α" and "β" are (possibly distinct) doxastic attitudes borne toward the propositions A_i. On the basis of this blueprint, we can generate a number of distinct bridge principles by varying the following parameters:

- Doxastic attitude: Does the principle govern full beliefs or credences (i.e., degrees of belief)?

- Constraint on antecedent of the conditional: How (if at all) is the antecedent to be constrained? Does the norm cover all logical consequences? Or perhaps only those that are known, believed, obvious, and so on?

- Deontic operator: Which deontic modal operator features in the requirement expressed in the consequent? Is it a claim about what the agent *ought*, has *permission*, or has (defeasible) *reason* to "do"?

- Scope of operator: Typically, the principle's consequent itself takes the form of a conditional—call it the embedded conditional. Ought the deontic operator O take wide scope with respect to it (i.e., $O(A \to B)$)? Ought it to take narrow scope ($A \to O(B)$)? Should it attach to both the consequent and the antecedent ($O(A) \to O(B)$)? Or should we think of the normative claim as involving a primitive, undecomposable conditional operator ($O(B \mid A)$)?

- Polarity: What is the polarity of the normative claim? Is it a positive demand that the agent have a particular attitude? Or is it a negative demand that the agent not have a particular attitude?

- Synchronic–diachronic: Are the principles synchronic, instructing us how the agent's doxastic state ought to be at any given moment in time? Or are they requirements about how one's beliefs should evolve over time?

A few examples may be useful.

1. If $A_1, \ldots, A_n \vDash C$, then S has reason to believe C if S believes each of the A_i.

2. If $A_1, \ldots, A_n \vDash C$, then S's credences ought to be such that[8] $cr(C) \geq \sum_{1 \leq i \leq n} cr(A_i) - (n-1)$.

3. If S knows that $A_1, \ldots, A_n \vDash C$, then S ought not simultaneously disbelieve C and believe each of the A_i.

4. If S believes that $A_1, \ldots, A_n \vDash C$, then, if S is permitted to believe each of the A_i, S is permitted to believe C.

Principles 1 and 2 have unconstrained antecedents. In the first, the normative claim in the consequent features the reason-operator, which takes narrow scope with respect to the consequent of the embedded conditional and governs full belief. The second principle is familiar from probability logic (Adams, 1998) and has been advanced as a bridge principle by Field (2009, 2015). It governs credences, and ought effectively takes wide scope in the consequent. Principle 3 differs in that it is restricted to known consequences and of negative polarity. It too employs the ought-operator, which takes wide scope over the embedded conditional, although it governs full belief. Principle 4 is relativized to a nonfactive attitude. It features the permission-operator, which attaches both to the antecedent and to the consequent of the embedded conditional and also governs full beliefs.

For simplicity, I am assuming that all principles are to be understood as synchronic principles. That is, principle 1 is to be read as

(Synchronic)   If S recognizes (at t) that $A_1, \ldots, A_n \vDash C$, then, if S believes each of the A_i at t, S has reason to believe C at t

as opposed to

(Diachronic)   If $A_1, \ldots, A_n \vDash C$, then, if S believes each of the A_i at t and t slightly precedes t', S has reason to believe C at t' (and from t' onward).

If there is an interesting connection between logic and rationality, there should be a viable bridge principle that articulates it. But what counts as a viable bridge principle? We can distinguish two approaches. On one approach (Field, 2009, 2015; MacFarlane, 2004; Milne, 2009; Steinberger, 2017b), we lay down various criteria of adequacy for our bridge principles. The principles that perform sufficiently well against these criteria are our contenders. The second approach—which I consider in section 6—proceeds from more fundamental epistemological

principles and seeks to derive a bridge principle from them (assuming such a principle is to be had).

5.   What Makes for a Viable Bridge Principle?

Let us begin by considering the former approach. The adequacy criteria appealed to in the literature derive in part from Harman's objections to potential bridge principles and in part from MacFarlane (2004). We can summarize them as follows:

Belief Revision: Suppose I believe both A and $A \rightarrow B$ (as well as accepting modus ponens). The mere fact that I have these beliefs and that I recognize them to jointly entail B does not normatively compel any particular attitude toward B on my part. In particular, it is not generally the case that I ought to, or have permission to, believe B. After all, B may be at odds with my evidence, and so it would be unreasonable of me to follow modus ponens slavishly by, as it were, "adding B to my belief box." The rational course of "action," rather, when B is untenable, is for me to relinquish my belief in at least one of my antecedent beliefs A and $A \rightarrow B$ on account of their unpalatable implications. Belief revision seems to tell against narrow-scope principles, at least against strict versions of such principles involving ought and permission. Similarly, because of the reflexivity of the consequence relation, these principles would imply that one ought to (or is permitted to) believe anything one in fact believes, which clearly seems problematic. Pinder (2017) defends a reasons-based narrow-scope principle (namely, principle 1 above) against such objections.[9]

Excessive Demands: Principles whose antecedents are unrestricted pose excessive demands on agents whose resources of time, computational power, stamina, and so on are limited. Take an ought-based narrow-scope principle according to which one's beliefs ought to be closed under logical consequence. Anyone who believes the axioms of Peano–Dedekind arithmetic ought to believe every last one of its theorems, even if a theorem's shortest proof has more steps than there are particles in the universe. But if the logical *ought* implies *can* (in the sense of what agents even remotely like us can do), such principles must be rejected.[10]

Clutter Avoidance: A related worry is this. Any of the propositions I believe entails an infinite number of propositions that are of no significance to me whatsoever. Not only do I not care about, say, the disjunction "Vienna is the capital of Austria or pigs can fly" entailed by my belief that Vienna is the capital of Austria, but it would be positively irrational for me to squander my meager cognitive resources on inferring trivial implications of my beliefs that are of no relevance to my goals.

Epistemic Paradoxes: Some maintain that there are various types of epistemic situations in which it is arguably not merely excusable for an agent to have logically incoherent beliefs but where such incoherence is permissible or even rationally mandated. The Preface Paradox arguably is a case in point (see Makinson, 1965). Here is a summary: suppose I am the author of a nontrivial nonfiction book. Having scrupulously checked the evidence for every one of my claims $A_1, \ldots, A_n$, I am highly confident (and believe) in each of them. I am also highly confident that at least one of my claims is fallacious. After all, I am fallible. Call the proposition that $\neg(A_1 \wedge \ldots \wedge A_n)$ the "preface proposition" (P). Clearly, my beliefs are inconsistent. Moreover, given reasonable assumptions, in believing P, I ought to disbelieve a straightforward logical consequence of my beliefs, namely, their conjunction $(A_1 \wedge \ldots \wedge A_n)$. The Preface Paradox presumably poses problems for all principles involving full belief (although credence-based principles may get around it).

The Strictness Test: At least when it comes to ordinary, readily recognizable logical implications leading to conclusions that the agent has reason to consider, the logical obligation should be strict: there is something amiss about an agent who endorses the premises and yet disbelieves the conclusion on account of stronger countervailing reasons (MacFarlane, 2004, p. 12).[11] The Strictness Test prima facie represents a strike against principles featuring the reason-operator, which countenances cases in which an agent believes the premises but disbelieves the conclusion on account of sufficient independent reasons for doing so.

The Priority Question: The attitudinal variants have a distinctive advantage when it comes to dealing with Excessive Demands worries. But relativizing one's logical obligations to, for example, one's believed or recognized logical consequences invites problems of its own, according to MacFarlane (2004): "We seek logical knowledge so that we will know how we ought to revise our beliefs: not just how we *will* be obligated to revise them when we acquire this logical knowledge, but how we are obligated to revise them even now, in our state of ignorance" (p. 12).

Logical Obtuseness: Suppose someone professes to believe A and B but refuses to take a stand on (neither believes nor disbelieves) the conjunction $A \wedge B$.

Intuitively, such a person is liable to criticism. Whereas the weaker (positive) reason-based principles fail to live up to the Strictness Test, they do not commit the sin of logical obtuseness since one at least has reason to believe, or not disbelieve, $A \wedge B$. Not so for principles with negative polarity. So long as the agent does not actively disbelieve $A \wedge B$, our negative bridge principles find no fault with cases like these. If this intuition carries any weight, negative principles may prove to be ultimately too weak (at least on their own).

There is little consensus over which bridge principle fares best in light of these desiderata (for an overview, see Steinberger, 2017b). One important sticking point is how to deal with the Preface Paradox and similar cases. MacFarlane (2004) argues that we must simply resign ourselves to the existence of an irresolvable normative conflict between the demands placed on us by logic and other epistemic norms. Field (2009) endorses principle 2, which constrains credences. The principle allows for one's credences in a consequence jointly entailed by all the propositions (e.g., their conjunction) to be very low, despite one's having high credence in each of the individual propositions. The principle is compatible with a broadly Bayesian view of credences, according to which a rational agent's credence function is (or is extendible to) a probability function. Another important issue is the question of the permissible level of idealization of a requirement of rationality. If our aim is to formulate principles of ideal rationality, Clutter Avoidance and Excessive Demands may matter less. But how, then, do these principles relate to ordinary agents like you and me (Harman, 1986)?

I argue (Steinberger, 2017a, 2019) that the approach as a whole is marred by our failure to distinguish three importantly different ways in which logic might be normative:

Directives provide first-personal guidance in the process of doxastic deliberation.

Evaluations serve as objective third-personal standards or ideals for classifying acts or states into correct and incorrect ones.

Appraisals serve as the basis for our (equally third-personal) criticisms of our epistemic peers and so underwrite our attributions of praise and blame.

To illustrate, consider the following act-utilitarian principle:

AU You ought to act in such a way as to maximize net happiness.

The principle might be an apt evaluative norm in that it serves as a metric for what is to count as a right action. Yet the norm is often of little help to an agent trying to figure out what to do. Typically, it will not be transparent to her which of the actions available to her maximize happiness. The norm offers the agent little by way of guidance and therefore is not fit to play the role of a directive.[12] Similarly, if our agent violates the evaluative utilitarian norm, she may nevertheless not be liable to criticism. Despite having violated the norm, she may have acted reasonably in light of how the situation presented itself to her. Conversely, she might have acted recklessly and yet, out of sheer luck, complied with the norm. In both cases, our appraisals and our evaluations come apart, and this is largely because our appraisals are, while our evaluations need not be, sensitive to the agent's perspective.

I argue (Steinberger, 2019) that this tripartite distinction reveals that the actors in this debate talk past one another on account of conceiving of the normative role of logic differently. Another advantage of the proposed analysis is this. The adequacy criteria are in tension with one another, leading participants in the debate to make unprincipled choices about which desiderata to discount in order to make a case for their favored principle(s). The threefold distinction has the virtue of explaining why our adequacy criteria are inconsistent and points us toward a philosophically well-motivated resolution. The criteria are inconsistent because they are motivated by the aforementioned incompatible conceptions of the normative role of logic. Once we distinguish between directives, evaluations, and appraisals, we find that each normative role is naturally associated with its own set of adequacy criteria, which form consistent subsets of our desiderata. Hence, for each normative role, there is a separate, well-defined, and—it is hoped—more tractable question as to whether logic is normative in that sense.

Even if my analysis is correct, though, it remains to be seen which (if any) of these normative roles could give rise to principles of rationality.

6. Logic and Epistemic Utility Theory

Let us now turn to the second approach. On this view, norms of epistemic rationality are to be derived from fundamental epistemic aims or values. Several candidates could be in the running: truth, knowledge, understanding, and so on. Also, we may ask whether we should be monists or pluralists about epistemic value.

Finally, given the value(s) at the center of our epistemology, what is its normative structure and how does it give rise to a theory of epistemic rationality? Rather than attempting a necessarily superficial survey of the options here, I propose to study a concrete proposal: epistemic utility theory (EUT). The prevalent version of EUT elects truth as its sole and fundamental epistemic value. What is more, the theory's value-theoretic superstructure is broadly consequentialist: any principle of epistemic rationality must earn its keep by promoting the aim of truth.

William James's famous slogan "Believe truth! Shun error!" serves as a good starting point.[13] While belief aims at the truth, our attempts at achieving it are subject to this double imperative. If my sole aim was the maximization of true belief without any concern for error, I might as well believe any proposition whatsoever. Conversely, if I was so cautious as to wish to avoid error at all costs, I would be best off suspending belief across the board. Our challenge, then, is to strike the optimal balance between maximizing true beliefs and avoiding false ones. EUT proposes to bring the tools of decision theory to bear on the problem.[14]

Let us consider the case of full belief. Suppose for simplicity that you are entertaining propositions belonging to a finite set $\mathscr{P}$. For each proposition in $\mathscr{P}$, the agent may believe (B), disbelieve (D), or suspend belief (S). Formally, we can represent the agent's various possible "choices" over $\mathscr{P}$ as belief functions $b: \mathscr{P} \to \{B, S, D\}$. Our belief functions are to be assessed based on their accuracy. Our standard of accuracy is the truth-value of the proposition at a possible world, where a world is simply represented by means of a consistent valuation function $w: \mathscr{P} \to \{t, f\}$.

In analogy with decision theory, we can conceive of each choice of a doxastic attitude relative to a world as producing a certain epistemic utility, which is represented by a numerical value. Epistemic utility can be represented by a further function that associates such a value to any attitude–truth-value pair:

$$eu: \quad \{B, D, S\} \times \{t, f\} \to (-\infty, \infty).$$

That is, for any proposition $A \in \mathscr{P}$, eu returns a score as a function of one's attitude toward it and the proposition's truth-value at the world. R is the score for getting it right, $-W$ is the penalty one incurs for getting it wrong, and suspending yields a neutral score. Hence,

$$eu(B, t) = eu(D, f) = R,$$
$$eu(S, t) = eu(S, f) = 0,$$
$$eu(B, f) = eu(D, t) = -W.$$

How are we to conceive of the relative values of R and W? There are three options:

- The *epistemic radical* values true belief to a higher degree than she disvalues false belief: $R > W$.
- The *epistemic centrist* values both to the same degree: $R = W$.
- The *epistemic conservative* disvalues false belief to a higher degree than she values true belief: $W > R$.

James believed one's stance toward the question to be a matter of intellectual temperament. There is, however, a prima facie case to be made for conservativeness. Suppose I flip a fair coin. Let p be the proposition that the coin lands heads. What attitude is it rational for me to adopt with respect to p? It is clear that in the absence of further information I should suspend. However, consider the following decision matrix:

p	$\neg p$	BB	BD	SS	BS
t	f	$R - W$	$2R$	0	$R+0$
f	t	$-W+R$	$-2W$	0	$-W+0$

(The remaining cases—DD, DB, SB, DS, SD—are strictly analogous and so can be omitted.) For the epistemic radical, the optimal choice is to believe p and disbelieve $\neg p$, or vice versa. But blindly going out on a limb in this way seems reckless. The centrist is no better off. She is indifferent between believing (or disbelieving) both p and its contradictory, believing p and disbelieving $\neg p$ (or vice versa), and suspending.[15] Conservatism, then, is our best option: the disvalue of believing falsely outstrips the value of believing truly. Since having contradictory beliefs always adds a net negative to one's score ($R - W$), one would be irrational to do so.

Having thus fixed the relative value of our rewards and penalties, we can determine the overall epistemic utility of a belief function at a world. Epistemic utility is generally assumed to be additive:[16] the overall epistemic utility of b, $EU(b)$, is

$$EU(b) = \sum_{A \in \mathscr{P}} eu(b(A), w(A)).$$

In general, of course, we are interested in the actual world and will do our best to choose the belief function with the greatest actual epistemic utility. However, we generally do not know which of our beliefs are accurate. The score of a belief function then hinges on facts about the world of which we are uncertain. What we can be certain of, though, is that a belief function that has less epistemic utility than another, *however* the world turns

out to be, can be eliminated. It would be plainly irrational to adopt it. Not only is such a belief function bound to be suboptimal, but I might have appreciated its suboptimality even in the absence of any knowledge about the actual world. Decision theory captures this form of irrationality in terms of the notion of dominance:

- A belief function b is *strictly dominated* by a belief function b' iff $EU(b', w) > EU(b, w)$ for all worlds w.
- A belief function b is *weakly dominated* by a belief function b' iff $EU(b', w) \geq EU(b, w)$ for all worlds w and there exists a world w' such that $EU(b', w') > EU(b, w')$.

Two brief remarks. First, one option's being dominated by another makes choosing it irrational only if the dominating option is not itself dominated. Consider this. A genie grants me a (single) cash reward: for any number n I care to specify, I receive exactly \$$n$. But of course, any choice of n is bound to be dominated. That is not to say, though, that there is no rational response (see Pettigrew, 2016a). Second, note that a belief function b's being dominated by another (undominated) function b' only shows that it would be irrational to opt for b. It says nothing about which belief function one should adopt (in particular, it does not in general recommend adopting b').

Following Easwaran (2016), we can say that a belief function b is *strongly coherent* just in case it is not even weakly dominated and that b is *weakly coherent* just in case it is not strongly dominated. The following rationality requirement thus falls right out of EUT's central commitments:

Strong Coherence: One ought to have strongly coherent beliefs.

Central for our purposes is the question of how logical principles of rationality relate to Strong Coherence and, more generally, whether they can be justified in the context of EUT.

First, let us observe that Strong Coherence requires that one believe all logical truths:

Logical Truth: If S's belief function is strongly coherent, then S believes all logical truths (in $\mathscr{P}$).

The following table illustrates why:

	B	S	D
A	R	0	$-W$
$\neg A$	$-W$	0	R

Suppose A is a logical truth. It follows that the second line in bold can be ignored because there is no logically possible world at which A is false. The score of one's belief function is thus determined on the basis of the first row of the table alone. Thus, any belief function b that suspends belief in, or disbelieves, a tautology will have less epistemic utility at every possible world than a belief function b' that agrees with b on all propositions aside from A, which it believes.

Next let us consider the following weak single-premise bridge principle. The principle is wide in scope and of negative polarity. Moreover, it is restricted to $\mathscr{P}$.

Single-Premise Closure: For all $A, C \in \mathscr{P}$, if $A \vDash C$, then S ought not both believe A and disbelieve C.

It turns out that that Strong Coherence entails Single-Premise Closure. Suppose belief function b violates Single-Premise Closure. That is, assume that $A, C \in \mathscr{P}$ and b believes A and disbelieves C. Then b is strictly dominated by the belief function b', which suspends on both propositions, as the following table illustrates (where, as above, the second line in bold, w_2, can be eliminated):

$A \vDash C$	A	C	BD	SS
w_1	t	t	$R - W$	0
w_2	**t**	**f**	**2R**	**0**
w_3	f	t	$-2W$	0
w_4	f	f	$-W + R$	0

The question is whether analogous results are to be had for the following two familiar requirements:

Multiple-Premise Closure: For all $\{A_1, \ldots, A_n, C\} \subseteq \mathscr{P}$, if $A_1, \ldots, A_n \vDash C$, then S ought not both believe all of the A_i and disbelieve C.

Consistency: If $\{A_1, \ldots, A_n\}$ is a logically inconsistent set, then S ought not believe each member of the set.

At the risk of ruining the suspense: Strong Coherence entails neither Multiple-Premise Closure nor Consistency. To see this, a small detour is necessary. First, let us introduce the notion of expected epistemic utility. The expected epistemic utility of a belief function b is the probability-weighted sum of its epistemic utilities across all possible worlds:

$$EEU_P(b) = \sum_{w \in W} P(w) EU(b, w).$$

The probability function P can be thought of as the subject's evidence or at least as partially determined by her evidence. The probability function might be interpreted as a credence function, as an evidential probability function, or as representing objective chances, depending on one's conception of evidence.

It can be shown that it is a sufficient condition for a belief function to be strongly coherent that there be a regular probability function relative to which it maximizes expected utility.[17] What is more, it can be shown that a belief function b has maximal expected utility just in case there exists a probability function P such that for any proposition A,[18]

- $b(A) = B$ iff $1 \geq P(A) \geq \dfrac{W}{R + W}$,

- $b(A) = S$ iff $\dfrac{W}{R + W} \geq P(A) \geq \dfrac{R}{R + W}$,

- $b(A) = D$ iff $\dfrac{R}{R + W} \geq P(A) \geq 0$.

It is noteworthy that if P is interpreted as the agent's credence function, this result yields a version of the so-called Lockean thesis, according to which fully (rationally) believing is (in a sense to be made precise) having credence in excess of a threshold. The threshold is set by the appropriate ratio between R and W.[19]

The upshot is that belief functions can maximize expected utility (and hence be strongly coherent) while satisfying neither Multiple-Premise Closure nor Consistency. The Preface Paradox can again serve as an example. Suppose our author entertains the propositions $A_1, \ldots, A_n$, which compose the body of her book, and that she has rational uniform credences with respect to these propositions. That is, her credence function cr is (or is extendible to) a probability function, and she has the same high credence, say, $cr(A_i) = .9$, with respect to each of the propositions. Assuming that $.9 \geq \dfrac{W}{R + W}$, the belief function b that believes all of the A_i maximizes expected utility with respect to cr and so is strongly coherent. However, provided that n is sufficiently large and that the author entertains $A_1 \wedge \ldots \wedge A_n$ and its negation, we get $cr(A_1 \wedge \ldots \wedge A_n) \leq \dfrac{R}{R + W}$, and so b, if it is to maximize expected utility, disbelieves $A_1 \wedge \ldots \wedge A_n$ (and believes its negation). The belief function b is then strongly coherent and yet violates both Multiple-Premise Closure and Consistency. We can resist this conclusion only if we ensure that $W \geq (n - 1)R$. Here n (we are assuming $n > 2$) is the number of propositions in question. EUT requires consistency and full closure just in case the disvalue of error exceeds the value of true belief by a factor of $n - 1$ (see Easwaran & Fitelson, 2015).

What are we to make of this? Here is an observation. Strong Coherence and Consistency are structurally very similar. Violations are, in both cases, a priori knowable indications that one's beliefs are less accurate than alternative ones. The difference comes down to a subtle quantifier shift: if b is an incoherent belief function, then there exists a belief function b' that has at least equal epistemic utility across all worlds and outperforms it in some. That is,

$$\exists b'(\forall w\ EU(b', w) \geq EU(b, w) \wedge$$
$$\exists w'\ EU(b', w') > EU(b, w')).$$

By contrast, if b is an inconsistent belief function, then, for every world w, there exists a belief function b' such that $EU(b', w) > EU(b, w)$, that is,

$$\forall w \exists b'\ EU(b', w) > EU(b, w).$$

According to EUT, the former violation spells irrationality, while the latter does not. But why should this minor difference be of such moment? If our aim really is the truth, then why should we content ourselves with maximizing epistemic utility in cases where, because our beliefs are inconsistent, some of them are bound to be false? Not only is maximizing expected epistemic utility no necessary condition for perfect accuracy, but it may sometimes positively discourage one from holding a true belief, because doing so would come with too high an opportunity cost: it would lower the expected epistemic utility of the belief function as a whole. Cases of such counterintuitive epistemic trade-offs have been the subject of much discussion.[20]

Many of these criticisms target the epistemic value theory undergirding EUT: that epistemic rationality is teleologically structured and that principles of rationality are justified only inasmuch as they further *expected* epistemic value (see Littlejohn, 2016).

Others question the assumed monism about epistemic value: is truth really the sole source of epistemic value? And even fellow monists may disagree about the nature of the fundamental aim. Why should truth be the North Star of epistemology as opposed to, for instance, knowledge?[21] Accidental true belief is not enough, they say. We want to get it right for the right reason. Finally, Leitgeb (2013, 2017) has argued for the possibility of providing a harmonious account of doxastic rationality according to which a probabilist approach to rational credence is compatible with a logical approach to rational full belief that incorporates full consistency and closure requirements.

Notes

1. Epistemic (or theoretical) rationality, here, is contrasted with practical rationality. Roughly, the former is concerned with what we ought to believe, and the latter is concerned with what we ought to do.

2. It might be thought that the two forms of coherence really are the same. Some, most notably Niko Kolodny (2007, 2008),

maintain that there is no separate demand that beliefs should cohere. To the extent that beliefs should indeed cohere, this is because beliefs that enjoy evidential support cannot fail but cohere in the right way. I will set Kolodny's challenge aside for the purposes of this discussion. Conversely, it might be thought that one's evidence is constituted only by one's beliefs. If so, a belief is evidentially supported precisely when it coheres with the entire body of relevant beliefs.

3. It is a matter of some controversy whether requirements of rationality have genuine normative force. While I allow myself to use formulations such as "norms of epistemic rationality," I do not mean to prejudge the issue.

4. For ease of exposition, I do not include function symbols, nor am I including identity among the logical constants.

5. In the case of atomic formulas of the form $P^1(t)$, we simply identify—for uniformity's sake—the "1-tuple" $\langle d \rangle$ with d. Hence, the extensions of one-place predicates continue to be sets of elements of D and not of 1-tuples of such objects.

6. For a much more detailed exposition, the interested reader may consult the excellent collaborative open-source textbook by Magnus and Button (2019) or Enderton (2001).

7. Henceforth, I use "logic" to mean FOL unless I explicitly indicate otherwise. Similarly, "⊨" designates the consequence relation of FOL.

8. Call $u(A)$ the uncertainty in A, defined as $u(A) = 1 - cr(A)$. We can then give the following tidier reformulation of the inequality: $u(C) \leq \sum_{1 \leq i \leq n} u(A_i)$.

9. In so doing, Pinder (2017) takes issue with an argument by Steinberger (2016) to the effect that appeals to the normativity of logic do not support a case for paraconsistent revisions of our logics (logics that reject the principle of explosion, according to which from an inconsistent set of propositions any proposition whatsoever logically follows).

10. It is for worries like these that Isaac Levi (2002) spoke of doxastic commitments, which do not demand the agent to believe but rather amount to something like a promise to believe. For similar issues in epistemic and doxastic logic, see chapter 5.1 by van Ditmarsch (this handbook).

11. MacFarlane takes inspiration from Broome (2000, p. 85).

12. One need not maintain that directives must always have transparent application conditions (see Srinivasan, 2015).

13. See Pettigrew (2016c) for a formal analysis of James's insights. For an alternative analysis, see Raidl and Spohn (2019).

14. The following exposition is inspired by Easwaran (2016) and Pettigrew (2017).

15. See also Easwaran and Fitelson (2015) as to why $W \leq R$ leads to counterintuitive consequences.

16. This assumption is not essential (see, e.g., Dorst, 2019).

17. A *regular* probability function, in the present context, is one that assigns every world a nonzero probability.

18. See Easwaran (2016) for details.

19. Dorst (2019) argues for a more sophisticated variable-threshold Lockean account, which he takes to provide a metaphysical reduction of full belief to credence. Easwaran (2016), by contrast, argues in favor of the primacy of full belief.

20. See Berker (2013), Carr (2017), Greaves (2013), and Littlejohn (2012, 2015) for criticisms along these lines as they pertain to EUT and epistemic consequentialism more broadly. See Pettigrew (2016b) for a response.

21. The go-to reference for much of the contemporary debate is Williamson (2000).

References

Adams, E. W. (1998). *A primer of probability logic.* Stanford, CA: CSLI Publications.

Alchourrón, C., Gärdenfors, P., & Makinson, D. (1985). On the logic of theory change: Partial meet contraction and revision functions. *Journal of Symbolic Logic, 50*(2), 510–530.

Berker, S. (2013). Epistemic teleology and the separateness of propositions. *Philosophical Review, 122*(3), 337–393.

Broome, J. (2000). Normative requirements. In J. Dancy (Ed.), *Normativity* (pp. 78–99). Oxford, England: Oxford University Press.

Carr, J. R. (2017). Epistemic utility theory and the aim of belief. *Philosophy and Phenomenological Research, 95*(3), 511–534.

Dorst, K. (2019). Lockeans maximize expected accuracy. *Mind, 128*(509), 175–211.

Easwaran, K. (2016). Dr. Truthlove, or, How I learned to stop worrying and love Bayesian probability. *Noûs, 50*(4), 816–853.

Easwaran, K., & Fitelson, B. (2015). Accuracy, coherence, and evidence. In T. Szabo Gendler & J. Hawthorne (Eds.), *Oxford studies in epistemology* (Vol. 5, pp. 61–96). Oxford, England: Oxford University Press.

Enderton, H. B. (2001). *A mathematical introduction to logic* (2nd ed.). San Diego, CA: Academic Press.

Field, H. (2009). What is the normative role of logic? *Proceedings of the Aristotelian Society, 83*, 251–268.

Field, H. (2015). What is logical validity? In C. R. Caret & O. T. Hjortland (Eds.), *Foundations of logical consequence* (pp. 33–70). Oxford, England: Oxford University Press.

Gentzen, G. (1969). Investigations into logical deduction. In M. Szabo (Ed.), *The collected papers of Gerhard Gentzen* (pp. 68–128). Amsterdam, Netherlands: North Holland. (Original work published 1934)

Greaves, H. (2013). Epistemic decision theory. *Mind, 122*(488), 915–952.

Harman, G. (1984). Logic and reasoning. *Synthese, 60*(1), 107–127.

Harman, G. (1986). *Change in view: Principles of reasoning.* Cambridge, MA: MIT Press.

Kolodny, N. (2007). How does coherence matter? *Proceedings of the Aristotelian Society, 107*(13), 229–263.

Kolodny, N. (2008). Why be disposed to be coherent? *Ethics, 118*(3), 437–463.

Leitgeb, H. (2013). The stability theory of belief. *Philosophical Review, 123*(2), 131–171.

Leitgeb, H. (2017). *The stability of belief: How rational belief coheres with probability.* Oxford, England: Oxford University Press.

Levi, I. (2002). Commitment and change of view. In J. L. Bermudez & A. Millar (Eds.), *Reason and nature: Essays in the theory of rationality* (pp. 209–32). Oxford, England: Clarendon Press.

Littlejohn, C. (2012). *Justification and the truth-connection.* Cambridge, England: Cambridge University Press.

Littlejohn, C. (2015). Who cares what you accurately believe? *Philosophical Perspectives, 29*(1), 217–248.

Littlejohn, C. (2016). The right in the good: A defense of teleological non-consequentialism in epistemology. In K. Ahlstrom-Vij & J. Dunn (Eds.), *Epistemic consequentialism* (pp. 23–47). Oxford, England: Oxford University Press.

MacFarlane, J. (2004). In what sense (if any) is logic normative for thought? Retrieved from https://www.johnmacfarlane.net/normativity_of_logic.pdf

Magnus, P. D., & Button, T. (with additions by A. Loftis & R. Trueman, updated by A. Thomas-Bolduc & R. Zach). (2019). *For all x: Calgary: An introduction to formal logic.* Retrieved from https://forallx.openlogicproject.org/

Makinson, D. C. (1965). The paradox of the preface. *Analysis, 25*(6), 205–207.

Milne, P. (2009). What is the normative role of logic? *Proceedings of the Aristotelian Society, 83*(1), 269–298.

Pettigrew, R. (2016a). *Accuracy and the laws of credence.* Oxford, England: Oxford University Press.

Pettigrew, R. (2016b). Making things right: The true consequences of epistemic consequentialism. In K. Ahlstrom-Vij & J. Dunn (Eds.), *Epistemic consequentialism* (pp. 220–239). Oxford, England: Oxford University Press.

Pettigrew, R. (2016c). Jamesian epistemology formalised: An explication of 'The will to believe.' *Episteme, 13*(3), 253–268.

Pettigrew, R. (2017). Epistemic utility and the normativity of logic. *Logos & Episteme, 8*(4), 455–492.

Pinder, M. (2017). A normative argument against explosion. *Thought, 6*(1), 61–70.

Raidl, E., & Spohn, W. (2020). An accuracy argument in favor of ranking theory. *Journal of Philosophical Logic, 49*, 283–313.

Spohn, W. (2012). *The laws of belief: Ranking theory and its philosophical applications*. Oxford, England: Oxford University Press.

Srinivasan, A. (2015). Normativity without Cartesian privilege. *Philosophical Issues, 25*(1), 273–299.

Steinberger, F. (2016). Explosion and the normativity of logic. *Mind, 125*(498), 383–419.

Steinberger, F. (2017a). Consequence and normative guidance. *Philosophy and Phenomenological Research, 98*(2), 306–328.

Steinberger, F. (2017b). The normative status of logic. In E. Zalta (Ed.), *The Stanford encyclopedia of philosophy*. Retrieved from https://plato.stanford.edu/archives/spr2017/entries/logic-normative/

Steinberger, F. (2019). Three ways in which logic might be normative. *Journal of Philosophy, 116*(1), 5–31.

Williamson, T. (2000). *Knowledge and its limits*. Oxford, England: Oxford University Press.

3.2 Natural Logic

David P. O'Brien

Summary

Declarative memory requires a predicate–argument structure to record propositions, and this is provided by a natural logic. A natural logic requires a representational format, some inference procedures, and a program for implementing the procedures and must attend to pragmatic influences. I show how inferences from a natural logic are integrated in reasoning with inferences from other practical processes, although a reasoner is not aware of the various origins of their inferences. I argue that other theories that eschew any role for a natural logic are not adequate without being integrated with a natural logic.

1. The Historical Shift in Attitudes about Rationality and Logic

The pre-Socratic Greeks introduced the idea that humans are rational, and after Aristotle, the basic Western tradition identified human rationality with logic—a perspective that dominated the Western perspective until the end of the 19th century, with only an occasional challenge by the likes of Bacon, Hume, or Nietzsche. Indeed, Boole (1854) treated logic as the study of human mental acts, so logic and human thinking were understood as more or less equivalent and part of the essence of human nature. The transition from the 19th to the 20th century, however, brought challenges to this view as psychology separated from philosophy after the founding of the new science late in the 19th century.

Although the most obviously popular claim that humans are not basically logical is found in Freud (e.g., 1914), work by logicians also encouraged the separation of logic from human thought. Philosophers like Frege and Russell, for example, argued for the separation of logic and epistemology from psychology (e.g., Gabbay & Woods, 2009; George, 1997; Passmore, 1994), and as professional logicians have focused on finding a foundation for mathematics in logic with their emphases on

complex proofs of soundness and completeness, they were not producing something inviting to someone seeking a model for ordinary human reasoning. At the same time, the newly established field of psychology was eager to divorce itself from philosophy (Boring, 1929; Mandler, 2007), seeking psychological explanations that were not imported from philosophical logic. So psychology turned away from logical rationality to rewards, punishments, reinforcers, instincts, defense mechanisms, and the like, and philosophers turned away from psychology to construct mathematical systems of logic that provided little of obvious interest psychologically. Henle (1962) noted, however, that these movements of drift between logic and psychology were based not on evidence but on a shift in attitude.

A shift in intellectual attitude is not an argument, and the fact that the assumption of human logical rationality had a two-and-a-half-millennia history should encourage us to reconsider its abandonment, keeping in mind that many different approaches have been proposed by logicians, and the lack of fit of any of them to psychology is not an argument that human reasoning lacks a logic. The stoic logic of Chrysippus, for example, was quite different from Aristotle's logic, and one might find things in Chrysippus that are psychologically interesting in ways that Aristotle's logic is not. The logics one finds in a contemporary standard-logic textbook, and in the professional journals for logicians with their truth tables and elaborate proofs of consistency and completeness, were not constructed to be models for human thought. To discover what kind of logic should be natural for humans, we need to think about why humans need a logic, what that logic would need to include, and what it would be unlikely to include.

2. Why We Should Expect a Natural Logic and What It Should Include

I begin with the fact that humans have a declarative memory, and it is axiomatic that to record something in declarative memory, a format must exist with which to record it (Braine & O'Brien, 1998; Fodor, 1981; O'Brien & Li, 2013). This format would need to keep track of entities and the properties that attach to those entities, and people hold propositional attitudes about what is expressed in the propositions that contain the entities and their properties, so the format must keep track of what is true and what is false so that people can assert, deny, suppose, question, attack, defend, believe, and doubt their propositions. Such a format is tantamount to a predicate–argument structure, so the approach described here begins with the assumption that the mind should include a logical format with which to record propositions. This logical format should bear some interesting relationship to the kinds of standard logics one finds in logic textbooks, although by no means would we expect it to be identical to any textbook logic, and we have referred to our approach as a natural logic or a mental logic as it is intended to provide an account of reasoning rather than to construct a formal system that is sound and complete. The task is different from that of a logician. Indeed, it begins by acknowledging that completeness of human logical reasoning has not been a consideration in human bioevolutionary development, although one would want their thinking to be sound in at least some meaningful way. The first task for a natural-logic theory is to describe the format, and once the format has been established, we need to address how inferences are drawn in the format, what sort of logical semantics is required, and how the natural logic integrates with inferences that stem from pragmatic concerns.

Standard-logic textbooks provide a language that includes symbols for connectives or operators (e.g., ∧, ∨, ¬, ⊃) and quantifiers (e.g., ∀, ∃), and such symbols correspond loosely to ordinary words of daily language (e.g., *and, or, not, if, all, some*). They also provide a truth-functional semantics in the form of truth tables, with *If p then q* given the truth-functional assignment of being true unless *p* is true and *q* is false—a definition that makes true such propositions as *I did not have a gun, but if I did, then I'm guilty as charged*, which few people would want to utter.

Propositions of this latter sort exemplify the paradoxes of material implication, and they provide a serious problem for an attempt to argue that the truth table for *if* can provide a sensible model for how people reason with *if*-propositions. Our natural-logic approach thus looks elsewhere to find the meanings of logic operators. Compare the following two propositions:

All the children who went to the zoo were in the third grade.

Some of the children who went to the zoo were in the third grade.

Suppose we discover that Álvaro went to the zoo, and we ask whether we can infer that Álvaro was in the third grade. The proposition beginning with *all* provides a basis for the inference that he was in the third grade, but the proposition beginning with *some* does not. Indeed, someone who does not understand which sentence sanctions the inference and which does not could not be said to understand the meaning of the words *all* and *some*. This illustrates how the natural-logic approach understands the logical semantics for its vocabulary. The meanings of the terms are given by the inferences that they sanction. The tradition of understanding semantics in terms of the inferences they sanction follows Wittgenstein (1958), Gentzen (1935/1964), Harman (1982), and Block (1998), among others, and has become known as a "conceptual-role" or "inferential-role semantics." This approach has been presented as a general theory for semantics across the lexicon, but our natural-logic approach makes no claims beyond the vocabulary of the natural logic, and for this part of the lexicon, a semantics based on the inferences that a term sanctions is straightforward because the schemas and reasoning program of natural-logic theory provide a simple and succinct description of what the forms are for those inferences.

I turn now to the question of the inference procedures, both because these procedures provide the logical semantics and, moreover, because an intelligent species needs inferential procedures that go beyond the information given (e.g., Bruner, 1997). Our hunter–gatherer ancestors would have benefited from an ability to make simple and immediate inferences. Imagine a gatherer in the forest who knows that the best locations for finding food at this time of the year are either next to the river or in the forest near the canyon. She encounters her cousin who has just returned from the riverbank and tells her that there is little to collect there this year as locusts have been there, so a simple and direct inference can be made that she is better served going to the forest near the canyon. Surely our ancestors needed to make such inferences to have survived and produced progeny, for it was their logical reasoning that kept them from needlessly visiting multiple sites in useless searches for food. This particular kind of inference can be captured by an inference schema for disjunctions, that is, alternatives, and the stoic logician Chrysippus attributed such a schema to dogs following a trail of scent as well as to humans. (For discussions of why immediate inference procedures are not provided by the way professional logicians deal with truth conditions, see Braine, 1978; Dopp, 1962; Fitch, 1973; Gentzen, 1935/1964; Lakoff, 1970.)

Simply having some inference-making schemas would not be sufficient: a reasoner would need some sort of reasoning program that applies the schemas to produce lines of reasoning. Further, given that our thoughts usually refer to practical concerns, an adequate theory of logical reasoning also would require an account of how the logical inferences are coordinated with inferences that refer to practical knowledge and practical interests. To exclude inferences that go beyond what is available on logic alone would be irrational on the part of investigators into rationality. An investigator of human reasoning should not interpret inferences made in laboratory reasoning tasks that do not follow from logic alone as irrational. Such inferences may well be quite rational even when not inferable on logical grounds alone.

This chapter presents the natural-logic theory of Braine and O'Brien (e.g., 1991, 1998) as a case study in how a natural logic can be developed. The presentation of the theory has three parts: the representational format (section 3), a set of reasoning schemas and a reasoning program that implements the schemas in constructing lines of reasoning (section 4), and a set of pragmatic principles that describe the interactions of the logical schemas with epistemic knowledge and practical goals (section 5).

3.   A Representational Format for a Natural Logic

Standard-logic textbooks first introduce the sentential level, which includes atomic propositions and connectives for negation, conjunction, disjunction, conditionals, and so on. A predicate level adds quantifiers and a predicate–argument structure, which can make it appear that reasoning at the predicate level is more difficult, but natural logic need have no such expectation.

Constructing a natural logic at the sentential level requires fewer decisions than does the predicate-logic level, providing only connectives but not quantifiers and the structure that quantifiers entail. The connectives in a natural logic are not identical to those in a standard logic. For example, "If p then q" is not equivalent to the material conditional of $p \supset q$, which as discussed earlier makes "If p then q" true whenever p is false. So the natural logic presented in Braine and O'Brien (1998) eschews

the symbols one finds in logic textbooks and represents sentential connectives with italicized English words: *if* for conditionals, *and* for conjunction, *or* for disjunction, and *not* for negation, although we understand that these words have additional pragmatic interpretations that go beyond their purely logical meanings, and in ordinary speech, words can convey meanings that differ from their ordinary surface meanings. A mother who tells a child, "Put your hand in the cookie jar and I'll send you to your room," is using *and* where one formally would expect the word *if*, and we would not think the proposition requires that "The hand is put in the cookie jar" and "The child will be sent to their room" are both true in order for the assertion to be true. Even though on the surface, the assertion uses a word usually reserved for conjunction, the meaning intended clearly is conditional. Meaning thus is not found in the surface structures of sentences but in the propositions that are held in a deeper language of thought. Linguistic input thus is translated into a language of thought from the language of speech, and then inferences made in the language of thought are translated back into the language of ordinary speech (see Braine & O'Brien, 1998; O'Brien & Li, 2013).

Constructing the representational format at the predicate-logic level requires that some decisions be made about the internal structure of propositions. Consider a universally quantified disjunction in a standard-logic textbook:

$(\forall x)(Px \vee Qx)$,   "For all x, x satisfies P or x satisfies Q,"

where the universal quantifier is placed outside the proposition that is being quantified, and the quantifier has scope over all instances of the bound variables within the parentheses, whatever the content predicates are to which the variables are attached.

A corresponding natural logic representation is:

S_1[All X] OR S_2[PRO-All X],
"All the X's satisfy S_1 or they satisfy S_2,"

where the scope of quantification is inside the expression and is specific to particular content domains. Quantificational scope in this example is marked by a PRO-notation, referring to the pronoun in the second clause that is an anaphoric reflex to the quantifier in the first clause. This choice about how to represent quantification reflects the use of pronouns that is typical in natural languages as a way to convey quantificational scope. Ioup (1975) surveyed 14 languages and found no evidence for the type of outside scope that is typical in standard-logic textbooks;

instead, quantifiers tend to be lexicalized inside noun phrases, as they are in the natural-logic representations, and quantification is bound to the specific content, as it tends to be in natural languages.

One needs to distinguish between alternative interpretations of the surface grammar of sentences like "All the boys are carrying a box," where one interpretation is that there is a single large box that is being carried by a group of boys, and a different interpretation is that each boy is carrying a different box from the other boys. The first interpretation can be captured by a representation of this sort:

CARRY BOX (BOYS),

and the second interpretation can be captured by one of this sort:

CARRY BOX (All BOYS).

Such sorts of collective sets are not at all unusual. Braine and O'Brien (1998) provided the example

"The prosecutors convinced the 12 members of the jury of the defendant's guilt,"

CONVINCED (PROSECUTORS) (All JURORS),

where the collective group of prosecutors convinced each individual jury member.

4. Inference Schemas and a Reasoning Program

Natural-logic theories propose that people construct lines of reasoning governed by inference schemas (e.g., Braine & O'Brien, 1991, 1998; Fitch, 1973; Gentzen, 1935/1964; Rips, 1994), and Braine and O'Brien emphasized the need to seek empirical evidence for guidance about which procedures used to construct inferences in lines of reasoning should be included. For example, Braine, Reiser, and Rumain (1984) found that college students tend not to accept "p or q" as following from p alone, even though it is a valid inference in standard logic and is included in the natural logic of Rips (1994); it was excluded on empirical grounds by Braine and O'Brien (1998).

Braine and O'Brien (1998, tables 6.1 and 6.2) presented a set of 14 schemas at the sentential level together with a reasoning program that applies the schemas to construct lines of reasoning. I describe these in some detail here to provide a sense of what is involved. Seven schemas are called "core schemas," and they are applied without restriction by the reasoning program whenever their conditions for application are met. For example (presented here in simplified form), when one holds "p

or q" and not-p together in working memory, one infers q, as did our gatherer described earlier who decided to go to the area of the canyon to procure food. Another core schema is modus ponens, which holds that when one considers "If p then q" and p jointly, one infers q.

Two schemas are called "feeder schemas," and their application is restricted so that they are applied only when their output can feed another inference. One of these infers "p and q" when both p and q are in working memory, and the other infers p alone (or infers q alone) from "p and q." The restriction is required because if left unrestricted, these feeder schemas could produce infinite loops. Also, people do not report their output when asked to write down everything that follows from some premises, yet such schemas seem to exist because people apply, for example, modus ponens to "If p and q then r" when they have both p and q separately.

The final four schemas include two that pertain to incompatibility and two that pertain to supposition. Supposition occurs when a proposition is treated as true for the purposes of reasoning with the supposed proposition, even when one does not know whether or not it is true. One of the supposition schemas combines with an incompatibility schema so that the supposition can be falsified as in a *reductio ad absurdum* argument. The other is a schema for conditional proof, which is used to infer propositions of the form "If p then q" or to evaluate whether "If p then q" follows from a set of premises. To infer a proposition of the form "If p then q," first suppose p. When a set of premises and background assumptions together with the supposition of p lead to the inference of q, this schema provides the inference of "If p then q."

The schemas are applied by a reasoning routine, which has two parts: a direct-reasoning routine (the DRR) that we claim is universally available and applied automatically, and some reasoning strategies that go beyond the DRR and are widely available among college students. The schema for conditional proof has a procedure in the DRR that applies when a line of reasoning is constructed to evaluate "If p then q": the proposition p is added to the premises as a suppositional premise, and reasoning proceeds toward q. When q is derived, "If p then q" is evaluated as true. When not-q is derived, "If p then q" is evaluated as false. The latter evaluation would not be made in standard logic with its material conditional—because p might be false, even when supposed as true, although both young children and adults seem to use this procedure (Braine & O'Brien, 1998, chapters 7 and 17).

Note that the combination of the schemas and the reasoning program enables precise predictions about what inferences people make, and some of these inferences differ from what would be predicted if natural reasoning were following standard textbook logic, like the evaluations of when conditional propositions are false just described. The combination of schemas and reasoning program also helps us understand the consistency of the schema for conditional proof and the modus ponens schema, which together provide the inferential semantics for *if*. Given that "If *p* then *q*" is derived by supposing *p* and then showing that this entails *q*, the introduction of *p* as a premise when one knows "If *p* then *q*" allows the conditional nature of *q* to be discharged so that *q* alone can be asserted. That is, modus ponens releases *q* from requiring the marking of *p* as suppositional when one knows that *p* is true, because the schema for conditional proof ensures that *q* is true when *p* is.

Reasoning under a supposition requires a constraint (see Braine & O'Brien, 1991). A sound line of reasoning can proceed only from true premises, and when a line of reasoning is done under a supposition, one needs to be sure that the propositions introduced under the supposition would still be true given the supposition. When it is not clear whether a proposition should be excluded under a counterfactual supposition, logic per se provides no clear way to resolve such cases, and interlocutors would need to negotiate about whether exclusion is required. Note, by the way, that the natural logic is able to deal with counterfactual conditionals without recourse to some special semantics that requires reference to a possible world that is most similar to the actual world. After all, what would the possible world be that is most close to the actual world when we consider the counterfactual "Philadelphia is north of New York City"? One would have to decide to move New York southward or to move Philadelphia northward, with all of the messy repercussions that such a decision would have.

For a detailed presentation of the complete set of predicate-level schemas and the reasoning program for their implementation, see table 11.3 in Braine and O'Brien (1998). I present the following schema for a universally quantified disjunction to illustrate the nature of the predicate-level schemas and to make the point that even though the notation is more complex than at the sentential level, the inferences are not necessarily more difficult.

$$S_1[\text{All } X] \text{ OR } S_2[\text{PRO-All } X]; [\alpha] \subseteq [X]; \text{NEG } S_2[\alpha] \therefore S_1[\alpha].$$

For example, "All of the boys from the village are at the river fishing or in the forest hunting. Uirá and Werá are boys from the village and they are not at the river fishing, so they are in the forest hunting." Unlike standard logic, which suggests that reasoning at the predicate level is intrinsically more difficult than at the sentential level, this example illustrates that intuitively, this is not the case in natural logic.

5. Pragmatic Principles

People do not reason only by applying natural-logic schemas. An intelligent person uses any information that is pertinent, and inferences thus can be made on the basis of many kinds of processes. A person should interpret words in plausible ways and not limit their interpretations to what a professional logician has in mind. *If*, for example, can invite additional sorts of interpretations (Geis & Zwicky, 1971)—for example, "If *p* then *q*" can invite interpretation as a biconditional, depending on context; if I am told that *if I paint the shed I will receive $50*, I expect that if I do not paint the shed, I will not receive the money. Natural logic does not constrain *or* as either exclusive or inclusive, but certain kinds of situations surely invite one interpretation over the other, which can expand the available inferences.

Story grammars or scripts provide additional inferences that a rational person would make. My knowledge about eating breakfast in a typical American diner tells me that I will not drink both tea and coffee, and this knowledge provided by a restaurant script can feed a natural-logic inference. Although a cognitive scientist is interested in the sources from which inferences are made, people are unlikely to attend to the varied sources that can lead to an inference or to distinguish purely logical inference from other sources. These things are not evidence of irrationality; in ordinary reasoning, inferences from various sources intertwine with inferences from natural logic to provide reasoning that meets practical concerns as lines of reasoning are constructed.

6. Empirical Support for the Braine–O'Brien Natural Logic

The most basic prediction of natural logic is that the inferences available on the DRR alone will be made routinely unless something complicates matters, and the evidence, generally from problems presenting neutral content, supports this prediction. Further, inferences that are valid in standard logic but not available on the schemas of natural logic are made far less often, as are inferences requiring strategies that go beyond what is available on the DRR alone. Finally, problem length per se does not predict problem solution, which depends for the most part on whether the required inferences are available in the DRR (Braine et al., 1984; Braine & O'Brien, 1998; see review in O'Brien, 2004).

Braine et al. (1984), with sentential problems, and Yang, Braine, and O'Brien (1998), with predicate-level problems, presented DRR problems and asked people to rate on a Likert-type scale the relative difficulty of each problem. Regression analyses provided weights for each schema. Problem difficulty was assumed to equal the sum of the difficulty weights for all of the schemas required to solve a problem. These weights were used to compute predicted difficulties for different problems that were presented to different people. The percentage of variance accounted for in correlations between the predicted and observed difficulties was 66% for sentential problems (55% with problem length partialed out) and 69% for predicate-level problems (56% with problem length partialed out). This indicates that people were not only solving these problems but also solving them using the natural-logic schemas.

When the schemas were embedded within story scenarios and readers were asked whether the natural-logic schemas were inferences or paraphrases, people showed that they did not know that they were making inferences at all, although inferences based on story scripts, grammars, and so forth were judged as being inferences rather than paraphrases (Lea, 1995; Lea, O'Brien, Fisch, Noveck, & Braine, 1990; O'Brien, Roazzi, Dias, & Soskova, 2007). Thus, the natural-logic inferences were made so easily that people did not think of them as requiring reasoning.

Braine et al. (1995) and O'Brien, Braine, and Yang (1994) presented problems with multiple premises and asked participants to write down all of their inferences in the order they made them. Consider the following problem with five premises from O'Brien et al. (1994) that referred to some letters written on a blackboard: (a) "N or P," (b) "Not N," (c) "If P then H," (d) "If H then Z," and (e) "Not both Z and S." Premises (a) and (b) trigger a logic schema to infer P, which then combines with (c) to infer H, which then combines with (d) to infer Z, which then combines with (e) to infer not-S. So, as each premise was read in turn, a schema was triggered that provided an inference. A matched problem presented the same premises but with the order reversed, so that no schema was triggered until all premises had been read and the information in the last two premises could trigger a schema. For both problems, the same inferences were made in the same order, demonstrating that the order of inferences was determined by availability of schemas. In summary, the predictions about which problems will be solved, about the relative difficulties

with which problems are perceived, and about the orders in which inferences are made are supported by the available data. Further, when the predictions differ from what would be predicted by a standard textbook logic, the data support the predictions of natural logic. Supporting evidence that the core inferences are available in children is presented in Braine and Rumain (1983) and Braine and O'Brien (1998).

7. Why Theories of Reasoning Need a Natural Logic

Proponents of several other theories have argued against any role for a natural logic. O'Brien and Manfrinati (2010) and O'Brien and Li (2013) have provided detailed discussions to counter their arguments against a natural logic. I concentrate here on whether those theories are adequate without including a natural logic.

Cheng and Holyoak (1985) introduced some content-specific inference rules for permissions and obligations that they proposed are inductively acquired, and Cosmides (1989) introduced some social-contract rules that she proposed have been acquired through our bioevolutionary history. Both Cheng and Holyoak (1985) and Cosmides (1989) argue that people do not use a natural logic but only content-dependent processes. Both theories rely on variants of Wason's selection task, comparing performance with universally quantified indicative conditionals to performance with sentences that should trigger either social-contract rules or pragmatic schemas. The problems required people to identify situations that could falsify the indicative conditionals or find rule violators of the pragmatic or social-contract rules, that is, instances of "p and not-q." Both theories reported more correct answers with their content, which they argued indicates that people do not reason with a natural logic but with their content-specific processes.

I address here whether these two theories provide an adequate account of human reasoning without including a natural logic, given that they specifically argue against any natural logic. (O'Brien & Manfrinati [2010] and O'Brien, Roazzi, Athias, & Brandão [2007] provide more extensive discussions of difficulties in interpreting the data for these content-specific theories.) The contention of both theories that reasoning relies not on a content-general natural logic but on some content-specific processes is based on very little evidence. They have investigated only a handful of conditionals that express permissions, obligations, or social contracts, and to support the notion that content-specific processes are sufficient to replace natural logic, they would need to provide a much larger set of types of content. I know of no metric with which to estimate the number of content areas that one would need to provide a general reasoning theory, but it must be immense, and I find it hard to imagine that either inductive or bioevolutionary methods would have provided such a number of content-dependent modules. Further, they would need to present evidence beyond a sole reliance on Wason's selection task.

O'Brien, Roazzi, Dias, Cantor, and Brooks (2004) and O'Brien (1995) discussed why data from Wason's selection task are difficult to interpret, and there is little discussion by critics of natural logic of exactly how one would solve the task using logic. The selection task actually is a meta-logical task when presented with indicative conditionals—it requires judging the conditions under which the truth of a conditional could be tested, so a logic inference alone would not be sufficient, whereas the task presented with pragmatic or social-contract conditionals and requiring only identification of rule violators is not a meta-logical task. No attention has been given to this difference by proponents of these theories. If these content-specific theories are to expand their focus, they will need to present problems other than the selection task.

Finally, these theories need to explain why *if*-sentences are used to convey these regulations. Let me illustrate the problem by pointing out an implicit difference in developmental expectations between the pragmatic-schemas approach and the natural-logic approach. If one assumes that the pragmatic rules are inductively acquired, as pragmatic-schemas theory does, then a content-general schema of the sort we describe would enter late in the process only, after many pragmatic schemas had been acquired, yet Braine and O'Brien (1998) present evidence from spontaneous speech that young children grasp the schemas for conditionals quite early.

Oaksford and Chater (e.g., 2010) proposed that the probability that someone will accept an inference from a conditional premise depends on their conditional belief in the conclusion given the premises and other background information. They make several arguments against natural logic, but their arguments, especially their arguments about defeasibility, stem from treating natural logic as equivalent to standard logic. More important is the question of whether Oaksford and Chater are able to account for the data without including a natural logic. Their approach depends on prior beliefs about the premises on which inferences rely, so the approach is mute when people receive premises about which they have no prior beliefs. The approach thus cannot address the large body of evidence in favor

on the natural-logic approach that refers to materials about which participants have no prior beliefs, like the size and colors of beads in an urn or letters written on an imaginary blackboard. Further, Oaksford and Chater have addressed only Wason's selection task and the four conditional syllogisms, and they need to explain the larger set of problems that we have investigated, which include conjunctions, negations, disjunctions, and conditional conclusions.

Evans and Over (2004) and Over and Evans (2003) proposed that people evaluate conditionals by imagining two situations, one in which both p and q are true and one in which p is true and q is false, and then assessing the probabilities of the two cases, and the probability of "If p then q" is assigned from the comparison of probabilities for the two models. So, "If p then q" is functionally equivalent to $P(q \mid p)$ computed from the probabilities assigned to the two models.

This proposal bears a clear resemblance to the natural-logic schema for conditional proof in that it begins with the supposition of p in constructing the two models in which p is true and then reasons toward q or not-q, but Evans and Over (2004) argued that because the natural-logic approach does not include either a possible-worlds semantics or a mental-models semantics, it cannot account for these probability judgments. Why one would need a possible-worlds or a mental-models semantics to judge these probabilities is not clear.

To illustrate that one does not need a separate logical semantics to make such judgments, imagine two boxes, a red box with 80 beads and a blue box with 20 beads. You are told that I have a bead in my hand and are asked to judge the probability that *if the bead came from one of these boxes, it came from the red box*. Using the natural-logic procedures for the conditional-proof schema, you begin by supposing that the bead came from one of these boxes, and you reason toward the proposition that it came from the red box. You discover that you cannot derive whether or not it came from the red box, so you know that the conditional cannot be shown to be true or false, but given that there are four times more balls in the red box and you are asked to judge the probability of the conditional, you interpret it as asking what is the probability that the ball came from the red box and respond that the probability is .80. No independent logical semantics of any other kind was required for someone to give such a response.

The theory of Evans and Over is not adequate to account for reasoning in general without a natural logic because their theory has not been developed beyond a single kind of judgment about conditionals and thus cannot provide an account of the larger set of findings concerning other connectives. It has the advantage over the content-dependent theories because it is not tied to a single content domain, and it obviously bears a close resemblance to the Braine–O'Brien natural-logic theory. Integrating their proposal with a natural logic thus seems eminently possible, and such an integration clearly needs to happen if their theory is to address a more general class of reasoning types, although the natural-logic theory seems already capable of making the judgments their theory describes.

I turn finally to mental-models theory (e.g., Johnson-Laird, 1983; Johnson-Laird & Byrne, 1991, 2002)—the only other theory to address the same broad class of reasoning problems that natural logic addresses—whose proponents consistently deny that there is any role in human reasoning for any natural logic. The models theory proposes that humans construct models that are adaptations of truth tables but represent not truth assignments but possible states of affairs. Because working memory is limited, people usually construct only some of the possible states of affairs. Although a complete representation of "If p then q" includes the three models corresponding to the material conditional, it usually is represented with

$[p]$ q
$\cdots$

where the ellipsis acts as a reminder that additional models could be constructed, and the brackets around p—called "exhaustivity markers"—constrain any such additional models so as not to include a token for p without a token for q.

If the minor premise for a modus ponens argument is added to the model for "If p then q," one gets

$[p]$ q; p; p q
$\cdots$

This set of models separates models for the two premises and then their combination with semicolons, and the conclusion, q, is "read off" from the final model and does not include p presumably because pragmatics precludes inclusion of categorical premises in a conclusion. In this way, they assert, one can derive a modus ponens conclusion with only the manipulation of models and no logic.

Precluding any logic from their theory, including the variables in a predicate–argument structure, is a significant loss. Johnson-Laird and Byrne (1993) wrote that although models do not contain variables, variables do occur "in the initial semantic representations" (p. 376),

that is, the representations from which they say their models are constructed.

Models for quantified propositions are almost identical to those for the sentential connective, and the complete model for "All P are Q" is

$[P]$ $[Q]$
$[P]$ $[Q]$
$[\sim P]$ $[Q]$
$[\sim P]$ $[\sim Q]$

Let's consider how models come from the premodels representations. Quantifiers provide "the raw material for a recursive loop that is used in building or manipulating a model. Thus, the universal quantifier 'all' elicits a recursion that deals with a set exhaustively, whereas the existential quantifier 'some' elicits a recursion that does not" (Johnson-Laird & Byrne, 1991, p. 178), which is illustrated with the example "All x's are equal to the sum of some y and some z," which is parsed to yield

(All x) (Some y) (Some z) ($x = y + z$).

A model, then, is constructed with an arbitrary value for the first variable term (x), another arbitrary value for the second (y), and a constrained value (the degrees of freedom being exhausted) for the third variable (z), looping over the equation several times, recording the output each time, resulting in a model like

[8 6] (1 6 4 2) (7 7 2 2 4 4),

where the two numbers in the first set are equal to some number in the second set plus some number in the third set. In this, Johnson-Laird and Byrne have succeeded in their goal of constructing a representation that has no logical variables or quantifiers, but at what cost?

They have told us repeatedly that conclusions are "read off" from final models, so what conclusion could be read off here? Consider the simplest logic inference—reiteration: $p \therefore p$. But here, the meaning of the initial representation prior to construction of the model is gone, and one has no way of retrieving "All x's are equal to the sum of some y and some z" from the final model. This is a steep cost to fulfill the goal of creating a reasoning theory that requires no logic inferences and no variables and only the manipulation of models. It's difficult to see that the procedure is analogous to the standard interpretation of quantifiers in the predicate calculus. The quantifiers and variables of the initial proposition have disappeared, and without them, one cannot retrieve the assertion from the model. Clearly, mathematical reasoning cannot be accomplished purely through the manipulation of models. How could one use this sort of procedure to prove "All natural numbers that end in zero are divisible by five" when one is limited to inclusion only of individual exemplars, square brackets to mark exhaustivity, and some ellipses to remind one that additional models might need to be constructed? Yet my students in undergraduate classes over many years have succeeded in understanding the necessary truth of this statement.

8. Conclusion

I have argued that an adequate account of human reasoning must include some predicate–argument structure and some inference processes for going beyond the information given and that other ways of thinking need to be integrated with the natural logic. Searching for counterexamples as one does with mental models would be a valuable skill to add to what is provided by natural logic, as would be the specialized searches for violators of pragmatic and social-contract rules. But to focus only on such specialized skills at the expense of the more general architecture provided by a natural logic would be to throw out the baby with the bathwater.

References

Block, N. (1998). Conceptual role semantics. In E. Craig (Ed.), *Routledge encyclopedia of philosophy* (Vol. 2, pp. 242–256). London, England: Routledge.

Boole, G. (1854). *An investigation of the laws of thought on which are founded the mathematical theories of logic and probabilities*. London, England: Walton and Maberly.

Boring, E. G. (1929). *A history of experimental psychology*. New York, NY: Century.

Braine, M. D. S. (1978). On the relation between the natural logic of reasoning and standard logic. *Psychological Review, 85*, 1–21.

Braine, M. D. S., & O'Brien, D. P. (1991). A theory of *if*: A lexical entry, reasoning program, and pragmatic principles. *Psychological Review, 98*, 182–203.

Braine, M. D. S., & O'Brien, D. P. (1998). *Mental logic*. Mahwah, NJ: Erlbaum.

Braine, M. D. S., O'Brien, D. P., Noveck, I. A., Samuels, M., Lea, R. B., Fisch, S. M., & Yang, Y. (1995). Predicting intermediate and multiple conclusions in propositional logic inference problems: Further evidence for a mental logic. *Journal of Experimental Psychology: General, 124*, 263–292.

Braine, M. D. S., Reiser, B. J., & Rumain, B. (1984). Some empirical justification for a theory of natural propositional logic.

In G. Bower (Ed.), *Psychology of learning and motivation: Vol. 18. Advances in research and theory* (pp. 313–371). New York, NY: Academic Press.

Braine, M. D. S., & Rumain, B. (1983). Logical reasoning. In J. H. Flavell & E. Markman (Eds.), *Handbook of child psychology: Vol. 3. Cognitive development* (pp. 263–339). New York, NY: Wiley.

Bruner, J. (1997). *The culture of education*. Cambridge, MA: Harvard University Press.

Cheng, P. W., & Holyoak, K. J. (1985). Pragmatic reasoning schemas. *Cognitive Psychology, 17*, 391–416.

Cosmides, L. (1989). The logic of social exchange: Has natural selection shaped how humans reason? Studies with the Wason selection task. *Cognition, 31*, 187–276.

Dopp, J. (1962). *Logiques construites par une méthode de déduction naturelle* [Logics built by a natural deduction method]. Louvain, Belgium: Nauwelaerts.

Evans, J. St. B. T., & Over, D. E. (2004). *If*. Cambridge, England: Cambridge University Press.

Fitch, F. B. (1973). Natural deduction rules for English. *Philosophical Studies, 24*, 89–104.

Fodor, J. A. (1981). *Representations: Philosophical essays on the foundations of cognitive science*. Cambridge, MA: Harvester Press.

Freud, S. (1914). *Psychopathology of everyday life*. London, England: T. Fisher Unwin.

Gabbay, D. M., & Woods, J. (Eds.). (2009). *Handbook of the history of logic: Vol. 5. Logic from Russell to Church*. Amsterdam, Netherlands: Elsevier.

Geis, M., & Zwicky, A. M. (1971). On invited inferences. *Linguistic Inquiry, 2*, 561–566.

Gentzen, G. (1964). Investigations into logical deduction. *American Philosophical Quarterly, 1*, 288–306. (Original work published 1935)

George, R. (1997). Psychologism in logic: Bacon to Bolzano. *Philosophy and Rhetoric, 30*, 213–242.

Harman, G. (1982). Conceptual role semantics. *Notre Dame Journal of Formal Logic, 23*(2), 242–256.

Henle, M. (1962). On the relation between logic and thinking. *Psychological Review, 69*(4), 366–378.

Ioup, G. (1975). Some quantifiers for universal scope. In J. Kimball (Ed.), *Syntax and semantics* (Vol. 4, pp. 37–58). New York, NY: Academic Press.

Johnson-Laird, P. N. (1983). *Mental models*. Cambridge, MA: Harvard University Press.

Johnson-Laird, P. N., & Byrne, R. M. J. (1991). *Deduction*. Hove, England: Erlbaum.

Johnson-Laird, P. N., & Byrne, R. M. J. (1993). Mental models or formal rules? *Behavioral and Brain Sciences, 16*, 368–376.

Johnson-Laird, P. N., & Byrne, R. M. J. (2002). Conditionals: A theory of meaning, pragmatics, and inference. *Psychological Review, 109*, 646–678.

Johnson-Laird, P. N., Byrne, R. M. J., & Schaeken, W. (1992). Propositional reasoning by model. *Psychological Review, 99*, 418–439.

Lakoff, G. (1970). Linguistics and natural logic. *Synthese, 22*, 151–271.

Lea, R. B. (1995). Online evidence for elaborative logical inference in text. *Journal of Experimental Psychology: Learning, Memory, and Cognition, 21*, 1469–1482.

Lea, R. B., O'Brien, D. P., Fisch, S. M., Noveck, I. A., & Braine, M. D. S. (1990). Predicting propositional logic inferences in text processing. *Journal of Memory and Language, 29*, 361–387.

Mandler, G. (2007). *A history of modern experimental psychology.* Cambridge, MA: MIT Press.

Oaksford, M., & Chater, N. (Eds.). (2010). *Cognition and conditionals: Probability and logic in human thinking.* Oxford, England: Oxford University Press.

O'Brien, D. P. (1995). Finding logic in human reasoning requires looking in the right places. In S. E. Newstead & J. St. B. T. Evans (Eds.), *Perspectives on thinking and reasoning: Essays in honor of Peter Wason* (pp. 189–216). Hove, England: Erlbaum.

O'Brien, D. P. (2004). Mental-logic theory: What it proposes, and reasons to take this proposal seriously. In J. Leighton & R. J. Sternberg (Eds.), *The nature of reasoning* (pp. 205–233). New York, NY: Cambridge University Press.

O'Brien, D. P., Braine, M. D. S., & Yang, Y. (1994). Propositional reasoning by model: Simple to refute in principle and in practice. *Psychological Review, 101*, 711–724.

O'Brien, D. P., & Li, S. (2013). Mental logic theory: A paradigmatic case for empirical research on the language of thought and inferential role semantics. *Journal of Foreign Languages, 36*, 27–41.

O'Brien, D., & Manfrinati, A. (2010). The mental logic theory of conditional propositions. In M. Oaksford & N. Chater (Eds.), *Cognition and conditionals: Probability and logic in human thinking* (pp. 39–54). Oxford, England: Oxford University Press.

O'Brien, D. P., Roazzi, A., Athias, R., & Brandão, M. C. (2007). What sorts of reasoning modules have been provided by evolution? Some experiments conducted among Tukano speakers in Brazilian Amazônia concerning reasoning about conditional propositions and about conditional probabilities. In M. Roberts (Ed.), *Integrating the mind: Domain general vs. domain specific processes in higher cognition* (pp. 59–81). Hove, England: Psychology Press.

O'Brien, D. P., Roazzi, A., Dias, M. G., Cantor, J. B., & Brooks, P. J. (2004). Violations, lies, broken promises, and just plain mistakes: The pragmatics of counterexamples, logical semantics, and the evaluation of conditional assertions, regulation, and promises in variants of Wason's selection task. In K. Manktelow & M. C. Chung (Eds.), *Psychology of reasoning: Theoretical and historical perspectives* (pp. 95–126). Hove, England: Psychology Press.

O'Brien, D. P., Roazzi, A., Dias, M. da G. B. B., & Soskova, J. (2007). Prevendo inferências lógico-predicativas no processamento de texto [Predicting predicate-logic inferences in text processing]. *Revista Interamericana de Psicología, 41*, 119–128.

Over, D., & Evans, J. St. B. T. (2003). The probability of conditionals: The psychological evidence. *Mind & Language, 18*, 340–358.

Passmore, J. (1994). *A hundred years of philosophy.* Harmondsworth, England: Penguin.

Rips, L. J. (1994). *The psychology of proof: Deductive reasoning in human thinking.* Cambridge, MA: MIT Press.

Wittgenstein, L. (1958). *Philosophical investigations.* Oxford, England: Basil Blackwell.

Yang, Y., Braine, M. D. S., & O'Brien, D. P. (1998). Some empirical justification of the mental predicate logic model. In M. D. S. Braine & D. P. O'Brien (Eds.), *Mental logic* (pp. 333–366). Mahwah, NJ: Erlbaum.

3.3 Psychological Theories of Syllogistic Reasoning

Sangeet Khemlani

Summary

Psychologists have studied syllogistic inferences for more than a century, because they can serve as a microcosm of human rationality. "Syllogisms" is a term that refers to a set of 64 reasoning arguments, each of which comprises two premises, such as, "All of the designers are women. Some of the women are not employees. What, if anything, follows?" People make systematic mistakes on such problems, and they appear to reason using different strategies. A meta-analysis showed that many existing theories fail to explain such patterns. To address the limitations of previous accounts, two recent theories synthesized both heuristic and deliberative processing. This chapter reviews both accounts and addresses their strengths. It concludes by arguing that if syllogistic reasoning serves as a sensible microcosm of rationality, the synthesized theories may provide directions on how to resolve broader conflicts that vex psychologists of reasoning and human thinking.

1. A Microcosm of Rationality

In 1908, the German scholar Gustav Störring published a 130-page manuscript detailing the results of the first known experiments on human reasoning. His main purpose in conducting them was to develop solutions to long-standing debates between logicians and philosophers, such as what people imagine when they reason. The studies worked like this: volunteers entered a dark room alone, sat down, and received a battery of deductive reasoning problems called "syllogisms," one after another. Störring recorded his observations of their verbal responses, reaction times, eye movements, gestures, and even their breathing patterns (Störring, 1908).

The research would likely be rejected were it to be submitted to any contemporary psychology journal. For one thing, Störring investigated only four participants. For another, he used an arbitrary experimental design,

and he failed to present any quantitative analysis of their behaviors except for a single table that listed averaged reaction times. But what Störring learned from his research was remarkable (see Clark, 1922; Knauff, 2013; Politzer, 2004). He noticed, for instance, that his volunteers were biased by the structure of the different reasoning problems: for some syllogisms, volunteers chose conclusions immediately, as though they could observe the answer directly. For other problems, they reported the sensation of *Nachdenken*, that is, "a feeling of deliberation." His volunteers appeared to adopt certain strategies as they carried out the task, and they seemed aware of their strategies well enough that they could articulate them. They reported that they used their imagination and mental imagery on many problems, and when probed further, they were able to depict that experience by sketching out corresponding diagrams. Perhaps the most important discovery was that Störring's volunteers had a limited ability to select logically *valid* conclusions, that is, conclusions that were true in all the situations in which the premises were true (cf. Jeffrey, 1981, p. 1)—for some problems, they produced correct answers, and for others, they made mistakes.

Scientists often make use of microcosms as a way of understanding broader phenomena. For instance, the geneticist Gregor Mendel examined pea plants to understand genetic inheritance; the entomologist Agostino Bassi studied silkworms to understand bacterial diseases. In his experiments on syllogisms, Störring had analyzed a feasible microcosm of human rationality. In the years that followed, syllogisms played an outsized role in educating contemporary researchers on the processes of thinking and reasoning. Many experiments investigated syllogistic reasoning in isolation, and many more used syllogistic reasoning as a stand-in for reasoning behavior more generally. For instance, Goel and colleagues ran a neuroimaging experiment in which they gave participants syllogistic reasoning problems with and without meaningful contents to discover that certain brain regions—such as temporal and frontal regions—systematically respond to semantic information (Goel, Buchel, Frith, & Dolan, 2000).

Perhaps syllogisms serve as an attractive microcosm of thinking behavior because of their simplicity and that there is only a finite number of them. Classical syllogisms (i.e., those investigated by Aristotle and Scholastic logicians) are reasoning arguments comprising multiple premises, such as

(1) All of the women are designers.

　　Some of the employees are not women.

　　What, if anything, follows?

These syllogisms contain a quantified noun phrase, such as "all of the women," and these quantifiers can be in one of four separate *moods*, that is, expressions comprising quantifiers and negations, as shown below:

All of the *a* are *b*.	(A*ab*)
Some of the *a* are *b*.	(I*ab*)
Some of the *a* are not *b*.	(O*ab*)
None of the *a* is *b*.	(E*ab*)

The parentheses indicate the abbreviation conventions adopted by Scholastic logicians (i.e., the 12th-century university scholars who gained access to Aristotle's works). Contemporary psychologists adopted those conventions, and we retain them here. Since syllogisms consist of two premises, the *terms* in the premises (e.g., "women," "designers," "employees") can occur in four different arrangements. These different arrangements are known as *figures*:

Figure 1	Figure 2	Figure 3	Figure 4
a – *b*	*b* – *a*	*a* – *b*	*b* – *a*
b – *c*	*c* – *b*	*c* – *b*	*b* – *c*

Other psychologists use different numbering systems for figures—and they sometimes include the conclusion as part of their numbering systems. Here we state the figures in terms of the premises only.

In sum, syllogisms concern 64 separate reasoning problems (4 moods of the first premise × 4 moods of the second premise × 4 separate figures). Many experiments on syllogisms focus on only these 64 problems, that is, they provide participants with the pairs of premises and then ask them to infer what follows from them. Typically, reasoners do not consider all the possible valid and invalid responses when they generate conclusions; they tend to describe just one or two. But across the problems overall, their conclusions can be classified into nine different structural patterns, as follows: "All of the *A*s are

*C*s" (abbreviated as A*ac*); "All of the *C*s are *A*s" (abbreviated as A*ca*); the responses that correspond to I*ac*, I*ca*, O*ac*, O*ca*, E*ac*, and E*ca*; and the response that no valid conclusion follows.

As Störring recognized, some syllogisms are quite easy, and others can be difficult. Consider this problem:

(2) Some of the designers are animators.

　　All of the animators are undergraduates.

　　What, if anything, follows?

Before reading further, how might you respond to the problem? If you inferred that some of the designers are undergraduates, you'd be correct. But now consider this problem:

(3) None of the designers are animators.

　　All of the animators are undergraduates.

　　What, if anything, follows?

This problem is significantly harder: the correct response is that some of the undergraduates are not designers. Why is (2) easy but (3) difficult? To answer the question, psychologists have run many studies on which inferences people conclude from the 64 syllogisms. Khemlani and Johnson-Laird (2012) compiled six of them together in a meta-analysis, which shows that the most common response to (2) is "Some of the designers are undergraduates" (see figure 3.3.1, left panel, row I*ab* A*bc*), and the most common response to (3) is "None of the designers are undergraduates" (see figure 3.3.1, left panel, row E*ab* A*bc*)—the response is an error, since (3) does not rule out the possibility that some of the designers are undergraduates.

In a typical study, responses to a syllogism vary from one individual to the next. Consider how people respond to (1), that is, figure 3.3.1, right panel, row A*ba* O*cb*: some draw an (erroneous) conclusion of the form O*ca* about half of the time; others make a different error and conclude I*ca*. Only about a fifth of university students (who typically serve as participants in studies on syllogisms) respond correctly that there is no valid conclusion. The psychologist's task is to explain the robust patterns of inference across the set of 64 problems. The difficulty for theorists is that reasoners approach the problems with different abilities and appear to develop different strategies (see table 3.3.1). The variability in reasoners' responses was enough to convince some theorists that the only way to understand how people reason syllogistically is to examine their individual differences (Stenning & Cox, 2006)—some reasoners generate correct

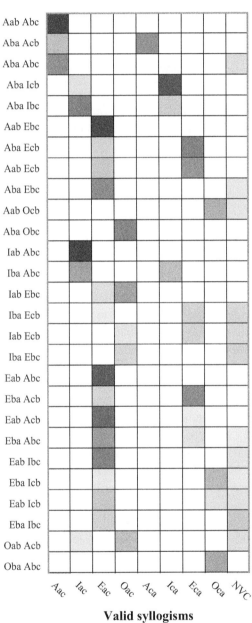

Valid syllogisms

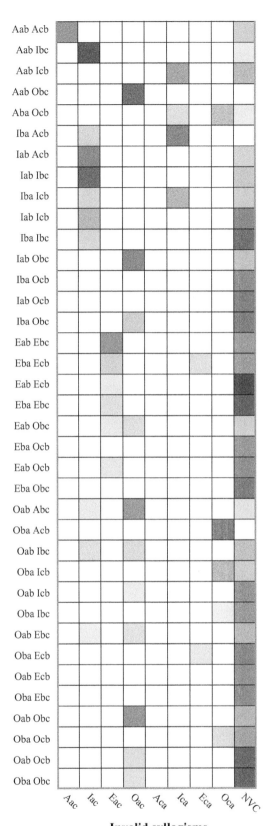

Invalid syllogisms

Figure 3.3.1

The percentages of responses to 64 syllogisms in the meta-analysis in Khemlani and Johnson-Laird (2012). Each of the 64 pairs of premises occurs in a row, and each of the possible responses occurs in a column. Abbreviations for premises are as follows: A*ac* = All of the *A* are *C*, I*ac* = Some of the *A* are *C*, E*ac* = None of the *A* is a *C*, O*ac* = Some of the *A* are not *C*, and NVC = No valid conclusion. The left panel denotes the 27 syllogisms with a valid definite conclusion, and the right panel denotes the 37 syllogisms without a valid definite conclusion. The grayscale in each cell indicates the proportion of corresponding conclusions (black = 100% and white = 16% or below). Hence, for the top-most valid syllogism, A*ab* A*bc*, nearly 100% of participants in the meta-analysis responded that A*ac* follows.

Table 3.3.1
Summary of robust sources of differences in syllogistic reasoning performance

Phenomenon	Description
Differences between problems	Some syllogisms are extremely difficult (1% correct solutions), and others are easy (90% correct conclusions). Differences in difficulty are reliable across studies.
Differences among individuals	Accuracy varies greatly: from one reasoner to another, between certain groups of university students, and as a function of intelligence and training.
The figural effect	The figure of a syllogism (i.e., the order in which terms are organized) affects the frequency and type of conclusions reasoners tend to draw.
The effect of contents	Individuals are more likely to accept believable conclusions and less likely to accept unbelievable ones, especially for invalid syllogisms. This is also referred to as the "belief bias" effect.
Developmental differences	Children are unable to draw sensible inferences from syllogisms until they can understand and produce quantifiers. The ability develops through adolescence up to adulthood.
Strategic differences	Reasoners acquire various strategies to solve syllogisms (e.g., which premise is interpreted first, how premises are interpreted, and how counterexamples are constructed).

answers for 85% of syllogisms, and others generate correct answers for only 15% of them (Johnson-Laird, 1983, pp. 118–119).

The study of syllogisms arguably resulted in a major challenge to human rationality. Early work by Woodworth and Sells (1935) suggested that, instead of reasoning their way through syllogisms, people were biased by the "atmosphere" created by the premises, which yielded a predisposition to accept a certain sort of conclusion. The atmosphere effect suggested wholesale irrationality: humans could diverge from normative reasoning behavior predictably and systematically (see chapter 2.3 by Johnson-Laird, this handbook). The result sparked a fascination with the extent to which human reasoning could be characterized as rational (see chapter 1.2 by Evans and chapter 3.1 by Steinberger, both in this handbook), and theorists began to devise accounts of the phenomena underlying syllogistic reasoning. The use of computational and formal tools helped some researchers implement psychological theories of the syllogism and test them against human data. As a result, after decades of research, nearly a dozen theories of the phenomenon had been proposed, and there existed a dire need to sort out the different theoretical proposals. Khemlani and Johnson-Laird (2012) surveyed existing psychological accounts of syllogistic reasoning to discover broad trends between them. The survey suggested that theories tended to fall into one of three groups: one group of theories explained syllogistic reasoning by appealing to sets of heuristics in how quantified statements were processed (e.g., Begg & Denny, 1969; Chater & Oaksford, 1999; Revlis, 1975; Wetherick & Gilhooly, 1995). For example, the so-called matching strategy

(Wetherick & Gilhooly, 1995) posited that for syllogisms such as

(4) Some of the designers are women.

Some of the women are employees.

What, if anything, follows?

people should conclude—erroneously—that "some of the designers are employees." The reason is because the conclusion matches the mood of the most "conservative" premise, that is, the premise that presupposes the existence of the fewest entities. And indeed, reasoners draw the predicted conclusion 61% of the time (see figure 3.3.1, right panel, row I*ab* I*bc*). But about a third of the time, they also accurately infer that "No valid conclusion" follows, and accounts based on heuristics have difficulty explaining the deliberative processes by which reasoners correct their mistakes (see also Ragni, Dames, Brand, & Riesterer, 2019). In order to account for deliberative reasoning, another group of psychological theories proposed that reasoners mentally simulate the situation described in the premises when they reason about syllogisms (Bucciarelli & Johnson-Laird, 1999; Guyote & Sternberg, 1981; Johnson-Laird & Steedman, 1978; Polk & Newell, 1995). The theories posited that mental simulations help explain both errors and correct responses: reasoners construct, and can make inferences from, initial simulations, but difficult syllogisms demand reasoners to consider alternative simulations (Johnson-Laird, 1983). A third group of theories assumed that syllogistic reasoning depends on mental proofs and rules of inference akin to those in formal logic (see, e.g., Braine & Rumain, 1983; Geurts, 2003; Politzer, 2007; Rips, 1994)—but such theories have systematic difficulty explaining how reasoners draw the conclusion that nothing follows from a set of premises, and so we presently address only the first two groups of theories.

Theories based on heuristics and theories based on deliberation both failed to explain many systematic patterns of syllogistic reasoning (Khemlani & Johnson-Laird, 2012). In retrospect, the debate between the two types of processing presents a false dichotomy. A robust theory of syllogistic inference needs to explain both heuristic and deliberative processing (see Evans & Stanovich, 2013; Johnson-Laird & Steedman, 1978). The most recent psychological accounts of syllogistic reasoning have sought to unify the two kinds of reasoning processes. This chapter reviews these recent theories and their computational implementations, and it summarizes their strengths and weaknesses. It also provides a broader perspective on how the theories address ongoing debates in the psychology of reasoning.

2. Unified Accounts of Syllogistic Reasoning

An ancient idea is that human thinking relies on two different systems: one fast, one slow. Peter Wason, with his students Philip Johnson-Laird and Jonathan Evans, proposed that reasoning processes should be construed in terms of two distinct, interreliant processes (Johnson-Laird & Wason, 1970; Wason & Evans, 1974). As Evans (2008, p. 263) notes, the dichotomy between heuristics and deliberation is closely related to dual processes because heuristics are thought to be a fast, shallow form of processing and deliberation is thought to be a slower, deeper form of processing. In practice, heuristics and deliberative thinking often occur sequentially, that is, a heuristic response is proposed and a deliberative process validates or falsifies it. More general accounts of dual processing are not committed to sequential processing—they permit that fast processes and slow processes can operate in parallel and interact with one another. The introduction noted that previous theories of syllogistic reasoning tended to account for one type of process over the other. It may be that previous theories were easier to formulate because it is difficult to anticipate the interactive effects of two interdependent processes. Yet, if it is indeed the case that human thinking depends on two inextricable processes, those theories were doomed to fail.

Two recent theories of syllogistic reasoning are unique in that they seek to model interactive processing. Both theories are built around the integrative idea that fast, heuristic processing is the result of a biased sampling procedure that can be formalized using probabilistic constraints and that slower, deliberative processing suggests

that reasoning depends on representations referred to as "mental models." The theory that people construct mental models when they reason originates from Johnson-Laird (1983; see also chapter 2.3 by Johnson-Laird, this handbook), who computationally developed earlier proposals that people build "small-scale models" of reality to anticipate events (Craik, 1943). Johnson-Laird's "model theory" posits that each mental model represents a distinct possibility or situation in the world. In other words, when reasoners draw inferences from syllogisms, they mentally simulate the situation referred to by the premises. The model theory predicts that problems that require reasoners to consider multiple mental models should be more difficult relative to those that require fewer models. As a result, models help explain reasoning difficulty in many domains (see, e.g., Johnson-Laird & Khemlani, 2013; Khemlani & Johnson-Laird, 2017). But, as Khemlani and Johnson-Laird (2012) show, previous implementations of the model theory tend to make overly liberal predictions of the kinds of syllogistic inferences people are likely to draw. Hence, the two latest theories of syllogistic reasoning add additional constraints that explain why reasoners are reticent to draw overly liberal conclusions from model-based representations. We describe each theory in turn.

2.1 The Probability Sampling Model

Masasi Hattori developed a recent account of syllogistic reasoning called the "probability sampling model" (PSM; Hattori, 2016). The account holds that reasoners interpret a set of syllogistic premises by constructing a prototypical representation of them (referred to as a "probability prototype model"). The probability prototype model uses circles to represent set-membership relations, and so it is closely related to, for example, Euler circles and Venn diagrams (see the meta-analysis in Khemlani & Johnson-Laird, 2012, which reviews other theories based on such diagrammatic systems, and also chapter 13.1 by Jamnik, this handbook). The various intersections of the circles in the prototype model denote different kinds of individuals. For example, a diagram of "All of the designers are women" would include a circle representing the set of designers embedded inside a circle representing the set of women. The intersections of the circles represent the different kinds of individuals consistent with the premise, for example, that one area would represent the designers who are also women, and another would represent women who are not designers. Hattori posits that reasoners annotate the different areas of the prototype model with information about the probability of their occurrence.

Once the probabilities are established, reasoners draw a finite set of random samples to construct a mental model, that is, a small set of tokens that denote the entities referred to in the premises. Hattori's adaptation of mental model theory is more restricted than previous accounts. For example, provided that the two kinds of areas established by a prototype model have equal probability, a sample mental model of "All of the designers are women" can be represented in the following diagram:

designer	woman
¬ designer	woman
designer	woman
¬ designer	woman

Each row of the diagram depicts the results of a random draw from the prototype model, and so the first row depicts a designer who is also a woman. The "¬" denotes the symbol for negation, and so the second row depicts a woman who is not a designer. The additional rows depict additional random draws from the prototype model. In the PSM, the establishment of a sample mental model is the central representation on which a unitary reasoning process operates. The algorithm works by applying a series of tests, one after another, to the sample mental model in order to generate a conclusion. Figure 3.3.2 provides a schematic of how the full theory works.

To test the theory's predictions, Hattori implemented the theory computationally and then ran simulations that compared the theory's predictions against eight separate data sets on syllogistic reasoning. He also compared the PSM's predictions against two other theories: Chater and Oaksford's probability heuristics model (Chater & Oaksford, 1999) and a parameterized version of mental model theory (Hattori's implementation of Johnson-Laird & Bara, 1984). His analyses show that the PSM matches the performance of both theories (Hattori, 2016, p. 308).

One major strength of the PSM is that it can explain how contents affect the kinds of inferences reasoners draw. For instance, consider the following problem:

(5) Some of the Frenchmen are wine-drinkers.

Some of the wine-drinkers are Italians.

What, if anything, follows?

A robust result from studies of syllogistic reasoning is that reasoners should be less apt to infer "Some of the Frenchmen are Italians," since they likely consider it either implausible or rare that a person is both French and Italian at the same time. Hattori's PSM can account for the effect as follows: the meanings of the premises

bias the way people construct the probability prototype model such that implausible areas of the representation are assigned low probabilities. Hattori implemented the process, and he described two studies that corroborate the predictions of his implementation (Hattori, 2016).

However, Hattori's theory is not without its limitations. One limitation is that the theory cannot explain how reasoners might draw inferences from more than two premises: in these cases, the two-dimensional probability sampling model may be unable to represent all of the possible different individuals. Another limitation is that under the PSM, deliberation occurs by applying a set of logical tests, one after another, to a sample mental model. The process is akin to the way heuristic processing operates in previous theories (e.g., Chater & Oaksford, 1999). No evidence at present suggests that reasoners apply such tests in a fixed, systematic order for each and every syllogism; indeed, the evidence suggests that reasoners tend to develop strategies over the course of a study on syllogisms, which would seem to conflict with the PSM. Moreover, running such tests in a fixed order is bound to produce a conclusion for any sample mental model—and so a clear consequence of the tests is that the PSM cannot account for why reasoners often spontaneously respond that "No valid conclusion" follows (Ragni et al., 2019). Other accounts based on sampling and constructing mental models suffer from similar issues (see, e.g., Tessler & Goodman, 2014). In general, it is too taxing for people to constantly and repeatedly apply the same set of tests to the representations they construct, and so the algorithm, although tractable, is limited and not cognitively plausible. In essence, the PSM does not explain the psychological processes that underlie how reasoners deliberate on syllogistic inferences.

The primary theoretical contribution of Hattori's (2016) probabilistic sampling model is that it integrates probabilistic machinery and sampling procedures with the construction of mental models. Another recent computational theory—mReasoner—likewise integrates probabilistic sampling with procedures with the construction of mental models.

2.2 mReasoner: A Unified Computational Implementation of Mental Model Theory

mReasoner (Khemlani & Johnson-Laird, 2021) is a unified computational implementation of mental model theory, which posits that reasoning depends on the construction and manipulation of mental models, that is, iconic simulations of possibilities (Bucciarelli & Johnson-Laird, 1999; Johnson-Laird, 1983; Johnson-Laird, Khemlani, &

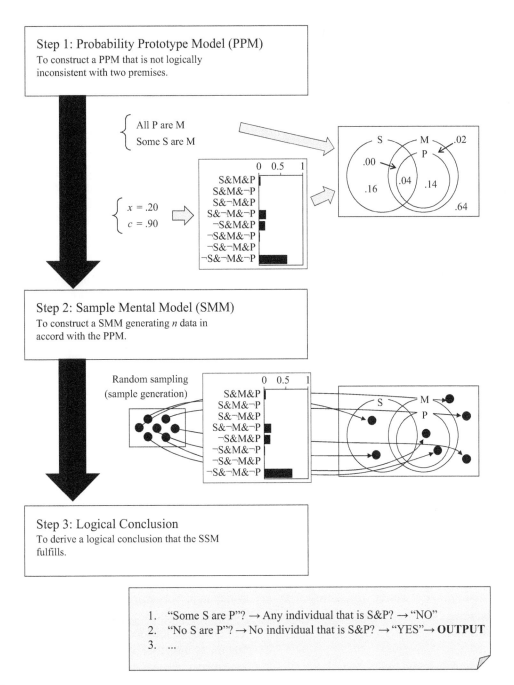

Figure 3.3.2
A schematic diagram of how Hattori's (2016) probability sampling model draws syllogistic conclusions. (Used with permission from Hattori, 2016.)

Goodwin, 2015a). The theory and its implementation are based on three fundamental assumptions:

- **Mental models represent possibilities:** a given assertion refers to a set of discrete possibilities that are observed or imagined (Johnson-Laird, 1983).

- **Iconicity and discreteness:** mental models are iconic, that is, their structures mirror the structures of what they represent (see Peirce, 1931–1958, vol. 4). Models

can also include abstract symbols, for example, the symbol for negation (Khemlani, Orenes, & Johnson-Laird, 2012). They are discrete in that they do not consist of continuous spaces, areas, and regions.

- **Dual processes:** reasoning, including syllogisms, is based on two interacting sets of processes. Rapid, heuristic reasoning occurs as a consequence of building and scanning a single model. Deliberative reasoning, by contrast, occurs as a result of revising the initial model to search for alternative models of heuristic conclusions (Khemlani & Johnson-Laird, 2013, 2016; Khemlani, Lotstein, Trafton, & Johnson-Laird, 2015).

The computational model makes syllogistic inferences by stochastically constructing a mental model directly from the premises. Hence, unlike the PSM, mReasoner eschews any intermediary representations (e.g., the probability prototype model). Two factors dictate how initial models are constructed: the size of a model (i.e., the maximum number of entities it represents) and the contents of a model. One parameter in the system controls the size of a mental model. It does so by basing the size on a sample drawn from a Poisson distribution. Another parameter governs the model's contents, which are drawn from the most common set of possibilities corresponding to a particular assertion (the canonical set) or else the complete set of possibilities consistent with the assertion. For example, in the case of "All of the designers are women," reasoners tend to consider only one canonical possibility: female designers. But the complete set of possibilities allows for women who are not designers, and a parameter in the system sets the probability of drawing from the complete set.

To illustrate how the system makes heuristic inferences, consider this syllogism:

(6) All of the designers are women.

Some of the women are not employees.

What, if anything, follows?

Suppose that the premises for (6) are input into mReasoner. The system may construct the following initial model:

designer	woman	
designer	woman	
designer	woman	employee
		employee

The diagram is similar to that of the sample mental model illustrated in the previous section: its rows denote separate individuals. To generate a heuristic conclusion, the system scans the model in the direction in which it was built. In the model above, for instance, the system builds tokens for "designers" first, "women" second, and "employees" third. Hence, the program draws an initial conclusion that interrelates "designers" to "employees" (e.g., some of the designers are not employees). This conclusion matches the preponderance of conclusions that reasoners spontaneously generate. For other sorts of syllogisms, the system draws initial intuitive

conclusions that interrelate "employees" to "designers," again depending on how the model was constructed.

Because the heuristic conclusion depends on just a single model, the system generates it quickly. But, as the example illustrates, the conclusion may be invalid. To correct the error, the program can call on a deliberative component to search for counterexamples to conclusions (Khemlani & Johnson-Laird, 2013). It operates by modifying the initial model, using a finite set of search strategies (Bucciarelli & Johnson-Laird, 1999). When the deliberative system is engaged, it can find a counterexample to the conclusion that some of the designers are not employees:

designer	woman	employee
designer	woman	employee
designer	woman	employee
	woman	

This model is one that represents the premises in (6) but falsifies the heuristic conclusion, and so the program responds that no valid conclusion holds. A separate parameter controls the probability that the deliberative system is engaged. Figure 3.3.3 provides a schematic diagram of the system's architecture.

mReasoner provides a close fit to the results from the data presented in the meta-analysis on syllogistic reasoning compared to alternative theories (Khemlani & Johnson-Laird, 2013). A major strength of the theory is that it is flexible enough to simulate reasoning about problems that include any number of premises, for example, one-premise "immediate" inferences (Khemlani et al., 2015), two-premise syllogisms (Khemlani & Johnson-Laird, 2016) and set-membership inferences (Khemlani, Lotstein, & Johnson-Laird, 2014), and more complex problems composed of three premises (Ragni, Khemlani, & Johnson-Laird, 2014). It does so by relying on a limited set of parameters that provide motivated constraints on how the theory's predictions vary (see figure 3.3.3). The flexibility of the system allows it to model individual differences and strategies in syllogistic reasoning (Bacon, Handley, & Newstead, 2003; Khemlani & Johnson-Laird, 2016; Ragni, Riesterer, Khemlani, & Johnson-Laird, 2018).

mReasoner has several limitations. One limitation is that at present, unlike the probability sampling model, mReasoner does not have machinery capable of explaining effects of content (i.e., the "belief bias" effect). In principle, the machinery is feasible, and a prototype system implements "semantic modulation" for sentential reasoning (see Khemlani, Byrne, & Johnson-Laird,

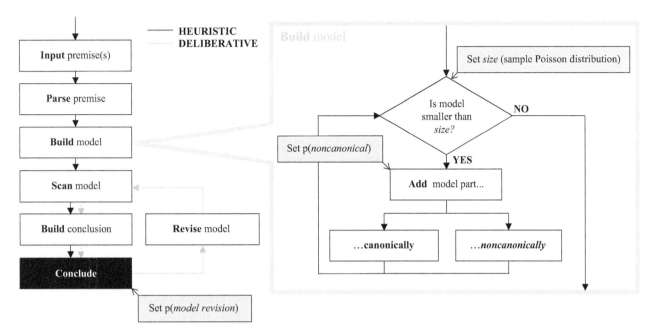

Figure 3.3.3

A schematic diagram of the program implemented in mReasoner for reasoning with mental models of quantified assertions such as "All the designers are women." White boxes denote procedures, gray boxes denote parameters, and the black box denotes the point at which the program generates a conclusion. The program integrates two separate reasoning pipelines: a fast, heuristic process (represented by solid arrows) that does not revise models and a slow, recursive, deliberative process (represented by dotted arrows) that revises initial models and draws conclusions from the set of alternative models. The diagram also highlights the stochastic algorithm that controls how models are built.

2018). But the system has not been applied to simulate belief bias effects. A second limitation of the theory is that, unlike theories based on Bayesian updating mechanisms (e.g., Chater & Oaksford, 1999; Oaksford & Chater, 2009), mReasoner does not explain how learning and reasoning interact with one another.

2.3 Integrated Models of Reasoning

The two newest theories of syllogistic reasoning described here were developed to achieve parsimony relative to previous theories. They share many commonalities: they both use sampling procedures to build up representations. They both result in the construction of mental models (i.e., mental simulations of possibilities). They're both implemented computationally, and so their limitations are clearly defined. Because of the clarity afforded by their computational implementations, both have been validated against numerous data sets—indeed, both accounts manage to explain far more data than any single previous account of syllogistic reasoning to date. And they predict many empirical phenomena beyond the narrow scope of the 64 syllogistic reasoning problems. As such, the theories may serve as promising foundations for future research.

3. General Discussion

Why have psychologists spent so much time and effort on understanding syllogistic reasoning? One reason is because Aristotle had made it central to his logic; logic was practically synonymous with syllogistic logic until Frege developed the predicate calculus, when logicians realized that syllogisms were but a small corner of the space of logical analysis. Nevertheless, syllogisms presented a feasible microcosm for study. By the middle of the 20th century, researchers began to recognize that people's syllogistic inferences could serve as a measure of their general ability to think rationally. Syllogisms were included in verbal reasoning tasks, entrance exams, and aptitude tests (e.g., Ekstrom, French, Harman, & Dermen, 1976; Nester & Colberg, 1984). Researchers of psychiatric disorders even began to use erroneous responses in syllogisms as a way to measure irrational thinking, for example, in patients with schizophrenia (Gottesman & Chapman, 1960; Williams, 1964), a practice that carries on into modern psychiatry (Mujica-Parodi, Malaspina, & Sackeim, 2000). Yet, no psychological account of syllogisms existed when Störring conducted his initial investigations into people's logical reasoning abilities. His primary goal was to resolve theoretical debates among philosophers and logicians; hence, it was necessary to develop a theory of syllogisms.

The generation of researchers who followed Störring developed theories that operated under the consensus that errors in reasoning were the result of fallible strategies. New theories continued to flourish late into the century, such that by its end, theorists had proposed a dozen different psychological accounts of syllogistic reasoning. As theories flourished, consensus among researchers dissipated. Khemlani and Johnson-Laird (2012) noted that the failure to arrive at a single, unified account was a catastrophe for the science of human thinking: "skeptics may well conclude that cognitive science has failed [as it] yields no consensus about a small, empirically tractable domain of reasoning."

The dozen theories of syllogistic reasoning cannot all be correct: they make inconsistent predictions and commitments. A trivial path forward might be to stitch together all of the theories' predictions by disregarding their inconsistencies; after all, people themselves can be inconsistent. Different people develop different reasoning strategies, and even the same person may use different strategies on different occasions (see, e.g., Ragni et al., 2018). Alas, an amalgamated account is unlikely to explain the robust patterns of responses depicted in figure 3.3.1. Worse still, it is unlikely to concern anything beyond the scope of the 64 syllogistic reasoning problems, and so it cannot provide much of a guide for how humans reason in general. A more productive way to resolve the impasse is to develop a new, unified theory of syllogistic reasoning, one that supersedes the old ones.

Hattori (2016) and Khemlani and Johnson-Laird (2013, 2016) took such an approach in their most recent proposals of reasoning by syllogism. Both proposals provide a framework for explaining (a) the interpretation and mental representation of syllogistic premises, (b) what the mind computes and how it carries out inferential tasks with such assertions, (c) the differences in difficulty from one inference to another and common errors, (d) how contents affect performance, and (e) individual differences in performance from one person to another. And their theories have been implemented in computer programs whose operations were used to simulate empirical data.

Perhaps the most fierce debate between theorists who study human reasoning is the challenge of integrating probabilistic and deductive inference (see, e.g., Baratgin et al., 2015; Evans & Over, 2013; Johnson-Laird, Khemlani, & Goodwin, 2015a, 2015b; Khemlani, 2018). Syllogistic reasoning stands at the nexus of the debate: people

have the ability to draw valid syllogistic inferences, that is, they have the ability to reason deductively. The result is evident in Störring's seminal studies and in numerous studies that followed. But reasoners' deductive abilities are limited; they often draw conclusions that are likely—but not guaranteed—to be true, and their own background knowledge biases their tendency to accept valid deductions. Some theorists sought to resolve the discrepancy by developing new probabilistic interpretations of validity (Chater & Oaksford, 1999; see also Evans, 2012; Over, 2009), by emphasizing those aspects of reasoning that are easiest to formulate with probabilistic machinery (e.g., Evans & Over, 2004), and by discounting the relative importance of deduction in daily thinking (e.g., Oaksford & Chater, 2002). But such treatments are controversial, even among theorists who argue that human rationality is fundamentally probabilistic (see Johnson-Laird et al., 2015a). As a result, arguments on how to integrate probability and deduction often result in deadlock. Just as the two theories of syllogistic reasoning reviewed in this chapter may help adjudicate between heuristic and deliberative processes, they may also represent ways to overcome larger deadlocks in debates on rationality: at their core, the theories advocate limited and biased sampling processes best characterized and formalized with probabilistic constraints. But the outputs of the sampling processes are discrete mental simulations—mental models—that help explain deductions, inductions, and errors in reasoning.

Acknowledgments

This research was supported by a grant from the Naval Research Laboratory to study deductive reasoning. I am grateful to my advisor and long-time collaborator, Phil Johnson-Laird, and I also thank Paul Bello, Gordon Briggs, Monica Bucciarelli, Ruth Byrne, Sam Glucksberg, Adele Goldberg, Geoff Goodwin, Hillary Harner, Tony Harrison, Laura Hiatt, Zach Horne, Laura Kelly, Markus Knauff, Joanna Korman, Max Lotstein, Marco Ragni, Nicolas Riesterer, and Greg Trafton for their intuitions and deliberations.

References

Bacon, A., Handley, S., & Newstead, S. (2003). Individual differences in strategies for syllogistic reasoning. *Thinking & Reasoning, 9*, 133–168.

Baratgin, J., Douven, I., Evans, J. St. B. T., Oaksford, M., Over, D., & Politzer, G. (2015). The new paradigm and mental models. *Trends in Cognitive Sciences, 19*, 547–548.

Begg, I., & Denny, J. P. (1969). Empirical reconciliation of atmosphere and conversion interpretations of syllogistic reasoning errors. *Journal of Experimental Psychology, 81*, 351–354.

Braine, M. D. S., & Rumain, B. (1983). Logical reasoning. In P. H. Mussen (Ed.), *Handbook of child psychology: Vol. 3. Cognitive development* (pp. 263–340). New York, NY: Wiley.

Bucciarelli, M., & Johnson-Laird, P. N. (1999). Strategies in syllogistic reasoning. *Cognitive Science, 23*, 247–303.

Chater, N., & Oaksford, M. (1999). The probability heuristics model of syllogistic reasoning. *Cognitive Psychology, 38*, 191–258.

Clark, R. S. (1922). An experimental study of silent thinking. *Archives of Psychology, 7*, 5–13.

Craik, K. J. W. (1943). *The nature of explanation.* Cambridge, England: Cambridge University Press.

Ekstrom, R. B., French, J. W., Harman, H. H., & Dermen, D. (1976). *Manual for kit of factor-referenced cognitive tests* (Vol. 102). Princeton, NJ: Educational Testing Service.

Evans, J. St. B. T. (2008). Dual-processing accounts of reasoning, judgment, and social cognition. *Annual Review of Psychology, 59*, 255–278.

Evans, J. St. B. T. (2012) Questions and challenges for the new psychology of reasoning. *Thinking & Reasoning, 18*, 5–31.

Evans, J. St. B. T., & Over, D. E. (2004). *If.* Oxford, England: Oxford University Press.

Evans, J. St. B. T., & Over, D. E. (2013). Reasoning to and from belief: Deduction and induction are still distinct. *Thinking & Reasoning, 19*, 267–283.

Evans, J. St. B. T., & Stanovich, K. E. (2013). Dual-process theories of higher cognition: Advancing the debate. *Perspectives on Psychological Science, 8*, 223–241.

Geurts, B. (2003). Reasoning with quantifiers. *Cognition, 86*(3), 223–251.

Goel, V., Buchel, C., Frith, C., & Dolan, R. J. (2000). Dissociation of mechanisms underlying syllogistic reasoning. *Neuroimage, 12*, 504–514.

Gottesman, L., & Chapman, L. J. (1960). Syllogistic reasoning errors in schizophrenia. *Journal of Consulting Psychology, 24*, 250–255.

Guyote, M. J., & Sternberg, R. J. (1981). A transitive-chain theory of syllogistic reasoning. *Cognitive Psychology, 13*(4), 461–525.

Hattori, M. (2016). Probabilistic representation in syllogistic reasoning: A theory to integrate mental models and heuristics. *Cognition, 157*, 296–320.

Jeffrey, R. (1981). *Formal logic: Its scope and limits* (2nd ed.). New York, NY: McGraw-Hill.

Johnson-Laird, P. N. (1983). *Mental models: Towards a cognitive science of language, inference, and consciousness.* Cambridge, MA: Harvard University Press.

Johnson-Laird, P. N., & Bara, B. G. (1984). Syllogistic inference. *Cognition, 16*, 1–61.

Johnson-Laird, P. N., & Khemlani, S. S. (2013). Toward a unified theory of reasoning. In B. Ross (Ed.), *The psychology of learning and motivation* (Vol. 59, pp. 1–42). Cambridge, MA: Academic Press.

Johnson-Laird, P. N., Khemlani, S., & Goodwin, G. P. (2015a). Logic, probability, and human reasoning. *Trends in Cognitive Sciences, 19*, 201–214.

Johnson-Laird, P. N., Khemlani, S., & Goodwin, G. P. (2015b). Response to Baratgin et al.: Mental models integrate probability and deduction. *Trends in Cognitive Sciences, 19*, 548–549.

Johnson-Laird, P. N., & Steedman, M. J. (1978). The psychology of syllogisms. *Cognitive Psychology, 10*, 64–99.

Johnson-Laird, P. N., & Wason, P. C. (1970). A theoretical analysis of insight into a reasoning task. *Cognitive Psychology, 1*, 134–148.

Khemlani, S. S. (2018). Reasoning. In S. L. Thompson-Schill (Ed.), *Stevens' handbook of experimental psychology and cognitive neuroscience* (4th ed., Vol. 3, pp. 385–427). Hoboken, NJ: Wiley.

Khemlani, S., Byrne, R. M. J., & Johnson-Laird, P. N. (2018). Facts and possibilities: A model-based theory of sentential reasoning. *Cognitive Science, 42*, 1887–1924.

Khemlani, S., & Johnson-Laird, P. N. (2012). Theories of the syllogism: A meta-analysis. *Psychological Bulletin, 138*, 427–457.

Khemlani, S., & Johnson-Laird, P. N. (2016). How people differ in syllogistic reasoning. In A. Papafragou, D. Grodner, D. Mirman, & J. Trueswell (Eds.), *Proceedings of the 38th Annual Conference of the Cognitive Science Society* (pp. 2165–2170). Austin, TX: Cognitive Science Society.

Khemlani, S., & Johnson-Laird, P. N. (2017). Illusions in reasoning. *Minds & Machines, 27*, 11–35.

Khemlani, S., & Johnson-Laird, P. N. (2021). Reasoning about properties: A computational theory. Manuscript in press at *Psychological Review*.

Khemlani, S., Lotstein, M., & Johnson-Laird, P. N. (2014). A mental model theory of set membership. In P. Bello, M. Guarini, M. McShane, & B. Scassellati (Eds.), *Proceedings of the 36th Annual Conference of the Cognitive Science Society* (pp. 2489–2494). Austin, TX: Cognitive Science Society.

Khemlani, S., Lotstein, M., Trafton, J. G., & Johnson-Laird, P. N. (2015). Immediate inferences from quantified assertions. *Quarterly Journal of Experimental Psychology, 68*, 2073–2096.

Khemlani, S., Orenes, I., & Johnson-Laird, P. N. (2012). Negation: A theory of its meaning, representation, and use. *Journal of Cognitive Psychology, 24*, 541–559.

Knauff, M. (2013). *Space to reason: A spatial theory of human thought.* Cambridge, MA: MIT Press.

Mujica-Parodi, L. R., Malaspina, D., & Sackeim, H. A. (2000). Logical processing, affect, and delusional thought in schizophrenia. *Harvard Review of Psychiatry, 8*, 73–83.

Nester, M. A., & Colberg, M. (1984). The effects of negation mode, syllogistic invalidity, and linguistic medium on the psychometric properties of deductive reasoning tests. *Applied Psychological Measurement, 8*, 71–79.

Oaksford, M., & Chater, N. (2002). Commonsense reasoning, logic and human rationality. In R. Elio (Ed.), *Common sense, reasoning and rationality* (pp. 174–214). Oxford, England: Oxford University Press.

Oaksford, M., & Chater, N. (2009). Précis of *Bayesian rationality: The probabilistic approach to human reasoning. Behavioral and Brain Sciences, 32*, 69–120.

Over, D. E. (2009). New paradigm psychology of reasoning. *Thinking & Reasoning, 15*, 431–438.

Peirce, C. S. (1931–1958). *Collected papers of Charles Sanders Peirce* (C. Hartshorne, P. Weiss, & A. Burks, Eds., 8 vols.). Cambridge, MA: Harvard University Press.

Politzer, G. (2004). Some precursors of current theories of syllogistic reasoning. In K. Manktelow & M. C. Chung (Eds.), *Psychology of reasoning: Theoretical and historical perspectives* (pp. 223–250). New York, NY: Psychology Press.

Politzer, G. (2007). The psychological reality of classical quantifier entailment properties. *Journal of Semantics, 24*(4), 331–343.

Polk, T. A., & Newell, A. (1995). Deduction as verbal reasoning. *Psychological Review, 102*, 533–566.

Ragni, M., Dames, H., Brand, D., & Riesterer, N. (2019). When does a reasoner respond: Nothing follows? In A. K. Goel, C. M. Seifert, & C. Freksa (Eds.), *Proceedings of the 41st Annual Conference of the Cognitive Science Society* (pp. 2640–2646). Montreal, Canada: Cognitive Science Society.

Ragni, M., Khemlani, S., & Johnson-Laird, P. N. (2014). The evaluation of the consistency of quantified assertions. *Memory & Cognition, 42*, 53–66.

Ragni, M., Riesterer, N., Khemlani, S., & Johnson-Laird, P. N. (2018). Individuals become more logical without feedback. In C. Kalish, M. Rau, T. Rogers, & J. Zhu (Eds.), *Proceedings of the 40th Annual Conference of the Cognitive Science Society* (pp. 2315–2320). Austin, TX: Cognitive Science Society.

Revlis, R. (1975). Two models of syllogistic reasoning: Feature selection and conversion. *Journal of Verbal Learning and Verbal Behavior, 14*, 180–195.

Rips, L. J. (1994). *The psychology of proof: Deductive reasoning in human thinking.* Cambridge, MA: MIT Press.

Stenning, K., & Cox, R. (2006). Reconnecting interpretation to reasoning through individual differences. *Quarterly Journal of Experimental Psychology, 59*, 1454–1483.

Störring, G. (1908). Experimentelle Untersuchungen über einfache Schlussprozesse [Experimental investigations of simple inference processes]. *Archiv für die gesamte Psychologie, 11*, 1–27.

Tessler, M. H., & Goodman, N. D. (2014). Some arguments are probably valid: Syllogistic reasoning as communication. In *Proceedings of the 36th Annual Meeting of the Cognitive Science Society* (pp. 1574–1579). Austin, TX: Cognitive Science Society.

Wason, P. C., & Evans, J. St. B. T. (1974). Dual processes in reasoning? *Cognition, 3*, 141–154.

Wetherick, N. E., & Gilhooly, K. J. (1995). 'Atmosphere', matching, and logic in syllogistic reasoning. *Current Psychology, 14*, 169–178.

Williams, E. B. (1964). Deductive reasoning in schizophrenia. *Journal of Abnormal and Social Psychology, 69*, 47–61.

Woodworth, R. S., & Sells, S. B. (1935). An atmosphere effect in formal syllogistic reasoning. *Journal of Experimental Psychology, 18*, 451–460.

Section 4   Probabilistic Reasoning

4.1   Subjective Probability and Its Dynamics

Alan Hájek and Julia Staffel

Summary

This chapter is a philosophical survey of some leading approaches in formal epistemology in the so-called
Bayesian tradition. According to them, a rational agent's
degrees of belief—*credences*—at a time are representable
with probability functions. We also canvas various further
putative "synchronic" rationality norms on credences.
We then consider "diachronic" norms that are thought
to constrain how credences should respond to evidence.
We discuss some of the main lines of recent debate and
conclude with some prospects for future research.

1.   Outline of the Chapter

You are certain that something exists, slightly less confident that your lottery ticket will lose, less confident
again that it will be cloudy tomorrow, still less confident
that your lottery ticket will win, and you have no confidence whatsoever that nothing exists. Traditional epistemology traffics mainly in the binary doxastic states
of *belief* and *knowledge*, and classical logic is thought
by many to provide norms for them (see chapter 3.1 by
Steinberger and chapter 5.1 by van Ditmarsch, both in
this handbook). But we also have a whole spectrum of
belief-like attitudes. Following Ramsey (1926/1990), we
may call them *degrees of belief* and regard them as being
answerable to a "logic of partial belief": probability theory.[1] This, in turn, gives that mathematical theory one of
its main interpretations: *subjective probability*.

The theory of subjective probability provides us
with a *normative, numerical* model of degrees of belief.
It uses precise numbers between 0 and 1 to represent
the degree to which someone believes something. Moreover, it formulates normative requirements that their
degrees of beliefs have to obey in order to be perfectly
rational. Hence, the philosophical investigation of subjective probability is less concerned with the question
of whether people are in fact rational—this question

is delegated to empirical psychologists (see chapter 4.5
by Chater & Oaksford, this handbook)—and more concerned with characterizing norms of ideal rationality
and reasoning. In this chapter, we survey a variety of
purported rational requirements and the philosophical
justifications that have been offered in their support—
both *synchronically* (for an agent at a given time) and
diachronically (for an agent at different times).

2.   The Probability Calculus

To understand subjective probability, we must first
understand *probability*. Kolmogorov's axiomatization
(1933/1950) remains the orthodox formalization of
probability—the so-called *probability calculus*.[2] We begin
with a "universal set" Ω regarded as the set of all possibilities of interest or of all possible outcomes of a given
random experiment. Probabilities are numerical values
assigned by a real-valued function P to certain subsets of
Ω—*events*—obeying the following axioms:

1. For all events X, $P(X) \geq 0$.   (Nonnegativity)
2. $P(\Omega) = 1$.   (Normalization)
3. For all disjoint events X and Y, $P(X \cup Y) = P(X) + P(Y)$.
   (Finite Additivity)

These axioms are often reformulated with events replaced
by sentences in a formal language. Normalization becomes

   $P(T) = 1$, where T is any tautology.

And Finite Additivity becomes

   for all logically incompatible sentences X and Y,
   $P(X \vee Y) = P(X) + P(Y)$.[3]

Conditional probability is probability *relative to*, or *given*,
some body of evidence or information. Following Kolmogorov, we symbolize "the probability of X, given Y"
as $P(X \mid Y)$ and define it as

$$P(X \mid Y) = \frac{P(X \cap Y)}{P(Y)}, \quad \text{if } P(Y) > 0.$$

For more on the probability calculus, see Hájek and Hitchcock (2016).[4]

This formalism is brought to life when we interpret "*P*." It might be regarded, for example, as representing *objective chances* in the world, independent of anyone's opinions; these might be understood as relative frequencies, as propensities, or as falling out of the best systematization of the universe—see Hájek (2019) for more discussion and references. However, according to the interpretation of interest here, "*P*" represents a rational agent's degrees of belief, also known as *degrees of confidence*, or *credences*. For example, your credence that a roll of a fair die lands an even number should be

$$P(2 \cup 4 \cup 6) = P(2) + P(4) + P(6) = \frac{1}{6} + \frac{1}{6} + \frac{1}{6} = \frac{1}{2}.$$

And your credence that it lands 6, *given* that it lands an even number, should be

$$\frac{\frac{1}{6}}{\frac{1}{2}} = \frac{1}{3}.$$

Going in the other direction, we might begin by assuming that a rational agent has various degrees of belief and ask how to represent them formally. The standard view among philosophers, economists, and others in the "Bayesian" tradition is that someone's degrees of belief should be representable by a probability function conforming to this calculus. This view is known as *probabilism*. This requires some analysis: *what are* degrees of belief? And it requires some justification: *why should* they obey these axioms? The justification will be tailored to the analysis.

3. Accounts of Subjective Probability, and Arguments for Probabilism

3.1 The Betting Account

The dominant approaches see degrees of belief as intimately tied to action and decision making. A behaviorist approach identifies degrees of confidence with betting behavior (e.g., de Finetti, 1937/1964). Its gist is as follows:

> Your probability of *E* is *x* iff *x* is the price for which you would either buy or sell a bet that pays a unit of money if *E* and that pays nothing otherwise.

There are various problems with this account. It assumes that there is a unique such price when there may be none or more than one. Moreover, the account is implausible when *E* is an event under your control. For example, you assign low probability to your impersonating a chicken today, but that may change if your doing so wins a $1 bet. (If $1 does not suffice, make it $1,000,000—raising

the further problem that your betting behavior is sensitive to the stakes involved, while your credences are not.) More generally, measuring your credences with bets may change what those credences are, as Ramsey observed. (Indeed, someone offering a high-stakes bet on a proposition may make you suspicious that they have inside knowledge about it!) And a given betting price may be jointly determined by many components—the value you attach to money, the thrill of gambling, a public show of bravado, and so on—only one of which is your credence.

3.2 The Dutch Book Argument

Despite its limitations, many authors regard the betting account as "fundamentally sound," in Ramsey's (1926/1990) words. As such, it is appealed to in one of the most important arguments for probabilism: the *Dutch Book argument*. A *Dutch book* is a set of bets bought or sold at such prices as to guarantee a net loss. We have the celebrated *Dutch book theorem*, first stated by Ramsey, and proven by de Finetti (1937/1964):

> If a set of (normalized) betting prices violates the probability calculus, then there is a Dutch book consisting of bets at those prices.

So, if your betting prices violate the probability calculus, you are susceptible to a Dutch book. However, the argument continues, if you are susceptible, then you are irrational. Hence, rationality requires your betting prices to obey the probability calculus. Identifying your credences with your betting prices, the argument concludes: rationality requires your credences to obey the probability calculus.

The argument is often personified with a canny bookie fleecing a hapless agent who has nonprobabilistic credences. For example, suppose you assign probability ½ to the Democrats winning the next election, probability ½ to the Republicans winning, but only probability 0.9 to either the Democrats or the Republicans winning. The bookie could sell you for 50 cents a bet that pays $1 if the Democrats win, sell you for 50 cents a bet that pays $1 if the Republicans win, and buy from you for 90 cents a bet that pays $1 if either of them win. Initially, you pay him a total of $1 and he pays you 90 cents, with you losing 10 cents upfront. Then, come what may, the bets that you hold pay exactly the same as the bet that he holds, so that your net loss remains the same.

Often neglected, but equally important, is the *converse* Dutch book theorem (Kemeny, 1955):

> If a set of betting prices (normalized to 1) *obeys* the probability calculus, then there is *no* Dutch book consisting of bets at those prices.

This dispels the concern that even so-called *coherent* agents, whose credences obey the probability calculus, might be susceptible to Dutch books. Not so: coherence inoculates us!

The Dutch book argument is often criticized for bringing out a *pragmatic* rather than an *epistemic* defect in an incoherent set of credences. Furthermore, some critics complain that an incoherent agent need not fall into the trap of being Dutch-booked—for example, you might just walk away instead of making the sure-loss package of bets. In response, Skyrms (1980), following Ramsey, interprets the argument as merely dramatizing an inconsistency in such an agent's evaluations—an epistemic defect. For more discussion, see Skyrms (1993), Vineberg (2001, 2016), and Hájek (2009).

3.3 Representation Theorems

We have noted some problems with the betting account of credences, which carry over to the Dutch book argument: if your credences come apart from your betting prices, a defect in the latter does not entail a defect in the former. Let's move on, then, to another account of credence and another kind of argument for probabilism that is based on it. We have Ramsey to thank for both.

He presupposes an operationalist attitude to defining such a theoretical term: "the degree of a belief . . . has no precise meaning unless we specify more exactly how it is to be measured" (Ramsey, 1926/1990, p. 167). His guiding idea is that probability "is a measurement of belief . . . *qua* basis of action" (Ramsey, 1926/1990, p. 171) and to use an agent's preferences among gambles to provide this measurement. He proposes a number of axioms governing such preferences. Some are "consistency" assumptions, such as transitivity, while others are structural, aimed at streamlining the mathematics. Together, they allow him to calibrate a *utility* scale for the agent, a measure of how desirable various gambles and their outcomes are. This, in turn, determines her credences as ratios of differences of utilities. He goes on to add further axioms that guarantee their obeying the probability calculus.

Savage (1954) similarly derives probabilities and utilities from preferences that are constrained by certain axioms. However, he is more concerned with the *normative* status of the axioms than Ramsey is: they constrain *rational* preferences (see also chapter 8.1 by Grüne-Yanoff, this handbook). Savage proves a *representation theorem*: for a given (representation of a) set of such preferences, there is a class of utility functions, unique up to positive linear transformations, and a unique probability function that jointly "represent" the preferences in the

following sense: the *expected utility* of an option ("act") is the weighted average of the utilities of its possible outcomes associated with a set of "states," the weights provided by the probabilities of the states. A utility function (chosen from such a class) and probability function are said to jointly *represent* an agent's preferences when she prefers A to B iff the expected utility of A is greater than that of B. Jeffrey (1965) offers an alternative axiomatization of rational preferences to Savage's, and Bolker proves another representation theorem (cited by Jeffrey), which typically does not yield a unique probability function.

Such representation theorems undergird an argument for probabilism: any rational agent has preferences that obey the axioms; as such, she is representable as an expected-utility maximizer, with weights interpreted as her credences. These credences obey Kolmogorov's axioms. However, some question the normative status of some of the preference axioms—they are there more to facilitate the mathematical representation. Others question the step from the rational requirement of the agent being *representable* probabilistically to the rational requirement of her credences *being* probabilities. For example, she may *also* be representable *non*probabilistically (see Zynda, 2000). Indeed, she may be representable in entirely implausible ways—for example, as a voodoo spirit maximizer (see Hájek, 2008).

Representation-theorem approaches to credences allow and even support an anti-realist stance regarding them: they are merely artifacts of a representation. (Stefánsson [2017] defends this view.) By contrast, a more realist understanding of credences fits well with our next arguments for probabilism. (See Eriksson & Hájek [2007] for further discussion of the metaphysics of credences, including their view that they should be regarded as primitive, irreducible to anything else; see also note 2 for references to approaches that take comparative probabilities as primitive.)

3.4 Accuracy Arguments

Belief is typically regarded as being a genuine mental state rather than merely a representational artifact. Moreover, it is often said to have a constitutive aim: *truth*. As such, a given belief may be evaluated on whether or not it achieves this aim—whether it is true or false (see also chapter 3.1 by Steinberger, this handbook). However, intermediate credences apparently *cannot* be said to be true or false. Are they nonetheless answerable to a truth norm?

A thinker may assign intermediate degrees of confidence to strike a balance between the competing aims of believing truths and avoiding belief in falsehoods. Ideally, the agent assigns full credence to all true propositions and zero credence to all false propositions. When she assigns intermediate credences, we can measure how close they are to the truth. A specific class of measures, called *scoring rules* or *accuracy measures*, has been developed for this purpose.

A popular measure is the Brier score: if the agent's credence in some event E is x, and E actually occurs, then the Brier score for this credence is $(1-x)^2$; if E doesn't actually occur, the score is $(0-x)^2$. We can compute the Brier score for an entire credence function at a particular world by summing the individual scores for the agent's credences at that world. The closer the score is to zero, the better, since a lower score indicates better accuracy or greater closeness to the truth. The Brier score has been put to practical use, for example, to score the accuracy of weather forecasters and political analysts. A feature that makes the Brier score, among other scoring rules, particularly suited for this purpose is that it is *strictly proper*: someone who is asked to report her credences expects to minimize the score she will receive by reporting her actual credences. By contrast, if we used an improper scoring rule, such as the absolute distance measure $|1-x|$ or $|0-x|$, to score the agent's credences' distance from the truth, she would sometimes be better off lying about her credences to improve her (own expectation of her) accuracy score.

Suppose we adopt the Brier score to measure accuracy. Then, if an agent has nonprobabilistic credences, there is an alternative probabilistic credence function that is more accurate in every possible world. Conversely, if she has coherent credences, there is no alternative probabilistic credence assignment that is more accurate in every possible world. Hence, all and only coherent credence functions avoid being *accuracy dominated*. A version of this mathematical result was first demonstrated by de Finetti (1974, pp. 87–91) and has since been developed further to show that it holds for additional suitable scoring rules besides the Brier score. The theorem is the central premise of the *accuracy dominance argument* for probabilism, which claims that having accuracy-dominated credences is a rational defect, which can only be avoided by having probabilistic credences (for developments of the theorem and the argument, see Joyce, 1998; Leitgeb & Pettigrew, 2010a, 2010b; Pettigrew, 2016).

The accuracy dominance argument is popular because it doesn't rely on pragmatic considerations—it is thought to be an argument for epistemic rationality that proceeds from purely epistemic premises. However, for the argument to be sound, we need to select a specific accuracy measure from the class of eligible strictly proper scoring rules (Bronfman, manuscript). This has presented something of a problem, since it is difficult to find compelling epistemic grounds for selecting a specific accuracy measure.[5]

4. Other Putative Synchronic Norms

4.1 Regularity

"If it can happen, then it has some probability of happening." Many authors have offered some version of this intuitive slogan as a constraint on credences (e.g., Carnap, 1950; Lewis, 1980; Shimony, 1955). More formally, *regularity* constraints have the form

If X is possible, then $C(X) > 0$,

where C represents the credences of some rational agent. There are various candidates for the sense of "possibility" here, but *logical*, *metaphysical*, and *epistemic* possibility are especially prominent.

There are both theoretical and pragmatic arguments in favor of regularity. Theoretical: Your credences should reflect your evidence; if your evidence does not decisively rule out X, so that X remains epistemically possible, your credence should not rule it out either. Furthermore, such an open mind regarding X is required for you to be able to learn it (Lewis, 1980). Pragmatic: If you assign probability 0 to a possible proposition X, then you are susceptible to a *semi-Dutch Book* (Shimony, 1955): you will consider fair a bet that you could lose and that could not possibly win you any money. According to the betting interpretation, you are prepared to pay \$1 for a \$1 bet on X. This seems irrational, since at best you will break even, and you could lose \$1. However, see Hájek (2012) for rebuttals of these arguments (and others).

Furthermore, there are both theoretical and pragmatic arguments *against* regularity. Theoretical: If the space of possibilities is sufficiently large, regularity is mathematically impossible. For example, if there are uncountably many possible outcomes, then a (real-valued) probability function cannot assign positive probabilities to all of them. Indeed, uncountably many of its assignments must be 0 (see Hájek, 2003). Pragmatic: Assuming that your gaining infinite expected utility is possible (as in Pascal's wager and the St. Petersburg paradox[6]), regularity requires you to assign positive probability to your gaining it. This, in turn, requires you to value this prospect infinitely, however small this probability—the same as you would if you had it for sure (see Hájek, 2012; for further opposition to regularity, see also Easwaran, 2014; Levi, 1989; T. Williamson, 2007).

4.2 The Principal Principle

We have mentioned that the "P" that Kolmogorov axiomatized might be interpreted objectively as *chance*. Lewis offers a further rationality constraint on credences, which codifies a certain kind of alignment between them and chances. Here is his classic statement of it (Lewis, 1980, p. 266):

> **The Principal Principle.** Let C be any reasonable initial credence function. Let t be any time. Let x be any real number in the unit interval. Let X be the proposition that the chance, at time t, of A equals x. Let E be any proposition compatible with X that is admissible at time t. Then

$$C(A/XE) = x.$$

An initial credence function is a "prior" probability function, representing one's credences before one receives *any* evidence. An admissible proposition is one "whose impact on credence about outcomes comes entirely by way of credence about the chances of those outcomes" (Lewis, 1980, p. 272). Typically, information about the past is admissible, whereas information about how the chance process turns out is inadmissible. An instance of the Principal Principle for a toss of a coin after noon is:

C(heads | *chance*$_{noon}$(heads) = ½, and another toss of the coin landed heads at 11:59 am) = ½.

Lewis's talk of an "initial" credence function may come as a surprise—it is not *your* credence function or that of any rational agent who has learned *anything*. Lewis (1994) presents the Principal Principle differently, speaking instead of "a rational credence function for someone whose evidence is limited to the past and present—that is, for anyone who doesn't have access to some very remarkable channels of information" (p. 483). This far more natural (nonequivalent) formulation is how the principle is usually understood. It is also surprising that the vast literature on the Principal Principle overlooks the change in formulation.[7] Pettigrew (2016) gives an accuracy-based justification for the principle, as well as a comprehensive overview and discussion of different formulations of it. For a Dutch book–style argument, see Pettigrew (2018).

4.3 The Reflection Principle

The Principal Principle is sometimes called an *expert principle* or a *deference principle*, because it requires rational thinkers to treat a source of probabilities—the chance function—like an expert and to defer to its assignments.

Another such principle is the *Reflection Principle* (van Fraassen, 1984), which enjoins thinkers to defer to their own future credences. Where t is a particular time, and $t + n$ is a later time,

$$C_t(A \mid C_{t+n}(A) = x) = x.$$

Informally stated: Given that at a later time your credence in some claim A will be x, your credence in A should already be x at the current time. For example, suppose you are certain that once you start your car tomorrow morning and hear the rattling of its old engine, you will be highly confident that the car needs servicing; then you should be just as confident *now* that the car will need servicing tomorrow.

Van Fraassen argued for the principle by showing that there is a Dutch book argument for it similar to the one for conditionalization (more on that below; for discussion, see also Briggs, 2009; Christensen, 1991; Mahtani, 2015). There is also an argument showing that obeying Reflection maximizes the expected accuracy of one's credences (Easwaran, 2013). However, it has been pointed out that, for the principle to be plausible, the range of situations in which it applies must be restricted. If you expect your future self to be either irrational or forgetful, then you should not defer to your future credences (Arntzenius, 2003; Christensen, 1991; Talbott, 1991). Van Fraassen (1989, 1999) offers a modified formulation of the principle to address these types of worries.

4.4 The Indifference Principle

Another putative norm of rationality is the *Indifference Principle*, which applies when we either have no evidence, or our evidence does not discriminate between the relevant possibilities. Here's a version of it:

> If there are n possibilities, and an agent has no evidence that discriminates between them, then the rational credence for her to assign to each possibility is $1/n$.

For example, suppose you are informed that you have been entered into some kind of lottery alongside 99 other participants. You know nothing about how the winner is determined. The Indifference Principle then advises that you should assign a 1/100 credence to winning.

The indifferent credence distribution is supported by minimax reasoning about the accuracy of one's credences: it minimizes the worst-case inaccuracy of one's credences (Pettigrew, 2016). It also minimizes *information* among all distributions over the possibilities. For a discrete probability distribution $P = (p_1, p_2, \ldots, p_n)$, the *information* of P is defined as

$$\sum_i p_i \log_2 p_i$$

(where $0 \log_2 0$ is regarded as 0). Equivalently, the indifferent distribution maximizes *entropy*, defined (by Shannon, 1948) as

$$-\sum_i p_i \log_2 p_i.$$

Entropy is regarded as a measure of the uncertainty of a distribution. A distribution that concentrates probability 1 on a single possibility is maximally informative and minimally uncertain; a uniform distribution is minimally informative and maximally uncertain. Entropy may also be generalized to continuous probability distributions, and the Indifference Principle may be generalized to the principle that the rational probability assignment, subject to a given constraint (e.g., background knowledge), is the distribution with maximum entropy that meets the constraint. Jaynes (2003) is a prominent advocate of this principle.

The Indifference Principle—and its generalization—is controversial, however, because it sometimes gives conflicting recommendations, made vivid in Bertrand's paradoxes (Bertrand, 1889). Van Fraassen (1989) illustrates this problem. Consider a factory that produces cubes with side lengths up to 1 foot. What should your credence be that a randomly selected cube has a side length up to ½ foot? The natural answer is of course ½. But this assumes that side length is the relevant quantity over which we should uniformly distribute our credence. We might just as well take the cubes' *volume* to be this quantity. The factory produces cubes with volumes up to 1 cubic foot. Then the question becomes how confident we should be that a randomly selected cube has a volume up to $\frac{1}{2}^3 = \frac{1}{8}$ cubic foot. Now the natural answer is ⅛! And we get yet another answer if we consider the area of the cube's faces instead. Since none of these ways of dividing up the possibilities into symmetrical alternatives stands out as privileged, the Indifference Principle cannot deliver a unique answer. It's a matter of debate whether this is a fatal objection to the principle. See Norton (2008), White (2009), and Novack (2010) for some defenses of the principle.

5. Diachronic Norms

5.1 Conditioning

Conditionalization or *conditioning* is the most widely endorsed rule that governs how a thinker should respond when she becomes certain of a new piece of evidence:

> When a thinker learns a piece of evidence E with certainty (and learns nothing in addition), then her new credence in any proposition A is equal to her old conditional credence $C(A|E)$.

For example, suppose you are 90% confident that your neighbor is home, given that his lights are on. Otherwise, you are only 50% confident that he is home. If you then see that his lights are on, your confidence that he is home should be 90%.

Both Dutch book and accuracy arguments have been offered in support of conditioning. The Dutch book is *diachronic*, because it requires that bets be placed at two different times—before and after learning the new evidence. Lewis (1999) shows that violations of conditionalization lead to Dutch book vulnerability. Skyrms (1987) provides the needed converse theorem: conditionalizers are immune to Dutch books (provided they don't commit any other Dutch-bookable offenses). Critics worry that diachronic Dutch book arguments are less compelling than synchronic ones. Christensen (1991) argues that Dutch-bookable inconsistencies between one's earlier and later self's credences are no more irrational than similarly Dutch-bookable disparities between one's own credences and one's spouse's credences. Taking the betting story literally, Levi (1987) argues that diachronic Dutch books are ineffective, since the agent would simply refuse the initial bets if she were aware of the bookie's strategy. Skyrms (1993) shows that once the decision faced by the agent is fully specified, she can be enticed to bet—and lose—even if she knows what the bookie is up to.

There are also variations of expected accuracy arguments for conditioning: relative to her initial credences, she will maximize her credences' expected accuracy upon learning whether E iff she conditionalizes (Easwaran, 2013; Greaves & Wallace, 2006; Leitgeb & Pettigrew, 2010b; Pettigrew, 2016; Schoenfield, 2017).

A concern about the accuracy arguments for conditioning is that they don't truly show that an agent who does otherwise is irrational. The credences C that recommend conditioning are those the agent accepted before she knew whether E, and the conditioning strategy recommends C only before it is settled whether E. Once she has learned whether E, she recognizes that C is outdated. Why should she update in a way that's recommended by credences that are then known to be defective (Pettigrew, 2016)? Moreover, as long as she adopts any coherent credences after learning whether E, any strictly proper inaccuracy measure will underwrite the result that the agent expects her current credences to be more accurate than any alternative credences (Easwaran, 2013). Pettigrew (2016) thus contends that these types of arguments can only show that rational agents must *plan* to conditionalize but not that they must follow through.

A different justification for conditioning comes from "minimal mutilation" considerations. Start with a credence function C, and impose the constraint that E is to be assigned probability 1 (and that it is the strongest such proposition). We want to move to the probability function *closest* to C that meets the constraint. There are various "distance" measures on the space of probability functions. A notable one is the *Kullback–Leibler divergence*, or *relative entropy*. For discrete distributions $P = (p_1, p_2, \ldots)$ and $Q = (q_1, q_2, \ldots)$, it is defined as

$$D(P, Q) = \sum_i p_i \log_2 \left(\frac{p_i}{q_i} \right).$$

It can be shown that $C(\cdot \mid E)$ is the function that minimizes entropy relative to C among all functions that meet the constraint. Other distance measures that underwrite conditioning in this way are the *Hellinger distance* and the *variation distance* (see Diaconis & Zabell, 1982). These results support conditioning insofar as they show that an agent who follows it changes her credences conservatively: she doesn't move away from her original credences any more than the evidence warrants, in a certain sense. However, one may ask why these particular distance measures are epistemically significant, since other distance measures deliver results that can disagree with conditioning. Moreover, one may ask why it's important to stay close to one's previous credences when they have become outdated.

5.2 Jeffrey Conditioning

One limitation of conditioning is that it applies only in cases where a thinker learns a piece of evidence with certainty. But sometimes we gather information without becoming certain of it. For example, suppose you glance at a tablecloth in poor light. Based on what you saw, you become more confident that the fabric is blue but not 100% confident. In this type of case, it is controversial whether there is any proposition that you have learned with certainty (although see Skyrms, 1987).

Jeffrey conditionalization, or *Jeffrey conditioning*, is an updating rule (named after Jeffrey, 1965) that can incorporate uncertain learning. Suppose $\{E_1, \ldots, E_n\}$ is a partition—a set of mutually exclusive and jointly exhaustive propositions—such that the agent's experience will change her credences in the elements of the partition from $C_{old}(E_i)$ to $C_{new}(E_i)$. Then, Jeffrey conditioning is this updating rule:

$$\text{for any } A, \ C_{new}(A) = \sum_i C_{old}(A \mid E_i) C_{new}(E_i).$$

In words: one's new credence in A is the weighted sum of one's old conditional credences $C_{old}(A \mid E_i)$, where the weights are one's new credences in the partition members. When the thinker becomes certain of a particular E_i,

Jeffrey conditioning reduces to standard conditioning. One interesting difference between standard and Jeffrey conditioning is that the former is commutative, while the latter is not. This means that when one receives more than one piece of evidence, standard conditioning always delivers the same credences in the end, regardless of the ordering of the updates. By contrast, with Jeffrey conditioning, the resulting credences need not be the same if the updates are made in a different order (Diaconis & Zabell, 1982; Levi, 1967).

There is currently no accuracy-based argument in favor of Jeffrey conditioning, but there is a diachronic Dutch book argument for it (Armendt, 1980; Skyrms, 1987). Additionally, it can be shown that the credence function that is arrived at by Jeffrey conditioning is the credence function *closest* to the agent's original credences that satisfies the constraint of assigning the values $C_{new}(E_i)$ to the members of the given partition, according to various "distance" measures (Kullback–Leibler, Hellinger, variation distance; see Diaconis & Zabell, 1982).

5.3 Relative Entropy, and More General Dynamics

The general problem for probability dynamics is this. Start with a credence function C, and impose a constraint K; to which credence function meeting K should C be revised? We have just seen solutions to this problem for the particular constraints that E be assigned credence 1 (conditioning) and that specified credences be assigned across a partition (Jeffrey conditioning). But we can imagine various other such constraints—for example, assign specified *conditional* credences across a partition or assign a probability in the interval $[x, y]$ to E. Any "distance" measure gives us a candidate for answering this general problem: move to the function closest to C, by that measure's lights, that satisfies K (when there is one). Relative entropy is an especially popular such measure (although van Fraassen [1981] raises his "Judy Benjamin problem" against it for a conditional-credence constraint).

6. Uniqueness

The putative norms of synchronic and diachronic rationality that we have specified so far don't generally single out a unique rational credence function for an agent at a time. Some philosophers, usually called *permissivists*, think that for any given body of evidence, there is more than one overall doxastic state a rational thinker could adopt on its basis (see, e.g., Douven, 2009; Kelly, 2013; Meacham, 2014; Schoenfield, 2014; for further references, see the excellent survey of the debate in Kopec &

Titelbaum, 2016). Proponents of *Uniqueness* disagree. Proponents of *Intra*personal Uniqueness argue that, for any given thinker who possesses a particular body of evidence, there is only one rational overall doxastic state this thinker can adopt in response to her evidence. Proponents of *inter*personal uniqueness argue for the even stronger thesis that there is only one rational response to any given body of evidence, the same for every thinker. (Early defenses of versions of Uniqueness can be found in Christensen, 2007; Feldman, 2007; White, 2005; and, more recently, in Dogramaci & Horowitz, 2016; Greco & Hedden, 2016.[8]) However, proponents of Uniqueness must admit that even if they are right, we probably won't ever know what these uniquely rational doxastic attitudes are.

7. Relaxing the Idealizations

The Bayesian theory of rational credences is highly idealized both in the manner in which it represents credences and in the normative demands it makes on them. Though we have so far granted that philosophers are mainly concerned with characterizing norms of *ideal* rationality, we may legitimately ask how that project relates to norms of rationality for *non-ideal* agents like us.

7.1   Imprecise Probabilities
One way in which the Bayesian view is idealized is in its representation of credences as being precisely point-valued. But realistically, many of our credences are not nearly so specific, and hence representing them this way is a distorting simplification of reality. Moreover, it has been argued that when our evidence is very unspecific, it would in fact be unfitting to form a precise credence on its basis (see, e.g., Joyce, 2005; Levi, 1985). For example, what is your credence that it's cloudy in Vladivostok? It seems hard to come up with a precise value. In response, it has been suggested that we represent credences as *interval-valued*. For example, one's credence that it's cloudy in Vladivostok might be represented as [0.1, 0.9]. More formally, and more generally, the entirety of someone's potentially imprecise credences could be represented as a set of precise credence functions that together capture the person's imprecise epistemic state.

However, this approach still has the problem that the intervals have artificially precise endpoints: where previously there was one precise value, now there are two! Furthermore, once people's credences are represented as sets of probability functions, the Bayesian norms of synchronic and diachronic rationality, and the arguments supporting them, have to be reformulated accordingly,

which is not always straightforward. There is also a substantial debate about how decision theory should work when we allow imprecise credences, with Elga (2010) raising a difficulty (see Rinard, 2015, for a response).

7.2   Forgetting/Information Loss
Another highly idealized assumption made by Bayesian representations of people's credences is that thinkers never forget or lose information. When an agent learns a piece of evidence with certainty, standard Bayesian models require that she remains certain of it. This is of course psychologically highly unrealistic, and it is also not obvious that forgetting or losing information is a rational defect. Titelbaum (2012) offers an account of how standard Bayesian updating rules need to be adapted in order to accommodate information loss (for important examples, see also Arntzenius, 2003) and self-locating credences (concerning who one is, where one is, and what time it is).

7.3   Human Limitations in Approximating Ideal Rationality
Even if we modify the standard Bayesian theory to allow for imprecise credences and information loss, the requirements of ideal rationality are still so demanding that humans cannot be expected to fully comply with them. Most obviously, the requirement that one must assign every tautology credence 1 exceeds the reasoning capacities of humans. Psychologists have also found evidence for other deviations from Bayesian rationality, such as our tendency not to change our credences enough in response to incoming evidence (Slovic & Lichtenstein, 1971; see also chapter 8.3 by Glöckner and chapter 8.5 by Hertwig & Kozyreva, both in this handbook). There is a major debate in psychology about how to interpret the relevant data and about whether such results demonstrate that humans are by and large rational (because they often come close to complying with Bayesian norms) or irrational (because they standardly deviate from them).

Once we grant that full compliance with Bayesian norms is excessively difficult for us to achieve, we may regard the norms as regulative ideals that we should approximate as much as is feasible. This raises two questions: First, what exactly does it mean to approximate ideal Bayesian rationality? Second, why is it beneficial to become less irrational, even if we can never reach the ideal?

A good way to measure how well someone's irrational credences approximate ideal rationality is to treat a credence function as a vector and to measure its distance from the closest vector that represents a rational credence function. For example, suppose someone has the following incoherent credences: $C(p) = 0.3$, $C(\sim p) = 0.4$, in vector notation: $\langle 0.3, 0.4 \rangle$. Different distance measures will identify different credence functions as closest. For example, we may use squared Euclidean distance as our distance measure, which is defined for vectors $X = \langle x_1, \ldots, x_n \rangle$ and $Y = \langle y_1, \ldots, y_n \rangle$ as

$$d(X,Y) = (x_1 - y_1)^2 + (x_2 - y_2)^2 + \cdots + (x_n - y_n)^2.$$

Then the closest coherent credence function is $C^*(p) = 0.45$, $C^*(\sim p) = 0.55$, and the distance between C and C^* is 0.045. But if we use other measures, such as absolute distance or the Kullback–Leibler divergence, we'll get different results (e.g., for the latter, the closest coherent credence function is $C^*(p) = 0.438$, $C^*(\sim p) = 0.572$). Which distance measure we use becomes important when we wonder why it is beneficial for thinkers to become more coherent, even if they can't become perfectly coherent. Suppose a thinker reduces the incoherence in her credence function by replacing her original incoherent credence function C_1 with a less incoherent credence function C_2 that is on the direct path toward the closest coherent credence function, as measured by squared Euclidean distance. C_2 is better than C_1 in at least two respects: C_2 is more accurate than C_1 in every possible world, according to the Brier score. Moreover, C_2 is vulnerable to a lower guaranteed loss from a Dutch book than C_1, provided we standardize the size of the bets the bookie is allowed to use. But what if we had used a different measure of closeness to coherence? For example, for the Kullback–Leibler measure, the accuracy result just mentioned holds only if we adopt the log-inaccuracy measure but not if we use the Brier score (see De Bona & Staffel, 2017, 2018; for earlier approaches to the problem, see also Schervish, Seidenfeld, & Kadane, 2002; Zynda, 1996). Hence, we can explain why it is good for thinkers to become less incoherent in specific ways, even if they can't ever have fully coherent credences: it gets them a greater portion of the accuracy benefits and the Dutch book–avoidance benefits that come with perfect coherence. However, these results crucially depend on finding the right combinations of distance measures, Dutch book, and inaccuracy measures.

7.4 Disagreement and Higher-Order Evidence

The fact that we can learn about our own imperfections also raises the question of how we should respond to information about our own ability to correctly assess evidence. Cases in which we disagree with others can provide us with such higher-order evidence. Suppose you and I have agreed to evenly split our lunch bill. We both do the calculations but get different answers.

One (or both) of us must have made a mistake (even by permissivist lights). What's the best repair strategy? *Conciliationists* maintain that we should both lower our credence that our respective answers are correct (Christensen, 2007). *Steadfasters*, by contrast, claim that if I've in fact assessed my evidence correctly, higher-order evidence of this type should not move me (Kelly, 2005).[9] Of course, not every disagreement indicates that someone has made a mistake. Different scenarios require different kinds of updates: if you disagree with me because you have more evidence, I should defer to you, but not if you have less evidence. If we disagree because we started from different permissible prior probabilities—before the evidence came in—perhaps we can both stick to our guns. If we find out we independently arrived at the same (high) degree of confidence in something, this might warrant raising our confidence even more (Lasonen-Aarnio, 2013). Claims about how rational agents should respond to disagreement and higher-order evidence thus yield further constraints on which prior probabilities are rationally permissible.

8. Conclusion

Formal epistemology has been one of the greatest growth areas in philosophy in the past couple of decades, and Bayesian approaches have been among the most dominant in formal epistemology. A very simple axiomatization codifies the putative synchronic norms on credences, and simple updating rules codify the putative diachronic norms on them. The result is an elegant theory that allows rigorous proofs of various claims that are taken to have epistemic significance. No wonder it has become so popular!

Moreover, there is still plenty more work to be done. Table 4.1.1 roughly represents the current state of play in Bayesian epistemology. The cells with checkmarks represent the putative norms that have already been given a particular type of putative justification (as far as we are aware at the time of writing). An empty cell can be empty for one of two reasons: either the missing justification is impossible, or it is just a matter of historical contingency that nobody has yet proved the requisite theorem; indeed, in some cases, perhaps nobody has even tried. The corresponding cells may thus be ripe for the taking. Moreover, when additional norms and justification strategies are proposed, the table itself will grow.

Then there are various other state-of-the-art topics that we have only been able to discuss briefly here: Uniqueness; whether credences may, or even must, be imprecise; the psychological reality (or otherwise) of Bayesianism

Table 4.1.1

Summary of putative justifications of norms

Putative justifications → Putative norms ↓	Dutch book	Accuracy	Representation theorem	Maximal entropy	Minimal distance	...
Probabilism	✔	✔	✔			
Regularity	✔					
Principal Principle	✔	✔				
Reflection Principle	✔	✔				
Indifference Principle		✔		✔		
Conditionalization	✔	✔			✔	
Jeffrey Conditionalization	✔				✔	
...						

and the extent to which humans are answerable to its norms; disagreement; and so on. And finally, there are still other "hot" topics that we have not even touched on: other putative norms; the relationship between quantitative credences and binary beliefs (which are a stock-in-trade of traditional epistemology); the relationship between credences and knowledge (likewise); quantitative versus comparative credences; Bayesianism versus other representations of uncertainty (Dempster–Shafer functions, ranking functions, belief revision theory, possibility theory, etc.); Bayesian decision theory—or rather, rival formulations of it; and more. We may thus be highly confident that Bayesian epistemology will remain a fertile research program for a long time to come!

Acknowledgments

We thank Arif Ahmed, Luc Bovens, Nicholas DiBella, Edward Elliott, Orri Stefánsson, Jeremy Strasser, Timothy Luke Williamson, James Willoughby, and the editors for very helpful comments.

Notes

1. It's an interesting question how we might go from the kinds of intuitive (nonnumerical) attributions of confidence with which we began to *numerical* degrees of belief. See, for example, Fishburn (1986) for discussion of the relationship between comparative and numerical probabilities. Stefánsson (2017) argues that comparative probabilities are primitive and psychologically real but that numerical probabilities are neither.

2. For alternative axiomatizations that take conditional probability to be primitive, see Popper (1959) and Rényi (1970).

3. Kolmogorov goes on to give an infinitary extension of the Finite Additivity axiom:

3′. (Countable Additivity) If $\{A_i\}$ is a countably infinite collection of (pairwise) disjoint events, then

$$P\left(\bigcup_{n=1}^{\infty} A_n\right) = \sum_{n=1}^{\infty} P(A_n).$$

However, the status of this axiom is more controversial (see, e.g., de Finetti, 1974), and we will not assume it in what follows. Moreover, a sentential reformulation of Countable Additivity requires some finessing, as classical logic admits only finite disjunctions.

4. It is a difficult question how conditional probabilities are related to probabilities of conditionals. We refer interested readers to chapter 4.4 by Pfeifer, chapter 4.6 by Oberauer and Pessach, and chapter 6.2 by Over and Cruz (all in this handbook), as well as to Hájek and Hall (1994) and Pfeifer and Douven (2014).

5. A critical discussion of another putative argument for probabilism, Cox's theorem, can be found in chapter 4.7 by Dubois and Prade (in this handbook).

6. There are unboundedly many references to both of these on the Internet. See also chapter 8.2 by Peterson (this handbook).

7. To be sure, the original Principal Principle holds as a constraint on rational agents more generally, assuming that they update by conditioning on evidence that is limited to the past and present—more on conditioning shortly. However, as we will see, the main arguments for conditioning have been questioned, as has conditioning itself. Lewis (1994) also offers a "New Principle" that explicitly supplants the original principle.

8. Carnap (1950) attempts to characterize a unique logical probability function and can thus also be seen as an early defender of Uniqueness. Various proponents of objective Bayesianism also fall into this camp (e.g., L. Williamson, 2010).

9. The literature on disagreement and higher-order evidence is vast. Matheson (2015) provides an overview of the debate. See also recent anthologies on the topic, such as Feldman and Warfield (2010) and Christensen and Lackey (2016).

References

Armendt, B. (1980). Is there a Dutch book argument for probability kinematics? *Philosophy of Science*, *47*(4), 583–588.

Arntzenius, F. (2003). Some problems for conditionalization and reflection. *Journal of Philosophy*, *100*(7), 356–370.

Bertrand, J. (1889). *Calcul des probabilités* [Calculus of probabilities]. Paris, France: Gauthier-Villars.

Briggs, R. (2009). Distorted reflection. *Philosophical Review*, *118*(1): 59–85.

Bronfman, A. (manuscript). *A gap in Joyce's argument for probabilism*. Unpublished manuscript.

Carnap, R. (1950). *Logical foundations of probability*. Chicago, IL: University of Chicago Press.

Christensen, D. (1991). Clever bookies and coherent beliefs. *Philosophical Review*, *100*(2), 229–247.

Christensen, D. (2007). Epistemology of disagreement: The good news. *Philosophical Review*, *116*(2), 187–217.

Christensen, D., & Lackey, J. (Eds.). (2016). *The epistemology of disagreement: New essays*. Oxford, England: Oxford University Press.

De Bona, G., & Staffel, J. (2017). Graded incoherence for accuracy-firsters. *Philosophy of Science*, *84*(2), 189–213.

De Bona, G., & Staffel, J. (2018). Why be (approximately) coherent? *Analysis*, *78*(3), 405–415.

de Finetti, B. (1964). Foresight: Its logical laws, its subjective sources. In H. E. Kyburg & H. E. Smokler (Eds.), *Studies in subjective probability* (pp. 100–110). New York, NY: Wiley. (Original work published in French in 1937)

de Finetti, B. (1974). *Theory of probability*. New York, NY: Wiley.

Diaconis, P., & Zabell, S. (1982). Updating subjective probability. *Journal of the American Statistical Association*, *77*(380), 822–830.

Dogramaci, S., & Horowitz, S. (2016). An argument for uniqueness about evidential support. *Philosophical Issues*, *26*(1), 130–147.

Douven, I. (2009). Uniqueness revisited. *American Philosophical Quarterly*, *46*(4), 347–361.

Easwaran, K. (2013). Expected accuracy supports conditionalization—and conglomerability and reflection. *Philosophy of Science*, *80*(1), 119–142.

Easwaran, K. (2014). Regularity and infinitesimal credences. *Philosophical Review*, *123*(1), 1–41.

Elga, A. (2010). Subjective probabilities should be sharp. *Philosophers' Imprint*, *10*(5), 1–11.

Eriksson, L., & Hájek, A. (2007). What are degrees of belief? *Studia Logica*, *86*(2), 183–213.

Feldman, R. (2007). Reasonable religious disagreements. In L. Antony (Ed.), *Philosophers without gods* (pp. 194–214). Oxford, England: Oxford University Press.

Feldman, R., & Warfield, T. A. (Eds.). (2010). *Disagreement*. Oxford, England: Oxford University Press.

Fishburn, P. C. (1986). The axioms of subjective probability. *Statistical Science*, *1*(3), 335–358.

Greaves, H., & Wallace, D. (2006). Justifying conditionalization: Conditionalization maximizes expected epistemic utility. *Mind*, *115*(459), 607–632.

Greco, D., & Hedden, B. (2016). Uniqueness and metaepistemology. *Journal of Philosophy*, *113*(8), 365–395.

Hájek, A. (2003). What conditional probability could not be. *Synthese*, *137*(3), 273–323.

Hájek, A. (2008). Arguments for—or against—probabilism? *British Journal for the Philosophy of Science*, *59*(4), 793–819.

Hájek, A. (2009). Dutch book arguments. In P. Anand, P. Pattanaik, & C. Puppe (Eds.), *The Oxford handbook of rational and social choice* (pp. 173–195). Oxford, England: Oxford University Press.

Hájek, A. (2012). Is strict coherence coherent? *Dialectica*, *66*(3), 411–424.

Hájek, A. (2019). Interpretations of probability. In E. N. Zalta (Ed.), *The Stanford encyclopedia of philosophy*. Retrieved from https://plato.stanford.edu/archives/fall2019/entries/probability-interpret/

Hájek, A., & Hall, N. (1994). The hypothesis of the conditional construal of conditional probability. In E. Eells & B. Skyrms (Eds.), *Probability and conditionals* (pp. 75–111). Cambridge, England: Cambridge University Press.

Hájek, A., & Hitchcock, C. (2016). Probability for everyone—even philosophers. In A. Hájek & C. Hitchcock (Eds.), *The Oxford handbook of probability and philosophy* (pp. 5–30). Oxford, England: Oxford University Press.

Jaynes, E. T. (2003). *Probability theory: The logic of science*. Cambridge, England: Cambridge University Press.

Jeffrey, R. C. (1965). *The logic of decision*. Chicago, IL: University of Chicago Press.

Joyce, J. M. (1998). A non-pragmatic vindication of probabilism. *Philosophy of Science*, *65*(4), 575–603.

Joyce, J. M. (2005). How probabilities reflect evidence. *Philosophical Perspectives*, *19*, 153–178.

Kelly, T. (2005). The epistemic significance of disagreement. *Oxford Studies in Epistemology*, *1*, 167–196.

Kelly, T. (2013). Evidence can be permissive. In M. Steup, J. Turri, & E. Sosa (Eds.), *Contemporary debates in epistemology* (2nd ed., pp. 298–311). Hoboken, NJ: Wiley.

Kemeny, J. G. (1955). Fair bets and inductive probabilities. *Journal of Symbolic Logic*, *20*(3), 263–273.

Kolmogorov, A. N. (1950). *Foundations of the theory of probability*. New York, NY: Chelsea. (Original work published in German as *Grundbegriffe der Wahrscheinlichkeitsrechnung* in 1933)

Kopec, M., & Titelbaum, M. (2016). The uniqueness thesis. *Philosophy Compass*, *11*(4), 189–200.

Lasonen-Aarnio, M. (2013). Disagreement and evidential attenuation. *Noûs*, *47*(4), 767–794.

Leitgeb, H., & Pettigrew, R. (2010a). An objective justification of Bayesianism I: Measuring inaccuracy. *Philosophy of Science*, *77*(2), 201–235.

Leitgeb, H., & Pettigrew, R. (2010b). An objective justification of Bayesianism II: The consequences of minimizing inaccuracy. *Philosophy of Science*, *77*(2), 236–272.

Levi, I. (1967). Probability kinematics. *British Journal for the Philosophy of Science*, *18*(3), 297–315.

Levi, I. (1985). Imprecision and indeterminacy in probability judgment. *Philosophy of Science*, *52*(3), 390–409.

Levi, I. (1987). The demons of decision. *The Monist*, *70*(2), 193–211.

Levi, I. (1989). Possibility and probability. *Erkenntnis*, *31*(2–3), 365–386.

Lewis, D. (1980). A subjectivist's guide to objective chance. In R. C. Jeffrey (Ed.), *Studies in inductive logic and probability* (Vol. II, pp. 83–132). Berkeley: University of California Press.

Lewis, D. (1994). Humean supervenience debugged. *Mind*, *103*(412), 473–490.

Lewis, D. (1999). Why conditionalize? In *Papers in metaphysics and epistemology* (pp. 403–407). Cambridge, England: Cambridge University Press.

Mahtani, A. (2015). Dutch books, coherence, and logical consistency. *Noûs*, *49*(3), 522–537.

Matheson, J. (2015). *The epistemic significance of disagreement*. New York, NY: Palgrave Macmillan.

Meacham, C. J. G. (2014). Impermissive Bayesianism. *Erkenntnis*, *79*(6)(Suppl.), 1185–1217.

Norton, J. D. (2008). Ignorance and indifference. *Philosophy of Science*, *75*(1), 45–68.

Novack, G. (2010). A defense of the principle of indifference. *Journal of Philosophical Logic*, *39*(6), 655–678.

Pettigrew, R. (2016). *Accuracy and the laws of chance*. Oxford, England: Oxford University Press.

Pettigrew, R. (2018, February 2). An almost-Dutch book argument for the Principal Principle. Retrieved from http://m-phi.blogspot.co.uk/2018/02/a-pragmatic-dutch-book-argument-for.html

Pfeifer, N., & Douven, I. (2014). Formal epistemology and the new paradigm psychology of reasoning. *Review of Philosophy and Psychology*, *5*(2), 199–221.

Popper, K. (1959). *The logic of scientific discovery*. New York, NY: Basic Books.

Ramsey, F. P. (1990). Truth and probability (original manuscript, 1926). In D. H. Mellor (Ed.), *Philosophical papers* (pp. 52–94). New York, NY: Cambridge University Press.

Rényi, A. (1970). *Foundations of probability*. San Francisco, CA: Holden-Day.

Rinard, S. (2015). A decision theory for imprecise probabilities. *Philosophers' Imprint*, *15*(7), 1–16.

Savage, L. J. (1954). *The foundations of statistics*. New York, NY: Wiley.

Schervish, M. J., Seidenfeld, T., & Kadane, J. B. (2002). Measuring incoherence. *Sankhya: The Indian Journal of Statistics*, *64*, 561–587.

Schoenfield, M. (2014). Permission to believe: Why permissivism is true and what it tells us about irrelevant influences on belief. *Noûs*, *48*(1), 193–218.

Schoenfield, M. (2017). Conditionalization does not (in general) maximize expected accuracy. *Mind*, *51*(4), 667–685.

Shannon, C. E. (1948). A mathematical theory of communication. *Bell System Technical Journal*, *27*(3), 379–423.

Shimony, A. (1955). Coherence and the axioms of confirmation. *Journal of Symbolic Logic*, *20*(1), 1–28.

Skyrms, B. (1980). Higher order degrees of belief. In D. H. Mellor (Ed.), *Prospects for pragmatism* (pp. 109–137). Cambridge, England: Cambridge University Press.

Skyrms, B. (1987). Dynamic coherence and probability kinematics. *Philosophy of Science*, *54*(1), 1–20.

Skyrms, B. (1993). A mistake in dynamic coherence arguments? *Philosophy of Science*, *60*(2), 320–328.

Slovic, P., & Lichtenstein, S. (1971). Comparison of Bayesian and regression approaches to the study of information processing in judgment. *Organizational Behavior and Human Performance*, *6*(6), 649–744.

Stefánsson, H. O. (2017). What is "real" in probabilism? *Australasian Journal of Philosophy*, *95*(3), 573–587.

Talbott, W. (1991). Two principles of Bayesian epistemology. *Philosophical Studies, 62*(2), 135–150.

Titelbaum, M. (2012). *Quitting certainties: A Bayesian framework modeling degrees of belief*. Oxford, England: Oxford University Press.

van Fraassen, B. C. (1981). A problem for relative information minimizers in probability kinematics. *British Journal for the Philosophy of Science, 32*, 375–379.

van Fraassen, B. C. (1984). Belief and the will. *Journal of Philosophy, 81*(5), 235–256.

van Fraassen, B. C. (1989). *Laws and symmetry*. Oxford, England: Oxford University Press.

van Fraassen, B. C. (1999). Conditionalization, a new argument for. *Topoi, 18*(2), 93–96.

Vineberg, S. (2001). The notion of consistency for partial belief. *Philosophical Studies, 102*(3), 281–296.

Vineberg, S. (2016). Dutch book arguments. In E. N. Zalta (Ed.), *The Stanford encyclopedia of philosophy*. Retrieved from https://plato.stanford.edu/archives/spr2016/entries/dutch-book/

White, R. (2005). Epistemic permissiveness. *Philosophical Perspectives, 19*, 445–459.

White, R. (2009). Evidential symmetry and mushy credence. In T. Szabó Gendler & J. Hawthorne (Eds.), *Oxford studies in epistemology* (pp. 161–186). Oxford, England: Oxford University Press.

Williamson, J. (2010). *In defense of objective Bayesianism*. Oxford, England: Oxford University Press.

Williamson, T. (2007). How probable is an infinite sequence of heads? *Analysis, 67*(3), 173–180.

Zynda, L. (1996). Coherence as an ideal of rationality. *Synthese, 109*(2), 175–216.

Zynda, L. (2000). Representation theorems and realism about degrees of belief. *Philosophy of Science, 67*, 45–69.

4.2 Bayes Nets and Rationality

Stephan Hartmann

Summary

Bayes nets are a powerful tool for researchers in statistics and artificial intelligence. This chapter demonstrates that they are also of much use for philosophers and psychologists interested in (Bayesian) rationality. To do so, we outline the general methodology of Bayes nets modeling in rationality research and illustrate it with several examples from the philosophy and psychology of reasoning and argumentation. Along the way, we discuss the normative foundations of Bayes nets modeling and address some of the methodological problems it raises.

1. The Virtues of Bayes Nets

Bayes nets (or "Bayesian networks") are a powerful tool for researchers in statistics and artificial intelligence. The reason for this consists in the many scientific virtues of Bayes nets. One of these virtues is their *representational power*: Bayes nets represent in an intuitive way the conditional independencies that hold between variables. Exploiting these conditional independencies, Bayes nets allow for a compact representation of a joint probability distribution over n variables: what would ordinarily, for binary variables without any additional constraints, require the specification of $2^n - 1$ parameters here needs significantly less. This has enormous practical advantages and hence initiated the development of probabilistic expert systems and other technical applications. Another virtue of Bayes nets is their *algorithmic power*: Bayes nets come with efficient algorithms to derive from the joint distribution whatever marginal or conditional probability one is interested in. Last but not least, the theory of Bayes nets is very elegant, and not much is needed to successfully apply it to new problems.

These are some of the reasons why Bayes nets have already found so many applications in various parts of science and engineering. The goal of this chapter is to demonstrate that Bayes nets are also of much use for philosophers and psychologists in the field of (Bayesian)

rationality. We will see that Bayes nets can be naturally integrated into the Bayesian framework and help solving problems that would otherwise be hard to address. In the following pages, we will outline the general methodology of Bayes nets modeling in rationality research and elaborate on the various functions of these models. The methodology will then be illustrated by analyzing a number of problems and questions from the philosophy and psychology of reasoning and argumentation.

The remainder of this chapter is organized as follows: section 2 provides a concise introduction to the theory of Bayes nets. Section 3 introduces Bayesian rationality. Here we distinguish between the general Bayesian framework and the models that are constructed within the framework to address a specific problem or question from rationality research. Bayes nets play a role in the latter but not in the former. We illustrate the methodology by examining a number of increasingly complex confirmation scenarios. Section 4 considers two examples from the philosophy and psychology of reasoning and argumentation in more detail. We will see that Bayes nets models help the rationality researcher to reconstruct certain reasoning and argumentation schemes and to identify possible holes in an argument (and to suggest a remedy). The section closes with the sketch of a general theory of Bayesian argumentation with a special focus on the role of indicative conditionals. It builds heavily on the use of Bayes nets. Section 5, finally, closes with a short outlook.

2. Bayes Nets in a Nutshell

A Bayes net organizes a set of variables into a *directed acyclic graph* (DAG). A DAG is a set of nodes and a set of arrows between those nodes. The only constraint is that there are no closed paths formed by following the arrows. A "root node" is a node with outgoing arrows only, a "parent" of a given node is a node from which an arrow points at the given node, and a "descendant" of a node is one that is pointed at by a corresponding arrow. Each node represents a propositional variable, which can take any number of mutually exclusive and jointly exhaustive values. To make a DAG into a Bayes net, one more step is required: we need to specify the prior probabilities for the variables in the root nodes and the conditional probabilities for the variables in all other nodes, given any combination of values of the variables in their respective parent nodes.

The arrows in a Bayes net carry information about the conditional independence relations between the variables in the DAG. This information is expressed by the *Parental Markov Condition* (PMC):

PMC A variable represented by a node in a Bayes net is independent of all variables represented by its non-descendant nodes, conditional on all variables represented by its parents.

Here is an illustration: consider the Bayes net in figure 4.2.1, involving the three propositional variables A, B, and C. Node C is a root node. It is the parent of A and B, and A and B are the children of C. Applying PMC, we find that $A \perp\!\!\!\perp B \mid C$ (read: A is independent of B given C). One also says that C *screens off* A from B.

Compare this Bayes net with the one in figure 4.2.2. Here B is the parent of C and C is the parent of A. Node A is a child of C and at the same time a descendant of B. Interestingly, applying PMC, we find that $A \perp\!\!\!\perp B \mid C$ also. The networks in figures 4.2.1 and 4.2.2 therefore represent the same conditional independence structure. That is, if a modeler has reason to assume that $A \perp\!\!\!\perp B \mid C$, then this conditional independence can be represented in a Bayes net in two ways. To single out one of them, further information is needed.

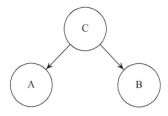

Figure 4.2.1
The "common cause" network.

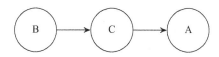

Figure 4.2.2
The chain network.

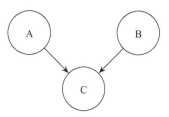

Figure 4.2.3
The collider (or "common effect") network.

Finally, let us consider the Bayes net in figure 4.2.3. Here the variables A and B are the parents of C. Applying PMC, we find that $A \perp\!\!\!\perp B$, that is, A is unconditionally independent of B (one could also write $A \perp\!\!\!\perp B \mid \varnothing$, where $\varnothing$ represents the empty set). Hence, the conditional independence structure instantiated in the Bayes net in figure 4.2.3 differs from the one in figures 4.2.1 and 4.2.2.

Conditional independence structures can be investigated in general, and there is a rich literature on the topic. Most important, conditional independence structures (i.e., the three-place relation $\cdot \perp\!\!\!\perp \cdot \mid \cdot$) satisfy the so-called *semi-graphoid axioms* (A. P. Dawid, 1979; Spohn, 1980), which can be used to derive new conditional independencies from already known ones.[1] This is important as it is easy to see that PMC does not allow us to identify *all* conditional independencies that hold in a DAG. Consider the Bayes net in figure 4.2.2. Here PMC implies only that $A \perp\!\!\!\perp B \mid C$. However, it also seems to be the case that $B \perp\!\!\!\perp A \mid C$, which does not follow from PMC. This conditional independence follows from $A \perp\!\!\!\perp B \mid C$ and the *symmetry axiom* (which is one of the semi-graphoid axioms). Applying the semi-graphoid axioms to find all conditional independencies is rather cumbersome, and so it would be helpful to have one criterion that identifies all conditional independencies in a DAG in a straightforward way. This is the *d*-separation criterion, which is explained in textbooks such as Darwiche (2014) and Pearl (1988).

While conditional independence structures can be studied abstractly, we are only interested in *probabilistic* conditional independence structures. For these, it is helpful to recall that two propositional variables A and B are probabilistically independent with respect to a probability measure P if and only if $P(A, B) = P(A)P(B)$ for all values of A and B. Equivalently, A and B are probabilistically independent with respect to a probability measure P if and only if $P(A \mid B) = P(A)$ for all values of A and B, where $P(A \mid B)$ stands for the conditional probability of A given B, which is defined as $P(A, B)/P(B)$ if $P(B) > 0$.[2] Generalizing this definition, two propositional variables A and B are probabilistically independent given the variable C with respect to a probability measure P if and only if $P(A, B \mid C) = P(A \mid C)P(B \mid C)$ for all values of A, B, and C. It is easy to see that this definition is equivalent to the following one: A and B are probabilistically independent given C with respect to a probability measure P iff $P(A \mid B, C) = P(A \mid C)$ for all values of A, B, and C. That is, once the value of C is known, learning the value of B does not change the probability of A (provided that all conditional probabilities are defined).

While conditional independencies can be read off from the network structure, a probability distribution defined over the DAG is needed to make specific probabilistic inferences. To do so, we express the joint probability distribution over a set of variables $A_1, \ldots, A_n$, which is organized into a DAG in terms of the prior probabilities of all root nodes and the conditional probabilities of all child nodes, given any combination of values of the variables in their respective parent nodes. Let $\mathrm{pa}(A_i)$ denote the set of parents of A_i. Then an application of the chain rule of the probability calculus yields the following expression ("the product rule") for the joint probability distribution over all variables:[3]

$$P(A_1, \ldots, A_n) = \prod_{i=1}^{n} P(A_i \mid \mathrm{pa}(A_i)) \tag{1}$$
$$= P(A_1 \mid \mathrm{pa}(A_1)) \cdot P(A_2 \mid \mathrm{pa}(A_2)) \cdots P(A_n \mid \mathrm{pa}(A_n)).$$

Let us illustrate the application of equation 1 with the DAG in figure 4.2.3. To make the DAG a Bayes net, we assume that all variables are binary (with their values represented by A, ¬A, etc.) and specify the (unconditional) probabilities of all root nodes, that is, $P(A) = a$ and $P(B) = b$, and the conditional probabilities of the child node given the four combinations of values of the two variables in its respective parent nodes, that is,

$$P(C \mid A, B) = \alpha, \qquad P(C \mid A, \neg B) = \beta,$$
$$P(C \mid \neg A, B) = \gamma, \qquad P(C \mid \neg A, \neg B) = \delta.$$

The product rule (i.e., equation 1) then allows us to compute whatever marginal or conditional probability we are interested in. For example, one easily sees that $P(A, B, C) = P(A)P(B)P(C \mid A, B) = ab\alpha$ and that $P(A, \neg B, C) = P(A)P(\neg B)P(C \mid A, \neg B) = a\bar{b}\beta$, where we have used the shorthand notation $\bar{x} := 1 - x$, which we will also use below.

Similarly, one can calculate $P(A \mid B, C)$ and $P(A \mid C)$ and show that the two probabilities are *not* identical. That is, A and B are unconditionally independent but conditionally dependent, given C. Fixing the value of the "common effect" variable C renders the "causes" A and B dependent.[4]

3. Bayesian Rationality

Bayesianism is the leading theory of uncertain reasoning.[5] Its starting point is the psychological truism that people believe contingent propositions such as "It will rain tomorrow" more or less strongly: they assign a certain *degree of belief* to a proposition. But what are rational degrees of belief? How can they be combined? And how should one change them if new evidence becomes available? To address these questions, we need a calculus for the representation of degrees of belief (i.e., a theory about the statics of rational belief), rules for changing them (i.e., a theory about the dynamics of rational belief), and a normative foundation for both.

Before moving on, it is useful to distinguish between the *Bayesian framework* and the *models* that are constructed within this framework. While the framework lays out the general features of Bayesian rationality and comes with a *normative foundation*, the models represent specific reasoning situations and help the researcher to tackle concrete problems. These models often involve Bayes nets.

3.1 The Bayesian Framework

Let us begin with the *static part* of Bayesianism. Here Bayesians identify degrees of belief with probabilities. As a consequence, the probability calculus puts specific constraints on the degrees of belief of an agent. For instance, if a rational agent assigns a degree of belief of .3 to the proposition "It will rain tomorrow," then this agent has to assign a degree of belief of .7 to the proposition "It will not rain tomorrow," as the latter proposition is the logical negation of the former and the probability of a proposition and its negation sum up to 1. More generally, a probability distribution P is defined over a Boolean algebra $\mathscr{B}$ of propositions, which comes with rules for the combination ($\wedge$ and $\vee$) and negation ($\neg$) of propositions. In a first step, the agent fixes the algebra, that is, she identifies all relevant propositions.[6] In the second step, the agent specifies a joint probability distribution over the algebra $\mathscr{B}$. As a result, the beliefs of the agent are *coherent*.

Turning to the *dynamic part*, Bayesians specify rules for changing ("updating") probabilities once new information becomes available. Let us assume, for example, that an agent has partial beliefs about the propositions A, B, and C. They are represented by propositional variables A, B, and C, and a prior probability distribution P is defined over them. The agent then learns that A is the case. That is, the new probability of A, that is, $P'(A)$, is 1. Here P' denotes the new ("posterior") probability distribution of the agent. So far, we only know the new value of the probability of A. But what are, for example, the new probabilities of B and C? And what is the full new joint probability distribution over A, B, and C? Bayesians argue that in this case, the agent should update her probability distribution according to the principle of Conditionalization ("Bayes' theorem"), that is, she should set

$$P'(X) = P(X \mid A)$$

for any proposition X in the algebra $\mathscr{B}$ under consideration.

But why should we identify degrees of belief with probabilities? And why should one update according to Conditionalization? That is, what is the *normative foundation* of the Bayesian framework? There are different ways to provide such a normative foundation, and it is a strength of Bayesianism that it is supported by a wide variety of such arguments. The two most popular types of arguments are pragmatic ("Dutch book arguments") and epistemic ("epistemic utility theory"). They are explained in chapter 4.1 by Hájek and Staffel (this handbook).

It is interesting to note that the dynamic part of Bayesianism (i.e., Conditionalization) can also be justified in a different way. To do so, we assume that the agent wants to be as conservative as possible with regard to changing her beliefs. That is, the agent is undogmatic and willing to modify her beliefs once new information comes in, but she wants to make sure that overall, her beliefs change as little as possible. This seems to be psychologically plausible and is also part of other theories of belief revision such as the AGM model, which is explained in chapter 5.2 by Rott (this handbook). More specifically, the principle of *Conservativity* demands that an agent who learns a new item of information make sure that the new probability distribution P' takes this new information into account as a constraint but also requests that P' differ as little as possible from the old (prior) probability distribution P.

Making this idea precise requires the specification of a measure of the distance between two probability distributions. It turns out that the most interesting and useful measures do not satisfy the axioms of a mathematical distance (i.e., of a metric space). For instance, the Kullback–Leibler divergence is not symmetrical and violates the triangle inequality. However, minimizing the Kullback–Leibler divergence yields Conditionalization if one takes into account the constraint that the probability of some proposition in the algebra shifts to 1 (see Diaconis & Zabell, 1982; Eva, Hartmann, & Rafiee Rad, 2020).

It is instructive to note an analogy to Newtonian mechanics here. Newtonian mechanics, too, provides a modeling framework. It specifies a static part (mass points, etc.) and a dynamic part (Newton's second law as a general dynamical law). Furthermore, there are justifications for both parts. What is more, the Newtonian framework comes with various assessment criteria and a (perhaps somewhat implicit) methodology for model construction (see Giere, 1990). This also holds for Bayesian modeling, which is the topic of the following subsection. Before that, however, we introduce Bayesian Confirmation Theory (BCT), which is the central philosophical application of Bayesianism. Its most general aspects are part of the Bayesian framework.

While qualitative confirmation theories formulate criteria that inform us whether or not a piece of evidence E confirms a hypothesis H (Sprenger, 2011), quantitative theories of confirmation (such as BCT) also tell us *how much* E confirms H. According to BCT, an agent starts with a subjective degree of belief that a certain hypothesis H is true—the *prior probability* $P(H)$ of the hypothesis. In the next step, a more or less expected piece of evidence, E, comes in. While E was uncertain before, it now becomes certain. To make sure that the beliefs of the agent remain coherent, the agent updates the probability of H and assigns a *posterior probability* $P'(H)$ to the hypothesis using Conditionalization, that is, $P'(H) = P(H \mid E)$. This can also be expressed as

$$P'(H) = \frac{P(H)}{P(H) + P(\neg H)x},$$

with the *likelihoods* $p := P(E \mid H)$ and $q := P(E \mid \neg H)$ and the *likelihood ratio* $x := q/p$. Now E *confirms* H iff the posterior probability $P'(H)$ (after learning E) is greater than the prior probability $P(H)$. Evidence E *disconfirms* (or "falsifies") H iff the posterior is smaller than the prior. If $P'(H) = P(H)$, then E is *irrelevant* for H. Equivalently, E confirms H iff the likelihood ratio $x < 1$, E disconfirms H iff $x > 1$, and E is irrelevant for H iff $x = 1$. For more on BCT, see Crupi (2016), Huber (2007), and chapter 4.3 by Merin (this handbook), which also discusses how to measure evidential relevance.

3.2 Bayesian Models

The general framework just described cannot be applied directly to concrete problems and questions. To do so, the agent has to specify (1) the relevant variables and (2) their relations. This may involve a considerable amount of modeling, as it isn't always clear which variables are the relevant ones.[7] Besides, different methodological values may conflict. One may, for example, favor a rather simple, intuitive, and understandable model. This can be achieved by taking into account only a small number of variables. Sometimes it is also possible to effectively combine various variables into one macro-variable. On the other hand, one might also want to account for the details of a reasoning scenario, and additional propositions might be needed for this. The modeler has to decide what to do. The same holds for the assumed relations between the variables. Are two variables really strictly (conditionally) independent? This may be controversial, and strict independencies are hard to come by.

However, once the modeler has decided what the relevant variables and their relations are, Bayes nets come in and help with the representation and the computations.[8] We will see that certain independence assumptions can do a lot of work and, if properly motivated, reduce the amount of subjectivity that one might complain about in a Bayesian model, due to the considerable freedom one otherwise has to fix a prior probability distribution. The structural constraints imposed in a Bayes net alone already contribute a great deal to the solution of a problem, and in the study of Bayesian rationality, probabilistic models often have Bayes nets as integral parts. Besides, once the Bayes net is fixed, the problem or question that prompted the construction of the model can be addressed by powerful mathematical machinery.

It has long been noted that models are an indispensable part of science.[9] They have important (pragmatic and epistemic) functions in the research process, and this also holds for the Bayes net models we are considering in this chapter. Here is a (nonexhaustive) list of the functions of Bayes net models in rationality research:

- Models help to *apply* the (Bayesian) theory. We mentioned already that not much follows from the general framework for a particular problem or question; it requires the model and the specification of details.
- Models help to *test* the (Bayesian) theory. Once the model is constructed, its consequences can be confronted with empirical data. Note, however, that Bayes net models are *normative models*, which raises additional problems (see Colyvan, 2013; Titelbaum, 2021).
- Models help to *solve* concrete problems, as many examples show (see also the illustrations in figure 4.2.4).
- Models *accommodate and explain* experimental data.
- Models help the researcher to *reconstruct and analyze* various reasoning and argumentation schemes and check their rationality.

- Bayes net models provide a *compact and intuitive representation* of a reasoning and argumentation situation and help with the bookkeeping of the variables and their relations.

3.3 Modeling Confirmation Scenarios

Large parts of the literature on BCT are only concerned with scenarios involving two variables—one representing the hypothesis (*H*) and the other representing the evidence (*E*). However, while much can be learned from focusing on simple scenarios (see, e.g., Earman, 1992; Howson & Urbach, 2006), it is clear that actual confirmation scenarios are much more complex and may raise new and intricate issues that do not show up in scenarios involving only two variables. If we want to reconstruct and analyze more complex confirmation scenarios, then Bayes nets prove to be extremely helpful. Without the use of Bayes nets, it is hard to keep track of the various dependencies and independencies and to properly evaluate what is going on, as the following example illustrates.

We consider the *testing of a hypothesis with partially reliable measurement instruments*. To begin with, we consider a situation where the evidence is uttered by a partially reliable information source. This information source could be a measurement instrument (which outputs a binary result, e.g., a positive or negative X-ray) or the testimony of a witness (e.g., in a murder case), again modeled as a binary variable, such as "I saw the suspect on the crime scene" or "I didn't see the suspect."

Such scenarios can be modeled by fixing the rates of false positives and false negatives of the information source. The rate of false positives (f_p) measures how often an instrument outputs "true" when in fact the hypothesis is false. Similarly, the rate of false negatives (f_n) measures how often an instrument outputs "false" when in fact the hypothesis is true. Ideally, both rates are zero, but in realistic cases, this is almost never the case. To proceed, it is easy to see that f_p and f_n are related to the likelihoods mentioned above: $f_p = P(E \mid \neg H) = q$ and $f_n = P(\neg E \mid H) = 1 - p$. While these likelihoods (or rates) are often available (e.g., in medical testing), this is not always the case (e.g., for witness reports). This suggests that one might want to construct more complex models for the information-gathering process of an agent.

Here is a simple example. Consider an agent who entertains the hypothesis H ("Paul is the murderer"). She then receives a witness report Rep (which is the evidence) to the effect that Paul indeed is the murderer. The agent then assumes—and this is a modeling assumption—that

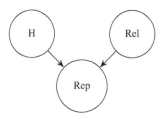

Figure 4.2.4
A Bayes net representing the test of a hypothesis with a partially reliable information source.

the witness is either reliable or not: if the witness is reliable, then she is fully reliable and therefore always tells the truth. That is, if Paul is the murderer, she reports that he is the murderer, and if Paul is not the murderer, she reports that Paul is not the murderer. However, if she is not reliable, then she is a *randomizer*, that is, she reports with a certain probability a (the so-called randomization parameter) that Paul is the murderer, independently of whether or not Paul is the murderer. Finally, the agent assumes that it is uncertain whether or not the witness is reliable and assigns a prior probability to the proposition Rel ("The witness is reliable").

Next we represent this situation with the ("collider") Bayes net in figure 4.2.4 and set $P(H) = h$, $P(\text{Rel}) = r$, and

$$P(\text{Rep} \mid \text{H, Rel}) = 1, \qquad P(\text{Rep} \mid \neg\text{H, Rel}) = 0,$$
$$P(\text{Rep} \mid \text{H}, \neg\text{Rel}) = a, \qquad P(\text{Rep} \mid \neg\text{H}, \neg\text{Rel}) = a.$$

With this, we calculate $P'(H) = P(H \mid \text{Rep})$, that is, the posterior probability of the hypothesis after receiving a positive witness report:

$$P'(H) = \frac{P(H, \text{Rep})}{P(\text{Rep})} = \frac{\sum_{Rel} P(H, Rel, \text{Rep})}{\sum_{H,Rel} P(H, Rel, \text{Rep})}$$
$$= \frac{\sum_{Rel} P(H)P(Rel)P(\text{Rep} \mid H, Rel)}{\sum_{H,Rel} P(H)P(Rel)P(\text{Rep} \mid H, Rel)} \qquad (2)$$
$$= \frac{h(r + a\bar{r})}{h(r + a\bar{r}) + \bar{h}a\bar{r}} = \frac{h(r + a\bar{r})}{hr + a\bar{r}}.$$

It is interesting to note that this model can be simulated by a two-variable model (using the variables H and Rep) with the corresponding likelihoods $p := P(\text{Rep} \mid H) = r + a\bar{r}$ and $q := P(\text{Rep} \mid \neg H) = a\bar{r}$. From this, one also sees that $p > q$ (for $r > 0$). Hence, as expected, Rep always confirms H. From equation 2, we obtain that $P'(H) = P(H)$ for $r = 0$. Likewise, for $r = 1$, we find that $P'(H) = 1$. This makes sense: if a perfectly reliable information source tells us that H is true, then H is true.

Because this three-variable model can be simulated by a two-variable model, it is not strictly necessary to study it in detail. However, the model allows us to reduce the likelihoods p and q to some parameters that are easier to grasp and interpret (namely, a and r). The model therefore provides (or suggests) a *mechanism* that generates

the likelihoods, but once these likelihoods are known, one can proceed with the two-variable model.

Things get more interesting when one studies more complex scenarios. Let us assume, for instance, that the agent receives two positive reports from two partially reliable information sources. We can then distinguish two scenarios and ask which of them provides more confirmation for the hypothesis in question. In the first scenario, the two reports are independent from each other. This is modeled by fixing two root nodes Rel_1 and Rel_2 (see figure 4.2.5). In the second scenario, we assume that the reports are dependent and model this by working with only one root node Rel, which is a parent of both Rep_1 and Rep_2 (see figure 4.2.6). One would expect that the first scenario provides more confirmation, ceteris paribus, as the two reports are independent. Here, the ceteris paribus clause makes sure that the priors of H and Rep (and of Rep_1 and Rep_2, respectively) are the same and that the randomization parameter is the same in both scenarios. But is this really the case? Do independent positive reports always result in more confirmation? A detailed analysis shows that the answer to this question is "no": the second scenario provides more confirmation if the values of a and r are sufficiently small.[10]

Extending and generalizing these ideas, Landes (2020) provides a systematic assessment of the *variety-of-evidence thesis*. This is the claim that more varied evidence confirms more strongly than less varied evidence. Landes

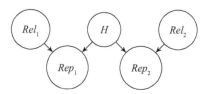

Figure 4.2.5
A Bayes net representing the test of a hypothesis with two independent instruments.

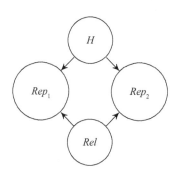

Figure 4.2.6
A Bayes net representing the test of a hypothesis with two dependent instruments.

explores in detail under which conditions the thesis holds. Further applications of the simple witness model described above can be found in the psychological literature (see, e.g., Hahn, Merdes, & von Sydow, 2018; Harris, Hahn, Madsen, & Hsu, 2016).

There are many other applications of Bayes nets in BCT. For instance, Dardashti, Hartmann, Thébault, and Winsberg (2019) investigate whether so-called analogue simulations can be used to confirm a theory. The (very rough) idea is this: A theory T predicts a phenomenon φ, but for certain (practical) reasons, the prediction cannot be directly tested. However, it turns out that there is another theory T', which is in important respects analogous to T and predicts the analogous phenomenon φ'. Fortunately, the prediction of φ' can be tested. The observation of φ' then confirms T', but does it also confirm T (as some authors argue for some cases)? Using the machinery of Bayes nets, Dardashti et al. (2019) show that this is indeed the case, given that certain conditions hold. Further applications concern the use of Bayes nets in legal reasoning. For example, Lagnado and collaborators (e.g., Connor Desai, Reimers, & Lagnado, 2016; Fenton, Neil, Yet, & Lagnado, in press) use Bayes nets to represent complex legal scenarios, demonstrating the power and the flexibility of the approach (see also chapter 11.4 by Prakken, this handbook).

4. Bayesian Reasoning and Argumentation

Confirmation theory and the analysis of confirmation scenarios are an important focus of Bayesianism. Bayesianism provides a clear criterion for when it is rational for an agent to claim that a piece of evidence supports a given hypothesis, and Bayes nets help to apply this theory to concrete cases so that the theory can be put to work. At this point, however, it is important to note that epistemic rationality comprises more than confirmation theory. In many cases of reasoning and argumentation, it is not immediately clear that such scenarios can be reconstructed (or represented) as confirmation scenarios. Interestingly, though, often it is possible. We illustrate this point with the sketch of a Bayesian account of argumentation (see also chapter 5.5 by Hahn & Collins and chapter 5.6 by Woods, both in this handbook).

The starting point of Bayesian argumentation is the observation that the premises and conclusions in typical real-life arguments are uncertain to the agent. We are more or less convinced of the premises of an argument, given the evidence we have for them (and our background knowledge), and this uncertainty transfers to the argument's conclusion. This holds whether the underlying argument scheme is valid or not. Hahn and Oaksford (2007) have observed that even so-called logical fallacies

(such as the argument scheme Denying the Antecedent) can be powerful in the sense that they can increase an agent's degree of belief in the conclusion of the argument. Hahn and Oaksford argue convincingly that it is not only the logical *structure* that is important when it comes to assessing the strength of an argument but also the *content* of the premises (see also Hahn & Hornikx, 2016). Generalizing from these insights, Eva and Hartmann (2018a) develop a general theory of Bayesian argumentation according to which "Argumentation is learning." This slogan connects argumentation with confirmation, and we will sketch the corresponding theory now. Some of its applications and further developments are discussed in the following subsections.

We consider an agent (agent 1) who entertains a set of propositions with a prior probability distribution P defined over it. This can be represented by a Bayes net. Now another agent (agent 2) wants to convince agent 1 of some proposition. She decides to do so in an indirect way by manipulating the beliefs of agent 1 about the premises of an argument. More specifically, agent 2 aims at getting agent 1 to increase her degree of belief in some of the premises. To make sure that her overall degrees of belief are coherent, agent 1 then updates on that new information, which leads to a new probability of the conclusion. Hence the slogan "Argumentation is learning." If the new probability of the conclusion is greater than the old one, then the argument has some force; if not, then not. Note that the possible logical relationship between the premises and the conclusion plays a role in the updating process. This has been noted long ago by Suppes (1966) and Adams (1996).

As an illustration, consider modus ponens, that is, the rule

$$A \to C$$
$$\underline{A}$$
$$C$$

and assume that the agent has a prior probability distribution P defined over the two variables. To reconstruct the argument in Bayesian terms, we assume that the agent learns the premises of the argument with certainty. We can then use Conditionalization to compute the new probability of C. Representing the conditional $A \to C$ by the corresponding material conditional $\neg A \vee C$ finally yields

$$P'(C) = P(C \mid A, \neg A \vee C) = \frac{P(A, C, \neg A \vee C)}{P(A, \neg A \vee C)} = 1.$$

Hence the argument succeeds. As a result, the agent increases the probability of the conclusion to 1 and thus becomes certain that C is true. In less ideal circumstances, for example, if the probability of the minor premise does not increase to 1 but to some value smaller than 1, one finds that the probability of the conclusion in a modus ponens argument increases but never reaches 1. Interestingly, for valid arguments, this always holds: if the conditional is learnt with certainty and the probability of the minor premise increases, then the probability of the conclusion of the argument increases. This is not the case for invalid arguments, which is a reason to prefer valid ones (see Adams, 1996). If the argument scheme is invalid, then it depends on the prior probability distribution of the agent whether or not the probability of the conclusion increases. We will come back to this in subsection 4.3. But before, let us illustrate Bayesian argumentation with two examples that also illustrate the importance of Bayes nets.

4.1 The No-Alternatives Argument

Consider the following argument:

A scientist entertains a theory H that satisfies several desirable conditions. Unfortunately, however, the theory cannot be tested empirically because it is mathematically too difficult to derive predictions from the theory or it is impossible to experimentally test the predictions of the theory. At this point, the scientist argues as follows: "Look, my colleagues and I tried hard to find an alternative to H that also satisfies the desirable conditions, but despite a lot of effort and brain power, we did not succeed. This supports the claim that H is true." That is, the scientist argues that an observation about the performance of the scientific community can provide a reason in favor of a scientific theory.

This is the No-Alternatives Argument (NAA). Quite recently, it has been put forward by defenders of string theory, which is a candidate for the most fundamental theory of physics that arguably provides a unified account of all four fundamental forces of nature (i.e., gravity, electromagnetism, and the weak and strong nuclear forces). Unfortunately, string theory lacks direct empirical confirmation, and no one expects this situation to change in the foreseeable future. This raises the question why scientists stick to a theory that is not (and perhaps never will be) confirmed by empirical data.

In his recent book *String Theory and the Scientific Method* (2013), R. Dawid suggests that the NAA is convincing, given certain conditions, and R. Dawid, Hartmann, and Sprenger (2015) provide a Bayesian analysis of the NAA. We present a slightly simplified version of their reconstruction and consider the following argument:

P$_1$: Theory H satisfies several desirable conditions.

P$_2$: Despite a lot of effort, the scientific community has not yet found an alternative to H that also satisfies these conditions.

C: Hence we have one good reason in favor of H.

Let F be the proposition "The scientific community has not yet found an alternative to H." One then has to show that F confirms H. Using BCT, we therefore have to show that $P'(\mathrm{H}) = P(\mathrm{H}\,|\,\mathrm{F}) > P(\mathrm{H})$. Note that F is neither a deductive nor an inductive consequence of H. Hence there cannot be a direct probabilistic dependence between the corresponding propositional variables. It is therefore natural to look for a third variable that facilitates the dependence. One possibility is a "common cause" structure with the two "effects" H and F and a so-far-unknown "common cause" variable Y that screens off H and F from each other. Such a structure seems appropriate as it renders H and F dependent if the agent does not know the value of Y, and H and F become independent once this value is known. But what could this variable be? We need to find another *active variable* about which the agent has beliefs and which is related to H and F in a simple and easy-to-justify way. (We certainly want to avoid that the result depends too much on the idiosyncrasies of the specific model.) R. Dawid et al. (2015) suggest the introduction of the multivalued variable Y, which has the following values:

Y$_i$: There are i distinct alternative theories that satisfy the desirable conditions.

Here, i runs from 0 to some maximal value N. Each Y$_i$ is a statement about the number of *existing* theories that satisfy the same conditions as H. It is easy to see that Y screens off F from H: Once we know the value of Y, learning F does not tell us anything new about the probability of H. To assess it, all that matters is that we know how many equally suitable candidate theories there are. Y facilitates the probabilistic dependence between F and H, since if there *is* only a small number of alternative theories, this would provide an explanation for why the scientists have not yet found any (i.e., it would explain F). In addition, if there are only a few alternative theories, this should also probabilistically affect our belief in the available theory. We finally arrive at the Bayes net depicted in figure 4.2.7 and formulate the following four plausible conditions (see unnumbered list below):

NAA1: $\forall i \geq 0,\ y_i := P(\mathrm{Y}_i) \in [0, 1)$.

NAA2: $h_i := P(\mathrm{H}\,|\,\mathrm{Y}_i)$ are monotonically decreasing in i.

NAA3: $f_i := P(\mathrm{F}\,|\,\mathrm{Y}_i)$ are monotonically decreasing in i.

NAA4: H and F are conditionally independent given Y, that is, $P(\mathrm{H}\,|\,\mathrm{F}, \mathrm{Y}_i) = P(\mathrm{H}\,|\,\mathrm{Y}_i)$.

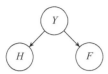

Figure 4.2.7
A Bayes net depicting the No-Alternatives Argument.

It can then be shown that the following theorem holds:

Theorem 1. Let P be a probability distribution satisfying the conditions NAA1–NAA4. Then F confirms H, that is, $P(H \mid F) > P(H)$, iff there exists a pair (i, j) with $i > j$ such that (1) $y_i y_j > 0$, (2) $f_i < f_j$, and (3) $h_i < h_j$.

This theorem allows for an assessment of the NAA. One may, for example, question NAA1. Are we really uncertain about the number of alternatives? Doesn't the *underdetermination thesis* in philosophy of science tell us that the number of alternatives to a given theory is always infinite? If this is so, then a rational agent should set $y_\infty = 1$ and the NAA does not go through. A defender of the NAA will therefore have to argue why y_∞ (and all other y_i, for that matter) should be assigned values smaller than 1. It has to be shown, then, that an agent can never be certain about the number of appropriate alternatives to a given theory.

Here is another issue. The NAA relies on how well F probes the variable Y. Note that there can be several complicating factors. For one, there might be an alternative explanation for why the scientific community has not yet found an alternative. For instance, R. Dawid et al. (2015) introduce an additional node D representing the difficulty of finding an alternative theory. The observation of F then may only confirm D (i.e., that it is very difficult to find an alternative theory). However, D is probabilistically independent of H, and hence the observation of F may not confirm the theory in question.

Note also that the NAA relies on it being possible to establish F in the first place. However, it is a nontrivial task to find agreement among the members of the scientific community about the existence or nonexistence of alternative theories. And even if there is agreement, this is only probative of the number of existing alternatives (i.e., the value of the variable Y), provided that the scientific community has attempted to explore the space of

alternative theories and has considered all problems one may encounter in this endeavor.

We conclude that the formal reconstruction of the NAA using Bayes nets helps the reasoner to put all assumptions on the table and highlights issues that need to be addressed to make the argument, if at all possible, convincing.

4.2 No Reason For Is a Reason Against

Next, consider the following argument:

You are interested in the question whether or not God exists. To address this question, you consider all arguments for the existence of God you can find in the literature. After a careful examination of them, you come to the conclusion that none of them is convincing. From *this* observation, you conclude that you have a reason *against* the existence of God because *no reason for is a reason against.*

This is the No Reason For Argument (NRF). While this reasoning may have some plausibility, it is not clear whether the argument is a good argument. To find out, we follow Eva and Hartmann (2018b) and provide a Bayesian reconstruction of the NRF, which proceeds analogously to the NAA and allows us to ask whether learning the premises of the argument increases the probability of its conclusion.

We consider the hypothesis H and introduce the proposition F, which says "I have not yet found a good argument in favor of H." Furthermore, let Y be a propositional variable whose values are the propositions Y_i: "There are exactly i good arguments in favor of H" ($i \geq 0$). It is plausible that an agent can be uncertain about the value of Y. Many rational agents would surely plead ignorance as to whether or not there are any undiscovered good arguments for the existence of God (and, if so, how many). Again, knowledge of the value of Y renders F independent of H: if I know that there are five good arguments for the existence of God, then the fact that I haven't yet found any one of them should be irrelevant to my belief in the existence of God. Furthermore, if I learn that there are more good arguments for God's existence than I previously thought, then that should raise my degree of belief in the existence of God. Finally, the more arguments there are for God's existence, the more likely it is that I find one. These considerations motivate the following conditions (see unnumbered list below):

NRF1: $\forall i \geq 0$, $y_i := P(Y_i) \in [0, 1)$.

NRF2: $h_i := P(H \mid Y_i)$ are monotonically increasing in i.

NRF3: $f_i := P(F \mid Y_i)$ are monotonically decreasing in i.

NRF4: H and F are conditionally independent on Y, that is, $P(H \mid Y_i, F) = P(H \mid Y_i)$.

NRF1–NRF4 are structurally near-identical to the conditions imposed on the NAA. The corresponding Bayes net model is also analogous. The only difference is that the h_i are now monotonically increasing in i. Accordingly, the proof for the following theorem is analogous to the proof of theorem 1.

Theorem 2. Let P be a probability distribution satisfying the conditions NRF1–NRF4. Then F disconfirms H, that is, $P(H|F) < P(H)$, iff there exists a pair (i,j) with $i > j$ such that (1) $y_i\,y_j > 0$, (2) $f_i > f_j$, and (3) $h_i < h_j$.

We contend that theorem 2 constitutes, under certain special circumstances, a full Bayesian vindication of the NRF argument. On the basis of this reconstruction, one can analyze how good specific NRF arguments are (for details, see Eva & Hartmann, 2018b).

4.3 Toward a General Theory of Reasoning and Argumentation

The examples given in the last two subsections illustrate the slogan "Argumentation is learning" mentioned above. Both examples involved the learning of a proposition, and the learning was modeled using Conditionalization. Note, however, that many argument schemes involve indicative conditionals of the form "If A, then C" as premises, which are notoriously difficult to deal with. In the introduction to this section, we represented an indicative conditional by the corresponding material conditional $\neg A \vee C$. This was convenient, as it allowed us to condition on it. However, the material conditional faces many problems, and it is even debated whether indicative conditionals can be modeled as propositions at all (for details, see Douven, 2018, and chapter 6.1 by Starr, this handbook). It is therefore advisable to look for a more general updating method that can be applied to both propositional and nonpropositional evidence. A general theory of Bayesian argumentation cannot work without such a method. To address this problem, Eva et al. (2020) develop the *distance-based approach to Bayesianism* that builds on the abovementioned principle of Conservativity. Learning the indicative conditional "If A, then C" from a perfectly reliable information source then suggests the constraint $P'(C|A) = 1$. The full posterior probability distribution P' follows by minimizing a suitable distance measure (such as the Kullback–Leibler divergence) between P' and P. Eva and Hartmann (2018a) apply this proposal to Bayesian argumentation and show how Bayes nets can be used to model, for example, disablers in an argument.

5. Outlook

In this chapter, we have introduced the theory of Bayes nets, shown how Bayes net models can be constructed within the Bayesian framework, and presented a number of examples from the philosophy and psychology of reasoning and argumentation that illustrate the power of the approach. However, Bayes nets have also been used in other areas of rationality research (such as the study of decision making), and there is no doubt that a whole range of further problems can be addressed with this methodology.

It is also worth noting that Bayes nets can be used with other theories of uncertainty as well. For example, Spohn (2012, chapter 7) shows that ranking theory satisfies the semi-graphoid axioms. Other authors investigate conditional independence structures for imprecise probabilities (see, e.g., Cozman, 2012; Halpern, 2003, chapter 4). Hence, nothing hinges on the Bayesian framework that we used in this chapter. At the same time, it is true that most applications of Bayes nets in rationality research studied so far presuppose the Bayesian framework, which is easy to use and has a solid normative foundation. It would be interesting to construct analogous models in other frameworks and to compare the resulting analyses with the corresponding Bayesian analyses (see also Colombo, Elkin, & Hartmann, in press). This will lead to further progress in rationality research.

Acknowledgments

Thanks to my collaborators Richard Dawid, Benjamin Eva, and Jan Sprenger and to the editors of this volume for their patience and valuable feedback.

Notes

1. For a textbook exposition, see Darwiche (2014, section 4.4) and Pearl (1988, section 3.1).

2. We follow the convention, adopted, for example, in Bovens and Hartmann (2003), to represent propositional variables in italics and their values in roman script. For instance, the variable A has the values A and $\neg$A. Here and in the remainder, we also use the shorthand notation $P(A, B)$ for $P(A \wedge B)$.

3. If A_j is a root node, then pa(A_j) is the empty set ($\emptyset$) and $P(A_j | pa(A_j)) = P(A_j)$.

4. We sometimes use causal language when talking about various Bayes nets. And indeed, causal intuitions help when it comes to construct a Bayes net model, which typically respects

the "causal direction." For instance, we always draw an arrow from the hypothesis variable to the corresponding evidence variable. Note, however, that all we need for the applications discussed in this chapter is that a Bayes net represents a joint probability distribution. Hence, the causal language is, strictly speaking, only of heuristic value here (see chapter 7.1 by Pearl, this handbook).

5. Section 5 of this volume surveys the various contenders. See also chapter 4.7 by Dubois and Prade (this handbook) and Halpern (2003). Oaksford and Chater (2007) discuss topics in the psychology of reasoning from a Bayesian point of view. Hájek and Hartmann (2010) and Hartmann and Sprenger (2010) survey Bayesian epistemology, and Sprenger and Hartmann (2019) investigate various topics in the philosophy of science from a Bayesian point of view.

6. "Relevant" means relevant for the specific problem or question the agent is interested in.

7. It would be very helpful to have an automated way to extract the relevant variables and their relations from potentially large data sets. For attempts in this direction, see Chalupka, Bischoff, Perona, and Eberhardt (2016).

8. The choice of the variables, and which relations between them one assumes, may also be suggested by considerations about which Bayes net provides a good representation.

9. See Frigg and Hartmann (2016/2020) for a survey of the corresponding philosophy of science literature.

10. For details and an explanation, see Bovens and Hartmann (2003).

References

Adams, E. W. (1996). *A primer of probability logic* (Synthese Library, Vol. 86). Boston, MA: Reidel.

Bovens, L., & Hartmann, S. (2003). *Bayesian epistemology*. Oxford, England: Clarendon Press.

Chalupka, K., Bischoff, T., Perona, P., & Eberhardt, F. (2016). Unsupervised discovery of El Niño using causal feature learning on microlevel climate data. In *Proceedings of the 32nd Conference on Uncertainty in Artificial Intelligence* (pp. 72–81). Arlington, VA: AUAI Press.

Colombo, M., Elkin, L., & Hartmann, S. (in press). Being realist about Bayes, and the predictive processing theory of mind. *British Journal for the Philosophy of Science*.

Colyvan, M. (2013). Idealisations in normative models. *Synthese, 190*, 1337–1350.

Connor Desai, S., Reimers, S., & Lagnado, D. (2016). Consistency and credibility in legal reasoning: A Bayesian network approach. In A. Papafragou, D. Grodner, D. Mirman, & J. C. Trueswell (Eds.), *Proceedings of the 38th Annual Meeting of the*

Cognitive Science Society (pp. 626–631). Austin, TX: Cognitive Science Society.

Cozman, F. G. (2012). Sets of probability distributions, independence, and convexity. *Synthese, 186*(2), 577–600.

Crupi, V. (2016). Confirmation. In E. N. Zalta (Ed.), *The Stanford encyclopedia of philosophy*. Retrieved from https://plato.stanford.edu/archives/spr2020/entries/confirmation/

Dardashti, R., Hartmann, S., Thébault, K., & Winsberg, E. (2019). Hawking radiation and analogue experiments: A Bayesian analysis. *Studies in History and Philosophy of Modern Physics, 67*, 1–11.

Darwiche, A. (2014). *Modeling and reasoning with Bayesian networks*. Cambridge, England: Cambridge University Press.

Dawid, A. P. (1979). Conditional independence in statistical theory. *Journal of the Royal Statistical Society, Series B, 41*, 1–31.

Dawid, R. (2013). *String theory and the scientific method*. Cambridge, England: Cambridge University Press.

Dawid, R., Hartmann, S., & Sprenger, J. (2015). The no alternatives argument. *British Journal for the Philosophy of Science, 66*(1), 213–234.

Diaconis, P., & Zabell, S. L. (1982). Updating subjective probability. *Journal of the American Statistical Association, 77*(380), 822–830.

Douven, I. (2018). *The epistemology of indicative conditionals: Formal and empirical approaches*. Cambridge, England: Cambridge University Press.

Earman, J. (1992). *Bayes or bust? A critical examination of Bayesian confirmation theory*. Cambridge, MA: MIT Press.

Eva, B., & Hartmann, S. (2018a). Bayesian argumentation and the value of logical validity. *Psychological Review, 125*(5), 806–821.

Eva, B., & Hartmann, S. (2018b). When no reason for is a reason against. *Analysis, 178*(3), 426–431.

Eva, B., Hartmann, S., & Rafiee Rad, S. (2020). Learning from conditionals. *Mind, 129*(514), 461–508.

Fenton, N., Neil, M., Yet, B., & Lagnado, D. (2020). Analyzing the Simonshaven case using Bayesian networks. *Topics in Cognitive Science, 12*(4), 1092–1114.

Frigg, R., & Hartmann, S. (2020). Models in science. In E. N. Zalta (Ed.), *The Stanford encyclopedia of philosophy*. Retrieved from https://plato.stanford.edu/archives/spr2020/entries/models-science/

Giere, R. N. (1990). *Explaining science: A cognitive approach*. Chicago, IL: University of Chicago Press.

Hahn, U., & Hornikx, J. (2016). A normative framework for argument quality: Argumentation schemes with a Bayesian foundation. *Synthese, 193*(6), 1833–1873.

Hahn, U., Merdes, C., & von Sydow, M. (2018). How good is your evidence and how would you know? *Topics in Cognitive Science, 10*(4), 660–678.

Hahn, U., & Oaksford, M. (2007). The rationality of informal argumentation: A Bayesian approach to reasoning fallacies. *Psychological Review, 114*(3), 704–732.

Hájek, A., & Hartmann, S. (2010). Bayesian epistemology. In J. Dancy, E. Sosa, & M. Steup (Eds.), *A companion to epistemology* (2nd ed., pp. 93–106). Oxford, England: Wiley-Blackwell.

Halpern, J. Y. (2003). *Reasoning about uncertainty.* Cambridge, MA: MIT Press.

Harris, A. J. L., Hahn, U., Madsen, J. K., & Hsu, A. S. (2016). The appeal to expert opinion: Quantitative support for a Bayesian network approach. *Cognitive Science, 40*(6), 1496–1533.

Hartmann, S., & Sprenger, J. (2010). Bayesian epistemology. In S. Bernecker & D. Pritchard (Eds.), *The Routledge companion to epistemology* (pp. 609–620). London, England: Routledge.

Howson, C., & Urbach, P. (2006). *Scientific reasoning: The Bayesian approach.* London, England: Open Court.

Huber, F. (2007). Confirmation and induction. In *Internet encyclopedia of philosophy.* Retrieved from https://iep.utm.edu/conf-ind/

Landes, J. (2020). Variety of evidence. *Erkenntnis, 85,* 183–223.

Oaksford, M., & N. Chater (2007). *Bayesian rationality: The probabilistic approach to human reasoning.* Oxford, England: Oxford University Press.

Pearl, J. (1988). *Probabilistic reasoning in intelligent systems.* San Francisco, CA: Morgan-Kaufmann.

Spohn, W. (1980). Stochastic independence, causal independence, and shieldability. *Journal of Philosophical Logic, 9,* 73–99.

Spohn, W. (2012). *The laws of belief: Ranking theory and its philosophical applications.* Oxford, England: Oxford University Press.

Sprenger, J. (2011). Hypothetico-deductive confirmation. *Philosophy Compass, 6*(7), 497–508.

Sprenger, J., & Hartmann, S. (2019). *Bayesian philosophy of science.* Oxford, England: Oxford University Press.

Suppes, P. (1966). Probabilistic inference and the concept of total evidence. In J. Hintikka & P. Suppes (Eds.), *Aspects of inductive logic* (pp. 49–65). Amsterdam, Netherlands: North Holland.

Titelbaum, M. G. (2021). Normative modeling. Retrieved from http://philsci-archive.pitt.edu/18670

4.3 Evidential Relevance

Arthur Merin

Summary

Signed evidential relevance among truth-valuable propositions has qualitative, comparative, and quantitative aspects. We explore the abstract theory of qualitative relevance quite fully, notably so the role of conditional irrelevance. Regarding ordinal and quantitative aspects of relevance, we offer a notational proposal and two extensions of the state of the art. First, we demonstrate task-related conditions for ordinal equivalence of all relevance measures satisfying a highly intuitive desideratum shared by the major relevance measures. This affords a useful degree of measure independence for many comparative relevance judgments. Second, we elicit relevance properties of disjunctive evidence to complement familiar facts about conjunctive evidence. The central role of relevance for reasoning is emphasized throughout.

1. Intuitions of Evidence in Argumentation

Will some proposition or statement, A, speak for the truth of another proposition, B, or against it? Questions of this kind are ubiquitous, and people answer them within the context of their general beliefs about the world. Example: Let A = "There are dark clouds in the sky today," B_1 = "It will rain today," B_2 = "It will be dry today," B_3 = "The Dow Jones will rise today." In typical belief contexts, A will speak for B_1, against B_2, and neither for nor against B_3. Similar relations obtain between statements about past events, as well as between possible experimental results and timeless scientific hypotheses.

Here is a way to reconstruct this kind of reasoning. We are uncertain about the truth values, True or False, of B_1, B_2, and B_3. We may be uncertain about the truth value of A. And we can distinguish, in principle, at least a handful of degrees of credence between certainty of truth and certainty of falsity. Now comes the principal ingredient of the reconstruction. If, being uncertain about A ("dark clouds"), we learnt for certain that A is true—just this much!—then our degrees of credence should move upward for B_1 ("rain"), go downward for B_2

("dry"), and remain unchanged for B_3 ("Dow up"). We need not actually become certain that B_1 is true. Neither need we become certain that B_2 is false. These predicaments are but special cases of "speaking for" and "speaking against."

Next suppose that our beliefs are as before, but that we learn that A ("dark clouds") is false; equivalently, $\neg A$ ("no dark clouds") is true. The suggestion here is as follows. Our credence in B_3 ("Dow up") would again remain unaffected. Our degree of credence for B_2 ("no rain") would go up. And our credence in B_1 ("rain") would go down. If so, learning that A is false must afford a basis for determining what A speaks for, against, and neither way. A will speak for B if $\neg A$ speaks against B.

In the diction of jurisprudence and of science, "speaking for" B means being *evidence for* B. "Speaking against" B means being *evidence against* B. In this evidential sense, we say that A is *positively relevant* ("*positive*") to B in a credal context or state *Cred* if and only if (iff) it speaks for B in *Cred*, *negatively relevant* ("*negative*") to B iff it speaks against B in *Cred*, evidentially *irrelevant* to B if it speaks neither for nor against B in *Cred*, and *relevant* iff not irrelevant.

Plausible as the idea may appear, it still remains vague and prone to apparent exceptions. Some framework for specifying it is needed to give it substance.

2. Evidential Relevance Explicated in Probability Theory

To turn intuitions about relevance into a theory, one interprets degrees of credence in A, B, and so on in a credal state, as credal probability values $P(A)$, $P(B)$, and so forth.[1]

Changes of credal state, from *Cred* to *Cred'*, are then represented by changes of probability measures, from P to P'. Any such change will yield credal reassessments $P'(A)$, $P'(B)$, and so on. However, it is one particular belief change regime that yields a workable explication of the intuitive notion of relevance. This regime is intimately associated with the concept of conditional probability, in terms of which relevance is standardly definable.

Definition 1. The *conditional probability* of A given B, notated $P(B \mid A)$, is defined by $P(B \mid A) =_{\mathrm{df}} P(AB)/P(A)$ whenever $P(A) \neq 0$, else undefined.[2]

We gloss it, roughly for now, as the probability that B would have if A were assumed. Recall now section 1. Suppose we learn A, no more, no less, and without any doubts or reservations. Then we should be in a new, "posterior" credal state, P', validating $P'(A) = 1$; $P'(\neg A) = 0$. Let our update be by *conditioning P on A*. Then the posterior state P' may be written P_A, and for all B, the posterior probability of B equals its conditional probability: $P_A(B) = P(B \mid A)$.

The probability-change explication of speaking "for" or "against" makes A positive to B in P iff $P_A(B) > P(B)$ and negative to B iff $P_A(B) < P(B)$. This yields

Definition 2. A is $\left\{ \begin{array}{c} positive \\ irrelevant \\ negative \end{array} \right\}$ to B (in P) iff

$$\left\{ \begin{array}{c} P(B \mid A) > P(B) \\ P(B \mid A) = P(B) \text{ or } P(A) = 0 \\ P(B \mid A) < P(B) \end{array} \right\}.$$

Multiplying the (in)equalities of definition 2 with $P(A)$ yields the equivalent definition 3, which we state with obvious abbreviations and using the standard symbol $(A \perp B)_P$ for "A is irrelevant to B in P":

Definition 3. $A \left\{ \begin{array}{c} pos \\ \perp \\ neg \end{array} \right\}_P B$ iff $P(AB) \gtreqless P(A) \cdot P(B)$.

Definition 3 depends on definition 2 for the underlying intuition. In return, it needs no completion clause and reveals the symmetry of relevance relations.

3. Properties of Qualitative Relevance and Extreme Positivity

Qualitative relevance of A to B, and of B to A, is specified only in terms of polarity. I present six of its properties in order of intuitive simplicity or importance.

(1) *Symmetry of qualitative relevance*: Explicated in probability theory, positive relevance, negative relevance, and irrelevance are *symmetric relations*:

Theorem 1. $A \left\{ \begin{array}{c} pos \\ \perp \\ neg \end{array} \right\}_P B$ iff $B \left\{ \begin{array}{c} pos \\ \perp \\ neg \end{array} \right\}_P A$.

The proof is obvious from definition 3. For instance, if A speaks for B, then B speaks for A, however weakly (intuitively).

(2) *Polarity reversal by negation*: Negating just one of A and B reverses relevance polarity and preserves irrelevance. Negating both of A and B leaves relevance polarity unchanged:

Theorem 2. $A \left\{ \begin{array}{c} pos \\ \perp \\ neg \end{array} \right\}_P B$ iff $\neg A \left\{ \begin{array}{c} neg \\ \perp \\ pos \end{array} \right\}_P B$ iff $A \left\{ \begin{array}{c} neg \\ \perp \\ pos \end{array} \right\}_P \neg B$ iff $\neg A \left\{ \begin{array}{c} pos \\ \perp \\ neg \end{array} \right\}_P \neg B$.

(3) *Certainty implies irrelevance*: Uncertainty is a necessary condition for relevance:

Theorem 3. A proposition A that has extreme probability, $P(A) = 0$ or $= 1$, is irrelevant, under P, to any other proposition and to itself.

For proof instantiate definition 3. To be relevant (or for anything to be relevant to it) in P, a proposition A must therefore be P-contingent, $0 < P(A) < 1$.

(4) *Extreme positive and negative relevance*: Classical logical entailment, $A \vdash B$, is transitive: $A \vdash B$ and $B \vdash C$ together imply $A \vdash C$. Positive relevance is *not* in general transitive. That is, $A \ pos_P B$ and $B \ pos_P C$ need not imply $A \ pos_P C$. An example will be given shortly. However, there are conditions under which positive relevance *is* transitive. One special case thereof engages "conclusive-evidence-for" as explicated in definition 4a:

Definition 4 (Carnap 1950, modified). (a) A is *extremely positive* to B iff $P(B) < P(B \mid A) = 1$. (b) A is *extremely negative* to B iff extremely positive to $\neg B$. We abbreviate (a) as $A \ expos_P B$ and (b) as $A \ exneg_P B$.

Clause (a) succinctly describes in a credal probability theory what is conveyed when someone asserts, without reservations, an indicative conditional "If A, then B." Of course, such an assertation can also be made with less certainty. Extreme positivity explicates what colloquial language mostly means by "entailment." The idea is that A is a compelling and, in this sense, *sufficient reason* for B. It is *sufficient* by $P(B \mid A) = 1$, which makes B a conditional certainty. It is a *reason* in virtue of $P(B) < P(B \mid A)$. Reasonhood has a reflex in our sense of evidential aesthetics: "If the Dow rises ($=A$), (then) 7 is a prime number ($=B$)" sounds nonsensical even when $P(A) > 0$ is uncontroversial, as normally it is, and $P(B \mid A) = 1$ must therefore hold because $P(B) = 1$ holds. Classical entailment, $A \vdash B$, is related to extreme relevance by

Lemma 1. If $0 < P(A), P(B) < 1$, and A entails B ($A \vdash B$), then $A \ expos_P B$.

Entailment, ⊢, is a transitive relation. So is its probabilistic counterpart, *P-entailment*, defined by $A \vdash_P B =_{df} P(A \rightarrow B) = 1$ (where → is material implication). Extreme positivity inherits the transitivity of ⊢ and ⊢$_P$:

Theorem 4. Extreme positive relevance is transitive: if $A \; expos_P \; B$ and $B \; expos_P \; C$, then $A \; expos_P \; C$.

(5) *Conditional irrelevance and conditional transitivity of positive relevance*: When positive relevance of A to B and of B to C is short of extreme, transitivity may fail. Here is an example: Assume $A \; expos_P \; B$, $C \; expos_P \; B$, and $P(AC) = 0$. These assumptions are easily satisfiable. For example, A = "Trump has won," B = "The U.S. has a new president," and C = "Clinton has won." Then $A \; pos_P \; B$, $B \; pos_P \; C$ (because of symmetry), yet $A \; exneg_P \; C$, hence $A \; neg_P \; C$.

Transitivity is, however, ensured by a supremely useful property. Relevance relations can be *conditional* on a proposition. On noting that $P(A) = P(A \mid \top)$ is a probability theorem (where ⊤ is any tautology), definition 3 becomes a special case of

Definition 5. A is *irrelevant* to (*independent* of) C conditionally on B under P, $(A \perp C \mid B)_P$, iff $P(AC \mid B) = P(A \mid B) \cdot P(C \mid B)$. When each of $(A \perp C \mid B)_P$ and $(A \perp C \mid \neg B)_P$ holds, we write $(A \perp C \mid \pm B)_P$.

We now have a sufficient condition for transitivity of positive relevance that need not be extreme:

Theorem 5. If A and B are irrelevant to one another conditionally on B and on $\neg B$, that is, if $(A \perp C \mid \pm B)_P$, then mutual positivity of A and B and mutual positivity of B and C imply mutual positivity of A and C.[3]

Condition $(A \perp C \mid \pm B)_P$ becomes more intuitable on realizing that the extreme-relevance configuration of $A \; expos_P \; B$ and $B \; expos_P \; C$ entails the antecedent of the following lemma and thus $(A \perp C \mid \pm B)_P$:

Lemma 2. If $P(B \mid A) = 1$, $P(C \mid B) = 1$, and $P(B) < 1$, then $(A \perp C \mid \pm B)_P$.

Hence, theorem 5 properly generalizes intuition-friendly theorem 4.

(6) *Conditional irrelevance and paradoxa of relevance*: Theorem 5 shows that $(A \perp C \mid \pm B)_P$ does not imply $(A \perp C)_P$. This counterintuitive fact constitutes an instance of a well-known statistics paradox. Suppose **B** and **¬B** specify complementary properties of members of a select population, say, "British" and "non-British." Let **A** stand for having been administered pharmaceutical Φ, and **C** for being cured of disease Δ. These properties and their complexes can be turned into propositions ($A, B, \neg B, C$, etc.), whose credal probabilities reflect statistical frequencies.

(Think of proposition D as "ι has property **D**," where ι is a randomly sampled individual.) Now suppose we investigate the relevance of A to C. Then theorem 5 implies that A may be positively relevant to C in the population as a whole, but irrelevant to C in each of subpopulations **B** and **¬B** (because, e.g., the drug Φ is actually inefficacious concerning Δ, but both the drug administration and the spontaneous remission rate are higher among British than among non-British citizens). This is an instance of "Simpson's Paradox" (see chapter 7.1 by Pearl, this handbook).

In return, issue-conditional independence $(A \perp C \mid \pm B)_P$ (referring to the issue whether B or $\neg B$) precludes otherwise possible, intuitively paradoxical relevance configurations such as the one where A and C are positive to B, but AC is irrelevant or negative to B (cf. Carnap, 1950, chapter 6). For Reichenbach (1956) shows, as a special case of a result for various combinations of relevance polarities:

Theorem 6. Suppose that $(A \perp C \mid \pm B)_P$ holds, A and B are mutually positive, and so are B and C. Then B and the conjunction AC are mutually positive.

Reichenbach's intended interpretations for theorems 5 and 6 had B as a *common cause* of A and C. Pearl (1988; see chapter 7.1 by Pearl, this handbook) surmised that medical diagnostic categories B (diseases, syndromes) are formed to make symptoms A and C conditionally independent given $\pm B$, by identifying B as their common cause.[4] A test of Pearl's thesis would thus be to search for diseases that exhibit Carnap-paradoxical symptomatologies. Anyone's credence in the thesis should increase if none were found. But why? Simply speaking, because finding a disease with a paradoxical symptomatology would be extremely negative to the thesis, which implies that finding none is positive to it. We now address this idea more generally.

4. An Application of Qualitative Relevance Reasoning

The most familiar applications of theorem 1 (symmetry) and theorem 2 (polarity reversal under negation) presuppose a sorting of propositions into two types: (1) "hypothesis" statements or "theories," labeled "H," and (2) observational "data reports" or "evidence statements," labeled "D" or "E." Suppose a hypothesis H having nonzero probability entails some data report D, which might be one possible outcome of an intended, future investigation. In symbols: $H \vdash D$. Then $P(D \mid H) = 1$ holds. Hence, provided $0 < P(D)$, $P(H) < 1$ holds, H is extremely positive, and thus positive, to D. By symmetry, $P(H \mid D) > P(H)$ must also hold. Suppose outcome D is then observed and P conditioned on it. Then $P_D(H) > P(H)$ must hold. H has become *more credible*. Alternatively, suppose outcome $\neg D$ is observed. By theorem 2, $P_{\neg D}(H) < P(H)$ holds, so H becomes *less credible*. This is what the relevance laws developed so far entail.

Complications arise when (a) D is already-known, "old" evidence to be explained, rather than predicted, and (b) P is required to reflect this knowledge. By theorem 3, such D must be P-irrelevant: $P_D(H) - P(H)$ (Glymour, 1980). A standard example has H instantiated by Einstein's theory of general relativity, which entailed D, a long-known and seemingly anomalous fact about the orbit of planet Mercury. This relationship was widely judged to increase confidence in H. Reconstructions of this judgment appeal to hypothetical, partly nonconditioning update histories (Jeffrey, 2004), introduce probabilities over primitive propositions about entailment or support (cf. Earman, 1992; Fitelson & Hartmann, 2015), or leave D slightly in doubt and choose a relevance measure (see below) that gives near-certain D an appreciable amount of relevance (cf. Hájek & Joyce, 2008). Working statisticians have, in this respect, an easier life. Faced with known data, they usually draw on ready-made probability models for conditional statistical frequencies of occurrence. This replaces commitment (b) by appropriate probability distributions for hypothetical predictions under repeatable conditions.[5]

5. Comparative Strength and Amount of Relevance

Consider propositions A_1 = "There are dark clouds," A_2 = "The weather forecast predicts rain," A_3 = "The weather forecast predicts dry conditions," and B = "It will rain." Imagine, as seems natural, that each of A_1 and A_2 would speak for B, while A_3 speaks against B. Now suppose that you wonder which of A_1 and A_2, if any, would speak more strongly for B than the other does. Or imagine that you learn A_1 and A_3 and wonder if A_1A_3 speaks for B, or against B, or does neither. Or imagine asking which of B ("It will rain") and B' ("It will be snowing") is more strongly spoken for by A_1 ("dark clouds") in a given credal state. Neither of these questions can usually be resolved by purely qualitative considerations.

Suppose, then, that A_1 and A_2 are each positive to B in credal state P. If we compare A_1 and A_2 by strength of relevance to B, the resultant ordering within the probabilistic explication will, in general, depend on measurements of strength. This fact already suffices to make measures of relevance a pertinent topic for

anyone interested in probabilistic relevance. Rankings may indeed depend on our choice of measuring function. This is much as could happen in ranking companies by asset value: rankings may differ according to the accounting convention used.

A further problem arises. If A and B are mutually relevant, A may be more strongly relevant to B than B is to A under one measure, less relevant under another, and equally relevant under a third. A helpful measurement notation should thus make the intended direction immediately clear, without tacitly relying on the classification of content as, say, evidence or hypothesis, which is often application dependent (see below). What motivates conventional assignments of relevance direction may then be freely investigated.

5.1 Notation and Terminology

In recent literature, degree of relevance is widely glossed as *"amount of '(incremental) confirmation'* of *'hypothesis'* H provided by *'evidence'* E under *probability measure P and measure of type C."* Notation is often *"$C_P(H, E)$,"* sometimes *"$C_P(E, H)$."* The $C_P(\cdot, \cdot)$ format affords no natural mnemonic of direction. True, its gloss will suggest a direction: *from* "evidence" *to* "hypothesis." Yet often enough we reason from hypotheses to possible data or find no warrant for the evidence/hypothesis type distinction. Thus, reliance on E and H abets premature identification of *structural argument position* with *content type*. Concomitantly, labels such as "confirmation," "corroboration," and "support" need a prefix "incremental" to avoid being misunderstood as designating conditional probability, $P(H \mid E)$. They also sound awkward with the prefix "negative." "Relevance" (Keynes, 1921; Carnap, 1950; Jeffrey, 2004) has no such problems.

All this speaks for a content-neutral approach to symbolization, at least alongside its common philosophy-of-science instantiation. I propose an iconic, signpost-like separator symbol for proposition arguments that indicates conventional measurement direction as the wording does in "relevance of A to B."

Definition 6 (Notation, Terminology). Let $R_T(A \succ B)_P$ designate—subject to conditions specified in definition 8—the *numerical amount of relevance* of proposition A to proposition B as assessed in terms of a relevance measure of type **T** under a credal probability measure P. Refer to the proposition arguments A and B preceding and following the signpost separator $\succ$ as being respectively in *source position* and in *target position*.

Source $\succ$ target direction affects numerical measurement values, at some point or other, for all relevance measures that do *not* have the property introduced in

Definition 7. A relevance measure $R_T(A \succ B)_P$ is *amount-symmetric* iff $R_T(B \succ A)_P = R_T(A \succ B)_P$ for all A, B, P.

Amount-symmetric R_T are, in this unambiguous sense, *nondirectional*.

5.2 Two Definitional Conditions on All Relevance Measures

Many distinct measures of relevance have been discussed in the literature (see, e.g., Eells & Fitelson, 2002; Crupi & Tentori, 2016). All measures validate

Definition 8. A function $R_T(A \succ B)_P$ mapping pairs $\langle P, \langle A, B \rangle \rangle$ consisting of a probability measure P and a pair $\langle A, B \rangle$ of propositions in the domain of P to numeric values is a *probabilistic relevance measure* iff it satisfies

(a) *Transparency*: $R_T(A \succ B)_P$ must be a smooth (i.e., differentiable) function of the probability distribution P over the atoms AB, $A \wedge \neg B$, $\neg A \wedge B$, $\neg A \wedge \neg B$ of the propositional algebra generated by A and B, and its value range must be a subinterval, proper or improper, of the extended real numbers $\mathbb{R}^* = \mathbb{R} \cup \{-\infty, \infty\}$.[6]

(b) *Quality Representation*: $R_T(A \succ B)_P$ must represent qualitative relevance polarity: its value range must contain an element $\iota := \iota_T$ such that

$$R_T(A \succ B)_P \gtreqless \iota_T \text{ iff } A \begin{Bmatrix} pos \\ \perp \\ neg \end{Bmatrix}_P B.$$

Measures in actual use have for neutral value either $\iota = 0$, the "neutral" identity element for addition, or $\iota = 1$, the multiplicative identity. Logarithmic transformations map multiplication to addition and thus multiplicative to additive identities.

In what follows, we mostly adopt the common format $R(E \succ H)$, which identifies the "source" proposition with "evidence" E and the "target" proposition with a "hypothesis" H. This makes for easy cross-reference to the confirmation literature, just as A and B do to the content-neutral mathematical literature.

5.3 The Big Three Measures of First Resort and Their Ordinal Equivalents

The "Big Three"—**D**, **Q**, and **L**—are the oldest and most commonly used relevance measures. Their formal definitions are given below. (**D**, **Q**, and **L** stand for "Difference," "Quotient," and "Likelihood.") Also given are the most useful and most common transforms for **Q** and **L**, namely, **QL** and **LL**.

$$\textbf{D:}\quad R_D(E \succ H)_P =_{df} \begin{cases} P(H|E) - P(H), & \text{if } P(E) \neq 0, \\ \qquad\qquad 0, & \text{else.} \end{cases}$$

$$\textbf{Q:}\quad R_Q(E \succ H)_P =_{df} \begin{cases} \dfrac{P(H|E)}{P(H)} = \dfrac{P(EH)}{P(E)P(H)} = \dfrac{P(E|H)}{P(E)}, & \text{if } P(E), P(H) \neq 0, \\ \qquad\qquad\qquad 1, & \text{else.} \end{cases}$$

$$\textbf{L:}\quad R_L(E \succ H)_P =_{df} \begin{cases} \dfrac{P(E|H)}{P(E|\neg H)}, & \text{if } P(H), P(\neg H), P(E|\neg H) \neq 0, \\ \infty, & \text{if } P(H), P(\neg H) \neq 0 \text{ and } P(E|\neg H) = 0 \neq P(E|H), \\ 1, & \text{else.} \end{cases}$$

$$\textbf{QL:}\quad R_{QL}(E \succ H)_P =_{df} \log R_Q(E \succ H)_P = \log P(EH) - \log P(E) - \log P(H).$$

$$\textbf{LL:}\quad R_{LL}(E \succ H)_P$$
$$=_{df} \begin{cases} \log R_L(E \succ H)_P = \log P(E|H) - \log P(E|\neg H), & \text{if } P(H), P(\neg H), P(E|H), P(E|\neg H) \neq 0, \\ \infty, & \text{if } P(E|\neg H) = 0 \neq P(E|H), \\ -\infty, & \text{if } P(E|H) = 0 \neq P(E|\neg H), \\ 0, & \text{else.} \end{cases}$$

Let me introduce two further measures to be mentioned below: **KO** and **Z**. Formal definitions for these are given at the bottom of this page.

Even more measures are discussed in the literature.

The "difference measure" R_D (Hugh MacColl's δ of 1897 or earlier; cf. Keynes, 1921, pp. 154–155) has irrelevance value $\iota = 0$ and range $(-1, 1)$. The "coefficient of influence" (W. E. Johnson, 1921 or earlier, published in Keynes, 1921) or "relevance quotient" Q^8 has $\iota = 1$ and range $[0, \infty)$. Its variant expression $\dfrac{P(EH)}{P(E)P(H)}$ reveals its amount symmetry: $R_Q(E \succ H)_P = R_Q(H \succ E)_P$. The "likelihood ratio" **L** has $\iota = 1$ and range $[0, \infty]$. Modulo logarithms, **L** was commended by C. S. Peirce, A. M. Turing, and I. J. Good. It has an information-theoretic claim to primacy, being both (i) sufficient and (ii) necessary to capture all the information in the data E that bears on the hypothesis bipartition $\{H, \neg H\}$. It conveys all of it (i) and any other sufficient statistic factors through it (ii) (cf. Savage, 1954, pp. 128–140).

The logarithmic **QL**-transform (Good, 1950) has $\iota = 0$ and range $[-\infty, \infty]$. The "log-likelihood ratio" **LL** has $\iota = 0$ and range $[-\infty, \infty]$ and is widely used in statistics. The "sigmoid" transform **KO** (Kemeny & Oppenheim, 1952) of **LL** has $\iota = 0$ and squashes the range to $[-1, 1]$. Finally, for the measure **Z** (Crupi & Tentori, 2013) the irrelevance value ι is 0 and the range is $[-1, 1]$. The various transforms' existence suffices to motivate a highly general concept.

Definition 9. Measure types **T** and **T′** are *ordinally equivalent* (**T** $\sim_\uparrow$ **T′**) iff, for all possible P and propositions A, B, C, D in the domain of P, $R_T(A \succ B)_P \gtreqless R_T(C \succ D)_P$ iff $R_{T'}(A \succ B)_P \gtreqless R_{T'}(C \succ D)_P$. Equivalently, **T** $\sim_\uparrow$ **T′** iff there exists some strictly monotonically increasing (isotonic) function f for which **T′** $= f($**T**$)$.[9]

Distinct, but ordinally equivalent, measures may nonetheless differ in convenience of range and, more important

for cognitive-engineering applications, in various algebraic properties.

The Big Three have appealing interpretations as additive or multiplicative operators. They map the *prior* credal value of a "target" proposition H to its value in the *posterior* credal state obtained by conditioning the prior P on the "source" proposition E. For **D** and **Q**, the credal values are probabilities. For **L** they are odds, $O(H)$ and $O(H \mid E)$, from which probabilities may be recomputed.[10]

$$P(H \mid E) = P(H) + R_D(E \succ H)_P, \tag{1}$$
$$P(H \mid E) = P(H) \cdot R_Q(E \succ H)_P, \tag{2}$$
$$\log P(H \mid E) = \log P(H) + R_{QL}(E \succ H)_P, \tag{2L}$$
$$O(H \mid E) = O(H) \cdot R_L(E \succ H)_P, \tag{3}$$
$$\log O(H \mid E) = \log O(H) + R_{LL}(E \succ H)_P. \tag{3L}$$

Observe that equations (2) and (3) express, respectively,

Bayes' Theorem: $P(H \mid E) = \dfrac{P(E \mid H)P(H)}{P(E)}$,

Bayes' Theorem in odds form:
$$\frac{P(H \mid E)}{P(\neg H \mid E)} = \frac{P(H)}{P(\neg H)} \cdot \frac{P(E \mid H)}{P(E \mid \neg H)}.$$

5.4 Axiomatic Characterization

A most illuminating characterization of the Big Three—although only up to ordinal equivalence—is given by Crupi, Chater, and Tentori (2013) and Crupi and Tentori (2016). Modulo labeling, they consider four axioms for measure types **T**:

Posterior Probability Decisiveness (POP-D): For all H, E_1, E_2, P: $R_T(E_1 \succ H)_P \gtreqless R_T(E_2 \succ H)_P$ iff $P(H \mid E_1) \gtreqless P(H \mid E_2)$.

Likelihood Decisiveness (LIK-D): For all E, H_1, H_2, P: $R_T(E \succ H_1)_P \gtreqless R_T(E \succ H_2)_P$ iff $P(E \mid H_1) \gtreqless P(E \mid H_2)$.

Modularity for Conditionally Independent Evidence (MCIE): For all H, E_1, E_2, P: $R_T(E_2 \succ H)_{P_{E_1}} = R_T(E_2 \succ H)_P$ whenever $(E_1 \perp E_2 \mid \pm H)_P$, where, as before, $P_E =_{df} P(\cdot \mid E)$, that is, P conditioned on E.

KO: $R_{KO}(E \succ H)_P$

$$=_{df} \begin{cases} \dfrac{P(E \mid H) - P(E \mid \neg H)}{P(E \mid H) + P(E \mid \neg H)} = \dfrac{P(EH) - P(E)P(H)}{P(H)P(E\neg H) + P(\neg H)P(EH)}, & \text{if } P(E), P(\neg E), P(H), P(\neg H) \neq 0, \\ \qquad\qquad\qquad\qquad\qquad 0, & \text{else.}^7 \end{cases}$$

Z: $R_Z(E \succ H)_P =_{df} \begin{cases} \dfrac{P(H \mid E) - P(H)}{1 - P(H)}, & \text{if } P(E) \neq 0, P(H) \neq 1, \text{ and } P(H \mid E) > P(H), \\ \dfrac{P(H \mid E) - P(H)}{P(H)}, & \text{if } P(E), P(H) \neq 0 \text{ and } P(H \mid E) \leq P(H), \\ \qquad\qquad 0, & \text{else.} \end{cases}$

Modularity for Disjoint Hypothesis Disjuncts (MDHD): For all E, H_1, H_2, P: if $P(H_1H_2) = 0$, then $R_\text{T}(E \succ H_1 \vee H_2)_P \gtreqless R_\text{T}(E \succ H_1)_P$ iff $P(H_2 \mid E) \gtreqless P(H_2)$.[11]

Then they derive a representation theorem:

Theorem 7 (Crupi et al., 2013; Crupi & Tentori, 2016). Let R_T be a real-valued function satisfying Transparency (definition 8a) with the possible exception of its smoothness requirement. Then R_T is identical to, or is an isotonic transformation of,

(i) R_D iff it validates POP-D and MDHD,

(ii) R_Q iff it validates POP-D and LIK-D,

(iii) R_L iff it validates POP-D and MCIE.

The challenging "if"-parts of theorem 7 yield ordinal equivalence classes $[\mathbf{T}] =_\text{df} \{\mathbf{T}' \mid \mathbf{T}' \sim_\uparrow \mathbf{T}\}$ of measures $\mathbf{T}$.[12] To determine specific measures **D**, **Q**, **L**, **QL**, **LL**, **KO**, and **Z**, further conditions are needed, among them "smoothness."

We now discuss the four axioms. Note first and importantly that POP-D and MCIE legislate for measurements comparing relevance of two "source/evidence" propositions E_1, E_2 to a given "target/hypothesis" H, while LIK-D and MDHD apply to comparisons of hypotheses H_1, H_2 for given E. Thus, POP-D and LIK-D do not compete and are indeed each validated by **Q**.

POP-D seems mandatory to most commentators. Its denial feels counterintuitive, and POP-D has also been argued to be necessary for enumerative induction based on the proportion of favorable and unfavorable outcomes of a series of structurally identical experiments (Steel, 2003). Only measures validating POP-D should thus be attractive as measures of first resort.[13]

LIK-D is often dubbed the "Law of Likelihood."[14] LIK-D feels less compelling than POP-D. However, intuitions on single axioms have limited normative value. Users who like POP-D and find direction-assignments unwarranted will adopt LIK-D, because POP-D and amount symmetry entail LIK-D. Users who value POP-D and prefer nonsymmetry at least for extreme relevance must allow violations of LIK-D.

MCIE, found in Good (1960) and Heckerman (1988), has its much desired special case in the respective conditional multiplicativity of **L** and additivity of **LL**. For, we have

Lemma 3: If $(E_1 \perp E_2 \mid \pm H)$ then $R_\text{L}(E_1E_2 \succ H)_P = R_\text{L}(E_1 \succ H)_P \cdot R_\text{L}(E_2 \succ H)_P$ and thus $R_\text{LL}(E_1E_2 \succ H)_P = R_\text{LL}(E_1 \succ H)_P + R_\text{LL}(E_2 \succ H)_P$.

These practical benefits of MCIE are unavailable for other isotonic transforms of **L** (e.g., **KO**).

MDHD looks sensible but faces an embarrassment (cf. Crupi & Tentori, 2016). Suppose E is extremely negative to H. Now weaken H to $H' = H \vee G$, by a disjunct G that is P-irrelevant to E and P-disjoint from H, that is, $P(GH) = 0$. Then $R_\text{D}(E \succ H)_P = R_\text{D}(E \succ H')_P$, even though E speaks conclusively against H but only inconclusively against H'. Since the **L**- and **Q**-measures do not face this problem, the property illustrates ordinal inequivalence of **D** to each of them.

5.5 Measure Dependence and Taskwise Independence

To what extent are purely ordinal assessments of comparative relevance measure dependent? The Big Three, **D**, **Q**, **L**, are known to be pairwise ordinally inequivalent. To localize the inequivalences, I offer the following results.

Theorem 8. For the task of ranking multiple "target" propositions H_j ($j = 1, \ldots, m \geq 2$) by strength $R(E \succ H_j)_P$ of relevance of a given "source" proposition E to each of them, measures **D**, **Q**, and **L** are pairwise ordinally inequivalent.

Theorem 9. For the task of ranking multiple "source" propositions E_i ($i = 1, \ldots, n \geq 2$) by strength $R(E_i \succ H)_P$ of relevance to a given "target" proposition H, all relevance measures validating POP-D (e.g., **D**, **Q**, **L**, **QL**, **LL**, **Z**) are pairwise ordinally equivalent.

Theorem 9 holds because POP-D prescribes an ordinal assessment schedule for any fixed target H and variable pair of source propositions $E_i, E_{i'}$. Indeed, theorem 9 and the nondirectionality of amount-symmetric **Q** (section 5.3) entail

Lemma 4. If POP-D is a soundness requirement on measures $R_\text{T}(E \succ H)_P$, then sound ordinal ranking of "source" propositions E_i ($i = 1, \ldots, n \geq 2$) by strength of relevance to a given "target" proposition H does not depend on the direction of measurement.

This lends credibility to the probabilistic explication of many judgments of comparative relevance that might otherwise seem critically measure dependent. Applications include solutions to Hempel's "Raven Paradox" (Hosiasson-Lindenbaum, 1940; more accessibly in Earman, 1992; Hájek & Joyce, 2008; Crupi, 2015). In abstract outline, this is the following predicament:

(1) H ("All ravens are black") is supported by the conjunctive proposition $A = A_1A_2$ (where $A_1 =$ "a is a raven" and $A_2 =$ "a is black"). Here "a" stands for a randomly encountered physical entity.

(2) A logical equivalent, G ("All nonblack things are nonravens"), of H is supported by conjunctive proposition $B = B_1B_2$ (where B_1 = "a is not black" and B_2 = "a is not a raven").

(3) G is entailed and thereby supported by C, "a is a white shoe."[15]

The problem arises because a plausible deductivist criterion of evidential support prescribes treating logical equivalents, here H and G, alike for support by evidence. Hence B (and, with mild further assumptions, C), in supporting G, should support H. But this is counterintuitive. Probabilistic proposals observe that B_1, "a is nonblack," is vastly more probable than A_1, "a is a raven," since there are vastly more nonblack things than ravens. Concomitantly, B will be vastly more probable than A. This pertinent observation and the resulting unsurprisingness of B make for its low relevance. To resolve the paradox formally, the known resolution proposals must also make some irrelevance assumptions. Hosiasson-Lindenbaum (1940, p. 138, ll. 5–8) assumes H to be probabilistically independent from each of A_1 and B_1. All known proposals use some familiar relevance measure or other to conclude that A and B are each positive to H, although B is positive only to a negligible degree. This explains the pretheoretical intuition that B is irrelevant to H, which extends to the irrelevance of C in the augmented story. Theorem 9 assures us that the suggestive ordinal result "B is less positive to G or H than A" holds under any possible measure satisfying POP-D.[16]

5.6 Features-in-Use of Various Relevance Measures

Measures contrast by various features that may be found desirable or undesirable, absolutely or in given uses, and normatively or descriptively. For instance, **D** must equally value prior-to-posterior shifts $0.2 \mapsto 0.4$, $0.5 \mapsto 0.7$, and $0.8 \mapsto 1$. By contrast, **L**- or **LL**-values for a target probability that increases from 0.999 to certainty, 1, must be (infinitely) higher than for an increase from 0.001 to 0.999. In this subsection, some properties affecting purely ordinal comparisons are discussed:

Extreme valuation of extreme relevance: Measures **L**, **Z**, and their ordinal equivalents attain the maximal (respectively, minimal) values of their ranges if and only if relevance is extremely positive (respectively, extremely negative). This property pleases anyone to whom conclusive argument is the strongest argument.

Amount symmetry: Measures **Q** and **QL** are amount-symmetric (section 5.1). So is the measure $R_C(E \succ H)_P =_{df} P(EH) - P(E) \cdot P(H)$ (Carnap, 1950), which, however, violates POP-D. Among measures that are not

amount-symmetric, **D** always makes the less probable of mutually relevant A and B more relevant by absolute amount to the more probable than vice versa. Formally: If not $(A \perp B)_P$, then $|R_D(A \succ B)_P| \gtreqless |R_D(B \succ A)_P|$ iff $P(A) \gtreqless P(B)$. **L** works likewise, except for extreme negativity.[17]

Intuitions on amount symmetry are strong only in the case of asymmetric extreme positivity. Example: Let A = "15 divides k," and B = "3 divides k," where k is only known to be a natural number ≥ 15. Evidently enough, A entails B, B does not entail A, and indeed $P(A) < P(B)$. We now have, for instance, $R_D(A \succ B)_P > R_D(B \succ A)_P$, in line with A's unilaterally extreme positivity to B. We also have $R_Q(A \succ B)_P = R_Q(B \succ A)_P$, which many will balk at, because it ignores the entailment asymmetry.

The next three features concern *conditional operation homomorphisms*, which are sometimes unconditional. Homomorphisms are structure-preserving mappings and are essential to predictability of many ordinal and all quantitative relevance properties of logical complexes. Here they map Boolean operations on propositions to equally simple, or at any rate to relatively simple, arithmetical operations on relevance values: inversion, addition, and weighted mean. Any relevance measure will afford some homomorphisms conditional on suitably narrow constraints, but most such constraints are not practically useful. Nor will all homomorphisms, conditional or otherwise, be desirable in practice. Let us look at three of them.

Relevance-functionality of negation: This refers to the inversion of relevance values r under negation: to $-r$ when $t_T = 0$, to $1/r$ when $t_T = 1$. It engages two distinct features: negation in target position and negation in source position. For source proposition E and target proposition H, target inversion means evidence E *always* speaks as strongly *for* hypothesis H as it speaks *against* $\neg H$. Professed opinion usually commends unconditional inversion in target position and recommends its denial for source position (cf. Eells & Fitelson, 2002). Measures **D**, **L**, and **Z** share this pair of positive and negative properties.

Conditional compositionality for conjunction: Lemma 3 (section 5.4) about **L** and **LL** exhibits the gold standard for useful conjunctive compositionality, in source position. The qualitative theorem 6 (section 3) presages its usefulness. No other familiar measures quite match this. For truly stark contrast, consider Carnap's **C**, which affords compositionality $R_C(AB \succ H)_P = R_C(A \succ H)_P + R_C(B \succ H)_P$ when $P(A \vee B) = 1$. This is neither practically useful nor intuitive.

Conditional compositionality for disjunction: Here **C** is very counterintuitive: $R_C(A \vee B \succ H)_P = R_C(A \succ H)_P + R_C(B \succ H)_P$ when $P(AB) = 0$. In the case $P(AB) = 0$, intuitive, useful compositionality situates the relevance that source proposition $A \vee B$ has for some target C *between* the disjuncts' relevances. The Big Three and their isotonic transforms meet this requirement. The disjunction's relevance is always some *convex combination* (i.e., a "mixture" or weighted mean) of the disjunct relevances.

Theorem 10.

(a) If $P(AB) = 0$, then $R_T(A \vee B \succ C)_P = \alpha R_T(A \succ C)_P + (1 - \alpha) R_T(B \succ C)_P$, where $0 \leq \alpha \leq 1$ for **T = D, Q** and, if neither A nor B is extremely relevant to C, also for **T = L**.

(b) The mixture weight α is $\dfrac{P(A)}{P(A) + P(B)}$ for **T = D, Q**, and it is $\dfrac{P(A \mid \neg C)}{P(A \mid \neg C) + P(B \mid \neg C)}$ for **T = L**.

The Big Three share this mixture property with the "expected utility" concept; see Savage (1954) and chapter 8.2 by Peterson (this handbook). This commonality makes apt the characterization of each measure as a "quasi-utility," a term that Good (1983), with a different motivation, had applied to **LL**. Quasi-utilities quantify the usefulness of "source" propositions toward promoting the "target" proposition in a credal context. Under this instrumental perspective, the relation between evidence and hypothesis is a special case of a more general rhetorical relationship. Establishment of a source proposition that is positively relevant to a target proposition is a resource toward the rhetorical end of establishing the target proposition.

6. Applications in Psychology

Besides uses in philosophy of science,[18] there are psychological applications of relevance reasoning. The best known among them involve quantitative or merely ordinal judgment tasks and have a common feature. Experimental subjects appear, in some contexts, to answer requests for posterior probability or odds judgments with replies that are consistent with their being covert relevance judgments. Prominent instances of this are the phenomenon of "prior probability (or base rate) neglect" and the "conjunction fallacy."[19]

The "neglect" phenomena, when they arise, may be seen as instances of Bayes' Theorem in odds form (section 5.3). In this instance, prior odds for H are tacitly set to 1 (i.e., assume equal probability of H and $\neg H$). These odds fail to reflect information supplied on the experimenter's instruction sheet that should warrant a skewed prior, say, with $P(H) \gg P(\neg H)$, before observation E impinges. Arithmetically, such information neglect will equate posterior odds with the likelihood ratio value for relevance.

The conjunction fallacy arose most famously in the following example from the late 1970s. A fictitious person, Linda, was introduced on the experimental instruction sheet with something like the following description, E: Linda, aged 31, is single, outspoken, and bright; has majored in philosophy; and has been demonstrating for equal opportunity causes. Participants were then requested to rank the probabilities of Linda being a bank teller (B), an active feminist (F), and an active feminist bank teller (BF)—presumably in the light of information E. A surprisingly large proportion of respondents ranked BF above B, which would contravene the probability law $P(BF \mid E) \leq P(B \mid E)$. Now, E was designed to be "representative" of F and "unrepresentative" of B. Many probabilists would explicate representativeness of a property Φ by the relevance of statements such as "Linda is Φ" to a reference description E, noting that F is positive to E and B negative. If one assumes that people responded to the experimenter's request with covert relevance judgments, the seemingly anomalous ranking is entirely feasible and indeed natural. See Tentori, Crupi, and Russo (2013) and Cevolani and Crupi (2015) for leads on relevance solutions of the conjunction fallacy.[20]

Finally, maximization of relevance (understood as increasing cognitive effects and reducing processing costs) has been conjectured in relevance theory to be a general principle structuring both cognition and communication. Relevance theory is popular in linguistics (Wilson & Sperber, 2004) but differs from the accounts of evidential relevance presented here. Empirically, this theory has been applied to the Wason selection task in Sperber, Cara, and Girotto (1995). Moreover, there is evidence that the measure delta-p = $P(H \mid E) - P(H \mid \neg E)$ (which is not discussed above) serves as an explication of participants' assessments of the relevance of E for H (see Skovgaard-Olsen, Singmann, & Klauer, 2016b). In Skovgaard-Olsen, Singmann, and Klauer (2016a), the relationship between this notion of relevance and indicative conditionals ("If A, then C") was investigated.

7. Conclusion

Our discussion of qualitative relevance relations has emphasized the importance of conditional independence assumptions. In practice, this means that one should seek out such independencies in one's domain of application.

Our discussion of relevance measures has shown the continued importance of conditional independence for several ordinal and compositionality properties of relevance. Another useful condition for compositionality, specifically of disjunctive evidence, was established for the major measures: the condition of mutual extreme negativity of disjuncts. More generally, our result on task-relative ordinal equivalence of the major measures provides an assurance that many ordinal relevance judgments are not unreasonably measure dependent.

If one must quantify relevance, the log-likelihood ratio **LL** is the default choice. It is the weakest sufficient statistics and has useful conditional additivity for conjunctive evidence. The relevance quotient **Q** and its logarithm **QL** are useful, for one, when relevance directionality is intrinsically unspecified. For special purposes, measures have been commended that may violate the desideratum POP-D discussed in section 5.4. Some philosophers indeed recommend measure pluralism to express diverse aspects of the relevance relation quantitatively. See Hájek and Joyce (2008) on both points.

Notes

Editorial note: **Arthur Merin died on May 24, 2019, shortly after finishing this chapter. Final minor revisions were undertaken by the editors.**

1. See chapter 4.1 by Hájek and Staffel (this handbook). To a good extent, the considerations of this chapter could be carried out in terms of other representations of degrees of credence; see chapter 4.7 by Dubois and Prade and chapter 5.3 by Kern-Isberner, Skovgaard-Olsen, and Spohn (both in this handbook).

2. The juxtaposition AB represents the conjunction of A and B.

3. $(A \perp C \mid \pm B)_P$ can be weakened to $(A \text{ not } neg \ C \mid \pm B)_P$ (W. E. Johnson; cf. Keynes, 1921). Theorem 5 follows from a result of Reichenbach (1956), which can be stated thus: If positive, neutral, and negative relevance polarities are represented by 1, 0, and -1, respectively, then, given $(A \perp C \mid \pm B)_P$, the relevance polarity of A to C is the arithmetical product of the relevance polarities of A to B and of B to C.

4. The specific benefit of conditional independence is relevance-additivity under lemma 3, section 5.4.

5. The language used in glosses is often counterfactual, particularly when referring to the "likelihood," $P(D \mid H)$, of H in view of D as the probability that D "would have had," given H. On likelihoods in statistics see, for example, Edwards (1972).

6. The label "Transparency" stands for "no unavoidable surprises." Hidden propositions and jumps would be surprises.

7. Thus, $R_{KO}(E \succ H)_P = \tanh[\frac{1}{2} \log_e R_{LL}(E \succ H)_P]$, that is, **KO** is just a transform of **LL**.

8. This is Carnap's 1950 name. Some authors call it "probability ratio measure" and use the label **R**, which, however, has distinct earlier referents.

9. A function $f: \mathbb{R} \to \mathbb{R}$ is *isotonic* iff $a < b$ implies $f(a) < f(b)$, for any $a, b \in \mathbb{R}$.

10. Reminder: $O(H) =_{df} P(H)/P(\neg H)$, $P(H) = O(H)/[O(H) + 1]$, and $O(H \mid E) =_{df} P(H \mid E)/P(\neg H \mid E)$.

11. Crupi and Tentori label POP-D "Final Probability," LIK-D "Law of Likelihood," MCIE "Modularity for Conditionally Independent Data," and MDHD "Disjunction of Alternative Hypotheses."

12. "Only if" parts for POP-D and LIK-D are in Steel (2003).

13. For other measures validating POP-D, among them **Z**, see Crupi and Tentori (2016). A uniqueness result for **Z** is presented in Crupi and Tentori (2013).

14. Unlike in statistics, the likelihood concept is extended by philosophers also to mutually compatible and nonexhaustive $\{H_i\}$.

15. (3) is an inessential embellishment that makes things vivid.

16. Applications of theorem 9 also include the generalization, to all measures validating POP-D, of purely ordinal results such as the following *proposition* (Merin, 1999): Let $A \stackrel{\vee}{\cdot} B =_{df} (A \vee B) \wedge \neg(A \wedge B)$ represent exclusive disjunction, and assume $(A \perp B \mid \pm H)$ and $1 = \iota_L < R_L(A \succ H)_P$, $R_L(B \succ H)_P < \infty$. Then $\iota_L < R_L(A \vee B \succ H)_P < \max[R_L(A \succ H)_P, \ R_L(B \succ H)_P] < R_L(AB \succ H)_P$, and $R_L(A \stackrel{\vee}{\cdot} B \succ H)_P < R_L(A \vee B \succ H)_P$, while none of $R_L(A \stackrel{\vee}{\cdot} B \succ H)_P \gtrless \iota_L$ are ruled out. The result shows, for one, that the defeasible "not both" implicature of assertions "*A* or *B*," which is commonly held to be inferred from the nonassertion of "*A* and *B*," can diminish the evidential relevance of inclusive "*A* or *B*" in issue-based discourse.

17. This is closely connected to what Roberto Festa calls "Matthew properties," referring to the biblical saying, "For onto every one that hath shall be given" (see Festa & Cevolani, 2016).

18. For more detail on these, see the surveys, some of them extensive, by Huber (2007), by Hájek and Joyce (2008), Crupi (2015), and Crupi and Tentori (2016).

19. See Kahneman (2011, chapters 15–16) on the two phenomena; see also chapter 1.2 by Evans and chapter 8.5 by Hertwig and Kozyreva (both in this handbook).

20. Another application of evidential relevance (see, e.g., Merin, 1999), which involves no cross-purposes in the interpretation of instructions, is to the theory of meaning for natural-language function words like "and," "or," "not," "but," "even," "also," "some," and "many."

References

Carnap, R. (1950). *Logical foundations of probability*. Chicago, IL: University of Chicago Press. 2nd ed. 1962.

Cevolani, G., & Crupi, V. (2015). Subtleties of naïve reasoning: Probability, confirmation, and verisimilitude in the Linda paradox. In M. Bianca & P. Piccari (Eds.), *Epistemology of ordinary knowledge* (pp. 211–230). Cambridge, England: Cambridge Scholars.

Crupi, V. (2015). Confirmation. In E. N. Zalta (Ed.), *The Stanford encyclopedia of philosophy*. Retrieved from https://plato.stanford.edu/archives/win2016/entries/confirmation/

Crupi, V., Chater, N., & Tentori, K. (2013). New axioms for probability and likelihood ratio measures. *British Journal for the Philosophy of Science, 64*, 189–204.

Crupi, V., & Tentori, K. (2013). Confirmation as partial entailment: A representation theorem in inductive logic. *Journal of Applied Logic, 11*, 364–372.

Crupi, V., & Tentori, K. (2016). Confirmation theory. In A. Hájek & C. Hitchcock (Eds.), *The Oxford handbook of probability and philosophy* (pp. 650–665). Oxford, England: Oxford University Press.

Earman, J. (1992). *Bayes or bust? A critical examination of Bayesian confirmation theory*. Cambridge, MA: MIT Press.

Edwards, A. W. F. (1972). *Likelihood*. Cambridge, England: Cambridge University Press.

Eells, E., & Fitelson, B. (2002). Symmetries and asymmetries in evidential support. *Philosophical Studies, 107*, 129–142.

Festa, R., & Cevolani, G. (2016). Unfolding the grammar of Bayesian confirmation: Likelihood and antilikelihood principles. *Philosophy of Science, 84*, 56–81.

Fitelson, B., & Hartmann, S. (2015). A new, Garber-style solution to the problem of old evidence. *Philosophy of Science, 82*, 712–717.

Glymour, C. (1980). *Theory and evidence*. Princeton, NJ: Princeton University Press.

Good, I. J. (1950). *Probability and the weighing of evidence*. London, England: Griffin.

Good, I. J. (1960). Weight of evidence, corroboration, explanatory power, information, and the utility of experiments. *Proceedings of the Royal Statistical Society B, 22*, 319–331.

Good, I. J. (1983). *Good thinking*. Minneapolis: University of Minnesota Press.

Hájek, A., & Joyce, J. M. (2008). Confirmation. In S. Psillos & M. Curd (Eds.), *The Routledge companion to philosophy of science* (pp. 115–128). Abingdon, England: Routledge.

Heckerman, D. E. (1988). An axiomatic framework for belief updates. In J. F. Lemmer & L. N. Kanal (Eds.), *Uncertainty in artificial intelligence* (Vol. 2, pp. 11–22). Amsterdam, Netherlands: North Holland.

Hosiasson-Lindenbaum, J. (1940). On confirmation. *Journal of Symbolic Logic, 5*, 133–148.

Huber, F. (2007). Confirmation and induction. *Internet encyclopedia of philosophy*. Retrieved from https://iep.utm.edu/conf-ind/

Jeffrey, R. C. (2004). *Subjective probability: The real thing*. Cambridge, England: Cambridge University Press.

Kahneman, D. (2011). *Thinking, fast and slow*. New York, NY: Farrar, Straus and Giroux.

Kemeny, J. G., & Oppenheim, P. (1952). Degree of factual support. *Philosophy of Science, 19*, 307–324.

Keynes, J. M. (1921). *A treatise on probability*. London, England: Macmillan.

Merin, A. (1999). Information, relevance and social decision-making: Some principles and results of decision-theoretic semantics. In L. S. Moss, J. Ginzburg, & M. De Rijke (Eds.), *Logic, language, and computation* (Vol. 2, pp. 179–221). Stanford, CA: CSLI Publications.

Pearl, J. (1988). *Probabilistic reasoning in intelligent systems*. San Mateo, CA: Morgan Kaufman.

Reichenbach, H. (1956). *The direction of time*. Berkeley: University of California Press.

Savage, L. J. (1954). *The foundations of statistics*. New York, NY: Wiley. Augmented ed. Dover Publications 1972.

Skovgaard-Olsen, N., Singmann, H., & Klauer, K. C. (2016a). The relevance effect and conditionals. *Cognition, 150*, 26–36.

Skovgaard-Olsen, N., Singmann, H., & Klauer, K. C. (2016b). Relevance and reason relations. *Cognitive Science, 41*(Suppl 5), 1202–1215.

Sperber, D., Cara, F., & Girotto, V. (1995). Relevance theory explains the selection task. *Cognition, 57*, 31–95.

Steel, D. (2003). A Bayesian way to make stopping rules matter. *Erkenntnis, 58*, 213–227.

Tentori, K., Crupi, V., & Russo, S. (2013). On the determinants of the conjunction fallacy: Probability vs. inductive confirmation. *Journal of Experimental Psychology: General, 142*, 235–255.

Wilson, D., & Sperber, D. (2004). Relevance theory. In L. R. Horn & G. L. Ward (Eds.), *The handbook of pragmatics* (pp. 607–632). Oxford, England: Blackwell.

4.4 Probability Logic

Niki Pfeifer

Summary

This chapter presents probability logic as a rationality framework for human reasoning under uncertainty. Selected formal-normative aspects of probability logic are discussed in the light of experimental evidence. Specifically, probability logic is characterized as a generalization of bivalent truth-functional propositional logic ("classical logic," for short), as being connexive, and as being nonmonotonic. The chapter discusses selected argument forms and associated uncertainty propagation rules. Throughout the chapter, the descriptive validity of probability logic is compared to classical logic, which was used as the gold standard of reference for assessing the rationality of human deductive reasoning in the 20th century.

1. Probability Logic Is a Generalization of Classical Logic

Probability logic as a rationality framework combines probabilistic reasoning with logical rule-based reasoning and studies formal properties of uncertain argument forms.

Among various approaches to probability logic (for overviews, see, e.g., Adams, 1975, 1998; Coletti & Scozzafava, 2002; Demey, Kooi, & Sack, 2019; Haenni, Romeijn, Wheeler, & Williamson, 2011; Hailperin, 1996), this chapter reviews selected formal-normative aspects of probability logic in the light of experimental evidence. The focus is on probability logic as a generalization of the classical propositional calculus ("classical logic," for short; for probabilistic generalizations of quantified statements, see, e.g., Hailperin, 2011; Pfeifer & Sanfilippo, 2017, 2019). The generalization is obtained (i) by the use of probability functions and (ii) by the introduction of the conditional event as a logical object, which is not expressible within classical logic. This generalization is currently most frequently investigated from a psychological point of view (see, e.g., Elqayam, Bonnefon, & Over,

2016; Oaksford & Chater, 2010; Pfeifer & Kleiter, 2009), and it is thus most suitable for discussing both empirical and normative aspects of probability logic as a rationality framework. The empirical focus is on investigating general patterns of human reasoning under uncertainty and not on data modeling.

Classical logic is bivalent as it deals with *true* and *false* as the two truth-values, which are assigned to propositional variables (denoted by uppercase letters *in italics*; for classical logic, see also chapter 3.1. by Steinberger, this handbook). Truth tables can be used to define logical connectives like *conjunction* ("A and B," denoted by "$A \& B$"), *disjunction* ("A or B," denoted by "$A \vee B$"), *negation* ("not-A," denoted by "$\sim A$"), or the *material conditional*, which is the disjunction $\sim A \vee B$. Probability logic generalizes classical logic by using the whole real-valued unit interval from 0 to 1 instead of just two truth-values: truth-value functions are generalized by probability functions. While classical logic is truth-functional, probability logic is only partially truth-functional as, usually, in the conclusion, even from point-valued premise probabilities, a probability interval is obtained. Probability functions can be used to formalize a real or an ideal agent's degrees of belief in propositions formed by logical connectives (e.g., in $A \& B$, $A \vee B$, or $\sim A$). Moreover, conditional probability functions can measure the degree of belief in a *conditional event*, that is, $p(B \mid A)$. The conditional event $B \mid A$ is a three-valued logical entity that is *true* if $A \& B$ is true, *false* if $A \& \sim B$ is true, and *void* (or undetermined) if $\sim A$ is true.[1] Since the conditional event cannot be expressed by a two-valued proposition, it is by definition not propositional and constitutes a further generalization of classical logic. Using the betting interpretation of probability, "true" means that the bet is won, "false" means that the bet is lost, and "void" means that the bet is called off (i.e., you get your money back).

Note that $p(A) = 0$ does not imply that A is logically impossible (i.e., a logical contradiction $\bot$). However, $p(\bot)$ is necessarily equal to zero. As it does not make sense to add $\perp$ to your stock of beliefs (or to bet on a conditional whose antecedent is $\perp$), it is obvious why "$A \mid \perp$" is undefined in the coherence approach.[2] The semantics of the conditional event matches the participants' responses in the truth table task (see, e.g., Wason & Johnson-Laird, 1972; see also chapter 4.6 by Oberauer & Pessach, this handbook): most people judge that (i) $A \& B$ confirms the conditional "If A, then B," (ii) $A \& \sim B$ disconfirms it, and (iii) $\sim A$ is irrelevant for "If A, then B." Under the material-conditional interpretation, one would expect that rational people judge that $\sim A$ confirms the conditional. As this expectation was violated, the response pattern (i)–(iii) was pejoratively called the "defective truth table." Within the rationality framework of probability logic, however, this response is perfectly rational, as it matches the semantics of the conditional event (see, e.g., Kleiter, Fugard, & Pfeifer, 2018; Over & Baratgin, 2017; Pfeifer & Tulkki, 2017b; see also chapter 6.2 by Over & Cruz, in this handbook).

2. Conditional Probability, Zero-Probability Antecedents, and Paradoxes

Traditionally, the conditional probability of B given A is defined by

(1) $p(B \mid A) =_{\text{def.}} p(A \& B)/p(A)$, if $p(A) > 0$.

Condition $p(A) > 0$ here serves to avoid fractions over zero. But what if $p(A) = 0$? Then the conditional probability is undefined or default assumptions about $p(B \mid A)$ are made. For example, some approaches suggest, by default, to equate $p(B \mid A)$ with 1 in this case (e.g., Adams, 1998, p. 57, footnote 5). However, this leads to wrong results, since then also $p(\sim B \mid A) = 1$, which violates the basic probabilistic principle $p(B \mid A) + p(\sim B \mid A) = 1$ (for a discussion, see Gilio, 2002). Moreover, from a practical point of view, if $p(B \mid A)$ is left undefined, counterintuitive consequences may follow. Consider, for example, the following "paradox of the material conditional":

(2) B; therefore, *if A, then B*.

Argument (2) consists of a premise B and a conditional as the conclusion. Under the material-conditional interpretation, (2) is *logically valid* (i.e., there is no model in which the premise set is true while the conclusion is false). However, natural-language instantiations can appear counterintuitive (substitute, for example, "The moon is made of green cheese" for A and "The sun will shine in Vienna" for B). This mismatch between the logical validity of (2) and its counterintuitive instantiations constitutes the paradox. From a logical point of view,

a logically valid argument remains logically valid whatever the instantiations are. If (2) is formalized in probability logic, however, the paradox is blocked when the conditional is represented by a conditional probability; then the corresponding argument is probabilistically noninformative, that is, for all probability values x:

(3) $p(B) = x$; therefore, $0 \le p(B \mid A) \le 1$ is coherent for all probability values x.

An argument is *probabilistically noninformative* if the premise probabilities do not constrain the probability of the conclusion. More technically, probabilistic noninformativeness means that for all coherent probability assessments of the premises, the tightest coherent probability bounds on the conclusion coincide with the unit interval $[0, 1]$.

What does "coherent" mean here? An assessment p on an arbitrary family **K** of conditional events is *coherent* if and only if, for any combination of bets on a finite subset of conditional events in **K**, it cannot happen that the values of the random gain,[3] when at least one bet is not called off, are all positive or all negative ("no Dutch book"; see also chapter 4.1 by Hájek & Staffel, this handbook). In the coherence-based approach, the avoidance of Dutch books is in the case of unconditional events equivalent to the solvability of a specific system of linear equations. This solvability reflects the existence of at least one probability distribution on a suitable partition of the constituents (i.e., the possible cases), which is compatible with the initial probability assessment. In geometrical terms, a probability assessment on n events can be represented by a prevision point **p** (i.e., a vector in $[0, 1]^n$) and the set of constituents by a set Q of binary points (i.e., of vectors in $\{0, 1\}^n$). Then, **p** is coherent if and only if **p** belongs to the convex hull of Q (de Finetti, 1974). In the case of conditional events, coherence amounts to the solvability of a suitable finite sequence of systems of linear equations (see, e.g., Biazzo & Gilio, 2000; Coletti & Scozzafava, 2002).

If the conditional in (2) is represented by the probability of the material conditional, the paradox is inherited. For all probability values x:

(4) $p(B) = x$; therefore, $x \le p(\sim A \vee B) \le 1$ is coherent.

Probability values strictly less than x in the conclusion of (4) are of course not coherent. Note that in general, (3) is probabilistically noninformative for all positive premise probabilities (i.e., $p(B) > 0$). For the extreme case $p(B) = 1$, when conditional probabilities are defined by (1), then $p(B \mid A) = 1$ or is undefined. This is obvious, since if $p(B) = 1$, then $p(A \& B) = p(A)$; therefore, by (1),

$p(B|A) = p(A \& B)/p(A) = p(A)/p(A) = 1$, provided $p(A) > 0$. If $p(A) = 0$, then $p(B|A)$ is undefined. This result is counterintuitive and does not match the experimental data: people interpret (3) as probabilistically noninformative, even in the case of $p(B) = 1$ (Pfeifer & Kleiter, 2011). However, in *coherence-based probability logic* (see, e.g., Coletti & Scozzafava, 2002; Gilio, Pfeifer, & Sanfilippo, 2016), where $p(B|A)$ is primitive and problems with zero-probability antecedents are avoided, $0 \le p(B|A) \le 1$ is coherent even in the extreme case $p(B) = 1$ (for a detailed proof, see Pfeifer, 2014). This example shows that the evaluation of the rationality of a probabilistic inference depends on whether the underlying probability concept allows for dealing with zero-probability antecedents or not. In the framework of coherence, the probabilistic noninformativeness of argument (3) holds for all probability values of the premises; however, for approaches that are based on (1), it holds only for positive probabilities.

3. From the Truth Table Task to the Probabilistic Truth Table Task

In the 20th century, the dominating rationality framework in the psychology of deductive reasoning was logic. Prominent examples are Braine and O'Brien's "mental logic" (1998; see also chapter 3.2 by O'Brien, this handbook), the "mental rule theory" by Rips (1994), and Johnson-Laird's (1983) theory of mental models (see chapter 2.3 by Johnson-Laird, this handbook). The rationality framework of the former two theories is derived from classical logical proof theory (which is syntactic and "rule-based"), whereas the latter one is based on logical model theory (which is semantic and "model-based"). According to these logic-based rationality frameworks, people are rational if they use logically valid rules of inference (like modus ponens) or if they build mental models that are inspired by truth tables. With the advent of the "new paradigm psychology of reasoning," which is characterized by using probabilistic rationality frameworks instead of logic, not only did the evaluation of the rationality of human inference change but also the task paradigms were adapted. The abovementioned truth table task, for example, became a *probabilistic* truth table task (PTTT, for short) to investigate how people interpret conditionals (Evans, Handley, & Over, 2003; Oberauer & Wilhelm, 2003; see also chapter 4.6 by Oberauer & Pessach, this handbook). From a probability-logical point of view, the PTTT presented the following premises to the participants:

(5) $p(A \& B) = x_1$, $p(A \& \sim B) = x_2$, $p(\sim A \& B) = x_3$, and $p(\sim A \& \sim B) = x_4$.

Then the participants were asked to infer their degree of belief in the conditional "If A, then B" based on the probabilistic information given in (5). The main experimental result obtained from this task was that most participants gave as their response values consistent with $x_1/(x_1+x_2)$, which corresponds to the conditional-probability interpretation of the conditional ($p(B|A)$; this is consistent with the "Ramsey test" as described in chapter 6.2 by Over & Cruz, this handbook). A significant minority responded with x_1, which corresponds to the conjunction interpretation of the conditional ($p(A \& B)$). Under the material-conditional interpretation, one would expect $x_1+x_3+x_4$ as the most frequent response type in this task. However, experimental evidence for this hypothesis was negligible. When the task was given several times to the same participants, among those who did not use conditional-probability responses in the first PTTT tasks, a "shift of interpretation" was observed: over the course of the experiment, these participants "shifted" to the conditional-probability interpretation. In the last tasks of the experiment, about 80% of the responses were consistent with the conditional-probability interpretation. This indicates that conditional probability is a key building block for a rationality framework for human reasoning about conditionals under uncertainty (see Fugard, Pfeifer, Mayerhofer, & Kleiter, 2011; Pfeifer, 2013).

In the PTTT, mostly indicative conditionals ("if–then" formulations) with "abstract" materials were used (like "If the figure shows a *square*, then the figure is *red*"). Interestingly, the finding that conditional probability is the best predictor for the data was also replicated for a larger variety of conditionals: causal conditionals ("If *cause*, then *effect*"), counterfactual conditionals ("If A were the case, then B would be the case"; Over, Hadjichristidis, Evans, Handley, & Sloman, 2007; Pfeifer & Stöckle-Schobel, 2015), and abductive conditionals ("If *effect*, then *cause*"; Pfeifer & Tulkki, 2017a).

The next sections explain why probability logic validates basic connexive principles and why it is nonmonotonic under the conditional-probability interpretation of the conditional.

4. Probability Logic Is Connexive

Connexive logics are motivated by the idea that there should be some connection between antecedents and consequents of conditionals in the sense that they should not contradict each other. Connexive logics are alternatives to (classical) logic as they are neither contained in, nor proper extensions of, it (for an overview, see Wansing, 2020). They include, for example, Aristotle's theses:

(AT1) $\sim$ (if $\sim A$, then A),

and

(AT2) $\sim$ (if A, then $\sim A$).

Under the material-conditional interpretation, (AT1) and (AT2) are contingent in logic (i.e., (AT1) is logically equivalent to $\sim A$ and (AT2) is logically equivalent to A). Thus, within logic, it is rational to say that it depends on the truth-value of A whether (AT1) and (AT2) are true. Indeed, (AT1) and (AT2) are not tautologies in logic. Experimental data suggest, however, that people believe that (AT1) and (AT2) must be true (Pfeifer, 2012). Probability logic allows for validating the rationality of (AT1) and (AT2). First, look at the conditionals (in terms of conditional probabilities): by coherence, $p(A\,|\sim A)$ and $p(\sim A\,|\,A)$ must be equal to zero. Second, the corresponding conditionals are negated by negating their consequents, that is, $p(\sim A\,|\sim A)$ and $p(\sim\sim A\,|\,A)$ $(= p(A\,|\,A))$, respectively. Since probability 1 is the only coherent assessment for $p(\sim A\,|\sim A)$ and $p(A\,|\,A)$, (AT1) and (AT2) are validated. This matches the experimental data (Pfeifer, 2012; Pfeifer & Tulkki, 2017b).

Boethius's theses are another instance of connexive principles. Like Aristotle's theses, Boethius's theses can be justified within probability logic (but not within classical logic). Boethius's theses are (the arrow denotes a conditional):

(BT1) $(A \to B) \to \sim(A \to \sim B)$

and

(BT2) $(A \to \sim B) \to \sim(A \to B)$.

Under the narrow-scope negation interpretation of negating conditionals,[4] the antecedent of (BT1) is interpreted in probability logic by $p(B\,|\,A) > x$ (for some threshold $x > .5$) and its consequent by $p(\sim\sim B\,|\,A) > x$. Since B is logically equivalent to $\sim\sim B$, (BT1) holds in probability logic. Analogously, (BT2) is validated in probability logic. Under the material-conditional interpretation, neither (BT1) nor (BT2) hold in general: (BT1) and (BT2) are both logically equivalent to A. Moreover, Abelard's first principle, which is another connexive principle, can be rationally justified in probability logic:

(AFP) $\sim((A \to B)\ \&\ (A \to \sim B))$.

Since, in general, $p(B\,|\,A) + p(\sim B\,|\,A) = 1$, it cannot be the case that both $p(B\,|\,A)$ and $p(\sim B\,|\,A)$ are "high" (i.e., at least strictly greater than .5). Therefore, (AFP) is

validated in probability logic. However, under the material-conditional interpretation, (AFP) is logically equivalent to A. Aristotle's and Boethius's theses and Abelard's first principle are intuitively plausible principles that hold in connexive and in probability logic but not in classical logic. Aristotle's theses have received strong experimental support (Pfeifer, 2012; Pfeifer & Tulkki, 2017b). For the other connexive principles, future empirical research is needed.

5.   Probability Logic Is Nonmonotonic

Nonmonotonic reasoning is about retracting conclusions in the light of new evidence. For example, from "If this animal is a bird (B), then this animal can fly (F)," one would not want to conclude "If this animal is a bird and a penguin ($B\,\&\,P$), then this animal can fly (F)." Classical logic, however, is monotonic: adding premises to a logically valid argument can only lead to an increase but never to a decrease of the conclusion set (for an overview, see, e.g., Antoniou, 1997; Strasser & Antonelli, 2019). Therefore, conclusions cannot be retracted in classical logic. Under the material-conditional interpretation, the abovementioned argument is logically valid. The argument form is called "monotonicity" (or "premise strengthening"):

(MON) $\sim B \vee F$ logically implies $\sim(B\,\&\,P) \vee F$.

In probability logic, however, the corresponding argument form is probabilistically noninformative, and monotonicity is therefore blocked, that is, for all probability values x:

(6) $p(F\,|\,B) = x$; therefore, $0 \le p(F\,|\,(B\,\&\,P)) \le 1$ is coherent.

Basic rationality principles for retracting conclusions in the light of new evidence are concentrated around System P (Kraus, Lehmann, & Magidor, 1990). The principles of System P are broadly considered to be minimal rationality requirements for any system of nonmonotonic reasoning. It is therefore a key system for reasoning in general. Various different semantics were developed for nonmonotonic reasoning systems, some of which are probability-logical (see, e.g., Adams, 1975; Gilio, 2002; Goldszmidt & Pearl, 1996; Hawthorne & Makinson, 2007; Schurz, 1997). Psychologically, Pfeifer and Kleiter (2005, 2006b, 2010) and Pfeifer and Tulkki (2017b) present experimental data supporting the descriptive validity of the coherence-based probability semantics of System P (Gilio, 2002; for experimental studies on the possibilistic semantics of System P, see, e.g., Benferhat, Bonnefon,

& Da Silva Neves, 2005; Da Silva Neves, Bonnefon, & Raufaste, 2002). In the coherence semantics, default conditionals like "If A, then *normally B*"[5] are interpreted as "high" coherent conditional probabilities that may also be imprecise (e.g., interval-valued probabilities). For each rule of System P, Gilio (2002) proved the probability propagation rules, which describe how the probabilities of the premises are propagated to the conclusion. As an example, consider the *and*-rule. For all probability intervals $[x_L, x_U]$ and $[y_L, y_U]$:

(AND) $x_L \leq p(B|A) \leq x_U$ and $y_L \leq p(C|A) \leq y_U$; therefore, $\max\{0, x_L + y_L - 1\} \leq p((B\&C)|A) \leq \min\{x_U, y_U\}$ is coherent.

It can easily be seen that even in cases where the premise probabilities are point-valued (i.e., $x_L = x_U$ and $y_L = y_U$), the probability of the conclusion is usually interval-valued. In the extreme case where the premise probabilities are equal to 1, the only coherent conclusion probability is also equal to 1.

Experimental data suggest that most people infer coherent interval responses in inference tasks corresponding to (AND). The majority of those people who violate coherence violate the lower bound (Pfeifer & Kleiter, 2005). These data speak also against the common misunderstanding that people are unable to perform probabilistic reasoning because of the high frequency of "conjunction fallacies" allegedly committed in Tversky and Kahneman's (1983) well-known Linda task. The conjunction fallacy consists in ranking the probability of a conjunction ($B\&C$) as higher than that of one of its conjuncts (C). In the context of (AND), this would mean that the *upper* probability bound on the conclusion is violated. In the experimental data on (AND), however, the people who violated the coherent interval violated the *lower* probability bound (Pfeifer & Kleiter, 2005). Even if the terms "and" and "probability" are mentioned, that does not mean that actual conjunction probabilities are investigated in the Linda task: the participants might not interpret the task as a task about conjunction probabilities.

Concerning the other rules of System P, a strong agreement between the participants' interval responses and the coherent intervals was observed (Pfeifer & Kleiter, 2005, 2006b, 2010; Pfeifer & Tulkki, 2017b). Moreover, the majority of the participants correctly understood that (6) and that contraposition (e.g., for all probability values x: if $p(B|A) = x$, then $0 \leq p(\sim A|\sim B) \leq 1$ is coherent) are probabilistically noninformative. This is interesting as these argument forms are logically valid under the material-conditional interpretation, but they cannot be validated without further assumptions in a nonmonotonic reasoning system. Adding monotonicity or contraposition to System P, for example, would make System P monotonic (which is undesirable of course). Transitivity is also probabilistically noninformative (i.e., for all probability values x and y: if $p(B|A) = x$ and $p(C|B) = y$, then $0 \leq p(C|A) \leq 1$ is coherent), and its addition to System P would make it monotonic. Experimental data suggest that people interpret the task material of Transitivity (presumably for pragmatic reasons) as Cumulative Transitivity (CUT) of System P (Pfeifer & Kleiter, 2006b, 2010): CUT changes the premises of Transitivity by conjunctively adding the antecedent of the first premise to the antecedent of the second premise. For all probability values x and y:

(CUT) $p(B|A) = x$, $p(C|(A\&B)) = y$; therefore, $xy \leq p(C|A) \leq xy + 1 - x$ is coherent (Gilio, 2002).

Note that the probability propagation rules of (CUT) coincide with those of the probabilistic modus ponens (Pfeifer & Kleiter, 2006a); for all probability values x and y:

(MP) $p(B) = x$, $p(C|B) = y$; therefore, $xy \leq p(C) \leq xy + 1 - x$ is coherent.

This close relationship between (CUT) and (MP) is explained by the fact that unconditional probabilities are defined in probability logic by the following principle:

(7) $p(A) =_{\text{def.}} p(A|\top)$, where "$\top$" denotes a logical tautology.

By replacing A with $\top$ in (CUT) and by (7), we obtain (MP). Modus ponens is one of the most frequently investigated argument forms in the psychology of reasoning. Its nonprobabilistic version is usually endorsed by most participants (Evans, Newstead, & Byrne, 1993). The clear majority of responses in tasks on probabilistic modus ponens support the predictions by probability logic (Pfeifer & Kleiter, 2009; Pfeifer & Tulkki, 2017b).

6.   Concluding Remarks

This chapter characterized probability logic as a generalization of logic and explained the importance of zero-probability antecedents. Probability logic is connexive and nonmonotonic. It is a powerful tool for investigating the rationality of reasoning in a unified and systematic way. Experimental studies support its descriptive validity. Future normative and descriptive research should include nested and compound conditionals. Probabilistic modus ponens and other argument forms, for example, were recently generalized to deal with nested conditionals and compounds of conditionals (Gilio, Pfeifer, & Sanfilippo, 2020; Sanfilippo, Pfeifer, & Gilio, 2017; Sanfilippo, Pfeifer, Over, & Gilio, 2018). Interestingly, the uncertainty propagation rules for modus ponens involving *nested* conditionals coincide with those of nonnested modus ponens (MP). For instance, consider the following nested instance of modus ponens: from "The cup breaks if dropped" ($D \rightarrow B$) and "If the cup breaks if dropped, then the cup is fragile" (($D \rightarrow B) \rightarrow F$) infer "The cup is fragile" (F). Here, the lower bound on the degree of belief in the conclusion F equals the product of the degrees of belief in the premises, and the upper bound on F equals the sum of the lower bound on F plus 1 minus the degree of belief in $D \rightarrow B$, which coincides with the uncertainty propagation rules of the nonnested (MP) (for details, see Sanfilippo et al., 2017). In this approach, which exploits the notions of conditional random quantities and conditional previsions, the law of import–export does not hold (Gilio & Sanfilippo, 2014; Sanfilippo, Gilio, Over, & Pfeifer, 2020), which is key to blocking Lewis's (1976) notorious triviality results. Lewis's triviality results show that sentences like $(D \rightarrow B) \rightarrow F$ must not be simply interpreted by $p(F|(B|D))$. Rather, for properly investigating such structures, a richer formal structure is required. Future work is needed to assess the psychological plausibility of this approach.

Finally, one might wonder why various nonclassical logics—which validate intuitively plausible rationality principles—were broadly neglected in the psychological literature, even if they were available already for decades. The reasons might be due to research traditions. This chapter proposed probability logic as a normatively and descriptively appealing rationality framework for human reasoning, which combines (i) the requirement of plausible *qualitative* logical principles (like nonmonotonicity and connexivity) with (ii) the expressibility of *quantitative* degrees of beliefs for investigating reasoning and argumentation under uncertainty.

Acknowledgments

The author thanks Angelo Gilio, Giuseppe Sanfilippo, Wolfgang Spohn, Hans Rott, and two anonymous referees for useful comments. He was supported by the DFG project PF 740/2-2 (part of the SPP 1516) and is currently supported by the BMBF project 01UL1906X.

Notes

1. Note that the conditional event must not be nested: neither A nor B in $B|A$ may contain occurrences of "|", because of Lewis's (1976) triviality results. For nested conditionals and logical operations among conditionals, more complex structures are needed to avoid triviality, which go beyond the scope of this chapter (see, e.g., the theory of "conditional random quantities"; Gilio, Pfeifer, & Sanfilippo, 2020; Gilio & Sanfilippo, 2014; Sanfilippo, Gilio, Over, & Pfeifer, 2020; Sanfilippo, Pfeifer, & Gilio, 2017; Sanfilippo, Pfeifer, Over, & Gilio, 2018).

2. Popper functions, however, allow for conditioning on contradictions (for a discussion, see Coletti, Scozzafava, & Vantaggi, 2001).

3. In betting terms, you evaluate the probability p of an event E on the understanding that, for each real number s, you are willing to pay ps with the proviso that you will receive either s or 0 according to whether E happens or does not happen, respectively. The random gain (or "net gain") G is the difference between what you receive and what you pay. Thus, $G = s - ps$, when E is true, and $G = -ps$, when E is false. For a probability assessment on a family of events **F**, the associated random gain is the sum of the random gains of each bet on all events in **F**.

4. For a wide-scope negation interpretation of negating conditionals in probability logic, see Gilio, Pfeifer, and Sanfilippo (2016) and Pfeifer and Sanfilippo (2017).

5. Here, the scope of "normally" is the whole conditional and not just the consequent B. In other words, this means "The conditional 'If A, then B' holds by default" or "Normally, B follows from A."

References

Adams, E. W. (1975). *The logic of conditionals*. Dordrecht, Netherlands: Reidel.

Adams, E. W. (1998). *A primer of probability logic*. Stanford, CA: CSLI.

Antoniou, G. (1997). *Nonmonotonic reasoning*. Cambridge, MA: MIT Press.

Benferhat, S., Bonnefon, J. F., & Da Silva Neves, R. (2005). An overview of possibilistic handling of default reasoning, with experimental studies. *Synthese, 146*(1–2), 53–70.

Biazzo, V., & Gilio, A. (2000). A generalization of the fundamental theorem of de Finetti for imprecise conditional probability assessments. *International Journal of Approximate Reasoning, 24*, 251–272.

Braine, M. D. S., & O'Brien, D. P. (Eds.). (1998). *Mental logic*. Mahwah, NJ: Erlbaum.

Coletti, G., & Scozzafava, R. (2002). *Probabilistic logic in a coherent setting*. Dordrecht, Netherlands: Kluwer.

Coletti, G., Scozzafava, R., & Vantaggi, B. (2001). Probabilistic reasoning as a general unifying tool. In S. Benferhat & P. Besnard (Eds.), *Symbolic and quantitative approaches to reasoning with* uncertainty (Lecture Notes in Artificial Intelligence, Vol. 2143, pp. 120–131). Cham, Switzerland: Springer.

Da Silva Neves, R., Bonnefon, J.-F., & Raufaste, E. (2002). An empirical test of patterns for nonmonotonic inference. *Annals of Mathematics and Artificial Intelligence, 34*(1–3), 107–130.

de Finetti, B. (1974). *Theory of probability.* New York, NY: Wiley.

Demey, L., Kooi, B., & Sack, J. (2019). Logic and probability. In E. N. Zalta (Ed.), *The Stanford encyclopedia of philosophy.* Retrieved from https://plato.stanford.edu/archives/sum2019/entries/logic-probability/

Elqayam, S., Bonnefon, J.-F., & Over, D. E. (Eds.). (2016). *New paradigm psychology of reasoning.* London: Routledge.

Evans, J. St. B. T., Handley, S. J., & Over, D. E. (2003). Conditionals and conditional probability. *Journal of Experimental Psychology: Learning, Memory, and Cognition, 29*(2), 321–335.

Evans, J. St. B. T., Newstead, S. E., & Byrne, R. M. J. (1993). *Human reasoning: The psychology of deduction.* Hove, England: Erlbaum.

Fugard, A. J. B., Pfeifer, N., Mayerhofer, B., & Kleiter, G. D. (2011). How people interpret conditionals: Shifts towards the conditional event. *Journal of Experimental Psychology: Learning, Memory, and Cognition, 37*(3), 635–648.

Gilio, A. (2002). Probabilistic reasoning under coherence in System P. *Annals of Mathematics and Artificial Intelligence, 34,* 5–34.

Gilio, A., Pfeifer, N., & Sanfilippo, G. (2016). Transitivity in coherence-based probability logic. *Journal of Applied Logic, 14,* 46–64.

Gilio, A., Pfeifer, N., & Sanfilippo, G. (2020). Probabilistic entailment and iterated conditionals. In S. Elqayam, I. Douven, J. St. B. T. Evans, & N. Cruz (Eds.), *Logic and uncertainty in the human mind: A tribute to David E. Over* (pp. 71–101). London, England: Routledge.

Gilio, A., & Sanfilippo, G. (2014). Conditional random quantities and compounds of conditionals. *Studia Logica, 102*(4), 709–729.

Goldszmidt, M., & Pearl, J. (1996). Qualitative probabilities for default reasoning, belief revision, and causal modeling. *Artificial Intelligence, 84,* 57–112.

Haenni, R., Romeijn, J.-W., Wheeler, G., & Williamson, J. (2011). *Probabilistic logics and probabilistic networks.* Dordrecht, Netherlands: Springer.

Hailperin, T. (1996). *Sentential probability logic.* Bethlehem, PA: Lehigh University Press.

Hailperin, T. (2011). *Logic with a probability semantics.* Bethlehem, PA: Lehigh University Press.

Hawthorne, J., & Makinson, D. (2007). The quantitative/qualitative watershed for rules of uncertain inference. *Studia Logica, 86,* 247–297.

Johnson-Laird, P. N. (1983). *Mental models: Towards a cognitive science of language, inference, and consciousness.* Cambridge, England: Cambridge University Press.

Kleiter, G. D., Fugard, A. J. B., & Pfeifer, N. (2018). A process model of the understanding of uncertain conditionals. *Thinking & Reasoning, 24*(3), 386–422.

Kraus, S., Lehmann, D., & Magidor, M. (1990). Nonmonotonic reasoning, preferential models and cumulative logics. *Artificial Intelligence, 44,* 167–207.

Lewis, D. (1976). Probabilities of conditionals and conditional probabilities. *Philosophical Review, 85,* 297–315.

Oaksford, M., & Chater, N. (Eds.). (2010). *Cognition and conditionals: Probability and logic in human thinking.* Oxford, England: Oxford University Press.

Oberauer, K., & Wilhelm, O. (2003). The meaning(s) of conditionals: Conditional probabilities, mental models and personal utilities. *Journal of Experimental Psychology: Learning, Memory, and Cognition, 29*(4), 680–693.

Over, D. E., & Baratgin, J. (2017). The "defective" truth table: Its past, present, and future. In N. Galbraith, E. Lucas, & D. E. Over (Eds.), *The thinking mind: A Festschrift for Ken Manktelow* (pp. 15–28). New York, NY: Routledge.

Over, D. E., Hadjichristidis, C., Evans, J. St. B. T., Handley, S. J., & Sloman, S. (2007). The probability of causal conditionals. *Cognitive Psychology, 54,* 62–97.

Pfeifer, N. (2012). Experiments on Aristotle's Thesis: Towards an experimental philosophy of conditionals. *The Monist, 95*(2), 223–240.

Pfeifer, N. (2013). The new psychology of reasoning: A mental probability logical perspective. *Thinking & Reasoning, 19*(3–4), 329–345.

Pfeifer, N. (2014). Reasoning about uncertain conditionals. *Studia Logica, 102*(4), 849–866.

Pfeifer, N., & Kleiter, G. D. (2005). Coherence and nonmonotonicity in human reasoning. *Synthese, 146*(1–2), 93–109.

Pfeifer, N., & Kleiter, G. D. (2006a). Inference in conditional probability logic. *Kybernetika, 42,* 391–404.

Pfeifer, N., & Kleiter, G. D. (2006b). Is human reasoning about nonmonotonic conditionals probabilistically coherent? In *Proceedings of the 7th Workshop on Uncertainty Processing, Mikulov* (pp. 138–150), http://www.utia.cas.cz/MTR/wupes06_papers

Pfeifer, N., & Kleiter, G. D. (2009). Framing human inference by coherence based probability logic. *Journal of Applied Logic, 7*(2), 206–217.

Pfeifer, N., & Kleiter, G. D. (2010). The conditional in mental probability logic. In M. Oaksford & N. Chater (Eds.), *Cognition and conditionals: Probability and logic in human thinking* (pp. 153–173). Oxford, England: Oxford University Press.

Pfeifer, N., & Kleiter, G. D. (2011). Uncertain deductive reasoning. In K. Manktelow, D. E. Over, & S. Elqayam (Eds.), *The science of reason: A Festschrift for Jonathan St. B. T. Evans* (pp. 145–166). Hove, England: Psychology Press.

Pfeifer, N., & Sanfilippo, G. (2017). Probabilistic squares and hexagons of opposition under coherence. *International Journal of Approximate Reasoning, 88*, 282–294.

Pfeifer, N., & Sanfilippo, G. (2019). Probability propagation in selected Aristotelian syllogisms. In G. Kern-Isberner & Z. Ognjanović (Eds.), *Symbolic and quantitative approaches to reasoning with uncertainty* (Lecture Notes in Computer Science, Vol. 11726, pp. 419–431). Cham, Switzerland: Springer.

Pfeifer, N., & Stöckle-Schobel, R. (2015). Uncertain conditionals and counterfactuals in (non-)causal settings. In G. Airenti, B. G. Bara, & G. Sandini (Eds.), *Proceedings of the EuroAsianPacific Joint Conference on Cognitive Science* (pp. 651–656). Aachen, Germany: CEUR Workshop Proceedings.

Pfeifer, N., & Tulkki, L. (2017a). Abductive, causal, and counterfactual conditionals under incomplete probabilistic knowledge. In G. Gunzelmann, A. Howes, T. Tenbrink, & E. Tavelaar (Eds.), *Proceedings of the 39th Annual Conference of the Cognitive Science Society* (pp. 2888–2893). Austin, TX: Cognitive Science Society.

Pfeifer, N., & Tulkki, L. (2017b). Conditionals, counterfactuals, and rational reasoning: An experimental study on basic principles. *Minds and Machines, 27*(1), 119–165.

Rips, L. J. (1994). *The psychology of proof: Deductive reasoning in human thinking.* Cambridge, MA: MIT Press.

Sanfilippo, G., Gilio, A., Over, D. E., & Pfeifer, N. (2020). Probabilities of conditionals and previsions of iterated conditionals. *International Journal of Approximate Reasoning, 121*, 150–173.

Sanfilippo, G., Pfeifer, N., & Gilio, A. (2017). Generalized probabilistic modus ponens. In A. Antonucci, L. Cholvy, & O. Papini (Eds.), *Symbolic and quantitative approaches to reasoning with uncertainty* (Lecture Notes in Computer Science, Vol. 10369, pp. 480–490). Cham, Switzerland: Springer.

Sanfilippo, G., Pfeifer, N., Over, D. E., & Gilio, A. (2018). Probabilistic inferences from conjoined to iterated conditionals. *International Journal of Approximate Reasoning, 93*, 103–118.

Schurz, G. (1997). Probabilistic default logic based on irrelevance and relevance assumptions. In D. Gabbay, R. Kruse, A. Nonnengart, & H. J. Ohlbach (Eds.), *Qualitative and quantitative practical reasoning* (Lecture Notes in Artificial Intelligence, Vol. 1244, pp. 536–553). Berlin, Germany: Springer.

Strasser, C., & Antonelli, G. A. (2019). Non-monotonic logic. In E. N. Zalta (Ed.), *The Stanford encyclopedia of philosophy.* Retrieved from https://plato.stanford.edu/archives/sum2019/entries/logic-nonmonotonic/

Tversky, A., & Kahneman, D. (1983). Extensional versus intuitive reasoning: The conjunction fallacy in probability judgment. *Psychological Review, 90*, 293–315.

Wansing, H. (2020). Connexive logic. In E. N. Zalta (Ed.), *The Stanford encyclopedia of philosophy.* Retrieved from https://plato.stanford.edu/archives/spr2020/entries/logic-connexive/

Wason, P. C., & Johnson-Laird, P. N. (1972). *Psychology of reasoning: Structure and content.* Cambridge, MA: Harvard University Press.

4.5 Bayesian Rationality in the Psychology of Reasoning

Nick Chater and Mike Oaksford

Summary

Bayesian models have become widespread throughout the cognitive and brain sciences, as providing natural ways to understand how the brain deals with uncertainty in perception, belief updating, decision making, and motor control. In this chapter, we ask how far the Bayesian approach helps psychologists understand perhaps the most direct expression of human rationality: explicit verbal reasoning. Historically, verbal reasoning has been modeled in the framework of mathematical logic; the "new paradigm" in the psychology of reasoning developed over the past 25 years has suggested that human reasoning and argumentation may often be better understood using probabilistic Bayesian methods.

1. The Psychology of Verbal Reasoning: A Brief History

Following in the footsteps of 20th-century analytic philosophy, the starting point for the psychology of verbal reasoning in the 1960s was the assumption that the standards of formal, binary logic should judge human reasoning.[1] Accordingly, researchers focused on how people reason with the so-called logical terms of language, most notably those that have natural counterparts in first-order logic: *and, or, not, if–then, all, some.* In a typical experiment, people would be given one or more premises (e.g., *Birds fly, Tweety is a bird*) and asked to generate, or evaluate, possible conclusions that may follow (e.g., *Tweety flies*). The assessment of specific patterns of logical reasoning was seen as central to a much wider project within cognitive science and artificial intelligence: viewing cognition as a whole as operating through logical reasoning over world knowledge encoded in an internal, logical, language of thought—a program encapsulated in the phrase "cognition is proof theory" (Fodor & Pylyshyn, 1988).

Since the mid-1990s, however, the psychology of verbal reasoning has begun to adopt what has become known as the "new paradigm" (Over, 2009), in which Bayesian probability theory, rather than logic, has been taken as the standard against which human performance should be compared (Oaksford & Chater, 1994). Moreover, the shift to a Bayesian viewpoint has also been associated with two connected changes: a focus on knowledge-rich real-world inferences, rather than narrow logical puzzles, and viewing verbal reasoning not as an isolated, internal cognitive process but as having an inherently social function, in argument and persuasion.

We begin with a brief sketch of logic-based accounts of reasoning and its difficulties before turning to probabilistic accounts of deduction, induction, abduction, and argumentation, concluding with brief reflections on how Bayesian approaches to verbal reasoning may connect with the wider program of Bayesian cognitive science.[2]

2. Logic-Based Approaches to Reasoning

In line with early research in the psychology of reasoning, let us take logic as our starting point and consider the modus ponens (MP) inference (see sentence display 1 below).

In classical logic, this inference is of course *valid*: the truth of the premises guarantees the truth of the conclusion, and it can readily be converted into an experiment: participants are instructed to assume the premises are true (whether they believe them or not) and to draw inferences from those premises alone. Notice that, from a Bayesian point of view, such reasoning may be highly unnatural: our subjective probabilities will, after all, typically be influenced by the entirety of our background

(1)	If I turn the hot tap, I get hot water	If p then q	(A)
	I turn the hot tap	p	(B)
Therefore:	I get hot water	q	(MP: A, B)

knowledge—we shall return to the implications of this observation below.

According to logic-based psychological approaches, verbal reasoning follows logical lines: either directly, through implementing a psychological version of the logician's proof theory (a putative mental logic; Rips, 1994; see chapter 3.3 by Khemlani, in this handbook), or indirectly, through implementing a psychological version of the logician's model theory (mental model theory [MMT], Johnson-Laird, 1983; see chapter 2.3 by Johnson-Laird, this handbook, and, for a recent critique of MMT, Oaksford, Over, & Cruz, 2019). But the appropriateness of any logic-based approach is thrown into question by the following observation. Suppose a friend says *I didn't get hot water this morning*; by the logical law of modus tollens (MT), we can infer *You didn't turn the hot tap*. But this is precisely the opposite of the correct conclusion: our friend is telling us that she didn't get hot water *despite turning the hot tap* (otherwise, the non-appearance of hot water would be too uninteresting to mention). So, in real discourse, we confidently infer that the hot tap was turned, in precise contradiction to the apparent recommendations of logic.

To explain how we reason in such cases requires recognizing, presumably using general principles of conversation (e.g., Grice, 1975; Sperber & Wilson, 1986), that the speaker is aiming to be relevant and informative. Moreover, what counts as relevant and informative depends not just on the given premises but on background knowledge (e.g., that a lack of hot water is sufficiently bothersome to be worth mentioning).

Note, too, that this example illustrates how real-world inferences routinely violate a fundamental law of classical treatments of the conditional. According to Strengthening of the Antecedent, *If p then q* implies *If p and r then q* for any *r*. Yet *If I turn the hot tap, I get hot water* does not imply *If I turn the hot tap and the heating is broken, I get hot water*. Indeed, quite the opposite! Unlike logical reasoning, everyday reasoning is *nonmonotonic*: conclusions can be overturned if more information is added (e.g., Oaksford & Chater, 1991).

3.   Deductive Reasoning with Propositions

Patterns of reasoning about the logical connectives, *and*, *or*, *not*, and, in particular, *if–then*, have been intensively studied experimentally and theoretically. One approach is probability logic (see chapter 4.4 by Pfeifer, this handbook), which seeks to generalize standard logic by requiring that reasoners respect not merely logical consistency but also probabilistic coherence: degrees of belief

associated with propositions should be consistent with probability theory (Coletti & Scozzafava, 2002; Pfeifer & Kleiter, 2009). Moreover, researchers in the new paradigm have explored a variety of related formal ideas ranging from so-called p-validity, Bayesian and Jeffrey conditionalization, and dynamic inference.

We begin by illustrating how probabilistic coherence applies to a generalization of the MP inference illustrated in (1), above:

$$(2) \qquad \text{If } p \text{ then } q \qquad \Pr(q \mid p) = a$$
$$\qquad\qquad p \qquad\qquad\qquad \Pr(p) = b$$
$$Therefore: \quad q \qquad\qquad \Pr(q) = [ab, ab + (1-b)]$$

Now, we no longer suppose that the premises are definitely true. Instead, each premise is assigned a probability (*a* and *b*, respectively), reflecting their degree of belief. Probabilistic coherence demands that the degree of belief in the conclusion must lie in the interval between (and including) the values ab and $ab + (1-b)$.[3]

What exactly does it mean to talk about the probability of a conditional $\Pr(\text{if } p \text{ then } q)$? The simplest approach is to equate this with the conditional probability $\Pr(q \mid p)$, an identification known as "the Equation" (Edgington, 1995). The conditional probability itself will depend on arbitrary background knowledge, using what is known as the Ramsey test, proposed by philosopher and mathematician Frank Ramsey (1929/1990). We add the antecedent (e.g., *The hot tap was turned*) of the conditional to our entire set of beliefs, adjust them to take account of this new information, and read off the probability of the conclusion (*There is hot water*) from this new set of beliefs. Notice, in particular, that from this point of view, reasoning depends not just on the given premises but on world knowledge in general.

Is the Equation appropriate from a psychological point of view? It has been confirmed in a large number of experiments (e.g., Evans, Handley, & Over, 2003; Oberauer & Wilhelm, 2003; Politzer, Over, & Baratgin, 2010), and experimental participants reject a crucial step[4] in philosophical arguments (Lewis, 1976) that appear to undermine the Equation by suggesting that it leads to an unacceptable triviality result (Douven & Verbrugge, 2013). Notice, too, that a conventional material-implication interpretation of the conditional (that *If p then q* is equivalent to *Not-p or q*) makes very different predictions, which are strongly at odds with participants' judgments.

How does probability coherence fare in explaining empirical data about how people reason with conditionals? Varying *a* and *b*, in the MP inference above, is seen reliably to yield probabilities of the conclusion that fall within the predicted coherence interval at above-chance levels (Cruz, Baratgin, Oaksford, & Over, 2015; Evans, Thompson, & Over, 2015; Pfeifer & Kleiter, 2009; Politzer & Baratgin, 2016; Singmann, Klauer, & Over, 2014). This is not true for all inference types, however, including modus tollens (if p then q; not-q; therefore not-p). Cruz et al. (2015) argue that this may stem from the complexity of the inference and show that probabilistic coherence is respected above chance for eight simple one-premise arguments (e.g., p or q; therefore if not-p then q).

Just as logical consistency has a parallel in probabilistic coherence, so logical validity has a corresponding probabilistic analogue, p-validity (Adams, 1998; Coletti & Scozzafava, 2002; Pfeifer & Kleiter, 2009). A p-valid argument is one for which it is not probabilistically coherent that the uncertainty of the conclusion $(1 - \Pr(conclusion))$ exceeds the sum of the uncertainties of the premises $(\sum_{i=1}^{n}(1 - \Pr(premise_i)))$.

For example, let us reconsider (2). Suppose, for example, $a = .9$ and $b = .5$, so that the sum of the premise uncertainties is $.1 + .5 = .6$. According to probabilistic coherence, the probability of the conclusion is in the interval [.45, .95], with corresponding uncertainties, therefore, ranging from .55 down to .05. So it is not probabilistically coherent that the uncertainty of the conclusion (in the interval [.05, .55]) exceeds the sum of the premise uncertainties, .6. Hence, the inference is p-valid (and, indeed, this will be true for MP quite generally).

P-validity mitigates all the more puzzling paradoxes that arise from the "material-conditional" (or what logicians prefer to call "material-implication") interpretation of *If p then q* used in standard first-order logic: that, as noted above, is equivalent to *Not-p or q*. According to this interpretation, numerous apparent paradoxes arise (see chapter 5.2 by Rott and chapter 6.1 by Starr, both in this handbook). For example, the premise *There is no life on Mars* appears to imply *If there is life on Mars, there are kangaroos on the moon*. Such inferences are not, though, p-valid (and, indeed, people do not endorse them; see Pfeifer & Kleiter, 2011).

Despite its theoretical interest, however, p-validity does not appear directly to conform to human reasoning judgments. Indeed, from a psychological point of view, reasoning seems oriented to determining how one should change one's beliefs in the light of new evidence and background knowledge (Harman, 1986), rather than determining the validity of arguments, whether validity is viewed from a logical or a probabilistic standpoint.

So far, we have considered the situation in which both premises are potentially uncertain. But what happens if one premise is learned to be true for certain? For example, suppose I am certain that you turned the hot tap, because I saw you do it or because I know you to be completely truthful. Then, $\Pr(p) = 1$ in (8), and the probability of the conclusion can directly be read off from the conditional probability of the conclusion, q, given p: $\Pr(q \mid p)$. This is an example of what is known as Bayesian conditionalization, and it yields a precise probability rather than an interval. This approach (combined with background assumptions concerning the probability of q given not-p) successfully captured early empirical data (Oaksford & Chater, 2007; Oaksford, Chater, & Larkin, 2000) on how people reason with MP, MT, and two logically invalid patterns of reasoning with conditionals (Denying the Antecedent [DA], where not-p is taken to imply not-q, and Affirming the Consequent [AC], where q is taken to imply p).

But suppose we are not completely certain that the hot water tap is turned on (i.e., what if $\Pr(p) < 1$)? A natural assumption from prior knowledge is that most taps, most of the time, are not being turned: tap-turning events are rare. So given evidence that the tap was turned may considerably raise an initial estimate, say, $\Pr_0(p) = .01$, to $\Pr_1(p) = .9$. In such cases, Jeffrey conditionalization (Jeffrey, 2004) provides a more general updating procedure (see chapter 4.1 by Hájek & Staffel, this handbook):

(10) $\Pr_1(q) = \Pr_0(q \mid p)\Pr_1(p) + \Pr_0(q \mid \neg p)(1 - \Pr_1(p))$.

Here, then, the probability of the conclusion $\Pr_1(q)$ will depend on prior assumptions about $\Pr_0(q \mid \neg p)$, so that belief updating depends on background knowledge even for modus ponens. Zhao and Osherson (2010; see also Hadjichristidis, Sloman, & Over, 2014) asked people to estimate the relevant probabilities and found that the probability of the conclusion aligns well with Jeffrey's rule.

We have so far assumed that changing the probability of one premise does not modify the probabilities of others. But this will not, of course, always be true, and in such cases, known as *dynamic inference*, reasoning is less straightforward (Adams, 1998; Gilio & Over, 2012; Oaksford & Chater, 2007, 2013).

Recall the example we outlined earlier, where we are informed *You didn't get hot water this morning*, but learning this will drastically change our belief in the conditional premise *If you turn the hot tap, you get hot water*. If it were not revised, we would simply conclude *You didn't turn the hot tap*. But the appropriate conclusion is that the probability of the conditional should drastically be revised downward: while heating systems normally work, learning *You didn't get hot water this morning* leads to the strong suspicion that your heating system is malfunctioning.

Such cases, where other premises modify, perhaps drastically, the conditional probability of the conditional premise, are said to violate invariance or rigidity between the initial ($\mathrm{Pr_0}$) and the revised distribution ($\mathrm{Pr_1}$): such violations imply that $\mathrm{Pr_1}(q \mid p) \neq \mathrm{Pr_0}(q \mid p)$. They are sometimes, but not always, observed (Zhao & Osherson, 2010), and allowing violations for MT (as in our example), as well as for AC and DA, improves fits to empirical data on conditional reasoning (Oaksford & Chater, 2007, 2013).

4. Deductive Reasoning with Quantifiers

So far, we have considered how people reason about combinations of, and relations between, propositions *p*, *q*, and so on. Researchers have also studied how people reason with the quantifiers *all*, *some*, *some-not*, and *none*, as well as with the so-called generalized quantifiers, *most* and *few*. For example, consider the syllogism

(3)   All beekeepers (*B*) are artists (*A*)

             Some chemists (*C*) are beekeepers

Therefore: Some chemists are artists

How can quantified inference be viewed in probabilistic terms? The Probability Heuristics Model (PHM; Chater & Oaksford, 1999) takes a direct approach, viewing quantified assertions as expressing probabilistic constraints: *All*: $\mathrm{Pr}(y \mid x) = 1$, *Some*: $\mathrm{Pr}(y \mid x) > 0$, *Some-not*: $\mathrm{Pr}(y \mid x) < 1$, *None*: $\mathrm{Pr}(y \mid x) = 0$, *Few*: $0 < \mathrm{Pr}(y \mid x) < \Delta$, *Most*: $1 - \Delta < \mathrm{Pr}(y \mid x) < 1$, for small Δ.

How these probabilities combine depends on the structural relationships between the premises—what is known in syllogistic logic as "figure" (i.e., the relative positions of the end terms, *A* and *C*, with respect to the middle term that links them, *B*). Each syllogistic figure can be represented by a dependency graph. For example, (3) has the following structure: *Chemists* → *beekeepers* → *artists*. The premises constrain the probabilities associated with the links in the graph. Here, for example, *Some chemists are beekeepers*, which applies to the first link in the graph and embodies the constraint that $\mathrm{Pr}(beekeeper \mid chemist) > 0$. Similarly, turning to the second link, *All beekeepers are artists*, corresponds to $\mathrm{Pr}(artist \mid beekeeper) = 1$. It is then possible to apply the notion of p-validity and to prove whether the conclusion *Some chemists are artists* follows (which in this case it does).

PHM does not predict reasoning behavior directly from p-validity, however (indeed, as we noted above, it seems that validity, whether probabilistic or logical, is not computed by reasoners, who are concerned instead with belief updating). Instead, PHM applies simple heuristics that reflect p-validity indirectly. The *min*-heuristic is defined over an "informativeness" ordering of the various quantified statements, such that *all* > *most* > *few* > *some* > *none* > *some-not* (based on Shannon surprisal: $I(s) = 1/\mathrm{Pr}(s)$; Shannon & Weaver, 1949). This ordering is justified by assuming *rarity*: most predicates apply only to small subsets of entities (i.e., there are far more nonartists than artists). The *min*-heuristic is that the quantifier in the conclusion should match the *least* informative quantifier in either of the premises. Thus, in (3), the conclusion should use the quantifier *some*. The *max*-heuristic is that the degree of confidence in the conclusion should depend on the expected informativeness of the quantifier in the *most* informative premises (here, *all*). Expected informativeness is ordered: *all* > *most* > *few* > *some* > *some-not* ≈ *none*, where the expectation is taken over the conclusions of the p-valid syllogisms for which that quantifier is the most informative premise (so p-validity enters indirectly into the analysis at this point).

PHM fits prior experiments using logical quantifiers that have been modeled using mental logic (Rips, 1994), with fewer parameters (Chater & Oaksford, 1999). Moreover, extending the approach to syllogisms with *most* and *few* led to data that corroborate the PHM account; these data remain outside the scope of any current logic-based account (Chater & Oaksford, 1999). Mental model theory (Johnson-Laird, 1983) provides an equally good account of experimental results that link memory span to syllogistic reasoning (Copeland & Radvansky, 2004), but PHM seems better able to capture data for syllogisms with many premises (Copeland, 2006). The patterns in syllogistic reasoning are complex and remain contested; one model comparison suggested that no current model of syllogistic reasoning is fully satisfactory (Khemlani & Johnson-Laird, 2012).

One promising recent approach is the Probabilistic Representation Model (PRM; Hattori, 2016), which combines aspects of PHM and mental model theory. PRM proposes that people construct, and reason with, a *probabilistic prototype model*, capturing the eight joint probabilities ranging over the terms, *A*, *B*, and *C*, and their negations (e.g., $\mathrm{Pr}(A, \neg B, C)$). PRM captures these eight probabilities using a simple probabilistic model. The premises may also imply that some of the joint probabilities will be zero. PRM then draws a random sample from the distribution, which will thereby automatically be consistent with the premises. For example, PRM might sample a person who is an *artist* and a *chemist* but not a *beekeeper*. A sequential application of the *min*-heuristic then determines the conclusion that is consistent with the *sample*. As in PHM, conclusion type is determined by the *min*-heuristic. In (3), *Some C are A* is a possible conclusion, if the sample contains at least one person who is both a *chemist* and an *artist*. If not, PRM checks for a conclusion with the next most informative quantifier and so on iteratively. If no conclusion is consistent with the sample, PRM generates "no valid conclusion." As in PHM, confidence in a conclusion is determined by the *max*-heuristic.

PRM models the empirical data slightly better than either PHM or mental model theory; it is also interesting that the best-fit sample size is six to seven items, roughly consistent with typical assumptions about working-memory capacity. PRM is also theoretically appealing, in that it assumes that reason operates through sampling from a probabilistic model, in a similar way to sampling models in other areas of cognition (e.g., Sanborn & Chater, 2016; Stewart, Chater, & Brown, 2006; Vul, Goodman, Griffiths, & Tenenbaum, 2014).

5. Inductive Reasoning

The psychology of verbal reasoning initially focused on deduction. Yet, from a Bayesian standpoint, reasoning is typically uncertain and knowledge-rich. Thus, Bayesian models are particularly well placed to deal with *inductive* verbal reasoning, where conclusions may extend, often very substantially, beyond the logical consequences of the premises. Inductive verbal reasoning also provides a bridge to the application of Bayesian models in perception, categorization, learning, and so on, which are widespread throughout the cognitive and brain sciences.

One of the best-studied areas of inductive verbal reasoning concerns property induction, such as

(4)   Blackbirds have property X

Therefore: Penguins/worms have property X

Such an induction cannot be a matter of pure logic. After all, any such inference will be logically invalid, as it is perfectly possible that the premise is true and the conclusion false. But also the difference between inferences concerning *penguins* and *worms* does not depend on logical terms but on the nature of the properties themselves. So, for example, the fact that penguins generally seem more similar to blackbirds than either is to worms helps explain why people often generalize properties from blackbirds to penguins more readily (e.g., concerning their anatomy and physiology).

Yet the picture is more complex—and crucially depends on the nature of the property and relevant background knowledge. So, for example, if the property is *being contaminated by a rare poison*, then people may suspect that worms and blackbirds may share this property and that blackbirds may ingest the poison by eating worms or through living in the same environment, while penguins live in a completely different ecosystem (e.g., Kemp & Tenenbaum, 2009).

A further natural question concerns the impact of multiple premises. Consider the following example:

(5)   Blackbirds have property X

               Ostriches have property X

Therefore: Penguins have property X

Here, the diversity of examples seems persuasive (e.g., Heit & Feeney, 2005). If blackbirds and *sparrows* have property X, we may suspect it is limited to garden birds; if blackbirds and *ostriches* do, we may suspect X is very widespread across birds as a whole, most likely including penguins. Getting such predictions right requires capturing complex knowledge, which may, for example, be modeled by hierarchical Bayesian models (Kemp & Tenenbaum, 2009).

Abductive reasoning involves "inference to the best explanation" given a set of sensory, linguistic, or scientific data (Harman, 1965). But how do we decide which explanation is best? One criterion, with a long intellectual history dating back to William of Ockham and beyond, is that we should prefer explanations that capture as much as possible with the fewest assumptions. Experimental work with adults and children indicates that people do indeed prefer explanatory simplicity (Lombrozo, 2016) but that they are also drawn to explanations that are as integrated as possible (Lombrozo & Vasilyeva, 2017). For example, people are more convinced by an explanation of weight loss and fatigue that does not explain each symptom independently (loss of appetite coincidentally combined with insomnia) but rather points to a single underlying cause (e.g., depression). Explanations are also preferred to the degree that they cover a diverse range of phenomena (Kim & Keil, 2003), and this is particularly true when people are prompted to pay attention to evidential diversity (Preston & Epley, 2005).

How are considerations such as simplicity and explanatory breadth connected to Bayesian explanations of reasoning? One direct connection is that a preference for simple explanations (where simplicity is judged by the length of description in a coding language) is equivalent to choosing the most probable explanation (according to an appropriate prior) (e.g., Chater, 1996). Moreover, Bayesian philosophy of science has attempted to recast a variety of explanatory virtues as corollaries of a Bayesian analysis (e.g., Bovens & Hartmann, 2003; Howson & Urbach, 1989), although whether this project is successful has been challenged (Douven & Schupbach, 2015; but see Wojtowicz & DeDeo, 2020).

So far, we have focused on cases where the premises from which reasoning proceeds are taken as given. But people are also active investigators into the world. This raises the question of how people should select information, or conduct experiments, that may be the basis for further reasoning. Initial work in the psychology of reasoning (e.g., Wason, 1968) presupposed that enquiry should seek only to falsify generalizations, on the principle that only falsification can give certainty (in the spirit of Popper's [1934/1959] falsificationist scientific methodology). But from a Bayesian point of view, both disconfirming and confirming evidence can lead us to update our beliefs. Thus, under some circumstances, attempts to confirm the hypothesis may be particularly fruitful.

Suppose we wish, for example, to investigate the somewhat fanciful hypothesis that eating blueberries turns your hair blue temporarily. According to the logic-based falsificationist search strategy, the sole objective is to find someone who has eaten blueberries and whose hair is nonblue and hence to search among blueberry eaters, on the one hand, and people with nonblue hair, on the other. Yet the latter search is likely to be futile—most people have nonblue hair and have also not recently eaten blueberries. On the other hand, it may well be worth asking the much smaller set of people with blue hair whether they have recently eaten blueberries; finding even a few of these might make one suspect that the rule has some validity.

From a Bayesian point of view, the objective of inquiry will often simply be to maximize the amount of information gained from sampling new evidence; of course, before gathering evidence, we cannot know how informative it will be. But Bayesian optimal data selection (Lindley, 1956) recommends that we maximize the *expected* amount of information gained (where the expectation is taken in the light of our probabilistic assumptions about the world). This is the starting point for a rational Bayesian explanation of what was previously viewed as "confirmation bias" in data acquisition tasks, including most notably Wason's (1968) card selection task (Oaksford & Chater, 1994). According to the Bayesian framework, the optimal strategy should depend on probabilistic background assumptions (e.g., the probability of having blue hair, in the example above), and these do, indeed, seem to moderate performance in data selection tasks.

From the standpoint of the logical paradigm, it is natural to see a profound division between deductive reasoning (guided by logical rules) and inductive reasoning (where conclusions extend beyond the logical consequences of the premises). But, as we have seen, both types of process can be modeled in a uniform way in a Bayesian framework. Indeed, it seems natural to see both types of reasoning as inherently uncertain and dependent on rich background knowledge, rather than purely on the verbal premises that are given in the problem. The Bayesian approach, therefore, seems to sit uneasily with the proposal that the mind combines a slow, logical reasoning system with fast associative processes for inductive reasoning (e.g., Heit & Rotello, 2010; Klauer, Beller, & Hütter, 2010; Rips, 2001; Stanovich, 2011). Interestingly, recent applications of signal detection theory to assess whether deductive and inductive reasoning are supported by separate processes or a single system appear to favor the latter (Stephens, Dunn, & Hayes, 2018).

6. Social Reasoning: Argumentation

The psychology of verbal reasoning has traditionally been pursued as if it were only concerned with the cognitive processes of isolated individuals. Yet verbal reasoning, like verbal behavior of all kinds, appears naturally to have a social role: as involved in communication, persuasion, and argumentation (Hahn & Oaksford, 2007; Mercier & Sperber, 2011, 2017). Indeed, understanding verbal reasoning in its communicative context may be crucial to understanding even the most elementary reasoning and to interpreting the significance of the statements over which reasoning occurs (Hilton, 1995). Moreover, the process of pragmatic enrichment may itself involve highly complex reasoning processes (Grice, 1975; Levinson, 2000; Sperber & Wilson, 1986), which are often of far greater sophistication than the patterns of reasoning (MP, MT, etc.) that are ostensibly being probed in experimental tasks. Recall that on hearing *I didn't get hot water this morning*, we immediately infer that the speaker turned the hot tap in the right direction, ran it for sufficiently long, and then discovered the unexpected lack of hot water. We assume, too, that she obtained cold, or perhaps lukewarm, water from the tap, rather than no water all—otherwise, some stronger statement (e.g., *There was no water at all this morning*) would have been given. Speaker and hearer have to do significant inferential work to align on an interpretation—and this process can misfire. It could be, for example, that the speaker is referring to the water at the gym, while the listener assumes they are referring to the water at home. The wider point is that verbal reasoning, as traditionally understood, presupposes that a great deal of "hidden" inferential work has already been carried out (Stenning & Cox, 2006).

Beyond questions of agreeing on the interpretation of what has been said, communication in general, and argumentation in particular, is invariably knowledge rich: whether new arguments or evidence will convince a person to change their beliefs can, in principle, depend on the entirety of their current knowledge. Argumentation is also local and interactive: interlocutors make a series of specific argumentative moves, which may be either accepted or else repelled by further moves, which may themselves be countered, and so on.

How far can a Bayesian approach extend from verbal reasoning to model argument more broadly? This has been a major focus of research for at least 15 years, focusing in particular on the question of whether so-called fallacies of informal argument, which are nonetheless widely deployed in real conversations, have a rational Bayesian foundation that may legitimately lead an audience to change their degrees of belief (Hahn & Oaksford, 2007; Oaksford & Hahn, 2004).

The range of fallacies that have been considered is broad, including arguments from ignorance, ad hominem arguments, and many more. Here, we focus on key examples: slippery-slope and circular arguments.

Consider two slippery-slope arguments (SSAs), both of which have been used in popular debate but differing in strength—from the patently ridiculous (6) to the more credible (7):

(6) If we allow gay marriage, then people will want to marry their pets.

(7) If voluntary euthanasia is legalized, then involuntary euthanasia will be.

From a Bayesian perspective, SSAs can be viewed as founded on a cost–benefit analysis (Corner, Hahn, & Oaksford, 2011; Hahn & Oaksford, 2007): do not take the relevant action (p: allow gay marriage) unless its benefit ($U(p)$) exceeds the expected cost of the consequence (q: allowing interspecies marriage) to which it might conceivably lead ($\Pr(q \mid p)U(q)$). Thus, SSAs are a species of warning (Bonnefon, 2009). A key difference between (6) and (7) is that the probabilities of the consequences seem very different: the chance of significantly raising the probability of interspecies marriage is infinitesimally small, whereas the chance of the extension of euthanasia to at least some nonvoluntary cases may be judged nontrivial (irrespective of how we evaluate either of these consequences).

Interestingly, SSAs can depend crucially on the degree to which our categories are flexible (Hahn & Oaksford, 2007). One reason that the jump from gay marriage to interspecies marriage seems so ludicrous is that these cases are judged to be highly dissimilar, and moreover,

the category of allowable marriages is itself quite rigid (it is not a matter of casual interpretation but is rooted in the law). On the other hand, different types of euthanasia may seem far more similar—and the category boundary (e.g., between quite what counts as voluntary or involuntary) may seem anything but rigid. More broadly, the similarity between p and q affects people's willingness both to assign them to the same category and to endorse the corresponding SSA (Corner et al., 2011; see also Rai & Holyoak, 2014, in the context of moral hypocrisy).

The "fallacy" of circularity (Hahn, 2011) is often viewed as problematic even though it is logically valid. To take the extreme case, *If p then p* is not objectionable because it succumbs to logical counterexamples—indeed, it is tautologically true. Instead, circular reasoning can seem argumentatively useless—it presupposes precisely what is to be proved. But how the presupposition operates can significantly affect how fallacious they appear to be, from the patently unconvincing (8) to the rather more compelling (9):

(8) God exists because the Bible says so and the Bible is the word of God

(9) Electrons exist because they make 3-cm tracks in a cloud chamber

A good argument, of any kind, should have the potential to change our beliefs (Harman, 1986). (9) is a type of inference to the best explanation: if we suppose the existence of electrons, that would explain (via a good deal of collateral knowledge about physical laws, the operation of cloud chambers, etc.) the 3-cm tracks; we can't think of any other compelling explanation for the observation of these tracks, so we infer that electrons do indeed exist. Thus, (9) concludes that electrons exist by assuming a theory that requires the existence of electrons and tracing its consequences, and so some nontrivial work has been done (although this work is "hidden" in the details of rival scientific theories and their predictions, which is not, of course, visible in the verbal statement of the argument). Indeed, (9) illustrates what appears to be a benign circularity, inevitable in abductive argument, whether concerning science or not: to explain an observation, we postulate an explanation that is then bolstered (or disconfirmed) to the degree that the explanation captures the observation of interest. (8) is, by contrast, much less compelling: there is no bolstering of the explanation by the observation—in essence, because the prediction is not a distinctive prediction. So, while rival physical theories will struggle to explain the observation of cloud chamber tracks, it is all too easy for any number of nontheological theories to explain why a religious text will claim divine authorship (these types of considerations can be captured in a hierarchical Bayesian model; Hahn, 2011; Hahn & Oaksford, 2007). Thus, the perceived strength of a circular argument depends not on its structure but on the credibility and number of possible alternative explanations (Hahn & Oaksford, 2007, experiments 1 and 2).

More broadly, from a Bayesian point of view, the study of verbal reasoning and the study of argumentation can be addressed using the same methods and are continuous, rather than corresponding to distinct domains. Indeed, despite the historical priority of the study of individual reasoning, it may be that communication and argumentation are more fundamental and that individual reasoning can only be understood in a social, communicative context (Hahn & Oaksford, 2007; Hilton, 1995; Mercier & Sperber, 2011, 2017; Oaksford & Chater, 2020).

7. Conclusion

Bayesian theories are now widespread in the psychology of verbal reasoning and argument, from what have traditionally been viewed as deductive reasoning tasks, to the study of induction, abduction, and argumentation. In each case, reasoning is presumed to be probabilistic and to depend not merely on the stated premises but typically on the totality of a person's beliefs. The purpose of reasoning of all kinds in daily life is, we have suggested, the updating of belief, rather than, for example, the evaluation of the validity of links between specific premises and conclusions (although this may be crucial in important special cases, such as checking mathematical proofs, legal arguments, and the predictions of scientific theories). Traditional logic-based models of reasoning, by contrast, typically focus on validity and reasoning from the information given. The complexities of semantic and pragmatic interpretation required to even start such reasoning, the relevance of unstated background knowledge, and the social and argumentative context have traditionally been viewed as rather difficult-to-eliminate confounding factors. But from a Bayesian point of view, these are not confounds but the very topic of interest. Indeed, more broadly, Bayesian approaches to verbal reasoning hold the promise of integration with the wider program of modeling world knowledge, social interaction, communication, and the basic cognitive mechanisms of perception, categorization, memory, and motor control within a single theoretical framework (Chater & Oaksford, 2008; Tenenbaum, Kemp, Griffiths, & Goodman, 2011).

Acknowledgments

N.C. was supported by the ESRC Network for Integrated Behavioural Science (grant ES/P008976/1).

Notes

1. As we have pointed out before (e.g., Oaksford & Chater, 2010), this starting point, especially for the conditional, contrasts starkly with the philosophy of logic and language (Nute, 1984), which abandoned the truth-functional view with the advent of the possible-worlds semantics for the conditional (Stalnaker, 1968).

2. We refer the reader to Oaksford and Chater (2020) for a more extensive review of the issues raised here.

3. An intuitive explanation of the origin of this interval is that $\Pr(q \mid \text{not-}p)$ can range between 0 and 1, the former yielding the lower bound and the latter the upper bound.

4. Douven and Verbrugge (2013) argue that the triviality arguments rely on the assumption that the conditional-forming operator is independent of belief states (van Fraassen, 1976). They present evidence that this is not the case because people do not judge the conditional probability of a conditional, that is, $\Pr(\text{if } p \text{ then } q \mid r)$, to be equal to the conditional probability $\Pr(q \mid p \& r)$.

References

Adams, E. W. (1998). *A primer of probability logic*. Stanford, CA: CSLI Publications.

Bonnefon, J.-F. (2009). A theory of utility conditionals: Paralogical reasoning from decision-theoretic leakage. *Psychological Review, 116*, 888–907.

Bovens, L., & Hartmann, S. (2003). *Bayesian epistemology*. Oxford, England: Oxford University Press.

Chater, N. (1996). Reconciling simplicity and likelihood principles in perceptual organization. *Psychological Review, 103*, 566–581.

Chater, N., & Oaksford, M. (1999). The probability heuristics model of syllogistic reasoning. *Cognitive Psychology, 38*, 191–258.

Chater, N., & Oaksford, M. (Eds.). (2008). *The probabilistic mind: Prospects for Bayesian cognitive science*. Oxford, England: Oxford University Press.

Coletti, G., & Scozzafava, R. (2002). *Probabilistic logic in a coherent setting*. Dordrecht, Netherlands: Kluwer.

Copeland, D. E. (2006). Theories of categorical reasoning and extended syllogisms. *Thinking & Reasoning, 12*, 379–412.

Copeland, D. E., & Radvansky, G. A. (2004). Working memory and syllogistic reasoning. *Quarterly Journal of Experimental Psychology A: Human Experimental Psychology, 57A*, 1437–1457.

Corner, A., Hahn, U., & Oaksford, M. (2011). The psychological mechanism of the slippery slope argument. *Journal of Memory and Language, 64*, 133–152.

Cruz, N., Baratgin, J., Oaksford, M., & Over, D. E. (2015). Bayesian reasoning with ifs and ands and ors. *Frontiers in Psychology, 6*, 192.

Douven, I., & Schupbach, J. N. (2015). The role of explanatory considerations in updating. *Cognition, 142*, 299–311.

Douven, I., & Verbrugge, S. (2013). The probabilities of conditionals revisited. *Cognitive Science, 37*, 711–730.

Edgington, D. (1995). On conditionals. *Mind, 104*, 235–329.

Evans, J. St. B. T., Handley, S. J., & Over, D. E. (2003). Conditionals and conditional probability. *Journal of Experimental Psychology: Learning, Memory, and Cognition, 29*, 321–335.

Evans, J. St. B. T., Thompson, V., & Over, D. E. (2015). Uncertain deduction and conditional reasoning. *Frontiers in Psychology, 6*, 398.

Fodor, J. A., & Pylyshyn, Z. W. (1988). Connectionism and cognitive architecture: A critical analysis. *Cognition, 28*, 3–71.

Gilio, A., & Over, D. (2012). The psychology of inferring conditionals from disjunctions: A probabilistic study. *Journal of Mathematical Psychology, 56*, 118–131.

Grice, H. P. (1975). Logic and conversation. In P. Cole & J. L. Morgan (Eds.), *Syntax and semantics: Vol. 3. Speech acts* (pp. 47–59). New York, NY: Academic Press.

Hadjichristidis, C., Sloman, S. A., & Over, D. E. (2014). Categorical induction from uncertain premises: Jeffrey's doesn't completely rule. *Thinking & Reasoning, 20*, 405–431.

Hahn, U. (2011). The problem of circularity in evidence, argument, and explanation. *Perspectives on Psychological Science, 6*, 172–182.

Hahn, U., & Oaksford, M. (2007). The rationality of informal argumentation: A Bayesian approach to reasoning fallacies. *Psychological Review, 114*, 704–732.

Harman, G. (1965). The inference to the best explanation. *Philosophical Review, 64*, 88–95.

Harman, G. (1986). *Change in view: Principles of reasoning*. Cambridge, MA: MIT Press.

Hattori, M. (2016). Probabilistic representation in syllogistic reasoning: A theory to integrate mental models and heuristics. *Cognition, 157*, 296–320.

Heit, E., & Feeney, A. (2005). Relations between premise similarity and inductive strength. *Psychonomic Bulletin and Review, 12*, 340–344.

Heit, E., & Rotello, C. M. (2010). Relations between inductive reasoning and deductive reasoning. *Journal of Experimental Psychology: Learning, Memory, and Cognition, 36*, 805–812.

Hilton, D. J. (1995). The social context of reasoning: Conversational inference and rational judgment. *Psychological Bulletin, 118*, 248–271.

Howson, C., & Urbach, P. (1989). *Scientific reasoning: The Bayesian approach*. LaSalle, IL: Open Court.

Jeffrey, R. C. (2004). *Subjective probability: The real thing*. Cambridge, England: Cambridge University Press.

Johnson-Laird, P. N. (1983). *Mental models*. Cambridge, England: Cambridge University Press.

Kemp, C., & Tenenbaum, J. B. (2009). Structured statistical models of inductive reasoning. *Psychological Review, 116*, 20–58.

Khemlani, S., & Johnson-Laird, P. N. (2012). Theories of the syllogism: A meta-analysis. *Psychological Bulletin, 138*, 427–457.

Kim, N. S., & Keil, F. C. (2003). From symptoms to causes: Diversity effects in diagnostic reasoning. *Memory & Cognition, 31*, 155–165.

Klauer, K., Beller, S., & Hütter, M. (2010). Conditional reasoning in context: A dual-source model of probabilistic inference. *Journal of Experimental Psychology: Learning, Memory, and Cognition, 36*, 298–323.

Levinson, S. C. (2000). *Presumptive meanings: The theory of generalized conversational implicature*. Cambridge, MA: MIT Press.

Lewis, D. K. (1976). Probabilities of conditionals and conditional probabilities. *Philosophical Review, 85*, 297–315.

Lindley, D. V. (1956). On a measure of the information provided by an experiment. *Annals of Mathematical Statistics, 27*(4), 986–1005.

Lombrozo, T. (2016). Explanatory preferences shape learning and inference. *Trends in Cognitive Sciences, 20*, 748–759.

Lombrozo, T., & Vasilyeva, N. (2017). Causal explanation. In M. Waldmann (Ed.), *Oxford handbook of causal reasoning* (pp. 415–432). Oxford, England: Oxford University Press.

Mercier, H., & Sperber, D. (2011). Why do humans reason? Arguments for an argumentative theory. *Behavioral and Brain Sciences, 34*, 57–74.

Mercier, H., & Sperber, D. (2017). *The enigma of reason*. Cambridge, MA: Harvard University Press.

Nute, D. (1984). Conditional logic. In D. Gabbay & F. Guenthner (Eds.), *Handbook of philosophical logic* (Vol. 2, pp. 387–439). Dordrecht, Netherlands: Reidel.

Oaksford, M., & Chater, N. (1991). Against logicist cognitive science. *Mind & Language, 6*, 1–38.

Oaksford, M., & Chater, N. (1994). A rational analysis of the selection task as optimal data selection. *Psychological Review, 101*, 608–631.

Oaksford, M., & Chater, N. (2007). *Bayesian rationality: The probabilistic approach to human reasoning*. Oxford, England: Oxford University Press.

Oaksford, M., & Chater, N. (2010). Causation and conditionals in the cognitive science of human reasoning. *Open Psychology Journal*, *3*, 105–118.

Oaksford, M., & Chater, N. (2013). Dynamic inference and everyday conditional reasoning in the new paradigm. *Thinking & Reasoning*, *19*, 346–379.

Oaksford, M., & Chater, N. (2020). New paradigms in the psychology of reasoning. *Annual Review of Psychology*, *71*, 305–330.

Oaksford, M., Chater, N., & Larkin, J. (2000). Probabilities and polarity biases in conditional inference. *Journal of Experimental Psychology: Learning, Memory, and Cognition*, *26*, 883–899.

Oaksford, M., & Hahn, U. (2004). A Bayesian analysis of the argument from ignorance. *Canadian Journal of Experimental Psychology*, *58*, 75–85.

Oaksford, M., Over, D. E., & Cruz, N. (2019). Paradigms, possibilities and probabilities: Comment on Hinterecker, Knauff, and Johnson-Laird (2016). *Journal of Experimental Psychology: Learning, Memory, and Cognition*, *45*, 288–297.

Oberauer, K., & Wilhelm, O. (2003). The meaning(s) of conditionals: Conditional probabilities, mental models, and personal utilities. *Journal of Experimental Psychology: Learning, Memory, and Cognition*, *29*, 680–693.

Over, D. E. (2009). New paradigm psychology of reasoning. *Thinking & Reasoning*, *15*, 431–438.

Pfeifer, N., & Kleiter, G. D. (2009). Framing human inference by coherence based probability logic. *Journal of Applied Logic*, *7*, 206–217.

Pfeifer, N., & Kleiter, G. D. (2011). Uncertain deductive reasoning. In K. Manktelow, D. E. Over, & S. Elqayam (Eds.), *The science of reason: A Festschrift for Jonathan St. B. T. Evans* (pp. 145–166). Hove, England: Psychology Press.

Politzer, G., & Baratgin, J. (2016). Deductive schemas with uncertain premises using qualitative probability expressions. *Thinking & Reasoning*, *22*, 78–98.

Politzer, G., Over, D. E., & Baratgin, J. (2010). Betting on conditionals. *Thinking & Reasoning*, *16*, 172–197.

Popper, K. (1959). *The logic of scientific discovery*. Abingdon, England: Routledge. (Original work published 1934)

Preston, J., & Epley, N. (2005). Explanations versus applications: The explanatory power of valuable beliefs. *Psychological Science*, *16*, 826–832.

Rai, T. S., & Holyoak, K. J. (2014). Rational hypocrisy: A Bayesian analysis based on informal argumentation and slippery slopes. *Cognitive Science*, *38*, 1456–1467.

Ramsey, F. P. (1990). General propositions and causality. In H. A. Mellor (Ed.), *Frank Ramsey: Philosophical Papers*. Cambridge, England: Cambridge University Press. (Original work published 1929)

Rips, L. J. (1994). *The psychology of proof: Deductive reasoning in human thinking*. Cambridge, MA: MIT Press.

Rips, L. J. (2001). Two kinds of reasoning. *Psychological Science*, *12*, 129–134.

Sanborn, A. N., & Chater, N. (2016). Bayesian brains without probabilities. *Trends in Cognitive Sciences*, *113*(33), E4764–E4766.

Shannon, C. E., & Weaver, W. (1949). *The mathematical theory of communication*. Urbana: University of Illinois Press.

Singmann, H., Klauer, K. C., & Over, D. E. (2014). New normative standards of conditional reasoning and the dual-source model. *Frontiers in Psychology*, *5*, 316.

Sperber, D., & Wilson. D. (1986). *Relevance: Communication and cognition*. Oxford, England: Blackwell.

Stalnaker, R. C. (1968). A theory of conditionals. In N. Rescher (Ed.), *Studies in logical theory* (pp. 98–112). Oxford, England: Blackwell.

Stanovich, K. E. (2011). *Rationality and the reflective mind*. Oxford, England: Oxford University Press.

Stenning, K., & Cox, R. (2006). Reconnecting interpretation to reasoning through individual differences. *Quarterly Journal of Experimental Psychology*, *59*, 1454–1483.

Stephens, R. G., Dunn, J. C., & Hayes, B. K. (2018). Are there two processes in reasoning? The dimensionality of inductive and deductive inferences. *Psychological Review*, *125*, 218–244.

Stewart, N., Chater, N., & Brown, G. D. A. (2006). Decision by sampling. *Cognitive Psychology*, *53*, 1–26.

Tenenbaum, J. B., Kemp, C., Griffiths, T. L., & Goodman, N. D. (2011). How to grow a mind: Statistics, structure, and abstraction. *Science*, *331*(6022), 1279–1285.

van Fraassen, B. C. (1976). Probabilities of conditionals. In W. L. Harper & C. A. Hooker (Eds.), *Foundations of probability theory, statistical inference, and statistical theories of science* (Vol. 1, pp. 261–301). Dordrecht, Netherlands: Reidel.

Vul, E., Goodman, N., Griffiths, T. L., & Tenenbaum, J. B. (2014). One and done? Optimal decisions from very few samples. *Cognitive Science*, *38*, 599–637.

Wason, P. C. (1968). Reasoning about a rule. *Quarterly Journal of Experimental Psychology*, *20*, 273–281.

Wojtowicz, Z., & DeDeo, S. (2020). From probability to consilience: How explanatory values implement Bayesian reasoning. *Trends in Cognitive Sciences*. https://doi.org/10.1016/j.tics.2020.09.013

Zhao, J., & Osherson, D. (2010). Updating beliefs in light of uncertain evidence: Descriptive assessment of Jeffrey's rule. *Thinking & Reasoning*, *16*, 288–307.

4.6 Probabilities and Conditionals

Klaus Oberauer and Danielle Pessach

Summary

What do conditionals have to do with probability? After a short introduction, the second section of this chapter will explain this connection and how it laid the groundwork for the "probabilistic turn" in the psychology of conditional reasoning. The rest of the chapter is devoted to two questions: (a) What is the *meaning* of conditionals, and (b) how do we *reason* with conditionals? We will answer these two questions from the perspective of probabilistic accounts of conditional reasoning. The third section is concerned with the probabilistic interpretation of the conditional, whereas the fourth section will cover current approaches to reasoning with (probability-)conditionals.

1. Conditionals and Rationality

Conditionals lie at the heart of human rationality. We use them to assert how states and events depend on each other—causally, logically, deontically, or in other ways (e.g., "If you want to learn more about conditionals, you should read this chapter"). We use them to express hypothetical thinking (e.g., "If we reduce CO_2 emissions now, we can still keep global warming below 2°C"). They figure centrally in two major systems for formalizing rationality, formal logic and the probability calculus: first, "if–then"—defined as the material conditional—was included as one of the junctors linking elementary propositions in propositional logic. Subsequently, the interpretation of "if–then" expressions as material conditionals has been criticized, leading to a "probabilistic turn" that links conditionals closely to conditional probabilities. This chapter reviews the arguments motivating this "probabilistic turn" and the related psychological research on how people interpret and reason with conditionals.

2. Why Turn to Probabilities?

The groundwork for the probabilistic turn was laid by the inaptitude of classical logic[1] to account for how people interpret conditionals, as well as for how they reason with them. Classical logic conceptualizes a conditional $A \to C$ (read as: "If A then C," where A is called the *antecedent* and C is called the *consequent*) as the *material conditional*. Whether a material conditional is true or false is defined by the truth-values of its constituents; it is therefore fully truth-functional. A truth function can be represented as a truth table, which maps the truth-value of a compound sentence such as "If A then C" as a function of the truth conditions of its components A and C. Table 4.6.1 shows the truth table for the material conditional. The material conditional is false when A is true and C is false and true in all other possible cases. This definition has consequences often characterized as paradoxical. From any true C, we can validly infer $A \to C$: from "Cats are mammals" we can infer "If cats are fish, then cats are mammals." Another paradox: from $\neg A$ ("not-A"; $\neg$ is used to denote a negation), we can validly infer $A \to C$. For example, "The moon is square" is clearly false; therefore, "If the moon is square, the president is a good person" and "If the moon is square, the president is a bad person" both follow. In the material-conditional account, $A \to C$ and $A \to \neg C$ can both be true.

Psychological research has shown that people do not understand conditionals as material conditionals. One possibility to assess how people understand $A \to C$ is to ask them to indicate for each possible combination of

Table 4.6.1

Truth table for the material conditional and responses in Johnson-Laird and Tagart (1969)

		Conditional ($A \to C$)	
A	C	MC	Responses
True	True	True	True
True	False	False	False
False	True	True	Irrelevant
False	False	True	Irrelevant

Note: MC = material conditional. Responses are participants' majority responses.

truth-values for A and C (i.e., for each line of the truth table) whether this combination renders the conditional true or false or is irrelevant to the truth of the conditional. This task is referred to as the *truth table task*. The usual pattern is displayed in table 4.6.1: participants tend to judge the false-antecedent cases as "irrelevant" rather than as cases rendering the conditional true (Evans, Handley, Neilens, & Over, 2008; Johnson-Laird & Tagart, 1969; Schroyens, 2010).

People also do not *reason* with conditionals as dictated by classical logic. In the psychology of reasoning, four simple inference forms—taking a conditional major premise and a categorical minor premise—are frequently studied. They are termed "modus ponens" (MP), "modus tollens" (MT), "affirmation of the consequent" (AC), and "denial of the antecedent" (DA). For example, MP in its abstract form is: from "If A then C" and A follows C. The full list of the four inferences is as follows:

$$\text{MP: } \frac{A \to C \quad A}{C} \qquad \text{AC: } \frac{A \to C \quad C}{A}$$

$$\text{MT: } \frac{A \to C \quad \neg C}{\neg A} \qquad \text{DA: } \frac{A \to C \quad \neg A}{\neg C}$$

Note: Statements above the horizontal lines are premises, below the line are conclusions.

While MP and MT are valid according to classical logic, AC and DA are not. However, in a conditional inference task (i.e., a task where participants have to judge the four inference forms as valid or not), people usually do not adhere to that norm: although almost everyone endorses MP as valid ($\sim 95\%$), the endorsement rate drops significantly for MT ($\sim 70\%$), and the so-called *fallacies* (AC and DA) are accepted as valid with rates well over 50% (e.g., Evans, Newstead, & Byrne, 1993; Evans, Over, & Manktelow, 1993; Schroyens & Schaeken, 2003).

The paradoxes of the material implication, its incompatibility with people's judgment (in the truth table task as well as in the conditional inference task), and the fact that everyday reasoning seldom deals with "true" and "false," but rather with uncertainty in different degrees, have led philosophers as well as psychologists to question the material conditional as an explication of the meaning of "if–then" statements. As an alternative, the "probability conditional" has been suggested (Adams, 1975; Bennett, 2003; Edgington, 1995).

3. Probabilistic Account for the Meaning of Conditionals: The Probability Conditional

The probability conditional is defined by "the Equation":

$$P(A \to C) = P(C \mid A).$$

The Equation defines the probability of the conditional as the conditional probability of the consequent given the antecedent. It was inspired by an account of how people assess conditionals, known as the *Ramsey test*. In probably one of the most famous footnotes in the history of philosophy, Ramsey wrote, "If two people are arguing 'If p will q?' and are both in doubt as to p, they are adding p hypothetically to their stock of knowledge and arguing on that basis about q. . . . We can say they are fixing their degrees of belief in q given p" (Ramsey, 1929/1990, p. 247). So, when we assess $A \to C$, we *suppose* A by adding it (hypothetically) to our beliefs about the world and assess, from that hypothetical belief state, how likely C is. The outcome of the Ramsey test is the conditional probability of C, given A, $P(C \mid A)$.

Before we continue to assess the merits and drawbacks of the probability conditional, some clarifications and specifications are necessary. First, when we speak of probabilities, we mean *subjective* probabilities, sometimes also called *degrees of belief*.

Second, while probabilities of conditionals equal conditional probabilities, they do not inherit all their (mathematical) features. The following two points should make this distinction clear: (1) conditional probabilities are mathematically defined by the Ratio formula: $P(C \mid A) = P(A \& C)/P(A)$. This does not apply to conditionals: for the purpose of determining $P(A \to C)$, $P(C \mid A)$ is defined as the outcome of the Ramsey test, that is, it is the result of assessing the probability of C on the supposition of A. Edgington (1995) gives the following example: we can be thinking about what to do next, A or B, considering what consequences those options would have and how likely they are. Obviously, we can do that without having fixed our degree of belief in A—after all, we are still deliberating whether to make A true. Hence, we can determine $P(C \mid A)$ without having a belief about $P(A)$ (Edgington, 1995, p. 266).

(2) Another consequence of the Ratio formula is that $P(C \mid A)$ is not defined when $P(A) = 0$. Stalnaker (1968) offers an extension of the Ramsey test that deals with cases where $P(A) = 0$ (i.e., counterfactual conditionals). If A is incompatible with our knowledge, we cannot simply add it to our stock of knowledge and still be coherent in what we believe. Stalnaker's extended Ramsey test requires that, before adding A, we make the minimally

necessary adjustments to our knowledge to make it compatible with A. However, this extended Ramsey test has been criticized, mostly because it is impossible to determine which adjustments to one's stock of knowledge are "minimal." Bennett (2003), among others, therefore maintains that when $P(A) = 0$, conditionals—with a few exceptions—are not defined and have no probability.

Now that the Equation is disambiguated, we can turn to its main implication: conditionals have no truth conditions.

3.1 A Triviality Result

The triviality result is Lewis's (1976) proof that there is no combination of cases in the truth table for which the joint probability systematically equals $P(C|A)$. Because any proposition conjoining A and C is defined by its truth-value, which in turn is a function of the truth-values of A and C, it follows that there is no proposition that satisfies the Equation (Lewis, 1976). Hence, if the meaning of a conditional such as "If A then C" is defined by the Equation, it cannot be a proposition (Adams, 1975; Bennett, 2003; Douven & Verbrugge, 2010; Edgington, 1995; Lewis, 1976). The consequences are far-reaching: if we accept the Equation, conditionals do not have truth conditions. This seems to be a high price to pay for an explanation of the conditional. For instance, it would not be meaningful to say that a conditional is true or false, and we could no longer characterize mental states such as believing or doubting a conditional as propositional attitudes (i.e., attitudes about a proposition). Why should we pay that price?

3.2 Arguments for the Probability Conditional

Let us first see how the probability conditional fares with regard to the objections against the material conditional. We saw above that the material conditional has some counterintuitive implications when the antecedent is false or when the consequent is true. Considering an antecedent that is certainly false (i.e., $P(A) = 0$), we can follow Bennett and regard the conditional as undefined or follow Stalnaker's extended Ramsey test and regard the conditional as more or less credible depending on how high $P(C|A)$ is after we have adjusted our beliefs to accommodate A. Either version saves us from having to endorse a conditional simply because its antecedent is false. Considering a case where the antecedent A is just very unlikely, but $P(A) > 0$, the Equation ensures that nothing paradoxical follows: $P(C|A)$ can still take on any value.

The same rationale applies to "true" consequents, when they actually are just very likely. It might well be that Peter always goes for a run before he goes to work, and those who know him think it very likely that he goes

for a run. However, "If Peter is injured, he goes for a run before he goes to work" is utterly unlikely. This example demonstrates that a high $P(C)$ does not imply a high $P(C|A)$, so that the probability conditional avoids the first paradox of the material conditional in most circumstances. Only when the consequent is a necessary truth, such that $P(C) = 1$, we are also forced to accept the conditional, because the conditional probability must be 1.[2]

Here comes another paradox of the material conditional: a material conditional "If A then C" is false if and only if A and $\neg C$, so from the falsity of "If A then C," we can infer that A and $\neg C$. Most people would agree that "If Peter is in London, then he is in France" is clearly false, regardless of whether Peter is in London or not. Yet the material-conditional account would require Peter to *be* in London for the conditional to be false. The probability conditional does not have that problem. In order to understand that, we have to clarify a consequence of the fact that conditionals have no truth conditions: when we ask, "What is the probability of $A \rightarrow C$?" we do *not* ask, "What is the probability that it is *true*?" A probability conditional can, of course, be true (when antecedent and consequent have obtained), and it can be false (when the antecedent has obtained but the consequent has not). Therefore, it *does* have a truth-value when the antecedent holds. However, in case the antecedent did not obtain (yet), it is neither true nor false; it is simply more or less likely. So one can restate the question how probable we think $A \rightarrow C$ is as how probable we think it is true *given that it has a truth-value*. The conditional has a truth-value when A is true, and in that case, it is true when $A \& C$ is true. Hence, the probability of $A \rightarrow C$, given that it has a truth-value, is $P(A \& C)/P(A)$, and that, of course, matches the conditional probability (Edgington, 1995). Turning back to our London–France example, we see that we have an extreme case: $P(C|A) = 0$. In this case, we can *say* that the conditional is "false" to express that our degree of belief is very low,[3] but in doing so, we do not assign the conditional a truth-value but refer to the conditional probability $P(C|A)$ instead (Jeffrey, 1991). We use "false" in a pleonastic sense, not referring to the value "false" in the truth table but to the zero probability of the conditional (Over & Baratgin, 2017). There are additional philosophical arguments in favor of the probability conditional (for a more detailed account, see Bennett, 2003). We now turn to the experimental evidence.

3.3 Empirical Evidence for the Probability Conditional

As mentioned above, participants in psychological experiments do not judge that false-antecedent cases render the conditional true but judge those cases to be irrelevant for the truth of the conditional (Johnson-Laird & Tagart, 1969). This is in line with the probabilistic account, because the fact that A is false has no bearing on $P(C|A)$. However, this finding can also be explained by an alternative theory that builds on the material conditional for its assumptions on how people represent conditional statements: mental model theory (Johnson-Laird & Byrne, 2002).[4] This theory is currently the main competitor to the probabilistic approach in the psychology of reasoning, and therefore we briefly introduce it next.

Mental model theory (MMT) assumes that the material conditional expresses the fully elaborated core meaning of conditionals. In addition, the theory specifies how conditionals are mentally represented and the psychological processes operating on them when people make judgments about conditionals or engage in conditional reasoning. MMT states that conditionals are represented as a set of mental models. A complete representation of a "basic" conditional (i.e., a conditional for which no contextual or knowledge factors modulate its core meaning) includes one model for each truth table case that makes the material conditional true (i.e., all cases that are possible, assuming the conditional is true):

$$A \quad C$$
$$\neg A \quad C$$
$$\neg A \quad \neg C$$

In this notation, each line is a mental model representing one possible state of the world. Note that only "true" rows of the truth table are represented as models, not the false ones (i.e., "$A \& \neg C$" is not represented because it would render the conditional false; Johnson-Laird & Byrne, 2002). MMT also states that usually people only represent the first model explicitly and add the others implicitly, leaving them unspecified:

$$A \quad C$$
$$[\ldots]$$

The three dots in square brackets are an "elliptical" representation of the unspecified cases. The full set of three models can be fleshed out from the initial representation, but that requires additional mental capacity and therefore is often not done.

These assumptions provide an explanation of why people tend to judge false-antecedent cases as "irrelevant" for the truth-value of conditionals in truth table cases: people fail to flesh out the full set of models and therefore cannot fit the $\neg A$-cases into any mental model of the conditional (Johnson-Laird & Byrne, 2002). However, this explanation has a problem: MMT also predicts—correctly—that people judge the conditional as "false" in case of $A \& \neg C$. Yet, this "false"

case of the truth table is not explicitly represented as a mental model, and as such, it has the same status as the $\neg A$-cases with respect to the initial, sparse representation in which only the $A \& C$–case is explicitly represented. Inferring that $A \& \neg C$ is incompatible with the conditional requires fleshing out the fully explicit set of three models (Johnson-Laird & Byrne, 2002). There is no mechanism that can explain how, when only the initial "$A \& C$" model is explicitly represented, $A \& \neg C$ can be distinguished from $\neg A \& C$ and $\neg A \& \neg C$. Therefore, in the truth table task, participants should be predicted to judge all three cases as irrelevant or all three as false.

The classic truth table task provides only indirect evidence for the probabilistic account of conditionals. A more direct test of the Equation would be to see whether participants really adhere to it in their judgments of probabilities.

Evans, Handley, and Over (2003) and Oberauer and Wilhelm (2003) examined whether participants' judgments about the probability of a conditional "If A then C" corresponded to the conditional probability $P(C|A)$ or to the probability according to the material-conditional account (i.e., $1 - P(A \& \neg C)$). Participants were tested on abstract conditionals concerning a set of cards that had, for example, colored letters printed on it (e.g., "If there is an A on the card, it is blue"). For each conditional, they were given the relevant frequencies of the four truth table cases (e.g., of the 100 cards, 50 had a blue A, 10 a red A, 20 a blue B, and 20 a red B printed on it). Participants were then asked to estimate the probability of the conditional. Both studies found strong evidence for the probabilistic account and against the material conditional: the probability of the conditional was highly correlated with $P(C|A)$, and a majority of participants was classified as adhering to the Equation.

Over, Hadjichristidis, Evans, Handley, and Sloman (2007) went a step further: instead of giving the relevant frequencies of the truth table cases, they used everyday conditionals and had participants judge the respective probabilities (with the restriction that they had to sum to 100%). Only the conditional probability (computed from the probability ratings for individual events) was correlated with the rated probability of the conditional. However, Evans and colleagues (2003; Evans, Handley, Neilens, & Over, 2007) as well as Oberauer and Wilhelm (2003) also found evidence of a conjunctive interpretation of conditionals: for a substantial subgroup of participants, $P(A \to C)$ was mostly influenced by $P(A \& C)$. This pattern of probability judgments is referred to as the "conjunctive pattern."

How can we explain the "conjunctive pattern" of probability judgments? Clearly, people do not think

$A \& C$ to be equal to $A \to C$, as a person equating the two would be utterly incapable of hypothetical thought.[5] It could be argued that the effect is task specific: in the tasks in which this pattern occurred, participants always had to judge how likely a conditional was true for a randomly drawn subset of cards. If there are participants who regard the conditional as true only if at least one of the cards was an "$A \& C$" card, this could explain the conjunctive pattern. Another possibility, suggested by Evans and colleagues (2003), is that this pattern reflects an incomplete Ramsey test. Participants set out to assess the probability of C on the supposition of A (i.e., once A is added to the stock of knowledge). Doing so requires one to compare the probability of resulting C-cases (i.e., $A \& C$) with that of the resulting $\neg C$-cases (i.e., $A \& C$). If participants stopped halfway during this procedure, omitting the computation of $P(A \& \neg C)$, that could explain why they come up with $P(A \& C)$ as the probability of $A \to C$. Oberauer, Geiger, Fischer, and Weidenfeld (2007) tested these two explanations but rejected both.

There are two other explanations for the conjunctive pattern: Edgington (2003) suggests that the wording of the question is ambiguous. As made clear above, the probability of a conditional is *not* the probability that it is true. Yet this is exactly what was asked in the experiments. So some participants could have ignored the "true" at the end of the sentence and answered with $P(C|A)$, while others took the question literally and hence responded with the only cases where the conditional is true, that is, the "$A \& C$"–cases. As far as we know, this hypothesis has not been tested directly; however, Oberauer and Wilhelm (2003) found no conjunctive pattern when participants were not required to answer with a probability estimate but simply had to decide whether a conditional was true or false. According to Edgington's explanation, this question should have elicited predominantly conjunctive responses, contrary to what was found. This leads us to the last attempt to explain the conjunctive pattern: the effect could be due to difficult arithmetic operations that are necessary to compute the conditional probabilities. There is some evidence favoring this explanation. As mentioned before, the conjunctive pattern disappeared when participants did not need to compute any probability but simply had to *assess* if their probability was high enough for them to accept the conditional as "true" (Oberauer & Wilhelm, 2003). Similarly, Over and colleagues (2007) did not find evidence of a conjunctive interpretation with their everyday conditionals. Although their participants had to give a numerical estimation of the probability of the conditional, they did not have to calculate them on arbitrary numbers provided by the experimenters but could use

their own, not necessarily explicit, estimation.[6] And last but not least, evidence supporting the arithmetic-load explanation comes from Fugard, Pfeifer, Mayerhofer, and Kleiter's study (2011). They used a facilitated-response paradigm: participants were asked about probabilities of a six-faced die landing a certain way in an "x out of y" format (e.g., "2 out of 6"), where participants could just reuse the frequencies given. For example, frequencies could be "Two sides have a yellow circle, two a yellow square, and two a red circle." The estimate of the probability of the conditional "If the die shows a yellow side, it shows a circle" only requires minimal arithmetic ability (namely, adding the two yellow cases). Fugard and colleagues found that (a) overall, approximately 73% of the participants judged the conditional according to the conditional probability, and (b) this percentage increased over the 70 trials each participant had to go through: 55% of the participants shifted to eventually adopt the conditional probability, most of which had adopted a conjunctive interpretation before.

Overall, the experimental evidence for the interpretation of our everyday conditional as the probability conditional is substantial (for further evidence stemming from negated conditionals, see, e.g., Handley, Evans, & Thompson, 2006). Recently, however, the probability conditional has been criticized for ignoring an important aspect of the meaning of conditionals: relevance.

3.4 Inferentialism

"If Peter is wearing a blue shirt today, the sea levels will rise." This conditional strikes most people as odd at least, because the antecedent has no apparent *relevance* for the consequent. Yet, under the material-conditional and probabilistic account, we are forced to accept it respectively as true or highly probable, simply because sea level rise is highly probable. What is wrong with this conditional is the missing semantic link between the antecedent and the consequent (and we therefore refer to it as a "missing-link conditional," in accordance with Douven, 2015). This is the main point of the critique of inferentialism (Douven, 2015; Krzyżanowska, 2015). Inferentialism states that a conditional is only true if there is a sufficiently strong argument from its antecedent, plus relevant background knowledge, to its consequent (Douven, 2015; Krzyżanowska, Wenmackers, & Douven, 2014). It requires an *inferential connection* between antecedent and consequent, which can be a deductive, abductive, or inductive one (Krzyżanowska et al., 2014).

Skovgaard-Olsen, Singmann, and Klauer (2016) tested this requirement experimentally. They found that participants' judgments were predicted almost perfectly by the Equation for regular conditionals, but this relation failed for missing-link conditionals. To be precise, the relation failed for some participants (who, irrespective of the conditional probability, assigned low probability to missing-link conditionals), while others still adhered to the conditional-probability account. This group pattern was confirmed by Skovgaard-Olsen, Kellen, Hahn, and Klauer (2017), who found that only a minority (17%) of participants adhered to the Equation for missing-link conditionals. Another approach to empirically test inferentialism against the probability conditional is offered by "and-to-if" inferences (also referred to as "centering": from "A & C" follows "If A then C"). This inference is valid according to both material- and probability-conditional accounts but invalid for an inferentialist. Although people indeed endorse centering (Cruz, Baratgin, Oaksford, & Over, 2015), that seems to be the case only with conditionals where the antecedent is *relevant* for the consequent and not with missing-link conditionals (Skovgaard-Olsen, Kellen, et al., 2017; Skovgaard-Olsen, Singmann, & Klauer, 2017). The main point of disagreement between proponents of the probability conditional and proponents of inferentialism is whether or not the effect of relevance is of a pragmatic or a semantic nature (i.e., whether or not it affects the meaning of the conditional), a question that is still hotly debated (see Douven, Elqayam, Singmann, & van Wijnbergen-Huitink, 2018; Krzyżanowska, Collins, & Hahn, 2017).

4. Probabilistic Accounts of Reasoning with Conditionals

As mentioned in the introductory section, if one accepts the material conditional as the meaning of conditionals, the usual pattern of people's evaluations of conditional inferences implies that people are extraordinarily bad at reasoning: the only inference they are really good at is modus ponens. Mental model theory (Johnson-Laird & Byrne, 2002), which accepts the material conditional as the normative standard, explains these findings by the failure of people to fully flesh out the set of mental models. MP and AC are endorsed if only the initial model (A & C) is represented. If participants manage to add a further model "¬A & C," they correctly endorse MP and reject AC and DA. To correctly endorse MT, they would have to further add the model "¬A & ¬C," because it is the only model of the conditional that is also compatible with the minor premise of the MT argument, "not C." Hence, if the fully explicit set of three models is fleshed out, participants accept the valid inferences (MP and MT) and reject the fallacies (AC and DA), which would be the normatively correct response by the material-conditional interpretation.

As we have seen in the previous section, there is strong evidence that people interpret conditional statements according to the probability conditional. How do probabilistic theories account for people's pattern of endorsement and rejection of the four simple conditional inferences? One approach is to understand inferences as Bayesian updating of our knowledge in light of the premises (Oaksford & Chater, 2007). This "purely Bayesian" approach interprets the four inferences as inferring the conditional probabilities of the conclusion, given the minor (categorical) premise. For instance, confronted with a DA argument of the form "If A then C; not C; therefore not A," reasoners are assumed to draw on their knowledge about C and A to assess the probability of "not A," given "not C." Oaksford, Chater, and Larkin (2000; see also Oaksford & Chater, 2001) were the first to propose this interpretation as a formal reasoning model. To model the person's knowledge, they require three parameters to capture the joint subjective probability distribution of A and C (i.e., the four truth table cases), from which the respective conditional probabilities can be computed: $e = 1 - P(C|A)$, referred to as the "exceptions parameter," $a = P(A)$ and $b = P(C)$. From this joint probability distribution, MP endorsement rate is predicted by $P(C|A)$, MT by $P(\neg A|\neg C)$, AC by $P(A|C)$, and DA by $P(\neg C|\neg A)$.

Oberauer (2006) compared the Oaksford–Chater model with a formalization of MMT and found that the probabilistic account did not fare well in explaining people's pattern of endorsement and rejection of conditional inferences. The problem for the Oaksford–Chater model is that to explain the high MP acceptance rate, it must estimate the exceptions parameter e to a very small value. This implies that the model also predicts high MT acceptance, contrary to the data. As a compromise, the model, when fit to data, has been found to underestimate endorsement of MP and overestimate endorsement of MT inferences (Oberauer, 2006; Schroyens & Schaeken, 2003). Oaksford and Chater (2007, 2013) presented a revision of their model in which they introduce a new exceptions parameter e', derived by updating e on a single occurrence of a counterexample to the conditional (i.e., an occurrence of an "A and ¬C" example). Whereas MP endorsement is still estimated by the "old" exceptions parameter e, estimation for all other inferences is based on the new parameter. This solves the MP–MT asymmetry problem, because MP and MT are now decoupled.

Oaksford and Chater (2007, 2013) justify the introduction of e' by arguing that the minor premises given in the context of inferences MT, AC, and DA provide information that is equivalent to presenting a counterexample to the conditional. Therefore, the endorsement rates of these inferences should be predicted with a different exceptions parameter. In our view, this justification is contrived: learning that "not A" or that "not C" does not provide reasons to question "If A then C." No amount of people who are not hit by a car, or of people who are not injured, should undermine our belief in the conditional "If a person is hit by a car, they will be injured." Therefore, although the revised model of Oaksford and Chater is better able to accommodate the data, it does so at the price of being less principled than the original model.

Another purely Bayesian approach, similar to the original Oaksford–Chater model, is based on Bayesian networks, a probabilistic graphical model that represents the dependencies between variables as a directed acyclic graph (Pearl, 2009). While the two approaches are similar, they are certainly not equivalent, but due to space restrictions, we cannot go into more detail here (for a detailed account, see chapter 4.2 by Hartmann and chapter 7.1 by Pearl in this handbook). One thing both approaches have in common is that they explain conditional reasoning as an inference from the minor premise to the conclusion.

But then, what is the role of the conditional (i.e., the major premise) in the inference task? Or put differently, what is the difference between inferring C from the two premises "If A then C" and A (i.e., the full MP argument) and inferring C from A alone (i.e., a reduced MP argument), as both are interpreted as estimating $P(C|A)$? Within the Bayesian framework, it is not clear how adding the conditional premise could play a role in knowledge updating. Normally, in Bayesian updating, additional premises are conditionalized on, that is, we assess $P(\text{conclusion} | \text{all premises})$. However, conditionalizing is not possible when the additional premise is itself a conditional probability. Oaksford and Chater (2007) propose that the effect of adding a conditional premise should be reflected in a decreased exceptions parameter $e = 1 - P(C|A)$, that is, the degree of belief in the sufficiency of the antecedent for the consequent increases. Hartmann and Rafiee Rad (2012) propose that instead of only increasing $P(C|A)$, all parameters capturing the joint probability distribution are adjusted so that the Kullback–Leibler distance between the old and the new distributions is minimized.

Klauer, Beller, and Hütter (2010) investigated the effect of the conditional on the inferences by presenting to participants both reduced inferences (i.e., without the major premise) and full inferences (i.e., the standard MP, MT, AC, and DA). The reduced inferences are viewed as an indication of how much the conclusion is endorsed based *only* on background knowledge. For example, when given the information that the streets are wet, most people would conclude that it probably has rained, even if they haven't been told "If it rains, the streets are wet." To model the effect of adding the conditional to the inference, Klauer and colleagues (2010) developed the dual-source model (DSM), a mixture model that implements Oaksford, Chater, and Larkin's (2000) probabilistic model to account for the reduced inferences (the "knowledge" component in the DSM terminology) but adds two components for the full inferences: (a) four form parameters, which capture to what extent the inference forms (i.e., MP, MT, AC, and DA) are seen as logically valid, and (b) a weighting parameter, which indicates how much weight is given to the form-based information relative to the knowledge-based information. Singmann, Klauer, and Beller (2016) compared the original and the revised Oaksford–Chater model (2000, 2007, 2013), the Bayes nets approach with Kullback–Leibler distance minimization (Hartmann & Rafiee Rad, 2012), and the DSM. They found that the DSM explains substantially more variance than the competing models and also confirmed a unique prediction of the DSM. Singmann and colleagues conclude that the effect of adding a conditional goes beyond updating the knowledge base (i.e., the joint probability distribution). Instead, they argue, adding the conditional to the minor premises provides participants with an inference form that is then evaluated in terms of logical soundness. It is important to note that logical soundness does *not* demand one to refer to bivalued logic: MP and MT are also *probabilistically* valid, whereas DA and AC are not (for an introduction to probability logic, see chapter 4.4 by Pfeifer, this handbook).

Further experiments provide support for the DSM by demonstrating that the form parameters can be selectively influenced by experimental variations of the inference form (e.g., to biconditional, "only if" vs. "if–then," or adding premises). Conversely, the form parameters do not change when the inference form is held constant (Klauer et al., 2010; Singmann et al., 2016). One limitation of the DSM is that, as a measurement model, it serves well to measure the effect of inference form but does not provide an explanation of *how* the validity of the inference forms is assessed.

Insight into what causes the effects of inference form could be gained by looking at a different class of theories, known as "dual-process" theories of reasoning. Various dual-process theories exist, even in the small world of conditional reasoning, but we will focus only on the general idea (for a detailed account, see chapter 2.5 by Klauer, this handbook). As the name suggests, they assume two processes, one (usually called Type 1) fast, automatic, and intuitive, the other (usually called Type 2) slower, controlled, and relying strongly on working-memory capacity. The dual-process framework thus naturally explains reasoning from the inference form in terms of the Type 2 process, whereas the knowledge component can be regarded as a Type 1 process. A candidate for a Type 2 process would be, for example, mental model theory, and indeed, Verschueren, Schaeken, and d'Ydewalle (2005) have proposed a dual-process model that integrates probabilistic inference as a Type 1 process and mental models as a Type 2 process. This model fared very well in a formal comparison of several competing models of conditional inferences (Oberauer, 2006).

But, as we have seen above, mental model theory builds on the material conditional, an assumption that is not supported by the empirical data, so we are left with a model that accounts for conditional inferences but is based on an interpretation of the conditional that is untenable. Geiger and Oberauer (2010) have sketched one potential way to keep the explanatory power of MMT for conditional inferences while avoiding the material-conditional interpretation of "if–then" premises. According to this proposal, the meaning of a conditional is not directly captured by a set of mental models that represent its truth conditions—rather, it is incorporated in procedural knowledge. A conditional statement "If A then C" is understood as an instruction to add an element C to every mental model of A in working memory. Conditional reasoning starts with building a mental model of the minor (categorical) premise. In case that model includes A, the procedure implementing the conditional adds C to it. For instance, when thinking through an MP argument, the reasoner first builds a mental model of A in working memory, representing the minor premise. The conditional premise instructs a procedure to augment this model, resulting in the model "$A \& C$," from which the conclusion "C" follows immediately. Thinking through an MT argument involves more steps: to represent the minor premise, the first model is "$\neg C$." The reasoner now tentatively augments this model by adding "A" and by adding "$\neg A$," effectively reasoning from the supposition that A, or the supposition that not A. The first augmentation results in a model "$A \& \neg C$," which triggers the procedure that incorporates the conditional premise. That procedure adds "C" to the model, in conflict with the "$\neg C$" element, and that conflict signals the contradiction between the supposition of A and the premises, which leads to elimination of the "$A \& \neg C$" model. The remaining "$\neg A \& \neg C$" model licenses the conclusion "not A."

5. Conclusion

Just rethink for a moment: if you had not known, would you have guessed that the probability of the conditional, judged by laypeople, is approximately equal to the conditional probability? On the one hand, this seems implausible, as calculating conditional probabilities is difficult. On the other hand, uncertainty pervades our knowledge of the world, and we have a rough sense of how likely certain events are to happen and how credible certain claims are. Probabilities are merely a way of expressing our degrees of certainty numerically. Moreover, a hallmark of human rationality is the consideration of hypothetical scenarios (Evans & Over, 2004): when we plan a course of action, we think about what is likely to happen if we do this or that. When we reflect on something that went wrong, we consider whether things would have turned out better if one or another unfortunate event had not happened.

Thinking about hypothetical scenarios does not require the ability to calculate conditional probabilities. It requires to form a representation of the hypothetical state—the supposition—and to use relevant knowledge to infer what is likely to happen, perhaps through a mental simulation (Hegarty, 2004). When people are asked how certain they are of their beliefs about what would happen, they can express their degree of certainty on a coarse scale of subjective probabilities; Khemlani, Lotstein, and Johnson-Laird (2015) have proposed a simple algorithm for arriving at such coarse probability estimates on the basis of mental models. Thinking about what happens, or would have happened, in hypothetical scenarios is an assessment of conditional probabilities: how probable is outcome C, given the hypothetical event or action A? Speaking about such hypothetical scenarios involves using conditionals (e.g., "If I go home early, my boss won't be pleased. I better stay another hour"). Therefore, a close link between subjective conditional probabilities and the meaning of a conditional statement appears plausible.

The probability conditional and the Equation have received compelling support from studying how people *understand* conditionals (Handley et al., 2006; Oberauer & Wilhelm, 2003; Over et al., 2007). At the same time, the question of how people *reason* with conditionals is still unresolved, both on the normative and the descriptive level. On the normative side, the purely Bayesian approaches have no rational procedure for updating knowledge, given conditional premises. Probability logic could be a possible answer to that problem, but that would mean abandoning the purely Bayesian approach to reasoning. On the descriptive side, people's rates of endorsement of conditional inferences under various conditions are explained only poorly by Bayesian updating models. Dual-source or dual-process models, in which Bayesian updating represents the knowledge-based process, appear more promising. These models require more work devoted to spelling out the form-based process in which reasoners use conditional premises to inform their inferences.

Notes

1. By "classical logic," we are referring to the standard bivalued logic, like, for example, first-order or propositional logic.

2. This has been the source of some criticism of the probability conditional, as we will see later on.

3. Obviously, the same rationale applies to "true" when our degree of belief is very high.

4. There have been various attempts at a substantial revision of mental model theory; see, for example, Barrouillet, Gauffroy, and Lecas (2008) and Johnson-Laird, Khemlani, and Goodwin (2015). For a discussion of the plausibility of these revisions, see Oberauer and Oaksford (2008) and Baratgin et al. (2015), respectively.

5. This person would, after hearing the warning "If you touch the wire, you will get an electric shock," answer "No, that is not true" because he does not touch the wire and is not shocked.

6. We do *not* suggest that people ask themselves first "What is the probability of $A \& C$?" and "What is the probability of $A \& \neg C$?" and then compute a conditional probability (see our remarks on the Ratio formula above) but that people can directly assess $P(C \mid A)$ if they have the relevant background knowledge, as is the case in Over and colleagues' experiment, which used everyday causal conditionals.

References

Adams, E. W. (1975). *The logic of conditionals: An application of probability to deductive logic.* Dordrecht, Netherlands: Reidel.

Baratgin, J., Douven, I., Evans, J. St.B.T., Oaksford, M., Over, D., & Politzer, G. (2015). The new paradigm and mental models. *Trends in Cognitive Sciences, 19*(10), 547–548.

Barrouillet, P., Gauffroy, C., & Lecas, J. F. (2008). Mental models and the suppositional account of conditionals. *Psychological Review.*

Bennett, J. (2003). *A philosophical guide to conditionals.* Oxford, England: Clarendon Press.

Cruz, N., Baratgin, J., Oaksford, M., & Over, D. E. (2015). Bayesian reasoning with ifs and ands and ors. *Frontiers in Psychology, 6,* 192.

Douven, I. (2015). How to account for the oddness of missing-link conditionals. *Synthese, 194,* 1541–1554.

Douven, I., Elqayam, S., Singmann, H., & van Wijnbergen-Huitink, J. (2018). Conditionals and inferential connections: A hypothetical inferential theory. *Cognitive Psychology, 101,* 50–81.

Douven, I., & Verbrugge, S. (2010). The Adams family. *Cognition, 117*(3), 302–318.

Edgington, D. (1995). On conditionals. *Mind, 104*(414), 235–329.

Edgington, D. (2003). What if? Questions about conditionals. *Mind & Language, 18*(4), 380–401.

Evans, J. St. B. T., Handley, S. J., Neilens, H., & Over, D. E. (2007). Thinking about conditionals: A study of individual differences. *Memory & Cognition, 35*(7), 1772–1784.

Evans, J. St. B. T., Handley, S. J., Neilens, H., & Over, D. (2008). Understanding causal conditionals: A study of individual differences. *Quarterly Journal of Experimental Psychology, 61*(9), 1291–1297.

Evans, J. St. B. T., Handley, S. J., & Over, D. E. (2003). Conditionals and conditional probability. *Journal of Experimental Psychology: Learning, Memory, and Cognition, 29*(2), 321–335.

Evans, J. St. B. T., Newstead, S. E., & Byrne, R. M. (1993). *Human reasoning: The psychology of deduction.* Hillsdale, NJ: Erlbaum.

Evans, J. St. B. T., & Over, D. E. (2004). *If.* Oxford, England: Oxford University Press.

Evans, J. St. B. T., Over, D. E., & Manktelow, K. I. (1993). Reasoning, decision making and rationality. *Cognition, 49*(1–2), 165–187.

Fugard, A. J. B., Pfeifer, N., Mayerhofer, B., & Kleiter, G. D. (2011). How people interpret conditionals: Shifts toward the conditional event. *Journal of Experimental Psychology: Learning, Memory, and Cognition, 37*(3), 635–648.

Geiger, S. M., & Oberauer, K. (2010). Towards a reconciliation of mental model theory and probabilistic theories of conditionals. In M. Oaksford & N. Chater (Eds.), *Cognition and conditionals: Probability and logic in human thinking* (pp. 289–307). Oxford, England: Oxford University Press.

Handley, S. J., Evans, J. St. B. T., & Thompson, V. A. (2006). The negated conditional: A litmus test for the suppositional conditional? *Journal of Experimental Psychology: Learning, Memory, and Cognition, 32*(3), 559–569.

Hartmann, S., & Rafiee Rad, S. (2012). *Updating on conditionals = Kullback–Leibler + causal structure.* Unpublished presentation. Retrieved from https://homepages.uni-regensburg.de/~pfn23853/rffcw/h.pdf

Hegarty, M. (2004). Mechanical reasoning by mental simulation. *Trends in Cognitive Sciences, 8*(6), 280–285.

Jeffrey, R. (1991). Matter-of-fact conditionals. *Proceedings of the Aristotelian Society, Supplementary Volumes, 65,* 161–183.

Johnson-Laird, P. N., & Byrne, R. M. J. (2002). Conditionals: A theory of meaning, pragmatics, and inference. *Psychological Review, 109*(4), 646–678.

Johnson-Laird, P. N., Khemlani, S. S., & Goodwin, G. P. (2015). Logic, probability, and human reasoning. *Trends in Cognitive Sciences.*

Johnson-Laird, P. N., & Tagart, J. (1969). How implication is understood. *American Journal of Psychology, 82*(3), 367–373.

Khemlani, S. S., Lotstein, M., & Johnson-Laird, P. N. (2015). Naive probability: Model-based estimates of unique events. *Cognitive Science, 39*(6), 1216–1258.

Klauer, K. C., Beller, S., & Hütter, M. (2010). Conditional reasoning in context: A dual-source model of probabilistic inference. *Journal of Experimental Psychology: Learning, Memory, and Cognition, 36*(2), 298–323.

Krzyżanowska, K. (2015). *Between "if" and "then": Towards an empirically informed philosophy of conditionals* (Unpublished doctoral dissertation). University of Groningen, Netherlands. Retrieved from http://www.rug.nl/research/portal/files/15864231/Title_and_contents_.pdf

Krzyżanowska, K., Collins, P. J., & Hahn, U. (2017). Between a conditional's antecedent and its consequent: Discourse coherence vs. probabilistic relevance. *Cognition, 164*(Suppl C), 199–205.

Krzyżanowska, K., Wenmackers, S., & Douven, I. (2014). Rethinking Gibbard's riverboat argument. *Studia Logica, 102*(4), 771–792.

Lewis, D. (1976). Probabilities of conditionals and conditional probabilities. In W. L. Harper, R. Stalnaker, & G. Pearce (Eds.), *Ifs* (pp. 129–147). Dordrecht, Netherlands: Springer.

Oaksford, M., & Chater, N. (2001). The probabilistic approach to human reasoning. *Trends in Cognitive Sciences, 5*(8), 349–357.

Oaksford, M., & Chater, N. (2007). *Bayesian rationality: The probabilistic approach to human reasoning.* Oxford, England: Oxford University Press.

Oaksford, M., & Chater, N. (2013). Dynamic inference and everyday conditional reasoning in the new paradigm. *Thinking & Reasoning, 19*(3), 346–379.

Oaksford, M., Chater, N., & Larkin, J. (2000). Probabilities and polarity biases in conditional inference. *Journal of Experimental Psychology: Learning, Memory, and Cognition, 26*(4), 883–899.

Oberauer, K. (2006). Reasoning with conditionals: A test of formal models of four theories. *Cognitive Psychology, 53*(3), 238–283.

Oberauer, K., Geiger, S. M., Fischer, K., & Weidenfeld, A. (2007). Two meanings of "if"? Individual differences in the interpretation of conditionals. *Quarterly Journal of Experimental Psychology, 60*(6), 790–819.

Oberauer, K., & Oaksford, M. (2008). What must a psychological theory of reasoning explain? Comment on Barrouillet, Gauffroy, and Lecas (2008). *Psychological Review, 115*(3), 773–778.

Oberauer, K., & Wilhelm, O. (2003). The meaning(s) of conditionals: Conditional probabilities, mental models, and personal utilities. *Journal of Experimental Psychology: Learning, Memory, and Cognition, 29*(4), 680–693.

Over, D. E., & Baratgin, J. (2017). The "defective" truth table: Its past, present, and future. In N. Galbraith, E. Lucas, & D. E. Over (Eds.), *The thinking mind: A Festschrift for Ken Manktelow* (pp. 15–28). London, England: Routledge.

Over, D. E., Hadjichristidis, C., Evans, J. St. B. T., Handley, S. J., & Sloman, S. A. (2007). The probability of causal conditionals. *Cognitive Psychology, 54*(1), 62–97.

Pearl, J. (2009). *Causality: Models, reasoning, and inference* (2nd ed.). Cambridge, England: Cambridge University Press. (Original work published 2000)

Ramsey, F. P. (1990). *Foundations of mathematics and other logical essays* (R. B. Braithwaite, Ed.). London, England: Routledge. (Original work published 1929)

Schroyens, W. (2010). A meta-analytic review of thinking about what is true, possible, and irrelevant in reasoning from or reasoning about conditional propositions. *European Journal of Cognitive Psychology, 22*(6), 897–921.

Schroyens, W., & Schaeken, W. (2003). A critique of Oaksford, Chater, and Larkin's (2000) conditional probability model of conditional reasoning. *Journal of Experimental Psychology: Learning, Memory, and Cognition, 29*(1), 140–149.

Singmann, H., Klauer, K. C., & Beller, S. (2016). Probabilistic conditional reasoning: Disentangling form and content with the dual-source model. *Cognitive Psychology, 88*, 61–87.

Skovgaard-Olsen, N., Kellen, D., Hahn, U., & Klauer, K. C. (2017). Conditionals, individual variation, and the scorekeeping task. In M. Gunzelmann, A. Howes, T. Tenbrink., & E. Davelaar (Eds.), *Proceedings of the 39th Annual Meeting of the Cognitive Science Society* (pp. 1084–1089). London, England: Cognitive Science Society. Retrieved from https://philpapers.org/rec/SKOCIV

Skovgaard-Olsen, N., Singmann, H., & Klauer, K. C. (2016). The relevance effect and conditionals. *Cognition, 150*, 26–36.

Skovgaard-Olsen, N., Singmann, H., & Klauer, K. C. (2017). Relevance and reason relations. *Cognitive Science, 41*, 1202–1215.

Stalnaker, R. C. (1968). A theory of conditionals. In W. L. Harper, R. Stalnaker, & G. Pearce (Eds.), *Ifs* (pp. 41–55). Dordrecht, Netherlands: Springer.

Verschueren, N., Schaeken, W., & d'Ydewalle, G. (2005). Everyday conditional reasoning: A working memory–dependent tradeoff between counterexample and likelihood use. *Memory & Cognition, 33*(1), 107–119.

4.7   Representing Belief: Beyond Probability and Logic

Didier Dubois and Henri Prade

Summary

This chapter surveys recent approaches to the representation of belief. There is a clash between the notion of degree of belief in the subjective probability tradition and the idea of certainty as accepted belief, often couched in the language of logic, especially modal logic. The attempt to consider degrees of certainty finds its origin in the works of Francis Bacon, often opposed to the one of Pascal. However, by reviewing some more recent trends in the representation of uncertainty, such as possibility theory, ranking theory, evidence theory, and imprecise probability, one may argue that these novel approaches try to bridge the gap between the two traditions, even if dropping some favorite properties on the way, such as the additivity of degrees of belief and the adjunction law for accepted beliefs.

1.   A Clash of Traditions

There are two traditions for representing belief, one based on probability, one based on modal logic. The probability tradition goes back to Pascal and other scientists of his time, who tried to measure our confidence in events (today we also speak of "propositions" in the logical sense). In those times, a distinction was made between so-called chances and probabilities (Shafer, 1978). The word "chance" was dedicated to repeatable events and measured via a combinatorial count of the frequencies of occurrence, assuming elementary events had equal chances to occur. The name "probability" was understood as a subjective notion, typically the confidence to be granted to a testimony. The latter view is still prevalent in important texts from the 18th century, such as d'Alembert and Diderot's encyclopedia.[1] The concept of subjective probability was to be taken up again in the 20th century by scholars like Ramsey (1926/1931) and de Finetti (1937). They tried to justify why degrees of belief should be additive. However, additivity can only be obtained by introducing severe restrictions on the

model of graded belief namely, it essentially prevents capturing the notion of ignorance. The reader is referred to the rather recent compilation by Huber and Schmidt-Petri (2009) for a survey of approaches to degrees of belief.

In contrast, the modal logic tradition was developed in the 20th century, even if modal logic per se goes back to medieval philosophy and even to ancient Greek philosophy. Hintikka (1962) proposed a so-called epistemic logic that tried to define a rational logical framework for the idea of knowledge and its articulation with an all-or-nothing notion of belief. Namely, knowledge is essentially true belief, while a simply believed proposition is not supposed to be actually true. In this setting, modeling ignorance is obvious as you just express that you believe neither a proposition nor its negation. Besides, one essential feature of this approach is that when you know that each of two propositions is true, you also know that their conjunction is true, which is sometimes called the *adjunction axiom* (Kyburg, 1997). This axiom is also supposed to hold for beliefs, with the understanding that modal beliefs represent *accepted beliefs*, that is, propositions taken for granted to the point of reasoning with them as if they were true.

There is a clash of intuitions between the two views, because for them to match, one needs to explain the connection between graded belief and belief *simpliciter*. One way is to add a threshold to graded beliefs and to retain propositions above this threshold, but it is then almost impossible to preserve the adjunction axiom if we model graded beliefs by probabilities. This is the point made by the lottery paradox (Kyburg, 1961), discussed later in this chapter.

The aim of this chapter is to provide an overview of belief representations that relate, and go beyond, subjective probability and epistemic logic, in the sense that some of them are graded versions of the epistemic logic approach and all of them give up the additivity assumption of probability theory. Most of these approaches have been developed only since the middle of the 20th century, with a sudden acceleration from 1975 on. This chapter should be of particular interest to psychologists, because they often stick to the alternative "logic vs. probability" and are not always aware of the fact that there is a large area beyond.

2.   Rationality Axioms for Set Functions Representing Belief

In this section, we survey mathematical models of belief and axioms that have been proposed to represent them. We consider the most general approaches and show that accepting the rule of adjunction considerably narrows the expressive power of representations of graded belief.

2.1   Degrees of Belief

We assume throughout the chapter that belief qualifies propositions, represented by subsets A, B, C, . . . of a set S of possible worlds or states of affairs. More precisely, A stands for the proposition that event A occurs or equivalently that some entity of interest, say, the result x of a measurement, lies in A. We do not distinguish between logically equivalent propositions. Quantifying (the possible lack of) belief is one instance of expressing uncertainty for an agent.

The most usual representation of uncertainty consists in assigning to each proposition or event A a number $g(A)$ in the unit interval $[0, 1]$. It evaluates the confidence (we deliberately use a more general term than "belief" or "plausibility") of the agent in the truth of the proposition "$x \in A$." This proposition can only be true or false by convention, even if the agent may ignore its truth-value. The following requirements sound natural:

$$g(\varnothing) = 0, \qquad g(S) = 1,$$

as you cannot have confidence in contradictions, and you should believe in tautologies. Monotonicity with respect to inclusion is also natural:

If $A \subseteq B$ then $g(A) \leq g(B)$.

Indeed, if A is more specific than B in the wide sense (in other words, A implies B), a rational agent should not be more confident in A than in B: all things being equal, the more imprecise a proposition, the more certain it is. Under these properties, the function g is sometimes called a *capacity* (after Choquet, 1953), sometimes a *fuzzy measure* (after Sugeno, 1977). In order to stick to the uncertainty framework, it was also called a *confidence measure* (Dubois & Prade, 1988). Such a set function represents the epistemic state of an agent, that is, what an agent knows or thinks he or she knows (irrespective of whether it is true or not).[2]

An important special case of a confidence measure is the probability measure $g = P$, which satisfies the *additivity* property:

If $A \cap B = \varnothing$, then $P(A \cup B) = P(A) + P(B)$.

Probability measures have very often been used to represent degrees of belief, hence with a subjective flavor, as opposed to limits of relative frequencies of repeatable events, which account for random phenomena. Contrary to frequencies, belief may be attached to unique events. However, the additivity axiom sounds much more natural when modeling frequencies than for degrees of belief. Actually, the latter were not supposed to be additive in pioneering works by Bernoulli and Lambert (Shafer, 1978).

To recover additivity, scientists of the mid-20th century such as de Finetti (1937) interpreted $P(A)$ as the amount of money the agent is ready to pay for buying a lottery ticket that earns \$1 if the proposition A is true. Moreover, it is assumed that the price is fair, that is, the agent would accept to sell this ticket under the same conditions for the same price, and wants to avoid sure loss. This interpretation of degrees of belief entails the additivity axiom. Hence, to represent degrees of belief, probability measures should be used. Such probabilities are called "subjective" (see chapter 4.1 by Hájek & Staffel, this handbook). Bayesian probabilists consider nonadditive degrees of belief irrational, since they incur sure money loss.

Remark: Supplying precise degrees of beliefs sounds cognitively very demanding. It seems more natural for an agent expressing his or her knowledge to only represent the relative strength of confidence in various propositions, rather than trying to force him or her to deliver numerical evaluations. It is indeed easier to assert that one proposition is more credible than another than to assess a particular degree of belief (whose meaning is not always simple to grasp) or even to guess a frequency for each of them. The idea of representing uncertainty by means of partial order relations over a set of events dates back to Ramsey (1926/1931), de Finetti (1937), and Koopman (1940). They tried to find an ordinal counterpart to subjective probabilities. Later, philosophers of logic such as David Lewis (1973) have considered other types of relations, including comparative possibilities in the framework of modal logic. We omit discussing this literature for the sake of brevity.

2.2   Accepted Belief

Alternatively to measuring degrees of belief, one may want to consider propositions as beliefs accepted by an agent, if the latter is ready to reason as if such believed propositions were true. In particular, if A and B are beliefs, the conjunction $A \cap B$ is also a belief (a debated issue). But the empty set should not be a belief. Moreover, if A is a belief, and $A \subseteq B$, then B is a belief. In other words, a set of beliefs should be consistent and deductively closed.

Consider a Boolean-valued set function N from 2^S to $\{0, 1\}$ (here, 2^S is the power set of S, i.e., the set of all subsets of S) such that $N(A) = 1$ if A is believed and $= 0$ otherwise. It is clear that N is a confidence measure (monotonic under inclusion) that satisfies the law of "minitivity":

$$N(A \cap B) = \min(N(A), N(B)).$$

This function is a special case of necessity measures (Dubois & Prade, 1988). It models the idea of certainty. It is easy to check that in the finite case, there exists a nonempty subset $E = \cap \{A: N(A) = 1\}$, and that $N(A) = 1$ if and only if $E \subseteq A$. The set E contains all states the agent does not consider impossible, given his or her beliefs, which is the simplest representation of an agent's epistemic state.[3] In other words, the function N describes propositions that can be proved from the epistemic state E.

2.3 Extracting Beliefs from Confidence Measures

In order to relate accepted beliefs to graded beliefs, one should extract accepted beliefs from a confidence function g. A natural way of proceeding is to define a belief as a proposition in which an agent has enough confidence. So we should define a positive belief threshold β such that A is a belief if and only if $g(A) \geq \beta > 0$. However, the closure of accepted beliefs under conjunction entails the following property:

Accepted belief postulate: *If $g(A) \geq \beta$ and $g(B) \geq \beta$, then $g(A \cap B) \geq \beta$.*

This requirement is very strong. As the set of accepted beliefs should not include the empty set, it is clear that one should have $\min(g(A), g(B)) < \beta$ if $A \cap B = \varnothing$. Worse, if the postulate holds for any positive threshold β, then it is clear that $g(A \cap B) \geq \min(g(A), g(B))$, but as g is monotonic under inclusion, it enforces the equality, which comes down to the statement that a capacity represents accepted beliefs if and only if $g(A \cap B) = \min(g(A), g(B))$, that is, g is a graded necessity measure, still denoted by N. Letting ι, a mapping from S to $[0, 1]$, be the function defined by $\iota(s) = N(S \setminus \{s\})$ (the degree of belief that the actual state of affairs is not s), it is clear that $N(A) = \min_{s \notin A} \iota(s)$. The value $1 - \iota(s)$ can be interpreted as the degree of plausibility $\pi(s)$ of state s, where π is the membership function of a fuzzy epistemic state, usually called a "possibility distribution" (Zadeh, 1978). The set function Π from 2^S to $[0, 1]$ with

$$\Pi(A) = 1 - N(A^c) = \max_{s \in A} \pi(s),$$

where A^c is the set-theoretical complement of A in S, represents the degree of plausibility of A, measuring to what extent A is not ruled out by the agent. This setting is the one of *possibility theory* (Dubois & Prade, 1988). This means that possibility theory accounts for the notion of accepted belief.

Possibility theory was proposed by Lotfi A. Zadeh (1978) in the late 1970s to represent uncertain pieces of information expressed by fuzzy linguistic statements and was later developed in an artificial intelligence perspective (Dubois & Prade, 1988, 1998). Formally speaking, the proposal is quite similar to the one made almost 30 years before by the economist George L. S. Shackle (1949), who advocated and developed a nonprobabilistic view of uncertainty based on the idea of degree of potential surprise. The degree of potential surprise attached to proposition A can be modeled as $N(A^c)$, namely, the more you believe A^c, the more surprising you find the occurrence of A.

A necessity-like function was explicitly used by L. Jonathan Cohen (1977) under the name "Baconian probability." It can be traced back to the English philosopher Francis Bacon, while probability was investigated by Pascal and followers, as pointed out by Cohen (1980). It is devoted to the idea of provability in contrast to probability, and it perfectly fits necessity measures. Especially if you can prove A with some confidence, you cannot at the same time claim you can prove its negation, which makes the condition $\min(g(A), g(A^c)) = 0$ natural. This condition is satisfied by necessity measures. So, the condition $N(A) > 0$ expresses that A is an accepted belief, the absolute value of $N(A)$ expressing the strength of acceptance. Such Baconian probabilities, viewed as grades of certainty, are claimed to be more natural than probabilities, for instance, for use in legal matters. Deciding whether someone is guilty cannot be done using statistics, nor can it be based on betting probabilities: you must prove guilt using convincing dedicated arguments.

About a decade later, in the late 1980s, Wolfgang Spohn (1988) introduced the notion of *ordinal conditional functions*, now called *ranking functions*, as a basis for a dynamic theory of epistemic states. Ranking functions κ are a variant of potential surprise, taking values in the nonnegative integers, that is, $N_\kappa(A^c) = 2^{-\kappa(A)}$ is a degree of potential surprise. The theory of ranking functions (see chapter 5.3 by Kern-Isberner, Skovgaard-Olsen, & Spohn, this handbook) can be developed in parallel with possibility theory,[4] even though they were independently devised.

Subjective probability has been justified by Leonard Savage (1954) from first principles in the setting of decisions under uncertainty. This result has been very important for popularizing probability as the natural way of representing degrees of belief. As it turns out, a similar approach has been carried out for possibility theory in the same act-based setting as the one of Savage, albeit assuming a finite state space S (for a detailed account, see Dubois, Prade, & Sabbadin, 2001). The approach leads to qualitative counterparts of expected utility and extends Wald's pessimistic and optimistic criteria to possibility distributions. They may be considered not discriminating enough and can be refined using special forms of expected utility that encode lexicographic refinements of min and max (leximin and leximax; Fargier & Sabbadin, 2005). For a survey, see Dubois, Fargier, Prade, and Sabbadin (2009); see also chapter 8.4 by Hill, this handbook.

2.4 Probability and Accepted Beliefs

It is clear that set functions representing accepted beliefs are at odds with probability measures supposed to capture rational degrees of belief. If we apply the threshold method to a probability function in order to recover accepted beliefs, then we fail to satisfy conjunctive closure since, however large the threshold β is,

$P(A) > \beta$ and $P(B) > \beta$ do not imply $P(A \cap B) > \beta$ for all A, B.

This point has been especially highlighted by Henry Kyburg (1961, 1997). He put forward the lottery paradox: if S contains the set of possible equiprobable outcomes of a chance game like buying a lottery ticket where only one wins, then, by making S big enough, we can make the probability of losing when betting on the chance outcome s, namely, $(|S| - 1)/|S|$, arbitrarily high. But if you buy all tickets, you are sure to win. This example questions the cogency of accepted belief understood as high-enough degree of belief. As a consequence, Kyburg proposed to give up the idea that accepted beliefs are deductively closed and constructed an appropriate logic accounting for this standpoint (Kyburg & Teng, 2012).

Another way out of the lottery paradox is suggested in Dubois, Fargier, and Prade (2004), namely, to restrict the set of probabilities. We look for probability measures that respect the adjunction rule, namely, for any context C and all A, B,

$P(A|C) > P(A^c|C)$ and $P(B|C) > P(B^c|C)$
$\qquad\qquad$ imply $P(A \cap B|C) > P(A^c \cup B^c|C)$.

This is the adjunction rule for $\beta = 0.5$. Such probability functions do exist, and they are called *big-stepped* probabilities (Benferhat, Dubois, & Prade, 1999) or also *atomic bound* probabilities (Snow, 1999); see also Leitgeb's (2014) stability theory. They have probability masses of

the form $p_1 > p_2 > \cdots > p_n$, where $p_i = P(\{s_i\})$, such that $p_i > p_{i+1} + p_{i+2} + \cdots + p_n$ for all $i < n$, that is, they are the discrete counterpart of exponential probability functions. The same kind of result can be obtained for any acceptance threshold. It suggests that the notion of accepted belief is more consistent with probability theory for distributions that are strongly biased toward some specific outcome (the opposite of the uniform distributions used in the lottery paradox), that is, those for which commonsense beliefs can be entertained.

3. Qualitative Possibility Theory: Reasoning with Defeasible Accepted Beliefs

As mentioned before, there are two different traditions for doxastic reasoning: modal logic, which captures accepted beliefs by modalities, and set functions, which use probability. The fact that there is a class of set functions that also captures accepted beliefs suggests a bridge between the two traditions (Dubois et al., 2004). Let us also mention a third tradition, which models partial ignorance using a truth-functional many-valued logic (e.g., Kleene logic, which expresses partial knowledge about atomic propositions only). This severely limits its representation power (Ciucci & Dubois, 2013). We claim that qualitative possibility and necessity functions, valued in a bounded chain, offer a unified framework for these traditions. In particular, in this approach, accepted beliefs are defeasible, and it has close connections with nonmonotonic reasoning, counterfactual reasoning, and belief revision (Dubois et al., 2004; see also chapter 5.2 by Rott and chapter 6.1 by Starr, both in this handbook).

3.1 Boolean Set Functions and Modalities: A Simple Epistemic Logic

Consider a propositional language $\mathscr{L}$ where well-formed formulas $p, q, \ldots$ encode propositions. In artificial intelligence, a consistent set of formulas in propositional logic is often called a *belief base* and, if deductively closed, a *belief set*. In the syntax of propositional logic, you can express the fact that a proposition is believed, but there is no way to express that it is *not* believed. All you can express is that its negation is believed. Introducing a belief modality $\Box$ prefixing p improves the situation as we can now distinguish between $\neg\Box p$ (p is not believed) and $\Box\neg p$ (its negation is believed). The standard approach to modeling accepted beliefs is indeed modal logic, after Hintikka (1962). Note that the meaning of the modality $\Box$ is described by the axioms ruling the logic. It can range from a very loose interpretation ("the agent has received information that p is true") to a very strong one ("the agent knows that p, in the sense of true belief that p").

The usual axioms of doxastic logic are those of the modal logic KD45 (see chapter 5.1 by van Ditmarsch, this handbook), which presupposes a modal language $\mathcal{M}$ that extends $\mathcal{L}$ and allows for nested modalities (the symbol $\Rightarrow$ represents material implication):

(PL) All axioms of propositional logics for M-formulas.
(K) $\Box(p \Rightarrow q) \Rightarrow (\Box p \Rightarrow \Box q)$
(D) $\Box p \Rightarrow \neg \Box \neg p$
(S4) $\Box p \Rightarrow \Box \Box p$
(S5) $\neg \Box p \Rightarrow \Box \neg \Box p$

The inference rules are modus ponens and necessitation (if p is a theorem, deduce $\Box p$). The semantics is in terms of accessibility relations $R \subseteq S \times S$. The satisfaction of $\Box p$ at a state s based on relation R is defined by $sR \subseteq [p]$, where $sR = \{s' : (s, s') \in R\}$ and $[p]$ denotes the set of states where p is true.

This approach can be simplified in order to relate doxastic logic with the representation of accepted belief from the previous section:

- KD45 uses a complex language, which we can restrict by putting the modality $\Box$ only before propositions in $\mathcal{L}$.

- Axioms S4 and S5, which are often called "positive" and "negative introspection," seem to be there basically to make complex formulas of the language $\mathcal{M}$ equivalent to simpler ones without nested modalities (even if they have a philosophical meaning).

It may sound counterintuitive to evaluate doxastic formulas on a real state of affairs; it is more natural to do it on epistemic or doxastic states (nonempty subsets of S).

The propositional language $\mathcal{L}_\Box$, whose atomic variables are of the form $\Box p$ ($p \in \mathcal{L}$), is the simplest language for an epistemic or doxastic logic. Note that it is disjoint from $\mathcal{L}$ in the sense that it cannot express objective formulas. Then we only keep the axioms K and D, and necessitation is modeled by an axiom saying that $\Box p$ is valid whenever p is a tautology of the propositional calculus. This system is called MEL (minimal epistemic logic) (Banerjee & Dubois, 2014).

The semantics of MEL is in terms of simple epistemic states $E \subseteq S$, and the satisfaction of $\Box p$ by E is then expressed by $E \subseteq [p]$, that is, p is true in all states that are not considered impossible for the agent, or alternatively, p is believed, that is, $N([p]) = 1$ for the Boolean necessity measure N equivalent to E. Indeed, axiom K ensures that $\Box(p \wedge q)$ is logically equivalent to $\Box p \wedge \Box q$, which is

the axiom of necessity measures. Axiom D expresses that $N(A) = 1$ implies $\Pi(A) = 1$. It ensures consistency (models of MEL are then nonempty subsets of S).

In other words, the logic MEL bridges the gap between epistemic logic (dropping the idea of introspection) and the representation of belief by means of set functions, identifying the latter with modalities. Especially the modality $\Diamond p = \neg \Box \neg p$ corresponds to possibility measures.

3.2 The Logic of Graded Acceptance

The logic of acceptance MEL can be extended to graded necessity functions N using a simple multimodal logic. If L is a finite totally ordered scale of necessity degrees (understood as describing a gradation in accepted beliefs), we can expand the modal language of the MEL logic to allow for several belief modalities denoted by $\Box^\lambda$, for $\lambda \in L$ with $\lambda > 0$, where the sentence $\Box^\lambda p$ encodes the statement $N([p]) \geq \lambda$, and the $\Box$-modality in MEL corresponds to $\Box^1$ (expressing full belief). The language of this logic is thus a propositional language $\mathcal{L}_\Box^L$ where atoms are of the form $\Box^\lambda p$, $p \in \mathcal{L}$, for $\lambda \in L$ with $\lambda > 0$. Note that the formula $\neg \Box^\lambda p$ stands for $N([p]) < \lambda$, which, due to the assumed finiteness of L, can be expressed as $\Pi([\neg p]) \geq 1 - s(\lambda)$, which is encoded as $\Diamond^{s(\lambda)} \neg p$, for the next lower value $s(\lambda)$ to λ in the scale L.

The sublogic obtained by fixing the value λ is a copy of the logic MEL (it satisfies axiom K, axiom D, and necessitation). There is also the weakening axiom: $\Box^\lambda p \Rightarrow \Box^\mu p$ if $\mu \leq \lambda$, and axiom D is valid in the stronger form $\Box^\lambda p \Rightarrow \Diamond^1 p$. The semantics is in terms of L-valued possibility distributions π representing gradual epistemic states, and π satisfies $\Box^\lambda p$ if and only if $N([p]) \geq \lambda$, where N is based on π. The soundness and completeness of this logic, called GPL (generalized possibilistic logic), have been proved (Dubois, Prade, & Schockaert, 2017).

This logic is very expressive and enables one to reason about ignorance and defeasible beliefs. As shown in the previous reference, it can encode several nonmonotonic formalisms, especially:

- The older standard possibilistic logic (Dubois, Lang, & Prade, 1994) is obtained by restricting the language $\mathcal{L}_\Box^L$ to conjunctions of atomic epistemic statements $\Box^\lambda p$, which are written (p, λ) in the original syntax. This logic is nonmonotonic (Benferhat, Dubois, & Prade, 1998).

- Conditional logics with statements of relative belief of the form $N(p) > N(q)$ can be encoded by GPL formulas of the form $\vee_{\lambda \neq 0}(\Box^\lambda p \wedge \neg \Box^\lambda q)$.

- The System P of Kraus, Lehmann, and Magidor (1990) uses nonmonotonic conditional statements $p \mathrel{|\!\sim} q$.

They express the plausible inference of q from p, which is modeled by the constraint $N(p \Rightarrow q) > N(p \Rightarrow \neg q)$.

- Answer-set programs can be expressed in GPL using a three-valued scale $L = \{1, \lambda, 0\}$, where $1 > \lambda > 0$, and requiring two necessity modalities: a strong one, $\square^1$, and a weak one, $\square^\lambda$.

In summary, GPL can be viewed as the logic of qualitative Baconian probabilities.

4. Nonadjunctive Settings for Rational Degrees of Belief

If we take it for granted that belief should come in degrees, and if, like Kyburg, we reject the adjunction rule, we are left with nonclassical logics where deduction is not closed under conjunction, like for instance the logic of risky knowledge (Kyburg & Teng, 2012). Or we give up the Boolean framework of logic altogether and concentrate on the properties of set functions that can model degrees of belief. The natural question is then whether degrees of belief should be additive at all. Fifty years ago, the answer was, Yes, of course. Since then, a number of proposals have emerged from which it follows that additivity should not be taken for granted. This section discusses reasons for questioning additive beliefs and focuses on some approaches to graded belief that are deliberately non-additive.

4.1 Can a Single Probability Distribution Capture Any Epistemic State?

The so-called Bayesian approach to subjective probability theory posits a uniqueness principle as a preamble to any kind of uncertainty modelling: Any state of knowledge is representable by a single probability distribution (see, for instance, Lindley, 1982). Note that indeed, if, following the fair bet procedure of de Finetti, an agent decides to directly assign subjective probabilities via buying prices to all possible outcomes in some game of chance, the coherence principle forces the agent to define a unique probability distribution.

Yet another mathematical attempt to justify probability theory as the only reasonable belief measure is the one by Richard T. Cox (1946). He relied on the Boolean structure of the set of events and on a number of postulates he considered compelling. Let $g(A \mid B) \in [0, 1]$ be a conditional degree of belief, A and B being events in a Boolean algebra, with $B \neq \varnothing$. The Cox axioms for conditional belief are as follows:

(i) $g(A \cap C \mid B) = F(g(A \mid C \cap B), g(C \mid B))$ (if $C \cap B \neq \varnothing$);

(ii) $g(A^c \mid B) = n(g(A \mid B))$ for $B \neq \varnothing$, where A^c is the complement of A in S;

(iii) The function F is supposed to be twice differentiable, with a continuous second derivative, while the function n is twice differentiable.

Cox claimed that, on such a basis, $g(A \mid B)$ must be isomorphic to a conditional probability measure.

This result has been repeated *ad nauseam* in the literature on artificial intelligence to justify probability combined with Bayes' rule as the only reasonable approach to represent and revise numerical degrees of belief (Horvitz, Heckerman, & Langlotz, 1986; Cheeseman, 1988; Jaynes, 2003). However, some reservations must be made. First, the original proof by Cox turned out to be faulty—see Paris (1994) for another proof, based on a weaker condition (iii) (it is enough that F be strictly monotonically increasing in each argument), but also on an additional technical density condition that requires an infinite setting. Moreover, Halpern (1999a,b) has shown that the result does not hold on finite spaces and that Cox's original conditions do not suffice to prove the result in the infinite setting. Independently of these technical issues, it should be noticed that postulate (i) sounds natural only if one takes the form of Bayes conditioning for granted; and postulate (ii) requires self-duality, which rules out representations of uncertainty due to partial ignorance, as seen later on. The above comments seriously weaken the alleged universality of Cox's results.

Applying the Bayesian credo as recalled above, justified via the avoidance of Dutch books, assuming fair prices for bets, or by obedience to Cox's axioms, forces the agent to use a single probability measure as the universal tool for representing uncertainty, whatever its source. This stance leads to serious representation difficulties already pointed out more than forty years ago (Shafer, 1976). For one, it means we give up making any difference between uncertainty due to incomplete information or ignorance, and uncertainty due to a purely random process, the next outcome of which cannot be predicted. One may indeed admit that additive degrees of beliefs are justified if they reflect extensive statistical evidence. But what if such information is not available?

Take the example of die tossing: The uniform probability assignment models the assumption that the die is fair. But if the agent assigns equal prices to bets assigned to all facets of the die, how can we interpret that? Is it because the agent is sure that the die is fair and its outcomes are driven by pure randomness (because, say, he or she could test it hundreds of times prior to placing the bets, or from counting cases)? Or is it because the agent who is given this die has just no idea whether the die is fair or not, and so has no reason to put more money on one facet than on another? Clearly the epistemic state of the agent is not the same in the first situation as in the second one. But the uniformly distributed probability function is mute about this issue and handles the two situations in the same way.

Next, the choice of a set of mutually exclusive outcomes depends on the chosen language, for example, the one used by the information source. However, several languages or points of view can co-exist for the same problem. Since there are several different possible representations of the state space, the probability assignment obtained from an agent will be language-dependent, especially in the case of ignorance: a uniform probability on one representation of the state space may conflict with a uniform one on another representation encoding of the same state space for the same problem, while in the case of ignorance this is the only available model. Shafer (1976) gives the following example: Consider the question of the existence of extraterrestrial life, about which the agent has no idea. If the variable v refers to the claim that life exists outside our planet ($v = $ li), or not ($v = \neg$li), then the agent proposes P_1(li) = $P_1(\neg$li) = ½ on $S_1 = \{$li, $\neg$li$\}$. However, it makes sense to interpret "life" as the disjunction between animal life (ali), and vegetable life only (vli), which leads to the state space $S_2 = \{$ali, vli, $\neg$li$\}$. The ignorant Bayesian agent is then bound to propose P_2(ali) = P_2(vli) = $P_2(\neg$li) = 1/3. As "li" is the disjunction of "ali" and "vli," the distributions P_1 and P_2 are not compatible with each other, while they are both supposed to represent ignorance. Another example comes from noticing that expressing ignorance about a real-valued quantity by means of a uniform distribution for $x \in [a, b]$, a positive interval, is not compatible with a uniform distribution on $y = \log(x) \in [\log(a), \log(b)]$, while the agent has the same level of ignorance about x and y.

Finally, Ellsberg's (1961) paradox (see chapter 8.2 by Peterson, this handbook) showed quite early that, when expressing preferences between gambles consisting in drawing balls from an urn the content of which is ill known, many experiments have shown that people tend to systematically violate Savage's axioms (especially the sure-thing principle), because they are pessimistic about rewarding events of unknown probability. One way of accounting for the results of these experiments is to give up the additivity of degrees of belief.

The above limitations of the expressive power of single probability distributions have motivated the emergence of other approaches to representing uncertainty. Some of them, as seen above, give up the numerical setting of degrees of belief and use ordinal or qualitative structures, like qualitative possibility theory. Another option is to tolerate incomplete information in the probabilistic approach, which leads to different mathematical models of various levels of generality. These are reviewed in the rest of this chapter.

4.2 Shafer Belief Functions and the Merging of Uncertain Testimonies

The theory of evidence by Glenn Shafer (1976) can be viewed as a specific interpretation of Dempster's (1967) upper-and-lower-probability framework for handling imprecise statistical information, or as the revival of the concept of probability (as opposed to chance) invented in the 17th century by Jakob Bernoulli, and later by Johann Heinrich Lambert, dealing with the problem of representing and merging unreliable testimonies (Shafer, 1978). Shafer's book is clearly in the latter tradition.

The main issue is first to model unreliable testimonies. Suppose that a witness claims that proposition E is true but the receiver only partially believes this statement, considering that the witness is reliable with probability p. So p can be viewed as the degree of belief of the receiver in proposition E due to the unreliability of the witness. The additivity issue is raised by the question, to what proposition should the complementary weight $1-p$ be assigned? The regular probabilist would assign it to the negated proposition E^c. But if the testimony E is interpreted by the receiver as its negation, it means that the latter thinks this witness lies, that is, that he or she says E when knowing it to be false. There is another option, namely that due to the incompetence of the witness, there is a probability $1-p$ that the testimony is just useless. In the latter case, the probability $1-p$ is assigned not to E^c, but to the whole of S, that is, to the state of ignorance. That is to say, with probability p the receiver knows that E is true and nothing else, and with probability $1-p$ he/she knows nothing, which is modeled by a basic assignment m from the power set 2^S to [0,1] such that $m(E) = p$ and $m(S) = 1-p$. A belief function that models such a simple unreliable testimony E is called a "simple support function."

More generally, consider a process whose outcomes are set-valued (i.e., imprecise) and uncertain (there is a probability value attached to each outcome). This is modeled by a more general basic assignment m from 2^S to [0, 1] such that $m(\emptyset) = 0$ and $\sum_{E \subseteq S} m(E) = 1$. Epistemic states E with $m(E) > 0$ are called "focal sets." The degree of belief $Bel(A)$ in a proposition A and its dual plausibility degree $Pl(A)$ are then defined by

$$Bel(A) = \sum\{m(E): E \subseteq A, E \neq \emptyset\};$$
$$Pl(A) = \sum\{m(E): E \cap A \neq \emptyset\}.$$

It is clear that the belief function *Bel* is non-additive, for example, $Bel(A) + Bel(A^c) \leq 1$, and the degree of plausibility is $Pl(A) = 1 - Bel(A^c)$. In the case of a simple support function, observe that when $A \neq S$, then $Bel(A) = p$ if $E \subseteq A$, and $Bel(A) = 0$ otherwise.

It is important to point out that belief functions generalize probabilities (recovered when for all E, $m(E) > 0$ implies that E is a singleton), Boolean necessity measures (recovered when $m(E) = 1$ for some epistemic state E), and also graded necessity measures (recovered when focal sets are all nested; an example of this is a simple support function). A degree of belief $Bel(A)$ clearly evaluates the probability of proving A from the available information. A plausibility degree $Pl(A)$ evaluates the probability that A is not logically incompatible with the available information. To use Cohen's (1977) terminology, belief functions join the probable and the provable, or bring Pascal and Bacon together. But doing so, belief functions are no longer additive, nor do they respect adjunction.

The major problem addressed by the 17th- and 18th-century pioneers is the merging of such testimonies. They proposed special cases of what is now known as Dempster's rule of combination: Let m_1 and m_2 be two mass assignments coming from independent sources. The result of the combination is a mass assignment $m_1 \otimes m_2$ defined by

for all $A \subseteq S$,
$(m_1 \otimes m_2)(A) = (1/K)\Sigma\{m_1(A_1) \cdot m_2(A_2): A = A_1 \cap A_2\}$,
where $K = \Sigma\{m_1(A_1) \cdot m_2(A_2): A_1 \cap A_2 \neq \varnothing\}$ and
$(m_1 \otimes m_2)(\varnothing) = 0$.

The assignment $m_1 \otimes m_2$ consists in intersecting any two overlapping focal sets, each coming from a distinct source, computing the probability of obtaining each subset A via such an intersection, and renormalizing the obtained mass assignment, as some pairs of focal sets may be conflicting. In the case of merging two simple support functions focusing on the same set E where $Bel_1(E) = m_1(E) = p_1$ and $Bel_2(E) = m_2(E) = p_2$, the resulting belief in E is $(Bel_1 \otimes Bel_2)(E) = p_1 + p_2 - p_1 p_2$, which leads to a reinforcement of the belief in E, a result already suggested by the pioneers of belief functions in the 17th and 18th century. This combination rule assumes that sources of information are independent, which makes the reinforcement effect plausible.

In Shafer's book, a major question was whether all belief functions can be expressed as the result of merging independent simple testimonies in the form of simple support functions. It turns out that only a subclass of belief functions, called *separable*, can be generated in this way. Later on, Smets (1995) tried to extend the notion of a simple support function so as to cover all belief functions. An extensive presentation of the theory of evidence as a theory of rational belief is proposed by Haenni (2009).

Finally, criteria for decision under uncertainty, where the latter is described by a belief function, are studied by Smets and Kennes (1994) and Jaffray (1989). The former propose to define a so-called *pignistic* probability measure from a belief function (generalizing Laplace's principle of insufficient reason), and apply the expected utility criterion. This probability measure coincides with the well-known Shapley value in game theory (Shapley, 1953) and in some sense projects the provable onto the probable. Jaffray (1989) proposes and axiomatizes an extension of Hurwicz's criterion.

4.3 Imprecise Probabilities, Desirability, and Generalized Betting Theory

The alternative approach to the modeling of degrees of belief consists in revisiting de Finetti's approach to subjective probability, dropping the constraint that the price proposed by a gambler for buying a lottery ticket should be fair. This view was pioneered by Cedric Smith (1961), Peter M. Williams (1975), and Robin Giles (1982), and more extensively developed by Peter Walley (1991).

In this approach, the agent offers buying prices for gambles. A gamble is a function f from S to the real line that expresses losses ($f(s) < 0$) or gains ($f(s) > 0$), depending on the actual state of affairs s. The gamble associated with a particular event is its characteristic function. The agent is not committed to selling such gambles at the same price as their buying price. The approach relies on so-called *desirable gambles* (Walley, 1991), which the agent would agree to buy for a positive price. The set of desirable gambles contains at least all positive gambles. Moreover, the sum of two desirable gambles is also considered desirable, and a desirable gamble remains desirable when multiplied by a positive constant. The maximal price at which the agent accepts to buy a gamble is the maximal value α such that $f - \alpha$ is desirable. It is called the *lower prevision* of a gamble f. It can be shown that given a set of gambles $\mathscr{G}$ and their lower previsions $LP(f)$, there is a convex set $\mathscr{P}_\mathscr{G}$ of probabilities, called a *credal set*, such that $LP(f) = \inf\{E_P(f): P \in \mathscr{P}_\mathscr{G}\}$ is the lower expectation of f_i according to $\mathscr{P}_\mathscr{G}$, for all $f \in \mathscr{G}$, where $E_P(f)$ is the expectation of f with respect to probability P. One important point is that any convex set of probabilities can be represented by lower previsions on some family of gambles.

In this setting, the upper prevision $UP(f)$ of a gamble f is provably equal to $-LP(-f)$. The upper prevision $UP(f)$ is thus the minimal selling price of f. If the credal set attached to a set of gambles and its lower previsions is empty, then the proposal is inconsistent and the agent incurs a sure loss after buying and resolving these gambles. Avoiding sure loss means that $UP(f) \geq LP(f)$ for all gambles f.

Moreover, due to the interaction between gambles, it may be that the consistent buying prices proposed by the agent for gambles $f \in \mathcal{G}$ are too low and could be raised without altering the credal set. A set of buying prices $bp(f)$ for $f \in \mathcal{G}$ is said to be *coherent* if and only if $LP(f) = bp(f)$ for all $f \in \mathcal{G}$. In other words, a set of buying prices for a set of gambles $\mathcal{G}$ is coherent if and only if for any $f \in \mathcal{G}$,

$$\inf\{E_P(f): E_P(f) \geq bp(f) \text{ for all } f \in \mathcal{G}\} = bp(f).$$

Under this approach the degree of belief in proposition A is a coherent lower probability $P_*(A) = \inf\{P(A): P \in \mathbb{P}\} = LP(1_A)$, the lower prevision of its characteristic function, where $\mathbb{P}$ is the credal set induced by the lower prevision $LP(f)$ on some gambles.

Some remarks are in order to position this approach with respect to other approaches to rational degrees of belief:

- The epistemic state of the agent is here represented by a credal set, but there is no ill-known probability inside. In particular, the interval $[P_*(A), P^*(A)]$, where $P^*(A) = 1 - P_*(A^c)$, is not supposed to contain an ill-known subjective (nor an objective) probability of A. Just as for belief functions, degrees of belief are precise and modelled by coherent lower probabilities.

- Mathematically, belief functions are a special case of coherent lower probabilities. They are super-additive set functions at any order, while lower probabilities from any credal set only satisfy the inequality $P_*(A \cup B) \geq P_*(A) + P_*(B)$ when A and B are disjoint. In particular, the mass function recomputed from P_* instead of *Bel* in (3) (called the Moebius transform of P_*) exists, and is unique but not necessarily positive (Chateauneuf & Jaffray, 1989).

- An attempt to justify belief functions as the only rational approach to degrees of belief under a betting framework in the style of Walley (and not as unreliable testimonies) was recently published by Kerkvliet and Meester (2018).

The gamble approach leads to a decision rule that is specific to the imprecise-probability setting, namely, a gamble f is preferred to a gamble g if and only the gamble $h = f - g$ is desirable, that is, if the lower expectation of this gamble with respect to the corresponding credal set $\mathbb{P}$ is positive. This yields a partial ordering on gambles. It

implies that $LP(f) \geq LP(g)$. The latter inequality is a decision rule that solves the Ellsberg paradox, in contrast with the decision rule $LP(f - g) \geq 0$ that satisfies Savage's sure-thing principle (see also chapter 8.2 by Peterson, this handbook).

4.4 Quantitative Possibility in the Setting of Imprecise Probability

It is natural to reconsider graded possibility and necessity measures in the setting of belief functions and imprecise probabilities. In fact they are at the crossroads of all non-additive approaches to uncertainty and may be interpreted in various ways:

- Necessity measures are a special case of belief functions. Their characteristic property is to have nested focal sets. In other words, they model coherent arguments in favor or disfavor of propositions. They are the only family of belief functions that obey the adjunction rule. Note that the weaker Baconian condition $\min(Bel(A), Bel(A^c)) = 0$ for all $A \subseteq S$ corresponds to overlapping (consistent) focal sets.

- As a consequence, necessity measures also stand for coherent lower probabilities. However, they correspond to a very cautious type of betting behavior, such that if the buying price for gambling on A is positive then the agent feels obliged to sell this gamble at the maximal price (Giles, 1982).

- One may borrow the operational semantics of the Bayesians to derive personal possibility and necessity degrees. If we adopt the framework of belief functions for representing an agent's knowledge and accept the idea that a belief function induces a pignistic probability for making decisions, then we may reverse this process. Given a subjective probability reflecting fair prices of gambles corresponding to random events, one may look for the least informative belief function that induces this subjective probability. It can be proved that it is always a necessity measure (Dubois, Prade, & Smets, 2008).

- Necessity functions induced on the unit interval by a suitable transformation of a Spohn ranking function (see chapter 5.3 by Kern-Isberner et al., this handbook) have nothing to do with lower probabilities. Basically, as shown in Spohn (1990), they are more closely related to powers of infinitesimal probabilities, for which the additivity axiom degenerates in the minitivity axiom.

- Yet another interpretation of possibility theory is in terms of likelihood. In statistical inference, given a parametric probabilistic model $P(\cdot | \theta)$ where $\theta \in \Theta$ is the parameter of the model, the probability $P(R | \theta)$

based on data set R is not the probability of θ based on R, only its likelihood. It represents a looser degree $lik(\theta) = P(R \mid \theta)$ of confidence in θ for the observer having received evidence R. Advocates of the likelihood approach (Edwards, 1972) refuse to attach prior probabilities to values of θ, basically because this quantity is not observable and is just a model artefact. Rather, it is natural to try and define the likelihood $lik(A)$ for any $A \subseteq \Theta$ from the values $lik(\theta)$ for $\theta \in A$. It has been shown that the only meaningful definition is $lik(A) = \max\{P(R \mid \theta): \theta \in A\}$ (Dubois, Moral, & Prade, 1997; Dubois, 2006). Hence, in the absence of prior probabilities, a likelihood function can be interpreted as a possibility measure. However, this kind of possibility measure is determined only up to a multiplicative constant, a specific feature that makes likelihood theory yet another kind of possibility theory.

5. Conclusion

This chapter has presented a survey of various approaches to the notion of belief, reflecting the progress made in the last 50 years. It seems that the frontal opposition between degrees of belief and accepted beliefs, that is, the Pascalian and Baconian traditions, may be alleviated to some extent if we give up the requirement that degrees of belief should be additive. There is a range of mathematical models standing between probability and modal logics, some of which retain the adjunction rule of Baconian probabilities. Some approaches blend the two traditions and are consistent with the requirement that you cannot at the same time believe in a proposition and believe in its negation. The Baconian tradition also touches upon the issue of formal argumentation, on which there is an abundant literature today (see Haenni, 2009, for its connection with Shafer belief functions). Argumentation can be viewed as a rational approach to handle inconsistency in reasoning due to conflicting pieces of information. One may argue that Baconian probabilities (in the form of, e.g., necessity functions, ranking functions, and the like) represent imprecise but conflict-free information, while ordinary probabilities capture precise but conflicting observations. The new theories of belief deal with both imprecise and conflicting information and seem to bridge the gap between the two traditions of belief representation. One may then consider belief in a more dynamical setting, where starting with more or less probable conflicting evidence one proceeds towards the provable via a suitable deliberation process involving argumentation.

Notes

1. See the entry "Probabilité," accredited to Benjamin de Langes de Lubières (1714–1790), in *Encyclopédie, ou Dictionnaire raisonné des sciences, des arts et des métiers*, by D. Diderot and J.-B. Le Rond d'Alembert (http://enccre.academie-sciences.fr/encyclopedie/page/v13-p403).

2. Halpern (2003) calls set functions of this kind "plausibility measures," not even assuming a total order on events (replacing [0, 1] by a partially ordered set). However, this terminology may lead to confusion with Shafer's older plausibility functions (see section 4.2).

3. Or doxastic state—we do not distinguish between the two in this chapter.

4. Up to the presence or not of technical assumptions like well-ordering in the infinite setting.

References

Banerjee, M., & Dubois, D. (2014). A simple logic for reasoning about incomplete knowledge. *International Journal of Approximate Reasoning, 55*, 639–653.

Benferhat, S., Dubois, D., & Prade, H. (1998). Practical handling of exception-tainted rules and independence information in possibilistic logic. *Applied Intelligence, 9*, 101–127.

Benferhat, S., Dubois, D., & Prade, H. (1999). Possibilistic and standard probabilistic semantics of conditional knowledge bases. *Journal of Logic and Computation, 9*, 873–895.

Chateauneuf, A., & Jaffray, J.-Y. (1989). Some characterizations of lower probabilities and other monotone capacities through the use of Möbius inversion. *Mathematical Social Sciences, 17*(3), 263–283.

Cheeseman, P. (1988). An inquiry into computer understanding (with comments and a rejoinder). *Computational Intelligence, 4*, 58–142.

Choquet, G. (1953). Théorie des capacités [Theory of capacities]. *Annales de l'Institut Fourier, 5*, 131–295.

Ciucci, D., & Dubois, D. (2013). A modal theorem-preserving translation of a class of three-valued logics of incomplete information. *Journal of Applied Non-Classical Logics, 23*(4), 321–352.

Cohen, L. J. (1977). *The probable and the provable*. Oxford, England: Clarendon Press.

Cohen, L. J. (1980). Some historical remarks on the Baconian conception of probability. *Journal of the History of Ideas, 41*, 219–231.

Cox, R. T. (1946). Probability, frequency, and reasonable expectation. *American Journal of Physics, 14*, 1–13.

de Finetti, B. (1937). La prévision: ses lois logiques, ses sources subjectives [Foresight: its logical laws, its subjective sources],

Annales de l'Institut Poincaré, 7, 1–68. In H. E. Kyburg Jr. & H. E. Smokler (1964, Eds.), *Studies in subjective probability* (pp. 53–118), New York, NY: Wiley.

Dempster, A. P. (1967). Upper and lower probabilities induced by a multivalued mapping. *Annals of Statistics, 28,* 325–339.

Dubois, D. (2006). Possibility theory and statistical reasoning. *Computational Statistics & Data Analysis, 51*(1), 47–69.

Dubois, D., Fargier, H., & Prade, H. (2004). Ordinal and probabilistic representations of acceptance. *Journal of Artificial Intelligence Research, 22,* 23–56.

Dubois, D., Fargier, H., Prade, H., & Sabbadin, R. (2009). A survey of qualitative decision rules under uncertainty. In D. Bouyssou, D. Dubois, H. Prade, & M. Pirlot (Eds.), *Decision-making process: Concepts and methods* (pp. 435–473). London, England: ISTE & Wiley.

Dubois, D., Lang, J., & Prade, H. (1994). Possibilistic logic. In D. M. Gabbay, C. J. Hogger, J. A. Robinson, & D. Nute (Eds.), *Handbook of logic in artificial intelligence and logic programming: Vol. 3. Nonmonotonic reasoning and uncertain reasoning* (pp. 439–513). Oxford, England: Oxford University Press.

Dubois, D., Moral, S., & Prade, H. (1997). A semantics for possibility theory based on likelihoods. *Journal of Mathematical Analysis and Applications, 205,* 359–380.

Dubois, D., & Prade, H. (1988). *Possibility theory: An approach to computerized processing of uncertainty.* New York, NY: Plenum Press.

Dubois, D., & Prade, H. (1998). Possibility theory: Qualitative and quantitative aspects. In D. M. Gabbay & P. Smets (Eds.), *Handbook of defeasible reasoning and uncertainty management systems: Vol. 1. Quantified representation of uncertainty and imprecision* (pp. 169–226). Dordrecht, Netherlands: Kluwer.

Dubois, D., Prade, H., & Sabbadin, R. (2001). Decision-theoretic foundations of qualitative possibility theory. *European Journal of Operational Research, 128*(3), 459–478.

Dubois, D., Prade, H., & Schockaert, S. (2017). Generalized possibilistic logic: Foundations and applications to qualitative reasoning about uncertainty. *Artificial Intelligence, 252,* 139–174.

Dubois, D., Prade, H., & Smets, P. (2008). A definition of subjective possibility. *International Journal of Approximate Reasoning, 48,* 352–364.

Edwards, A. W. F. (1972). *Likelihood.* Cambridge, England: Cambridge University Press.

Ellsberg, D. (1961). Risk, ambiguity, and the Savage axioms. *Quarterly Journal of Economics, 75*(4), 643–669.

Fargier, H., & Sabbadin, R. (2005). Qualitative decision under uncertainty: Back to expected utility. *Artificial Intelligence, 164*(1–2), 245–280.

Giles, R. (1982). Foundations for a theory of possibility. In M. Gupta & E. Sanchez (Eds.), *Fuzzy information and decision processes* (pp. 183–195). Amsterdam, Netherlands: North-Holland.

Haenni, R. (2009). Non-additive degrees of belief. In F. Huber & C. Schmidt-Petri (Eds.), *Degrees of belief* (pp. 121–160). Berlin, Germany: Springer.

Halpern, J. Y. (1999a). A counterexample to theorems of Cox and Fine. *Journal of Artificial Intelligence Research, 10,* 67–85.

Halpern, J. Y. (1999b). Technical addendum: Cox's theorem revisited. *Journal of Artificial Intelligence Research, 11,* 429–435.

Halpern, J. Y. (2003). *Reasoning about uncertainty.* Cambridge, MA: MIT Press.

Hintikka, J. (1962). *Knowledge and belief.* Ithaca, NY: Cornell University Press.

Horvitz, E. J., Heckerman, D. E., & Langlotz, C. P. (1986). A framework for comparing alternative formalisms for plausible reasoning. In T. Kehler (Ed.), *Proceedings, AAAI-86: Fifth national conference on artificial intelligence, August 11–15, 1986* (Vol. 1, pp. 210–214). San Mateo, CA: Morgan Kaufmann.

Huber, F., & Schmidt-Petri, C. (Eds.). (2009). *Degrees of belief.* Berlin, Germany: Springer.

Jaffray, J. Y. (1989). Linear utility theory for belief functions. *Operations Research Letters, 8,* 107–112.

Jaynes, E. T. (2003). *Probability theory: The logic of science.* Cambridge, England: Cambridge University Press.

Kerkvliet, T., & Meester, R. (2018). A behavioral interpretation of belief functions. *Journal of Theoretical Probability, 31,* 2112–2128.

Koopman, B. O. (1940). The bases of probability. *Bulletin of the American Mathematical Society, 46,* 763–774.

Kraus, S., Lehmann, D., & Magidor, M. (1990). Nonmonotonic reasoning, preferential models and cumulative logics. *Artificial Intelligence, 44,* 167–207.

Kyburg, H. (1961). *Probability and the logic of rational belief.* Middletown, CT: Wesleyan University Press.

Kyburg, H. (1997). The rule of adjunction and reasonable inference. *Journal of Philosophy, 94,* 109–125.

Kyburg, H., & Teng, C. M. (2012). The logic of risky knowledge, reprised. *International Journal of Approximate Reasoning, 53,* 274–285.

Leitgeb, H. (2014). The stability theory of belief. *Philosophical Review, 123*(2), 131–171.

Lewis, D. K. (1973). *Counterfactuals.* London, England: Basil Blackwell (2nd ed.). Worcester, England: Billing.

Lindley, D. V. (1982). Scoring rules and the inevitability of probability. *International Statistics Review, 50,* 1–26.

Paris, J. B. (1994). *The uncertain reasoner's companion: A mathematical perspective*. Cambridge, England: Cambridge University Press.

Ramsey, F. P. (1931). Truth and probability. In F. P. Ramsey, *The foundations of mathematics and other logical essays* (R. B. Braithwaite, Ed., pp. 156–198). London, England: Kegan, Paul, Trench, Trubner. (Original work 1926)

Savage, L. J. (1954). *The foundations of statistics*. New York, NY: Wiley.

Shackle, G. L. S. (1949). *Expectation in economics*. Cambridge, England: Cambridge University Press.

Shafer, G. (1976). *A mathematical theory of evidence*. Princeton, NJ: Princeton University Press.

Shafer, G. (1978). Non-additive probabilities in the work of Bernoulli and Lambert. *Archive for History of Exact Sciences, 19*(4), 309–370.

Shapley, L. S. (1953). A value for *n*-person games. In H. W. Kuhn & A. W. Tucker (Eds.), *Contributions to the theory of games* (Vol. II, pp. 307–317). Princeton, NJ: Princeton University Press.

Smets, P. (1995). The canonical decomposition of a weighted belief. In C. S. Mellish (Ed.), *Proceedings of the 14th International Joint Conference on Artificial Intelligence* (pp. 1896–1901). San Mateo, CA: Morgan Kaufmann.

Smets, P., & Kennes, R. (1994). The transferable belief model. *Artificial Intelligence, 66*, 191–234.

Smith, C. A. B. (1961). Consistency in statistical inference and decision. *Journal of the Royal Statistical Society B, 23*, 1–37.

Snow, P. (1999). Diverse confidence levels in a probabilistic semantics for conditional logics. *Artificial Intelligence, 113*, 269–279.

Spohn, W. (1988). Ordinal conditional functions: A dynamic theory of epistemic states. In W. L. Harper & B. Skyrms (Eds.), *Causation in decision, belief change, and statistics* (Vol. 2, pp. 105–134). Dordrecht, Netherlands: Kluwer.

Spohn, W. (1990). A general, nonprobabilistic theory of inductive reasoning. In R. D. Shachter, T. S. Levitt, L. N. Kanal, & J. F. Lemmer (Eds.), *Uncertainty in artificial intelligence* (Vol. 4, pp. 149–158). Amsterdam, Netherlands: North Holland.

Sugeno, M. (1977). Fuzzy measures and fuzzy integrals—A survey. In M. M. Gupta, G. N. Saridis, & B. R. Gaines (Eds.), *Fuzzy automata and decision processes* (pp. 89–102). Amsterdam, Netherlands: North Holland.

Walley, P. (1991). *Statistical reasoning with imprecise probabilities* (Monographs on Applied Statistics and Probability 42). New York, NY: Chapman and Hall.

Williams, P. M. (2007). Notes on conditional previsions. *International Journal of Approximate Reasoning, 44*, 366–383 (unpublished research report, 1975).

Zadeh, L. A. (1978). Fuzzy sets as a basis for a theory of possibility. *Fuzzy Sets and Systems, 1*, 3–28.

Section 5 Belief Revision, Defeasible Reasoning, and Argumentation Theory

5.1   Doxastic and Epistemic Logic

Hans van Ditmarsch

Summary

We present modal-logical semantics for knowledge and belief, alternative semantics for knowledge, and also systems with interaction between knowledge and belief. We then present common knowledge and common belief, as well as distributed knowledge. We conclude with topics involving change of knowledge and belief: public announcements, unsuccessful updates, knowability, growth of common knowledge, and resolving distributed knowledge.

1.   Knowledge

1.1   Knowledge and Ignorance

Let p stand for the proposition "Paneer is rose water." This is true. Amir, however, is uncertain whether it is true (maybe paneer is soft cheese?). If it is true, he considers it possible that it is true and that it is false. If false, he considers it possible that it is false and that it is true. "Considers possible" induces a binary *indistinguishability* relation between states of the world that is therefore obviously reflexive and symmetric and that is also transitive (see below for exceptions involving vagueness). It is an equivalence relation. Figure 5.1.1b depicts the situation.

If Amir and his friend Bala are both ignorant whether p, and they know this about each other (and so on), then a simple model for that situation again consists of two states for the different truth values of p and is such that both Amir and Bala cannot distinguish them: figure 5.1.1b.

In that situation, both know what the other is uncertain about. It is easy to conceive of scenarios where this is not the case, for example, when Bala considers it possible that Amir knows p, that he knows $\neg p$, and that he is ignorant about p: figure 5.1.1d. This is known as *higher-order uncertainty*.

Models for higher-order uncertainty can be arbitrarily large. Suppose that Amir and Bala both know a number

and that the other's number is one more or one less: the *consecutive numbers riddle* (Littlewood, 1953). An infinite chain of possibilities results: figure 5.1.1f (p now stands for "the number is even").

1.2   The Modal Logic of Knowledge

The language of epistemic logic is that of propositional logic expanded with inductive clauses $K_a p$ for "Agent a knows proposition p." The dual modality $\langle K \rangle_a p$ can be defined by abbreviation as $\neg K_a \neg p$ and stands for "Agent a considers p possible." The language of epistemic logic is interpreted on pointed Kripke models. A *Kripke model* consists of a domain of abstract objects called "states" or "worlds," a set of binary relations R_a (one for each agent a) on this domain, and valuations V_p for atomic propositions p, specifying for each state whether atom p is true or false there. (In this simplified presentation, the names of states are their value of atom p, and we also use p for other than atomic propositions.) A *pointed* Kripke model has a designated "actual" state. Formula $K_a p$ (for any proposition p) is true in state s of a model iff p is true in all states t of the model that are R_a-accessible from s. On models interpreting knowledge, the accessibility relations are assumed to be equivalence relations. We will list and discuss the principles of knowledge below in view of how rational they are. The logic of knowledge is called **S5**.

1.3   Belief

A difference between knowledge and belief is that knowledge is supposed to be correct, whereas belief may be mistaken. But a belief's being true does not make it knowledge. The difference between knowledge and belief is a notoriously difficult question. It also involves evidence and justification, as well as Gettier examples (Gettier, 1963). It will not be addressed here.

The correctness of knowledge is formalized in the axiom $K_a p \rightarrow p$ (**T**), which characterizes the reflexivity of R_a. Reflexivity guarantees nonemptyness of the relation. Removing this axiom from the logic allows an accessibility relation R_a to be empty, in which case beliefs are

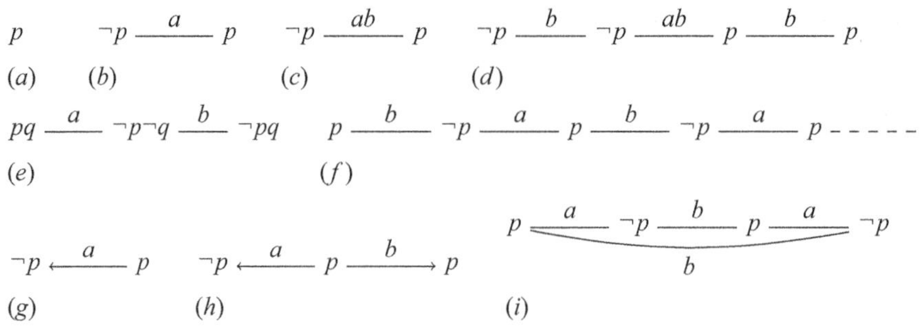

p $\neg p \xrightarrow{\,a\,} p$ $\neg p \xrightarrow{\,ab\,} p$ $\neg p \xrightarrow{\,b\,} \neg p \xrightarrow{\,ab\,} p \xrightarrow{\,b\,} p$

(a)   (b)   (c)   (d)

$pq \xrightarrow{\,a\,} \neg p\neg q \xrightarrow{\,b\,} \neg pq$ $p \xrightarrow{\,b\,} \neg p \xrightarrow{\,a\,} p \xrightarrow{\,b\,} \neg p \xrightarrow{\,a\,} p \; -----$

(e)   (f)

$\neg p \xleftarrow{\,a\,} p$ $\neg p \xleftarrow{\,a\,} p \xrightarrow{\,b\,} p$ $p \xrightleftharpoons{\,a\,} \neg p \xrightarrow{\,b\,} p \xrightarrow{\,a\,} \neg p$

(g)   (h)   (i)

Figure 5.1.1

Models for knowledge and belief. States are named with the valuations of atoms. Visual conventions for relations R_a are transitivity, symmetry *except for a-labeled edges with arrows at end* and reflexivity *except for states from which lead a-labeled edges with arrows at end*. These conventions are sufficient to display **S5** models and **KD45** models unambiguously for a given number of agents.

inconsistent. A more austere way of preserving consistency but removing correctness is merely to require that *you consider possible what you believe*. In the logical language, instead of an inductive clause for knowledge, we now have an inductive clause $B_a p$ for "Agent a believes p." Formally, the above requirement is then $B_a p \to \langle B \rangle_a p$ (**D**). The logic of belief is called **KD45**.

Typical examples are when p is true but agent a believes it to be false (figure 5.1.1g), and where p is true, a mistakenly believes that a and b believe it to be false, whereas b believes that a and b correctly believe it to be true (figure 5.1.1h).

Knowledge and belief in these modal **S5** and **KD45** incarnations have become standard since Hintikka (1962) but have roots going back to medieval and classical times.

1.4 Rationality and Knowledge

We now present the principles of knowledge by way of questioning their rationality. Except for the correctness of knowledge, they are also the principles of belief. There are no principles for multiagent interaction. We therefore omit the index a in K_a.

T $Kp \to p$.

What you know is true (factivity or correctness of knowledge).

If this fails, why not call the notion "belief"? Some approaches still call it knowledge: if the set of counterexamples, which may or may not include what is actually the case, is sufficiently *small* (according to some formal notion of small and large, such as measure theory; Ben-David & Ben-Eliyahu, 1994), in other words, if the counterexamples are insignificant. Or it is the *fallible knowledge* of Özgün (2017, chapter 5), which ties the notion of knowledge only to justification and evidence.

4 $Kp \to KKp$.

You know what you know (positive introspection).

You have to make pairs of matching color out of the heap of socks coming out of the washing. Two socks may have the same color, and the next pair, and the next pair . . . But you are likely to end up with a last pair of socks of different color. The "same-color" relation is not transitive. Knowledge of socks' colors is *vague* (van Deemter, 2010). A different kind of attack emphasizes that rationality is bounded. Birds know how to fly, but they do not *know* that they know how to fly. Higher-order cognition in animals is (also) addressed in Verbrugge (2009).

5 $\langle K \rangle p \to K \langle K \rangle p$.

You know what you do not know (negative introspection).

Negative introspection is most often attacked: I am unaware of much that I do not know. So I do not know what I do not know. Negative introspection is justifiable if the relevant facts that agents may be uncertain about are known: the closed-world assumption. The **S5** notion of knowledge is considered suitable for describing the cognitive architecture of artificial agents. Well known is the *interpreted system* (Fagin, Halpern, Moses, & Vardi, 1995) consisting of a set of global states composed in turn of the local states of individual agents/processors. An agent cannot distinguish global states with the same local state value: **S5** knowledge.

Various alternative modal logics of knowledge weaken negative introspection. The logic **S4** is the system without **5**. Building a ladder from **S4** to **S5** are the systems **S4.2**, **S4.3**, and **S4.4**, obtained by adding the principles below to **S4**, for which (including **5**) we also give corresponding relational properties. The **.4** property is known as "remote symmetry."

.2 $\langle K\rangle Kp \rightarrow K\langle K\rangle p$ $\forall xyz\,(Rxy \wedge Rxz \rightarrow \exists w\,(Ryw \wedge Rzw))$

.3 $K(p \rightarrow \langle K\rangle q) \vee K(q \rightarrow \langle K\rangle p)$ $\forall xyz\,(Rxy \wedge Rxz \rightarrow Ryx \vee Rzy)$

.4 $p \rightarrow (\langle K\rangle p \rightarrow K\langle K\rangle p)$ $\forall xyz\,(Rxy \wedge Ryz \rightarrow Rzy)$

5 $\langle K\rangle p \rightarrow K\langle K\rangle p$ $\forall xyz\,(Rxy \wedge Rxz \rightarrow Ryz)$

Negative introspection has also been criticized because in combination with factivity, it implies $p \rightarrow K\langle K\rangle p$ (this characterizes symmetry; Hintikka, 1962; Stalnaker, 2005). This direct connection between objective truth and subjective truth is considered unacceptable: the agent should not be "allowed" to create knowledge out of thin air.

In logics combining knowledge and awareness (Fagin & Halpern, 1988; such logics are often known as logics for bounded rationality), the failure of negative introspection may even *define* unawareness. You are *unaware of p* if you don't know p and you also don't *know* that you don't know p: $\neg Kp \wedge \neg K\neg Kp$. This is known as the Modica–Rustichini definition of unawareness (Modica & Rustichini, 1994).

$K(p \rightarrow q) \rightarrow (Kp \rightarrow Kq)$

Deductive closure under the scope of knowledge.

This principle and the next principle of "necessitation" are together often known as those embodying logical *omniscience* and thus *full rationality*. Logics satisfying weaker notions, of which we will give two examples, are associated with *bounded rationality*.

In neighborhood structures (Pacuit, 2017), a formula is known in world w if there is a *neighborhood* of w (a designated subset typically containing w) that is the extension of the formula (i.e., the set of worlds where the formula holds). Let $p \rightarrow q$ and p have such extensions. Their intersection is contained in the extension of q. If the latter is not a neighborhood, $p \rightarrow q$ and p are both known, but not q.

In certain logics of knowledge and awareness (Fagin & Halpern, 1988; Halpern & Pucella, 2011), you know p if p is a member of the (fixed) set of "knowable" formulas in every accessible state; $p \rightarrow q$ and p may always be in that set, but not q. In that case, again, $p \rightarrow q$ and p are both known, but not q.

p implies Kp.

Necessitation of knowledge (i.e., if p is *valid*, then Kp is valid).

In neighborhood semantics, instead of necessitation, the weaker rule of *monotony* may hold: "$p \rightarrow q$ implies $Kp \rightarrow Kq$." In logics of knowledge and awareness, necessitation may not hold either: $p \vee \neg p$ is valid, but $K(p \vee \neg p)$ is invalid if the agent is unaware of p.

1.5 Combining Knowledge and Belief

Principles that are typically part of systems combining knowledge and belief, and where the main difference between the two is that beliefs, but not knowledge, may be incorrect, are $Kp \rightarrow Bp$ (knowledge implies belief) and $Bp \rightarrow BKp$ (you believe to know what you believe).

The logics **KD45** and **S5** do not combine very well. If apart from $Bp \rightarrow BKp$ one also has $Bp \rightarrow KBp$, then belief trivializes to knowledge: $Bp \leftrightarrow Kp$ (van der Hoek, 1993).

But **KD45** and **S4.4** combine well. In **S4.4**, belief is definable in terms of knowledge, namely, as $Bp \leftrightarrow \langle K\rangle Kp$, as was observed by Lenzen (1978). In this combination, knowledge is also defined by belief, namely, as $Kp \leftrightarrow p \wedge Bp$. Halpern, Samet, and Segev (2009) call this *explicit* definability. They also show that in the **S5**–**KD45** combination, knowledge is not explicitly definable in terms of belief, but only implicitly.

1.6 Knowing Whether

You know *whether p*, formally $Kw\,p$, if you know that p or you know that $\neg p$—more generally, if all accessible states have the same value for p. The principles of "knowing whether" are different from the principles of knowledge, for example, $Kw\,p \leftrightarrow Kw\,\neg p$. For the expressivity of such epistemic logics, this does not matter, as "knowing that" and "knowing whether" are interdefinable: $Kp \leftrightarrow p \wedge Kw\,p$, and $Kw\,p \leftrightarrow Kp \vee K\neg p$. In modal logics without the principle $Kp \rightarrow p$, such as **KD45**, knowledge may no longer be definable in terms of knowing whether. Indeed, the models of figure 5.1.1a and figure 5.1.1g are indistinguishable in the p states; in the former, Amir believes that p, and in the latter, Amir believes that $\neg p$; therefore, in either case, Amir "believes whether p" (in natural language this is infelicitous). Knowing-whether logics are founded in logics of contingency (Montgomery & Routley, 1966). Related notions such as knowing value, knowing how, and so on are already discussed in Hintikka (1962). A recent survey is Wang (2018).

2. Common Knowledge and Belief, and Distributed Knowledge

When celebrating Christmas with your young cousin, you should not spoil the fun by telling her that Santa Claus is actually dressed-up Uncle Ben. In fact, she knows this. So you both know this. But it is not common knowledge. Only when she yells, "But it is Uncle Ben," does it become common knowledge. Lewis (1969) gave the example of knowing that you have to drive on the left-hand side of the road: you know it, the car approaching you on this narrow road knows it, but you would feel very unsafe if it were all you know. However, it is common knowledge that you have to drive on the left-hand side. But how can you have obtained common knowledge with the total stranger in the other car? This is absurd. This sort of common knowledge is a *convention*: you are supposed to know. If you cause an accident, you cannot claim ignorance.

2.1 The Modal Logic of Common Knowledge

To the logic of knowledge, we add operators Cp for "The agents have common knowledge of p." The dual operator is $\langle C \rangle p$. The formula $\langle C \rangle p$ is true in a state of a Kripke model iff p is true at the end of any finite path of accessibility/indistinguishability links involving any agent. This accessibility is described by the arbitrary iteration of the union of all accessibility relations: let R_C be the transitive closure of the union of all relations R_a, then Cp is true in state s if p is true in all states t that are R_C-accessible. Similarly, we add to the logic of belief modalities $CB\,p$ for common belief of p.

Common knowledge has the properties of knowledge, but common belief does not have the properties of belief. It is easy to see that $Cp \rightarrow p$, $Cp \rightarrow CCp$, and $\langle C \rangle p \rightarrow C\langle C \rangle p$. But for common belief, negative introspection does not hold. Figure 5.1.1g demonstrates that $\neg CBp \rightarrow CB\,\neg CBp$ is false.

For example, in figure 5.1.1f, it is common knowledge that Bala does not know whether p. The logic of common knowledge is indeed more expressive than the logic of knowledge. This is proved by comparing the set of all finite-length chains (with leftmost actual state) consisting of iterated paired b–a links with that same set plus the infinite chain as in figure 5.1.1f. The formula $\langle C \rangle \langle K \rangle_b p$ can distinguish the former from the latter set, but no formula in the logic of knowledge can distinguish these sets, as it cannot look into the chains beyond the finite modal depth of that formula.

The history of common knowledge comes with the names Friedell (1969) and Lewis (1969). Van Ditmarsch, van Eijck, and Verbrugge (2009) give a historical survey in the form of a Socratic dialogue.

2.2 Relativized Common Knowledge

Temporal logic with the binary "until" modality is more expressive than temporal logic with the unary "going to be" (always in the future) modality. Similarly, epistemic logic with the binary so-called *relativized common knowledge* modality is more expressive than that with the unary common knowledge modality (Kooi & van Benthem, 2004). In a given state s, the agents have common

knowledge of p relative to q (notation $C_q p$) iff p is true in all states t that are reachable by finite R_C-paths satisfying q. Figure 5.1.1e illustrates the difference between relativization and a "mere" antecedent in an implication: in the pq state, $C_q p$ is true, but $C(q \rightarrow p)$ is false.

2.3 Distributed Knowledge

Suppose that Amir knows that p and that Bala knows that $p \rightarrow q$. Then together they know that q: if they were to communicate, they could make q common knowledge. Seen from a static perspective: if they were the same agent, then that agent could deduce q from p and $p \rightarrow q$. That agent only considers a world possible if Amir and Bala both consider it possible. We write Dp for "The agents have distributed knowledge of p"; Dp is true if p is true in all states accessible by the intersection of all accessibility relations R_a. The notion of distributed belief goes back to Hayek (1945) and Halpern and Moses (1990).

The logic with distributed knowledge is more expressive than the logic without: figure 5.1.1c and figure 5.1.1i cannot be distinguished in the logic with merely K_a and K_b (the models are bisimilar), but in the former, Dp is false in the p state, whereas in the latter, it is true in the p states.

A survey of logics of knowledge, belief, common knowledge, and distributed knowledge is van Ditmarsch, Halpern, van der Hoek, and Kooi (2015, chapter 1). Goldblatt (2003) presents modal logics in their historical context.

3. Change of Knowledge and Belief

The formation of knowledge and belief is crucial to give meaning to the concepts of knowledge and belief, and even more so for common knowledge and belief (see also chapter 2.6 by van Benthem, Liu, & Smets, this handbook).

3.1 Unsuccessful Update

Moore (1942) observed the incoherence of the statement "I went to the movies last Tuesday and I don't believe that." This form, $p \wedge \neg Kp$ (belief can be treated similarly), became later known as the *Moore sentence*. Assuming that you believe what you say, we get $K(p \wedge \neg Kp)$. Using the discussed properties of knowledge, this is equivalent to $Kp \wedge K\neg Kp$ and thus to $Kp \wedge \neg Kp$, which is contradictory. Now consider Bala informing Amir that Amir does not know that p is true; by conversational implicature, this is formalized as $p \wedge \neg K_a p$, which is not contradictory. (Hintikka [1962] devotes a whole chapter to such phenomena.) However, Bala cannot truthfully inform Amir *twice* in this way. After informing Amir of $p \wedge \neg K_a p$, true at the moment of utterance, Amir has learnt that p, that is, $K_a p$, so $p \wedge \neg K_a p$ is now false.

We can semantically verify this in *public announcement logic* (Plaza, 1989). A public announcement is a novel piece of information that is considered reliable (it is assumed true) and is simultaneously observed/incorporated by all agents. The result of the public announcement of p is the restriction of the Kripke model to the subdomain satisfying p. Now consider figure 5.1.1b: in the p state, $p \wedge \neg K_a p$ is true. In the $\neg p$ state, it is not. This announcement therefore induces the transition of figure 5.1.1b to figure 5.1.1a, the singleton-p state, in which obviously $K_a p$ is true. The dynamic interpretation of the Moore sentence as an *unsuccessful update* is due to Gerbrandy (2007; see also van Ditmarsch & Kooi, 2006). Furthermore, the Moore sentence demonstrates that announcing a proposition need not make it common knowledge.

3.2 Growth of Common Knowledge

The other problem with common knowledge is that it does not exist. This is very clear in the case of complete strangers in approaching cars having "common knowledge" of the traffic rules but less clear in other cases. In order to obtain common knowledge, special observation conditions need to apply. The agents must simultaneously observe the event and must also have joint awareness of their observing this event. Such conditions are fulfilled in a group of people facing each other wherein one of them is addressing the group: the observation is of a sound under conditions of mutual eye contact. This explains the term "public announcement." Loopholes can always be found: you may not have been paying attention, you could not hear what was being said, and so on. But you would then later need an excuse if you did not act upon the information: you were *supposed* to have heard.

The directness of speech in company should be contrasted with separating the sending and receiving of messages, after which growth of common knowledge is impossible (Halpern & Moses, 1990).

3.3 Knowability

In figure 5.1.1b, we can make two announcements, assuming the truth of p: the trivial announcement or the announcement of p. So, there is an announcement after which $K_a p$, and there is an announcement after which (still) $\neg K_a p$. Consider a logic with *quantification* over announcements: let $\blacklozenge K p$ mean that there is an announcement after which p is known. Inspired by Fine (1970), such a logic was proposed in Balbiani et al. (2008). We can also read $\blacklozenge K p$ as "p is knowable," such that $p \rightarrow \blacklozenge K p$ means that all truths are knowable. This is not valid. We have already seen that $p \wedge \neg K p$ is not knowable. Knowability (Fitch, 1963) is indeed interpreted in

such dynamic terms in van Benthem (2004) and Balbiani et al. (2008). A surprising result is that $\blacklozenge(K p \vee K \neg p)$ (i.e., $\blacklozenge K w p$) is valid (van Ditmarsch, van der Hoek, & Iliev, 2012).

3.4 Changing Knowledge and Belief Simultaneously

We can also change **S5** knowledge and **KD45** belief simultaneously. Plausibility models are Kripke models with equivalence relations R_a to encode knowledge, but where the states in each epistemic equivalence class are ordered according to plausibility. The agent believes p iff p is true in all most plausible states (the order has to be a well-preorder, which guarantees that most plausible states do exist). This induces a doxastic relation R'_a to interpret belief, contained in R_a. A public announcement restricts the epistemic relations to the novel domain. It is then determined anew what the most plausible states are, and this determines the new doxastic relation.

Figure 5.1.1b and figure 5.1.1g jointly represent that Amir does not know whether p but believes that $\neg p$. After public announcement of p, the single remaining state is p: figure 5.1.1a. This p state has now become the most plausible state. So figure 5.1.1a also depicts the doxastic relation. Amir now knows and believes that p. He has revised his knowledge and his belief (see Baltag & Smets, 2008; van Benthem, 2007; van Ditmarsch, 2005; a precursor is Spohn, 1988. These matters are also discussed in chapter 5.2 by Rott and chapter 5.3 by Kern-Isberner, Skovgaard-Olsen, & Spohn, both in this handbook).

3.5 Resolving Distributed Knowledge

Not all distributed knowledge can become common knowledge. Let us give two very different examples. In the p state of figure 5.1.1b, it is distributed knowledge that p is true, and agent a does not know this: we have that $K_b(p \wedge \neg K_a p)$, and thus $D(p \wedge \neg K_a p)$. But this cannot become common knowledge: if b informs a of p, p is common knowledge, not $p \wedge \neg K_a p$. Second, consider figure 5.1.1i: as we have already seen, p is distributed knowledge between a and b. But communication between a and b cannot make this common knowledge, because figure 5.1.1i and figure 5.1.1b encode the same information in epistemic logic (they are bisimilar). Making distributed knowledge common knowledge is called *resolving* distributed knowledge by Ågotnes and Wáng (2017).

Acknowledgments

Hans van Ditmarsch is also affiliated with IMSc (Chennai, India) and IRISA (Rennes, France). Reviewers are kindly thanked for their comments.

References

Ågotnes, T., & Wáng, Y. N. (2017). Resolving distributed knowledge. *Artificial Intelligence, 252*, 1–21.

Balbiani, P., Baltag, A., van Ditmarsch, H., Herzig, A., Hoshi, T., & de Lima, T. (2008). "Knowable" as "known after an announcement." *Review of Symbolic Logic, 1*(3), 305–334.

Baltag, A., & Smets, S. (2008). A qualitative theory of dynamic interactive belief revision. In G. Bonanno, W. van der Hoek, & M. Wooldridge (Eds.), *Logic and the foundations of game and decision theory (LOFT 7)* (Texts in Logic and Games, Vol. 3, pp. 13–60). Amsterdam, Netherlands: Amsterdam University Press.

Ben-David, S., & Ben-Eliyahu, R. (1994). A modal logic for subjective default reasoning. In *Proceedings, Ninth Annual IEEE Symposium on Logic in Computer Science* (pp. 477–486). IEEE Computer Society Press.

Fagin, R., & Halpern, J. Y. (1988). Belief, awareness, and limited reasoning. *Artificial Intelligence, 34*(1), 39–76.

Fagin, R., Halpern, J. Y., Moses, Y., & Vardi, M. Y. (1995). *Reasoning about knowledge*. Cambridge, MA: MIT Press.

Fine, K. (1970). Propositional quantifiers in modal logic. *Theoria, 36*(3), 336–346.

Fitch, F. B. (1963). A logical analysis of some value concepts. *Journal of Symbolic Logic, 28*(2), 135–142.

Friedell, M. F. (1969). On the structure of shared awareness. *Behavioral Science, 14*, 28–39.

Gerbrandy, J. (2007). The surprise examination in dynamic epistemic logic. *Synthese, 155*(1), 21–33.

Gettier, E. (1963). Is justified true belief knowledge? *Analysis, 23*, 121–123.

Goldblatt, R. (2003). Mathematical modal logic: A view of its evolution. *Journal of Applied Logic, 1*(5–6), 309–392.

Halpern, J. Y., & Moses, Y. (1990). Knowledge and common knowledge in a distributed environment. *Journal of the ACM, 37*(3), 549–587.

Halpern, J. Y., & Pucella, R. (2011). Dealing with logical omniscience: Expressiveness and pragmatics. *Artificial Intelligence, 175*(1), 220–235.

Halpern, J. Y., Samet, D., & Segev, E. (2009). Defining knowledge in terms of belief: The modal logic perspective. *Review of Symbolic Logic, 2*(3), 469–487.

Hayek, F. (1945). The use of knowledge in society. *American Economic Review, 35*, 519–530.

Hintikka, J. (1962). *Knowledge and belief: An introduction to the logic of the two notions*. Ithaca, NY: Cornell University Press.

Kooi, B., & van Benthem, J. (2004). Reduction axioms for epistemic actions. In R. A. Schmidt, I. Pratt-Hartmann, M.

Reynolds, & H. Wansing (Eds.), *AiML-2004: Advances in modal logic* (Technical Report Series No. UMCS-04-9-1, pp. 197–211). Manchester, England: University of Manchester.

Lenzen, W. (1978). Recent work in epistemic logic. *Acta Philosophica Fennica, 30*(2), 1–219.

Lewis, D. K. (1969). *Convention: A philosophical study*. Cambridge, MA: Harvard University Press.

Littlewood, J. E. (1953). *A mathematician's miscellany*. London, England: Methuen.

Modica, S., & Rustichini, A. (1994). Awareness and partitional information structures. *Theory and Decision, 37*, 107–124.

Montgomery, H., & Routley, R. (1966). Contingency and non-contingency bases for normal modal logics. *Logique et Analyse, 9*, 318–328.

Moore, G. E. (1942). A reply to my critics. In P. Schilpp (Ed.), *The philosophy of G. E. Moore* (pp. 535–677). Evanston, IL: Northwestern University.

Özgün, A. (2017). *Evidence in epistemic logic: A topological perspective* (Doctoral thesis, University of Lorraine, Nancy, France).

Pacuit, E. (2017). *Neighborhood semantics for modal logic*. Cham, Switzerland: Springer.

Plaza, J. (1989). Logics of public communications. In M. L. Emrich, M. S. Pfeifer, M. Hadzikadic, & Z. W. Ras (Eds.), *Proceedings of the 4th International Symposium on Methodologies for Intelligent Systems (ISMIS 1989): Poster session program* (pp. 201–216). Charlotte, NC: Oak Ridge National Laboratory.

Spohn, W. (1988). Ordinal conditional functions: A dynamic theory of epistemic states. In W. L. Harper & B. Skyrms (Eds.), *Causation in decision, belief change, and statistics* (Vol. II, pp. 105–134). Dordrecht, Netherlands: Kluwer.

Stalnaker, R. (2005). On logics of knowledge and belief. *Philosophical Studies, 128*(1), 169–199.

van Benthem, J. (2004). What one may come to know. *Analysis, 64*(2), 95–105.

van Benthem, J. (2007). Dynamic logic of belief revision. *Journal of Applied Non-Classical Logics, 17*(2), 129–155.

van Deemter, K. (2010). *Not exactly: In praise of vagueness*. Oxford, England: Oxford University Press.

van der Hoek, W. (1993). Systems for knowledge and beliefs. *Journal of Logic and Computation, 3*(2), 173–195.

van Ditmarsch, H. (2005). Prolegomena to dynamic logic for belief revision. *Synthese, 147*, 229–275.

van Ditmarsch, H., Halpern, J. Y., van der Hoek, W., & Kooi, B. (Eds.). (2015). *Handbook of epistemic logic*. London, England: College Publications.

van Ditmarsch, H., & Kooi, B. (2006). The secret of my success. *Synthese, 151*, 201–232.

van Ditmarsch, H., van der Hoek, W., & Iliev, P. (2012). Everything is knowable—how to get to know *whether* a proposition is true. *Theoria, 78*(2), 93–114.

van Ditmarsch, H., van Eijck, J., & Verbrugge, R. (2009). Common knowledge and common belief. In J. van Eijck & R. Verbrugge (Eds.), *Discourses on social software* (pp. 99–122). Amsterdam, Netherlands: Amsterdam University Press.

Verbrugge, R. (2009). Logic and social cognition: The facts matter, and so do computational models. *Journal of Philosophical Logic, 38*(6), 649–680.

Wang, Y. (2018). Beyond knowing that: A new generation of epistemic logics. In H. van Ditmarsch & G. Sandu (Eds.), *Jaakko Hintikka on knowledge and game-theoretical semantics* (Outstanding Contributions to Logic, Vol. 12, pp. 499–533). Cham, Switzerland: Springer.

5.2 Belief Revision

Hans Rott

Summary

This chapter gives an introduction to the problem of rational belief change and its formal modeling in qualitative logical theories. It first presents an outline of the approach based on rationality postulates and of the most important constructive approaches within the classical AGM model of belief revision. Then it gives the basic ideas of extensions of the classical model to iterated belief revision, to multiple revision and to two-dimensional belief change, to belief merging and to belief updates, as well as to default reasoning.

1. The Problem of Belief Revision

The term "belief revision" refers to the change of a system of beliefs or opinions in response to new information, particularly to information that is inconsistent with this system. In the 1960s and 1970s, the works of Isaac Levi and William Harper prepared the ground for systematic studies of belief change processes. In the 1980s, a seminal research paradigm of belief revision was established by Carlos Alchourrón, Peter Gärdenfors, and David Makinson (also known as "AGM"), who discovered a common structure in the logic of normative systems and of the logic of counterfactual conditionals (see chapter 6.1 by Starr, this handbook)—areas that turned out to be structurally related and were subsequently merged into the field of belief revision (also known as "theory change"). Belief revision theories are different from descriptive psychological investigations (see chapter 5.4 by Gazzo Castañeda & Knauff, this handbook) and investigations of theory dynamics as conducted in the philosophy of science in that they aim at providing postulates and construction recipes for *rational* processes of belief change.

The problem of belief revision is best illustrated by an example. Suppose you believe that

(a) The magician's swords are real swords. $\quad s$

(b) Lisa, the magician's assistant, is in the box. $\quad b$

(c) If Lisa is in the box, she cannot escape the swords. $\quad b \to \neg e$

(d) If the magician's swords are real swords $\quad s \wedge \neg e \to d$ and Lisa cannot escape them, she will die.

Like any person capable of good reasoning, you conclude from these premises that

(e) Lisa will die. $\quad d$

At the end of the show, however, Lisa climbs out of the box without a scratch. This is a surprise to you. You receive new information $\neg d$ that contradicts your initial premises (a)–(d). Thinking about the matter, you realize that given your initial premises, logic in a way compelled you to believe d. Now that you have found out, luckily, that d is not true after all, and you want to add $\neg d$ to your stock of beliefs, there are only two options: either your logical reasoning that delivered the conclusion d was faulty, or (at least) one of the premises (a)–(d) was false. Assuming that your logic has been flawless and you want to accept the new piece of information $\neg d$, you have to give up (at least) one of your premises. Logic, however, is completely silent about *which* of the premises to give up. Something else has to step in that provides you with guidance how to proceed. Belief revision theories specify what this "something else" is and how it helps you to resolve such problems in the transformation of belief states.

In general, you will not want to give up *all* of your beliefs in order to accommodate the new information $\neg d$ and avoid becoming inconsistent. This radical move would mean an unnecessary loss of valuable information. So when giving up the belief d, you have to decide which of your reasons for holding d to retain and which to retract. You have to *choose* between retracting either s or b or $b \to \neg e$ or $s \wedge \neg e \to d$. You may, for instance, decide that you were too rash in endorsing $b \to \neg e$. Lisa might twist her body into a corner of the box so as to avoid being impaled by the swords.

2.   The Representation of Belief

Any theory of the rational formation and transformation of belief states consists of (1) a static part describing the states of belief that a person may be in and (2) a dynamic part describing how belief states should change in response to external input. The static picture constrains the dynamic picture. Assuming that there are such things as plain beliefs, belief revision theorists need to decide how to represent *beliefs* (what a person believes) and *belief states* (the mental states that believers are in).

This chapter considers qualitative models in which belief states don't carry any numerical information and reasoners accept new information without numerical qualification. There are no numbers representing degrees of "certainty," "security," "plausibility," and so on, and belief is always plain belief. In quantitative theories, on the other hand, reasoners do not, or in any case do not have to, accept new information *simpliciter* but can accept it with a certain degree of certainty or plausibility. The best-known quantitative modeling of belief is the probabilistic one, but there are alternative numerical methods of belief change using *ranking functions* (Spohn, 2012; similar ideas are part of possibility theory as summarized by Dubois & Prade, 2015; cf. chapter 4.7 by Dubois & Prade and chapter 5.3 by Kern-Isberner, Skovgaard-Olsen, & Spohn, both in this handbook). Quantitative approaches are more expressive than qualitative methods. Their additional expressive power is a substantial advantage (it is essential, for instance, for expressing significant independence relations between beliefs), but there is a price to be paid for it: often it is not clear where the numbers come from and what exactly they are supposed to mean. While in probability theory, numerical values may be explicated in terms of betting quotients, the meaning of the numbers in ranking functions is less clear (but see Spohn, 2012, chapter 8).

Belief revision theories have traditionally represented beliefs as sentences or propositions (i.e., roughly, contents of sentences). We will focus on sentential models that typically work with a language **L** closed under applications of the Boolean operators of negation, conjunction, disjunction, and material implication.

There is an important contrast in modern epistemology between foundationalist and coherentist theories of the justification of belief. With a grain of salt, this contrast is mirrored in belief revision theories. *Foundationalists* in belief revision assume that there is a set of basic beliefs that are somehow given and may be thought of as justified in a direct, noninferential way. All other beliefs are justified only insofar as they can be derived from the stock of basic beliefs. In more formal terms, such theories presume that there is a *belief base* H of distinguished sentences that have, epistemologically speaking, an independent standing. Another factor to be decided in a formal modeling of belief states is the logic governing the language **L**. A *logic* or *consequence operation* is a function Cn that assigns to any belief base H the set $Cn(H)$ of all logical consequences of H. It is commonly assumed that Cn is Tarskian and includes classical propositional logic.

We call H a *base for* the belief set K if and only if $K = Cn(H)$. *Belief sets* are sets K of sentences in **L** that are closed under Cn, that is, $K = Cn(K)$. If A is an element of K, then A is *believed* (*held to be true*) by a reasoner whose belief set is K. If the negation ¬A is an element of K, then A is *rejected in K*. Of course, in general, there are also sentences A that are neither believed nor rejected but on which the reasoner suspends judgment. A belief set can be viewed as a theory that is a partial description of the world.

A belief derived from a belief base is justified by those elements of the base that are used in its derivation. Like other foundationalist theories in epistemology, the modeling using belief bases suggests that propositions should not be accepted as beliefs unless they are positively justified. Reasoners are supposed to keep track of the justifications for their beliefs. In contrast, *coherence theories* hold that reasoners need not keep track of the pedigrees of their beliefs. They should instead focus on the logical or inferential structure of the beliefs—what matters for the rationality of a belief is how it coheres with the other beliefs that are accepted in the current belief state. For the coherentist, there is no designated set of basic beliefs: each belief depends in some way or other on there being certain other beliefs that support it.

As models of explicit (active, occurrent) belief, *bases* are psychologically more realistic since they are usually finite entities. On the other hand, *belief set* dynamics offers a competence model that helps us to understand what people ought to do ideally if they were not bounded by limited logical or computational reasoning capabilities. There are also interpretations that make sense of the requirement of logical closure even for less-than-ideal reasoners: a belief set may be taken to represent the set of beliefs a reasoner is *committed* to (Levi, 1997) or the set of beliefs *ascribed* to a reasoner. More thoughts on logical closure can be found in chapter 3.1 by Steinberger (this handbook).

3.   Kinds of Belief Change

If belief states are represented by sets of sentences, there can only be three doxastic attitudes: a sentence can be accepted, rejected, or neither. We distinguish two basic types of belief change: a sentence A can be *inserted* into a belief set (turned into a belief) or *deleted* from a belief set (turned into a nonbelief). The former is called a *revision* by A, denoted by $K*A$, and the latter a *contraction* with respect to A, denoted by $K \div A$.

An *expansion* $K + A$ of a belief set K by a new piece of information A is simply formed by set-theoretic addition and subsequent logical closure: $K + A = Cn(K \cup \{A\})$. An expansion thus defined is closed under logical consequence, and it is consistent as long as A is consistent with K. But what if A is inconsistent with K? According to the classical rule of *ex falso quodlibet*, a single inconsistency entails any arbitrary sentence, so there is only one inconsistent set closed under Cn, namely, the set L of *all* sentences in the language $\mathbf{L}$. Thus, the expansion operation is useless in the belief-contravening case. If consistency is to be preserved, belief-contravening changes require choices in retractions that cannot be made on the basis of set theory and logic alone.

This specific problem of belief-contravening revision was illustrated by our introductory example. From a logical point of view, there were several ways of constructing the revision when accommodating $\neg d$. There is no purely logical reason for making one choice rather than another among the sentences eligible for retraction. The reasoner needs additional information concerning these sentences and a well-defined method for constructing revisions that takes into account a number of coherence constraints (see section 4.1). Such a method is represented formally by a *revision function* $*$ associated with the reasoner's belief set K that can take any sentence A and return as value the revised belief set $K*A$.[1]

The contraction process faces parallel problems. The belief set K of our initial example contains the premises s, b, $b \rightarrow \neg e$, and $s \wedge \neg e \rightarrow d$ together with all their logical consequences (among which is d). Suppose that you want to contract K with respect to d. Of course, the target sentence d itself must be deleted from K when forming the contracted belief set $K \div d$. But this is not enough. At least one of the four premises must be given up as well in order to prevent d from being rederived through Cn. Again, there is no purely logical reason for making one choice rather than another. Deleting d from your belief set poses quite the same problems as adding $\neg d$ to it.

We have now formed an idea that the problem of belief revision (by $\neg d$) is closely related to the problem of belief contraction (with respect to d). In parallel with revision functions, one can introduce the concept of a *contraction function* $\div$ associated with the reasoner's belief set K that takes sentences A and returns as values the

contracted belief sets $K \div A$. While for belief-contravening revision, the problem of maintaining consistency is most palpable, in belief contraction, there is no problem with consistency: shrinking one's set of beliefs can never introduce an inconsistency. But instead, the problem of logical closure makes itself felt rather acutely. The logical interaction between an updated "data base" and its derived consequences may be seen as the ultimate source of the problem of belief change.

4. Two Strategies for Characterizing Rational Changes of Belief

When tackling the problem of belief revision in more concrete detail, two general strategies have been followed. First, one can write down a list of desiderata that an appropriate belief revision function should fulfill. That is, the standards for revision and contraction functions can be laid down in the form of *rationality postulates*. Second, one can present *explicit constructions* for rational changes of belief. The solution to the problem of belief revision will not be complete unless we know how to define and compute appropriate revision and contraction functions for a given belief state. We will see that constructions of belief changes make essential use of "doxastic preference relations" (or similar structures on which the necessary choices may be based).

4.1 The Strategy Based on Rationality Postulates
In their seminal study, Alchourrón, Gärdenfors, and Makinson (1985) proposed a set of general postulates for the rational revision of belief sets. They are now usually called the *AGM postulates*:

($*1$) $K*A$ is closed under Cn.
($*2$) $A \in K*A$.
($*3$) $K*A \subseteq K + A$.
($*4$) If $\neg A \notin K$, then $K + A \subseteq K*A$.
($*5$) If A is consistent, so is $K*A$.
($*6$) If $Cn(A) = Cn(B)$, then $K*A = K*B$.
($*7$) $K*(A \wedge B) \subseteq (K*A) + B$.
($*8$) If $\neg B \notin K*A$, then $(K*A) + B \subseteq K*(A \wedge B)$.

We can think of this set as coming in four pairs. Postulates ($*2$) and ($*6$) concern the *input*: the input should be accepted, and it is its content, not its syntactic form, that matters. The other three pairs encode three different ideas of coherence. Postulates ($*1$) and ($*5$) represent a *static* notion of coherence very much like the idea of a reflective equilibrium. They concern the beliefs held by a person at a certain point in time. Consistency is the major driving force behind belief revision theories, and it is a minimal notion of static coherence.[2] Closure is a more demanding idea of static coherence. Postulates (*3) and (*4) formulate a *dynamic* notion of coherence. They refer to a short sequence of belief states, namely, the transition from a state to its successor state—but only when the new piece of information A is consistent with the belief set K: in this case, the revision is formed by set-theoretically adding A to K and then taking the logical closure. Dynamic coherence in this sense instantiates a (restricted) maxim of conservatism or, in different terms, of informational economy, minimal change, minimum mutilation (Quine), or cognitive inertia. Prior beliefs should not be abandoned, and new beliefs should not be adopted, beyond necessity. Minimal change has frequently been advertised as one of the principal ideas motivating belief revision theories, but this claim is controversial (Rott, 2000). Postulates (*7) and (*8) encode a *dispositional* notion of coherence referring to potential transitions from a single belief state to various possible successor states. More specifically, (*7) and (*8) relate the change by a conjunction to the change by one of the conjoined sentences. Dispositional coherence in belief revision turned out to be closely related to the coherence of choices as studied in rational choice theory. While static coherence is essentially a logical notion, dynamic and dispositional coherence are notions relating to two different concepts of economic behavior.[3] But notice that all postulates except (*2) refer to the underlying logic Cn.

Isaac Levi advanced the thesis that any rational belief revision occasioned by a new piece of information A is decomposable into two successive steps, namely, an elimination of $\neg A$, followed by an expansion by A (which produces no inconsistency, thanks to the antecedent contraction step). This suggestion, which is in accordance with the intuitive discussion of our introductory example, has become known as the *Levi identity*:

(L) $K * A = (K \div \neg A) + A$.

Many authors have endorsed (L) and sought to reduce the problem of belief revision to the problem of belief contraction. But there is also a converse equation defining contraction in terms of revision. This is the so-called *Harper identity*, which is, in a rather precise sense, the converse of the Levi identity:

(H) $K \div A = K \cap (K * \neg A)$.

According to (H), the contraction of a belief set K with respect to A is that part of K that will survive K's revision by $\neg A$. Using equations (L) and (H) as bridge principles, a set of eight *AGM postulates* for the rational *contraction*

of belief sets is equivalent to the set of postulates for belief revision. Here are just the most interesting ones:

(÷7) $K \div A \cap K \div B \subseteq K \div (A \wedge B)$.
(÷8)   If $A \notin K \div (A \wedge B)$, then $K \div (A \wedge B) \subseteq K \div A$.

The conjunction postulates (÷7) and (÷8) compare contractions of K with respect to a conjunction $A \wedge B$ with contractions with respect to the conjoined sentences A and B. (÷7) states that any sentence that is contained both in the contraction of K with respect to A and in the contraction of K with respect to B is contained in the contraction of K with respect to $A \wedge B$. As a kind of converse, if the contraction with respect to $A \wedge B$ eliminates A, then any sentence contained in the contraction with respect to $A \wedge B$ is also contained in the contraction with respect to A. This is natural if eliminating $A \wedge B$ is taken to mean eliminating A or eliminating B (or eliminating both, if A and B are equally entrenched).

AGM called the first six postulates for revisions and contractions the *basic* postulates and the last two ones *supplementary* postulates. The latter pair of postulates entails that belief change operations can be construed as being guided by some well-behaved doxastic preference relation (cf. section 4.2). In many contexts, however, postulates (*7) and (*8) have turned out to be too strong. Interesting weakenings, for example, are the following:

(*7c)   If $B \in K * A$, then $K * (A \wedge B) \subseteq K * A$.
(*8c)   If $B \in K * A$, then $K * A \subseteq K * (A \wedge B)$.

Taken together, these conditions say that if B is contained in $K * A$, then $K * A$ is equal to $K * (A \wedge B)$. The "c" in "(*7c)" and "(*8c)" means "cumulative," due to the fact that there are corresponding postulates for nonmonotonic logics that are characteristic for *cumulative reasoning* (cf. section 6.4).

4.2   The Constructive Strategy

We have already seen that logical postulates are not sufficient to determine a particular change of K with respect to some proposition A. The construction of belief changes has to draw on extra-logical factors. In some constructions, the syntactic structure of the base for K is itself taken to encode the relevant information (Hansson, 1991; Lewis, 1981; Nebel, 1992). Since the 1990s, most researchers have assumed that these extra-logical factors are *doxastic preference relations* (*plausibility relations*) over sentences or over possible worlds and that such relations are part of the reasoner's belief state. Others have suggested that external, objective criteria (like objective probability, informational content, truthlikeness, or objective similarity between possible worlds) should play the key role in fixing the revision-guiding structures.

AGM explored three methods for the construction of the contractions.[4] The first method defines $K \div A$ as the intersection of the best maximal subsets of K not implying A, where what is best is determined by a doxastic preference relation (*partial meet contraction*). This method was later shown to be intimately related to a semantic modeling using possible worlds (Grove, 1988; Katsuno & Mendelzon, 1991). Here, $K \div A$ is defined as the set of sentences in K that are true in all maximally "plausible" worlds that do not satisfy A, where what is most plausible is determined by a system of spheres of possible worlds centered on the worlds satisfying K. The second method keeps in $K \div A$ precisely those propositions of K that are "safe" in all minimal subsets of K that imply A, where again what is safe is determined by a doxastic preference relation (*safe contraction*). The third method keeps in $K \div A$, roughly speaking, those propositions that are doxastically "well entrenched," where entrenchment relations are yet another kind of doxastic preferences (*entrenchment-based contraction*). Importantly, the doxastic preference relations used in each of these constructions are independent of the particular sentence to be contracted.

It is natural to ask how the postulates strategy and the constructions strategy relate to one another. Assuming that doxastic preferences meet a number of formal requirements (like transitivity and completeness), AGM and their followers were able to prove various *representation theorems* to the effect that a contraction operation for a belief set K satisfies the rationality postulates for contraction if and only if it can be (re)constructed by any of the standard construction recipes. Revisions are taken care of with the help of the Levi identity. As a corollary, the three standard methods are thereby proven equivalent.[5]

While the idea of informational economy has played a rather modest role in AGM-style theorizing of belief revision, ideas from economics have always formed an essential part of belief revision theories through the systematic study of the structure and application of preference relations to cognitive choice problems. Varying (usually weakening) the formal requirements for the relevant preference relations results in variations (usually weakenings) of the postulates encoding dispositional coherence. Some natural relaxations of the requirements for preference relations (including the renunciation of completeness) lead to the postulates (∗7c) and (∗8c) mentioned above. Rationality constraints for choices (as studied by leading economists like P. Samuelson,

K. Arrow, and A. Sen; see chapter 8.1 by Grüne-Yanoff, this handbook) can be applied in quite a straightforward way to choice problems involved in belief change processes—choices concerning least plausible beliefs or concerning most plausible possible worlds—and were thus shown to "translate" into rationality postulates for belief revision (Rott, 2001). Seen from this perspective, the problem of belief revision is just an instantiation of the general problem of rational choice, with a few additional constraints placed by the background logic Cn. While the formation and transformation of belief sets would usually be classed as a problem of theoretical reason, the making of choices belongs to the realm of practical reason (cf. chapter 2.2 by Wedgwood, this handbook). However, the interpretation of this connection is far from clear. Should we take the use of choice functions as indicating that rational believers are free to decide to believe or disbelieve a proposition according to their preferences? Are they free to adopt the doxastic preferences they wish?[6]

5. Iterated Changes of Belief

There is a broad consensus about how to use selection structures in one-shot belief revision: take the AGM solution. However, iterated belief change or, equivalently, the change of selection structures exhibits substantial ambiguities (Darwiche & Pearl, 1997). Even within the narrow confines of simple qualitative theories (no numbers, only propositional inputs), there are quite a few models that have been proposed. They typically adopt the AGM postulates for one-shot belief revision and add one rationality postulate taking care of iterations. We look at five important proposals.

The first model uses one and the same selection structure for all possible belief sets. This structure does not represent the reasoner's current belief state but is *external* to it (Areces & Becher, 2001). In contrast, the other four models identify the reasoner's belief state with her selection structure. The second model is maximally *conservative*: the selection structure is changed so that the input just gets accepted, but the changes made are absolutely minimal, and as a consequence, the newly acquired information is very easily lost in further belief changes (Boutilier, 1996). The third model, *restrained* revision, is close to the conservative one but gives the input somewhat more impact (Booth & Meyer, 2006). The fourth model is *moderate* in that it recommends giving high priority to the incoming information, but it does not wipe out all plausibility distinctions between worlds or situations that do not comply with it (Nayak, 1994). The fifth model is *radical* in that the input is given uncompromising priority over the previous beliefs: any world or situation that does not satisfy the input is considered maximally implausible after the revision (Segerberg, 1998). Each of these models is well motivated by a possible-worlds semantics and can be axiomatically characterized by a single axiom taking care of the iteration case: it equates $(K*A)*B$ with $K*(A \wedge B)$, if B is consistent with $K*A$ (external and conservative revision); if B is consistent with $K*A$ and A is consistent with $K*B$ (restrained revision); if B is consistent with A (moderate revision); or in any case (radical revision). And it equates $(K*A)*B$ with $K*B$ otherwise (exception: with $L*B$ for external revision[7]). The belief set resulting from any finite number of revision steps can be generated inductively by repeated applications of the respective axiom. We can see that according to each model, two successive revision steps can be reduced to a one-shot revision, mostly by means of some characteristic case distinction (but this kind of reducibility has recently been challenged by Booth & Chandler, 2017). Each of these competing models for iterated belief change is adequate for some cases, but it is easy to come up with examples showing that they have drawbacks in other cases. Unfortunately, no *methodology* has been put forward yet that would tell us when it is rational to use which model.

6. Further Developments

The AGM model is a classic by now that has in many ways been extended and revised.

6.1 Extensions of Belief Revision

Multiple revisions are occasioned by sets of sentences that have to be accepted or withdrawn simultaneously (Delgrande & Jin, 2012; Fuhrmann & Hansson, 1994). In *nonprioritized belief revision*, a reasoner does not invariably incorporate the new information but first decides in the light of her current belief state whether or not she wants to revise her beliefs at all (Hansson, 1997). *Two-dimensional belief revision* is a method that changes beliefs in response to imperatives of the form "Accept A with a degree of plausibility that at least equals that of B" (Cantwell, 1997; Fermé & Rott, 2004; Rott, 2012). Put differently, this belief change operation takes two arguments, an *input sentence A* and a *reference sentence B*, or the input is of the form "$B \leq A$." This method of changing belief states can be used for belief revision (with respect to the input sentence) as well as for belief contraction (with respect to the reference sentence). Two-dimensional belief revision is much more flexible than the methods of iterated belief revision mentioned in

section 5. For instance, beliefs can be *lowered*, and nonbeliefs can be *raised*, in doxastic status while still remaining beliefs and nonbeliefs, respectively.

6.2 Belief Merging

The field of *belief merging* or *belief fusion* is concerned with the rational aggregation of two or more belief states (Baral, Kraus, Minker, & Subrahmanian, 1992; Konieczny & Pino Pérez, 2002; Nayak, 1994; Pigozzi, 2015). In many variants of merging, negotiation procedures or aggregation processes similar to those known from social choice theory find application. While belief merging should be thought of as primarily combining belief states of several agents, Konieczny and Pino Pérez (2002) also make room for propositional "inputs" in the original AGM sense (they call them "integrity constraints"). They show that the merging postulates can be modified or supplemented in a wide variety of interesting ways, provide a constructive approach, and give appropriate representation theorems for basic as well as for supplemented models.

6.3 Updates

Katsuno and Mendelzon (1992) discovered an important distinction between belief revisions and belief updates. *Revisions* are prompted by new information about an unchanging world, while *updates* are prompted by information about changes in the world. The difference is most conspicuous in the case where the new piece of information A is consistent with the belief set K. Revisions by A make a person eliminate from the set of worlds compatible with K those worlds that do not satisfy A. If you learn that Lisa is the owner of the theater, with the understanding that she has owned it for some time, then you simply discard all possible worlds in which she is not the owner (this is in accordance with dynamic coherence). In contrast, if you learn that Lisa is the owner of the theater with the understanding that she has just bought it, then you don't drop any of your possible worlds but modify each of them individually by "letting Lisa become" the owner of the theater. An update by A makes a person consider, for each world compatible with K, how this world would develop if A were brought about in it. A formal mark of such updates is that they violate postulate (*4) and instead satisfy the following monotonicity condition:

(*M) If $K_1 \subseteq K_2$, then $K_1 * A \subseteq K_2 * A$. (*-Monotonicity)

6.4 Default Inferences as Expectation Revision

Given the information that Tweety is a bird, you will typically conclude that Tweety can fly. But given the information that Tweety is a penguin, which of course implies that Tweety is a bird, you will *not* infer that Tweety can fly. In commonsense reasoning, an extension of the set of premises not only *gains* information but often *loses* some conclusions that were drawn on the basis of the smaller premise set. Everyday inferences thus do not conform to the rules of classical logic but exhibit patterns of *defeasible* or *nonmonotonic* reasoning. Most conspicuously, they violate the following rule of ordinary (Tarskian) logics:

(M) If $H \subseteq H'$, then $Inf(H) \subseteq Inf(H')$. (*Inf-Monotonicity*)

Gärdenfors and Makinson (1994) suggested that the set of defeasible consequences *Inf(H)* of a finite set of premises $H = \{A_1, \ldots, A_n\}$ can be regarded as the theory that results from the revision of some set of *background assumptions*, *defaults*, or *expectations E* by the conjunction of the premise set *H*. They proposed the following equation:

(I) $Inf(H) = E * (A_1 \wedge \ldots \wedge A_n)$.

For the inference operation *Inf* thus defined, adding information can defeat previously drawn conclusions, just as when we enlarge the singleton premise set containing only "Tweety is a bird" (A_1) with the premise "Tweety is a penguin" (A_2). Given our expectations, the input A_1 makes us conclude that Tweety can fly, whereas conjoining the input A_2 will make us draw the opposite conclusion (note that, given our background knowledge, $A_1 \wedge A_2$ is equivalent to A_2 alone).

Defeasible or nonmonotonic inference operations are quite irregular as compared to Tarskian logics. The failure of (M) is indeed incisive. However, commonsense reasoning nevertheless appears to leave intact quite a number of important classical inference patterns (Brewka, Marek, & Truszczyński, 2011; Kraus, Lehmann, & Magidor, 1990; Makinson, 1994, 2005). Gärdenfors and Makinson (1994) proved correspondences of the basic and full sets of rationality postulates for belief revision and appropriate sets of rationality postulates for default reasoning, and numerous other correspondences are detailed in Rott (2001). Results such as these suggest that belief change and defeasible reasoning are indeed just "two sides of the same coin" (Gärdenfors, 1991).

7. Conclusion

Research on rational belief change originally started out in philosophy and is now most actively pursued in computer science, but it is an essentially interdisciplinary undertaking. One of the tasks ahead is to develop a methodology for when to apply which of the many models on offer. Another task is to close the gap between idealized models of perfectly rational reasoners and the actual facts about the limited logical and computational capacities of human beings. But whether we view it from a normative or an empirical perspective, the formation and transformation of beliefs is a problem that raises its head just about everywhere.

Notes

1. Many authors present revision functions as two-place functions taking a pair of a belief set and a sentence and returning a new belief set. This way of putting things appears misleading to me, since one and the same belief set may well be coupled with different belief revision strategies, so * as acting on pairs $\langle K, A \rangle$ cannot be functional in the K argument. For more details, see Rott (1999).

2. The consistency requirement can be seen as marking off the enterprise of belief revision from that of paraconsistent logic, which shares the aim of explicating rational deliberation in the face of contradictory information. A logic or inference operation is *paraconsistent* if it does not satisfy the classical rule of *ex falso quodlibet* (i.e., if not everything is derivable from a classically inconsistent set of premises). Wassermann (2011), Girard and Tanaka (2016), and Testa, Coniglio, and Ribeiro (2017) argue that one should *combine* belief revision and paraconsistent logics.

3. See Rott (2003). The thesis that recipes for belief revision ought to be derived from decision-theoretic principles has been championed by Isaac Levi at least since Levi (1991).

4. For more detailed overviews, see Gärdenfors and Rott (1995), Hansson (1999), and Fermé and Hansson (2018).

5. A great variety of representation theorems can be found in Gärdenfors (1988), Hansson (1999), Bochman (2001), Rott (2001), and Fermé and Hansson (2018).

6. This raises the question of doxastic voluntarism, which is linked with belief revision theory in Rott (2017).

7. "External" methods based on global (i.e., belief-state independent) similarities or distances between possible worlds behave differently (cf. Lehmann, Magidor, & Schlechta, 2001).

References

Alchourrón, C., Gärdenfors, P., & Makinson, D. (1985). On the logic of theory change: Partial meet contraction and revision functions. *Journal of Symbolic Logic, 50*, 510–530.

Areces, C., & Becher, V. (2001). Iterable AGM functions. In M.-A. Williams & H. Rott (Eds.), *Frontiers in belief revision* (pp. 261–277). Dordrecht, Netherlands: Kluwer.

Baral, C., Kraus, S., Minker, J., & Subrahmanian, V. S. (1992). Combining multiple knowledge bases consisting of first-order theories. *Computational Intelligence, 8*, 45–71.

Bochman, A. (2001). *A logical theory of nonmonotonic inference and belief change.* Berlin, Germany: Springer.

Booth, R., & Chandler, J. (2017). The irreducibility of iterated to single revision. *Journal of Philosophical Logic, 46,* 405–418.

Booth, R., & Meyer, T. (2006). Admissible and restrained revision. *Journal of Artificial Intelligence Research, 26,* 127–151.

Boutilier, C. (1996). Iterated revision and minimal change of conditional beliefs. *Journal of Philosophical Logic, 25,* 263–305.

Brewka, G., Marek, V., & Truszczyński, M. (Eds.). (2011). *Nonmonotonic reasoning: Essays celebrating its 30th anniversary.* London, England: College Publications.

Cantwell, J. (1997). On the logic of small changes in hypertheories. *Theoria, 63,* 54–89.

Darwiche, A., & Pearl, J. (1997). On the logic of iterated belief revision. *Artificial Intelligence, 89,* 1–29.

Delgrande, J., & Jin, Y. (2012). Parallel belief revision: Revising by sets of formulas. *Artificial Intelligence, 176,* 2223–2245.

Dubois, D., & Prade, H. (2015). Possibility theory and its applications: Where do we stand? In J. Kacprzyk & W. Pedrycz (Eds.), *Springer handbook of computational intelligence* (pp. 31–60). Berlin, Germany: Springer.

Fermé, E., & Hansson, S. O. (2018). *Belief change: Introduction and overview.* Berlin, Germany: Springer.

Fermé, E., & Rott, H. (2004). Revision by comparison. *Artificial Intelligence, 157,* 5–47.

Fuhrmann, A., & Hansson, S. O. (1994). A survey of multiple contractions. *Journal of Logic, Language and Information, 3,* 39–76.

Gärdenfors, P. (1988). *Knowledge in flux: Modeling the dynamics of epistemic states.* Cambridge, MA: MIT Press.

Gärdenfors, P. (1991). Belief revision and nonmonotonic logic: Two sides of the same coin? In J. van Eijck (Ed.), *Logics in AI: JELIA 1990* (Lecture Notes in Computer Science (Lecture Notes in Artificial Intelligence), vol. 478). Berlin, Germany: Springer.

Gärdenfors, P., & Makinson, D. (1994). Nonmonotonic inference based on expectations. *Artificial Intelligence, 65,* 197–245.

Gärdenfors, P., & Rott, H. (1995). Belief revision. In D. M. Gabbay, C. J. Hogger, & J. A. Robinson (Eds.), *Handbook of logic in artificial intelligence and logic programming: Vol. IV. Epistemic and temporal reasoning* (pp. 35–132). Oxford, England: Oxford University Press.

Girard, P., & Tanaka, K. (2016). Paraconsistent dynamics. *Synthese, 193,* 1–14.

Grove, A. (1988). Two modellings for theory change. *Journal of Philosophical Logic, 17,* 157–70.

Hansson, S. O. (1991). *Belief base dynamics* (Unpublished doctoral dissertation). Uppsala University, Uppsala, Sweden.

Hansson, S. O. (1999). *A textbook of belief dynamics: Theory change and database updating.* Dordrecht, Netherlands: Kluwer.

Hansson, S. O. (Ed.). (1997). Non-prioritized belief revision [Special issue]. *Theoria, 63.*

Katsuno, H., & Mendelzon, A. O. (1991). Propositional knowledge base revision and minimal change. *Artificial Intelligence, 52,* 263–294.

Katsuno, H., & Mendelzon, A. O. (1992). On the difference between updating a knowledge base and revising it. In P. Gärdenfors (Ed.), *Belief revision* (pp. 183–203). Cambridge, England: Cambridge University Press.

Konieczny, S., & Pino Pérez, R. (2002). Merging information under constraints: A logical framework. *Journal of Logic and Computation, 12,* 773–808.

Kraus, S., Lehmann, D., & Magidor, M. (1990). Nonmonotonic reasoning, preferential models and cumulative logics. *Artificial Intelligence, 44,* 167–207.

Lehmann, D., Magidor, M., & Schlechta, K. (2001). Distance semantics for belief revision. *Journal of Symbolic Logic, 66,* 295–317.

Levi, I. (1991). *The fixation of belief and its undoing.* Cambridge, England: Cambridge University Press.

Levi, I. (1997). The logic of full belief. In I. Levi, *The covenant of reason: Rationality and the commitments of thought* (pp. 40–69). Cambridge, England: Cambridge University Press.

Lewis, D. (1981). Ordering semantics and premise semantics for counterfactuals. *Journal of Philosophical Logic, 19,* 217–234.

Makinson, D. (1994). General patterns in nonmonotonic reasoning. In D. M. Gabbay, C. J. Hogger, & J. A. Robinson (Eds.), *Handbook of logic in artificial intelligence and logic programming: Vol. III. Nonmonotonic reasoning and uncertain reasoning* (pp. 35–110). Oxford, England: Oxford University Press.

Makinson, D. (2005). *Bridges from classical to nonmonotonic logic.* London, England: King's College Publications.

Nayak, A. C. (1994). Iterated belief change based on epistemic entrenchment. *Erkenntnis, 41,* 353–390.

Nebel, B. (1992). Syntax-based approaches to belief revision. In P. Gärdenfors (Ed.), *Belief revision* (pp. 52–88). Cambridge, England: Cambridge University Press.

Pigozzi, G. (2015). Belief merging and judgment aggregation. In E. N. Zalta (Ed.), *The Stanford encyclopedia of philosophy.* Retrieved from https://plato.stanford.edu/archives/win2016/entries/belief-merging/

Rott, H. (1999). Coherence and conservatism in the dynamics of belief. Part I: Finding the right framework. *Erkenntnis, 50*(2–3), 387–412.

Rott, H. (2000). Two dogmas of belief revision. *Journal of Philosophy, 97,* 503–522.

Rott, H. (2001). *Change, choice and inference: A study of belief revision and nonmonotonic reasoning.* Oxford, England: Oxford University Press.

Rott, H. (2003). Economics and economy in the theory of belief revision. In V. F. Hendricks, K. F. Jørgensen, & S. A. Pedersen (Eds.), *Knowledge contributors* (pp. 57–86). Dordrecht, Netherlands: Kluwer.

Rott, H. (2012). Bounded revision: Two-dimensional belief change between conservative and moderate revision. *Journal of Philosophical Logic, 41,* 173–200.

Rott, H. (2017). Negative doxastic voluntarism and the concept of belief. *Synthese, 194,* 2695–2720.

Segerberg, K. (1998). Irrevocable belief revision in dynamic doxastic logic. *Notre Dame Journal of Formal Logic, 39,* 287–306.

Spohn, W. (2012). *The laws of belief: Ranking theory and its philosophical applications.* Oxford, England: Oxford University Press.

Testa, R. R., Coniglio, M. E., & Ribeiro, M. M. (2017). AGM-like paraconsistent belief change. *Logic Journal of the IGPL, 25,* 632–672.

Wassermann, R. (2011). On AGM for non-classical logics. *Journal of Philosophical Logic, 40,* 271–294.

5.3 Ranking Theory

Gabriele Kern-Isberner, Niels Skovgaard-Olsen, and Wolfgang Spohn

Summary

Ranking theory is one of the salient formal representations of doxastic states. It differs from others in being able to represent both belief in a proposition (i.e., taking it to be true) and degrees of belief (i.e., beliefs as more or less firm) and thus to generally account for the dynamics of these beliefs. It does so on the basis of fundamental and compelling rationality postulates and is hence one way of explicating the rational structure of doxastic states. Thereby, it provides foundations for accounts of defeasible or nonmonotonic reasoning. It has widespread applications in philosophy, it proves to be highly useful in artificial intelligence, and it has started to find applications as a model of reasoning in psychology.

1. Belief and Degrees of Belief

Epistemic or doxastic attitudes[1] representing how the world is like come in degrees, whether you call them degrees of belief, of uncertainty, or of plausibility (there are more terms around). There are various accounts of those degrees, amply presented in this handbook.[2] The interests in those accounts are manifold: philosophers are concerned with the rational nature of those degrees, artificial intelligence (AI) researchers are interested in their computational feasibility, psychologists deal with their actual manifestations, and all sides argue about how well they are suited to model human reasoning.

However, we also have the notion of belief *simpliciter*. Related notions are those of acceptance or judgment. These are indeed the more basic notions when it comes to truth, to truly representing the world. Beliefs can be true, but degrees of belief cannot. The latter rather relate to action (see chapter 8.2 by Peterson, this handbook). Accounts of degrees of belief invariably have great difficulties in doing justice to this fundamental point. There is a questionable tendency to take degrees of belief as basic and to belittle those difficulties.

So we need to theoretically account for belief *simpliciter*. The first attempt was doxastic logic (see chapter 5.1 by van Ditmarsch, this handbook). However, it is static and lacks a dynamic perspective. This perspective has been unfolded in belief revision theory (see chapter 5.2 by Rott, this handbook). However, this theory has problems with iterated belief revision, which is required for a complete dynamic account.

Ranking theory promises both to represent belief and degrees of belief *and* to provide a complete dynamics for both. These features give it a prominent place in the spectrum of possible theories. It was first presented in English in Spohn (1988) and fully developed in Spohn (2012). Easy access is provided in Spohn (2009). Its far-reaching applications in philosophy of science, epistemology, and even normative reasoning may be found, for example, in Spohn (2012, 2015, 2020). There is no place here to go into any of them.

Below, we present the basics of the theory in section 2 and its dynamic aspects in section 3. Section 4 is comparative. Section 5 gives a short introduction to the theory's relevance for artificial intelligence, and section 6 explains how it can be put to use in psychology.

2. The Basics of Ranking Theory

Grammatically, "believe" is a transitive verb. In the phrase "*a* believes that *p*," "*a*" refers to a (human) subject and "that *p*" seems to be the object. What does "that *p*" stand for? That is, what are the objects of belief? This is a difficult and most confusing issue extensively discussed in philosophy (under the rubric "propositions"; see, e.g., McGrath, 2012). Here, we cut short the issue, as is usual in formal epistemology, by saying in a noncommittal way that "that *p*" stands for the proposition expressed by "*p*," where that proposition is its truth condition, the set of possibilities or possible worlds in which *p* obtains or "*p*" is true.

Hence, we simply assume a set W of (mutually exclusive and jointly exhaustive) possibilities. These may be coarse-grained and refer only to a few things of interest; they need not consist in entire possible worlds. Each subset of W, that is, each element of the power set $P(W)$ of W, is a *proposition*.

Now, the basic representation of a belief state is simply as the set of propositions believed or taken to be true in that state, its *belief set*. Traditionally, a belief set $\mathcal{B} \subseteq \mathcal{P}(W)$ has to satisfy two rationality requirements: $\mathcal{B}$ must be *consistent*, that is, $\bigcap \mathcal{B} \neq \varnothing$, and $\mathcal{B}$ must be *deductively closed*, that is, if $\bigcap \mathcal{B} \subseteq A$, then $A \in \mathcal{B}$.

These two rationality requirements may seem entirely obvious. The rationale of deductive logic is to check what we must not believe and what we are committed to believe. Note, however, that deductive closure is lost when we identify belief with subjective probability above a certain threshold: it easily happens that the probabilities of two propositions are above the threshold, while that of their conjunction is below. Thus, the lottery and the preface paradox and the general desire to stick to a probabilistic representation of belief have led to a contestation of these requirements (see, e.g., Christensen, 2004). Here we stick to them as absolutely basic (see chapter 3.1 by Steinberger, this handbook). Of course, these requirements can be maintained only under a dispositional understanding of belief; occurrent thought cannot be deductively closed (see also section 6 of the introduction, this handbook).

The notion of a belief set is static. However, belief sets continuously change, and we must account for how they change (or should rationally change). We cannot do so on a qualitative level. In those changes, we often give up old beliefs and replace them by new ones, and then we give up less well-entrenched beliefs and keep better-entrenched ones (see chapter 5.2 by Rott, this handbook). Roughly, this calls for some entrenchment order or, indeed, for some kind of degrees of belief measuring the strength of entrenchment. Here, ranking theory commences. Let us start with some brief formal explanations.

Definition 1: κ is a *negative ranking function* for W iff κ is a function from $\mathcal{P}(W)$ into the set of natural numbers plus infinity ∞ such that for all $A, B \subseteq W$,

(1) $\kappa(W) = 0$ and $\kappa(\varnothing) = \infty$,

(2) $\kappa(A \vee B) = \min\{\kappa(A), \kappa(B)\}$ (the *law of disjunction*).[3]

The basic interpretation is that κ expresses degrees of *disbelief* (whence the qualification "negative"). If $\kappa(A) = 0$, A is not disbelieved at all. This allows that $\kappa(\neg A) = 0$ as well (where $\neg A$ is the negation of A); in that case, we have indifference or suspension of judgment regarding A. If $\kappa(A) > 0$, A is disbelieved or taken to be false, and

the more so the larger $\kappa(A)$. So, positive *belief* in A is expressed by disbelief in $\neg A$, that is, by $\kappa(\neg A) > 0$ (which implies that $\kappa(A) = 0$).

This interpretation explains axioms (1) and (2): axiom (1) says that the tautology W is not disbelieved, and hence the contradiction $\varnothing$ is not believed. This entails that beliefs are consistent according to κ. Axiom (1) moreover says that the contradiction is indeed maximally disbelieved. And axiom (2) states that you cannot disbelieve a disjunction less strongly than its disjuncts. This entails in particular that if you believe two conjuncts, you also believe their conjunction. Hence, beliefs are deductively closed according to κ. In other words, the belief set $\mathcal{B} = \{A \mid \kappa(\neg A) > 0\}$ associated with κ satisfies the two basic rationality requirements. Note, moreover, that (1) and (2) entail

(3) either $\kappa(A) = 0$ or $\kappa(\neg A) = 0$ (or both)   (the *law of negation*),

that is, you cannot (dis)believe A and $\neg A$ at once.

For an illustration, consider Tweety. Tweety has, or fails to have, each of three properties: being a bird (B), being a penguin (P), and being able to fly (F). This opens up eight possibilities. Suppose you have no idea who or what Tweety is, but somehow you do not think that it might be a penguin. Then your negative ranks for the eight possibilities (which determine the ranks for all other propositions) may be as given in table 5.3.1 (chosen in some plausible way—but see below how the numbers may be justified).

In this case, the strongest proposition you believe is that Tweety is *either* no penguin and no bird ($\neg B \wedge \neg P$) *or* a flying bird and no penguin ($F \wedge B \neg P$); all other possibilities are disbelieved. Hence, you neither believe that Tweety is a bird nor that it is not a bird. You are also indifferent concerning its ability to fly. But you believe, for example, that if Tweety is a bird, it is not a penguin and can fly ($\neg B \vee (\neg P \wedge F)$), and that if Tweety is a penguin, it can fly ($\neg P \vee F$)—each if–then taken as material implication. Surely you believe the latter only because you believe that Tweety is not a penguin in the first place. The large ranks in the last column indicate your strong disbelief in penguins *not* being birds. This may suffice as a first illustration.

Table 5.3.1

Prior beliefs about Tweety

κ	$B \wedge \neg P$	$B \wedge P$	$\neg B \wedge \neg P$	$\neg B \wedge P$
F	0	4	0	11
$\neg F$	2	1	0	8

We will see the reasons for starting with negative ranks. But of course, we can also introduce the positive counterpart to κ by defining β to be a *positive ranking function* iff there is a negative ranking function κ such that $\beta(A) = \kappa(\neg A)$ for all propositions A. The function β represents *degrees of belief*. Of course, (1) and (2) translate directly into axioms for β.

We may as well represent degrees of belief and degrees of disbelief in a single function by defining τ to be a *two-sided ranking function* iff there is a negative ranking function κ and the corresponding positive ranking function β such that $\tau(A) = \beta(A) - \kappa(A) = \kappa(\neg A) - \kappa(A)$ for all propositions A. Thus, we have $\tau(A) > 0$, < 0, or $= 0$ according to whether A is believed, disbelieved, or neither in τ. Therefore, this is perhaps the most intuitive notion. However, the mathematics is best done in terms of negative ranking functions. It is clear, though, that the three functions are interdefinable.

There is an important interpretational degree of freedom that we have not yet noticed. So far, we said that belief in A is represented by $\kappa(\neg A) = \beta(A) = \tau(A) > 0$. However, we may often find it useful to raise the threshold for belief, as we do informally in asking, "Do you *really* believe A?" That is, we may as well say that belief in A is only represented by $\kappa(\neg A) = \beta(A) = \tau(A) > z$ for some $z \geq 0$. This seems to be a natural move: belief is vague. Where does it commence, and when does it cease? And this vagueness seems well represented by that parameter z. This move at the same time enlarges the range of suspension of judgment to the interval from $-z$ to z. The remarkable point about axioms (1) and (2) is that they guarantee belief sets to be consistent and deductively closed, however we choose the threshold z. They are indeed equivalent to this general guarantee.

3. Conditional Ranks, Reasons, and the Dynamics of Ranks

So far, we have sketched only the static part of ranking theory. However, we mentioned that the numeric ranks are essentially used to account for the dynamics of belief; they are not just to represent greater and lesser firmness of (dis)belief. To achieve this, the crucial notion is that of conditional ranks:

Definition 2: Let κ be a negative ranking function for W and $\kappa(A) < \infty$. Then the *conditional rank* of B given A is defined as $\kappa(B \mid A) = \kappa(A \wedge B) - \kappa(A)$.

We might rewrite this definition as

(4) $\kappa(A \wedge B) = \kappa(A) + \kappa(B \mid A)$ (the *law of conjunction*).

This is highly intuitive. For, what is your degree of disbelief in $A \wedge B$? One way for $A \wedge B$ to be false is that A is false; this contributes $\kappa(A)$ to that degree. However, if A is true, B must be false, and this adds $\kappa(B \mid A)$.

It immediately follows for all propositions A and B with $\kappa(A) < \infty$ that

(5) $\kappa(B \mid A) = 0$ or $\kappa(\neg B \mid A) = 0$ (the *conditional law of negation*).

This law says that even conditional belief must be consistent. If both $\kappa(B \mid A)$ and $\kappa(\neg B \mid A)$ were > 0, both B and $\neg B$ would be (dis)believed given A, and this must be excluded, as long as the condition A itself is considered possible. Indeed, given definition 2 and axiom (1), we could axiomatize ranking theory also by (5) instead of (2). Hence, the only substantial assumption written into ranking functions is conditional consistency.

Axioms (1) and (2) did not refer to any *cardinal* properties of ranking functions. However, the definition of conditional ranks involves arithmetical operations and thus presupposes a cardinal understanding of ranks. We will see below how this may be justified. We hasten to add that one could as well define positive conditional ranks by $\beta(B \mid A) = \kappa(\neg B \mid A)$ and two-sided conditional ranks by $\tau(B \mid A) = \kappa(\neg B \mid A) - \kappa(B \mid A)$.

As an illustration, consider again table 5.3.1 and the conditional beliefs contained therein. We can see that precisely the (material) if–then propositions non-vacuously held true correspond to conditional beliefs. According to the κ specified, you believe, for example, that Tweety can fly given it is a bird (since $\kappa(\neg F \mid B) = 1$), and also given it is a bird but not a penguin (since $\kappa(\neg F \mid B \wedge \neg P) = 2$) and that Tweety cannot fly given it is a penguin (since $\kappa(F \mid P) = 3$). Hence, your vacuous belief in the material implication "If Tweety is a penguin, it can fly" does not amount to a corresponding conditional belief. In other words, "if . . . then . . ." expresses conditional belief rather than material implication, as suggested by the Ramsey test for conditionals (see also chapter 6.1 by Starr, this handbook).

A first fundamental application of conditional ranks lies in the notion of an *epistemic reason*, which is at the center of the entire handbook. It is very natural to say that A is a reason for B iff A speaks in favor of B or confirms B, if A makes B more plausible or less implausible, or if B is more credible or less incredible given A than given $\neg A$. This explanation works for any conception of conditional degrees of belief. In a probabilistic interpretation, it amounts to Carnap's notion of incremental confirmation or positive relevance (Carnap, 1950/1962), which is basic for confirmation theory.[4] Ranking-theoretically, it leads to

Definition 3: A is a *reason for* B relative to the negative ranking function κ or the associated two-sided ranking function τ iff $\tau(B|A) > \tau(B|\neg A)$.

We can show that if A is a deductive reason for B, that is, if $A \subseteq B$, then A is also a reason for B according to definition 3 (given $\kappa(A)$, $\kappa(\neg B) < \infty$). Clearly, this definition, depending entirely on the subject's doxastic state, provides only a subjectively relativized notion of a reason. Philosophers strive for a more objective notion of a reason,[5] perhaps because they take objective deductive reasons as a paradigm. In our view, the extent to which a more objective notion may be reached is a philosophically fundamental, alas very open, issue (see Spohn, 2018).

On this account, reasons can take four significant forms, depending on whether $\tau(B|A)$ and $\tau(B|\neg A)$ are positive or negative. For example, A is a *sufficient* reason for B iff B is believed given A and not believed given $\neg A$. We suggest that this is indeed the core meaning of the term "sufficient reason," although it is often used differently.

Moreover, we may define that B is (doxastically) *irrelevant* to or *independent* of A if neither A nor $\neg A$ is a reason for B. On this basis, a theory of (conditional) independence can be developed in ranking terms in wide-ranging analogy to the probabilistic theory. For instance, the theory of Bayes nets as developed by Pearl (1988) (see also chapter 4.2 by Hartmann, this handbook) could equally well be stated in terms of ranking theory (see Spohn, 1988; Goldszmidt & Pearl, 1996).

A second fundamental application of conditional ranks lies in the *dynamics* of beliefs and ranks. As in probability theory, we may say that, upon learning evidence E, we should simply move to the degrees of belief conditional on E. Thereby, though, the evidence E acquires maximal certainty, that is, probability 1 or positive rank ∞. This seems too restrictive. In general, evidence may be (slightly) uncertain, and our rules for doxastic change through evidence or learning—we do not attend to changes caused in other ways, like forgetting—should take account of this. In ranking theory, this is achieved by two principles: first, conditional ranks given the evidence E and given its negation $\neg E$ are not changed by the evidence itself—how could it change them?—and second, the evidence E does not become maximally certain but improves its position by n ranks, where n is a free parameter characterizing the specific information process.[6] These two assumptions suffice to uniquely determine the kinematics of ranking functions (i.e., ranking-theoretic conditionalization).

In order to see how this works, look again at our Tweety example: suppose you learn in some way, and accept with firmness 2, that Tweety is a bird. Thus, you shift up $\neg B$-possibilities by 2 and keep constant the rank differences within B and within $\neg B$. This results in the posterior ranking function κ' given in table 5.3.2.

In κ', you believe that Tweety is a bird able to fly but not a penguin; you still neglect this possibility. So, in κ', you believe more than in κ; in belief revision theory (cf. chapter 5.2 by Rott, this handbook), this would be called a "belief expansion."

Next, to your surprise, you tentatively learn and accept, say, with firmness 1 that Tweety is indeed a penguin. This results in another ranking function κ'', which shifts all P-possibilities down by 1 and all $\neg P$-possibilities up by 1, so that P is indeed believed with firmness 1 (i.e., $\kappa''(\neg P) = 1$; see table 5.3.3).

So, you have changed your mind and believe in κ'' that Tweety is a penguin bird that cannot fly. In belief revision theory, this would be called a "belief revision." Obviously, belief contraction (cf. chapter 5.2 by Rott, this handbook), where you simply give up a belief previously held without replacing it by a new one, can also be modeled by ranking-theoretic conditionalization. The example already demonstrates that this rule of belief change can be iteratively applied ad libitum.

An important application of ranking-theoretic conditionalization is that it delivers a measurement procedure for ranks that justifies the cardinality of ranks. This procedure refers to iterated belief contraction. Its point is this: if your iterated contractions behave as prescribed by ranking theory,[7] then this behavior uniquely determines your ranking function up to a multiplicative constant. That is, your ranks can thereby be measured on a ratio scale (see Hild & Spohn, 2008). The consequences of the fact that ranks are measured only on a ratio scale await investigation. They imply, for example, a problem analogous to the problem of the interpersonal comparison of utilities.

Table 5.3.2

Learning that Tweety is a bird

κ'	$B \wedge \neg P$	$B \wedge P$	$\neg B \wedge \neg P$	$\neg B \wedge P$
F	0	4	2	13
$\neg F$	2	1	2	10

Table 5.3.3

Learning that Tweety is a penguin

κ''	$B \wedge \neg P$	$B \wedge P$	$\neg B \wedge \neg P$	$\neg B \wedge P$
F	1	3	3	12
$\neg F$	3	0	3	9

4. Comparisons

The formal structure defined by axioms (1) and (2) has been called *Baconian probability* by Cohen (1980). Its first clear appearance is in the functions of potential surprise developed by Shackle (1949). The structure is also hidden in Rescher (1964) and is clearly found in Cohen's own work (1970, 1977). The crucial formal advance of ranking theory lies in the definition of conditional ranks, which cannot be found in these works and makes the theory a properly cardinal one.

Belief revision theory was precisely about the dynamics of belief. However, it only conceived of entrenchment *orders*. And within their ordinal framework, there was no clear solution of the problem of iterated belief revision (see chapter 5.2 by Rott, this handbook).

Possibility theory, building on early work by Zadeh (1978) and developed by Dubois and Prade (see chapter 4.7 by Dubois & Prade, this handbook), is in fact equivalent to ranking theory; axiom (2) is the characteristic property of possibility measures. However, the interpretation of those measures was intentionally left open, leaving considerable formal uncertainty as to how to conceive of conditional degrees of possibility.

The theory of Dempster–Shafer (DS) belief functions, as developed in Shafer (1976), seems to be a far more general theory (see again chapter 4.7 by Dubois & Prade, this handbook), which comprises probability theory and also ranking theory as a special case. Shafer (1976, chapter 10) defines so-called consonant belief functions, which appear to be equivalent to negative ranking functions. However, their respective dynamic behaviors diverge, a fact that prevents reduction of ranking theory to DS theory (see Spohn, 2012, section 11.9).

Of course, the largest comparative issue is how ranking theory relates to probability theory. A comparison of their axioms and their form of conditionalization suggests translating the sum of probabilities into the minimum of ranks and the product and the quotient of probabilities into the addition and the subtraction of ranks, respectively. This translation works only for *negative* ranks; that's why they provide the formally preferred version of ranking theory. And it explains why very many things that can be done with probability theory also work for ranking theory in a meaningful way.

However, the translation does not justify conceiving ranks in probabilistic terms. As mentioned in section 2, belief in *A* cannot be probabilistically represented by $P(A) = 1 - \varepsilon$ if one sticks to the consistency and deductive closure of belief sets. The relation of probability and belief is hotly debated in philosophy, without a clear

solution emerging (see, e.g., the proposals of Leitgeb, 2017, and Raidl & Skovgaard-Olsen, 2017). Therefore, our attitude has always been to independently develop ranking theory as a theory of belief.

5. Ranking Functions in Artificial Intelligence

Besides probability theory and logic, ranking functions are among the most popular formalisms used for knowledge representation[8] and reasoning (KRR), and their popularity is still increasing because they provide a very versatile framework for many central operations in KRR, as already sections 2 to 4 pointed out. Most importantly, ranking functions are a convenient common basic tool for nonmonotonic reasoning and belief revision. Belief revision already has been explained in more detail in section 3, and nonmonotonic reasoning also deals with belief dynamics insofar as conclusions may be given up when new information arrives (so, the consequence relation is not monotonic, as in classical logic). Both fields emerged in the 1980s (partly) as a reaction to the incapability of classical logic to handle problems in everyday life that intelligent systems like robots were expected to tackle. Knowledge or belief about the world is usually uncertain, and the world is always changing. Therefore, AI systems built upon classical logics failed. So-called preferential models (Makinson, 1989) provide an important semantics for nonmonotonic logics. Their basic idea is to order worlds according to normality and to focus on the minimal ones (i.e., the most plausible ones) for reasoning. Likewise, AGM belief revision theory (see chapter 5.2 by Rott, this handbook) needs orderings of worlds to become effective. For both fields, ranking functions offer a quite perfect technical tool that also complies nicely with the intuitions behind the techniques. Moreover, they can also evaluate conditionals and are an attractive qualitative counterpart to probabilities (see section 3).

Judea Pearl was probably the first renowned AI scientist to make use of ranking functions; his famous *system Z* (Pearl, 1990) is based on them. He has continuously emphasized the commonsense structural qualities of probabilities and developed ranking functions as an interesting qualitative counterpart to probabilities. He set up his system Z as an "ultimate system of nonmonotonic reasoning" in terms of ranking functions. To date, it is one of the best and most convenient approaches to implement high-quality nonmonotonic reasoning.

Consequently, ranking functions are deeply connected with nonmonotonic and uncertain reasoning and with belief change, which are core topics in KRR. Many researchers make use of them in one way or another even if they rely on more general frameworks. Darwiche and Pearl (1997) presented general postulates for the iterated revision of general epistemic states but illustrated their account with ranking functions. So did Jin and Thielscher (2007) and Delgrande and Jin (2012) when they devised novel postulates for iterated and multiple revision. Interestingly, the independence properties for advanced belief revision that were proposed in those studies can be related to independence with respect to ranking functions (see Spohn, 2012, chapter 7) in analogy with probabilistic independence (see Kern-Isberner & Huvermann, 2017).

Indeed, as suggested in section 3, ranking functions are particularly well suited for iterated belief change because they can easily be modified in accordance with AGM theory, returning new ranking functions that are readily available for a subsequent change operation. The main AGM operations are revision (adopting a belief) and contraction (giving up a belief), related by Levi and Harper identities (see chapter 5.2 by Rott, this handbook). In ranking theory, the connections between these operations are even deeper, since (iterated) contraction is just a special kind of (iterated) ranking conditionalization. Indeed, the results of Kern-Isberner, Bock, Sauerwald, and Beierle (2017) show that iterated revision and contraction can be performed by a common methodology.

Continuing on that, and beyond the practicality and diversity of ranking functions, it is crucial to understand that they are not just a pragmatically good choice but indeed allow for deep theoretical foundations of approaches to reasoning. It is the ease and naturalness with which they can handle conditionals—very similar to probabilities—that make them an excellent formal tool for modeling reasoning. Given that conditionals are, on the one hand, crucial entities for nonmonotonic and commonsense reasoning and belief change, and, on the other hand, formal entities fully accessible to conditional logics, this capability provides a key feature for logic-based approaches connecting nonmonotonic logics and belief change theories with commonsense and general human reasoning. More precisely, conditional ranks give meaning to differences between degrees in belief when observing A versus $A \wedge B$ (see the law of conjunction in section 3), and the examples of belief change given in section 3 illustrate nicely how it is easily possible to preserve these differences under change when using ranking functions. This property has been elaborated as a *principle of conditional preservation* in Kern-Isberner (2004), giving rise to defining c-representations and c-revisions (all belief changes shown in section 3 are c-revisions). C-representations are c-revisions starting

from a uniform ranking function and allowing for reasoning from conditional belief bases. Ranking theory is one of the few formal frameworks (probability theory is another) that is rich and expressive enough to allow such a precise formalization of conditional preservation that supports both belief change and inductive reasoning as a common methodology.

6. Ranking Theory in Psychology

Potentially, ranking theory has applications in many areas of psychology. To illustrate, psychological research on belief revision has been carried out under the inspiration of AGM theory and probabilistic updating (Baratgin & Politzer, 2010; Oaksford & Chater, 2013; Wolf, Rieger, & Knauff, 2012; see also chapter 5.4 by Gazzo Castañeda & Knauff, this handbook). Such work could be extended by ranking theory, given that one of its central motivations was to represent a notion of full belief, in contrast to probabilistic update mechanisms, while improving upon AGM theory to allow for iterative revisions. However, as Colombo, Elkin, and Hartmann (2018) note, probabilistic Bayesian approaches are currently enjoying a boom in cognitive science to the neglect of alternative formal frameworks, like ranking theory.

As said, possibility theory is mathematically equivalent to ranking theory, yet differs fundamentally in its intended interpretation. In Da Silva Neves, Bonnefon, and Raufaste (2002) and Benferhat, Bonnefon, and Da Silva Neves (2005), possibility theory was subjected to empirical testing. In Da Silva Neves et al. (2002), a direct route was chosen by testing whether participants' possibility judgments satisfy the rationality postulates codified in System P (see chapter 4.4 by Pfeifer, this handbook) augmented by *rational monotonicity* ($A \mid\!\sim C$ and not $A \mid\!\sim \neg B$; therefore $A \wedge B \mid\!\sim C$),[9] in case their responses violated *monotonicity* ($A \mid\!\sim C$; therefore $A \wedge B \mid\!\sim C$). Interestingly, no such direct test of ranking theory based on the participants' judgments of disbelief or implausibility has yet been made.

What exists is the following: Isberner and Kern-Isberner (2017) investigated whether belief revision with ranking functions could retrodict findings of temporal delay when processing implausible information in tasks where plausibility judgments would interfere with the task constraints (a finding known as the "epistemic Stroop effect"). A guiding assumption underlying this work is that ranking theory can be used to represent the situational model that participants construct during language comprehension. Moreover, Eichhorn, Kern-Isberner, and Ragni (2018) proposed a conditional-logical model based on ranking functions that allows for the elaboration of plausible background knowledge.

In Ragni, Eichhorn, Bock, Kern-Isberner, and Tse (2017), it was investigated whether ranking theory could retrodict the suppression effect (Byrne, 1989), where endorsement rates of classically valid modus ponens (MP) ($A \rightarrow B, A$; therefore B) and modus tollens (MT) ($A \rightarrow B, \neg B$; therefore $\neg A$) are decreased when further premises indicating possible defeaters are presented. To illustrate: Normally, inferring "Lisa will study late in the library" from the premises "Lisa has an essay to write" and "If Lisa has an essay to write, then Lisa will study late in the library" would be seen as unproblematic. But if the additional premise "If the library is open, then Lisa will study late in the library" becomes available, then participants are much more reluctant to draw this inference. However, as Ragni et al. (2017) show, the inference mechanism exploiting ranking functions does not in itself retrodict the suppression effect. To do so, further assumptions about the underlying knowledge base instantiated in long-term memory need to be made. In addition, Ragni et al. (2017) present experiments that test ranking theory's ability to predict the participants' reasoning with MP and MT once keywords indicating nonmonotonicity such as "normally" are inserted. Characteristic of this line of research is that c-representations are used to inductively infer ranking functions that satisfy the constraints set by an assumed knowledge base.

Moreover, the account of conditionals in Spohn (2013, 2015) has inspired a series of experiments. Spohn (2013, 2015) outlines a number of expressive roles of conditionals that go beyond the Ramsey test, which merely takes conditionals to express conditional beliefs. For instance, conditionals may express reason relations as specified in definition 3. In Skovgaard-Olsen (2014), a logistic regression model was presented both to formulate predictions for the participants' evaluations of the conclusions of MP, MT, AC (affirming the consequent: $A \rightarrow B, B$; therefore A), and DA (denying the antecedent: $A \rightarrow B, \neg A$; therefore $\neg B$) inferences and to suggest a solution to the problem of updating based on conditional information. In Skovgaard-Olsen, Singmann, and Klauer (2016), a reason-relation reading of the conditional was contrasted experimentally with the Ramsey test in participants' probability and acceptability evaluations of indicative conditionals, and support was obtained for the reason-relation reading. In Skovgaard-Olsen, Kellen, Hahn, and Klauer (2019), it was found that there are patterns of individual variation in these results. In Skovgaard-Olsen, Singmann, and Klauer (2017), participants' perceived relevance and reason-relation evaluations were investigated, and some

first evidence for the explications of reason relations and perceived relevance introduced above was obtained. However, characteristic of this latter line of research is that it adopts an indirect route to testing ideas from ranking theory by exploiting its extension to conditionals in Spohn (2013, 2015) and its parallels to probability theory (see also Henrion, Provan, Del Favero, & Sanders, 1994).

Presently, paradigms operationalizing degrees of beliefs as probabilities are much better established than tasks using ranking functions. This is so despite the fact that the arithmetical operations of ranking theory require much less computational effort than those of probability theory. For instance, it is well known that participants exhibit difficulties in properly integrating information about base rates of the rarity of a given disease in evaluating how likely it is that a person has the disease given a positive test result. Yet, interestingly, Juslin, Nilsson, Winman, and Lindskog (2011) find that the notorious base-rate neglect could be reduced when participants are given the tasks in a logarithmic format. Juslin et al. (2011) therefore conjecture that a linear, additive integration of information becomes more intuitive in the absence of access to overriding analytic rules that use a multiplicative format, like probabilities.

Nevertheless, there is a lack of direct experimental investigations of ranking functions. This is perhaps due to the following challenges:

(i) Negative ranking functions are useful for conducting proofs. But it initially presents a conceptual challenge to think in terms of disbelief in negations of propositions as a way of representing full beliefs. Two-sided ranking functions solve this problem. But they come at the cost of having different rules applying to the negative and positive ranges of the scale.

(ii) An operationalization of ranking functions in terms of iterated contractions exists (see above), but it has not received the same kind of experimental implementation as the operationalization of probabilities in terms of betting quotients.

(iii) Since negative ranks take natural numbers from 0 to infinity, there is no natural way of nonarbitrarily dividing the scale into regions of ascending degrees of disbelief other than a region of zero disbelief, a region of above-zero disbelief, and a region of maximal disbelief. In contrast, the crude division of the probability scale of real numbers between 0 and 1 into decimal regions enables participants and experimenters to make qualitative differentiations between low degrees of belief (e.g., [0.0, 0.3)), medium degrees of belief (e.g., [0.3, 0.7]), and strong degrees of belief (e.g., (0.7, 1.0]).

If challenges such as these can be overcome in future work, ranking theory potentially has a lot to offer to psychology and to cognitive science more generally.

Notes

1. Strictly speaking, "epistemic" only refers to knowledge, although it is often used more widely. Because we will talk only about belief, we prefer to use "doxastic" throughout. See also chapter 5.1 by van Ditmarsch (this handbook).

2. See chapter 4.1 by Hájek and Staffel, chapter 4.5 by Chater and Oaksford, chapter 4.7 by Dubois and Prade, chapter 8.3 by Glöckner, and chapter 8.4 by Hill (all in this handbook).

3. We use the notation of propositional logic, not the set-theoretic one. No confusion can arise thereby.

4. See chapter 4.3 by Merin (this handbook). Sections 1–4 of that chapter work just as well with ranking theory.

5. See also chapter 2.1 by Broome, chapter 2.2 by Wedgwood, and chapter 12.2 by Smith (all in this handbook).

6. This is completely analogous to Jeffrey conditionalization in probability theory (see Jeffrey, 1983, chapter 11; chapter 4.1 by Hájek & Staffel, this handbook). Shenoy (1991) has contributed an important variant.

7. In fact, we need no more than twofold nonvacuous contractions.

8. In AI, the distinction between knowledge and belief is usually quite vague.

9. Here, $\vdash\!\!\!\sim$ represents nonmonotonic or default entailment.

References

Baratgin, J., & Politzer, G. (2010). Updating: A psychologically basic situation of probability revision. *Thinking & Reasoning, 16*, 253–287.

Benferhat, S., Bonnefon, J. F., & Da Silva Neves, R. (2005). An overview of possibilistic handling of default reasoning, with experimental studies. *Synthese, 146*, 53–70.

Byrne, R. M. (1989). Suppressing valid inferences with conditionals. *Cognition, 31*, 61–83.

Carnap, R. (1962). *Logical foundations of probability.* Chicago, IL: University of Chicago Press. (Original work published 1950)

Christensen, D. (2004). *Putting logic in its place: Formal constraints on rational belief.* Oxford, England: Oxford University Press.

Cohen, L. J. (1970). *The implications of induction.* London, England: Methuen.

Cohen, L. J. (1977). *The probable and the provable.* Oxford, England: Oxford University Press.

Cohen, L. J. (1980). Some historical remarks on the Baconian conception of probability. *Journal of the History of Ideas, 41*, 219–231.

Colombo, M., Elkin, L., & Hartmann, S. (2018). Being realist about Bayes, and the predictive processing theory of mind. *British Journal for the Philosophy of Science*, doi: 10.1093/bjps/axy059.

Darwiche, A., & Pearl, J. (1997). On the logic of iterated belief revision. *Artificial Intelligence, 89*, 1–29.

Da Silva Neves, R., Bonnefon, J.-F., & Raufaste, E. (2002). An empirical test of patterns for nonmonotonic inference. *Annals of Mathematics and Artificial Intelligence, 34*, 107–130.

Delgrande, J., & Jin, Y. (2012). Parallel belief revision: Revising by sets of formulas. *Artificial Intelligence, 176*, 2223–2245.

Eichhorn, C., Kern-Isberner, G., & Ragni, M. (2018). Rational inference patterns based on conditional logic. In S. A. McIlraith & K. Q. Weinberger (Eds.), *Proceedings of the Thirty-Second AAAI Conference on Artificial Intelligence (AAAI-18)* (pp. 1827–1834). Palo Alto, CA: AAAI Press.

Goldszmidt, M., & Pearl, J. (1996). Qualitative probabilities for default reasoning, belief revision, and causal modeling. *Artificial Intelligence, 84*, 57–112.

Henrion, M., Provan, G., Del Favero, B., & Sanders, G. (1994). An experimental comparison of numerical and qualitative probabilistic reasoning. In R. Lopez de Mantaras & D. Poole (Eds.), *Uncertainty in Artificial Intelligence: Proceedings of the Tenth Conference* (pp. 319–326). Burlington, MA: Morgan Kaufmann.

Hild, M., & Spohn, W. (2008). The measurement of ranks and the laws of iterated contraction. *Artificial Intelligence, 172*, 1195–1218.

Isberner, M.-B., & Kern-Isberner, G. (2017). Plausible reasoning and plausibility monitoring in language comprehension. *International Journal of Approximate Reasoning, 88*, 53–71.

Jeffrey, R. C. (1983). *The logic of decision.* Chicago, IL: University of Chicago Press.

Jin, Y., & Thielscher, M. (2007). Iterated belief revision, revised. *Artificial Intelligence, 171*, 1–18.

Juslin, P., Nilsson, H., Winman, A., & Lindskog, M. (2011). Reducing cognitive biases in probabilistic reasoning by the use of logarithm formats. *Cognition, 120*, 248–267.

Kern-Isberner, G. (2004). A thorough axiomatization of a principle of conditional preservation in belief revision. *Annals of Mathematics and Artificial Intelligence, 40*, 127–164.

Kern-Isberner, G., Bock, T., Sauerwald, K., & Beierle, C. (2017). Iterated contraction of propositions and conditionals under the principle of conditional preservation. *EPiC Series in Computing, 50*, 78–92.

Kern-Isberner, G., & Huvermann, D. (2017). What kind of independence do we need for multiple iterated belief change? *Journal of Applied Logic, 22*, 91–119.

Leitgeb, H. (2017). *The stability of belief: How rational belief coheres with probability*. Oxford, England: Oxford University Press.

Makinson, D. (1989). General theory of cumulative inference. In M. Reinfrank, J. de Kleer, M. L. Ginsberg, & E. Sandewall (Eds.), *Non-monotonic reasoning* (pp. 1–18). Berlin, Germany: Springer.

McGrath, M. (2012). Propositions. In E. N. Zalta (Ed.), *The Stanford encyclopedia of philosophy*. Retrieved from https://plato.stanford.edu/archives/spr2014/entries/propositions/

Oaksford, M., & Chater, N. (2013). Dynamic inference and everyday conditional reasoning in the new paradigm. *Thinking & Reasoning, 19*(3), 346–379.

Pearl, J. (1988). *Probabilistic reasoning in intelligent systems: Networks of plausible inference*. San Mateo, CA: Morgan Kaufmann.

Pearl, J. (1990). System Z: A natural ordering of defaults with tractable applications to nonmonotonic reasoning. In R. Parikh (Ed.), *Theoretical Aspects of Reasoning about Knowledge: Proceedings of the Third Conference (TARK 1990)* (pp. 121–135). San Francisco, CA: Morgan Kaufmann.

Ragni, M., Eichhorn, C., Bock, T., Kern-Isberner, G., & Tse, A. P. P. (2017). Formal nonmonotonic theories and properties of human defeasible reasoning. *Minds and Machines, 27*, 79–117.

Raidl, E., & Skovgaard-Olsen, N. (2017). Bridging ranking theory and the stability theory of belief. *Journal of Philosophical Logic, 46*, 577–609.

Rescher, N. (1964). *Hypothetical reasoning*. Amsterdam, Netherlands: North-Holland.

Shackle, G. L. S. (1949). *Expectation in economics*. Cambridge, England: Cambridge University Press. (2nd edition 1952, 2012)

Shafer, G. (1976). *A mathematical theory of evidence*. Princeton, NJ: Princeton University Press.

Shenoy, P. P. (1991). On Spohn's rule for revision of beliefs. *International Journal of Approximate Reasoning, 5*, 149–181.

Skovgaard-Olsen, N. (2014). *Making ranking theory useful for psychology of reasoning* (Unpublished doctoral dissertation). University of Konstanz, Konstanz, Germany. Retrieved from http://nbn-resolving.de/urn:nbn:de:bsz:352-0-262692

Skovgaard-Olsen, N., Kellen, D., Hahn, U., & Klauer, K. C. (2019). Norm conflicts and conditionals. *Psychological Review, 126*(5), 611–633.

Skovgaard-Olsen, N., Singmann, H., & Klauer, K. C. (2016). The relevance effect and conditionals. *Cognition, 150*, 26–36.

Skovgaard-Olsen, N., Singmann, H., & Klauer, K. C. (2017). Relevance and reason relations. *Cognitive Science, 41*, 1202–1215.

Spohn, W. (1988). Ordinal conditional functions: A dynamic theory of epistemic states. In W. L. Harper & B. Skyrms (Eds.), *Causation in decision, belief change, and statistics* (pp. 105–134). Dordrecht, Netherlands: Springer.

Spohn, W. (2009). A survey of ranking theory. In F. Huber & C. Schmidt-Petri (Eds.), *Degrees of belief* (pp. 185–228). Dordrecht, Netherlands: Springer.

Spohn, W. (2012). *The laws of belief: Ranking theory and its philosophical applications*. Oxford, England: Oxford University Press.

Spohn, W. (2013). A ranking-theoretic approach to conditionals. *Cognitive Science, 37*, 1074–1106.

Spohn, W. (2015). Conditionals: A unifying ranking-theoretic perspective. *Philosopher's Imprint, 15*, 1–30.

Spohn, W. (2018). Epistemic justification: Its subjective and its objective ways. *Synthese, 195*, 3837–3856.

Spohn, W. (2020). Defeasible normative reasoning. *Synthese, 197*, 1391–1428.

Wolf, A. G., Rieger, S., & Knauff, M. (2012). The effects of source trustworthiness and inference type on human belief revision. *Thinking & Reasoning, 18*, 417–440.

Zadeh, L. A. (1978). Fuzzy sets as a basis for a theory of possibility. *Fuzzy Sets and Systems, 1*, 3–28.

5.4 Defeasible Reasoning and Belief Revision in Psychology

Lupita Estefania Gazzo Castañeda and Markus Knauff

Summary

How do people change their mind in the light of new information? In defeasible reasoning, people withdraw previously drawn *inferences* (conclusions) in light of new evidence. In belief revision, people abandon prior *beliefs* (premises) to accommodate new information. Cognitive research aims at developing *descriptive* theories of these cognitive functions, without making strong *normative* claims. However, such normative claims are more or less implicit in many cognitive theories of human reasoning. This chapter summarizes the experimental state of the art in the field and then presents the most important psychological theories of defeasible reasoning and belief revision. We also discuss whether both cognitive processes are really so different from a cognitive (descriptive) point of view. Both cognitive processes seem to rely on the availability of disablers, the facts that prevent people from believing the conclusions or premises of a reasoning problem.

1. Defeasible Reasoning and Belief Revision in Human Reasoning

A fundamental prerequisite of rationality is that humans are willing to change their mind in light of new evidence. For instance, suppose that Jack has learned that if he goes to bed late, then he will be tired the next day. So, if he is going to bed late, he can conclude that he will be tired the next day. But if he finds out that the next day is a holiday, then it makes sense for Jack to change this conclusion. He can conclude that he will not be tired because he is able to sleep in. That is, new information can make people reject previous conclusions that would have been drawn otherwise. Sometimes, however, the information we get is more than an exception to a rule. New information can also conflict with our beliefs. Imagine, for instance, that Jack usually has problems falling asleep. So the psychotherapist tells him that if he meditates in the evening, he will fall asleep more easily.

Following the advice, Jack meditates in the evening. But he is still not able to fall asleep. This is not what he expected. Therefore, Jack has to revise his prior beliefs and conclude either that the psychotherapist was wrong or that he meditated inappropriately.

These two cases of reasoning are called defeasible reasoning and belief revision, respectively. In *defeasible reasoning*, people withdraw previously drawn conclusions in light of new evidence. Jack withdraws the conclusion of being tired after going to bed late because he knows that he is able to sleep in. This does not mean that he no longer believes that going late to bed makes him tired. In fact, in defeasible reasoning, one still believes in the prior propositions (basic beliefs), but one knows that in light of specific circumstances, some conclusions (derived beliefs) are not suitable anymore. This is different in *belief revision*, in which new information is in conflict with one's prior beliefs. The fact that Jack is not able to fall asleep is inconsistent with what the psychotherapist told him. This inconsistency between new evidence and prior beliefs can only be solved by revising these prior beliefs. In other words, in belief revision, it is not the conclusion that is changed; rather, one has to change prior beliefs in order to reach consistency. In the psychology of reasoning, these prior beliefs are often given by a set of premises. The conclusion is usually the belief that follows from these premises.

From a formal point of view, defeasible reasoning and belief revision are *nonmonotonic*, that is, conclusions can be retracted based on further evidence. Psychologically, that means that a person abandons accepted consequences from previous beliefs (Elio & Pelletier, 1997). The investigation of nonmonotonicity has its origins in philosophy and artificial intelligence (AI), in which many *normative* theories have been developed to describe how intelligent systems can rationally operate with incomplete information in changing worlds (for an overview, see Antoniou, 1997). In psychology, however, the investigation of defeasible reasoning and belief revision is still young. For a long time, psychologists concentrated only on deduction, which is monotonic, that is, a conclusion has to follow necessarily from its premises and cannot be altered by additional information. However, as our initial examples show, it is also sensible to change conclusions in light of new evidence. Because of that, psychologists have also started to develop several *descriptive* theories of human nonmonotonic reasoning and the psychological processes behind defeasible reasoning and belief revision. The aim of this chapter is to give an overview of this psychological research.

2. The Empirical Investigation of Defeasible Reasoning

Defeasibility is a central property of human reasoning. In everyday life, people are often forced to reason with uncertain and incomplete information and thus must often change previously drawn conclusions in light of new evidence. Most of the psychological research on defeasible reasoning focused on *conditional inferences* and the circumstances under which the conclusions from such inferences are withdrawn. There are two main experimental paradigms: the overt and the covert paradigm (Politzer & Bonnefon, 2006).

2.1 Defeasibility in the Overt Paradigm

The overt paradigm was introduced by Rumain, Connell, and Braine (1983) but made popular by Byrne (1989). There are usually two experimental conditions. In the simple condition, participants have to solve inferences consisting of a conditional premise of the form "If p, then q," a categorical premise containing p, q, or their negations and a conclusion that has to be evaluated. An example:

If Jack goes to bed late, then Jack will be tired.

Jack goes to bed late.

Jack will be tired.

In this case, the conclusion is true because this is the valid modus ponens (MP) inference: if p (the antecedent) is true, then q (the consequent) is also true. Another valid inference is modus tollens (MT: if p then q; not-q; therefore, not-p). By contrast, the other two inferences, *affirmation of the consequent* (AC: if p then q; q; therefore, p) and *denial of the antecedent* (DA: if p then q; not-p; therefore, not-q), are fallacies in classical logic because p is only sufficient, but not necessary, for q. Participants typically have no problem in recognizing MP inferences as logically valid, and they do it also for MT inferences, although to a lesser extent. The results regarding AC and DA are mixed and highly dependent on the task demands (Evans, Newstead, & Byrne, 1993).

In the extended condition, though, an additional premise is added to the inference. This additional premise can contain either a disabler or an alternative (Cummins, 1995; Cummins, Lubart, Alksnis, & Rist, 1991). *Disablers* are additional preconditions that prevent p from leading to q. They are thus relevant for MP and MT and can be presented either as concrete facts (e.g., Jack is able to sleep in) or as conditionals. An example:

If Jack goes to bed late, then he will be tired.

If Jack is able to sleep in, then he will not be tired.

Jack goes to bed late.

Jack will be tired.

Alternatives, by contrast, are alternative reasons for q that are not p. They are relevant for AC and DA inferences because they make it clear that p is not necessary for q. Alternatives can be also phrased either as facts (e.g., Jack has a disrupted sleep) or as conditionals. For example:

If Jack goes to bed late, then he will be tired.

If Jack has a disrupted sleep, then he will be tired.

Jack does not go to bed late.

Jack will not be tired.

The defeasibility of human reasoning is shown by comparing the acceptance rates for conclusions in the simple and the extended condition. This comparison usually shows that disablers and alternatives affect inferences differently. When disablers are presented, participants accept fewer MP and MT conclusions than without disablers (e.g., Byrne, 1989; Byrne, Espino, & Santamaria, 1999; De Neys, Schaeken, & d'Ydewalle, 2003b; Stevenson & Over, 1995). When alternatives are presented, participants accept fewer AC and DA conclusions than without alternatives (e.g., Byrne, 1989; De Neys et al., 2003b; Rumain et al., 1983).

2.2 Defeasibility in the Covert Paradigm

The covert paradigm was used for the first time by Cummins (Cummins et al., 1991). In the covert paradigm, alternatives and disablers are not presented explicitly as additional premises but are just implicitly present in the premises. For this, a first group of participants is confronted with a set of conditionals and asked to generate as many disablers and alternatives as they can. For example:

If Jack goes to bed late, then he will be tired.

Jack goes to bed late, but is not tired (or: Jack is tired, but he did not go to bed late).

Please write down as many factors as you can that could make this situation possible.

Depending on how many disablers and alternatives participants generate, the conditionals are subdivided in those having (1) many disablers and many alternatives, (2) many disablers and few alternatives, (3) few disablers and many alternatives, and (4) few disablers and few alternatives. These conditionals are then embedded in MP, MT, AC, and DA inferences and presented to a new group of participants who have to indicate how sure they are that the conclusion can be drawn. The second group of participants never sees the alternatives and disablers. Nevertheless, the acceptability of the conclusion is a function of the number of disablers and alternatives. Participants accept fewer MP and MT conclusions when conditionals have many instead of few disablers, and accept fewer AC and DA conclusions when conditionals have many instead of few alternatives (e.g., Cummins, 1995; De Neys, Schaeken, & d'Ydewalle, 2003a, 2003b; Gazzo Castañeda & Knauff, 2018).

2.3 The Importance of Content and Knowledge

One of the main reasons why human reasoning is defeasible stems from people's background knowledge and their prior beliefs about the content of conditionals (De Neys et al., 2003a, 2003b; Dieussaert, De Neys, & Schaeken, 2005; Evans & Over, 2004; Johnson-Laird & Byrne, 2002; Oaksford & Chater, 2007). Indeed, the relation between knowledge, beliefs, and content is not trivial (see chapter 13.3 by Knauff, this handbook). However, here we simply use the term "knowledge" in the sense of true or justified belief.

As shown by the findings with the covert paradigm, defeasibility does not require an explicit presentation of new evidence; instead, people activate their beliefs about the content of conditionals even if no explicit disablers or alternatives are presented (De Neys et al., 2003b; Vadeboncoeur & Markovits, 1999). Therefore, prior beliefs are an important prerequisite for defeasible reasoning. Without background knowledge (in the sense of true beliefs), people cannot know which information is a disabler or an alternative that can affect inferences. Experimentally, this has been shown in studies comparing participants' acceptance of conclusions for problems with unfamiliar and familiar content (Cummins, 1995; Gazzo Castañeda & Knauff, 2020; Markovits, 1986), by comparing the weighting of domain-specific disablers by laypeople and experts (Gazzo Castañeda & Knauff, 2016), and recently by varying the familiarity of the person or object in a conditional (Gazzo Castañeda & Knauff, 2019). In all these studies, the results are consistent: when participants do not have domain knowledge, they also make more MP, MT, AC, and/or DA inferences. In other words,

they do not know the circumstances that could defeat the conclusions. It is therefore argued that the retrieval of disablers and alternatives affects the *perceived* sufficiency and necessity relations between p and q: disablers make the antecedent less sufficient for the consequent so that fewer MP and MT conclusions are accepted, and alternatives make the antecedent less necessary for the consequent so that fewer AC and DA conclusions are accepted (Cummins, 1995; Thompson, 1994, 1995).

Another factor related to background knowledge that is important for defeasible reasoning is the associative strength between the premises and the disablers in memory. Some disablers are more strongly associated with one's prior knowledge of how to *prevent q* from happening, and some alternatives are more strongly associated with one's prior knowledge of how to *cause q* (De Neys et al., 2003a; Quinn & Markovits, 1998). For instance, in our initial example "If Jack goes to bed late, then he will be tired," the disabler "Jack is able to sleep in" might be a stronger disabler than "Jack takes a cold morning shower." Therefore, both the number and the associative strength of disablers and alternatives affect people's acceptance of conclusions. The more of these strongly associated disablers or alternatives exist, the more probable it is that at least one will be retrieved, and the more probable it is that a conclusion will be withdrawn (De Neys et al., 2003a, 2003b; Vadeboncoeur & Markovits, 1999). Initially, an "all-or-nothing phenomenon" was assumed (De Neys et al., 2003b, p. 582): as soon as a person retrieves one disabling condition or alternative, the corresponding conclusion is rejected (cf. Vadeboncoeur & Markovits, 1999). However, in a series of experiments, De Neys and colleagues (2003b) showed that the consideration of disablers and alternatives is gradual: every additional disabler or alternative has an impact on the degree to which people accept a conclusion. Therefore, people's acceptance of the conclusion is related in a monotonic fashion to the number of known disablers and alternatives.

Besides knowledge about the number and strength of disablers and alternatives, also their frequency of occurrence is important. Geiger and Oberauer (2007)[1] noticed that it is important to distinguish between the number of different disablers and the frequency with which they are thought to occur (i.e., p & not-q cases). Sometimes, there are many disablers for one conditional, but their frequency of occurrence is low (and vice versa). For instance, for the conditional "If you open the fridge, then the light inside goes on," there are many disablers (e.g., broken fridge, broken light bulb, no electricity), but their overall occurrence is low (usually, the light goes on when the fridge is opened). When both factors are disentangled, the frequency of exceptions seems to predict people's withdrawal of conclusions more accurately than the number of disablers (Geiger & Oberauer, 2007). Defeasibility in human reasoning is thus not a phenomenon in which a conclusion is either accepted or rejected categorically. Instead, the withdrawal of conclusions happens by degrees and depends on how much one knows about the circumstances that may affect the conditional relationship.

2.4   The Importance of the Experimental Context

How the context influences defeasible reasoning has not received as much attention as content-related factors. However, contextual factors are nonetheless interesting as they can moderate the degree to which participants consider potential disablers and alternatives. For instance, Vadeboncoeur and Markovits (1999) showed that people's automatic activation of background knowledge about disablers can be inhibited by instructions emphasizing the logical necessity of conditional inferences. Furthermore, Markovits, Lortie Forgues, and Brunet (2010) showed that people's tendency to consider the frequency of exceptions can be inhibited by the presentation of a dichotomous response format. They showed that only when scaled response formats were available did information about the frequency of exceptions affect the acceptance of conclusions. But when there was only a dichotomous response format, then people rejected the conclusion already in light of a single disabler. Finally, the consideration of disablers and alternatives can also be moderated by the introduction of words questioning the believability of the conditional or the disabler. George (1997), for instance, conducted experiments where conditional statements were either presented traditionally as "if p then q" or with an additional "very probable" or "not very probable" before the conditional. He found that the certainty of the conditional statement influenced participants' belief in the conclusion. Stevenson and Over (1995) elicited similar effects by introducing doubt in the disabler. Participants were confronted with conditionals such as "If John goes fishing, he will have a fish supper" and disablers such as "If John catches a fish, he will have a fish supper." However, when participants were told, for example, that John is always lucky, they rejected fewer conclusions.

3.   The Empirical Investigation of Belief Revision

In defeasible reasoning, people change their conclusions (derived beliefs) in light of new evidence. In belief revision, people instead have to make changes in their system of prior beliefs. This system of beliefs can be rather complex and interconnected, as we know from research in AI and philosophy. Psychologists are usually more pragmatic and often just treat the set of premises in an inference problem as the set of prior beliefs. Although this is a rather strong simplification, it is justified in many experimental settings in the psychological laboratory. Belief revision is then based on the assumption that a set of premises (beliefs) has to be consistent. Therefore, in our meditation example, Jack has to revise his beliefs: knowing that he did not fall asleep, he cannot believe that meditating makes him sleep. He has to find a way to reconcile his beliefs with what he observed. But which belief will he abandon? In AI and philosophy, two main accounts of which belief should be rejected have been proposed: minimal change (Harman, 1986) and epistemic entrenchment (Gärdenfors, 1988). According to the principle of *minimal change*, one should in case of inconsistencies make just enough changes in the belief set as are necessary to reach consistency. The principle of *epistemic entrenchment* arises from the idea that some beliefs are less entrenched than others and are therefore more easily given up in cases of conflict. Minimal change and epistemic entrenchment are well established in many formal theories of belief revision (e.g., Alchourrón, Gärdenfors, & Makinson, 1985; Gärdenfors, 1988; see chapter 5.2 by Rott and chapter 5.3 by Kern-Isberner, Skovgaard-Olsen, & Spohn, both in this handbook). However, from a psychological point of view, it is not clear how minimal change and epistemic entrenchment are implemented cognitively (cf. Elio & Pelletier, 1997). Therefore, many psychological studies have been carried out to understand the psychological processes behind these concepts and belief revision.

3.1   Experimental Findings

Psychological experiments on belief revision usually follow the same structure: participants are confronted with problems consisting of two (or more) premises and an additional fact. This additional fact, however, is inconsistent with the premises. Participants are then told that the additional fact is true but that the premises are uncertain. The task of the participants is to decide which of the two premises they believe less or abandon. One of the premises usually contains a general statement, such as a conditional (e.g., "If p then q") or a universally quantified statement (e.g., "All As are Bs" or "No As are Bs"). Let us illustrate this with an example from conditional reasoning:

If you meditate in the evening, then you fall asleep more easily.

You meditate in the evening.

But: You do not fall asleep more easily.

You will have noticed that the premises are the same as in MP. The additional fact, however, is not the conclusion that follows from the MP inference. It is even inconsistent with the set of premises. Participants thus have to indicate which premise they abandon: the conditional or the categorical one. In a similar way, belief revision can also be tested with MT:

If you meditate in the evening, then you fall asleep more easily.

You did not fall asleep more easily.

But: You meditated in the evening.

Again, participants have to decide whether they want to discard the conditional or the categorical premise.

The experimental findings on belief revision are heterogeneous. The first psychological studies on belief revision were conducted with quantified statements. Participants had to choose between revising premises with a universal quantifier (e.g., "All vertebrates have a backbone") or with particular facts (e.g., "This amoeba is not a vertebrate"). They consistently chose to revise the particular fact rather than the premise with the universal quantifier (e.g., Revlin, Cate, & Rouss, 2001; Revlis & Hayes, 1972; Revlis, Lipkin, & Hayes, 1971). The authors explained their findings by arguing that statements with universal quantifiers can be seen as general laws and are therefore more entrenched and believed more. However, years later, other studies had opposite results. In an influential study, Elio and Pelletier (1997) created conditional inference problems and found that the conditional premise—which is also a general law—is abandoned more readily than the categorical premise. This effect was more pronounced for MP than for MT. The authors argue that the conditional is revised more often because if–then "regularities" describe interdependences between particular facts and are therefore more uncertain (Elio & Pelletier, 1997). Elio and Pelletier's findings have been replicated several times (e.g., Byrne & Walsh, 2005, experiment 1; Elio, 1997, experiment 1; Politzer & Carles, 2001).

One explanation for these conflicting findings is the premises' content. Studies finding a preference for believing the conditional or universal premise used content-rich material, for which the reasoner has background knowledge. By contrast, studies finding a preference for *disbelieving* the conditional or universal premise used unfamiliar content, for which the reasoner does not have

prior knowledge. This was tested by Byrne and Walsh (2005). In two experiments, they varied the reasoner's familiarity with the conditional premise by creating conditionals for which the reasoner had no knowledge ("If the ruin was inhabited by Pings it had a forcefield surrounding it") or had knowledge ("If water was poured on the campfire the fire went out"). After presenting to the participants the unquestionable fact, they had to write down which premise they revised. When the content was unfamiliar, participants preferred to revise the conditional premise. But when the content was familiar, participants preferred to revise the categorical premise. Similarly, in another study, Politzer and Carles (2001) varied the plausibility of conditionals and found that more plausible conditionals were revised less often than less plausible ones. Correspondingly, Wolf, Rieger, and Knauff (2012) also found that conditional premises uttered by less trustworthy sources are revised more often than those uttered by more trustworthy ones. Along these lines, the number of disablers affects belief revision too. Elio (1997) constructed belief revision problems with conditionals that differed in their number of disablers (including the ones from Cummins et al., 1991). She found that the conditional premise was believed less, and abandoned more often, for conditionals with many disablers compared to conditionals with few disablers.

Overall, these studies show that human belief revision depends on the content of the premises and the trustworthiness of the information source. Revlis's assumption that premises containing general rules, such as conditionals, are more entrenched because they are lawlike applies only as long as this general rule is known by, and plausible for, the reasoner. When the conditional or universal premise is unfamiliar instead, then reasoners have no problems in revising such beliefs. In other words, entrenchment depends on the knowledge base of the reasoner (Byrne & Walsh, 2005; Politzer & Carles, 2001). However, this content-dependent approach cannot explain all revision choices. For instance, Elio and Pelletier (1997) found that the preference for revising the conditional premise instead of the categorical one was more pronounced for MP than for MT. This effect has also been replicated several times (e.g., Khemlani & Johnson-Laird, 2011; Revlin et al., 2001). There are thus additional, inference-dependent processes involved in belief revision. We present some explanations for this MP–MT asymmetry in the next section.

3.2 Spatial Belief Revision

The empirical findings on belief revision we have presented so far were concerned with conditionals. This is understandable because "if–then" statements are closely related to our beliefs about the world. Moreover, the experimental paradigm follows the philosophical tradition in which an inconsistency arises between an additional fact and the conclusion that can be derived from a set of premises. But in everyday life, people are often forced to change other kinds of beliefs. For instance, when following directions or trying to orient themselves in foreign cities, people also have to change beliefs about spatial location when they are lost.

Our group was the first that explored belief revision processes in the domain of spatial reasoning. In spatial reasoning, people infer new spatial relations between objects in space from information that is already given (Knauff, 2013; Ragni & Knauff, 2013). But what happens if new information disagrees with the prior information about the objects' locations in space? In our belief revision paradigm, participants get two or more premises describing the spatial relations between a set of objects. Afterward, an additional fact is presented that conflicts with the spatial arrangement consistent with the premises. Participants have to decide which objects they relocate in order to reach consistency. For example:

A is to the left of *B*.

B is to the left of *C*.

But: *A* is to the right of *C*.

The two premises allow the construction of the linear spatial arrangement *A–B–C*. However, the contradictory fact says that *A* is to the right of *C*. There are two ways to solve this inconsistency: either moving *A* to the right of *C* or moving *C* to the left of *A*. Following the psychological and linguistic tradition on the functional asymmetry between arguments, *A* is called the *to-be-located object* (LO) and *C* the *reference object* (RO) (Landau & Jackendoff, 1993). The empirical findings show that participants prefer to relocate the LO instead of the RO (Knauff, Bucher, Krumnack, & Nejasmic, 2013), although this effect can be modulated by properties of the objects such as size and movability (Nejasmic, Bucher, & Knauff, 2015).

4. Psychological Theories of Defeasible Reasoning and Belief Revision

So far, we have described how defeasible reasoning and belief revision are investigated in psychology and given an overview of the main experimental findings. Now we want to give an overview of the current psychological theories of defeasible reasoning and belief revision. This will, hopefully, facilitate the understanding of the

relatively complex (and not always consistent) pattern of effects and provide insights into the cognitive processes behind defeasible reasoning and belief revision.

4.1 The Theory of Mental Models

The theory of mental models (Johnson-Laird, 2010; Johnson-Laird & Byrne, 1991; see chapter 2.3 by Johnson-Laird, in this handbook) describes human reasoning as a semantic process where people use the meaning of premises to construct internal mental representations—the models—of what would be the case if the premises were true. These mental models are inspected and evaluated in order to find information that is not already given in the premises. The construction of mental models follows the principle of truth, meaning that only what is true is represented. Therefore, following the postulates of classical logic, a conditional "If *p* then *q*" allows the construction of the following mental models:

$$
\begin{array}{cc}
p & q \\
\text{not-}p & q \\
\text{not-}p & \text{not-}q
\end{array}
$$

A conclusion is considered *true* if it follows necessarily from the premises. A conclusion is considered *false* if there is a counterexample to it, that is, a model in which the premises are true but the conclusion is false. For instance, in the models above, we can see that if *p* is true, *q* is also true. There is no model in which *p* is true but *q* is false. Therefore, the MP conclusion is true. Along the same lines, also the validity of MT and the invalidity of AC and DA can be shown.

Furthermore, the model theory considers the restrictions from working memory. It is postulated that people do not consider all mental models from the beginning. Instead, they only construct one mental model, the *explicit* mental model *p* & *q*, which represents the information explicitly provided in the conditional:

$$
\begin{array}{cc}
p & q \\
\ldots &
\end{array}
$$

All other possible models (not-*p* and *q*, and not-*p* and not-*q*) are only represented implicitly, denoted by the three dots. Only if required by the task are these implicit models fleshed out into fully explicit models. Several experiments have shown that people prefer to construct specific models (*preferred mental models*) while ignoring other possibilities that are also consistent with the premises (*neglected mental models*) (Jahn, Knauff, & Johnson-Laird, 2007; Knauff, 2013). Moreover, the sequence of models generated is not random but follows a specific, cognitively efficient model variation mechanism (Ragni

& Knauff, 2013). This theory has recently been fundamentally revised and extended to account for ordinary people's reasoning beyond deduction (Hinterecker, Knauff, & Johnson-Laird, 2016; Johnson-Laird, Khemlani, & Goodwin, 2015b).

Defeasible reasoning Mental model theory accounts for defeasible reasoning by assuming that reasoners introduce disablers as additional antecedents (Byrne et al., 1999; Johnson-Laird & Byrne, 2002). Imagine, for instance, that a reasoner is confronted with the conditional "If Jack goes to bed late, then he will be tired" and considers the disabler that he can sleep in. This results in the following mental model:

bed late   sleeping in   tired

. . .

Because reasoners know that after sleeping in, people are usually not tired, they will additionally construct the following models, based on general knowledge:

sleeping in	not tired
not sleeping in	tired
not sleeping in	not tired

Thus, when people have to decide whether q follows from p, they combine the models from background knowledge with the models of the premises. However, the combination of these models would yield a contradiction:

bed late   sleeping in   tired   not tired

Therefore, Johnson-Laird and Byrne (2002) propose the principle of *pragmatic modulation*: in such cases, the model from prior knowledge has priority, allowing reasoners to conclude that, if Jack goes to bed late but is able to sleep in, he will not be tired. The model theory thus describes the consideration of disablers in terms of counterexample search in general knowledge. The impact of the frequency of exceptions is accounted for by assuming that the probability of an event can be inferred from the proportion of models in which the event occurs (e.g., Johnson-Laird, 2001; Johnson-Laird, Legrenzi, Girotto, Legrenzi, & Caverni, 1999). For instance, when reasoning about the conditional "If Jack goes to bed late, then he is tired," reasoners may construct the following mental models, containing the chances that q actually follows from p:

bed late	tired	3
bed late	not tired	1
not bed late	tired	2
not bed late	not tired	4

In this case, there are three chances that Jack is tired out of four chances that he goes to bed late (see also Johnson-Laird, Khemlani, & Goodwin, 2015a).

Belief revision According to mental model theory, belief revision happens in three steps (Johnson-Laird, Girotto, & Legrenzi, 2004; Legrenzi & Johnson-Laird, 2005): first, participants try to construct a mental model where all premises are satisfied. If this is not possible, participants infer that the set of premises is inconsistent. Second, participants try to solve the inconsistency. If the contradicting fact only conflicts with a single premise, then this premise is disbelieved. However, if the contradicting fact conflicts with more premises, then people will revise the premise whose model is in conflict with the contradictory fact or that fails to represent the fact. This process is called the *mismatch principle* (Johnson-Laird et al., 2004). For instance, in the case of MP, people are confronted with the conditional "If p then q," the categorical premise p, and the contradictory fact not-q. This contradictory fact is clearly in conflict with the explicit model from the conditional premise (p q). Therefore, the mismatch principle predicts that, for MP, the conditional premise will be abandoned. In MT, by contrast, people are confronted with the conditional "If p then q," the categorical premise not-q, and the contradictory fact p. In this case, p is not in conflict with the explicit model and thus the categorical premise will be abandoned instead of the conditional. This explains why the conditional premise is more often disbelieved in MP than MT. Finally, after belief revision, participants search for a *diagnostic explanation* that resolves the inconsistency they observed. For instance, in our example about the failed meditation, such an explanation could be "It was the wrong meditation and therefore he did not fall asleep" (see Legrenzi & Johnson-Laird, 2005).

4.2 Probabilistic Theories

Probabilistic theories assume Bayesian probability theory as the right norm for human reasoning (Evans, 2012; Oaksford & Chater, 2007; see chapter 4.5 by Chater & Oaksford and chapter 4.6 by Oberauer & Pessach, both in this handbook). Proponents of probabilistic theories of reasoning assume that people treat conditionals probabilistically by understanding the probability of a conditional, $P(\text{if } p \text{ then } q)$, as the conditional probability of q given p, $P(q \mid p)$—a relationship known as "the Equation" (Edgington, 1995; Evans & Over, 2004). The conditional probability can be calculated by performing the Ramsey test (Ramsey, 1929/1990; see also Evans, Handley, & Over, 2003). According to the Ramsey test, people first suppose that p holds. On the basis of this hypothetical belief, people then estimate the probability of q. For instance, in the case of "If Jack goes to bed late, then he will be tired," reasoners first assume that Jack goes to bed late. Then they start thinking about how probable it is that a person actually feels tired after going to bed late ($P(p \& q)$) and how probable it is that this is not the case ($P(p \& \text{not-}q)$). The higher $P(p \& q)$ is relative to $P(p \& \text{not-}q)$, the higher is the conditional probability $P(q \mid p)$ (Evans & Over, 2004; Evans et al., 2003). And the higher this conditional probability, the more probable it is that an MP conclusion is accepted (Oaksford, Chater, & Larkin, 2000).

Defeasible reasoning The notion of conditional probabilities captures defeasible reasoning directly. In fact, nonmonotonic and defeasible reasoning were the purposes probabilistic theories were developed for in the first place (Oaksford & Chater, 2013). All information that influences the conditional probability affects people's acceptance of conclusions. Therefore, disablers affect reasoning: disablers lower the conditional probability $P(q \mid p)$, and as a consequence, also fewer MP and MT conclusions are drawn (Weidenfeld, Oberauer, & Hörnig, 2005). In a similar way, alternatives affect inferences too: alternatives lower the conditional probability $P(p \mid q)$, and therefore fewer AC and DA conclusions are drawn (Verschueren, Schaeken, & d'Ydewalle, 2005). Further, given that probabilistic theories do not assume mental representations of concrete p-, q-, not-p-, or not-q-cases, they can explain why in addition to the number of disablers, also the overall frequency of exceptions affects inferences. Probabilistic approaches do not require people to construct concrete representations from general knowledge to be aware that p can happen without q. Instead, the effect of disablers is operationalized by the probability $P(p \& \text{not-}q)$, which is directly related to the frequency of exceptions.

Belief revision On the normative side, there are many accounts that explain belief change, for instance, through Bayesian conditionalization. However, from a psychological point of view, these accounts may be problematic (see Oaksford & Chater, 2013). Therefore, further research is needed to understand how people change their probability distributions after learning new information. For instance, Wolf, Rieger, and Knauff (2012) provided an account of how probabilities could describe the asymmetry in belief revision for MP and MT. The main assumption is that the contradictory fact changes the probabilities of the premises. For instance, in the case of belief revision after MP, the fact not-q is presented as true; therefore, $P(\text{not-}q) = 1$. This implies that $P(q) = 0$, and consequently, $P(q \mid p) = P(\text{if } p \text{ then } q) = 0$. However, in the case of belief revision after MT, the fact p is presented as true; therefore, $P(p) = 1$. Consequently, $P(q \mid p) = P(\text{if } p \text{ then } q) = q$. The comparison of these probabilities shows

why the conditional premise is often believed less for MP than MT.

The influence of plausibility and trustworthiness on belief revision is described by probabilistic theories through the role of long-term memory. The main idea is that long-term memory about the content of a conditionals affects the corresponding conditional probability (Oaksford & Chater, 2013). Similarly, also knowledge about plausibility and trustworthiness is part of our knowledge system. Therefore, participants give lower probabilities to implausible conditionals and to conditionals uttered by untrustworthy sources. As a consequence, these are believed less and abandoned more readily (cf. Wolf et al., 2012).

4.3 Dual-Process Theories

Dual-process theories propose that people use *two kinds of thinking* or *modes of processing* that are employed depending on factors such as tasks, time constraints, or cognitive capacity (Evans, 2008; Evans & Stanovich, 2013; see chapter 2.4 by Fiedler, Prager, & McCaughey, this handbook). People can reason heuristically, fast, and effortlessly (known as "System 1" or "Type 1" processes), and they can reason analytically, slowly, and dependently on working memory (known as "System 2" or "Type 2" processes) (e.g., Evans, 2008; Kahneman & Frederick, 2005). It is argued that System 1 serves as a default system, whose responses can be either overridden or supported by System 2 (Evans & Stanovich, 2013).

Defeasible reasoning As we have shown, disablers and alternatives can affect reasoning in two ways: either as single instances or through the frequency of p & not-q- and not-p & q-cases. Verschueren et al. (2005) argue that both aspects are important and can be unified in a dual-process approach. They argue that the consideration of single disablers and alternatives is an analytic process—as proposed by mental model theory. In contrast, the consideration of the frequency information is considered a heuristic process, in which the overall likelihood of q given p and of p given q matters—as proposed by the probabilistic theories. Verschueren et al.'s (2005) dual-process model has been tested experimentally and corroborated several times (e.g., Markovits, Brisson, & de Chantal, 2015, 2017). As predicted by dual-process theories, fast responses are best explained by the likelihood of q given p (for MP and MT inferences) or the likelihood of p given q (for AC and DA inferences), slow responses are best described by the availability of disablers or alternatives, and conclusions based on likelihoods can be overridden by specific disablers and alternatives.

Belief revision A dual-process approach to belief revision can be found in Wolf and Knauff (2008). In their experiments, they varied the probability of the conditional premise (0%, 50%, or 100%). They found that when the conditional premise was highly probable or highly improbable, participants used this probability information as a heuristic for belief revision: they believed the conditional premise more for highly probable conditionals but believed the categorical premise more for highly improbable conditionals. However, when the probability of the conditional was around 50%, people reasoned more analytically, and the mismatch principle from mental model theory predicted belief revision choices: the conditional premise was believed more for MT than for MP.

5. Defeasible Reasoning or Belief Revision?

So far, we have treated defeasible reasoning and belief revision separately. The main reason is that both functions are often discussed within different theoretical (normative) frameworks and communities. In defeasible reasoning, conclusions are withdrawn in light of new evidence, while in belief revision, the new evidence makes people abandon previous beliefs (in the psychology lab typically operationalized as premises). However, a germane question is whether defeasible reasoning and belief revision do indeed rely on distinct *cognitive* processes—a crucial descriptive question from a psychological point of view. In fact, the psychological literature on defeasible reasoning and belief revision shows that both depend on the content of the premises. In defeasible reasoning, many disablers and unbelievable content increase people's willingness to reject conclusions from conditional premises. In a similar way, in belief revision, conditionals with many disablers and implausible content are revised more often than those with fewer disablers or plausible content. These similarities were noticed by Dieussaert et al. (2005). They conducted a belief revision experiment using conditionals with many and with few disablers and asked the participants twice about their belief in the conditional premise: before presenting the contradicting fact and after. They found no significant differences in the believability ratings before and after the introduction of the inconsistency. This shows that people's disbelief in the conditional premise was not the outcome of a process of belief revision but a product of people's consideration of disablers. Dieussaert and colleagues therefore argue that people's withdrawal of conclusions in conditional reasoning and in belief revision are "two sides of the same

coin" (Dieussaert et al., 2005, p. 29; see also Gärdenfors, 1990).

At first sight, the findings by Dieussaert et al. (2005) seem counterintuitive. In belief revision, participants abandon premises in the search for consistency, while in defeasible reasoning, there is no inconsistency at all. Nevertheless, there is evidence suggesting that this distinction is only a result of differing experimental paradigms. For instance, Walsh and Johnson-Laird (2009) showed that when participants are asked to explain inconsistencies, they rarely just negate the conditional premise but rather explain inconsistencies by naming disablers that question the causal link between antecedent and consequent (see also Johnson-Laird et al., 2004). In a later experiment, Khemlani and Johnson-Laird (2011) showed that in case of inconsistencies, participants prefer to give *explanations* instead of direct refutations of the premises. That is, participants seem to resolve inconsistencies by finding disablers, without the need of a revision in the strict sense. Similarly, also Politzer and Carles (2001) found that when participants can choose between completely abandoning one of the premises and only casting doubt on them, they prefer to cast doubt. These findings suggest that under everyday circumstances, people do not spontaneously abandon existing beliefs but instead try to find explanations for inconsistencies in the set of beliefs. This has consequences for the relation between defeasible reasoning and belief revision. An example:

If you meditate in the evening, then you fall asleep more easily.

You meditate in the evening.

But: You do not fall asleep more easily.

In a belief revision paradigm, participants have to choose which premise they reject. However, in everyday life, participants would rather try to explain the inconsistency by the introduction of a disabler:

If you meditate in the evening, then you fall asleep more easily.

If it is the wrong meditation, then you do not fall asleep more easily.

You meditate in the evening.

You do not fall asleep more easily.

Now the inference resembles one of defeasible reasoning: by introducing the disabler, the formally inconsistent fact turns into the "conclusion" that participants would have accepted in a defeasible reasoning paradigm. The search for explanations plays a vital role in these cognitive processes, albeit one that is not so frequently considered in normative accounts of defeasible reasoning and belief revision.

6. Concluding Remarks and Further Perspectives

Many psychological studies help us to understand human defeasible reasoning and belief revision. Still, there is need for further research. For instance, the principle of truth of mental model theory has difficulties in capturing people's consideration of *p* & not-*q*-cases during reasoning (Geiger & Oberauer, 2010). Similarly, probabilistic theories are only computational theories (according to Marr's, 1982, terminology) that do not describe the cognitive processes behind reasoning (Oaksford & Chater, 2003). However, cognitive theories of reasoning must explain not only *what* is computed but also *how* it is computed in the human mind (Hinterecker, Knauff, & Johnson-Laird, 2019; see the introductory chapter by Knauff & Spohn, this handbook). Dual-process theories have been criticized for not describing qualitative differences but only quantitative differences in reasoning (Osman, 2013) and for just exemplifying "the backwards development from precise theories to surrogates" (Gigerenzer, 2011, p. 739; but see chapter 1.2 by Evans, this handbook).

What remains clear from our review is that still more research is needed to understand the cognitive processes that underlie human defeasible reasoning and belief revision. Here, more collaboration between psychology, philosophy, and AI could be helpful. There are many normative and formal theories on defeasible reasoning (e.g., Kraus, Lehmann, & Magidor, 1990; Spohn, 2012; see chapter 5.3 by Kern-Isberner, Skovgaard-Olsen, & Spohn, this handbook) and belief revision (e.g., Alchourrón et al., 1985; see chapter 5.2 by Rott, in this handbook), which can be helpful in cognitive psychology, too. At the same time, however, these formal theories should also consider psychological findings and adapt their formalisms to the actual cognition and behavior of real people. We are well aware of the different views on this *is–ought problem* (see Knauff and Spohn's introduction to this handbook). Yet, it is obvious that there are still too few links between the normative and the descriptive work on defeasibility and belief revision. On the one hand, many of the formalisms from philosophy and AI have not been tested regarding their capability to model actual human reasoning (cf. Ragni, Eichhorn, Bock, Kern-Isberner, & Ping Ping Tse, 2017). On the other hand, psychological research has neglected nonmonotonicity for a long time. Therefore, psychologists have often prematurely considered people irrational, just because they did not follow the norms of classical logic. In fact, many of the psychological studies we have reported here do not even use the technical term "defeasible reasoning," although they have investigated the circumstances under which it is rational to withdraw otherwise valid conclusions. To avoid such dangers of disciplinary division, we need a more interdisciplinary approach. Ideally, this research combines the strengths of psychology, philosophy, and AI to obtain a broader understanding of how a cognitive system, be it real or abstract, human or machine, can flexibly respond to new information and therefore change its belief system in a rational way.

Note

1. Geiger and Oberauer (2007) only investigated the relation between the number of disablers and the frequency of exceptions. However, it is also conceivable that the frequency of alternatives affects inferences.

References

Alchourrón, C. E., Gärdenfors, P., & Makinson, D. (1985). On the logic of theory change: Partial meet contraction and revision functions. *Journal of Symbolic Logic, 50*, 510–530.

Antoniou, G. (1997). *Nonmonotonic reasoning.* Cambridge, MA: MIT Press.

Byrne, R. M. J. (1989). Suppressing valid inferences with conditionals. *Cognition, 31*, 61–83.

Byrne, R. M. J., Espino, O., & Santamaria, C. (1999). Counterexamples and the suppression of inferences. *Journal of Memory and Language, 40*, 347–373.

Byrne, R. M. J., & Walsh, C. R. (2005). Resolving contradictions. In V. Girotto & P. N. Johnson-Laird (Eds.), *The shape of reason: Essays in honour of Paolo Legrenzi* (pp. 117–122). Hove, England: Psychology Press.

Cummins, D. D. (1995). Naive theories and causal deduction. *Memory & Cognition, 23*, 646–658.

Cummins, D. D., Lubart, T., Alksnis, O., & Rist, R. (1991). Conditional reasoning and causation. *Memory & Cognition, 19*, 274–282.

De Neys, W., Schaeken, W., & d'Ydewalle, G. (2003a). Causal conditional reasoning and strength of association: The disabling condition case. *European Journal of Cognitive Psychology, 15*, 161–176.

De Neys, W., Schaeken, W., & d'Ydewalle, G. (2003b). Inference suppression and semantic memory retrieval: Every counterexample counts. *Memory & Cognition, 31*, 581–595.

Dieussaert, K., De Neys, W., & Schaeken, W. (2005). Suppression and belief revision, two sides of the same coin? *Psychologica Belgica, 45*, 29–46.

Edgington, D. (1995). On conditionals. *Mind, 104*, 235–329.

Elio, R. (1997). What to believe when inferences are contradicted: The impact of knowledge type and inference rule. In M. Shafto & P. Langley (Eds.), *Proceedings of the Nineteenth Annual Conference of the Cognitive Science Society* (pp. 211–216). Mahwah, NJ: Erlbaum.

Elio, R., & Pelletier, F. J. (1997). Belief change as propositional update. *Cognitive Science, 21*, 419–460.

Evans, J. St. B. T. (2008). Dual-processing accounts of reasoning, judgment, and social cognition. *Annual Review of Psychology, 59*, 255–278.

Evans, J. St. B. T. (2012). Questions and challenges for the new psychology of reasoning. *Thinking & Reasoning, 18*, 5–31.

Evans, J. St. B. T., Handley, S. J., & Over, D. E. (2003). Conditionals and conditional probability. *Journal of Experimental Psychology: Learning, Memory, and Cognition, 29*, 321–355.

Evans, J. St. B. T., Newstead, S. E., & Byrne, R. (1993). *Human reasoning: The psychology of deduction.* Hove, England: Erlbaum.

Evans, J. St. B. T., & Over, D. E. (2004). *If.* Oxford, England: Oxford University Press.

Evans, J. St. B. T., & Stanovich, K. E. (2013). Dual-process theories of higher cognition: Advancing the debate. *Perspectives on Psychological Science, 8*, 223–241.

Gärdenfors, P. (1988). *Knowledge in flux.* Cambridge, MA: MIT Press.

Gärdenfors, P. (1990). Belief revision and nonmonotonic logic: Two sides of the same coin? In L. Carlucci Aiello (Ed.), *Proceedings of the Ninth European Conference on Artificial Intelligence* (pp. 768–773). London, England: Pitman Publishing.

Gazzo Castañeda, L. E., & Knauff, M. (2016). Defeasible reasoning with legal conditionals. *Memory & Cognition, 44*, 499–517.

Gazzo Castañeda, L. E., & Knauff, M. (2018). Quantifying disablers in reasoning with universal and existential rules. *Thinking & Reasoning, 24*, 344–365.

Gazzo Castañeda, L. E., & Knauff, M. (2019). The specificity of terms affects conditional reasoning. *Thinking & Reasoning, 25*, 72–93.

Gazzo Castañeda, L. E., & Knauff, M. (2020). Everyday reasoning with unfamiliar conditionals. *Thinking & Reasoning.* doi:10.1080/13546783.2020.1823478

Geiger, S. M., & Oberauer, K. (2007). Reasoning with conditionals: Does every counterexample count? It's frequency that counts. *Memory & Cognition, 35*, 2060–2074.

Geiger, S. M., & Oberauer, K. (2010). Towards a reconciliation of mental model theory and probabilistic theories of conditionals.

In M. Oaksford & N. Chater (Eds.), *Cognition and conditionals: Probability and logic in human thinking* (pp. 289–307). Oxford, England: Oxford University Press.

George, C. (1997). Reasoning from uncertain premises. *Thinking & Reasoning, 3*, 161–189.

Gigerenzer, G. (2011). Personal reflections on theory and psychology. *Theory & Psychology, 20*, 733–743.

Harman, G. (1986). *Change in view.* Cambridge, MA: MIT Press.

Hinterecker, T., Knauff, M., & Johnson-Laird, P. N. (2016). Modality, probability, and mental models. *Journal of Experimental Psychology: Learning, Memory, and Cognition, 42*, 1606–1620.

Hinterecker, T., Knauff, M., & Johnson-Laird, P. N. (2019). How to infer possibilities: A reply to Oaksford et al. (2018). *Journal of Experimental Psychology: Learning, Memory, and Cognition, 45*(2), 298–301.

Jahn, G., Knauff, M., & Johnson-Laird, P. N. (2007). Preferred mental models in reasoning about spatial relations. *Memory & Cognition, 35*, 2075–2087.

Johnson-Laird, P. N. (2001). Mental models and deduction. *Trends in Cognitive Sciences, 5*, 434–442.

Johnson-Laird, P. N. (2010). Mental models and human reasoning. *Proceedings of the National Academy of Sciences, 107*, 18243–18250.

Johnson-Laird, P. N., & Byrne, R. M. J. (1991). *Deduction.* Hove, England: Erlbaum.

Johnson-Laird, P. N., & Byrne, R. M. J. (2002). Conditionals: A theory of meaning, pragmatics, and inference. *Psychological Review, 109*, 636–678.

Johnson-Laird, P. N., Girotto, V., & Legrenzi, P. (2004). Reasoning from inconsistency to consistency. *Psychological Review, 111*, 640–661.

Johnson-Laird, P. N., Khemlani, S. S., & Goodwin, G. P. (2015a). Logic, probability, and human reasoning. *Trends in Cognitive Science, 19*, 201–214.

Johnson-Laird, P. N., Khemlani, S. S., & Goodwin, G. P. (2015b). Response to Baratgin et al.: Mental models integrate probability and deduction. *Trends in Cognitive Sciences, 19*, 548–549.

Johnson-Laird, P. N., Legrenzi, P., Girotto, V., Legrenzi, M. S., & Caverni, J.-P. (1999). Naive probability: A mental model theory of extensional reasoning. *Psychological Review, 106*, 62–88.

Kahneman, D., & Frederick, S. (2005). A model of heuristic judgment. In K. Holyoak & R. G. Morrison (Eds.), *The Cambridge handbook of thinking and reasoning* (pp. 267–294). Cambridge, England: Cambridge University Press.

Khemlani, S., & Johnson-Laird, P. N. (2011). The need to explain. *Quarterly Journal of Experimental Psychology, 64*, 2276–2288.

Knauff, M. (2013). *Space to reason: A spatial theory of human thought.* Cambridge, MA: MIT Press.

Knauff, M., Bucher, L., Krumnack, A., & Nejasmic, J. (2013). Spatial belief revision. *Journal of Cognitive Psychology, 25,* 147–156.

Kraus, S., Lehmann, D., & Magidor, M. (1990). Nonmonotonic reasoning, preferential models and cumulative logics. *Artificial Intelligence, 44,* 167–207.

Landau, B., & Jackendoff, R. (1993). "What" and "where" in spatial language and spatial cognition. *Behavioral and Brain Sciences, 16,* 217–265.

Legrenzi, P., & Johnson-Laird, P. N. (2005). The evaluation of diagnostic explanations for inconsistencies. *Psychologica Belgica, 45,* 19–28.

Markovits, H. (1986). Familiarity effects on conditional reasoning. *Journal of Educational Psychology, 78,* 492–494.

Markovits, H., Brisson, J., & de Chantal, P.-L. (2015). Additional evidence for a dual-strategy model of reasoning: Probabilistic reasoning is more invariant than reasoning about logical validity. *Memory & Cognition, 43,* 1208–1215.

Markovits, H., Brisson, J., & de Chantal, P.-L. (2017). Logical reasoning versus information processing in the dual-strategy model of reasoning. *Journal of Experimental Psychology: Learning, Memory, and Cognition, 43,* 72–80.

Markovits, H., Lortie Forgues, H., & Brunet, M.-L. (2010). Conditional reasoning, frequency of counterexamples, and the effect of response modality. *Memory & Cognition, 38,* 485–492.

Marr, D. (1982). *Vision.* San Francisco, CA: Freeman.

Nejasmic, J., Bucher, L., & Knauff, M. (2015). Grounded spatial belief revision. *Acta Psychologica, 157,* 144–154.

Oaksford, M., & Chater, N. (2003). Conditional probability and the cognitive science of conditional reasoning. *Mind & Language, 18,* 359–379.

Oaksford, M., & Chater, N. (2007). *Bayesian rationality: The probabilistic approach to human reasoning.* Oxford, England: Oxford University Press.

Oaksford, M., & Chater, N. (2013). Dynamic inference and everyday conditional reasoning in the new paradigm. *Thinking & Reasoning, 19,* 346–379.

Oaksford, M., Chater, N., & Larkin, J. (2000). Probabilities and polarity biases in conditional inference. *Journal of Experimental Psychology: Learning, Memory, and Cognition, 26,* 883–899.

Osman, M. (2013). A case study: Dual-process theories of higher cognition—Commentary on Evans & Stanovich (2013). *Perspectives on Psychological Science, 8,* 248–252.

Politzer, G., & Bonnefon, J.-F. (2006). Two varieties of conditionals and two kinds of defeaters help reveal two fundamental types of reasoning. *Mind & Language, 21,* 484–503.

Politzer, G., & Carles, L. (2001). Belief revision and uncertain reasoning. *Thinking & Reasoning, 7,* 217–234.

Quinn, S., & Markovits, H. (1998). Conditional reasoning, causality and the structure of semantic memory: Strength of association as a predictive factor for content effects. *Cognition, 68,* B93–B101.

Ragni, M., Eichhorn, C., Bock, T., Kern-Isberner, G., & Ping Ping Tse, A. (2017). Formal nonmonotonic theories and properties of human defeasible reasoning. *Minds and Machines, 27,* 79–117.

Ragni, M., & Knauff, M. (2013). A theory and a computational model of spatial reasoning with preferred mental models. *Psychological Review, 120,* 561–588.

Ramsey, F. P. (1990). General propositions and causality (original manuscript, 1929). In D. H. Mellor (Ed.), *Philosophical papers* (pp. 145–163). Cambridge, England: Cambridge University Press.

Revlin, R., Cate, C. L., & Rouss, T. S. (2001). Reasoning counterfactually: Combining and rending. *Memory & Cognition, 29,* 1196–1208.

Revlis, R., & Hayes, J. R. (1972). The primacy of generalities in hypothetical reasoning. *Cognitive Psychology, 3,* 268–290.

Revlis, R., Lipkin, S. G., & Hayes, J. R. (1971). The importance of universal quantifiers in a hypothetical reasoning task. *Journal of Verbal Learning and Verbal Behaviour, 10,* 86–91.

Rumain, B., Connell, J., & Braine, M. D. (1983). Conversational comprehension processes are responsible for reasoning fallacies in children as well as adults: *If* is not the biconditional. *Developmental Psychology, 19,* 471–481.

Spohn, W. (2012). *The laws of belief: Ranking theory and its philosophical applications.* Oxford, England: Oxford University Press.

Stevenson, R. J., & Over, D. E. (1995). Deduction from uncertain premises. *Quarterly Journal of Experimental Psychology, 48A,* 613–643.

Thompson, V. A. (1994). Interpretational factors in conditional reasoning. *Memory & Cognition, 22,* 742–758.

Thompson, V. A. (1995). Conditional reasoning: The necessary and sufficient conditions. *Canadian Journal of Experimental Psychology, 49,* 1–58.

Vadeboncoeur, I., & Markovits, H. (1999). The effect of instructions and information retrieval on accepting the premises in a conditional reasoning task. *Thinking & Reasoning, 5,* 97–113.

Verschueren, N., Schaeken, W., & d'Ydewalle, G. (2005). A dual-process specification of causal conditional reasoning. *Thinking & Reasoning, 11,* 239–278.

Walsh, C. R., & Johnson-Laird, P. N. (2009). Changing your mind. *Memory & Cognition, 37,* 624–631.

Weidenfeld, A., Oberauer, K., & Hörnig, R. (2005). Causal and noncausal conditionals: An integrated model of interpretation and reasoning. *Quarterly Journal of Experimental Psychology, 58A,* 1479–1513.

Wolf, A. G., & Knauff, M. (2008). The strategy behind belief revision: A matter of judging probability or the use of mental models? In B. C. Love, K. McRae, & V. M. Sloutsky (Eds.), *Proceedings of the 30th Annual Conference of the Cognitive Science Society* (pp. 831–836). Austin, TX: Cognitive Science Society.

Wolf, A. G., Rieger, S., & Knauff, M. (2012). The effects of source trustworthiness and inference type on human belief revision. *Thinking & Reasoning, 18,* 417–440.

5.5 Argumentation Theory

Ulrike Hahn and Peter Collins

Summary

Unlike persuasion, the study of "rational argument" asks what convinces and what *should* convince (see, e.g., Hahn & Oaksford, 2012). It combines normative and descriptive concerns, both within theory development and the study of behavior. In this chapter, we show the centrality of norms, discuss how norms have shaped empirical research, and highlight how descriptive work has informed normative developments, showing the tight coupling between argumentation, reasoning, and rationality. Finally, we identify future directions for research.

1. What Are Arguments; Why Study Them?

We distinguish two relevant senses of "argument": arguments as objects and arguments as a kind of social process. Arguments-as-objects are sets of statements: claims or conclusions supported by premises.[1] The strength of support varies. Arguments can be deductive: the conclusion follows necessarily; if one believes the premises, one denies the conclusion on pain of inconsistency. Arguments can also be inductive or abductive: in either case, the inference is ampliative in the sense that the conclusion may be false even if the premises are true. Often such inference is cast as defeasible (presumptive): an argument that "may be strong enough to provide evidence to warrant rational acceptance of its conclusion, given that its premises are acceptable" (Walton, Reed, & Macagno, 2008, p. 1). But acceptance of the conclusion is tentative: it may be overturned by further information (Walton et al., 2008). It is subject to debate to what extent these different types require different normative frameworks. In the remainder, we use the term "inductive" broadly to refer to ampliative arguments, setting aside (unless otherwise indicated) differences between statistical generalizations ("inductive" in a narrower sense), arguments from evidence to causes (e.g., "abduction"), and defeasible arguments.

Arguments-as-processes are dialogues in a social context: dialectical exchanges, with discussants proposing and opposing claims, say, to convince someone of the truth of some claim or persuade someone to perform some action (for a typology, see Walton, 2008). In recent decades, much argumentation research has centered, explicitly or implicitly, on the extent to which arguments-as-objects can be understood without the wider vantage point of arguments-as-processes.

Psychology has tended to focus on arguments-as-objects. In particular, the psychology of reasoning has investigated how people derive, or endorse, conclusions, typically of deductive and inductive arguments (for a survey, see Manktelow, 2012). In mainstream psychology of reasoning, studies have given relatively little attention to social and conversational contexts (although see Evans, Neilens, Handley, & Over, 2008; Girotto, Kemmelmeier, Sperber, & van der Henst, 2001; Mercier & Sperber, 2011, 2017; Sperber, Cara, & Girotto, 1995; Stevenson & Over, 2001). Yet, if one closely considers how people change their beliefs upon hearing certain premises, one must eventually consider the contexts in which such premises might be uttered. For example, if one tries to understand why participants deviate from the logical prescriptions of modus tollens, one soon sees the need to consider why a particular premise might be uttered in real life (see, e.g., Hahn & Oaksford, 2012).

An undue focus on arguments-as-processes suffers complementary difficulties. Historically, it was argued that so-called fallacies of argumentation—such as "circular arguments" or "question-begging"—could only be adequately understood as fallacious by considering them as argumentative "moves" in a dialectical context (see, e.g., Hahn, 2011, and references therein). Hence, a circular argument such as "God exists, because God exists" is viewed as fallacious because it violates a procedural commitment such as a requirement to meet one's "burden of proof." This renders the argument weak even though it is actually deductively valid and, in that sense, strong. However, such explanation by procedural appeal to the burden of proof is, arguably, itself question-begging, because the argument can only fail to meet that burden of proof *because* it is weak. This, however, means that some independent measure of the strength of the argument's content is required. In other words, the procedural failure presupposes the weakness it is trying to explain (Hahn, 2011; Hahn & Oaksford, 2007).

Any satisfactory account of argument thus ultimately requires consideration of the properties of both arguments-as-objects and arguments-as-processes. In other words, to understand when an argument should convince, we need two distinct, but interconnected, sets of norms.

2. Norms

"Good argumentation" is widely recognized as a desirable goal, for instance, by education systems (Crowell & Kuhn, 2014; Kuhn & Crowell, 2011). In defining "goodness," one can usefully distinguish two types of norm (Hahn & Oaksford, 2012): first, procedural norms, aimed at regulating discussants' behavior to achieve consensus and, second, epistemic norms, aimed at regulating structure and content to achieve truth or justified belief.

Procedural accounts have drawn inspiration from natural-language pragmatics (Walton, 2008). Understood abstractly, arguments comprise the following procedures: proponents secure commitments from their opponents to advance their own case and exploit weaknesses in their opponents' argument to undermine their opponents' case (Walton, 1989). An influential example is pragmadialectics, which posits different stages of argumentative dialogues and a set of rules to regulate behavior at each stage (van Eemeren & Grootendorst, 1984, 1992, 1995, 2004). An intuitive rule is the "burden of proof" rule, according to which proponents are expected to offer evidence to support their standpoints.

Argument content plays a deeper role in epistemic norms. Some epistemic norms are familiar from the psychology of reasoning. A central normative system is classical logic, which enforces consistency through its definition of logical validity: that the truth of the premises guarantees the truth of the conclusion. A "good argument," from the perspective of classical logic, is a valid argument. This does not, however, suffice for ampliative argument, leaving the majority of everyday informal argument unaccounted for.

Probability theory generalizes classical logic and adds to logical consistency the notion of probabilistic coherence. This has been exploited in the body of work known as "Bayesian Argumentation" (Hahn & Oaksford, 2007;

Zenker, 2012). A good argument, from this perspective, is one in which the premises mandate an increase in the probability (degree of belief) we assign to the conclusion. This framework formally explains so-called fallacies of argumentation (such as circular arguments, mentioned above; see Hahn & Oaksford, 2006, 2007; on the catalogue of fallacies, see, e.g., Woods, Irvine, & Walton, 2004), thus filling a long-standing theoretical "gap" (Hamblin, 1970). A probabilistic approach can be extended to include utilities, invoking decision theory to capture considerations of argument strength (see, e.g., Corner, Hahn, & Oaksford, 2011; Evans et al., 2008).

In the literature, there have been tensions between procedural and epistemic approaches, illustrated by Walton et al.'s (2008) critical discussion of probabilities as a model for argumentation and by Bhatia and Oaksford's (2015) test of competing procedural and probabilistic predictions for ad hominem arguments. Nevertheless, argumentation is a complex phenomenon, and different normative systems might be valuable for different facets of argumentation. Uniquely, procedural rules, for instance, advise participants how to behave to secure consensus and can be understood as ethical prescriptions (Christmann, Mischo, & Flender, 2000; Schreier, Groeben, & Christmann, 1995). Ultimately, the two systems are complementary, not least because procedural rules will likely have to appeal to epistemic norms for their foundation (Corner & Hahn, 2013).

Presently, it is not clear that procedural rules have deep normative force (Corner & Hahn, 2011; Collins & Hahn, 2016; Walton et al., 2008). There are, however, deep justifications for the probabilistic approach to argumentation: mathematical arguments for the optimality of Bayesian inference (Leitgeb & Pettigrew, 2010a, 2010b; Pettigrew, 2016; Rosenkrantz, 1992) and its relationship with belief accuracy or truth. Ultimately, at least one factor in an appropriate foundation for procedural rules is likely to be that they *maximize opportunities for truth to emerge* (although it is not necessarily the only factor; see recent considerations of virtue epistemology in argument, which blur lines between content and process when it comes to quality; Aberdein, 2010; Paglieri, 2015). Normative systems may help to resolve disputes (van Eemeren, Garssen, & Meuffels, 2009) and resolve them in accuracy-maximizing ways. Detailed elaborations of such foundations have yet to be provided.

3. Empirical Evidence

Similarly intertwined when it comes to argumentation are normative and empirical considerations: the conceptual justification of norms for argumentation means that data on argumentation can speak to human rationality, about which there is considerable debate in cognitive science. At the same time, probing people's adherence to putative normative systems governing argumentation means probing the conventional validity of these systems, that is, examining how far they underpin the argumentative behavior of typical discussants, untrained in analysis or evaluation of argumentation (Schellens, Šorm, Timmers, & Hoeken, 2017; van Eemeren et al., 2009). We next sketch current evidence on how people assess arguments and how those data bear on the validity of normative systems.

One source of evidence is naturalistic data: the conceptual analysis of real-world arguments. Procedural-rules theorists have analyzed extended arguments (van Eemeren & Houtlosser, 1999). So, too, have Bayesians, a notable example being a Bayesian reconstruction of the entire body of evidence in one of the United States' most famous court cases (Kadane & Schum, 1996). The argumentation-scheme literature has compiled systems of generalized "types" of arguments that form the basic components of much of everyday informal argument. A wide range of different schemes has been proposed (Garssen, 1997; Hastings, 1962; Kienpointner, 1992; Perelman & Olbrechts-Tyteca, 1969; Schellens, 1985; Walton et al., 2008), with over 50 different schemes distinguished by some authors. Some aspects of those typologies may arguably be redundant, but these schemes are not yet exhaustive, for example, in the domain of causal argument (Hahn, Bluhm, & Zenker, 2017). These (generalized) descriptive data have themselves been subjected to normative analysis from various viewpoints (e.g., Walton et al., 2008), including a probabilistic perspective (Hahn & Hornikx, 2016). Such qualitative data suggest the usefulness of procedural and epistemic normative systems.

More controlled data come from cognitive-psychological experiments. These experiments tend to manipulate properties of arguments and have participants assess the reasonableness, persuasiveness, or convincingness of the arguments on Likert-style scales. In the procedural-rules tradition, van Eemeren et al. (2009) have found evidence for the conventional validity of procedural rules. In one study, participants read arguments containing legitimate attacks on a person, disguised fallacious ad hominem attacks, and openly abusive ad hominem attacks. Participants distinguished among these conditions in judging reasonableness (similarly, in the "fairness rules" formulation, see Christmann et al., 2000; Schreier et al., 1995). In the argumentation-schemes tradition, Schellens et al.

(2017) found that participants were sensitive to different argument parameters for a range of schemes, including arguments from authority, analogy, and example. Participants also invoked such parameters in open-ended questioning.

A larger literature has tested whether the Bayesian approach to argumentation has conventional validity, finding considerable support. The approach is to identify probabilistic parameters, manipulate them experimentally, and assess whether participants are sensitive to them. Arguments are typically set in dialogues. Take arguments from ignorance. As Hahn and Oaksford (2007) show, these arguments can be modeled analogously to drug testing. For example:

(1) Drug A is toxic because a toxic effect has been observed.

(2) Drug A is not toxic because no toxic effect has been observed.

Argument (2) is an argument from ignorance. Reconstructed probabilistically, the strength of (1) and (2) can be modeled with Bayes' rule. The crucial parameters are sensitivity (P(Effect | Toxicity)) and specificity (P(NoEffect | NoToxicity)). In several tasks, participants were influenced by these parameters.

Participants show the influence of probabilistic parameters in other common arguments. For instance, a Bayesian analysis of slippery-slope arguments—"If you allow gay marriage, people will soon want to marry their pets"—predicts influence of both probabilities and utilities, that is, P(Outcome | Action) and U(Outcome) (see Hahn & Oaksford, 2007). Experimental evidence suggests such influence and links it to cognitive mechanisms for categorization (Corner et al., 2011). Similarly, Bayesian models have been applied to arguments involving sources. Inspired by models of testimony, these models account well for participants' assessments of ad hominem arguments and arguments from expertise (Harris, Hahn, Madsen, & Hsu, 2016; Harris, Hsu, & Madsen, 2012).

Thus far, there is evidence that participants, at a group level, endorse procedural rules, argumentation schemes, and probabilistic norms. That they endorse probabilistic norms can also be taken as evidence that people approximate behaviors that have deep conceptual justification.

4. Future Directions

The psychology of argumentation has yet to fully capitalize on a wide range of possible cognitive-psychological methods for exploring argumentation. Two promising areas are exploration of individual differences and intervention studies to improve argumentation. As we will see below, there is already suggestive, but limited, evidence. The limitation arises from the fact that we do not currently have reliable, theory-neutral measures of argumentation skills. Such measures would be challenging to develop but would yield rich rewards.

We turn, first, to the study of individual differences. Such studies allow us to probe conventional validity in more detail, asking whether norms are endorsed to the same extent by all participants and what participant characteristics drive differences in endorsement. Perhaps more interestingly, studies could also offer evidence for or against certain norms, when paired with a reliable measure of argumentation skills. In principle, adherence to certain norms, or related cognitive skills, should correlate with higher scores on measures of argument quality.

Existing research offers suggestive evidence. Argumentation skills have largely been assessed using procedural norms: for instance, the number of counterarguments and rebuttals might be counted (Nussbaum & Asterhan, 2016) without deep assessment of the content (although see Kuhn, 1991, for some more epistemic criteria, albeit with unclear validity). Compared with such norms, few people attain proficiency (for discussion, see Asterhan & Schwarz, 2016). Individual characteristics such as level of education (Kuhn, 1991) or epistemological belief (Kuhn, Cheney, & Weinstock, 2000; Kuhn, Wang, & Li, 2010) predict success to some degree. For instance, there are "multiplists," who view facts as freely chosen opinions and tend not to engage in argumentation at all, but there are also "evaluativists," who view facts as judgments that can be evaluated against standards of evidence (Kuhn et al., 2000; Kuhn et al., 2010). This literature could be enriched by considering other norms and a theory-neutral definition of skill.

Another potentially rich source of evidence is intervention studies: studies that set out to improve argumentative skills. These studies can offer data on conventional validity through their control groups. As with individual-difference studies, the data are more suggestive than decisive. A growing literature demonstrates that intervention can increase adherence to procedural norms, in particular, increase the frequency of attempts to make claims, justify them, respond to other arguments, and compare competing arguments (Crowell & Kuhn, 2011, 2014). Successful interventions range from an intensive and complex three-year program with children (Crowell & Kuhn, 2014; Kuhn & Crowell, 2011) to an hourlong intervention with undergraduate students (Zavala & Kuhn, 2017). Perhaps the most striking evidence comes from studies that demonstrate transfer from argumentation tasks to other tasks, such as from training in mathematical argumentation to scores on standardized tests of mathematics and reading (Nussbaum & Asterhan, 2016). Such assessments are theory neutral regarding argumentation. Thus far, there do not seem to be equivalent studies testing interventions based on argumentation schemes or Bayesian norms. But there is no reason to consider such interventions impossible, and they would pave the way for contrastive studies.

Strikingly, the empirical literature does not seem to have asked whether particular norms help resolve disputes. While this question is sometimes considered conceptual (van Eemeren et al., 2009), it seems straightforward to imagine its empirical study. There is already a large literature on group judgment and decision making, exploring both how people interact to reach a single judgment or decision (for a review, see Kerr & Tindale, 2004) and the extent to which argument improves performance (e.g., Mercier, Trouche, Yama, Heintz, & Girotto, 2015). Such methods could be adapted to test whether compliance with sets of norms improves the likelihood or efficiency of arriving at a consensus. The consensus answers could be assessed in the usual way, to test whether these norms increase success in tracking the truth.

5. Conclusions

The psychology of argumentation offers valuable, if limited, data. There is evidence that people implicitly endorse both procedural and epistemic norms for argumentation. Argumentation seems rational in the sense that it is a considered, rule-governed behavior. Moreover, data suggest that (at least some) people approximate probabilistic norms when evaluating arguments—behavior that has deeper conceptual justification and a stronger claim on rationality.

Nevertheless, much remains to be done. Probabilistic norms, in particular, lend themselves to constrained models, which offer substantial advantages over existing approaches, where much is left unspecified, rendering the models hard to falsify (for discussion, see Schellens et al., 2017). Theories and models of argumentation can, likewise, link pragmatic reasoning to specific forms of argument, offering increased specificity to the study of natural-language pragmatics (Macagno & Walton, 2013).

In sum, we have considered the merits of studying the psychology of argumentation, its connections with the psychology of reasoning, and the central feature they both share: the interdependence of descriptive and normative considerations.

Note

1. Sometimes, "arguments-as-objects" may also be used to refer to just the premises or "reasons" themselves. It should be clear from the context which use is intended.

References

Aberdein, A. (2010). Virtue in argument. *Argumentation, 24*(2), 165–179.

Asterhan, C. S. C., & Schwarz, B. B. (2016). Argumentation for learning: Well-trodden paths and unexplored territories. *Educational Psychologist, 51*(2), 164–187.

Bhatia, J.-S., & Oaksford, M. (2015). Discounting testimony with the argument ad hominem and a Bayesian congruent prior model. *Journal of Experimental Psychology: Learning, Memory, and Cognition, 41*(5), 1548–1559.

Christmann, U., Mischo, C., & Flender, J. (2000). Argumentational integrity: A training program for dealing with unfair argumentative contributions. *Argumentation, 14*(4), 339–360.

Collins, P. J., & Hahn, U. (2016). Arguments and their sources. In F. Paglieri, L. Bonelli, & S. Felletti (Eds.), *The psychology of argument: Cognitive approaches to argumentation and persuasion* (pp. 129–149). London, England: College Publications.

Corner, A., & Hahn, U. (2013). Normative theories of argumentation: Are some norms better than others? *Synthese, 190*(16), 3579–3610.

Corner, A., Hahn, U., & Oaksford, M. (2011). The psychological mechanism of the slippery slope argument. *Journal of Memory and Language, 64*(2), 133–152.

Crowell, A., & Kuhn, D. (2014). Developing dialogic argumentation skills: A 3-year intervention study. *Journal of Cognition and Development, 15*(2), 363–381.

Evans, J. St. B. T., Neilens, H., Handley, S. J., & Over, D. E. (2008). When can we say 'if'? *Cognition, 108*(1), 100–116.

Garssen, B. J. (1997). *Argumentatieschema's in pragma-dialectisch perspectief: Een theoretisch en empirisch onderzoek* [Argumentation schemes in the pragma-dialectical perspective: A theoretical and empirical inquiry]. Amsterdam, Netherlands: IFOTT.

Girotto, V., Kemmelmeier, M., Sperber, D., & van der Henst, J.-B. (2001). Inept reasoners or pragmatic virtuosos? Relevance and the deontic selection task. *Cognition, 81*(2), B69–B76.

Hahn, U. (2011). The problem of circularity in evidence, argument, and explanation. *Perspectives on Psychological Science, 6*(2), 172–182.

Hahn, U., Bluhm, R., & Zenker, F. (2017). Causal argument. In M. R. Waldmann (Ed.), *The Oxford handbook of causal reasoning* (pp. 475–494). Oxford, England: Oxford University Press.

Hahn, U., & Hornikx, J. (2016). A normative framework for argument quality: Argumentation schemes with a Bayesian foundation. *Synthese, 193*(6), 1833–1873.

Hahn, U., & Oaksford, M. (2006). A Bayesian approach to informal reasoning fallacies. *Synthese, 152*, 207–223.

Hahn, U., & Oaksford, M. (2007). The rationality of informal argumentation: A Bayesian approach to reasoning fallacies. *Psychological Review, 114*(3), 704–732.

Hahn, U., & Oaksford, M. (2012). Rational argument. In K. J. Holyoak & R. G. Morrison (Eds.), *The Oxford handbook of thinking and reasoning* (pp. 277–298). Oxford, England: Oxford University Press.

Hamblin, C. L. (1970). *Fallacies*. London, England: Methuen.

Harris, A. J. L., Hahn, U., Madsen, J. K., & Hsu, A. S. (2016). The appeal to expert opinion: Quantitative support for a Bayesian network approach. *Cognitive Science, 40*(6), 1496–1533.

Harris, A. J. L., Hsu, A. S., & Madsen, J. K. (2012). Because Hitler did it! Quantitative tests of Bayesian argumentation using ad hominem. *Thinking & Reasoning, 18*(3), 311–343.

Hastings, A. C. (1962). *A reformulation of the modes of reasoning in argumentation* (Unpublished doctoral dissertation). Northwestern University, Evanston, IL.

Kadane, J. B., & Schum, D. A. (1996). *A probabilistic analysis of the Sacco and Vanzetti evidence*. Mahwah, NJ: Wiley.

Kerr, N. L., & Tindale, R. S. (2004). Group performance and decision making. *Annual Review of Psychology, 55*(1), 623–655.

Kienpointner, M. (1992). *Alltagslogik: Struktur und Funktion von Argumentationsmustern* [Everyday logic: The structure and function of argumentation patterns]. Stuttgart-Bad Cannstatt, Germany: Frommann.

Kuhn, D. (1991). *The skills of argument*. Cambridge, England: Cambridge University Press.

Kuhn, D., Cheney, R., & Weinstock, M. (2000). The development of epistemological understanding. *Cognitive Development, 15*(3), 309–328.

Kuhn, D., & Crowell, A. (2011). Dialogic argumentation as a vehicle for developing young adolescents' thinking. *Psychological Science, 22*(4), 545–552.

Kuhn, D., Wang, Y., & Li, H. (2010). Why argue? Developing understanding of the purposes and values of argumentive discourse. *Discourse Processes, 48*(1), 26–49.

Leitgeb, H., & Pettigrew, R. (2010a). An objective justification of Bayesianism I: Measuring inaccuracy. *Philosophy of Science, 77*(2), 201–235.

Leitgeb, H., & Pettigrew, R. (2010b). An objective justification of Bayesianism II: The consequences of minimizing inaccuracy. *Philosophy of Science, 77*(2), 236–272.

Macagno, F., & Walton, D. (2013). Implicatures as forms of argument. In A. Capone, F. L. Piparo, & M. Carapezza (Eds.), *Perspectives on pragmatics and philosophy* (pp. 203–225). Dordrecht, Netherlands: Springer.

Manktelow, K. (2012). *Thinking and reasoning: An introduction to the psychology of reason, judgment and decision making*. Hove, England: Psychology Press.

Mercier, H., & Sperber, D. (2011). Why do humans reason? Arguments for an argumentative theory. *Behavioral and Brain Sciences, 34*(2), 57–74.

Mercier, H., & Sperber, D. (2017). *The enigma of reason*. Cambridge, MA: Harvard University Press.

Mercier, H., Trouche, E., Yama, H., Heintz, C., & Girotto, V. (2015). Experts and laymen grossly underestimate the benefits of argumentation for reasoning. *Thinking & Reasoning, 21*(3), 341–355.

Nussbaum, E. M., & Asterhan, C. S. C. (2016). The psychology of transfer from classroom argumentation. In F. Paglieri, L. Bonelli, & S. Felletti (Eds.), *The psychology of argument: Cognitive approaches to argumentation and persuasion* (pp. 407–422). London, England: College Publications.

Paglieri, F. (2015). Bogency and goodacies: On argument quality in virtue argumentation theory. *Informal Logic, 35*(1), 65–87.

Perelman, C., & Olbrechts-Tyteca, L. (1969). *The new rhetoric: A treatise on argumentation* (J. Wilkinson & P. Weaver, Trans.). Notre Dame, IN: University of Notre Dame Press.

Pettigrew, R. (2016). *Accuracy and the laws of credence*. Oxford, England: Oxford University Press.

Rosenkrantz, R. D. (1992). The justification of induction. *Philosophy of Science, 15*, 527–539.

Schellens, P. J. (1985). *Redelijke argumenten: Een onderzoek naar normen voor kritische lezers* [Reasonable arguments: A study of norms for critical readers]. Dordrecht, Netherlands: Foris.

Schellens, P. J., Šorm, E., Timmers, R., & Hoeken, H. (2017). Laypeople's evaluation of arguments: Are criteria for argument quality scheme-specific? *Argumentation, 31*(4), 681–703.

Schreier, M., Groeben, N., & Christmann, U. (1995). "That's not fair!" Argumentational integrity as an ethics of argumentative communication. *Argumentation, 9*(2), 267–289.

Sperber, D., Cara, F., & Girotto, V. (1995). Relevance theory explains the selection task. *Cognition, 57*(1), 31–95.

Stevenson, R. J., & Over, D. E. (2001). Reasoning from uncertain premises: Effects of expertise and conversational context. *Thinking & Reasoning, 7*(4), 367–390.

van Eemeren, F. H., Garssen, B., & Meuffels, B. (2009). *Fallacies and judgments of reasonableness: Empirical research concerning the pragma-dialectical discussion rules*. Dordrecht, Netherlands: Springer.

van Eemeren, F. H., & Grootendorst, R. (1984). *Speech acts in argumentative discussions: A theoretical model for the analysis of discussions directed toward solving conflicts of opinion*. Dordrecht, Netherlands: Floris Press.

van Eemeren, F. H., & Grootendorst, R. (1992). *Argumentation, communication, and fallacies: A pragma-dialectical perspective*. Mahwah, NJ: Erlbaum.

van Eemeren, F. H., & Grootendorst, R. (1995). The pragma-dialectical approach to fallacies. In H. V. Hansen & R. C. Pinto (Eds.), *Fallacies: Classical and contemporary readings* (pp. 130–144). Philadelphia: Pennsylvania State University Press.

van Eemeren, F. H., & Grootendorst, R. (2004). *A systematic theory of argumentation: The pragma-dialectical approach*. Cambridge, England: Cambridge University Press.

van Eemeren, F. H., & Houtlosser, P. (1999). Strategic manoeuvring in argumentative discourse. *Discourse Studies, 1*(4), 479–497.

Walton, D. N. (1989). Dialogue theory for critical thinking. *Argumentation, 3*(2), 169–184.

Walton, D. N. (2008). *Informal logic: A pragmatic approach* (2nd ed.). Cambridge, England: Cambridge University Press.

Walton, D. N., Reed, C., & Macagno, F. (2008). *Argumentation schemes*. Cambridge, England: Cambridge University Press.

Woods, J., Irvine, A. D., & Walton, D. N. (2004). *Argument: Critical thinking, logic and the fallacies*. London, England: Pearson/Prentice Hall.

Zavala, J., & Kuhn, D. (2017). Solitary discourse is a productive activity. *Psychological Science, 28*(5), 578–586.

Zenker, F. (2012). *Bayesian argumentation: The practical side of probability*. New York, NY: Springer Science & Business Media.

5.6 Reasoning and Argumentation

John Woods

Summary

Drawing on the work of informal logicians, speech communication scholars, rhetoricians, critical-thinking theorists, empirical and theoretical linguists, cognitive and experimental psychologists, dialogue theorists and dialectical logicians, and, more recently, computer scientists and artificial intelligence (AI) theorists, argumentation theory is a large and growing network of enquiry into natural-language argument and its tie to reason-giving and inference. This entry concentrates on theoretical developments from the late 1950s to the present day. My remarks are organized under the following headings: (1) Argument, (2) Reasoning, (3) Toulmin and the Rhetorical Turn, (4) Hamblin Contra the Logicians, (5) The Revival of Fallacy Theory, (6) Informal Logic Contra the Textbooks, (7) Pragma-Dialectics, and (8) More Recent Developments. The coverage is not complete. For want of space, I omit developments in speech communication, rhetoric, and critical thinking, concerning which see Blair (2019).

1. Argument

Logic was founded as a theory of argument. Aristotle's syllogistic logic was the first nonmonotonic, relevant, and paraconsistent logic, and some decent approximation to an intuitionist one, each of which today is termed "nonclassical" (Woods, 2015/2018). Aristotle was also the originator of a metalogic in which he produced a near-perfect proof of the semi-decidability of validity in syllogistically formulated arguments[1] (Corcoran, 1972). The point of mentioning this here is that in most of the approaches to argument under review here, decidability proofs are neither sought nor possible.

Aristotle also placed the analysis of fallacies—*paralogismoi*—on the front burner of logical theory. In 1879 and following, however, logic changed course. Its new purpose was to provide secure foundations for number theory in a newly created second-order quantificational system in which fallacy theory had no place[2] (Frege, 1884/1950).

As we have it now, argumentation theory is an investigation of how in various contexts conclusions are arrived at from premises. Conclusions take different forms—about what is true or false, probable or improbable, plausible or implausible, persuasive or unpersuasive, about what is best to do, most prudent to do, and so on. Although human conclusion-drawing need not be argumentally structured, it is widely believed that premiss–conclusion reasoning can insightfully be *modeled* as premiss–conclusion argument (see, e.g., Mercier & Sperber, 2017).

Two further distinctions call for attention: the first recognizes three different ways in which the consequence-relation manifests itself—consequence-*having* (or entailment), consequence-*spotting* (or recognition), and consequence-*drawing* (or inference). Having is a binary relation between premises and conclusion. Spotting and drawing require *agents* and hence are ternary relations (Woods, 2013/2014). It is broadly accepted that a theory that works well for entailment also works well for inference. In fact, this is seriously mistaken (Harman, 1970).

While the Trinitarian character of consequence has received little recognition in the literature, our second distinction, that between an argument *as-such* and an argument *in-use*, is at least implicitly grasped. An argument as-such is a finite sequence of statements whose terminal member, the conclusion, is a consequence of the other members, its premises. Arguments in-use are what happens to arguments as-such when goal-directed agents engage them in support of their various ends. Logics of arguments as-such concentrate on the consequence-having relation (Hitchcock, 2017, chapter 7; Woods, 2011). Logics for consequence-spotting and for consequence-drawing are agent centered, goal directed, and (frequently) dynamically interactive.

The word "argument" is an open-textured term, subject to the loose constraints of family resemblance, hence a term that precludes definition by way of necessary and sufficient conditions. The extension of that term is large and sprawling, stretching from conversationally expressed differences of opinion to counsels' closing summations at trial, newspaper editorials and political cartoons, partisan placards, World Trade Organization negotiations, and nightly *à table* screaming matches between Bill and Sue. Use of the word "argumentation" adds no clarity to this already fuzzy assortment. This is just to say that "argumentation" is no improvement as a synonym of "argument." Even so, in some uses, argumentation is a *theory* of argument, and in others, it is a *theory of* theories of argument (Blair, 2012, pp. 198–199).

Rhetorical studies range more broadly, often with insightful probes into the give-and-take of arguments at their least civilized. Theorists of argument are responsive to varying standards: they can aim for normative accuracy. They can aim for descriptive accuracy. Sometimes they aim for both. A descriptively accurate theory of arguments of kind K is an objectively true account of how K-arguments are made and how they play out in real life. A normatively accurate theory of arguments of kind K is an account of what it takes for a well-made K-argument to satisfy what are usually implicit norms of goodness. It would be seriously wrong to assume that these hoped-for goals are only the ones we see listed in the introductory textbooks—deductive validity and inductive strength. A glaring omission is the logic of abductive reasoning (Gabbay & Woods, 2005).

Argument kinds make for a loose and somewhat fuzzy taxonomy, neither exclusive nor exhaustive, and with lots of cross-border traffic. In none of its subdisciplines is there anything resembling full coverage of the extension of the word "argument." In most cases, there are features of it that even the better accounts ignore or downplay. In the dominant approaches, the theorist picks what he takes to be a manageable subset of the word's capacious denotation, with as little offence as possible to the word's also large connotation. This is a perfectly reasonable thing to do and an instructive reminder of the rule not to bite off more than one can chew. Still, subset choices are at some risk of selection bias, which might, in turn, incline a theorist to presume that the theory of his own bite of the pie is in some ways canonical.

2. Reasoning

There is a question about arguing and reasoning. Are they One or Many? On the One view, arguing is just reasoning out loud, whereas reasoning is just arguing with oneself privately. The Many view is supported by the fact that young human beings routinely draw inferences before acquiring the linguistic means for making arguments. Reasoning is typically a search for conclusions from premises at hand, whereas arguing is often a search for premises to support some already hoped-for conclusion. Robert Pinto (1995) sees an argument as an invitation to draw an inference, that is, to adjust the invitee's belief-state in accordance with how its premises stand to its conclusion and he, the invitee, stands to its premisses. While Sue can invite Bill to draw the conclusion she sees as following from her proffered premises, Bill is often able to draw his own inferences without invitation and also without the need of having to make the case to himself for drawing it. It would be an overreach to dismiss these solo inferences as inherently defective or subpar. Indeed, the frequency of argumentally *un*invited inferences considerably exceeds the frequency of invited ones (Woods, 2013/2014).

Another still unsettled question are the relations between good reasoning, good reason-giving, and the exercise of *rationality*. On one finding, a reasoner is reasoning well when (or to extent that) he is reasoning in the way that a rational agent does (or would). Likewise, someone is good at giving good reasons for something when (or to the extent that) she does so in the way that a rational agent does (or would) in those same circumstances. Especially pinching is the issue of how to understand situations in which a theory's rules of rationality are *systematically* disobliged by the reasoning-behavior of actual human beings. Does massive discompliance discredit the theory, or can the theory save itself by invoking its normative authority over human behavior (see chapter 2.5 by Klauer, this handbook)?

There are two different ways of approaching the normativity question, not only in the theory of argument but in all the sciences of both empirically discernible and normatively assessable human behavior. Some hold that normative rules are what the theorists themselves take the *common intuitive* view of them to be (Johnson, 2000). Others hold that rational performance is how the *ideally rational* agent would perform. Typical of these are Bayesian epistemologists, classical decision theorists, and neoclassical economists (see chapter 4.5 by Chater & Oaksford and chapter 9.4 by Dhami & al-Nowaihi, both in this handbook). Floating in these precincts are ideal norms to which no human being could approximate in any finite degree: one provides that the ideally rational agent closes his beliefs under truth-preserving consequence. Another provides that when an ideally rational decision is taken, the decider has full command of all information of material significance. Neoclassical economists postulate that an ideally rational economist is subject to the infinite divisibility of her subjective utilities, largely because this enables economics to harness the mathematical firepower of the calculus. In each of these cases, it is put about that any real-life human agent is performing rationally to the extent that he approximates to the ideal norms. So conceived of, the norms are accurate descriptions of how the ideally rational agent performs and an inaccurate description of how we ourselves perform. Even so, the story goes, the empirically false postulates are *normatively* binding upon us. This is not a well-received doctrine in by far most of the sectors of argumentation theory that are reviewed in this note. Of course, this is not to deny the approximable standards of goodness implicit in cognitive practice. For a good review, see the Introduction to this volume by Knauff and Spohn.

3. Toulmin and the Rhetorical Turn

In 1958, persuasion-arguments had a flourishing home in rhetoric but had largely disappeared from logic's theoretical halls.[3] This would start to change when Toulmin became aware of speech communication studies and informal logicians became aware of Toulmin.[4] In the textbooks of the time, natural-language arguments were mainly conceived of in the as-such way. Logic's task was to provide mapping rules[5] from subsets of real-life arguments to structures of entities constructed from elements of an uninterpreted formal language. Called "logical forms," the idea was that if a property of interest for a natural-language argument—validity, say—could be shown to have a formal counterpart instantiated by the original argument's logical form, then the formal property's natural-language counterpart would also be instantiated there.[6] This is problematic: it is easily shown that there is no strict one-to-one correspondence between valid natural-language arguments and arguments having a valid logical form in some or other logistic system.[7] Toulmin made a twofold departure in *The Uses of Argument.* He abandoned the mapping-to-logical-form test for properties of logical interest, and he abandoned the preoccupation with argument as-such and turned toward arguments in-use.

Toulmin is best known for the model of argument that bears his name. Its core idea is that arguing is case-making and that cases are made by human beings for the entertainment and assessment of other human beings. When an argument is under way, its would-be conclusion is a "claim," a proposition on behalf of which a case is being made and whose having been made is a condition of its meriting acceptance. Claims (C) are supported by "grounds" (G), which take the form of data, facts, or evidence. Grounds answer challenges claims. A "warrant" (W) is a statement that in turn endorses the support rendered by the ground. Claims, grounds, and warrants are necessary conditions for something's being an argument. The next three parameters are discretionary, depending on context. A "backing" (B) lends support to a warrant if it's in doubt. A "rebuttal" (R) is a hedging device, indicating that in default of contrary indications, the backing is secure. A "qualifier" (Q) modifies an arguer's confidence at any given point. When these conditions are met, the first three always and the next three as may be, the arguer has made an argument for his claim. Toulmin's own example of an argument that fits his model can be set out as follows: "Nigel is a British citizen" (C), "Nigel was born in Bermuda" (G), "A person born in Bermuda is a legal citizen of Britain" (W), "Nigel trained as a barrister in London, specializing in citizenship, so he knows that a person born in Bermuda is a legal citizen of Britain" (R). "Unless, that is to say, he has betrayed Britain and spied for another country." Q is a discretionary prefix, attachable as context might indicate.

It is puzzling why warrants don't serve as premises in the arguments they support. Toulmin concedes that Bermudians are British citizens and Nigel is a Bermudian. He allows that it follows that Nigel is a British citizen, but it follows from "Nigel was born in Bermuda" alone but *not* in the absence of the fact that Bermudans are British citizens. Although Toulmin didn't adequately dig into what brings this to pass, he was on to something important. Not everything necessary for an argument's logical success is suitable for use as a premise (Carroll, 1895; Ryle, 1950) or, as in some cases, even capable of being *formulated* as one.[8] The reason why is that background information is often implicit and tacit (Polanyi, 1966; Woods, 2021).

4. Hamblin contra the Logicians

Toulmin was disappointed in the quality of the introductory logic textbooks at midcentury. Twelve years later, the Australian philosopher Charles Hamblin (1970) turned the disappointment into a *scandale.* He scorned the fallacy chapters of the leading texts of the day for their silly and unrealistic examples. The textbooks recycled what their predecessors had to say of the fallacies, as they in turn recycled what their own predecessors said, resulting in a dog's breakfast "so incoherent that we have every reason to look for some enlightenment at its historical source" (Hamblin, 1970, p. 50). Hamblin follows his own advice with admirable effect. His book *Fallacies* traces the history of fallacy theory from its inception in antiquity, revealing its presence in every succeeding era until its post-Fregean fade-out in the last fifth of the 19th century. The survey shows the adaptability of Aristotelian themes to the freshly arising insights of his successors. *Fallacies* betrays no impatience with formal logic. Hamblin was drawn to a formalized dialectical model of argument and devoted two chapters to it. In so doing, he clearly caught a wave. Formal dialectic was on its way to being a growth industry (see, e.g., Henkin, 1961; Hintikka, 1968; Lorenz, 1961; Lorenzen, 1960; Lorenzen & Lorenz, 1978; Rescher, 1977).

Fallacies had a second and more serious indictment to file: the reason there is no scholarship or depth in those pallid fallacy chapters is that logicians had betrayed their heritage. Aristotle invented their subject and placed the fallacies project at its dead center. But as Hamblin dolefully notes, "We have no *theory* of fallacy at all, in the sense in which we have theories of correct reasoning or inference" (Hamblin, 1970, p. 11). *Fallacies* was a *cri de cœur* for logicians to reattach the fallacies project to logic. Unlike Toulmin, Hamblin wasn't reaching out to the speech communication community. Hamblin was calling upon logicians to produce a thick theory of our mistaking something bad for something good, attended by the means to stop doing it. It is a given that an intellectual instrument purpose-built for the logicism of Frege, Whitehead, and Russell couldn't be of much use in the analysis of *paralogismoi*.[9] A logician whose only professional interest lay in the foundations of mathematics isn't likely to have much working capital or appetite for fallacy theory. Hamblin is perfectly aware of this. His point is that securing the foundations of arithmetic shouldn't be the sole business of logic.

One of the merits of Hamblin's indispensable book is the care with which it plots the historical course of fallacy theory, and thereby of logic itself, from the 4th century B.C. to the present day. Its author also found space for a chapter on the Indian tradition in logic. A further virtue is the respect it shows medieval contributions to logic, from which Hamblin draws considerable inspiration for his own advances in formal dialectic, in his chapter 8. As already mentioned, Hamblin is in the vanguard of a more general revival of formal dialogue and dialectical logic.[10]

It cannot be said that logicians flocked to Hamblin's challenge and still less that he himself produced the theory he called for. I say this notwithstanding the rightful interest that chapter 8 has attracted to the Hamblin Game H. All the same, from that day to this, not a single article on fallacies has yet appeared in the *Journal of Symbolic Logic*.

5. The Revival of Fallacy Theory

In 1972, there appeared the first of a series of papers by John Woods and Douglas Walton on the fallacies that would extend to the mid-1980s. Woods was a philosophical logician working on the modal character of the entailment relation, and Walton, his former student at the University of Toronto, was an action theorist. In 1971, they were briefly in San Francisco, where they resolved to reserve the next two years to clear up the fallacies and then get back to their other business. Forty-seven years later, they are both still writing about fallacies. Woods and Walton accepted Hamblin's insistence that fallacy theory is the proper business of logic. For each fallacy, they would seek an existing theoretical framework and adapt it to model the logical structure of the fallacy at hand. Theirs was a benign and not uncommon parasitism, mining the tried and true for some not originally intended good. *Fallacies: Selected Papers 1972–1982* (Woods & Walton, 1989/2007) makes selective use of intuitionist, modal, and relatedness logics; causal and epistemic logics; and erotetic (interrogative), deontic, and dialectic logics. The book's chapter 10 shows how a cumulative model of enquiry based on the semantics of Kripke's intuitionist logic can block circular arguments.[11] Chapter 8 appropriates Burge's (1977) aggregate theory for the analysis of the composition fallacy. Scattered throughout the book are adaptations of numerous frameworks for dialogue games, some of their own making.[12] Overall, the Woods–Walton Approach (as it came to be called) embodies the maxim that if the foot of a fallacy fits the shoe of a suitably retrofitted logic, then the fallacy should wear it. It is a pluralistic approach eschewing any notion that a beautiful treatment of some one fallacy could be made canonical for how to treat them all. Some critics, notably Rob Grootendorst, protested that on the Woods–Walton Approach (WWA), every fallacy has its own logic. Just so!

In chapter 1 of Woods and Walton's *Selected Papers*, first published in 1972, it was shown that the truth conditions on the logical consequence relation cannot serve as reliable rules of deductive inference.[13] A running theme of the WWA is that the informal fallacies, if fallacies they be, are not inherently fallacious in virtue of syntactic form. Whether an argument or piece of reasoning is fallacious is always affected by contextual particularities, such as the goal-directedness of its parts. What is true of the logics of the WWA is true of informal logics across the board: they are practically oriented logics. What the WWA also had, some of the others would reject as overconfidence in the assumption that philosophical treatments of the fallacies are best achieved under the gravitational pull of successful preexisting theories.[14]

The Woods–Walton papers were not by any means the only ones of note to appear in the close aftermath of 1970.[15] For a while, the WWA dominated by force of numbers, if not always the press of merit. In time, it would be clear that the greater value of the WWA was the stimulus it provided for further work on fallacious reasoning, some at rather substantial levels.[16] Not the least of these developments are Walton's 40 or so books either directly on the fallacies or significantly related to them. Since the early days, each of the WWA parties has parted company with their shared model, what with Walton's circumspect expansions of the pragma-dialectic model (see below) and Woods's move to a naturalized logic in company with a causal-response epistemological reliabilism (Woods, 2013/2014; chapter 3.2 by O'Brien, this handbook).

6. Informal Logic contra the Textbooks

Toulmin didn't like the textbooks of the 1950s, and Hamblin abjured the ones of the 1970s. The two leading texts of those times were Irving Copi's *Introduction to Logic* (1953) and *An Introduction to Logic and Scientific Method*, by Morris Cohen and Ernest Nagel (1934). Neither deserved the scorn that would be heaped upon them, notwithstanding the slightness of their fallacies coverage. In 1971, something consequential happened: Howard Kahane published *Logic and Contemporary Rhetoric*. Spurred by the student-power turbulence of the day, the standard textbooks were scorned for their irrelevance to the current upheavals. Kahane's book reflected two facts welcomed by its new readers: one was that mainstream symbolic logic had strayed too far from the give-and-take of good and bad reasoning on the ground. The other was that rhetoricians had stayed closer to home and offered a broader, more varied, and more appealing assortment of argument strategies. The relevance revolution was not confined to introductory logic texts,[17] but it was the spur that motivated teachers of introductory courses to write new and more responsive ones. The now-spurned old primers would be placed on their *Index Librorum Prohibitorum*.[18] Correspondingly, teachers of introductory logic courses were motivated to push back against what they rightly saw entirely as the expropriation of logic by mathematicians. The two developments were joined at the hip, one to "relevantize" the textbooks, the other to "demathematicize" logic and restore it to its former primacy in philosophy. Jointly, the two developments marked the birth of late-20th-century informal logic.

Beginning with their *Logical Self-Defence* in 1977, Ralph Johnson, J. Anthony Blair, and their colleagues established at the University of Windsor (Ontario) the organizational and communicational infrastructure for informal logic (IL). In so doing, they displayed an enduring steadfastness and remarkable networking skills. Windsor is home base for triennial meetings of the Ontario Society for the Study of Argumentation and also the place of publication of *Informal Logic*, the journal of record for that subject, and its related monograph series, *Windsor Studies in Argumentation*. Informal logicians have never seen fit to band together and issue manifestos. Windsor is head office infrastructurally but not doctrinally. Perhaps the better part of what explains this openness is that informal logicians tend rather more to agree on what they *don't* like and rather less on what they *do* like.

One point of near-universal agreement is IL's hostility to the suggestion conveyed by the textbooks of the day to the effect that first-order mathematical logic gives the best way to represent and evaluate deductively valid inferences as they occur in real life (Govier, 1987; Scriven, 1980). It is strange that on this very point, there is scant reference to Harman (1970). One can also see in the dissatisfactions of 1970 and onward a reaction to the entry of *formal semantics*, in the manner of Tarski, to the philosophy of language. Since language is the principal medium of argument, it is only reasonable to expect that its philosophical theorists would have something insightful to say about it. To put it succinctly, informal logicians were resisting the idea that the best way to construct a philosophical theory of real-life argument is to craft it in the way that Tarski did for the natural-language truth-predicate in "The Concept of Truth in Formalized Languages" (1935/1983), on one reading of which the natural-language predicate "true" comes out as transfinitely ambiguous. In their determination *not* to produce this sort of semantics for the concept of argument, informal logicians do themselves credit in respecting the realities of human speech.

It is virtually impossible to produce a nontrivial overview of the characteristic substance of informal logic, or in what its intellectual distinctiveness lies. Some writers concentrate on the premissory end of argument, others on the conclusional end, and others still on the premiss–conclusion link. Beyond the truisms, there is little by way of solid consensus. For example, Johnson's and Blair's criteria for argumental goodness—premiss-acceptability, premiss–conclusion relevance, and sufficiency of premisses for conclusion-drawing—have had a long play. But, as they say, the devil is in the details, concerning which there are interesting differences of opinion:[19] some informalists see themselves as normative theorists, while others are more interested in descriptive accuracy. Some think that informal logic won't flourish unless it stays on top of developments in epistemology.[20] Others insist that the target of good argument is interparty concordance, not knowledge, even if truth might sometimes be a contingent collateral benefit of it.[21] Further unsettled questions are whether argument and inference are essentially the same, whether argumentation theories should be empirically sensitive, whether arguments are inherently (or even mainly) dialectical—that is, attack-and-defend contests—and whether courses in informal logic enhance one's critical-thinking skills (Blair, 2019). Some emphasize the importance of taking metadialogues and meta-arguments seriously into account (Finocchiaro, 2013; Krabbe, 2003). However, this is not to say that the informal logic idea is beyond informative articulation (see, e.g., Hitchcock, 2007; Johnson, 1996).

7. Pragma-Dialectics

If informal logic carries a nominal Canadian identity, pragma-dialectics (PD) has a more substantial Dutch one. It began with van Eemeren and Grootendorst (1984) as a simplified version of the sort of theory advanced in Barth and Krabbe (1982), a difficult and mathematically sophisticated book entitled *From Axiom to Dialogue*. The pragma-dialectical book—*Speech Acts in Argumentative Discussion*—reflected the authors' confidence that conflict-resolution arguments could successfully be modeled without the encumbrance of complex mathematicized formalisms. The counterpart city to Windsor is Amsterdam: it is the conference home of the International Society for the Study of Argumentation (ISSA) and of the journal of record *Argumentation* and its related monograph series. People who work in informal logic tend to have more arrows in their quiver than the Amsterdamers do. While many informal logicians write about other things,[22] pragma-dialectics is the sole preoccupation of those who espouse it. While it is true that there is no party line in *Argumentation*—the Amsterdam journal—the same is not true of its practitioners. I have sometimes heard it bruited in Amsterdam that there is a reason for this difference: the pragma-dialectical approach has produced a canonical theory, and the informal logic approach has not. It takes at least two conditions for a theory to achieve canonical status: it must

stand tall in the marketplace of ideas, and it must have a large, durable, and approving readership. There are certainly widely acclaimed individuals in the IL community, and some of their views have had more staying power than others. True, but the PD theory is a *house-brand*, and nothing in IL is quite that.

The PD theory has two main parts: the first is a normative theory of critical discussions furnished by an ideal model of them. The second part of the PD theory incorporates, under the heading "strategic maneuvering," features derived from rhetoric and designed to make critical discussions persuasive. I'll sketch the first and most central part of the theory, beginning with the 10 critical discussion rules. The first thing to note is a change in the meaning of "fallacy." A fallacy is now conceived of as *any* violation of the discussion rules.

Paraphrasing van Eemeren, Grootendorst, and Snoeck Henkemans (2002), the rules can be paraphrased as follows:

- Parties are free to advance standpoints and also to challenge them.
- The burden of proof should lie on him who has advanced the standpoint.
- Attack- and defend-arguments must be on topic.
- Parties must not deny their own unexpressed premisses or misattribute an unexpressed premiss to the other.
- Parties must not use premises not accepted by the other party or deny a premiss accepted by their interlocutors.
- The only conclusive defense of a challenged standpoint is by way of an appropriate argumentation scheme correctly applied.
- An argument is conclusive only if it is valid or can be made so by the explicitization of implicit premises.
- If a defense fails, it must be withdrawn, and if it succeeds, the challenge must be retracted.
- Moreover, parties must not speak in an unclear or confusing manner and must try to interpret one another fairly.

The rules themselves jointly constitute a standpoint. It is disputed whether it satisfies its own rule 9.

The PD approach has its critics, of course. Here are some of the main misgivings: (1) It is not clear whether the rules biconditionally define the PD's own concept of "critical discussion." (2) If they do, it is not clear whether, or to what extent, the PD model is actually *instantiated* in conversationally expressed differences of opinion in real life. (3) If the theory is empirically inadequate, it is not clear whether it could be saved. (4) Could it be saved by its status in an *ideal model*? (5) Could it be saved by the ideal model's *normative authority*? (6) The deductivist validity condition is problematic: *any* invalid argument can be validated by adding as a premiss a conditional whose antecedents are its other premisses and whose conclusion is its conclusion. (7) Finally, how well does the model "travel"? Concerning this last question, it is put about that the PD model is effective in the analysis and evaluation of collective-bargaining treaty negotiation, parliamentary debates, political arguments, arguments about health policy, and visual argumentation. I am bound to say in reply that, however questions (1) to (6) are to be answered, in all good conscience the answer to (7) must be that it doesn't travel especially well.

Like all those in the reference class of this note, invocations of idealized normative models leave two critical chores undone: one is to specify the conditions under which an account of something of a kind *K* is a "model" of its behavioral instantiations. Second, what makes the model an "idealization" of *K*'s behavioral instantiations? Wherein lies its "normative authority" to bind in our own behavioral instantiations of *K*? For a thorough and carefully balanced view of these and related matters, one could do no better than consult Blair (2012, chapter 20).

This leaves the question, Why does the PD approach have such appeal? A pair of answers come to mind. The less serious one is this: it is said in Windsor and other like places that the appeal of the PD approach lies in its *easiness*. The more serious answer is that the PD approach's underlying pragma-dialectical character reflects some key truths about how human beings advance their views in reason-giving ways, for the consideration of other humans who think otherwise. If Blair is right, the difficulties with the PD approach might derive from its own *theoretical handling* of the underlying idea. If so, it could be that its supporters are rather more approving of the idea than of the theory.

8. More Recent Developments

Developments of importance include the expanding migration of argumentation theory to the analysis and evaluation of legal reasoning (Feteris, 1999; Prakken, 1997; Sartor, 1995; Walton, 2002; Woods, 2015/2018; Zenker, Dahlman, Bååth, & Sarwar, 2015), a renewed interest in argument by mathematical logicians (Barringer, Gabbay, & Woods, 2012; Gabbay, 2013; van Benthem, 2011), an enlarging rapprochement between logic and psychology (Oaksford & Chater, 2007; Paglieri, Bonelli, & Felletti, 2016; chapter 5.5 by Hahn &

Collins and chapter 5.4 by Gazzo Castañeda & Knauff, both in this handbook), and the hook-up between the argumentation schemes literature, on the one hand, and software engineering and AI, on the other. We also see in this alliance an interesting overlap with legal argumentation (Verheij, 2003): initially introduced by Perelman and Olbrechts-Tyteca (1958) and Hastings (1962), an argumentation scheme is a pattern of reasoning in everyday kinds of argument, adjoined to which are questions designed to facilitate the evaluation of arguments thus configured. For example, are the premises accurate? Are there some unexpressed but load-bearing premises? Might there be some allowable exceptions?[23] Under what assumption would the scheme in question intelligibly be construed as a successful form of argument? Would a reasonable answer to that question make it defeasibly reasonable to regard the assumption as provisionally true?[24]

Argumentation Schemes for Presumptive Reasoning (Walton, 1996) reveals a preoccupation with presumption. Walton tends to identify presumptive reasoning with defeasible reasoning. Some say that neither is, strictly speaking, a species of the genus "reasoning." A piece of reasoning is defeasible just because it doesn't close the world and therefore lies open to the possibility of defeat or weakening. Presumption—some say—is no kind of reasoning at all but rather a reasoner's disposition to view defeating conditions as surprising. In many of these patterns of fallacious argument, the questions they raise are thought to be helpful in distinguishing the fallacious from the nonfallacious. For Walton (2013), "a fallacy is an argument, a pattern of argumentation, or something that purports to be an argument, that falls short of some standard of correctness as used in a conversational context and poses a serious obstacle to the realization of the goal of the dialogue" (p. 215). In essence, his is the PD conception. Walton's further purpose was to show how argument schemes can help students in analyzing and judging arguments. In Walton (1996), there are 29 such schemes, but in Walton, Reed, and Macagno (2008), the list has swollen to 96, a daunting concordance for a student to keep on top of but no chore at all for a computer. It is ample evidence of Walton's penchant for a pluralism that gives a multitude of different ways for something to be a fallacy, no fewer than the number of a scheme's rules times 96.

Computer modeling presents all theorists of empirically discernible and normatively assessable behavior with interesting challenges. If the theorist in question seeks computer models of his account, he'll have to formulate them in a way that enables the software engineer to get hold of them. The argumentation schemes approach has had substantial appeal for AI engineers whose goal is to model them in ways that assist with "argument-mining" and multiagent analysis. This is brought about by feeding in answers to the questions raised by a scheme as further premisses of defeasible arguments now made valid by the additions. Notwithstanding the formidable reach of computer modeling, it is virtually impossible to bring off until data sets are suitably massaged for computational grasp. In so saying, we are back where we started. What wears the trousers here? The data or the mathematical models that seek them out?

I close with a remark from Christopher Tindale's Introduction to Blair (2012, p. xiv). Although specifically about Walton, it applies equally to the argument schemes approach in general:

> In the review of Walton he asks some fundamental questions of argumentation schemes: from where do they come, and from where do they derive their probative force? The origins of the Walton list had not been clear, and their cogency had been taken for granted by a number of theorists. But not by Blair. In similar fashion, he challenges the critical questions that accompany the schemes. What motivates these questions, and how do we know when a list is complete?

Acknowledgments

Thanks to Tony Blair and Erik Krabbe for helpful information and constructive suggestions and especially to my patient editor Wolfgang Spohn for his help in shrinking this note to within striking distance of the required word count. Blair was one of this submission's reviewers; he offered three suggestions, all of them adopted for the better.

Notes

1. A property is semi-decidable in a theory just in case there is a finitary algorithmic procedure for determining that something has it if indeed it does have it.

2. Roughly speaking a theory is of first order just in case its sole objects of reference are individuals. In a second-order theory, both individuals and properties or classes are objects of reference.

3. A still-influential work of that same year is Perelman and Olbrechts-Tyteca (1958).

4. Toulmin (1958), with Brockriede and Ehninger (1960) responding and Toulmin, Rieke, and Janik (1979) responding in turn.

5. Also mistakenly called "rules of *translation*."

6. Think also of entailment, deducibility, consistency, provability, equivalence, and interdeducibility.

7. Consider, for example, the valid argument, "The shirt is red, so the shirt is colored." There is no logistic system in which it maps to a valid form.

8. For example, is memory or background information helpful, even when not decomposable into well-individuated propositionally expressible units?

9. Logicism is the doctrine according to which every theorem of arithmetic is provable from the laws of symbolic logic.

10. In some of the more highly formalized ones, decidability and semi-decidability proofs are an intelligible, and sometimes even attainable, objective (van Benthem, 2011).

11. Cumulative models preclude the retraction of answers in a question-and-answer game. The preclusion originated in Aristotle's treatment of refutation arguments.

12. For example, Circle Games, DD games, and the so-called Woods–Walton fragment.

13. *In medias res*, they were made aware that the same point had been made in Harman (1970).

14. See, for example, Rescher (1976).

15. See, for example, Massey (1975), Broyles (1975), Barker (1976), and Mackenzie (1979).

16. Later works of note include Walton (1984), Govier (1987), van Eemeren and Grootendorst (1992), Hansen and Pinto (1995), Johnson (2000), and Tindale (2007). See also the papers in Hitchcock (2017), Blair (2012), and Wohlrapp (2014). Also important is Plantin (2018).

17. The student-power movement of the time rewrote university governance policies throughout the Western world.

18. Oddly enough, the dismissal didn't stop informal logicians, especially in the early decades, from mining the indexed books to motivate their theories. As one critic observed, "Pretty low-hanging fruit!"

19. For premiss adequacy, see Freeman (2005). For relevance, see Gabbay and Woods (2003) and Walton (2004). For premiss sufficiency, see Woods (2013/2014, chapter 7).

20. Plumping for the justified-true-belief model of knowledge are Biro and Siegel (1992) and Freeman (2005). Plumping for the causal-response model of reliabilism is Woods (2013/2014). See also chapter 3.2 by O'Brien (this handbook).

21. See here van Eemeren and Grootendorst (2004). Essential counterreading is Blair (2012, chapter 20).

22. For example, Finocchiaro on Galileo, Siegel on scientific relativism, Walton on courage, Barth on Quisling, Govier on peace, and Woods on bioethics.

23. Schemes are prefigured as *topica* in Aristotle's monograph of that same name.

24. See also Hitchcock (2017, chapter 14), Rawman and Reed (2009), and Wyner (2016).

References

Barker, J. A. (1976). The fallacy of begging the question. *Dialogue, 15*, 241–255.

Barringer, H., Gabbay, D. M., & Woods, J. (2012). Temporal argumentation networks. *Argument & Computation, 2–3*, 143–202.

Barth, E. M., & Krabbe, E. C. (1982). *From axiom to dialogue: A philosophical study of logics and argumentation.* Berlin, Germany: De Gruyter.

Biro, J., & Siegel, H. (1992). Normativity, argumentation and an epistemic theory of fallacies. In F. H. van Eemeren, R. Grootendorst, J. A. Blair, & Willard, C. A. (Eds.), *Proceedings of the Second International Conference on Argumentation* (pp. 330–337). Amsterdam, Netherlands: SicSat.

Blair, J. A. (2012). *Groundwork in the theory of argumentation: Selected papers of J. Anthony Blair.* Dordrecht, Netherlands: Springer.

Blair, J. A. (Ed.). (2019). *Studies in critical thinking.* Windsor, Ontario, Canada: WSIA (Windsor Studies in Argumentation).

Broyles, J. E. (1975). The fallacies of composition and division. *Philosophy & Rhetoric, 8*, 108–113.

Burge, T. (1977). A theory of aggregates. *Noûs, 11*, 97–118.

Carroll, L. (1895). What the Tortoise said to Achilles. *Mind, 4*, 278–280.

Cohen, M. R., & Nagel, E. (1934). *An introduction to logic and scientific method.* New York, NY: Harcourt, Brace.

Copi, I. M. (1953). *Introduction to logic.* New York, NY: Macmillan.

Corcoran, J. (1972). Completeness of an ancient logic. *Journal of Symbolic Logic, 37*, 696–702.

Feteris, E. T. (1999). *Fundamentals of legal argumentation.* Dordrecht, Netherlands: Springer.

Finocchiaro, M. A. (2013). *Meta-argumentation: An approach to logic and argumentation theory* (Studies in Logic, Vol. 42). London, England: College Publications.

Freeman, J. P. (2005). *Acceptable premises: An epistemic approach to an informal logic problem.* New York, NY: Cambridge University Press.

Frege, G. (1950). *The foundations of arithmetic: A logico-mathematical enquiry into the concept of number* (J. L. Austin, Trans.). Oxford, England: Blackwell. (Original work published 1884)

Gabbay, D. M. (2013). *Meta-logical investigations in argumentation networks* (Studies in Logic, Vol. 44). London, England: College Publications.

Gabbay, D. M., & Woods, J. (2003). *A practical logic of cognitive systems: Vol. 1. Agenda relevance: A study in formal pragmatics.* Amsterdam, Netherlands: North Holland.

Gabbay, D. M., & Woods, J. (2005). *A practical logic of cognitive systems: Vol. 2. The reach of abduction: Insight and trial.* Amsterdam, Netherlands: North-Holland.

Govier, T. (1987). *Problems in argument analysis and evaluation.* Dordrecht, Netherlands: Foris.

Hamblin, C. L. (1970). *Fallacies.* London, England: Methuen.

Hansen, H. V., & Pinto, R. C. (1995). *Fallacies: Classical and contemporary readings.* University Park: Pennsylvania State University Press.

Harman, G. H. (1970). Induction: A discussion of the relevance of the theory of knowledge to the theory of induction (with a digression to the effect that neither deductive logic nor the probability calculus has anything to do with inference). In M. Swain (Ed.), *Induction, acceptance, and rational belief* (pp. 83–99). Dordrecht, Netherlands: Reidel.

Hastings, A. C. (1962). *A reformulation of the modes of reasoning in argumentation* (Unpublished doctoral dissertation). Northwestern University, Evanston, IL.

Henkin, L. (1961). Some remarks on infinitely long formulas. In *Infinitistic methods: Proceedings of the Symposium on Foundations of Mathematics, Warsaw, 2–9 September 1959* (pp. 167–183). Oxford, England: Pergamon Press.

Hintikka, J. (1968). Language games for quantifiers. In N. Rescher (Ed.), *Studies in logical theory* (pp. 46–72). Oxford, England: Blackwell.

Hitchcock, D. (2007). Informal logic and the concept of argument. In D. Jacquette (Ed.), *Philosophy of logic*, Vol. 5 of D. M. Gabbay & J. Woods (Eds.), *Handbook of the philosophy of science* (pp. 101–129). Amsterdam, Netherlands: Elsevier.

Hitchcock, D. (2017). *On reasoning and argument: Essays in informal logic and on critical thinking.* Cham, Switzerland: Springer.

Johnson, R. H. (1996). *The rise of informal logic: Essays on argumentation, critical thinking, reasoning and politics.* Newport News, VA: Vale Press.

Johnson, R. H. (2000). *Manifest rationality: A pragmatic theory of argument.* Mahwah, NJ: Erlbaum.

Johnson, R. H., & Blair, J. A. (1977). *Logical self-defense.* Toronto, Ontario, Canada: McGraw-Hill.

Kahane, H. (1971). *Logic and contemporary rhetoric.* Belmont, CA: Wadsworth.

Krabbe, E. C. W. (2003). Metadialogues. In F. H. van Eemeren, J. A. Blair, C. A. Willard, & A. F. Snoeck Henkemans (Eds.), *Anyone who has a view* (pp. 641–644). Dordrecht, Netherlands: Kluwer.

Lorenz, K. (1961). *Arithmetik und Logik als Spiele* [Arithmetic and logic as games] (Unpublished doctoral dissertation). Christian-Albrechts-Universität Kiel, Germany.

Lorenzen, P. (1960). Logik und Agon [Logic and agon]. In *Atti del XII Congresso Internazionale di Filosofia (Venezia, 12–18 Settembre 1958)* (Vol. 4, pp. 187–194). Florence, Italy: Sansoni.

Lorenzen, P., & Lorenz, K. (1978). *Dialogische Logik* [Dialogical logic]. Darmstadt, Germany: Wissenschaftliche Buchgesellschaft.

Mackenzie, J. D. (1979). Question-begging in non-cumulative systems. *Journal of Philosophical Logic, 8*, 117–133.

Massey, G. J. (1975). Are there good arguments that bad arguments are bad? *Philosophy in Context, 4*, 61–77.

Mercier, H., & Sperber, D. (2017). *Enigmas of reason*. Cambridge, MA: Harvard University Press.

Oaksford, M., & Chater, N. (2007). *Bayesian rationality: The probabilistic approach to human reasoning*. Oxford, England: Oxford University Press.

Paglieri, F., Bonelli, L., & Felletti, S. (Eds.). (2016). *The psychology of argument: Cognitive approaches to argumentation and persuasion*. London, England: College Publications.

Perelman, C., & Olbrechts-Tyteca, L. (1958). *La nouvelle rhétorique: Traité de l'argumentation* [The new rhetoric: Treatise on argumentation]. Paris: Presses Universitaires de France.

Pinto, R. C. (1995). *Argument, inference and dialectic*. Dordrecht, Netherlands: Kluwer.

Plantin, C. (2018). *Dictionary of argumentation*. London, England: College Publications.

Polanyi, M. (1966). *The tacit dimension*. Garden City, NY: Doubleday.

Prakken, H. (1997). *Logical tools for modelling legal argument: A study of defeasible reasoning in law*. Dordrecht, Netherlands: Kluwer.

Rahwan, I., & Reed, C. (2009). The argument interchange format. In I. Rahwan & G. R. Simari (Eds.), *Argumentation in artificial intelligence* (pp. 383–402). Boston, MA: Springer.

Rescher, N. (1976). *Plausible reasoning*. Assen, Netherlands: Van Gorcum.

Rescher, N. (1977). *Dialectics: A controversy-oriented approach to the theory of knowledge*. Albany: State University of New York Press.

Ryle, G. (1950). 'If,' 'so,' and 'because.' In M. Black (Ed.), *Philosophical analysis* (pp. 323–340). Ithaca, NY: Cornell University Press.

Sartor, G. (1995). Defeasibility in legal reasoning. In Z. Bankowski, I. White, & U. Hahn (Eds.), *Informatics and the foundations of legal reasoning* (Law and Philosophy Library, Vol. 21, pp. 119–157). Dordrecht, Netherlands: Springer.

Scriven, M. J. (1980). The philosophical and practical significance of informal logic. In J. A. Blair & R. H. Johnson (Eds.), *Informal logic, the first international symposium* (pp. 147–160). Inverness, CA: Edgepress.

Tarski, A. (1983). The concept of truth in formalized languages. In A. Tarski, *Logic, semantics, metamathematics: Papers from 1923 to 1938* (J. Corcoran, Ed.; 2nd ed., pp. 152–278). Indianapolis, IN: Hackett. (Original work published 1935)

Tindale, C. W. (2007). *Fallacies and argument appraisal*. Cambridge, England: Cambridge University Press.

Toulmin, S. E. (1958). *The uses of argument*. Cambridge, England: Cambridge University Press.

Toulmin, S. E., Rieke, R., & Janik, A. (1979). *An introduction to reasoning*. New York, NY: Macmillan.

van Benthem, J. (2011). *Logical dynamics of information and interaction*. New York, NY: Cambridge University Press.

van Eemeren, F. H., & Grootendorst, R. (1984). *Speech acts in argumentative discussions: A theoretical model for the analysis of discussions directed towards solving conflicts of opinion*. Dordrecht, Netherlands: Foris.

van Eemeren, F. H., & Grootendorst, R. (1992). *Argumentation, communication and fallacies: A pragma-dialectical perspective*. Hillsdale, NJ: Erlbaum.

van Eemeren, F. H., & Grootendorst, R. (2004). *A systematic theory of argumentation: The pragma-dialectical approach*. Cambridge, England: Cambridge University Press.

van Eemeren, F. H., Grootendorst, R., & Snoeck Henkemans, A. F. (2002). *Argumentation: Analysis, evaluation, presentation*. Mahwah, NJ: Erlbaum.

Verheij, B. (2003). Dialectical argumentation with argumentation schemes: An approach to legal logic. *Artificial Intelligence and Law, 11*, 167–195.

Walton, D. N. (1984). *Logical dialogue games and fallacies*. Lanham, MD: University Press of America.

Walton, D. N. (1996). *Argumentation schemes for presumptive reasoning*. Mahwah, NJ: Erlbaum.

Walton, D. N. (2002). *Legal argumentation and evidence*. University Park: Pennsylvania State University Press.

Walton, D. (2004). *Relevance in argumentation*. Mahwah, NJ: Erlbaum.

Walton, D. (2013). *Methods of argument*. New York, NY: Cambridge University Press.

Walton, D., Reed, C., & Macagno, F. (2008). *Argumentation schemes*. Cambridge, England: Cambridge University Press.

Wohlrapp, H. R. (2014). *The concept of argument: A philosophical foundation*. Dordrecht, Netherlands: Springer. (Original work published 2008)

Woods, J. (2011). Whither consequence? *Informal Logic, 31*, 318–343.

Woods, J. (2013). *Errors of reasoning: Naturalizing the logic of inference* (Studies in Logic, Vol. 45). London, England: College Publications. (Reprinted with corrections in 2014)

Woods, J. (2014). *Aristotle's earlier logic* (2nd rev. ed.). Paris, France: ISTE/Hermes Science. (Original work published 2001)

Woods, J. (2015). *Is legal reasoning irrational? An introduction to the epistemology of law* (Law and Society, Vol. 2). London, England: College Publications. (Reprinted with corrections in 2018)

Woods, J. (2021). Four grades of ignorance-involvement and how they nourish the cognitive economy. *Synthese, 198*, 3339–3368.

Woods, J., & Walton, D. (2007). *Fallacies: Selected papers 1972–1982*. London, England: College Publications 2007. (Original work published 1989)

Wyner, A. (2016). A functional perspective on argumentation schemes. *Argument and Computation, 7*, 113–133.

Zenker, F., Dahlman, C., Bååth, R., & Sarwar, F. (2015). Giving reasons *pro et contra* as a debiasing technique in legal decision making. In D. Mohammed & M. Lewinski (Eds.), *Proceedings of the First European Conference on Argumentation* (pp. 807–820). London, England: College Publications.

Section 6 Conditional and Counterfactual Reasoning

6.1 Conditional and Counterfactual Logic

William B. Starr

Summary

A logic aims to capture certain rational constraints on reasoning in a precise formal language that has a precise semantics. For conditionals, this aim has produced a vast and mathematically sophisticated literature that could easily be the topic of a whole volume of this size. In this chapter, the focus will be on covering the major logical analyses developed by philosophers, the key issues that motivate them, and their connection to views developed in artificial intelligence, linguistics, and psychology.

1. Analyzing Conditionals

Research on conditionals makes use of a few key terms and notation:

Conditional A sentence of the form "If A then B," and its variants.

Antecedent The A component of a conditional.

Consequent The B component of a conditional.

Notation: "If A then B" is represented in logics as $A \to B$, where A and B are logical representations of two sentences A and B.

Conditionals are typically divided into two broad classes:[1]

Indicative Conditionals For example, "If Maya sang, then Nelson danced."

Counterfactual Conditionals For example, "If Maya had sung, then Nelson would have danced."

A crucial dividing line between indicatives and counterfactuals is that counterfactuals can be used felicitously to talk about situations where the antecedent is contrary-to-fact (or thought to be false; Stalnaker, 1975; Veltman, 1986). This contrast is evident in (1) and (2):[2]

(1) Maya has definitely never sung. #If Maya sang, then Nelson danced.

(2) Maya has definitely never sung. If Maya had sung, then Nelson would have danced.

Corresponding indicative and counterfactual conditions can also differ in truth-value (D. K. Lewis, 1973, p. 3) or at least in their justification (Adams, 1975). Assuming Oswald was a lone shooter, the indicative (3) is straightforwardly true and justified, while the counterfactual (4) is false, or at least unjustified:

(3) If Oswald didn't kill Kennedy, someone else did.

(4) If Oswald hadn't killed Kennedy, someone else would've.

It is worth emphasizing that the term "counterfactual" should not be taken too literally, as it does not mean that a sentence of this form *must* have a contrary-to-fact antecedent (Anderson, 1951). For example, the counterfactual (5) is used as part of an argument that the antecedent is true:

(5) If there had been intensive agriculture in the pre-Columbian Americas, the natural environment would have been impacted in specific ways. That is exactly what we find in many watersheds.

A logic of conditionals typically aims to say which arguments involving $A \to B$ are deductively valid.[3]

Deductive Validity An argument with premises $P_1, \ldots, P_n$ and conclusion C is *deductively valid* just in case it is impossible for the premises to be true while the conclusion is false. *Notation*: $P_1, \ldots, P_n \vDash C$.

However, as discussed in section 4, some instead follow Adams (1975) and focus on inductive support: whether the premises being true, or highly probable, makes the conclusion *highly probable*. Either of these approaches counts as pursuing a *semantic* approach to the logic of conditionals. The semantic approach is well suited to capturing the pervasive *context sensitivity* of conditional reasoning, which will be a prominent theme of this chapter.

Section 2 will outline the major challenges for a logic of conditionals and explain why classical truth-functional logic is not adequate. This will motivate section 3, which discusses three different analyses that draw on tools from modal logic: strict-conditional analyses, similarity analyses, and restrictor analyses. This section will also outline how these analyses have been developed using a different approach to semantics: dynamic semantics. In section 4, the chapter will turn to analyses that rely instead on the tools of probability theory.[4]

2. Logic, Conditionals, and Context

How can we systematically specify what the world must be like if a given conditional is true and thereby capture patterns of valid deductive arguments involving it? It turns out to be rather difficult to answer this question using the tools of classical logic. Seeing why will help identify the key challenges for a logic of conditionals.

The logical semantics developed by Frege, Tarski, and Carnap provided useful analyses of English connectives like *and*, *or*, and *not* using the Boolean truth-functional connectives $\wedge$, $\vee$, and $\neg$. In truth-functional semantics (see chapter 3.1 by Steinberger, this handbook), the meaning of a sentence is identified with its truth-value True (1) or False (0), and the meaning of a connective is identified with a function from one or more truth-values to another, as depicted in table 6.1.1. The best truth-functional approximation of "If . . . then . . ." is the material conditional $\supset$. $A \supset B$ is false when A is true and B false, and it is true otherwise (making it equivalent to $\neg A \vee B$). This analysis is employed in most introductory logic textbooks because it captures three key logical features of conditionals:[5]

Modus Ponens $A \rightarrow B, A \vDash B$. (If a conditional and its antecedent are true, its consequent must be true.)

No Affirming the Antecedent $A \rightarrow B \nvDash A$. (A conditional can be true even when its antecedent is not.)

No Affirming the Consequent $A \rightarrow B \nvDash B$. (A conditional can be true even when its consequent is not.)

The intuitive appeal of modus ponens is clear in (6a). The appeal of the inferences in (6b) and (6c) can

be indirectly confirmed by showing that a conditional can be consistent with the negation of its antecedent or consequent.

(6) a. If Maya sang, then Nelson danced. Maya did sing. Therefore, Nelson danced.

b. If Maya sang, then Nelson danced. But Maya did not sing.

c. If Maya sang, then Nelson danced. But Nelson didn't dance.

To see that modus ponens is valid for $A \supset B$, note that when $A \supset B$ is true and A is true, B must be true—as rows 1, 3, and 4 of table 6.1.1 show. Row 4 shows that it is possible for $A \supset B$ to be true without either A or B being true.

Despite having some attractive features, the material conditional analysis is widely taken to be incorrect.[6] There are particular problems for analyzing English conditionals as $\supset$. But there are also general problems with *any* truth-functional analysis. Let us consider the more general problems first.

Many counterfactuals have false antecedents and consequents, but some are true and some false. For example, (7a) is false—given that Joplin's "Mercedes Benz" was a critique of consumerism—and (7b) is true:

(7) a. If Janis Joplin were alive today, she would drive a Mercedes Benz.

b. If Janis Joplin were alive today, she would metabolize food.

This is not possible on a truth-functional analysis: the truth-values of antecedent and consequent determine a unique truth-value for the whole sentence.

With a bit of care, a similar point can be made about indicatives: suppose a standard die has been tossed, but you do not know which side has landed face up. Then (8a) is intuitively true, while (8b) is false:

(8) a. If the die came up 2, it came up even.

b. If the die came up 1, it came up even.

This intuition persists even when you get to see that the die came up 3: it would seem that (8a) is true and (8b) false even though both have a false antecedent and consequent.

Another kind of problem for truth-functional analyses centers on the context sensitivity of conditionals. The basic observation is that the truth-value of a conditional can vary from one context of use to another, even when the truth-values of the antecedent and consequent stay the same across the two contexts.

Table 6.1.1
Negation ($\neg$), conjunction ($\wedge$), disjunction ($\vee$), and material conditional ($\supset$) as functions of True (1)/False (0).

A	B	$\neg A$	$A \wedge B$	$A \vee B$	$A \supset B$
1	1	0	1	1	1
1	0	0	0	1	0
0	1	1	0	1	1
0	0	1	0	0	1

This is clearest with counterfactuals. Quine (1982, chapter 3) voiced skepticism that any semantic analysis of counterfactuals was possible by highlighting puzzling pairs like (9):

(9) a. If Caesar had been in charge [in Korea], he would have used the atom bomb.

   b. If Caesar had been in charge [in Korea], he would have used catapults.

But D. K. Lewis (1973, p. 67) took these examples to show that the truth-conditions of counterfactuals are *context sensitive*. The antecedents and consequents of (9a) and (9b) are all false, but in some conversational contexts, (9a) seems true and (9b) false, and in other conversational contexts, (9a) seems false and (9b) true. Consider evaluating (9a) and (9b) in a context where we have explicitly discussed and agreed that Caesar was, first and foremost, a ruthless military leader. In such a context, (9a) seems like a true thing to say, while (9b) seems false. By contrast, consider a context where we have explicitly discussed and agreed that Caesar was, first and foremost, a technologically conservative military leader. Then (9b) seems like a true thing to say, while (9a) seems false.

Similar examples illustrate the same point for indicative conditionals. Suppose we have mutually established that a die in our possession has 3 on every side except for one, which has 2. The die has been tossed, but we do not know how it came up. Then (10) seems true:

(10) If the die came up even, it came up 2.

Suppose we then find out that the die came up 3. It seems that (10) was still a true thing to say. However, consider a more ordinary context where we have mutually established that our die is a standard one. The die has been tossed, but we do not know how it came up. It would be false to say (10). Now suppose we later find out that the die came up 3. It still seems like (10) is false. This shows that the truth-value of (10) can vary from context to context, even when the truth-values of its antecedent and consequent are held fixed.

These general problems for a truth-functional analysis are compounded by particular weaknesses in the material conditional analysis. This is particularly clear for counterfactuals. As rows 3 and 4 of table 6.1.1 make clear, $A \supset B$ is true any time A is false. This means that the material conditional validates the following logical principle:

Material Antecedent $\neg A \vDash A \supset B.$

But this means that *all* genuinely contrary-to-fact conditionals are true on a material conditional analysis. So

not only is (11a) incorrectly predicted to be true, but *both* (11a) and (11b) are predicted to be true despite the fact that they seem to be contradictory:

(11) a. If the Earth hadn't existed, the moon would have existed.

   b. If the Earth hadn't existed, the moon wouldn't have existed.

For indicatives, this problem surfaces in another way: recall that indicatives are not felicitous to use when their antecedent has been explicitly denied. This is quite puzzling if that pattern of use is actually a valid form of argument.

Another major problem for the material conditional analysis stems from the fact that it validates

Material Negation $\neg(A \supset B) \vDash A.$

But neither of the following are compelling arguments for the existence of God, despite having (plausibly) true premises:

(12) a. It's not true that if God exists, he's a turnip. Therefore, God exists.

   b. It's not true that if God had existed, he would be a turnip. Therefore, God exists.

But the shortcomings of this analysis are instructive, as they establish some clear criteria for a more successful analysis:

Non-Truth-Functionality The truth-value of a conditional is not determined by the truth-values of its antecedent and consequent.

Context Dependence The truth-value of a conditional depends on certain features of the context in which it is used.

Logical Constraints Conditionals do not obey Material Antecedent or Material Negation.

This is the starting point for analyses that appeal to *possible worlds*.

3. Conditionals and Possible Worlds

Table 6.1.1 helps make salient a crucial assumption of truth-functional semantics: the truth-value of a complex sentence is determined only by the truth-values of its component sentences *in that row*. Once the truth-values of A and B have been settled, the truth-value of $A \wedge B$ has been settled. But table 6.1.1 also makes salient an alternative approach: what if the truth-value of a complex sentence is determined by the *distribution* of its component sentences' truth-values across a number of rows?

Pursuing this alternative analysis requires clarifying exactly what "a row" is and settling on a particular account of which distributions matter. Following the tradition in modal logic, one can think of the rows as *possible worlds* (Kripke, 1963). The two basic analyses surveyed in this section—strict-conditional analyses ("strict analyses"; section 3.1) and similarity analyses (section 3.2)—amount to different views about which distributions of truth-values across worlds make conditionals true. Sections 3.3 and 3.4 discuss two resources for extending these two kinds of analyses. Before exploring the analyses in detail, it is useful to state their shared assumptions and their main difference.

Strict and similarity analyses both employ the idea of *possible worlds w*, which are simply ways the world could be or could have been. They are treated as primitive points in the set of all possible worlds W. They play a crucial role in assigning truth-values to sentences: a sentence A is only true given a possible world w, but since w is genuinely possible, it cannot be the case that both A and $\neg A$ are true at w. Consider, then, a language with just three atomic sentences A, B, C, namely, "Maya ate apples," "Maya ate bananas," and "Maya ate cherries." At least eight possible worlds are then needed, as listed in table 6.1.2:[7] strict and similarity analyses both aim to say when a conditional like $A \rightarrow B$ is true in a given world w and do so on the basis of whether certain A-worlds *relevant* to w are also B-worlds. But they differ in terms of how they determine this set of worlds relevant to w.

Different methods for selecting relevant worlds produce a distinctive logical difference between strict and similarity analyses. This logical difference is easiest to illustrate with the following principle:

Antecedent Strengthening $A \rightarrow C \vDash (A \wedge B) \rightarrow C$.

Goodman (1947) observes that this does not hold for counterfactuals, since (13a) is true and (13b) false:

Table 6.1.2
Possible worlds for A, B, C

	A	B	C
w_1	1	1	1
w_2	1	1	0
w_3	1	0	1
w_4	1	0	0
w_5	0	1	1
w_6	0	1	0
w_7	0	0	1
w_8	0	0	0

(13) a. If I had struck this match, it would have lit.

   $S > L$

   b. If I had struck this match and done so in a room without oxygen, it would have lit.

   $(S \wedge \neg O) > L$

D. K. Lewis (1973, p. 10) dramatized this counterexample by considering "Sobel sequences" such as (14). Surprisingly, adding more information to the antecedent repeatedly flips the truth-value of the counterfactual.

(14) a. If I had shirked my duty, no harm would have ensued.

   $I > \neg H$

   b. Though, if I had shirked my duty and you had too, harm would have ensued.

   $(I \wedge U) > H$

   c. Yet, if I had shirked my duty, you had shirked your duty, and a third person had done more than their duty, then no harm would have ensued.

   $(I \wedge U \wedge T) > \neg H$

   $\vdots$

For indicatives, (15a) may be true when (15b) is not:

(15) a. If Maya sang at the party, then Nelson danced at the party.

   b. If Maya sang at the party and Nelson wasn't there, then Nelson danced at the party.

Similarity analyses predict that antecedent strengthening is invalid while strict analyses predict that it is valid. Strict analyses address the counterexamples above *pragmatically*: the conclusions of the arguments are false, but that is because pragmatic mechanisms force the conclusion to be interpreted relative to a different set of worlds than the premises. In doing so, these analyses appeal to a feature already highlighted above: conditionals' context sensitivity.

The general pattern, of which antecedent strengthening is an instance, is:

Antecedent Monotonicity If $B \vDash A$ then $A \rightarrow C \vDash B \rightarrow C$.

Another instance of this is:

Simplification of Disjunctive Antecedents (SDA) $(A \vee B) \rightarrow C \vDash (A \rightarrow C) \wedge (B \rightarrow C)$.

Antecedent monotonicity also leads (indirectly) to (Starr, 2019, section 2.1):

Transitivity $A \rightarrow B$, $B \rightarrow C \vDash A \rightarrow C$.

Contraposition $A \rightarrow B \vDash \neg B \rightarrow \neg A$.

Counterexamples similar to those above have been presented by similarity theorists to transitivity, contraposition, and SDA (see, e.g., D. K. Lewis, 1973; McKay & van Inwagen, 1977; Stalnaker, 1968). Strict theorists have also aimed to address these examples pragmatically (e.g., Warmbrōd, 1981). So the key issue dividing strict and similarity issues is the following:

Pragmatic or Semantic Nonmonotonicity? Is the non-monotonicity of conditional antecedents best explained semantically or pragmatically?

- *Strict Theorists*: pragmatically (e.g., Gillies, 2007, 2009; Warmbrōd, 1981)

- *Similarity Theorists*: semantically (e.g., D. K. Lewis, 1973; Stalnaker, 1968)

This chapter will not weigh in on this question, instead presenting the basics of these two approaches. Antecedent strengthening will be the focus for concreteness, but it is just a representative of antecedent monotonicity more generally.

3.1 Strict-Conditional Analyses

Strict-conditional analyses began with the basic idea that $A \rightarrow B$ is true just in case all of the A-worlds are B-worlds (Peirce, 1896). This has been refined using a crucial tool from the semantics of modal logic: *accessibility functions* R. The function R takes a world w and returns the set $R(w)$ of worlds that are accessible from, or relevant to, w (Kripke, 1963).[8]

Strict Conditional $A \Rightarrow B$ is true in w, given R, just in case every A-world in $R(w)$ is a B-world.

To illustrate this definition, consider a particular world w_7 and a particular accessibility function R_1 where $R_1(w_7)$ corresponds to the set of worlds bolded in table 6.1.3. $A \Rightarrow B$ comes out true in this case: every bold world where A is 1 is a world where C is 1. $A \Rightarrow C$ would come out false on a slightly different accessibility function R_2, which is just like R_1 except it includes w_2: a world where A is 1 and C is 0. R_2 in fact shows why antecedent negation does not hold for the strict conditional: $\neg A$ is true in w_7 and yet $A \Rightarrow C$ is false in w_7, relative to R_2. Similarly, this is also a counterexample to material negation since $\neg(A \Rightarrow C)$ is true in w_7, relative to R_2, and A is false in w_7. These count as counterexamples given the following definition of validity.

Modal Validity $P_1, \ldots, P_n \vDash C$ just in case for every w and R, if $P_1, \ldots, P_n$ are true in w, relative to R, then C is true in w, relative to R.

According to this definition, antecedent strengthening is valid, an issue to which we will return. Modus

Table 6.1.3

Possible worlds for A, B, C; worlds in $R_1(w_7)$ in bold

	A	B	C
w_1	1	1	1
w_2	1	1	0
w_3	1	0	1
w_4	1	0	0
w_5	0	1	1
w_6	0	1	0
w_7	0	0	1
w_8	0	0	0

ponens is valid for the strict conditional if one requires that accessibility functions satisfy *reflexivity*: for all w, w is in $R(w)$. Accordingly, the strict conditional addresses the main *logical* problems faced by the material conditional analysis. As discussed at the outset of this section, the strict analysis also captures non-truth-functionality: the truth-value of $A \Rightarrow B$ depends not just on the actual truth-values of A and B but on their truth-values across all the worlds in $R(w)$.

The context sensitivity of conditionals is central to contemporary strict analyses, particularly when responding to examples like (13), (14), and (15). These examples seem to show that antecedent strengthening is not valid for conditionals, contrary to the strict analysis. While early work on the logic of strict conditionals did not include this component, subsequent work like Warmbrōd (1981) has done so by treating $R(w)$ as one aspect of context that changes as utterances take place.

Warmbrōd (1981) proposes that the accessibility function $R(w)$ corresponds to the background suppositions of the conversationalists and will change as the conversation unfolds. In particular, Warmbrōd (1981) proposes that $R(w)$ also changes as the result of conditionals being asserted. Warmbrōd (1981) notes that *trivial* strict conditionals are not pragmatically useful in conversation. A strict conditional $A \Rightarrow C$ is trivial just in case A is inconsistent with $R(w)$—there is no A-world in $R(w)$. Asserting $A \Rightarrow C$ in such a context does not provide any information: every conditional of the form $A \Rightarrow X$ is true. But Warmbrōd (1981) proposes that conversationalists adopt a pragmatic rule of charitable interpretation to make sense of why the speaker asserted $A \Rightarrow C$ rather than a different conditional:

(P) If the antecedent A of a conditional is itself consistent, then there must be at least one A-world in $R(w)$.

Given this, $R(w)$ can change as a result of asserting a conditional. This part of the view is central to explaining away counterexamples to antecedent monotonic validities, like antecedent strengthening.

Consider again this counterexample to antecedent strengthening:

(13) a. If I had struck this match, it would have lit.

    b. If I had struck this match and done so in a room without oxygen, it would have lit.

On a strict analysis, if (13a) is true in a world w, relative to R_0, then $R_0(w)$ must exclude any worlds where the match is struck but there is no oxygen in the room. However, if (13b) is interpreted against R_0, the antecedent will be inconsistent with $R_0(w)$ and so express a trivial strict conditional. According to Warmbrōd (1981), interpreting (13b) according to (P) forces the conversationalist to adopt a new, modified accessibility function $R_1(w)$ where the presence of oxygen is no longer assumed. If this is right, then (13) is not really a counterexample to the validity of antecedent strengthening: the premise is true relative to R_0, and the conclusion is only false relative to R_1. Warmbrōd (1981) and others develop versions of this account to address all of the proposed counterexamples to antecedent-monotonic patterns.[9] These accounts do have an important limitation: they do not capture nested conditionals and do not actually predict how $R(w)$ evolves to satisfy (P). Von Fintel (2001), Gillies (2007, 2009), and Willer (2017) offer accounts that remove these limitations, using the tools of dynamic semantics covered in section 3.4.

3.2 Similarity Analyses

The basic idea of similarity analyses is that $A \to C$ is true in w when C is true in all of the A-worlds most similar to w (D. K. Lewis, 1973; Stalnaker, 1968; Stalnaker & Thomason, 1970). One way of precisely formulating this appeals to a selection function f, which takes a world w and a proposition p and returns the set of p-worlds most similar to w: $f(w, p)$.[10] This is used to define the truth-conditions of a similarity-based conditional, notated ">," as follows:[11]

Similarity Analysis $A > C$ is true in w, relative to f, just in case every world in $f(w, A)$ is a C-world.

While a strict analysis assumes a *single* set of relevant worlds $R(w)$ for all conditionals, f can select a *different* set of worlds for each different antecedent. For example, it is perfectly possible for $f(w, A \wedge B)$ to contain worlds that are not in $f(w, A)$, even though all $A \wedge B$-worlds are A-worlds. Different similarity analyses propose different constraints on f. The candidate constraints are given in table 6.1.4.[12] To make clear that this permits $f(w, A \wedge B)$ to contain

worlds not in $f(w, A)$ and how this invalidates antecedent strengthening, let's consider a concrete example.

Consider the worlds and selection function f_1 in table 6.1.5. $A > C$ is true in w_6, relative to f_1, since A and C are true in w_3. But $(A \wedge B) > C$ is false in w_6, relative to f_1, since $A \wedge B$ is true and C false in w_2. It is easily verified that f_1 satisfies all four constraints in table 6.1.4. This should highlight the key difference between strict and similarity analyses: strict analyses assume a fixed set of relevant worlds for all antecedents, while similarity analyses allow the set of relevant worlds to vary from antecedent to antecedent—even among logically related antecedents. It is also worth highlighting a point that will matter later: nothing in the formal analysis requires $f(w, p)$ to hold fixed facts of w, even if they are unrelated to p. This is clear with $f_1(A, w_6) = \{w_3\}$, which does not preserve the fact that B is true in w_6 and that B's truth may be unrelated to A being false in w_6.

The constraints on selection functions listed in table 6.1.5 are partly motivated by our intuitive concept of similarity but also by logical considerations. Success enforces that the set of most similar p-worlds to w are in fact p-worlds. But the other constraints correspond to certain logical validities—see Starr (2019) for a thorough discussion of this. For the purposes of this chapter, only

Table 6.1.4

Candidate constraints on selection functions; $p, q \subseteq W$ and $w \in W$

(a) $f(w, p) \subseteq p$	success
(b) $f(w, p) = \{w\}$, if $w \in p$	strong centering
(c) if $f(w, p) \subseteq q$ and $f(w, q) \subseteq p$ then $f(w, p) = f(w, q)$	uniformity
(d) $f(w, p)$ contains *at most one* world	uniqueness

Table 6.1.5

Possible worlds for A, B, and C; $f_1(A, w_6)$ in bold, $f_1(A \wedge B, w_6)$ underlined

	A	B	C
w_1	1	1	1
$\underline{w_2}$	$\underline{1}$	$\underline{1}$	$\underline{0}$
$\mathbf{w_3}$	**1**	**0**	**1**
w_4	1	0	0
w_5	0	1	1
w_6	0	1	0
w_7	0	0	1
w_8	0	0	0

strong centering will matter, as it ensures that similarity conditionals validate modus ponens.[13]

Material antecedent and material negation are invalid for the similarity conditional for the same reasons that they were invalid for the strict conditional. The falsity of A in w does not ensure that the A-worlds most similar to w are B-worlds. So $\neg A \nvDash A > B$. $A > B$ can be false in w when A is false in w, but in one of the A-worlds most similar to w, B is false. So $\neg(A > B) \nvDash A$.

The context sensitivity of conditionals can be captured on similarity analyses by highlighting the fact that judgments of similarity are themselves context dependent (D. K. Lewis, 1973). As D. K. Lewis (1973, p. 67) details, different contexts can make salient different properties of the things we are talking about, and this affects what counts as a similar world to our own. This is illustrated with the pair in (9), discussed back in section 2:

(9) a. If Caesar had been in charge [in Korea], he would have used the atom bomb.

 b. If Caesar had been in charge [in Korea], he would have used catapults.

Consider a context where Caesar's brutality is made salient. It will then be held fixed when determining which worlds where Caesar was in charge in Korea count as most similar to our own. As a result, (9a) will come out true and (9b) false. Other contexts will have the opposite effect. Quine's (1982, chapter 3) claim that there is no fact of the matter whether these counterfactuals are true comes from failing to embed them in natural contexts. Subsequent work such as K. S. Lewis (2018) and Ippolito (2016) develops this idea.

Many similarity theorists explicitly limited the analysis to counterfactuals (e.g., D. K. Lewis, 1973). But Stalnaker (1975) applied the analysis to indicative conditionals by saying that indicatives and counterfactuals differ in how they are context sensitive. Recall that indicatives and counterfactuals can differ in truth-value. Setting aside conspiracy theories, (3) is true and (4) false:

(3) If Oswald didn't kill Kennedy, someone else did.

(4) If Oswald hadn't killed Kennedy, someone else would've.

Stalnaker (1975) explains this contrast in terms of a general account of how assertion works. The mutual assumptions of the conversationalists can be modeled as a set of worlds c called the *context set*—the set of worlds compatible with what everyone is assuming and with what everyone is assuming that everyone is assuming and so on. When a proposition p is asserted and accepted, then the new context set is $c \cap p$—the set of all worlds in the previous context set and compatible with p.

Stalnaker (1975) Analysis of Indicatives For an indicative conditional $A > B$, if $w \in c$, then $f(w, A) \subseteq c$.

When evaluated in a world compatible with the context set, the most similar antecedent worlds must also be within the context set.

This predicts (3) to be true, since it requires that its interpretation hold fixed the fact that Kennedy was killed—that proposition is part of the context set against which (3) is asserted. By contrast, (4) does not require that to be held fixed and so allows the most similar antecedent worlds to be ones where Kennedy was not killed at all. On this analysis, all conditionals are context sensitive, but indicative conditionals are specifically sensitive to the context set. This goes some way in explaining the observation that indicative conditionals are only felicitous when their antecedent has not been explicitly ruled out—recall (1) and (2) from section 1.

3.3 Restrictor Analyses

The restrictor analysis of conditionals originates with D. K. Lewis (1975) and Kratzer (1981, 1986) and argues for a dramatic change in the logical analysis of conditionals. It has been assumed that conditionals have a logical form like $A \to B$ and that their analysis must proceed by finding the right meaning for $\to$. But restrictor analyses contend that this is wrong. An analysis of conditionals should begin with modal adverbs like *must*, *would*, *might*, and *probably* that occur in what is normally thought to be the consequents of conditionals. A conditional like "If Maya sang, Nelson probably danced" should be thought of primarily as a sentence of the form *Probably*(*D*). All the *if*-clause does is restrict the domain of worlds over which *probably* quantifies. While "Nelson probably danced" says that Nelson danced in most of the worlds, "If Maya sang, Nelson probably danced" says that Nelson danced in most of the worlds where Maya sang.

D. K. Lewis (1975) and Kratzer (1986) argue for a restrictor analysis by observing that no uniform contribution can be assigned to $A \to B$ that captures the different meanings that conditionals have when different modal adverbs occur in the consequent. This argument will not be summarized here because restrictor analyses are not actually competitors to strict and similarity analyses—they simply specify a different form those two analyses can take: as Kratzer (1991, p. 649) details, material conditionals, strict conditionals, and similarity conditionals can all be modeled within a restrictor analysis.

A restrictor analysis provides a resource for extending the empirical coverage of strict and similarity analyses: while the conditionals considered so far involve a necessity modal of some sort, a more general analysis is needed.[14] Further, a restrictor analysis provides a more flexible theory that could, in principle, provide an account on which conditionals are sometimes strict and sometimes similarity conditionals. For more on this approach, see Kratzer (2012).

3.4 Dynamic Analyses

This chapter has assumed that the meaning of a sentence corresponds to its truth-conditions: the set of worlds in which it is true. Formally, this means that the semantics is specified as a function $[\![A]\!]$ that maps sentences A of the formal language to subsets of W (the set of all possible worlds). Logic followed suit: it requires that valid arguments preserve truth. But there is another dynamic perspective on meaning: it is the characteristic way a sentence changes a context in which it is uttered (Groenendijk & Stokhof, 1991; Heim, 1982; Kamp, 1981; Veltman, 1996).

On Veltman's (1996) approach, a dynamic semantics is specified as a function $[A]$ that maps one information state s (a subset of W) to another information state s'. This is written $s[A] = s'$, and it is said that s' is the result of updating s with A. The meaning of A resides in the difference between the two states: one prior to the use of A and one posterior to the use of A. A expresses truth-conditional information about worlds by eliminating worlds from s. But it can also express a global property of the information s that is quite different than distinguishing between worlds.

For example, the sentence $Might(A)$ tests whether s is consistent with A. If not, s' is reduced to $\varnothing$. But if it is, then s' is left as it is. On this view, $Might(A)$ expresses a property of the information state s, without treating that property as something that distinguishes *worlds* in s from each other. Logic follows suit: valid arguments preserve information rather than truth. More specifically, updating with the conclusion after updating with the premises provides no additional information:

Dynamic Validity $P_1, \ldots, P_n \vDash C$ just in case for any information state s, $s[P_1] \ldots [P_n] = s[P_1] \ldots [P_n][C]$.

Both strict and similarity analyses have drawn inspiration from this dynamic approach.

Von Fintel (2001) and Gillies (2007) develop dynamic strict analyses of counterfactuals and argue that they can better explain ordering effects. Among them are *reverse Sobel sequences*, which are simply the reversal of

the sequences like (14) presented by D. K. Lewis (1973, p. 10). The important observation is that reversing these sequences is not felicitous:

(14) a. If I had shirked my duty and you had too, harm would have ensued.

    b. #If I had shirked my duty, no harm would have ensued.

Von Fintel (2001) and Gillies (2007) observe that similarity analyses render sequences like (14) semantically consistent. Their theories predict this infelicity by providing a systematic theory of how counterfactuals update context.[15] These analyses involve richer models of context than Veltman (1996).

Gillies (2004, 2009) develops a dynamic strict analysis of indicative conditionals on which $s[A \to B]$ tests that all the A-worlds in s are C-worlds. If the test is passed, s stays as it is. If it is failed, s is reduced to $\varnothing$. This analysis, as well as the dynamic definition of validity, navigates a number of tricky issues in conditional logic. Gillies (2004) uses it to diffuse counterexamples to modus ponens (McGee, 1985). Willer (2012) extends this solution to counterexamples to modus ponens presented by Kolodny and MacFarlane (2010). Stojnić (2016) integrates a dynamic strict analysis with a theory of discourse coherence and modal anaphora to address counterexamples to modus ponens and modus tollens. Gillies (2009) uses a dynamic strict analysis to counter a number of arguments (e.g., Edgington, 1995; Gibbard, 1981), which say that it is not possible to assign truth-conditions to indicative conditionals that are stronger than a material conditional and weaker than a classical strict conditional and capture the *import–export equivalence* between $A \to (B \to C)$ and $(A \wedge B) \to C$. Nondynamic similarity theories invalidate import–export but offer no evidence in its favor.

Starr (2014) proposes that the general meaning of conditionals is that given by Gillies (2009) but that counterfactuals contain an operator in their antecedent that allows them to talk about the most similar antecedent worlds. This operator is argued for on the basis of work in linguistics such as Iatridou (2000). Starr (2014) argues that this dynamic strict analysis of indicatives and dynamic similarity analysis of counterfactuals solves a number of outstanding problems. Willer (2017) has developed a related account and argues that only a dynamic strict analysis can explain why indicative versions of (14) are felicitous. This remains an active area of research.

3.5 Discussion

Both strict and similarity analyses can overcome the basic shortcomings of the material conditional analysis. They validate modus ponens while invalidating material antecedent and material negation, and they are not truth-functional. They also leave room for the context sensitivity of conditionals. The main debate between them is whether nonmonotonic inferences like antecedent strengthening should be rendered invalid (similarity analyses) or whether the proposed counterexamples involve changes in context (strict analyses). This is a subtle and ongoing debate.

There is, however, a more pressing challenge to both of these analyses. To explain how a given conditional like (17) expresses a true proposition, a similarity analysis must specify which particular conception of similarity informs it.

(17) If my computer were off, the screen would be blank.

Of course, the strict analysis is in the same position. It cannot predict the truth of (17) without specifying a particular accessibility relation. In turn, the same question arises: on what basis do ordinary speakers determine some worlds to be accessible and others not? Theories like those discussed above do not directly address this question, as they are primarily concerned with the logic of conditionals. But a wealth of examples illustrates that there are systematic generalizations about how sentences are judged to be true, and it is not clear that strict or similarity analyses are well positioned to capture this. Readers interested in this next turn in the debate are directed to Starr (2019, section 2.5).

4. Conditionals and Probability

Conditional reasoning involves uncertainty and evaluating the consequences of the world being different than it actually is. The study of probability has provided a suite of tools designed for just these purposes. It is no accident, then, that many philosophers, psychologists, and computer scientists have drawn on these tools to analyze the meaning and logic of conditionals. The seminal proposal here comes from Adams (1975), which analyzes conditionals in terms of conditional probability. This basic idea, as well as its development by philosophers (Edgington, 1995) and psychologists (Evans & Over, 2004), is surveyed in section 4.1. Recently, research in artificial intelligence (AI), philosophy, psychology, and linguistics has appealed to a related probabilistic tool: Bayesian networks. Section 4.2 will survey those accounts.

4.1 Conditional Probability

The seminal proposal from Adams (1975) is this:

Adams' Thesis The assertability of "*B* if *A*" is proportional to $P(B \mid A)$, where *P* is a probability function representing the agent's subjective credences.

Probabilities are real numbers between 0 and 1 assigned to propositional variables *A*, *B*, *C*, . . . Adams takes these probabilities to reflect an agent's subjective credence; for example, $P(A) = 0.6$ reflects that they think *A* is slightly more likely than not to be true.[16] $P(B \mid A)$ is the credence in *B* conditional on *A* being true and is defined as follows:

Conditional Probability $P(B \mid A) = \dfrac{P(A \wedge B)}{P(B)}$.

Surprisingly, the logic of conditionals based on conditional probability developed by Adams (1975) is more or less the same as the similarity analyses pursued by Stalnaker (1968) and D. K. Lewis (1973). In particular, it invalidates antecedent-monotonic patterns, along with material negation and material antecedent. Its major difference from similarity analyses comes in its account of context sensitivity and non-truth-functionality.

This can be illustrated with examples discussed earlier. Recall the context of (8). You know a standard die has been tossed, but you do not know which side has landed face up. (8a) seems like a good assertion to make, while (8b) does not:

(8) a. If the die came up 2, it came up even.

   b. If the die came up 1, it came up even.

When you get to see that the die came up 3, it still seems right to say that (8a) was a justified assertion even if you would not now *assert* that conditional.[17] The same goes for (8b). Both intuitions are captured on a conditional probability analysis. The conditional probability of the die coming up even given that it came up 2 is 1, while the conditional probability of the die coming up even given that it came up 1 is 0. (8) was an example of non-truth-functionality, and the conditional probability analysis captures this: instead of the truth of a conditional being determined by the truth of its parts, the probability of a conditional is being determined by the probability of its parts. As it turns out, this view is even more radical. A conditional probability $P(B \mid A)$ cannot be modeled as the probability that a conditional proposition $B \mid A$ is true (D. K. Lewis, 1976). So $P(B \mid A)$ should not be thought of as the probability that a proposition is true. For this reason, many philosophers articulate a conditional probability analysis as holding that conditionals do not express truth-evaluable propositions but merely a ratio of credences in truth-evaluable propositions (Edgington, 1995).

The context sensitivity of conditionals is reflected in the core idea of the conditional probability analysis: conditionals express credences. As those credences change, so will the ratios between them. Consider again (10) and its context: we have mutually established that a die in our possession has 3 on every side except for one, which has 2. The die has been tossed, but we do not know how it came up. Then (10) seems like a justified assertion:

(10) If the die came up even, it came up 2.

Indeed, the conditional probability of the die coming up 2, given that it came up even, is 1. In this context, our credence that the die came up any even number other than 2 is 0, so $P(Even \wedge Two) = P(Two)$ and $\frac{P(Two)}{P(Two)} = 1$. But when we believe the die to be a standard one, the assertion of (10) will be quite unjustified. Believing the die to be standard amounts to these credences: $P(Even) = \frac{1}{2}, P(Two) = \frac{1}{6}$, and $P(Even \wedge Two) = \frac{1}{6}$. So, $P(Even \mid Two) = \frac{\frac{1}{6}}{\frac{1}{2}} = \frac{1}{3}$.

Since conditional probabilities are only defined when the antecedent is assigned a nonzero probability, Adams' thesis is of limited use for counterfactuals. Further, indicative and counterfactual pairs often differ in their assertability, for example, (3) and (4). To address this, Adams (1976) proposed the *prior probability* analysis of counterfactuals:

Adams' Prior Probability Analysis The assertability of "*B* would have been, if *A* had been" is proportional to $P_0(B \mid A)$, where P_0 is the agent's credence prior to learning that *A* is false.

Consider a counterfactual variant of (10) uttered in the context where the die is known to have 3 on five sides and 2 on one, and we have just learned that the die came up 3.

(18) If the die had come up even, it would have come up 2.

Prior to learning that the die came up 3, and so didn't come up even, $P(Even \mid Two) = 1$, as just discussed. So the prior probability analysis correctly predicts this utterance of (18) to be a perfect assertion. As with the indicative, this would change if we believed the die to be 12-sided instead.

There are other cases, however, where Adams' prior probability analysis makes incorrect predictions. Edgington (2004, p. 21) discusses these cases and concludes that

the analysis requires an amendment: P_0 may also reflect any facts the agent learns after they learn that the antecedent is false, provided that those facts are causally independent of the antecedent. Kvart (1986) integrates causal information into a different objective conditional probability analysis.

More recently, this focus on causality and probability has inspired a different approach to counterfactuals. Many using probabilistic tools now favor modeling causal information in terms of Bayesian networks (Pearl, 2009). It is then possible to formulate a semantics for counterfactuals directly in terms of Bayesian networks. As section 4.2 will explain, Bayesian networks have significant advantages to a standard probabilistic representation when trying to formulate a computationally tractable representation of an agent's knowledge about the world.

4.2 Bayesian Networks

When an agent's credence in *B* is the same as their credence in *B* given *A* and *B* given $\neg A$, *B* is probabilistically independent of *A*:

Probabilistic Independence *B* is probabilistically independent of *A* just in case $P(B) = P(B \mid A) = P(B \mid \neg A)$.

Bayesian networks are built on the mathematical insight that it is possible to represent an agent's credences by representing only the conditional probabilities of dependent variables and the probabilities of independent variables. This mathematical insight has been taken to be of great importance to artificial intelligence and cognitive science. It makes probabilistic representations of agents' beliefs computationally tractable.[18] But it also stores immensely useful information. It facilitates counterfactual reasoning, reasoning about actions, and explanatory reasoning.

An agent's knowledge about a system containing eight variables could be represented by the directed acyclic graph and system of structural equations between those variables in figure 6.1.1. While the arrows mark relations of probabilistic dependence, the equations characterize the nature of the dependence; for example, "$H = F \vee G$" means that the value of H is determined by the value of $F \vee G$ (but not vice versa). Consider just the three rightmost nodes of figure 6.1.1. They are an appropriate representation for an agent who has credences about three propositions, and their probabilistic dependencies correspond to the indicative conditionals (19a) and (19b). Let us further suppose they have the unconditional credence corresponding to (19c).

(19) a. If both Fran and Greta attend, Harriet attends.

    b. If either Fran or Greta don't attend, Harriet won't attend.

    c. Fran attended, Greta did not, and so Harriet did not.

    d. If Greta had attended, Harriet would have attended.

The counterfactual (19d) seems true in this scenario. Pearl's (2009, chapter 7) analysis can capture this:

Interventionism Evaluate $G > H$ relative to a Bayesian network by removing any incoming arrows to G, setting its value to 1, and projecting this change forward through the remaining network. If H is 1 in the resulting network, $G > H$ is true; otherwise, it's false.

On this method, one first intervenes on G: remove the arrow coming in to G and the equation $G = E$, and replace it with $G = 1$. One then solves for H using the equation $H = F \wedge G$. Since intervention does not affect the value of F, it remains 1. So, it follows that $H = 1$ and that the counterfactual is true. Pearl (2009, chapter 7) shows that the logic of interventionist counterfactuals is very close to similarity analyses (D. K. Lewis, 1973; Stalnaker, 1968)—hence my use of ">" here.

Unlike similarity analyses, Bayesian networks provide explicit models of the knowledge that makes counterfactuals true. This allows them to better navigate the numerous counterexamples to the similarity analysis surveyed in Starr (2019, section 2.5) and provide an explicit theory of how counterfactuals are context sensitive. To be sure, interventionism has limitations and faces a number of counterexamples.[19] But there is now a burgeoning interdisciplinary literature refining interventionism (Kaufmann, 2013; Lucas & Kemp, 2015; Schulz, 2007, 2011) and pursuing alternatives also based on Bayesian networks (Hiddleston, 2005; Rips, 2010). This is a very active area of research.

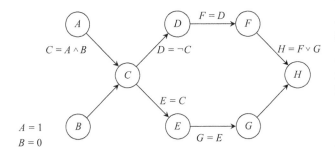

Figure 6.1.1
Bayesian network and structural equations.

5.   Conclusion

Recent work on the logic of conditionals maintains that they have three key properties: they are non-truth-functional, they are context sensitive, and their antecedents are interpreted nonmonotonically. Certain core validities, like modus ponens, and invalidities, like material negation and material antecedent, have been captured alongside these key properties. Possible-worlds analyses have come in two basic varieties: strict analyses and similarity analyses. These two varieties have been augmented in various ways using restrictor analyses of modality and dynamic semantics.

Accounting for the particular contextual features that fix the truth-conditions of conditionals remains a challenge for these approaches. Probabilistic analyses present a promising option here but are still very much in development.

Notes

1. While commonly assumed, this division is debated (e.g., Dudman, 1988). Others prefer the categories of indicative and subjunctive conditionals (e.g., Declerck & Reed, 2001, p. 99; von Fintel, 1999). For an overview of these issues, see Starr (2019). For a dedicated survey of subjunctive conditionals, see von Fintel (2012). For indicatives, see Gillies (2012) and Edgington (2014).

2. Here, the "#" is used in the descriptive convention of linguistics, where it indicates the native speaker judgment that the sentences are grammatical but can't be used in this context.

3. This chapter will assume that the if–then structure in indicative and counterfactual conditionals should have the same logical analysis and that their difference should be explained in terms of their different morphology. Not all philosophers have shared this assumption (e.g., D. K. Lewis, 1973), but it remains the default one for good reasons. See Starr (2019) for further discussion.

4. For a more exhaustive and formal survey of conditional logics, see Arló-Costa, Egré, and Rott (2007/2019).

5. As is often remarked in those textbooks, it is the only truth function that captures these features.

6. While Grice (1989) attempted a pragmatic defense, it has never been satisfactorily extended to the problems presented here.

7. For technical reasons not discussed here, it is generally assumed in modal logic that there are more possible worlds than unique combinations of atomic truth-values.

8. Kripke (1963), as well as most work in modal logic, actually uses accessibility *relations* $R(w, w')$. But accessibility functions

simplify the presentation and are definable in terms of accessibility relations: $R(w) = \{ w' \mid R(w, w') \}$.

9. They also address the "paradoxes of strict implication," for example, $\neg \Diamond A \vDash A \Rightarrow B$.

10. See D. K. Lewis (1973, section 2.7) for the various formulations. The set selection formulation makes the limit assumption: A-worlds do not get indefinitely more and more similar to w. D. K. Lewis (1973) rejected this assumption, but it will merely simplify exposition here. See Starr (2019) for discussion of the limit assumption.

11. $f(w, A)$ is shorthand for $f(w, \llbracket A \rrbracket)$, where $\llbracket A \rrbracket$ is the set of worlds in which A is true.

12. Pollock (1976) adopts (a) and (b), D. K. Lewis (1973) and Nute (1975) adopt (a)–(c), and Stalnaker (1968) adopts (a)–(d).

13. In fact, weak centering suffices: $w \in f(w, p)$ if $w \in p$.

14. Where there is no overt modal, Kratzer (1986) proposes that conditionals involve a covert epistemic necessity modal.

15. Moss (2012) and K. S. Lewis (2018) are similarity analyses that explain these data pragmatically.

16. Probabilities are taken to obey the Kolmogorov axioms. See Hájek (2019) for details.

17. As noted in section 1, indicatives are infelicitous when their antecedent is known to be false. This is expected on Adams's analysis, since $P(B \mid A)$ is undefined when $P(A) = 0$.

18. A complete description of an agent's credences involves a joint probability distribution over all Boolean combinations of the variables. For a system with 8 variables, this requires storing $2^8 = 256$ probability values, while the Bayesian network would require only 18—one conditional probability for each Boolean combination of the parent variables and one for each of the independent variables. See Sloman (2005, chapter 4) and Pearl (2009, chapter 1) for details.

19. It does not apply to conditionals with logically complex antecedents or consequents. This limitation is addressed by Briggs (2012), who also axiomatizes the logic and compares it to D. K. Lewis (1973) and Stalnaker (1968)—significantly extending the analysis and results in Pearl (2009, chapter 7). For counterexamples, see Hiddleston (2005) and Rips (2010).

References

Adams, E. W. (1975). *The logic of conditionals*. Dordrecht, Netherlands: Reidel.

Adams, E. W. (1976). Prior probabilities and counterfactual conditionals. In W. Harper & C. Hooker (Eds.), *Foundations of probability theory, statistical inference, and statistical theories of science* (Vol. 6a of The University of Western Ontario Series in Philosophy of Science, pp. 1–21). Dordrecht, Netherlands: Springer.

Anderson, A. R. (1951). A note on subjunctive and counterfactual conditionals. *Analysis, 12*(2), 35–38.

Arló-Costa, H., Egré, P., & Rott, H. (2019). The logic of conditionals. In E. N. Zalta (Ed.), *The Stanford encyclopedia of philosophy*. Retrieved from https://plato.stanford.edu/archives/sum2019/entries/logic-conditionals/

Briggs, R. A. (2012). Interventionist counterfactuals. *Philosophical Studies, 160*(1), 139–166.

Declerck, R., & Reed, S. (2001). *Conditionals: A comprehensive empirical analysis* (Vol. 37 of Topics in English Linguistics). New York, NY: De Gruyter Mouton.

Dudman, V. H. (1988). Indicative and subjunctive. *Analysis, 48*(3), 113–122.

Edgington, D. (1995). On conditionals. *Mind, New Series, 104*(413), 235–329.

Edgington, D. (2004). Counterfactuals and the benefit of hindsight. In P. Dowe & P. Noordhof (Eds.), *Cause and chance: Causation in an indeterministic world* (pp. 12–27). London, England: Routledge.

Edgington, D. (2014). Indicative conditionals. In E. N. Zalta (Ed.), *The Stanford encyclopedia of philosophy*. Retrieved from https://plato.stanford.edu/archives/win2014/entries/conditionals/

Evans, J. St. B. T., & Over, D. (2004). *If*. Oxford, England: Oxford University Press.

Gibbard, A. (1981). Two recent theories of conditionals. In W. L. Harper, R. C. Stalnaker, & G. Pearce (Eds.), *Ifs: Conditionals, beliefs, decision, chance, time* (pp. 211–247). Dordrecht, Netherlands: Reidel.

Gillies, A. S. (2004). Epistemic conditionals and conditional epistemics. *Noûs, 38*(4), 585–616.

Gillies, A. S. (2007). Counterfactual scorekeeping. *Linguistics and Philosophy, 30*(3), 329–360.

Gillies, A. S. (2009). On truth-conditions for *if* (but not quite only *if*). *Philosophical Review, 118*(3), 325–349.

Gillies, A. S. (2012). Indicative conditionals. In G. Russell & D. G. Fara (Eds.), *The Routledge companion to philosophy of language* (pp. 449–465). New York, NY: Routledge.

Goodman, N. (1947). The problem of counterfactual conditionals. *Journal of Philosophy, 44*, 113–118.

Grice, H. P. (1989). Indicative conditionals. In *Studies in the way of words* (pp. 58–85). Cambridge, MA: Harvard University Press.

Groenendijk, J., & Stokhof, M. (1991). Dynamic predicate logic. *Linguistics and Philosophy, 14*(1), 39–100.

Hájek, A. (2019). Interpretations of probability. In E. N. Zalta (Ed.), *The Stanford encyclopedia of philosophy*. Retrieved from https://plato.stanford.edu/archives/fall2019/entries/probability-interpret/

Heim, I. R. (1982). *The semantics of definite and indefinite noun phrases* (Unpublished doctoral dissertation). Retrieved from http://citeseerx.ist.psu.edu/viewdoc/download?doi=10.1.1.473 .1996&rep=rep1&type=pdf

Hiddleston, E. (2005). A causal theory of counterfactuals. *Noûs, 39*(4), 632–657.

Iatridou, S. (2000). The grammatical ingredients of counterfactuality. *Linguistic Inquiry, 31*(2), 231–270.

Ippolito, M. (2016). How similar is similar enough? *Semantics and Pragmatics, 9*(6), 1–60.

Kamp, H. (1981). A theory of truth and semantic representation. In J. A. Groenendijk, T. Janssen, & M. Stokhof (Eds.), *Formal methods in the study of language* (pp. 277–322). Dordrecht, Netherlands: Foris.

Kaufmann, S. (2013). Causal premise semantics. *Cognitive Science, 37*(6), 1136–1170.

Kolodny, N., & MacFarlane, J. (2010). Ifs and oughts. *Journal of Philosophy, 107*(3), 115–143.

Kratzer, A. (1981). The notional category of modality. In H.-J. Eikmeyer & H. Rieser (Eds.), *Words, worlds, and contexts: New approaches in word semantics* (pp. 38–74). Berlin, Germany: De Gruyter.

Kratzer, A. (1986). Conditionals. In A. M. Farley, P. T. Farley, & K.-E. McCullough (Eds.), *Proceedings of the 22nd Regional Meeting of the Chicago Linguistic Society* (pp. 1–15). Chicago, IL: University of Chicago.

Kratzer, A. (1991). Modality. In A. von Stechow & D. Wunderlich (Eds.), *Semantics: An international handbook of contemporary research* (pp. 639–650). Berlin, Germany: De Gruyter Mouton.

Kratzer, A. (2012). *Modals and conditionals: New and revised perspectives*. New York, NY: Oxford University Press.

Kripke, S. A. (1963). Semantical analysis of modal logic I: Normal modal propositional calculi. *Zeitschrift für Mathematische Logik und Grundlagen der Mathematik, 9*, 67–96.

Kvart, I. (1986). *A theory of counterfactuals*. Indianapolis, IN: Hackett.

Lewis, D. K. (1973). *Counterfactuals*. Cambridge, MA: Harvard University Press.

Lewis, D. K. (1975). Adverbs of quantification. In E. L. Keenan (Ed.), *Formal semantics of natural language* (pp. 3–15). Cambridge, England: Cambridge University Press.

Lewis, D. K. (1976). Probabilities of conditionals and conditional probabilities. *Philosophical Review, 85*(3), 297–315.

Lewis, K. S. (2018). Counterfactual discourse in context. *Noûs, 52*(3), 481–507.

Lucas, C. G., & Kemp, C. (2015). An improved probabilistic account of counterfactual reasoning. *Psychological Review, 122*(4), 700–734.

McGee, V. (1985). A counterexample to modus ponens. *Journal of Philosophy, 82*(9), 462–471.

McKay, T. J., & van Inwagen, P. (1977). Counterfactuals with disjunctive antecedents. *Philosophical Studies, 31*, 353–356.

Moss, S. (2012). On the pragmatics of counterfactuals. *Noûs, 46*(3), 561–586.

Nute, D. (1975). Counterfactuals. *Notre Dame Journal of Formal Logic, 16*(4), 476–482.

Pearl, J. (2009). *Causality: Models, reasoning, and inference* (2nd ed.). Cambridge, England: Cambridge University Press.

Peirce, C. S. (1896). The regenerated logic. *The Monist, 7*(1), 19–40.

Pollock, J. L. (1976). *Subjunctive reasoning*. Dordrecht, Netherlands: Reidel.

Quine, W. V. O. (1982). *Methods of logic* (4th ed.). Cambridge, MA: Harvard University Press.

Rips, L. J. (2010). Two causal theories of counterfactual conditionals. *Cognitive Science, 34*(2), 175–221.

Schulz, K. (2007). *Minimal models in semantics and pragmatics: Free choice, exhaustivity, and conditionals* (Unpublished doctoral dissertation). Retrieved from https://pure.uva.nl/ws/files /4368825/52782_Schulz.pdf

Schulz, K. (2011). "If you'd wiggled A, then B would've changed": Causality and counterfactual conditionals. *Synthese, 179*, 239–251.

Sloman, S. (2005). *Causal models: How people think about the world and its alternatives*. Oxford, England: Oxford University Press.

Stalnaker, R. (1968). A theory of conditionals. In N. Rescher (Ed.), *Studies in logical theory* (pp. 98–112). Oxford, England: Blackwell.

Stalnaker, R. (1975). Indicative conditionals. *Philosophia, 5*, 269–286.

Stalnaker, R. C., & Thomason, R. H. (1970). A semantic analysis of conditional logic. *Theoria, 36*, 23–42.

Starr, W. B. (2014). A uniform theory of conditionals. *Journal of Philosophical Logic, 43*(6), 1019–1064.

Starr, W. B. (2019). Counterfactuals. In E. N. Zalta (Ed.), *Stanford encyclopedia of philosophy*. Retrieved from https://plato .stanford.edu/archives/fall2019/entries/counterfactuals/

Stojnić, U. (2016). One's *modus ponens*: Modality, coherence and logic. *Philosophy and Phenomenological Research, 95*(1), 167–214.

Veltman, F. (1986). Data semantics and the pragmatics of indicative conditionals. In E. C. Traugott, A. ter Meulen, J. S. Reilly, & C. A. Ferguson (Eds.), *On conditionals* (pp. 147–168). Cambridge, England: Cambridge University Press.

Veltman, F. (1996). Defaults in update semantics. *Journal of Philosophical Logic, 25*(3), 221–261.

von Fintel, K. (1999). The presupposition of subjunctive conditionals. In U. Sauerland & O. Percus (Eds.), *The interpretive tract: Working papers in syntax and semantics* (MIT Working Papers in Linguistics, Vol. 25, pp. 29–44). Cambridge: Massachusetts Institute of Technology, Department of Linguistics.

von Fintel, K. (2001). Counterfactuals in a dynamic context. In M. Kenstowicz (Ed.), *Ken Hale: A life in language* (pp. 123–152). Cambridge, MA: MIT Press.

von Fintel, K. (2012). Subjunctive conditionals. In G. Russell & D. G. Fara (Eds.), *The Routledge companion to philosophy of language* (pp. 466–477). New York, NY: Routledge.

Warmbrōd, K. (1981). An indexical theory of conditionals. *Dialogue, Canadian Philosophical Review, 20*(4), 644–664.

Willer, M. (2012). A remark on iffy oughts. *Journal of Philosophy, 109*(7), 449–461.

Willer, M. (2017). Lessons from Sobel sequences. *Semantics and Pragmatics, 10*. doi:10.3765/sp.10.4

6.2 The Suppositional Theory of Conditionals and Rationality

David E. Over and Nicole Cruz

Summary

In a suppositional theory, a conditional, *if A then B*, is interpreted as *B, supposing A*, and its probability, *P(if A then B)*, is the conditional probability of *B* given *A*, *P(B|A)*. Conditionals are ubiquitous in reasoning and essential to it. There is strong experimental support for the hypothesis that *P(if A then B) = P(B|A)* in people's judgments, and this finding opens up the possibility of giving a unified Bayesian account of human rationality, reasoning, and decision making. Probability theory can be used to define validity in reasoning, and coherence intervals can be specified for the conclusions of inferences. It is rational, by Bayesian standards, to conform to these intervals, and by this measure, people are often rational but sometimes commit fallacies and have biases. Investigating the coherence and incoherence in people's reasoning will give us a deeper understanding of its rationality.

1. Conditionals and Supposing

The suppositional theory of conditionals (Edgington, 2014) is a contribution to the Bayesian analysis of human rationality and its understanding of subjective probability judgments as expressing degrees of belief. To state the theory informally, it treats a natural-language indicative conditional, *if A then B*, as equivalent to *B, supposing A*, and it identifies the subjective probability of the conditional, *P(if A then B)*, with the probability of *B* given *A*, so that *P(if A then B) = P(B|A)*. This identity, which originates in foundational work on subjective probability theory (de Finetti, 1936/1995, 1937/1964; Ramsey, 1926/1990, 1929/1990), has such deep implications for a Bayesian account of rationality and reasoning that it has been called *the Equation* (Edgington, 1995).

Given the Equation for the natural-language conditional, the normative standard for rational reasoning with it is Bayesian probability theory (chapter 4.5 by Chater & Oaksford, this handbook; Oaksford & Chater,

2007; Over, 2020; Over & Cruz, 2018). In the psychology of reasoning, the Equation becomes *the conditional probability hypothesis*: that people's probability judgments will conform to *P(if A then B) = P(B|A)*. A suppositional theory of the conditional, as we will interpret it, has both normative and descriptive aspects. It endorses the Equation as setting the correct normative standard for people's conditional reasoning and implies the descriptive conditional probability hypothesis. Supporters of the theory aim to confirm the hypothesis in experimental studies and to investigate how far people comply with the standards of probability theory in their conditional reasoning. Suppositional theory does not imply this compliance will be perfect. Even if people judge that *P(if A then B) = P(B|A)*, cognitive limitations will sometimes result in biases and fallacious inferences. However, suppositional theory does propose that people's conditional reasoning will be more accurately described by probability theory than by binary extensional logic (Elqayam & Over, 2013).

Reasoning from *A* to *B* (supplemented with background beliefs) can be "summed up" with a conditional assertion *if A then B*, and assertions of *if A then B* can be supported by reasoning from *A* to *B* (plus background information). Thus, any research program implying that *P(if A then B) = P(B|A)*, as a normative principle and as broadly descriptively adequate, places conditional probability at the center of its account of human reasoning. This identity opens up the possibility of integrating the study of reasoning with that of judgment and decision making, which depends so much on conditional probability, so producing an overall account, normative and descriptive, of human rationality (for developments along these lines, see Oaksford & Chater, 2007, 2020; for the debate in psychology about human rationality, see chapter 1.2 by Evans, this handbook).

2. The Conditional Probability Hypothesis

Psychological research has highly corroborated the conditional probability hypothesis as descriptive of people's probability judgments across a wide range of conditionals (Evans, Handley, Neilens, & Over, 2007; Evans, Handley, & Over, 2003; Fugard, Pfeifer, Mayerhofer, & Kleiter, 2011; Oberauer & Wilhelm, 2003; chapter 4.6 by Oberauer & Pessach, this handbook; Over, Hadjichristidis, Evans, Handley, & Sloman, 2007; Pfeifer & Kleiter, 2005; Singmann, Klauer, & Over, 2014). Its intuitive appeal can be appreciated by considering these examples:

(1) If the fair coin is tossed (*F*), then the result will be Heads (*H*).
(2) If the double-headed coin is tossed (*D*), then the result will be Tails (*T*).

The above are *singular conditionals*. They are about specific events: the next toss of the fair, or the double-headed, coin. A *general* conditional is about a whole domain of objects, as in "If an animal is a bird, then it can fly." There are clearly counterexamples to this conditional in its domain, there being birds that cannot fly. A singular conditional with a similar topic would be, "If the next animal we see is a bird, then it will be able to fly." This conditional could be highly probable, depending on where we are in the world, and could turn out to be true. We will focus on singular conditionals like (1) and (2) in this chapter (see Cruz & Oberauer, 2014, on general conditionals).

Intuitively, the probability of (1), $P(if\ F\ then\ H)$, is .5, and the probability of (2), $P(if\ D\ then\ T)$, is 0. These are the correct judgments by the Equation, and those that we would predict using the conditional probability hypothesis, since $P(H|F) = .5$ and $P(T|D) = 0$. These judgments are not, however, correct and predictable if the natural-language conditional is interpreted as the *material conditional* of binary extensional logic, $P(if\ A\ then\ B) = P(A \supset B) = P(not\text{-}A\ or\ B)$. The material conditional $A \supset B$ is logically equivalent to the disjunction *not-A or B*, and similarly, $not\text{-}A \supset B$ is logically equivalent (*not-not-A* becoming *A*) to *A or B*. However, the identification of $P(if\ A\ then\ B)$ with $P(not\text{-}A\ or\ B)$ is most counterintuitive, and the same can be said for identifying $P(if\ not\text{-}A\ then\ B)$ with $P(A\ or\ B)$. Assume the coins referred to in (1) and (2) are valuable ancient Roman examples in a locked museum display case. Then the probabilities that they will be tossed, $P(F)$ and $P(D)$, are very low, making $P(not\text{-}F)$ and $P(not\text{-}D)$, and so $P(not\text{-}F\ or\ H)$ and $P(not\text{-}D\ or\ T)$, very high. But it is absurd to claim that the probabilities of (1) and (2) are very high merely because these coins are highly unlikely to be tossed, and it is difficult to see how rational reasoning and rational decision making, with natural-language conditionals *if A then B*, can be given a unified treatment when $P(if\ A\ then\ B) = P(A \supset B) = P(not\text{-}A\ or\ B)$.

The logical problem with identifying, for instance, $P(if\ F\ then\ H)$ and $P(not\text{-}F\ or\ H)$ is that it is valid to infer *not-F or H* from *not-F*, which entails that $P(not\text{-}F) \leq P(not\text{-}F\ or\ H)$, forcing $P(not\text{-}F\ or\ H)$ to be high whenever $P(not\text{-}F)$ is high. A truth table can be used in classical logic to prove that it is logically valid to infer *A or B* from *A*, an inference termed *or-introduction*. Table 6.2.1 is the truth table for *A or B*, and it clearly shows that *A or B* is true when *A* is true, making the inference from *A* to *A or B* valid, with the further normative result that $P(A) \leq P(A\ or\ B)$. Table 6.2.2 is the truth table for the material conditional $A \supset B$ or, equivalently, *not-A or B*, which validly follows from *not-A*. But to claim that it is valid to infer a natural-language conditional *if A then B* from *not-A* is one of the "paradoxes" of the material conditional (Edgington, 2014; Evans & Over, 2004). Its validity entails that $P(not\text{-}A) \leq P(if\ A\ then\ B)$, which produces the highly counterintuitive results we have seen in the example above, and people have been found to reject the validity of this inference in their probability judgments (Cruz, Over, & Oaksford, 2017; Pfeifer & Kleiter, 2011; Politzer & Baratgin, 2016).

The only major theory in the psychology of reasoning that closely related (until recently) people's representation of the natural-language conditional to the material conditional was *mental model theory*, with the result that the paradoxes of the material conditional were "valid" in this theory (Johnson-Laird & Byrne,

Table 6.2.1

The truth table for the disjunction *A or B*

A	B	A or B
1	1	1
1	0	1
0	1	1
0	0	0

Note: 1 = true; 0 = false.

Table 6.2.2

The truth table for the material conditional $A \supset B$, equivalent to *not-A or B*

A	B	A⊃B
1	1	1
1	0	0
0	1	1
0	0	1

Note: 1 = true; 0 = false.

1991, pp. 73–74). But counterintuitive implications like those described above have led to a radical revision of mental model theory, in which the paradoxes are no longer considered valid (Johnson-Laird, Khemlani, & Goodwin, 2015). The revision is in a state of development (chapter 2.3 by Johnson-Laird, this handbook), and we will not consider it here (but see Baratgin et al., 2015; Oaksford, Over, & Cruz, 2019; Over, 2020; Over & Cruz, 2018).

Support for the conditional probability hypothesis also disconfirms the claim (Johnson-Laird & Byrne, 1991, p. 74) that a natural-language conditional can be given the same truth table as the material conditional (table 6.2.2). This table entails that $P(A \supset B) = P(not\text{-}A$ $or\ B)$, but $P(B \mid A) = P(not\text{-}A\ or\ B)$ only in certain extreme cases (Gilio & Over, 2012), for example, when $P(A) = P(B) = 1$. Four possibilities are displayed in table 6.2.2: $A \& B$, $A \& not\text{-}B$, $not\text{-}A \& B$, and $not\text{-}A \& not\text{-}B$. As an example, assume we judge that these possible states have the same probability: $P(A \& B) = P(A \& not\text{-}B) = P(not\text{-}A \& B) = P(not\text{-}A \& not\text{-}B) = .25$. We now clearly see that $P(B \mid A) = P(A \& B)/P(A) = .5$, but $P(not\text{-}A$ $or\ B) = P(A \& B) + P(not\text{-}A \& B) + P(not\text{-}A \& not\text{-}B) = .75$. More informally, the problem is that $not\text{-}A\ or\ B$ makes a disjunctive statement about the $not\text{-}A$-possibilities, with $not\text{-}A\ or\ B$ true in those possibilities, and so the probability of these $not\text{-}A$-cases contributes to $P(not\text{-}A\ or\ B)$ but not to $P(B \mid A)$. Lewis (1976) proved a much more general result. He considered conditionals like those in Lewis (1973) and Stalnaker (1968). We might call these Lewis-style conditionals and symbolize them with $if\ A$ $then_L\ B$, which means that B holds in the closest possible world (or worlds) in which A holds. In this way, if $A\ then_L\ B$ can hold in a $not\text{-}A$-world, and that implies that $P(if\ A\ then_L\ B)$ can be affected by $P(not\text{-}A)$. Because of that, Lewis was able to prove that $P(if\ A\ then_L\ B)$ cannot generally be identified with $P(B \mid A)$, which depends only on A-worlds.

3. The Ramsey Test and the de Finetti and Jeffrey Tables

A suppositional conditional is not a material conditional or a Lewis-style conditional (Edgington, 2014). In suppositional theory, $if\ A\ then\ B$ means B, supposing A, which does not make any statement about, or depend upon, $not\text{-}A$-possibilities. We can make the identification that $P(if\ A\ then\ B) = P(B \mid A)$ in a suppositional theory, and so this conditional does satisfy the Equation. It can simply be called a *suppositional conditional*, but it has also been termed a *probability conditional* (Adams, 1998) and a

conditional event (de Finetti, 1936/1995, 1937/1964). We can assess $P(if\ A\ then\ B)$ for a suppositional conditional by using the *Ramsey test*, which was first described by Ramsey (1929/1990) and later extended by Stalnaker (1968). In this psychological process, people make a judgment about $P(if\ A\ then\ B)$ by supposing A, while making minimal changes to preserve (or restore) consistency in their beliefs, and then judging their degree of belief in B under this supposition, yielding their subjective conditional probability for B given A, $P(B \mid A)$. The Ramsey test has had a major impact in philosophy, psychology, and other parts of cognitive science (Edgington, 1995; Evans & Over, 2004; Oaksford & Chater, 2007; Pearl, 2013).

We can explain the Ramsey test more fully with an example: assume we are in a car and trying to make a rational decision about which route to take home one particular afternoon. We are discussing the following two conditionals relevant to this decision:

(3) If we take the short city route (S), it will be a quick journey (Q).

(4) If we take the long countryside route (L), it will be an enjoyable journey (E).

The short city route home is an uninteresting drive, but the long countryside route can be a pleasant drive. Of course, it is not certain that the city route will be quick and the countryside route enjoyable. A number of uncertain factors can affect these outcomes: traffic jams, changing weather conditions, and accidents. For a rational decision, we must assess both $P(if\ S\ then\ Q)$ and $P(if$ $L\ then\ E)$, and we can do that via the Ramsey test and the Equation. We might make the first Ramsey test on (3), by supposing S and making a judgment about how likely Q is to follow, and the second on (4), by supposing L and making a judgment about how likely E is to follow. Perhaps in both tests, we judge that traffic jams and so on are unlikely on this particular day, and so we end the tests by judging that $P(Q \mid S)$ and $P(E \mid L)$ are both high, giving us high confidence in (3) and (4). A rational decision would then depend on our relative desires, that particular afternoon, for a quick drive as opposed to a pleasant one.

Let us say that, after some pondering, we decide that afternoon to take the longer but more pleasant drive, and we set off on the long countryside route. Now our probability judgments change, and we become certain that S is false: $P(not\text{-}S)$ becomes 1. In this case, Ramsey (1929/1990) said that we would lose interest in (3), which he described as becoming "void," and de Finetti (1936/1995, 1937/1964) also took this view and used "void" in this way. We would not use (3) when we were on the countryside route and still less an indicative conditional beginning, "If we are on the city route." We could, though, take an interest in and assert a counterfactual:

(5) If we had taken the short city route home, it would have been a quick journey.

Counterfactuals are a topic of considerable interest in cognitive, social, and developmental psychology. For instance, Kahneman and Miller (1986) established the interesting result that, when the city route is our usual way home, we would feel more regret when we had an accident on the countryside route than when we had an accident on the city route. Our relatively intense regret after an accident on the countryside route and our high confidence in counterfactual (5), which would come from a Ramsey test on (5) similar to the one on (3), would be closely linked. This regret could affect, rationally or irrationally, our future decision making about which route to take. With space limitations, we will, however, continue to focus on indicative conditionals, like (1)–(4), in the rest of this chapter (but see Over, 2017; Over & Cruz, 2019).

In de Finetti's account of indicative conditionals, *if A then B* is true when *A* and *B* hold, false when *A* holds and *B* does not, and "void" when *A* does not hold. These three values can be represented in what has been called a *de Finetti table*, and we display this as table 6.2.3. In traditional research on truth tables in the psychology of reasoning, participants produce a so-called defective table that arguably matches table 6.2.3 and confirms de Finetti's analysis (Baratgin, Over, & Politzer, 2013; Over & Baratgin, 2017; Politzer, Over, & Baratgin, 2010). As we have already noted, Ramsey also used "void" for an indicative conditional with an antecedent known to be false. He added that we lose interest in these indicative conditionals "except as a question about what follows from certain laws or hypotheses" (Ramsey, 1929/1990, p. 155). This suggests that conditionals that follow from logical laws, such as *if A & B then B*, are not void but rather certainly true (Baratgin et al., 2013), and the Ramsey test itself implies that, in general, a conditional with a false antecedent is only void in not having a *factual* truth value, and that it has its conditional probability as a subjective value. A table in which *if A then B* has the value $P(B|A)$ when *A* is false has been called a *Jeffrey table* (after Jeffrey, 1991), and it is found, explicitly or implicitly, in contemporary followers of de Finetti and Ramsey (Cruz & Oberauer, 2014; Over & Baratgin, 2017; Over & Cruz, 2018; Pfeifer & Kleiter, 2009). Table 6.2.4 is the Jeffrey table for *if A then B*.

Table 6.2.3

The de Finetti table for *if A then B*

A	B	if A then B
1	1	1
1	0	0
0	1	V
0	0	V

Note: 1 = true; 0 = false; V = void.

Table 6.2.4

The Jeffrey table for *if A then B*

A	B	if A then B
1	1	1
1	0	0
0	1	$P(B|A)$
0	0	$P(B|A)$

Note: 1 = true, 0 = false; $P(B|A)$ = the subjective conditional probability of *B* given *A*.

4. Rational Reasoning from Degrees of Belief

The suppositional theory of conditionals employs the Ramsey test and de Finetti and Jeffrey tables to give a probabilistic account of conditionals. It rests on the foundational work of Ramsey and de Finetti on conditionals and, more generally, on their identification of people's probability judgments with their degrees of belief. With its focus on degrees of belief, suppositional theory is a prime example of the new Bayesian paradigm in the psychology of reasoning (Elqayam & Over, 2013). Contributions to the traditional paradigm were primarily concerned with inferences from premises that were supposed to be *assumed* true, which meant that these were, in effect, taken to be certain, and the relevant normative standard was binary extensional logic (Johnson-Laird & Byrne, 1991; Over & Cruz, 2018). However, most human inferences, in ordinary affairs and science, are from uncertain beliefs or from hypotheses that, though plausible, are not certain. This uncertainty has to be taken into account in reasoning to arrive at a rational degree of confidence in a conclusion.

For instance, after we decide to take the countryside route home and set off on it, we can use the valid inference form of modus ponens (MP) to infer that we will have an enjoyable journey from (4) as an additional premise. More formally, we would use MP to infer *E* from *if L then E* as the major premise and *L* as the minor premise. We would not, though, treat (4) as certain or simply assume it. Possible *disabling* conditions, like a sudden change in the weather, could make (4) false. By the Ramsey test, (4) has a probability that is equal to $P(E|L)$, which is high but is not 1, with 1 representing certainty.

But let us consider first the simpler case of a one-premise inference, rather than a two-premise inference, like MP. In their classic article on conjunction, Tversky and Kahneman (1983) pointed out that, because it is logically valid to infer B from $A \& B$, people should judge that $P(B) \geq P(A \& B)$. Tversky and Kahneman relied on a necessary relation between logical validity and *coherent* probability judgments, which conform to the rules of probability theory. To be incoherent by violating the principle that $P(B) \geq P(A \& B)$ in one's probability judgments is to commit what Tversky and Kahneman called the *conjunction fallacy*.

The coherence of probability judgments can be used to define validity. We can say that a one-premise inference is probabilistically valid, *p-valid*, if and only if the probability of the conclusion can never be coherently lower than the probability of the premise. For an inference with two or more premises, we first define the uncertainty of a statement as 1 minus its probability. For example, when $P(A) = .6$, the uncertainty of A is .4. Then an inference is *p-valid* if and only if, for all coherent probability assignments, the uncertainty of the conclusion is never greater than the sum of the uncertainties of the premises (Adams, 1998). A p-valid inference never increases the uncertainty present in its premises, and it differs from a strong inductive argument in precisely this respect (Evans & Over, 2013).

The one-premise inference from $A \& B$ to B is known as *&-elimination* in logic, and we can ask what rational degree of confidence we should have in B when there is some uncertainty about $A \& B$. The answer given above is that our degree of confidence should be such that $P(B) \geq P(A \& B)$. More formally, for $P(A \& B) = x$, the *coherence interval* for this inference is $[x, 1]$, meaning that $P(B)$ should lie between x and 1 inclusive.

Coherence intervals can also be derived for inferences with more than one premise. The coherence interval for MP follows from the Equation and the total probability theorem of probability theory. Consider (4) again, inferring E from *if L then E* and L. By the total probability theorem,

(6) $P(E) = P(L)P(E|L) + P(not\text{-}L)P(E|not\text{-}L)$.

When we definitely decide to take the long countryside route home, we judge $P(L) = 1$, and we can infer from

(6) that our confidence that we will have an enjoyable journey is $P(E|L)$, which we might assess as .8 since there is some modest probability of a disabling condition that could make our journey unpleasant. Other people might agree that $P(E|L) = .8$, but not knowing what we have decided, perhaps set $P(L) = .5$. Using (6), they can then compute $P(L)P(E|L) = (.5)(.8) = .4$. When $P(L) = .5$, $P(not\text{-}L)$ is of course also .5. The last part of equation (6), $P(E|not\text{-}L)$, might be unknown. But by probability theory, it has to lie between 0 and 1. When it is 0, $P(E) = P(L)P(E|L) = .4$. And when it is 1, $P(E) = P(L)P(E|L) + P(not\text{-}L) = .9$. To be coherent, the probability of the conclusion of MP for these people must then lie in the interval $[.4, .9]$, that is, between .4 and .9.

The derivations of coherence intervals for inferences (Evans, Thompson, & Over, 2015; chapter 4.4 by Pfeifer, this handbook; Pfeifer & Kleiter, 2009) yield a normative theory for the rational degrees of confidence people should have in the conclusions they infer from uncertain premises. To fall outside the coherence interval for an inference is to be incoherent and exposed to a *Dutch book* (i.e., a bet that will be lost no matter which possible outcome occurs). Dutch book arguments are the foundation of the normative Bayesian theory of coherence and rationality (Howson & Urbach, 2005).

People are not always coherent in their degrees of belief, but it is quite plausible that explicit inference would help them to be more rational at this basic level, for otherwise what benefit can come from the time and effort that go into explicit inference? Studies have shown that people are sometimes more coherent in the degrees of belief that they have in the conclusions of explicit inferences, but not necessarily (Cruz, Baratgin, Oaksford, & Over, 2015; Cruz, Over, Oaksford, & Baratgin, 2016; Cruz et al., 2017; Evans et al., 2015; Singmann et al., 2014). A striking example of when explicit inference does not facilitate coherence is a context in which the conjunction fallacy occurs (Tversky & Kahneman, 1983). In this context, people can incoherently judge $P(B) < P(A \& B)$ even when they explicitly infer B from $A \& B$ (Cruz et al., 2015) and even though their judgments for the same inference are coherent when using neutral contents (Cruz et al., 2015; Politzer & Baratgin, 2016). This finding underlines the robustness of the fallacy and the importance of accounting for it (Costello & Watts, 2014; Tentori, Crupi, & Russo, 2013).

A further important inference form for the psychology of reasoning is *centering*. One-premise centering is sometimes called *conjunctive sufficiency*: it is the inference from $A \& B$ as a single premise to *if A then B* as a conclusion. Two-premise centering is the inference from *A* and *B* as separate premises to *if A then B* as a conclusion. Centering is a valid inference for a wide range of conditionals, including the material conditional, Lewis-style conditionals, and the suppositional conditional, and there is evidence that people conform to it in their degrees of belief (Cruz et al., 2015; Cruz et al., 2016; Pfeifer & Tulkki, 2017; Politzer & Baratgin, 2016). In contrast, the inference is invalid in *inferentialist* theories of the conditional (Douven, 2015, 2016; Douven, Elqayam, Singmann, & van Wijnbergen-Huitink, 2018). In semantic versions of these theories, *if A then B* means that there is a deductive, inductive, or abductive link between *A* and *B*, and *if A then B* can only be true when there is such a link between *A* and *B*. Supporters of this position argue that centering will be rejected for *missing-link* conditionals. These are conditionals *if A then B* with no relation between *A* and *B*, and the claim is that these conditionals do not follow from the truth of *A* & *B*. The following is an example of a missing-link conditional:

(7) If it is sunny in Canberra (*C*) today, then it is raining in London (*N*).

Supposing that *C* and *N* are both true, and so *C* & *N* is true, inferentialists would argue that (7) does not follow. It certainly could be pragmatically very odd in a discussion with someone to begin with *C* and *N* as premises, for the purpose of explicitly inferring (7) as a conclusion. On the other hand, if participants in an experiment are given a missing-link conditional *if A then B* first, followed by confirmation of *A* and *B*, they hold that *if A then B* is true (Skovgaard-Olsen, Kellen, Krahl, & Klauer, 2017, figure 1). Indeed, if we were to assert (7) first in a discussion, however irrelevantly or oddly, and were to place a bet on it, we would surely claim to have stated a truth and to have won our bet when *C* and *L* turned out later to be true. This claim is in accordance with the de Finetti and Jeffrey tables for *if A then B* and a bet on it, and it has been strongly supported empirically in truth table tasks (see also Evans et al., 2003; Oberauer & Wilhelm, 2003), but it goes against the assumptions of inferentialism.

On the other hand, it is hard to see how (7), or any other missing-link conditional, can be used in rational decision making. When there is no epistemic relation between *A* and *B*, making *A* and *B* independent, $P(B|A) = P(B)$, and *A* cannot be a reason for doing *B*. There is evidence that in some contexts, people conform to the Equation more often when $P(B|A) > P(B|not\text{-}A)$, suggesting an interpretation of *if A then B* in which *A* raises the probability of *B* (Skovgaard-Olsen, Singmann, & Klauer, 2016), but there is also evidence *against* this prediction (Oberauer,

Weidenfeld, & Fischer, 2007; Over et al., 2007; Singmann et al., 2014). Cruz et al. (2016) find that the lack of an epistemic relation between *A* and *B* affects probability judgments not only about conditionals but also about conjunctions *A* & *B* and disjunctions *A or B*. This implies that a missing-link effect on probability judgments is not specific to conditionals, suggesting that it is pragmatic and not semantic. An effect that is not specific to conditionals cannot be used to distinguish the meaning of conditionals from that of other statement types. The findings of Cruz et al. (2016) also suggest that having a common topic of discourse between *A* and *B* is more important for the pragmatic acceptability of statements than a specific epistemic relation (see, on this, Over & Cruz, 2018; for a related account of why the computation of subjective probabilities for compound events is more difficult when *A* and *B* come from different domains, see Sanborn & Chater, 2016). The uselessness of missing-link conditionals for rational decision making does not have to be a matter of semantic meaning.

5. Rational Belief Updating

Suppositional reasoning is fundamental to rational decision making and to the belief revision and updating that goes with it. In scientific Bayesian inference, we might assign a low prior probability to a hypothesis *H* at time 1, making $P_1(H)$ low, and infer (via Bayes' theorem) a high probability of *H* supposing we get an finding *F* in an experiment, making $P_1(H|F)$ high. Assume we now run an experiment at time 2 and learn that *F* holds (and nothing else relevant). We can then use the new information about *F* to update our degree of belief in *H* at time 2, $P_2(H) = P_1(H|F)$, in an inference termed *(strict) conditionalization* (Howson & Urbach, 2005).

In a suppositional theory, we can identify this Bayesian process with conditional reasoning over time, in which we begin with $P_1(H)$ low and $P_1(if\ F\ then\ H)$ high, and then after we run the experiment and find that $P_2(F)$ holds, we can infer that $P_2(H)$ is high using *dynamic MP* (Oaksford & Chater, 2013): inferring high confidence in *H* from high confidence in *if F then H* after learning *F* as a result of acquiring new information. Bayesian inference like this is often an essential part of ordinary rational decision making.

A little more formally, suppose we begin at time 1 with a degree of confidence in *H*, say $P_1(H) = .5$, and in *H* given *F*, say $P_1(H|F) = P(if\ F\ then\ H) = .8$. At a later time 2, after acquiring new information, we judge that $P_2(F) = 1$, and we can then infer that $P_2(H) = .8$. When we are less sure of *F* at time 2, say $P_2(F) = .7$, we can use

Jeffrey conditionalization (Jeffrey, 1983) to infer what our confidence in H should be at time 2:

(8) $P_2(H) = P_2(F)P_1(H \mid F) + P_2(not\text{-}F)P_1(H \mid not\text{-}F)$.

Jeffrey conditionalization is derived from the total probability theorem, an instance of which is (6). Both it and (strict) conditionalization depend on the assumption that the conditional probabilities are *invariant* (Oaksford & Chater, 2013), which means that $P_1(H \mid F) = P_2(H \mid F)$ and $P_1(H \mid not\text{-}F) = P_2(H \mid not\text{-}F)$. People do not always conform to Jeffrey conditionalization (Hadjichristidis, Sloman, & Over, 2014; Zhao & Osherson, 2010), perhaps because they tend to focus on the component— either $P_2(F)P_1(H \mid F)$ or $P_2(not\text{-}F)P_1(H \mid not\text{-}F)$—that has the higher probability and do not take into account the less probable one.

The problem of making a precise decision about $P_1(H \mid not\text{-}F)$ can be bypassed by considering the coherence interval: the minimum value of $P_1(H \mid not\text{-}F)$ is 0, and the maximum value is 1, allowing us to derive, as above, a coherence interval for Jeffrey conditionalization, justified when invariance holds. How far people's dynamic reasoning and belief updating are coherent, and their subjective conditional probabilities invariant, are important questions for future research.

6. Conclusion

In suppositional theory, a natural-language indicative conditional, *if A then B*, is equivalent to *B, supposing A*, and its probability is the subjective conditional probability of its consequent given its antecedent: $P(if\ A\ then\ B) = P(B \mid A)$. This theory is a central part of a fully Bayesian approach to the psychology of reasoning and of judgment and decision making, which aims to unify these two subject areas and to study how degrees of belief are revised and updated in human reasoning, as well as how this process affects human decision making. The underlying standard of rationality in this approach, for both reasoning and decision making, is coherence, which means consistency with probability theory. People are sometimes coherent and sometimes incoherent in their reasoning and decision making, and unifying the two subject areas should lead to deeper understanding of why this is so.

References

Adams, E. W. (1998). *A primer of probability logic*. Stanford, CA: CSLI Publications.

Baratgin, J., Douven, I., Evans, J. St. B. T., Oaksford, M., Over, D. E., & Politzer, G. (2015). The new paradigm and mental models. *Trends in Cognitive Sciences, 19*(10), 547–548.

Baratgin, J., Over, D. E., & Politzer, G. (2013). Uncertainty and de Finetti tables. *Thinking & Reasoning, 19*, 308–328.

Costello, F., & Watts, P. (2014). Surprisingly rational: Probability theory plus noise explains biases in judgment. *Psychological Review, 121*(3), 463–480.

Cruz, N., Baratgin, J., Oaksford, M., & Over, D. E. (2015). Bayesian reasoning with ifs and ands and ors. *Frontiers in Psychology, 6*, 192.

Cruz, N., & Oberauer, K. (2014). Comparing the meanings of "if" and "all." *Memory & Cognition, 42*, 1345–1356.

Cruz, N., Over, D., & Oaksford, M. (2017). The elusive oddness of or-introduction. In G. Gunzelmann, A. Howes, T. Tenbrink, & E. Davelaar (Eds.), *Proceedings of the 39th Annual Conference of the Cognitive Science Society* (pp. 259–264). Austin, TX: Cognitive Science Society.

Cruz, N., Over, D., Oaksford, M., & Baratgin, J. (2016). Centering and the meaning of conditionals. In A. Papafragou, D. Grodner, D. Mirman, & J. C. Trueswell (Eds.), *Proceedings of the 38th Annual Conference of the Cognitive Science Society* (pp. 1104–1109). Austin, TX: Cognitive Science Society.

de Finetti, B. (1995). The logic of probability (R. B. Angell, Trans.). *Philosophical Studies, 77*, 181–190. (Original work published 1936)

de Finetti, B. (1964). Foresight: Its logical laws, its subjective sources. In H. E. Kyburg & H. E. Smokler (Eds.), *Studies in subjective probability* (pp. 55–118). New York, NY: Wiley. (Original work published 1937)

Douven, I. (2015). How to account for the oddness of missing-link conditionals. *Synthese, 194*(5), 1–14.

Douven, I. (2016). *The epistemology of indicative conditionals*. Cambridge, England: Cambridge University Press.

Douven, I., Elqayam, S., Singmann, H., & van Wijnbergen-Huitink, J. (2018). Conditionals and inferential connections: A hypothetical inferential theory. *Cognitive Psychology, 101*, 50–81.

Edgington, D. (1995). On conditionals. *Mind, 104*, 235–329.

Edgington, D. (2014). Indicative conditionals. In E. N. Zalta (Ed.), *The Stanford encyclopedia of philosophy*. Retrieved from http://plato.stanford.edu/archives/win2014/entries/conditionals/

Elqayam, S., & Over, D. E. (2013). New paradigm psychology of reasoning: An introduction to the special issue edited by Elqayam, Bonnefon, and Over. *Thinking & Reasoning, 19*, 249–265.

Evans, J. St. B. T., Handley, S., Neilens, H., & Over, D. E. (2007). Thinking about conditionals: A study of individual differences. *Memory & Cognition, 35*, 1772–1784.

Evans, J. St. B. T., Handley, S. J., & Over, D. E. (2003). Conditionals and conditional probability. *Journal of Experimental Psychology: Learning, Memory, and Cognition, 29*, 321–335.

Evans, J. St. B. T., & Over, D. E. (2004). *If*. Oxford, England: Oxford University Press.

Evans, J. St. B. T., & Over, D. E. (2013). Reasoning to and from belief: Deduction and induction are still distinct. *Thinking & Reasoning, 19*, 267–283.

Evans, J. St. B. T., Thompson, V. A., & Over, D. E. (2015). Uncertain deduction and conditional reasoning. *Frontiers in Psychology, 6*, 398.

Fugard, A. J. B., Pfeifer, N., Mayerhofer, B., & Kleiter, G. D. (2011). How people interpret conditionals: Shifts toward the conditional event. *Journal of Experimental Psychology: Learning, Memory, and Cognition, 37*, 635–648.

Gilio, A., & Over, D. E. (2012). The psychology of inferring conditionals from disjunctions: A probabilistic study. *Journal of Mathematical Psychology, 56*, 118–131.

Hadjichristidis, C., Sloman, S. A., & Over, D. E. (2014). Categorical induction from uncertain premises: Jeffrey's doesn't completely rule. *Thinking & Reasoning, 20*(4), 405–431.

Howson, C., & Urbach, P. (2005). *Scientific reasoning: The Bayesian approach* (3rd ed.). Chicago, IL: Open Court.

Jeffrey, R. C. (1983). *The logic of decision* (2nd ed.). Chicago, IL: University of Chicago Press.

Jeffrey, R. (1991). Matter-of-fact conditionals I. *Proceedings of the Aristotelian Society, 65*, 161–183.

Johnson-Laird, P. N., & Byrne, R. M. J. (1991). *Deduction*. Hove, England: Erlbaum.

Johnson-Laird, P. N., Khemlani, S., & Goodwin, G. P. (2015). Logic, probability, and human reasoning. *Trends in Cognitive Science, 19*, 201–214.

Kahneman, D., & Miller, D. T. (1986). Norm theory: Comparing reality to its alternatives. *Psychological Review, 93*, 136–156.

Lewis, D. (1973). *Counterfactuals*. Cambridge, MA: Harvard University Press.

Lewis, D. (1976). Probabilities of conditionals and conditional probabilities. *Philosophical Review, 85*, 297–315.

Oaksford, M., & Chater, N. (2007). *Bayesian rationality: The probabilistic approach to human reasoning*. Oxford, England: Oxford University Press.

Oaksford, M., & Chater, N. (2013). Dynamic inference and everyday conditional reasoning in the new paradigm. *Thinking & Reasoning, 19*, 346–379.

Oaksford, M., & Chater, N. (2020). New paradigms in the psychology of reasoning. *Annual Review of Psychology, 71*, 305–330.

Oaksford, M., Over, D. E., & Cruz, N. (2019). Paradigms, possibilities, and probabilities: Comment on Hinterecker et al. (2016). *Journal of Experimental Psychology: Learning, Memory, and Cognition, 45*, 288–297.

Oberauer, K., Weidenfeld, A., & Fischer, K. (2007). What makes us believe a conditional? The roles of covariation and causality. *Thinking & Reasoning, 13*(4), 340–369.

Oberauer, K., & Wilhelm, O. (2003). The meaning(s) of conditionals: Conditional probabilities, mental models, and personal utilities. *Journal of Experimental Psychology: Learning, Memory, and Cognition, 29*, 680–693.

Over, D. E. (2017). Causation and the probability of causal conditionals. In M. Waldmann (Ed.), *The Oxford handbook of causal reasoning* (pp. 307–325). Oxford, England: Oxford University Press.

Over, D. E. (2020). The development of the new paradigm in the psychology of reasoning. In S. Elqayam, I. Douven, J. St. B. T. Evans, & N. Cruz (Eds.), *Logic and uncertainty in the human mind* (pp. 243–263). Abingdon, England: Routledge.

Over, D. E., & Baratgin, J. (2017). The "defective" truth table: Its past, present, and future. In N. Galbraith, D. E. Over, & E. Lucas (Eds.), *The thinking mind: The use of thinking in everyday life* (pp. 15–28). Hove, England: Psychology Press.

Over, D. E., & Cruz, N. (2018). Probabilistic accounts of conditional reasoning. In L. J. Ball & V. A. Thompson (Eds.), *International handbook of thinking and reasoning* (pp. 434–450). Hove, England: Psychology Press.

Over, D. E., & Cruz, N. (2019). Philosophy and the psychology of conditional reasoning. In A. Aberdein & M. Inglis (Eds.), *Advances in experimental philosophy of logic and mathematics* (pp. 225–249). London, England: Bloomsbury Academic.

Over, D. E., Hadjichristidis, C., Evans, J. St. B. T., Handley, S. J., & Sloman, S. A. (2007). The probability of causal conditionals. *Cognitive Psychology, 54*, 62–97.

Pearl, J. (2013). Structural counterfactuals: A brief introduction. *Cognitive Science, 37*, 977–985.

Pfeifer, N., & Kleiter, G. D. (2005). Coherence and nonmonotonicity in human reasoning, *Synthese, 146*(1–2), 93–109.

Pfeifer, N., & Kleiter, G. D. (2009). Framing human inference by coherence based probability logic. *Journal of Applied Logic, 7*, 206–217.

Pfeifer, N., & Kleiter, G. D. (2011). Uncertain deductive reasoning. In K. Manktelow, D. E. Over, & S. Elqayam (Eds.), *The science of reason: A Festschrift for Jonathan St B. T. Evans* (pp. 145–166). Hove, England: Psychology Press.

Pfeifer, N., & Tulkki, L. (2017). Conditionals, counterfactuals, and rational reasoning: An experimental study on basic principles. *Minds and Machines, 27*(1), 119–165.

Politzer, G., & Baratgin, J. (2016). Deductive schemas with uncertain premises using qualitative probability expressions. *Thinking & Reasoning, 22*, 78–98.

Politzer, G., Over, D. E., & Baratgin, J. (2010). Betting on conditionals. *Thinking & Reasoning, 16*, 172–197.

Ramsey, F. P. (1990). General propositions and causality (original manuscript, 1929). In D. H. Mellor (Ed.), *Philosophical papers* (pp. 145–163). Cambridge, England: Cambridge University Press.

Ramsey, F. P. (1990). Truth and probability (original manuscript, 1926). In D. H. Mellor (Ed.), *Philosophical papers* (pp. 52–94). Cambridge, England: Cambridge University Press.

Sanborn, A. N., & Chater, N. (2016). Bayesian brains without probabilities. *Trends in Cognitive Sciences, 20*(12), 883–893.

Singmann, H., Klauer, K. C., & Over, D. (2014). New normative standards of conditional reasoning and the dual-source model. *Frontiers in Psychology, 5,* 16.

Skovgaard-Olsen, N., Kellen, D., Krahl, H., & Klauer, K. C. (2017). Relevance differently affects the truth, acceptability, and probability evaluations of "and," "but," "therefore," and "if–then." *Thinking & Reasoning, 23,* 449–482.

Skovgaard-Olsen, N., Singmann, H., & Klauer, K. C. (2016). The relevance effect and conditionals. *Cognition, 150,* 26–36.

Stalnaker, R. (1968). A theory of conditionals. In N. Rescher (Ed.), *Studies in logical theory* (pp. 98–112). Oxford, England: Blackwell.

Tentori, K., Crupi, V., & Russo, S. (2013). On the determinants of the conjunction fallacy: Probability versus inductive confirmation. *Journal of Experimental Psychology: General, 142,* 235–255.

Tversky, A., & Kahneman, D. (1983). Extensional versus intuitive reasoning: The conjunction fallacy in probability judgment. *Psychological Review, 90,* 293–315.

Zhao, J., & Osherson, D. (2010). Updating beliefs in light of uncertain evidence: Descriptive assessment of Jeffrey's rule. *Thinking & Reasoning, 16*(4), 288–307.

6.3 Conditional and Counterfactual Reasoning

Ruth M. J. Byrne and Orlando Espino

Summary

We consider a set of questions about conditionals and counterfactuals that have led to experimental discoveries able to distinguish between competing explanations of human reasoning. The first part of the chapter focuses on reasoning with factual conditionals, such as "If she made a good painting, she got a treat," and addresses two related questions: (1) Do people think about what is possible or what is probable when they understand a conditional? and (2) When do people infer a conditional relation? The second part of the chapter focuses on reasoning with counterfactual conditionals, such as "If she had done everything right, she would have passed her driving test," and also addresses two questions: (1) Do people make similar inferences from factual and counterfactual conditionals? and (2) Do people construct "embodied" mental representations of counterfactuals? The discoveries made from experimental investigations of these questions provide insights into how the mind accomplishes reasoning.

1. Conditionals and Counterfactuals

Conditional inferences are at the very center of everyday human reasoning. Even the youngest child who believes "If I make a good painting, I will get a treat," on hearing that her painting is good, will expect a treat. People make conditional inferences often in everyday life. Some seem immediate and obvious. Other conditional inferences require people to mull over them, to consider different possibilities. An office worker who knows "If I leave the office at 5 p.m., I will get home in time for dinner," on finding himself still at his desk as the clock strikes 5 p.m., may conclude that he will not get home on time—or, alternatively, he may begin to think about other routes to ensure he makes it. And people make inferences even more readily from counterfactual conditionals, about what once was possible but is so no longer. A learner driver who believes "If I had done everything right, I would have passed my driving test," reflecting on the fact that she did not pass, will conclude readily that she must have done something wrong. Inferences from counterfactuals occupy a special place at the center of conditional reasoning.

Conditionals and counterfactuals provide a unique window onto the human mind. They have been the subject of many intriguing and often conflicting analyses, especially in philosophy (e.g., Adams, 1975; Lewis, 1973; Stalnaker, 1968) and psychology (e.g., Evans & Over, 2004; Johnson-Laird & Byrne, 2002; Oaksford & Chater, 2007; Sloman & Lagnado, 2005). There have been many striking experimental discoveries about conditional inference (for a review, see Nickerson, 2015) and about counterfactual inference (for a review, see Byrne, 2016). Our goal in this chapter is to consider just a few of the crucial questions about them that have led to important findings since they first became the subject of intensive psychological study in the 1960s and 1970s. The discoveries provide important insights not only into how the mind accomplishes reasoning but also how the mind supports other sorts of thinking, including thinking that may seem quite remote from reasoning, such as creative imagination. People often create "if only . . ." thoughts about how the past could have turned out differently and "what if . . ." thoughts about how the future could be different (e.g., Roese & Epstude, 2017; Walsh & Byrne, 2004). Remarkably, the creation of such imagined alternatives to reality appears to depend on the same cognitive processes that support conditional and counterfactual reasoning (Byrne, 2005).

2. Conditional Reasoning

The frequency with which people make some conditional inferences is very high. Almost all participants in experiments, when they are told "If he wrote a good essay, he got a high mark" and "He wrote a good essay," make the modus ponens inference "He got a good mark" (see Nickerson, 2015). The frequency of other conditional inferences varies. Usually about half to two thirds of participants, when they are told "He got a high mark," make the affirmation of the consequent inference "He wrote a good essay" (see Schroyens, Schaeken, & d'Ydewalle, 2001). Often about half of participants, when they hear "He did not get a high mark," make the modus tollens inference "He did not write a good essay" whereas the other half say that nothing follows. Similarly, when they hear "He did not write a good essay" about half to two thirds of participants make the denial of the antecedent inference, "He did not get a good mark" (see table 6.3.1).

How the human mind carries out reasoning is contended, and there is as yet no agreement on the cognitive processes that underlie conditional inference. An early set of theories proposed that people access a mental logic that consists of abstract inference rules, such as "If p then q; p; therefore q," and they construct a mental derivation of a conclusion, akin to a logical proof (e.g., Braine & O'Brien, 1998; Rips, 1994). On this sort of theory, people make modus ponens readily because it corresponds directly to an inference rule in their mental logic; they have difficulty with modus tollens because there is no rule in their mental logic corresponding to it, and they must instead construct a mental derivation using related rules. Early psychological studies of conditional reasoning, influenced by long-standing philosophical analyses of the validity of inferences, assumed propositional and predicate logics as the normative standard against which to compare human reasoning (e.g., Jeffrey, 1981). Another early set of theories proposed that the mind contains reasoning modules that consist of domain-sensitive inference rules, specialized for content such as obligation and permission, or social regulations of costs and benefits (e.g., Cheng & Holyoak, 1985; Cosmides, 1989). Domain-sensitive rule theories limited their explanations of conditional reasoning to performance in the Wason selection task (see Ragni, Kola, & Johnson-Laird, 2018). They proposed that people make inferences akin to modus ponens readily because the schemas for most domains contain an inference rule

corresponding to it, for example, "If the action is to be taken, the precondition must be met"; they make inferences akin to modus tollens less often because few schemas contain an inference rule for it. These two sorts of theories—abstract and domain-sensitive inference rule theories—have few active champions in contemporary research on reasoning.

Nowadays, two other theories are hotly debated, one based on Bayesian probability and the other on mental models. The probabilist approach proposes that people make conditional inferences by relying on Bayesian probability calculations—different versions have been based on suppositions (e.g., Evans & Over, 2004), probabilistic logic (e.g., Oaksford & Chater, 2007), and causal Bayesian networks (e.g., Lucas & Kemp, 2015; Sloman & Lagnado, 2005). They propose that people make inferences from a conditional such as "If there are apples, there are oranges" by assessing conditional probability (the probability of an event occurring given that another event has already occurred). People compare the probability that there are oranges given there are apples, $P(B|A)$, to the probability that there are no oranges given there are apples, $P(\text{not-}B|A)$, and they do not think about the situations in which there are no apples (Evans, Handley, & Over, 2003). They make the modus ponens inference if their prior beliefs indicate $P(B|A)$ is greater than $P(\text{not-}B|A)$; they have difficulty with the modus tollens inference because they have not thought about the probability of situations in which there are no apples. On this view, probability logic is viewed as the normative standard against which to compare human reasoning (e.g., Cruz, Baratgin, Oaksford, & Over, 2015).

In contrast, the mental model theory proposes that the cognitive processes that underlie reasoning construct iconic simulations that correspond to the way the world would be if the assertion was true (e.g., Byrne & Johnson-Laird, 2009; Johnson-Laird & Byrne, 1991). People understand the conditional by envisaging a single possibility at the outset, "There are apples and there are oranges" (Johnson-Laird & Byrne, 2002). If need be, they can "flesh out" their models to think about other

Table 6.3.1
Four inferences from the conditional "If he wrote a good essay, he got a high mark" (If A then B)

Inference	Minor premise	Conclusion	Form
Modus ponens	He wrote a good essay	He got a high mark	$A \therefore B$
Affirm consequent	He got a high mark	He wrote a good essay	$B \therefore A$
Modus tollens	He did not get a high mark	He did not write a good essay	Not-$B \therefore$ not-A
Deny antecedent	He did not write a good essay	He did not get a high mark	Not-$A \therefore$ not-B

possibilities that are consistent with the conditional, such as "There are no apples and no oranges." The possibilities are conjunctive, that is, it is possible that there are apples and there are oranges, and it is possible that there are no apples and there are no oranges (e.g., Khemlani, Byrne, & Johnson-Laird, 2018). People interpret conditionals in many different ways—and so conditionals cannot be reduced to the truth-functional material implication of propositional logic (e.g., Johnson-Laird & Byrne, 2002). Instead, the normative standard against which to compare human reasoning is the semantic principle that an inference is valid if there are no counterexamples, that is, it is impossible for the premises to be true and the conclusion false (e.g., Johnson-Laird, Khemlani, & Goodwin, 2015). People make the modus ponens inference readily because it corresponds to the single possibility they have thought about at the outset; they find modus tollens more difficult because it requires them to flesh out their models to be more explicit.

Each of the theories predicts the difference between modus ponens and modus tollens, which they were designed to explain, but they make different predictions for other inferences (see Byrne & Johnson-Laird, 2009, for some examples). Difficulties in comparing the two sorts of theories arise because some probabilist accounts lack an algorithmic-level specification of the mental representations and cognitive processes that the human mind relies on to reason (see Oaksford, Over, & Cruz, 2019). Nonetheless, two crucial questions have led to discoveries that help to distinguish between the theories.

2.1 Do People Think about What Is Possible or What Is Probable?

According to the mental model theory, when people understand "if," they think about what is possible (Johnson-Laird & Byrne, 2002). They understand the conditional "If there are apples, there are oranges" by envisaging initially just a single possibility, in line with a principle of parsimony, because of working-memory constraints. Their models are small-scale dynamic simulations of the imagined possibility, but for convenience, they can be represented here in the following simple diagram:

apples   oranges

. . .

The three dots in the diagram indicate that there may be other possibilities, and people can "flesh out" their models to think about some of them, for example,

apples   oranges
no apples   no oranges

Reasoners come to many different interpretations of "if," and some of them flesh out their initial representation to correspond to this *biconditional* interpretation, as if the conditional asserted "If and only if there are apples, then there are oranges." Others flesh out their models to include a further possibility:

apples   oranges
no apples   no oranges
no apples   oranges

which corresponds to a *conditional* interpretation of "if" (see table 6.3.2). In line with a principle of truth, people do not tend to think about the possibilities that are ruled out by an assertion. The conditional rules out as impossible one situation, the one in which there are apples and there are no oranges. The biconditional rules out as impossible two situations, the one in which there are apples and there are no oranges and another in which there are no apples and there are oranges. People do not tend to think about these impossibilities.

On the probabilist account, people understand "if" by assessing conditional probability (e.g., Evans & Over, 2004; Oaksford & Chater, 2007). To do so, they carry out a *Ramsey test* (after Ramsey, 1929/1990), in which they hypothetically suppose the "if" clause to be the case, make changes to their beliefs as needed, and assess, on this basis, the probability of the "then" clause. They compare the strength of their prior belief that there are apples and there are oranges to the strength of their prior belief that there are apples and there are no oranges (Evans et al., 2003). In other words, they think about the probability of a situation that is true according to the conditional and compare it to a situation the conditional rules out as false. To work out the likelihood

Table 6.3.2
The situations people think about for a conditional "If there are apples, then there are oranges" and a biconditional "If and only if there are apples, then there are oranges" according to two theories

	Mental models	Probability and prior beliefs
Conditional	apples and oranges no apples and no oranges no apples, and oranges	apples and oranges apples and no oranges
Biconditional	apples and oranges no apples and no oranges	apples and oranges apples and no oranges no apples, and oranges

that if there are apples, then there are oranges, people must think not only about what is possible if the conditional is true: there are apples and oranges; they must also think about what is not possible if the conditional is true: there are apples but no oranges. Moreover, they do not think about their beliefs concerning what happens when there are no apples, because they consider such instances irrelevant (Evans et al., 2003).

Recent empirical studies about judgments of truth and probability test predictions derived from these competing accounts (e.g., Byrne & Johnson-Laird, 2019; Cruz et al., 2015; Elqayam & Over, 2013; Goodwin, 2014; Hinterecker, Knauff, & Johnson-Laird, 2016; Pfeifer & Tulkki, 2017). I will illustrate just one of the discoveries. Consider the following short narrative:

> Carmen went shopping to the market. When she looked at the poster there, she saw it said, "If there are apples, there are oranges." When Carmen looked at the shelves, she saw that there were apples and there were oranges. Carmen checked her list of purchases.

When participants in experiments read a conditional such as "If there are apples, there are oranges" in this short narrative, they are able to read the subsequent conjunction "There were apples and there were oranges" very quickly, in just under a second and a half, on average (Espino, Santamaría, & Byrne, 2009). The conditional "primes" them to read the affirmative conjunction quickly. A biconditional such as "If and only if there are apples, there are oranges" also primes them to read the affirmative conjunction just as quickly compared to the conditional. Now consider instead the following narrative:

> Carmen went shopping to the market. When she looked at the poster there, she saw it said, "If there are apples, there are oranges." When Carmen looked at the shelves, she saw that there were no apples and there were oranges. Carmen checked her list of purchases.

Participants read the conjunction with the negated antecedent—"There were no apples and there were oranges"—more quickly when they were primed by a conditional compared to a biconditional (see figure 6.3.1). Why does the conditional prime participants to read the conjunction with the negative antecedent, "There were no apples and there were oranges," more quickly compared to the biconditional?

The mental model theory explains the experimental result readily: people mentally represent what is possible, not what is impossible. They read the conjunction "There were no apples and there were oranges" more quickly when it is primed by a conditional, because it is one of the true possibilities consistent with the conditional, whereas they take longer to read the conjunction when it is primed by the biconditional, because it is a false possibility ruled out by the biconditional (Espino et al., 2009).

The experimental result is more challenging to explain on the probabilist account. People do not think about their beliefs concerning what happens when the "if" clause is not true (Evans et al., 2003). So the theory predicts no difference in reading times for "There are no apples and there are oranges" when primed by a conditional versus a biconditional. In fact, the suppositional version even predicts a difference in the opposite direction. A biconditional is understood as two related conditionals, "If there are apples, there are oranges, and if there are oranges, there are apples" (Evans et al., 2003; see also Handley, Evans, & Thompson, 2006). People must compare their belief that there are apples and there are oranges to their belief that there are apples and there are no oranges for the first conditional, and in addition, they must compare their belief that there are oranges and there are apples to their belief that there are oranges and there are no apples for the second conditional. Thus, for a biconditional, they think about three prior beliefs, rather than just two (see table 6.3.2). The theory must predict that people will read the conjunction "There are no apples and there are oranges" *more* quickly when it is primed by the biconditional than the conditional, the opposite of the experimental result (Espino et al., 2009).

We can conclude that when people understand a conditional, their initial thoughts are about what is possible, not what is impossible. We can also conclude that when they understand a conditional, their initial thoughts are not about what is probable (since that would require them to have thought about what is impossible as well as what is possible).

2.2 When Do People Infer a Conditional Relation?

Many studies examine how people infer a causal relationship from the observation of objects and events (for reviews, see Johnson-Laird & Khemlani, 2017; Oaksford & Chater, 2017). But people can infer a conditional relationship even in non-causal situations. Suppose you know "There are roses in the garden or there are lilies or both." Can you infer "If there are no roses there are lilies"? About half of participants in experiments make the inference (Espino & Byrne, 2013). It occurs even when the assertions contain negated clauses, for example, when people are told "There are no daisies in the garden or there are no buttercups or both," they infer "If there are daisies, there are no buttercups" (see figure 6.3.2).

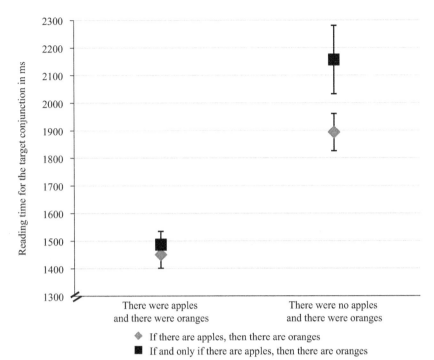

Figure 6.3.1

Reading times for conjunctions primed by conditionals or biconditionals (adapted from Espino et al., 2009, experiment 2). Error bars are standard error of the mean.

The discovery that when people hear about disjunctive alternatives, they make an inference of a conditional relation between them, is intriguing. Why would they do so?

According to the mental model theory, the inference is a reasonable one to make (Johnson-Laird & Byrne, 2002; see also Ormerod & Richardson, 2003). The disjunction "There are roses or lilies or both" is consistent with three possibilities:

roses	no lilies
no roses	lilies
roses	lilies

And the conditional "If there are no roses there are lilies" is consistent with the same three possibilities:

no roses	lilies
roses	no lilies
roses	lilies

Because there are no possibilities that the two assertions do not have in common, the inference from the disjunction to the conditional is valid—there are no counterexamples to it. Of course, the inference is difficult, requiring the comparison of multiple possibilities for each assertion, which is likely to exceed most people's working-memory capacity (see also Murray & Byrne, 2005). Unsurprisingly, only about half of participants make the inference (see figure 6.3.2). Instead, many of them make a revealing error: they choose the conclusion

that matches their initial model of the conditional (see Espino & Byrne, 2013).

Crucially, according to the mental model theory, the inference between the disjunction and the conditional is valid in either direction, from "or" to "if" and from "if" to "or." For example, suppose you know instead "If there are no tulips in the garden, there are daffodils." Can you infer "There are tulips or daffodils or both"? About one third of participants make the inference. Tellingly, in most cases, participants make the inference from "or" to "if" more often than the one from "if" to "or," as figure 6.3.2 shows.

The difference is revealing because it challenges the predictions of probabilist theories (Oberauer, Geiger, & Fischer, 2011; Over, Evans, & Elqayam, 2010). According to probabilist theories, the inference from "or" to "if"—the one that people make most often—is not valid; only the one from "if" to "or" is valid. A probabilistically valid (p-valid) inference is one in which the likelihood of the conclusion is not lower than the likelihood of the premise. The inference from "or" to "if" is not p-valid because the likelihood of the conditional can be *lower* than the likelihood of the disjunction. The equation for the general relation between the probability of a disjunction "*A* or *B*," $P(A \text{ or } B)$, and the probability of a conditional "If not-*A*, *B*," $P(B \mid \text{not-}A)$, is as follows (as given in Over et al., 2010, p. 142):

$$P(A \text{ or } B) = P(A) + P(B \mid \text{not-}A) - P(A)P(B \mid \text{not-}A)$$

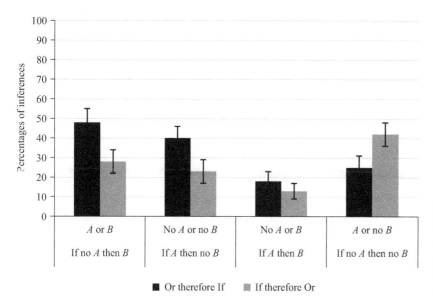

Figure 6.3.2
Percentages of inferences from a disjunction to a conditional and from a conditional to a disjunction (adapted from Espino & Byrne, 2013, experiment 1b). Error bars are standard error of the mean.

Hence, the inference from "or" to "if" is not p-valid (see also Oberauer et al., 2011). In contrast, the inference from "if" to "or" *is* p-valid because the likelihood of the disjunction cannot be *smaller* than the likelihood of the conditional. The relation is shown by the expansion of the probability of the conditional (as provided in Oberauer et al., 2011, p. 97), given by $P(B \mid \text{not-}A)$:

$$P(A) \times P(B \mid \text{not-}A) + P(\text{not-}A) \times P(B \mid \text{not-}A)$$

and the probability of the disjunction:

$$P(A) + P(\text{not-}A) \times P(B \mid \text{not-}A)$$

People judge the probability of a conditional premise to be higher than, or the same as, the probability of a disjunctive conclusion for "if" to "or" inferences, and they judge the probability of a conditional conclusion to be higher than, or the same as, the probability of a disjunctive premise for "or" to "if" inferences (Cruz et al., 2015; see also Gilio & Over, 2012). But nonetheless, as figure 6.3.2 shows, they make the inference from "or" to "if" more often than the inference from "if" to "or," even though the former is p-invalid and the latter is p-valid. The experimental result is the opposite of what would be expected if people understood conditionals by calculating likelihood, and it runs counter to the predictions of probabilist theories.

We can conclude that people are willing to infer a conditional relation from alternative possibilities. They infer a conditional from a disjunction because they think about the possibilities that are consistent with each assertion and appreciate that they are the same. Although the inference is valid in both directions in that there are no counterexample possibilities, people

may be more inclined to make the inference from the disjunction to the conditional because the disjunction appears to assert categorically that something is the case, in contrast to the conditional, which appears to assert a hypothetical relation.

These two key discoveries shed light on the nature of the mental algorithms that underlie conditional inference. Further light is shed by discoveries about inferences from counterfactual conditionals.

3. Counterfactual Reasoning

A counterfactual such as "If she had done everything right, she would have passed her driving test" can seem to mean something very different from its factual counterpart, "If she did everything right, she passed her driving test." The counterfactual seems to convey that its antecedent is not the case—"She did not do everything right"—and it seems to convey that its consequent is not the case either—"She did not pass her driving test." Hence, assessing the truth of a counterfactual poses unique logical problems—since every counterfactual has a false antecedent, every counterfactual must be true on a traditional truth-functional account of conditionals, according to which a conditional is true if its antecedent is false or its consequent is true (see Nickerson, 2015). The development of possible-worlds semantics in modal logics led to several alternative analyses of counterfactuals (e.g., Lewis, 1973; Stalnaker, 1968). Their study in the psychology of reasoning has also had a significant impact in the past 20 years since they have come under increasing scrutiny in experiments. Two research questions, and the discoveries they led to, illustrate the novel contribution of the study of counterfactuals to understanding the reasoning mind.

3.1 Do People Make Similar Inferences from Factual and Counterfactual Conditionals?

From a conditional in the indicative mood—which is often called a *factual* conditional—such as, "If Paolo was in Venice, then Marco was in Padua," when people are told, "Marco was not in Padua," only about half of them make the modus tollens inference, "Paolo was not in Venice," and the remainder tend to say that nothing follows. Strikingly, inferences from a conditional in the subjunctive mood—which is often called a *counterfactual* conditional—such as, "If Paolo had been in Venice, then Marco would have been in Padua," show a different pattern. When people know "Marco was not in Padua," most of them readily make the modus tollens inference "Paolo was not in Venice." The otherwise complex inference is made easily from the counterfactual (Byrne & Tasso 1999), as figure 6.3.3 shows.

Moreover, people continue to make the modus ponens inference, from "Paolo was in Venice" to "Marco was in

Padua," as readily from the counterfactual as from the factual conditional. Similar effects occur for the *denial of the antecedent* inference, which is made more often from the counterfactual than the factual conditional, and the *affirmation of the consequent* inference, which is made equally from both sorts of conditionals.

This "counterfactual inference effect" is robust and has been studied extensively (for a review, see Byrne, 2017). It occurs not only for locational content but also for causal and definitional counterfactuals (Thompson & Byrne, 2002) and various everyday scenarios (Frosch & Byrne, 2012). It has been examined for inducements such as promises and threats (Egan & Byrne, 2012) and deontic content such as obligations (Quelhas & Byrne, 2003). Why are otherwise difficult inferences easy from a counterfactual?

The discovery helps distinguish alternative explanations of human reasoning. According to the mental model theory, people understand a counterfactual such as "If there had been apples there would have been oranges" by thinking about two possibilities from the outset. They mentally represent the conjecture mentioned in the conditional, "There are apples and there are oranges," and

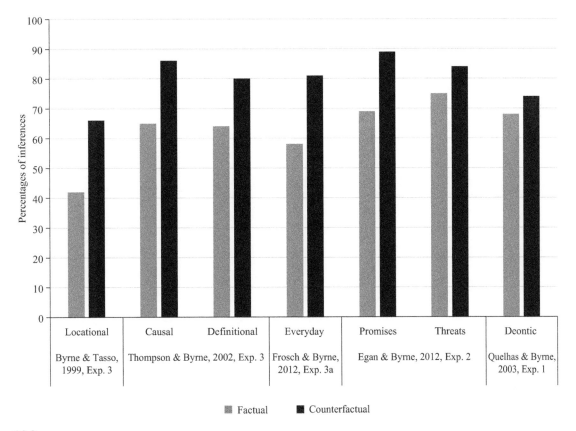

Figure 6.3.3
The percentages of modus tollens inferences made from factual and counterfactual conditionals for different contents (adapted from Byrne & Tasso, 1999; Egan & Byrne, 2012; Frosch & Byrne, 2012; Quelhas & Byrne, 2003; Thompson & Byrne, 2002).

they also mentally represent the presupposed or known facts, "There are no apples and there are no oranges." They keep track of the epistemic status of their models, as corresponding to a counterfactual possibility, or the facts (e.g., Johnson-Laird & Byrne, 1991):

Counterfactual: apples   oranges
Facts: no apples   no oranges

. . .

The theory explains a number of curious findings about counterfactuals. For example, when people hear the counterfactual, they mistakenly remember that they were told there were no apples and no oranges (Fillenbaum, 1974). They judge that someone who utters the counterfactual means to imply there were no apples and no oranges (Thompson & Byrne, 2002). Strikingly, they read the conjunction "There were no apples and there were no oranges" more quickly when they are primed by the counterfactual compared to the factual conditional, but they read the conjunction "There were apples and there were oranges" equally quickly when they are primed by either conditional, as figure 6.3.4 shows (Santamaría, Espino, & Byrne, 2005; see also de Vega, Urrutia, & Riffo, 2007). They look at images corresponding to "no apples and no oranges" more often for the counterfactual than the factual conditional, as shown by eye-tracking studies (Orenes, García-Madruga, Gómez-Veiga, Espino,

& Byrne, 2019; see also Ferguson & Sanford, 2008; Nieuwland & Martin, 2012). The dual meaning of counterfactuals has been corroborated by studies that rely on measures of brain activity such as event-related potential (ERP) measures and functional magnetic resonance imaging (fMRI) measures (e.g., Ferguson, Sanford, & Leuthold, 2008; Kulakova, Aichhorn, Schurz, Kronbichler, & Perner, 2013; Van Hoeck et al., 2013).

The proposal that people think of two possibilities from the outset for the counterfactual predicts the finding that they readily make the modus tollens inference (Byrne & Tasso, 1999). When they hear, "There are no oranges," they do not have to flesh out their models, as they do for a factual conditional. They have already represented a possibility in which there are no oranges. They can eliminate the model corresponding to the counterfactual conjecture and conclude, "There are no apples." Similarly, they make the modus ponens inference as readily from the counterfactual as the factual conditional because when they hear, "There are apples," they can update their models to eliminate the model corresponding to the presupposed facts and note instead that the facts are that there are apples and oranges. They can then conclude, "There are oranges."

Of course, the fully fleshed out models for the counterfactual are the same as those for the factual conditional, even though the epistemic status of the models

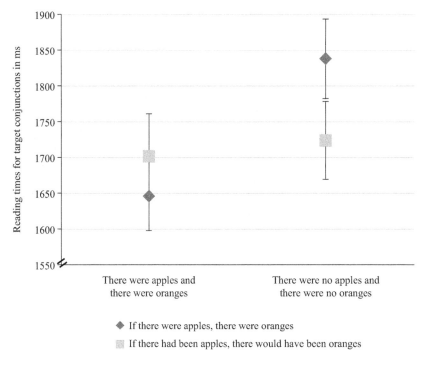

Figure 6.3.4
Reading times for conjunctions primed by factual or counterfactual conditionals (adapted from Santamaría et al., 2005, experiment 1). Error bars are standard error of the mean.

is different, for example, on a conditional interpretation of "if":

Counterfactual:	apples	oranges
Facts:	no apples	no oranges
Counterfactual:	no apples	oranges

Even though counterfactuals can seem to mean something very different from factual conditionals, their semantics runs in parallel (e.g., Byrne & Johnson-Laird, 2019). A counterfactual such as "If Oswald hadn't shot Kennedy, then someone else would have" seems debatable, whereas the corresponding factual one, "If Oswald didn't shoot Kennedy, then someone else did," seems true (Adams, 1970). But the comparison is misleading (Byrne & Johnson-Laird, 2019). The parallel to Adams's counterfactual should be "If Oswald hasn't shot Kennedy, then someone else will," which is also debatable. And the counterfactual parallel to his factual conditional, which assumes Kennedy has been shot, should be, "If Oswald hadn't been the person who shot Kennedy, then someone else must have been," which indeed seems true (see Byrne & Johnson-Laird, 2019).

One probabilist approach proposes that people understand a counterfactual by assessing its likelihood, which at the present time is the same as the probability of the factual conditional at a previous time (Over et al., 2007). People think only about one situation, in line with a singularity principle (Evans, 2007, p. 74). They combine their prior belief in the conditional probability of the counterfactual, that is, the likelihood of the counterfactual consequent "There would have been oranges" given the counterfactual antecedent "if there had been apples," with their prior beliefs in each of the implied facts, "There were no apples" and "There were no oranges," to arrive at a single assessment of their belief in the counterfactual (Over et al., 2007). Hence, it is challenging for this suppositional theory to explain the findings that people make modus tollens more often from counterfactuals than factual conditionals and that they appear to have ready access not only to the counterfactual conjecture but also to the presupposed facts.

Another probabilist theory is based on causal Bayes nets (e.g., Pearl, 2013). People construct causal models, for example,

$$ \text{(A)} \longrightarrow \text{(B)} \longrightarrow \text{(C)} $$

in which nodes represent causes and effects, arrows indicate causal direction, and a conditional probability table at each node gives the probability that a node is present or absent (e.g., Lucas & Kemp, 2015; Sloman & Lagnado, 2005). Changes to the causal model are made by counterfactual "interventions" on a node (e.g., imagining that B is absent deletes links into B). A counterfactual's probability relies on the "do" intervention operator, $P(c \mid b. \, do(\neg b))$, that is, the conditional probability of c, given that b was observed, but counterfactually removed (e.g., Dehghani, Iliev, & Kaufmann, 2012; Lagnado, Gerstenberg, & Zultan, 2013; Meder, Hagmayer, & Waldmann, 2009). The account predicts the *opposite* of the counterfactual inference effect, namely, that people *refrain* from modus tollens when counterfactuals describe the prevention of a cause (e.g., Sloman & Lagnado, 2005). For example, participants tend to make the modus tollens inference for a factual question, "Suppose A was observed to not be moving, would B still be moving?" but not for a counterfactual one, "Suppose A were prevented from moving, would B still be moving?" (Sloman & Lagnado, 2005). A challenge for the causal model theory is that the counterfactual inference effect is observed for conditionals that convey not only causal relations but also a wide variety of other sorts of conditional relation, as figure 6.3.3 shows.

Of course, not all conditionals in the subjunctive mood are counterfactuals (e.g., Dudman, 1988). So-called Anderson conditionals (e.g., "If he had taken arsenic, he would have exactly these symptoms"; Adams, 1975) make use of the subjunctive mood, but their antecedents may be indicatives in disguise (e.g., "If he in fact took arsenic, he would have exactly these symptoms"). And other counterfactuals, such as counterpossibles (counterfactuals with impossible antecedents), for example, "If Hobbes had squared the circle, sick children in the mountains of South America at the time would have cared" (see Berto & Jago, 2018; Williamson, 2018), have received scant psychological attention.

3.2 Do People Construct "Embodied" Mental Representations of Counterfactuals?

There is abundant evidence that when people understand a counterfactual such as "If the flowers had been roses, the trees would have been orange trees," they think about two possibilities, the conjecture, "There were roses and orange trees," and its opposite, the presupposed facts. But how are the presupposed facts mentally represented? They could be represented explicitly through the use of negation, "There are no roses and there are no orange trees," or they could be represented by alternates, "There are poppies and there are apple trees." Which sorts of mental representations do people rely on?

The idea that the presupposed facts are represented as alternates is derived from the theory that the meaning of concepts, such as "roses," is *embodied*, in a modality-specific, experientially based representation grounded in character-istics of sensory or motor processes (e.g., Barsalou, Simmons, Barbey, & Wilson, 2003). The embodied proposal extends even to abstract concepts such as negation (e.g., Glenberg, Robertson, Jansen, & Johnson-Glenberg, 1999). In an embodied system, a negation such as "There is not a rose" is represented by an alternate, such as "There is a poppy," and it may require several steps, for example, first representing "There is a rose" and then inhibiting the representation (e.g., Kaup, Lüdtke, & Zwaan, 2006; Mayo, Schul, & Burnstein, 2004).

In an experimental test of the embodied idea, participants were told that in a garden, the flowers are roses or poppies and the trees are orange trees or apple trees, and they then read a counterfactual, "If the flowers had been roses, the trees would have been orange trees." When they were told, "The trees were not orange trees," most of them tended to infer, "The flowers were poppies" (Espino & Byrne, 2018). They rarely said, "The flowers were not roses." In other words, they tended to infer what *is* the case, from information about what is *not* the case. This "inference-to-alternates effect" is a robust strategy that occurs in many different situations.

One situation in which the inference-to-alternates effect does *not* occur is revealing about the source of

the tendency. Participants were told that the flowers are roses or poppies *or lilies* and the trees are orange trees or apple trees, that is, the context for flowers was a multiple one rather than a binary one. Now when they read a counterfactual "If the flowers had been roses, the trees would have been orange trees," and "The trees were not orange trees," many of them said, "The flowers were not roses," as figure 6.3.5 shows.

Why do people exhibit an inference-to-alternates effect in a binary context but not in a multiple context? On the embodied view, the presupposed facts for the counterfactual are represented as alternates, in both sorts of context. It explains the observation of an inference-to-alternates effect in the binary context, but it makes the wrong prediction for the multiple context. When people are told, "The trees are not orange trees," they should conclude, "There are poppies or lilies." But they do not do so.

According to the mental model theory, the mental representation of negation is as iconic as possible, but it can include symbols, such as a propositional-like tag "no" or some other annotation to capture negation (e.g., Johnson-Laird & Byrne, 2002; Orenes, Beltrán, & Santamaría, 2014). In a binary context, people simulate the presupposed facts by thinking about alternates:

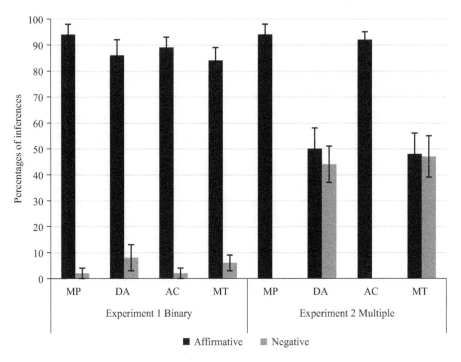

Figure 6.3.5
Percentages of modus ponens (MP), denial of the antecedent (DA), affirmation of the consequent (AC), and modus tollens (MT) inferences from counterfactuals in a binary or multiple context phrased as affirmative, e.g., "The flowers were poppies," or as negative, e.g., "The flowers were not roses" (adapted from Espino & Byrne, 2018). Error bars are standard error of the mean.

Counterfactual: roses   oranges
Facts: poppies   apples

. . .

When they are told, "There are no oranges," they eliminate the first model and conclude, based on the second model, "There are poppies." Hence, the inference-to-alternates effect is observed. But in a multiple context, the number of alternates they would have to consider is likely to exceed working memory:

Counterfactual: roses   oranges
Facts: poppies   apples
                     lilies   apples

. . .

Instead they can switch to models that are annotated with propositional-like tags or symbols:

Counterfactual: roses   oranges
Facts: no roses   no oranges

. . .

Hence, when they are told, "There are no oranges," they conclude, based on the second model, "There are no roses," that is, the inference-to-alternates effect is eliminated. We can conclude that how people represent the presupposed facts depends on the context. People tend to mentally simulate the presupposed facts by considering alternates but they can switch strategy to a symbolic annotation when the possibilities exceed working-memory constraints.

These two key discoveries about inferences from counterfactuals shed further light on the nature of the mental algorithms that underlie conditional inference.

4. Conclusion

Conditional inferences are at the very center of everyday human reasoning, and counterfactual inferences occupy a special place there. Conditionals and counterfactuals provide a window onto the nature of mental simulations and algorithms in the human mind. In this chapter, we have considered some important discoveries that help distinguish explanations of conditional and counterfactual reasoning. When people understand, and reason about, a *factual* conditional, they think about what the conditional indicates is possible, not what it rules out as impossible. When they hear about alternative possibilities, they infer a conditional relation between them. The findings indicate that people think about what is possible, not what is probable. When they understand, and reason about, a *counterfactual* conditional, they are able to make complex inferences very readily. They make

inferences about what *is* the case, even when they are told what is *not* the case. The findings indicate that people think about dual possibilities for a counterfactual—not only the counterfactual conjecture but also the presupposed facts—and they construct iconic mental simulations that can include symbolic annotations. These discoveries about conditional and counterfactual reasoning provide important insights into how the mind accomplishes reasoning by imagining possibilities of different sorts.

Acknowledgments

We thank our colleagues, Phil Johnson-Laird and Mark Keane, for helpful discussions related to this chapter.

References

Adams, E. W. (1970). Subjunctive and indicative conditionals. *Foundations of Language, 6,* 89–94.

Adams, E. W. (1975). *The logic of conditionals.* Dordrecht, Netherlands: Reidel.

Barsalou, L. W., Simmons, W. K., Barbey, A. K., & Wilson, C. D. (2003). Grounding conceptual knowledge in modality-specific systems. *Trends in Cognitive Science, 7,* 84–91.

Berto, F., & Jago, M. (2018). Impossible worlds. In E. N. Zalta (Ed.) *The Stanford encyclopedia of philosophy.* Retrieved from https://plato.stanford.edu/archives/fall2018/entries/impossible-worlds/

Braine, M. D. S., & O'Brien, D. P. (Eds.). (1998). *Mental logic.* Mahwah, NJ: Erlbaum.

Byrne, R. M. J. (2005). *The rational imagination: How people create alternatives to reality.* Cambridge, MA: MIT Press.

Byrne, R. M. J. (2016). Counterfactual thought. *Annual Review of Psychology, 67,* 135–157.

Byrne, R. M. J. (2017). Counterfactual thinking: From logic to morality. *Current Directions in Psychological Science, 26*(4), 314–322.

Byrne, R. M. J., & Johnson-Laird, P. N. (2009). 'If' and the problems of conditional reasoning. *Trends in Cognitive Sciences, 13,* 282–287.

Byrne, R. M. J., & Johnson-Laird, P. N. (2020). *If* and *or:* Real and counterfactual possibilities in their truth and probability. *Journal of Experimental Psychology: Learning, Memory, and Cognition, 46*(4), 760–780.

Byrne, R. M. J., & Tasso, A. (1999). Deductive reasoning with factual, possible, and counterfactual conditionals. *Memory & Cognition, 27,* 726–740.

Cheng, P. W., & Holyoak, K. J. (1985). Pragmatic reasoning schemas. *Cognitive Psychology*, *17*(4), 391–416.

Cosmides, L. (1989). The logic of social exchange: Has natural selection shaped how humans reason? Studies with the Wason selection task. *Cognition*, *31*(3), 187–276.

Cruz, N., Baratgin, J., Oaksford, M., & Over, D. E. (2015). Bayesian reasoning with ifs and ands and ors. *Frontiers in Psychology*, *6*, 192.

Dehghani, M., Ilicv, R., & Kaufmann, S. (2012). Causal explanation and fact mutability in counterfactual reasoning. *Mind & Language*, *27*(1), 55–85.

de Vega, M., Urrutia, M., & Riffo, B. (2007). Cancelling updating in the comprehension of counterfactuals embedded in narratives. *Memory & Cognition*, *35*, 1410–1421.

Dudman, V. H. (1988). Indicative and subjunctive. *Analysis*, *48*, 113–122.

Egan, S. M., & Byrne, R. M. J. (2012). Inferences from counterfactual threats and promises. *Experimental Psychology*, *59*(4), 227–235.

Elqayam, S., & Over, D. E. (2013). New paradigm psychology of reasoning: An introduction to the special issue edited by Elqayam, Bonnefon, and Over. *Thinking & Reasoning*, *19*, 249–265.

Espino, O., & Byrne, R. M. J. (2013). The compatibility heuristic in non-categorical hypothetical reasoning: Inferences between conditionals and disjunctions. *Cognitive Psychology*, *67*(3), 98–129.

Espino, O., & Byrne, R. M. J. (2018). Thinking about the opposite of what is said: Counterfactual conditionals and symbolic or alternate simulations of negation. *Cognitive Science*, *42*(8), 2459–2501.

Espino, O., Santamaría, C., & Byrne, R. M. J. (2009). People think about what is true for conditionals, not what is false: Only true possibilities prime the comprehension of "if." *Quarterly Journal of Experimental Psychology*, *62*, 1072–1078.

Evans, J. St. B. T. (2007). *Hypothetical thinking: Dual processes in reasoning and Judgement*. Hove, England: Psychology Press.

Evans, J. St. B. T., Handley, S. J., & Over, D. E. (2003). Conditionals and conditional probability. *Journal of Experimental Psychology: Learning, Memory, and Cognition*, *29*(2), 321–335.

Evans, J. St. B. T., & Over, D. E. (2004). *If*. Oxford, England: Oxford University Press.

Ferguson, H. J., & Sanford, A. J. (2008). Anomalies in real and counterfactual worlds: An eye-movement investigation. *Journal of Memory and Language*, *58*, 609–626.

Ferguson, H. J., Sanford, A. J., & Leuthold, H. (2008). Eye-movements and ERPs reveal the time course of processing negation and remitting counterfactual worlds. *Brain Research*, *1236*, 113–125.

Fillenbaum, S. (1974). Information amplified: Memory for counterfactual conditionals. *Journal of Experimental Psychology*, *102*(1), 44–49.

Frosch, C. A., & Byrne, R. M. J. (2012). Causal conditionals and counterfactuals. *Acta Psychologica*, *141*(1), 54–66.

Gilio, A., & Over, D. (2012). The psychology of inferring conditionals from disjunctions: A probabilistic study. *Journal of Mathematical Psychology*, *56*(2), 118–131.

Glenberg, A. M., Robertson, D. A., Jansen, J. L., & Johnson-Glenberg, M. C. (1999). Not propositions. *Cognitive Systems Research*, *1*(1), 19–33.

Goodwin, G. P. (2014). Is the basic conditional probabilistic? *Journal of Experimental Psychology: General*, *143*, 1214–1241.

Handley, S., Evans, J. St. B. T., & Thompson, V. (2006). The negated conditional: A litmus test for the suppositional conditional? *Journal of Experimental Psychology: Learning, Memory, and Cognition*, *32*, 559–569.

Hinterecker, T., Knauff, M., & Johnson-Laird, P. N. (2016). Modality, probability, and mental models. *Journal of Experimental Psychology: Learning, Memory, and Cognition*, *42*(10), 1606–1620.

Jeffrey, R. C. (1981). *Formal logic*. New York, NY: McGraw-Hill.

Johnson-Laird, P. N., & Byrne, R. M. J. (1991). *Deduction*. Hillsdale, NJ: Erlbaum.

Johnson-Laird, P. N., & Byrne, R. M. J. (2002). Conditionals: A theory of meaning, pragmatics, and inference. *Psychological Review*, *109*, 646–678.

Johnson-Laird, P. N., & Khemlani, S. S. (2017). Mental models and causation. In M. R. Waldmann (Ed.), *The Oxford handbook of causal reasoning* (pp. 169–188). Oxford, England: Oxford University Press.

Johnson-Laird, P. N., Khemlani, S. S., & Goodwin, G. P. (2015). Logic, probability, and human reasoning. *Trends in Cognitive Sciences*, *19*, 201–214.

Kaup, B., Lüdtke, J., & Zwaan, R. A. (2006). Processing negated sentences with contradictory predicates: Is a door that is not open mentally closed? *Journal of Pragmatics*, *38*(7), 1033–1050.

Khemlani, S. S, Byrne, R. M. J., & Johnson-Laird, P. N. (2018). Facts and possibilities: A model theory of sentential reasoning. *Cognitive Science*, *42*(6), 1887–1924.

Kulakova, E., Aichhorn, M., Schurz, M., Kronbichler, M., & Perner, J. (2013). Processing counterfactual and hypothetical conditionals: An fMRI investigation. *NeuroImage*, *72*, 265–271.

Lagnado, D. A., Gerstenberg, T., & Zultan, R. (2013). Causal responsibility and counterfactuals. *Cognitive Science*, *37*(6), 1036–1073.

Lewis, D. (1973). *Counterfactuals*. Oxford, England: Blackwell.

Lucas, C. G., & Kemp, C. (2015). An improved probabilistic account of counterfactual reasoning. *Psychological Review, 122*, 700–734.

Mayo, R., Schul, Y., & Burnstein, E. (2004). "I am not guilty" vs "I am innocent": Successful negation may depend on the schema used for its encoding. *Journal of Experimental Social Psychology, 40*(4), 433–449.

Meder, B., Hagmayer, Y., & Waldmann, M. R. (2009). The role of learning data in causal reasoning about observations and interventions. *Memory & Cognition, 37*(3), 249–264.

Murray, M. A., & Byrne, R. M. J. (2005). Attention and working memory in insight problem solving. In B. G. Bara, L. Barsalou, & M. Bucciarelli (Eds.), *Proceedings of the Twenty-Seventh Annual Conference of the Cognitive Science Society* (pp. 1571–1576). Mahwah, NJ: Erlbaum.

Nickerson, R. (2015). *Conditional reasoning*. Oxford, England: Oxford University Press.

Nieuwland, M. S., & Martin, A. E. (2012). If the real world were irrelevant, so to speak: The role of propositional truth-value in counterfactual sentence comprehension. *Cognition, 122*, 102–109.

Oaksford, M., & Chater, N. (2007). *Bayesian rationality: The probabilistic approach to human reasoning*. Oxford, England: Oxford University Press.

Oaksford, M., & Chater, N. (2017). Causal models and conditional reasoning. In M. R. Waldmann (Ed.), *The Oxford handbook of causal reasoning* (pp. 327–346). Oxford, England: Oxford University Press.

Oaksford, M., Over, D., & Cruz, N. (2019). Paradigms, possibilities, and probabilities: Comment on Hinterecker, Knauff, and Johnson-Laird (2016). *Journal of Experimental Psychology: Learning, Memory, and Cognition, 45*(2), 288–297.

Oberauer, K., Geiger, S., & Fischer, K. (2011). Conditionals and disjunctions. In K. Manktelow, D. Over, & S. Elqayam (Eds.), *The science of reason: A Festschrift for Jonathan St B. T. Evans* (pp. 93–118). Hove, England: Psychology Press.

Orenes, I., Beltrán, D., & Santamaría, C. (2014). How negation is understood: Evidence from the visual world paradigm. *Journal of Memory and Language, 74*, 36–45.

Orenes, I., García-Madruga, J. A., Gómez-Veiga, I., Espino, O., & Byrne, R. M. J. (2019). The comprehension of counterfactual conditionals: Evidence from eye-tracking in the visual world paradigm. *Frontiers in Psychology, 10*, 1172.

Ormerod, T., & Richardson, J. (2003). On the generation and evaluation of inferences from single premises. *Memory & Cognition, 31*(3), 467–478.

Over, D. E., Evans, J. St. B. T., & Elqayam, S. (2010). Conditionals and non-constructive reasoning. In M. Oaksford & N.

Chater (Eds.), *Cognition and conditionals* (pp. 135–151). Oxford, England: Oxford University Press.

Over, D. E., Hadjichristidis, C., Evans, J. St. B. T., Handley, S. J., & Sloman, S. A. (2007). The probability of causal conditionals. *Cognitive Psychology, 54*, 62–97.

Pearl, J. (2013). Structural counterfactuals: A brief introduction. *Cognitive Science, 37*, 977–985.

Pfeifer, N., & Tulkki, L. (2017). Conditionals, counterfactuals, and rational reasoning: An experimental study on basic principles. *Minds and Machines, 27*, 119–165.

Quelhas, A. C., & Byrne, R. M. J. (2003). Reasoning with deontic and counterfactual conditionals. *Thinking & Reasoning, 9*, 43–66.

Ragni, M., Kola, I., & Johnson-Laird, P. N. (2018). On selecting evidence to test hypotheses: A theory of selection tasks. *Psychological Bulletin, 144*(8), 779–796.

Ramsey, F. P. (1990). General propositions and causality (original manuscript 1929). In D. H. Mellor (Ed.), *Philosophical papers* (pp. 145–163). London, England: Humanities Press.

Rips, L. J. (1994). *The psychology of proof*. Cambridge, MA: MIT Press.

Roese, N. J., & Epstude, K. (2017). The functional theory of counterfactual thinking: New evidence, new challenges, new insights. *Advances in Experimental Social Psychology, 56*, 1–79.

Santamaría, C., Espino, O. & Byrne, R. M. J. (2005). Counterfactual and semifactual conditionals prime alternative possibilities. *Journal of Experimental Psychology: Learning, Memory, and Cognition, 31*, 1149–1154.

Schroyens, W. J., Schaeken, W., & d'Ydewalle, G. (2001). The processing of negations in conditional reasoning: A meta-analytic case study in mental model and/or mental logic theory. *Thinking & Reasoning, 7*(2), 121–172.

Sloman, S. A., & Lagnado, D. A. (2005). Do we "do"? *Cognitive Science, 29*, 5–39.

Stalnaker, R. C. (1968). A theory of conditionals. In N. Rescher (Ed.), *Studies in logical theory* (pp. 98–112). Oxford, England: Blackwell.

Thompson, V., & Byrne, R. M. J. (2002). Reasoning counterfactually: Making inferences about things that didn't happen. *Journal of Experimental Psychology: Learning, Memory, and Cognition, 28*, 1154–1170.

Van Hoeck, N., Ma, N., Ampe, L., Baetens, K., Vandekerckhove, M., & Van Overwalle, F. (2013). Counterfactual thinking: An fMRI study on changing the past for a better future. *Social Cognitive Affective Neuroscience, 8*, 556–564.

Walsh, C. R., & Byrne, R. M. J. (2004). Counterfactual thinking: The temporal order effect. *Memory & Cognition, 32*, 369–378.

Williamson, T. (2018). Counterpossibles. *Topoi, 37*, 357–368.

6.4 Utility Conditionals

Jean-François Bonnefon

Summary

Utility conditionals are statements like "If you update this app, your phone will crash" or "If I let you do this, my boss will fire me," where antecedents, consequents, or both are actions or outcomes that are desirable or undesirable for some agents. Because they describe desirable and undesirable states of the world, utility conditionals support peculiar inferences that draw on both theoretical rationality and practical rationality. They offer a convenient device for studying how people reason about the beliefs, preferences, and decisions of others. As a consequence, they are a natural point of contact between disparate fields such as theory of mind, Bayesian approaches to reasoning, and behavioral economics.

1. Three Ways We Reason with Utility Conditionals

Utility conditionals (Bonnefon, 2009) are statements of the form "if p, q" where p or q (or both) are actions that increase or decrease the utility (in the economic sense) of some agents. For example:

(1) a. If you wash the car, I'll let you borrow it tonight.

    b. If you update this app, your phone will crash.

    c. If you testify against me, you will have an accident.

    d. If I let you do this, my boss will fire me.

    e. If Peyton moves to Paris, Alex will be unhappy.

    f. If I study instead of partying, I will get good grades.

Your washing the car is useful to me, you would not like your phone to crash, having accidents is bad, I don't want to be fired, Alex would be unhappy to be unhappy, and having good grades is good. Thus, all these examples are utility conditionals.

Some utility conditionals are speech acts, statements that perform a social function, for which we have a label: conditional (1a) is a *promise*, and conditional (1b) is a *warning*. Other utility conditionals are speech acts for which we do not have as clear a label: conditional

(1c) would be something like a veiled threat, and conditional (1d) is something like an apologetic interdiction. Other utility conditionals yet, such as (1e) and (1f), do not seem to be speech acts at all.

One line of research about utility conditionals is concerned with how people sort them out into categories such as promises, threats, tips, and warnings (these categories are also of interest for deontic reasoning; see chapter 11.2 by Elqayam, this handbook). For example, in a promise such as conditional (1a), the consequent of the conditional is an event under the control of the speaker, which has positive valence for the listener. Change the valence to something negative, and you get a threat; change the event such that it is no longer under the control of the speaker, and you get a tip; change both, as in conditional (1b), and you get a warning (Evans, 2005; López-Rousseau & Ketelaar, 2004, 2006).

The frontiers between these four categories can be blurry, though. For example, conditional (1c) has the

Figure 6.4.1

Distance between promises, threats, tips, and warnings along the axes of valence and control. Some "warnings" are really veiled threats, and some threats are issued as "promises," perhaps for dramatic effect.

properties of a warning but sounds like a threat. And indeed, eye-tracking studies have shown that readers are not too perturbed when such a statement is referred to as a threat (Wray, Wood, Haigh, & Stewart, 2016)—suggesting that the speaker's explicit control over the realization of q is not a strong constraint for distinguishing between threats and warnings. Altogether, other eye-tracking studies (Haigh, Ferguson, & Stewart, 2014; Haigh, Stewart, Wood, & Connell, 2011; Stewart, Haigh, & Ferguson, 2013; Wood, Haigh, & Stewart, 2016) paint a picture not unlike figure 6.4.1: the frontiers around threats are blurry (both warnings and promises can be disguised or "indirect" threats), but tips are clearly identifiable as conditionals "if p, q" where q is something positive for the listener over which the speaker has no control.

Regardless of the way we sort utility conditionals into linguistic or pragmatic categories, there are three ways to reason about them, which involve three different forms or mixtures of rationality.

The first way, which I will not address in this chapter because it only concerns a subset of utility conditionals, is to consider their use as persuasion tools. Indeed, some utility conditionals can be used as *arguments* for or against a given course of action or policy (Schellens & De Jong, 2004). More precisely, these utility conditionals are equivalent to *arguments from consequences*, in which the good or bad consequences of some action are pointed out in order to argue that this action should or should not be taken (Bonnefon, 2016). The persuasive use of utility conditionals appeals to practical rationality. Indeed, a core question within this field of research is to assess whether people are convinced by "good" arguments—arguments that would convince a rational agent who wants actions to maximize expected utility (Corner & Hahn, 2007; Corner, Hahn, & Oaksford, 2011; Feteris, 2002; Hahn & Oaksford, 2006, 2007; Hoeken, Šorm, & Schellens, 2014; Hoeken, Timmers, & Schellens, 2012; Thompson, Evans, & Handley, 2005).

The second way one can reason about utility conditionals is to use them to derive standard syllogistic inferences (modus ponens, modus tollens, etc.), for example,

(2) a. If Peyton moves to Paris, Alex will be unhappy.

   b. Alex is happy.

   c. Therefore, Peyton has not moved to Paris.

This form of reasoning about utility conditionals primarily draws on theoretical rationality, that is, it mostly aims to maintain belief consistency. We will see, though, that practical rationality can sneak into this kind of reasoning, because what we believe of the practical rationality

of others can affect the confidence we have in utility conditionals and thus the confidence we have in the conclusions of syllogisms that use them as main premises. For example, consider

(3) a. Lucy, if you don't eat your salad, you will be grounded for 5 years.

   b. Lucy does not eat her salad.

   c. Therefore, Lucy is grounded for 5 years.

As I will discuss in section 2, the reason why this conclusion sounds doubtful has to do with the lack of credibility of conditional (3a), and this lack of credibility derives from our assumptions about whether it would be practically rational to enforce the threat expressed in conditional (3a).

The third way one can reason about utility conditionals is to use them to derive nonstandard inferences about what agents want, believe, and do. For example:

(4) a. If Alex leaves early, she will be fired.

   b. Therefore, Alex won't leave early.

I call this inference "nonstandard" because it is not licensed by a logical reading of (4a). Indeed, there are no logical grounds to infer not-p from "if p, q." Instead, the inference that Alex won't leave early is based on the assumption that Alex is a practically rational agent who would not want to be fired and thus will not take an action that would get her fired. As I will discuss in section 3, such nonstandard inferences can concern actions, as in (4b), but also preferences and beliefs.

2.   Standard Inferences

For the sake of simplicity, I will focus in this section on the most basic conditional inference, modus ponens (for other phenomena, see, e.g., Couto, Quelhas, & Byrne, 2017; Demeure, Bonnefon, & Raufaste, 2009; Hilton, Kemmelmeier, & Bonnefon, 2005). Given a utility conditional "if p, q" and the observation that p is true, how likely are we to conclude that q is true?

First, we need to observe that some utility conditionals are more credible than others. For example, I have already discussed the distinction between promises and tips. When speakers make promises, they usually are in a position to fulfill these promises:

(5) If you tidy up your room, I'll make your favorite dessert.

But when speakers give tips, they cannot actually guarantee that the tip will actually result in the expected consequence:

(6) If you work hard, your teacher will give you a good grade.

Accordingly, we can expect reasoners to be relatively confident that promises will be carried out but less confident in the consequences of tips. As a consequence, reasoners are more likely to carry out modus ponens inferences from promises than from tips (Couto et al., 2017; Evans, Neilens, Handley, & Over, 2008), and *mutatis mutandis*, they are more likely to carry out modus ponens inferences from threats than from warnings.

Even promises and threats, though, can be more or less credible. Compare, for example, the following two conditionals:

(7) a. If you don't eat your salad, you will be grounded tonight.
    b. If you don't eat your salad, you will be grounded for 5 years.

Threat (7b) is so disproportionate to the offense that it loses its credibility (Amgoud, Bonnefon, & Prade, 2005; López-Rousseau, Diesendruck, & Benozio, 2011; Verbrugge, Dieussaert, Schaeken, & Van Belle, 2004). Parents who would unleash such punishment onto their children for mere peccadilloes would soon find themselves out of options—not to mention the effort that would be required from the parents to actually maintain the punishment for 5 years. Assuming that the parents have some modicum of practical rationality, it seems unlikely that they would actually carry out the threat.

Just like the vast majority of conditionals, utility conditionals have exceptions, that is, circumstances that weaken the inferential connection from their antecedent to their consequent. For example, Gazzo Castañeda and Knauff (2016b) have investigated "if crime, then punishment" utility conditionals, such as

(8) a. If a person downloads music from the Internet without allowance, then this person will be punished for breaching the copyright law.
    b. If a person kills another human, then this person will be punished for manslaughter.

Because reasoners find it easier to imagine exceptions to (8a) than to (8b), they are less likely to derive modus ponens inferences from (8a) than from (8b). In other words, utility conditionals behave just like regular conditionals in this respect. However, the magnitude of the crime (i.e., the disutility inflicted by whomever took the action in the antecedent clause) affects modus ponens performance when the modal "will" is replaced with the modal "should" (Gazzo Castañeda & Knauff, 2016a):

(9) a. If a person gains admission to an event without paying, then this person should be punished for obtaining benefits by devious means.
    b. If a person maltreats a minor in their charge, then this person should be punished for maltreatment of wards.

Reasoners take into account exceptions and exculpatory circumstances when they apply modus ponens to (9a), but they ignore these exceptions and exculpatory circumstances when dealing with (9b), because of the moral outrage they feel about maltreatment. In other words, the utility component of the conditional moderates the effect of exceptions on the application of modus ponens (see also Gazzo Castañeda, Richter, & Knauff, 2016).

3. Nonstandard Inferences

The most important feature of utility conditionals is that they support nonstandard inferences that are based on assumptions of practical rationality. Consider this example:

(10) If Alex writes an application, she'll get the award.

Conditional (10) suggests that Alex will write an application, based on three assumptions: that she wants the award enough to take the time to write an application, that she knows she'll get the award if she does, and that she's a rational agent who takes actions that increase her net utility. People are very well attuned to this calculus (Bonnefon & Hilton, 2004; Bonnefon & Sloman, 2013; Elqayam, Thompson, Wilkinson, Evans, & Over, 2015; Evans et al., 2008; Jara-Ettinger, Hyowon, Schulz, & Tenenbaum, 2016). Given a utility conditional "if p, q" where p is an action of an agent and q a positive (or negative, respectively) consequence of that action, reasoners will infer that the agent will or will not (and even *should* or should not) do q. They are more likely to make that inference when the magnitude of q increases, they are less likely to make that inference when the probability of q given p decreases, and they are less likely to make that inference when they can think of a way to obtain q that is less costly than p. Furthermore, eye-tracking evidence suggests that they make these inferences very quickly, in real time, as they read the conditional (Haigh & Bonnefon, 2015a, 2015b).

Not only can people predict actions based on utility conditionals, but they can also make inferences about the beliefs and preferences of other agents:

(11) a. If Brenda uses her loyalty card, she will save $20.
     b. Brenda does not use her loyalty card.

Conditional (11a) would usually encourage the inference that Brenda uses her loyalty card. Given that this inference is falsified by (11b), reasoners look for a reason to explain the behavior of Brenda. Their expectation that Brenda would use the loyalty card was based on at least four assumptions: (i) Brenda cares about saving $20, (ii) she knows she can save the $20 by using her loyalty card, (iii) she's a rational agent, and (iv) she is able to use her loyalty card. When they have no reason to believe that (iii) and (iv) may be the case, they decide to revise their assumptions about Brenda's beliefs or preferences: that is, they conclude *either* that Brenda did not know about the opportunity to save money *or* that she could not be bothered to save $20 (Bonnefon, Girotto, & Legrenzi, 2012).

Utility conditionals can express more complex situations than just one action and its consequences for the agent that takes it. Consider, for example,

(12) If you hurt me, I will help her.

This conditional encapsulates a complex, three-agent situation: if *a* hurts *b*, *b* will help *c*. Is *a* going to hurt *b*? Reasoners try to make sense of such a complex situation by fitting it into an easier template, such as a social contract in which people trade favors or exchange threats (Bonnefon, 2012; Bonnefon, Haigh, & Stewart, 2013; see also chapter 10.6 by Cosmides & Tooby, this handbook). More precisely, they hesitate between a good-for-good interpretation (for some reason, *b* wants to be hurt by *a*, and *b* wants *a* to help *c*; therefore, *a* will hurt *b*) and a bad-for-bad interpretation (*a* does not want *b* to help *c*, so *b* is threatening to help *c* if *a* hurts *b*; therefore, *a* will not hurt *b*).

Reasoners make similar template-fitting inferences about the preferences of agents when they do not have intuitions about these preferences. Consider the following example, where "tymping" and "yorbing" are nonverbs whose interpretation is open:

(13) If Peyton tymps Jesse, then Jesse will yorb Peyton.

When told that Jesse enjoys being tymped, reasoners apply a social-contract template and infer that Peyton likes to be yorbed (and *mutatis mutandis* if told that Jesse does not enjoy being yorbed). Reasoners can even revise their interpretation of nonambiguous verbs such as "hurt":

(14) If Peyton hurts Jesse, then Jesse will zim Peyton.

Here reasoners spontaneously assume that Peyton would not like to be zimmed. But if told that Peyton would actually like to be zimmed, they revise their assumptions about the preferences of Jesse, coming to the conclusion that Jesse likes to be hurt (for other such template effects, see Bonnefon et al., 2013).

4. Future Directions

Utility conditionals bridge a gap between research on reasoning and research on decision making, between theoretical rationality and practical rationality. They offer a convenient device for studying how people reason about the beliefs, preferences, and decisions of others. As a consequence, they are a natural point of contact between disparate fields such as theory of mind, Bayesian approaches to reasoning, moral judgment (see chapter 12.3 by Wiegmann & Sauer, this handbook), and behavioral economics (De Vito & Bonnefon, 2014; Jara-Ettinger et al., 2016). The greater challenge ahead will be to tackle what economists call "other-regarding preferences," that is, preferences over the outcomes of other individuals (Cooper & Kagel, 2015). There is currently no well-defined theory about how people weight the utilities of others when making their decisions and least of all a theory of how people think *others* weight the utilities of others when making their decisions. Utility conditionals offer a useful tool to explore this uncharted territory and to discover the naïve axioms of reasoners' folk theory of decision.

References

Amgoud, L., Bonnefon, J.-F., & Prade, H. (2005). An argumentation-based approach for multiple criteria decision. *Lecture Notes in Computer Science, 3571*, 269–280.

Bonnefon, J.-F. (2009). A theory of utility conditionals: Paralogical reasoning from decision-theoretic leakage. *Psychological Review, 116*, 888–907.

Bonnefon, J.-F. (2012). Utility conditionals as consequential arguments: A random sampling experiment. *Thinking & Reasoning, 18*, 379–393.

Bonnefon, J.-F. (2016). Predicting behavior on the basis of arguments from consequences. In F. Paglieri, L. Bonelli, & S. Felletti (Eds.), *The psychology of argument: Cognitive approaches to argumentation and persuasion* (pp. 1–15). London, England: College Publications.

Bonnefon, J.-F., Girotto, V., & Legrenzi, P. (2012). The psychology of reasoning about preferences and unconsequential decisions. *Synthese, 187*, 27–41.

Bonnefon, J.-F., Haigh, M., & Stewart, A. J. (2013). Utility templates for the interpretation of conditional statements. *Journal of Memory and Language, 68*, 350–361.

Bonnefon, J.-F., & Hilton, D. J. (2004). Consequential conditionals: Invited and suppressed inferences from valued outcomes. *Journal of Experimental Psychology: Learning, Memory, and Cognition, 30*, 28–37.

Bonnefon, J.-F., & Sloman, S. A. (2013). The causal structure of utility conditionals. *Cognitive Science, 37*, 193–209.

Cooper, D. J., & Kagel, J. H. (2015). Other-regarding preferences: A selective survey of experimental results. In J. H. Kagel & A. E. Roth (Eds.), *The handbook of experimental economics* (Vol. 2, pp. 217–288). Princeton, NJ: Princeton University Press.

Corner, A., & Hahn, U. (2007). Evaluating the meta-slope: Is there a slippery slope argument against slippery slope arguments? *Argumentation, 21*, 349–359.

Corner, A., Hahn, U., & Oaksford, M. (2011). The psychological mechanism of the slippery slope argument. *Journal of Memory and Language, 64*, 153–170.

Couto, M., Quelhas, A. C., & Byrne, R. M. J. (2017). Advice conditionals about tips and warnings: Interpretations and inferences. *Journal of Cognitive Psychology, 29*, 364–380.

Demeure, V., Bonnefon, J.-F., & Raufaste, E. (2009). Politeness and conditional reasoning: Interpersonal cues to the indirect suppression of deductive inferences. *Journal of Experimental Psychology: Learning, Memory, & Cognition, 35*, 260–266.

De Vito, S., & Bonnefon, J.-F. (2014). People believe each other to be selfish hedonic maximizers. *Psychonomic Bulletin & Review, 21*, 1331–1338.

Elqayam, S., Thompson, V. A., Wilkinson, M. R., Evans, J. St. B. T., & Over, D. E. (2015). Deontic introduction: A theory of inference from is to ought. *Journal of Experimental Psychology: General, 41*, 1516–1532.

Evans, J. St. B. T. (2005). The social and communicative function of conditional statements. *Mind & Society, 4*, 97–113.

Evans, J. St. B. T., Neilens, H., Handley, S. J., & Over, D. E. (2008). When can we say 'if'? *Cognition, 108*, 100–116.

Feteris, E. T. (2002). A pragma-dialectical approach of the analysis and evaluation of pragmatic argumentation in a legal context. *Argumentation, 16*, 349–367.

Gazzo Castañeda, L. E., & Knauff, M. (2016a). Defeasible reasoning with legal conditionals. *Memory & Cognition, 44*, 499–517.

Gazzo Castañeda, L. E., & Knauff, M. (2016b). When *will* is not the same as *should*: The role of modals in reasoning with legal conditionals. *Quarterly Journal of Experimental Psychology, 69*, 1480–1497.

Gazzo Castañeda, L. E., Richter, B., & Knauff, M. (2016). Negativity bias in defeasible reasoning. *Thinking & Reasoning, 22*, 209–220.

Hahn, U., & Oaksford, M. (2006). A Bayesian approach to informal argument fallacies. *Synthese, 152*, 207–236.

Hahn, U., & Oaksford, M. (2007). The rationality of informal argumentation: A Bayesian approach to argument fallacies. *Psychological Review, 114*, 704–732.

Haigh, M., & Bonnefon, J.-F. (2015a). Conditional sentences create a blind spot in theory of mind during narrative comprehension. *Acta Psychologica, 160*, 194–201.

Haigh, M., & Bonnefon, J.-F. (2015b). Eye movements reveal how readers infer intentions from the beliefs and desires of others. *Experimental Psychology, 62*, 206–213.

Haigh, M., Ferguson, H. J., & Stewart, A. J. (2014). An eye-tracking investigation into readers' sensitivity to actual versus expected utility in the comprehension of conditionals. *Quarterly Journal of Experimental Psychology, 67*, 166–185.

Haigh, M., Stewart, A. J., Wood, J. S., & Connell, L. (2011). Conditional advice and inducements: Are readers sensitive to implicit speech acts during comprehension? *Acta Psychologica, 136*, 419–424.

Hilton, D. J., Kemmelmeier, M., & Bonnefon, J.-F. (2005). Putting *if*s to work: Goal-based relevance in conditional directives. *Journal of Experimental Psychology: General, 135*, 388–405.

Hoeken, H., Šorm, E., & Schellens, P. J. (2014). Arguing about the likelihood of consequences: Laypeople's criteria to distinguish strong arguments from weak ones. *Thinking & Reasoning, 20*, 77–98.

Hoeken, H., Timmers, R., & Schellens, P. J. (2012). Arguing about desirable consequences: What constitutes a convincing argument? *Thinking & Reasoning, 18*, 394–416.

Jara-Ettinger, J., Hyowon, G., Schulz, L. E., & Tenenbaum, J. B. (2016). The naïve utility calculus: Computational principles underlying commonsense psychology. *Trends in Cognitive Sciences, 20*, 589–604.

López-Rousseau, A., Diesendruck, G., & Benozio, A. (2011). My kingdom for a horse: On incredible promises and unpersuasive warnings. *Pragmatics and Cognition, 19*, 399–421.

López-Rousseau, A., & Ketelaar, T. (2004). "If . . .": Satisficing algorithms for mapping conditional statements onto social domains. *European Journal of Cognitive Psychology, 16*, 807–823.

López-Rousseau, A., & Ketelaar, T. (2006). Juliet: If they do see thee, they will murder thee: A satisficing algorithm for pragmatic conditionals. *Mind & Society, 5*, 71–77.

Schellens, P. J., & De Jong, M. (2004). Argumentation schemes in persuasive brochures. *Argumentation, 18*, 295–323.

Stewart, A. J., Haigh, M., & Ferguson, H. J. (2013). Sensitivity to speaker control in the online comprehension of conditional tips and promises: An eye-tracking study. *Journal of Experimental Psychology: Learning, Memory & Cognition, 39*, 1022–1036.

Thompson, V. A., Evans, J. St. B. T., & Handley, S. J. (2005). Persuading and dissuading by conditional argument. *Journal of Memory and Language, 53,* 238–257.

Verbrugge, S., Dieussaert, K., Schaeken, W., & Van Belle, W. (2004). Promise is debt, threat another matter: The effect of credibility on the interpretation of conditional promises and threats. *Canadian Journal of Experimental Psychology, 58,* 106–112.

Wood, J. S., Haigh, M., & Stewart, A. J. (2016). This isn't a promise, it's a threat. *Experimental Psychology, 63,* 89–97.

Wray, H., Wood, J. S., Haigh, M., & Stewart, A. J. (2016). Threats may be negative promises (but warnings are more than negative tips). *Journal of Cognitive Psychology, 28,* 593–600.

Section 7   Causal and Diagnostic Reasoning

7.1 Causal and Counterfactual Inference

Judea Pearl

Summary

All accounts of rationality presuppose knowledge of how actions affect the state of the world and how the world would have changed had alternative actions been taken. The chapter presents a framework called the structural causal model (SCM), which operationalizes this knowledge and explicates how it can be derived from both theories and data. In particular, it shows how counterfactuals are computed and how they can be embedded in a calculus that solves critical problems in the empirical sciences.

1. Actions, Physical and Metaphysical

If the options available to an agent are specified in terms of their immediate consequences, as in "make him laugh," "paint the wall red," "raise taxes," or, in general, $do(X = x)$, where x is a value that variable X can take, then a rational agent is instructed to maximize the expected utility over all options x,

$$EU(x) = \sum_y P_x(y)U(y). \tag{1}$$

Here, $U(y)$ stands for the utility of outcome $Y = y$, and $P_x(y)$—the focus of this chapter—stands for the (subjective) probability that outcome $Y = y$ would prevail had action $do(X = x)$ been performed so as to establish condition $X = x$.

It has long been recognized that Bayesian conditionalization, that is, $P_x(y) = P(y \mid x)$, is inappropriate for serving in equation (1), for it leads to paradoxical results of several kinds (see Pearl, 2000a, pp. 108–109; Skyrms, 1980). For example, patients would avoid going to the doctor to reduce the probability that they are seriously ill, barometers would be manipulated to reduce the chance of storms, doctors would recommend a drug to male and female patients but not to patients with undisclosed gender, and so on. Yet the question of what function should substitute for $P_x(y)$, despite decades of thoughtful debates (Cartwright, 1983; Harper, Stalnaker, & Pearce,

1981; Jeffrey, 1965), seems to still baffle philosophers in the 21st century (Arló-Costa, 2007; Weirich, 2008/2016; Woodward, 2003). Modern discussions over evidential versus causal decision theory (chapter 8.2 by Peterson, this handbook) echo these debates.

Most studies of rationality have dealt with the utility function $U(y)$, its behavior under various shades of uncertainty, and the adequacy of the expectation operator in equation (1). Relatively little has been said about the probability $P_x(y)$ that governs outcomes $Y = y$ when an action $do(X = x)$ is contemplated. Yet regardless of what criterion one adopts for rational behavior, it must incorporate knowledge of how our actions affect the world. We must therefore define the function $P_x(y)$ and explicate the process by which it is assessed or inferred, be it from empirical data or from world knowledge. We must also ask what mental representation and thought processes would permit a rational agent to combine world knowledge with empirical observations and compute $P_x(y)$.

Guided by ideas from structural econometrics (Haavelmo, 1943; Spirtes, Glymour, & Scheines, 1993; Strotz & Wold, 1960), I have explored a conditioning operator called $do(x)$ (Pearl, 1995) that captures the intent of $P_x(y)$ by simulating an intervention in a causal model of interdependent variables (Pearl, 2009b).

The idea is simple. To model an action $do(X = x)$, one performs a "mini-surgery" on the causal model, that is, a minimal change necessary for establishing the antecedent $X = x$, while leaving the rest of the model intact. This calls for removing the mechanism (i.e., equation) that nominally assigns values to variable X and replacing it with a new equation, $X = x$, that enforces the intent of the specified action. This mini-surgery (not unlike Lewis's "little miracle") makes precise the idea of using a "minimal deviation from actuality" to define counterfactuals.

One important feature of this formulation is that the postintervention probability, $P(y \mid do(x))$, can be derived from preinterventional probabilities provided one possesses a diagrammatic representation of the processes that govern variables in the domain (Pearl, 2000a; Spirtes, Glymour, & Scheines, 2001). Specifically, the post-intervention probability reads:[1]

$$P(x, y, z \mid do(X = x^*)) = \begin{cases} P(x, y, z)/P(x \mid z), & \text{if } x = x^*, \\ 0, & \text{if } x \neq x^*. \end{cases} \quad (2)$$

Here, z stands for any realization of the set Z of "past" variables, y is any realization of the set Y of "future" variables, and "past" and "future" refer to the occurrence of the action event $X = x^*$.[2]

This feature, to be further discussed in section 2, is perhaps the key for the popularity of graph-theoretical methods in causal inference applications. It states that the effects of policies and interventions can be predicted without knowledge of the functional relationships (or mechanisms) among X, Y, and Z. The preinterventional probability and a few qualitative features of the model (e.g., variable ordering) are sufficient for determining the postintervention probabilities as in equation (2).

The philosophical literature spawned a totally different perspective on the probability function $P_x(y)$ in equation (1). In a famous letter to David Lewis, Robert Stalnaker (1972/1981) suggested to replace conditional probabilities with probabilities of conditionals, that is, $P_x(y) = P(x > y)$, where "$x > y$" stands for the counterfactual conditional "Y would be y if X were x" (see chapter 6.1 by Starr, this handbook). Using a "closest-worlds" semantics, Lewis (1973) defined $P(x > y)$ using a probability-revision operation called "imaging," in which probability mass "shifts" from worlds to worlds, governed by a measure of "similarity." Whereas Bayes conditioning $P(y \mid x)$ transfers the entire probability mass from worlds excluded by $X = x$ to all remaining worlds, in proportion to the latter's prior probabilities $P(\cdot)$, imaging works differently: each excluded world w transfers its mass individually to a select set $S_x(w)$ of worlds that are considered "closest" to w among those satisfying $X = x$. Joyce (1999) used the "\"-symbol, as in "$P(y \setminus x)$," to denote the probability resulting from such an imaging process and derived a formula for $P(y \setminus x)$ in terms of the selection function $S_x(w)$.

In Pearl (2000a, p. 73), I have shown that the transformation defined by the do-operator, equation (2), can be interpreted as an imaging-type mass transfer if the following two provisions are met:

Provision 1: The choice of "similarity" measure is not arbitrary; worlds with equal histories should be considered equally similar to any given world.

Provision 2: The redistribution of weight within each selection set $S_x(w)$ is not arbitrary either; equally similar worlds should receive mass in proportion to their prior probabilities.

This tie-breaking rule is similar in spirit to the Bayesian policy and permits us to generalize equation (2) to disjunctive actions, as in "exercise at least 30 minutes daily" or "paint the wall either green or purple" (Pearl, 2017).

The theory that emerges from the do-operator (equation (2)) offers several conceptual and operational advantages over Lewis's closest-world semantics. First, it does not rest on a metaphysical notion of "similarity," which may be different from person to person and thus could not explain the uniformity with which people interpret causal utterances. Instead, causal relations are defined in terms of our scientific understanding of how variables interact with one another (to be explicated in section 2). Second, it offers a plausible resolution of the "mental representation" puzzle: how do humans represent "possible worlds" in their minds and compute the closest one, when the number of possibilities is far beyond the capacity of the human brain? Any credible theory of rationality must account for the astonishing ease with which humans comprehend, derive, and communicate counterfactual information. Finally, it results in practical algorithms for solving some of the most critical and difficult causal problems that have challenged data analysts and experimental researchers in the past century (see Pearl & Mackenzie, 2018, for an extensive historical account). I call this theory the structural causal model (SCM).

In the rest of the chapter, we will focus on the properties of SCM and explicate how it can be used to define counterfactuals (section 2), to control confounding and predict the effect of interventions and policies (section 3), to define and estimate direct and indirect effects (section 4), and, finally, to ensure generalizability of empirical results across diverse environments (section 5).

2. Counterfactuals and SCM

At the center of the structural theory of causation lies a "structural model," M, consisting of two sets of variables, U and V, and a set F of functions that determine or simulate how values are assigned to each variable $V_i \in V$. Thus, for example, the equation

$$v_i = f_i(v, u)$$

describes a physical process by which variable V_i is *assigned* the value $v_i = f_i(v, u)$ in response to the current values, v and u, of all variables in V and U. Formally, the triplet $\langle U, V, F \rangle$ defines an SCM, and the diagram that captures the relationships among the variables is called the *causal graph G* (of M). The variables in U are considered "exogenous," namely, background conditions for which no explanatory mechanism is encoded in model M. Every instantiation $U = u$ of the exogenous variables uniquely determines the values of all variables in V, and hence, if we assign a probability $P(u)$ to U, it defines a probability function $P(v)$ on V. The vector $U = u$ can also be interpreted as an experimental "unit," which can stand for an individual subject, agricultural lot, or time of day, since it describes all factors needed to make V a deterministic function of U.

The basic counterfactual entity in structural models is the sentence "Y would be y had X been x in unit (or situation) $U = u$," denoted $Y_x(u) = y$. Letting M_x stand for a modified version of M, with the equation(s) of set X replaced by $X = x$, the formal definition of the counterfactual $Y_x(u)$ reads

$$Y_x(u) = Y_{M_x}(u). \tag{3}$$

In words, the counterfactual $Y_x(u)$ in model M is defined as the solution for Y in the "modified" submodel M_x. Galles and Pearl (1998) and Halpern (1998) have given a complete axiomatization of structural counterfactuals, embracing both recursive and nonrecursive models (see also Pearl, 2009b, chapter 7).[3]

Since the distribution $P(u)$ induces a well-defined probability on the counterfactual event $Y_x = y$, it also defines a joint distribution on all Boolean combinations of such events, for instance, "$Y_x = y$ & $Z_{x'} = z$," which may appear contradictory, if $x \neq x'$. For example, to answer retrospective questions, such as whether Y would be y_1 if X were x_1, given that in fact Y is y_0 and X is x_0, we need to compute the conditional probability $P(Y_{x_1} = y_1 \mid Y = y_0, X = x_0)$, which is well defined once we know the forms of the structural equations and the distribution of the exogenous variables in the model.

In general, the probability of the counterfactual sentence $P(Y_x = y \mid e)$, where e is any propositional evidence, can be computed by the following three-step process (Pearl, 2009b, p. 207):

Step 1 (abduction): Update the probability $P(u)$ to obtain $P(u \mid e)$.

Step 2 (action): Replace the equations determining the variables in set X by $X = x$.

Step 3 (prediction): Use the modified model to compute the probability of $Y = y$.

In temporal metaphors, step 1 explains the past (U) in light of the current evidence e, step 2 bends the course of history (minimally) to comply with the hypothetical antecedent $X = x$, and finally, step 3 predicts the future (Y) based on our new understanding of the past and our newly established condition, $X = x$.

2.1 Example: Computing Counterfactuals in Linear SCM

We illustrate the working of this three-step algorithm using a linear structural equation model, depicted by the graph in figure 7.1.1.

To motivate the analysis, let X stand for the level of assistance (or "treatment") given to a student, Z stands for the amount of time the student spends studying, and Y, the outcome, stands for the student's performance on the exam. The algebraic version of this model takes the form of the following equations:

$$x = \varepsilon_1,$$
$$z = \beta x + \varepsilon_2,$$
$$y = \alpha x + \gamma z + \varepsilon_3.$$

The coefficients α, β, and γ are called "structural coefficients," to be distinguished from regression coefficients, and represent direct causal effects of the corresponding variables. Under appropriate assumptions, say that the error terms ε_1, ε_2, and ε_3 are mutually independent, the structural coefficients can be estimated from data. Our task, however, is not to estimate causal effects but to answer counterfactual questions taking the model as given.

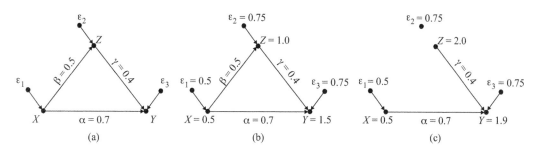

Figure 7.1.1
Structural models used for answering a counterfactual question about an individual $u = (\varepsilon_1, \varepsilon_2, \varepsilon_3)$. (a) The generic model. (b) The u-specific model. (c) The modified model necessary to accommodate the antecedent $Z = 2$ of the counterfactual question Q_1.

Let us consider a student named Joe, for whom we measure $X = 0.5$, $Z = 1$, $Y = 1.5$, and about whom we ask a counterfactual question:

Q_1: What would Joe's score have been had he doubled his study time?

Using our subscript notation, this question amounts to evaluating $Y_{Z=2}(u)$, with u standing for the distinctive characteristics of Joe, namely, $u = (\varepsilon_1, \varepsilon_2, \varepsilon_3)$, as inferred from the observed data $\{X = 0.5, Z = 1, Y = 1.5\}$.

Following the algorithm above, the answer to this question is obtained in three steps:

1. Use the data to compute the exogenous factors $\varepsilon_1, \varepsilon_2, \varepsilon_3$. (These are the invariant characteristics of unit u and do not change by interventions or counterfactual hypothesizing.) In our model, we get (see figure 7.1.1b):

$$\varepsilon_1 = 0.5$$
$$\varepsilon_2 = 1 - 0.5 \times 0.5 = 0.75,$$
$$\varepsilon_3 = 1.5 - 0.5 \times 0.7 - 1 \times 0.4 = 0.75.$$

2. Modify the model, to form $M_{Z=2}$, in which Z is set to 2 and all arrows to Z are removed (figure 7.1.1c).

3. Compute the value of Y in the mutilated model formed in step 2, which gives

$$Y_{Z=2} = 0.5 \times 0.7 + 2.0 \times 0.4 + 0.75 = 1.90.$$

We can thus conclude that, had he doubled his study time, Joe's score would have been 1.90 instead of 1.5. This example illustrates the need to modify the original model (figure 7.1.1a), in which the combination $X = 1$, $\varepsilon_2 = 0.75$, $Z = 2.0$ constitutes a contradiction.

2.2 The Two Principles of Causal Inference

Before describing specific applications of the structural theory, it will be useful to summarize its implications in the form of two "principles," from which all other results follow:

Principle 1: The law of structural counterfactuals

Principle 2: The law of structural independence

The first principle is described in equation (3) and instructs us how to compute counterfactuals and their probabilities from a structural model. This, together with principle 2, will allow us (section 3) to determine what assumptions one must make about reality in order to infer probabilities of counterfactuals from either experimental or passive observations.

Principle 2 defines how structural features of the model entail dependencies in the data. Remarkably, regardless of the functional form of the equations in the model and

regardless of the distribution of the exogenous variables U, if the latter are mutually independent and the model is recursive, the distribution $P(v)$ of the endogenous variables must obey certain conditional independence relations, stated roughly as follows: whenever sets X and Y are "separated" by a set Z in the graph, X is independent of Y given Z (Verma & Pearl, 1988). This "separation" condition, called "d-separation" (Geiger, Verma, & Pearl, 1990; Pearl, 2000a, pp. 16–18), constitutes the link between the causal assumptions encoded in the causal graph (in the form of missing arrows) and the observed data. It is defined formally as follows:

Definition 1 (d-separation). A set S of nodes is said to *block* a path p if either

1. p contains at least one arrow-emitting node that is in S, or

2. p contains at least one *collider* (i.e., a node obtaining head-to-head arrows) that is outside S and has no descendant in S.

If S blocks *all* paths from set X to set Y, it is said to "d-separate X and Y," and then variables X and Y are independent given S, written $X \perp\!\!\!\perp Y \mid S$.[4]

D-separation implies conditional independencies for every distribution $P(v)$ that is compatible with the causal assumptions embedded in the diagram. To illustrate, the diagram in figure 7.1.2a implies $Z_1 \perp\!\!\!\perp Y \mid (X, Z_3, W_2)$, because the conditioning set $S = \{X, Z_3, W_2\}$ blocks all paths between Z_1 and Y. The set $S = \{X, Z_3, W_3\}$, however, leaves the path (Z_1, Z_3, Z_2, W_2, Y) unblocked (by virtue of the collider at Z_3), and so, the independence $Z_1 \perp\!\!\!\perp Y \mid (X, Z_3, W_3)$ is not implied by the diagram.

3. Intervention, Identification, and Causal Calculus

To maximize the expectation defined by equation (1), a central problem for any rational agent is that of inferring the probability $P_x(y)$ from empirical data. In the context of social or medical policy making, this amounts

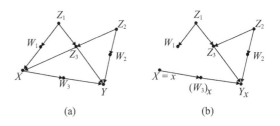

(a)   (b)

Figure 7.1.2

Illustrating the intervention $do(X = x)$. (a) The original model M. (b) The intervention submodel M_x and the counterfactual Y_x.

to estimating the interventional probability $P(y \mid do(x))$, which is defined, using the counterfactual Y_x, as[5]

$$P(y \mid do(x)) \triangleq P(Y_x = y). \tag{4}$$

Given a model M, the effect of an intervention $X = do(x)$ can be predicted from the submodel M_x as shown in figure 7.1.2. Figure 7.1.2b illustrates the submodel M_x created by the atomic intervention $do(x)$; it sets the value of X to x and thus removes the influence of W_1 and Z_3 on X. We similarly define the result of *conditional interventions* by

$$P(y \mid do(x), z) \triangleq P(y,z \mid do(x)) / P(z \mid do(x))$$
$$= P(Y_x = y \mid Z_x = z). \tag{5}$$

$P(y \mid do(x), z)$ captures the z-specific effect of X on Y, that is, the effect of setting X to x among those units only for which $Z = z$.

A second important question concerns *identification* in partially specified models: given a set A of qualitative causal assumptions, as embodied in the structure of the causal graph, can the controlled (postintervention) distribution, $P(y \mid do(x))$, be estimated from the available data that are governed by the preintervention distribution $P(z, x, y)$? In linear parametric settings, the question of identification reduces to asking whether some model parameter, β, has a unique solution in terms of the parameters of P (say the population covariance matrix). In the nonparametric formulation, the notion of "has a unique solution" does not directly apply since quantities such as $Q = P(y \mid do(x))$ have no parametric signature and are defined procedurally by a symbolic operation on the causal model M, as in figure 7.1.2b. The following definition captures the requirement that Q be estimable from the data:

Definition 2 (Identifiability; Pearl, 2000a, p. 77). A causal query Q is *identifiable from* data compatible with a causal graph G, if for any two (fully specified) models M_1 and M_2 that satisfy the assumptions in G, we have

$$P_1(v) = P_2(v) \implies Q(M_1) = Q(M_2). \tag{6}$$

In words, equality in the probabilities $P_1(v)$ and $P_2(v)$ induced by models M_1 and M_2, respectively, entails equality in the answers that these two models give to query Q. When this happens, Q depends on P only and should therefore be expressible in terms of the parameters of P.

When a query Q is given in the form of a *do*-expression, for example, $Q = P(y \mid do(x), z)$, its identifiability can be decided systematically using an algebraic procedure known as the "*do*-calculus" (Pearl, 1995). It consists of three inference rules that permit us to equate interventional and observational distributions whenever certain d-separation conditions hold in the causal diagram G.

3.1 The Rules of the Do-Calculus

Let X, Y, Z, and W be arbitrary disjoint sets of nodes in a causal directed acyclical graph (DAG) G. We denote by $G_{\overline{X}}$ the graph obtained by deleting from G all arrows pointing to nodes in X. Likewise, we denote by $G_{\underline{X}}$ the graph obtained by deleting from G all arrows emerging from nodes in X. To represent the deletion of both X-incoming and Z-outgoing arrows, we use the notation $G_{\overline{X}\underline{Z}}$.

The following three rules are valid for every interventional distribution compatible with G:

Rule 1 (Insertion/deletion of observations):

$$P(y \mid do(x), z, w) = P(y \mid do(x), w)$$
$$\text{if } (Y \perp\!\!\!\perp Z \mid X, W)_{G_{\overline{X}}}. \tag{7}$$

Rule 2 (Action/observation exchange):

$$P(y \mid do(x), do(z), w) = P(y \mid do(x), z, w)$$
$$\text{if } (Y \perp\!\!\!\perp Z \mid X, W)_{G_{\overline{X}\underline{Z}}}. \tag{8}$$

Rule 3 (Insertion/deletion of actions):

$$P(y \mid do(x), do(z), w) = P(y \mid do(x), w)$$
$$\text{if } (Y \perp\!\!\!\perp Z \mid X, W)_{G_{\overline{X}\,\overline{Z(W)}}}, \tag{9}$$

where $Z(W)$ is the set of Z-nodes that are not ancestors of any W-node in $G_{\overline{X}}$.

To establish identifiability of a causal query Q, one needs to repeatedly apply the rules of the *do*-calculus to Q, until an expression is obtained that no longer contains a *do*-operator;[6] this renders it estimable from nonexperimental data. The *do*-calculus was proven to be complete for queries in the form $Q = P(y \mid do(x), z)$ (Huang & Valtorta, 2006; Shpitser & Pearl, 2006), which means that if Q cannot be reduced to probabilities of observables by repeated application of these three rules, then such a reduction does not exist, that is, the query is not estimable from observational studies without strengthening the assumptions.

3.2 Covariate Selection: The Backdoor Criterion

One of the most powerful results emerging from the *do*-calculus is a method of identifying a set of variables that, if measured, would permit us to predict the effect of action from passive observation. This set of variables coincides with the set Z of equation (2), which we called "past" in section 1 and will now receive a formal characterization in definition 3.

Consider an observational study in which we wish to find the effect of some treatment (X) on a certain outcome (Y), and assume that the factors deemed relevant to the problem are structured as in figure 7.1.2a; some are affecting the outcome, some are affecting the treatment, and some are affecting both treatment and response. Some of these factors may be unmeasurable, such as genetic trait or lifestyle, while others are measurable, such as gender, age, and salary level. Our problem is to select a subset of these factors for measurement and adjustment such that if we compare treated versus untreated subjects having the same values of the selected factors, we get the correct treatment effect in that subpopulation of subjects. Such a set of factors is called a "sufficient set" or a set "appropriate for adjustment" (see Greenland, Pearl, & Robins, 1999; Pearl, 2000b, 2009a). The following criterion, named "backdoor" (Pearl, 1993), provides a graphical method of selecting such a set of factors for adjustment:

> **Definition 3** (Admissible sets—the backdoor criterion). A set S is *admissible* (or "sufficient") for estimating the causal effect of X on Y if two conditions hold:
>
> 1. No element of S is a descendant of X.
> 2. The elements of S "block" all "backdoor" paths from X to Y—namely, all paths that end with an arrow pointing to X.

Based on this criterion, we see, for example, that in figure 7.1.2, the sets $\{Z_1, Z_2, Z_3\}$, $\{Z_1, Z_3\}$, $\{W_1, Z_3\}$, and $\{W_2, Z_3\}$ are each sufficient for adjustment, because each blocks all backdoor paths between X and Y. The set $\{Z_3\}$, however, is not sufficient for adjustment because it does not block the path $X \leftarrow W_1 \leftarrow Z_1 \rightarrow Z_3 \leftarrow Z_2 \rightarrow W_2 \rightarrow Y$.

The intuition behind the backdoor criterion is as follows. The backdoor paths in the diagram carry spurious associations from X to Y, while the paths directed along the arrows from X to Y carry causative associations. Blocking the former paths (by conditioning on S) ensures that the measured association between X and Y is purely causal, namely, it correctly represents the target quantity: the causal effect of X on Y. Conditions for relaxing restriction 1 are given in Pearl (2009b, p. 338), Pearl and Paz (2014), and Shpitser, VanderWeele, and Robins (2010).[7]

The implication of finding a sufficient set, S, is that stratifying on S is guaranteed to remove all confounding bias relative to the causal effect of X on Y. In other words, it renders the causal effect of X on Y identifiable, via the *adjustment formula*[8]

$$P(Y = y \mid do(X = x))$$
$$= \sum_s P(Y = y \mid X = x, S = s)P(S = s). \quad (10)$$

Since all factors on the right-hand side of the equation are estimable (e.g., by regression) from preinterventional data, the causal effect can likewise be estimated from such data without bias. Note that equation (2) is a special case of equation (10), where S is chosen to include all variables preceding X in the causal order. Moreover, the backdoor criterion implies the independence $X \perp\!\!\!\perp Y_x \mid S$,

also known as "conditional ignorability" (Rosenbaum & Rubin, 1983), and provides therefore the scientific basis for most inferences in the potential-response framework.

The backdoor criterion allows us to write equation (10) by inspection, after selecting a sufficient set, S, from the diagram. The selection criterion can be applied systematically to diagrams of any size and shape, thus freeing analysts from judging whether "X is conditionally ignorable given S," a formidable mental task required in the potential-response framework. The criterion also enables the analyst to search for an optimal set of covariates— namely, a set S that minimizes measurement cost or sampling variability (Tian, Paz, & Pearl, 1998).

> **Theorem 1** (identification of interventional expressions). Given a causal graph G containing both measured and unmeasured variables, the consistent estimability of any expression of the form
>
> $$Q = P(y_1, y_2, \ldots, y_m \mid do(x_1, x_2, \ldots, x_n), z_1, z_2, \ldots, z_k)$$
>
> can be decided in polynomial time. If Q is estimable, then its estimand can be derived in polynomial time. Furthermore, the algorithm is complete.

The results stated in theorem 1 were developed in several stages over the past 20 years (Pearl, 1993, 1995; Shpitser & Pearl, 2006; Tian & Pearl, 2002). Bareinboim and Pearl (2012a) extended the identifiability of Q to combinations of observational and experimental studies.

It is important to note at this point that the *do*-operator can be used not merely for fixing a variable at a predetermined value x but also for analyzing "soft interventions." For example, the effect of additive interventions, such as "administer 5 mg of insulin to a given patient," can be estimated using the *do*-calculus (Pearl, Glymour, & Jewell, 2016, p. 109). Likewise, the effects of stochastic interventions (e.g., "change the frequency with which this patient receives a drug") can be estimated by a method based on the *do*-operator (Pearl, 2009b, p. 113). The versatility of the *do*-operator is further discussed in Pearl (2009b, section 11.4).

4. Mediation Analysis

Mediation analysis aims to uncover causal pathways along which changes are transmitted from causes to effects. Interest in mediation analysis stems from both scientific and practical considerations. Scientifically, mediation tells us "how nature works," and practically it enables us to predict behavior under a rich variety of conditions and policy interventions. For example, in coping with the age-old problem of gender discrimination

(Bickel, Hammel, & O'Connell, 1975; Goldberger, 1984), a policy maker may be interested in assessing the extent to which gender disparity in hiring can be reduced by making hiring decisions gender-blind, compared with eliminating gender inequality in education or job qualifications. The former concerns the "direct effect" of gender on hiring while the latter concerns the "indirect effect" or the effect *mediated* via job qualification.

The role that mediation analysis plays in rational decision making revolves around the richer set of options that emerges from understanding "how the world works." For example, the option of using mosquito nets was not considered by decision makers when malaria was believed to be caused by *mal'aria* ("bad air"). It was fairly rational in those days to use breathing masks in swampy areas. It is hardly rational today, given the overwhelming evidence about the mediating effect of the *Anopheles* mosquito. The logic of properly accounting for empirical data in one's belief system is an aspect of rational behavior that is gaining increased attention among researchers (Pearl, 2013).

The structural model for a typical mediation problem takes the form

$$t = f_T(u_T), \quad m = f_M(t, u_M), \quad y = f_Y(t, m, u_Y), \quad (11)$$

where T (treatment), M (mediator), and Y (outcome) are discrete or continuous random variables; f_T, f_M, and f_Y are arbitrary functions; and U_T, U_M, and U_Y represent, respectively, omitted factors that influence T, M, and Y. In figure 7.1.3a, the omitted factors are assumed to be arbitrarily distributed but mutually independent. In figure 7.1.3b, the dashed arcs connecting U_T and U_M (as well as U_M and U_T) encode the understanding that the factors in question may be dependent.

4.1 Natural Direct and Indirect Effects

Using the structural model of equation (11), four types of effects can be defined for the transition from $T = 0$ to $T = 1$:[9]

Total effect:

$$\begin{aligned} TE &= E[f_Y[1, f_M(1, u_M), u_Y] - f_Y[0, f_M(0, u_M), u_Y]] \\ &= E[Y_1 - Y_0] \\ &= E[Y \mid do(T=1)] - E[Y \mid do(T=0)]. \end{aligned} \quad (12)$$

TE measures the expected increase in Y as the treatment changes from $T = 0$ to $T = 1$, while the mediator is allowed to track the change in T as dictated by the function f_M.

Controlled direct effect:

$$\begin{aligned} CDE(m) &= E\{f_Y[1, M=m, u_Y] - f_Y[0, M=m, u_Y]\} \\ &= E[Y_{1,m} - Y_{0,m}] \\ &= E[Y \mid do(T=1, M=m)] - E[Y \mid do(T=0, M=m)]. \end{aligned} \quad (13)$$

CDE measures the expected increase in Y as the treatment changes from $T = 0$ to $T = 1$, while the mediator is set to a specified level $M = m$ uniformly over the entire population.

Natural direct effect:[10]

$$\begin{aligned} NDE &= E\{f_Y[1, f_M(0, u_M), u_T] - f_Y[0, f_M(0, u_M), u_T]\} \\ &= E[Y_{1, M_0} - Y_{0, M_0}]. \end{aligned} \quad (14)$$

NDE measures the expected increase in Y as the treatment changes from $T = 0$ to $T = 1$, while the mediator is set to whatever value it *would have attained* (for each individual) prior to the change, that is, under $T = 0$.

Natural indirect effect:

$$\begin{aligned} NIE &= E\{f_Y[0, f_M(1, u_M), u_Y] - f_Y[0, f_M(0, u_M), u_Y]\} \\ &= E[Y_{0, M_1} - Y_{0, M_0}]. \end{aligned} \quad (15)$$

NIE measures the expected increase in Y when the treatment is held constant, at $T = 0$, and M changes to whatever value it would have attained (for each individual) under $T = 1$. It captures, therefore, the portion of the effect that can be explained by mediation alone while disabling the capacity of Y to respond to X.

We note that, in general, the total effect can be decomposed as

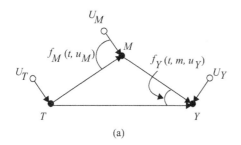 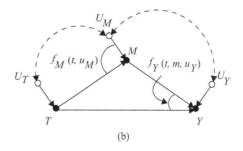

Figure 7.1.3

(a) The basic nonparametric mediation model, with no confounding. (b) A confounded mediation model in which dependence exists between U_M and (U_T, U_Y).

$$TE = NDE - NIE_r, \tag{16}$$

where NIE_r stands for the natural indirect effect under the reverse transition, from $T = 1$ to $T = 0$. This implies that NIE is identifiable whenever NDE and TE are identifiable. In linear systems, where reversal of transitions amounts to negating the signs of their effects, we have the standard additive formula, $TE = NDE + NIE$.

We further note that TE and $CDE(m)$ are *do*-expressions and can, therefore, be estimated from experimental data. Not so NDE and NIE; both are counterfactual expressions that cannot be reduced to *do*-expressions. The reason is simple: there is no way to disable the direct effect by intervening on any variable in the model. The counterfactual language permits us to circumvent this difficulty by (figuratively) changing T to affect M while feeding Y the prior value of T.

Since theorem 1 assures us that the identifiability of any *do*-expression can be determined by an effective algorithm, TE and $CDE(m)$ can be identified by those algorithms. NDE and NIE, however, require special analysis, given in the next subsection.

4.2 Sufficient Conditions for Identifying Natural Effects

The following is a set of assumptions or conditions, marked A-1 to A-4, that are sufficient for identifying both direct and indirect natural effects. Each condition is communicated using the causal diagram.

There exists a set W of measured covariates such that

A-1 No member of W is a descendant of T.

A-2 W blocks all backdoor paths from M to Y (not traversing $X \to M$ and $X \to Y$).

A-3 The W-specific effect of T on M is identifiable (using theorem 1 and possibly using experiments or auxiliary variables).

A-4 The W-specific joint effect of $\{T, M\}$ on Y is identifiable (using theorem 1 and possibly using experiments or auxiliary variables).

Theorem 2 (identification of natural effects). When conditions A-1 and A-2 hold, the natural direct effect is experimentally identifiable and is given by

$$
\begin{aligned}
NDE = \sum_m \sum_w &[E(Y \mid do(T=1, M=m), W=w) \\
&- E(Y \mid do(T=0, M=m), W=w)] \\
&\times P(M=m \mid do(T=0), W=w) \times P(W=w).
\end{aligned}
\tag{17}
$$

The identifiability of the *do*-expressions in equation (17) is guaranteed by conditions A-3 and A-4 and can be determined by theorem 1.

In the nonconfounding case (figure 7.1.3a), NDE reduces to

$$
\begin{aligned}
NDE = \sum_m &[E(Y \mid T=1, M=m) - E(Y \mid T=0, M=m)] \\
&\times P(M=m \mid T=0),
\end{aligned}
\tag{18}
$$

which came to be known as the *mediation formula* (Pearl, 2012).

Shpitser (2013) further provides complete algorithms for identifying natural direct and indirect effects and extends these results to path-specific effects with multiple treatments and multiple outcomes.

5. External Validity and Transportability

To support the choice of optimal actions on the basis of nonexperimental data, the role of the *do*-calculus is to remove the *do*-operator from the query expression. We now discuss a totally different application, to decide if experimental findings from environment π can be transported to a new, potentially different environment π^*, in which only passive observations can be performed. This problem, labeled "transportability" in Pearl and Bareinboim (2011), is at the heart of every scientific investigation since, invariably, experiments performed in one environment (or population) are intended to be used elsewhere, where conditions may differ.

To formalize problems of this sort, a graphical representation called [a] "selection diagram" was devised (figure 7.1.4), which encodes knowledge about differences and commonalities between populations. A selection diagram is a causal diagram annotated with new variables, called "S-nodes," which point to the mechanisms where discrepancies between the two populations are suspected to be located. The task of deciding if transportability is feasible now reduces to the syntactic problem of separating (using the *do*-calculus) the *do*-operator from the S-variables in the query expression $P(y \mid do(x), z, s)$. In effect, this separation renders the disparities irrelevant to what we learn in the experimental setup.

Theorem 3 (Pearl & Bareinboim, 2011). Let D be the selection diagram characterizing two populations, π and π^*, and S a set of selection variables in D. The relation $R = P^*(y \mid do(x), z)$ is transportable from π and π^* if and only if the expression $P(y \mid do(x), z, s)$ is reducible, using the rules of the *do*-calculus, to an expression in which S appears only as a conditioning variable in *do*-free terms.

While theorem 3 does not specify the sequence of rules leading to the needed reduction (if such exists), a complete and effective graphical procedure was devised

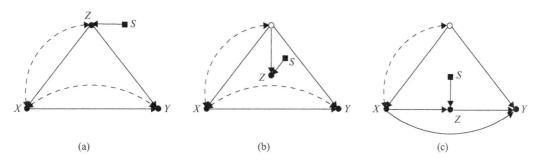

Figure 7.1.4
Selection diagrams depicting differences in populations. (a) The two populations differ in age distributions. (b) The populations differ in how reading skills (Z) depend on age (an unmeasured variable, represented by the hollow circle), and the age distributions are the same. (c) The populations differ in how Z depends on X. Dashed arcs (e.g., $X \leftarrow\!-\!-\!\rightarrow Y$) represent the presence of latent variables affecting both X and Y.

by Bareinboim and Pearl (2012b), which also synthesizes a *transport formula* whenever possible. Each transport formula determines what informations need to be extracted from the experimental and observational studies and how they ought to be combined to yield an unbiased estimate of the relation $R = P(y\,|\,do(x),s)$ in the target population π^*. For example, the transport formulas induced by the three models in figure 7.1.4 are given by

$$P(y\,|\,do(x),s) = \sum_z P(y\,|\,do(x), z)P(z\,|\,s), \qquad \text{(a)}$$

$$P(y\,|\,do(x),s) = P(y\,|\,do(x)), \qquad \text{(b)}$$

$$P(y\,|\,do(x),s) = \sum_z P(y\,|\,do(x),z)P(z\,|\,x,s). \qquad \text{(c)}$$

Each of these formulas satisfies theorem 3, and each describes a different procedure of pooling information from π and π^*.

For example, (c) states that to estimate the causal effect of X on Y in the target population π^*, $P(y\,|\,do(x),z,s)$, we must estimate the z-specific effect $P(y\,|\,do(x),z)$ in the source population π and average it over z, weighted by $P(z\,|\,x,s)$, that is, the conditional probability $P(z\,|\,x)$ estimated at the target population π^*. The derivation of this formula follows by writing

$$P(y\,|\,do(x),s) = \sum_z P(y\,|\,do(x), z, s)P(z\,|\,do(x),s)$$

and noting that rule 1 of the *do*-calculus authorizes the removal of s from the first term (since $Y \perp\!\!\!\perp S\,|\,Z$ holds in $G_{\bar{x}}$), and rule 2 authorizes the replacement of $do(x)$ with x in the second term (since the independence $Z \perp\!\!\!\perp X$ holds in $G_{\underline{x}}$).

A generalization of transportability theory to multiple environments has led to a method called "data fusion" (Bareinboim & Pearl, 2016), aimed at combining results from many experimental and observational studies, each conducted on a different population and under a different set of conditions, so as to synthesize an aggregate measure of effect size in yet another environment, different from the rest. This fusion problem has received enormous attention in the health and social sciences, where it is typically handled inadequately by a statistical method called "meta-analysis," which "averages out" differences instead of rectifying them.

Using multiple "selection diagrams" to encode commonalities among studies, Bareinboim and Pearl (2013) "synthesized" an estimator that is guaranteed to provide an unbiased estimate of the desired quantity based on information that each study shares with the target environment. Remarkably, a consistent estimator may be constructed from multiple sources even in cases where it is not constructible from any one source in isolation.

Theorem 4 (Bareinboim & Pearl, 2013).

- Nonparametric transportability of experimental findings from multiple environments can be determined in polynomial time, provided suspected differences are encoded in selection diagrams.

- When transportability is feasible, a transport formula can be derived in polynomial time that specifies what information needs to be extracted from each environment to synthesize a consistent estimate for the target environment.

- The algorithm is complete, that is, when it fails, transportability is infeasible.

Another problem that falls under the data fusion umbrella is that of "selection bias" (Bareinboim, Tian, & Pearl, 2014), which requires a generalization from a subpopulation selected for a study to the population at large, the target of the intended policy.

Selection bias is induced by preferential selection of units for observation, usually governed by unknown factors, thus rendering the data no longer representative of the environment (or population) of interest. Selection bias represents a major obstacle to valid causal and statistical inferences. It cannot be removed by randomized experiments and can rarely be detected in either experimental or observational studies.[11] For instance, in a typical study of the effect of training programs on earnings, subjects achieving higher incomes tend to report their earnings more frequently than those who earn less. The data-gathering process in this case will reflect this distortion in the sample proportions, and since the sample is no longer a faithful representation of the population, biased estimates will be produced regardless of how many samples were collected. Our ability to eliminate such bias by analytical means thus provides a major opportunity to the empirical sciences.

6. Conclusions

Rational decisions demand rational assessments of the likely consequences of one's actions. This chapter offers a formal and normative account of how such assessments should be shaped by empirical observations and by prior world knowledge. The account is based on modern research in causal inference, which extends beyond probability and statistics and is becoming, in my opinion, an integral part of the theory of rationality.

One of the crowning achievements of modern work on causality has been to formalize counterfactual reasoning within a structure-based representation, the very representation researchers use to encode scientific knowledge. We showed that every structural equation model determines the truth value of every counterfactual sentence. Therefore, we can determine analytically if the probability of a counterfactual sentence is estimable from experimental or observational studies or a combination thereof.

This enables us to infer the behavior of specific individuals, identified by a distinct set of characteristics, as well as the average behavior of populations, identified by preintervention features or postintervention response. Additionally, this formalization leads to a calculus of actions that resolves some of the most daunting problems in the empirical sciences. These include, among others, the control of confounding, the evaluation of interventional policies, the assessment of direct and indirect effects, and the generalization of empirical results across heterogeneous environments. The same calculus can be leveraged to generate rational explanations for action recommended or actions taken in the past.

Acknowledgments

I am grateful to the coeditor, Wolfgang Spohn, for inviting me to participate in this handbook, for offering helpful comments on the first version of this chapter, and for pointing me to his earlier papers (Spohn, 1978, 1983), in which several key ideas on causal decision theory first appeared.

This research was supported in parts by grants from the International Business Machines Corporation (IBM) (#A1771928), the National Science Foundation (#IIS-1302448, #IIS-1527490, and #IIS-1704932), and the Office of Naval Research (#N00014-17-12091).

Notes

1. The relation between P_x and P takes a variety of equivalent forms, including the backdoor formula, truncated factorization, adjustment for direct causes, or the inverse probability weighting shown in equation (2) (Pearl, 2000a, pp. 72–73). The latter form is the easiest to describe without appealing to graphical notation. But see equation (10) in section 3.1 for a more general formula and definition 3 for a formal definition of the set Z.

2. I will use "future" and "past" figuratively; "affected" and "unaffected" (by X) are more accurate technically (i.e., descendants and nondescendants of X, in graph-theoretical terminology). The derivation of equation (2) requires that processes be organized recursively (avoiding feedback loops); more intricate formulas apply to nonrecursive models. See Pearl (2009b, pp. 72–73) or Spirtes, Glymour, & Scheines (2001) for a simple derivation of this and equivalent formulas. Equation (2) has also been anticipated in Spohn (1978, sections 3.3 and 5.2).

3. The structural definition of counterfactuals given in equation (3) was first introduced in Balke and Pearl (1995).

4. By a "path," we mean a sequence of consecutive edges in the graph regardless of direction. See Pearl (2009b, p. 335) for a gentle introduction to d-separation and its proof. In linear models, the independencies implied by d-separation are valid for nonrecursive models as well.

5. An alternative definition of $do(x)$, invoking population averages only, is given in Pearl (2009b, p. 24).

6. Such derivations are illustrated in graphical details in Pearl (2009b, p. 87).

7. In particular, the criterion devised by Pearl and Paz (2014) simply adds to condition 2 of definition 3 the requirement that X and its nondescendants (in Z) separate its descendants (in Z) from Y.

8. Summations should be replaced by integration when applied to continuous variables, as in Imai, Keele, and Yamamoto (2010).

9. Generalizations to arbitrary reference points, say from $T = t$ to $T = t'$, are straightforward. These definitions apply at the population levels; the unit-level effects are given by the expressions under the expectation. All expectations are taken over the factors

U_M and U_Y. Note that in this section, we use parenthetical notation for counterfactuals, replacing the subscript notation used in sections 2 and 3.

10. Natural direct and indirect effects were conceptualized in Robins and Greenland (1992) and were formalized using equations (14) and (15) in Pearl (2001).

11. Remarkably, selection bias can be detected by combining experimental and observational studies, if certain coherence inequalities are violated (Pearl, 2009b, p. 294).

References

Arló-Costa, H. (2007). The logic of conditionals. In E. N. Zalta (Ed.), *The Stanford encyclopedia of philosophy*. Retrieved from http://plato.stanford.edu/archives/sum2019/entries/logic-conditionals/

Balke, A., & Pearl, J. (1995). Counterfactuals and policy analysis in structural models. In P. Besnard & S. Hank (Eds.), *Uncertainty in Artificial Intelligence: Proceedings of the Eleventh Conference (1995)* (pp. 11–18). San Francisco, CA: Morgan Kaufmann.

Bareinboim, E., & Pearl, J. (2012a). Causal inference by surrogate experiments: z-identifiability. In N. de Freitas & K. P. Murphy (Eds.), *Proceedings of the Twenty-Eighth Conference on Uncertainty in Artificial Intelligence* (pp. 113–120). Corvallis, OR: AUAI Press.

Bareinboim, E., & Pearl, J. (2012b). Transportability of causal effects: Completeness results. In J. Hoffman & B. Selman (Eds.), *Proceedings of the Twenty-Sixth AAAI Conference on Artificial Intelligence* (pp. 698–704). Menlo Park, CA: AAAI Press.

Bareinboim, E., & Pearl, J. (2013). Meta-transportability of causal effects: A formal approach. In C. M. Carvalho & P. Ravikumar (Eds.), *Proceedings of the Sixteenth International Conference on Artificial Intelligence and Statistics* (JMLR Workshop and Conference Proceedings, Vol. 31, pp. 135–143). Scottsdale, AZ: PMLR.

Bareinboim, E., & Pearl, J. (2016). Causal inference and the data-fusion problem. *Proceedings of the National Academy of Sciences, 113*, 7345–7352.

Bareinboim, E., Tian, J., & Pearl, J. (2014). Recovering from selection bias in causal and statistical inference. In C. E. Brodley & P. Stone (Eds.), *Proceedings of the Twenty-Eighth AAAI Conference on Artificial Intelligence* (pp. 2410–2416). Palo Alto, CA: AAAI Press.

Bickel, P. J., Hammel, E. A., & O'Connell, J. W. (1975). Sex bias in graduate admissions: Data from Berkeley. *Science, 187*, 398–404.

Cartwright, N. (1983). *How the laws of physics lie*. Oxford, England: Clarendon Press.

Galles, D., & Pearl, J. (1998). An axiomatic characterization of causal counterfactuals. *Foundations of Science, 3*(1), 151–182.

Geiger, D., Verma, T., & Pearl, J. (1990). d-separation: From theorems to algorithms. In M. Henrion, R. D. Shachter, L. N. Kanal, & J. F. Lemmer (Eds.), *Uncertainty in AI* (Vol. 5, pp. 139–148). Amsterdam, Netherlands: North-Holland.

Goldberger, A. S. (1984). Reverse regression and salary discrimination. *Journal of Human Resources, 19*(3), 293–318.

Greenland, S., Pearl, J., & Robins, J. (1999). Causal diagrams for epidemiologic research. *Epidemiology, 10*(1), 37–48.

Haavelmo, T. (1943). The statistical implications of a system of simultaneous equations. *Econometrica, 11*, 1–12.

Halpern, J. (1998). Axiomatizing causal reasoning. In G. Cooper & S. Moral (Eds.), *Proceedings of the Fourteenth Conference on Uncertainty in Artificial Intelligence* (pp. 202–210). San Francisco, CA: Morgan Kaufmann.

Harper, W. L., Stalnaker, R., & Pearce, G. (1981). *Ifs*. Dordrecht, Netherlands: Reidel.

Huang, Y., & Valtorta, M. (2006). Pearl's calculus of intervention is complete. In R. Dechter & T. Richardson (Eds.), *Proceedings of the Twenty-Second Conference on Uncertainty in Artificial Intelligence* (pp. 217–224). Corvallis, OR: AUAI Press.

Imai, K., Keele, L., & Yamamoto, T. (2010). Identification, inference, and sensitivity analysis for causal mediation effects. *Statistical Science, 25*(1), 51–71.

Jeffrey, R. C. (1965). *The logic of decision*. New York, NY: McGraw-Hill.

Joyce, J. (1999). *The foundations of causal decision theory*. Cambridge, England: Cambridge University Press.

Lewis, D. (1973). Counterfactuals and comparative possibility. *Journal of Philosophical Logic, 2*(4), 418–446.

Pearl, J. (1993). Comment: Graphical models, causality, and intervention. *Statistical Science, 8*(3), 266–269.

Pearl, J. (1995). Causal diagrams for empirical research. *Biometrika, 82*(4), 669–710.

Pearl, J. (2000a). *Causality: Models, reasoning, and inference*. New York, NY: Cambridge University Press.

Pearl, J. (2000b). Comment on A. P. Dawid's "Causal inference without counterfactuals." *Journal of the American Statistical Association, 95*(450), 428–431.

Pearl, J. (2001). Direct and indirect effects. In J. S. Breese & D. Koller (Eds.), *Proceedings of the Seventeenth Conference on Uncertainty in Artificial Intelligence* (pp. 411–420). San Francisco, CA: Morgan Kaufmann.

Pearl, J. (2009a). Causal inference in statistics: An overview. *Statistics Surveys, 3*, 96–146.

Pearl, J. (2009b). *Causality: Models, reasoning, and inference* (2nd ed.). New York, NY: Cambridge University Press.

Pearl, J. (2012). The causal mediation formula—A guide to the assessment of pathways and mechanisms. *Prevention Science, 13*, 426–436.

Pearl, J. (2013). The curse of free-will and the paradox of inevitable regret. *Journal of Causal Inference, 1*, 255–257.

Pearl, J. (2017). Physical and metaphysical counterfactuals: Evaluating disjunctive actions. *Journal of Causal Inference, 5*(2), 20170018, eISSN 2193–3685.

Pearl, J., & Bareinboim, E. (2011). Transportability of causal and statistical relations: A formal approach. In W. Burgard & D. Roth (Eds.), *Proceedings of the Twenty-Fifth Conference on Artificial Intelligence (AAAI-11)* (pp. 247–254). Menlo Park, CA: AAAI Press. Available at https://ftp.cs.ucla.edu/pub/stat_ser/r372-aaai-corrected-reprint.pdf

Pearl, J., Glymour, M., & Jewell, N. (2016). *Causal inference in statistics: A primer.* New York, NY: Wiley.

Pearl, J., & Mackenzie, D. (2018). *The book of why: The new science of cause and effect.* New York, NY: Basic Books.

Pearl, J., & Paz, A. (2014). Confounding equivalence in causal inference. *Journal of Causal Inference, 2*, 75–93.

Robins, J., & Greenland, S. (1992). Identifiability and exchangeability for direct and indirect effects. *Epidemiology, 3*(2), 143–155.

Rosenbaum, P. R., & Rubin, D. B. (1983). The central role of propensity score in observational studies for causal effects. *Biometrika, 70*, 41–55.

Shpitser, I. (2013). Counterfactual graphical models for longitudinal mediation analysis with unobserved confounding. *Cognitive Science, 37*(6), 1011–1035.

Shpitser, I., & Pearl, J. (2006). Identification of conditional interventional distributions. In R. Dechter & T. Richardson (Eds.), *Proceedings of the Twenty-Second Conference on Uncertainty in Artificial Intelligence* (pp. 437–444). Corvallis, OR: AUAI Press.

Shpitser, I., VanderWeele, T., & Robins, J. M. (2010). On the validity of covariate adjustment for estimating causal effects. In P. Grünwald & P. Spirtes (Eds.), *Proceedings of the Twenty-Sixth Conference on Uncertainty in Artificial Intelligence* (pp. 527–536). Corvallis, OR: AUAI Press.

Skyrms, B. (1980). *Causal necessity.* New Haven, CT: Yale University Press.

Spirtes, P., Glymour, C., & Scheines, R. (1993). *Causation, prediction, and search.* New York, NY: Springer.

Spirtes, P., Glymour, C., & Scheines, R. (2001). *Causation, prediction, and search* (2nd ed.). Cambridge, MA: MIT Press.

Spohn, W. (1978). *Grundlagen der Entscheidungstheorie* [Foundations of decision theory]. Kronberg/Taunus, Germany: Scriptor.

Spohn, W. (1983). *Eine Theorie der Kausalität* (unpublished Habilitationsschrift) [A theory of causality]. Munich, Germany: Ludwig Maximilian University.

Stalnaker, R. (1981). Letter to David Lewis. In W. Harper, R. Stalnaker, & G. Pearce, *Ifs* (pp. 151–152). Dordrecht, Netherlands: Reidel. (Original work published 1972)

Strotz, R. H., & Wold, H. O. A. (1960). Recursive versus nonrecursive systems: An attempt at synthesis. *Econometrica, 28*, 417–427.

Tian, J., Paz, A., & Pearl, J. (1998). *Finding minimal separating sets* (Tech. Rep. No. R-254). Los Angeles: University of California.

Tian, J., & Pearl, J. (2002). A general identification condition for causal effects. In R. Dechter, M. Kearns, & R. S. Sutton (Eds.), *Proceedings of the Eighteenth National Conference on Artificial Intelligence* (pp. 567–573). Menlo Park, CA: AAAI Press/MIT Press.

Verma, T., & Pearl, J. (1988). Causal networks: Semantics and expressiveness. In R. D. Shachter, T. S. Levitt, L. N. Kanal, & J. F. Lemmer (Eds.), *Proceedings of the Fourth Workshop on Uncertainty in Artificial Intelligence* (pp. 352–359). Mountain View, CA: AUAI Press.

Weirich, P. (2016). Causal decision theory. In E. N. Zalta (Ed.), *The Stanford encyclopedia of philosophy.* Retrieved from http://plato.stanford.edu/archives/win2016/entries/decision-causal/

Woodward, J. (2003). *Making things happen: A theory of causal explanation.* Oxford, England: Oxford University Press.

7.2   The Rationality of Everyday Causal Cognition

Michael R. Waldmann

Summary

Normative and descriptive theories of causal cognition are tightly connected. Theories of causal cognition proposed in psychology have in most cases precursors in philosophy and other normative disciplines. For example, associative theories, causal Bayes net theories, and dispositional theories of causal cognition all are inspired by positions developed in philosophy. The competition between normative accounts is therefore mirrored in psychology. Findings that seem rational within one normative account may appear deficient within a competing theory. Consistency with a specific normative account is, however, only one criterion to assess rationality; to assess rationality, also correspondence of causal representations to causal relations in the world has been used. The present chapter will selectively review research, focusing on causal reasoning and learning, to demonstrate the tight coupling between normative and descriptive theories of everyday causal cognition.

1.   Normative and Descriptive Theories

Causal cognition belongs to our most central cognitive competencies. Without causal knowledge, we would not be able to make predictions and diagnoses, generate explanations, design artifacts, solve problems, or plan and act. The ubiquity of causal cognition has attracted scientists from various disciplines to this topic. Philosophers have studied causality for centuries, but more recently, the topic has also motivated research in the fields of psychology, artificial intelligence, anthropology, biology, economics, physics, and statistics (among others). Causality is a genuinely interdisciplinary topic attracting both researchers interested in developing *normative methods* of causal analysis and researchers pursuing the *descriptive* goal of understanding how humans (in different age groups) and nonhuman animals reason about causal relations (for overviews, see Waldmann, 2017; Waldmann & Hagmayer, 2013).

Yet until recently, causal reasoning has been curiously absent from mainstream cognitive psychology. Although there has been some limited work in more specific areas, such as developmental or social psychology, it did not play a significant role in fundamental theories of human cognition. To date, textbooks on cognitive psychology do not contain chapters on causal reasoning. One reason for this neglect is that psychology was for many decades dominated by the view that cognitive mechanisms, such as associative learning, are general and can be specified independently from the domains to which they are applied. In many areas that addressed knowledge acquisition, domain-general similarity-based theories dominated, which neglected the role of causal knowledge (e.g., categorization, induction, conditional reasoning, learning). The situation has slowly changed in the past two decades, with more and more research devoted to the particular characteristics of reasoning and learning about causal systems. In fact, the relevance of causal knowledge has now been acknowledged in virtually all fields of higher-level cognition (see Waldmann, 2017). The breadth of this literature makes it necessary to restrict the focus in the present chapter, which will primarily discuss general theoretical frameworks and empirical studies on causal reasoning and learning.

Theories of causal cognition proposed in psychology have in most cases precursors in philosophy and other normative disciplines. It is not an accident that the goals of normative and descriptive theories of causality overlap. Both scientists and laypeople intend their causal theories to be correct. Causal claims are therefore generally associated with normative force (see Spohn, 2002; Waldmann, 2011). The common goals of normative and descriptive accounts of causal cognition may be the reason why psychologists often turn to normative theories as an inspiration for psychological theories. One of the most recent examples are *causal Bayes nets*, which were first developed in philosophy and engineering (see Pearl, 1988, 2000; Spirtes, Glymour, & Scheines, 2001; Spohn, 2012) but have also been adopted by psychologists as models of everyday causal reasoning (for reviews, see Rottman, 2017; Rottman & Hastie, 2014; Waldmann, 2017; Waldmann & Hagmayer, 2013; see also chapter 7.1 by Pearl, this handbook).

Although scientists and laypeople have common goals, it is also plausible to expect differences in their methodological approaches and the resulting types of representations of causal knowledge. Causal domains can fundamentally differ, so that a method that has been developed to acquire knowledge in economics or sociology, for example, differs from methods employed in chemistry or physics. Causal Bayes nets (see chapter 4.2 by Hartmann, this handbook), which were intended as general models of causation, are more often used in the social sciences than in physics. Similarly, laypeople use causal knowledge in various everyday domains, including intuitive psychology, biology, and sociology. Again, although there are attempts to develop abstract theories of causality, such as Bayes nets, to model knowledge in these different areas, there are also more specific theories that focus on unique domain properties (e.g., intuitive physics or belief–desire psychology). Moreover, everyday knowledge is often rudimentary, erroneous, and driven by unsubstantiated beliefs (see Rozenblit & Keil, 2002; Sloman & Fernbach, 2017). Nevertheless, causal knowledge is probably adequate more often than not; otherwise, it would not be adaptive.

Another difference between normative and descriptive approaches is that philosophers and scientists interested in methodology generally try to develop a uniform coherent account based on few fundamental principles. By contrast, laypeople are often *satisficers* (Simon, 1956): they use tools that work for a given problem, but they are rarely interested in achieving coherence and consistency (see Arkes, Gigerenzer, & Hertwig, 2016; see also chapter 8.5 by Hertwig & Kozyreva, this handbook).

2.   How Rational Is Everyday Causal Cognition?

Given that most psychological theories of causal cognition have been inspired by normative accounts, most research addresses, at least implicitly, the question of whether everyday causal cognition conforms to the normative principles developed in philosophy, statistics, or machine learning. However, the normative evaluation of causal cognition is complicated by the fact that there is no unique normative account of causation that is universally accepted. The competition within the normative discourse is reflected in psychological theories. For example, associative theories, which can be traced back to the philosopher David Hume (1748/1977), model causation

as statistical covariation, whereas causal Bayes nets view associative approaches as deficient and argue that they neglect important specific features of causality. A further complexity is added by the fact that even within each of these frameworks, different competing normative and descriptive theories have been developed. Thus, depending on the normative view of the researcher, findings can be seen either as confirming or as violating norms (see the following sections for examples).

When researchers discover deviations from initially held normative accounts, different strategies are pursued. Sometimes the findings lead to modifications of the theory with the primary goal of maximizing *descriptive* adequacy (e.g., different versions of associative theories). At other times, the strategy is to develop a modified *normative* theory by adding features that had previously been neglected (e.g., sensitivity to uncertainty in Bayesian theories). These attempts seek to be both descriptively adequate and normative.

Whereas the debates about which norm is appropriate address rationality in terms of coherence with specific norms, another perspective on rationality focuses on whether causal cognition corresponds to objective relations in the world (the correspondence view of rationality; Arkes et al., 2016; see also chapter 8.5 by Hertwig & Kozyreva, this handbook). Thus, one important question is whether people tend to distort observations and whether possible distortions can be rationalized as adaptive. There is evidence showing that learners are often poor in correctly encoding probability information (Rottman & Hastie, 2014) and tend to over- or underestimate covariations, especially when these are zero (for a review, see Waldmann & Hagmayer, 2013). There have been attempts to argue that some of these distortions may actually be rational because they reflect the integration of prior knowledge, but not all cases can be explained this way.

3.   Frameworks of Causal Cognition

The plurality of philosophical theories of causality is also reflected in psychological approaches. Various frameworks differ in how they model causality and causal cognition (see also Waldmann & Mayrhofer, 2016). I will use the term "framework" to describe classes of theories that use substantially different theoretical concepts to capture causality. Frameworks also differ in the tasks they are trying to model. In this section, I will describe different frameworks applied in causal reasoning research. The main distinguishing features of frameworks are the proposed causal *relata* (that is, the type of entities that enter causal relations) and the way causal *relations* are modeled.

3.1 The Dependency Framework

The dependency view of causation is adopted by several psychological theories, which share central framework assumptions but otherwise often compete. Theories that can be subsumed under the dependency framework include associative theories (see Le Pelley, Griffiths, & Beesley, 2017), covariation theories (e.g., Cheng & Novick, 1992; Perales, Catena, Cándido, & Maldonado, 2017), power probabilistic contrast (PC) theory (Cheng, 1997), causal model theories (e.g., Gopnik et al., 2004; Rehder, 2017a, 2017b; Rottman, 2017; Sloman, 2005; Waldmann & Hagmayer, 2001; Waldmann & Holyoak, 1992; Waldmann, Holyoak, & Fratianne, 1995), and Bayesian inference theories (Griffiths & Tenenbaum, 2005, 2009; Lu, Yuille, Liljeholm, Cheng, & Holyoak, 2008; Meder, Mayrhofer, & Waldmann, 2014).

According to dependency theories, a variable C is a cause of variable E if E statistically depends upon C. There is an extensive debate in philosophy about the proper *causal relata* in dependency theories (e.g., events, propositions, facts, properties, or states of affairs; see Ehring, 2009; Spohn, 2012). *Causal relations* are often graphically depicted by causal arrows directed from cause to effect (see figure 7.2.1). For example, a causal model could be postulated that uses the binary variables representing the effect forest fire (present vs. absent) and as potential causes a lit match (e.g., dropped by an arsonist vs. not dropped) and a lightning (present vs. absent) (see the common-effect model in figure 7.2.1a). The dependencies, then, encode a set of hypothetical situations consistent with the causal model. Causal models can also express mechanism knowledge as chains or networks of mediating variables.

Representative research topics

Associative theories One early popular class of models of causal cognition are *associative theories*, which mainly focus on learning. Inspired by the philosopher David Hume (1748/1977), the key claim is that causal learning is based on associating repeatedly observed spatiotemporally contiguous event pairs. How exactly such associations arise, and which covariation measures they approximate, is a matter of great debate between competing associative theories (see Le Pelley et al., 2017; Perales et al., 2017).

Whereas Hume additionally assumed that causes and effects that enter associations are ordered in the temporal order of their occurrence in the world (i.e., causes precede their effects), psychological theories of association focus on the temporal order of learning cues and outcomes, which can refer either to causes or to effects. For example, for a physician, a cue could be an observed symptom and the outcome the diagnosis of a disease. In this case, the cue refers to an effect of its cause, the disease.

Associative theories versus causal model theory Whereas in associative theories, weights are acquired in the cue–outcome direction regardless of whether cues represent causes or effects, a crucial feature of causal models is that they contain arrows that represent causal directionality regardless of the order in which information about the variables is acquired (see figure 7.2.1). This crucial difference led to a debate in the 1990s about whether people are sensitive to causal directionality and use causal model representations or whether they disregard this crucial feature of causality and acquire associative knowledge in the cue–outcome direction, analogous to multiple regression analysis in statistics. Thus, this debate was about which normative principle people apply in everyday causal cognition. Waldmann and Holyoak (1992) presented evidence supporting the causal model perspective (see also Waldmann, 1996, 2000; Waldmann et al., 1995). In the meantime, much more evidence for causal model representations has been collected (e.g., Gopnik et al., 2004; Rehder, 2017a, 2017b; Sloman, 2005), although there is also some evidence showing that in complex learning tasks, learners sometimes fall back to simpler associative representations (see, e.g., López, Cobos, & Caño, 2005).

Estimating causal strength Another debate related to a normative discussion about the proper norm concerns how causal strength parameters should be estimated. Most associative theories assume that causal strength is equal to the observed covariation. A popular measure for single cause–effect relations, used both by rule-based theories and by associative theories such as the Rescorla–Wagner rule (Rescorla & Wagner, 1972), is *delta-p*, which is the difference between the probability of an effect when its cause is present and the probability of the effect when its cause is present and the probability of the effect

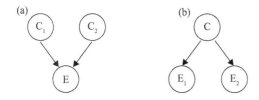

Figure 7.2.1
Two examples of causal Bayes nets: (a) common-effect model and (b) common-cause model.

when its cause is absent. Conceptually, this measure views causes as *difference makers*.

Although *delta-p* is an accepted covariation measure, Cheng (1997) has questioned its normative and psychological suitability as a measure of causal strength. Associative theories add up multiple associative weights when multiple causes are present, which may lead to the nonsensical prediction that an effect will occur with a probability higher than 1. Thus, possible ceiling effects are not accounted for. Another example of a problematic ceiling effect are cases in which the probability of the effect in the absence of the cause is already at maximum (i.e., 1) so that the cause cannot further increase the probability of the effect. Concluding from such an observation that the cause has zero strength seems counterintuitive.

Cheng (1997) has therefore suggested a new measure of causal strength, *causal power*, which represents the probability with which a cause, acting alone, generates or prevents a target effect. Given that causes are never observed alone, causal power is a theoretical quantity that needs to be inferred on the basis of observed data. Cheng's theory uses *delta-p* along with default background assumptions about the underlying causal model (e.g., independence of target cause and unobserved alternative causes) to infer the causal power parameter.

Cheng's (1997) theory is not the only theory of causal strength estimation that claims to be both normative and descriptively adequate. Cheng's goal was to develop a theory that generates point estimates of causal power. By contrast, Griffiths and Tenenbaum (2005) used the framework of Bayesian inference theory and analyzed causal inference in the context of their *causal support model*, which incorporates uncertainty of parameter estimates by considering distributions of parameters. Whereas Cheng's (1997) focus were judgments of causal strength, which were modeled as parameter estimation tasks, the causal support model focuses on the assessment of the likelihood of the presence of a causal link between a target cause C and an effect E. Griffiths and Tenenbaum argue that this is actually the question learners try to answer when asked to assess causal strength. The causal support model aims to contrast the evidence in favor of a causal model in which a link exists between C and E with a causal model in which these two variables are independent and E is only influenced by background causes. One notable feature of the model is that it can rationally explain why people sometimes offer positive numerical estimates when they have observed contingencies that are actually at zero.

Observing versus intervening A unique feature that sets causal Bayes nets apart from associative theories is their ability to predict the outcomes of interventions based on observational knowledge alone (Pearl, 2000; Spirtes et al., 2001; see also chapter 7.1 by Pearl, this handbook). Assume, for example, that we have observed that atmospheric pressure, barometer readings, and weather covary and that we have reason to believe that atmospheric pressure is the common cause of the two effects. Both causal Bayes nets and associative theories would allow us to make observational predictions between the three events. If we observe, for example, high barometer readings, we are licensed to predict good weather. However, without further observations, causal Bayes nets also allow us to correctly predict that intervening on the barometer (for example, by shaking it) and thus changing the reading would not alter the weather. According to Bayes nets accounts, we would represent such an action as the result of a *do*-operator (Pearl, 2000) that is the sole cause of the new reading, whereas the previous cause, atmospheric pressure, is now causally disconnected from the barometer. In the common-cause model, this would amount to removing the arrow between the atmospheric pressure variable and the barometer variable ("graph surgery"). Using this method, more complex planning situations can also be modeled (see Pearl, 2000). Associative theories, by contrast, can only predict that intervening on the barometer leads to outcomes that are different from those in situations in which the barometer is simply observed, when the learner has actually intervened in the past and observed the associated outcome (i.e., interventional learning); these theories cannot make correct interventional predictions based on observations alone.

The distinction between observing and intervening represents an interesting test case to distinguish between competing theories within the dependency framework. Therefore, a number of psychologists have investigated whether human subjects are capable of deriving interventional predictions from mere observations (Meder, Hagmayer, & Waldmann, 2008, 2009; Sloman & Lagnado, 2005; Waldmann & Hagmayer, 2005). The general finding in these studies is that participants proved capable of deriving interventional predictions from observations, although in more complex tasks, some participants seemed to confuse observational with interventional knowledge. Moreover, there are also some preliminary findings suggesting that rats, too, are capable of deriving interventional predictions from observations (Blaisdell, Sawa, Leising, & Waldmann, 2006; Leising, Wong, Waldmann, & Blaisdell, 2008).

Modeling causal queries Most studies on causal reasoning and learning have addressed how people make *predictive* inferences (from causes to effects) or assessments of causal strength. Causal knowledge representations, however, also allow for other types of queries. More recently, psychologists have developed computational models of various types of queries and have empirically tested these models. These studies are again a test case for normative theories that model how causal queries *should* be answered. For example, a set of studies on diagnostic reasoning by Meder et al. (2014) shows how deviations from previously held norms can lead to a modified normative theory that is also descriptively adequate (for overviews, see Meder & Mayrhofer, 2017; chapter 7.3 by Meder & Mayrhofer, this handbook).

Meder et al. (2014) discovered that participants who were asked to make a diagnostic judgment did not simply report the empirical probability of the cause given the effect, which has been considered the standard normative measure of diagnostic probability, but decrease the diagnostic estimates when causal strength in the predictive direction is low. Meder and colleagues showed that this behavior is in fact rational under a Bayesian inference view. They developed a modified Bayesian model that is not only sensitive to the observed diagnostic probability but also incorporates the degree of uncertainty about the underlying causal model in the judgments.

Other studies have addressed the question of how people make *singular* causal judgments (Cheng & Novick, 2005; Stephan & Waldmann, 2018; Stephan, Mayrhofer, & Waldmann, 2020). For example, it may have been observed that Peter, a heavy smoker, has contracted lung disease. It is known that generally heavy smokers tend to contract lung disease later in their lives. A singular query might in this case ask whether Peter's smoking actually caused *his* lung disease, in contrast to the possibility that in his case, the observed events are just a coincidence. The new model, which is again derived from normative accounts, tries to explain how people respond to such queries.

Acquisition of causal model knowledge One of the major engineering achievements of Bayes net research in computer science are algorithms that allow machines to induce causal networks from covariation data using minimal prior knowledge (e.g., constraint-based algorithms, Bayesian inference methods; Pearl, 2000; Spirtes et al., 2001). In psychology, there has been a debate whether these algorithms are plausible models of human learning (see Gopnik et al., 2004; Griffiths & Tenenbaum, 2009). One problem is that these algorithms typically require data sets of a size that clearly surpasses what human learners can process. Also, it is unclear whether humans are capable of making, without the aid of computers, the required precise estimates about conditional independence relations. Nevertheless, some psychologists have tried to test whether human subjects are capable of inducing causal models from covariation alone. In general, relatively simple tasks were presented in which a restricted set of causal models with a small number of variables (typically 3) were offered as candidates between which subjects were asked to choose. Nevertheless, performance was generally poor (Steyvers, Tenenbaum, Wagenmakers, & Blum, 2003). Some studies have shown that allowing subjects to intervene helps a little bit (Bramley, Lagnado, & Speekenbrink, 2015; Gopnik et al., 2004; Steyvers et al., 2003). The general conclusion from these findings is that in principle, people can induce causal models from covariation data, but performance is poor unless learning conditions are extremely favorable. This is an example of research showing deviations from what normative models can achieve.

The implausibility of domain-general algorithms of causal structure induction has led Waldmann (1996) to propose the view that people generally use prior knowledge about the structure of causal models to guide learning in a top-down fashion ("knowledge-based causal induction"; see also Lagnado, Waldmann, Hagmayer, & Sloman, 2007). Various cues, including temporal order, interventions, instructions, and prior knowledge, can guide the initial hypotheses about causal structure. Numerous studies have shown that learners are capable of using these cues and are able to prioritize them in cases of conflict.

Griffiths and Tenenbaum (2009) have proposed *hierarchical* Bayesian inference strategies to model knowledge- or theory-based causal induction. The central assumption is that probabilistic inference is carried out at multiple levels of abstraction, which influence each other and are updated simultaneously. In causal learning, these levels include the data, alternative causal models, and the theory level, which may, for example, encode physical domain knowledge. Hierarchical Bayesian models speed up learning because the range of alternative hypotheses being considered is constrained by both the data and the theory level.

Whereas theories combining top-down knowledge with bottom-up learning have initially focused on fairly abstract causal knowledge, such as knowledge about causal directionality, more recent accounts have incorporated specific domain knowledge. For example, Waldmann (2007) has presented evidence showing that domain knowledge about different types of interactions of physical quantities influences the functional form of how multiple causes of a common effect are assumed to combine (see also Griffiths & Tenenbaum, 2009).

Following up on earlier work demonstrating the role of temporal assumptions about cause–effect latencies (Buehner & May, 2003; Hagmayer & Waldmann, 2002), more sophisticated models of the influence of time on causal induction have been proposed (Bramley, Gerstenberg, Mayrhofer, & Lagnado, 2018; Stephan et al., 2020). Other approaches, postulating an integration of intuitive Newtonian physics with probabilistic inference in causal models ("noisy Newton"), were offered by Gerstenberg and Tenenbaum (2017) and Sanborn, Mansinghka, and Griffiths (2013; see also Bramley, Gerstenberg, Tenenbaum, & Gureckis, 2018).

Violations of Bayes net predictions Research about causal Bayes nets has initially found a surprisingly good fit between their theoretical assumptions and human causal inference (see Rottman & Hastie, 2014). However, more recently, deviations have been discovered, which led to attempts to modify the theory.

Diagnostic reasoning was already discussed above as an example of how deviations from the predictions of normative theories may lead to a revised account that claims to be a better theory of diagnostic reasoning, both descriptively and normatively. Another popular topic questioning Bayes nets as a psychological account are apparent Markov violations, first discovered by Rehder and Burnett (2005). They showed, for example, that subjects who were presented with a common-cause model (see figure 7.2.1b) and then were asked to infer the probability of an effect given its cause were influenced by how many other effects of this cause were present or absent. *Prima facie*, this finding violates the Markov constraint, which is assumed to be a central feature of Bayes nets. According to this constraint, the requested cause–effect inference should be independent of whether collateral effects are present or absent.

Whereas some philosophers use this kind of evidence as a demonstration of the inadequacy of Bayes nets (e.g., Cartwright, 2001), cognitive psychologists chose a similar strategy as Meder et al. (2014) and tried to show that Markov violations are only apparent violations of the normative account. They proposed extended Bayes net models with hidden variables, which explain the apparent Markov violations within normative Bayes nets that honor the Markov condition. Various theories have been proposed to motivate these extensions, for example, by postulating additional mechanism knowledge

(Park & Sloman, 2013) or intuitions about dispositions (Mayrhofer & Waldmann, 2015). Other accounts combine Bayes net representations with associative processes (Rehder, 2014; Rehder & Waldmann, 2017).

3.2 The Disposition Framework

A completely different view answers the question of why an observed causal relation between events holds by focusing on the participants involved in the causal interaction. An often-studied example are the two animated colliding billiard balls in Michotte's (1963) task.

The normative status of dispositional theories can be debated. Some specific accounts were motivated by Aristotelian philosophy, which for centuries was viewed as rational but has been demonstrated to contradict modern physics (White, 2006). More recently, however, versions of dispositional theories have been developed that claim that these theories provide a better account of rational scientific research than competing views in philosophy of science (Cartwright & Pemberton, 2013; Mumford & Anjum, 2011).

One important difference between dependency and dispositional theories concerns the *causal relata*. Whereas dependency theories focus on variables that, for instance, encode the presence or absence of events, dispositional theories use *objects* as primary entities. These objects may be humans or nonhuman entities. The dispositions of objects can be static (e.g., the solubility of sugar) or they can be transient and dynamic such as the sudden exertion of force when moving an object. Causal relations between events arise when objects are placed in specific situational contexts allowing them to express their dispositions or powers. Thus, dispositional theories are looking for deeper explanations of observed dependencies underlying the observed covariation. One can view this as a focus on underlying mechanisms; however, the mechanisms have different properties from mechanisms modeled within the dependency framework (e.g., as chains or networks of variables; see Stephan, Tentori, Pighin, & Waldmann, 2021).

Dispositional theories can refer to very specific dispositions (e.g., of aspirin; see Mumford & Anjum, 2011), but the accounts most popular in psychology typically distinguish between just two types of objects, often called *causal agents* and *causal patients*. A popular theory, which was initially developed in linguistic semantics, is *force dynamics* (see Gärdenfors, 2014; Talmy, 1988). For example, Gärdenfors (2014) analyzes causal scenarios as interactions between a causal agent, endowed with a force, and a causal patient, which can be an animate or inanimate, concrete or abstract object. The patient can carry a counterforce resisting the action of the agent. Forces are primarily physical, but they can be extended metaphorically to social or mental forces (e.g., threats, commands, persuasions). Force dynamics has been used in linguistics to characterize verb semantics and argument structure. For example, in "Peter hits Mary," "hit" has two arguments, one describing an agent (Peter), the other the patient (Mary).

Wolff (2007) has developed psychological variants of force dynamics (see also Wolff, Barbey, & Hausknecht, 2010; for an overview, see Wolff & Thorstad, 2017). His force theory states that people evaluate configurations of forces attached to what Wolff calls "affectors" (i.e., agents) and "patients," which may vary in direction and degree with respect to an end state.

Representative research topics Numerous studies on force dynamics have been conducted in the context of linguistic semantics. My focus here will be on selected findings from psychology.

Understanding causal terms Wolff (2007) used force theory to analyze the meaning of abstract causal concepts, such as *cause*, *prevent*, *enable*, and *despite* (see also Wolff & Song, 2003). For example, when we say, "High winds caused the man to move toward the bench," we mean that the patient (the man) had no tendency to move toward the bench, the affector (i.e., agent; the wind) acted against the patient, and the resultant of the forces acting on the patient was directed toward the result of moving toward the bench. This constellation of forces describes how we understand the term *cause*. Related analyses were proposed for the other abstract causal terms. Wolff and colleagues tested their theory, for example, by presenting visual scenes to subjects depicting stationary or moving people or objects and asking subjects to choose between alternative causal terms (see Wolff, 2007; Wolff & Thorstad, 2017).

Initially, the theory was applied to individual causal relations between agents and patients. Later, the theory was extended to more complex scenarios, such as causal chains or complex preventive relations, such as double preventions. For example, consider somebody hitting a pole that prevents a tent from falling. The action (hitting the pole) in this case prevents a preventer (the pole keeping the tent upright), which leads to the falling of the tent. Similarly, causation by omission can be modeled by considering the forces that are in play when an action is not executed (see Wolff et al., 2010; Wolff & Barbey, 2015).

Causal asymmetry Another example of a dispositional account (although different from force dynamics) is

White's (2006) theory. Many of White's studies have focused on the Michotte (1963) task, in which subjects observe two-dimensionally rendered animated moving balls. In a *launching* scenario, for example, object X, a ball, moves toward a resting object Y, another ball, and touches it. At this moment, object X stops and object Y begins moving, eliciting a causal impression. White (2009) has shown that subjects tend to view the launching ball (X) as the agent and the launched ball (Y) as the patient (or "cause" and "effect object," in his terminology). Subjects typically describe the launching scene as an event in which "X launched Y" instead of using the equally valid description that "Y stopped X." Moreover, force estimates for X tend to be higher than force estimates for Y. According to White, both findings are indicators of the underlying dispositional distinction between the two types of objects. Causal asymmetry contradicts Newtonian physics because the physical force on object Y exerted by object X is equal in magnitude (but opposite in direction) to that on object X exerted by object Y. From a Newtonian perspective, the collision is perfectly symmetric, and both descriptions (i.e., "X launched Y" and "Y stopped X") should be equally appropriate (see also Mayrhofer & Waldmann, 2014, 2016, for follow-up studies).

White (2006, 2009) has proposed a dispositional theory of causal asymmetry that links perceived scenes to stored representations of sensomotoric experiences of our actions on objects. According to White, we experience our own agency along with its associated forces during the course of our ontogenetic development. When we perceive a scene, we compare the movements of the objects with these stored representations. We tend to overestimate the force of the causal agent (i.e., cause object) relative to the counterforce of the manipulated patient (i.e., effect object) because this asymmetry corresponds to our experience of resistance that is overcome by our action.

4. Conclusion

Despite many attempts to develop a unitary theory of causal reasoning, this review of the literature shows that fundamentally different frameworks compete. Each framework has philosophical precursors that provide an initial normative basis, but within each framework, these accounts have been further developed with the goal to provide both descriptively more adequate theories and, at least in some cases, also improved normative accounts. Given that both the descriptive and the normative theories developed within and across the frameworks compete, no uniform normative or descriptive theory of causal cognition has emerged.

The frameworks differ in terms of the causal relata they employ and the way causal relations are construed. These differences make them more or less suitable for modeling different tasks. For example, dependency theories are particularly good at modeling how people predict events within complex causal models, whereas dispositional theories are typically applied to linguistic phrases or visual scenarios involving a limited number of participants. These specializations may be the reason why the different frameworks rarely compete directly with each other. Most of the debates in the causal cognition literature concern competing theories *within* each framework. There are some limited attempts to cross the aisle and develop unitary frameworks that also apply to the tasks of the competitor (Cheng, 1993; Wolff, 2014), but these attempts have so far not been widely adopted by the research community.

As a consequence, other proposals have been presented, which try to address the issue of competing frameworks. One position is *causal pluralism*, which accepts that different tasks may be modeled best by different types of theories (e.g., Lombrozo, 2010). Another approach is to develop *hybrid* theories that focus on how different kinds of causal representations interact in specific tasks. An example of a hybrid theory that combines dependency with dispositional representations has been presented by Mayrhofer and Waldmann (2015; see also Waldmann & Mayrhofer, 2016). They claim that verbal instructions, whose causal content can be best modeled by dispositional accounts, influence how causal Bayes nets are construed that are used to make causal inferences. These are just initial attempts. Future research will have to address in greater detail whether unitary, pluralistic, or hybrid accounts are best suited to model causal cognition.

Acknowledgments

This research was supported by Deutsche Forschungsgemeinschaft (DFG) Grant WA 621/24-1. I am grateful to Simon Stephan for helpful comments.

References

Arkes, H. R., Gigerenzer, G., & Hertwig, R. (2016). How bad is incoherence? *Decision, 3*, 20–39.

Blaisdell, A. P., Sawa, K., Leising, K. J., & Waldmann, M. R. (2006). Causal reasoning in rats. *Science, 311*, 1020–1022.

Bramley, N. R., Gerstenberg, T., Mayrhofer, R., & Lagnado, D. A. (2018). Time in causal structure learning. *Journal of Experimental Psychology: Learning, Memory, and Cognition, 44*, 1880–1910.

Bramley, N. R., Gerstenberg, T., Tenenbaum, J. B., & Gureckis, T. M. (2018). Intuitive experimentation in the physical world. *Cognitive Psychology, 105*, 9–38.

Bramley, N. R., Lagnado, D. A., & Speekenbrink, M. (2015). Conservative forgetful scholars: How people learn causal structure through interventions. *Journal of Experimental Psychology: Learning, Memory, and Cognition, 41*, 708–731.

Buehner, M. J., & May, J. (2003). Rethinking temporal contiguity and the judgement of causality: Effects of prior knowledge, experience, and reinforcement procedure. *Quarterly Journal of Experimental Psychology Section A, 56*, 865–890.

Cartwright, N. (2001). What is wrong with Bayes nets? *The Monist, 84*, 242–264.

Cartwright, N., & Pemberton, J. M. (2013). Aristotelian powers: Without them, what would modern science do? In R. Groff & J. Greco (Eds.), *Powers and capacities in philosophy: The new Aristotelianism* (pp. 93–112). New York, NY: Routledge.

Cheng, P. W. (1993). Separating causal laws from casual facts: Pressing the limits of statistical relevance. In D. L. Medin (Ed.), *The psychology of learning and motivation* (Vol. 30, pp. 215–264). New York, NY: Academic Press.

Cheng, P. W. (1997). From covariation to causation: A causal power theory. *Psychological Review, 104*, 367–405.

Cheng, P. W., & Novick, L. R. (1992). Covariation in natural causal induction. *Psychological Review, 99*, 365–382.

Cheng, P. W., & Novick, L. R. (2005). Constraints and nonconstraints in causal learning: Reply to White (2005) and to Luhmann and Ahn (2005). *Psychological Review, 112*, 694–706.

Ehring, D. (2009). Causal relata. In H. Beebee, C. Hitchcock, & P. Menzies (Eds.), *The Oxford handbook of causation* (pp. 387–413). Oxford, England: Oxford University Press.

Gärdenfors, P. (2014). *The geometry of meaning: Semantics based on conceptual spaces*. Cambridge, MA: MIT Press.

Gerstenberg, T., & Tenenbaum, J. (2017). Intuitive theories. In M. R. Waldmann (Ed.), *The Oxford handbook of causal reasoning* (pp. 515–547). New York, NY: Oxford University Press.

Gopnik, A., Glymour, C., Sobel, D. M., Schulz, L. E., Kushnir, T., & Danks, D. (2004). A theory of causal learning in children: Causal maps and Bayes nets. *Psychological Review, 111*, 1–30.

Griffiths, T. L., & Tenenbaum, J. B. (2005). Structure and strength in causal induction. *Cognitive Psychology, 51*, 354–384.

Griffiths, T. L., & Tenenbaum, J. B. (2009). Theory-based causal induction. *Psychological Review, 116*, 661–716.

Hagmayer, Y., & Waldmann, M. R. (2002). How temporal assumptions influence causal judgments. *Memory & Cognition, 30*, 1128–1137.

Hume, D. (1977). *An enquiry concerning human understanding.* Indianapolis, IN: Hackett. (Original work published 1748)

Lagnado, D. A., Waldmann, M. R., Hagmayer, Y., & Sloman, S. A. (2007). Beyond covariation: Cues to causal structure. In A. Gopnik & L. E. Schultz (Eds.), *Causal learning: Psychology, philosophy, and computation* (pp. 154–172). Oxford, England: Oxford University Press.

Leising, K. J., Wong, J., Waldmann, M. R., & Blaisdell, A. P. (2008). The special status of actions in causal reasoning in rats. *Journal of Experimental Psychology: General, 137*, 514–527.

Le Pelley, M. E., Griffiths, O., & Beesley, T. (2017). Associative accounts of causal cognition. In M. R. Waldmann (Ed.), *The Oxford handbook of causal reasoning* (pp. 13–28). New York, NY: Oxford University Press.

Lombrozo, T. (2010). Causal-explanatory pluralism: How intentions, functions, and mechanisms influence causal ascriptions. *Cognitive Psychology, 61*, 303–332.

López, F. J., Cobos, P. L., & Caño, A. (2005). Associative and causal reasoning accounts of causal induction: Symmetries and asymmetries in predictive and diagnostic inferences. *Memory & Cognition, 33*, 1388–1398.

Lu, H., Yuille, A. L., Liljeholm, M., Cheng, P. W., & Holyoak, K. J. (2008). Bayesian generic priors for causal learning. *Psychological Review, 115*, 955–984.

Mayrhofer, R., & Waldmann, M. R. (2014). Indicators of causal agency in physical interactions: The role of the prior context. *Cognition, 132*, 485–490.

Mayrhofer, R., & Waldmann, M. R. (2015). Agents and causes: Dispositional intuitions as a guide to causal structure. *Cognitive Science, 39*, 65–95.

Mayrhofer, R., & Waldmann, M. R. (2016). Causal agency and the perception of force. *Psychonomic Bulletin & Review, 23*, 789–796.

Meder, B., Hagmayer, Y., & Waldmann, M. R. (2008). Inferring interventional predictions from observational learning data. *Psychonomic Bulletin & Review, 15*, 75–80.

Meder, B., Hagmayer, Y., & Waldmann, M. R. (2009). The role of learning data in causal reasoning about observations and interventions. *Memory & Cognition, 37*, 249–264.

Meder, B., & Mayrhofer, R. (2017). Diagnostic reasoning. In M. R. Waldmann (Ed.), *The Oxford handbook of causal reasoning* (pp. 433–457). New York, NY: Oxford University Press.

Meder, B., Mayrhofer, R., & Waldmann, M. R. (2014). Structure induction in diagnostic causal reasoning. *Psychological Review, 121*, 277–301.

Michotte, A. E. (1963). *The perception of causality.* New York, NY: Basic Books.

Mumford, S., & Anjum, R. L. (2011). *Getting causes from powers.* New York, NY: Oxford University Press.

Park, J., & Sloman, S. A. (2013). Mechanistic beliefs determine adherence to the Markov property in causal reasoning. *Cognitive Psychology, 67*, 186–216.

Pearl, J. (1988). *Probabilistic reasoning in intelligent systems: Networks of plausible inference.* San Mateo, CA: Morgan Kaufmann.

Pearl, J. (2000). *Causality: Models, reasoning, and inference.* Cambridge, England: Cambridge University Press.

Perales, J., Catena, A., Cándido, A., & Maldonado, A. (2017). Rules of causal judgment: Mapping statistical information onto causal beliefs. In M. R. Waldmann (Ed.), *The Oxford handbook of causal reasoning* (pp. 29–51). New York, NY: Oxford University Press.

Rehder, B. (2014). Independence and dependence in human causal reasoning. *Cognitive Psychology, 72*, 54–107.

Rehder, B. (2017a). Categories as causal models: Categorization. In M. R. Waldmann (Ed.), *The Oxford handbook of causal reasoning* (pp. 347–375). New York, NY: Oxford University Press.

Rehder, B. (2017b). Categories as causal models: Induction. In M. R. Waldmann (Ed.), *The Oxford handbook of causal reasoning* (pp. 377–413). New York, NY: Oxford University Press.

Rehder, B., & Burnett, R. (2005). Feature inference and the causal structure of categories. *Cognitive Psychology, 50*, 264–314.

Rehder, B., & Waldmann, M. R. (2017). Failures of explaining away and screening off in described versus experienced causal learning scenarios. *Memory & Cognition, 45*, 245–260.

Rescorla, R. A., & Wagner, A. R. (1972). A theory of Pavlovian conditioning: Variations in the effectiveness of reinforcement and non-reinforcement. In A. H. Black & W. F. Prokasy (Eds.), *Classical conditioning II: Current research and theory* (pp. 64–99). New York, NY: Appleton-Century-Crofts.

Rottman, B. M. (2017). The acquisition and use of causal structure knowledge. In M. R. Waldmann (Ed.), *The Oxford handbook of causal reasoning* (pp. 85–114). New York, NY: Oxford University Press.

Rottman, B. M., & Hastie, R. (2014). Reasoning about causal relationships: Inferences on causal networks. *Psychological Bulletin, 140*, 109–139.

Rozenblit, L., & Keil, F. C. (2002). The misunderstood limits of folk science: An illusion of explanatory depth. *Cognitive Science, 26*, 521–562.

Sanborn, A. N., Mansinghka, V. K., & Griffiths, T. L. (2013). Reconciling intuitive physics and Newtonian mechanics for colliding objects. *Psychological Review, 120*, 411–437.

Simon, H. A. (1956). Rational choice and the structure of the environment. *Psychological Review, 63*, 129–138.

Sloman, S. (2005). *Causal models: How people think about the world and its alternatives.* New York, NY: Oxford University Press.

Sloman, S., & Fernbach, P. (2017). *The knowledge illusion: Why we never think alone*. New York, NY: Riverhead Books.

Sloman, S. A., & Lagnado, D. A. (2005). Do we "do"? *Cognitive Science, 29*, 5–39.

Spirtes, P., Glymour, C., & Scheines, R. (2001). *Causation, prediction and search*. New York, NY: Springer.

Spohn, W. (2002). The many facets of the theory of rationality. *Croatian Journal of Philosophy, 2*, 247–262.

Spohn, W. (2012). *The laws of belief: Ranking theory and its philosophical applications*. Oxford, England: Oxford University Press.

Stephan, S., Mayrhofer, R., & Waldmann, M. R. (2020). Time and singular causation—a computational model. *Cognitive Science, 44*, e12871.

Stephan, S., Tentori, K., Pighin, S., & Waldmann, M. R. (2021). Interpolating causal mechanisms: The paradox of knowing more. *Journal of Experimental Psychology: General*. Advance online publication.

Stephan, S., & Waldmann, M. R. (2018). Preemption in singular causation judgments: A computational model. *Topics in Cognitive Science, 10*, 242–257.

Steyvers, M., Tenenbaum, J. B., Wagenmakers, E.-J., & Blum, B. (2003). Inferring causal networks from observations and interventions. *Cognitive Science, 27*, 453–489.

Talmy, L. (1988). Force dynamics in language and cognition. *Cognitive Science, 12*, 49–100.

Waldmann, M. R. (1996). Knowledge-based causal induction. In D. R. Shanks, K. J. Holyoak, & D. L. Medin (Eds.), *The psychology of learning and motivation: Vol. 34. Causal learning* (pp. 47–88). San Diego, CA: Academic Press.

Waldmann, M. R. (2000). Competition among causes but not effects in predictive and diagnostic learning. *Journal of Experimental Psychology: Learning, Memory, and Cognition, 26*, 53–76.

Waldmann, M. R. (2007). Combining versus analyzing multiple causes: How domain assumptions and task context affect integration rules. *Cognitive Science, 31*, 233–256.

Waldmann, M. R. (2011). Neurath's ship: The constitutive relation between normative and descriptive theories of rationality. *Behavioral and Brain Sciences, 34*, 273–274.

Waldmann, M. R. (Ed.). (2017). *The Oxford handbook of causal reasoning*. New York, NY: Oxford University Press.

Waldmann, M. R., & Hagmayer, Y. (2001). Estimating causal strength: The role of structural knowledge and processing effort. *Cognition, 82*, 27–58.

Waldmann, M. R., & Hagmayer, Y. (2005). Seeing vs. doing: Two modes of accessing causal knowledge. *Journal of Experimental Psychology: Learning, Memory, and Cognition, 31*, 216–227.

Waldmann, M. R., & Hagmayer, Y. (2013). Causal reasoning. In D. Reisberg (Ed.), *The Oxford handbook of cognitive psychology* (pp. 733–752). New York, NY: Oxford University Press.

Waldmann, M. R., & Holyoak, K. J. (1992). Predictive and diagnostic learning within causal models: Asymmetries in cue competition. *Journal of Experimental Psychology: General, 121*, 222–236.

Waldmann, M. R., Holyoak, K. J., & Fratianne, A. (1995). Causal models and the acquisition of category structure. *Journal of Experimental Psychology: General, 124*, 181–206.

Waldmann, M. R., & Mayrhofer, R. (2016). Hybrid causal representations. In B. Ross (Ed.), *The psychology of learning and motivation* (Vol. 65, pp. 85–127). New York, NY: Academic Press.

White, P. A. (2006). The causal asymmetry. *Psychological Review, 113*, 132–147.

White, P. A. (2009). Perception of forces exerted by objects in collision events. *Psychological Review, 116*, 580–601.

Wolff, P. (2007). Representing causation. *Journal of Experimental Psychology: General, 136*, 82–111.

Wolff, P. (2014). Causal pluralism and force dynamics. In B. Copley & F. Martin (Eds.), *Causation in grammatical structures* (pp. 100–118). New York, NY: Oxford University Press.

Wolff, P., & Barbey, A. K. (2015). Causal reasoning with forces. *Frontiers in Human Neuroscience, 9*, 1–21.

Wolff, P., Barbey, A. K., & Hausknecht, M. (2010). For want of a nail: How absences cause events. *Journal of Experimental Psychology: General, 139*, 191–221.

Wolff, P., & Song, G. (2003). Models of causation and the semantics of causal verbs. *Cognitive Psychology, 47*, 276–332.

Wolff, P., & Thorstad, R. (2017). Force dynamics. In M. R. Waldmann (Ed.), *The Oxford handbook of causal reasoning* (pp. 147–167). New York, NY: Oxford University Press.

7.3 Diagnostic Causal Reasoning

Björn Meder and Ralf Mayrhofer

Summary

Rational theories of diagnostic reasoning assume that the reasoner's goal is to infer the conditional probability of a cause given an effect from the available data. Typically, diagnostic reasoning is modeled within a statistical inference framework, with Bayes' rule applied to the obtained covariation information serving as the normative standard. This chapter analyzes diagnostic reasoning from the perspective of causal induction, using the framework of causal Bayes net theory to instantiate different accounts of rational diagnostic reasoning. These approaches elucidate the relevant kinds of inputs, computations, and outputs by differentiating between parametric causal models and observable contingency information. A particularly interesting feature of these accounts is that they can include predictions that systematically deviate from the traditional, purely statistical norm. The analyses highlight key issues for constructing a rational theory of diagnostic reasoning and the experimental study of human rationality.

1. Reasoning from Effect to Cause

Diagnostic causal reasoning pertains to inferences from effect to cause, such as reasoning from symptoms to diseases in medical diagnosis. Probabilistic diagnostic inferences can be considered a special case of Bayesian inference, where beliefs about unobserved states of the world (the cause events) are updated in light of observed data (the effect events). The question to what extent people can make appropriate diagnostic inferences has been central to debates on human rationality, with probability theory, and Bayes' rule in particular, serving as a normative or descriptive reference point (Gigerenzer, 1996; Kahneman & Tversky, 1996). Recent advances in causal modeling have provided new insights, with respect to both long-standing norms of rationality and the descriptive adequacy of models of human diagnostic reasoning. *Causal Bayes net theory*, that is, probabilistic inference over causal graphical models, provides a formal framework for representing causal dependencies and modeling different kinds of probabilistic causal inferences (Pearl, 2000; Spirtes, Glymour, & Scheines, 1993; Spohn, 1976/1978, as cited in Spohn, 2001; see also chapter 4.2 by Hartmann, this handbook). The framework explicates the relations between observed covariation information and an underlying causal model that generates the data. Importantly, causal Bayes net theory has the expressive power to instantiate different rational models of diagnostic inference, thereby contesting the idea of a single normative benchmark for evaluating the rationality of human diagnostic reasoning (Meder & Mayrhofer, 2017b; Meder, Mayrhofer, & Waldmann, 2014; Waldmann, Cheng, Hagmayer, & Blaisdell, 2008). This chapter highlights the ways in which a causal analysis of diagnostic reasoning can inform issues of theoretical rationality and guide empirical research.

2. Rational Models of Diagnostic Reasoning

Different models of diagnostic reasoning have been postulated that serve the dual purpose of providing normative standards and constituting candidate models of human cognition. Common to these approaches is the goal to infer the conditional probability of a cause given an effect. They critically differ, however, with respect to their assumptions regarding the relation between unobservable causal structures and observable covariation information, as well as the involved representations and computations.

2.1 Diagnostic Reasoning as Statistical Inference

The most elemental form of diagnostic reasoning involves inferences from a binary (present vs. absent) effect event E to a binary cause event C. For instance, the cause event C could be a disease, with c and $\neg c$ denoting the presence and absence, respectively, of the disease, and the effect event E could be a particular symptom, with e and $\neg e$ denoting the presence and absence, respectively, of

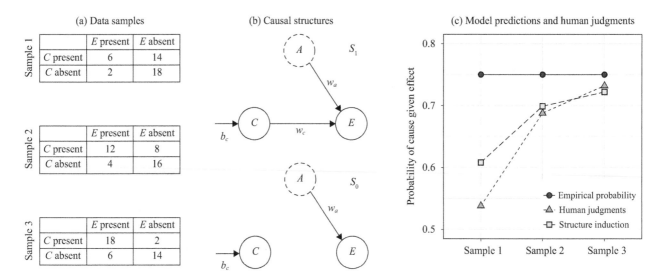

Figure 7.3.1

(a) Three joint frequency distributions over cause C and effect E. In each data sample, the diagnostic probability of cause C given effect E is 0.75. (b) Alternative causal structures that may have generated the data. According to structure S_1, there is a causal relation between C and E, as well as background causes A that can cause E independently of C. According to structure S_0, candidate cause C and effect E are independent; any observed empirical contingency is merely coincidental and caused by background causes A. (c) Model predictions and empirical results. The empirical probability of cause given effect is 0.75 in all three data samples. If uncertainty about possible causal structures is taken into account, different diagnostic probabilities result, depending on the extent to which the data warrant the existence of a causal relation (i.e., supports S_1 over S_0). Mean human judgments are not invariant across the three data sets, suggesting that people are sensitive to uncertainty about causal structure (Meder et al., 2014, experiment 2).

the symptom. Here, we consider the situation where the diagnostic inference is based on a sample of observed data. For instance, a doctor may have data on how many patients with the disease have a particular symptom, how many without the disease have the symptom, and so on. Given this covariation information, the diagnostic probability of cause given effect, $P(c \mid e)$, can be inferred using Bayes' rule:

$$P(c \mid e) = \frac{P(e \mid c) \cdot P(c)}{P(e \mid c) \cdot P(c) + P(e \mid \neg c) \cdot P(\neg c)} \qquad (1)$$
$$= \frac{P(e \mid c) \cdot P(c)}{P(e)},$$

where $P(c)$ denotes the prior probability (base rate) of the cause, $P(e \mid c)$ is the likelihood of the effect given that the cause is present, and $P(e \mid \neg c)$ is the likelihood of the effect if the cause is absent. Consider the data sample shown in figure 7.3.1a (top). According to these empirical frequencies, $P(c) = 20/40 = 0.5$, $P(e \mid c) = 6/20 = 0.3$, and $P(e \mid \neg c) = 2/20 = 0.1$. Plugging these values into equation (1) yields the diagnostic probability $P(c \mid e) = 0.75$.

Whether people reason in line with Bayes' rule has been pivotal to disputes about human rationality. Studies in the 1950s and 1960s indicated that subjects updated their beliefs to a lesser extent than prescribed by Bayes'

rule (Edwards, 1968; Phillips & Edwards, 1966). By and large, though, the experimental findings were considered evidence for sound probabilistic reasoning, giving rise to the metaphor of "man as intuitive statistician" (Peterson & Beach, 1967). Researchers in the heuristics-and-biases program, however, came to a rather different conclusion, arguing that typically people's diagnostic inferences are not in line with classic Bayesian norms (e.g., Kahneman, Slovic, & Tversky, 1982; Tversky & Kahneman, 1974; for a critical review, see Koehler, 1996). The discrepancy has led researchers to more precisely characterize the conditions under which people can make appropriate diagnostic inferences, for instance, when probabilistic information is conveyed in particular frequency formats (Gigerenzer & Hoffrage, 1995; McDowell & Jacobs, 2017; Meder & Gigerenzer, 2014).

The conditional probability of cause given effect derived by applying Bayes' rule to verbally described probabilities or sample data has been endorsed by many researchers as the normative standard, even when they otherwise debate whether and under what conditions people can solve such tasks. Next, we analyze diagnostic reasoning from the perspective of inductive causal inferences, using causal Bayes net theory to implement different candidate models for a rational account of diagnostic reasoning. These approaches differentiate between parametric causal models and observable contingency information and can therefore lead to very different predictions than accounts that derive the conditional probability of cause given effect directly from the observed empirical data without any reference to an underlying causal model.

2.2 Diagnostic Reasoning as Causal Inference

One goal of rational agents is to acquire knowledge about the causal structure of noisy environments, in order to support prediction, diagnosis, and control (Waldmann, 2017; chapter 7.2 by Waldmann, this handbook). In this view, diagnostic inferences operate on representations that preserve the directionality of causal relations (as opposed to undirected statistical or associative relations). This distinguishes the account from Bayes' rule, which is applicable to arbitrary statistically related events and does not make any reference to possible causal relations that may underlie the observed data.

The standard causal model for situations involving a binary cause event and a binary effect event is structure S_1 in figure 7.3.1b. This graph states that there is a causal relation between C and E, as well as an amalgam of unobserved background causes A that occur independently of C and can also generate E. Cheng (1997; see also Glymour, 2003; Griffiths & Tenenbaum, 2005; Novick & Cheng, 2004) showed that the generative *causal power* of a cause—the unobservable probability with which C produces E—can be estimated according to

$$w_c = \frac{P(e \mid c) - P(e \mid \neg c)}{1 - P(e \mid \neg c)}, \qquad (2)$$

where w_c is the causal power of C with respect to E. The diagnostic probability of cause given effect can be inferred from the parameterized causal structure:

$$
\begin{aligned}
P(c \mid e) &= \frac{P(e \mid c) \cdot P(c)}{P(e \mid c) \cdot P(c) + P(e \mid \neg c) \cdot P(\neg c)} \\
&= \frac{w_c b_c + w_a b_c - w_c w_a b_c}{w_c b_c + w_a - w_c w_a b_c},
\end{aligned} \qquad (3)
$$

where w_c is the causal power of C with respect to E, b_c denotes the base rate (prior probability) of the cause, and w_a denotes the strength of the background cause A.

The quantities used to parameterize the causal structure are typically single-point estimates derived directly from the empirical frequencies (i.e., maximum likelihood estimates of $P(c)$, $P(e \mid c)$, $P(e \mid \neg c)$, and, hence, w_c; see, e.g., Cheng, 1997). In this case, the conditional probability of cause given effect derived from the parameterized causal

graph exactly corresponds to the values obtained from directly applying Bayes' rule to the empirical frequencies, although diagnostic reasoning takes place on the causal rather than the data level (Meder et al., 2014).[1] This is not necessarily the case, though, if uncertainty about causal parameters or alternative generative causal models is incorporated in the diagnostic inference process.

2.3 Diagnostic Reasoning: Parameter Uncertainty and Causal Structure Uncertainty

Analyzing diagnostic reasoning from the perspective of causal induction enables consideration of different kinds of uncertainty in the inference process. *Parameter uncertainty* arises if estimates are derived from limited and potentially noisy data. Formally, this type of uncertainty can be modeled by using probability *distributions* for representing causal parameters, rather than single-point estimates, with the parameter distributions being updated in light of the data using Bayesian inference (Lu, Yuille, Liljeholm, Cheng, & Holyoak, 2008). Importantly, the posterior distributions quantify and represent uncertainty explicitly (e.g., through the variance of the distributions).

Another type of uncertainty is *structure uncertainty*, which pertains to possible causal models that may underlie the observed data. The *structure induction model of diagnostic reasoning* (Meder et al., 2014) takes this into account by considering two causal graphs (figure 7.3.1b): instead of using only the default structure S_1, the model also considers an alternative structure S_0 according to which there is *no* causal relation between C and E. The intuition behind this is that an observed contingency between C and E may not be indicative of a causal relation but merely coincidental, and rational agents should take this into account. The data are then used to estimate the causal structures' parameters through Bayesian inference, and under each parameterized structure, an estimate of the diagnostic probability is derived. Under S_1, assuming uniform prior distributions over the parameters, the computed probability will approximate the empirically observed $P(c \mid e)$. Structure S_0, by contrast, states that C and E are independent events; therefore, observing E does not provide diagnostic evidence for C (i.e., $P(c \mid e) = P(c)$). The structure induction model computes how likely each of the two structures is, given the data, and forms a weighted average of the entailed diagnostic probabilities, with the resulting estimate taking into account both parameter and structure uncertainty.

Depending on the relative probability of S_0 and S_1, the inferred diagnostic probability can strongly diverge from the empirical probability. Figure 7.3.1c illustrates this for the three data sets shown in figure 7.3.1a: in all three data sets, the empirical probability is $P(c \mid e) = 0.75$; therefore, a rational agent might conclude that the diagnostic probability of cause given effect is the same in all three situations. Consideration of alternative causal structures leads to very different inferences. For instance, for sample 1, the structure induction model entails a diagnostic probability of 0.61, much lower than the empirical probability of 0.75. This discrepancy arises from the fact that the contingency between cause and effect is relatively weak, so that structures S_0 and S_1 are almost equally likely to have generated the data. For samples 2 and 3, the discrepancy is smaller, as these data indicate a stronger contingency between C and E, making it more likely that S_1 is indeed the true generating model.

Interestingly, human diagnostic judgments mirror structure uncertainty: judgments strongly vary, although the empirical probability of cause given effect is identical across the three data sets (figure 7.3.1c; for details, see Meder et al., 2014). From the perspective of the classic statistical inference perspective, this response pattern looks irrational. By contrast, viewed as resulting from a causal inference strategy that is adapted to the uncertainties of the world outside the laboratory, the judgments reflect rational reasoning.

3. Rational Models of Causal Attribution

Theories of diagnostic reasoning typically assume that the goal of the reasoner is to infer the conditional probability of cause given effect. In many diagnostic reasoning scenarios, however, it is judgments of the *causal responsibility* (or *causal attribution*) that are of interest, that is, judgments of the probability that a candidate cause brought about the effect. For instance, instead of merely assessing the probability of a particular genetic disposition's being present, a doctor may want to find out whether that disposition is the cause of a patient's symptoms. This quantity is different from the diagnostic probability of cause given effect. For instance, if there is no causal relation (i.e., $P(e \mid c) = P(e \mid \neg c)$ and, therefore, $w_c = 0$), it holds that $P(c \mid e) = P(c)$. But, intuitively, if there is no causal dependency, then the probability that C produced E is zero.

Whereas a purely statistical account lacks the expressive power to model judgments of causal attribution, they can be formalized within a causal modeling approach (Cheng & Novick, 2005). Let $c \rightarrow e$ denote that the presence of effect E is generated by the presence of cause C. The query of whether the occurrence of effect E can be attributed to cause C translates to determining

the conditional probability $P(c \rightarrow e \mid e)$. Given parameterized structure S_1, this quantity can be computed as follows (for details and for further measures of causal responsibility, see Cheng & Novick, 2005):

$$\begin{aligned} P(c \rightarrow e \mid e) &= \frac{P(e \mid c \rightarrow e) \cdot P(c \rightarrow e)}{P(e)} \\ &= \frac{b_c w_c}{b_c w_c + w_a - b_c w_c w_a}. \end{aligned} \tag{4}$$

In its original formulation, parameters b_c, w_a, and w_c are maximum likelihood point estimates directly derived from the empirical data.[2] Bayesian variants of the model (Holyoak, Lee, & Lu, 2010) can incorporate parameter uncertainty, and the computations can also be incorporated into the structure induction model of diagnostic reasoning, in which case estimates of causal responsibility take into account both parameter and structure uncertainty (Meder et al., 2014). This approach has been successfully used to account for people's judgments of causal attribution about singular cases (Stephan & Waldmann, 2016, 2018).

4. Diagnostic Reasoning in Complex Causal Networks

Real-world scenarios typically involve complex causal networks relating multiple causes and multiple effects, such as diagnostic reasoning with multiple symptoms and multiple possible diseases. In the causal Bayes net framework, the factorization of the joint probability distribution over the considered variables is determined by the causal relations in the graph. This follows from applying the *causal Markov condition*, according to which the state of any variable in the graph is a function only of its direct causes, rendering it independent of all other variables except its direct and indirect effects (Hausman & Woodward, 1999; for a critique, see Cartwright, 1989, 1999). For instance, in the common-effect model shown in figure 7.3.2a, causes A and C are unconditionally independent but dependent conditional on their common effect B. In the common-cause model (figure 7.3.2b), A and C are two effects of a common cause B. If the Markov condition holds, this implies that A and C are unconditionally dependent but independent conditional on their common cause B (B "screens off" the correlation between A and C; Reichenbach, 1956). Similarly, in the causal-chain model (figure 7.3.2c), A and C are unconditionally dependent but conditionally independent given B.

The causality-based factorization implies particular dependence and independence relations among the variables, which facilitate and constrain diagnostic inferences

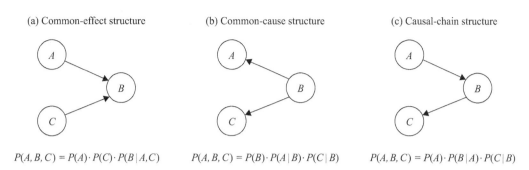

(a) Common-effect structure   (b) Common-cause structure   (c) Causal-chain structure

$$P(A,B,C) = P(A) \cdot P(C) \cdot P(B \,|\, A,C) \qquad P(A,B,C) = P(B) \cdot P(A \,|\, B) \cdot P(C \,|\, B) \qquad P(A,B,C) = P(A) \cdot P(B \,|\, A) \cdot P(C \,|\, B)$$

Figure 7.3.2
Basic causal structures comprising three variables A, B, and C. Applying the causal Markov condition to each causal graph entails a different factorization of the joint probability distribution $P(A,B,C)$ over domain variables A, B, and C.

across causal networks. For instance, when reasoning diagnostically in common-effect structures, *explaining away* is an intriguing inference pattern (Morris & Larrick, 1995; Pearl, 1988). Explaining away occurs when conditioning not only on the effect but also on the known presence of an alternative cause. Consider figure 7.3.2a and assume B is a symptom (e.g., fever) with two independent, not mutually exclusive, causes A and C (e.g., a virus and a bacterial infection). With respect to cause A, explaining away corresponds to the inequality $P(a \,|\, b) > P(a \,|\, b, c)$. Knowing that a patient has a fever will raise the probability of her being infected with the virus, but learning additionally that she has a bacterial infection will *lower* the probability of the virus to some extent. In other words, knowing that not only effect B but also the alternative cause C is present will explain away some of the diagnostic evidence provided by B with respect to A (for an overview, see Khemlani & Oppenheimer, 2011).

Another kind of diagnostic reasoning involves inferences from multiple effects (e.g., different symptoms) to an underlying cause (e.g., a disease), constituting a common-cause structure (figure 7.3.2b; Meder & Mayrhofer, 2017a). According to this structure, effects A and C are unconditionally dependent (e.g., lung cancer and yellow teeth correlate because of their common cause, smoking) but independent conditional on their common cause B. This property strongly simplifies diagnostic inferences, because the number of estimates required to parameterize the causal structure is greatly reduced. In particular, the joint likelihood of the effects given the cause can be computed as the product of the individual likelihoods of the effects given the cause. However, research also indicates that people's judgments do not always honor the causal Markov condition, suggesting important pathways for further investigating the rationality of human reasoning (Mayrhofer & Waldmann, 2016; Park & Sloman, 2013; Rottman & Hastie, 2016).

5. Toward a Rational Theory of Diagnostic Reasoning

Models of diagnostic reasoning fundamentally differ with respect to their assumptions, representations, and involved computations. As a consequence, they can make strongly diverging predictions in many situations. What are the implications for constructing a rational theory of diagnostic inference, and what does this mean for the experimental study of human rationality?

On the theoretical level, analyzing diagnostic reasoning from the perspective of inductive causal inference provides new insights into an old problem. Causal Bayes net theories have the expressive power to distinguish between the data level (i.e., covariation information) and parametric causal models that could underlie the observations. A causal modeling framework also enables formalizing diagnostic inferences that do not pertain to the conditional probability of cause given effect, such as judgments of causal responsibility. Some of the accounts that can be implemented within this framework are isomorphic to the classic, purely statistical model in the sense that the inferred diagnostic probability of cause given effect is identical to applying Bayes' rule to the empirical probabilities. Other models make very different predictions about what would constitute a rational solution to the inference problem, for instance, when incorporating uncertainty about alternative causal structures into the judgment process. As a result, inferences that look irrational from the perspective of one model (e.g., a statistical account operating on the data level) would be considered rational from the perspective of another model (e.g., a causal inference account that considers structure uncertainty).

This also raises critical methodological issues for the experimental study of human rationality, which all too often involves comparing human behavior to a single, supposedly normative, yardstick. Rarely, if ever, will there be a single rational theory for a particular phenomenon (Cohen, 1981; Gigerenzer, 1996; cf. chapter 2.4 by Fiedler, Prager, & McCaughey, this handbook). First, different frameworks exist that can—and should—be used to construct rational theories, including probability theory (Anderson, 1990; chapter 4.5 by Chater & Oaksford, this handbook), logic-based theories (Ragni & Knauff, 2013), ranking theory (Spohn, 2012; chapter 5.3 by Kern-Isberner, Skovgaard-Olsen, & Spohn, this handbook), and theories of bounded and ecological rationality (Chase, Hertwig, & Gigerenzer, 1998; chapter 8.5 by Hertwig & Kozyreva, this handbook). Comparing different types of models can provide insights that could not be gained when restricting the analysis to one particular theoretical viewpoint. Moreover, within a particular methodological framework, different models can be implemented and defended as rational, challenging the common approach of comparing human behavior to a single, supposedly normative, standard (Meder et al., 2014). In the case of elemental diagnostic reasoning, all the models discussed in this chapter rely on some form of probabilistic inference, but their scope, assumptions, and predictions strongly differ. From the perspective of the behavioral sciences, these models should be considered candidate theories, not standards, of human behavior (McKenzie, 2003). In this view, the influence runs both ways: rational theories can inform empirical research, but if there is a stable behavioral pattern that is inconsistent with a particular rational model, one should also revise one's beliefs about the appropriateness of the presumed normative yardstick.

Acknowledgments

This research was supported by grants ME 3717/2-2 and MA 6545/1-2 from the Deutsche Forschungsgemeinschaft (DFG) as part of the priority program "New Frameworks of Rationality" (SPP 1516). We thank Anita Todd for editing the manuscript.

Notes

1. Consider sample 1 in figure 7.3.1a (top). The empirical probabilities derived from the frequency data are $P(c) = 0.5$, $P(e\,|\,c) = 0.3$, and $P(e\,|\,\neg c) = 0.1$. Accordingly, $w_c = 0.22$ (equation (2)), with $P(c)$ and $P(e\,|\,\neg c)$ serving as estimates for b_c and w_a, respectively. Plugging these values into equation (3) yields $P(c\,|\,e) = 0.75$; the same value results when applying Bayes' rule to the empirical probabilities (equation (1)).

2. Consider again sample 1 in figure 7.3.1a (top). Using these data to parameterize structure S_1 with maximum likelihood

point estimates yields $b_c = 0.5$, $w_c = 0.22$, and $w_a = 0.1$. Plugging these values into equation (3) yields $P(c\,|\,e) = 0.75$, whereas the estimate for the probability that C produced E derived from equation (4) yields $P(c \to e\,|\,e) = 0.56$.

References

Anderson, J. R. (1990). *The adaptive character of thought.* Hillsdale, NJ: Erlbaum.

Cartwright, N. (1989). *Nature's capacities and their measurement.* Oxford, England: Oxford University Press.

Cartwright, N. (1999). Causal diversity and the Markov condition. *Synthese, 121,* 3–27.

Chase, V. M., Hertwig, R., & Gigerenzer, G. (1998). Visions of rationality. *Trends in Cognitive Sciences, 2,* 206–214.

Cheng, P. W. (1997). From covariation to causation: A causal power theory. *Psychological Review, 104,* 367–405.

Cheng, P. W., & Novick, L. R. (2005). Constraints and non-constraints in causal reasoning: Reply to White (2005) and to Luhmann & Ahn (2005). *Psychological Review, 112,* 694–707.

Cohen, L. J. (1981). Can human irrationality be experimentally demonstrated? *Behavioral and Brain Sciences, 4,* 317–331.

Edwards, W. (1968). Conservatism in human information processing. In B. Kleinmuntz (Ed.), *Formal representation of human judgment* (pp. 17–52). New York, NY: Wiley.

Gigerenzer, G. (1996). On narrow norms and vague heuristics: A reply to Kahneman and Tversky (1996). *Psychological Review, 103,* 592–596.

Gigerenzer, G., & Hoffrage, U. (1995). How to improve Bayesian reasoning without instruction: Frequency formats. *Psychological Review, 102,* 684–704.

Glymour, C. (2003). Learning, prediction and causal Bayes nets. *Trends in Cognitive Science, 7,* 43–47.

Griffiths, T. L., & Tenenbaum, J. B. (2005). Structure and strength in causal induction. *Cognitive Psychology, 51,* 334–384.

Hausman, D. M., & Woodward, J. (1999). Independence, invariance and the causal Markov condition. *British Journal for the Philosophy of Science, 50,* 521–583.

Holyoak, K. J., Lee, H. S., & Lu, H. (2010). Analogical and category-based inference: A theoretical integration with Bayesian causal models. *Journal of Experimental Psychology: General, 139,* 702–727.

Kahneman, D., Slovic, P., & Tversky, A. (Eds.). (1982) *Judgment under uncertainty: Heuristics and biases.* New York, NY: Cambridge University Press.

Kahneman, D., & Tversky, A. (1996). On the reality of cognitive illusions: A reply to Gigerenzer's critique. *Psychological Review, 103,* 582–591.

Khemlani, S. S., & Oppenheimer, D. M. (2011). When one model casts doubt on another: A levels-of-analysis approach to causal discounting. *Psychological Bulletin, 137*, 195–210.

Koehler, J. J. (1996). The base rate fallacy reconsidered: Descriptive, normative and methodological challenges. *Behavioral and Brain Sciences, 19*, 1–54.

Lu, H., Yuille, A. L., Liljeholm, M., Cheng, P. W., & Holyoak, K. J. (2008). Bayesian generic priors for causal learning. *Psychological Review, 115*, 955–984.

Mayrhofer, R., & Waldmann, M. R. (2016). Sufficiency and necessity assumptions in causal structure induction. *Cognitive Science, 40*, 2137–2150.

McDowell, M., & Jacobs, P. (2017). Meta-analysis of the effect of natural frequencies on Bayesian reasoning. *Psychological Bulletin, 143*, 1273–1312.

McKenzie, C. R. M. (2003). Rational models as theories—not standards—of behavior. *Trends in Cognitive Sciences, 7*, 403–406.

Meder, B., & Gigerenzer, G. (2014). Statistical thinking: No one left behind. In E. J. Chernoff & B. Sriraman (Eds.), *Advances in mathematics education: Probabilistic thinking: Presenting plural perspectives* (pp. 127–148). Dordrecht, Netherlands: Springer.

Meder, B., & Mayrhofer, R. (2017a). Diagnostic causal reasoning with verbal information. *Cognitive Psychology, 96*, 54–84.

Meder, B., & Mayrhofer, R. (2017b). Diagnostic reasoning. In M. R. Waldmann (Ed.), *The Oxford handbook of causal reasoning* (pp. 433–458). New York, NY: Oxford University Press.

Meder, B., Mayrhofer, R., & Waldmann, M. R. (2014). Structure induction in diagnostic causal reasoning. *Psychological Review, 121*, 277–301.

Morris, M. W., & Larrick, R. P. (1995). When one cause casts doubt on another: A normative analysis of discounting in causal attribution. *Psychological Review, 102*, 331–355.

Novick, L. R., & Cheng, P. W. (2004). Assessing interactive causal power. *Psychological Review, 111*, 455–485.

Oaksford, M., & Chater, N. (1994). A rational analysis of the selection task as optimal data selection. *Psychological Review, 101*, 608–631.

Park, J., & Sloman, S. (2013). Mechanistic beliefs determine adherence to the Markov property in causal reasoning. *Cognitive Psychology, 67*, 186–216.

Pearl, J. (1988). *Probabilistic reasoning in intelligent systems*. San Francisco, CA: Morgan-Kaufmann.

Pearl, J. (2000). *Causality: Models, reasoning and inference*. Cambridge, England: Cambridge University Press.

Peterson, C. R., & Beach, L. R. (1967). Man as intuitive statistician. *Psychological Bulletin, 68*, 29–46.

Phillips, L. D., & Edwards, W. (1966). Conservatism in a simple probability inference task. *Journal of Experimental Psychology, 72*, 346–354.

Ragni, M., & Knauff, M. (2013). A theory and a computational model of spatial reasoning with preferred mental models. *Psychological Review, 120*, 561–588.

Reichenbach, H. (1956). *The direction of time*. Berkeley: University of California Press.

Rottman, B. M., & Hastie, R. (2016). Do people reason rationally about causally related events? Markov violations, weak inferences, and failures of explaining away. *Cognitive Psychology, 87*, 88–134.

Spirtes, P., Glymour, C., & Scheines, P. (1993). *Causation, prediction, and search*. New York, NY: Springer.

Spohn, W. (1978). *Grundlagen der Entscheidungstheorie* [Foundations of decision theory]. Kronberg/Taunus, Germany: Scriptor.

Spohn, W. (2001). Bayesian nets are all there is to causal dependence. In M. C. Galavotti, P. Suppes, & D. Costantini (Eds.), *Stochastic dependence and causality* (pp. 157–172). Stanford, CA: CSLI Publications.

Spohn, W. (2012). *The laws of belief: Ranking theory and its philosophical applications*. Oxford, England: Oxford University Press.

Stephan, S., & Waldmann, M. R. (2016). Answering causal queries about singular cases. In A. Papafragou, D. Grodner, D. Mirman, & J. C. Trueswell (Eds.), *Proceedings of the 38th Annual Conference of the Cognitive Science Society* (pp. 2795–2801). Austin, TX: Cognitive Science Society.

Stephan, S., & Waldmann, M. R. (2018). Preemption in singular causation judgments: A computational model. *Topics in Cognitive Science, 10*, 242–257.

Tversky, A., & Kahneman, D. (1974). Judgment under uncertainty: Heuristics and biases. *Science, 185*, 1124–1131.

Waldmann, M. R. (Ed.). (2017). *The Oxford handbook of causal reasoning*. New York, NY: Oxford University Press.

Waldmann, M. R., Cheng, P. W., Hagmayer, Y., & Blaisdell, A. P. (2008). Causal learning in rats and humans: A minimal rational model. In N. Chater & M. Oaksford (Eds.), *The probabilistic mind: Prospects for Bayesian cognitive science* (pp. 453–484). Oxford, England: Oxford University Press.

Part III Practical Rationality

8.1   Preferences and Utility Functions

Till Grüne-Yanoff

Summary

This chapter discusses the notions of preference and utility in relation to individual judgment and decision making. The first two sections sketch the formal properties of preferences—in particular, completeness and transitivity—and of utility functions—in particular, continuity, differentiability, diminishing marginal utility, and risk attitudes. Whether these properties should be considered as conditions of practical rationality is also discussed in section 2. The third section focuses on the dominant interpretations of the formal preference and utility notions, in particular, distinguishing between mental and behavioral interpretations. The fourth section discusses relations between total and partial preferences, and the last section presents some models of rational preference change.

1.   Preference Relations

The most common way to represent preferences is as a binary relation between two relata. This is how preference is typically conceptualized both in the social sciences and in philosophy. The properties that this relation is assumed to have are commonly described in a formalized language.[1] The preferences studied in preference logic are usually the preferences of rational individuals, but preference logic is also used in psychology and behavioral economics, where the emphasis is on actual preferences as revealed in behavior.

1.1   Concepts and Notation

The basic formal preference concept consists of a binary relation $\succcurlyeq$ on a set of alternatives A. This relation is most commonly used to denote the objectives of a decision maker as a *comparative evaluation*. That is, $x \succcurlyeq y$ compares two members x and y of A, describing x to be *at least as good as y*. This comparison might represent subjective desires of an individual but also her evaluative judgments. Furthermore, it might also represent evaluative judgments of

social agents (e.g., groups, companies, government institutions) or intersubjective evaluations that do not coincide with the subjective attitudes of any particular agent (e.g., certain types of moral judgments). Finally, preference relations are also used to represent choice patterns.

The logical properties of this comparative relation do not seem to differ between the cases where they correspond to what we usually call "preferences" and the cases where they do not. The term "preference logic" is therefore standardly used to analyze the basic properties of these relations irrespective of their interpretations.

From the relation $\succcurlyeq$, two further relations on A can be derived. The *strict preference* relation $\succ$ is defined by

$$x \succ y \quad \text{if and only if} \quad x \succcurlyeq y \text{ and not } y \succcurlyeq x \qquad (1)$$

and reads as "x is strictly preferred to y," "x is better than y," or "y is worse than x." The *indifference* relation is defined by

$$x \sim y \quad \text{if and only if} \quad x \succcurlyeq y \text{ and } y \succcurlyeq x \qquad (2)$$

and reads as "x is indifferent to y" or "x is equal in value to y" (von Wright, 1963).

The above-described relations relate members of A, the set of alternatives. Most broadly, one can characterize A as a set of propositions (Jeffrey, 1965/1983). Economists typically consider A to consist of vectors of consumption goods (Debreu, 1959). In most applications, members of A are assumed to be mutually exclusive (i.e., none of them is compatible with, or included in, any of the others). Preferences over a set of mutually exclusive relata are referred to as *exclusionary preferences* (S. O. Hansson, 2001).

The properties discussed in this section all concern such exclusionary preferences. In practice, people also have preferences between relata that are *not* mutually exclusive. These are called *combinative preferences* (S. O. Hansson, 2001).

An important kind of preferences are *preferences over certain outcomes*, where the relata are conjunctions of propositions that are interpreted as describing specific states of the world. Another kind are *preferences over lotteries*, where the relata are mutually exclusive and jointly exhaustive disjuncts of propositions that are interpreted as alternative states of the world, each of which might be realized with a certain probability. Many relata of actual preferences do not satisfy either of these two conditions—in particular, the relevant disjuncts might be only partly identified, thus making it difficult to describe the relata as well-formed formulas. Such relata feature in decision making under *deep uncertainty* (S. O. Hansson, 1996).

1.2 Completeness

In most applications of preference logic, it is taken for granted that the following property, called "completeness" or "connectedness," should be satisfied:

$$x \succeq y \lor y \succeq x, \quad \text{for all } x, y \in A, \tag{3}$$

that is, $\succeq$ is assumed to connect every member of A to every other member.

Completeness is assumed in many applications, not least in economics. Bayesian decision theory is a case in point. The Bayesian decision maker is assumed to make her choices in accordance with a complete preference ordering over the available options (Savage, 1954/1972, p. 18). However, in many everyday cases, we do not have, and do not need, complete preferences. Consider a person who has to choose between three objects *a*, *b*, and *c*. If she knows that she prefers *a* to the others, she does not have to make up her mind about the relative ranking among *b* and *c*. In such cases, preference *incompleteness* might be rationally justifiable.

In many practical contexts, people might exhibit incomplete preferences, even if not rationally justified. This raises the question how incompleteness can be resolved, which might happen in different ways. First, if the reason for preference incompleteness is lack of knowledge or reflection of one's own desires or judgments, then through observation, introspection, logical inference, or some other means of discovery, the missing preference relations might be discovered. In that case, the incompleteness is resolved in exactly one way, by arriving at exactly one complete preference relation. Second, incompleteness may be resolvable in several different ways. In this case, one's own desires or judgments on these issues are genuinely undetermined, so that the missing preferences cannot be discovered but must be *constructed* (Slovic, 1995). Finally, incompleteness may be irresolvable. Even with all the necessary resources available for preference discovery or construction, a person might be unable to say which she prefers. This last case of preference incompleteness is often also discussed as "preference *incommensurability*" (Chang, 1997).

1.3 Transitivity

By far the most discussed logical property of preferences is *transitivity*:

$$x \succeq y \land y \succeq z \to x \succeq z, \quad \text{for all } x, y, z \in X. \tag{4}$$

That is, if *x* is weakly preferred to *y*, and *y* weakly preferred to *z*, then *x* is weakly preferred to *z*. Experiments suggest that *in*transitive preferences are quite common (e.g., Tversky, 1969). Nevertheless, there is a strong tradition, not least in economic applications, to regard $\succeq$-transitivity as a necessary prerequisite of rationality.

The most famous argument in favor of preference transitivity is the *money-pump argument*. Its basic idea is that transitivity ensures against practical losses (Davidson, McKinsey, & Suppes, 1955; Ramsey, 1928/1950). Consider a situation where a stamp collector has cyclic preferences with respect to three stamps, denoted *x*, *y*, and *z*. He prefers *x* to *y*, *y* to *z*, and *z* to *x*. He is willing to pay 10 cents every time he exchanges a stamp for one he prefers. Now he enters a stamp shop with stamp *x*. The stamp dealer offers him to trade in *x* for *z* if he pays 10 cents. He accepts the deal. Next, the dealer offers him to trade in *z* for *y*, again for a 10-cent fee, which he accepts. Finally, the dealer offers him to trade in *y* for *x*, again for a 10-cent fee, which he accepts. The collector now has obtained the same stamp with which he entered the shop but is 30 cents poorer. Unless he revises his preferences such that they satisfy (4), he will accept further trades from the dealer until he runs out of funds. The money-pump argument thus claims to show that transitivity is a necessary condition in order to avoid negative practical results.

Various arguments have been raised against the money-pump justification of transitivity. First, it has been argued that it relies on particular conditions, which must be in place for intransitive preferences to be exploited in this way. In particular, it must be assumed that the thus-pumped individual cannot take recourse to some strategy that avoids his being pumped—for example, through precommitment or resolution (McClennen, 1990). Second, it has been argued that the money-pump argument itself must make relatively strong consistency assumptions, which undermine its generality (Cubitt & Sugden, 2001).

Another argument for the normative appropriateness of preference transitivity suggests that transitivity is constitutive of the meaning of preference. Drawing an analogy to length measurement, Davidson (1976/1980) asks, "If length is not transitive, what does it mean to use a number to measure length at all? We could find or invent an answer, but unless or until we do, we must strive to interpret 'longer than' so that it comes out transitive. Similarly for 'preferred to'" (p. 273). Violating transitivity,

Davidson claims, thus undermines the very meaning of preferring one option over others.

Yet another argument rests on the importance of preferences for choice. When agents choose simultaneously from all the elements of an alternative set, then preferences should be choice-guiding. They should have such a structure that they can be used to guide their choice among the elements of that set. But when choosing, for example, from $\{x, y, z\}$, a preference relation $\succ$ such that $x \succ y \succ z \succ x$ does not guide choice at all: any or none of the alternatives should be chosen according to $\succ$. The transitivity of preference, it is therefore suggested, is a necessary condition for a meaningful connection between preferences and choice. A critic, however, can point out that preferences are important even when they cannot guide choices. Take, for example, preferences over lottery outcomes: these are real preferences, regardless of the fact that one cannot *choose* between lottery outcomes. Furthermore, the indifference relation does not satisfy choice guidance either. That does not make it irrational to be indifferent between alternatives. Finally, the necessary criteria for choice guidance are much weaker than transitivity (S. O. Hansson, 2001, pp. 23–25). Thus, choice guidance can be an argument for the normative appropriateness of transitivity only under certain restrictions, if at all (for further discussion, see Anand, 1993).

All of these arguments for transitivity remain controversial. In addition, examples have been offered that suggest intransitive preferences to be rational at least sometimes. One important class of such counterexamples is derived from the classic *sorites paradox*. Perhaps the most famous example applied to preference transitivity is Quinn's self-torturer. Here a device has been implanted into the body of a person. The device has 1,001 settings, from 0 (off) to 1,000. Each increase leads to a negligible increase in pain. Each week, the self-torturer "has only two options—to stay put or to advance the dial one setting. But he may advance only one step each week, and he may never retreat. At each advance he gets \$10,000." In this way he may "eventually reach settings that will be so painful that he would then gladly relinquish his fortune and return to 0" (Quinn, 1990, p. 79). The self-torturer thus reveals intransitive preferences. But Quinn argued that these preferences are *not* irrational, because the self-torturer has good reasons to turn the dial, yet also has good reasons to want to revert to the original position. Much depends here on what might constitute good reasons to turn the dial. Some have argued that the self-torturer's inability to distinguish between two adjacent dial settings might constitute such a reason (Luce, 1956). Others have sought to explain the intransitive preferences as the result of a heuristic decision procedure (Voorhoeve & Binmore, 2006). It therefore seems that, ultimately, the rationality judgment depends on the process through which these preferences are generated, and here considerable ambiguity remains.

Another important class of counterexamples to transitivity is derived from the classic *Condorcet paradox*. A simple example here is Schumm's Christmas ornaments (Schumm, 1987), involving boxes containing three balls, colored red, blue, and green, respectively; they are represented by the vectors $\langle R1, G1, B1 \rangle$, $\langle R2, G2, B2 \rangle$, and $\langle R3, G3, B3 \rangle$. A chooser might strictly prefer box 1 to box 2, since they contain (to her) equally attractive blue and green balls, but the red ball of box 1 is more attractive than that of box 2. She prefers box 2 to box 3, since they are equal but for the green ball of box 2, which is more attractive than that of box 3. And finally, she prefers box 3 to box 1, since they are equal but for the blue ball of box 3, which is more attractive than that of box 1. Thus,

$$R1 \succ R2 \sim R3 \sim R1,$$

$$G1 \sim G2 \succ G3 \sim G1,$$

$$B1 \sim B2 \sim B3 \succ B1; \text{ and}$$

$$\langle R1, G1, B1 \rangle \succ \langle R2, G2, B2 \rangle \succ \langle R3, G3, B3 \rangle \succ \langle R1, G1, B1 \rangle.$$

The described situation yields a preference cycle, which contradicts transitivity of strict preference—yet arguably, each of these preferences is rational.

1.4 Order Typology

The completeness and transitivity of $\succeq$, in combination with the definitions of $\succ$ and $\sim$, imply the following properties (Mas-Colell, Whinston, & Green, 1995, p. 7):

$\succ$ is *irreflexive*: $x \succ x$ never holds. (5)

$\succ$ is *asymmetric*: $x \succ y \rightarrow \neg(z \succ x)$ for all $x, y \in A$. (6)

$\succ$ is *transitive*: $x \succ y \wedge y \succ z \rightarrow x \succ z$ for all $x, y, z \in A$. (7)

$\succ$ is *acyclical*: there is no series $x_1, \ldots, x_n$
of alternatives such that $x_1 \succ \ldots \succ x_n \succ x_1$. (8)

$\sim$ is *reflexive*: $x \sim x$ for all $x \in A$. (9)

$\sim$ is *symmetric*: $x \sim y \rightarrow y \sim x$ for all $x, y \in A$. (10)

$\sim$ is *transitive*: $x \sim y \wedge y \sim z \rightarrow x \sim z$
for all $x, y, z \in A$. (11)

IP-transitivity: $x \sim y \wedge y \succ z \rightarrow x \succ z$
for all $x, y, z \in A$. (12)

PI-transitivity: $x \succ y \wedge y \sim z \rightarrow x \succ z$
for all $x, y, z \in A$. (13)

Weak connectivity of $\succeq$: if $x \neq y$, then
$x \succeq y \vee y \succeq x$ for all $x, y, z \in A$. (14)

All of these properties individually or in conjunction constitute a weakening of the completeness and transitivity of $\succeq$ (with the exception of those conjunctions that imply completeness or transitivity). Given the various counterarguments discussed above, some authors have suggested replacing completeness and transitivity with some of their weaker implications.

Sometimes, the following property is added to the above list, even though it is not implied by the completeness and transitivity of $\succeq$:

$$\textit{Antisymmetry of } \succeq: \quad x \succeq y \wedge y \succeq x \rightarrow x = y. \quad (15)$$

Preference orderings of varying strength can be characterized by combinations of the above properties. Some of their common names are given in table 8.1.1 (Debreu, 1959; Sen, 1970).

2. Utility Functions

Another way to represent value comparisons or choice patterns is via a real-valued function $u: A \rightarrow \mathbb{R}$, which maps the members of A into the real numbers. Of two alternatives x and y, we say "x is strictly preferred to y," "x is better than y," or "y is worse than x" if and only if $u(x) > u(y)$, and "x is indifferent to y" or "x is equal in value to y" if and only if $u(x) = u(y)$.

2.1 Ordinal Scale Utility Functions

Under this interpretation, there is an obvious connection to the relational representation of preferences. Specifically,

$$x \succeq y \quad \text{if and only if} \quad u(x) \geq u(y) \quad (16)$$
(Ordinal Representation).

Any function u that assigns a larger number to x than to y will work as such a representation. Consequently,

the function u can be replaced with any function u' as long as u' is a *positive monotone transformation* of u. As this transformation property is the defining characteristic of ordinal scales, we call this an *ordinal preference representation*.

However, a preference relation $\succeq$ can be ordinally represented by a utility function *only* if it is transitive and complete (von Neumann & Morgenstern, 1947). An *incomplete* preference ordering has an ordinal representation that satisfies

$$\text{if } x \succeq y \text{ then } u(x) \geq u(y). \quad (17)$$

The converse is obviously not true. However, incomplete preferences have been represented as *sets* of utility functions (Aumann, 1962).

Not every transitive and complete preference relation $\succeq$ can be ordinally represented by a utility function. Consider, for example, a lexicographic preference over goods bundles $(x, y) \in \mathbb{R} \times \mathbb{R}$ with

$$(x_1, y_1) \succeq (x_2, y_2) \quad \text{iff} \quad x_1 \geq x_2 \text{ or } (x_1 = x_2 \text{ and } y_1 \geq y_2).$$

Unless $x_i \neq x_j$ for all goods bundles (x_i, y_i), (x_j, y_j) with $i \neq j$, such a relation $\succeq$ cannot be represented by a utility function, although $\succeq$ is transitive and complete (Debreu, 1954).

A sufficient condition for a transitive and complete preference relation $\succeq$ to be ordinally representable is the continuity of $\succeq$:

$\succeq$ is *continuous* iff for all $x \in X$, the *upper contour set of x*, $\{y \in X: y \succeq x\}$, and the *lower contour set of x*, $\{y \in X: x \succeq y\}$, are *closed*, that is, they include their boundaries.

Lexicographic preferences like the above example are not continuous. Every continuous, transitive, and complete preference relation $\succeq$ is ordinally representable by a utility function u (Mas-Colell et al., 1995, pp. 46–47).

In the social sciences, utility functions are typically analyzed with the tools of maximization under constraints. This requires that the thus-analyzed functions are *twice differentiable*. However, not every continuous, transitive, and complete preference relation $\succeq$ is representable by a differentiable utility function. Consider, for example, the preference $(x_1, y_1) \succeq (x_2, y_2)$ iff $\min\{x_1, y_1\} \geq \min\{x_2, y_2\}$ (Leontief, 1941).

Lexicographic and Leontief preferences appear at least sometimes to be fully rational. For example, it seems rational to lexicographically prefer subsistence goods to luxury goods when living at subsistence levels. Similarly, it seems rational to have Leontief preferences for perfectly complementary goods, like left and right shoes. Therefore, standard utility functions *cannot* ordinally represent *all* rational preferences.

Table 8.1.1

An (incomplete) overview of ordering terminologies

	Properties	Name(s)
1.	Reflexive, transitive	*Preorder, quasi-order*
2.	Reflexive, transitive, antisymmetric	*Partial order*
3.	Irreflexive, transitive	*Strict partial order*
4.	Reflexive, transitive, complete	*Total preorder, complete quasi-ordering, weak ordering*
5.	Reflexive, transitive, complete, antisymmetric	*Chain, linear ordering, complete ordering*
6.	Asymmetric, transitive, weakly connected	*Strict total order, strong ordering*

2.2 Interval-Scale Utility Functions

Although utility functions cannot represent all preferences, an ordinal-scale functional representation of $\succeq$ represents *nothing but* information contained in $\succeq$. In other words, all information contained in an ordinal utility function can be represented by a (transitive and complete) ordering $\succeq$. This is not the case if we interpret utility functions *cardinally*—that is, either as an *interval scale* or as a *ratio scale*. A utility function as an interval scale allows for degrees of difference between preferences, such that, for example, the utility assignments $u(x) = 12$, $u(y) = 6$, and $u(z) = 3$ represent the information that the difference in preference intensity between x and y is twice as large as the difference between y and z. Ratio-scale utility functions are quite rare; they show up, for example, in prospect theory (see chapter 8.3 by Glöckner, this handbook) and in experienced-utility measures (see section 3.1).

Interval differences are particularly important for analyzing preferences over lotteries under the *expected utility* model. Lotteries are ordered tuples of outcome–probability pairs that are mutually exclusive and jointly exhaustive. For example, $[x, p; y, q; z]$ is the lottery that gives outcome x with probability p, y with probability q, and z with probability $1 - p - q$. Under the expected utility model, the utility of a lottery is the sum of the utilities of its outcomes, weighted by their probabilities (von Neumann & Morgenstern, 1947):

$$u([x_1, p_1; x_2, p_2; \ldots ; x_n, p_n]) = \sum_{i=1}^{n} p_i \times u(x_i). \qquad (18)$$

In the example, then, $u([x, p; y, q; z]) = p \times u(x) + q \times u(y) + (1 - p - q) \times u(z)$. This is a very powerful tool for analyzing preferences over lotteries, making them comparable to preferences over certain outcomes. But it requires an interval utility function. For example, a comparison of lottery $[x, p; y, q; z]$ to certain outcome y depends not only on whether $x \succ y \succ z$ but also on whether the difference in preference intensity between x and y, weighted by p, is larger than the difference between y and z, weighted by $1 - p - q$. Such information cannot be provided by an ordinal scale but requires an interval-scale interpretation of u.

Interval-scale utilities are represented by sets of functions, where each member u of the set is a positive linear transformation of another member u': $u(x) = a \times u'(x) + b$, where $a > 0$. An interval-scale representation of a preference relation $\succeq$ is thus a proper subset of an ordinal-scale representation of $\succeq$. For such a representation to exist, and to be unique up to positive linear transformation, however, $\succeq$ must satisfy certain properties. Different approaches to these representations include Ramsey (1928), von Neumann and Morgenstern (1947),

Savage (1954/1972), Jeffrey (1965/1983), and Fishburn (1970). There are substantial differences between these approaches and their respective assumptions, but the rationality of at least some of the required preference properties in each of these approaches is controversial. Thus, in addition to the above-discussed limitations of ordinal utility representations, the interval scale further restricts which rational preferences utility functions can represent. For more detail, see chapter 8.2 by Peterson on decision theory (in this handbook).

Interval utility scales represent differences in intensities of preference. This allows describing two abstract properties: the relation between utility and quantities of goods as well as the relation between utility and uncertainty.

When evaluating a quantifiable good or service, the *marginal utility* is the value added by the last increment of that good or service. For example, if the value of x sheep is $u(x) = 15$ and that of $x + 1$ sheep is $u(x + 1) = 17$, then the marginal utility of adding one sheep to the x previous ones is 2. The so-called *law of diminishing marginal utility* (also known as a "Gossen's first law"; Gossen, 1854/1983) states that marginal utilities decrease, ceteris paribus, as additional amounts of a good or service are added to available resources. One striking implication of this principle is that any good, however much valued in its first unit, will eventually be valued less than even the first unit of a good that is valued rather little. This insight solves the *paradox of value* (famously presented by Smith, 1776/1904): water commands a lower market price than diamonds, despite its being on the whole more useful, in terms of survival, than diamonds. Why? Because most people are well supplied with water and thus value an additional unit very little. By contrast, diamonds are scarce, and many people value the first unit of diamonds higher than the nth unit of water.

When evaluating lotteries, *risk attitudes* express evaluations of the respective uncertainties of the alternatives. For example, a lottery [$100, 0.5; $0] has an expected value of $50; those being indifferent between the lottery and $50 for certain are called *risk neutral*, those preferring the certain outcome *risk averse*, and those preferring the lottery *risk loving*. It is widely agreed that moderate risk aversion is perfectly rational at least in some cases—for example, when someone with few resources prefers a certain gain of $100,000 to an even gamble between receiving $200,000 or nothing (Arrow, 1971).

The standard von Neumann–Morgenstern (vNM) expected utility model allows for the representation of such risk attitudes but only by equating them with changing marginal utilities. By definition (18), the utility of lottery [$100, 0.5; $0] is the probability-weighted sum of its outcomes; it can differ from the utility of $50 for certain only if the utilities of outcomes are not linear with the monetary amounts. Consequently, general risk aversion in the vNM model is identical to diminishing marginal utility.

This is conceptually and empirically problematic. Conceptually, evaluations of quantities and of uncertainty seem distinct. For example, Bengt Hansson (1988) imagines a professional gambler who turns down an offer to trade a single copy of a book he likes for an even-chance gamble between receiving no copy of the book and three copies. According to the vNM model, the gambler must be risk averse. Yet the gambler might reject such an analysis: being a professional gambler, he has habituated himself to being risk neutral. The reason he turns down the gamble, he says, is simply that the second and third copies are of almost no worth to him. Thus, he exhibits diminishing marginal utility for additional books but no risk aversion.

Furthermore, the vNM account cannot represent various empirically observed choice patterns, whose interpretation in terms of risk preferences seems normatively legitimate. For example, a person rejecting a small-stakes gamble (e.g., [$100, 0.5; –$100]) can be represented in the vNM model only with a diminishing utility function for money, which would entail that the same person would reject very reasonable large-stakes gambles (e.g., [$10^{10}, 0.5; –$100]) (Rabin, 2000). Other examples concern choice patterns exhibited in the famous Allais and Ellsberg paradoxes, which cannot be represented by vNM utilities but are often considered expressions of rational attitudes toward risks (see chapter 8.2 by Peterson, this handbook).

The root of these problems seems to be the identification of risk attitudes with nonlinear evaluation of quantity of goods in the vNM model.[2] Various solutions to these problems thus propose to separate the two. One approach is to replace (18) with a risk-weighted expected utility model (Buchak, 2013; Quiggin, 1982), where the agent's risk function r weighs the probabilities p in a nonlinear fashion:

$$u([x_1, p_1; x_2, p_2; \ldots ; x_n, p_n]) = \sum_{i=1}^{n} r(p_i) \times u(x_i). \quad (19)$$

In particular, risk aversion would be represented by a convex risk function (e.g., $r(p) = p^2$). This representation, however, syntactically separates risk attitudes from evaluations. Critics have argued that risk attitudes are a kind of desire and therefore propose to model them in the Jeffrey framework as desirabilities over chance propositions instead (Stefánsson & Bradley, 2019).

3. What Do Preferences and Utilities Represent?

The above two sections concerned formal properties of representational tools. Although these properties of course impose constraints, it turns out that these constraints are not sufficient to uniquely determine what preferences and utilities represent. Instead, multiple coherent positions have been developed, and lively controversies in philosophy and the social sciences continue between them. I will distinguish three broad categories: (1) those that consider utilities as representing mental states—like pleasure or happiness—and consider preferences as derived from utilities, (2) those that consider preferences as representing mental states—like comparative likings or judgments—and derive utilities from preferences, and (3) those that consider preferences as representing choice patterns and derive utilities from preferences.

3.1 Utility as Representing Mental States

Although earlier authors had proposed the maximization of happiness as a moral principle, is it only with Bentham (1789/1988) that happiness is explicitly identified as the presence of pleasure and the absence of pain that people feel as the consequence of action. Bentham argued that happiness can be measured in its intensity, duration, uncertainty, and remoteness and proposed to call this aggregate measure *utility*.

Bentham proposed utility in this sense only as a normative criterion of what should be considered morally good. Early marginalist economists like Gossen (1854/1983) and Jevons (1871) instead aimed to explain social phenomena, based on the motives of (representative) human individuals, and they used the utility concept for that purpose.[3] In Jevons's (1871) words, they "attempted to treat Economy as a Calculus of Pleasure and Pain." Furthermore, they created the utility function as a cardinal representation of mental states. In this framework, the notion of preference, to the extent that it was used at all, was merely derived from hedonistic utility.

While this perspective has been largely replaced in 20th-century social science, it has recently seen something of a revival in Daniel Kahneman's theory of *experienced utility*. Kahneman assumes that people at every moment experience what he terms *instant utility*—meaning pleasure and/or pain. This utility has quantity and valence, with a neutral point on the boundary between desirable and undesirable, between pleasure and pain, and can be measured on a ratio scale. By integrating instant utility over a period, the *total utility* for that period can be computed. Given that on this approach, utility can be measured independently, utility maximization becomes an empirical, testable hypothesis. To the extent that utility maximization is seen as a criterion of rationality, the experienced-utility approach claims to offer an empirical test for the rationality of individuals in particular domains of behavior (Kahneman, Wakker, & Sarin, 1997).

3.2 Preferences as Representing Mental States

In the social sciences, the preference concept became important for explanatory and predictive purposes with Irving Fisher's (1892/1961) and Vilfredo Pareto's (1906/2014) methodological criticisms of hedonic cardinal utility. Pareto argued that because an accurate measurement procedure for cardinal hedonic utility was not available, social scientists should constrain themselves to merely ordinal comparisons (Bruni & Guala, 2001). This argument turned preference into a fundamental notion of the social sciences, replacing (hedonic) utility.

Economists in the 1930s (Hicks & Allen, 1934) radicalized Pareto's idea and argued that cardinal utility should be excluded in order to purge economics of psychological hedonism. However, their concept of preference retained psychological content: people were assumed to act purposefully and therefore to have preferences that really constitute mental evaluations, rather than being ex post rationalizations of behavior (Lewin, 1996).

Based on (18) and stated preferences over lotteries with known probabilities, von Neumann and Morgenstern devised a measurement of interval utility. Later, Savage (1954/1972) and Jeffrey (1965/1983) devised simultaneous measurements of utilities and probabilities from stated preferences. These new concepts of utility, however, were very different from the older hedonic concept: here the preference concept is basic and the cardinal utility function merely derived.[4]

Because utility is derived from preferences using representation theorems, the assumptions necessary for these theorems are built into the utility representation. Utility representations of this kind thus already assume rationality properties like transitivity, independence, and continuity. It would therefore be pointless to try to test whether such utility representations meet these rationality criteria (cf. Davidson's arguments discussed in section 2.3).

3.3 Preferences as Representing Choice Patterns

In economics, the revealed-preference approach defined preference in terms of choice. Historically, this approach developed out of the pursuit of behavioristic foundations for economic theories—that is, the attempt to eliminate the mental-preference interpretation altogether. Specifically, the downward slope of the demand function—one of the central features of microeconomics—could be

derived only from properties of marginal utilities, yet these properties were difficult to measure with the mentalistic Allan–Hicks utility concept. Consequently, Paul Samuelson in 1938 proposed to "start anew . . . dropping off the last vestiges of the utility analysis" (Samuelson, 1938, pp. 61–62).

The basic idea was to assume that *choice* satisfied certain rationality postulates and then define a *revealed preference* relation $x \succeq_R y$ between two goods bundles x and y as the choice of x in conditions where either x or y could have been chosen.[5] In other words, if an individual at current prices can afford both purchasing x and purchasing y, given her budget, and she chooses to purchase x, then she is said to revealed-prefer x to y. Based on the above definitions (1) and (2), revealed indifference and revealed strict preferences can be derived from $\succeq_R$. Furthermore, by determining revealed preferences over relevant lotteries and applying one of the expected utility frameworks discussed in section 2.2, these preferences can be represented on an interval utility scale. Note, however, that in this case, the utility function does not represent any mental content at all but only recorded choice patterns.

Although this approach was highly influential at the time, not all economists followed Samuelson in this radical proposal—so that today, the mentalist and the revealed-preference interpretation of preferences are often both presented in textbooks (e.g., Mas-Colell et al., 1995). Indeed, it might be the case that Samuelson himself later changed his mind, shifting from the *definition* of preference in terms of choice to a *measurement* of preference based on choice data (Hands, 2014). Nevertheless, there are many economists today who defend the behaviorist interpretation, for example, by denying that preferences represent the causes of choice (Binmore, 2009) and by insisting on "mindless economics" (Gul & Pesendorfer, 2008). Concurrently, there is an ongoing discussion among philosophers whether the current concept of preference used by economists is a separate theoretical concept—defined against the common use as capturing choice patterns—or whether it indeed is based on the mental, "folk-theoretic" notion (Mäki, 2000; Ross, 2014).

Critics of the behaviorist interpretation have pointed out that the common "folk" explanations of action require both a motivational component (represented by, e.g., preferences or utility) and an epistemic component (i.e., beliefs represented by, e.g., probabilities). The revealed-preference account neglects this epistemic component and therefore cannot provide folk explanations (Hausman, 2012; List & Dietrich, 2016; Sen, 1993). While the defenders of revealed preference might concede as much (see above), Hausman further argues that the folk-psychological account already provides a lexical definition of preferences and that it is counterproductive to craft a stipulative definition of preferences that is supposed to replace this folk notion in economics.

However, defenders of revealed preference further argue that it is necessary to distinguish between those agents who indeed have preferences as states of minds (e.g., humans and maybe higher animals) and those agents who do not (e.g., machines, plants, or institutions). The former category might choose on the basis of their mentalistically interpreted preferences, and hence the above-discussed effort can aim at eliciting the preferences on which their choices are based. The latter category, despite their lack of states of mind, might nevertheless exhibit behavior that can be interpreted as rational choice. In those cases, one can only speak of preferences reconstructed from choice, without claiming that these preferences describe mental states at all (Gul & Pesendorfer, 2008; Ross, 2014).

4. Preferences Combination

A preference relation $\succeq$ might be constructed from a vector of preferences $\langle \succeq_1, \ldots, \succeq_n \rangle$. Such a relationship has a number of different interpretations. For example, on an *intrapersonal* interpretation, $\succeq$ might be the *total preference* of an individual—that is, an overall comparison of two relata, taking all relevant considerations into account. The items of $\langle \succeq_1, \ldots, \succeq_n \rangle$ might be the partial preferences of this individual—for example, desirable characteristics of economic goods (Lancaster, 1966) or different reasons that one may have to prefer one of the options to another (Dietrich & List, 2013; Pettit, 1991). For example, $\succeq_1$ might represent evaluations of the respective health benefits of different foodstuffs, while $\succeq_2$ might represent the comparative sustainability of their production.

An *interpersonal* interpretation, in contrast, would interpret $\langle \succeq_1, \ldots, \succeq_n \rangle$ as the collection of total preferences of individuals $1, \ldots, n$ in a group and $\succeq$ as the group preference. For example, the managing board of a company might consist of n members, each of whom has a preference for a certain course of action. The final decision of which action to take might be interpreted as the expression of the group preference $\succeq$.

Some authors have argued that for intrapersonal cases, total preferences are always the result of an aggregation of partial preferences (e.g., Hausman, 2012). These authors assume that a total preference relation is uniquely determined by the partial preference relations through a process of aggregation.

Other authors reject the idea that total preferences are uniquely derivable from partial preferences. Instead, they claim that total preferences are constructed at the moment of elicitation and thus influenced by contexts and framings of the elicitation procedure that are not encoded in preexisting partial preferences (Payne, Bettman, & Johnson, 1993). Total preferences seem to be influenced by direct affective responses that are independent of cognitive processes (Zajonc, 1980). For instance, food preferences seem to be partly determined by habituation and are therefore difficult to explain as the outcome of a process exclusively based on well-behaved partial preferences. According to this view, partial preferences are in many cases ex post rationalizations of total preferences rather than the basis from which total preferences are derived.

If total preferences are the result of an aggregation of partial preferences, the question arises what this aggregation process might be. Two kinds of approaches can be distinguished. First, one could seek to develop *intra*personal aggregation by following models of social choice that describe *inter*personal aggregation of preferences (Arrow, 1963; Sen, 1970). However, such models do not allow for trade-offs between partial preferences. That is, if a particular x is much better than a y in one dimension k, the preference $y \succ x$ might nevertheless be possible because the strength of y in other dimensions over x outweighs dimension k. This approach requires (i) a cardinal measure of preference intensity and (ii) the comparability of these preference intensities across dimensions (Keeney & Raiffa, 1993).

5. Preference Change

Preferences sometimes change. This poses difficult predictive and explanatory challenges for social scientists, many of which remain currently unsolved. In this section, I will instead focus on the question of what might constitute *rational preference change*, discussing three approaches: temporal-discounting models, consistency-restoring preference revision models, and doxastic preference change (for a broader overview, see Grüne-Yanoff & Hansson, 2009).

5.1 Temporal-Discounting Models
Sometimes preferences change because the temporal distance between the time of evaluation and the realization of the evaluated event changes (see chapter 10.4 by Raub on rational choice, this handbook). For example, many people consider a tedious task to be performed in three months' time less bad than the same task to be performed

now, even if all background conditions are equal. As this is an intensity change, preferences must be represented on an interval scale. The standard model then represents the utility $u(x,t)$ of an alternative x at time t as equal to the time-independent utility of x (which one might think of as the utility of x if it were realized now), discounted by the delay:

$$u(x,t) = u(x)/(1+r)^t, \tag{20}$$

where r is the discount rate. This is the *exponential discounting model*, proposed by Samuelson (1937), which still dominates economic analysis. It is a fact that people at least sometimes do not discount in agreement with the exponential model. Various nonexponential models have been proposed, of which the *hyperbolic discounting model* is the most prominent.

While it is obvious that the behavior of many people can only be described by assuming some kind of time-discounting, the question remains whether such discounting is rational and justifiable. At least two questions need to be distinguished—first, whether discounting utility of a prospect merely for its distance in time is ever justified, particularly whether steep discount rates can be justified—irrespective of what the particular form of the discounting function is. Critics argue that one should want one's life, as a whole, to go as well as possible and that counting some parts of life more than others interferes with this goal (Pigou, 1920; Ramsey, 1928; Rawls, 1971). According to this view, it is irrational to prefer a smaller immediate good to a greater future good, because now and later are equal parts of one life. Choosing the smaller good or the greater bad makes one's life, as a whole, turn out worse: "Rationality requires an impartial concern for all parts of our life. The mere difference of location in time, of something's being earlier or later, is not a rational ground for having more or less regard for it" (Rawls, 1971, p. 293). Critics of temporal discounting often attribute apparent departures from temporal neutrality to a cognitive illusion, which causes people to see future pleasures or pains in some diminished form.

Against the temporal neutrality of preferences, some have argued that there is no enduring, irreducible entity over time to whom all future utility can be ascribed; they deny that all parts of one's future are equally parts of oneself (Parfit, 1984). They argue, instead, that a person is a succession of overlapping selves related to varying degrees by memories, physical continuities, similarities of character and interests, and so on. On this view, it may be just as rational to discount one's "own" future preferences as to discount the preferences of another, distinct individual, because the divisions between the stages of one's life may be as "deep" as the distinctions between individuals.

The second question concerns the particular form of discounting. One of the central properties of Samuelson's exponential model is that it is *order-preserving*. That is, an individual who prefers $(y,t+i)$ to (x,t) at some time previous to t will also prefer $(y,t+i)$ to (x,t) at time t. The only difference between these two evaluations is the increase in utility for both $(y,t+i)$ and (x,t). The exponential model ensures that these intensities increase proportionally, so that the preference ranking of options never changes.

Many people experience temporal preference changes that contradict this order-preserving property: they start out preferring $(y,t+i)$ to (x,t) at some earlier time—perhaps because y, when compared to x at the same time, is so much better. But as they approach t, and hence the realization of x, they experience a *preference reversal*: near to and at t, they suddenly prefer (x,t) to $(y,t+i)$. For a simple example, consider a person who prefers one apple today to two apples tomorrow but (today) prefers two apples in 51 days to one apple in 50 days. Although this is a plausible preference pattern, it is incompatible with the exponential model. It can, however, be accounted for in a model with a declining discount rate. Pioneered by Ainslie (1992), psychologists and behavioral economists have therefore proposed to replace Samuelson's exponential-discounting model with a model of *hyperbolic discounting*. The hyperbolic model discounts the future consumption with a parameter inversely proportional to the delay of consumption and hence can represent preference reversals like the above (for an overview, see Grüne-Yanoff, 2015).

Choosing according to hyperbolic-discounted utility leads to periodic abandonment of previous consumption plans in favor of new ones, which would be in turn abandoned in the next period. Such patterns of time-inconsistent choice are often related to failures of *self-control*, and hyperbolic discounting is therefore often seen as an expression of irrational preferences (Strotz, 1956).

However, some authors have argued that hyperbolically discounted preferences are adaptive for environments in which the uncertainty of obtaining a distant reward increases nonlinearly with the temporal distance (Sozou, 1998). In the light of these results, hyperbolic discounting might be rational at least in some contexts. Against this, others have argued that it might be possible to separate these considerations of uncertainty from pure time preferences (Andreoni & Sprenger, 2012) and that nonexponential discounting of pure time preferences remains a sign of irrationality.

5.2 Consistency-Restoring Preference Revision Models

If an agent forms a specific preference as the result of some experience, further changes in her overall preference state are often necessary to regain consistency. Using tools from belief revision (see chapter 5.2 by Rott on belief revision, this handbook), preference change has been modeled as an adjustment to such inputs (Grüne-Yanoff, 2013; S. O. Hansson, 1995; Liu, 2011). Changes in preference are triggered by inputs that are represented by sentences expressing new preference patterns.

For example, if a subject grows tired of her previous favorite brand of mustard, x, and starts to like brand z better, then this will be represented by a change with the sentence "z is better than x," in formal language: $z \succ x$, as an input. However, a change in which the previous preference $x \succ z$ is substituted by the new preference $z \succ x$ can happen in different ways. For instance, there may be a third brand, y, that was previously placed between x and z in the preference ordering. The instruction to make the new preference relation satisfy $z \succ x$ does not tell us where y should be placed in the new ordering. The new ordering may, for instance, be either $z \succ x \succ y$ or $z \succ y \succ x$. One way to deal with this is to include additional information in the input, for instance, specifying which element(s) of the alternative-set should be moved while the others keep their previous positions. In my example, if only z is going to be moved, then the outcome should satisfy $z \succ x \succ y$. These and other considerations make it necessary to modify the standard model of belief change in order to accommodate the subject matter of preferences.

5.3 Doxastic Preference Change

Preferences also change as a consequence of belief changes. Two kinds of beliefs are especially important here. The first is the belief that the presence of state x would make a desired state y more likely—a rise in probability of y given x produces a rise in the desirability of x, and vice versa. The second kind of belief relevant for doxastic preference change concerns prospects that influence the preference for other prospects without being probabilistically related. For example, one's preference for winning a trip to Florida in the lottery will crucially depend on one's belief about the weather there during the specified travel time, even though these two prospects are probabilistically unrelated. More generally, if $x \wedge y$ is preferred to $x \wedge \neg y$, with x and y probabilistically not correlated, then a rise in the probability of x will result in a rise in the desirability of y (even if it does not affect the probability of y), and vice versa.

Jeffrey (1977) provided a simple model of preference change as the consequence of an agent's coming to believe a proposition A to be true. His model incorporates both kinds of belief relevant for doxastic preference change. The basic idea is that $\langle u, P \rangle$ represents the unconditional preference $\succeq$ if P is the probability distribution based on the agent's actual information. The *conditional preference ordering* $\succeq_A$, in contrast, is represented by the tuple $\langle u, P_A \rangle$, where P_A is the probability distribution based on the counterfactual scenario that the agent accepts proposition A as true. That is, the agent imagines that if he changed his whole belief system from P to P_A, then he would have the preference relation $\succeq_A$ as represented by $\langle u, P_A \rangle$ (for more discussion of conditional preferences, see Bradley, 2005; Joyce, 1999; Luce & Krantz, 1971).

Notes

1. The study of these properties can be traced back to Book III of Aristotle's Topics. Since the early 20th century, several philosophers have studied the structure of preferences with logical tools. In 1957 and 1963, respectively, Sören Halldén and Georg Henrik von Wright proposed the first complete systems of preference logic (Halldén, 1957; von Wright, 1963).

2. This identification has a long tradition, going back to Daniel Bernoulli's solution to the St. Petersburg paradox (Bernoulli, 1738/1954).

3. Jevons discussed the effect of character, gender, and race on the utility function and avoided dealing with the implied heterogeneity of utility functions by invoking such a representative individual (White, 1994).

4. Defenders of the Benthamite conception call this new vNM representation *decision utility*, to distinguish it from their *experienced utility*.

5. The two main assumptions are the *weak axiom of revealed preferences* (WARP) and the *strong axiom of revealed preferences* (SARP). WARP says that if x is chosen when y is available (in some alternative-set A), then there must not be another alternative-set A' containing both alternatives for which y is chosen and x is not. SARP says that if from a set A_1 of alternatives, x is chosen when y and z are available, and if in some other set of alternatives, A_2, y is chosen while z is available, then there can be no set of alternatives containing alternatives x and z for which z is chosen and x is not. (SARP says this for chains of any length.)

References

Ainslie, G. (1992). *Picoeconomics: The strategic interaction of successive motivational states within the person.* Cambridge, England: Cambridge University Press.

Anand, P. (1993). The philosophy of intransitive preference. *Economic Journal, 103*, 337–346.

Andreoni, J., & Sprenger, C. (2012). Risk preferences are not time preferences. *American Economic Review, 102*(7), 3357–3376.

Arrow, K. J. (1963). *Social choice and individual values* (2nd ed.). New Haven, CT: Yale University Press. (Original work published 1951)

Arrow, K. J. (1971). The theory of risk aversion. In *Essays in the theory of risk-bearing* (pp. 90–120). Chicago, IL: Markham.

Aumann, R. J. (1962). Utility theory without the completeness axiom. *Econometrica, 30*(3), 445–462.

Bentham, J. (1988). *An introduction to the principles of morals and legislation.* Buffalo, NY: Prometheus. (Original work published 1789)

Bernoulli, D. (1954). Exposition of a new theory on the measurement of risk (L. Sommer, Trans.). *Econometrica 22*(1), 22–36. (Original work published 1738)

Binmore, K. (2009). *Rational decisions.* Princeton, NJ: Princeton University Press.

Bradley, R. (2005). Radical probabilism and mental kinematics. *Philosophy of Science, 72,* 342–364.

Bruni, L., & Guala, F. (2001). Vilfredo Pareto and the epistemological foundations of choice theory. *History of Political Economy, 33,* 21–49.

Buchak, L. (2013). *Risk and rationality.* Oxford, England: Oxford University Press.

Chang, R. (Ed.). (1997). *Incommensurability, incomparability, and practical reason.* Cambridge, England: Cambridge University Press.

Cubitt, R. P., & Sugden, R. (2001). On money pumps. *Games and Economic Behavior, 37*(1), 121–160.

Davidson, D. (1980). Hempel on explaining action. In *Essays on actions and events* (pp. 261–275). Oxford, England: Oxford University Press. (Original work published 1976)

Davidson, D., McKinsey, J. C. C., & Suppes, P. (1955). Outlines of a formal theory of value, I. *Philosophy of Science, 22,* 140–160.

Debreu, G. (1954). Representation of a preference ordering by a numerical function. In R. M. Thrall, C. H. Coombs, & R. L. Davis (Eds.), *Decision processes* (pp. 159–166). New York, NY: Wiley.

Debreu, G. (1959). *Theory of value: An axiomatic analysis of economic equilibrium.* New Haven, CT: Yale University Press.

Dietrich, F., & List, C. (2013). A reason-based theory of rational choice. *Noûs, 47*(1), 104–134.

Fishburn, P. C. (1970). *Utility theory for decision making.* New York, NY: Wiley.

Fisher, I. (1961). *Mathematical investigations in the theory of value and prices.* New Haven, CT: Yale University Press. (Original work published 1892)

Gossen, H. H. (1983). *Die Entwicklung der Gesetze des menschlichen Verkehrs und der daraus fließenden Regeln für menschliches Handeln* [The laws of human relations and the rules of human action derived therefrom]. Boston, MA: MIT Press. (Original work published 1854)

Grüne-Yanoff, T. (2013). Preference change and conservatism. *Synthese, 190*(14), 2623–2641.

Grüne-Yanoff, T. (2015). Models of temporal discounting 1937–2000: An interdisciplinary exchange between economics and psychology. *Science in Context, 28*(4), 675–713.

Grüne-Yanoff, T., & Hansson, S. O. (Eds.). (2009). *Preference change: Approaches from philosophy, economics and psychology* (Theory and Decision Library). Dordrecht, Netherlands: Springer.

Gul, F., & Pesendorfer, W. (2008). The case for mindless economics. In A. Caplin & A. Schotter (Eds.), *The foundations of positive and normative economics: A handbook* (pp. 3–42). Oxford, England: Oxford University Press.

Halldén, S. (1957). *On the logic of "better."* Lund, Sweden: Library of Theoria.

Hands, D. W. (2014). Paul Samuelson and revealed preference theory. *History of Political Economy, 46,* 85–116.

Hansson, B. (1988). Risk aversion as a problem of conjoint measurement. In P. Gärdenfors & N.-E. Sahlin (Eds.), *Decision, probability, and utility: Selected readings* (pp. 136–158). Cambridge, England: Cambridge University Press.

Hansson, S. O. (1995). Changes in preference. *Theory and Decision, 38,* 1–28.

Hansson, S. O. (1996). Decision making under great uncertainty. *Philosophy of the Social Sciences, 26*(3), 369–386.

Hansson, S. O. (2001). *The structure of values and norms.* Cambridge, England: Cambridge University Press.

Hausman, D. M. (2012). *Preference, value, choice, and welfare.* New York, NY: Cambridge University Press.

Hicks, J. R., & Allen, R. G. D. (1934). A reconsideration of the theory of value. Part I. *Economica, 1*(1), 52–76.

Jeffrey, R. C. (1977). A note on the kinematics of preference. *Erkenntnis, 11,* 135–141.

Jeffrey, R. C. (1983). *The logic of decision* (2nd ed.). Chicago, IL: University of Chicago Press. (Original work published 1965)

Jevons, W. S. (1871). *The theory of political economy.* London, England: Macmillan.

Joyce, J. M. (1999). *Foundations of causal decision theory.* Cambridge, England: Cambridge University Press.

Kahneman, D., Wakker, P. P., & Sarin, R. (1997). Back to Bentham? Explorations of experienced utility. *Quarterly Journal of Economics, 112*(2), 375–406.

Keeney, R. L., & Raiffa, H. (1993). *Decisions with multiple objectives: Preferences and value tradeoffs*. Cambridge, England: Cambridge University Press.

Lancaster, K. J. (1966). A new approach to consumer theory. *Journal of Political Economy, 74*(2), 132–157.

Leontief, W. W. (1941). *The structure of American economy, 1919–1929: An empirical application of equilibrium analysis*. Cambridge, MA: Harvard University Press.

Lewin, S. B. (1996). Economics and psychology: Lessons for our own day from the early twentieth century. *Journal of Economic Literature, 34*(3), 1293–1323.

List, C., & Dietrich, F. (2016). Mentalism versus behaviourism in economics: A philosophy-of-science perspective. *Economics & Philosophy, 32*(2), 249–281.

Liu, F. (2011). *Reasoning about preference dynamics*. Dordrecht, Netherlands: Springer.

Luce, R. D. (1956). Semiorders and a theory of utility discrimination. *Econometrica, 24*, 178–191.

Luce, R. D., & Krantz, D. H. (1971). Conditional expected utility. *Econometrica, 39*, 253–271.

Mäki, U. (2000). Reclaiming relevant realism. *Journal of Economic Methodology, 7*(1), 109–125.

Mas-Colell, A., Whinston, M. D., & Green, J. R. (1995). *Microeconomic theory*. New York, NY: Oxford University Press.

McClennen, E. F. (1990). *Rationality and dynamic choice*. Cambridge, England: Cambridge University Press.

Pareto, V. (2014). *Manual of political economy: A critical and variorum edition*. Oxford, England: Oxford University Press. (Original work published 1906)

Parfit, D. (1984). *Reasons and persons*. Oxford, England: Oxford University Press.

Payne, J. W., Bettman, J. R., & Johnson, E. J. (1993). *The adaptive decision maker*. Cambridge, England: Cambridge University Press.

Pettit, P. (1991). Decision theory and folk psychology. In M. Bacharach & S. Hurley (Eds.), *Foundations of decision theory* (pp. 147–175). Cambridge, MA: Blackwell.

Pigou, A. C. (1920). *The economics of welfare*. London, England: Macmillan.

Quiggin, J. (1982). A theory of anticipated utility. *Journal of Economic Behavior and Organization, 3*(5), 323–343.

Quinn, W. S. (1990). The puzzle of the self-torturer. *Philosophical Studies, 59*, 79–90.

Rabin, M. (2000). Risk aversion and expected utility theory: A calibration theorem. *Econometrica, 68*(5), 1281–1292.

Ramsey, F. P. (1928). A mathematical theory of saving. *Economic Journal, 38*, 543–559.

Ramsey, F. P. (1950). Truth and probability (original manuscript, 1928). In R. B. Braithwaite (Ed.), *The foundations of mathematics and other logical essays* (pp. 156–198). London, England: Routledge & Kegan Paul.

Rawls, J. (1971). *A theory of justice*. Oxford, England: Oxford University Press.

Ross, D. (2014). *Philosophy of economics*. New York, NY: Palgrave Macmillan.

Samuelson, P. A. (1937). A note on measurement of utility. *Review of Economic Studies, 4*, 155–161.

Samuelson, P. A. (1938). A note on the pure theory of consumer's behaviour. *Economica, 5*(17), 61–71.

Savage, L. J. (1972). *The foundations of statistics* (2nd ed.). New York, NY: Dover. (Original work published 1954)

Schumm, G. F. (1987). Transitivity, preference and indifference. *Philosophical Studies, 52*, 435–437.

Sen, A. (1970). *Collective choice and social welfare*. San Francisco, CA: Holden Day.

Sen, A. (1993). Internal consistency of choice. *Econometrica, 61*(3), 495–521.

Slovic, P. (1995). The construction of preference. *American Psychologist, 50*(5), 364–371.

Smith, A. (1904). *An inquiry into the nature and causes of the wealth of nations*. London, England: Methuen. (Original work published 1776)

Sozou, P. D. (1998). On hyperbolic discounting and uncertain hazard rates. *Proceedings of the Royal Society of London, Series B: Biological Sciences, 265*(1409), 2015–2020.

Stefánsson, H. O., & Bradley, R. (2019). What is risk aversion? *British Journal for the Philosophy of Science, 70*(1), 77–102.

Strotz, R. H. (1956). Myopia and inconsistency in dynamic utility maximization. *Review of Economic Studies, 23*(3), 165–180.

Tversky, A. (1969). Intransitivity of preferences. *Psychological Review, 76*, 31–48.

von Neumann, J., & Morgenstern, O. (1947). *Theory of games and economic behavior* (2nd ed.). Princeton, NJ: Princeton University Press.

von Wright, G. H. (1963). *The logic of preference*. Edinburgh, Scotland: Edinburgh University Press.

Voorhoeve, A., & Binmore, K. (2006). Transitivity, the sorites paradox, and similarity-based decision-making. *Erkenntnis, 64*(1), 101–114.

White, M. V. (1994). Bridging the natural and the social: Science and character in Jevons' political economy. *Economic Inquiry, 32*, 429–444.

Zajonc, R. B. (1980). Feeling and thinking: Preferences need no inferences. *American Psychologist, 35*(2), 151–175.

8.2 Standard Decision Theory

Martin Peterson

Summary

What is it rational to do in a choice situation, given the limited and sometimes unreliable information available to an agent? The dominant view among decision theorists is that rational agents act in accordance with the principle of maximizing expected utility. This chapter surveys the debate on the expected utility principle and discusses the most significant objections raised in the literature.

1. Right versus Rational Decisions

Decision theory is a theory of practical rationality that seeks to answer the following question: what is it rational to do in a choice situation, given the limited and sometimes unreliable information available to an agent? Decision theorists distinguish between *right* decisions and *rational* ones. If you play the National Lottery and win, your decision was right (because the outcome was optimal), but if the odds of winning were as long as they usually are (and the act of playing the lottery did not in itself give you great pleasure), the decision was irrational. In light of the information available to you, you had more reason to abstain from playing than for doing so: the value of hitting the jackpot did not outweigh the low probability of winning. The task decision theorists set themselves is to articulate and defend this line of thought. How, exactly, should information about an act's possible outcomes guide our decisions? This chapter surveys some of the most influential answers and discusses a number of significant objections raised in the literature.

A fundamental assumption accepted by more or less all decision theorists is that the notion of rationality described by decision theory is a means–ends notion in which the rationality of a decision depends on the agent's beliefs and desires (see chapter 2.1 by Broome, this handbook). Decision theorists have little to say about what beliefs and desires are, but they assume that beliefs and desires can vary in strength. It is widely believed that the agent's degree of belief in something

can be represented by a subjective probability function (see section 4 of this chapter and chapter 4.1 by Hájek & Staffel, this handbook) and that desires can be represented by a utility function. The stronger you believe something and the more you desire it, the higher is your subjective probability and utility.

2. The Principle of Maximizing Expected Utility

Pascal (1670/1967), Nicolaus Bernoulli (in a letter from 1713), and Cramer (1728/1975) propose that an act is rational if and only if its *expected utility* is at least as high as that of every alternative act.[1] This has been the dominant view among decision theorists ever since, but there is significant disagreement about how the expected utility principle should be interpreted and justified. The expected utility of an act with n possible outcomes $x_i, \ldots, x_n$ is

$$\sum_{i=1}^{n} p_i \cdot u(x_i), \qquad (1)$$

where p_i is the probability of outcome x_i and $u(x_i)$ the utility of x_i (see chapter 8.1 by Grüne-Yanoff, this handbook). If the number of possible outcomes x is infinite, the expected utility is $\int_{x \in X} [p(x) \cdot u(x)] dx$.

3. The Law of Large Numbers

A possible reason for accepting the principle of maximizing expected utility (or expected value more generally) is that agents who do so will almost certainly be better off in the long run. This insight is formalized in a mathematical theorem known as the law of large numbers:[2] if a random experiment for which the expected value is E_x is repeated n times, then the probability that the average value $\bar{X}$ of the experiment differs from E_x by more than some small number ε converges to 0 as the number of experiments n goes to infinity, and this holds true for every positive $\varepsilon > 0$ no matter how small:

$$\lim_{n \to \infty} p(|\bar{X} - E_x| > \varepsilon) = 0. \qquad (2)$$

This mathematical insight might be of some importance for agents who face similar choice situations a large number of times. However, many decisions are unique. Claims about what *would* happen if one were to make decisions that one will *not actually* make therefore seem irrelevant. Would it, for instance, be rational for you (given your interests and financial circumstances) to study at university X rather than university Y? Should you buy the house offered for sale for $400,000, or should you look for a cheaper house? Would it be rational to quit your job to sail around the world? When answering these questions, it seems irrelevant what would happen if you were to face similar choices a million times, because you *know* that you will only face each choice a small number of times. Keynes (1923/2000) expressed this insight in its most brutal form: "In the long run we are all dead" (p. 80).

4. Axiomatic Analyses of the Expected Utility Principle

4.1 Von Neumann and Morgenstern

Several authors have developed axiomatic analyses of the expected utility principle, which are sometimes interpreted as normative arguments for accepting the principle. See, for example, Ramsey (1926/1931), von Neumann and Morgenstern (1947), Savage (1954/1972), and Jeffrey (1983).[3] None of these axiomatizations entail that rational agents will make their decisions by consciously multiplying probabilities and utilities. The axioms merely entail that agents forming preferences *in accordance with* the axioms can be described *as if* they were maximizing expected utility. As Blackburn (1998) puts it, these representation and uniqueness theorems provide us with a "grid for imposing interpretation: a mathematical structure, designed to render processes of deliberation mathematically tractable, whatever those processes are" (p. 135).

Von Neumann and Morgenstern focus on decision problems in which probabilities can be defined exogenously, meaning that the agent somehow knows the probability of every event relevant to her decision. In this section, we will use a slightly different notation than that used in chapter 8.1 by Grüne-Yanoff (this handbook). Recall that von Neumann and Morgenstern conceive of acts as lotteries: xpy is the lottery that yields x with probability p and y with probability $1-p$ (and this holds regardless of whether x and y are basic prizes or new lotteries). The possible outcomes x or y are either basic prizes (such as money or a ticket to the opera) or lotteries over basic prizes. The relation $\succcurlyeq$ is a binary preference relation over the set of all possible lotteries L. Von Neumann and Morgenstern propose that agents are rationally required to form preferences over lotteries that meet the following structural requirements, for all lotteries and basic prizes x, y, z and all probabilities p and q:

vNM 1 (Weak Order) $\succcurlyeq$ is a complete, reflexive, and transitive relation on L.

vNM 2 (Independence) $x \succcurlyeq y$ if and only if $xpz \succcurlyeq ypz$.

vNM 3 (Continuity)   If $x \succcurlyeq y \succcurlyeq z$, then there exist some p and q such that $xpz \succcurlyeq y \succcurlyeq xqz$.

The independence axiom is the most controversial of the three axioms. In table 8.2.1, it seems reasonable to prefer lottery x to y, because this will give you $1 million for sure. However, the independence axiom entails that if you prefer x to y, then you must also prefer xpz to ypz. This is counterintuitive. It seems perfectly reasonable to prefer ypz to xpz, because there is no certain outcome here. The .10 probability of winning $5 million instead of $0 might very well be considered more attractive than a slightly larger probability (.11) of winning a much smaller amount ($1 million). For discussions of the ordering and continuity axioms, see chapter 8.1 by Grüne-Yanoff (this handbook).

Von Neumann and Morgenstern's theorem states that whenever vNM 1–3 are satisfied, there exists a real-valued function $u(x)$ that takes a lottery in L as its argument and returns a real number between 0 and 1, which has the following properties:

(1) $x \succcurlyeq y$ if and only if $u(x) \geq u(y)$.

(2) $u(xpy) = p \cdot u(x) + (1-p) \cdot u(y)$.

(3) For every other function u' satisfying (1) and (2), there exist some constants $c > 0$ and d such that $u' = cu + d$.

Properties (1) and (2) constitute the representation part: the agent assigns higher utility to outcomes she judges to be preferable, and the utility of a lottery equals the expected utility of its outcomes. Property (3) is the uniqueness part: any other function u' that satisfies the first two properties

Table 8.2.1
The independence axiom

	Ticket 1	Ticket 2–11	
Lottery x	$1 million	$1 million	
Lottery y	$0	$5 million	

	Ticket 1	Ticket 2–11	Ticket 12–100
Lottery xpz	$1 million	$1 million	$0
Lottery ypz	$0	$5 million	$0

is a linear transformation of u, meaning that utility is measured on an interval scale (see chapter 8.1).

A common objection to normative interpretations of von Neumann and Morgenstern's result is that it puts the cart before the horse (for an overview, see Peterson, 2008, section 2.5). The agent acting in accordance with these axioms does not prefer an act *because* its expected utility is optimal. She can be described *as if* she were acting from the expected utility principle, but we have no reason to believe that the utility function we ascribe to the agent is equivalent to the function that guides her choices or that such a function exists. The problem is thus that the theory is not action-guiding. The ordering axiom explicitly requires that agents have complete preferences over *all possible* lotteries (acts), including very complex lotteries (acts), such as the ones mentioned earlier: Would it be rational for you to study at university X rather than university Y? Should you buy the house offered for sale for $400,000, or should you wait and hope the price drops to $350,000? Should you quit your job to sail around the world? Agents who have complete preferences over all possible lotteries hardly need advice from decision theorists.

The standard response to this objection is that the axioms could be action-guiding in an indirect sense. They can be treated as "inference rules" that can help nonideal agents revise rationally impermissible preferences. If you, say, discover that your preferences are incomplete, or intransitive, then the axioms tell you how to overcome this deficiency: revise your preferences, in whatever way you wish, so they meet the structural requirements imposed by the axioms. However, this type of indirect action guidance seems somewhat unhelpful. If you are trying to figure out if it is rational for you to study at university X or university Y or whether to offer $400,000 for a house you like, and the decision theorist is merely able to tell you that "anything goes" *as long as the axioms are fulfilled*, then decision theory will not be of much help to you.

4.2 Savage

The axiomatizations proposed by Ramsey (1926/1931), Savage (1954/1972), and Jeffrey (1983) differ from von Neumann and Morgenstern's in that probabilities are derived from the axioms themselves. In these axiomatizations, the probability function is not an exogenously defined function. Furthermore, since the axioms regulate the agent's preferences, we can interpret probabilities derived from these axioms as *subjective* probabilities, or credences (see chapter 4.1 by Hájek & Staffel, this handbook). To say that the probability is .90 that it will rain tomorrow means that you believe to a fairly strong

degree that it will rain tomorrow: your credence makes you willing to pay something worth .90 units of utility for a bet in which you win 1 unit if it rains tomorrow and nothing otherwise.

Savage conceives of acts as functions from a set of possible states of the world to outcomes. The act of not bringing the umbrella to the football game is a function in which the two possible states are "rain" and "no rain," and the possible outcomes are "the agent gets wet" and "the agent does not get wet," depending on which state turns out to be the true state of the world.

Some of the axioms proposed by Savage resemble those proposed by von Neumann and Morgenstern. This includes the assumption that the agent's preferences satisfy the weak ordering axiom, which raises the same issue about putting the cart before the horse discussed earlier. Another axiom borrowed from von Neumann–Morgenstern (and, ultimately, from Ramsey, 1926/1931) is the so-called *sure-thing principle*. This is Savage's name for a much-discussed version of the independence axiom. In Savage's terminology, two acts f and g *agree* with each other in a set S of states if and only if $f(s) = g(s)$ for all $s \in S$. In this terminology, the sure-thing principle can be formulated as follows:

If f and g, and f' and g', are such that

1. in $\neg S$, the complement of S, f agrees with g, and f' agrees with g',

2. in S, f agrees with f', and g agrees with g', and

3. $f \succeq g$,

then $f' \succeq g'$.

Table 8.2.2 is helpful for highlighting the similarities with von Neumann and Morgenstern's independence axiom. Imagine that you are offered a choice among four lotteries, each of which has 100 tickets. It seems rationally permissible to prefer f to g but g' to f' (for the reason mentioned in the discussion of the independence axiom), but the sure-thing principle controversially entails that if you prefer f to g, then you must prefer f' to g'.

Table 8.2.2

The sure-thing principle

	S		$\neg S$
	Ticket no. 1	Ticket nos. 2–11	Ticket nos. 12–100
Act f	$1 million	$1 million	$1 million
Act g	$0	$5 million	$1 million
Act f'	$1 million	$1 million	$0
Act g'	$0	$5 million	$0

One of the axioms in Savage's theory, which does not figure in von Neumann and Morgenstern's (in which the probability function is defined exogenously), is designed to test whether the agent considers two states to be equally probable. Savage's own formulation of this axiom is quite complex, but the following explanation offered by Broome (1990) is more helpful:

> Savage . . . needs to test whether two events, say E and F, are equally probable. He does this by taking a pair of outcomes, say A and B, that are known not to be indifferent. He forms a gamble $(A, E; B, F)$ in which A comes about in event E and B in F. And he forms the opposite gamble $(B, E; A, F)$. The events are equally probable if and only if these gambles are indifferent. Savage's Postulate 4 says that this test will deliver the same answer whatever pair of non-indifferent outcomes A and B are used. (p. 483)

Broome notes that a problem with this axiom is that it seems to assume what it is supposed to show: the assumption that the agent will be indifferent between the gambles only if they are judged to be equally probable *presupposes* that probabilities and utilities are aggregated in accordance with the expected utility principle.

Savage's uniqueness and representation theorem is similar to von Neumann and Morgenstern's, except that the probability function is derived from the axioms. The theorem guarantees the existence of a utility function that is unique up to positive linear transformations and a probability function such that one act is preferred to another just in case its expected utility exceeds that of the latter.

4.3   Bolker–Jeffrey

The theory proposed in Jeffrey (1965) was axiomatized by Bolker (1967) and presented in a slightly modified form in the second edition of Jeffrey's book (Jeffrey, 1983). The Bolker–Jeffrey theory differs from von Neumann and Morgenstern's and Savage's in several important respects.

In the Bolker–Jeffrey theory, it is assumed that the agent has preferences over *propositions*, such as "I have a refreshing swim with friends," rather than lotteries (von Neumann & Morgenstern, 1947), or functions from states to outcomes (Savage, 1954/1972). Utilities and probabilities are also assigned to propositions, meaning that the Bolker–Jeffrey theory is more economical from an ontological point of view. Another philosophical advantage is that the use of propositions enables the agent to retain her original beliefs about the world throughout the deliberative process. There is no need for the agent to imagine complex lotteries that require the existence of causal processes she does not believe in. The agent can continue to believe whatever she believes, without adding any new beliefs or modifying old ones.

In the Bolker–Jeffrey theory, it is assumed that preferences over the Boolean algebra B (with the zero removed; contradictions have no place in the preference ordering) satisfy the weak ordering axiom, are continuous, and satisfy the following two axioms:

(Averaging)   If a, b in B are contraries (i.e., cannot both be true), then
$a \succ b$ implies $a \succ (a \vee b) \succ b$, and
$a \sim b$ implies $a \sim (a \vee b) \sim b$.

(Impartiality)   If a, b, c in B are pairwise contraries, and $a \sim b$ but not $a \sim c$, and $(a \vee c) \sim (b \vee c)$, then for every d in B that is contrary to a and b, $(a \vee d) \sim (b \vee d)$.

The averaging axiom is somewhat similar to, but weaker than, the independence axiom. It requires that the agent ranks a disjunction of two propositions somewhere between the two disjuncts. This may seem reasonable. However, a possible reason for ranking a disjunction above or below the disjuncts could be that the uncertainty introduced by the disjunction sometimes has a positive or negative effect on the value of the proposition. Imagine, for instance, that you prefer the proposition "My partner loves me" to "My partner breaks up with me tomorrow." If so, it does not seem irrational to rank the disjunction of these two propositions lower than "My partner breaks up with me tomorrow." Not knowing for sure if your relationship is over might be worse than knowing that it is. If you learn that your partner will break up with you tomorrow, you can adjust your plans for the future and move on.

The impartiality axiom resembles the equiprobability axiom in Savage's theory. The intuition it seeks to capture is similar to that summarized by Broome in the quote in section 4.2: to test whether the agent considers two propositions to be equally probable, we can ask whether the agent is indifferent between $a \vee c$ and $b \vee c$ whenever she is indifferent between a and b. If the agent, for instance, is indifferent between a and b but prefers $a \vee c$ over $b \vee c$, then she does not consider a and b to be equally probable (as long as she maximizes expected utility; more about this soon). The impartiality axiom states that this holds true for every third proposition c.

Broome (1990) points out that the impartiality axiom assumes what it is supposed to show, just as axiom 4 in Savage's theory. If the agent prefers a to b, then in case she maximizes expected utility, the fact that she prefers $a \vee c$ over $b \vee c$ entails that she considers a to be more probable than b. However, if she aggregates probabilities and utilities in some other way, she has no reason to accept this axiom. It is not entirely satisfactory to justify the expected utility principle by appealing to an axiom that will be attractive primarily to those who accept the expected utility principle.

4.4   Other Axiomatizations

A radically different approach to the expected utility principle is to start with preferences over certain outcomes (basic prizes), as well as beliefs about the world, and then generate preferences over risky acts from this ex ante information. The ex ante approach steers clear of the objection that traditional axiomatizations put the cart before the horse. The drawback is that it seems somewhat optimistic to think that ordinary agents have access to exogenously defined utility and probability functions (see Peterson, 2004).

5.   Pragmatic Arguments

Faced with counterexamples to some of the key axioms of expected utility theory, decision theorists have attempted to offer further warrant for the axioms proposed by von Neumann and Morgenstern, Savage, and others by formulating *pragmatic arguments* (see, e.g., McClennen, 1990). A pragmatic argument is a device that seeks to show that anyone who violates a purported rationality requirement can be placed in a situation in which she has to act contrary to her own preference.

Donald Davidson, J. C. C. McKinsey, and Patrick Suppes (1955) propose the following pragmatic argument in defense of the transitivity condition, which is key to the assumption that rational preferences form a weak order:

> Mr. S. is offered his choice of three jobs by a cynical department head (never mind what department): He can be a full professor with a salary of $5,000 (alternative *a*), an associate professor at $5,500 (alternative *b*) or an assistant professor at $6,000 (alternative *c*). Mr. S. reasons as follows: $a \succ b$ since the advantage in kudos outweighs the small difference in salary; $b \succ c$ for the same reason; $c \succ a$ since the difference in salary is now enough to outweigh a matter of rank. . . . It is clear that the set of Mr. S.'s preferences makes a rational choice impossible, for whichever alternative he chooses there will be another alternative which is preferred to it. (Davidson et al., 1955, p. 145)

This argument is sometimes presented in a *diachronic* form, that is, as a series of choices to be made at different points in time. Imagine, for instance, that you sign the contract for the assistant professorship with a salary of $6,000 (alternative *c*). Right after you do so, the cynical department head reminds you that you prefer *b* to *c*. He offers you to swap if you pay him a small administrative

fee, say $.01. Because you prefer *b* minus the small fee to *c*, you accept his offer. However, right after you sign the contract for job *b*, he offers you to swap *b* for *a* if you pay a small fee, and once you have paid the fee, he offers you to pay him for swapping again, and so on. The cyclical structure of your preferences forces you to keep paying small amounts for swapping over and over again. After 100 million swaps, you have lost $1 million but gained nothing, which seems to indicate that your preferences are irrational. This type of pragmatic argument is known as a *money-pump argument*.

Schick (1986) claims that a rational agent with cyclical preferences will "see which way the wind is blowing . . . [and] seeing what is in store for him, he may well reject the offer and thus stop the pump" (p. 117). This is, however, a problematic response. Schick's argument can at best cast doubt on the diachronic money pump sketched above, but it does not affect the original *synchronic* version proposed by Davidson et al. (1955). As stressed by Gustafsson (2013, p. 462), Davidson et al.'s point is that "it is irrational to choose an alternative to which another alternative is preferred." This is irrational even if the agent makes only *one* such choice. There is no need to consider diachronic versions of the argument in which the agent has to take into account what will happen in the future.

Another problem with money-pump arguments for transitivity is that they do *not* show that *every possible* violation of transitivity will lead to a sure loss. Preferences can violate the transitivity condition in many different ways, and not every violation forces the agent to act against her own preference. Imagine, for instance, that you strictly prefer *a* to *b*, *b* to *c*, but are indifferent between *a* and *c*. With these preferences, you would be rationally obliged to pay a small amount for swapping *c* for *b* and then pay again for swapping *b* for *a*. However, once you have acquired *a*, you have no reason to continue swapping. Since you are indifferent between *a* and *c*, you may keep *a*, which stops the pump. By keeping *a*, you do not choose an alternative to which another is preferred. However, because you are indifferent between *a* and *c*, your preference violates the transitivity condition.

The worry outlined above can be further analyzed by distinguishing between strong ("forcing") and weak ("nonforcing") pragmatic arguments (see, e.g., Peterson, 2015). In a *strong* pragmatic argument, the agent's preferences *guarantee* that a rational agent will choose an alternative to which another is preferred. In a *weak* pragmatic argument, the agent's preferences *permit* the rational agent to choose an alternative to which another is preferred. If you, for instance, strictly prefer *a* to *b*, and *b* to *c*, but are indifferent between *a* and *c*, then you are rationally *permitted* to swap over and over again, each time paying a small fee when you swap to a strictly preferred option. However, you would not be rationally *obliged* to keep swapping once you reach *a*. Some scholars think that weak pumps are strong enough: by merely showing that your preferences *permit* you to be money-pumped, we can conclude that your preferences are irrational. Defenders of strong pumps insist that for a pragmatic argument to be successful, we have to establish that the agent is rationally *obliged* to be money-pumped.

The distinction between weak and strong pumps is also relevant for assessing pragmatic arguments for the assumption that rational preferences are complete. It is reasonable to think that if you have no preference between *a* and *c*, then you are *rationally permitted*, but not *rationally obliged*, to swap. Hence, if you strictly prefer *a* to *b*, and *b* to *c*, you would be vulnerable to a weak, but not a strong, money pump.

Some authors have argued that we can justify the independence axiom (see section 4.1) by appealing to pragmatic considerations. Figure 8.2.1 summarizes a sequential choice problem with two choice nodes (boxes) and three chance nodes (circles). At the first choice node (the leftmost box), the agent chooses between going "up" and "down." If the agent goes up and event ¬E occurs (the probability of which is 89/100), she wins $0. However, if event E occurs (the probability for this is 11/100), the agent faces a new choice, in which she can choose to either walk away with $1 million or accept a lottery in which she wins

$5 million with probability 10/11 and $0 with probability 1/11. If the agent goes down at the first choice node, she faces a lottery in which the outcomes and probabilities are exactly as if she were to decide at the outset to go up at both the first and second choice nodes, except that a small bonus ε has been added to each possible outcome.

The outcomes and probabilities in figure 8.2.1 are analogous to those in table 8.2.1. Note that

x: $1 million;
y: $5 million with probability 10/11, otherwise $0;
xpz: $1 million with probability 11/100, otherwise $0;
ypz: $5 million with probability 10/100, otherwise $0.

To construct a pragmatic argument for the independence axiom, suppose the agent's preferences are as follows:

$$x \succ y, \tag{1}$$

$$ypz \succ xpz. \tag{2}$$

For the reason explained in section 4.1, these preferences violate the independence axiom. To construct the argument, we also assume that the small bonus ε in figure 8.2.1 does not reverse the agent's preference between *ypz* and *xpz*. Given that the bonus is small enough, this is not a controversial assumption:

$$ypz \succ xpz + \varepsilon \succ xpz. \tag{3}$$

Because the agent prefers *ypz* to *xpz*, per (2), she will go up at the first choice node in figure 8.2.1 with the intention to go down at the second if event E occurs. However, if the agent reaches the second choice node, she will go up, since she prefers *x* to *y*, per (1). According to the agent's preferences, $1 million for sure is worth more to her than the risky gamble in which she wins $5 million with probability 10/11. However, this means that she is willing to forsake a small but certain bonus. If the agent goes down at the first choice node, she would obtain the same outcome *plus* the bonus ε. From a pragmatic point of view, agents who act in accordance with (1)–(3) thus forsake sure gains, compared to what they could have achieved had they instead revised their preferences in accordance with the independence axiom.

Is this a convincing pragmatic argument for the independence axiom? To start with, note that the argument only purports to show that *one of many possible violations* of the independence axiom causes the agent to forsake the small bonus ε. The argument does not purport to show that *every possible* violation has the same pragmatic effect. Suppose, for instance, that the agent prefers *x* to *y* but is indifferent, or has no preference, between *ypz* and *xpz*. If this is the case, she is at most vulnerable to a weak pragmatic argument but not to a strong one.

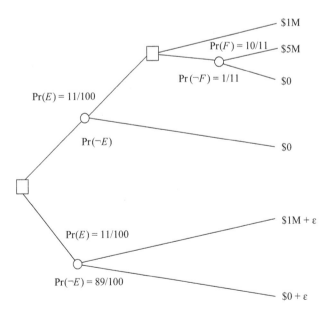

Figure 8.2.1
A decision tree (after Rabinowicz, 1995).

A more fundamental concern is that the argument presupposes a theory of sequential decision-making few people find plausible. In essence, the problem is that the agent who goes up at the first choice node with the intention of going down at the second is overly *myopic* (shortsighted). Schick's (1986, p. 117) observation that a rational agent will see "what is in store for him" seems equally applicable here as in the money-pump argument for transitivity mentioned earlier. According to Schick, a *sophisticated* agent facing a sequential choice problem should reason backward: if the agent were to reach the second choice node, she would prefer *x* to *y*, and she knows this already at the first choice node. Therefore, there are only two options open to her: she can (i) go down at the first choice node or (ii) go up at the first choice node with the intention to go up at the second. The third alternative, up followed by down, is not a genuine alternative. It should thus be deleted from the list of options. Therefore, the example in figure 8.2.1 fails to show that agents who violate the independence axiom will act contrary to their own preference.

Rabinowicz (1995) argues that it is possible to overcome this objection: the sophisticated approach does not make the agent immune to pragmatic arguments. In figure 8.2.2, the sophisticated agent with the preferences described in (1) and (2) would reason backward just as before: she prefers $1 million for sure (i.e., *x*) to $5 million with a probability of 10/11 (*y*), so she knows already at the first choice node that she would go up at the second choice node. At the first node, she

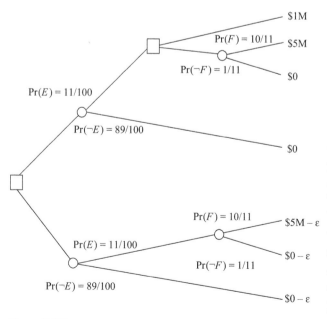

Figure 8.2.2
A decision tree for sophisticated agents (after Rabinowicz, 1995).

thus faces a choice between (i) going down and (ii) going up with the intention of going up at the second choice node. Because the agent prefers *ypz* to *xpz*, and the subtraction of ε from the outcomes of *ypz* does not alter her preference, she will go down at the first node. However, according to Rabinowicz, down is dominated by the plan to go up at the first node and down at the second. From a pragmatic point of view, the sophisticated agent is thus acting contrary to her own preference. She prefers *ypz* to *ypz* − ε, but the sophisticated approach forces her to choose *ypz* − ε. Advocates of the sophisticated approach respond to this that the dominating plan (up, down) is not feasible and can therefore be ignored.

McClennen (1990) rejects the myopic and sophisticated approaches and proposes what he calls a *resolute* approach. Resolute agents adopt a plan at the first choice node for how to act at all later choice nodes and then stick to the initial plan as long as no new information is received. The plan will typically be conditional: "Go up at the second node if event *E* occurs" and so on. Resolute agents who violate the independence axiom cannot be money-pumped, but a common objection is that resolute choice presupposes perfect self-control (for a discussion, see, e.g., Peterson & Vallentyne, 2018).

6. Causal versus Evidential Decision Theory

Decision theorists disagree on how rational agents should respond to information about causal processes. The *locus classicus* for this debate is a thought experiment first discussed in print by Nozick (1969), who attributes it to the physicist William Newcomb.

Imagine a supercomputer that is very good at predicting people's choices by scanning their brains. You are asked to choose between two boxes. The first box is transparent, and you can see that it contains exactly $1,000. The other box is not transparent. It contains either $0 or $1 million, depending on how the supercomputer predicts you will choose. The operator, whom you have no reason to distrust, tells you that (i) if the supercomputer predicts that you will take only the first box, then the operator will put $1 million in the second box, but (ii) if the supercomputer predicts that you will take both boxes, then the operator will put $0 in the second box. The supercomputer's prediction is highly accurate but not infallible. Of the 1,000 predictions the computer has made so far, the computer made correct predictions 990 times.

You are now asked to choose between the following options:

(a) Take box 1 ($1,000) and box 2 (either $0 or $1 million).

(b) Take only box 2 (either $0 or $1 million).

Note that the supercomputer *first* predicts how you will choose; *then* the operator adjusts the content in box 2 by putting either $1 million or $0 in the box; and *finally*, after these events have already happened, you make your choice. The temporal order is important because it guarantees that your choice *does not causally influence* the content of box 2.

Causal decision theory (CDT) holds that the only rational option is to take both boxes. Because the operator has already put $1 million or $0 in the second box when it is time for you to choose, the decision you make will not affect the content of the second box. If you take both boxes, you will thus get $1,000 in addition to whatever is in the second box. Another way to put this is to say that taking two boxes *dominates* taking one box: no matter what the world is like ($0 or $1 million in the second box), you will be at least as well or better off if you take both boxes (for a detailed defense of CDT, see, e.g., Joyce, 1999).

Advocates of evidential decision theory (EDT) take issue with what they consider to be an overly narrow analysis offered by causal decision theorists. Their point is that the analysis offered by causal decision theorist does not accurately reflect the relevant information available to the agent. In Newcomb's problem, you know that the probability that the supercomputer's prediction is correct is 99%. This holds true even if you take only one box. The choice you make thus *counts as evidence* for what the supercomputer has predicted you will do. The only rational option according to advocates of EDT is to calculate the expected utility such that probabilities are conditional on which act is performed (see table 8.2.3).

To simplify the calculation, we assume that the agent's utility for money is linear. (Note that we could easily modify the prizes to fit any utility function the agent may have.) It can then be easily verified that the expected utility of taking only the second box exceeds that of taking both boxes, meaning that this is the only rationally permissible option.

Which theory do we have most reason to accept? Egan (2007) argues that the following example shows that CDT sometimes has unacceptable implications:

Paul is debating whether to press the "kill all psychopaths" button. It would, he thinks, be much better to live in a world with no psychopaths. Unfortunately, Paul is quite confident that only a psychopath would press such a button. Paul very strongly prefers living in a world *with* psychopaths to dying. Should Paul press the button? (p. 97)

Joyce (2012), an influential advocate of CDT, argues that "CDT does not recommend shooting as the uniquely rational act. . . . Agents who think their way through [this example] from the perspective of CDT will wind up being correctly indifferent between shooting and refraining once they have taken *all their causally relevant information in to account*" (p. 125). It is an open question whether this is a satisfactory response to Egan-style examples. Many people seem convinced that it is not rationally permissible to shoot or press the button (see also Ahmed, 2014, and chapter 7.1 by Pearl, this handbook).

7. Paradoxes and Puzzles

This section reviews some paradoxes and puzzles not mentioned in the previous sections.

7.1 Allais's Paradox

The point of departure for Allais's (1953) paradox, which is not a paradox in a strict philosophical sense but rather an unintuitive implication of the expected utility principle, is the four acts f, g, f', and g' listed in table 8.2.2 in section 4.2. To state the paradox, we calculate the *arithmetic difference* in expected utility between acts f and g, as well as between f' and g'. Since the agent's utility of money is unknown, we write

$$u(f) - u(g) =$$
$$= u(1\text{M}) - [.01 \cdot u(0\text{M}) + .1 \cdot u(5\text{M}) + .89 \cdot u(1\text{M})]$$
$$= .11 \cdot u(1\text{M}) - [.01 \cdot u(0) + .1 \cdot u(5\text{M})],$$
$$u(f') - u(g') =$$
$$= [.11 \cdot u(1\text{M}) + .89 \cdot u(0)] - [.9 \cdot u(0\text{M}) + .1 \cdot u(5\text{M})]$$
$$= .11 \cdot u(1\text{M}) - [.01 \cdot u(0) + .1 \cdot u(5\text{M})].$$

As noted in section 4.2, it seems reasonable to prefer f to g but g' to f', even though this violates Savage's sure-thing principle. However, the difference in expected utility between f and g is exactly the same as that between f'

Table 8.2.3

Newcomb's problem

	$1 million in second box		$0 in second box	
Take second box only	$1 million	(prob. .99)	$0	(prob. .01)
Take both boxes	$1 million + $1,000	(prob. .01)	$1,000	(prob. .99)

and g' regardless of what the agent's utility of money is. Therefore, whenever f is preferred to g, the agent will also prefer f' to g', regardless of the agent's utility of money. This challenge to the sure-thing principle is thus a challenge to the principle of maximizing expected utility. Regardless of how much or little you care about money, you are bound to violate the expected utility principle if you prefer f to g but g' to f'. The "paradox" consists in the fact that there is no way of assigning utility to money that renders the expected utility principle compatible with a preference for f over g and a preference for g' over f'.

The standard response to Allais's paradox is to point out that the description of the outcomes in table 8.2.2 is incomplete. It does not take into account that you will be *disappointed* if you choose act g and do not become a millionaire. A more accurate representation would be the one in table 8.2.4.

Although we do not know exactly what the agent's utility of "$0 and disappointment" is, it seems reasonable to assume that it is less than the utility of $0. Therefore, the difference in expected utility between f and g is not the same as that between f' and g'.

7.2   The St. Petersburg Paradox

The St. Petersburg paradox is one of the oldest and most well-known problems in decision theory. It is not a paradox in a strict philosophical sense (just like the other problems discussed here) but merely a counterintuitive implication of the expected utility principle.

Imagine that a fair coin is tossed n times until it lands heads up for the first time, after which you receive a prize worth 2^n units on your personal utility scale. If, for instance, the coin lands heads on the first toss, you win 2 units, and if you get to toss it five times, you win $2^5 = 32$ units, and so on. How much would a rational agent be willing to pay for an opportunity to playing this game?

The expected utility of the St. Petersburg game is

$$\frac{1}{2} \cdot 2 + \frac{1}{4} \cdot 4 + \frac{1}{8} \cdot 8 + \cdots = 1 + 1 + 1 + \cdots = \infty.$$

Table 8.2.4

Allais's paradox

	Ticket 1	Ticket 2–11	Ticket 12–100
Act f	$1 million	$1 million	$1 million
Act g	$0 and disappointment	$5 million	$1 million
Act f'	$1 million	$1 million	$0
Act g'	$0	$5 million	$0

Hence, a rational agent would, according to the expected utility principle, be willing to sacrifice *all* her assets for the opportunity to play this game once. But this seems absurd. The most likely outcome of the game is that the agent wins 2, 4, or 8 units of utility. Can a remote possibility of winning a huge amount really compensate for a very likely loss of all one's assets?

In Bernoulli's (1738/1954) original formulation of the St. Petersburg paradox, the prizes were gold coins rather than units of utility. The concept of utility was invented by Cramer (1728/1975) as a way of explaining why a rational agent is not required to sacrifice all her assets for the opportunity to play the game. If the agent's marginal value (utility) of gold is decreasing, the expected utility of the gamble is finite. However, although the notion of decreasing marginal utility is still important in many economic analyses, modern philosophers are aware that the paradox can be reformulated in the manner mentioned above, in which the agent wins 2^n units of *utility* rather than some units of gold. This reformulation of the paradox of course presupposes that utility is unbounded (Peterson, 2020).

7.3   The Pasadena Puzzle

The gist of the Pasadena Puzzle is that some gambles seem to have no expected utility at all. Before we state the paradox, it is helpful to recall that some infinite series have no unique sum. Consider, for instance, the alternating harmonic series $\frac{(-1)^{n-1}}{n}$. We can write this as $1 - \frac{1}{2} + \frac{1}{3} - \frac{1}{4} + \frac{1}{5} \pm \cdots$. The sum of this series, added up in this order, is $\ln 2$. However, the Riemann rearrangement theorem entails that the sum of the alternating harmonic series (and other conditionally convergent series) will depend on the *order* in which the terms are summed up. In fact, the terms can be rearranged in a permutation such that the alternating harmonic series converges to any given value, including $+\infty$ and $-\infty$.

To turn this somewhat esoteric mathematical fact into an objection to the expected utility principle, Nover and Hájek (2004) ask us to imagine that you toss a fair coin n times until it lands heads up for the first time. If n is odd, you win $2^n/n$ units of value, but if n is even, you have to *pay* $2^n/n$. Do we have any reason to think that the expected value of the Pasadena gamble is $\sum_n \frac{(-1)^{n-1}}{n} = 1 - \frac{1}{2} + \frac{1}{3} - \frac{1}{4} + \frac{1}{5} \pm \cdots = \ln 2$? Recall that we could also sum up the terms in some other order that would make the series converge to any real number we want. Suppose, for instance, that we write down each term of the sum on separate pieces of paper and drop all of them on the floor before we sum up the terms. Why not just sum up the terms in the order we happen to pick them up from the floor?

Easwaran (2008) responds to this challenge by introducing a distinction between weak and strong expected value. He shows that the Pasadena gamble's (only) weak expected value is ln 2. However, Lauwers and Vallentyne (2016) remind us that not all gambles have any weak expected value. Their suggestion is that we should rather think of the entire interval of all possible values the series can take as its value. However, a problem with this proposal is that it is not clear how we ought to choose between something worth, say, [−∞, +∞] and 0. The interval [−∞, +∞] is clearly not worth as much as 0, because if it were, we could obviously reduce the value of the series to a point value, which would make the interval value redundant.

8. Conclusion

There is persistent disagreement about how the principle of maximizing expected utility should be interpreted and justified. None of the positions reviewed in this chapter appears to be fully satisfactory. However, despite all these problems, it seems likely that the expected utility principle will continue to dominate debates in decision theory for the years to come. There are few, if any, promising alternatives to the expected utility principle. Buchak's (2013) recent modification of the principle is worth considering, but the decision rule she proposes makes the agent vulnerable to pragmatic objections similar to those mentioned in section 5, as she points out herself. Kahneman and Tversky's (1979) prospect theory offers reasonable descriptive predictions, but no one thinks their theory would fare better than the expected utility principle from a normative perspective (see also chapter 8.4 by Hill, this handbook).

Notes

1. Pascal and Bernoulli discuss the principle of maximizing expected *monetary value*. Cramer's contribution was to clearly state the distinction between monetary value and utility. (He used the term "moral value.")

2. The law of large numbers comes in two versions, the "weak" and the "strong" law of large numbers. The version presented here is the weak version.

3. Note that the first edition of von Neumann and Morgenstern's book, published in 1944, does not include their axiomatization of the expected utility principle. The axioms first appeared in an appendix to the second edition published in 1947.

References

Ahmed, A. (2014). *Evidence, decision and causality*. Cambridge, England: Cambridge University Press.

Allais, M. (1953). Le comportement de l'homme rationnel devant le risque: Critique des postulats et axiomes de l'école Américaine [Rational man's behavior in the presence of risk: Critique of the postulates and axioms of the American school]. *Econometrica, 21*, 503–546.

Bernoulli, D. (1954). Specimen theoriae novae de mensura sortis [Exposition of a new theory on the measurement of risk]. *Econometrica, 22*, 23–36. (Original work published 1738)

Blackburn, S. (1998). *Ruling passions*. Oxford, England: Oxford University Press.

Bolker, E. D. (1967). A simultaneous axiomatization of utility and subjective probability. *Philosophy of Science, 34*, 333–340.

Broome, J. (1990). Bolker–Jeffrey expected utility theory and axiomatic utilitarianism. *Review of Economic Studies, 57*, 477–502.

Buchak, L. (2013). *Risk and rationality*. Oxford, England: Oxford University Press.

Cramer, G. (1975). Letter from Cramer to Nicolas Bernoulli. London, 21 May 1728. In B. L. van der Waerden (Ed.), *Die Werke von Jakob Bernoulli: Vol. 3. Wahrscheinlichkeitsrechnung* [The works of Jakob Bernoulli: Vol. 3. Probability calculus]. Basel, Switzerland: Birkhäuser. (Original work published 1728)

Davidson, D., McKinsey, J. C. C., & Suppes, P. (1955). Outlines of a formal theory of value, I. *Philosophy of Science, 22*, 140–160.

Easwaran, K. (2008). Strong and weak expectations. *Mind, 117*, 633–641.

Egan, A. (2007). Some counterexamples to causal decision theory. *Philosophical Review, 116*, 93–114.

Gustafsson, J. E. (2013). The irrelevance of the diachronic money-pump argument for acyclicity. *Journal of Philosophy, 110*, 460–464.

Jeffrey, R. C. (1965). *The logic of decision*. Chicago, IL: University of Chicago Press.

Jeffrey, R. C. (1983). *The logic of decision* (2nd ed.). Chicago, IL: University of Chicago Press.

Joyce, J. M. (1999). *The foundations of causal decision theory*. Cambridge, England: Cambridge University Press.

Joyce, J. M. (2012). Regret and instability in causal decision theory. *Synthese, 187*, 123–145.

Kahneman, D., & Tversky, A. (1979). Prospect theory: An analysis of decisions under risk. *Econometrica, 47*, 263–291.

Keynes, J. M. (2000). *A tract on monetary reform*. Amherst, NY: Prometheus Books. (Original work published 1923)

Lauwers, L., & Vallentyne, P. (2016). Decision theory without finite standard expected value. *Economics and Philosophy, 32*(3), 383–407.

McClennen, E. F. (1990). *Rationality and dynamic choice.* Cambridge, England: Cambridge University Press.

Nover, H., & Hájek, A. (2004). Vexing expectations. *Mind, 113,* 237–249.

Nozick, R. (1969). Newcomb's problem and two principles of choice. In N. Rescher (Ed.), *Essays in honor of Carl G. Hempel: A tribute on the occasion of his sixty-fifth birthday* (pp. 114–146). Dordrecht, Netherlands: Reidel.

Pascal, B. (1967). *Pensées.* New York, NY: Modern Library. (Original work published 1670)

Peterson, M. (2004). From outcomes to acts: A non-standard axiomatization of the expected utility principle. *Journal of Philosophical Logic, 33,* 361–378.

Peterson, M. (2008). *Non-Bayesian decision theory: Beliefs and desires as reasons for action.* Springer.

Peterson, M. (2015). Prospectism and the weak money pump argument. *Theory and Decision, 78,* 451–456.

Peterson, M. (2020). The St. Petersburg paradox. In E. N. Zalta (Ed.), *The Stanford encyclopedia of philosophy.* Retrieved from https://plato.stanford.edu/archives/fall2020/entries/paradox -stpetersburg/

Peterson, M., & Vallentyne, P. (2018). Self-prediction and self-control. In J. Bermudez (Ed.), *Self-control, decision theory, and rationality* (pp. 48–71). Cambridge, England: Cambridge University Press.

Rabinowicz, W. (1995). To have one's cake and eat it, too: Sequential choices and expected-utility violations. *Journal of Philosophy, 92,* 586–620.

Rabinowicz, W. (2000). Money pump with foresight. In M. J. Almeida (Ed.), *Imperceptible harms and benefits* (pp. 123–154). Dordrecht, Netherlands: Kluwer.

Ramsey, F. P. (1931). Truth and probability (original manuscript, 1926). In R. B. Braithwaite (Ed.), *The foundations of mathematics and other logical essays* (pp. 156–198). London, England: Kegan Paul.

Savage, L. J. (1972). *The foundations of statistics* (2nd ed.). Dover, England: Wiley. (Original work published 1954)

Schick, F. (1986). Dutch bookies and money pumps. *Journal of Philosophy, 83,* 112–119.

von Neumann, J., & Morgenstern, O. (1947). *Theory of games and economic behavior* (2nd ed.). Princeton, NJ: Princeton University Press.

8.3 Prospect Theory

Andreas Glöckner

Summary

Prospect theory is one of the most influential descriptive theories of risky choice. In this chapter, the core features of the theory are summarized and the theory is contrasted against theoretical standards of rationality as defined by expected utility theory. Developments of the theory, critical debates, and recent findings are presented. Empirical evidence overall supports prospect theory as a descriptive model of risky choice. The theory can qualitatively account for multiple deviations from theoretical rationality, and it predicts simple risky choices from given information better than heuristics and other competing models. Findings particularly concerning more complex risky choices and decisions from experiences, however, also show limitations of the theory. Prospect theory is mainly considered an as-if model that predicts choices only. Findings concerning the cognitive processes underlying risky choice are mixed and warrant further investigation.

1. Prospect Theory: A Descriptive Theory of Risky Choice

In everyday decisions, individuals often have to select between (more or less) risky options for which outcomes realize with particular probabilities. Here, "risk" refers to the fact that one of several possible states of the world can realize, but it is uncertain which. These states of the world lead to outcomes that differ concerning value or utility for the decision maker. Risky decision can in principle concern any kind of objects, and prospects can range from highly complex and important decisions, such as job selection or company investments, to consumer decisions for any kind of products (e.g., insurances, lottery tickets, health states, food).

According to typical standards of theoretical rationality as, for example, the economic approaches (Becker, 1976; von Neumann & Morgenstern, 1947), such decisions should only be influenced by the subjective utilities of outcomes and the probabilities of states of the

world to realize (for details, see chapter 8.2 by Peterson, this handbook). Typical empirical investigations of risky decisions reduce complexity by focusing on exactly these factors. Research thus often abstracts from all further context factors and focuses on the selection between gambles with outcome and probability information provided. An example would be the choice between gambles A and B for which A pays €4 with certainty and B pays €10 with probability .50 and €0 otherwise. Prospect theory (Kahneman & Tversky, 1979; Tversky & Kahneman, 1992) is the most prominent descriptive theory for decisions between monetary gambles and also for risky choice outside the lab (Camerer, 2005; Kahneman, 2003; Rabin, 2003).

Of particular interest are regularities captured by prospect theory that deviate from the predictions of a strict interpretation of theoretical rationality as introduced above. In this chapter, I review prospect theory and recent findings and discuss relations between prospect theory and rationality.

2. Core Features of Prospect Theory and Important Findings

Prospect theory contains four core features: (i) reference dependence, (ii) diminishing sensitivity of outcomes relative to the reference point, (iii) loss aversion, and (iv) a nonlinear probability weighting function with overweighting of small probabilities and underweighting of large probabilities (figure 8.3.1; more detailed explanations for all four features below).

Specifically, the theory predicts (i) "that the carriers of value are changes in wealth or welfare, rather than final states" (Kahneman & Tversky, 1979, p. 277). This means that individuals perceive outcomes (or final wealth states) not in absolute terms but as changes relative to a reference point. The subjective utility of an option (which Kahneman and Tversky refer to as "value," V) is a function of the gains and losses relative to the reference points and their probabilities. The exact way in which outcomes are transformed into subjective

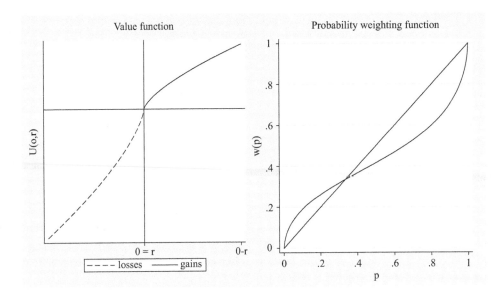

Figure 8.3.1

Prospect theory functions (examples) for transforming outcomes *o* into utilities *U(o,r)*, taking into account a reference point *r* (left), and probabilities *p* into decision weights *w(p)* (right).

utilities is described by a value function with a kink at the reference point (figure 8.3.1, left). It is assumed that outcomes are considered and transformed in isolation, instead of being integrated with other outcomes or probabilities first.

Prospect theory assumes (ii) a concave value function in the gains and a convex value function in the losses, reflecting the principle of diminishing sensitivity of differences from the reference point (figure 8.3.1, left). From this functional form, it follows that individuals should be risk averse in the gain domain and risk seeking in the loss domain.[1]

The theory predicts (iii) that people are loss averse, which means that they are more averse to *losses* relative to their reference level than they are attracted to same-sized gains. This is reflected by the steeper value function in the losses as compared to the gains in figure 8.3.1 (left).

Finally, prospect theory predicts that (iv) outcomes are not weighted by their objective probabilities but by decision weights, which (according to more recent versions of the theory) follow an inverse S-shaped function (i.e., decision weights regress to the mean for extreme probabilities: figure 8.3.1, right; Tversky & Kahneman, 1992). Outcomes with low probabilities receive increased weight (i.e., decision weights are higher than the respective probabilities), whereas outcomes with high probabilities receive reduced weight (i.e., decision weights are lower than the respective probabilities).[2]

Prospect theory can account for a multitude of findings concerning risky choices in the field (for a review,

see Camerer, 2005) and also in the lab. Overweighting of small probabilities, for example, can explain why people buy insurances and lottery tickets although their price is higher than their expected value and/or expected utility. Loss aversion can explain why stock returns are too high relative to bond returns (equity premium) and why people demand higher prices for selling a product they own as compared to their willingness to pay for the same product (endowment effects). Diminishing sensitivity can explain why people sell winning stocks too early (risk aversion in the gain domain) and hold losing stocks for too long (risk seeking in the loss domain) and why they shift to long shots in horse betting at the end of the day.

One of the most influential demonstrations of deviations from rationality predicted by prospect theory are framing effects (Tversky & Kahneman, 1981). In typical framing studies, choice problems are presented in different contexts (frames), while holding the normatively relevant factors (probabilities and outcomes) of the options constant. In the classic Asian disease study, for example, substantial shifts in preferences resulted from presenting two programs to cure an Asian disease either in a loss frame (by mentioning the persons who would be killed) or in a gain frame (by mentioning the remaining persons who would survive). In the loss frame, most people were risk seeking and preferred the risky program, a program in which all people could be killed with some probability. In the gain frame, in contrast, individuals were risk averse and preferred the safe program that assured the survival of some people (Tversky

& Kahneman, 1981). The very robust finding (e.g., Klein et al., 2014) that a mere reframing of what were objectively the same options influences choice, constitutes a violation of invariance that "cannot be defended as normative" (Kahneman, 2003, p. 163).

3.   Normative Theories of Risky Choice and Prospect Theory

Descriptive theories of decision making, such as prospect theory, aim to describe and predict how people make decisions. Normative theories, in contrast, describe how decisions *should* be made and define a standard of rationality against which such descriptive models are typically evaluated. Normative theories have to be defined with respect to the goals that a decision maker aims to achieve. Not taking all goals of a person into account can thus lead to wrong standards (cf. chapter 9.4 by Dhami & al-Nowaihi, this handbook).

3.1   Standard Rational Models: Expected Value and Expected Utility Theory

If the decision maker's only goal is to maximize the average (or sum) of payoffs over an infinite number of decisions, then she should choose the option that maximizes expected value (*EV*). Here, *EV* is defined as the sum of lottery outcomes (payoffs) o_i multiplied by their probabilities p_i for all possible outcomes i: $EV = \Sigma_i p_i o_i$. In the above example, according to *EV*, the risky gamble B ($EV = 0.5 \times €10 = €5$) should be chosen over the safe gamble A ($EV = €4$).

This picture has to be qualified for at least three reasons: utility might not be linearly related to monetary outcomes, individuals may have further or different goals than maximizing the average of payoffs, and decisions are not infinitely repeated. All three issues have been (or at least can be) taken into account in expected utility (EU) theory to develop a normative model for risky choice that goes beyond *EV*.

According to one common implementation of EU theory for monetary gambles, individuals choose the option that maximizes $EU = \Sigma_i p_i o_i^\alpha$, with p_i being the probability of outcome o_i to realize and $0 < \alpha < 1$. In the gamble example introduced above, for a person with a reasonable sensitivity to outcomes (often also referred to as "risk aversion") of $\alpha = 0.7$, choosing the safe gamble A ($EU = 2.64$) over the risky gamble B ($EU = 2.51$) in a one-time decision is rational according to this implementation of EU theory. The form of this *EU* value function is concave, which was originally suggested by Bernoulli (1738/1954) to account for the St. Petersburg paradox.[3] The *EU* value function

considered here is equivalent to the value function of prospect theory when considering the gain domain only (figure 8.3.1, left, upper-right sector). Hence, EU theory assumes a diminishing marginal utility of money. That is, raising the wealth level of an individual by a particular amount (e.g., €1,000) leads to a larger change in utility than adding the same amount once again (e.g., adding another €1,000).

Importantly, in contrast to prospect theory, EU theory does not assume that utility is reference-dependent. Outcomes are assessed relative to a wealth level of zero. As a consequence, according to EU theory, there is no qualitative differentiation between gains and losses, and they should be treated the same (i.e., no reflection of the utility curve at the reference point and therefore no risk seeking in the losses; no loss aversion). Finally, EU theory assumes that probabilities are used without any transformation in weights. The core difference between EV and EU theory is that the latter assumes a value function that transforms outcomes into utilities (in this case by raising them to the power of α).

Individuals often not only want to maximize the sum of outcomes over a long or infinite number of gambles but also have further goals. In some situations, for example, it can be rational to be short-sighted and to think about the current situation in isolation, not taking into account potentially higher future outcomes. It will be reasonable for individuals to avoid falling below a minimum wealth level in order to keep a decent life, and short-sightedness might be extremely rational for an individual to ensure survival. More complex further goals are possible too. Individuals want to maximize joint outcomes, maximize the difference in wealth compared to others (Van Lange, 1999), or avoid sticking out from the crowd. These further goals can, in principle, be accounted for in EU theory by incorporating them (e.g., as further preferences) in individuals' utility functions (see chapter 9.4 by Dhami & al-Nowaihi, this handbook).

3.2   Prospect Theory and Rationality

Tversky and Kahneman (1981) used EU theory as a normative (rational) standard for risky choice by stating, "When faced with a choice, a rational decision-maker will prefer the prospect that offers the highest expected utility" (p. 453). They then demonstrated that behavior systematically deviates from these standards. Prospect theory, a descriptive decision theory, aims to account for individuals' actual choice behavior, which often follows, but in some cases systematically deviates from, these rational standards. The theory does not aim at providing new standards of rationality; quite the contrary, the description and prediction of deviations from these standards lie at the heart of the theory.

4. Theory Development and Versions of Prospect Theory

The original version of prospect theory (Kahneman & Tversky, 1979) contained two phases. In the "editing phase," a preliminary analysis of the offered prospects is conducted. The editing often yields a simpler representation of the prospects, for example, by discarding shared outcomes, combining probabilities of equal outcomes, or coding outcomes as gains and losses. In the subsequent "evaluation phase," the edited prospects are evaluated, and the option with the highest utility is chosen. The original version did not contain a formal description of the probability weighting function. Cumulative prospect theory, which was introduced in 1992, included fully specified functions for outcomes and probability weighting, as depicted in figure 8.3.1 (Tversky & Kahneman, 1992). Furthermore, the specification of the probability weighting function included a specification of outcome-rank-dependent probability weights and a cumulation principle to ensure that probability weights add up to 1. Therefore, probability weights are influenced by the size of the outcome they are attached to relative to the other outcomes.[4] Hence, cumulative prospect theory predicts not only nonlinear probability weighting but also that the same outcome with the same probability receives different weights, depending on the size of the other outcomes, which is in conflict with normative models such as EU theory. In the version suggested in 1992, cumulative prospect theory contains five parameters:

- α: the sensitivity to outcomes in the gains (it determines the curvature of the value function in the gains; figure 8.3.1, left, upper-right sector),
- β: sensitivity to outcomes in the losses (it determines the curvature of the value function in the losses, which is additionally scaled by λ; figure 8.3.1, left, lower-left sector),
- λ: loss aversion (it captures the relative increase in sensitivity to losses, that is, the factor by which losses loom larger than equal gains, when assuming $\alpha = \beta$),
- γ: probability weighting in gains (it determines the shape of the probability weighting function for gains, that is, the sensitivity to probabilities in gains),
- δ: probability weighting in losses (it determines the shape of the probability weighting function for losses, that is, the sensitivity to probabilities in losses).

Since the exact formalization of cumulative prospect theory is not of particular interest for the current chapter, it is not provided here, and I refer the reader to Tversky and Kahneman (1992) for a description of the full formalization of the theory.

Further extensions concerned, for example, modifications to the probability weighting function (for an overview, see Stott, 2006); the introduction of a two-stage model of prospect theory, taking into account in a first step probability estimation in decisions under uncertainty (i.e., without explicit probability information available) based on support theory (Fox & Tversky, 1998); and an extension that allows reference points to be uncertain too (i.e., lotteries as reference points; Schmidt, Starmer, & Sugden, 2008). Overall, prospect theory inspired several extensions and theoretical developments, but the core features generally remained (for an excellent overview, see Wakker, 2010).

5. Empirical Findings

The capability of prospect theory to account for risky choice behavior has been critically investigated using various methodological approaches. A full review would be beyond the scope of this chapter. I will exemplify some of the methodologies, debates, and crucial findings.

5.1 Predictive Ability

In the original paper, Kahneman and Tversky (1979) presented results for 12 selected decision problems in which the majority of participants chose the option that was predicted by prospect theory and contrary to EU theory. However, these problems were explicitly constructed to disprove EU theory. It remained unclear whether prospect theory is indeed a good theory for predicting choices also in a less selective set of decision problems. Furthermore, it has been argued that cumulative prospect theory (due to its free parameters) might only be able to fit data post hoc but that simpler models, such as the priority heuristic,[5] might be better at predicting choice, due to the overfitting of prospect theory (Brandstätter, Gigerenzer, & Hertwig, 2006).

We put this question to a direct test in an adversarial collaboration with a proponent of the heuristic perspective (Glöckner & Pachur, 2012). Participants made incentivized decisions in a large and diverse set of gamble problems at two points in time. Importantly, the set of gamble problems was intensely discussed a priori to avoid giving any model a strategic advantage. The competing models included various versions of prospect theory, EU theory, EV theory, and all existing simple heuristics that we were aware of. All models were fit to the data at time 1 to predict choices at time 2. The results showed that neither priority heuristic nor any of the other heuristics—or their adaptive use as specified in some models—could predict choices nearly as well as even a simple EV theory. EV theory in turn predicted significantly worse than EU theory. The best predictions by far were achieved by cumulative prospect theory. Among different versions, the model without any free parameters using Kahneman and Tversky's parameter values from 1992 did fairly well, but individual-level fitting improved predictive ability further and close to the maximum level possible given limited reliability in choices.

Overall, this and many further investigations, including analyses in applied settings (Camerer, 2005), show that prospect theory is a model well suited for describing risky choice behavior. Given this excellent performance as a descriptive theory, prospect theory seems particularly relevant as a starting point for reflecting about extended standards of practical rationality. Prominent alternative models seem less relevant for deriving such extended standards, considering their typically worse performance as descriptive models. For example, given that no single heuristic (simplified decision strategy) and no model of adaptive strategy usage exists that can predict behavior in the most basic gambling paradigm nearly as well as prospect theory (Glöckner & Pachur, 2012), such models might be less useful for deriving extended standards of rationality (but see Gigerenzer & Brighton, 2009, for a different position).

As a limitation, however, it remains unclear whether this superiority of cumulative prospect theory in predicting simple choices between two lotteries with two outcomes each and explicitly provided probabilities generalizes to more complex risky choice tasks, for which heuristics might be more prevalent (see below).

5.2 Critical Property Testing

A second approach to test cumulative prospect theory against alternative models is to investigate critical properties of the theory that are predicted under any set of parameters. Birnbaum conducted intense empirical investigations following this approach, particularly taking into account more complex lotteries with three outcomes. He demonstrated multiple critical choice anomalies that prospect theory could not account for but alternatively suggested models such as the transfer of attention exchange model (Birnbaum, 2006, 2008). Birnbaum showed that splitting the most (least) attractive branches of a gamble makes them more (less) attractive and can lead to preference reversals that prospect theory cannot account for.

For example, most individuals prefer gamble A: .85 to win $100 and .15 to win $50, over gamble B: .95 to win $100 and .05 to win $7 (Birnbaum, 2008, Table 1). This (majority) preference, however, reverses if the less attractive outcome is split for gamble A (.85, $100; .10, $50; .05, $50) and the more attractive outcome is split for gamble B (.85, $100; .10, $100; .05, $7), although the games remain substantially equivalent.

Similarly, it was shown that in gambles with multiple winning and losing outcomes, choices are more in line with a p-win heuristic (i.e., choose the option that maximizes the overall probability of winning outcomes) and could not be accounted for well by prospect theory (Payne, 2005). These investigations show limitations of prospect theory concerning its ability to account for more complex decision tasks (i.e., ones with more than two outcomes). Given the higher task complexity, it is psychologically plausible that behavior is less in line with prospect theory in more complex tasks due to limitations of deliberate cognitive capacity. Further investigations are needed to test whether the average predictive ability of prospect theory in a broad set of more complex tasks falls below the performance of competing theories.

5.3 Decisions from Experience

The standard paradigm for investigating prospect theory relies on decisions from descriptions, that is, decisions between gambles with explicitly provided probabilities (see examples above). Many everyday decisions, however, involve decisions from experience, in which probabilities are not explicitly provided but have to be inferred. This has led to a large literature on ambiguity (unknown probabilities) in the economic literature (Ellsberg, 1961), surveyed by Trautmann and van de Kuilen (2015) and Wakker (2010). I will focus on another stream of literature, initiated by Erev, Hertwig, and colleagues (Barron & Erev, 2003; Hertwig, Barron, Weber, & Erev, 2004). These authors showed substantial differences between decisions from experience and decisions from description, the so-called description–experience gap. In typical studies on decisions from experience (Hertwig et al., 2004), individuals are allowed to sample outcomes as long as they wish, before making a consequential choice.[6] For example, an individual might sample five outcomes from gamble A (€4, €0, €4, €4, €4) and five outcomes from gamble B (€3, €3, €3, €3, €3) and then decide to stop search and select gamble A. Hertwig et al. (2004) showed, for example, that for a choice between gamble A: .8, €4; .2, €0, and gamble B: €3, 1.0 (i.e., with certainty), most individuals choose gamble B in decisions from descriptions but gamble A in decisions from experience.

One important factor that contributes to this difference in choice behaviors are sampling biases. Of particular importance here is the bias that results from the fact that rare events are more likely to be underrepresented in small samples due to the skewness of the binomial distribution. That is, small samples have a higher likelihood of containing too *few* rare events than of containing too many.[7]

It has been controversially debated whether a new theory has to be developed for decisions from experience (Hertwig & Erev, 2009) or whether choices can be explained by sampling biases plus prospect theory (Fox & Hadar, 2006). While the former perspective assumes that cognitive processes might be entirely different in decisions from experience, proponents of the latter perspective assume that individuals merely sample outcomes to estimate probabilities. These estimates are, however, influenced by sampling biases, which in turn lead to different choices in experience-based decisions.

A large-scale prediction competition (Erev et al., 2010) showed that a psychologically not very plausible "ensemble" (mixture) of models predicted choices in experience-based decision making best. Several further process models were developed and tested (Barron & Erev, 2003; Gonzalez & Dutt, 2011; Lejarraga, Dutt, & Gonzalez, 2012). A more recent model comparison, taking into account the data from the Erev prediction competition and additional data (Glöckner, Hilbig, Henninger, & Fiedler, 2016), provided support for prospect theory as a model for decisions from experience too. It was found that a two-stage model (cf. Fox & Hadar, 2006; Fox & Tversky, 1998), assuming (i) probability estimations that are influenced by sampling biases and (ii) an application of prospect theory based on these probability estimations, could explain choice behavior very well and better than important alternative process models (Gonzalez & Dutt, 2011; Lejarraga et al., 2012).

Furthermore, it was shown that the initially assumed direction of the description–experience gap, a qualitative difference in that rare events are overweighted in description but underweighted in experience, does not hold universally (Glöckner et al., 2016; Kellen, Pachur, & Hertwig, 2016). The assumed gap is highly problem dependent, as confirmed by a large-scale meta-analysis (Wulff, Mergenthaler-Canseco, & Hertwig, 2018). Specifically, after accounting for sampling bias[8] and when taking into account a more representative sample of tasks, the description–experience gap reverses (Glöckner et al., 2016; Kellen et al., 2016). That is, when considering the outcomes that people have truly experienced by sampling, they overweight small probabilities even more strongly in decisions from experience than in decisions from

description. This finding is in line with much research showing that smaller probabilities tend to be overestimated due to a universal regression-to-the-mean effect (K. Fiedler & Unkelbach, 2014).

Overall, this research highlights that sampling biases and also further (partially task-dependent) factors have to be taken into account in modeling decisions from experience. Given the complexity of the findings (for an excellent summary, see Wulff et al., 2018), it is up to further research whether adaptations to cumulative prospect theory are sufficient to account for decisions from experience or whether qualitatively different models have to be developed. In a first approximation, a model based on sampling bias and cumulative prospect theory does surprisingly well also in predicting decisions from experience (in the sampling paradigm).

One important distinction in experience-based decision making concerns the differentiation between the sampling paradigm (discussed so far) and a feedback paradigm. In the feedback paradigm, each sampled outcome from a gamble is paid, whereas in the sampling paradigm, only the outcome of the final choice is paid.

Considerable differences between the two paradigms, as well as between them and the standard decision-from-description paradigm, indicate that dependent on the requirements of the situation, people do not focus on the current choice task only. Instead, individuals seem to adapt—at least partially—to whether adopting a narrower or a broader perspective, by taking into account the outcomes of further decisions, is appropriate. This should be considered in future theoretical developments.

5.4 Process Measures

Aside from the analysis of choice behavior, further measures of, for example, information search and/or response time are used to test hypotheses concerning the underlying cognitive processes (Schulte-Mecklenbeck, Kühberger, & Johnson, 2019). Various such process measures have been applied to risky choice to determine whether and how prospect theory might be implemented computationally and whether it is a reasonable model at all. Similarly to the classic bounded rationality critique of EU theory (Simon, 1955), it is unreasonable to assume that individuals deliberately apply complex multiplication operations to implement prospect theory. Hence, prospect theory is usually considered an as-if model instead of a process model.

Classic studies on processes in risky choice (Payne, Bettman, & Johnson, 1988) used the computer-based information search paradigm Mouselab. In Mouselab, individuals can reveal pieces of information (outcomes and probabilities) concerning multiple complex lotteries by moving the mouse to the respective information boxes. Payne et al. (1988) have demonstrated that in complex risky choice paradigms (multiple options with multiple outcomes each), manipulations of the environmental structure and time pressure induce changes in information search. This indicates that individuals also change their decision strategies. Payne et al. show that these changes are adaptive to the structure of the environment, in that individuals tend to change to strategies that are more successful in the respective environments.

The findings by Payne et al. preclude that (all) individuals search for all pieces of information and use prospect theory to evaluate them. Results can be better explained by assuming that individuals are adaptive decision makers and switch, for example, from more complex to simpler and less effortful strategies (i.e., simple heuristics that ignore large parts of the information) dependent on task demands (Payne, Bettman, & Johnson, 1993). These results converge with the findings from critical property testing (see section 5.2) indicating that prospect theory's ability to account for choices in more complex tasks might be limited.

Eye-tracking studies provide further insights concerning the potential processes underlying risky choices in less complex tasks. In these studies (S. Fiedler & Glöckner, 2012; Franco-Watkins & Johnson, 2011; Glöckner & Herbold, 2011), it was repeatedly shown that individuals (i) look up all pieces of information, (ii) mainly search information *within* gambles (instead of comparing between gambles), and (iii) show choices that are in line with a compensatory integration of outcomes and probabilities. Thereby, (iv) response times are too short to allow for an integration of outcomes and probabilities by deliberate calculation. Also, (v) individuals look up information with short fixations, while deliberate calculations are accompanied by long fixations. These findings are hard to reconcile with the assumption that—even for simple risky choice tasks—weighted sums of (transformed) outcomes and probabilities are deliberately calculated. The processes might be better accounted for by evidence accumulation models (Busemeyer & Townsend, 1993; Usher & McClelland, 2004) or parallel constraint satisfaction models (Glöckner & Herbold, 2011; Holyoak & Simon, 1999), which allow approximating weighted integration of probabilities and outcomes by psychologically plausible processes.

6. Summary and Future Directions

Prospect theory is arguably the most influential descriptive model for risky choice. Prospect theory assumes (i) reference dependence, (ii) diminishing sensitivity of outcomes relative to the reference point, (iii) loss aversion, and (iv) a nonlinear probability weighting function with overweighting of small probabilities and underweighting of large probabilities. Prospect theory predicts multiple deviations from expected utility theory, the standard normative model of theoretical rationality. These predictions have received ample support in the lab, and the theory can explain deviations from theoretical rationality that are observed in the field.

As a descriptive model of choice, prospect theory does not aim to define new standards of rationality. Quite the contrary, Kahneman and Tversky developed the theory to (among other things) account for deviations from these standards, such as violations of the invariance principle as demonstrated by framing effects.

A review of recent findings and debates indicated that prospect theory is still one of the best models for predicting simple decisions from descriptions. For complex tasks, the evidence is more mixed. In risky choice between options with multiple outcomes, evidence from critical property testing and process tracing converges to show that individuals rely on different, potentially more heuristic processes. For decision from experience, sampling biases in probability estimation and subsequent application of prospect theory can account for choice behavior surprisingly well. Still, in both areas, more research is needed to clarify the issues and to develop and comparatively test improved models.

Process measures for simple decisions from descriptions indicate that individuals do *not* apply prospect theory by deliberately calculating weighted sums of (transformed) outcomes and probabilities—even if choices are in line with the theory's predictions. This highlights the status of prospect theory as an as-if model. To integrate (transformed) outcomes and probabilities, individuals seem to rely on qualitatively different processes, which also have to be explored in future research.

Notes

1. When taking into account the value function and the probability weighting function (described below) simultaneously, a more complex fourfold pattern of risk attitudes emerges (i.e., risk aversion for gains and risk seeking for losses of high probability, as well as risk seeking for gains and risk aversion for losses of low probability; Tversky & Kahneman, 1992).

2. According to cumulative prospect theory (Tversky & Kahneman, 1992), decision weights are furthermore dependent on the rank of the outcome, and it is assured that they add up to 1 (details below and in note 4).

3. The paradox is that individuals are not willing to pay a particularly high amount of money for playing the following St. Petersburg game, which has an infinitely large *EV* (for a detailed discussion, see also chapter 8.2 by Peterson, this handbook): a coin is flipped repeatedly until heads appears. If, on the first coin flip, tails appears, the person receives €1 (otherwise zero). The payment amount doubles with each further coin flip landing on tails. When heads appears, the game ceases and the player receives the amount determined at the previous coin flip (i.e., only this one amount is paid). See also section 7.2 of chapter 8.2 by Peterson (this handbook).

4. This is realized by outcomes being sorted by magnitude separately for gains and losses. The decision weight for the highest outcome in the gains (and the lowest outcome in the losses) is directly determined according to the probability weighting function w (figure 8.3.1, right), whereas the weights of the subsequent outcomes are determined as differences in weights. For example, consider a gamble with three outcomes O_1: .20, €20; O_2: .40, €10; O_3: .40, €5. The probability weight of O_1 is the weight $w(p_1)$ with $p_1 = .20$ being transformed according to function w. The weight for O_2 is the difference between the weights for $w(p_1 + p_2)$ with $p_1 + p_2 = .60$ and $w(p_1)$. The weight for O_3 is the difference between weights $w(p_1 + p_2 + p_3)$ with $p_1 + p_2 + p_3 = 1$ and $w(p_1 + p_2)$. Note that in this example, the decision weight for O_2 will typically be lower than for O_3, although both outcomes have the same probability of .40.

5. According to the priority heuristic, individuals apply a stepwise noncompensatory comparison of specific attributes of gambles, namely, (i) worst outcomes, (ii) probabilities of worst outcomes, and (iii) best outcomes with particular difference thresholds.

6. This describes the sampling paradigm. In the also-often-used feedback paradigm, the outcome of each draw is paid (Barron & Erev, 2003). Differences are discussed at the end of this section.

7. In our example, the likelihood of the rare event from gamble *A* (.2, €0) to be underrepresented (i.e., experiencing €0 in *less* than 2 out of 10 samples) is .38, while the likelihood of it being overrepresented in the sample (i.e., experiencing €0 in *more* than 2 out of 10 samples) is .32, according to the binomial distribution.

8. Realized, in this case, by considering only the outcomes that people had indeed experienced and not the hypothetical probabilities of the gambles.

References

Barron, G., & Erev, I. (2003). Small feedback-based decisions and their limited correspondence to description-based decisions. *Journal of Behavioral Decision Making, 16*(3), 215–233.

Becker, G. S. (1976). *The economic approach to human behavior.* Chicago, IL: University of Chicago Press.

Bernoulli, D. (1954). Exposition of a new theory on the measurement of risk. *Econometrica, 22,* 23–36. (Original work published 1738)

Birnbaum, M. H. (2006). Evidence against prospect theories in gambles with positive, negative, and mixed consequences. *Journal of Economic Psychology, 27*(6), 737–761.

Birnbaum, M. H. (2008). New paradoxes of risky decision making. *Psychological Review, 115*(2), 463–501.

Brandstätter, E., Gigerenzer, G., & Hertwig, R. (2006). The priority heuristic: Making choices without trade-offs. *Psychological Review, 113*(2), 409–432.

Busemeyer, J. R., & Townsend, J. T. (1993). Decision field theory: A dynamic-cognitive approach to decision making in an uncertain environment. *Psychological Review, 100*(3), 432–459.

Camerer, C. F. (2005). Prospect theory in the wild: Evidence from the field. In M. H. Bazerman (Ed.), *Negotiation, decision making and conflict management* (Vols. 1–3, pp. 575–588). Northampton, MA: Elgar.

Ellsberg, D. (1961). Risk, ambiguity, and the Savage axioms. *Quarterly Journal of Economics, 75*(4), 643–669.

Erev, I., Ert, E., Roth, A. E., Haruvy, E., Herzog, S. M., Hau, R., . . . Lebiere, C. (2010). A choice prediction competition: Choices from experience and from description. *Journal of Behavioral Decision Making, 23*(1), 15–47.

Fiedler, S., & Glöckner, A. (2012). The dynamics of decision making in risky choice: An eye-tracking analysis. *Frontiers in Psychology, 3,* 335.

Fiedler, K., & Unkelbach, C. (2014). Regressive judgment: Implications of a universal property of the empirical world. *Current Directions in Psychological Science, 23*(5), 361–367.

Fox, C. R., & Hadar, L. (2006). "Decisions from experience" = sampling error + prospect theory: Reconsidering Hertwig, Barron, Weber & Erev (2004). *Judgment and Decision Making, 1*(2), 159–161.

Fox, C. R., & Tversky, A. (1998). A belief-based account of decision under uncertainty. *Management Science, 44*(7), 879–895.

Franco-Watkins, A. M., & Johnson, J. G. (2011). Applying the decision moving window to risky choice: Comparison of eye-tracking and mousetracing methods. *Judgment and Decision Making, 6*(8), 740–749.

Gigerenzer, G., & Brighton, H. (2009). Homo heuristicus: Why biased minds make better inferences. *Topics in Cognitive Sciences, 1,* 107–143.

Glöckner, A., & Herbold, A.-K. (2011). An eye-tracking study on information processing in risky decisions: Evidence for compensatory strategies based on automatic processes. *Journal of Behavioral Decision Making, 24,* 71–98.

Glöckner, A., Hilbig, B. E., Henninger, F., & Fiedler, S. (2016). The reversed description–experience gap: Disentangling sources of presentation format effects in risky choice. *Journal of Experimental Psychology: General, 145*(4), 486–508.

Glöckner, A., & Pachur, T. (2012). Cognitive models of risky choice: Parameter stability and predictive accuracy of prospect theory. *Cognition, 123*(1), 21–32.

Gonzalez, C., & Dutt, V. (2011). Instance-based learning: Integrating sampling and repeated decisions from experience. *Psychological Review, 118*(4), 523–551.

Hertwig, R., Barron, G., Weber, E. U., & Erev, I. (2004). Decisions from experience and the effect of rare events in risky choice. *Psychological Science, 15*(8), 534–539.

Hertwig, R., & Erev, I. (2009). The description–experience gap in risky choice. *Trends in Cognitive Sciences, 13*(12), 517–523.

Holyoak, K. J., & Simon, D. (1999). Bidirectional reasoning in decision making by constraint satisfaction. *Journal of Experimental Psychology: General, 128*(1), 3–31.

Kahneman, D. (2003). A psychological perspective on economics. *American Economic Review, 93*(2), 162–168.

Kahneman, D., & Tversky, A. (1979). Prospect theory: An analysis of decision under risk. *Econometrica, 47*, 263–292.

Kellen, D., Pachur, T., & Hertwig, R. (2016). How (in)variant are subjective representations of described and experienced risk and rewards? *Cognition, 157*, 126–138.

Klein, R. A., Ratliff, K. A., Vianello, M., Adams, R. B., Jr., Bahník, Š., Bernstein, M. J., . . . Nosek, B. A. (2014). Investigating variation in replicability: A "many labs" replication project. *Social Psychology, 45*(3), 142–152.

Lejarraga, T., Dutt, V., & Gonzalez, C. (2012). Instance-based learning: A general model of repeated binary choice. *Journal of Behavioral Decision Making, 25*(2), 143–153.

Payne, J. W. (2005). It is whether you win or lose: The importance of the overall probabilities of winning or losing in risky choice. *Journal of Risk and Uncertainty, 30*(1), 5–19.

Payne, J. W., Bettman, J. R., & Johnson, E. J. (1988). Adaptive strategy selection in decision making. *Journal of Experimental Psychology: Learning, Memory, and Cognition, 14*(3), 534–552.

Payne, J. W., Bettman, J. R., & Johnson, E. J. (1993). *The adaptive decision maker*. Cambridge, England: Cambridge University Press.

Rabin, M. (2003). The Nobel Memorial Prize for Daniel Kahneman. *Scandinavian Journal of Economics, 105*(2), 157–180.

Schmidt, U., Starmer, C., & Sugden, R. (2008). Third-generation prospect theory. *Journal of Risk and Uncertainty, 36*(3), 203–223.

Schulte-Mecklenbeck, M., Kühberger, A., & Johnson, J. G. (Eds.). (2019). *A handbook of process tracing methods* (2nd ed.). New York, NY: Routledge.

Simon, H. A. (1955). A behavioral model of rational choice. *Quarterly Journal of Economics, 69*(1), 99–118.

Stott, H. P. (2006). Cumulative prospect theory's functional menagerie. *Journal of Risk and Uncertainty, 32*(2), 101–130.

Trautmann, S. T., & van de Kuilen, G. (2015). Ambiguity attitudes. In G. Keren & G. Wu (Eds.), *The Wiley Blackwell handbook of judgment and decision making* (Vol. 1, pp. 89–116). Oxford, England: Blackwell.

Tversky, A., & Kahneman, D. (1981). The framing of decisions and the psychology of choice. *Science, 211*(4481), 453–458.

Tversky, A., & Kahneman, D. (1992). Advances in prospect theory: Cumulative representation of uncertainty. *Journal of Risk and Uncertainty, 5*, 297–323.

Usher, M., & McClelland, J. L. (2004). Loss aversion and inhibition in dynamical models of multialternative choice. *Psychological Review, 111*(3), 757–769.

Van Lange, P. A. M. (1999). The pursuit of joint outcomes and equality in outcomes: An integrative model of social value orientation. *Journal of Personality and Social Psychology, 77*(2), 337–349.

von Neumann, J., & Morgenstern, O. (1947). *Theory of games and economic behavior*. Princeton, NJ: Princeton University Press.

Wakker, P. P. (2010). *Prospect theory: For risk and ambiguity*. Cambridge, England: Cambridge University Press.

Wulff, D. U., Mergenthaler-Canseco, M., & Hertwig, R. (2018). A meta-analytic review of two modes of learning and the description–experience gap. *Psychological Bulletin, 144*(2), 140–176.

8.4 Decision under Uncertainty

Brian Hill

Summary

A series of famous examples casts doubt on the standard, Bayesian account of belief and decision in situations of considerable uncertainty. They have spawned a significant literature in economics and, to a lesser extent, philosophy. This chapter surveys some of this literature, with an emphasis on the normative issue of rational decision.

1.   The Challenge of Uncertainty

Imagine that you are faced with two urns, each containing only black and white balls. For one of the urns (the unknown urn), that is all you know; for the other (the known urn), you have counted the balls and know that exactly half are black. What would you say if asked for the probability that the next ball drawn from the known urn was black? And what about the unknown urn? And, if you had to place a bet on the next ball drawn from one of the urns being black, would you bet on the known or the unknown urn? And what if the bets were on the next ball being white?

This example crops up in several disciplines. Keynes (1921/2004) uses it to motivate his notion of *weight of evidence*: "It is evident that in either case the probability of drawing a white ball is ½, but that the weight of the argument in favor of this conclusion is greater [for the known urn]" (p. 75). Replacing direct observation of the composition of the known urn with sampling yields something close to Popper's "paradox of ideal evidence." In economics, the example is associated with Ellsberg (1961) and has spawned a significant literature on *decision under uncertainty*[1] over the past 30 years.

Its importance lies in the challenge it poses to the standard account of rational belief and decision, Bayesianism (see chapter 8.2 by Peterson, this handbook), and in particular to

- the **Bayesian thesis about rational belief**: It can be represented by a function assigning a single number

(between 0 and 1) to each proposition or event, which satisfies the laws of probability;

- and the **Bayesian thesis about rational decision**: The chosen action in any decision is that which maximizes the *expected utility* or *desirability* on the basis of the agent's beliefs.

Although not formulated as such, Keynes's and Popper's points appear to challenge the first thesis: representing belief by probabilities cannot, allegedly, capture the weight of evidence supporting a belief. As often noted, this argument is not watertight: there are several differences between one's Bayesian probabilities concerning the two urns, including one's beliefs about their composition or the robustness of beliefs to further observations (e.g., Joyce, 2005).

By contrast, Ellsberg's version involves decision. He observed a tendency (borne out in subsequent experiments; Camerer & Weber, 1992) to prefer betting on the ball drawn from the known urn over betting on the one drawn from the unknown urn, whatever the color—a pattern of behavior that has come to be known as *uncertainty aversion* or *ambiguity aversion*. (The field often qualifies the unknown urn as involving *ambiguity* or *Knightian uncertainty*.[2]) Such preferences conflict with the Bayesian theses. A Bayesian would choose to bet on black from the known urn (call this event B_k) over black in the unknown urn (B_u) only if $p(B_k) > p(B_u)$, where p represents her beliefs. However, she would have the same preference over bets on white only if $p(W_k) > p(W_u)$. Clearly, no probability function p can satisfy both inequalities. So the Bayesian must condemn these so-called Ellsberg preferences as irrational. Some find this drastic. This opens up the question of what more permissive account should or can replace it.

Although distinct, the points concerning belief representation and decision are intimately related. The difference in your ignorance about the two urns—or in the "weight of evidence" in the two cases—seems to justify a preference between them (Ellsberg, 1961; Levi, 1986; Gilboa, Postlewaite, & Schmeidler, 2009). Indeed, Ellsberg's point about decision adds a twist to the debate about weight of evidence: whatever ways there are of capturing something like weight in the framework of Bayesian probability, they remain irrelevant for choice. Indeed, the two criticisms are perhaps strongest when combined: *Bayesianism*, it appears, *reserves no role for the weight of evidence (or similar factors) in choice.*

This survey will present some of the main responses to this challenge, particularly in the economic field of decision under uncertainty, but also in philosophy. While the focus will be on decision, the relationship with belief representation means that this issue cannot be ignored. Indeed, it will structure the survey: different belief representations (replacing Bayesian probabilities) naturally require different, although not unrelated, decision rules (replacing expected utility). The emphasis will be solely on rational decision, although the reader should bear in mind that some proposals were developed with other goals in mind (e.g., tractability for economic modeling or descriptive accuracy). For ease of exposition, we shall largely eschew technicalities, at times abstracting liberally from precise formulations to focus on the gist. For details, discussions of dimensions other than the normative one, or more on the relevant economic literature, the reader is referred to excellent and more complete existing surveys such as Gilboa and Marinacci (2013) and Machina and Siniscalchi (2014).

Table 8.4.1
Terminology and definitions

Terminology	Explanation
S	The set of states of the world (state space)
X	The set of consequences (consequence space)
f, g, etc.	Acts (objects of choice): functions from S to X
E, F, etc.	Events: subsets of S
Δ	The set of probability measures over the state space S (probability space)
$EU_p f$	The expected utility of act f calculated with probability measure $p \in \Delta$ and utility function u: $EU_p f = \sum_{s \in S} p(s)u(f(s))$
$1, 0, x, \dots$	Consequence yielding utility value $1, 0, x, \dots$
$1_E 0$	Bet yielding consequence **1** if event E occurs and consequence **0** otherwise
The functional V represents the preference relation $\succcurlyeq$.	For every pair of acts f, g: $f \succcurlyeq g$ if and only if $V(f) \geq V(g)$.

Table 8.4.1 lists the main terminology and definitions used throughout the chapter, which are standard in the economic branch of decision theory (although little depends on the use of this framework). For simplicity, we assume a fixed utility function throughout the chapter and use **1**, **0**, **x** to refer to consequences yielding utility 1, 0, x, respectively, according to this function. We assume sets to be finite, as far as possible (so as to use sums instead of integrals). Some notions are illustrated on the previous example in table 8.4.2. The columns correspond to the states of the world, each specifying the resolution of all relevant uncertainty (the color of the ball drawn from each urn). The rows correspond to the acts—in this case, bets yielding 1 unit of utility if won and 0 otherwise. The entries in the table are the consequences obtained under each act and in each state (e.g., if you bet on B_k and the state B_k & B_u realizes, you obtain a consequence worth 1 utility unit).

2. Multiple Priors

A popular reaction to the opening example focuses on the *precision* of the Bayesian representation of beliefs. In the unknown urn, there is no evidence to justify a particular call on the probability of black on the next draw. Yet, by insisting that a precise probability value must be assigned to this event, Bayesianism has—the thought goes—no way of expressing this ignorance. A natural move is thus to allow imprecision in valuations, by using *sets* of probability measures as representations of belief. Bayesianism amounts to the special case where the agent's set is a singleton. Such a representation has been discussed and defended under names such as *credal sets* in philosophy (Levi, 1986; Joyce, 2010) and *imprecise probabilities* in statistics (Walley, 1991); in economics, one speaks of *multiple priors* or *sets of priors*.

How could or should one decide on the basis of a set $\mathcal{C}$ of priors? One of the earliest theories in economics— *maximin Expected Utility* or *maximin-EU*, developed by Gilboa and Schmeidler (1989)[3]—looks at the worst-case expected utility across the set; that is, it considers the representation of preferences by

$$\min_{p \in \mathcal{C}} EU_p f. \tag{1}$$

The rule is *cautious* or *pessimistic*, insofar as it bases preferences on the worst the act can do, according to the probability measures in the set $\mathcal{C}$. As such, it can straightforwardly account for the Ellsberg preferences. (The interested reader may verify this using the set $\mathcal{C}^* = \{p \in \Delta: 0.3 \leq p(B_u) \leq 0.7; p(B_k) = 0.5\}$.)

Table 8.4.2

Ellsberg choices

	B_k & B_u (Black ball drawn from both urns)	B_k & W_u (Black drawn from known urn, white from unknown urn)	W_k & B_u	W_k & W_u
$1_{B_k}0$ (bet on B_k)	1	1	0	0
$1_{B_u}0$ (bet on B_u)	1	0	1	0
$1_{W_k}0$ (bet on W_k)	0	0	1	1
$1_{W_u}0$ (bet on W_u)	0	1	0	1

One purportedly restrictive aspect of this rule is the focus on the worst case; an apparently less extreme alternative is the *α-maximin-EU* or *Hurwicz criterion*, along the lines suggested by Hurwicz (1951; Jaffray, 1989):

$$\alpha \min_{p \in \mathcal{C}} EU_p f + (1 - \alpha) \max_{p \in \mathcal{C}} EU_p f, \qquad (2)$$

where α is a number between 0 and 1. This rule contains maximin-EU as a special case (where $\alpha = 1$) but goes beyond the arguably extreme case of total caution by taking into consideration both the worst and the best case. Other generalizations in this direction allow the α to depend more or less strongly on the act f being evaluated (Ghirardato, Maccheroni, & Marinacci, 2004).

Another approach focuses on what an agent can say "for sure" on the basis of her set of priors. Under a popular proposal, a preference for one act over another is formed only if all probability measures in the set "agree" that the former has higher expected utility. In other words, for all acts $f, g, f \succcurlyeq g$ if and only if

$$EU_p f \geq EU_p g, \qquad \text{for all } p \in \mathcal{C}. \qquad (3)$$

This *unanimity rule* has been defended in economics (Bewley, 1986), in statistics under the name *maximality* (Walley, 1991), as well as by several philosophers (S. Bradley & Steele, 2016).[4] Note that it does not order certain pairs of acts: the reader is invited to check, for instance, that it does not order the Ellsberg bets under the set of priors $\mathcal{C}^*$ above. Related rules include Levi's (1986) E-admissibility, which picks out acts that are best according to at least one probability measure in the set.

Beyond intuitive remarks about their reasonableness, how can decision rules such as these be evaluated on normative grounds? Two potentially relevant families of results have been developed in the field; we shall illustrate them on maximin-EU.

2.1 Evaluation 1: Implications for Choice

One way of evaluating the normative credentials of a decision rule is in terms of its *implications for choice*. If it leads to unpalatable choices, such as choosing to obtain a sure loss, then this provides good reason for skepticism; if its consequences for choice seem sensible, then this may provide arguments in its favor. Any particular choice results from the combination of the decision rule (such as (1)) and the agent's attitudes (the set of priors and utility function). To focus evaluation on the rule, decision theory considers the implications it has *no matter* the agent's attitudes. For instance, the maximin-EU rule generates transitive preferences (see below) for any $\mathcal{C}$ and u used. The central results in decision under uncertainty—*representation theorems*—fully characterize the implications of the decision rule for choice. That is, they provide necessary and sufficient conditions on preferences—called *axioms*—for there to be some specification of the agent's attitudes that represents them according to the rule. Figure 8.4.1 presents the general schema on maximin-EU (for details, see Gilboa, 2009; Gilboa & Marinacci, 2013).

These results help pinpoint the properties of preferences that distinguish decision rules from one another. For instance, the Bayesian expected utility rule satisfies (see chapter 8.1 by Grüne-Yanoff and chapter 8.2 by Peterson, both in this handbook):[5]

Preferences $\succcurlyeq$ satisfy a (particular) set of axioms	if and only if	there exists a set of priors $\mathcal{C}$ and utility function u such that $\succcurlyeq$ is represented by (1) calculated with $\mathcal{C}$ and u.

Moreover, the representing $\mathcal{C}$ and u are suitably unique.

Figure 8.4.1

Representation theorem.

Weak Order (WO): The preference relation is transitive (for all f, g, h, $f \succsim g$ and $g \succsim h$ imply $f \succsim h$) and complete (for all f, g, $f \succsim g$ or $g \succsim f$).

Sure Thing Principle (STP): The preference across any pair of acts is independent of what the acts yield on events where they agree.

For instance, in table 8.4.2, since $1_{B_k}0$ and $1_{B_u}0$ agree on the event that the balls drawn from the two urns are the same color (i.e., the first and fourth states), STP implies that preferences over these acts should be independent of their consequences on this event. The same goes for $1_{W_k}0$ and $1_{W_u}0$. However, since on the complement event (i.e., the event consisting of the second and third states), $1_{B_k}0$ coincides with $1_{W_u}0$ and $1_{B_u}0$ coincides with $1_{W_k}0$, STP implies that $1_{B_k}0 \succsim 1_{B_u}0$ if and only if $1_{W_u}0 \succsim 1_{W_k}0$. Hence, the Ellsberg preferences (for $1_{B_k}0$ over $1_{B_u}0$ and for $1_{W_u}0$ over $1_{W_u}0$) violate STP.

In terms of choice, the maximin-EU and unanimity rules take complementary approaches: the former weakens STP while retaining WO; the latter drops the completeness clause of WO while holding on to STP. Various arguments have been proposed, in both the philosophical and economic literatures, that violating one or other of these axioms leads to unsavory consequences, particularly in *dynamic* contexts. They have been used by some to argue for, say, dropping WO rather than STP (S. Bradley & Steele, 2016; Seidenfeld, 1988). They have also been used to criticize any divergence from Bayesianism—and hence all of the approaches discussed in this survey—as irrational (Hammond, 1988; Al-Najjar & Weinstein, 2009; Elga, 2010).[6] This is currently the main battleground between Bayesianism and its critics and deserves a survey in itself. Note that even if such arguments are correct about the weaknesses of non-Bayesian approaches in dynamic contexts, these need to be traded off against their strengths in dealing with the role of evidence in choice. Some have claimed that the latter outweigh the former (Gilboa et al., 2009).

Apart from dropping STP, maximin-EU retains two axioms satisfied by expected utility:

P4: The preference for a bet on E over a bet on F with the same stakes is independent of the stakes.

Uncertainty Aversion (UA): For any pair of disjoint events E, F with $1_E 0 \sim 1_F 0$, $\frac{1}{2}_{E \cup F} 0 \succsim 1_E 0$.

Preferences over bets on events—for a bet on B_k over a bet on B_u (table 8.4.2)—reflect agents' "willingness to bet" and are typically considered to be related to their beliefs (Savage, 1954). P4 states that the relative willingness to bet on different events is independent of the stakes involved—that is, of the consequences of winning or losing the bet, assuming that these are the same for both bets.[7] Uncertainty Aversion translates the caution of maximin-EU. $\frac{1}{2}_{E \cup F} 0$ "hedges" the uncertainty involved in the bets on E and F: the uncertainty is "halved," insofar as the payoff depends on whether E or F realizes, not which one does, but the winnings are halved as well. For an agent who dislikes uncertainty, such hedging can only be attractive, and hence the hedge is (weakly) preferred to the initial bets. This axiom is often taken as the property characterizing *uncertainty* (or *ambiguity*) *aversion*. However, it seems that people are not universally ambiguity averse (Wakker, 2010); for instance, in a version of the running example with 10 possible colors rather than 2, there is a tendency to bet on the unknown urn. Were such preferences to be deemed reasonable, this would be a blow for the rational credentials of maximin-EU. The α-maximin-EU rule retains P4 but drops UA and hence can accommodate these preferences.

2.2 Evaluation 2: Beliefs, Tastes, and Uncertainty Attitudes

Another criterion for evaluating a decision rule concerns its capacity to neatly separate beliefs from tastes or values. Consider the following oft-cited criticism of maximin-EU. In our running example, you know nothing about the composition of the unknown urn, so any probability measure is possible; the set $\mathcal{C}_0 = \{p \in \Delta: 0 \leq p(B_u) \leq 1, p(B_k) = 0.5\}$ captures this. However, according to maximin-EU with this set, a bet on the unknown urn is considered worse than a bet on black from an urn known to contain 1 black ball and 99 white ones. Such caution seems extreme.

This objection relies on two assumptions: first, that the set of priors involved in the decision rule represents beliefs and, second, that the beliefs should perfectly match available information. Concerning the latter assumption, the set $\mathcal{C}_0$ may reflect the objectively available information, but it does not capture a Bayesian agent's beliefs, for these must be precise; so why should things be different for non-Bayesians? Indeed, the set of priors is often understood to represent the agent's state of belief, which incorporates, but may go beyond, the "objective" information (see also R. Bradley, 2017). Indeed, many economists subscribe to the revealed-preference paradigm (see chapter 8.1 by Grüne-Yanoff, this handbook) and take the set of priors to be the uniquely determined set specified in the appropriate representation theorem (figure 8.4.1).

Turning to the first assumption, suppose that $\mathcal{C}_1 = \{p \in \Delta: 0.4 \leq p(B_u) \leq 0.6\}$ represents your preferences (according to (1)), so you do not hold the preference claimed in the objection.[8] Since this set goes beyond the "objective" information, one can ask why you are using it. Is it because you have further information, or an inclination to "believe" in the principle of insufficient reason? Or is it instead because you are not so cautious as to use $\mathcal{C}_0$ but rather have a higher tolerance of uncertainty? These two possibilities are radically different: the former concerns beliefs (and their formation), whereas the latter relates to values or tastes for bearing uncertainty. Decision theory has developed formal tools that can shed light on such questions.

The central concept is comparative uncertainty aversion: between two agents, which (if any) is *more* uncertainty averse? The following is a widely accepted behavioral definition of the concept:

Comparative Uncertainty Aversion. Agent 1 is *more uncertainty averse* than agent 2 if, whenever agent 1 weakly prefers an act *f* over a sure *c*, then so does agent 2.

If Ann prefers an uncertain act *f* to a sure (utility) amount *c*, then she is not so averse to uncertainty as to consider bearing the uncertainty involved in *f* to be worse than getting the "non-uncertain" *c*. If Bob exhibits the same preference in all such cases, then he is not that averse to uncertainty either, so he is (weakly) less uncertainty averse than Ann.

Influenced by the analogy with risk attitude, economists have generally considered uncertainty attitude as a taste (for bearing uncertainty) and been interested in its consequences for, say, investment decisions. In particular, they have characterized comparative uncertainty aversion in terms of the parameters of decision models, with results such as that in figure 8.4.2 for maximin-EU (Ghirardato & Marinacci, 2002). It tells us that differences in uncertainty attitude correspond to differences in the set of priors. The fact that a comparison of tastes or values—as uncertainty attitudes are understood to be—translates to a difference in the set of priors casts doubt on the claim that this set reflects *only* beliefs. Indeed, the conclusion often drawn is that there is no clean interpretation of the set of priors: it reflects aspects of both belief and uncertainty attitude

(Klibanoff, Marinacci, & Mukerji, 2005). The maximin-EU account, it seems, does not have the resources to determine whether and to what extent the set $\mathcal{C}_1$ reflects enhanced beliefs or uncertainty tolerance.

These interpretational subtleties signal a perhaps more severe departure from the Bayesian benchmark than first imagined. Bayesianism vaunts a clear separation of beliefs (captured by the probability measure) and tastes (reflected in the utility function); maximin-EU apparently does not. To the extent that such a separation is desirable, this could bode ill for its normative credentials. For instance, in public decisions (e.g., concerning climate policy), it is customary for one group to supply the factual judgments (e.g., climate experts) and another to provide the values (e.g., society or its representatives); without the separation of beliefs and tastes, this division of labor is compromised. While current arguments against value-free science tend to focus on difficulties faced *in practice* (see Douglas, 2009, and chapter 14.2 by Bueter, this handbook), using sets of priors to report uncertainty risks ruling out such a possibility *in principle*. Moreover, the separation allows questions of theoretical rationality (e.g., learning) to be treated independently of those of practical rationality (e.g., decision; see chapter 2.2 by Wedgwood, this handbook). But if an agent's set of priors reflects not only the evidence acquired but also her uncertainty attitude, then both will play a role in the formation of such sets; any theory of belief update will thus also have to incorporate value considerations related to uncertainty tolerance. To date, work on belief updating for sets of priors does not seem to have grappled with this issue.

2.3 Across Multiple Prior Rules

In comparing the various rules for choosing on the basis of sets of priors, the belief–taste separation issue seems not to favor any particular one but rather concerns the multiple prior representation in general: results à la figure 8.4.2 for the unanimity rule suggest that it suffers from similar problems.[9] As concerns implications for choice, while some dispute the normative credentials of the preferences promoted by rules retaining WO (such as maximin-EU), few dispute the preference orderings yielded by the unanimity rule: if it recommends against

For agents 1 and 2 with the same utility function, agent 1 is more uncertainty averse than agent 2 — if and only if — agent 2's set of priors in a subset of agent 1's set of priors.

Figure 8.4.2
Comparative Uncertainty Aversion.

an act as being worse than another for all probability measures, then it would seem like a bad idea to choose it. However, as noted previously, the rule may remain silent on some comparisons between acts; the issue is what to do in these cases.

Some suggest that there is nothing more to be said about rational decision: the unanimity rule provides all the guidance there is, and in cases where it is silent, there is no more guidance to be had (S. Bradley & Steele, 2016). Others invoke "mechanisms" that are specific to such cases of "indecision," for instance, choosing a (contextually provided) status quo option, taking a deferral option, or choosing at random. Some have suggested "picking" a precise probability in the set of priors, or "sharpening," for the purposes of decision (Joyce, 2010). Such a procedure needs to be carried out in a coherent way across decisions, to avoid agents making chains of decisions that yield sure losses. Rules such as maximin-EU can be thought of as providing principles for "picking" a probability: it always chooses a probability measure that evaluates the act as badly as possible among those in the set. Although the relevant probability measure differs according to the act evaluated (the measure used to evaluate a bet on white may not be relevant for a bet on black), the representation theorem tells us that the rule is invulnerable to the aforementioned problems. Indeed, each of the rules discussed above satisfying WO can be thought of as one way of "complementing" the unanimity approach in situations where it remains silent. The possibility of complementing approaches violating WO—seen as a strong rational base—by invoking considerations such as "security"—which may not enjoy the same interpersonally valid rational credentials—has been discussed by Levi (1986). Gilboa, Maccheroni, Marinacci, and Schmeidler (2010) invoke a similar intuition, distinguishing "objectively" from "subjectively rational" preferences, represented according to the unanimity and maximin-EU rules, respectively. They provide axioms characterizing the maximin-EU "completion" of given unanimity preferences.

3. Nonadditive Probabilities

Another reaction to Bayesianism's apparent difficulties in accounting for evidence focuses on the additivity of probability functions. A range of proposals, including Dempster–Shafer belief functions, possibility functions, and Shackle's degrees of surprise (see chapter 4.7 by Dubois & Prade, this handbook), employ real-valued functions assigning a number in [0, 1] to each event and satisfying a mild monotonicity condition: bigger events

do not get a lower value. Such (monotonic real-valued set) functions are called *capacities* (other terms used, in various literatures, include "fuzzy," "confidence," and "plausibility measures").

A first-pass decision rule involving capacities would keep the expected utility formula (table 8.4.1) but with capacities in the place of probabilities. It has long been known that such a rule has very unattractive behavioral properties: any decision maker using it will strictly prefer some dominated act (i.e., one that does worse in all states than another available act; see Quiggin, 1982; Wakker, 2010). Basically, the main decision rule under uncertainty using capacities that avoids such problems is the "Choquet Expected Utility" or "CEU" rule proposed by Schmeidler (1989), which evaluates an act f according to

$$\sum_{x_i} v(\{s: u(f(s)) \geq x_i\})[x_i - x_{i+1}], \qquad (4)$$

where v is the capacity, u is the utility function, and x_i are the utility values of consequences of f, in decreasing order. Beyond being the main rule suitable for the sorts of belief representation mentioned above, it has proved popular descriptively (see section 5 below), partly because of its implications for choice: the rule does not assume UA (although it does satisfy P4), which, as noted, is sometimes violated.

However, UA is satisfied whenever the capacity is convex, that is, whenever $v(E \cup F) + v(E \cap F) \geq v(E) + v(F)$ for all events E, F. In this case, the Choquet integral coincides with the maximin-EU evaluation using the set of probability measures dominating the capacity (called the *core*), that is, $\{p \in \Delta: \text{for all } E \subseteq S, p(E) \geq v(E)\}$. Since several nonadditive probability representations—such as Dempster–Shafer belief functions—are convex capacities, this means that the decision rule for them is (equivalent to) maximin-EU. So the previous remarks also hold for them. More generally, results à la figure 8.4.2 indicate that the capacity in the CEU rule reflects attitude to uncertainty (as well as, potentially, beliefs), so the separation issue remains problematic.

4. Confidence and Other Second-Order Representations

A third approach focuses on a purported multidimensional character of beliefs: to use Keynesian vocabulary, beyond the *balance* of evidence supporting a probabilistic judgment, there is also its *weight*. The thought that both of these "dimensions" are involved in the representation of beliefs dates back at least as far as Peirce (1878): "to express the proper state of our belief, not one number but two are requisite, the first depending on the inferred probability, the second on the amount of knowledge on which that probability is based." Similar distinctions have more recently played a prominent role in cognitive psychology (Griffin & Tversky, 1992). To capture this idea, representations generally employ some structure over the probability space Δ.

A simple example is a (weak) order over the probability space or, equivalently, a nested family of sets of probability measures, denoted Ξ (figure 8.4.3). R. Bradley (2017) and Hill (2013, 2019) argue that such a structure can capture an agent's *confidence in her beliefs*. Each set in the family represents the beliefs or probability judgments held at a given level of confidence: the probability judgments that hold for all measures in the set are those that the agent holds with (at least) the corresponding amount of confidence. For instance, an agent represented according to figure 8.4.3 has high confidence in the probability judgment of 0.5 for B_k but only low confidence in the probability judgment of 0.5 for B_u. Larger sets in the family (for which fewer judgments hold) correspond to higher levels of confidence. Conceptually, confidence in a probability judgment is related to the evidence underpinning it (R. Bradley, 2017), so this approach relates to the aforementioned tradition. Technically, the representation is simply the "system of spheres" representation from belief revision and conditional logic (see chapter 5.2 by Rott and chapter 6.1 by Starr, both in this handbook), applied over the probability space rather than the state space. Belief representations in this spirit have been proposed by Gärdenfors and Sahlin (1982) and Nau (1992), although they assume that the second-order structure is cardinal: a number is assigned to each probability measure. Naturally, every cardinal structure over the probability space induces the ordinal structure described above (by ordering probability measures according to the value assigned). Similar uncertainty representations have also been promoted in econometrics (Manski, 2013).

Like multiple priors (section 2), several decision rules operate with this representation of beliefs. A notable family encapsulates the arguably reasonable maxim: *the higher the stakes involved in the decision, the more confidence is required in a belief for it to play a role.* They involve a function D assigning to each decision a confidence level—formally, a set in the nested family Ξ representing the agent's beliefs—and evaluate acts on the basis of the set picked out as appropriate for the decision by D. For each multiple-prior decision rule (section 2), there is a "confidence version," involving the same rule but allowing the set of priors involved to vary according to the decision, and in particular the stakes involved in it. For instance, the maximin-EU member of this family represents preferences according to

$$\min_{p \in D(f)} EU_p f. \tag{5}$$

The main difference for choice with respect to maximin-EU lies in P4, which directly clashes with the intuition that, for different stakes, different levels of confidence may be appropriate and hence different beliefs may inform the decision, leading to potentially different willingness to bet. Decision rule (5) satisfies a weakening of this axiom, which allows the willingness to bet to change with the stakes, in line with the aforementioned maxim. Unlike maximin-EU, results à la figure 8.4.2 suggest a clean separation between beliefs and values (Hill, 2013): the family of sets Ξ represents the beliefs while the function D represents the agent's uncertainty aversion, or taste for choosing on the basis of limited confidence. Drawing on similar results for the confidence version of the unanimity rule, Hill (2019) argues that these two points—the mild yet motivated divergence from multiple prior models in terms of choice implications, and the clean separation of beliefs and tastes—are general properties of the approach.

As noted previously, any cardinal representation of confidence in beliefs can be used with this sort of decision procedure, by "forgetting" the numbers and using only the order. However, other decision rules have been developed that make specific use of the cardinal structure. Prominent ones are the *variational preferences* rule (Maccheroni, Marinacci, & Rustichini, 2006), which represents preferences by

$$\min_{p \in \Delta}(EU_p f + c(p)), \tag{6}$$

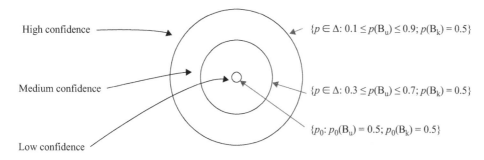

Figure 8.4.3
Confidence in beliefs.

where c is a real-valued function on Δ, and the *confidence preferences* rule (Chateauneuf & Faro, 2009),

$$\min_{p \in \Delta} \frac{1}{\varphi(p)} EU_p f, \tag{7}$$

where φ is a function from Δ to $[0, 1]$. The former rule is motivated by an important literature on robustness in macroeconomics: a central model developed there (Hansen & Sargent, 2008) is a special case. The latter was motivated by and technically related to the literature on fuzzy sets: φ is a fuzzy set of probability measures. Both approaches diverge from maximin-EU by weakening P4, without motivating the divergence with a maxim similar to that behind (5). However, these models do not cleanly separate beliefs and tastes: results à la figure 8.4.2 suggest that c and φ capture uncertainty attitudes and hence cannot be thought of as pure representations of belief.

5. Probabilities after All

Finally, some strive to retain the probabilistic representation of beliefs. Inspired by work on decision under risk, one approach assumes that agents assign a precise probability $p(E)$ to each event E but that in choice, these probability values are "deformed" by a *weighting function* w, so that the decision weight attached to the event is $w(p(E))$. Since the composition of w and p is a capacity, this approach, popular in prospect theory (Wakker, 2010; see chapter 8.3 by Glöckner, this handbook), uses the CEU rule (section 3). A different approach reverts to a second-order probability μ over the space of first-order probabilities Δ. To the extent that this is a second-order structure, it can be thought of as related to the confidence approaches in section 4 (Marinacci, 2015).

The main challenge for these approaches lies in the incompatibility between the Ellsberg preferences and the property of *probabilistic sophistication*, which, roughly, states that all the decision-relevant information about an event is summarized in the probability assigned to it (Machina & Schmeidler, 1992). For instance, in the Ellsberg example, if the agent assigns probability 0.5 to both B_k and B_u, as is often deemed reasonable, then the deformed weight $w(0.5)$ assigned to the two events is the same, and the CEU rule predicts indifference between the bets. Likewise, a second-order probability μ generates a "reduced" probability $\Sigma p\mu(p)$, which can be used in the context of the expected utility rule; doing so is equivalent to the following evaluation, which applies expected utility at both levels:

$$\int_\Delta EU_p f \, d\mu(p) \tag{8}$$

However, as noted at the outset, expected utility cannot accommodate the Ellsberg preferences.

Faced with this situation, a common reply is to treat (the probabilities of) different events differently in decision according to the type of event, and more specifically the *source of uncertainty*, to which it belongs. The Ellsberg known and unknown urns are different sources of uncertainty, and so, the idea goes, the same probability assignment with respect to events from the different sources can be treated differently in decision.

Source dependence was introduced in the behavioral literature (Tversky & Fox, 1995) and continues to play a central role in current developments of one of the most important descriptive theories of decision, prospect theory (Abdellaoui, Baillon, Placido, & Wakker, 2011; Wakker, 2010; see chapter 8.3 by Glöckner, this handbook). Recent versions incorporate source preference by using the CEU rule with weighting functions that depend on the source of uncertainty, so probabilities about events concerning the known urn are weighted differently from those concerning the unknown urn.

For second-order probabilities, the following *smooth ambiguity* decision rule has been proposed (Klibanoff et al., 2005):[10]

$$\int_\Delta \varphi(EU_p f) \, d\mu(p), \tag{9}$$

with φ a real-valued function on utility values. The transformation function φ translates the difference in attitude to first-order and second-order uncertainty; interpreting the former as "physical uncertainty" and the latter as "model uncertainty," it thus reflects different attitudes to these two sources of uncertainty (Marinacci, 2015). In particular, whenever φ is nonlinear, the attitudes differ, and the rule can accommodate Ellsberg preferences. This is one of the prominent models in the literature admitting a clean separation of beliefs from uncertainty attitudes: the second-order probability μ can be understood as a representation of the agent's state of belief, whereas φ reflects her uncertainty attitude (Klibanoff et al., 2005). This separation, combined with the tractability and familiarity of the largely Bayesian framework, has contributed to its increasing popularity in economic modeling.

While some approaches invoking source dependence, particularly in the behavioral literature, have plainly descriptive aims, others seem to harbor normative ambitions (Marinacci, 2015) and hence call for an appraisal of their rational credentials. Criticisms have focused on their resolution of the challenge of reserving a role for weight of evidence in decision (section 1). They insist that the Bayesian representation captures all relevant aspects of belief but introduce another element (the φ in (9)) to account for the role apparently played by weight of evidence. Some have suggested that this tension—in particular, the use of a parameter representing uncertainty attitude (φ) to reflect a characteristic of evidence, which is a factor pertaining to belief—is damaging for the rational credentials of approaches invoking source dependence (Hill, 2019).

Notes

1. We follow economists in using "decision under uncertainty" to refer to cases where the probabilities of the various outcomes are not provided, as opposed to "decision under risk," where they are given.

2. After the notion of uncertainty introduced by Knight (1921).

3. Similar decision rules have been defended in philosophy (Gärdenfors & Sahlin, 1982) and developed in robust statistics, where they are called "Γ-maximin" (Berger, 1985).

4. Different versions vary according to whether the inequalities are weak (as in (3)) or strict.

5. Throughout, we retain the standard names for axioms used in the literature. For further details, the reader is referred to Gilboa (2009) and Gilboa and Marinacci (2013), for example.

6. While these arguments focus on decision, they are related to "dilation" arguments that focus specifically on learning under imprecise probabilities (Seidenfeld & Wasserman, 1993; Joyce, 2010).

7. This axiom, which was introduced by Savage (1954), whose nomenclature we adopt, is related to Gilboa and Schmeidler's C-Independence (1989).

8. Applying the maximin-EU rule with this set of priors evaluates a bet on black from the unknown urn using the worst-case probability for black—0.4—and this is better than the probability of getting black from an urn with one black ball out of 100.

9. For α-maximin-EU, despite some indications of belief–taste separation (Ghirardato, Maccheroni, & Marinacci, 2004), there are reasons for skepticism (Klibanoff, Marinacci, & Mukerji, 2005). It is complicated by the interdependence between the representing set of priors and the α (the uniqueness is weaker than in figure 8.4.1), which, without controversial assumptions, is unconducive to a clean separation.

10. For related approaches, see Ergin and Gul (2009), Nau (2006), and Seo (2009).

References

Abdellaoui, M., Baillon, A., Placido, L., & Wakker, P. P. (2011). The rich domain of uncertainty: Source functions and their experimental implementation. *American Economic Review, 101*(2), 695–723.

Al-Najjar, N. I., & Weinstein, J. (2009). The ambiguity aversion literature: A critical assessment. *Economics & Philosophy, 25,* 249–284.

Berger, J. O. (1985). *Statistical decision theory and Bayesian analysis.* New York, NY: Springer.

Bewley, T. F. (1986). Knightian decision theory. Part I. *Decisions in Economics and Finance, 25*(2), 79–110.

Bradley, R. (2017). *Decision theory with a human face.* Cambridge, England: Cambridge University Press.

Bradley, S., & Steele, K. (2016). Can free evidence be bad? Value of information for the imprecise probabilist. *Philosophy of Science, 83*(1), 1–28.

Camerer, C., & Weber, M. (1992). Recent developments in modeling preferences: Uncertainty and ambiguity. *Journal of Risk and Uncertainty, 5*(4), 325–370.

Chateauneuf, A., & Faro, J. H. (2009). Ambiguity through confidence functions. *Journal of Mathematical Economics, 45,* 535–558.

Douglas, H. E. (2009). *Science, policy, and the value-free ideal.* Pittsburgh, PA: University of Pittsburgh Press.

Elga, A. (2010). Subjective probabilities should be sharp. *Philosopher's Imprint, 10.* http://hdl.handle.net/2027/spo.3521354.0010.005

Ellsberg, D. (1961). Risk, ambiguity, and the Savage axioms. *Quarterly Journal of Economics, 75,* 643–669.

Ergin, H., & Gul, F. (2009). A theory of subjective compound lotteries. *Journal of Economic Theory, 144*(3), 899–929.

Gärdenfors, P., & Sahlin, N.-E. (1982). Unreliable probabilities, risk taking, and decision making. *Synthese, 53*(3), 361–386.

Ghirardato, P., Maccheroni, F., & Marinacci, M. (2004). Differentiating ambiguity and ambiguity attitude. *Journal of Economic Theory, 118*(2), 133–173.

Ghirardato, P., & Marinacci, M. (2002). Ambiguity made precise: A comparative foundation. *Journal of Economic Theory, 102*(2), 251–289.

Gilboa, I. (2009). *Theory of decision under uncertainty.* Cambridge, England: Cambridge University Press.

Gilboa, I., Maccheroni, F., Marinacci, M., & Schmeidler, D. (2010). Objective and subjective rationality in a multiple prior model. *Econometrica, 78*(2), 755–770.

Gilboa, I., & Marinacci, M. (2013). Ambiguity and the Bayesian paradigm. In D. Acemoglu, M. Arellano, & E. Dekel (Eds.), *Advances in economics and econometrics: Vol. 1. Economic theory: Tenth world congress* (Econometric Society Monographs, Vol. 49, pp. 179–242). Cambridge, England: Cambridge University Press.

Gilboa, I., Postlewaite, A., & Schmeidler, D. (2009). Is it always rational to satisfy Savage's axioms? *Economics & Philosophy, 25*(3), 285–296.

Gilboa, I., & Schmeidler, D. (1989). Maxmin expected utility with non-unique prior. *Journal of Mathematical Economics, 18*(2), 141–153.

Griffin, D., & Tversky, A. (1992). The weighing of evidence and the determinants of confidence. *Cognitive Psychology, 24*(3), 411–435.

Hammond, P. J. (1988). Consequentialist foundations for expected utility theory. *Theory and Decision, 25*, 25–78.

Hansen, L. P., & Sargent, T. J. (2008). *Robustness*. Princeton, NJ: Princeton University Press.

Hill, B. (2013). Confidence and decision. *Games and Economic Behavior, 82*, 675–692.

Hill, B. (2019). Confidence in beliefs and rational decision making. *Economics & Philosophy, 35*(2), 223–258.

Hurwicz, L. (1951). Some specification problems and applications to econometric models. *Econometrica, 19*(3), 343–344.

Jaffray, J.-Y. (1989). Linear utility theory for belief functions. *Operations Research Letters, 8*, 107–112.

Joyce, J. M. (2005). How probabilities reflect evidence. *Philosophical Perspectives, 19*, 153–178

Joyce, J. M. (2010). A defense of imprecise credences in inference and decision making. *Philosophical Perspectives, 24*, 281–323

Keynes, J. M. (2004). *A treatise on probability*. Mineola, NY: Dover. (Original work published 1921)

Klibanoff, P., Marinacci, M., & Mukerji, S. (2005). A smooth model of decision making under ambiguity. *Econometrica, 73*(6), 1849–1892.

Knight, F. H. (1921). *Risk, uncertainty, and profit*. Boston, MA: Houghton Mifflin.

Levi, I. (1986). *Hard choices: Decision making under unresolved conflict*. Cambridge, England: Cambridge University Press.

Maccheroni, F., Marinacci, M., & Rustichini, A. (2006). Ambiguity aversion, robustness, and the variational representation of preferences. *Econometrica, 74*(6), 1447–1498.

Machina, M. J., & Schmeidler, D. (1992). A more robust definition of subjective probability. *Econometrica, 60*(4), 745–780.

Machina, M. J., & Siniscalchi, M. (2014). Ambiguity and ambiguity aversion. In M. J. Machina & W. K. Vicusi (Eds.), *Handbook of the economics of risk and uncertainty* (Vol. 1, pp. 729–807). Amsterdam, Netherlands: North-Holland.

Manski, C. F. (2013). *Public policy in an uncertain world: Analysis and decisions*. Cambridge, MA: Harvard University Press.

Marinacci, M. (2015). Model uncertainty. *Journal of the European Economic Association, 13*(6), 1022–1100.

Nau, R. F. (1992). Indeterminate probabilities on finite sets. *Annals of Statistics, 20*(4), 1737–1767.

Nau, R. F. (2006). Uncertainty aversion with second-order utilities and probabilities. *Management Science, 52*, 136–145.

Peirce, C. S. (1878). The probability of induction. *Popular Science Monthly, 12*(705), 83.

Quiggin, J. (1982). A theory of anticipated utility. *Journal of Economic Behavior & Organization, 3*(4), 323–343.

Savage, L. J. (1954). *The foundations of statistics*. New York, NY: Dover.

Schmeidler, D. (1989). Subjective probability and expected utility without additivity. *Econometrica, 57*, 571–587.

Seidenfeld, T. (1988). Decision theory without "independence" or without "ordering." *Economics & Philosophy, 4*, 267–290.

Seidenfeld, T., & Wasserman, L. (1993). Dilation for sets of probabilities. *Annals of Statistics, 21*(3), 1139–1154.

Seo, K. (2009). Ambiguity and second-order belief. *Econometrica, 77*, 1575–1605.

Tversky, A., & Fox, C. R. (1995). Weighing risk and uncertainty. *Psychological Review, 102*, 269–283.

Wakker, P. P. (2010). *Prospect theory for risk and ambiguity*. Cambridge, England: Cambridge University Press.

Walley, P. (1991). *Statistical reasoning with imprecise probabilities* (Monographs on Statistics and Applied Probability, Vol. 42). London, England: Chapman and Hall.

8.5 Bounded Rationality: A Vision of Rational Choice in the Real World

Ralph Hertwig and Anastasia Kozyreva

Summary

This chapter focuses on the conception of bounded rationality introduced by Herbert Simon as an alternative to the perfect rationality of the omniscient homo economicus, then further developed in psychology and economics. Bounded rationality is a principle stating that real-world rational cognition is limited by bounds on time, computational power, foresight, and knowledge; it is also a theory of what it means to be rational given limited human cognition and an uncertain and complex world. We outline the foundations of bounded rationality in Simon's work and explore interpretations of bounded rationality in three research programs: in economics, the optimization-under-constraints program; in psychology, the heuristics-and-biases program and the program on ecological rationality.

> It is time to take account ... of the empirical limits on human rationality, of its finiteness in comparison with the complexities of the world with which it must cope. (Simon, 1957b, p. 198)

> Broadly stated, the task is to replace the global rationality of economic man with the kind of rational behavior that is compatible with the access to information and the computational capacities that are actually possessed by organisms ... in the kinds of environments in which such organisms exist. (Simon, 1955, p. 99)

1. Foundations of Bounded Rationality

1.1 From Metaphors to Models of Choice

If there was any period in human history during which the Enlightenment ideals of rationality and humanity were at risk of being shattered completely, it was the 20th century. Paradoxically, this century, abounding with the horrors of human irrationality and unprecedented self-destruction, was also the most fruitful period for advancing both formal normative theories of rational choice and psychologically realistic models of reasonable behavior. Herbert

Simon's work on bounded rationality was one such 20th-century advancement. It established a new requirement for theorizing about human choice: the need to understand how real people, not agents with supreme foresight and unlimited computational skills, make decisions when faced with limited time and knowledge. The time was ripe to comprehend how people achieve their goals in the face of a staggering environmental complexity that surpasses the limits of the computational resources of any human being or any information-processing system.

Simon introduced the concept of bounded rationality in the mid-1950s to describe a new theoretical and empirical approach to studies of choice and rational behavior. First, he defined bounded rationality as a principle that emphasizes the limits on computational capacities of agents in the real world who have only limited time, knowledge, foresight, and cognitive resources:

> The principle of bounded rationality: The capacity of the human mind for formulating and solving complex problems is very small compared with the size of the problems whose solution is required for objectively rational behavior in the real world—or even for a reasonable approximation to such objective rationality. (Simon, 1957b, p. 198)

He thereby positioned bounded rationality as an alternative to the concepts of perfect or substantive rationality rooted in classical rational choice theory that were prominent in the neoclassical economics of his day (more on this juxtaposition in section 1.2). The idea of cognitive limitations and impractical normative standards in the real world was already present in his early research on decision making within organizations (Simon, 1947/1957a). Although Simon initially embraced the normative standards of rational choice theory, showing that real people fall short of these standards,[1] in later works, he progressed toward a positive theory of bounded rationality (most notably beginning with Simon, 1955, 1956). This extension from principle to positive theory implies that bounded rationality is not merely a simplification of perfect or substantive rationality but a descriptive and prescriptive model of actual choice behavior in the real world. In fact, Simon's ultimate endeavor was to reconstruct and reinvent the "theory of the rational" (Simon, 1957b, p. 200). On this view, bounded rationality represents a new vision of what it means to be rational given both the inescapable uncertainty and complexity of the world and the limitations of the human cognitive system. Both the principle of bounded rationality and the vision of bounded rationality as a new descriptive and prescriptive theory of rational choice imply that understanding (and eventually predicting) the choice behavior of real humans is only possible when psychological processes are empirically analyzed. This study of what Simon called "procedural rationality" ultimately amounts to a theory of the efficient computational procedures by which agents (individual human beings, organizations, or computers) arrive at good solutions to the problems they face (Simon, 1976).

Simon went on to develop his approach to bounded rationality across a diverse set of scientific disciplines such as economics, cognitive psychology, and artificial intelligence, adding new levels to his concept over time. One of the most important developments concerned his inquiry into an adaptive, or ecological, level of rationality. Specifically, he stressed that the essence of rational behavior consists in how an organism can adapt in order to achieve its goals under the constraints of its environment and its own cognitive limitations. These two dimensions of rational behavior—cognitive and environmental—gave rise to the scissors metaphor that encapsulates bounded rationality's theoretical core: "Human rational behavior (and the rational behavior of all physical symbol systems) is shaped by a scissors whose two blades are the structure of task environments and the computational capabilities of the actor" (Simon, 1990, p. 7).

The scissors metaphor was not the only one Simon used to portray human choice and choice environments. As he recalled in his autobiography, he saw the labyrinth as another powerful symbol for decision making. Simon (1991) envisioned decision making "in terms of successive choices along a branching path" (p. 86) rather than as an optimizing process. The first step in his search to turn his metaphors of bounded rationality into realistic models of rational behavior was to criticize the narrow and unrealistic conception of rationality found in neoclassical economics.

1.2 The Limits of the Olympian Model of Rationality

Simon (1983) opposed what he called the "Olympian model" of rationality (p. 19), which encompasses concepts such as perfect or substantive rationality, (subjective) expected utility theory, and the unfailingly rational

homo economicus.[2] In order to satisfy its normative demands, this model requires a decision maker with unlimited cognitive resources who can form mental models representing, in a collectively exhaustive and mutually exclusive way, all future relevant states of the world. Simon argued that while rational choice theory was worthy of a home in Platonic heaven, it was simply out of place in the real world (Simon, 1983, p. 13). In "A Behavioral Model of Rational Choice" (Simon, 1955), he spelled out the unrealistic assumptions of perfectness and logical omniscience that the Olympian approach to rationality made about the agent:

> If we examine closely the "classical" concepts of rationality outlined above, we see immediately what severe demands they make upon the choosing organism. The organism must be able to attach definite pay-offs (or at least a definite range of pay-offs) to each possible outcome. This, of course, involves also the ability to specify the exact nature of outcomes—there is no room in the scheme for "unanticipated consequences." The pay-offs must be completely ordered—it must always be possible to specify, in a consistent way, that one outcome is better than, as good as, or worse than any other. And, if the certainty or probabilistic rules are employed, either the outcomes of particular alternatives must be known with certainty, or at least it must be possible to attach definite probabilities to outcomes. (pp. 103–104)

In actual human choice, such demanding requirements can rarely be met. Outside what Savage (1954) called "small worlds"—highly simplified environments such as monetary gambles—real people cannot live up to the ideal of making decisions by specifying all possible outcomes, assigning probabilities and values to each, and then maximizing the expected payoffs. It would be a misrepresentation, however, to assume that people cannot live up to this decision-making ideal due to cognitive destitution or a mere lack of skill; Simon (1972) credited the *irreducible uncertainty* inherent in any human reckoning about the future as the reason people did not act according to the Olympian model.

Consider firefighters. Like emergency room doctors, police officers, and soldiers, firefighters face enormous time pressure, high stakes, and inescapable uncertainty. They do not have the luxury of mentally generating all possible courses of actions, specifying their respective outcomes, and then evaluating them. Not only do firefighters face epistemic uncertainty, but they also face the aleatory uncertainty[3] inherent in their environment. The Olympian model thus seems to be an obscenely lofty conception of decision making. Professional experts do not make decisions by comparing all possible options but rather by generating a single good course of action from the start (Klein, 1998, p. 17). By drawing on a repertoire of past situations compiled from real and virtual experience, firefighters and their ilk quickly identify a plausible course of action and play it out mentally to test whether it could work in the current situation. If success seems plausible, it will be implemented; otherwise, it will be modified or replaced by the next-best option until a "good enough" alternative is found. This is an example of a fast and intuitive problem-solving performance—one for which the expert cannot necessarily describe in detail the underlying reasoning processes. Moreover, the key to experts' success lies in the way they approach and analyze their environments. Returning to Simon's scissors, one blade—the environment—provides the cues that are in turn matched by the second blade—the mind—which draws on its resources (e.g., memory) to find a suitable decision strategy. On this view, "intuition is nothing more and nothing less than recognition" (Simon, 1992, p. 155).

Next to irreducible uncertainty and time pressure, there is another key reason that the Olympian model prospers in Platonic heaven but withers in earthly reality: computational intractability. Consider chess, a game Simon studied extensively (Chase & Simon, 1973). Chess offers a choice set of about 30 legal moves (a number that stays more or less constant until the end of the game) and a time horizon of about 40 moves per player until one party concedes; 30^{80} possible sequences (about 10^{118}) follow from the original position (see also Shannon, 1950). No human or computer can generate and evaluate all these consequences. Even Deep Blue, the IBM computer that beat chess world champion Garry Kasparov in 1997 and that could examine some 300 million possible moves per second, had to evaluate positions through "a high-performance alpha-beta search that expands a vast search tree by using a large number of clever heuristics and domain-specific adaptations" (Silver et al., 2018, p. 1). Relative to many real-world decisions, however, chess is a piece of cake; indeed, "the problem space associated with the game of chess is very much smaller than the space associated with the game of life" (Simon, 1976, p. 72). In many social interactions, the rules are not as well defined as they are in chess, the set of possible actions is much vaster, there are more than two parties involved, and goals are unclear or, even worse, in conflict, thus compounding the intractability problem with additional complexity. If chess is computationally intractable and thus beyond the reach of Olympian optimization, so are many of the decisions people face in the real world.

In other words, even though the classical understanding of rationality as set out in Olympian models is theoretically coherent and even appealing, it profoundly misrepresents the reality of decision making: the set of decisions where Platonic heaven and earthly reality intersect is very narrow.[4] For Simon (1989), the fundamental question for the study of bounded rationality was, therefore, "How do human beings reason when the conditions for rationality postulated by the model of neoclassical economics are not met?" (p. 377). This question has an important implication. Classical concepts of rationality not only fail to describe what people actually do but also fail to offer procedural advice about how to find solutions to many real-world problems. The study of boundedly rational decision making, on the other hand, is ultimately concerned with both descriptive and prescriptive questions—a point to which we return later.

1.3 Principles of Bounded Rationality

Firefighters face limits on time, information, and certainty—yet they still make decisions, as do emergency room doctors and chess players. How do reasonable people decide when optimization is out of reach, and how can scientists examine this behavior? In order to show how Simon answers this question, we distinguish four principles underlying his investigation of bounded rationality.

The first principle calls for a behaviorally informed theory of choice, which should take into account and specify what conditions govern human behavior in different real-world situations. In Simon's (1992) words, "The study of the behavior of an adaptive system like the human mind is not a logical study of optimization but an empirical study of the side conditions that place limits on the approach to the optimum" (p. 157). Failing to understand not just the "design product (the alternative finally chosen) but the design process as well" (p. 156) compromises behavioral science's ability to explain and predict behavior. This postulate was a bone of contention between Simon and Milton Friedman, who fiercely defended classical economic theory by arguing that "'complete realism' is clearly unattainable, and the question whether a theory is realistic 'enough' can be settled only by seeing whether it yields predictions that are good enough for the purpose at hand" (Friedman, 1953, p. 41). Simon, however, insisted that a realistic account of the cognitive processes was required to successfully predict human behavior (see also Simon, 1976, on procedural rationality).

A second principle is that realistic models will be models of "approximate methods" (Simon, 1990, p. 6). "Approximate" here should not be interpreted to mean that these methods will produce inferior solutions or lack cognitive sophistication. Simple methods, for instance, can exploit evolved cognitive, visual, and motoric capacities that can be complex and demanding but are nevertheless easy for the mind to execute, such as pattern recognition, emotions (see Hanoch, 2002), tracking motion in space, and the ability to adaptively forget (Schooler & Hertwig, 2005). By exploiting the mind's sophisticated capacities, approximate methods can remain computationally slim and operate under constraints on time and information.

A third principle of bounded rationality is that when optimization is out of reach, people *satisfice*—they look for good-enough answers. Satisficing does not mean merely taking a good-enough alternative and giving up on the best possible alternative. Rather, satisficing is "the process of finding alternatives by heuristic search with the use of a stop rule based upon adjustable aspirations" (Simon, 1982, p. 323). In natural situations, alternatives are often encountered sequentially (e.g., potential mating partners; Miller & Todd, 1998), and the number of available alternatives is almost always too large to be exhaustively explored; furthermore, alternatives may vanish if not chosen immediately. Under these circumstances, a limited search is implemented and the point at which a search will be terminated is determined by an aspiration level—"a goal variable that must be reached or surpassed by a satisfactory decision alternative" (Selten, 2001, pp. 13–14). Aspiration levels can be adjusted depending on one's success in finding satisfying options. Note that not every boundedly rational behavior requires satisficing. Rather, a range of heuristic processes can be employed to reach good decisions in both strategic games and games against nature (Hertwig & Herzog, 2009; see also section 2.3 of this chapter).

A final principle of the empirical study of bounded rationality requires taking into account the properties of an organism's environment. In *The Sciences of the Artificial* (1969/1996), Simon argued that an organism's behavior results not merely from its inherent characteristics but also from the structures of its surroundings. Taking the path of an ant on a beach as an example, Simon claimed that an "ant, viewed as a behavior system, is quite simple. The apparent complexity of its behavior over time is largely a reflection of the complexity of the environment" (p. 52). An ant's environment may be an obstacle course, forcing the ant to repeatedly change direction around pebbles, rocks, and puddles (figure 8.5.1). An observer who saw only the ant's laborious path, without the properties of the environment, may be tempted to attribute substantial complexity to the cognitive mechanisms underlying the ant's behavior. Yet, if the ant's behavior is understood as the result of its interaction with its environment, the explanation may be much more mundane. The mechanisms producing the complex path may be simple rules

such as, "If the left vision field sensors for obstacles lights up, turn to the right, and vice versa." Even though actual ants might be guided by a multitude of other factors as well, this example reveals that the study of bounded rationality must include the study of the structural properties of the surroundings in which organisms, whether ants or humans, make decisions. Explaining complex behaviors exclusively in cognitive terms risks misattributing—or worse, profoundly misconstruing—the causes of the complexity.

Having clarified the basic principles behind Simon's view, we now turn to various research programs in psychology and economics that claim to have further developed the theory of bounded rationality. As we will see, these programs are quite diverse and give varying weight to the theoretical and methodological notions and principles reviewed above.

2. Interpretations of Bounded Rationality

There are at least three main research programs aiming to interpret bounded rationality and develop it further (see also Gigerenzer & Selten, 2001): the economic research program of *optimization under constraints*, the approach in psychology known as the *heuristics-and-biases program* that led to the foundation of behavioral economics (and behavioral law), and a further program in psychology known as *ecological rationality* or the study of *fast-and-frugal heuristics*.

2.1 Optimization under Constraints
Simon's candid and fundamental criticism of neoclassical economics and its view of global and omniscient

Figure 8.5.1
An ant walking on a beach illustrates that a behavior (here, the path the ant takes) that might appear complex to an external observer is actually a function of the interaction between the organism and its environment.

rationality was generally not met with enthusiasm by economists. However, in an attempt to reconcile neo-classical economics with behavioral and cognitive limits, George Stigler, a leading figure of the Chicago school of economics, proposed a notion often described as "optimization under constraints" (Stigler, 1961).[5] As a former student of Frank Knight, Stigler (1961) was mindful of the role of uncertainty in human choice, stating that "our understanding of economic life will be incomplete if we do not systematically take account of the cold winds of ignorance" (p. 224).

In order to incorporate ignorance—or, to use a less pejorative term, incomplete knowledge—in the process of choice, Stigler (1961) focused on one key idea of bounded rationality: the concept of limited search (and, by extension, the respect for real decision makers' finite resources). The method for determining when to stop searching is interpreted in terms of a cost–benefit analysis. On this view, conditions of limited search and less-than-perfect information appear to easily square with models of optimization. Specifically, the models in this class assume that a stopping rule optimizes with respect to the relevant currency (e.g., time, computation, money). Thus, when the cost of searching for new alternatives or other pieces of information exceeds the benefits of further exploration, the search will be terminated. For example, when looking for a new car, a buyer explores the market until the expected marginal costs (i.e., time and effort) of looking for new options exceed the expected marginal benefits (i.e., finding a car that more closely meets the buyer's preferences). Although this conception of psychologically realistic decision making apparently reconciles bounded rationality with optimization (see also Sargent, 1993), the cost–benefit analysis imposes the "burden of estimating the expected marginal return of search and the opportunity cost" (Simon, 1982, p. 296). Indeed, optimization under constraints requires the same amount of information and even more knowledge and computation (e.g., determining the costs of continuing search, such as search's opportunity costs; Vriend, 1996) compared to neoclassical optimization models. Moreover, as Simon (1979) noted, it "poured the search theory back into the old bottle of utility maximization" (p. 503) without paying due attention to a key element of the theory of bounded rationality, namely, that choice can be made with incomplete information and without carrying out optimization procedures.

2.2 Heuristics and Biases

The theory of bounded rationality is meant to be a behaviorally and psychologically grounded approach to decision making. One of the research programs that has contributed significantly to developing this approach is the work instigated by Amos Tversky and Daniel Kahneman. Their heuristics-and-biases approach has mapped the impact of cognitive limitations on people's judgments and decisions and documented a large catalog of systematic deviations from norms of rationality, drawn from probability theory, statistics, and axioms of rational choice (Kahneman, Slovic, & Tversky, 1982; Tversky & Kahneman, 1974). Kahneman interpreted bounded rationality in terms of behavior that diverges from such norms and, by extension, optimality. In this spirit, he claimed, "Our research attempted to obtain a map of bounded rationality, by exploring the systematic biases that separate the beliefs that people have and the choices they make from the optimal beliefs and choices assumed in rational-agent models" (Kahneman, 2003, p. 1449).

This statement both highlights the key normative assumption of the heuristics-and-biases program and outlines the contours of its map of bounded rationality. Rational-agent models (Olympian models, according to Simon) as well as their axioms and assumptions (e.g., Bayesian probability updating) are explicitly acknowledged as the benchmarks against which people's judgments should be compared. This means that the decision problems in question are ones in which an optimal or normative solution exists. By extension, bounded rationality is not so much a map of human choice in the world of the unknown and the uncertain as it is an empirical collection of decisional deficiencies in problems with assumed normative solutions. This is also evident in prospect theory, the most influential descriptive alternative choice theory to expected utility theory (Kahneman & Tversky, 1979; Tversky & Kahneman, 1992). Indeed, as Kahneman emphasized, their aim was to "assemble the minimal set of modifications of expected utility theory that would provide a descriptive account of . . . choices between simple monetary gambles" (Kahneman, 2000, p. x). The theory's goal was to explain a wide range of violations of axioms and predictions of expected utility theory, thus accepting the normative force of the theory. Prospect theory enriches the expected utility framework by introducing a few psychological concepts (e.g., nonlinear probability weighting, reference points, loss aversion; for a detailed account of prospect theory, see chapter 8.3 by Glöckner, this handbook). Yet its scope, at least in its original form, is confined to the world of risk (Savage's small worlds) with known possible outcomes and without the possibility of surprises. As Simon repeatedly stressed, these are not the conditions under which individuals and organizations make decisions most of the time.

Outside of the domain of risky choice, the heuristics-and-biases program has invoked a small set of heuristics, or mental shortcuts, to explain the myriad errors—also called "cognitive illusions"—that people commit in their inferences and decisions. Specifically, the program's argument is that "people rely on a limited number of heuristic principles which reduce the complex task of assessing probabilities and predicting values to simpler judgmental operations. In general these heuristics are quite useful, but sometimes they lead to severe and systematic errors" (Tversky & Kahneman, 1974, p. 1124).

Three key heuristics—representativeness, availability, and anchoring and adjustment—have been proposed to account for a wide range of cognitive illusions such as the base-rate fallacy, the overconfidence bias, the conjunction fallacy, the hot-hand fallacy, and the failure to appreciate regression to the mean or the law of large numbers. Two theoretical premises buttress these heuristics. One is that of attribute substitution, such as similarity-for-probability substitution, which is the mechanism behind the representativeness heuristic (Kahneman & Frederick, 2002). It describes the idea that a heuristic evaluates a target attribute of a judgment object (e.g., the likelihood that X is a Y or a Z) by substituting it with a property of the object that comes more readily to mind (e.g., the degree to which X resembles a Y or a Z). This occurs when a person who learns that Linda is an outspoken woman concerned with social justice judges that it is more likely that Linda is a "feminist bank teller" than—as is necessarily at least as likely—a "bank teller," because they believe Linda's characteristics are representative of a feminist (Tversky & Kahneman, 1983). Error-prone attribute substitution occurs when three conditions are satisfied: first, the target attribute (e.g., probability) is cognitively relatively inaccessible; second, the substituting attribute (e.g., similarity) is highly accessible; and, third, the mind's "controller" does not reject the substitution.

This controlling entity relates to the second theoretical premise key to Kahneman's (2011) view of bounded rationality—namely, that two, admittedly fictitious, cognitive systems make up the human mind. System 1 is automatic, effortless, and associative. Reasoning errors are typically the product of this speedy system and its propensity to propose the solution that most readily comes to mind. The awareness that an error may have occurred requires the more reflective and effortful System 2. This system acts like a teacher identifying a student's mistake and rectifying the process that produced it. But System 2 also tires easily. Too often, instead of analyzing the proposals of System 1, System 2 is content with the solutions

offered, thus leaving decision makers unaware that they are about to commit an error.

The heuristics-and-biases interpretation of bounded rationality in terms of a map of the systematic deviations between people's behavior and norms of rationality has been highly influential within psychology and its neighboring fields and has even prompted new fields of research such as behavioral economics (e.g., Thaler, 2016) and behavioral law (e.g., Sunstein, 2000). However, it has also provoked a number of critical objections. One is that the standards against which human reasoning is compared in this program are narrow and predominantly content-free coherence norms (Arkes, Gigerenzer, & Hertwig, 2016). Another is that the identified cognitive illusions are by no means as impervious to change (e.g., via different formats for representing probabilistic information; see, e.g., Hoffrage, Lindsey, Hertwig, & Gigerenzer, 2000) as has often been claimed. Finally, it has been demonstrated that at least some behaviors seemingly based on cognitive illusions are not best understood as manifestations of human irrationality; instead, they may be the product of evolved and adaptive decision rules. These rules, however, are not adapted to experimentalists' simplified environments and so are wrongly interpreted as irrational behavior (Fawcett et al., 2014; see also chapter 10.6 by Cosmides & Tooby, this handbook, for an evolutionary perspective on human reasoning).

Interestingly, this is a development already predicted by Simon (1957b), who wrote, "I believe that the return swing of the pendulum will begin, that we will begin to interpret as rational and reasonable many facets of human behavior that we now explain in terms of affect" (p. 200).

2.3 Ecological Rationality

The third influential interpretation of bounded rationality is ecological rationality and the study of fast-and-frugal heuristics.[6] This approach aims to understand boundedly rational decision mechanisms based on how they match the statistical structures of choice environments. Following Simon's emphasis on understanding and modeling rational behavior as shaped by both cognitive and environmental structures, research on ecological rationality explores the adaptive toolbox of simple heuristics that human minds have developed through individual, cultural, or evolutionary learning (see Gigerenzer, Todd, & ABC Research Group, 1999; Hertwig, Hoffrage, & ABC Research Group, 2013; Hertwig, Pleskac, Pachur, & the Center for Adaptive Rationality, 2019; Todd, Gigerenzer, & ABC Research Group, 2012). Simple heuristics, understood as the boundedly rational mind's main tools in games against nature and social games (Hertwig & Herzog, 2009), are defined as follows: "A heuristic is a strategy that ignores part of the information, with the goal of making decisions more quickly, frugally, and/or accurately than more complex methods" (Gigerenzer & Gaissmaier, 2011, p. 454).

Proposing a new "theory of the rational" (Simon, 1957b, p. 200), ecological rationality strives to model rational behavior in terms of correspondence norms (Hammond, 2000). Correspondence norms represent a measure of cognitive success in achieving one's goals in the world—a consequentialist interpretation of rationality in cognition (Kozyreva & Hertwig, 2021; Schurz & Hertwig, 2019; see also chapter 1.3 by Schurz, this handbook). Depending on the environment, cognitive success could be measured in terms of criteria such as accuracy, speed, frugality, robustness, or accountability. An important methodological principle in research on ecological rationality is comparatively testing heuristic strategies against informationally and computationally complex strategies. Ideally, the success of each strategy should be measured against that of others and across a reference class of environments (for examples of such comparative tests, see Gigerenzer & Brighton, 2009; Spiliopoulos & Hertwig, 2020). An important insight stemming from tests involving a set of strategies and environments is that heuristics do not result in good or bad performance in and of themselves; a heuristic's performance is relative to the structure of the environment in which it is employed. If a heuristic is able to exploit the structure at hand, the heuristic can be surprisingly accurate. Similarly, it can yield dismal performance when facing environment structures that do not match its architecture or its assumptions about the environment. Just as the ant's complex path cannot be understood apart from the environment in which it unfolded, cognitive tools and their ecological rationality cannot be divorced from the environmental structures they may exploit.

Research on ecological rationality has produced three key discoveries pertaining to the understanding of bounded rationality. First, bounded rationality does not mean that simplicity in search, information integration, and decision rules will inevitably result in second-rate performance. In fact, less is often more. Second, bounded rationality requires an ongoing examination of the environmental structures that support heuristics' performance. To "describe, predict and explain the behavior of a system of bounded rationality, we must both construct a theory of the system's processes and describe the environments to which it is adapting" (Simon, 1990, p. 7). One could go even further. Rather than cataloging piecemeal descriptions of environments, one could harness existing theories of environments that shed light on the

factors and dynamics that amplify, attenuate, or even wipe out the environmental structure in question. Based on such a theory, one would then examine how the environmental processes interact with theories of the system's process—for an illustration of this approach, see an analysis of the risk–reward structure using the foraging theory of ideal free distribution in Pleskac, Hertwig, Leuker, and Conradt (2019). Third, heuristics can enable good performance in a wide range of domains (e.g., judgment, choice, inference, classification) in which people face uncertainty, lack of information, and time pressure.

In briefly reviewing the last discovery relating to the often surprising success of heuristics, we start with the question of why people recruit boundedly rational heuristics to begin with. The classical explanation is that people resort to heuristics in a kind of compromise, saving cognitive effort at the cost of accuracy (Shah & Oppenheimer, 2008). Humans and other animals rely on heuristics because searching for and processing information can be taxing, and heuristics offer relief by trading reduced accuracy for faster, more frugal cognition. This accuracy–effort trade-off—variously conceptualized as a rational and beneficial trade-off (Payne, Bettman, & Johnson, 1993) or as one that is constrained by capacity limitations and paid for in terms of reasoning mistakes (the heuristics-and-biases view)—has been touted as a potentially universal law of cognition. But there is an alternative explanation for why people employ heuristics: less computation and less information can result in performance that is as good as, or even better than, that of computationally and informationally complex strategies. In a wide range of studies, often involving computer simulations and analytical work, the ecological rationality program has demonstrated numerous instances of such less-is-more effects (Gigerenzer, Hertwig, & Pachur, 2011; Spiliopoulos & Hertwig, 2020).

One way to understand these effects is through the distinction between prediction and fitting. *Prediction* occurs when the data have not yet been observed and a model, whether heuristic or complex in nature, with fixed parameter values is employed to predict future events. *Fitting*, in contrast, occurs when the data have already been observed and the parameters of a model are estimated in order to maximize the fit between the data and the model's behavior. Generally, the more free parameters a model has, the better its fit. This rule, however, does not hold for predictions. When parameters need to be estimated from small or unreliable samples, the function between predictive accuracy and model flexibility (e.g., the number of free parameters) is typically inversely U-shaped. This means that models that are too simple (e.g., have too few parameters) or too complex can fail in prediction (Pitt, Myung, & Zhang, 2002). Heuristics—at least some—can occupy the sweet spot between too little and too much complexity.

The distinction between prediction and fitting is related to another theoretical construct relevant to understanding why heuristics can perform so well: the bias–variance framework (exploited in machine learning; Geman, Bienenstock, & Doursat, 1992). Heuristics can succeed because they smartly trade off two components of prediction error: *bias* (how well, on average, the model can agree with the ground truth) and *variance* (the variation around this average). Heuristics tend to have higher bias but lower variance than more complex models (with more adjustable parameters), which explains why heuristics perform relatively better when making predictions from small environmental samples (Gigerenzer & Brighton, 2009; Katsikopoulos, Schooler, & Hertwig, 2010). Variance is a substantial source of error when information about the environment is sparse, thus exposing more complex models to the risk of overfitting by virtue of the flexibility granted by their parameters. Heuristics, however, are less flexible and thus less likely to overfit, which gives them the chance to outperform more complex models when knowledge about the environment is incomplete and uncertainty is high.

Still another approach to understanding heuristics' ecological rationality is in terms of the match between environmental structures (i.e., statistical properties that reflect patterns of information distribution in the given ecology) and cognitive strategies (Hogarth & Karelaia, 2006, 2007). For instance, one notable property of an environment is the presence of a *noncompensatory* cue. This is an attribute that has a much higher correlation with the ground truth than all other attributes combined, which means it cannot be outweighed (compensated) by any combination of less valid cues or attributes (for details and other environmental structures, see Şimşek, 2013). Noncompensatory cues can be exploited by lexicographic heuristics such as Take-the-Best (Gigerenzer & Goldstein, 1996) and LEX (Payne et al., 1993) that order attributes according to importance and process them sequentially.

Finally, another significant contribution of ecological rationality explores the ways in which heuristics can be engineered to foster better decisions in uncertain circumstances. As we have seen, the heuristics-and-biases program has focused on a small set of heuristics and refrained from proposing heuristics to help people make better decisions. This is consistent with the view that heuristics cannot compete with, let alone outperform, more complex models. In contrast, the ecological rationality program has suggested and inspired new heuristics (often in terms of fast-and-frugal trees; Martignon, Katsikopoulos, & Woike, 2008) for a wide range of professional uses, including geographic profiling, prescribing antibiotics, and allocating marketing resources (see Gigerenzer et al., 2011). This ongoing work suggests that the study of boundedly rational heuristics not only examines which heuristics people use but also addresses profound prescriptive concerns, thereby providing the basis for boosting people's competences by giving them smart and easy-to-use tools and decision aids (Hertwig & Grüne-Yanoff, 2017; the boosting approach to public policy making is an alternative to the nudging approach that is rooted in the heuristics-and-biases program; Thaler & Sunstein, 2008).

3. Conclusion

Having already inspired groundbreaking investigations into a map of cognitive illusions and an adaptive toolbox of ecologically rational heuristics, the exploration of bounded rationality has nevertheless in many regards only just begun. Great challenges are ahead, including constructing an encompassing theory-based taxonomy of environmental structures that decision makers perceive and the heuristics that exploit them. Another largely neglected research topic is how simple heuristics, by interacting with environment structures as well as with the heuristics of other decision makers, give rise to complexity in the environment (see, e.g., Hertwig, Davis, & Sulloway, 2002; for a related concern, see Schelling, 2006). In the years to come, the study of bounded rationality should further explore and detail the content of the adaptive toolbox beyond simple heuristics, investigating other cognitive tools that people—both as individuals and in groups—use to handle uncertainty (see Hertwig et al., 2019). Perhaps the most ambitious challenge of all is to uncover the extent to which integrating the sciences and methodologies of heuristics, search, and learning (Hertwig, 2015; Lejarraga & Hertwig, in press) can offer a unified framework for the study of the boundedly rational mind. In a future of vast complexity, informational affluence, and potential for surprise, the need for a realistic vision of rational choice will only grow—and with it the scientific exploration of bounded rationality.

Acknowledgments

We are grateful to Max Albert, Hartmut Kliemt, and Gregory Wheeler for their comments and suggestions. We thank Deb Ain for editing the manuscript.

Notes

1. In the second edition of *Administrative Behavior*, Simon (1947/1957) wrote, "Administrative theory is peculiarly the theory of intended and bounded rationality—of the behavior of human beings who *satisfice* because they have not the wits to *maximize*" (p. xxiv). He later asserted, "In *Administrative Behavior*, bounded rationality is largely characterized as a residual category—rationality is bounded when it falls short of omniscience" (Simon, 1979, p. 502).

2. Simon employed several terms throughout his writings to define this theory, including "global rationality," "perfect rationality," "objective rationality," "substantive rationality," "full rationality," "homo economicus," and "Olympian model of rationality." The target of these descriptions was the dominant classical rational choice theory, including, most notably, expected utility and subjective expected utility theories. In rational choice theory, norms of coherence and axioms of expected utility theory (e.g., transitivity, completeness) are taken as benchmarks of rational decision making. This approach, sometimes described as the standard picture of rationality (Stein, 1996), remains the most commonly used normative approach to rationality in decision sciences.

3. The distinction between aleatory and epistemic uncertainty (or objective and subjective uncertainty) stems from the duality inherent in the modern concept of probability, which encompasses both epistemic probability (subjective degrees of belief) and aleatory probability (stable frequencies displayed by chance devices; Hacking, 1975/2006). In a similar vein, epistemic uncertainty refers to incomplete knowledge or information, whereas aleatory uncertainty stems from the statistical properties of the environment, which exist independently of a person's knowledge (Kozyreva & Hertwig, 2019).

4. In his Nobel Prize lecture, Simon (1979) stated, "A strong positive case for replacing the classical theory by a model of bounded rationality begins to emerge when we examine situations involving decision making under uncertainty and imperfect competition. These situations the classical theory was never designed to handle, and has never handled satisfactorily" (p. 497).

5. The idea of optimization under constraints was originally outlined by Good in 1952 when he observed that agents in the real world seek to minimize costs involved in obtaining information and making decisions (in terms of both time and effort; see also Wheeler, 2018/2019).

6. This use of the term "ecological rationality" is not to be confused with the terminologically similar but conceptually different concept developed by Vernon Smith (2003).

References

Arkes, H. R., Gigerenzer, G., & Hertwig, R. (2016). How bad is incoherence? *Decision*, 3(1), 20–39.

Chase, W. G., & Simon, H. A. (1973). Perception in chess. *Cognitive Psychology*, 4(1), 55–81.

Fawcett, T. W., Fallenstein, B., Higginson, A. D., Houston, A. I., Mallpress, D. E., Trimmer, P. C., & McNamara, J. M. (2014). The evolution of decision rules in complex environments. *Trends in Cognitive Sciences*, 18(3), 153–161.

Friedman, M. (1953). *Essays in positive economics*. Chicago, IL: University of Chicago Press.

Geman, S., Bienenstock, E., & Doursat, R. (1992). Neural networks and the bias/variance dilemma. *Neural Computation*, 4(1), 1–58.

Gigerenzer, G., & Brighton, H. (2009). Homo heuristicus: Why biased minds make better inferences. *Topics in Cognitive Science*, 1(1), 107–143.

Gigerenzer, G., & Gaissmaier, W. (2011). Heuristic decision making. *Annual Review of Psychology*, 62, 451–482.

Gigerenzer, G., & Goldstein, D. G. (1996). Reasoning the fast and frugal way: Models of bounded rationality. *Psychological Review*, 103(4), 650–669.

Gigerenzer, G., Hertwig, R., & Pachur, T. (2011). *Heuristics: The foundations of adaptive behavior*. Oxford, England: Oxford University Press.

Gigerenzer, G., & Selten, R. (Eds.). (2001). *Bounded rationality: The adaptive toolbox*. Cambridge, MA: MIT Press.

Gigerenzer, G., Todd, P. M., & ABC Research Group. (1999). *Simple heuristics that make us smart*. Oxford, England: Oxford University Press.

Good, I. J. (1992). Rational decisions. In S. Kotz & N. L. Johnson (Eds.), *Breakthroughs in statistics: Foundations and basic theory* (pp. 365–377). New York, NY: Springer. (Original work published 1952)

Hacking, I. (2006). *The emergence of probability: A philosophical study of early ideas about probability, induction and statistical inference* (2nd ed.). Cambridge, England: Cambridge University Press. (Original work published 1975)

Hammond, K. R. (2000). Coherence and correspondence theories in judgment and decision making. In T. Connolly, H. A. Arkes, & K. R. Hammond (Eds.), *Judgment and decision making: An interdisciplinary reader* (pp. 53–65). Cambridge, England: Cambridge University Press.

Hanoch, Y. (2002). "Neither an angel nor an ant": Emotion as an aid to bounded rationality. *Journal of Economic Psychology*, 23(1), 1–25.

Hertwig, R. (2015). Decisions from experience. In G. Keren & G. Wu (Eds.), *The Wiley Blackwell handbook of judgment and decision making* (Vol. 1, pp. 240–267). Chichester, England: Wiley Blackwell.

Hertwig, R., Davis, J. N., & Sulloway, F. J. (2002). Parental investment: How an equity motive can produce inequality. *Psychological Bulletin, 128*(5), 728–745.

Hertwig, R., & Grüne-Yanoff, T. (2017). Nudging and boosting: Steering or empowering good decisions. *Perspectives on Psychological Science, 12*(6), 973–986.

Hertwig, R., & Herzog, S. M. (2009). Fast and frugal heuristics: Tools of social rationality. *Social Cognition, 27*(5), 661–698.

Hertwig, R., Hoffrage, U., & ABC Research Group. (2013). *Simple heuristics in a social world.* Oxford, England: Oxford University Press.

Hertwig, R., Pleskac, T. J., Pachur, T., & Center for Adaptive Rationality. (2019). *Taming uncertainty.* Cambridge, MA: MIT Press.

Hoffrage, U., Lindsey, S., Hertwig, R., & Gigerenzer, G. (2000). Communicating statistical information. *Science, 290*(5500), 2261–2262.

Hogarth, R. M., & Karelaia, N. (2006). Regions of rationality: Maps for bounded agents. *Decision Analysis, 3*(3), 124–144.

Hogarth, R. M., & Karelaia, N. (2007). Heuristic and linear models of judgment: Matching rules and environments. *Psychological Review, 114*(3), 733–758.

Kahneman, D. (2000). Preface. In D. Kahneman & A. Tversky (Eds.), *Choices, values, and frames* (pp. ix–xvii). Cambridge, England: Cambridge University Press.

Kahneman, D. (2003). Maps of bounded rationality: Psychology for behavioral economics. *American Economic Review, 93*(5), 1449–1475.

Kahneman, D. (2011). *Thinking, fast and slow.* London, England: Penguin Books.

Kahneman, D., & Frederick, S. (2002). Representativeness revisited: Attribute substitution in intuitive judgment. In T. Gilovich, D. Griffin, & D. Kahneman (Eds.), *Heuristics and biases: The psychology of intuitive judgment* (pp. 49–81). Cambridge, England: Cambridge University Press.

Kahneman, D., Slovic, P., & Tversky, A. (Eds.). (1982). *Judgment under uncertainty: Heuristics and biases.* Cambridge, England: Cambridge University Press.

Kahneman, D., & Tversky, A. (1979). Prospect theory: An analysis of decision under risk. *Econometrica: Journal of the Econometric Society, 47*(2), 263–291.

Katsikopoulos, K. V., Schooler, L. J., & Hertwig, R. (2010). The robust beauty of ordinary information. *Psychological Review, 117*(4), 1259–1266.

Klein, G. A. (1998). *Sources of power: How people make decisions.* Cambridge, MA: MIT Press.

Kozyreva, A., & Hertwig, R. (2021). The interpretation of uncertainty in ecological rationality. *Synthese, 198*, 1517–1547. doi:10.1007/s11229-019-02140-w

Lejarraga, T., & Hertwig, R. (in press). How experimental methods shaped views on human competence and rationality. *Psychological Bulletin.*

Martignon, L., Katsikopoulos, K. V., & Woike, J. (2008). Categorization with limited resources: A family of simple heuristics. *Journal of Mathematical Psychology, 52*(6), 352–361.

Miller, G. F., & Todd, P. M. (1998). Mate choice turns cognitive. *Trends in Cognitive Sciences, 2*(5), 190–198.

Payne, J. W., Bettman, J. R., & Johnson, E. J. (1993). *The adaptive decision maker.* Cambridge, England: Cambridge University Press.

Pitt, M. A., Myung, I. J., & Zhang, S. (2002). Toward a method of selecting among computational models of cognition. *Psychological Review, 109*(3), 472–491.

Pleskac, T. J., Hertwig, R., Leuker, C., & Conradt, L. (2019). Using risk–reward structures to reckon with uncertainty. In R. Hertwig, T. Pleskac, T. Pachur, & the Center for Adaptive Rationality (Eds.), *Taming uncertainty* (pp. 51–70). Cambridge, MA: MIT Press.

Sargent, T. J. (1993). *Bounded rationality in macroeconomics: The Arne Ryde memorial lectures.* Oxford, England: Clarendon Press.

Savage, L. J. (1954). *The foundations of statistics.* New York, NY: Wiley.

Schelling, T. C. (2006). *Micromotives and macrobehavior.* New York, NY: Norton.

Schooler, L. J., & Hertwig, R. (2005). How forgetting aids heuristic inference. *Psychological Review, 112*(3), 610–628.

Schurz, G., & Hertwig, R. (2019). Cognitive success: A consequentialist account of rationality in cognition. *Topics in Cognitive Science, 11*(1), 7–36.

Selten, R. (2001). What is bounded rationality? In G. Gigerenzer & R. Selten (Eds.), *Bounded rationality: The adaptive toolbox* (pp. 13–36). Cambridge, MA: MIT Press.

Shah, A. K., & Oppenheimer, D. M. (2008). Heuristics made easy: An effort-reduction framework. *Psychological Bulletin, 134*(2), 207–222.

Shannon, C. E. (1950). Programming a computer for playing chess. *The London, Edinburgh, and Dublin Philosophical Magazine and Journal of Science, 41*(314), 256–275.

Silver, D., Hubert, T., Schrittwieser, J., Antonoglou, I., Lai, M., Guez, A., . . . Hassabis, D. (2018). A general reinforcement learning algorithm that masters chess, shogi, and Go through self-play. *Science, 362*(6419), 1140–1144.

Simon, H. A. (1955). A behavioral model of rational choice. *Quarterly Journal of Economics, 69*(1), 99–118.

Simon, H. A. (1956). Rational choice and the structure of the environment. *Psychological Review, 63*(2), 129–138.

Simon, H. A. (1957a). *Administrative behavior* (2nd ed.). New York, NY: Macmillan. (Original work published 1947)

Simon, H. A. (1957b). *Models of man, social and rational: Mathematical essays on rational human behavior in a social setting*. New York, NY: Wiley.

Simon, H. A. (1972). Theories of bounded rationality. *Decision and Organization*, *1*(1), 161–176.

Simon, H. A. (1976). From substantive to procedural rationality. In T. J. Kastelein, S. K. Kuipers, W. A. Nijenhuis, & G. R. Wagenaar (Eds.), *25 years of economic theory: Retrospect and prospect* (pp. 65–86). Boston, MA: Springer.

Simon, H. A. (1979). Rational decision making in business organizations. *American Economic Review*, *69*(4), 493–513.

Simon, H. A. (1982). *Models of bounded rationality: Empirically grounded economic reason* (Vol. 3). Cambridge, MA: MIT Press.

Simon, H. A. (1983). *Reason in human affairs*. Stanford, CA: Stanford University Press.

Simon, H. A. (1989). The scientist as problem solver. In D. Klahr & K. Kotovsky (Eds.), *Complex information processing: The impact of Herbert A. Simon* (pp. 373–398). Hillsdale, NJ: Erlbaum.

Simon, H. A. (1990). Invariants of human behavior. *Annual Review of Psychology*, *41*(1), 1–20.

Simon, H. A. (1991). *Models of my life*. New York, NY: Basic Books.

Simon, H. A. (1992). What is an "explanation" of behavior? *Psychological Science*, *3*(3), 150–161.

Simon, H. A. (1996). *The sciences of the artificial* (3rd ed.). Cambridge, MA: MIT Press. (Original work published 1969)

Şimşek, Ö. (2013). Linear decision rule as aspiration for simple decision heuristics. *Advances in Neural Information Processing Systems*, *26*, 2904–2912.

Smith, V. L. (2003). Constructivist and ecological rationality in economics. *American Economic Review*, *93*(3), 465–508.

Spiliopoulos, L., & Hertwig, R. (2020). A map of ecologically rational heuristics for uncertain strategic worlds. *Psychological Review*, *127*(2), 245–280.

Stein, E. (1996). *Without good reason: The rationality debate in philosophy and cognitive science*. Oxford, England: Clarendon Press.

Stigler, G. J. (1961). The economics of information. *Journal of Political Economy*, *69*(3), 213–225.

Sunstein, C. R. (Ed.). (2000). *Behavioral law and economics*. Cambridge, England: Cambridge University Press.

Thaler, R. H. (2016). Behavioral economics: Past, present, and future. *American Economic Review*, *106*(7), 1577–1600.

Thaler, R. H., & Sunstein, C. R. (2008). *Nudge: Improving decisions about health, wealth and happiness*. New York, NY: Simon & Schuster.

Todd, P. M., Gigerenzer, G., & ABC Research Group. (2012). *Ecological rationality: Intelligence in the world*. Oxford, England: Oxford University Press.

Tversky, A., & Kahneman, D. (1974). Judgment under uncertainty: Heuristics and biases. *Science*, *185*(4157), 1124–1131.

Tversky, A., & Kahneman, D. (1983). Extensional versus intuitive reasoning: The conjunction fallacy in probability judgment. *Psychological Review*, *90*(4), 293–315.

Tversky, A., & Kahneman, D. (1992). Advances in prospect theory: Cumulative representation of uncertainty. *Journal of Risk and Uncertainty*, *5*(4), 297–323.

Vriend, N. J. (1996). Rational behavior and economic theory. *Journal of Economic Behavior & Organization*, *29*(2), 263–285.

Wheeler, G. (2019). Bounded rationality. In E. N. Zalta (Ed.), *The Stanford encyclopedia of philosophy*. Retrieved from https://plato.stanford.edu/archives/fall2019/entries/bounded-rationality/

8.6 Reasoning, Rationality, and Metacognition

Valerie A. Thompson, Shira Elqayam, and Rakefet Ackerman

Summary

In this chapter, we explore how the study of metacognition advances our understanding of human rationality. "Metacognition" refers to the processes by which we monitor and control our ongoing cognitive activities. These monitoring processes manifest as feelings on the continuum between certainty and uncertainty, which then are the stimulus for control processes (e.g., a feeling of uncertainty in an answer is a stimulus to rethink that answer, search online). Because feelings of (un)certainty are based on heuristic cues, such as the ease with which an answer comes to mind, they can mislead our control processes and produce unwarranted levels of confidence in our decisions. We also argue that metacognitive principles can help us to understand why people are not the cognitive misers that they are assumed to be and why they continue to waste cognitive resources on difficult or unsolvable problems.

1. Reasoning, Rationality, and Metacognition

Here, we explore how understanding metacognitive processes can shed light on issues of rationality. At first glance, the connection between metacognition and rationality may seem straightforward. Metacognition, after all, is typically understood to be "thinking about one's thinking," a process that seems integral to achieving rational decisions.[1] However, most researchers are less concerned with this self-reflective stance and more concerned with the ubiquitous processes by which we monitor and control our ongoing cognitions. By "monitoring," we mean the processes that keep track of how well we are doing; control processes include changing strategies, maintaining course, or giving up.

Monitoring and control again sound like high-level, reflective processes, and there is no doubt that they can be. In most ordinary instances, however, they operate passively (Koriat, 2007), in much the same way as a thermostat passively monitors the temperature of a room.

The associated control process would be to send a signal to the furnace to start producing warm air when the temperature drops below a certain threshold (Thompson, 2016). Examples include the following:

1) You meet someone whose face you recognize but whose name you are not certain of. You decide not to address her by name.

2) You order a desk from a furniture store and it arrives, unassembled. Your first thought is "I can't put that together." You then look at the instructions, reconsider, and begin assembling. Halfway through, you realize that your first judgment was right and call a friend to help.

3) You are working on writing a manuscript and are dissatisfied with the progress. Even though the words accurately convey your thoughts, you think that it does not have that certain ring that will convince readers about the points you wish to make. You decide to take a break and make a fresh start in the morning.

In each of these examples, a set of cognitive processes is unfolding, involving memory retrieval, solving a problem, writing a manuscript, and a second set of processes layered on top, which we refer to as "monitoring and control processes." The monitoring processes are assumed to be continuous and, as above, to manifest primarily as a sense of (un)certainty about how well things are proceeding. In this chapter, we will primarily be concerned with monitoring and control processes associated with reasoning, problem solving, and decision making, an area of study termed "Meta-Reasoning" (Ackerman & Thompson, 2015, 2017).

The meta-reasoning framework posits many different kinds of monitoring judgments. A *judgment of solvability* is an initial assessment of whether a problem is solvable and whether one personally is capable of solving it, as the self-assembly example in number 2 illustrates. The *feeling of rightness* (FOR) accompanies an initial, intuitive answer to a problem, as illustrated by example 1.

Examples 2 and 3 illustrate the ongoing nature of monitoring and the assessment of intermediate confidence. At the end of a process, one is left with a final sense of confidence or feeling of error that one has successfully or unsuccessfully completed the task at hand.

The reason these monitoring judgments matter is their link to control processes: feelings of certainty and uncertainty are the arbiters of action (Thompson, 2016). A feeling of certainty is a signal that all is well and to carry on with the present course of action. Uncertainty signals the need to reassess. For example, reasoning problems that cue a strong FOR to an initial answer receive less reanalysis than problems that cue a low FOR, whereas a low FOR is a cue to look for alternative answers (Thompson, Prowse Turner, & Pennycook, 2011); similarly, low intermediate confidence is a signal to give up (Ackerman, 2014) or seek additional resources, as in the self-assembly example, and initial judgments of solvability (example 2) predict persistence in problem solving (Lauterman & Ackerman, 2019).

The trouble is that these monitoring judgments are usually made implicitly, on the basis of heuristic cues (Koriat, 1997). Heuristic cues can be based on general self-perceptions (e.g., "I am no good at DIY projects"), global information about the task (e.g., "These instructions seem particularly opaque"), and characteristics of individual items (see Ackerman, 2019, for a review). One item-based cue is fluency, or the ease with which an answer comes to mind. Fluency is a cue to confidence and FOR (Ackerman & Zalmanov, 2012; Thompson, Evans, & Campbell, 2013), regardless of accuracy. That is, answers that come to mind fluently are experienced positively (Thompson & Morsanyi, 2012) and lead to a sense of certainty, as does familiarity (Reder & Ritter, 1992).

The implications for human rationality are clear: because the monitoring processes can send misleading signals, control processes may not operate in an optimal manner. That is, one might believe that a problem is solvable, leading one to persist in trying to solve it, even when it is not (Lauterman & Ackerman, 2019). Similarly, one might have a strong FOR in a suboptimal answer and thus fail to think further; conversely, one might have a weak FOR in a good solution and thus spend unnecessary time thinking it over (Thompson et al., 2011; Thompson & Johnson, 2014). Indeed, one of the reasons that people commit errors on traditional heuristics-and-biases tasks is that the first answer tends to come to mind fluently and is therefore held with high confidence (Thompson et al., 2013). As a result, one may fail to reflect further on the answer and give an answer based on beliefs, stereotypes, or familiarity,

when the correct answer requires one to think logically or probabilistically.

In this chapter, we consider the advantages and drawbacks of the proper allocation of deliberate thinking and the role of normative theories, such as logic and probability, as markers of rationality. As stated earlier, the goal of this chapter is to consider how understanding metacognitive processes can advance our understanding of human rationality. In doing so, we examine how metareasoning and metacognition shed new, and sometimes surprising, light on rationality, when considered from a variety of perspectives. In addition to normative correctness, we consider issues related to pragmatic rationality, bounded rationality, and epistemic rationality.

2. Overconfidence and Epistemic Rationality

Epistemic rationality is usually understood to mean having true beliefs about the world (i.e., believing true propositions and disbelieving false ones).[2] In this section, we demonstrate how miscalibrated judgments of confidence play a role in acquiring and maintaining false beliefs about the world and the manner in which one evaluates one's knowledge of the world.

The *feeling of knowing* is one type of metacognitive judgment we make in evaluating our own knowledge. For example, one may be quite sure that one will ultimately be able to remember the capital city of Uruguay, even if it currently eludes recall. These feelings are often inferences that rely on cues such as the amount of information that comes to mind (Koriat, 1993), for example, it starts with M, is on the coast, and so on. Dunning (2011) argued that these cues underlie a set of phenomena he termed "meta-ignorance": an ignorance about the things one does not know about. Examples of meta-ignorance include *overclaiming*, in which people claim to have knowledge about topics that actually do not exist. This tendency is especially pronounced in areas in which people claim to be highly knowledgeable (Atir, Rosenzweig, & Dunning, 2015). For example, participants who rated themselves to be knowledgeable about personal finance were more likely to claim knowledge about fictitious financial constructs, such as "fixed-rate deduction." Another example is the *illusion of explanatory depth*, in which people believe that they understand how common objects, such as toilets and toasters, work, when in fact they do not (Fernbach, Rogers, Fox, & Sloman, 2013; Rozenblit & Keil, 2002). It appears that our familiarity and experience with these objects creates a misplaced sense of understanding. Dunning (2011) describes numerous instances of situations, ranging from the trivial to the calamitous, in which people demonstrate that they are unaware of how little they know or how poorly they understand a topic or situation, leading them to sometimes dangerous actions (e.g., rewiring the electrical system in one's house). In such cases, people have a metacognitive experience feeling that leads them to believe that they know more than they actually do.

A third example is the *Dunning–Kruger effect*, in which people demonstrate more confidence in their responses than their performance warrants (Kruger & Dunning, 1999; see also Fischhoff, Slovic, & Lichtenstein, 1977). Of particular relevance to the rationality debate is the demonstration that this effect extends to tasks designed to measure reasoning biases (Pennycook, Ross, Koehler, & Fugelsang, 2017). In one study, participants completed the *cognitive reflection test* (Frederick, 2005), which consists of three items such as the following:

> A ball and a bat together cost $1.10. The bat costs $1.00 more than the ball. How much does the ball cost?

About two-thirds of respondents mistakenly answer "10¢," when a bit of basic algebra would indicate that the correct answer is "5¢." The answer "10¢" comes easily to mind, possibly because people miss or misinterpret the "more than" (Mata, Ferreira, Voss, & Kollei, 2017). As such, the results of this test are commonly interpreted as a measure of people's tendency to respond intuitively rather than reflectively (Toplak, West, & Stanovich, 2011). Pennycook et al. (2017) observed a typical Dunning–Kruger effect: those who performed well tended to underestimate their performance, whereas those who performed most poorly believed they did three times better than they actually did (for related findings, see Mata, Ferreira, & Sherman, 2013). The poor performers are therefore "doubly cursed" (Kruger & Dunning, 1999): not only do they perform poorly, but they are unaware of it and therefore are not in a position to ameliorate their performance.

There are two other relevant points to make about confidence and overconfidence. First, evidence has emerged showing that confidence has a trait-like property, in that people who are confident (or overconfident) about their performance in one domain also tend to be so in other domains (Jackson & Kleitman, 2014; Stankov, Kleitman, & Jackson, 2014). Second, as argued above, confidence is the arbiter of action. In this case, confidence has been found to predict decision-making style (Jackson & Kleitman, 2014; Jackson, Kleitman, Stankov, & Howie, 2016). To test this hypothesis, reasoners were asked whether they wanted to take action on each decision, for example, by submitting an answer (i.e., to count as part of the total

score) or administering a treatment for a fictitious disease. It was found that confident reasoners take actions that are congruent with the decision they made, regardless of whether it was accurate. Consequently, those who are overconfident make errors of commission (i.e., they act when they should not). Conversely, those who are underconfident make errors of omission (i.e., fail to act when they are correct).

Up to this point, we have discussed ways in which metacognition is a source of epistemic irrationality, by producing monitoring judgments, based on unreliable cues, that lead to overconfidence. A more optimistic view comes from work on consistency and on metacognitive coherence. For the latter, read: regularity and predictability; it means the sense of fit between environmental cues and our knowledge and expectations (Topolinski, 2015, 2018).[3]

We start with work showing that metacognitive judgments are sensitive to semantic coherence, that is, they are sensitive to the rules that govern associations between entries in our mental lexicon[4] (Sweklej, Balas, Pochwatko, & Godlewska, 2014; for a related view, see Betsch & Glöckner, 2010). Much of the work in support of this claim is based on a variation of Mednick and Mednick's (1967) Remote Associates Test. The experimental paradigm involves presenting participants with word triads. The researchers defined triads to be coherent when the three words were associated with a fourth word (e.g., "struck," "beam," and "light" are associated with "moon"). For incoherent triads, there is no common remote associate (e.g., "car," "tiger," and "cream"). Semantic coherence, in this context, is a type of fluency cue—it triggers a feeling of processing ease (or, if lacking, difficulty). Bolte and Goschke (2005) observed that participants are able to identify semantic coherence of triads at above-chance rates, even when they were required to make such judgments after a brief presentation (namely, in 1.5 seconds; for a review, see Topolinski, 2015; but see Ackerman & Beller, 2017).

Although it is not usual to think of them in such a way, the judgments of semantic coherence used in this line of research are akin to metacognitive monitoring judgments done regarding one's own cognitive performance (Thompson, 2014). We know that people can make judgments regarding their knowledge under similar time constraints (Reder & Ritter, 1992). Moreover, both types of judgments make an inference about a mental state, based on the heuristic cues that generate feeling of fluency (Topolinski, 2015). In Ackerman and Thompson's (2015, 2017) framework, both are similar to judgments of solvability, which, as described above, are prospective judgments about whether the participant would be able to solve the problem at hand (see examples in Ackerman & Beller, 2017; Lauterman & Ackerman, 2019).

While consistency and coherence are different aspects of metacognitive judgment, they both have to do with regularity and predictability versus conflict. Evidence for the role of consistency in metacognitive judgments can be found in the so-called *conflict detection effect* (De Neys, 2014). Many classic reasoning problems are designed to be tricky, in that they pit two conflicting answers against each other, as in the bat and ball problem above. People's confidence in their answers to such problems is lower than it is to control versions in which there is no trick (De Neys, Cromheeke, & Osman, 2011; Thompson et al., 2011; Thompson & Johnson, 2014). That is, when such problems generate a feeling of disfluent processing, by permitting two conflicting answers, it is reflected in people's confidence judgments.

Indeed, Koriat (2012) argued that consistency is the major determinant of confidence and also underlies fluency. His *self-consistency model* (SCM) defines self-consistency as a mnemonic cue that captures agreement among a variety of considerations, including task performance and knowledge. The SCM applies in cases where there are two alternatives from which to choose. To decide, people are assumed to gather/retrieve information about those two alternatives. Confidence in the decision is determined by the consistency of the information that favors the preferred alternative: choices that have many pieces of information favoring them, and few drawbacks, engender confidence and will be made relatively more fluently than those that are less consistent.

3. Normative, Practical, and Bounded Rationality in Light of Metacognitive Research

So far, the literature we reviewed resonates well with the *heuristics and biases* tradition in the psychology of decision making (e.g., Gilovich, Griffin, & Kahneman, 2002; and see chapter 2.4 by Fiedler, Prager, & McCaughey, this handbook). The core idea of heuristics and biases is *meliorist* (a term we borrow from Stanovich, 1999): people draw on heuristic mental shortcuts, which in everyday life often work well enough, but in many contexts can be spectacularly off the mark, leading to biased reasoning and decision making. This approach highlights the notorious *normative–descriptive gap*: it holds that there is a set of norms that ought to be met (such as those of the probability calculus or classical logic), but actual behavior falls short of them.

However, there is more to rationality than meeting normative rules (or "rationality$_2$," as Evans & Over, 1996,

labeled it). Rationality can also be gauged by practical or instrumental standards ("rationality$_1$," in the Evans and Over nomenclature): does the thought or decision or act lead to achieving one's goals? Behavior that is normatively irrational can often be argued to be pragmatically rational and vice versa. The term often mentioned in this context is "bounded rationality" (Simon, 1982; and see chapter 8.5 by Hertwig & Kozyreva, this handbook), an idea that turns out to be surprisingly metacognitive. Simon highlights the inherent cognitive and biological limitations on human thinking: we do not have unbounded working memory or unlimited attentional resources; we do not live forever. Some normative solutions to rational puzzles are computationally intractable in that they would require more than the lifetime of the universe to be computed, never mind a human life span. Thus, with human rationality necessarily bounded by these limitations, the rational approach is, rather than to try and find the slippery, often intractable, optimal solutions, to scale down to solutions that are just good enough: to *satisfice*, in Simon's terminology.

Seen through the lens of metacognitive research, satisficing is a type of stopping rule (see Ackerman & Thompson, 2015, 2017): it tells us when we can stop searching for a better response, because the one we have is good enough. Nonetheless, the concept of satisficing on its own is rather underspecified. All it tells us is that people abort the search prior to achieving a normatively valid solution, but it does not tell us the details of the underlying psychological mechanisms, for example, that conceptual fluency (ease of processing) might trigger aborting the search regardless of how close to optimal the solution is. The unique insight from metacognitive research is that satisficing is not some sort of downgraded normative processing, where stopping depends on some quantitatively partial fulfilment of normative parameters. Rather, stopping processing depends on cues that are *qualitatively different* from normative parameters of the solution and only happen to (modestly) correlate with them (see Ackerman, 2014).

4. Meta-Reasoning and the Rationality of Persistence

4.1 Are Humans Cognitive Misers?
One surprising insight that metacognitive research on reasoning and decision making can afford us on human rationality regards the so-called *cognitive miser hypothesis*. Originally proposed in the context of social cognition (Fiske & Taylor, 1984, 1991), the cognitive miser hypothesis suggests that people are mental Harpagons, stingily doling out cognitive resources like the miser in

Molière's play reluctantly giving up his gold. For example, people are capable of overcoming social stereotyping to treat people as individuals rather than cast types, but this demands mental effort. Although people possess the requisite mental resources, they are often loath to invest them, thus succumbing to stereotyping due to mental laziness. The cognitive miser hypothesis was later adopted into the rationality debate by the meliorist tradition (Evans & Stanovich, 2013; Stanovich, 2009), where it transmuted into the more specific idea that biases (and therefore irrationality) result from defaulting into less effortful processing.

How sound is the cognitive miser hypothesis? There are several presuppositions here: (1) people tend to stop searching for solutions prematurely, even though (2) they have the necessary cognitive resources to complete the search, and (3) this leads to bias and normative irrationality. It follows that (4) investing more cognitive effort should correlate positively with being normatively rational. The problem is that each one of these theses is suspect. Theses (2) and (3) have both been challenged before (for thesis (2), see, e.g., Elqayam, 2012; for thesis (3), see Gigerenzer, Todd, & ABC Research Group, 1999). Where meta-reasoning research can provide novel insight is mainly with (1) and (4). Contrary to widespread consensus, people often *fail* to satisfice, and any correlation between cognitive investment and normative responding is weak at best (for a review, see Ackerman, 2014).

4.2 Thinking in Vain

Bounded rationality and satisficing have long been a source of contention within the great rationality debate. Within the heuristics and biases (aka meliorist) tradition, the term "bounded rationality" is used almost as a synonym for "error." In contrast, the Panglossian tradition (again borrowing a term from Stanovich, 1999), particularly in the literature dealing with the fast-and-frugal heuristics approach (Gigerenzer et al., 1999), sees bounded rationality as the epitome of human rationality (see also chapter 8.5 by Hertwig & Kozyreva, this handbook). In this chapter, we take no side in that particular debate, except to note that both sides share the presupposition that people do indeed satisfice—which is exactly what metacognitive research calls into question. There are numerous cases, we learn from metacognitive research, in which the opposite is true: people systematically *waste* cognitive effort.

Nelson and Leonesio (1988) called this *labor in vain*. Metacognitive research shows that when people can self-pace their own learning or problem solving, they invest the longest time in the most difficult items, those for

which they have little chance of success (Ackerman, 2014; Koriat, Ma'ayan, & Nussinson, 2006; Undorf & Ackerman, 2017). They do so even when they themselves judge their chances of success with these items to be slim and even when they are offered incentives for accuracy (in fact, incentives for accuracy exacerbate, rather than mitigate, labor in vain). These participants over- rather than underinvest cognitive effort, making them cognitive *wasters* rather than cognitive *misers*. The labor-in-vain hypothesis is diametrically opposed to the cognitive miser hypothesis. Importantly, the weight of the metacognitive evidence favors labor in vain.

A major source of evidence comes from Ackerman's (2014) *diminishing criterion model* (DCM). It shows that although reasoners do adjust their levels of aspiration—and, as a consequence, their rate of satisficing—to changing circumstances, the updating lags far behind diminishing resources. In Ackerman's (2014) experiment 3, for example, participants solved the easiest items in about 10 seconds with over 90% confidence, whereas the most difficult items took about 90 seconds, and even then, participants' confidence was not much higher than 20%. Thus, it took a long time for the penny to drop and for participants to stop trying for a better (i.e., more confidently held) solution. An important component in the DCM is the time limit—the maximum amount of time the respondent is willing to invest in each task item. Recent studies have shown across a large variety of tasks, including memorizing, various problem-solving tasks, and real-life web searches, that people limit the amount of time that they spend considering their response, with the aim of not wasting time on items they expect to make little progress on. However, it is nonetheless the case that the items on which the most time is spent have the lowest rates of success (Ackerman, Yom-Tov, & Torgovitsky, 2020; Undorf & Ackerman, 2017). Thus, it is not always the case that reasoners resort to satisficing, and even when they do, they sometimes fail to satisfice sufficiently, even by their own lights, when they judge their chances of success to be slim.

The labor-in-vain effect has also been observed in people's information-gathering strategies. This research takes place in the fast-and-frugal heuristics research tradition, where it is posited that people solve complex problems using simple heuristics (Gigerenzer et al., 1999). Nonetheless, it has been observed that to make a decision, people often gather more information than specified by the heuristic. A common task asks people to decide between a pair of alternatives, such as which stock to invest in. They are allowed to gather information about each of the options prior to reaching a decision. People often continue to gather information even when they already have enough information to discriminate between options (see Bröder & Newell, 2008; for reviews, see Hilbig, 2010). Indeed, people will pay for that information, even if it is not helpful (Newell, Weston, & Shanks, 2003). This suggests that having additional information, regardless of quality, may help people reach a confidence threshold where they are comfortable with their decision.

4.3 Cognitive Investment and Normative Responding

We now turn to the idea that there should be a positive relation between investment of cognitive effort and normatively correct responding. The main evidence against this idea comes from Thompson's two-response paradigm (Thompson et al., 2011). The core method in this paradigm consists of presenting participants with (typically) classic reasoning, decision-making, or moral judgment problems and asking them to provide a first quick, intuitive response, followed by a second, more considered response. Each of the two responses is also accompanied by rating on a metacognitive confidence scale. For the first response, the judgment is the feeling of rightness (FOR) mentioned above. For the second response, the judgment reflects a final judgment of confidence. The typical finding is that people seldom change their response from the first to the second response, but they are more likely to do so when metacognitive cues trigger low FOR—typically as a result of processing disfluency. As noted above, these cues only marginally coincide with the normatively correct response.

The cognitive miser hypothesis would suggest that in the rare cases that people change their mind between the first and the second response, it should be more often in the direction of a normatively correct response than the opposite. Actual findings suggest that the connection is much more complex. For example, people change from a normative to a nonnormative solution almost as often as the other way around. While there often is an increase in normatively correct responses over time, it is typically small (e.g., Bago & De Neys, 2019; Shynkaruk & Thompson, 2006; Thompson et al., 2011; Thompson & Johnson, 2014). Thus, investing more time and thought does not necessarily lead to more normative solutions.

Similarly, Stupple, Pitchford, Ball, Hunt, and Steel (2017) found only a small positive correlation between latencies (a potential proxy for cognitive effort) and correct solutions for the cognitive reflection test (Frederick, 2005), which aims to measure the ability to provide normatively correct responses against the lure of intuitively compelling but incorrect responses. Even more interesting are item-level results. For the bat and ball problem

(which requires mental arithmetic to solve correctly), the cognitive miser hypothesis was supported: latencies positively predicted the percentage of correct responses, and correct responses took longest. However, on the lily pad problem (which requires little mental arithmetic and is more in the nature of an insight item), latencies actually *negatively* predicted normatively correct responses. For the latter, intuitively compelling erroneous responses were fastest, but nonintuitive erroneous responses were slowest—slower than correct responses, in fact. This suggests that reasoners were investing time and effort, just to end up with the wrong answers.

Taken together, the findings from Thompson et al. (2011) and from Stupple et al. (2017) seem to hint at an inverted-U-shaped relation between normative correctness and cognitive resources. Beyond a certain point, additional cognitive resources have little role to play and might in fact turn out to be detrimental to normative responding. If this is correct, we can outline the boundaries of the cognitive miser hypothesis, which seems to have limited scope; beyond that, evidence supports the labor-in-vain hypothesis.

4.4 Rationality and Further Deliberation

Last, we turn to the contribution of meta-reasoning research to a little-explored but significant issue in rationality: information gathering. Classic Bayesian decision theory provides a normative model for selecting between alternative options but is silent on how an option set is assembled in the first place. When do we stop looking for alternatives? Suppose you are looking for a second medical opinion. Does it make sense to look for a third opinion? A fourth? Bounded rationality tells us that eventually people will abort the search and that they are likely to do so prematurely, but this principle is underspecified. Douven (2002) outlined a philosophical model of the rationality of further deliberation, suggesting that in any given search, the rationality of searching further is determined by four parameters:

1. How satisfactory does the agent find the existing options? Ceteris paribus, the better the existing options, the lower the value of further search.

2. What is the agent's estimate, or prediction, of her chance of success in finding a better alternative? Ceteris paribus, the stronger the prediction of success, the higher the value of further search.

3. What is the cost of further search? Ceteris paribus, the higher the cost, the lower the rationality of further search.

4. What is the time limit for the search?

These options can all be operationalized in metacognitive terms (Ackerman, Douven, Elqayam, & Teodorescu, 2020). The first thing to note is that parameter (1) is closely related to Thompson's feeling of rightness, the metacognitive confidence measure. The higher the FOR, the more satisfactory the agent will find the existing options and the less inclined they will be to search further. Evidence for this comes from the two-response paradigm and from reasoners' reluctance to change responses accompanied by high FOR. Parameter (2), prediction of success, might be akin to judgment of solvability (Ackerman & Thompson, 2017).

An important insight that comes from meta-reasoning research is that parameters (3) and (4)—that is, the cost of further search and time limit—are not independent: in Ackerman's (2014) diminishing criterion model, time itself is a cost: as time goes by, participants are increasingly willing to settle for less ambitious goals, in which they have lower confidence.

5. Conclusions

In sum, we have argued that understanding metacognitive processes can advance our understanding of epistemic, pragmatic, and bounded rationality. Because cues to confidence may be unrelated to the accuracy or adequacy of decisions, we are often overconfident in our reasoning performance and lay claim to knowledge about fictitious concepts. We recast Simon's (1982) notion of satisficing as a metacognitive stopping rule, demonstrating that in many circumstances, people fail to satisfice, even when they have enough information to make an informed choice. Thus, rather than being cognitive misers, people tend to labor in vain, spending precious cognitive resources on difficult or unsolvable problems.

Notes

The first two authors contributed equally to the preparation of the chapter.

1. Nelson and Narens (1990) borrowed the "meta" from Hilbert's "metamathematics" and Carnap's "metalanguage." Nelson (1996) anchors the term in Tarski's philosophy of truth, where "object language" denotes referents that are not linguistic (e.g., chair, happiness), while "meta-language" (and "meta-meta-language," and so on) refers to linguistic terms such as "truth." Thus, "Snow is white" is in the object language, whereas "It is true that snow is white" is in the meta-language. The "meta" in "metacognition" similarly refers to a referential level of processing: the content of thought is the object level and the monitoring of that content is the meta-level.

2. In philosophy, the role of truth-conduciveness in epistemic rationality is moot, particularly the relation between epistemic and practical rationality. As Stich (1999) argued, the conception of truth in philosophy is too fragmented to allow a single normative standard. In psychology of reasoning, Evans (2014) advocated a view of epistemic rationality as subservient to pragmatic rationality. For a recent defense of the connection between truth-conduciveness and pragmatic rationality, see Schurz (2014) and Schurz and Hertwig (2019), although we note that Schurz concedes that this view is somewhat undermined by placebo effects (beneficial false beliefs). Here we stick to the classic notion of epistemic rationality based on truth-conduciveness, as this is the view that dominates most of the metacognitive literature, but the reader is invited to keep this caveat in mind.

3. The metacognitive conception of coherence is therefore distinct from the Bayesian one in than it does not rely on conformity to the probability calculus.

4. The term "semantic coherence" is borrowed from linguistics, where it means the extent to which words adhere to a word formation rule, so that they are predictable given the rule (Aronoff, 1976, p. 38).

References

Ackerman, R. (2014). The diminishing criterion model for metacognitive regulation of time investment. *Journal of Experimental Psychology: General*, *143*(3), 1349–1368.

Ackerman, R. (2019). Heuristic cues for meta-reasoning judgments: Review and methodology. *Psychological Topics*, *28*(1), 1–20.

Ackerman, R., & Beller, Y. (2017). Shared and distinct cue utilization for metacognitive judgements during reasoning and memorisation. *Thinking & Reasoning*, *23*(4), 376–408.

Ackerman, R., Douven, I., Elqayam, S., & Teodorescu, K. (2020). Satisficing, meta-reasoning, and the rationality of further deliberation. In S. Elqayam, I. Douven, J. St. B. T. Evans, & N. Cruz (Eds.), *Logic and uncertainty in the human mind: A tribute to David Over* (pp. 10–26). London, England: Routledge.

Ackerman, R., & Thompson, V. A. (2015). Meta-reasoning: What can we learn from meta-memory? In A. Feeney & V. A. Thompson (Eds.), *Reasoning as memory* (pp. 164–178). London, England: Psychology Press.

Ackerman, R., & Thompson, V. A. (2017). Meta-reasoning: Monitoring and control of thinking and reasoning. *Trends in Cognitive Science*, *21*(8), 607–617.

Ackerman, R., Yom-Tov, E., & Torgovitsky, I. (2020). Using confidence and consensuality to predict time invested in problem solving and in real-life web searching. *Cognition*. Advance online publication. doi.org/10.1016/j.cognition.2020.104248

Ackerman, R., & Zalmanov, H. (2012). The persistence of the fluency–confidence association in problem-solving. *Psychonomic Bulletin and Review, 19*, 1187–1192.

Aronoff, M. (1976). *Word formation in generative grammar.* Cambridge, MA: MIT Press.

Atir, S., Rosenzweig, E., & Dunning (2015). When knowledge knows no bound: Self-perceived expertise predicts claims of impossible knowledge. *Psychological Science, 26*(8), 1295–1303.

Bago, B., & De Neys, W. (2019). The smart system 1: Evidence for the intuitive nature of correct responding on the bat-and-ball problem. *Thinking & Reasoning, 25*(3), 257–299.

Betsch, T., & Glöckner, A. (2010). Intuition in judgment and decision making: Extensive thinking without effort. *Psychological Inquiry, 21*, 1–16.

Bolte, A., & Goschke, T. (2005). On the speed of intuition: Intuitive judgments of semantic coherence under different response deadlines. *Memory & Cognition, 33*(7), 1248–1255.

Bröder, A., & Newell, B. R. (2008). Challenging some common beliefs: Empirical work within the adaptive toolbox metaphor. *Judgment and Decision Making, 3*, 205–214.

De Neys, W. (2014). Conflict detection, dual processes, and logical intuitions: Some clarifications. *Thinking & Reasoning, 20*, 169–187.

De Neys, W., Cromheeke, S., & Osman, M. (2011). Biased but in doubt: Conflict and decision confidence. *PLoS ONE, 6*, e15954.

Douven, I. (2002). Decision theory and the rationality of further deliberation. *Economics & Philosophy, 18*(2), 303–328.

Dunning, D. (2011). The Dunning–Kruger effect: On being ignorant of one's own ignorance. In J. Olson & M. P. Zanna (Eds.), *Advances in experimental social psychology* (Vol. 44, pp. 247–296). New York, NY: Elsevier.

Elqayam, S. (2012). Grounded rationality: Descriptivism in epistemic context. *Synthese, 189*(1), 39–49.

Evans, J. St. B. T. (2014). Two minds rationality. *Thinking & Reasoning, 20*(2), 129–146.

Evans, J. St. B. T., & Over, D. E. (1996). *Rationality and reasoning.* Hove, England: Psychology Press.

Evans, J. St. B. T., & Stanovich, K. E. (2013). Dual-process theories of higher cognition: Advancing the debate. *Perspectives on Psychological Science, 8*(3), 223–241.

Fernbach, P. M., Rogers, T., Fox, C. R., & Sloman, S. A. (2013). Political extremism is supported by an illusion of understanding. *Psychological Science, 24*, 939–946.

Fischhoff, B., Slovic, P., & Lichtenstein, S. (1977). Knowing with certainty: The appropriateness of extreme confidence. *Journal of Experimental Psychology: Human Perception and Performance, 3*, 552–564.

Fiske, S. T., & Taylor, S. E. (1984). *Social cognition.* New York, NY: McGraw-Hill.

Fiske, S. T., & Taylor, S. E. (1991). *Social cognition: From brains to culture* (2nd ed.). Los Angeles, CA: Sage.

Frederick, S. (2005). Cognitive reflection and decision making. *Journal of Economic Perspectives, 19*(4), 25–42.

Gigerenzer, G., Todd, P. M., & ABC Research Group. (1999). *Simple heuristics that make us smart.* Oxford, England: Oxford University Press.

Gilovich, T., Griffin, D., & Kahneman, D. (2002). *Heuristics and biases: The psychology of intuitive judgement.* Cambridge, England: Cambridge University Press.

Hilbig, B. E. (2010). Reconsidering "evidence" for fast-and-frugal heuristics. *Psychonomic Bulletin & Review, 17*, 923–930.

Jackson, S. A., & Kleitman, S. (2014). Individual differences in decision-making and confidence: Capturing decision tendencies in a fictitious medical test. *Metacognition and Learning, 9*, 25–49.

Jackson, S. A., Kleitman, S., Stankov, L., & Howie, P. (2016). Individual differences in decision making depend on cognitive abilities, monitoring and control. *Journal of Behavioral Decision Making, 30*, 209–223.

Koriat, A. (1993). How do we know that we know? The accessibility model of the feeling of knowing. *Psychological Review, 100*(4), 609–639.

Koriat, A. (1997). Monitoring one's own knowledge during study: A cue-utilization approach to judgments of learning. *Journal of Experimental Psychology: General, 126*(4), 349–370.

Koriat, A. (2007). Metacognition and consciousness. In P. D. Zelazo, M. Moscovitch, & E. Thompson (Eds.), *The Cambridge handbook of consciousness* (pp. 289–325). Cambridge, England: Cambridge University Press.

Koriat, A. (2012). The self-consistency model of subjective confidence. *Psychological Review, 119*(1), 80–113.

Koriat, A., Ma'ayan, H., & Nussinson, R. (2006). The intricate relationships between monitoring and control in metacognition: Lessons for the cause-and-effect relation between subjective experience and behavior. *Journal of Experimental Psychology: General, 135*(1), 36–69.

Kruger, J., & Dunning, D. (1999). Unskilled and unaware of it: How difficulties in recognizing one's own incompetence lead to inflated self-assessments. *Journal of Personality and Social Psychology, 77*, 1121–1134.

Lauterman, T., & Ackerman, R. (2019). Initial judgment of solvability in non-verbal problems—a predictor of solving processes. *Metacognition and Learning, 14*(3), 365–383.

Mata, A., Ferreira, M. B., & Sherman, S. J. (2013). The metacognitive advantage of deliberative thinkers: A dual-process perspective on overconfidence. *Journal of Personality and Social Psychology, 105*(3), 353–373.

Mata, A., Ferreira, M. B., Voss, A., & Kollei, T. (2017). Seeing the conflict: An attentional account of reasoning errors. *Psychonomic Bulletin & Review, 24,* 1980–1986.

Mednick, S. A., & Mednick, M. T. S. (1967). *Examiner's manual, Remote Associates Test: College and adult forms 1 and 2.* Boston, MA: Houghton Mifflin.

Nelson, T. O. (1996). Consciousness and metacognition. *American Psychologist, 51*(2), 102–116.

Nelson, T. O., & Leonesio, R. J. (1988). Allocation of self-paced study time and the "labor-in-vain effect." *Journal of Experimental Psychology: Learning, Memory, and Cognition, 14*(4), 676–686.

Nelson, T. O., & Narens, L. (1990). Metamemory: A theoretical framework and new findings. In G. Bower (Ed.), *The psychology of learning and motivation: Advances in research and theory* (Vol. 26, pp. 125–173). San Diego, CA: Academic Press.

Newell, B. R., Weston, N. J., & Shanks, D. R. (2003). Empirical tests of a fast-and-frugal heuristic: Not everyone "takes-the-best." *Organizational Behavior and Human Decision Processes, 91,* 82–96.

Pennycook, G., Ross, R. M., Koehler, D. J., & Fugelsang, J. A. (2017). Dunning–Kruger effects in reasoning: Theoretical implications of the failure to recognize incompetence. *Psychonomic Bulletin & Review, 24,* 1474–1484.

Reder, L. M., & Ritter, F. E. (1992). What determines initial feeling of knowing? Familiarity with question terms, not with the answer. *Journal of Experimental Psychology: Learning, Memory, and Cognition, 18,* 435–451.

Rozenblit, L., & Keil, F. C. (2002). The misunderstood limits of folk science: An illusion of explanatory depth. *Cognitive Science, 26,* 521–562.

Schurz, G. (2014). Cognitive success: Instrumental justifications of normative systems of reasoning. *Frontiers in Psychology, 5.* doi:10.3389/fpsyg.2014.00625

Schurz, G., & Hertwig, R. (2019). Cognitive success: A consequentialist account of rationality in cognition. *Topics in Cognitive Science, 11*(1), 7–36.

Shynkaruk, J. M., & Thompson, V. A. (2006). Confidence and accuracy in deductive reasoning. *Memory & Cognition, 34,* 619–632.

Simon, H. A. (1982). *Models of bounded rationality.* Cambridge, MA: MIT Press.

Stankov, L., Kleitman, S., & Jackson, S. A. (2014). Measures of the trait of confidence. In G. J. Boyle, D. H. Saklofske, & G. Matthews (Eds.), *Measures of personality and social psychological constructs* (pp. 158–189). Amsterdam, Netherlands: Academic Press.

Stanovich, K. E. (1999). *Who is rational? Studies of individual differences in reasoning.* Mahwah, NJ: Erlbaum.

Stanovich, K. E. (2009). Distinguishing the reflective, algorithmic, and autonomous minds: Is it time for a tri-process theory? In J. St. B. T. Evans & K. Frankish (Eds.), *In two minds: Dual processes and beyond* (pp. 55–88). Oxford, England: Oxford University Press.

Stich, S. P. (1990). *The fragmentation of reason: Preface to a pragmatic theory of cognitive evaluation.* Cambridge, MA: MIT Press.

Stupple, E. J. N., Pitchford, M., Ball, L. J., Hunt, T. E., & Steel, R. (2017). Slower is not always better: Response-time evidence clarifies the limited role of miserly information processing in the Cognitive Reflection Test. *PLoS ONE, 12*(11), e0186404.

Sweklej, J., Balas, R., Pochwatko, G., & Godlewska, M. (2014). Intuitive (in)coherence judgments are guided by processing fluency, mood, and affect. *Psychological Research, 78,* 141–149.

Thompson, V. A. (2014). What intuitions are . . . and are not. *Psychology of Learning and Motivation, 60,* 35–75.

Thompson, V. A. (2016). Certainty and action. In N. Galbraith, E. Lucas, & D. Over (Eds.), *The thinking mind: A Festschrift for Ken Manktelow* (pp. 80–96). Oxford, England: Routledge.

Thompson, V. A., Evans, J. St. B. T., & Campbell, J. I. C. (2013). Matching bias on the selection task: It's fast and it feels good. *Thinking & Reasoning, 19,* 431–452.

Thompson, V. A., & Johnson, S. C. (2014). Conflict, metacognition, and analytic thinking. *Thinking & Reasoning, 20*(2), 215–244.

Thompson, V. A., & Morsanyi, K. (2012). Analytic thinking: Do you feel like it? *Mind & Society, 11,* 93–105.

Thompson, V. A., Prowse Turner, J., & Pennycook, G. (2011). Intuition, metacognition, and reason. *Cognitive Psychology, 63,* 107–140.

Toplak, M., West, R. F., & Stanovich, K. (2011). The cognitive reflection test as a predictor of performance on heuristics-and-biases tasks. *Memory & Cognition, 39,* 1275–1289.

Topolinski, S. (2015). Intuition: Introducing affect into cognition. In A. Feeney & V. A. Thompson (Eds.), *Reasoning as memory* (pp. 146–163). London, England: Psychology Press.

Topolinski, S. (2018). The sense of coherence: How intuition guides reasoning and thinking. In L. J. Ball & V. A. Thompson (Eds.), *The Routledge international handbook of thinking and reasoning* (pp. 559–574). London, England: Routledge.

Undorf, M., & Ackerman, R. (2017). The puzzle of study time allocation for the most challenging items. *Psychonomic Bulletin & Review, 24*(6), 2003–2011.

Section 9   Game Theory

9.1 Classical Game Theory

Max Albert and Hartmut Kliemt

Summary

Classical game theory extends rational choice theory to interactions among several actors. This chapter introduces the formal language of classical noncooperative game theory in a self-contained way. Emphasis on the extensive form of games and its presuppositions clarifies game theory's scope and limits as a modern *lingua franca* of the behavioral sciences. The discussion of rational behavior in games centers on the equilibrium concept and its most important refinement, subgame perfection. A final section relates this discussion to the interpretation of classical game theory in terms of "reasoning about knowledge."

1. Outline of the Chapter

"Classical game theory" extends rational choice theory to interactions among several actors. We use the term to distinguish our topic from related but distinct fields like epistemic, evolutionary, and behavioral game theory. We focus on noncooperative game theory, which is more fundamental than cooperative game theory (Harsanyi & Selten, 1988, chapter 1). Our presentation is mostly ahistorical; for historical accounts, see Weintraub (1992) and Leonard (2010). References were primarily selected to provide convenient entry points to the literature. For in-depth coverage of mathematical aspects, we refer readers to Fudenberg and Tirole (1991) and Maschler, Solan, and Zamir (2013).

Section 2 approaches games as mathematical structures. Section 3 focuses on applying game theory with the aim of explaining or predicting interactions among several actors. Section 4 explains, and comments on, the tension within applied game theory between the interpretation we present and an alternative interpretation in terms of "reasoning about knowledge."

2. Game-Theoretic Language

2.1 The Rules of a Game

Primitive terms of mathematical game theory are "player," "move," "payoff," and "probability." The use of these terms is restricted by certain formal requirements but otherwise unspecified. Depending on the application, a player can be an individual person, a group of persons (say, a firm), or even a bacterium. There are applications where a single person is modeled as a set of different players or where a player stands for logical operations. Probabilities satisfy the Kolmogorov axioms (see chapter 4.1 by Hájek & Staffel, this handbook), but their interpretation is left open. Payoffs are real numbers, but it remains unspecified what these numbers represent. For ease of exposition, we nevertheless speak of players "making," "selecting," or "choosing" moves and "receiving" or "earning" payoffs; moreover, we cast a female as our generic player and as players 1 and 3, as well as a male as player 2.

The definition of a game as a mathematical structure, often called "the rules of the game," specifies all moves that are possible in the game, implying that rule-breaking moves cannot exist. A "play of the game" is a sequence of moves, each made by a specific player. Each possible play of a game comes with a "payoff profile," that is, an assignment of one payoff to each player.

The rules of a game are summarized in the game's "extensive form," also called "game tree." Figure 9.1.1a, for instance, describes a game that begins with player 1 choosing between moves A and B. If she chooses A, player 2 must choose between W and X; if player 1 chooses B, player 2 must choose between moves Y and Z. After moves X, Y, or Z of player 2, the game ends. After move W, player 1 must choose between moves C and D, after which the game ends. A "path" is a succession of nodes and branches. Each complete path through the tree, for instance, A–W–D or B–Y (plus the relevant nodes), represents a possible play of the game. The associated payoff profile is displayed at the end of the path; thus, A–W–D is associated with the payoff profile $(4, 6)$, that is, player 1 earns 4 and player 2 earns 6.

All game trees we consider begin with a unique initial node (indicated by a circle) and end with a finite number of final nodes represented by payoff profiles. Bullets represent intermediate nodes. Branches connect each node—except the initial node—with its immediate

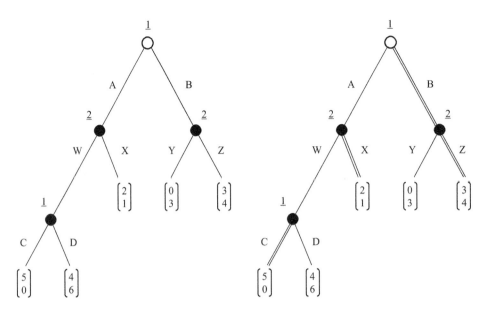

Figure 9.1.1

(a) A strictly sequential game. The pure strategies of player 1 are *AC*, *BC*, *AD*, and *BD*. Those of player 2 are *WY*, *XY*, *WZ*, and *XZ*.
(b) The pure-strategy profile (*BC*, *XZ*), illustrated by the highlighted moves, implies the path *B–Z* with associated payoff profile (3, 4).

predecessor. Each node of a game tree is reached from the initial node through a unique path. The games we consider are finite in the sense that there are finitely many complete paths, each of finite length.

Each nonfinal node is assigned to one specific player. Names of players are numbers appearing (underscored) close to their nodes. There may, however, exist passive players without nodes who receive payoffs but cannot move. The branches emerging from a player's node correspond to different moves available to the player at the node. Names of players' possible moves are capital letters appearing at the respective branches.

The path reaching a node corresponds to the history of play up to the node; therefore, this path is also called the node's "history." A complete path ends with the payoff profile associated with the corresponding play of the game. Payoff profiles list payoffs in the order of the players' names, from 1 to *n*.

A situation in a game where a player must move is called "information set." In figure 9.1.1a, all information sets are singleton sets containing exactly one node; in this case, distinguishing between nodes and information sets is inessential. Each singleton information set, or node, of a player comes with its own set of moves. Thus, player 2 chooses among *W* and *X* at his left node and between *Y* and *Z* at his right node; his total number of different moves is 4.

This is different in figure 9.1.2a, where the two nodes of player 2 belong to the same information set (as indicated by the broken-line connection). The two moves

possible at this two-element information set, *W* and *X*, are displayed at each of the two nodes; however, the two branches labeled *W* stand for a single move, as do the two branches labeled *X*. In total, player 2 has only two different moves (in contrast to figure 9.1.1a, where he has four). In applications, information sets are used to describe situations where a player is unable to observe (or remember, or notice in some other way) what had happened before she was called upon to move. The game, then, begins with player 1 choosing between *A* and *B*, followed by player 2 choosing between *W* and *X*. In case of *A* and *W*, player 1 chooses for a second time, between *C* and *D*.

Generally, each node of a player belongs to exactly one information set; any number of nodes of a player can belong to the same information set. At each of her information sets, a player has at least two possible moves, of which exactly one must be chosen if the play of the game reaches (a node of) the information set. The possible moves at an information set are displayed at every node of that information set, with their names indicating which moves at the different nodes are the same moves (so that exchanging names at one node but not at the others leads to a different game).

In a "simultaneous" game, each player has at most one information set, and each complete path runs through all information sets (as in figure 9.1.3). A nonsimultaneous game is often called "sequential," although it need not be "strictly sequential" in the sense that each information set is a singleton set (as in figure 9.1.1). Most games are neither simultaneous nor strictly sequential

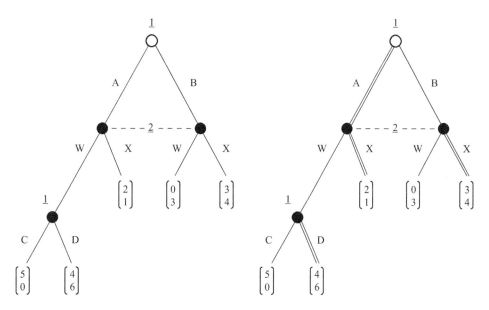

Figure 9.1.2
(a) A game that is neither strictly sequential nor simultaneous. Player 2 has only two pure strategies, W and X, since both his nodes belong to the same information set, meaning that he must choose the same move, W or X, at both nodes. (b) The pure-strategy profile (AD, X), illustrated by the highlighted moves, implies the path A–X with associated payoff profile $(2, 1)$.

(see, e.g., figures 9.1.2, 9.1.4, and 9.1.5). Strictly sequential games are often (rather confusingly) called "games of perfect information."

2.2 Strategies and Equilibrium

A pure strategy of a player is a function assigning to each of her information sets exactly one of the possible moves. In figure 9.1.1, each player has two information sets with two possible moves each; therefore, each player has four (2×2) pure strategies. We denote pure strategies by a sequence of letters spelling out the moves selected at each information set (in the order from left to right and top to bottom). The four pure strategies of player 1, then, are AC, AD, BC, and BD, while player 2's four strategies are WY, WZ, XY, and XZ.

A pure-strategy profile is a tuple of pure strategies, one for each active player, again in the order of players' names. In figure 9.1.1, a pure-strategy profile determines a unique complete path through the game tree, that is, a unique play of the game and, therefore, a unique payoff profile. For instance, the profile (BC, XZ) implies the path B–X with payoff profile $(3, 4)$ (see figure 9.1.1b). Not each move selected by the strategies occurs in the play of the game generated by (BC, XZ): player 2 is not called upon to move at his left-hand node; nevertheless, his strategy selects X. Moreover, player 1's own choice of B prevents that she is called upon to move at her second node, where her strategy selects C. Yet, distinguishing between strategies like BD and BC turns out to be relevant.

In figure 9.1.2, player 2 has only two pure strategies, called, like the moves, "W" and "X" (accepting a minor ambiguity in the notation). Figure 9.1.2b illustrates the strategy profile (AD, X). In contrast to figure 9.1.1, player 2 has no pure strategy that discriminates, through its assignment of moves, between the different histories preceding his two nodes. Nonsingleton information sets, then, decrease the number of pure strategies.

The "strategic (normal) form" of a game is a list of all pure-strategy profiles together with the resulting payoff profiles. For two-player games, the strategic form is written in the form of a cross-tabulation. Table 9.1.1 displays the strategic forms of the games of figures 9.1.1 and 9.1.2. Strategies differing only at nodes that are out of reach due to the strategies themselves (like BC and BD in figure 9.1.1) generate identical entries in the strategic form. Different games can have the same strategic form.

Strategies can be stochastic. Let $\{E_1, \ldots, E_n\}$ be a finite set. A "mixture" of the set's elements, written as $p_1 \circ E_1 \oplus p_2 \circ E_2 \oplus \cdots \oplus p_n \circ E_n$, is a probability distribution assigning to each E_k a probability $p_k \geq 0$ with $\sum_k p_k = 1$. Probability-0 elements can be dropped; the degenerate cases $1 \circ E_k$ are identified with E_k. Mixtures of elements from different sets are always stochastically independent.

A "mixed strategy" of a player is a mixture of her pure strategies. A "behavioral strategy" of a player is a function assigning to each of her information sets a mixture of the moves at that information set, called a "mixed move." Mixed and behavioral strategies include the pure

Table 9.1.1

The strategic forms of the games of figure 9.1.1 (left-hand side) and figure 9.1.2 (right-hand side)

		2						2	
		WY	WZ	XY	XZ			W	X
	AC	$\underline{5}, 0$	$\underline{5}, 0$	$\underline{2}, \underline{1}$	$2, \underline{1}$		AC	$\underline{5}, 0$	$\underline{2}, \underline{1}$
	AD	$4, \underline{6}$	$4, \underline{6}$	$\underline{2}, 1$	$2, 1$		AD	$4, \underline{6}$	$\underline{2}, 1$
1	BC	$0, 3$	$3, \underline{4}$	$0, 3$	$\underline{3}, \underline{4}$	1	BC	$0, 3$	$\underline{3}, \underline{4}$
	BD	$0, 3$	$3, \underline{4}$	$0, 3$	$\underline{3}, \underline{4}$		BD	$0, 3$	$\underline{3}, \underline{4}$

Note: Underscored payoffs indicate best replies to the other player's pure strategies. A strategy profile leading to a cell where all payoffs are underscored—for example, (*AC*, *XY*) on the left—is an equilibrium.

strategies as degenerate cases. Further extensions of the strategy set, such as correlated (i.e., stochastically dependent) strategies (Maschler et al., 2013, chapter 8), go beyond classical game theory.

The payoff profile associated with a profile of mixed or behavioral strategies is, in general, a profile of expected (values of) payoffs (see also section 2.4 below). Subsequently, the term "payoff" covers "expected payoff"; we add "expected" only for special emphasis.

Exogenous stochastic events are represented by a player without payoffs (usually called "Nature"), playing, at each of its nodes, a specific mixed move. Although this extension has many important applications (see, e.g., Fudenberg & Tirole, 1991, parts III & IV, on so-called Bayesian games), we can disregard it for present purposes.

A "solution concept" selects, for each game in some class of games and given strategy sets (pure, mixed, or behavioral), a set of strategy profiles as solutions. The basic solution concept of classical game theory is the Nash equilibrium (subsequently, just "equilibrium").

An "equilibrium" is a strategy profile where the strategy of each player is a best reply to the "complementary strategy profile," that is, the tuple of strategies of the other players (Nash, 1950). "Best reply" of a player means: no other strategy of the player achieves a higher payoff—or expected payoff, if relevant—against the complementary strategy profile. In an equilibrium, then, each player's strategy maximizes the player's payoff given the other players' strategies; no player's payoff can be increased by changing only the player's own strategy while leaving the complementary strategy profile unchanged.

We focus on equilibria of two-player games since it usually does not matter whether the complementary strategy profile appearing in the definition of equilibrium involves one or several players.

2.3　Pure-Strategy Equilibrium

Figure 9.1.3 shows the general form and several examples of simultaneous games with two players where each player has two moves (simultaneous 2×2 games). We find the pure-strategy equilibria of these and all other finite games by underscoring, in the strategic form, those payoffs of each player that are associated with best replies to the other player's pure strategies. A pure-strategy equilibrium is associated with a cell where all payoffs are underscored (see table 9.1.1 and figures 9.1.3 and 9.1.4).

In most games, a player's equilibrium strategy cannot be determined independently of the equilibrium strategies of the other players. The Prisoner's Dilemma (figure 9.1.3a) is the most prominent exception: player 1's strategy *U* is strictly dominated by her strategy *D*, that is, *D* yields a higher payoff than *U*, no matter which (pure or mixed) strategy is played by player 2. Similarly, player 2's strategy *L* is strictly dominated by his strategy *R*. Therefore, the only equilibrium is (*D*, *R*).

An equilibrium is "strict" if and only if (iff) each player has exactly one best reply to the complementary strategy profile. With the exception of (*U*, *L*) in figure 9.1.3d, the equilibria shown in figure 9.1.3 are strict.

A pure-strategy equilibrium in a game without moves of Nature selects a unique play of the game, that is, a unique complete path through the game tree, called an "equilibrium path." In an equilibrium with mixing (by Nature or by conventional players), the term "equilibrium path" denotes the set of complete paths with positive probability.

2.4　Mixed-Strategy Equilibrium

As shown by Matching Pennies (MP; figure 9.1.3b), there are games without pure-strategy equilibria. Yet, all finite games have at least one mixed-strategy equilibrium (Nash, 1950).

A profile of mixed strategies yields a probability distribution on the set of all payoff profiles and, thus, a profile of expected payoffs. For the general game of figure 9.1.3, we find that the mixed-strategy profile ($p \circ U \oplus 1 - p \circ D$, $q \circ L \oplus 1 - q \circ R$) yields payoffs $pqx_1 + p(1-q)x_2 + (1-p)qx_3 + (1-p)(1-q)x_4$ for player 1 and $pqy_1 + p(1-q)y_2 + (1-p)qy_3 + (1-p)(1-q)y_4$ for player 2.

The definition of equilibrium remains unchanged. In MP, for instance, we find the unique mixed-strategy equilibrium ($\frac{1}{2} \circ U \oplus \frac{1}{2} \circ D$, $\frac{1}{2} \circ L \oplus \frac{1}{2} \circ R$) with payoff profile (0, 0).

Mixing maximizes a player's payoff iff all pure strategies played with positive probability are best replies to the complementary strategy and, therefore, yield the same payoff. This insight has three important consequences, which we spell out for simultaneous 2×2 games but which hold, *mutatis mutandis*, for other games

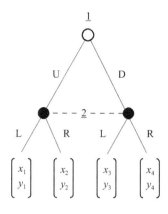

3, 3	0, $\underline{4}$	
$\underline{4}$, 0	$\underline{1}$, $\underline{1}$	

(a) Prisoner's Dilemma: $x_3 > x_1 > x_4 > x_2$, $y_2 > y_1 > y_4 > y_3$.

$\underline{1}$, -1	-1, $\underline{1}$	
-1, $\underline{1}$	$\underline{1}$, -1	

(b) Matching Pennies (or Inspection Game): $x_1 = x_4 > x_2 = x_3$, $y_2 = y_3 > y_1 = y_4$.

$\underline{4}$, $\underline{2}$	0, 0	
0, 0	$\underline{2}$, $\underline{4}$	

(c) Battle of the Sexes: $x_1 > x_4 > x_3 = x_2$, $y_4 > y_1 > y_3 = y_2$.

$\underline{4}$, $\underline{4}$	0, 3	
$\underline{4}$, 0	$\underline{1}$, $\underline{1}$	

(d) Nameless game.

Figure 9.1.3
Below the general form, extensive and strategic, of a 2×2 simultaneous game, three well-known named games (general structure indicated) and a nameless game are displayed.

where mixing occurs in equilibrium. First, no equilibrium where mixing occurs is strict. Second, if both players mix in equilibrium, one player's mixed strategy must equalize the expected payoffs of the *other* player's pure strategies (indicating how to compute the equilibrium). Third, small changes in one player's payoffs affect only the *other* player's equilibrium strategy. For instance, modifying MP by adding 1 to player 1's payoff in (U,L) leads to the new equilibrium ($\frac{1}{2} \circ U \oplus \frac{1}{2} \circ D$, $\frac{2}{5} \circ L \oplus \frac{3}{5} \circ R$).

Pure-strategy equilibria are special mixed-strategy equilibria. Accordingly, "Battle of the Sexes" (figure 9.1.3c) has three mixed-strategy equilibria: (U,L), (D,R), and ($\frac{2}{3} \circ U \oplus \frac{1}{3} \circ D$, $\frac{1}{3} \circ L \oplus \frac{2}{3} \circ R$).

2.5 Subgame Perfection

Most games have several equilibria. So-called refinements of the equilibrium concept select subsets of these equilibria. We consider only the most important refinement, subgame perfection (Selten, 1965, 1975).

Subgame perfection is concerned with moves at information sets off the equilibrium path (i.e., information sets reached with probability 0 in equilibrium). Since these moves do not affect the players' equilibrium payoffs, they can fail to maximize payoffs *from the perspective of the information set where they are taken*. Examples are player 1's move D and player 2's move X in the equilibrium (BD, XZ) in figure 9.1.1 (not highlighted; see also table 9.1.1). Subgame perfection aims at eliminating equilibria containing such non-payoff-maximizing moves.

Payoff maximization from the perspective of an information set requires that subsequent moves and, in the case of non-singleton sets, preceding moves are taken into account. Consider the equilibrium (XU, R) in figure 9.1.4. All moves except the initial move X of player 1

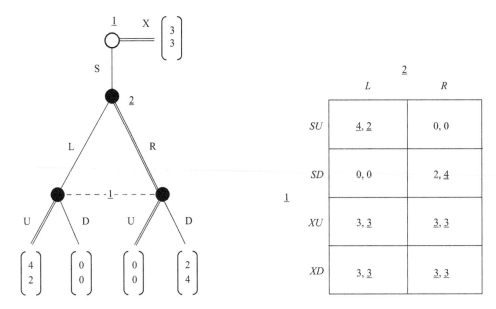

Figure 9.1.4
Player 1 decides whether to start (S) a subgame (type "Battle of the Sexes") or to exit (E) instead. The strategic form shows three pure-strategy equilibria; only (XU, R) (highlighted) is not subgame-perfect. An additional subgame-perfect equilibrium is (XM, N) with $M = \frac{2}{3} \circ U \oplus \frac{1}{3} \circ D$ and $N = \frac{1}{3} \circ L \oplus \frac{2}{3} \circ R$.

are off the equilibrium path. Given player 1's subsequent move U, player 2's move R does not maximize his payoff. Given player 2's preceding move R, player 1's move U does not maximize her payoff.

Subgame perfection addresses this interdependency problem by considering the section of the game beginning at player 2's node as a separate game, called a "subgame," requiring that the equilibrium of the complete game be consistent with some equilibrium of the subgame. The subgame has three equilibria, (U, L), (D, R), and ($\frac{2}{3} \circ U \oplus \frac{1}{3} \circ D$, $\frac{1}{3} \circ L \oplus \frac{2}{3} \circ R$), with payoff profiles $(4, 2)$, $(2, 4)$, and $(\frac{4}{3}, \frac{4}{3})$, respectively. In all three, both players' information sets are on the equilibrium path, implying payoff maximization by both.

In the complete game of figure 9.1.4, strategy S maximizes player 1's payoff iff her payoff in the subgame is at least 3. Accordingly, there are three subgame-perfect equilibria of the complete game: (XD, R), (SU, L), and a behavioral-strategy equilibrium (XN, M) with $N = \frac{2}{3} \circ U \oplus \frac{1}{3} \circ D$ and $M = \frac{1}{3} \circ L \oplus \frac{2}{3} \circ R$. The equilibrium ($XU, R$), in contrast, is not subgame perfect since (U, R) is not an equilibrium of the subgame.

Generally, a section of a game tree is a subgame iff it starts with a singleton information set, contains all successors to its starting node down to the final nodes with the payoffs, and contains no incomplete information sets (i.e., any non-singleton information set is either completely included or completely excluded). The complete game is an (improper) subgame of itself. A subgame

considered in isolation, then, satisfies the definition of a game.

A strategy profile is a subgame-perfect equilibrium of a game iff, for all subgames, the profile's restriction to the subgame is an equilibrium of the subgame. Subgame-perfect equilibria are, in general, equilibria in behavioral strategies because mixed strategies cannot describe mixed moves at information sets that are, according to the strategy itself, reached with probability 0 (Fudenberg & Tirole, 1991, pp. 87–88, esp. figure 3.10). Every finite game has at least one subgame-perfect equilibrium in behavioral strategies (Selten, 1965).

In simultaneous and other games without proper subgames (figures 9.1.3 and 9.1.5), each equilibrium is subgame perfect by definition. In all other games, subgame-perfect equilibria are found by (a generalized version of) "backward induction." This solution method begins with finding one equilibrium for each of the "smallest" subgames (those that have no proper subgames themselves). It then goes on to more and more encompassing subgames, always preserving the equilibria of already solved subgames when solving the more encompassing ones (as in figure 9.1.4's game above).

In strictly sequential games, each nonfinal node is also the first node of a subgame. The smallest subgames begin at the last player nodes and have only one active player. In figure 9.1.1b, backward induction first selects C by player 1 and Z by player 2 at the two last nodes, then X at the left node of player 2, and, finally, B at the initial node. Thus, the unique backward-induction solution is (BC, XZ). This solution is an equilibrium (see table 9.1.1), and its restriction to any proper subgame is an equilibrium of the subgame—for instance, (C, X) for the subgame beginning at player 2's left node. Therefore, the backward-induction solution is a subgame-perfect equilibrium.

In figure 9.1.2, the only proper subgame begins at the second node of player 1 and has a single equilibrium where player 1 chooses C. Of the two pure-strategy equilibria of this game (see table 9.1.1), then, only (BC, X) is subgame perfect.

Yet, subgame perfection still allows for non-payoff-maximizing moves. Figure 9.1.5 illustrates two strategy profiles of Selten's (1975) "Horse Game." Both are subgame-perfect equilibria: no player can achieve a higher payoff by unilaterally changing her strategy, and there are no proper subgames. Yet, in figure 9.1.5a, the move of player 2 obviously fails to maximize his payoff from the perspective of his information set.

The problem is that subgame perfection addresses the issue of non-payoff-maximizing moves only indirectly, by considering subgames, which do not exist in the Horse Game. This suggests a more direct approach: require that each player at each information set maximize her payoffs. In the case of singleton information sets, this works. However, consider the equilibrium in figure 9.1.5b, where player 3's information set is not reached. Since this is a non-singleton information set, identification of a payoff-maximizing move would require an appeal to preceding moves reaching the information set—but there are none. Mathematically, payoff maximization at player 3's information set requires a probability distribution on the set of nodes, but any such distribution is arbitrary. In more complicated games, a similar problem occurs if

several preceding moves reach a non-singleton information set off the equilibrium path. These problems provide the starting point for competing refinements beyond subgame perfection (e.g., Maschler et al., 2013, chapter 7).

3. Applying Game Theory

In applied game theory, game-theoretical models are used to explain and predict behavior. A model is a theory together with the description of a situation, real or hypothetical, to which the theory is applied in order to deduce predictions or explananda (Bunge, 1973, pp. 97–99).

We restrict attention to human behavior. In classical game theory, the relevant theory of human behavior is rational choice theory (RCT; cf. Sugden, 1991, and chapter 10.4 by Raub, this handbook), which is applied to situations described by a game tree. Players are individual persons. Payoffs are utilities representing players' preferences. Probabilities are either physical probabilities known to all players or players' subjective degrees of belief. In the latter case, the probabilities are, quite arbitrarily (Sugden, 1991, p. 768), assumed to be the same for all players (the "common prior assumption").

3.1 Rationality and Preferences

Applied game theory assumes that players are rational actors in the sense of RCT, know the game tree, and have true beliefs about the strategies played by the other players.

According to RCT, an actor is rational in the sense that she has, for any set X of mutually exclusive and jointly exhaustive options, a complete preference ordering described by a complete and transitive relation $\succeq$ on X, called "weak preference" (see chapter 8.1 by Grüne-Yanoff, this handbook). For $a, b \in X$, $a \succeq b$ means either

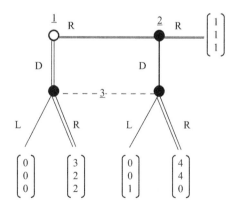
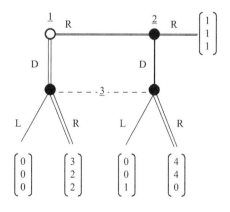

Figure 9.1.5

(a) Selten's Horse Game with equilibrium (D, R, R) highlighted. Player 2 does not maximize his payoffs from the perspective of his information set. (b) Selten's Horse Game with equilibrium (R, R, L) highlighted. It is unclear what it would mean for player 3 to maximize her payoffs from the perspective of her information set.

that she strictly prefers a to b (symbolically, $a \succ b$) or that she is indifferent between the two (symbolically, $a \sim b$). A real-valued function u defined on X is said to represent the preference relation (and is called "utility function") iff, for all $a, b \in X$, $u(a) \geq u(b)$ implies $a \succeq b$ and vice versa. This ensures that $u(a) > u(b)$ iff $a \succ b$, and $u(a) = u(b)$ iff $a \sim b$.

Under the standard ("revealed-preference") interpretation, preferences only summarize behavioral dispositions: $a \succ b$ means that the actor would not choose b if she had to choose from a subset B of X with $a, b \in B$. This subjunctive conditional exhausts the meaning of the strict preference relation. It implies that the actor would always maximize her utility: she would choose $c \in B$ only if $u(c) = \max_{x \in B} u(x)$. If B contains several utility-maximizing options, the theory implies that the actor would choose one of them; apart from that, her behavior would be indeterminate.

This interpretation implies that RCT considers neither decision making nor deliberation or reasoning as mental processes; it is only concerned with the resulting choices. Neither utilities nor preferences are reasons for, or causes of, choices; they just summarize hypothetical choices, that is, behavioral dispositions.

Preferences themselves may result from all kinds of motivations—materialistic or nonmaterialistic, selfish or unselfish, and so on—but they are "all-inclusive": all factors influencing choices are already taken into account (Binmore, 1994, pp. 104–109). Although subjective well-being may motivate an actor's choices, her preferences may reflect other factors as well (e.g., moral considerations). Therefore, RCT is not wedded to the assumption that an actor strictly preferring a to b is subjectively better off if she gets a than if she gets b (Hausman, 2012, pp. 81–83).

In addition to some theory of motivation, predictions or explanations also require hypotheses (lacking in RCT) that specify the conditions under which players are rational and have true beliefs about the situation and other players' strategies—for instance, shared cultural background (Meyerson, 2009) or experience and sufficiently high stakes (Binmore, 2007, chapter 1). Given these possible specifications, RCT is not a single theory but a family of distinct theories relying on a common mathematical language. Applied game theory inherits this feature.

3.2 Preferences in Games

In game theory, actors are players who choose moves at information sets. This has two implications. First, the objects of choice are moves at information sets, not strategies. If a player has several information sets, she cannot choose a strategy in a single act of choice. Each information set, whether reached during play or not, requires its own (possibly hypothetical) choice act.

Second, a player's utilities attach to complete paths, which typically contain moves by other players. Therefore, the objects of preferences, namely, complete paths, differ from the objects of choice, namely, moves at information sets. As such, this is not problematic: given beliefs about other players' strategies, choosing moves at information sets amounts to choosing among complete paths (see section 3.5 below). However, it becomes a problem if the question of *who* chooses matters to the players: in these cases, preferences over complete paths defined in terms of hypothetical choices may not exist (Hausman, 2012, pp. 31–33).

Consider, for instance, a game where Eve must accept (A) or decline (D) Adam's marriage proposal. Assume that Adam wishes her to accept; *in this sense*, he prefers the A-path to the D-path. However, this is not the same as saying that, hypothetically, Adam would choose the A-path over the D-path *if it were his choice*. If, plausibly, the distinction between "the X-path chosen by Eve" and "the X-path chosen by Adam" matters to Adam, he cannot have preferences *in the revealed-preference sense* over paths chosen by Eve.

If, in contrast to this example, the objects of a player's desires are, physically or in the player's mind, separable from the paths (e.g., monetary gains), it does not matter who chooses. Preferences over complete paths derive from preferences defined in terms of hypothetical choices over these separable objects. This separability assumption is sometimes called "consequentialism" (Hausman, 2012, pp. 42–45).

3.3 Uncertainty and Beliefs

Extending RCT to "choice under uncertainty" (see also chapter 8.2 by Peterson, this handbook) requires a consequentialist distinction between actions, states, and consequences that are separable from actions and states (see table 9.1.2). The elements of each set (of actions, of states, and of consequences) are mutually exclusive and jointly exhaustive. An actor's set of actions describes the options among which she must choose. The set of states describes all aspects of the situation not under her control. States differ in those aspects about which she is uncertain. If an action is taken in some state, one of the consequences obtains.

Von Neumann and Morgenstern's (2004) theory of choice under risk assumes that a physical probability distribution on the set of states is given and known to the actor. Each action, then, implies a probability distribution on the set of consequences, called a "lottery." If the actor has a complete preference ordering on the set of all conceivable lotteries, and if her preferences satisfy several further axioms, there exists a real-valued utility function on the set of consequences such that the lotteries' expected utilities represent the actor's preferences over actions.

Yet, if a player has nonconsequentialist motivations, her preferences may violate the axioms of the theory, and an expected utility representation may not exist (Bradley & Stefánsson, 2017). Examples are von Neumann and Morgenstern's (2004, pp. 629–630) "utility of gambling" and certain procedural-fairness motivations (Diamond, 1967; Verbeek, 2001).

In Savage's (1954) theory of choice under radical uncertainty, no physical probabilities are given; the probabilities are subjective. Both the utility function on the set of consequences and the subjective probabilities are derived from a preference ordering on the set of all conceivable actions (considered as functions assigning consequences to states). Again, the (subjective) expected utilities of the actions represent the actor's preferences.

In a strict revealed-preference interpretation of Savage's theory, the expected utilities of actions just summarize the actor's behavioral dispositions. Usually, however, the subjective probability of a state is interpreted as a personal degree of belief that the state obtains. In contrast to the utilities of the consequences, beliefs are mental states.

This interpretation of beliefs as mental states, on the one hand, and utilities of complete paths as expression of behavioral dispositions, on the other hand, is typical of classical game theory, whether or not beliefs are explicitly introduced as subjective probabilities.

3.4 Information, Causality, Time, and Memory

The game tree describes the information flow in plays of the game. Players called upon to move can observe at

Table 9.1.2

Decision problem with four actions A_i, three states S_j, and five consequences C_k

		States		
		S_1	S_2	S_3
Actions	A_1	C_1	C_1	C_1
	A_2	C_1	C_2	C_3
	A_3	C_5	C_2	C_1
	A_4	C_4	C_2	C_1

Note: Uncertainty about the states implies uncertainty about the consequences of actions.

which information set they are, but, except for singleton information sets, not at which node. Since the transmission of information requires a causal link, the game tree describes at least some aspects of the causal relations relevant in the game. However, Newcomb's problem (recast as a game) shows that some conceivable causal structures cannot be described with the help of a game tree (Albert & Heiner, 2003; Binmore, 1994, pp. 242–256).

The flow of information also describes the flow of time: if a player observes a move, the move must have occurred earlier. Where no observations take place (as, e.g., in simultaneous games), the game tree is consistent with any timing.

Applied game theory mostly assumes "perfect recall": players forget neither their own moves nor anything they have known or observed before (Fudenberg & Tirole, 1991, p. 81). But models of imperfect recall are possible.

In figure 9.1.6a, player 1 forgets her initial move. It seems natural to assume that, at her second information set, she behaves as if she were an independent third player with player 1's preferences. The resulting equilibria—for instance, (L, DX, L) (three-player notation) or (LL, DX) (two-player notation)—assume that, as the third player, she has correct beliefs about her first-player strategy, although she forgot what she did in that role (which is not implausible with respect to routine behavior).

In figure 9.1.6b, player 1 does not know whether she has already moved ("absentmindedness"; Piccione & Rubinstein, 1997). Her equilibrium beliefs are the subject matter of philosophy's "sleeping beauty problem" (Cisewski, Kadane, Schervish, Seidenfeld, & Stern, 2016).

3.5 Equilibrium and Beliefs about Other Players

A player's behavioral dispositions in a game are described by a behavioral strategy. A behavioral strategy is a complete contingent plan of action, that is, a list of subjunctive conditionals stating, for each information set of the player, which (possibly mixed) move the player would choose at this information set. According to RCT, a player would choose a move only if the move maximized her expected utility given her beliefs at the relevant information set.

The concept of equilibrium among rational players originates from economics; traditionally, it involves the assumption that everybody's beliefs are correct. If one assumes that *all* beliefs *everywhere* in the game are correct, RCT implies that all moves, those actually taken in the game as well as the hypothetical moves off the equilibrium path, maximize the relevant player's expected utility from the perspective of the information set where they are supposed to be made.

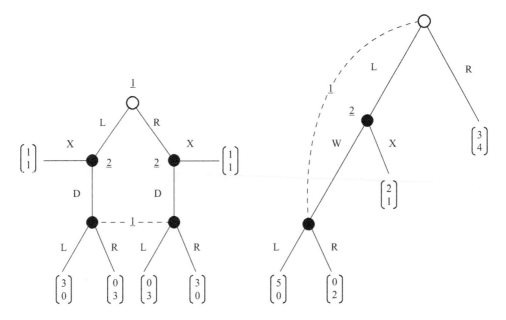

Figure 9.1.6
(a) At her second information set, player 1 has forgotten her initial move. (b) Player 1 does not know whether she moved before ("absentmindedness").

The game of Blackmail in figure 9.1.7 illustrates the basic argument. Eve (player 1) knows some damaging fact about Adam (player 2). She can keep silent (S) or, by threatening to reveal the fact, try to extort money from him (X). If Eve chooses X, Adam can give in and pay (P) or report Eve to the police (R), in which case she goes to jail but the damaging fact is revealed, to Adam's distress. In the strategic form, we find two pure-strategy equilibria, (S,R) and (X,P).

Players' utilities indicate how, given the opportunity, they would choose among complete paths. If Eve chose X, Adam could choose between complete paths X–R and X–P; preferring X–P, he would choose P. Since Eve has correct beliefs about what Adam would do and prefers the complete path X–P to the complete path S, she chooses X. Therefore, players play in accordance with the subgame-perfect equilibrium (X,P).

More generally, a player's beliefs are conditional on information sets, that is, they take into account what the player knows at the information sets. Players anticipate their own beliefs and, therefore, their own moves. Even if a player believes that she would never reach a certain information set, her beliefs about how other players would move after this information set had been reached are the beliefs she would have if this—in the player's eyes, impossible—event happened. The requirement that beliefs are conditional on information sets and correct, even in the case of counterfactual considerations, implies that a player would not, upon encountering an

unexpected move by another player, revise her beliefs about subsequent moves by this or any other player.

Consider, for instance, the unique subgame-perfect equilibrium (BC, XZ) in figure 9.1.1b. The belief of player 2 at his left node that player 1 would play C at her last node already takes into account that player 1 would have to have played the unexpected move A in order for player 2 to find himself at his left node.

The whiff of inconsistency accompanying backward induction (e.g., Sugden, 1991, pp. 771–774) derives from the assumption that players derive their correct beliefs about other players' strategies from the belief that the other players are rational—a belief that might turn out to be false if an unexpected move occurred. In our presentation, however, we do not, and need not, refer to beliefs in the rationality of other players. It might be argued that such beliefs follow from the assumption that players know the extensive form of the game, which, after all, contains players' utilities, whose interpretation is tied to RCT. Yet, one need not assume that players know other players' utilities. It suffices to assume instead that each player has correct beliefs about the strategies played by the other players. This weaker assumption avoids any potential inconsistencies from the outset.

In many games, players' beliefs about the strategies played by the other players are sufficient to determine their utility-maximizing moves from the perspective of each information set. Not so in the equilibrium (R,R,L) of figure 9.1.5b. Player 3's correct belief that players 1 and

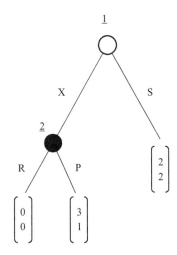
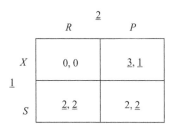

Figure 9.1.7

The game of Blackmail. The strategic form shows two equilibria, (X, P) and (S, R). Only (X, P) is subgame-perfect.

2 would both play R implies radical uncertainty for the hypothetical situation where, unexpectedly, she finds herself called upon to move. In order to decide what she would do, player 3 must form beliefs about the alternative histories that, counterfactually, might have led to this situation.

Traditionally, RCT resolves radical uncertainty by assuming arbitrary beliefs in the form of a subjective probability distribution. This is a first step toward further refinements justifying and, often, restricting equilibrium beliefs.

One instructive example of such a refinement is "forward induction" (Maschler et al., 2013, p. 261). In figure 9.1.4, let player 2 know that player 1 would never play her—strictly dominated—strategies SD or SM (where $M = \frac{2}{3} \circ U \oplus \frac{1}{3} \circ D$). If player 2 found himself called upon to move, he would not believe, then, that player 1 would play either D or M: since player 1 must have had played S before, D or M would imply the strategies SD or SM, respectively, which are ruled out. Therefore, player 2 would not consider playing either R or $N = \frac{1}{3} \circ L \oplus \frac{2}{3} \circ R$ (his best replies to SD and SM, respectively). But then, only one subgame-perfect equilibrium remains: (SU, L).

Equilibria where players are supposed to mix among strategies or moves lead to another problem with beliefs (cf., e.g., Sugden, 1991, pp. 765–768). In Matching Pennies (figure 9.1.3b), game theory predicts the mixed-strategy equilibrium $(\frac{1}{2} \circ U \oplus \frac{1}{2} \circ D, \frac{1}{2} \circ L \oplus \frac{1}{2} \circ R)$, while RCT says that the players' behaviors are indeterminate because they are indifferent between their pure strategies. This contradiction is resolved if one assumes that (a) each player's behavior is indeterminate so that the other player faces radical uncertainty, and (b) each player assigns subjective

probabilities $\frac{1}{2}$ to the other player's choices. Thus, the probabilities are interpreted as subjective beliefs without physical correlates. Yet, while internally consistent, this interpretation clashes with empirical applications where the same probabilities appear as physical probabilities of actions.

The same problem occurs in nonstrict pure-strategy equilibria like (U, L) in figure 9.1.3d, where player 1 is indifferent but player 2 nevertheless expects a specific move.

3.6 Credibility, Commitment, and the Explicitness Requirement

The strategic form represents a game as if players could choose strategies. However, as already explained, players can choose only moves. A player's strategy is a collection of subjunctive conditionals about her choices at her information sets.

The example of Blackmail (figure 9.1.7) illustrates an important consequence of this fact. Considered as such a subjunctive conditional, Adam's *strategy R* in Blackmail is an implicit threat to Eve: "I would play R if you played X." This threat, however, is not credible: Adam would not make the *move R* at this information set. The equilibrium (S, R) in Blackmail assumes that Eve has false beliefs about what Adam would do.

One might wonder whether Adam, to deter Eve, could commit himself to report her if she blackmailed him. However, in marked contrast to cooperative game theory, noncooperative game theory requires games to be interpreted under the proviso that only those choices exist that are *explicitly* modeled as possible moves (Harsanyi & Selten, 1988, chapter 1). As it stands, the model of Blackmail contains no information set where Adam could choose to commit himself. This leaves us with only two modeling options: either the model remains as is, expressing our belief that, in the situation we wish to model, a commitment option does not exist, or, if we believe otherwise, we must explicitly introduce a commitment option into the model.

Commitments can take several forms (Güth & Kliemt, 2007); for simplicity, we only consider removing unwanted options (proverbially, "burning one's bridges"). Consider an extended version of Blackmail (figure 9.1.8) where Adam can choose whether to commit himself in this sense to report Eve if she blackmails him and where Eve can observe whether he commits. In this game, the original game of Blackmail appears as a subgame reached only if Adam does not commit himself (move N). In the subgame reached after committing himself (move C), he has no choice between R and P; if Eve chose to blackmail him, she would automatically be reported to the police.

Noncooperative game theory is silent on the question of whether, or how, commitment is in fact possible. For instance, when revisiting Newcomb's problem, Spohn (2012) allows for (observable) commitment as a purely mental process, an exertion of willpower. In a game-theoretic model based on this hypothesis, however, such a commitment must explicitly appear as a move in the game tree (Binmore, 1994, pp. 165–167, 173–182). It is one of the merits of the explicitness requirement of noncooperative game theory that it forces us to represent all factual assumptions explicitly.

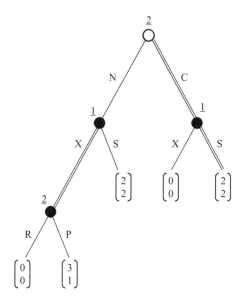

Figure 9.1.8
The extended game of Blackmail with initial commitment possibility for player 2. The only subgame-perfect equilibrium (highlighted) is (CP, XS).

The explicitness requirement and the assumption that preferences are all-inclusive play an important role in discussions concerning the Prisoner's Dilemma (figure 9.1.3a). Many contributors found it hard to swallow the conclusion that rational players play (D,R) instead of (U,L), which both players would prefer. However, in order to avoid this conclusion, one must change either the utilities or the structure of the game tree such that a different game results (Binmore, 1994, pp. 104–109).

4. Game Theory and "Reasoning about Knowledge"

The question of how players might come to have correct beliefs about other players' strategies becomes particularly relevant if one considers game theory not only as predicting what players will do but also as advising players what they should do in the light of the theory itself. The condition that all potential advisees believe in the predictions/prescriptions of a theory and can find no way to do better by deviating unilaterally implies equilibrium recommendations and equilibrium play (Dacey, 1976).

This insight instigated a tradition of game theory as "reasoning about knowledge." Von Neumann and Morgenstern (2004) already searched for a theory of rational play that would imply correct beliefs for rational players of unlimited intellectual capacities who know and accept the theory. Those who still work in this tradition replace knowledge of the theory by "common knowledge" of the game tree, that is, all players know the tree, all players know that all players know it, and so on, through all stages of meta-knowledge. Game theory becomes logical reasoning about common knowledge, including common knowledge of the theory of rational play (Fagin, Halpern, Moses, & Vardi, 1995).

This logico-philosophical rather than psychological "reasoning about knowledge" approach to interactive decision making appeals to game theorists (notably, economists) to the present day. Identifying equilibrium strategies by logical reasoning apparently keeps their discipline separate from psychology. However, their claim that game theory can, in this way, *explain* human interactions seems far-fetched and unlikely to survive the contact with reality. For this reason, Selten (1999, p. 303), while endorsing game theory for its philosophical interest, referred to it as "rationology," likening it to theology rather than science.

Yet, as our presentation emphasizes, classical game theory is not inevitably moored to the "reasoning about knowledge" approach and its most recent version, epistemic game theory (see chapter 9.2 by Perea, this handbook). Applications can often rely on simple psychological hypotheses about the circumstances under which players may be rational and have correct beliefs about the situation and other players' strategies. Evolutionary game theory (see chapter 9.3 by Alexander, this handbook) shows simple mechanisms by which even players with low cognitive abilities learn to play sophisticated equilibria. Behavioral economics (see chapter 9.4 by Dhami & al-Nowaihi, this handbook) supplies not only new hypotheses about players' motivations but also alternatives to RCT as the basic theory of choice. It is a testimony of its intellectual achievement that many of these new developments still make use of, or at least take their inspiration from, classical game theory.

Acknowledgments

For helpful comments and suggestions, we thank Volker Gadenne, Werner Güth, Sebastian Krügel, Fabian Meckl, Wolfgang Spohn, and an anonymous referee.

References

Albert, M., & Heiner, R. A. (2003). An indirect-evolution approach to Newcomb's problem. *Homo Oeconomicus, 20*, 161–194.

Binmore, K. G. (1994). *Playing fair: Game theory and the social contract* (Vol. I). Cambridge, MA: MIT Press.

Binmore, K. G. (2007). *Does game theory work?* Cambridge, MA: MIT Press.

Bradley, R., & Stefánsson, H. O. (2017). Counterfactual desirability. *British Journal for the Philosophy of Science, 68*, 485–533.

Bunge, M. (1973). *Method, model and matter*. Dordrecht, Netherlands: Reidel.

Cisewski, J., Kadane, J. B., Schervish, M. J., Seidenfeld, T., & Stern, R. (2016). Sleeping beauty's credences. *Philosophy of Science, 83*, 324–347.

Dacey, R. (1976). Theory absorption and the testability of economic theory. *Zeitschrift für Nationalökonomie, 36*, 247–267.

Diamond, P. A. (1967). Cardinal welfare, individualistic ethics, and interpersonal comparison of utility: Comment. *Journal of Political Economy, 75*, 765–766.

Fagin, R., Halpern, J. Y., Moses, Y., & Vardi, M. Y. (1995). *Reasoning about knowledge*. Cambridge, MA: MIT Press.

Fudenberg, D., & Tirole, J. (1991). *Game theory*. Cambridge, MA: MIT Press.

Güth, W., & Kliemt, H. (2007). The rationality of rational fools. In F. Peter & H. B. Schmidt (Eds.), *Rationality and commitment* (pp. 124–149). Oxford, England: Oxford University Press.

Harsanyi, J. C., & Selten, R. (1988). *A general theory of equilibrium selection in games*. Cambridge, MA: MIT Press.

Hausman, D. M. (2012). *Preference, value, choice, and welfare*. Cambridge, England: Cambridge University Press.

Leonard, R. (2010). *Von Neumann, Morgenstern, and the creation of game theory: From chess to social science 1900–1960*. Cambridge, England: Cambridge University Press.

Maschler, M., Solan, E., & Zamir, S. (2013). *Game theory*. Cambridge, England: Cambridge University Press.

Meyerson, R. B. (2009). Learning from Schelling's *Strategy of conflict*. *Journal of Economic Literature, 47*, 1109–1125.

Nash, J. F. (1950). Equilibrium points in *n*-person games. *PNAS, 36*, 48–49.

Piccione, M., & Rubinstein, A. (1997). On the interpretation of decision problems with imperfect recall. *Games and Economic Behavior, 20*, 3–24.

Savage, L. J. (1954). *The foundations of statistics*. New York, NY: Wiley.

Selten, R. (1965). Spieltheoretische Behandlung eines Oligopolmodells mit Nachfrageträgheit [Game-theoretic treatment of an oligopoly model with demand inertia]. *Zeitschrift für die gesamte Staatswissenschaft, 121*, 301–324, 667–689.

Selten, R. (1975). Reexamination of the perfectness concept for equilibrium points in extensive games. *International Journal of Game Theory, 4*, 25–55.

Selten, R. (1999). Response to Shepsle and Laitin. In J. Alt, M. Levi, & E. Ostrom (Eds.), *Competition and cooperation: Conversations with Nobelists about economics and political science* (pp. 303–308). New York, NY: Russell Sage Foundation.

Spohn, W. (2012). Reversing 30 years of discussion: Why causal decision theorists should one-box. *Synthese, 187*, 95–122.

Sugden, R. (1991). Rational choice: A survey of contributions from economics and philosophy. *Economic Journal, 101*, 751–785.

Verbeek, B. (2001). Consequentialism, rationality and the relevant description of outcomes. *Economics & Philosophy, 17*, 181–205.

von Neumann, J., & Morgenstern, O. (2004). *Theory of games and economic behavior* (60th anniversary ed.). Princeton, NJ: Princeton University Press.

Weintraub, E. R. (Ed.). (1992). *Toward a history of game theory*. Durham, NC: Duke University Press.

9.2 Epistemic Game Theory

Andrés Perea

Summary

In this chapter, we review some of the most important ideas, concepts, and results in epistemic game theory, with a focus on the central idea of *common belief in rationality*. We start by showing how belief hierarchies can be encoded by means of epistemic models with types and how this encoding can be used to formally define common belief in rationality. We next indicate how the induced choices can be characterized by a recursive elimination procedure and how the concept relates to Nash equilibrium. Finally, we investigate how the idea of common belief in rationality can be extended to dynamic games by looking at several plausible ways in which players may revise their beliefs.

1. From Classical to Epistemic Game Theory

Classical game theory, as explored in chapter 9.1 by Albert and Kliemt (in this handbook), was pioneered by the seminal work of von Neumann (1928/1959), von Neumann and Morgenstern (1944), and Nash (1950, 1951). It presents a series of concepts that select, for every game and each of the players in that game, a set of possible choices.

In *epistemic game theory*, we concentrate on the *beliefs* that motivate these choices. These may be beliefs about the possible choices of the opponents (*first-order* beliefs) but also beliefs about the *beliefs* of others (*higher-order* beliefs). Putting these first-order and higher-order beliefs together leads to *belief hierarchies*—the language of epistemic game theory. The aim of epistemic game theory is to impose reasonable conditions on such belief hierarchies and to explore the *behavioral consequences* resulting from these conditions.

As with many disciplines in science, it is difficult to say when epistemic game theory really started off. Morgenstern (1935/1976), more than 80 years ago, already stressed the importance of belief hierarchies in economic analysis, but it took a long time before belief hierarchies structurally entered the analysis of human behavior in economic systems and games. A possible reason for this long delay lies in the complexity of a belief hierarchy. Despite being a very natural object, it is quite difficult to work with because it involves *infinitely many layers*.

The purpose of this chapter is to provide an overview of some of the most important ideas and results in epistemic game theory, with a focus on the central concept of *common belief in rationality*. The outline is as follows: in section 2, we show how infinite belief hierarchies in static games can conveniently be encoded by means of epistemic models with types and use it in section 3 to formally define common belief in rationality. In section 4, we present a recursive elimination procedure that characterizes the choices that can be rationally made under common belief in rationality. In section 5, we discuss the epistemic gap between common belief in rationality and the famous notion of Nash equilibrium. In section 6, finally, we discuss how the idea of common belief in rationality can be extended to dynamic games.

For a more comprehensive overview of epistemic game theory, the reader is referred to the overview paper by Brandenburger (2007), the textbook by Perea (2012), the handbook chapter of Dekel and Siniscalchi (2015), the encyclopedia entry by Pacuit and Roy (2015), and the book by Battigalli, Friedenberg, and Siniscalchi (in press).

2. Belief Hierarchies and Types

The central idea in epistemic game theory is that of *common belief in rationality*. Informally, it states that you do not only choose rationally yourself but also believe that your opponents will choose rationally, that your opponents believe that the other players will choose rationally, and so on. Most other reasoning concepts in epistemic game theory may be viewed as refinements, or variants, of common belief in rationality. The intuitive idea of common belief in rationality is already present in Spohn (1982) and in the concept of rationalizability (Bernheim, 1984; Pearce, 1984), although the latter two papers do not formally define the notion.

An important question is how the idea of common belief in rationality can be defined formally. Consider a *finite static game*[1] $G = (C_i, u_i)_{i \in I}$, where I is the finite set of players, C_i the finite set of choices for player i, and $u_i: \times_{j \in I} C_j \to \mathbb{R}$ player i's utility function. It is assumed that all these ingredients—the set of players, the sets of choices, and the utility functions—are *commonly believed* among the players. Moreover, here and in the rest of this chapter, we restrict to noncooperative games. When we say that player i believes in the opponents' rationality, we mean that player i believes that every opponent j chooses optimally, given what player i believes that player j believes about his opponents' choices. For this to be formally defined, we need to specify i's belief about j's choice—a *first-order* belief—together with i's belief about j's belief about his opponents' choices, which is a *second-order* belief. Similarly, to formally define that player i believes that player j believes in his opponents' rationality, we need to additionally specify the belief that i holds about the belief that j holds about the belief that every opponent k holds about the other players' choices, which is a *third-order* belief. By continuing in this fashion, we can form, for any given player i, an infinite string originating in a first-order belief about the opponents' choices, a second-order belief about the opponents' first-order beliefs, a third-order belief about the opponents' second-order beliefs, and so on. Such infinite strings of beliefs are called *belief hierarchies*. They constitute the central concept of the language of epistemic game theory.

In view of the fact that belief hierarchies are *infinite* strings, making it hard to write them down explicitly, epistemic game theorists typically encode them in an easy and compact way as *types* in the sense of Harsanyi (1967–1968). The main idea is as follows: in a belief hierarchy, player i holds, for every opponent j, a belief about j's choice, j's first-order belief, j's second-order belief, and so on. That is, a belief hierarchy for player i specifies, for every opponent j, a belief about j's choice and j's belief hierarchy. If we replace the words "belief hierarchy" by "type" and formalize beliefs by probability distributions, we obtain the following definition of an epistemic model with types:

Definition 1 (Epistemic Model with Types). Consider a finite static game $G = (C_i, u_i)_{i \in I}$. A *finite epistemic model* for G is a tuple $M = (T_i, b_i)_{i \in I}$, where T_i is the finite set of types for player i, and $b_i: T_i \to \Delta(C_{-i} \times T_{-i})$ is i's belief mapping, which assigns to every type $t_i \in T_i$ a probabilistic belief $b_i(t_i) \in \Delta(C_{-i} \times T_{-i})$ on the choice–type combinations of i's opponents.

In this definition, we have used the following pieces of notation: for every finite set X, we denote by $\Delta(X)$ the set

of probability distributions on X. By $C_{-i} \times T_{-i} := \times_{j \neq i} (C_j \times T_j)$, we denote the set of choice–type combinations for i's opponents.

A finite epistemic model may be viewed as a convenient way to encode belief hierarchies in a finite manner. Indeed, for every type in an epistemic model, we may derive the full belief hierarchy it induces.

To see how this works, consider the two-player game in table 9.2.1, where player 1's choices are in the rows and player 2's choices are in the columns, together with an epistemic model in table 9.2.2.

The superscript of a type always specifies the choice that is optimal for that particular type. This will be shown later. The expression $b_1(t_1^c) = (0.6) \cdot (e, t_2^e) + (0.4) \cdot (f, t_2^g)$ means that type t_1^c assigns probability 0.6 to the event that player 2 chooses e and is of type t_2^e, as well as assigns probability 0.4 to the event that player 2 chooses f and is of type t_2^g. On the other hand, $b_1(t_1^a) = (g, t_2^g)$ means that type t_1^a assigns probability 1 to the event that player 2 chooses g and is of type t_2^g.

Consider the type t_1^b. As t_1^b believes that, with probability 1, player 2 chooses e and is of type t_1^e, the induced first-order belief is that player 1 believes that, with probability 1, player 2 chooses e. Moreover, as player 2's type t_2^e has the belief $(0.6) \cdot (c, t_1^c) + (0.4) \cdot (d, t_1^a)$ about player 1, player 2's type t_2^e assigns probability 0.6 to player 1 choosing c and probability 0.4 to player 1 choosing d. Hence, the second-order belief induced by type t_1^b is that player 1 assigns probability 1 to the event that player 2 assigns probability 0.6 to player 1 choosing c and probability 0.4 to player 1 choosing d. In a similar fashion,

Table 9.2.1

A two-player game

	e	f	g	h
a	0,0	4,1	4,4	4,3
b	3,2	0,0	3,4	3,3
c	2,2	2,1	0,0	2,3
d	1,2	1,1	1,4	0,0

Table 9.2.2

An epistemic model for the game in table 9.2.1

Types	$T_1 = \{t_1^a, t_1^b, t_1^c\}, T_2 = \{t_2^e, t_2^g, t_2^h\}$
Beliefs for player 1	$b_1(t_1^a) = (g, t_2^g)$
	$b_1(t_1^b) = (e, t_2^e)$
	$b_1(t_1^c) = (0.6) \cdot (e, t_2^e) + (0.4) \cdot (f, t_2^g)$
Beliefs for player 2	$b_2(t_2^e) = (0.6) \cdot (c, t_1^c) + (0.4) \cdot (d, t_1^a)$
	$b_2(t_2^g) = (a, t_1^a)$
	$b_2(t_2^h) = (c, t_1^c)$

we can derive the higher-order beliefs, and hence the full belief hierarchy, for the type t_1^b and for all the other types in the epistemic model.

In the game-theoretic literature, people often use *infinite* instead of finite epistemic models, because they wish to work with models that encode *all possible* belief hierarchies. Such exhaustive models are also called *terminal* type structures. That terminal type structures exist for every finite static game—something that is far from obvious—has been shown by Armbruster and Böge (1979), Böge and Eisele (1979), Mertens and Zamir (1985), and Brandenburger and Dekel (1993). For this chapter, we have chosen to work with finite epistemic models instead for two reasons. First, finite epistemic models are easier to work with than terminal type structures, since no advanced measure-theoretic or topological machinery is needed. Moreover, as we will see, this choice does not affect the main results we discuss.

The game-theoretic literature also uses alternative ways of encoding belief hierarchies, such as Kripke structures (Kripke, 1963) and Aumann structures (Aumann, 1974, 1976). The first is the predominant model in the logical and philosophical literature, whereas the latter is often used by economists. Both models use *states* instead of types and assign to every state and every player i a choice for player i, together with a belief for player i about the states. In a similar way as above, one can then derive from such a structure a belief hierarchy for every player at every state. In this chapter, we have chosen to encode belief hierarchies by means of types, but the whole chapter could have been written by using Kripke structures or Aumann structures instead.

3. Common Belief in Rationality

In the previous section, we have seen that belief hierarchies can be encoded by means of epistemic models with types. This now enables us to provide a formal definition of common belief in rationality, starting from the first layer of common belief in rationality, which states that player i believes that every opponent chooses rationally.

To express this event within the formalism of epistemic models with types, we must first define what it means for a choice to be optimal for a type. Consider an epistemic model $M = (T_i, b_i)_{i \in I}$ for a static game $G = (C_i, u_i)_{i \in I}$, a type $t_i \in T_i$, and a choice $c_i \in C_i$. Then,

$$u_i(c_i, t_i) := \sum_{(c_{-i}, t_{-i}) \in C_{-i} \times T_{-i}} b_i(t_i)(c_{-i}, t_{-i}) \cdot u_i(c_i, c_{-i})$$

denotes the *expected utility* for type t_i of choosing c_i. We say that choice c_i is *optimal* for type t_i if $u_i(c_i, t_i) \geq u_i(c_i', t_i)$ for all $c_i' \in C_i$. In the epistemic model of table 9.2.2, it can

be verified that a is optimal for the type t_1^a, b is optimal for the type t_1^b, and c is optimal for the type t_1^c. Similarly, e is optimal for player 2's type t_2^e, g is optimal for the type t_2^g, and h is optimal for the type t_2^h.

Remember that a type t_i holds a probabilistic belief $b_i(t_i)$ on the opponents' choice–type combinations. For a type t_i to believe in the opponents' rationality means that $b_i(t_i)$ must only assign positive probability to opponents' choice–type pairs where the choice is optimal for the type.

Definition 2 (Belief in the Opponents' Rationality). Consider a finite epistemic model $M = (T_i, b_i)_{i \in I}$ for a finite static game $G = (C_i, u_i)_{i \in I}$. A type $t_i \in T_i$ *believes in the opponents' rationality* if $b_i(t_i)((c_j, t_j)_{j \neq i}) > 0$ only if, for every opponent $j \neq i$, choice c_j is optimal for type t_j.

In the epistemic model of table 9.2.2, it can be verified that types t_1^a, t_1^b, t_2^g, and t_2^h believe in the opponents' rationality, but the other two types do not. Indeed, the type t_1^c for player 1 assigns positive probability to player 2's choice–type pair (f, t_2^g), where f is not optimal for the type t_2^g, and hence t_1^c does not believe in player 2's rationality. Similarly, player 2's type t_2^e assigns positive probability to player 1's choice–type pair (d, t_1^a), where d is not optimal for t_1^a, and hence t_2^e does not believe in player 1's rationality.

With the definition of belief in the opponents' rationality at hand, we can now recursively define k-fold belief in rationality for all $k \geq 1$, which finally enables us to formalize common belief in rationality.

Definition 3 (Common Belief in Rationality). Consider a finite epistemic model $M = (T_i, b_i)_{i \in I}$ for a finite static game $G = (C_i, u_i)_{i \in I}$.

(Induction start) A type $t_i \in T_i$ expresses onefold belief in rationality if t_i believes in the opponents' rationality.

(Induction step) For $k > 1$, a type $t_i \in T_i$ expresses k-fold belief in rationality if $b_i(t_i)((c_j, t_j)_{j \neq i}) > 0$ only if, for every opponent $j \neq i$, type t_j expresses $(k-1)$-fold belief in rationality.

A type $t_i \in T_i$ expresses *common belief in rationality* if t_i expresses k-fold belief in rationality for every $k \geq 1$.

Hence, twofold belief in rationality entails that a type only assigns positive probability to opponents' types that express onefold belief in rationality. In other words, the player believes that every opponent believes in his opponents' rationality. Similarly, threefold belief in rationality corresponds to the event that a player believes that his opponents believe that their opponents believe in their opponents' rationality, and so on.

Within a finite static game $G = (C_i, u_i)_{i \in I}$, we say that player i can *rationally choose $c_i \in C_i$ under common belief in rationality* if there is a finite epistemic model $M = (T_i, b_i)_{i \in I}$ and a type $t_i \in T_i$ such that t_i expresses common belief in rationality and c_i is optimal for t_i. That is, choice c_i can be supported by some belief hierarchy that expresses common belief in rationality.

In the epistemic model of table 9.2.2, it can be verified that types t_1^c and t_2^e do not express onefold belief in rationality. Indeed, type t_1^c assigns positive probability to the choice–type pair (f, t_2^g), where f is not optimal for the type t_2^g and similarly for type t_2^e. Next, types t_1^b and t_2^h express onefold but not twofold belief in rationality. To see why, note that type t_1^b believes that player 2 is of type t_2^e, which does not express onefold belief in rationality, and similarly for type t_2^h. Finally, types t_1^a and t_2^g express common belief in rationality. Indeed, type t_1^a believes that player 2 is of type t_2^g, whereas type t_2^g believes that player 1 is of type t_1^a. As both t_1^a and t_2^g express onefold belief in rationality, it can inductively be shown that t_1^a expresses k-fold belief in rationality for all k and hence expresses common belief in rationality, and similarly for type t_2^g. Consequently, player 1 can rationally choose a and player 2 can rationally choose g under common belief in rationality.

4. Recursive Procedure

Suppose that in a given static game, we are interested in the choices that the players can rationally make under common belief in rationality. Is there an easy method to find these choices, without having to resort to epistemic models with types? That is the question that will be addressed in this section.

The key to answering this question is lemma 3 in Pearce (1984), which we will reproduce below. To state the lemma formally, we need the following definitions. Consider a finite static game $G = (C_i, u_i)_{i \in I}$, a choice c_i, and a belief $b_i \in \Delta(C_{-i})$ about the opponents' choices. Then,

$$u_i(c_i, b_i) := \sum_{c_{-i} \in C_{-i}} b_i(c_{-i}) \cdot u_i(c_i, c_{-i})$$

denotes the expected utility of choice c_i under the belief b_i. Choice c_i is called *optimal* in G for the belief b_i if $u_i(c_i, b_i) \geq u_i(c_i', b_i)$ for all $c_i' \in C_i$. Choice c_i is called *strictly dominated* in G if there is some randomization $r_i \in \Delta(C_i)$ such that

$$u_i(c_i, c_{-i}) < \sum_{c_i' \in C_i} r_i(c_i') \cdot u_i(c_i', c_{-i}) \text{ for all } c_{-i} \in C_{-i}.$$

In the literature, such randomizations $r_i \in \Delta(C_i)$ are typically called *mixed strategies* or *randomized choices*, and they are often interpreted as real objects of choice for player i.

In this chapter, however, we assume that players do not randomize when making a decision, and these randomizations r_i are merely used as an auxiliary device to characterize choices that are optimal for some belief. The reason is that players are assumed to be expected utility maximizers, and hence a player can never increase his expected utility by randomizing over his choices.

Lemma 3 in Pearce (1984) can now be stated as follows:

Lemma 1 (Pearce, 1984). Consider a finite static game $G = (C_i, u_i)_{i \in I}$ and a choice $c_i \in C_i$. Then, there is a belief $b_i \in \Delta(C_{-i})$ such that c_i is optimal in G for b_i if, and only if, c_i is not strictly dominated in G.

This lemma can be used to characterize the choices a player can rationally make if he believes in his opponents' rationality. Let G^1 be the reduced game that remains if we eliminate, for every player, the choices that are strictly dominated in G. For a player to believe in the opponents' rationality thus means, by lemma 1, that his belief is fully concentrated on opponents' choices in G^1. By applying lemma 1 to the reduced game G^1, we thus conclude that, for every player, the choices he can rationally make if he believes in the opponents' rationality are exactly the choices in G^1 that are not strictly dominated in G^1. That is, these are the choices that survive *two rounds* of elimination of strictly dominated choices. In a similar vein, it can be shown that the choices that can rationally be made if a player believes in his opponents' rationality, and believes that his opponents believe in their opponents' rationality (i.e., if he expresses up to *two*fold belief in rationality), are exactly the choices that survive *three rounds* of elimination of strictly dominated choices. By continuing in this fashion, we arrive at the following elimination procedure:

Definition 4 (Iterated Elimination of Strictly Dominated Choices). Consider a finite static game $G = (C_i, u_i)_{i \in I}$.

(Induction start) Let $G^0 := G$ be the full game.
(Induction step) For every $k > 0$, let G^k be the reduced game that remains if we eliminate from G^{k-1} all choices that are strictly dominated in G^{k-1}.

A choice $c_i \in C_i$ survives iterated elimination of strictly dominated choices if c_i is in G^k for all $k > 0$.

By the argument above, we thus see that G^2 contains exactly those choices that can rationally be made if a player believes in the opponents' rationality. By iterating this argument, we conclude that, for every $k \geq 2$, the k-fold reduced game G^k contains exactly those choices that can rationally be made under some belief hierarchy that expresses up to $(k-1)$-fold belief in rationality. This argument already appears in Spohn (1982). In particular, the choices that survive the full procedure will be exactly those choices that can rationally be made under common belief in rationality. This leads to the following central result, which is based on theorems 5.2 and 5.3 in Tan and Werlang (1988), and which Brandenburger (2014) has called the "fundamental theorem of epistemic game theory." Brandenburger and Dekel (1987) offer in their proposition 2.1 a similar result, characterizing common belief in rationality by "best reply sets" instead of an elimination procedure.

Theorem 1 (Fundamental Theorem of Epistemic Game Theory). Consider a finite static game $G = (C_i, u_i)_{i \in I}$ and a choice $c_i \in C_i$. Then, c_i can rationally be made under common belief in rationality if, and only if, c_i survives iterated elimination of strictly dominated choices.

The fundamental theorem would remain unaffected if we were to use terminal type structures (hence, *infinite* epistemic models) instead of *finite* epistemic models to define common belief in rationality. To illustrate the procedure of iterated elimination of strictly dominated choices and the theorem above, consider the game G from table 9.2.1. In the full game G, it is easily verified that player 1's choice d is strictly dominated by the randomization that assigns probability 0.5 to his choices a and b and that player 2's choice f is strictly dominated by the randomization that assigns probability 0.5 to his choices g and h. No other choices are strictly dominated. Hence, G^1 is the game obtained after eliminating the choices d and f. Within the onefold reduced game G^1, player 1's choice c is strictly dominated by b (or, rather, the randomization that assigns probability 1 to b), and player 2's choice e is strictly dominated by h. Hence, G^2 is the game obtained from G^1 after eliminating the choices c and e. Finally, within G^2, player 1's choice b is strictly dominated by a, and player 2's choice h is strictly dominated by g. As such, only the choices a and g survive iterated elimination of strictly dominated choices, and hence, by theorem 1, these are the only choices that can rationally be made under common belief in rationality.

5. Nash Equilibrium

For many decades, the concept of Nash equilibrium (Nash, 1950, 1951) has dominated the classical approach to game theory, inspiring many refinements such as perfect equilibrium (Selten, 1975) and proper equilibrium (Myerson, 1978) for static games, as well as subgame-perfect equilibrium (Selten, 1965) and sequential equilibrium (Kreps

& Wilson, 1982) for dynamic games. See chapter 9.1 by Albert and Kliemt (this handbook) for a discussion of the latter two concepts. However, for a long time, it remained unclear what epistemic conditions are needed for players to choose in accordance with Nash equilibrium. The purpose of this section is to investigate Nash equilibrium from an epistemic perspective and to link it to the conditions of common belief in rationality that we have explored so far. Let us start by giving the definition of Nash equilibrium. See also chapter 9.1, where it is called *mixed-strategy equilibrium*.

Definition 5 (Nash Equilibrium). Consider a finite static game $G = (C_i, u_i)_{i \in I}$. A Nash equilibrium in G is a tuple $(\sigma_i)_{i \in I}$ where $\sigma_i \in \Delta(C_i)$ for every player i, such that $\sigma_i(c_i) > 0$ only if

$$\sum_{c_{-i} = (c_j)_{j \neq i} \in C_{-i}} \left(\prod_{j \neq i} \sigma_j(c_j) \right) \cdot u_i(c_i, c_{-i}) \geq$$
$$\sum_{c_{-i} = (c_j)_{j \neq i} \in C_{-i}} \left(\prod_{j \neq i} \sigma_j(c_j) \right) \cdot u_i(c_i', c_{-i})$$

for all $c_i' \in C_i$.

In other words, a Nash equilibrium is a tuple of probability distributions on choices such that a choice only receives positive probability if it is optimal against the probability distributions on the opponents' choices. Traditionally, these probability distributions σ_i have been interpreted as conscious randomizations, or *mixed strategies*, by the players. A more recent approach, adopted by Spohn (1982), Aumann and Brandenburger (1995), and other authors, is to interpret σ_i as the (common) probabilistic belief that i's opponents have about i's choice, and this is also the interpretation we use here.

A Nash equilibrium $(\sigma_i)_{i \in I}$ induces, in a natural way, a belief hierarchy for player i in which his (first-order) belief about the opponents' choices is given by $(\sigma_j)_{j \neq i}$, his (second-order) belief about j's belief about his opponents' choices is given by $(\sigma_k)_{k \neq j}$, and so on. Such belief hierarchies are called *simple* in Perea (2012). Moreover, this belief hierarchy can be shown to express common belief in rationality, relying on the optimality conditions in a Nash equilibrium. To see this, consider the belief hierarchy for player i induced by a Nash equilibrium $(\sigma_i)_{i \in I}$. Then, player i only assigns positive probability to a choice c_j of player j if $\sigma_j(c_j) > 0$. By the optimality condition of Nash equilibrium, this is only the case if c_j is optimal against $(\sigma_k)_{k \neq j}$, which is what player i believes that player j believes about his opponents' choices. Altogether, we see that player i only assigns positive probability to c_j if c_j is optimal for player j, given what player i believes that player j believes about his opponents' choices. That is, with this belief hierarchy, player i believes in j's rationality. In a similar vein, it can be shown that with this belief hierarchy, induced by a Nash equilibrium, player i also believes that every opponent j believes in his opponents' rationality, and so on. Hence, every Nash equilibrium induces, for every player, a belief hierarchy that expresses common belief in rationality. We can thus say that Nash equilibrium implies common belief in rationality.

But is the other direction also true? Does common belief in rationality necessarily lead to Nash equilibrium? The answer, as we will see, is *no*. Consider the two-player game in table 9.2.3, which is similar to the game of figure 1 in Bernheim (1984).

It may be verified that all three choices can rationally be made under common belief in rationality. However, there is only one Nash equilibrium (σ_1, σ_2) in this game, where σ_1 assigns probability 1 to c and σ_2 assigns probability 1 to f. Hence, in this example, Nash equilibrium imposes more restrictions than just common belief in rationality. But what are these extra restrictions?

To see this most clearly, consider the epistemic model, together with its graphical representation, in figure 9.2.1.

It may be verified that all types in the epistemic model express common belief in rationality. Moreover, the respective superscripts of the types indicate the choice that is optimal for that type. Remember that only the choices c and f are supported by a Nash equilibrium in this game.

Consider the type t_1^a that supports the choice a—a choice that is not supported by a Nash equilibrium. The induced belief hierarchy states that, on the one hand, player 1 believes that player 2 chooses e but, on the other hand, believes that 2 believes that 1 believes that 2 chooses d. That is, player 1 believes that player 2 is *incorrect* about 1's first-order belief. The same can be said about his type t_1^b. In contrast, the type t_1^c that supports the Nash equilibrium choice c induces a belief hierarchy in which 1 believes that 2 is correct about 1's first-order belief.

It turns out that in two-player games, this *correct beliefs assumption*—that is, that a player believes that his opponent is correct about his first-order belief—is exactly what separates common belief in rationality

from Nash equilibrium. This is reflected in Spohn's (1982) theorem on page 253, and Aumann and Brandenburger's (1995) theorem A, which both state that in two-player games, mutual belief in rationality, together with mutual belief in the actual first-order beliefs, leads to Nash equilibrium. Here, mutual belief in rationality means that player 1 believes in 2's rationality, and player 2 believes in 1's rationality. Similarly, mutual belief in the actual first-order beliefs means that player 1 is correct about 2's first-order belief, and player 2 is correct about player 1's first-order belief. From a one-person perspective (in which conditions are imposed on the belief hierarchy of a *single* player i) the Spohn–Aumann–Brandenburger conditions thus state that player i believes that j is rational, believes that j believes that i is rational, that i believes that j is correct about i's first-order belief, and that i believes that j believes that i is correct about j's first-order belief. In particular, Spohn, Aumann, and Brandenburger show that the *first two layers* of common belief in rationality, together with the correct beliefs assumptions above, are enough to imply Nash equilibrium. Not all layers of common belief in rationality are needed. Polak (1999) shows, however, that if mutual belief in the actual first-order beliefs is strengthened to *common* belief in the actual first-order beliefs, then the Spohn–Aumann–Brandenburger conditions would imply common belief in rationality. Other epistemic foundations for Nash equilibrium in two-player games, which in some way or another involve the correct beliefs assumption above, can be found in Tan and Werlang (1988), Brandenburger and Dekel (1989), Asheim (2006), and Perea (2007a). As the reasonability of the correct beliefs assumption can be debated—after all, why should an opponent be correct about your first-order belief?—these papers implicitly point at the problematic assumptions underlying Nash equilibrium.

For more than two players, the above conditions are no longer enough to characterize Nash equilibrium. For such games, Nash equilibrium additionally implies that i's belief about j's choice must be stochastically independent from i's belief about k's choice and that i's belief about j's belief about k's choice must be the same as i's belief about k's choice. The first property follows from the fact that in a Nash equilibrium $(\sigma_i)_{i \in I}$, the belief of i about the opponents' choices is given by the independent probability distributions $(\sigma_j)_{j \neq i}$, whereas the second condition is implied by the property that i's belief about j's belief about k's choice and i's belief about k's choice are both given by σ_k. These two conditions are not implied by common belief in rationality, and hence the gap between Nash equilibrium and common belief in rationality is

Table 9.2.3

A two-player game

	d	e	f
a	0,3	3,0	0,2
b	3,0	0,3	0,2
c	2,0	2,0	2,2

Types	$T_1 = \{t_1^a, t_1^b, t_1^c\}$, $T_2 = \{t_2^d, t_2^e, t_2^f\}$
Beliefs for player 1	$b_1(t_1^a) = (e, t_2^e)$
	$b_1(t_1^b) = (d, t_2^d)$
	$b_1(t_1^c) = (f, t_2^f)$
Beliefs for player 2	$b_2(t_2^d) = (a, t_1^a)$
	$b_2(t_2^e) = (b, t_1^b)$
	$b_2(t_2^f) = (c, t_1^c)$

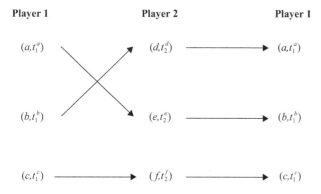

Player 1   **Player 2**   **Player 1**

(a, t_1^a) (d, t_2^d) ⟶ (a, t_1^a)

(b, t_1^b) (e, t_2^e) ⟶ (b, t_1^b)

(c, t_1^c) ⟶ (f, t_2^f) ⟶ (c, t_1^c)

Figure 9.2.1

Epistemic model for the game in table 9.2.3 and a graphical representation.

even bigger in games with more than two players. Epistemic foundations for Nash equilibrium in games with more than two players can be found in Brandenburger and Dekel (1987), Aumann and Brandenburger (1995), Perea (2007a), Barelli (2009), and Bach and Tsakas (2014).

6. Dynamic Games

So far, we have been exploring *static* games, where all players only make one choice, and players choose in complete ignorance of the opponents' choices. We now investigate how the idea of common belief in rationality can be translated to *dynamic games*. In a dynamic game, players may choose one after the other, may choose more than once, and may fully or partially observe what the opponents have done in the past when it is their turn to move. As a consequence, a player may need to *revise* his belief about the opponents when he discovers that his previous belief has been contradicted by some of the opponents' past choices. As an illustration, consider the game from figure 9.2.2, which is based on Reny (1992).

If player 1 believes that player 2 would rationally choose *g* at his last move, then he would choose *a* at the beginning. Common belief in rationality thus seems to suggest that player 2 should initially believe that player 1 chooses *a*. However, when it is player 2's turn to move, this initial belief has been contradicted by player 1's past play, and hence player 2 must revise his belief about player 1. But how? As we will see, there are at least two plausible ways for player 2 to revise his belief.

One option is to interpret player 1's past move *b* as a *mistake*, yet at the same time maintain the belief that player 1 would choose rationally at his second move, as well as the belief that player 1 believes that player 2 would rationally choose *g* at his second move. In that case, player 2 would believe, upon observing *b*, that player 1 would choose *e* at this second move, and therefore player 2 would choose *c*. This type of reasoning, in which the players are free to interpret "surprising" past moves as mistakes but believe that the opponents will choose rationally in the future, believe that the opponents always believe that their opponents will choose rationally in the future, and so on, is called *backward induction reasoning* and is formally captured by the concept of *common belief in future rationality* (Perea, 2014). Similar lines of reasoning are present in Penta (2015), Baltag, Smets, and Zvesper (2009), and the concept of *sequential rationalizability* (Dekel, Fudenberg, & Levine, 1999, 2002; Asheim & Perea, 2005). Backward induction reasoning is also implicitly present in the backward induction procedure (for a survey of the various epistemic foundations for backward induction, see Perea, 2007b) and the equilibrium concepts of subgame-perfect equilibrium (Selten, 1965) and sequential equilibrium (Kreps & Wilson, 1982; for a formal statement, see Perea & Predtetchinski, 2019).

Another option for player 2, after observing the "surprising" move *b*, is to interpret *b* as a conscious, optimal choice for player 1. However, this is only possible if player 2 believes that player 1 would choose *f* afterward and if player 2 believes that player 1 assigns a high probability to player 2 making the suboptimal choice *h* at his second move. Consequently, player 2 would choose *d* and, in case he is asked to make a second move, choose *g*. This type of reasoning, where a player, whenever possible, tries to interpret "surprising" past choices as conscious, optimal choices, is called *forward induction reasoning*. It can be formalized by the epistemic condition of *strong belief in the opponents' rationality* (Battigalli & Siniscalchi, 2002), which states that a player, whenever possible, must believe that his opponents are implementing optimal strategies.[2] The concepts that most closely implement this type of reasoning are *explicable equilibrium*

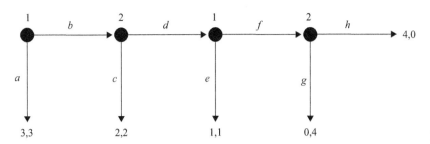

Figure 9.2.2
Reny's game.

(Reny, 1992) and *extensive-form rationalizability* (Pearce, 1984; Battigalli, 1997), where the latter has epistemically been characterized by *common strong belief in rationality* in Battigalli and Siniscalchi (2002). See also Battigalli and Friedenberg (2012), in which forward induction with exogenous restrictions on the players' beliefs is studied.

As the example above illustrates, backward and forward induction reasoning may lead to different strategy choices. Indeed, player 2 chooses *c* under backward induction reasoning but would choose (*d,g*) under forward induction reasoning. However, both types of reasoning lead to the same outcome, which is the terminal history following *a*. Battigalli (1997) has shown that the latter is always true in dynamic games with perfect information without relevant ties by proving that in every such game, the forward induction concept of extensive-form rationalizability always uniquely leads to the backward induction *outcome*. This result is remarkable, as forward induction and backward induction represent two completely different lines of reasoning. The connection between these two lines of reasoning in general dynamic games is one of the many intriguing problems in epistemic game theory that need further exploration.

Notes

1. Such a static game may also correspond to the strategic form of a dynamic game, as explained in chapter 9.1 by Albert and Kliemt (this handbook).

2. Strong belief in rationality is very similar to *assumption of rationality* in static games, which has been used by Brandenburger, Friedenberg, and Keisler (2008) to epistemically characterize the iterated elimination of weakly dominated choices.

References

Armbruster, W., & Böge, W. (1979). Bayesian game theory. In O. Moeschlin & D. Pallaschke (Eds.), *Game theory and related topics* (pp. 17–28). Amsterdam, Netherlands: North-Holland.

Asheim, G. B. (2006). *The consistent preferences approach to deductive reasoning in games* (Theory and Decision Library). Dordrecht, Netherlands: Springer.

Asheim, G. B., & Perea, A. (2005). Sequential and quasi-perfect rationalizability in extensive games. *Games and Economic Behavior, 53*, 15–42.

Aumann, R. J. (1974). Subjectivity and correlation in randomized strategies. *Journal of Mathematical Economics, 1*, 67–96.

Aumann, R. J. (1976). Agreeing to disagree. *Annals of Statistics, 4*, 1236–1239.

Aumann, R., & Brandenburger, A. (1995). Epistemic conditions for Nash equilibrium. *Econometrica, 63*, 1161–1180.

Bach, C. W., & Tsakas, E. (2014). Pairwise epistemic conditions for Nash equilibrium. *Games and Economic Behavior, 85*, 48–59.

Baltag, A., Smets, S., & Zvesper, J. A. (2009). Keep "hoping" for rationality: A solution to the backward induction paradox. *Synthese, 169*, 301–333.

Barelli, P. (2009). Consistency of beliefs and epistemic conditions for Nash and correlated equilibrium. *Games and Economic Behavior, 67*, 363–375.

Battigalli, P. (1997). On rationalizability in extensive games. *Journal of Economic Theory, 74*, 40–61.

Battigalli, P., & Friedenberg, A. (2012). Forward induction reasoning revisited. *Theoretical Economics, 7*, 57–98.

Battigalli, P., Friedenberg, A., & Siniscalchi, M. (in press). *Epistemic game theory: Reasoning about strategic uncertainty*.

Battigalli, P., & Siniscalchi, M. (2002). Strong belief and forward induction reasoning. *Journal of Economic Theory, 106*, 356–391.

Bernheim, B. D. (1984). Rationalizable strategic behavior. *Econometrica, 52*, 1007–1028.

Böge, W., & Eisele, T. (1979). On solutions of Bayesian games. *International Journal of Game Theory, 8*, 193–215.

Brandenburger, A. (2007). The power of paradox: Some recent developments in interactive epistemology. *International Journal of Game Theory, 35*, 465–492.

Brandenburger, A. (Ed.). (2014). *The language of game theory: Putting epistemics into the mathematics of games* (World Scientific Series in Economic Theory, Vol. 5). Singapore: World Scientific.

Brandenburger, A., & Dekel, E. (1987). Rationalizability and correlated equilibria. *Econometrica, 55*, 1391–1402.

Brandenburger, A., & Dekel, E. (1989). The role of common knowledge assumptions in game theory. In F. Hahn (Ed.), *The economics of missing markets, information and games* (pp. 46–61). Oxford, England: Oxford University Press.

Brandenburger, A., & Dekel, E. (1993). Hierarchies of beliefs and common knowledge. *Journal of Economic Theory, 59,* 189–198.

Brandenburger, A., Friedenberg, A., & Keisler, H. J. (2008). Admissibility in games. *Econometrica, 76,* 307–352.

Dekel, E., Fudenberg, D., & Levine, D. K. (1999). Payoff information and self-confirming equilibrium. *Journal of Economic Theory, 89,* 165–185.

Dekel, E., Fudenberg, D., & Levine, D. K. (2002). Subjective uncertainty over behavior strategies: A correction. *Journal of Economic Theory, 104,* 473–478.

Dekel, E., & Siniscalchi, M. (2015). Epistemic game theory. In P. Young & S. Zamir (Eds.), *Handbook of game theory with economic applications* (Vol. 4, pp. 619–702). Amsterdam, Netherlands: North-Holland.

Harsanyi, J. C. (1967–1968). Games with incomplete information played by "Bayesian" players, I–III, *Management Science, 14,* 159–182, 320–334, 486–502.

Kreps, D. M., & Wilson, R. (1982). Sequential equilibria. *Econometrica, 50,* 863–894.

Kripke, S. (1963). A semantical analysis of modal logic I: Normal modal propositional calculi. *Zeitschrift für mathematische Logik und Grundlagen der Mathematik, 9,* 67–96.

Mertens, J.-F., & Zamir, S. (1985). Formulation of Bayesian analysis for games with incomplete information. *International Journal of Game Theory, 14,* 1–29.

Morgenstern, O. (1976). Perfect foresight and economic equilibrium. In A. Schotter (Ed.), *Selected economic writings of Oskar Morgenstern* (pp. 169–183). New York: New York University Press. (Original work published 1935)

Myerson, R. B. (1978). Refinements of the Nash equilibrium concept. *International Journal of Game Theory, 7,* 73–80.

Nash, J. F., Jr. (1950). Equilibrium points in *n*-person games. *Proceedings of the National Academy of Sciences of the United States of America, 36,* 48–49.

Nash, J. F. (1951). Non-cooperative games. *Annals of Mathematics, 54,* 286–295.

Pacuit, E., & Roy, O. (2015). Epistemic foundations of game theory. In E. N. Zalta (Ed.), *The Stanford encyclopedia of philosophy.* Retrieved from https://plato.stanford.edu/archives/spr2015/entries/epistemic-game/

Pearce, D. G. (1984). Rationalizable strategic behavior and the problem of perfection. *Econometrica, 52,* 1029–1050.

Penta, A. (2015). Robust dynamic implementation. *Journal of Economic Theory, 160,* 280–316.

Perea, A. (2007a). A one-person doxastic characterization of Nash strategies. *Synthese, 158,* 251–271.

Perea, A. (2007b). Epistemic foundations for backward induction: An overview. In J. van Benthem, D. Gabbay, & B. Löwe (Eds.), *Interactive logic: Proceedings of the 7th Augustus de Morgan Workshop, London* (Texts in Logic and Games, Vol. 1, pp. 159–193). Amsterdam, Netherlands: Amsterdam University Press.

Perea, A. (2012). *Epistemic game theory: Reasoning and choice.* Cambridge, England: Cambridge University Press.

Perea, A. (2014). Belief in the opponents' future rationality. *Games and Economic Behavior, 83,* 231–254.

Perea, A., & Predtetchinski, A. (2019). An epistemic approach to stochastic games. *International Journal of Game Theory, 48,* 181–203.

Polak, B. (1999). Epistemic conditions for Nash equilibrium, and common knowledge of rationality. *Econometrica, 67,* 673–676.

Reny, P. J. (1992). Backward induction, normal form perfection and explicable equilibria. *Econometrica, 60,* 627–649.

Selten, R. (1965). Spieltheoretische Behandlung eines Oligopolmodells mit Nachfragezeit [Game-theoretic treatment of an oligopoly model with demand time]. *Zeitschrift für die gesamte Staatswissenschaft, 121,* 301–324, 667–689.

Selten, R. (1975). Reexamination of the perfectness concept for equilibrium points in extensive games. *International Journal of Game Theory, 4,* 25–55.

Spohn, W. (1982). How to make sense of game theory. In W. Stegmüller, W. Balzer, & W. Spohn (Eds.), *Philosophy of economics: Proceedings, Munich, July 1981* (pp. 239–270). Berlin, Germany: Springer.

Tan, T. C.-C., & Werlang, S. R. da C. (1988). The Bayesian foundations of solution concepts of games. *Journal of Economic Theory, 45,* 370–391.

von Neumann, J. (1959). Zur Theorie der Gesellschaftsspiele. *Mathematische Annalen, 100,* 295–320. Translated by S. Bargmann as "On the theory of games of strategy (S. Bargmann, Trans.). In A. W. Tucker & R. D. Luce (Eds.), *Contributions to the theory of games* (Vol. IV, pp. 13–43). Princeton, NJ: Princeton University Press. (Original work published 1928)

von Neumann, J., & Morgenstern, O. (1944). *Theory of games and economic behavior.* Princeton, NJ: Princeton University Press.

9.3 Evolutionary Game Theory

J. McKenzie Alexander

Summary

Evolutionary game theory analyzes the outcomes of interactions between boundedly rational individuals in strategic settings. Originally developed in the field of population biology, economists and other social scientists became interested in evolutionary game theory because it provided a natural way of incorporating dynamics into game theory. Given the wide variety of different evolutionary dynamics and models, a number of interesting and surprising results have been proven. Under some dynamics, both weakly and strictly dominated strategies can persist in the limit. And, although it is typically the case that population states corresponding to Nash equilibria are fixed points of the evolutionary dynamics, some dynamics allow for fixed points that do *not* correspond to Nash equilibria. The connection between evolutionary dynamics and the traditional solution concept of game theory—the Nash equilibrium—is thus found to be subtle and complex.

1. The Origins of Evolutionary Game Theory

Evolutionary game theory emerged in the 1970s from the realization that frequency-dependent fitness introduced a strategic aspect to evolution, even in contexts where none of the agents were traditionally "rational." The first work of modern evolutionary game theory is generally understood to be Maynard Smith and Price (1973), in which the concept of the *evolutionarily stable strategy* (ESS) was introduced.[1] Yet one can see the beginnings of game-theoretic approaches to the study of evolution in the work of R. A. Fisher, who, in *The Genetical Theory of Natural Selection* (1930), attempted to explain the approximate equality of the mammalian sex ratio using arguments that were essentially game-theoretic in nature.

After its introduction in population biology, evolutionary game theory became of interest to economists and social scientists more generally. Initially, it was hoped that, because the concept of an evolutionarily stable strategy provided a proper strengthening of the Nash equilibrium solution concept, evolutionary game theory could offer insight into the equilibrium selection problem: when multiple Nash equilibria exist, which one rational players would settle upon. In addition, the fact that evolutionary game theory provided dynamical models of quasi-rational deliberation was seen to address a lacuna in the traditional theory of games, which largely provided a static analysis of strategic problems. In *Theory of Games and Economic Behavior*, von Neumann and Morgenstern wrote,

> We repeat most emphatically that our theory is thoroughly static. A dynamic theory would unquestionably be more complete and therefore preferable. But there is ample evidence from other branches of science that it is futile to try to build one as long as the static side is not thoroughly understood. (§ 4.8.2)

The "most [emphatic]" reminder of the von Neumann and Morgenstern theory being static was needed because, although the extensive-form representation of games does make explicit the temporal relation between the choice nodes of players, individual strategies were represented atemporally, where a strategy specified the choice of a player, in advance, at all possible choice nodes.

Over time, it was recognized that evolutionary game theory was less successful in delivering on these aims than initially hoped. The equilibrium selection problem reappeared in the form of competing notions of evolutionary stability, and the plurality of dynamical models yielded varying results regarding the long-term survival of strategies. For example, whereas some concepts of evolutionary stability rule out even weakly dominated strategies, in the dynamical setting, it turns out that both weakly *and* strictly dominated strategies can persist for some families of dynamics. In addition, some families of dynamics—particularly imitative dynamics—have the property that not every fixed point of the dynamics corresponds to a Nash equilibrium of the underlying game.

Since its inception, evolutionary game theory has morphed into a more general study of the population dynamics of boundedly rational agents (see chapter 8.5 by Hertwig & Kozyreva, this handbook), with the once-central concept of an evolutionarily stable strategy, and its subsequent refinements, no longer in central place. The family of models developed provides useful mathematical tools for modeling and understanding the behavior of quasi-rational individuals interacting in strategic contexts, but care needs to be exercised to ensure that the model and the target system are aligned before drawing conclusions.

2. Game-Theoretic Fundamentals

Consider a game played by a group of N individuals, where each player i has a set S_i of *pure strategies*. Each strategy $s \in S_i$ can be thought of as an action available to player i; the complete set of actions taken by all players determines the payoffs each player receives. In traditional game theory, it often proves useful to allow a player to choose a strategy at random, so as to keep other individuals uncertain as to what she will do; such strategies are called *mixed strategies* and correspond to probability distributions over the set of pure strategies. The set of mixed strategies for player i is denoted Δ_i, and a particular mixed strategy for player i will be denoted $\sigma_i \in \Delta_i$. Mixed strategies, while indispensable for many important theoretical results in traditional game theory (see chapter 9.1 by Albert & Kliemt, this handbook), such as Nash's theorem that every finite n-player game has at least one equilibrium, play a less central role in evolutionary game theory.

If $\vec{\sigma} = (\sigma_1, \ldots, \sigma_N)$ denotes the particular strategies used by each player, also known as a *strategy profile*, the payoff to player i is given by $\pi_i(\vec{\sigma})$, where π_i is the *payoff function* for that player. In two-player games, we will use the notation "$\pi_i(\sigma|\mu)$" to denote the fitness of player i when σ is played against μ. A two-player game is *symmetric* when, for all strategies σ, μ, the payoff of σ played against μ does not depend on the identity of the player; for such games, the payoff function will simply be written "π."

The fundamental solution concept for traditional game theory is the *Nash equilibrium*, defined as follows:

Definition. A strategy profile $\sigma^* = (\sigma_1^*, \ldots, \sigma_n^*)$ is a Nash equilibrium if and only if for each player i and all strategies $\sigma_i \in \Delta_i$,

$$\pi_i(\sigma_i^*, \sigma_{-i}^*) \geq \pi_i(\sigma_i, \sigma_{-i}^*),$$

where "$(\sigma_i, \sigma_{-i}^*)$" denotes the strategy profile identical to σ^* except that the strategy for player i is replaced by σ_i.

A Nash equilibrium is a set of mutual best responses for each player, where no individual can improve their payoff by adopting an alternative strategy. This property makes it a natural minimal solution concept for traditional game theory. All games with a finite number of players and finitely many strategies have at least one Nash equilibrium.

3. Evolutionarily Stable Strategies

To see that the concept of a Nash equilibrium is too *weak* for capturing the idea of evolutionary stability, consider the game shown in figure 9.3.1a. That game has two Nash equilibria in pure strategies: (S_1, S_1) and (S_2, S_2). But imagine this game played in a population where individuals are paired at random and fitness corresponds to expected payoff. Informally, it is clear that a population consisting entirely of the S_2 type would not be evolutionarily stable. In such a population, a novel S_1 mutant receives the same payoff as the incumbent population and thus does not incur a fitness disadvantage. This means an S_1 mutant can invade the population. If two S_1 mutants appear, the payoff of an S_1–S_1 interaction exceeds that of an S_2–S_2 interaction, which means the S_1 type has an expected fitness greater than the average of the population and thus would be expected to increase in number, potentially even driving the S_2 type to extinction.[2] The problem stems from the fact that a Nash equilibrium only requires that deviant play *does no better* than the equilibrium strategy, while allowing for the possibility that deviant play may *do as well as* the equilibrium strategy.

A *strict Nash equilibrium* is one where any deviation from the equilibrium profile leaves players worse off. To see that the concept of a strict Nash equilibrium is too *strong* to capture the idea of evolutionary stability, consider the game in figure 9.3.1b, the Hawk–Dove of Maynard Smith and Price (1973).[3] In this game, two individuals compete over a resource with value V. The strategies individuals may adopt are "Hawk," which escalates to the point of fighting, if necessary, and "Dove," which makes a show of escalating initially but then backs down just before the point of fighting. Given this, if a Hawk meets a Dove, the Hawk gets the entire resource. If two Doves meet, they share the resource. If two Hawks meet, they escalate to the point of fighting until one is injured, with cost $-C$, and retreats.[4] If $V < C$, this game has no pure-strategy Nash equilibrium and one mixed-strategy Nash equilibrium where both individuals play Hawk with probability V/C. Denote this mixed strategy by σ^*. From the fundamental theorem of mixed-strategy Nash equilibria (see Gintis, 2009), it can be shown that

$$\pi(\text{Hawk}|\sigma^*) = \pi(\text{Dove}|\sigma^*) = \pi(\sigma^*|\sigma^*).$$

Therefore, for any other mixed strategy $\mu \neq \sigma^*$, it is the case that $\pi(\mu|\sigma^*) = \pi(\sigma^*|\sigma^*)$, and so σ^* is not a strict

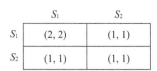

	S_1	S_2
S_1	(2, 2)	(1, 1)
S_2	(1, 1)	(1, 1)

(a) A Nash equilibrium allows for evolutionary drift, and hence may not be evolutionarily stable.

	Hawk	Dove
Hawk	$\left(\dfrac{V-C}{2}, \dfrac{V-C}{2}\right)$	$(V, 0)$
Dove	$(0, V)$	$\left(\dfrac{V}{2}, \dfrac{V}{2}\right)$

(b) A game whose Nash equilibrium is not strict when $V < C$, yet is evolutionarily stable.

Figure 9.3.1

Why the concept of a Nash equilibrium is too weak to capture the idea of evolutionary stability.

Nash equilibrium strategy. However, unlike the game in figure 9.3.1a, here it can be shown that $\pi(\sigma^* \mid \mu) > \pi(\mu \mid \mu)$. This means that the mutant strategy μ is at a fitness disadvantage in that it does less well *when it plays against itself* than the incumbent strategy σ^* does when it plays against μ. This means that the mutant strategy μ would be unable to invade and, hence, the incumbent strategy σ is evolutionarily stable.

These considerations led Maynard Smith (1982) to propose the following:

Definition. A strategy σ is an *evolutionarily stable strategy* (ESS) if and only if for all strategies $\mu \neq \sigma$, it is the case that

either $\pi(\sigma \mid \sigma) > \pi(\mu \mid \sigma)$,

or $\pi(\sigma \mid \sigma) = \pi(\mu \mid \sigma)$ and $\pi(\sigma \mid \mu) > \pi(\mu \mid \mu)$.

The first clause states that every strict Nash equilibrium strategy is an ESS; the second clause states that, if an ESS is not a strict best reply to itself, then the ESS must be a strict best reply to all *other* strategies. Although it is true that all strict Nash equilibrium strategies are evolutionarily stable, the converse does not hold. The second-order best-reply condition means that an ESS provides a real strengthening of the Nash equilibrium solution concept.

An alternative definition of an evolutionarily stable strategy, first introduced by Taylor and Jonker (1978) and used by Weibull (1995) and others, is as follows. Let σ and μ be strategies, and let $\varepsilon\mu + (1 - \varepsilon)\sigma$ for $\varepsilon > 0$ denote the strategy, which ε of the time plays μ and $1 - \varepsilon$ of the time plays σ. For small ε, this can be thought of as either a small "tremble" around the base strategy σ or, equivalently, as the strategy employed by a randomly chosen opponent in a population of individuals where almost all play σ but a small number of mutants adopt μ.

Definition. A strategy σ is an *evolutionarily stable strategy* if and only if for every strategy $\mu \neq \sigma$, there exists a $0 < \varepsilon_\mu < 1$ such that

$\pi(\sigma \mid \varepsilon\mu + (1 - \varepsilon)\sigma) > \pi(\mu \mid \varepsilon\mu + (1 - \varepsilon)\sigma)$

for all $0 < \varepsilon < \varepsilon_\mu$.

It is easily shown that this definition, which can be interpreted as a mathematically precise formulation of the definition given by Maynard Smith and Price (1973), is equivalent to the one proposed by Maynard Smith (1982).

The above definition of an ESS states that the expected payoff of an ESS in a mixed population consisting of the ESS and a small number of mutants is strictly greater than the expected payoff of the mutant in that same mixed population. However, Hofbauer, Schuster, and Sigmund (1979) give the following characterization of an ESS, which not only has the benefit of being conceptually clearer but also proves to be more useful when we move to considering evolutionary dynamics in the next section:

Definition. A strategy σ is *locally superior* if there exists an ε-neighborhood around σ such that, for all strategies $\mu \neq \sigma$ in that neighborhood, it is the case that $\pi(\sigma \mid \mu) > \pi(\mu \mid \mu)$.

From this, one can then prove the following:

Theorem 1. A strategy σ is an ESS if and only if σ is locally superior.

There are a number of theorems one can prove about ESSs. For example:

Theorem 2. No weakly dominated strategy is an ESS.

This is of interest because, as shown in figure 9.3.2, some weakly dominated strategies constitute Pareto-optimal Nash equilibria. In such cases, it can be rational to play a weakly dominated strategy even though it is not evolutionarily stable.

The *support* of a mixed strategy σ is the set of pure strategies to which σ assigns nonzero probability. One can show that if μ and σ are ESSs with identical supports, then $\mu = \sigma$. From this, it immediately follows that

	S_1	S_2
S_1	(1, 1)	(100, 0)
S_2	(0, 100)	(100, 100)

Figure 9.3.2

The fact that no weakly dominated strategy is an ESS means that some Pareto-optimal Nash equilibria are excluded, as shown here.

	Rock	Paper	Scissors
Rock	(0, 0)	(−1, 1)	(1, −1)
Paper	(1, −1)	(0, 0)	(−1, 1)
Scissors	(−1, 1)	(1, −1)	(0, 0)

Figure 9.3.3
The game of Rock–Paper–Scissors, which has no ESS.

Theorem 3. The number of ESSs is finite (and possibly zero).

The latter half of the above theorem is easily proven by considering the well-known game of Rock–Paper–Scissors, shown in figure 9.3.3. The only Nash equilibrium is in mixed strategies and assigns equal probability to all three pure strategies. The zero-sum nature of the game means that neither of the two conditions in the definition of an ESS are met, and so this game has no evolutionarily stable strategy.

4. Dynamics for Evolutionary Games

The main shortcoming of the evolutionarily-stable-strategy concept is that it, like that of a Nash equilibrium, is *static*. It would be useful if we could say something about the behavior of a population in an out-of-equilibrium environment, especially in games where multiple ESSs exist or where no ESS exists. For this reason, we now turn to dynamical models of evolutionary games. As we shall see, there is an imperfect match between static and dynamical notions of evolutionary stability, and although there are some stability concepts that possess a useful degree of generality, considerable variation exists in the behavior of a population between continuous and discrete models, as well as between unstructured and structured population models.

To begin, let $S = \{S_1, \ldots, S_n\}$ denote the set of pure strategies and let $p_i(t) \in [0, 1]$ denote the proportion of the population using strategy i at time t. It is assumed that no player employs a mixed strategy and that $\sum_{i=1}^{n} p_i(t) = 1$. Let Δ_n denote the set of all possible frequency distributions of strategies over the population. If the only thing that matters, from an evolutionary point of view, is the frequency with which a strategy is used in the population, then $\vec{p}(t) = \langle p_1(t), \ldots, p_n(t) \rangle \in \Delta_n$ represents the total state of the population at time t. A game, in this context, is a function $W: \Delta_n \to \mathbb{R}^n$, which

maps states of the population to a vector of expected payoffs—which will typically be thought of as expected fitnesses—for each pure strategy. In what follows, we use $W_i(\vec{p}(t))$ to denote the expected payoff to strategy i when the population is in state $\vec{p}$ at time t.

Sandholm (2010a, 2010b) provides a useful framework for thinking about evolutionary dynamics, demonstrating how continuous dynamics at the population level can be derived from particular learning rules—also known as *revision protocols*—used by the individuals comprising the population. This in effect provides the "microfoundations for existing dynamics" (Sandholm, 2010b, p. 31), the evolutionary game theory analogue of providing microfoundations for macroeconomics.

Sandholm models a revision protocol as a map from the current expected payoffs of each strategy in the population, as well as its present composition, to a transition matrix expressing the rate at which strategies switch from one type to another. More formally, a revision protocol is a map, generally assumed to be Lipschitz-continuous, as follows:

$$\rho : \mathbb{R}^n \times \Delta_n \to \mathbb{R}_+^{n \times n}.$$

Given a particular vector of expected payoffs $\pi \in \mathbb{R}^n$ and a population state $\vec{p} \in \Delta_n$, the *ij*th entry of $\rho(\pi, \vec{p})$, written $\rho_{ij}(\pi, \vec{p})$, is the rate at which individuals switch from strategy i to strategy j.

Once a revision protocol is specified, the evolutionary dynamics for the population are derived by substituting the revision protocol into the following schema of differential equations defined on Δ_n (see equation 1 below). For the precise mathematical details, see Sandholm (2010b).

Much of the early literature in evolutionary game theory focused on the *replicator dynamics*, the first dynamical model of evolutionary games, proposed by Taylor and Jonker (1978). The core idea behind the replicator dynamics is that the proportion of the population that follows a given strategy increases if the fitness of that strategy is higher than the average fitness of the population, which is denoted by $\phi(\vec{p})$. The average fitness of the population is just the weighted average of the individual-strategy fitnesses, with the proportion of the population following the respective strategies used as weights:

$$\phi(\vec{p}) = \sum_{i \in S} p_i \cdot W_i(\vec{p}).$$

The *continuous replicator dynamics* is the system of differential equations given by

$$\frac{dp_i}{dt} = \left(\begin{array}{c} \text{Rate at which the population} \\ \text{switches } \textit{to} \text{ strategy } i \end{array} \right) - \left(\begin{array}{c} \text{Rate at which the population} \\ \text{switches } \textit{from} \text{ strategy } i \end{array} \right). \tag{1}$$

$$\frac{dp_i}{dt} = p_i \cdot [W_i(\vec{p}) - \phi(\vec{p})], \quad \text{for } i \in S.$$

Note that if $p_i(0) > 0$, then $p_i(t) > 0$ for all t. This is of course compatible with $p_i(t) \to 0$ as $t \to \infty$; the point is that, under the replicator dynamics, strategies can be driven to extinction in the limit, but they cannot actually be driven out in finite time. A related point is that the replicator dynamics cannot *introduce* strategies into the population if they are originally absent: if $p_i(0) = 0$, then $p_i(t) = 0$ for all t.

In the simplest case, if population members interact at random and the probability of interacting with strategy j equals $p_j(t)$, then the fitness of strategy i is $W_i(\vec{p}(t)) = \sum_j p_j(t) \cdot \pi(S_i | S_j)$. This is the continuous *linear* replicator dynamics (henceforth, simply "the replicator dynamics").

Schlag (1998) showed that the replicator dynamics can be derived from the following revision protocol: suppose that each player in the population selects someone at random and compares their individual payoff with that of the person selected. If the person selected received a higher payoff, then the player adopts the strategy of the person selected with a probability proportional to the difference in payoffs. That is, if $\vec{\pi}$ denotes the vector of expected payoffs for the population in state $\vec{p}$, then $\rho_{ij}(\vec{\pi}, \vec{p}) = p_j \cdot [\pi_j - \pi_i]_+$, where $[x]_+ = \max(x, 0)$. The replicator dynamics, then, is the population-level description of the outcome when people *imitate with probability proportional to success*.

At this point, a number of interesting differences already emerge between the static ESS concept and the outcomes

of dynamical models: no weakly dominated strategy can be an ESS, but the replicator dynamics does not necessarily eliminate weakly dominated strategies. Indeed, in the case of the game of figure 9.3.4, almost all evolutionary trajectories converge to a state in which some proportion of the population follows a weakly dominated strategy.

Although the replicator dynamics need not eliminate weakly dominated strategies, Akin (1980) showed that, in most cases, the replicator dynamics will eliminate *strictly* dominated strategies:

Theorem 4. Let $\vec{p} \in \Delta_n$ be a point in the interior of the simplex. If S_i is strictly dominated, then $\lim_{t \to \infty} p_i(t) = 0$ under the replicator dynamics.

The requirement that all strategies must be represented in the initial state is easily seen to be necessary: consider the prisoner's dilemma, and suppose that the population begins in the state where everyone follows the strategy Cooperate. Although Cooperate is strictly dominated by Defect, the fact that the replicator dynamics cannot introduce absent strategies means that the strictly dominated state of All-Cooperate will endure.

Although the replicator dynamics generally eliminates strictly dominated strategies, the same is not true for other types of evolutionary dynamics. Indeed, Hofbauer and Sandholm (2011) show that many families of evolutionary dynamics have the property of not eliminating strictly dominated strategies. This result shows how classical conceptions of rational behavior, at the individual level, can become decoupled from population-level

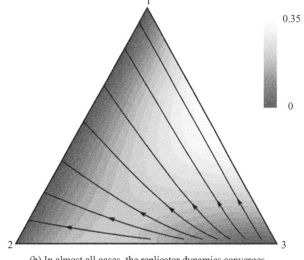

	S_1	S_2	S_3
S_1	(1, 1)	(1, 1)	(1, 0)
S_2	(1, 1)	(1, 1)	(0, 0)
S_3	(0, 1)	(0, 0)	(0, 0)

(a) A game where strategy S_2 is weakly dominated by S_1.

(b) In almost all cases, the replicator dynamics converges to a polymorphic state featuring a mix of both S_1 and S_2.

Figure 9.3.4

Weakly dominated strategies may not be eliminated by the replicator dynamics. Shading indicates the relative speed of the evolutionary trajectories.

outcomes. In what follows, two additional evolutionary dynamics are introduced, which—despite the plausible nature of the underlying revision protocol—do not, in general, eliminate strictly dominated strategies.

As we've seen, the replicator dynamics can be derived from a revision protocol that imitates strategies with probability proportional to success. One might wonder whether imitation is a reliable guide to future payoffs. Imitation certainly suffers from an inability to introduce new strategies into the population. Perhaps it would be better to adopt a strategy only taking into account its performance compared to aggregate properties of the overall population. One such revision protocol, which yields an interesting evolutionary dynamic, is the following: suppose that the rate at which players switch to strategy j does not depend on what strategy they currently employ, or what their expected payoff is in the current population state, but rather only on whether the expected payoff of strategy j exceeds the average payoff of the population. Sandholm (2010a) shows that when

$$\rho_{ij}(W(\vec{p}), \vec{p}) = [W_j(\vec{p}) - \phi(\vec{p})]_+, \text{ for all } j \in S,$$

the evolutionary dynamic obtained is the Brown–Nash–von Neumann (BNN) dynamic (Brown & von Neumann, 1950). This dynamic belongs to a more general class of dynamics known as *separable excess payoff dynamics*. The BNN dynamic takes the following form:

$$\frac{dp_i}{dt} = [W_i(\vec{p}) - \phi(\vec{p})]_+ - p_i \sum_{j \in S}[W_j(\vec{p}) - \phi(\vec{p})]_+, \text{ for all } i \in S.$$

One interesting difference between the BNN dynamic and the replicator dynamic is that the BNN dynamic can introduce *new* strategies into the population. Figure 9.3.5 shows this for the game Bad Rock–Paper–Scissors. The phase diagram shows the evolutionary trajectories for three different initial population states, each beginning at vertices of the simplex. The BNN dynamic moves the population along each of the respective faces in the direction of the dominating strategy. At some point, there is a fitness advantage to playing the third, unrepresented, strategy, and so the dynamics moves into the interior of the simplex and converges to an orbit.

The revision protocol that yields the BNN dynamics compares alternative strategies to the average payoff of the population. In a sense, this is a strange motivation because the average payoff of the population, in most cases, does not correspond to the payoff obtained by any particular strategy. As an alternative, consider the revision protocol that selects a new strategy by comparing the expected payout of alternative strategies to the expected payout of one's current strategy, where only those strategies with higher expected payoffs have a nonzero chance of being adopted. That is,

$$\rho_{ij}(W(\vec{p}, \vec{p})) = [W_j(\vec{p}) - W_i(\vec{p})]_+.$$

This revision protocol generates an evolutionary dynamic first discussed by Smith (1984) and hence is known as the *Smith dynamic*. This dynamic is a member of a more general family of evolutionary dynamics

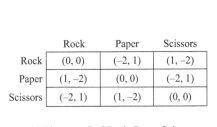

(a) The game Bad Rock–Paper–Scissors.

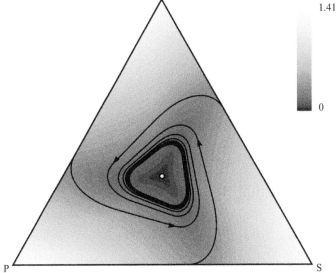

Figure 9.3.5
The Brown–Nash–von Neumann dynamic can introduce new strategies into the population that are not initially present.

known as *impartial pairwise comparison dynamics*. (It is "impartial" in that the switch rate does not depend upon the player's current strategy i.) The Smith dynamic takes the following form:

$$\frac{dp_i}{dt} = \sum_{j \in S} p_j \cdot [W_i(\vec{p}) - W_j(\vec{p})]_+ - p_i \sum_{j \in S} [W_j(\vec{p}) - W_i(\vec{p})]_+.$$

Both the Smith dynamic and the BNN dynamic allow strictly dominated strategies to persist in the limit. To see this, consider the game of figure 9.3.6a. Here, the game of Bad Rock–Paper–Scissors is augmented with a fourth strategy: a "feeble twin." The feeble twin strategy is essentially the same as Scissors except that all of its payoffs are decremented by an additional amount $\varepsilon > 0$. Figure 9.3.6b shows how, under the Smith dynamic, populations beginning at the four vertices of the simplex (states such as All-Rock, All-Paper, etc.) follow trajectories such that all four strategies are played with positive probability, *including* the strictly dominated feeble twin strategy.

The phenomenon illustrated in figure 9.3.6 is a specific instance of the following general result:

Theorem 5 (Hofbauer & Sandholm, 2011). For all evolutionary dynamics that belong to either the family of impartial pairwise comparison dynamics or the family of separable excess payoff dynamics, there exists a game such that, from most initial conditions, there is a strictly dominated strategy played by a proportion of the population that is bounded away from 0 and exceeds ⅙ infinitely often as time approaches infinity.

One interesting point about the result of theorem 5 is that both the BNN dynamics and the Smith dynamics are generated by revision protocols that are not obviously implausible. Thus, irrational outcomes at the population level—the survival of strictly dominated strategies—can be generated by revision protocols that, at the individual level, may appear sensible.

All of the above three dynamics are *single-population* models, in that they assume the underlying game is

	Rock	Paper	Scissors	Twin
Rock	(0, 0)	(−2, 1)	(1, −2)	(1, −2−ε)
Paper	(1, −2)	(0, 0)	(−2, 1)	(−2, 1−ε)
Scissors	(−2, 1)	(1, −1)	(0, 0)	(0, −ε)
Twin	(−2−ε, 1)	(1−ε, −1)	(−ε, 0)	(−ε, −ε)

(a) In this version of Bad Rock–Paper–Scissors, the fourth strategy Twin is the same as Scissors except that all of its payoffs are decreased by ε.

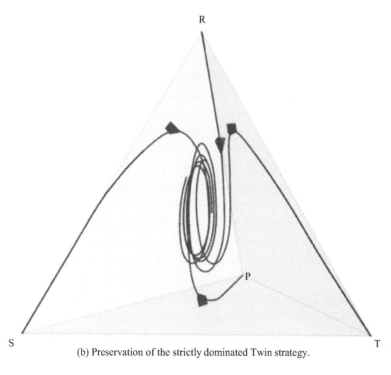

(b) Preservation of the strictly dominated Twin strategy.

Figure 9.3.6
The Smith dynamic allows strictly dominated strategies to be preserved.

symmetric. Since there is no way of distinguishing row and column players, it must be the case that $\pi(S_i | S_j) = \pi(S_j | S_i)$. Asymmetric games are typically analyzed using *multipopulation* models, one population for Row and another population for Column, which give rise to a coupled system of differential equations. Suppose, for example, that the base game is the two-player asymmetric game *Battle of the Sexes*, shown in figure 9.3.7a. The two-population replicator dynamics treats the Row and Column players as being drawn from separate subpopulations, where the expected fitness of a Row player depends on the distribution of strategies found in the Column subpopulation, and vice versa, as indicated in figure 9.3.7b. (The superscripts R and C

denote the Row and Column subpopulation, respectively.) In this case, the replicator dynamics takes the following form (again, the explicit time dependencies are suppressed for clarity):

$$\frac{dp_i^R}{dt} = p_i^R \cdot [W_i^R(\vec{p}^C) - \phi(\vec{p}^R)], \quad \text{for } i = 1, 2,$$

and

$$\frac{dp_i^C}{dt} = p_i^C \cdot [W_i^C(\vec{p}^R) - \phi(\vec{p}^C)], \quad \text{for } i = 1, 2.$$

For the payoff matrix given in figure 9.3.7a, the specific system of coupled differential equations is obtained by substituting the following values into the above:

	Column	
	S_1	S_2
S_1	(1, 2)	(0, 0)
S_2	(0, 0)	(2, 1)

(a) Battle of the Sexes.

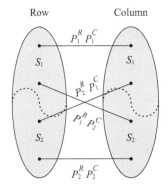

(b) A two-population model with the interaction probabilities indicated.

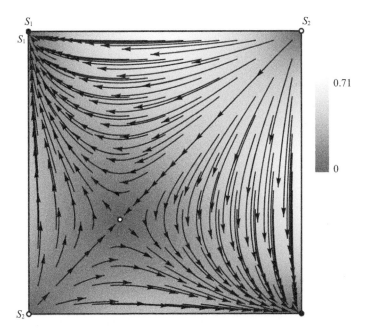

(c) Phase diagram for the two-population, continuous linear replicator dynamics for Battle of the Sexes. The two stable rest points are $\langle (1, 0), (1, 0) \rangle$ and $\langle (0, 1), (0, 1) \rangle$, denoted by solid circles; unstable rest points are indicated by unfilled circles.

Figure 9.3.7

A two-population version of the replicator dynamics.

$$W_1^R(\vec{p}^C) = p_1^C, \qquad\qquad W_1^C(\vec{p}^R) = 2p_1^R,$$
$$W_2^R(\vec{p}^C) = 2p_2^C, \qquad\qquad W_2^C(\vec{p}^R) = p_2^R,$$
$$\phi(\vec{p}^R) = p_1^R p_1^C + 2p_2^R p_2^C, \qquad \phi(\vec{p}^C) = 2p_1^R p_1^C + p_2^R p_2^C.$$

Figure 9.3.7c illustrates the phase diagram for Battle of the Sexes under the two-population replicator dynamics. Cressman (2003) provides a detailed investigation of multipopulation models with reference to extensive-form games.

5.   Stability

One natural question to ask is what connection—if any—exists between rest points of the various dynamical systems under consideration and the Nash equilibria (and/or evolutionarily stable strategies) of the underlying game. As a preliminary, an important conceptual distinction needs to be made: the various dynamics considered model the evolutionary flows over population states, where a population state is a *distribution of strategies* over the individual players. In contrast, a Nash equilibrium strategy (or an evolutionarily stable strategy) is a *probability distribution* over pure strategies adopted by a player. These are very different entities. But, as we will see, as long as we are careful about the conceptual differences, important connections exist.

Inspection of figure 9.3.7c suggests one possible connection. The underlying game has three Nash equilibria: $\langle(1,0), (1,0)\rangle$, $\langle(0,1), (0,1)\rangle$, and $\langle\tfrac{1}{3},\tfrac{2}{3}\rangle,\langle\tfrac{2}{3},\tfrac{1}{3}\rangle$. Of the three Nash equilibria, the first two feature evolutionarily stable strategies. If all three of these Nash equilibria are interpreted as *population states*, all correspond to rest points of the dynamics, with the two states corresponding to evolutionarily stable strategies being stable points of the dynamics. (The exact meaning of "stability" will be specified below.) However, the fact that additional rest points exist shows that, at least for the replicator dynamics, it is not the case that the rest points correspond to Nash equilibria.[5] A precise statement of the relationship between the two concepts requires introducing some definitions from the theory of dynamical systems.

A *rest point* of a dynamical system (also known as a *fixed point*) is a point $\vec{p}$ such that $dp_i/dt = 0$ for all $i = 1, \ldots, n$. A stability concept provides a characterization of the behavior of a dynamical system in the local region around a fixed point. There are several natural stability concepts used in evolutionary game theory.

Definition. Let $\vec{p} \in \Delta_n$ be a rest point of the replicator dynamics. Then $\vec{p}$ is said to be *Lyapunov-stable* if for every $\varepsilon > 0$, there exists a $\delta > 0$ such that $|\vec{p}(0) - \vec{p}| < \delta$ implies $|\vec{p}(t) - \vec{p}| < \varepsilon$ for all $t \geq 0$. That is,

every trajectory that is sufficiently close to the fixed point $\vec{p}$ (i.e., no more than δ away) remains close to $\vec{p}$ (i.e., no more than ε away) for all future times.

The intuition behind Lyapunov stability is that there is not a local "push" away from the rest point. Another intuition about stability is that there is a local "pull" *toward* the rest point. This motivates the following:

Definition. Let $\vec{p}$ be a fixed point of the replicator dynamics. Then $\vec{p}$ is *asymptotically stable* if $\vec{p}$ is Lyapunov-stable and, in addition, there exists a $\delta > 0$ such that if $|\vec{p}(0) - \vec{p}| < \delta$, then $\lim_{t \to \infty} \vec{p}(t) = \vec{p}$.

Asymptotic stability captures one aspect of evolutionary stability: that the population is not easily displaced from the rest point corresponding to a particular distribution of strategies. A displacement from an asymptotically stable rest point, provided it is not too great, will return to the rest point in the limit.

Definition. If $\vec{p}$ is Lyapunov-stable but not asymptotically stable, then $\vec{p}$ is said to be *neutrally stable*.

In light of the Hofbauer et al. (1979) result that an evolutionarily stable strategy is one that is locally superior, we can introduce an analogous definition in the context of evolutionary games. In particular, Hofbauer and Sigmund (2002) propose the following:

Definition. Let $\vec{p} \in \Delta_n$, and let A be the payoff matrix for the underlying evolutionary game. Then $\vec{p}$ is a *Nash equilibrium* if, for all $\vec{q} \in \Delta_n$, it is the case that

$$\vec{p} \cdot A\vec{p} \geq \vec{q} \cdot A\vec{p}.$$

The state $\vec{p}$ is an *evolutionarily stable state* if, for all $\vec{q} \neq \vec{p}$ in a neighborhood around $\vec{p}$,

$$\vec{p} \cdot A\vec{q} \geq \vec{q} \cdot A\vec{q}.$$

Note that, despite the mathematical similarity between the various definitions, the term "evolutionarily stable *state*" is used to stress the difference between the population and the strategy interpretation. But, in effect, an evolutionarily stable state is the distribution corresponding to an evolutionarily stable strategy interpreted as the frequency with which pure strategies appear in a population. For example, in the case of the Hawk–Dove game, the evolutionarily stable *strategy* is the mixed strategy $\sigma = \tfrac{V}{C}$ Hawk $+ (1 - \tfrac{V}{C})$ Dove, and the evolutionarily stable *state* is the population where $\tfrac{V}{C}$ of the population follows Hawk and $1 - \tfrac{V}{C}$ follows Dove.

Returning to the game of Rock–Paper–Scissors from figure 9.3.3, the phase diagram for the replicator dynamics is shown in figure 9.3.8. The rest point corresponding to the single Nash equilibrium of the game, $\langle\tfrac{1}{3},\tfrac{1}{3},\tfrac{1}{3}\rangle$, although not an evolutionarily stable state, is neutrally stable. Any small displacement from the rest point in the center of the simplex will result in the population cycling around the rest point in a stable orbit.

These two observations provide an illustration of the following theorem, proven by Hofbauer and Sigmund (2002):

Theorem 6. Let A be the payoff matrix for a symmetric two-player game. Then

If $\vec{p} \in \Delta_n$ is a Nash equilibrium of A, then $\vec{p}$ is a rest point of the replicator dynamics;

if $\vec{p}$ is Lyapunov-stable under the replicator dynamics, then it is a Nash equilibrium of the underlying game.

One can also prove (see Hofbauer & Sigmund, 2002) that the following connection holds between evolutionarily stable states and asymptotic stability, for the replicator dynamics:

Theorem 7. Let A be the payoff matrix for a symmetric two-player game. If $\vec{p} \in \Delta_n$ is an evolutionarily stable state for A, then $\vec{p}$ is asymptotically stable under the replicator dynamics with A as the underlying game.

It turns out that if one adopts a slightly stronger stability concept than that of an evolutionarily stable state, the above result for the replicator dynamics can be extended to multiple families of dynamics. Taylor and Jonker (1978) define a *regular ESS* as follows:

Definition. Let $S = \{1, \ldots, n\}$ be a set of pure strategies for a symmetric two-player game with payoff matrix A. Let σ be a Nash equilibrium for A. Then σ is a *regular ESS* if $\pi(i \mid \sigma) < \pi(\sigma \mid \sigma)$ whenever $i \notin \mathrm{supp}(\sigma)$, and $\vec{x} \cdot A\vec{x} < 0$ whenever $\mathrm{supp}(x) \subset \mathrm{supp}(\sigma)$, $\vec{x} \neq \vec{0}$ and $\Sigma x_i = 0$.

The concept of a regular ESS is a slight strengthening of the ESS concept. Every regular ESS is an ESS, but not

conversely. However, Taylor and Jonker note that *most* ESSs are regular, and if π is a payoff function for which a nonregular ESS exists, one can find another payoff function π', arbitrarily close to π, that only has regular ESSs. Intuitively, a regular ESS is an ESS σ where the support of σ contains *all* of the pure strategies that are best responses to σ.

One can then prove the following:

Theorem 8 (Hofbauer & Sandholm, 2011). Let σ be a regular ESS for an evolutionary game G. Then, σ is asymptotically stable under

- any impartial pairwise comparison dynamic for G;
- any separable excess payoff dynamic for G;
- the best response dynamic for G.

While the generality of this result is noteworthy, it must be stressed that what this establishes is a validation of "the use of regular ESS as a blanket sufficient condition for local stability under evolutionary game dynamics" (Sandholm, 2010b). Local stability is not necessarily a good guide to the expected behavior of a population in an evolutionary game. Consider, for example, the game shown in figure 9.3.9: although S_3 is a regular ESS and is asymptotically stable, the *basin of attraction* for S_3—that is, the set of all points that converge to S_3 as $t \to \infty$—can be made arbitrarily small by adjusting the value of ε accordingly.

Furthermore, it can also be the case that the basin of attraction for a population state is the entire interior of the simplex space, even though the population state is *not* asymptotically stable. Figure 9.3.10 illustrates a game where, under the replicator dynamics, the long-term behavior is such that the population will always converge to A. Yet A is not asymptotically stable, as small

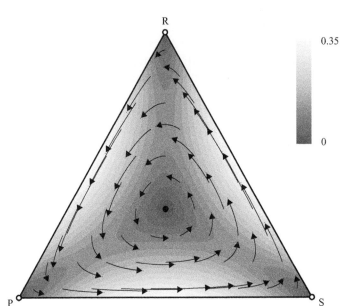

Figure 9.3.8

For the game of Rock–Paper–Scissors, the population state in which all three strategies are equally represented is a neutrally stable state of the replicator dynamics.

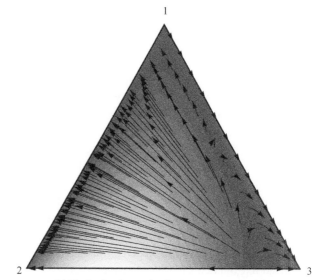

	S_1	S_2	S_3
S_1	$(0, 0)$	$(1, 0)$	$(\varepsilon, 0)$
S_2	$(0, 1)$	$(1, 1)$	$(-0, 0)$
S_3	$(0, \varepsilon)$	$(0, 0)$	$(2\varepsilon, 2\varepsilon)$

(a) The payoff matrix.

(b) The basin of attraction for S_3 appears on the right of the simplex and can be made arbitrarily small by adjusting the value of ε accordingly. Nevertheless, S_3 is a regular ESS and is asymptotically stable.

Figure 9.3.9
An ESS can have an arbitrarily small basin of attraction.

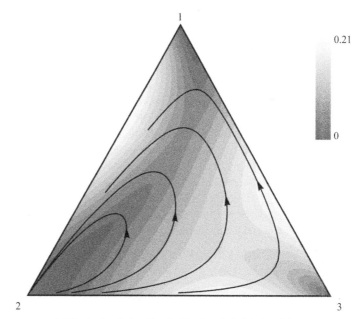

	S_1	S_2	S_3
S_1	$(0, 0)$	$(0, 1)$	$(2, 0)$
S_2	$(1, 0)$	$(0, 0)$	$(0, 0)$
S_3	$(0, 2)$	$(0, 0)$	$(1, 1)$

(a) The payoff matrix.

(b) The basin of attraction for S_2 is the whole interior of the simplex, even though S_2 is not asymptotically stable. Small displacements from S_2 result in the population following a lengthy trajectory, which carries it very far away from the all-S_2 state before eventually returning. (The trajectories ultimately converge to the all-S_2 point; the trajectories appear to stop in the middle of the simplex due to numeric limits on the differential equation solver.)

Figure 9.3.10
A population state can be the expected evolutionary outcome without being asymptotically stable.

displacements may result in the population being carried very far away before eventually returning via a circuitous route.

Evolutionary game theory provides a great variety of tools for modeling the boundedly rational behavior of populations in strategic contexts. Yet, as we have seen, the evolutionary outcomes may not always be rational: strictly dominated strategies may be played with positive probability, as figure 9.3.7b shows. The fact that a population state is a (regular) evolutionarily stable state does not guarantee that it is, in any way, the expected or even likely evolutionary outcome, as seen in figure 9.3.9. And the long-term convergence behavior of a population can also be decoupled from asymptotic stability, as in figure 9.3.10. With great variety comes great complexity and hence the need to exercise care when drawing inferences about expected outcomes.

Acknowledgments

I thank William Sandholm, Emin Dokumaci, and Francisco Franchetti for their open-source software suite Dynamo, which was used to create many figures in this article (see Franchetti & Sandholm, 2013).

Notes

1. It should be noted that, in his collection of essays *On Evolution*, published in 1972, Maynard Smith gives the same definition of an ESS in the essay "Game Theory and the Evolution of Fighting." But he clearly notes in the introduction to the volume that "I would probably not have had the idea for this essay if I had not seen an unpublished manuscript on the evolution of fighting by Dr George Price. . . . Unfortunately, Dr Price is better at having ideas than at publishing them. The best I can do therefore is to acknowledge that if there is anything in the idea, the credit should go to Dr Price and not to me."

2. Whether this would, in fact, happen depends on specific features of the dynamics.

3. In their 1973 paper, the game is called the "Hawk–Mouse" game as a result of George Price's objection to using the term "Dove," due to its religious connotations.

4. It is assumed that the losing Hawk is selected at random with equal probability, which is why the payoffs in the upper-left cell of the payoff matrix take the form that they do.

5. The two additional rest points are where the entire Row population follows S_2 and the Column population follow S_1, and vice versa. These are unstable rest points, but they exist because the replicator dynamics cannot introduce strategies into the population if they are originally absent.

References

Akin, E. (1980). Domination or equilibrium. *Mathematical Biosciences, 50*(3–4), 239–250.

Brown, G. W., & von Neumann, J. (1950). Solutions of games by differential equations. In H. W. Kuhn & A. W. Tucker (Eds.), *Contributions to the theory of games* (Vol. 1, Annals of Mathematics Studies, Nr. 24, pp. 73–80). Princeton, NJ: Princeton University Press.

Cressman, R. (2003). *Evolutionary dynamics and extensive form games*. Cambridge, MA: MIT Press.

Fisher, R. A. (1930). *The genetical theory of natural selection*. Oxford, England: Oxford Clarendon Press.

Franchetti, F., & Sandholm, W. H. (2013). An introduction to Dynamo: Diagrams for evolutionary game dynamics. *Biological Theory, 8*, 167–178.

Gintis, H. (2009). *Game theory evolving* (2nd ed.). Princeton, NJ: Princeton University Press.

Hofbauer, J., & Sandholm, W. H. (2011). Survival of dominated strategies under evolutionary dynamics. *Theoretical Economics, 6*(3), 341–377.

Hofbauer, J., Schuster, P., & Sigmund, K. (1979). A note on evolutionary stable strategies and game dynamics. *Journal of Theoretical Biology, 81*, 609–612.

Hofbauer, J., & Sigmund, K. (2002). *Evolutionary games and population dynamics*. Cambridge, England: Cambridge University Press.

Maynard Smith, J. (1972). *On evolution*. Edinburgh, Scotland: Edinburgh University Press.

Maynard Smith, J. (1982). *Evolution and the theory of games*. Cambridge, England: Cambridge University Press.

Maynard Smith, J., & Price, G. R. (1973). The logic of animal conflict. *Nature, 246*, 15–18.

Sandholm, W. H. (2010a). Local stability under evolutionary game dynamics. *Theoretical Economics, 5*(1), 27–50.

Sandholm, W. H. (2010b). *Population games and evolutionary dynamics*. Cambridge, MA: MIT Press.

Schlag, K. H. (1998). Why imitate, and if so, how? A boundedly rational approach to multi-armed bandits. *Journal of Economic Theory, 78*(1), 130–156.

Smith, M. J. (1984). The stability of a dynamic model of traffic assignment—An application of a method of Lyapunov. *Transportation Science, 18*(3), 245–252.

Taylor, P. D., & Jonker, L. B. (1978). Evolutionary stable strategies and game dynamics. *Mathematical Biosciences, 40*(1–2), 145–156.

von Neumann, J., & Morgenstern, O. (1944). *Theory of games and economic behavior*. Princeton, NJ: Princeton University Press.

Weibull, J. W. (1995). *Evolutionary game theory*. Cambridge, MA: MIT Press.

9.4 Rationality in Economics: Theory and Evidence

Sanjit Dhami and Ali al-Nowaihi

Summary

We examine the various senses in which economists use the term "rationality" in various domains such as certainty, risk and uncertainty, time-discounting, and strategic interaction. Although the rationality definitions in economics are precise and axiomatic, there has been much misunderstanding about them. For this reason, we also outline some of the commonly drawn implications and auxiliary assumptions and separate these from the axiomatic definitions. Finally, we confront the rationality axioms and the implications with the empirical evidence, drawing on insights from the exciting new field of behavioral economics.

1. Outline of the Chapter

Rationality has diverse meanings in the social sciences and in common discourse. The dominant paradigm in economics, *neoclassical economics*, defines rationality in a precise way. This forms the basis of a coherent body of economic theory that allows for sharp and testable predictions. However, economists typically justify their assumption of rationality on grounds of plausibility and tractability rather than on the only admissible grounds in science: conformity with the empirical evidence.

The plan of this essay is as follows. Section 2 specifies the meaning of rationality in economics. Section 3 considers some auxiliary assumptions that are either regarded as part of a broader concept of rationality or are sometimes mistakenly believed to be consequences of rationality as defined in section 2. Finally, section 4 pits the neoclassical view of rationality against the evidence that has emerged over the past several decades from the exciting and fastest-growing field in economics: behavioral economics. It is impossible to cite all the papers in such a short essay, so the reader is alerted to suitable surveys that contain other relevant references.

2. Rationality in Economics

This section explains the precise nature of rationality in economics. In each case, the primitives are the preferences that individuals have over appropriately defined sets of objects.

2.1 The Assumption of Rationality under Certainty

Suppose that an individual has a set $X \subset \mathbb{R}^n$ of choices, where $\mathbf{x} \in X$ may be interpreted as a bundle of n goods or services. Let $\succeq$ be a binary preference relation: for $\mathbf{x}, \mathbf{y} \in X$, $\mathbf{x} \succeq \mathbf{y}$ means that $\mathbf{x}$ is considered "at least as good as" $\mathbf{y}$.

Definition 1. The individual is *rational*, or equivalently, the preference relation $\succeq$ is *rational*, if the following hold:

For all $\mathbf{x}, \mathbf{y} \in X$, we have either $\mathbf{x} \succeq \mathbf{y}$ or $\mathbf{y} \succeq \mathbf{x}$. (Completeness)

Let $\mathbf{x}, \mathbf{y}, \mathbf{z} \in X$ with $\mathbf{x} \succeq \mathbf{y}$ and $\mathbf{y} \succeq \mathbf{z}$. Then $\mathbf{x} \succeq \mathbf{z}$. (Transitivity)

Definition 2. If $\mathbf{x} \succeq \mathbf{y}$, then we say that $\mathbf{x}$ is *at least as good as* $\mathbf{y}$. If $\mathbf{x} \succeq \mathbf{y}$ and $\mathbf{y} \succeq \mathbf{x}$, then we say that $\mathbf{x}$ is *indifferent to* $\mathbf{y}$, and we write this as $\mathbf{x} \sim \mathbf{y}$. On the other hand, if $\mathbf{x} \succeq \mathbf{y}$ but it is not the case that $\mathbf{y} \succeq \mathbf{x}$, then we say that $\mathbf{x}$ is *strictly preferred to* $\mathbf{y}$ and we write $\mathbf{x} \succ \mathbf{y}$.

The *assumption of rationality* states that a decision maker is rational in the sense of definition 1, that is, a decision maker has a preference relation $\succeq$ on the set X of choices such that pairwise comparisons between any two alternatives can be made (completeness) and preferences are transitive.

Note that a violation of rationality (definition 1) in itself does not entail a violation of any law of logic. There is a widespread consensus among economists that the assumption of rationality is plausible. However, the crucial question is whether it conforms to human behavior. We examine the evidence in section 4.

Also, note that the definition of rationality, definition 1, can be stated for *any* nonempty set X, not just $X \subset \mathbb{R}^n$. For example, X could be a set of lotteries (see subsection 2.2, below) or consumption streams (see subsection 2.3, below). However, restricting X to be a subset of $\mathbb{R}^n$ allows us to introduce a useful continuity property (without explicitly defining a topology on X).

Definition 3 (Continuity). The preference relation $\geq$ is *continuous* if, for each $\mathbf{y} \in X$, both the sets $\{\mathbf{x} \in X: \mathbf{y} \geq \mathbf{x}\}$ and $\{\mathbf{z} \in X: \mathbf{z} \geq \mathbf{y}\}$ are closed.

A *utility function* is simply a useful, and equivalent, alternative way of expressing the preference relation $\geq$:

Definition 4 (Utility Function). A utility function $u: X \to \mathbb{R}$ *represents* the preference relation $\geq$, if for $\mathbf{x}, \mathbf{y} \in X$, $\mathbf{x} \geq \mathbf{y}$ if, and only if, $u(\mathbf{x}) \geq u(\mathbf{y})$.

But does such a utility function exist?

Proposition 1. Suppose that the preference relation $\geq$ on X is complete, transitive, and continuous. Then there exists a continuous utility function u that represents $\geq$.

The converse of proposition 1 also holds:

Proposition 2. If a continuous function u represents the preference relation $\geq$ on X, then $\geq$ must be complete, transitive, and continuous.

It is often more convenient to work with a utility function than with preferences.

2.2 Rationality under Risk

Risk is pervasive in most economic problems. Let $X = \{x_1, x_2, \ldots, x_n\}$ be a fixed finite set of real-valued outcomes such that $x_1 \leq x_2 \leq \ldots \leq x_n$.[1]

A *simple lottery*, or *simple gamble*, L, is a shorthand way of capturing risk:

$$L = (x_1, p_1; x_2, p_2; \ldots; x_n, p_n), \tag{1}$$

where $p_1, \ldots, p_n$ are the respective probabilities corresponding to the outcomes $x_1, \ldots, x_n$ such that $p_i \in [0, 1]$ and $\sum_{i=1}^n p_i = 1$. The lottery $(x_i, 1)$ denotes an outcome x_i received with certainty; this is the situation considered in subsection 2.1. Basically, a *compound lottery* is a lottery whose outcomes are themselves lotteries rather than real numbers.

Let $\mathscr{L}$ be the set of all lotteries over X (simple and compound). Which lottery will the decision maker choose?

The most widely used theory under risk in economics is *expected utility theory* (EU). Suppose that the decision maker is endowed with a binary preference relation over $\mathscr{L}$ denoted by $\geq$. So, for $L_1, L_2 \in \mathscr{L}$, $L_2 \geq L_1$ means that the lottery L_2 is "at least as good as" the lottery L_1.

Consider the following axioms on behavior under risk:

Axiom 1 (Order). Order requires the following two conditions:

For all lotteries L_1, L_2, either $L_2 \geq L_1$ or $L_1 \geq L_2$. (Completeness)

For all lotteries L_1, L_2, L_3: if $L_3 \geq L_2$ and $L_2 \geq L_1$, then $L_3 \geq L_1$. (Transitivity)

Two further binary relations between lotteries can be defined in terms of $\geq$:

$L_1 \sim L_2 \Leftrightarrow L_2 \geq L_1$ and $L_1 \geq L_2$; (Indifference)

$L_2 \succ L_1 \Leftrightarrow$ it is not the case that $L_1 \geq L_2$. (Strict preference)

Axiom 2 (Best and Worst). $x_n \succ x_1$ (i.e., $(x_n, 1) \succ (x_1, 1)$).

Axiom 3 (Continuity). For each lottery L, there is a $p \in [0, 1]$ such that $L \sim (x_1, 1 - p; x_n, p)$.

Axiom 4 (Independence). For all lotteries L_1, L_2, L, and all $p \in [0, 1]$, $L_2 \geq L_1 \Leftrightarrow (L_2, p; L, 1 - p) \geq (L_1, p; L, 1 - p)$.

The axioms of completeness, transitivity, and continuity introduced above are analogous to those introduced in subsection 2.1. From axiom 2, under certainty, a decision maker strictly prefers the highest to the lowest outcome. The continuity property in axiom 3 asserts that every lottery (no matter how complex) is equivalent to a simple lottery obtained by probabilistically mixing the highest outcome, x_n, and the lowest outcome, x_1. The independence axiom, axiom 4, requires that if a lottery L_2 is preferred to a lottery L_1, then mixing each with a third lottery L (with the same mixing probability p) does not alter the preference.

The next definition gives "rationality" a precise meaning under risk:

Definition 5 (Axioms of Rationality under Risk). The axioms *order, best and worst, continuity, independence,* and *reduction* are collectively termed the *axioms of rationality under risk.*[2]

Proposition 3 (von Neumann–Morgenstern Expected Utility Representation). Suppose that the binary relation $\geq$ on the set $\mathscr{L}$ of lotteries satisfies the axioms of rationality under risk (definition 5). Then there is a von Neumann–Morgenstern utility function $EU: \mathscr{L} \to \mathbb{R}$ that represents $\geq$, that is, $L_2 \geq L_1$ if, and only if, $EU(L_2) \geq EU(L_1)$. This function takes the form

$$EU(L) = \sum_{i=1}^n p_i u(x_i), \tag{2}$$

where $u(x_i)$ is a real number assigned to the outcome x_i, also called the *utility* of x_i.

The converse of proposition 3 also holds. This framework can be extended to uncertainty (where probabilities are subjective, rather than objective; Dhami, 2016, section 1.3; Savage, 1954).

Finally, it is assumed that all probabilities satisfy the Kolmogorov (1950) axioms of probability.

2.3 Rationality in Choices Made over Time

Let $Z \subset \mathbb{R}$ be the set of possible outcomes and let $\Gamma \subset \mathbb{R}_+$ be the set of corresponding times at which these outcomes are realized. Elements of the set Z are denoted by z_j ($j = 1, 2, \ldots$) and elements of the set Γ by t_j ($j = 1, 2, \ldots$). Denote by (z_j, t_j) an *outcome–time pair*, which *guarantees* the outcome z_j at time t_j. Let $\geq$ denote the binary preference ordering "at least as good as" on the Cartesian product $X = Z \times \Gamma$. Notice that we have now redefined the set X from subsection 2.1 as outcome–time pairs. Also, let $\succ$ and $\sim$ denote, respectively, the binary relations "strictly preferred to" and "indifferent to" on the set $X = Z \times \Gamma$. The following five axioms B1–B5 from Fishburn and Rubinstein (1982) are typically called the *axioms of rationality for time-discounting*:

Definition 6 (Axioms of Rationality for Time-Discounting). Let $z_m, z_n \in Z$ and $t_p, t_q \in \Gamma$.

B1. *Ordering*: $\geq$ is complete and transitive.

B2. *Monotonicity*: If $z_m > z_n$, then $(z_m, t) \succ (z_n, t)$ for any $t \in \Gamma$ (larger rewards are preferred, keeping time fixed).

B3. *Impatience*: Let $t_q < t_p$ and $z \in Z$ with $z > 0$. Then, $(z, t_q) \succ (z, t_p)$ (the same outcome is preferred at an earlier date). Also $(0, t_q) \sim (0, t_p)$ (no time preference for zero rewards).

B4. *Continuity*: Fix some outcome–time pair $(z_j, t_j) \in Z \times \Gamma$. Then $\{(z,t) : (z,t) \geq (z_j, t_j)\}$ and $\{(z,t) : (z_j, t_j) \geq (z,t)\}$ are closed.

B5. *Stationarity*: If $(z_m, t_q) \sim (z_n, t_p)$, then $(z_m, t_q + t) \sim (z_n, t_p + t)$ for any $t \in \Gamma$ (indifference between two outcomes depends only on the length of the time difference between the two).

We have already encountered the analogues of axioms B1 and B4 in subsection 2.1. Axioms B2 and B3 are eminently reasonable. Axiom B5 is crucial to producing the additive separability under exponential discounting. Fishburn and Rubinstein (1982) write, "However, we know of no persuasive argument for stationarity as a psychologically viable assumption" (p. 681).

The utility representation of the *axioms of rationality for time-discounting* is given next:

Proposition 4 (Fishburn & Rubinstein, 1982, p. 682). If B1–B5 hold, then for any $\delta \in (0, 1)$, there exists a continuous increasing real-valued function u defined on the domain Z such that

(i) For all (z_m, t_q), (z_n, t_p) in $Z \times \Gamma$, $(z_m, t_q) \geq (z_n, t_p)$ if, and only if,

$$\delta^{t_q} u(z_m) \geq \delta^{t_p} u(z_n).$$

(ii) The outcome–time pair $(0, t)$ is treated as $\delta^t u(0) = 0$.

(iii) Given δ, u is unique up to multiplication by a positive constant.

The *axioms of rationality for time-discounting* are testable; the evidence is considered in section 4.

In the light of proposition 4, the resulting model is called the *exponentially discounted utility model* (EDU).

2.4 Rationality in Game Theory

Game theory deals with strategic situations in which individuals mutually influence the payoffs of others through their actions and beliefs. Rationality plays a key role in the development of the central concepts in game theory. These are already developed in chapter 9.1 by Albert and Kliemt, chapter 9.2 by Perea, and chapter 9.3 by Alexander (all in this handbook), so this discussion is omitted. The chapter assumes a knowledge of the basic equilibrium concepts (see, e.g., Fudenberg & Tirole, 1991).

In the simplest setting of a *static game of complete information*, a Nash equilibrium places no restrictions on the beliefs of players. Such belief restrictions are dealt with in the *epistemic foundations of a Nash equilibrium*, and they turn out to be extremely demanding and unlikely to be met in practice. Suppose that the following conditions hold: mutual knowledge of the payoffs, mutual knowledge of rationality of players, common knowledge of the beliefs (or conjectures) each player has about the others, and common priors. Then, Aumann and Brandenburger (1995) showed, the common conjectures about what other players will do are in agreement and constitute a Nash equilibrium of the game. Polak (1999) showed that if mutual knowledge of the payoffs is strengthened to *common* knowledge of payoffs, then the Aumann–Brandenburger conditions imply common knowledge of rationality. Recent research offers an unsettled view of the epistemic foundations (Gintis, 2009).

If the predictions of a Nash equilibrium and its refinements are violated by the empirical evidence, then so are the epistemic conditions. We briefly summarize the empirical evidence in section 4.

A Nash equilibrium is often justified as the outcome of a learning process. However, such processes do not guarantee convergence of play to a Nash equilibrium, and when convergence does take place, there is no guarantee that the underlying learning process is plausible (Fudenberg & Tirole, 1991, pp. 25–29). Even when convergence does take place under a plausible learning process, there is no guarantee that it does so in a reasonable amount of time that is representative of real-world interactions.

3.   Some Auxiliary Assumptions

In this section, we briefly outline some auxiliary assumptions that are commonly used in economics in addition to the assumption of rationality, given in section 2. These assumptions are sometimes mistakenly believed to be consequences of the rationality assumptions of section 2. At other times, they are regarded as part of a more restricted concept of rationality.

3.1   Individuals Have Self-Regarding Preferences

Individual i is said to have *other-regarding preferences* (or *social preferences*) if the individual cares about own-consumption, x_i, and the consumption of others, x_{-i}, that is, individual i's utility function is $u_i(x_i, x_{-i})$. In definitions 1 and 4 and propositions 1 and 2, preferences are purely *self-regarding* since the utility of any individual i only depends on own-consumption; that is, instead of $u_i(x_i, x_{-i})$, we could as well just talk about $u_i(x_i)$. This has led to a common narrative that rationality implies self-regarding preferences.

This, however, is not correct. One may define the initial set X in any way one wishes to. For instance, for individual i, elements $x \in X$ may be appropriately defined as pairs of the form (x_i, x_{-i}). One can then proceed, as in section 2, to derive a utility function with other-regarding preferences. However, in actual practice, neoclassical economics has not taken this direction.[3]

All this has changed with the advent of behavioral economics, in which other-regarding preferences play a central role. We now know a great deal about the nature of other-regarding preferences through lab, field, and neuroeconomic evidence. Behavioral models of other-regarding preferences have made several precise and confirmed predictions. Experiments have played a key role in the precise identification of other-regarding preferences. Experimental games such as the dictator game, the ultimatum game, the public goods game, the trust game, and the gift-exchange game, among many others, have allowed stringent tests of competing theories. For surveys, the interested reader may consult Fehr and Schmidt (2006) and Dhami (2016, part 2).

3.2   There Is No Role for Emotions in Decision Making

Rationality in economics has also been taken to imply cold, emotionless deliberation. This does appear to be an indirect implication of definitions 1, 5, and 6. None of the axioms explicitly takes account of emotional states. In contrast, there has been growing interest in modeling emotions in behavioral economics.

In principle, one could modify the classical framework by introducing a set M of emotional states and, for any $m \in M$, index the preference relation $\succsim$ in section 2 by m to get $\succsim_m$; this applies to certainty, to uncertainty, and to the time dimension. This allows one's preferences to be dependent on emotions. In an emotional state, one may make erratic or inconsistent decisions. In anger, one could signal credibly one's own future intended actions. *Anxiety* is an important emotion that has been used to explain the equity-premium puzzle in finance (Caplin & Leahy, 2001). *Projection bias*, the tendency to project our current emotional states to future states of the world, explains why those awaiting kidney transplants have relatively lower reported well-being as compared to a healthy control group, yet after the transplant, both groups report similar well-being (Loewenstein, O'Donoghue, & Rabin, 2003). For a large number of other applications, and for the relevant evidence, the interested reader can consult Dhami (2016, part 6). Behavioral game theory has used emotions to successfully explain diverse economic phenomena under the rubric of *psychological game theory* (Dhami, 2016, section 13.5).

3.3   Perfect Attention, Unlimited Computing Power, and No Misperceptions

The choice set X, whether it be the space of outcomes, lotteries, or outcome–time pairs, can be infinite (section 2). Thus, the rationality assumption requires individuals to make a large number of pairwise comparisons in order to choose the most preferred or maximal element in the set X. In addition, transitivity must not be violated. This may require enormous computation power, memory, and attention. In contrast, the evidence shows that people are subject to inattention, misperception, and limited computation powers (see section 4 below).

Consider the following typical example of what underlies almost all areas of economics:

Example 1. A self-regarding individual lives for T time periods and cares only about the dated stream of his consumption, that is, $c_1, c_2, \ldots, c_T$. At time t, $t = 1$, $2, \ldots, T$, he inherits a stock of savings, S_{t-1}, from the previous period. This gives him the interest payment rS_{t-1}, where $r \geq 0$ is the return on one-period bonds. In addition, he has income y_t from other sources (e.g., labor income). Thus, his total period-t income is $y_t + rS_{t-1}$. This is spent on consumption c_t, and the balance, $s_t = y_t + rS_{t-1} - c_t$, is saved. Thus, $S_t = S_{t-1} + s_t$. Let his utility from period-t consumption be $u(c_t)$ and his subjective discount factor be $\delta \in (0, 1)$. His objective at time t is to choose $c_t, c_{t+1}, \ldots, c_T$ so as to maximize the present value of his stream of utilities $\sum_{i=t}^{i=T} \delta^{t-i}u(c_i)$. At the time of his death, in period T, he cannot be in debt, so we impose the condition $S_T \geq 0$.

The problem formulated in example 1 can be given the following mathematical form that we refer to as "problem-T":

For each time $t = 1, 2, \ldots, T$, choose $c_t, c_{t+1}, \ldots, c_T$ so as to maximize

$$\sum_{i=t}^{i=T} \delta^{t-i}u(c_i) \quad \text{(objective function)}$$

subject to the following conditions:

Initial condition:

S_{t-1} given.

Dynamic equation:

$S_i = y_i + (1+r)S_{i-1} - c_i, \quad$ for $i = t, t+1, \ldots, T.$

Terminal condition: $S_T \geq 0.$

Problem-T is a simple example of a *dynamic optimization problem*. It is an *optimization problem* because we have an objective function we want to maximize. It is a *dynamic problem* because excessive consumption (say) in the early periods will not leave enough money for adequate consumption in the final periods. The initial state of the consumer is his initial stock of savings, S_0. His final state is zero savings (because it is suboptimal to leave unused savings). The control variables are the consumption levels.

The solution method for problem-T is by *backward induction*. The consumer should start with the final period T. Then for *every* possible value in the stock of savings, S_{T-1}, inherited from the penultimate period, $T-1$, find the optimal level of consumption by choosing c_T so as to maximize $u(c_T)$ subject to $y_T + (1+r)S_{T-1} - c_T = 0$. Then move back in time and repeat for period $T-1$. The consumer should continue until reaching the first period, $t = 1$, and then substitute for the initial value for savings, S_0. This will then determine the optimal consumption levels $c_1^*, c_2^*, \ldots, c_T^*$. There are, however, three problems that we now turn to:

1. Although backward induction is conceptually simple, it requires a huge number of calculations. For example, consider a 40-year horizon and a period length of one year (so $T = 40$). Suppose that the income y_t at each possible time t can take two possible values, high

and low, and corresponding to this, we have just two possible optimal levels of consumption, low and high. Depending on the values of consumption, income and initial wealth, this could require the consumer to carry out up to $2^{40} > 10^{12}$ (one trillion) calculations.

2. Even when the problem is simplified, say, for a three-year horizon, so that the total number of calculations is, at most, $2^3 = 8$, experimental subjects do not use backward induction (see Dhami, 2016, part 4).

3. Any real-life situation is far more complex than example 1. For example, individuals might be uncertain of the time of death, future incomes, and future rates of return. Bequests might have to be calculated optimally. Decisions might be taken in hot, and unanticipated, emotional states, and capital markets might be imperfect.

The maintained assumption in neoclassical economics is that the man on the street can solve these hard problems literally, in his head, in an instant (analytically or numerically) or behave in a manner *as if* he can solve the problem. However, when tested, the assumptions and predictions of many leading neoclassical models fail (see section 4).

Furthermore, in many real-life situations, one might be distracted by other pressing issues (limited time, information overload, deadlines, marital discord, emotional lows) that make the supposed optimization exercise even more challenging. Yet, in real life, individuals routinely make complex decisions in the presence of uncertainty. So how do they do it? Could it be that they do not use the sort of optimization that economists believe they engage in and use instead simple rules of thumb and various boundedly rational methods? See subsection 4.5, below, and Dhami (2016, part 7).

3.4   We Can Ignore Irrationality in Financial and Other Markets

In economics, financial markets are typically held out to be the epitome of efficiency. The dominant tradition in finance argues that we can ignore any irrationality (in the sense considered in section 2) in financial markets. This argument is based on two assumptions but with little empirical support (Dhami, 2016, chapter 21). First, although irrational individuals, if there are any, may make errors, these errors will cancel out in the aggregate. Second, irrational market participants (or *noise traders*), if there are any, will be driven out of the market by the actions of rational market participants (or *arbitrageurs*).

The assumption of Bayesian rationality in finance is central to the *efficient markets hypothesis* (EMH), which is the cornerstone of modern finance.[4] The EMH asserts that the price of an asset at any point in time equals the present discounted value of the stream of future incomes from that asset (i.e., its fundamental value). The evidence, however, is that humans, even professionals, are poor at Bayesian updating, and many assets can deviate for long stretches of time from their fundamental values (Dhami, 2016, sections 19.7 and 19.18 and chapter 21).

4. The Evidence on Rationality in Economics

In this section, we briefly consider the evidence on the rationality assumptions of section 2 and the auxiliary assumptions of section 3.

4.1 Evidence on the Basic Assumptions of Rationality

The most basic definition of rationality (definition 1) requires that preferences be complete and transitive. Both conditions are violated in experiments. Iyengar and Lepper (2000) exposed subjects to either a limited-choice condition (6 varieties of jams) or an intensive-choice condition (24 varieties of jams). There was no difference in the percentage of jams sampled in both conditions. However, nearly 30% of the consumers in the limited-choice condition purchased a jam jar. The corresponding number in the intensive-choice condition was only 3%. This suggests that individuals may find it more difficult to make choices when there are too many options; this may violate completeness.

The most famous of the violations of transitivity occurs in the experiments on *preference reversals*; this is a problem in the domain of risk. Consider two probabilities $p < P$, two outcomes $z < Z$, and the following pair of lotteries:

$$P\text{-bet} = (0, 1 - P; z, P), \qquad \$\text{-bet} = (0, 1 - p; Z, p).$$

The P-bet has a higher probability, P, of winning, and the $-bet has a higher prize, Z. The decision maker is given the following two tasks:

1. Direct choice between the P-bet and the $-bet.
2. Assign *certainty equivalents* to the P-bet and the $-bet; these are denoted, respectively, by C_P and $C_\$$.[5]

The typical empirical finding, much replicated, is that the P-bet is chosen over the $-bet in the first task and $C_P < C_\$$ in the second task (Lichtenstein & Slovic, 2006). Transitivity fails because $\$\text{-bet} \sim C_\$ \succ C_P \sim P\text{-bet} \succ \$\text{-bet}$ (where $\succ$ denotes strict preference).

Unlimited attention is implicit in the assumption of rationality (definition 1). However, there is growing interest in the implications of *limited attention*, which reflects limited cognitive abilities of humans. Here are some findings from this literature (Dhami, 2016, section 19.17): consumers pay less attention to taxes that are not listed on price stickers but added at cash registers; displaying information about taxes on price stickers reduces sales. Individuals pay relatively less attention to electronic toll payments, allowing operators to levy higher tolls as compared to, say, cash paid at manual toll collections, which make the payment salient. Car sales and car prices are sensitive to integer multiples of the 10,000-mile threshold, so there is too large a gap in prices of cars with mileage readings of 9,999 and 10,000 miles because consumers pay too much attention to the leftmost digit. Limited attention is also increasingly used to explain suboptimal economic decisions of the poor, whose attention is diverted by the basic problems of food, shelter, and clean water, a form of *cognitive taxes* that reduce *mental bandwidth* (Datta & Mullainathan, 2014; World Development Report, 2015). There are now promising theoretical attempts to formalize limited attention in which consumers pay differential attention to various attributes of a good (Gabaix, 2014).

Despite this evidence, the common consensus in economics, and in behavioral economics, is to maintain a minimum degree of rationality as dictated by definition 1. A compromise could be to retain rationality in the sense of definition 1, but only within a suitably narrow frame or context (Gintis, 2009, 2017). However, it is not clear if all evidence that contradicts complete and transitive preferences could be explained by invoking different contexts and frames.

4.2 Evidence on Rationality in Risk and Uncertainty

The leading example of a violation of the independence axiom is the Allais paradox (Allais, 1953; Dhami, 2016, section 1.5; see also chapter 8.2 by Peterson, this handbook). It is now well recognized that *nonlinear probability weighting* (NLPW) explains the Allais paradox. On the other hand, EU requires that probabilities enter into the utility function linearly.

The evidence for NLPW has grown enormously over the years (Fehr-Duda & Epper, 2012). NLPW plays a central role in the modern alternatives to EU, such as *rank-dependent utility* (RDU) due to Quiggin (1982) and *prospect theory* (PT) due to Kahneman and Tversky (1979) and Tversky and Kahneman (1992). PT has been enormously successful in explaining the evidence, particularly where both EU and RDU fail. PT has been successfully incorporated into formal economic models in many areas of economics; for surveys, see Kahneman and Tversky (2000) and Dhami (2016, part 1); see also chapter 8.3 by Glöckner (this handbook).

These are by no means the only refutations of EU.[6] Other refutations include the following (Dhami, 2016, chapter 3): unreasonable attitudes toward risk, the endowment effect, anomalies of tax evasion, the efficacy of goal-setting behavior, backward-bending labor supply curve of taxi drivers, the equity-premium puzzle, contract choice and renegotiation, and the pricing of assets and the skewness of asset returns. Furthermore, close genetic relatives of humans (e.g., primates such as capuchin monkeys) exhibit similar violations of EU, and their behavior supports PT. The implication is that preferences supporting PT were hardwired into the common ancestors of humans and capuchin monkeys.

4.3 Evidence on Rationality in Time-Discounting

The utility representation of rational time preferences takes the form of the EDU (proposition 4, recall subsection 2.3). Under EDU, the discount factor δ is constant. Under EDU, the same discount factor δ applies independently of (1) the magnitude of an outcome (absence of *magnitude effect*) and (2) the sign, positive or negative, of an outcome (absence of *gain–loss asymmetry*). Furthermore, preferences are independent of the shapes of consumption profiles.

Strong empirical evidence refutes each of these features of the EDU model. Behavioral models of time-discounting have been proposed that are able to account for all of these violations. Consider the following violations of the EDU model reported in Thaler (1981), which have been often replicated (see Dhami, 2016, chapter 9, for the updated evidence):

1. *Magnitude effect*: Subjects reported being indifferent to $15 received now and $60 in a year's time, $250 received now and $350 received in a year's time, and $3,000 received now and $4,000 in a year's time. The implied respective annual discount factors are 0.25, 0.71, and 0.75; these are not identical with those required by EDU. Thus, larger amounts are discounted less.

2. *Gain–loss asymmetry*: The discount factors for losses turned out to be larger than those for gains, so losses are discounted less than gains. This could be because losses are more salient than gains.

3. *Common difference effect*: When asked how much money in one month, one year, and 10 years, respectively, would make them indifferent to receiving $15 now, the median response of subjects was $20, $50, and $100, respectively. The respective implied annual discount factors are 3%, 30%, and 83%. Hence, the shorter the horizon, the lower the implied annual discount factor (greater impatience); in contrast, under EDU, the annual discount rate factor stays constant.

The common difference effect suggests a taste for immediate gratification. A major thrust of the behavioral economics literature is to explain this phenomenon. A tractable and popular alternative to EDU is *quasi-hyperbolic discounting* (Laibson, 1997; Phelps & Pollak, 1968) in which time preferences are represented by[7]

$$U(c_0,\ldots,c_T) = u(c_0) + \beta \sum_{t=1}^{T} \delta^t u(c_t), \quad 0 < \beta < 1. \quad (3)$$

In (3), except for the presence of $\beta \in (0, 1)$, the utility function is identical to the EDU model. The factor β shrinks the value of future lifetime utility relative to current utility, thereby creating an additional *present-bias for current consumption*, or a taste for immediate gratification. It also leads to a pattern of increasing discount rates as we approach the present that potentially explain the common difference effect.

A taste for immediate gratification creates self-control problems. Furthermore, people have imperfect awareness about their future self-control problems; hence, they often take inadequate preventive measures. This also leads to the *time inconsistency problem* (i.e., choices made now about a future date t need not be optimal when the individual reoptimizes at time t). Classically, in economics, time-inconsistent choices were considered irrational (and are impossible under EDU), but evidence suggests that they often occur.

The quasi-hyperbolic framework resolves several problems that are unexplained by the axioms of rationality for time-discounting (definition 6). Some of these are as follows (for details, see Dhami, 2016, chapter 11): Why does consumption track income so closely? Why do individuals undersave for retirement? Why is there a sharp drop in consumption at retirement? Why do individuals hold illiquid assets and credit card debt simultaneously? What causes addictions? Why do some smokers pay more to buy smaller packs of cigarettes or support sales taxes on cigarettes? Why do some people buy annual gym memberships when they could save money by paying on a pay-as-you-go basis?

4.4 Evidence on Rationality in Strategic Interaction

Classical game theory (CGT) makes precise and testable predictions (see chapter 9.1 by Albert & Kliemt, this handbook). However, behavior in the early rounds of games and in games unfamiliar to the players is often inconsistent with the predictions. Learning and experience may or may not produce data consistent with the predictions (Camerer, 2003; Dhami, 2016, parts 4 and 5). People make many important decisions only a few times in their lives and experience limited learning (e.g., the choice of a university degree, a marriage partner, a house, consumer durables, or a pension plan). Firms also make many important decisions rarely (capital restructuring, mergers, sunk costs in major machinery and equipment, and choice of a new product).

Even in decisions that are taken frequently, the environment is ever changing and uncertain and hence always novel. Thus, arguably the choices made in the early rounds of an experiment are vital in understanding strategic human behavior, yet these choices do not typically conform to the predictions of CGT. Such data are, in many cases, much better explained by *behavioral models of game theory* such as *level-k models*, *quantal response equilibrium*, *analogy-based equilibrium*, *cursed equilibrium*, *team reasoning*, and *evidential reasoning* (for details, see Dhami, 2016, chapter 13).

We summarize some of the evidence here. It is impractical to cite all the references in such a short chapter; see the book-length treatments in Camerer (2003) and Dhami (2016, part 4) for the details.

1. In games requiring more than two steps of iterated deletion of dominated strategies, a majority of the players violate the predictions of CGT.

2. Extensive-form games and their equivalent normal-form representations often elicit different behavioral responses.

3. The backward induction prediction of CGT in dynamic games of full information is refuted by the evidence, even when subjects highly trained in backward induction (e.g., chess grandmasters) play centipede games.

4. Even when the simplest mechanisms are tested in experiments (e.g., the Abreu–Matsushima and the Glazer–Perry mechanisms, which are dominance-solvable), the results are unsupportive of CGT. This casts doubt on the implementability of more complex mechanisms.

5. Subjects play mixed strategies in experiments but not in the proportions predicted by a *mixed-strategy Nash equilibrium*. There is some positive evidence from sports contexts, but there are many confounding factors in such experiments, and it is not clear if such motor skills have external validity for other economic contexts (e.g., choice of pension plans).

6. In games with multiple equilibria, none of the selection principles in CGT, such as *payoff dominance* or *risk dominance*, fully account for the data. The equilibria are strongly *history dependent*. *Preplay communication* enhances coordination.

7. The CGT predictions of alternating offers bargaining games are not supported by the evidence. When players reject an offer, they often make counteroffers that give them an even lower share (disadvantageous counteroffers). The predictions of bargaining under one-sided asymmetric information in CGT are not supported, but there is surprising and unexplained support for the predictions of two-sided asymmetric information under CGT. Under unstructured bargaining, which is more realistic, factors unimportant in CGT, such as *self-serving bias*, explain *bargaining impasse* under full information.

8. The pattern of searches and lookups in Rubinstein's alternating offers bargaining game using the MOUSE-LAB software does not support the predictions of CGT.

9. In signaling games, there is no support for the predictions of CGT beyond the Cho–Kreps *intuitive criterion*.

10. There is evidence that several emotions, such as guilt-aversion, surprise-seeking, anger, and inferring intentions behind the actions of others, directly enter into the utility functions of players. This runs counter to CGT and forms the subject matter of the promising approach in *psychological game theory*.

4.5 Heuristics and Biases: An Alternative to Optimization

As noted in subsections 2.2 and 3.3, the dominant view in economics is that people are *fully rational Bayesian decision makers*. By this, it is meant that they use unbounded rationality, have unlimited attention and computing abilities, optimize using all appropriate mathematical and statistical tools, and obey the laws of standard (Kolmogorov) probability theory (not just Bayes' rule). Tversky and Kahneman (1971, 1974) put this view to the test, and this has spawned an enormous literature, called the *heuristics and biases* research program (see also chapter 2.4 by Fiedler, Prager, & McCaughey and chapter 8.5 by Hertwig & Kozyreva, both in this handbook).

The evidence showed that individuals are far from fully rational Bayesian decision makers, but their behavior is not random either. Individuals do not take account of all available information, do not correctly use Bayes' law and other laws of statistics, and eschew mathematical optimization in favor of simple *rules of thumb* or *heuristics*. Most heuristics are *fast*, in terms of the computation time required, and *frugal* in the use of information. Since these heuristics do not optimize, their performance is usually suboptimal. Even statistically sophisticated researchers and experts typically rely on these judgment heuristics.[8] Gerd Gigerenzer and collaborators have shown that when there is uncertainty about the optimal solution, the performance of heuristics compares favorably with more complex mathematical or statistical methods (see Dhami, 2016, section 19.15).

When individuals use the *representativeness heuristic*, they compare the likeness of even small samples with features of the population distribution. Hence, they assign too high a probability that small samples share the properties of large samples and behave "as if" they obeyed a *law of small numbers*, rather than the statistically correct *law of large numbers*. Thus, when asked to produce random sequences, the constructed sequences show up too much *negative autocorrelation*. This creates the *gambler's fallacy* (reluctance to bet on previously winning numbers). The converse, a belief in *positive autocorrelation* in outcomes when there is none, is known as the *hot hands fallacy* (basketball and football players who score heavily in a game are often perceived to be on a hot streak even when there is none). The implications for financial markets (and generally for economics) are profound, and this provides a much better explanation of behavior in financial markets (Dhami, 2016, part 7).

Other heuristics that are increasingly used in economics and finance to explain human behavior include the *availability heuristic*, the *anchoring heuristic*, *base rate neglect* in using Bayes' law, *conservatism* or *underweighting of the likelihood of a sample*, *hindsight bias*, *confirmation bias*, and *regression to the mean*.[9] In particular, the failure of Bayes' law is problematic for all areas of economic theory because Bayesian rationality lies at the heart of modern economic theory.

It is often claimed that presenting information in a *frequency format* rather than in a *probability format* eliminates biases arising from using heuristics (Cosmides & Tooby, 1996; Gigerenzer & Hoffrage, 1995). While this is true for some heuristics, the case for the efficacy of the frequency format is vastly overstated (Dhami, 2016, chapter 19). Further, most real-world economic information is arguably presented in a probability format.

Cognitive limitations force individuals to group different events into separate categories and to treat all events in the same category in an identical manner. This insight plays a central role in new behavioral equilibrium concepts, such as *analogy-based equilibrium* and *cursed equilibrium*, as well as for understanding *persuasion* and *advertising* (Dhami, 2016, sections 13.7 and 19.12).

People form *mental models* of the world that can take the form of beliefs about causal relations, social identities, categorizing disparate information into coarse categories, and social worldviews (see chapter 2.3 by Johnson-Laird, this handbook). Mental models are pervasive, help

economize on cognitive costs, are transmitted from generation to generation, are typically inertial once formed, and can both assist and deter optimal decisions, depending on their accuracy (Datta & Mullainathan, 2014; World Development Report, 2015). Humans are hardwired with some mental models, while in other cases, culture plays an important role in their formation. This provides another alternative to optimization in making decisions (Dhami, 2016, section 19.3).

Another alternative to Bayesian rationality is Herbert Simon's seminal approach to bounded rationality (Simon, 1979; see chapter 8.5 by Hertwig & Kozyreva, this handbook). In a direct reference to optimization, Simon (1979) famously notes, "But there are no direct observations that individuals or firms do actually equate marginal costs and revenues." The Nobel Prize winner Reinhard Selten (2001) draws attention to the fact that economic problems are *NP-complete*, that is, the number of steps in the required solution grows exponentially with the size of the problem (e.g., the traveling salesman problem).[10] Even for apparently simple problems of this sort, the computing time required to solve the problem is very high. It then becomes a leap of faith that the man on the street can solve this problem or act "as if" he could. Yet, neoclassical economics does make that leap of faith.

Herbert Simon proposes that instead of being Bayesian reasoners, individuals *satisfice*, that is, they have an *aspiration level* and adjust gradually in its direction through a sequence of steps (Simon, 1979, 2000; Selten, 2001). Empirical evidence is supportive of the theory (Caplin, Dean, & Martin, 2011): subjects appear to search sequentially and stop searching when their aspiration level is achieved. This can demonstrate how the cooperative outcome could be achieved as a solution to the prisoner's dilemma game (Karandikar, Mookherjee, Debraj, & Vega-Redondo, 1998).

Finally, the approach of Gerd Gigerenzer and colleagues on fast and frugal heuristics is another worthwhile alternative to Bayesian rationality (see Dhami, 2016, sections 19.14 and 19.15; Gigerenzer & Hoffrage, 1995). They show that heuristics can often outperform selected optimization methods, but in many cases, the optimization benchmark in these problems is far from clear. In some cases, these heuristics are derived from evolutionary adaptation, but mostly they are provided by the experimenter.

4.6 Evidence on the Efficiency of Financial Markets

The efficient markets hypothesis, EMH (see section 3.4), is rejected by a very large body of empirical evidence. Each of the following empirical findings rejects EMH:[11] prices of substitute assets differ consistently; there is gradual, rather than instantaneous, flow of information in the stock market; asset prices exhibit momentum at the short horizon and mean reversion at the long horizon; asset bubbles, particularly in the housing market, are well documented; stock markets crash even when there is "no news"; and asset prices exhibit underreaction and overreaction to new news. Indeed, a mistaken belief in EMH can be disastrous for financial markets, as demonstrated by the decision of the chairman of the Federal Reserve, Alan Greenspan, to not intervene in financial markets before the crash of 2007, believing that markets are efficient.

The refutation of both assumptions in the opening paragraph of subsection 3.4 and the discrediting of EMH on empirical grounds have led to the fast-growing discipline of *behavioral finance*. Behavioral finance recognizes that individuals are boundedly rational and follow a range of judgment heuristics that can give biased solutions (subsection 4.5). Contrary to the assertion in classical finance, the errors of noise traders often aggregate in the same direction.

In the event of a departure of prices from fundamental values and a strong belief on the part of noise traders that this mispricing will persist, rational investors will bet on the mispricing getting worse. Hence, the normal forces bringing prices back in line with fundamental values might not work, and we might observe bubbles on asset prices, or extended periods during which prices depart from fundamental values. The presence of institutional investors does not solve the problem either, because the deposits they receive from small investors depend on their past returns (*performance-based arbitrage*). Hence, they may liquidate their loss-making assets prematurely before the end of the year to demonstrate high returns, even when they ought to have held these assets longer.

There is also a growing literature on corporate finance that applies the behavioral evidence on humans to CEOs of companies (Dhami, 2016, section 21.7). For instance, it has been demonstrated that overconfident CEOs may undertake mergers even when there is no economic case for doing so. This might well explain why so many mergers fail.

5. Conclusions

Neoclassical economics, which is still the mainstream, does well to propose precise and testable assumptions and rigorously derive their logical conclusions. However, the evidence (section 4) is not consistent with the basic assumptions of rationality (section 2) and the auxiliary assumptions (section 3). Building on empirical evidence and insights from cognitive psychology, behavioral economics has made great strides in building a new economics that is as rigorous as neoclassical economics but in better conformity with the evidence.

Acknowledgments

We are grateful for valuable comments from the editor and two referees. However, limitations of space did not allow us to implement all of their suggestions.

Notes

1. For ease of exposition, we restrict ourselves to a fixed finite set X of real-valued outcomes. Extensions to vector-valued outcomes are possible and to sets X that are not necessarily finite. See, for example, Kreps (1990), who also makes a strong case for exploring alternatives to expected utility theory.

2. The reduction-of-compound-lotteries axiom is not formally stated above. For a formal definition, see Dhami (2016, p. 86).

3. To be sure, there are some models of *keeping up with the Joneses* and *snob consumption*, but these have played only a fringe role in modern economics.

4. Bayesian rationality augments the rationality assumptions from section 2 with the assumption that individuals update their beliefs when new information arrives, by using Bayes' rule.

5. The certainty equivalent is a guaranteed sum of money that makes the decision maker indifferent between accepting the money or the lottery. Thus, $\text{-bet} \sim C_\$ and $C_P \sim P$-bet, where $\sim$ denotes indifference.

6. The interested reader can peruse Kahneman and Tversky (2000), Starmer (2000), and Dhami (2016) for the details.

7. See Dhami (2016, chapter 10) for alternative models that explain the common difference effect.

8. For the claims in this paragraph, see Gilovich, Griffin, and Kahneman (2002); Kahneman (2003, 2011); Tetlock (2006); and Dhami (2016, part 7).

9. For a discussion of these heuristics, evidence, incorporation into new theoretical models, and economic applications, see Dhami (2016, chapter 19).

10. Suppose that a salesman is provided with a list of cities and the distances between each pair of cities. The objective, starting with a home city, is to find the shortest possible route such that the salesman visits each city exactly once and returns to the home city.

11. For the claims made in this section and the relevant references, the interested reader can consult the surveys in Shleifer (2000) and Dhami (2016, chapter 21).

References

Allais, M. (1953). Le comportement de l'homme rationnel devant le risque: Critique des postulats et axiomes de l'école Américaine [Rational behavior under risk: Criticism of the postulates and axioms of the American School]. *Econometrica, 21*, 503–546.

Aumann, R. J., & Brandenburger, A. (1995). Epistemic conditions for Nash equilibrium. *Econometrica, 63*, 1161–1180.

Camerer, C. F. (2003). *Behavioral game theory: Experiments in strategic interaction*. Princeton, NJ: Princeton University Press.

Caplin, A., Dean, M., & Martin, D. (2011). Search and satisficing. *American Economic Review, 101*(7), 2899–2922.

Caplin, A., & Leahy, J. (2001). Psychological expected utility and anticipatory feelings. *Quarterly Journal of Economics, 116*, 55–79.

Cosmides, L., & Tooby, J. (1996). Are humans good intuitive statisticians after all? Rethinking some conclusions from the literature on judgment under uncertainty. *Cognition, 58*, 1–73.

Datta, S., & Mullainathan, S. (2014). Behavioral design: A new approach to development policy. *Review of Income and Wealth, 60*(1), 7–35.

Dhami, S. (2016). *Foundations of behavioral economic analysis*. Oxford, England: Oxford University Press.

Fehr, E., & Schmidt, K. M. (2006). The economics of fairness, reciprocity and altruism: Experimental evidence and new theories. In S.-C. Kolm & J. M. Ythier (Eds.), *Handbook of the economics of giving, altruism and reciprocity* (Vol. 1, pp. 615–691). Amsterdam, Netherlands: Elsevier.

Fehr-Duda, H., & Epper, T. (2012). Probability and risk: Foundations and economic implications of probability-dependent risk preferences. *Annual Review of Economics, 4*, 567–593.

Fishburn, P. C., & Rubinstein, A. (1982). Time preference. *International Economic Review, 23*(3), 677–694.

Fudenberg, D., & Tirole, J. (1991). *Game theory*. Cambridge, MA: MIT Press.

Gabaix, X. (2014). A sparsity-based model of bounded rationality. *Quarterly Journal of Economics, 129*(4), 1661–1710.

Gigerenzer, G., & Hoffrage, U. (1995). How to improve Bayesian reasoning without instruction: Frequency formats. *Psychological Review, 102*, 684–704.

Gilovich, T., Griffin, D. W., & Kahneman, D. (Eds.). (2002). *Heuristics and biases: The psychology of intuitive judgment*. New York, NY: Cambridge University Press.

Gintis, H. (2009). *The bounds of reason: Game theory and the unification of the behavioral sciences*. Princeton, NJ: Princeton University Press.

Gintis, H. (2017). *Individuality and entanglement: The moral and material bases of social life*. Princeton, NJ: Princeton University Press.

Iyengar, S. S., & Lepper, M. R. (2000). When choice is demotivating: Can one desire too much of a good thing? *Journal of Personality and Social Psychology, 79*(6), 995–1006.

Kahneman, D. (2003). Maps of bounded rationality: Psychology of behavioral economics. *American Economic Review, 93*, 1449–1475.

Kahneman, D. (2011). *Thinking, fast and slow*. New York, NY: Farrar, Straus and Giroux.

Kahneman, D., & Tversky, A. (1979). Prospect theory: An analysis of decision under risk. *Econometrica, 47*(2), 263–291.

Kahneman, D., & Tversky, A. (2000). *Choices, values and frames*. Cambridge, England: Cambridge University Press.

Karandikar, R., Mookherjee, D., Debraj, R., & Vega-Redondo, F. (1998). Evolving aspirations and cooperation. *Journal of Economic Theory, 80*(2), 292–331.

Kolmogorov, A. N. (1950). *Foundations of the theory of probability*. New York, NY: Chelsea. (Original work published in German 1933)

Kreps, D. M. (1990). *A course in microeconomic theory*. New York, NY: Harvester Wheatsheaf.

Laibson, D. (1997). Golden eggs and hyperbolic discounting. *Quarterly Journal of Economics, 112*(2), 443–478.

Lichtenstein, S., & Slovic, P. (Eds.). (2006). *The construction of preference*. New York, NY: Cambridge University Press.

Loewenstein, G., O'Donoghue, T., & Rabin, M. (2003). Projection bias in predicting future utility. *Quarterly Journal of Economics, 118*, 1209–1248.

Phelps, E. S., & Pollak, R. A. (1968). On second best national savings and game equilibrium growth. *Review of Economic Studies, 35*(2), 185–199.

Polak, B. (1999). Epistemic conditions for Nash equilibrium, and common knowledge of rationality. *Econometrica, 67*, 673–676.

Quiggin, J. (1982). A theory of anticipated utility. *Journal of Economic Behavior and Organization, 3*, 323–343.

Savage, L. J. (1954). *The foundations of statistics*. New York, NY: Wiley.

Selten, R. (2001). What is bounded rationality? In G. Gigerenzer & R. Selten (Eds.), *Bounded rationality: The adaptive toolbox* (Dahlem Workshop Report, pp. 1–12). Cambridge, MA: MIT Press.

Shleifer, A. (2000). *Inefficient markets: An introduction to behavioral finance*. Oxford, England: Oxford University Press.

Simon, H. A. (1979). Rational decision making in business organizations (Nobel Memorial Lecture, 8 December, 1978). *American Economic Review, 69*(4), 493–513.

Simon, H. A. (2000). Bounded rationality in social science: Today and tomorrow. *Mind & Society, 1*(1), 25–39.

Starmer, C. (2000). Developments in non-expected utility theory: The hunt for a descriptive theory of choice under risk. *Journal of Economic Literature, 38*, 332–382.

Tetlock, P. E. (2006). *Expert political judgment: How good is it? How can we know?* Princeton, NJ: Princeton University Press.

Thaler, R. H. (1981). Some empirical evidence on dynamic consistency. *Economics Letters, 8*, 201–207.

Tversky, A., & Kahneman, D. (1971). Belief in the law of small numbers. *Psychological Bulletin, 76*, 105–110.

Tversky, A., & Kahneman, D. (1974). Judgement under uncertainty: Heuristics and biases. *Science, 185*(4157), 1124–1131.

Tversky, A., & Kahneman, D. (1992). Advances in prospect theory: Cumulative representation of uncertainty. *Journal of Risk and Uncertainty, 5*(4), 297–323.

World Development Report. (2015). *Mind, society, and behavior.* Washington, DC: World Bank.

Section 10 Aspects of Social Rationality

10.1 Social Epistemology

Franz Dietrich and Kai Spiekermann

Summary

Social epistemology studies knowledge in social contexts. We distinguish two types, depending on the holder of knowledge. Epistemology *in* groups studies individuals learning from others, while epistemology *of* groups studies groups as knowers, literally or metaphorically. Group knowledge can emerge explicitly, through aggregation procedures like voting, or implicitly, through institutions like deliberation or prediction markets. In the truth-tracking paradigm, group beliefs or actions aim at "correctness"—in virtue of facts that are empirical or normative, real or constructed, objective or only intersubjective, universal or relative, and so on. Procedures and institutions are evaluated by epistemic performance: Are they truth-conducive? Do groups become "wiser" than their members? This chapter reviews several procedures and institutions, discussing epistemic successes and failures. Jury theorems provide formal arguments for epistemic success, but only some have defensible premises. It will be argued that larger groups can outperform smaller groups, yet without becoming infallible.

1. Scope and Problems of Social Epistemology

Social epistemology is the branch of epistemology that studies knowledge in social contexts. This review first sets the stage by introducing clarifications, distinctions, and applications (section 1). It then discusses aggregation procedures (section 2) and institutional arrangements (section 3) that generate collective outcomes that "track the truth" or lead to epistemic failures (section 4).

1.1 Epistemology in Groups versus Epistemology of Groups

A first distinction pertains to the knowledge holder, who is either a group member or the group as a whole. Epistemology *in* groups might be regarded as a species of individual epistemology, although its focus lies in how individual knowledge depends on social inputs;

typical questions are rational responses to testimony and peer disagreement and "irrational" or subpersonal responses such as belief contagion through social media. Epistemology *of* groups ascribes knowledge to the group; let us talk here of *social* knowledge or belief (List & Pettit, 2011; see also Goldman & O'Connor, 2019; Schmitt, 1999). The meaning and status of social knowledge or belief is controversial. The group in question can be more or less structured and cohesive, with implications for whether the group qualifies as an *agent* and for what it *means* for the group to believe or know something, ranging from literal to purely metaphorical meanings. Arguably, a committee, firm, or state (if treated as an agent) can believe something literally, while a prediction market or the random group participating in a survey can believe something only metaphorically. List (2014) distinguishes between three types of social beliefs: *aggregate* beliefs are mere summaries of beliefs of group members; *common* beliefs are ultimately individual beliefs held by each group member, with a common awareness of these beliefs; and *corporate* beliefs are beliefs of the group in a literal sense, which presupposes that the group qualifies as an agent. Theoretical distinctions aside, it is evident that we routinely invoke social knowledge or beliefs, for instance, when saying that "we know" something or that "prediction markets knew" that Obama would become president of the United States. In a formal analysis, our distinction between epistemology *in* groups and *of* groups is sharp: the former is studied using belief revision models, the latter is studied using aggregation models. In this review, we focus on epistemology *of* groups, but for other aspects of social epistemology, see Goldman and O'Connor (2019), Goldman and Whitcomb (2011), Goldman (1999), and Schmitt (1999).[1]

1.2 Narrow versus Broad Social Epistemology

Narrowly construed, social epistemology addresses only social knowledge or belief: it addresses how a legal court learns whether the defendant is guilty, not how it decides whether to convict him; how a community of physicists discovers a law, not how it decides whether to perform an experiment; and so on. Broadly construed, however, social epistemology addresses social decisions in a wide sense, covering not only beliefs but also, for instance, actions: courts passing sentences, scientists choosing experiments, parliaments passing laws, governments setting goals, commissions working out ethical standards, and so forth. For such a social decision to be an epistemic matter, it must have the property of being *correct or incorrect* in virtue of some fact or truth. A court sentence may be an epistemic matter, its being correct or incorrect depending on whether the crime has been committed and merits the sentence. Selecting Oscar winners *might* be an epistemic matter. One reason for adopting this broad notion of social epistemology could be that correct decisions often are, or can be rationalized as, the result of beliefs. This chapter uses "(social) decision" in the widest sense, to include, for instance, social beliefs and social actions.[2] To be an object of social epistemology, broadly construed, the decision must, however, be correctness-apt.[3]

1.3   The Truth to Be Tracked

One can distinguish between three different types of facts that make social beliefs or social decisions more broadly correct or incorrect: logical, empirical, and normative facts. For example:

- The mathematical community forms mathematical beliefs. Correctness depends only on logical facts.

- The monetary policy committee of a central bank predicts whether inflation will rise. Correctness depends on empirical facts, besides logical facts.

- A group of doctors decides whether to shift funds from medical research to patient treatment. Correctness depends on normative and empirical facts, besides logical facts.

Social epistemology with normative facts is controversial, since the existence and the nature of normative facts are debatable.[4] But epistemology about normative facts should not be dismissed, because truth-tracking is meaningful regardless of whether normative facts are real or constructed (e.g., socially constructed), objective or at least intersubjective (within the group), universal or relativistic (as in cultural relativism), and natural or non-natural. The fact—normative or other—should, however, be suitably stable and procedure-independent. If the fact is of a constructed kind, then its existence and nature should not emerge from the social procedure but precede that procedure; otherwise, there is no truth to be tracked, but one to be constructed.

1.4   Truth-Tracking and Rationality

Rationality, understood here as internal coherence (i.e., formal rather than substantive rationality), is necessary but not sufficient for correctness. Beliefs can be perfectly consistent yet false, betterness judgments can be transitive yet misguided, and so on.[5] Rationality is thus only one step toward correctness. But this step is already very difficult to take when the holder of the beliefs (judgments, evaluations, and so on) is a group rather than an individual. Importantly, rationality of the group members does not guarantee rationality of the group, as we know from many aggregation paradoxes, such as Condorcet's voting paradox and the discursive dilemma, and from general impossibility results such as Arrow's Theorem (see section 2.1 and also chapter 10.2 by Schmid, this handbook). The sheer difficulty of achieving social rationality might explain why much of social choice theory focuses on social rationality, setting aside correctness. This limited focus does not place social choice theory outside the social-epistemological endeavor, however, as rationality is necessary for correctness. The interaction between individual rationality, social rationality, and truth-tracking is complex. For example, irrational individual belief revision can hinder individual truth-tracking while leading to a fruitful epistemic diversity that promotes social truth-tracking (see section 3.2).

1.5   Social Knowledge, Justification, and Reliability

Most well-known problems about individual knowledge carry over to social knowledge. For instance, it is controversial whether knowledge is justified true belief, a thesis threatened by "Gettier cases," in which a subject (for us: a group) forms justified true beliefs by sheer coincidence (Gettier, 1963). It is also controversial whether justification to hold a belief comes from sufficient evidence or from a reliable procedure (see Goldman & Beddor, 2016, on evidentialism versus reliabilism). Despite their familiarity, such "knowledge problems" can take a distinctive form for *social* knowledge, notably because groups use different procedures than individuals to acquire beliefs (e.g., voting procedures). Nonetheless, this chapter shall set aside the question of what exactly social knowledge means and instead zoom in on how social beliefs and decisions more broadly can be formed and track the truth.

1.6   Social Epistemology Applied to Epistemic Democracy

Social epistemology is partly a positive theory of which institutions and procedures *do* track the truth. By contrast, the core claim of *epistemic* justifications of democracy is normative: institutions and procedures *should* track the truth to be legitimate (Estlund, 2008). By contrast, *proceduralist* justifications of democracy maintain that institutions or procedures are not legitimate because they track any independent truth but because they are procedurally fair, which often means that voters should have an equal say and alternatives should get an equal chance. The difference is fundamental: if there is an Oracle of Delphi that always tells the truth, then the procedure of blindly implementing the oracle's recommendation for society is epistemically good (it tracks the truth) but procedurally bad (it is totally undemocratic). The debate between epistemic and procedural theories of democracy is ongoing (for a hybrid position, see Peter, 2007). For epistemic theories of democracy, social epistemology is a highly relevant enterprise.

2.   Aggregation Procedures to Track the Truth

This section turns to prominent procedures generating social beliefs (section 2.1) or other truth-tracking decisions (section 2.2) and then sketches how *jury theorems* help establish the truth-conduciveness of such procedures (section 2.3).

2.1   Forming Social Beliefs or Judgments
Aggregating beliefs on a single proposition A group of individuals is interested in whether a given proposition p is true. For instance, an economic panel is interested in whether inflation will rise, or a jury in whether the defendant has committed the crime. Each group member holds a belief about p, in the binary form of "yes" or "no." An aggregation procedure takes the individual beliefs about p as input and returns a social belief about p as output. Social abstentions are allowed: society can say "yes," or say "no," or abstain (but for simplicity, individuals cannot abstain). Here are some examples. *Majority rule* makes society believe what the majority believes (and abstain in case of a tie). *Asymmetric supermajority rules* make social belief in p harder to achieve than social disbelief by requiring a supermajority support for a social "yes" while otherwise opting for a social "no." *Symmetric supermajority rules* make social "yes" and "no" equally hard to achieve by requiring the same supermajority support for each. This often results in social abstention, especially for high thresholds. Extreme cases of (asymmetric or symmetric) supermajority rules are (asymmetric or symmetric) *unanimity rules*, in which the supermajority threshold is a unanimity threshold. Later, in section 2.3, we analyze how likely it is that resulting social beliefs track the truth.

Aggregating beliefs about multiple propositions: Judgment aggregation Many groups are interested simultaneously in

the truth of different propositions, something referred to as a *judgment-aggregation problem* (e.g., Dietrich, 2007; List & Pettit, 2002). For instance, a court may need collective beliefs about three propositions: the defendant has committed a certain act (p), such an act breaks the law (q), and the defendant has broken the law (r). A judgment-aggregation problem does not simply reduce to several belief-aggregation problems about a single proposition each, because the propositions in question are typically logically interconnected. If in our court example, the court comes to believe p and q, then it must believe r. Disbelieving r would be logically inconsistent; abstaining on r (neither believing nor disbelieving r) would be deductively unclosed. The trouble is that voting on each relevant proposition in isolation often generates inconsistent and deductively unclosed social beliefs: even if each member of the court holds consistent and complete (and thus deductively closed) beliefs, it may happen that a majority believes p, another majority believes q, and yet a majority disbelieves r. This phenomenon is referred to as the *doctrinal paradox* (Kornhauser & Sager, 1986) or the *discursive dilemma* (Pettit, 2001). The discursive dilemma generalizes far beyond propositionwise majority voting. General impossibility theorems establish that, as soon as propositions are sufficiently interconnected, there exist *no* propositionwise judgment-aggregation procedures that are well behaved in certain senses; see in particular List and Pettit's (2002) theorem and the Arrow-type theorem for judgment aggregation (Dietrich & List, 2007; Dokow & Holzman, 2010; both building on results of Klaus Nehring and Clemens Puppe later published in Nehring & Puppe, 2010). So, whether a proposition is socially believed must depend not only on the individual beliefs about *this* proposition but also on the individual beliefs about other propositions. This holistic nature of social beliefs makes social beliefs a more interesting and less transparent concept. The most famous holistic procedure to form social beliefs is the *premise-based procedure* (e.g., Dietrich & Mongin, 2010; Pettit, 2001). In our court example, this procedure determines the social belief on the propositions p and q (the *premise propositions*) through a majority vote on each of these propositions, thereafter logically deducing the social belief about r (the *conclusion proposition*). So r is socially believed if and only if p and q are both socially believed. This is holistic since the social belief about r is no longer determined only by the individual beliefs about r, *potentially* overruling a majority belief about r. But the premise-based procedure comes with its own problems; for instance, it presupposes that we can prioritize certain (premise) propositions over other (conclusion) propositions. Rival holistic procedures are distance-based rules (e.g., Miller & Osherson, 2008), sequential rules (e.g., Dietrich, 2015; List, 2004), relevance-based rules (Dietrich, 2015), and "approximate majoritarian" rules (Nehring, Pivato, & Puppe, 2014).

Probabilistic opinion pooling If we adopt the Bayesian paradigm, social beliefs should come in degrees rather than in binary yes/no form. What probabilities should society assign to propositions, given the probabilities assigned by the individuals? This is the so-called *opinion-pooling* problem—the probabilistic counterpart of the judgment-aggregation problem (for reviews, see Dietrich & List, 2016; Genest & Zidek, 1986). Once beliefs are probabilistic rather than binary, coherence of beliefs consists in respecting probabilistic principles like additivity, rather than logical principles like deductive closure. The picture reverses entirely: propositionwise ("local") aggregation of beliefs no longer runs into trouble. The social probability of any proposition can simply be the average (or a weighted average) of the individual probabilities, which guarantees coherent social beliefs as long as individual beliefs are coherent. Such "linear averaging" procedures have been characterized axiomatically by McConway (1981), Wagner (1982), Mongin (1995), and Dietrich and List (2017). Although social beliefs produced by linear averaging are coherent, they are not fully Bayesian. Bayesianism is usually taken to require two things: holding beliefs in probabilistic form ("probabilism") and revising beliefs through Bayes' rule ("conditionalization"), as explained in chapter 4.1 by Hájek and Staffel (this handbook). Social beliefs generated by linear averaging violate the second requirement: they fail to be revised via Bayes' rule. Why? Assume some proposition is learnt, so that every individual revises his or her beliefs (via Bayes' rule, assuming that individuals are Bayesian). If at any time the current social beliefs are the average of the current individual beliefs, then the post-information social beliefs usually differ from the initial social beliefs updated via Bayes' rule. In short, this social belief revision is non-Bayesian. Opinion pooling through *geometric* averaging repairs this flaw, that is, produces dynamically rational social beliefs. Here, social probabilities of worlds are obtained by a possibly weighted *geometric* average of individual probabilities (followed by a simple normalization ensuring that the total probability of worlds is 1). Geometric opinion pooling is in fact the *only* way to guarantee "Bayesian" group beliefs of a suitably well-behaved sort, as has only recently been shown (Dietrich, 2019; Russell, Hawthorne, & Buchak, 2015). Yet geometric pooling has a different flaw, which linear pooling avoids: social probabilities are not robust to refining or

coarsening the algebra of propositions considered, making social beliefs description-sensitive.

Aggregating evaluations—absolute versus ordinal approach Certain objects (e.g., wines, political candidates, holiday destinations) must be evaluated in terms of some criterion (e.g., moral value, well-being, aesthetic value, or size). Let this be an epistemic problem: evaluations express *beliefs* about facts of some kind—which is plausible for some evaluations (e.g., of size) and more controversial for others (e.g., of well-being). A first question is whether value is measured in absolute or ordinal terms. Consider the set X of objects evaluated. *Ordinal* evaluations are captured by a binary relation $\succeq$ on X, where "$x \succeq y$" means that x is at least as valuable as, or ranks weakly above, y (with respect to the relevant criterion). *Absolute* evaluations are captured by a function assigning to each object in X a value or rank from a set V of possible values or ranks, for example, from {very good, good, . . .} (moral evaluation) or {beautiful, ugly, . . .} (aesthetic evaluation) or {large, medium, . . .} (size evaluation) or a set of numbers (numerical evaluations, e.g., of size). What should the *social* evaluations of objects be, given the evaluations by the group members?

- In the ordinal case, this problem is structurally the notorious preference-aggregation problem, reinterpreting preference relations as value–judgment relations. One of many proposals is to use pairwise majority voting: object x is socially ranked over object y if and only if more individuals rank x over y than y over x. This procedure can lead to social cycles: some object x is majority-ranked over another, y, which is majority-ranked over another, z, which is majority-ranked over x (Condorcet's *voting paradox*). It generalizes into Arrow's impossibility theorem (Arrow, 1963): very roughly, not just pairwise majority voting but *any* pairwise aggregation procedure is flawed. This impossibility can be escaped by using some "holistic" rather than pairwise aggregation rule, for instance, the Borda rule or a distance-based rule. A different escape is to replace ordinal by absolute evaluations.

- The picture is brighter when aggregating *absolute* evaluations—an aggregation problem called the *social grading problem* by Balinski and Laraki (2011). In case of just two possible values—"approved" and "nonapproved"—the most natural procedure is "approval voting": an object is socially approved if the number of individuals approving it is at least as high as for each other object, and socially nonapproved otherwise (Brams & Fishburn, 2007). Balinski and Laraki generalize this procedure to any set of values linearly ranked from "highest" to "lowest," for example, {good, medium, bad}, where "good" ranks above "medium," which ranks above "bad." To socially evaluate an object x, first order the individuals such that the first individual evaluates x at least as highly as the second, the second at least as highly as the third, and so on. The social value of x is the value assigned to x by the middle individual in that order (assuming an odd number of individuals for simplicity). So, if in a wine evaluation among three judges, some wine receives the evaluations (exquisite, exquisite, drinkable), then this wine is socially evaluated as exquisite. This so-called *majority-judgment rule* (one might have called it the "median rule") has several appealing features (see Balinski & Laraki, 2011) but remains controversial within social choice theory with its ordinalist tradition.

Aggregating evaluations can be regarded as a special case of judgment aggregation by using propositions of type "x ranks over y" (for options x and y in X), in the ordinal case, or "x has value or rank v" (for options x in X and values v in V), in the cardinal case.

2.2 Making Social Decisions

The procedures discussed above generate social beliefs of different sorts. This section turns to procedures for making other social decisions.[6] To stay within social epistemology (broadly construed), the decision in question is assumed to track some truth, in particular, can be "correct" or "incorrect" in some procedure-independent sense (see sections 1.2, 1.3, and 1.6). For example, a group might have to choose a member from a set K of social alternatives. K could contain just two alternatives, as when a court decides whether to convict or acquit the defendant. K could instead contain many alternatives, as when a court decides between different sentences. When using *plurality rule*, each individual votes for exactly one alternative in K, and society chooses the alternative receiving the highest number of votes (or one such alternative in case of a tie). Plurality rule reduces to simple majority rule in the two-alternative case. In the many-alternative case, plurality rule can lead to problems: suppose alternative k^* in K is the "correct" alternative. Plurality rule will normally fail to select k^* if K contains many alternatives similar to k^* ("clones" of k^*), because k^* will tend to lose votes to its clones. Worse, plurality will usually not even select one of the "approximately correct" clones of k^*, because these clones will themselves tend to lose votes to similar alternatives. One response is to replace plurality rule with *approval voting* (Brams & Fishburn, 2007), in which individuals

can vote for any number of alternatives and the most often approved alternative wins; here the correct alternative will not tend to lose votes to its clones because individuals can approve many alternatives. Another response is to base the social choice on how each individual ranks alternatives from "best" to "worst." Here we aggregate individual rankings into a socially winning alternative (or set of winning alternatives, if ties are permitted). This aggregation problem differs structurally from the preference-aggregation problem mentioned above, in that social outputs are winning alternatives rather than rankings of alternatives. Procedures with such outputs are called *social-choice* rules rather than *preference-aggregation* rules. Nonetheless, the logical difficulties surrounding preference aggregation—illustrated by Condorcet's voting paradox and culminating in Arrow's (1963) theorem—reemerge for social-choice procedures. For instance, there may not exist any "Condorcet winner," an alternative that is majority-ranked over each other alternative—a problem closely related to Condorcet's voting paradox.

2.3 Jury Theorems: Formal Truth-Tracking Arguments

Do aggregation procedures like those just introduced succeed in tracking the truth? Jury theorems provide formal "wisdom of crowds" arguments, to the effect that appropriate procedures—typically majoritarian procedures—tend to generate correct social beliefs or other social decisions. Jury theorems can be powerful instruments—but they can also convey a misleading message when applied wrongly, because their optimistic conclusions may rely on misguided assumptions. The simplest jury theorem is Condorcet's (1785) jury theorem. It assumes a majority vote between two social alternatives of which exactly one is "correct" or "better," for instance, whether or not to convict or acquit a defendant. Jury theorems address the effect of increasing the size of the group. They typically conclude that "groups are wise" in the sense that one or both of two controversial hypotheses holds:

- *The growing-reliability hypothesis*: Larger groups are better truth-trackers. That is, they are more likely to reach the correct outcome (in most jury theorems: by majority) than smaller groups or single individuals.

- *The infallibility hypothesis*: Huge groups are infallible truth-trackers. That is, the likelihood of a correct outcome tends to full certainty as the group becomes larger and larger (Dietrich & Spiekermann, 2020).

Different jury theorems differ in which of the two conclusions they reach and which premises they rest on. For instance, Condorcet's theorem reaches both conclusions, based on two simple premises: an *independence* assumption, whereby voters have independent probabilities of voting for the correct alternative, and a *competence* assumption, whereby these correct-voting probabilities exceed ½ and are the same for each voter. The infallibility hypothesis then follows easily using the law of large numbers and the growing-reliability hypothesis follows from a more sophisticated combinatorial argument. The infallibility hypothesis, however, is overly optimistic and has left many with the (correct) impression that "something" must be wrong with those jury theorems that reach that conclusion, although there is confusion about the source of the problem. Some blame Condorcet's unrealistic competence assumption. However, the implausible infallibility conclusion remains if competence can vary across individuals as long as it exceeds ½ *on average* (Dietrich, 2008; Owen, Grofman, & Feld, 1989). The real problem is the independence assumption. Although this assumption can also be weakened *without* losing infallibility (e.g., Pivato, 2017, and the literature reviewed in Dietrich & Spiekermann, 2020), *plausible* weakenings of independence make infallibility collapse (Dietrich, 2008; Dietrich & List, 2004).

An important source of independence failure are *common causes* affecting voters: common evidence, common paradigms or perspectives, and even noninformational common causes such as room temperature (Dietrich & Spiekermann, 2013). The limited nature of available evidence places objective bounds on the reliability of majority judgments, which cannot be miraculously overcome by including more and more voters exposed to the same limited evidence. But it would be hasty to dismiss jury theorems. Although infallibility is unrealistic, the growing-reliability conclusion *can* be saved: that conclusion is reached by a jury theorem that revises Condorcet's independence and competence assumptions in defensible ways (Dietrich & Spiekermann, 2013, 2020). In other work, Condorcet's jury theorem has been extended beyond binary social decisions, such as choices between many alternatives via plurality voting (List & Goodin, 2001). In fact, there is a jury theorem for almost any standard voting rule, including approval voting, distance-based rules, the Borda rule, and other scoring rules (Pivato, 2017). The question is thus not whether jury theorems "exist" to defend a given aggregation rule but whether those theorems use acceptable premises.[7]

3.   Institutions beyond Mere Aggregation

Besides voting procedures, there are many other institutional and social processes for promoting social knowledge and correct social decisions. This section gives three important examples: deliberation, distributed search, and prediction markets.

3.1   Deliberation

The procedures discussed so far emphasize information aggregation. Deliberation, by contrast, emphasizes the exchange of reasons, prior to or instead of aggregation. Is deliberation truth-conducive? Several epistemic effects are at work: (i) deliberation can make private information or reasons public and allow participants to incorporate information and reasons of others; (ii) critical exchanges can eliminate bad reasons or inconsistencies and make good reasons and consistent viewpoints stand out more; (iii) deliberation might eliminate biases and reduce the influence of opinion leaders and other common causes; (iv) while deliberating, the decision problem can evolve, as new options come on the agenda and existing issues are reframed (Goodin & Spiekermann, 2018, chapter 9); and (v) deliberation can induce a meta-consensus about the structure of the decision space (Dryzek & List, 2003; cf. Dietrich & List, 2010; List, 2002).

Extensive empirical research on "deliberative polls" suggests many benefits of deliberation (e.g., Fishkin, 2018, and much of his other work). It is particularly striking that opportunities for citizens to deliberate with decision makers increase political knowledge (Esterling, Neblo, & Lazer, 2011), although it is hard to disentangle whether this is an effect of the deliberative process or of opportunities for influence. However, there are also warnings that deliberation can lead to increased polarization (Sunstein, 2002). A good overview of the deliberation literature is Bächtiger, Dryzek, Mansbridge, and Warren (2018); Landemore (2017) is helpful for epistemic aspects.

3.2   Distributed Search

Some problems are best tackled by dividing epistemic labor. Scientific research, for example, progresses not because all scientists work on the same problems, with the same frameworks, theories, evidence, and so on, but because of a competitive diversity of approaches, with incentive structures rewarding originality while ensuring that resources are not too concentrated. In a similar vein, some have argued that federalism is an epistemically advantageous political system because different political subunits can search for and try out different solutions to a problem, an advantage that Judge Brandeis referred to as the "laboratory" of federalism (*New State Ice Co. v. Liebmann*, 1932). Distributed search is particularly suitable when (i) the search domain is large (e.g., many rival theories could be tested); (ii) identifying correct choices requires effort, for example, testing a theory (to an extent sufficient for acceptance or rejection) is an expensive and lengthy process; and (iii) search in different locations causes useful by-products (e.g., new discoveries).

In distributed search, individual rationality need not be conducive to social rationality. In scientific research, for example, a bad incentive system would render it individually rational for each scientist to pursue research in novel areas receiving most attention and funding. However, that will likely lead to inefficient allocation of resources while valuable subfields are left unexplored, hampering overall scientific progress. A socially rational division of scientific labor would promote more diversity, more replication, and work on important projects with less imminent rewards. The influential work by Kitcher (1993) and Strevens (2003) shows how a mix of incentive schemes helps modern science manage this tension, often with success. The importance of diversity and the tension between individual and social rationality have been modeled in different ways (e.g., Hong & Page, 2012; Weisberg & Muldoon, 2009; Zollman, 2010; see also chapter 14.1 by Andersen & Andersen, this handbook). The theoretical results suggest that a socially irrational division of epistemic labor can emerge from individual rationality.

3.3 Prediction Markets
Prediction markets are financial markets from which predictions of contingent events, such as election outcomes, emerge. The securities traded on these markets are futures on the occurrence of these events. The mechanisms behind prediction markets are aggregation and incentivization (Mann, 2016; Wolfers & Zitzewitz, 2004). We have already seen that aggregation through voting can track the truth. The clever addition of prediction markets is the incentive created by gains when the future is under- or overpriced. The competitive setting motivates participants to try to outwit others by finding more information. The prediction market is "successful" when it predicts—more precisely, attaches high price (probability) to—events that eventually happen. Prediction markets have often successfully predicted political events (see, e.g., the Iowa Electronic Markets[8]), business events, sports events, and even scientific events (Arrow et al., 2008; Dreber et al., 2015). However, they are not infallible, mostly because of inefficient markets, systematic mistakes of traders, and insufficient or misleading information. For example, the Iowa Prediction Markets wrongly predicted a "no" in the 2016 UK "Brexit" referendum and a Clinton victory over Trump in the 2016 US presidential election.[9]

Prediction markets track the truth in an interestingly different way: while standard aggregation procedures transform explicit individual judgments into an explicit group judgment or other decision, prediction markets transform entirely behavioral inputs (buying and selling) into market prices from which implicit, but precise, group judgments can be read off. Although individual judgments are revealed only indirectly through acts of buying or selling, they are elicited reliably: while voting gives incentives to misrepresent, prediction markets do not normally provide such opportunities because counter-judgmental trading leads to losses.

4. Some Sources of Social Epistemic Failure

This chapter has already shown several reasons why groups can fail to form correct beliefs or other correct decisions: its members can fall prey to biases, be influenced by opinion leaders, or follow misleading evidence. This section briefly reviews some further sources of social epistemic failure. Most epistemic failures begin as individual failures, pertaining to epistemology *in* groups. But since individual beliefs shape group beliefs, individual failures become social failures and may even get amplified through deliberation or aggregation. So they also pertain to epistemology *of* groups, the main focus of this chapter. An exception is strategic voting, which involves no false individual beliefs but may lead to social epistemic failures.

4.1 Strategic Voting
Even if voters care for nothing but correctness of social beliefs or decisions more broadly, they can have strategic incentives to misrepresent their views—which is surprising given the absence of conflicting goals. For instance, a juror may vote for convicting the defendant, *although* her private information suggests innocence, as she reasons strategically that this makes the jury's (majority or supermajority) decision more likely correct, despite making her own vote less likely correct (see Austen-Smith & Banks, 1996; for details of the strategic rationale, see Dietrich & Spiekermann, 2020). So individual and social rationality can conflict even when, unlike in a prisoner's dilemma or tragedy of the commons, individual and social interests are perfectly aligned.

It is, however, important to put things into perspective. First, it is debatable whether such strategizing is socially harmful: while strategic voters systematically ignore private information (leading to underinformed social outcomes), they do so in order to improve social outcomes. The harmfulness of strategizing depends on the models and assumptions used. Second, voters often have perfectly rational grounds for voting truthfully, as they may care about expressing their own view instead of (or in addition to) caring about correctness of the voting outcome, a phenomenon usually ignored in the strategic voting literature with its consequentialist orientation (Dietrich & Spiekermann, 2020). Paradoxically, voters stop strategizing because they *stop* (not *begin*) sharing the same goal of correct outcomes.

4.2 Social Pressure

A potent cause of epistemic failure is the desire to conform with perceived expectations or norms, to avoid conflict or cognitive dissonance, to avoid mistakes in public, or to please peers. For example, group deliberation is likely to overemphasize information that was already widely accepted prior to deliberation and underemphasize information that is sparsely spread and yet crucial for a correct decision. Such an underemphasis can happen not just because there are fewer people who hold the information in the first place but also because those who hold it stay silent, predicting, perhaps correctly, that the information will be controversial (Gigone & Hastie, 1993; Stasser & Titus, 2003). Related effects of social pressure are the amplification of individual errors in deliberation and the possible enforcement of extreme positions—two deliberation failures (Sunstein & Hastie, 2015a, 2015b).

4.3 Motivated Cognition

The desire to fit into an identity group or comply with social norms can bias the acquisition and cognitive processing of factual information in the first place (e.g., Spiekermann & Weiss, 2016). For example, Kahan et al. (2012) describe how different political-cultural backgrounds create different pressures to seek or avoid information about climate change and to process such information in biased ways. Kahan et al.'s left-liberal subjects believe that climate change is a problem, right-libertarians much less so. Scientific or numeracy skills further increase this divergence, suggesting that better understanding leads to even more biased processing.

4.4 Epistemic Injustice

Someone suffers epistemic injustice if she is "wronged specifically in her capacity as a knower" (Fricker, 2007, p. 20). Fricker distinguishes between testimonial and hermeneutic injustice. *Testimonial* injustice is experienced if testimony is discounted because of an identity prejudice; for example, the evidence of a person of minority background is dismissed *because* of their background. *Hermeneutic* injustice can arise if oppressed individuals or groups do not have adequate concepts to conceive of, make sense of, or communicate their experiences. For example,

Fricker explains that victims of sexual harassment found it difficult to understand their experience and explain it to others before the term "sexual harassment" was coined. Testimonial and hermeneutic injustice undermine the ability of individuals and ultimately the group to generate correct beliefs or other correct decisions.

4.5 Epistemic Skepticism and Nihilism

A more fundamental threat to epistemic success is the refusal to take a truth-seeking attitude. *Epistemic skepticism* is the belief that there is no truth to be tracked or that access to the truth is impossible. It can be distinguished from *epistemic nihilism*, the view that truth does not matter, regardless of whether or not it exists and is accessible. Whereas skepticism is a respectable epistemological position, nihilism is an epistemologically and morally dubious attitude. However, genuine nihilists will by definition not care about whether their attitude is defensible, which makes nihilism particularly immune to arguments and hard to fight against. Harry Frankfurt (2005) has suggested that the lack of care for the truth is one aspect of "bullshitting," a lack of commitment to the truth and epistemic norms of truthfulness (cf. Cassam, 2019, especially chapter 4). The so-called fake news phenomenon seems fueled by epistemic nihilism, although the debate about this is still in its infancy (e.g., Goodin & Spiekermann, 2018, chapter 21; Mukerji, 2018; cf. Rini, 2017). The skepticism or nihilism of individuals can scale up to the group level, undermining the truth-tracking ability of media organizations, public forums, or the political system as a whole.

Acknowledgments

We thank Wolfgang Spohn and two anonymous referees for very helpful suggestions. We also acknowledge financial support by the French National Research Agency (ANR) through three grants (ANR-17-CE26-0003, ANR-16-FRAL-0010, and ANR-17-EURE-0001).

Notes

1. This volume touches upon other issues of formal epistemology in chapter 2.6 by van Benthem, Liu, and Smets; chapter 5.2 by Rott; chapter 5.3 by Kern-Isberner, Skovgaard-Olsen, and Spohn; chapter 5.5 by Hahn and Collins; and chapter 5.6 by Woods.

2. By using "social decision" broadly to encompass social beliefs we do not assume that group beliefs (let alone individual beliefs) always are under direct voluntary control. We set aside whether groups have such control, and what it would mean.

3. If groups are construed as intentional agents, they also hold non-belief attitudes like desires, preferences, or intentions. This

immediately raises the question of whether holding such other attitudes is also "correct" or "incorrect" in some relevant sense (possibly just for certain attitudes, e.g., for Scanlon's judgment-sensitive attitudes; Scanlon, 1998, p. 20). The correctness of holding, say, an intention might depend on whether its content is morally permissible. For instance, it might be correct for a firm to intend to maximize shareholder value if maximizing it is morally permissible. Like social beliefs and correctness-apt actions, correctness-apt social attitudes might be regarded as a matter of social epistemology (broadly construed). Whether or not one adopts this inclusive view, some of the technical machinery of social epistemology—like jury theorems—can be used to study whether group attitudes become correct.

4. A hybrid position is to deny normative facts but to grant that normative judgments can be correct or incorrect as acts (on other grounds than the truth or falseness of their contents). This is still compatible with social epistemology broadly construed (cf. section 1.2).

5. What (formal) rationality means precisely depends on the type of belief (or other attitude or decision) considered and can be fleshed out differently. Binary yes–no beliefs are often required to be consistent and deductively closed. Probabilistic beliefs are often required to satisfy Kolmogorov's axioms. Ordinal betterness judgments over some choice options are often required to be transitive and perhaps complete.

6. The social beliefs or judgments discussed in section 2.1 often already generate social actions: society can choose what it believes to be best (ordinally) or what it approves or rates highest (absolutely) or what is recommended by its judgments, and so on. The current section, however, investigates procedures that generate social actions *directly*, rather than generating beliefs or judgments *supporting* actions.

7. While standard jury theorems address the epistemic performance of a given procedure (usually majority voting), one may alternatively search for the epistemically optimal procedure, thereby accounting for differential competence across individuals (Ben-Yashar & Nitzan, 1997; Dietrich, 2006).

8. Available at https://iemweb.biz.uiowa.edu (last accessed April 17, 2019).

9. The low probabilities attached to Trump or Brexit victories were, strictly speaking, not mistakes. They merely showed that prediction markets, like many pollsters, underestimated support for Trump and Brexit.

References

Arrow, K. J. (1963). *Social choice and individual values* (2nd rev. ed.). New York, NY: Wiley.

Arrow, K. J., Forsythe, R., Gorham, M., Hahn, R., Hanson, R., Ledyard, J. O., . . . Zitzewitz, E. (2008). The promise of prediction markets. *Science, 320*(5878), 877–878.

Austen-Smith, D., & Banks, J. S. (1996). Information aggregation, rationality, and the Condorcet jury theorem. *American Political Science Review, 90*(1), 34–45.

Bächtiger, A., Dryzek, J. S., Mansbridge, J., & Warren, M. E. (Eds.). (2018). *The Oxford handbook of deliberative democracy.* Oxford, England: Oxford University Press.

Balinski, M., & Laraki, R. (2011). *Majority judgment: Measuring, ranking, and electing.* Cambridge, MA: MIT Press.

Ben-Yashar, R. C., & Nitzan, S. I. (1997). The optimal decision rule for fixed-size committees in dichotomous choice situations: The general result. *International Economic Review, 38*(1), 175–186.

Brams, S., & Fishburn, P. C. (2007). *Approval voting* (2nd ed.). New York, NY: Springer Science & Business Media.

Cassam, Q. (2019). *Vices of the mind: From the intellectual to the political.* Oxford, England: Oxford University Press.

Condorcet, A., Marquis de. (1785). *Essai sur l'application de l'analyse à la probabilité des décisions rendues à la pluralité des voix* [An essay on the application of probability theory to plurality decision-making]. Paris, France: Imprimerie Royale.

Dietrich, F. (2006). General representation of epistemically optimal procedures. *Social Choice and Welfare, 26*(2), 263–283.

Dietrich, F. (2007). A generalised model of judgment aggregation. *Social Choice and Welfare, 28*(4), 529–565.

Dietrich, F. (2008). The premises of Condorcet's jury theorem are not simultaneously justified. *Episteme, 5*(1), 56–73.

Dietrich, F. (2015). Aggregation theory and the relevance of some issues to others. *Journal of Economic Theory, 160*, 463–493.

Dietrich, F. (2019). A theory of Bayesian groups. *Noûs, 53*, 708–736.

Dietrich, F., & List, C. (2004). A model of jury decisions where all jurors have the same evidence. *Synthese, 142*, 175–202.

Dietrich, F., & List, C. (2007). Arrow's theorem in judgment aggregation. *Social Choice and Welfare, 29*(1), 19–33.

Dietrich, F., & List, C. (2010). Majority voting on restricted domains. *Journal of Economic Theory, 145*(2), 512–543.

Dietrich, F., & List, C. (2016). Probabilistic opinion pooling. In A. Hájek & C. Hitchcock (Eds.), *The Oxford handbook of probability and philosophy* (pp. 519–543). Oxford, England: Oxford University Press.

Dietrich, F., & List, C. (2017). Probabilistic opinion pooling generalized. Part one: General agendas. *Social Choice and Welfare, 48*(4), 747–786.

Dietrich, F., & Mongin, P. (2010). The premiss-based approach to judgment aggregation. *Journal of Economic Theory, 145*(2), 562–582.

Dietrich, F., & Spiekermann, K. (2013). Epistemic democracy with defensible premises. *Economics & Philosophy*, *29*(1), 87–120.

Dietrich, F., & Spiekermann, K. (2020). Jury theorems. In M. Fricker, P. J. Graham, D. Henderson, & N. J. L. L. Pedersen (Eds.), *The Routledge handbook of social epistemology* (pp. 386–396). New York, NY: Routledge.

Dokow, E., & Holzman, R. (2010). Aggregation of binary evaluations. *Journal of Economic Theory*, *145*(2), 495–511.

Dreber, A., Pfeiffer, T., Almenberg, J., Isaksson, S., Wilson, B., Chen, Y., . . . Johannesson, M. (2015). Using prediction markets to estimate the reproducibility of scientific research. *Proceedings of the National Academy of Sciences*, *112*(50), 15343–15347.

Dryzek, J., & List, C. (2003). Social choice theory and deliberative democracy: A reconciliation. *British Journal of Political Science*, *33*(1), 1–28.

Esterling, K. M., Neblo, M. A., & Lazer, D. M. J. (2011). Means, motive, and opportunity in becoming informed about politics: A deliberative field experiment with members of Congress and their constituents. *Public Opinion Quarterly*, *75*(3), 483–503.

Estlund, D. M. (2008). *Democratic authority: A philosophical framework*. Princeton, NJ: Princeton University Press.

Fishkin, J. S. (2018). *Democracy when the people are thinking: Revitalizing our politics through public deliberation*. Oxford, England: Oxford University Press.

Frankfurt, H. G. (2005). *On bullshit*. Princeton, NJ: Princeton University Press.

Fricker, M. (2007). *Epistemic injustice: Power and the ethics of knowing*. Oxford, England: Oxford University Press.

Genest, C., & Zidek, J. V. (1986). Combining probability distributions: A critique and an annotated bibliography. *Statistical Science*, *1*(1), 114–135.

Gettier, E. L. (1963). Is justified true belief knowledge? *Analysis*, *23*(6), 121–123.

Gigone, D., & Hastie, R. (1993). The common knowledge effect: Information sharing and group judgment. *Journal of Personality and Social Psychology*, *65*(5), 959–974.

Goldman, A. I. (1999). *Knowledge in a social world*. Oxford, England: Clarendon Press.

Goldman, A., & Beddor, B. (2016). Reliabilist epistemology. In E. N. Zalta (Ed.), *The Stanford encyclopedia of philosophy*. Retrieved from https://plato.stanford.edu/archives/win2016/entries/reliabilism/

Goldman, A., & O'Connor, C. (2019). Social epistemology. In E. N. Zalta (Ed.), *The Stanford encyclopedia of philosophy*. Retrieved from https://plato.stanford.edu/archives/fall2019/entries/epistemology-social/

Goldman, A. I., & Whitcomb, D. (Eds.). (2011). *Social epistemology: Essential readings*. Oxford, England: Oxford University Press.

Goodin, R. E., & Spiekermann, K. (2018). *An epistemic theory of democracy*. Oxford, England: Oxford University Press.

Hong, L., & Page, S. E. (2012). Some microfoundations of collective wisdom. In H. Landemore & J. Elster (Eds.), *Collective wisdom: Principles and mechanisms* (pp. 56–71). Cambridge, England: Cambridge University Press.

Kahan, D. M., Peters, E., Wittlin, M., Slovic, P., Ouellette, L. L., Braman, D., & Mandel, G. (2012). The polarizing impact of science literacy and numeracy on perceived climate change risks. *Nature Climate Change*, *2*(10), 732–735.

Kitcher, P. (1993). *The advancement of science: Science without legend, objectivity without illusions*. Oxford, England: Oxford University Press.

Kornhauser, L. A., & Sager, L. G. (1986). Unpacking the Court. *Yale Law Journal*, *96*, 82–117.

Landemore, H. (2017). Beyond the fact of disagreement? The epistemic turn in deliberative democracy. *Social Epistemology*, *31*(3), 277–295.

List, C. (2002). Two concepts of agreement. *The Good Society*, *11*(1), 72–79.

List, C. (2004). A model of path-dependence in decisions over multiple propositions. *American Political Science Review*, *98*(3), 495–513.

List, C. (2014). Three kinds of collective attitudes. *Erkenntnis*, *79*(9), 1601–1622.

List, C., & Goodin, R. E. (2001). Epistemic democracy: Generalizing the Condorcet jury theorem. *Journal of Political Philosophy*, *9*(3), 277–306.

List, C., & Pettit, P. (2002). Aggregating sets of judgments: An impossibility result. *Economics & Philosophy*, *18*(1), 89–110.

List, C., & Pettit, P. (2011). *Group agency: The possibility, design, and status of corporate agents*. Oxford, England: Oxford University Press.

Mann, A. (2016). The power of prediction markets. *Nature News*, *538*(7625), 308–310.

McConway, K. J. (1981). Marginalization and linear opinion pools. *Journal of the American Statistical Association*, *76*(374), 410–414.

Miller, M. K., & Osherson, D. (2008). Methods for distance-based judgment aggregation. *Social Choice and Welfare*, *32*(4), 575–601.

Mongin, P. (1995). Consistent Bayesian aggregation. *Journal of Economic Theory*, *66*(2), 313–351.

Mukerji, N. (2018). What is fake news? *Ergo: An Open Access Journal of Philosophy*, *5*, 923–946.

Nehring, K., Pivato, M., & Puppe, C. (2014). The Condorcet set: Majority voting over interconnected propositions. *Journal of Economic Theory*, *151*, 268–303.

Nehring, K., & Puppe, C. (2010). Abstract Arrowian aggregation. *Journal of Economic Theory*, *145*(2), 467–494.

New State Ice Co. v. Liebmann, 285 U.S. 262 (1932).

Owen, G., Grofman, B., & Feld, S. L. (1989). Proving a distribution-free generalization of the Condorcet jury theorem. *Mathematical Social Sciences*, *17*, 1–16.

Peter, F. (2007). Democratic legitimacy and proceduralist social epistemology. *Politics, Philosophy & Economics*, *6*(3), 329–353.

Pettit, P. (2001). Deliberative democracy and the discursive dilemma. *Philosophical Issues*, *11*(1), 268–299.

Pivato, M. (2017). Epistemic democracy with correlated voters. *Journal of Mathematical Economics*, *72*, 51–69.

Rini, R. (2017). Fake news and partisan epistemology. *Kennedy Institute of Ethics Journal*, *27*(2), E-43–E-64.

Russell, J. S., Hawthorne, J., & Buchak, L. (2015). Groupthink. *Philosophical Studies*, *172*(5), 1287–1309.

Scanlon, T. M. (1998). *What we owe to each other*. Cambridge, MA: Belknap Press of Harvard University Press.

Schmitt, F. (1999). Social epistemology. In J. Greco & E. Sosa (Eds.), *The Blackwell guide to epistemology* (pp. 354–382). Malden, MA: Blackwell.

Spiekermann, K., & Weiss, A. (2016). Objective and subjective compliance: A norm-based explanation of 'moral wiggle room.' *Games & Economic Behavior*, *96*, 170–183.

Stasser, G., & Titus, W. (2003). Hidden profiles: A brief history. *Psychological Inquiry*, *14*(3–4), 304–313.

Strevens, M. (2003). The role of the priority rule in science. *Journal of Philosophy*, *100*(2), 55–79.

Sunstein, C. R. (2002). The law of group polarization. *Journal of Political Philosophy*, *10*(2), 175–195.

Sunstein, C. R., & Hastie, R. (2015a). Garbage in, garbage out? Some micro sources of macro errors. *Journal of Institutional Economics*, *11*(3), 561–583.

Sunstein, C. R., & Hastie, R. (2015b). *Wiser: Getting beyond groupthink to make groups smarter*. Boston, MA: Harvard Business Review Press.

Wagner, C. (1982). Allocation, Lehrer models, and the consensus of probabilities. *Theory and Decision*, *14*(2), 207–220.

Weisberg, M., & Muldoon, R. (2009). Epistemic landscapes and the division of cognitive labor. *Philosophy of Science*, *76*(2), 225–252.

Wolfers, J., & Zitzewitz, E. (2004). Prediction markets. *Journal of Economic Perspectives*, *18*(2), 107–126.

Zollman, K. J. S. (2010). The epistemic benefit of transient diversity. *Erkenntnis*, *72*(1), 17–35.

10.2 Collective Rationality

Hans Bernhard Schmid

Summary

"Collective rationality" is used in a variety of ways in the relevant literature, two of which stand out: (a) the rationality of cooperation and coordination and (b) corporate rationality (or the rationality of group agents). These two conceptions are markedly different from each other in the issues they concern and in the concepts of collectivity and rationality they involve. Upon closer inspection, however, it turns out that from different sides, they converge on the same issue, namely, the capacity for rational joint action.

1. Rational Joint Action, Cooperation, and Corporation

In the first of the two versions in which collective rationality is currently discussed, it is a relational feature of choices of individuals. The paradigmatic case of this sense of collective rationality is mutual cooperation in a two-person prisoner's dilemma. Individual choices are collectively rational if they realize a mutually cooperative outcome (an outcome that is better for both than mutual defection), even though individual strategic reasoning recommends mutual defection. Defection is individually rational (in a strategic sense) because from the point of view of each of the participating individuals, choosing to defect realizes a better state no matter whether the other cooperates or defects. Among strategically rational individuals, mutual defection is thus the expected outcome—which seems a bit foolish given that they could have realized an outcome that is better from both participants' points of view. Collective rationality in this sense is thus defined in contradistinction against (a strategic conception of) individual rationality. Conversely, there is an air of individual irrationality about individual choices that are collectively rational, even though in terms of efficiency, the participants are ultimately better off through mutual cooperation. Collective rationality in this sense is a matter of the rationality of cooperation (although, as will be argued below, *coordination* might be

a better paradigm to study this first conception of collective rationality). The rationality of cooperation and coordination matters because it explains the provision of common and public goods (cooperation) and the role of conventions in action (coordination).

In the second salient sense of the word, collective rationality is an attribute of collective or group agents, their attitudes, and their actions. It is the feature in virtue of which collective or group agents (such as political parties, universities, or business corporations) are consistent in their attitudes and actions over more or less complex interdependent decisions. In this second sense, collective rationality is defined in analogy to (rather than in contradistinction against) individual rationality. Just as rational individual agency needs some degree of consistency across issues, it seems that at least some collectives need to ensure that their collective views and decisions are mutually consistent too. The problem of collective rationality in this second sense, as discussed in the recent literature, is that it is not guaranteed by the rationality of individuals. No plausible fixed aggregation procedure robustly ensures the rationality of collective decisions among rational individuals. Achieving collective rationality is thus an extra achievement necessary for organized social life—the "collectivization of reason," as it is sometimes called in the literature. This second conception of collective rationality matters insofar as the credibility and trustworthiness of group agents such as institutions and organizations depend on their consistency in their attitudes and actions.

These two conceptions of collective rationality—the rationality of coordination and cooperation, as well as corporate rationality—thus seem to be markedly different in their targets and approaches. Perhaps more important, they radically differ both with regard to the idea of rationality as well as with regard to the notion of collectivity they entail. The rationality of cooperation between individuals is a matter of *efficiency*, and it involves collectivity in the sense that it is a *relational feature of participant individual choices*. By contrast, the rationality of collective agents (or corporate rationality) is a matter of *consistency*, and it involves collectivity in terms of *agency that is collective rather than individual*. A further distinction is that while the rationality of cooperation is defined in *contradistinction* from individual rationality, corporate rationality is defined in *analogy* to individual rationality.

After overviews over each of these two debates (sections 2 and 3), this chapter makes a suggestion concerning how, despite their differences, they might hang together (section 4). The claim is that the feature that rationalizes cooperation (and coordination) between individuals is also the basic force that drives toward coherence between group attitudes. Both conceptions thus converge on the capacity to think and act together, from a joint point of view, and to do so consistently and effectively. This capacity is currently mostly discussed under labels such as "collective intentionality."

2. The Rationality of Coordination and Cooperation

In the first sense of the word, it is sometimes claimed in the literature that human social activity cannot be based on individual rationality alone but has to be complemented with—or perhaps partially replaced by—collective rationality. The relation between individual and collective rationality is thought to be difficult because the two can—and often do—come into conflict with each other. The basic problem has been noticed at least since Hobbes (cf. Gauthier, 1979; with a bit of good will, one could perhaps reconstruct a passage in Plato's *Republic* [359a] along similar lines). In a game-theoretic framework (cf. chapter 9.1 by Albert & Kliemt, this handbook), the issue at stake has been at the focus of social theory since the 1950s under the name of "the prisoner's dilemma" (Kuhn, 2017; an alternative—and equally correct—spelling is "prisoners' dilemma").

The prisoner's dilemma (PD) is a situation of interdependent choices between two available options (choices, or strategies), *c* and *d*, where in the case of two participants *A* and *B*, the outcome if *A* *d*'s and *B* *c*'s realizes best what *A* values and worst what *B* values (and vice versa for *B*'s *d*'ing and *A*'s *c*'ing), and where mutual *c*'ing realizes what *A* and *B* value better than mutual *d*'ing. The PD does not presuppose selfish preferences or values. It can arise between altruists, too (cf. Tilley, 1991), although it is usually illustrated with the example of self-interested (and mutually disinterested) participants and has its most important applications in this context. The reason why the system of evaluations that is the PD is so important in the debate is that it captures the basic structure

of the problem of common and public goods—that is, goods that are nonexcludable and therefore invite free-riding (the "tragedy of the commons"). The dilemma, as it is usually conceived of in the literature, is "a conflict between individual and collective rationality" (Orbell & Wilson, 1979, p. 411). It is claimed that in situations such as the PD, "individually rational" choice results in "collective irrationality" (e.g., Sorensen, 2004, p. 265). In this sense, collective rationality is thus a property of a set of individual choices that are such that the individuals reap the benefits of mutual cooperation in the face of temptation to defect. Because of this structure of incentives, there is, however, an air of irrationality about cooperative choices. Standard strategic rationality demands of individuals that if there is an option that realizes a better outcome (measured by one's own desires or values) no matter what the relevant others do, one should choose that strategy. The cooperative choice is strictly dominated by the defective choice—to choose the cooperative strategy realizes a worse outcome no matter what the other does. This recommends defection in PDs as the rational choice (and since this is true for both participants, mutual defection is the only Nash equilibrium).

It is hard to accept from an intuitive point of view, however, that it should be rational for agents with common knowledge of rationality as well as common knowledge of the payoff structure that they should knowingly realize a combination of choices that will make each of them worse off than another available outcome. To put this point in the relevant jargon: the problem is that mutual defection is the only Pareto-inefficient outcome of a PD. It seems counterintuitive to say that it is rational to be inefficient. Agents whose rationality is only strategic, but foreseeably inefficient, thus appear to be "rational fools" (Sen, 1977).

Although not all relevant authors use the term "collective rationality" in this context, the project of safeguarding (combinations of) cooperative individual choices in PDs from the allegation of irrationality has featured prominently in the agenda of social theory and social philosophy, especially since the 1970s (for an extended discussion see Townley, 2008, chapter 9). The reason why this is an important issue is that there are limits to aligning individual strategic rationality with the production of public goods via institutional design. To some degree, social order depends on individual cooperative-mindedness, and among likeminded people, cooperative-mindedness is efficient—and is thus not simply irrational, although the rationality involved here is somewhat in tension with practical rationality of the standard individual (strategic) kind.

Most prominent among the more recent attempts is Paul Weirich's (2009) conception of collective rationality. Weirich defends participant cooperative choices from the allegation of irrationality by arguing that self-support rather than utility maximization is the standard to which it should be held; a combination of cooperative choices is self-supporting. In this vein, Weirich argues that it is a sufficient condition for the rationality of a combination of choices (a "group act," in Weirich's terms) that its individual members' acts are rational (in his revised sense).

Another conception of collective rationality in this sense is Christopher McMahon's (2001). According to his view, it is (collectively) rational to contribute to a "cooperative scheme" if the "outcome of the scheme, when one's contribution is added to the others that will actually be made, exceeds the value to one of the noncooperative outcome" (McMahon, 2001, pp. 21–22).

This conception of collective rationality is, in turn, similar to David Gauthier's (1986) classical concept of constrained maximization (cf. chapter 12.1 by Fehige & Wessels, this handbook, as well as a considerable number of similar attempts to vindicate the rationality of cooperative choices). However, Gauthier avoids using the term "collective rationality." Indeed, it is far from obvious why the rationality of cooperative choices, as conceived in this debate, should be *collective* rather than individual. After all, it is the *individuals* who profit from mutual cooperation (or suffer from mutual defection), not some collective entity such as "the team." In this regard, one might even think that rather than a *dilemma* (i.e., a choice between equally problematic alternatives), the PD is really a *paradox* (i.e., a self-contradiction; an example for this interpretation is Davis, 1977) or perhaps a *tragedy* (a contradiction between intention and effect; cf. Tilley, 1994). Strategic rationality (that is aimed at maximizing expected utility) recommends a choice of which it is known that it fails to maximize utility among strategically rational individuals. The paradox of the PD is thus, in this view, that "so-called irrational players . . . will both fare much better than so-called rational ones" (Luce & Raiffa, 1957, p. 96; the fact that conversely, "rational" players fare worse than irrational ones might be called the tragedy of the PD).

Although the efficiency of mutual cooperation is certainly a good ground to claim *some sort* of rationality for the respective choices, considered in combination, the sense in which the rationality of cooperation is collective rather than individual certainly needs clarification. From this perspective, rather than pitting collective rationality against individual rationality, the PD seems

to separate efficiency from strategy dominance, and it is not obvious how exactly the rationality of the efficient choice involves collectivity other than in the rather obvious sense that it takes at least two to realize a mutually cooperative result.

In this regard, the debate on *coordination* is illuminating. While cooperation in PDs requires the participant to choose against their strategic self-interest, there is no such conflict in pure coordination problems (for a discussion of coordination and cooperation in game theory and behavioral economics see Butler, 2012). In situations of pure coordination, there is no dominant strategy for the participants; what is rational for them to do, individually, depends on what they expect the others to choose (of whom it is known that they are in the same situation). Where none of the available equilibria is better for all of the participants, the problem is usually very effectively regulated by conventions (or salience) in real life, and although it seems intuitively obvious that it is rational for the participants to solve coordination problems by choosing the conventional option, it is impossible to sustain this intuition in terms of individual strategic reasoning (where mutual expectations are interdependent rather than fixed or convergent). Classical rational choice theory cannot recommend the choice of the "salient" strategy as the one rational choice. Even more counterintuitive is the recommendation of classical rational choice theory in the case where one coordination equilibrium is better for all than the other(s) (such as in the Hi–Lo game, where the choice of "Hi" by both participants realizes a better result for both than the choice of "Lo" by both partners; "Lo"/"Lo" is still better for both than "Hi"/"Lo" or "Lo"/"Hi"). In this case, it seems intuitively very clear that common knowledge of rationality and of the utilities rationalizes the according choice (i.e., "Hi"). Choosing "Lo" would be irrational. However, this intuition is, again, not sustained by classical rational choice theory. (Especially counterintuitive is the mixed-strategy equilibrium strategy with a higher probability of the strategy that has the worse coordination equilibrium as a potential outcome, i.e., "Lo." The mixed-strategy equilibrium thus equalizes the expected utilities of the two strategies.)

The general structure of interdependent expectations in situations of coordination ("multiple contingency") leads such social theorists as Talcott Parsons and Niklas Luhmann to think that social science should be framed in systems-theoretic terms rather than in terms of rational agency—a strike through the Gordian knot of rationalizing coordination that seems to have lost a great deal of its appeal in social science over the past couple of decades

(Schmid, 2011; for the use of game theory in the social sciences, cf. chapter 10.4 by Raub, this handbook). In recent action and decision theory, the issue has been taken up, and a series of theories of the rationality of coordinated choices has been developed. In this context, the *collectivity* of the rationality in question has received particular attention. Most prominently, Robert Sugden (1993) and Michael Bacharach (2006) have suggested that the rationality of coordinated choices is collective in the sense of a certain kind, or mode, of reasoning. Rather than reasoning strategically from their individual points of view, the participants engage in *team reasoning*. They recognize situations of coordination (such as the paradigmatic Hi–Lo game) as situations in which "the team"—"we," from the agents' perspective—has to choose, and this rationalizes their contribution to the best collective choice. The way this matters to the prisoner's dilemma is that aversion against exploiting others (or being exploited by them) transforms a PD into a Hi–Lo game with mutual cooperation as the better equilibrium (Gold & Sugden, 2007, pp. 288ff.). The way in which this relates to the open question of the collectivity of the collective rationality in rational team reasoning is that it is "as a team" (rather than "as individuals") that the participants reason. As Susan Hurley (2003) puts it in her pithy title, "The limits of individualism are not the limits of rationality." It is widely accepted in the relevant literature that this should not be taken as saying that it is simply another rational choice for individuals to identify with teams and reason from a team perspective, as this would lead back into the problem of interdependent expectations (it is rational *if* the others choose to identify with the team too, etc.). At this point, Elizabeth Anderson's "priority of identity" principle is relevant: before the question of rational choice can be answered, the question of the identity of the agent has to be settled, and very few agents are "merely individuals" but rather members of a wide variety of groups (Anderson, 2001, p. 30). However, the claim that the—singular or plural—identity of the chooser is always a matter of a subpersonal framing process that is beyond the scope of reasoning is equally dissatisfying as the "rational identity choice" view. Clearly, we do sometimes reason about from which point of view to approach a decision, such as in the case of role conflicts, and it is not clear that we should take a "purely individual" point of view as the default.

3.   The Rationality of Corporations

In a second and apparently quite different sense of the word, "collective rationality" is used in the literature for a particular relation between beliefs, desires, and perhaps intentions had by a collection, or group. Collective rationality is the feature in virtue of which there are "consistent and complete group attitudes towards the propositions on the agenda" (List & Pettit, 2011, p. 49; cf. chapter 10.1 by Dietrich & Spiekermann, this handbook). In this sense, the very idea of collective rationality is predicated on the assumption that besides individuals, collectives, too, can have an agenda, that is, that they, too, can be epistemic and practical agents (and be usefully treated as such in social science) and thus be rational. This assumption has received increasing support in the recent literature, most notably from Christian List and Philip Pettit (2011). This marks a decided departure from the widespread older methodological precept in social science and social philosophy to treat only individuals (and not collectives) as agents and thus as (potentially) rational.

On this line, collective rationality is approached as an *aggregation problem*: how can individual judgments (preferences, beliefs, or desires) be aggregated in such a way as to yield a *consistent* collective standpoint across interdependent issues? It turns out on this line of research that no simple nondictatorial aggregation rule ensures collective rationality in a robust way (i.e., across a larger number of interdependent decisions)—and that if we manage to be robustly rational (consistent in our joint position) as the teams and organizations we are, we cannot achieve this through mechanical application of simple judgment aggregation procedures (e.g., majority voting) alone.

This second concept of collective rationality is best introduced indirectly. For in the 20th-century literature in economics and other social sciences, collective rationality—in the sense of rational group agency—has repeatedly and prominently been declared to be neither possible nor even desirable. A particularly clear statement to this effect is in the chapter on individual and collective rationality in James M. Buchanan's (1962/1999) *Calculus of Consent*:

> It is difficult to understand why group decisions should be directed toward the achievement of any specific end or goal. Under the individualistic postulates, group decisions represent outcomes of certain agreed-upon rules for choice after the separate individual choices are fed into the process. There seems to be no reason why we should expect these final outcomes to exhibit any sense of order which might, under certain definitions of rationality, be said to reflect rational social action. Nor is there reason to suggest that rationality, even if it could be achieved through appropriate modification of the rules, would be "desirable." Rational social action, in this sense, would seem to be neither a positive prediction of the results that might emerge from group

activity nor a normative criterion against which decision-making rules may be "socially" ordered. (pp. 31–32)

In this view, the idea of collective rationality appears simply as an "illegitimate transfer from the individual to society" (Arrow, 1963, p. 120; cf. Suzumura, 1983, p. 94). According to Buchanan's "individualistic postulate," only individuals have a will and values and make choices. Why deny collectives such features and abilities? The worry behind the individualistic position seems to be that collective agency demotes the participating individuals from their place as proper agents to mere "organs" of the collective organism (cf. Buchanan, 1962/1999, pp. 11ff.). This worry is intimately connected to the history of methodological individualism (Heath, 2015; Udehn, 2001). Buchanan's views on collective rationality thus reflect Max Weber's (1922/1968) canonical statements and echo Joseph Alois Schumpeter's (1908/2010, chapter 6) view—with the remarkable difference that both of these classical methodological individualists saw the "individualistic principle" (Schumpeter) as a *methodological* precept for the particular type or branch of social science they envisaged, and they explicitly allowed for conceptions of "societal values" (Schumpeter) or "collective agents" (Weber) for other social sciences.

The most influential challenge to Buchanan's ban on collective rationality comes from Kenneth J. Arrow: "Collective rationality in the social choice mechanism is not . . . an illegitimate transfer from the individual to society, but an important attribute of a genuinely democratic system capable of full adaptation to varying environments" (Arrow, 1963, p. 120). In particular, collective rationality is the remedy against "democratic paralysis"—"a failure to act due not to a desire for inaction but an inability to agree on the proper action" (Arrow, 1963, p. 120). Thus, contrary to Buchanan's view, collective rationality may well appear as desirable, and very desirable indeed. But is it possible?

The way the issue has historically been approached is in terms of *aggregation* of individual preferences to a social preference, and the debate revolved around Condorcet's famous paradox (see Gehrlein, 2006) as the central challenge to collective rationality from an aggregative point of view. Assume a group consisting of A, B, and C. A's preference order is $x > y > z$, B's is $y > z > x$, and C's is $z > x > y$. Each individual member's preferences are complete and transitive. However, if they vote pairwise over the alternatives, it comes out that there is a majority of two for $x > y$, a majority of two for $y > z$, but also a majority of two for $z > x$. Thus, the group's preferences are intransitive: the preference order is circular.

Their predicament, as a group, is thus such that based on their aggregation procedure according to the majority rule, they cannot agree on one option, but neither are they indifferent between the options. A further reason to think that this makes the group irrational is that they can now be money-pumped (cf. chapter 8.1 by Grüne-Yanoff, this handbook): give them x for free, then offer them z, which they prefer over x and are thus willing to pay for, then make them pay for y, then x, then z again, and so on and so forth until they are broke. It seems intuitively irrational to have a structure of preferences that makes one vulnerable to such exploitation.

Kenneth J. Arrow's (1963) famous *impossibility theorem*—which can be seen as a generalization of the Condorcet theorem (List, 2011)—shows that collective rationality is incompatible with other plausible conditions of collective choice: unrestricted domain (there is a unique and complete ranking of collective choices no matter what the individual preferences are that go into the decision process), non-dictatorship (no individual has the power to always determine the group's choice), independence of irrelevant alternatives (the social preference between x and y should depend only on the individuals' preferences over x and y), and Pareto efficiency (if every participant individual prefers x over y, there is a collective preference of x over y). Many authors have searched for—and suggested—potential escape routes. Most prominent among these is Amartya Sen's (1969) proposal of a weaker conception of collective rationality; however, this condition turned out to contain an oligarchy (see Blair, Bordes, Kelly, & Suzumura, 1976).

While Arrow's impossibility theorem concerns preference aggregation, *judgment aggregation* is at the center of the more recent debate on the desirability and (im)possibility of collective rationality. This development might be seen as a generalization (cf. Nehring, 2003), as a shift of focus (depending on whether or not one interprets preferences as judgments—a preference of x over y can be interpreted as the judgment that x is better than [or preferable to] y), or perhaps as not making much of a difference at all (if one interprets epistemic judgments as preferences over what to believe). Again, the point is that groups and other collective bodies often have to arrive at judgments concerning what is the case and what should be done, and they have to show some degree of consistency in their judgment in order to function as the group or collective body they are.

For this new wave in the discussion on social choice, the *discursive dilemma* plays a similar role as did Condorcet's paradox for the Arrovian line and has largely replaced the latter in the recent debate. The main point, however, is the same: the combination of majority voting rule and rationality of the participant individuals does not ensure collective rationality. Philip Pettit's (2003) example for the discursive dilemma is a political party of three members that is unanimously committed to the view that deficit spending is not a good idea. Now a series of policy decisions comes up. The party first votes on whether or not taxes should be increased—two of the three members vote "no," so this is the party view. Then the issue of defense spending comes up—two of three members vote for an increase, one for a reduction, making "increase" the party line. Then the issue of other government spending comes up—again, increase has a two-thirds pro majority, with only one member voting for reduction. Given the party's view on deficit spending, the party is thus inconsistent. However, the participant individuals may well be individually consistent and thus rational: one of the votes for an increase of defense spending and other spending comes from the member who voted for a tax increase, while the second vote for increased defense spending comes from a member who voted for a reduction of other spending, and the second vote for an increase of other spending comes from a member who judged that defense spending should be reduced, so that each of the participants individually respects the incompatibility of increased spending, non-increase of taxes, and rejection of deficit spending.

Thus, the discursive dilemma again shows that the combination of individual rationality and majority voting does not ensure collective rationality. It is controversially discussed in the literature on the discursive dilemma, however, how much collective rationality really matters. For example, in the context of the question of democratic legitimacy, Fabienne Peter (2009) distinguishes reason-responsiveness from rational consistency (cf. chapter 2.1 by Broome, this handbook) and argues that the latter is not in itself necessary for legitimacy:

> Is it really the case that, as a democratic collective, we ought to satisfy some requirements of collective rationality? Does the binding force of legitimate decision hinge on whether or not these decisions satisfy requirements of rationality? . . . In light of the distinction between reasons and rationality, it is possible to see that to deny that democratic legitimacy includes a rationality requirement does not entail that the demand to respect a democratic decision as legitimate does not depend on reasons. It is only to deny that there is independent normative force in the fact that collective decisions satisfy some conditions of consistency. (pp. 144–145)

While this seems plausible as far as the question of democratic legitimacy is concerned, an issue of credibility and trustworthiness comes up when inconsistency is not plausibly and robustly precluded. Philip Pettit thus argues for the desirability (and indeed necessity) of collective rationality with his example of a political party: "the party cannot tolerate collective inconsistency, because that would make it a laughing-stock among its followers and in the electorate at large; it could no longer claim to be seriously committed to its alleged purpose" (Pettit, 2003, p. 178). Examples for the importance of (this sense of) collective rationality include cases where groups are trusted (be it by their members or by outsiders, e.g., as testifiers); in such cases, it does indeed seem plausible to assume that groups need to achieve, at the collective level, the sort of robust coherence that characterizes rational individual agency. For a group testimony that p to function as such, there has to be trust in collective rationality. One might object, however, that any such argument presupposes what is claimed to be shown, that is, that group agency matters (for an instructive discussion, cf. Roth, 2014).

In an Arrovian vein, List and Pettit (2011) show that collective rationality is incompatible with the counterparts of Arrow's conditions for preference aggregation (cf. List, 2011; cf. chapter 10.1 by Dietrich & Spiekermann, this handbook). The upshot is—very roughly—this: if the participants can have a rational view on everything that is at stake, and if all the individuals' views are to count for the collective judgment, there is no simple aggregation rule that robustly ensures collective rationality. This severely limits the prospects for the sort of coherence that seems necessary to take a group seriously as a rational agent. (Similar to Arrow's theorem, one might object that the simple majority rule works rather well in most cases of collective decision making and that the problem with the simple majority rule arises only in the special case of interdependent decisions; however, a plausible counterargument is that trusting a group requires some modal robustness and that a group that cannot ensure collective rationality in the complicated case cannot be trusted in simple cases.)

The debate on social choice has thus again come to disagree with Buchanan's "individualistic principle": it is argued that collective rationality is desirable after all. However, the result so far is only grist to Buchanan's mill in that it appears that under the stated conditions, no application of a simple aggregation rule will robustly ensure collective rationality, even if all the individual inputs fed into the process are rational.

One way to interpret this result is that collective rationality cannot be conceived of in terms of an aggregation function. Collective rationality is not a matter of organizational design. Collective rationality cannot be achieved reliably by feeding rational individual inputs into a mechanism that yields a collective attitude. Admittedly, some groups do need such organizational mechanisms—think of modern democracy. (And it is no coincidence that modern social choice theory originates from investigations of voting procedures in mass society.) But what the impossibility results have shown might just be that thinking of collective rationality on this paradigm is problematic or even wrong-headed. Perhaps such organizational measures as voting procedures are attempts to emulate collective rationality under unfavorable circumstances. Smaller groups typically resort to votes only if in the process of deliberation dissent proves to be persistent; in many other cases, groups work out a joint view in the same way they might jointly prepare a meal—not by mechanically assembling whatever the participants happen to bring along but by working out a joint plan and a division of tasks. Thinking about collective rationality on the model of voting mechanisms might be mistaking the emulation for the original.

Indeed, in the last chapter of their book on *Group Agency* (2011, chapter 9), List and Pettit sketch a "behavioral" alternative to the "organizational" route out of the aggregation problem. Roughly speaking, the view is that for groups to get their act together, they have to self-identify with the team in a way that is "immediate" in the sense that it leaves no "identification gap." An identification gap opens up whenever members think of their group in third-personal terms rather than adopting the first-person-plural perspective (e.g., by my thinking of my professional group in terms of "the philosophy department" rather than in terms of *us*). If, from my perspective, the consistency problem is "the philosophy department's," it is not *my* problem. Thus conceived, the group is another agent, over and above the heads of its members, and collective irrationality seems to be a problem only if I am, for some additional reason, bothered by it. From a first-person-plural perspective, however, inconsistency is *our* problem, and it involves me in a way that does not run through any extra reason for me to be bothered by it.

4.   Collective Intentionality

Even though these two conceptions of collective rationality differ in the sort of practical problems they concern as well as in the notions of rationality and collectivity involved in each of the conceptions, they point toward one and the same issue: the two solutions to the two different problems of collective rationality converge on the topic of thinking and acting as a team. The sort of reasoning behind collective rationality appears to be collective in a way in which the participants are

self-identified as members that is not in need of further individual motivational or evidential support. But how could individuals ever be "blindly" identified with their teams in such a way that their group's interests and consistency requirements should motivate (and perhaps even justify) their contribution without an extra individual motivation?

One way to address this issue is via the first-person-plural perspective. In recent philosophical research, this has mostly been investigated under the label *collective intentionality*, that is, "the power of minds to be jointly directed at objects, matters of fact, states of affairs, goals, or values" (Schweikard & Schmid, 2013). Although philosophers disagree over how to analyze collective intentional states, the first-person-plural perspective figures prominently in this debate. From the members' points of view, their team's attitude (or lack thereof, or inconsistency thereof) is not just some other agent's business, which concerns them only as external constraints on their own actions. Rather, it is *theirs*, collectively. Put in first-personal terms, that is why it concerns *me*, as a member, in a way that is different from a third-party interest or consistency problem, and much closer to the way in which *my own* interests and consistency problems concern me. The way I am committed by my interests, and bothered by my inconsistencies, is via the particular way I know them, which is self-knowledge, self-consciousness, or self-awareness. One way to cash out the capacity for joint thought and action further is thus in terms of plural (prereflective) self-awareness (Schmid, 2014, 2018). The way in which team reasoners and self-identified members know, or feel, to be reasoning together self-commits to a shared goal in the same way in which an individual preference commits one to an individual goal, although the cases are not identical. In the individual case, an attitude of the form "*x* realizes best what I value, but why should I do it?" makes little rational sense (although it may occur in the case of weakness of will): judging that an alternative is best is to be committed to choosing it in the same way as judging that something is true is to believe it; there is, under normal circumstances, no room for questions concerning whether or not a proposition should be believed after its truth has been settled. In the collective case, however, judgment and rational choice seem to come apart, to some degree. To judge, as a team, that *x* is best *for us* leaves some room open for rational questioning whether or not *I* should contribute (one might need some assurance from others before deciding to pull one's weight). It should be noted, however, that this plural case is still very different from a third-personal case.

"Our" judgment commits me to a different degree than my own, but in contrast to an outsider's judgment, it *does* commit me. If I decide to act against what we, together, judge to be best, I certainly owe an explanation, while the converse is not true: our judgment provides a *pro tanto* reason for my contribution (cf. Gilbert, 2013). After all, the view in question is *ours*, and that implicates me—it is first-personal, although not in the singular form, but in the plural. Thus, our judgment self-commits me (I need no reason to be committed by a judgment besides knowing it to be ours) in the same general way as my own judgment self-commits me just in virtue of my self-awareness of it.

This also bears on the question of consistency, and again, the analogies and disanalogies to the singular case are instructive. Our individual minds contain inconsistent attitudes aplenty, but becoming *aware* of them (in the right first-personal way) is being committed to sorting them out. I can perhaps *have* incompossible intentions, but as soon as I become first-personally *aware* of them, they lose their status as practical commitments, and the issue concerning what to do is reopened. Plural self-awareness works in the same way: it is in virtue of our first-person-plural knowledge of attitudes as *ours* that we cannot be unbothered by their potential inconsistencies but have to make up our (plural) mind about what to think and what to do. In this way, plural (prereflective) self-awareness constitutes a *collective* consistency requirement in the same way as singular (prereflective) self-awareness constitutes an *individual* consistency requirement. In this way, collective self-awareness makes it not only desirable but possible, too, to be effective and consistent as the teams we are.

References

Anderson, E. (2001). Unstrapping the straitjacket of "preference": A comment on Amartya Sen's contributions to philosophy and economics. *Economics & Philosophy, 17*, 21–38.

Arrow, K. J. (1963). *Social choice and individual values.* New York, NY: Wiley.

Bacharach, M. (2006). *Beyond individual choice: Teams and frames in game theory.* Princeton, NJ: Princeton University Press.

Blair, D. H., Bordes, G., Kelly, J. S., & Suzumura, K. (1976). Impossibility theorems without collective rationality. *Journal of Economic Theory, 13*, 361–379.

Buchanan, J. M. (1999). *The calculus of consent.* Indianapolis, IN: Liberty Fund. (Original work published 1962)

Butler, D. J. (2012). A choice for "me" or for "us"? Using we-reasoning to predict cooperation and coordination in games. *Theory and Decision, 73*, 53–76.

Davis, L. H. (1977). Prisoners, paradox, and rationality. *American Philosophical Quarterly, 14*(4), 319–327.

Gauthier, D. (1979). *The logic of Leviathan: The moral and political theory of Thomas Hobbes.* Oxford, England: Clarendon Press.

Gauthier, D. (1986). *Morals by agreement.* Oxford, England: Clarendon Press.

Gehrlein, W. V. (2006). *Condorcet's paradox.* Berlin, Germany: Springer.

Gilbert, M. (2013). *Joint commitment: How we make the social world.* Oxford, England: Oxford University Press.

Gold, N., & Sugden, R. (2007). Theories of team agency. In F. Peter & H. B. Schmid (Eds.), *Rationality and commitment* (pp. 280–312). Oxford, England: Oxford University Press.

Heath, J. (2015). Methodological individualism. In E. N. Zalta (Ed.), *The Stanford encyclopedia of philosophy.* Retrieved from https://plato.stanford.edu/archives/spr2015/entries/methodological-individualism/

Hurley, S. (2003). The limits of individualism are not the limits of rationality. *Behavioral and Brain Sciences, 26*(2), 164–165.

Kuhn, S. (2017). Prisoner's Dilemma. In E. N. Zalta (Ed.), *The Stanford encyclopedia of philosophy.* Retrieved from https://plato.stanford.edu/archives/spr2017/entries/prisoner-dilemma/

List, C. (2011). The logical space of democracy. *Philosophy & Public Affairs, 39*(3), 262–297.

List, C., & Pettit, P. (2011). *Group agency.* Oxford, England: Oxford University Press.

Luce, R. D., & Raiffa, H. (1957). *Games and decisions: Introduction and critical survey.* New York, NY: Wiley.

McMahon, C. (2001). *Collective rationality and collective reasoning.* Cambridge, England: Cambridge University Press.

Nehring, K. (2003). Arrow's theorem as a corollary. *Economic Letters, 80*, 379–382.

Orbell, J. M., & Wilson, L. A., II (1979). Institutional solutions to the N-prisoners' dilemma. *American Political Science Review, 72*(2), 411–421.

Peter, F. (2009). Political legitimacy without collective rationality. In B. de Bruin & C. F. Zurn (Eds.), *New waves in political philosophy* (pp. 143–157). New York, NY: Palgrave.

Pettit, P. (2003). Groups with minds of their own. In F. Schmitt (Ed.), *Socializing metaphysics* (pp. 167–195). New York, NY: Rowman & Littlefield.

Roth, A. S. (2014). Indispensability, the discursive dilemma, and groups with minds of their own. In S. R. Chant, F. Hindriks, & G. Preyer (Eds.), *From individual to collective intentionality: New essays* (pp. 137–162). Oxford, England: Oxford University Press.

Schmid, H. B. (2011). The idiocy of strategic reasoning. *Analyse & Kritik, 1*, 35–56.

Schmid, H. B. (2014). Plural self-awareness. *Phenomenology and the Cognitive Sciences, 13*, 7–24.

Schmid, H. B. (2018). The subject of "we intend." *Phenomenology and the Cognitive Sciences, 17*, 231–243.

Schumpeter, J. A. (2010). *The nature and essence of economic theory*. New Brunswick, NJ: Transaction Publishers. (Original work published 1908)

Schweikard, D. P., & Schmid, H. B. (2013). Collective intentionality. In E. N. Zalta (Ed.), *The Stanford encyclopedia of philosophy*. Retrieved from https://plato.stanford.edu/archives/sum2013/entries/collective-intentionality/

Sen, A. K. (1969). Quasi-transitivity, rational choice and collective decisions. *Review of Economic Studies, 36*(3), 381–393.

Sen, A. K. (1977). Rational fools. *Philosophy & Public Affairs, 6*, 317–344.

Sorensen, R. (2004). Paradoxes of rationality. In A. R. Mele & P. Rawling (Eds.), *The Oxford handbook of rationality* (pp. 257–277). Oxford, England: Oxford University Press.

Sugden, R. (1933). Thinking as a team: Towards an explanation of nonselfish behavior. *Social Philosophy & Policy, 10*, 69–89.

Suzumura, K. (1983). *Rational choice, collective decisions, and social welfare*. Cambridge, England: Cambridge University Press.

Tilley, J. (1991). Altruism and the prisoner's dilemma. *Australasian Journal of Philosophy, 69*(3), 246–287.

Tilley, J. (1994). Accounting for the "tragedy" in the Prisoner's Dilemma. *Synthese, 99*(2), 251–276.

Townley, B. (2008). *Reason's neglect: Rationality and organizing*. Oxford, England: Oxford University Press.

Udehn, L. (2001). *Methodological individualism: Background, history and meaning*. London, England: Routledge.

Weber, M. (1968). *Economy and society*. Berkeley: University of California Press. (Original work published 1922)

Weirich, P. (2009). *Collective rationality: Equilibrium in cooperative games*. Oxford, England: Oxford University Press.

10.3 Rationality in Communication

Georg Meggle

Summary

Communicative actions are actions aiming at being understood. This is the common ground of many different theories of communication. Explicating what is entailed in this smallest common denominator is the first and foremost task of a general theory of communicative actions. This chapter covers three different kinds of communicative theory: an individualistic theory of communication following ideas of Paul Grice (section 1) and two collectivist theories of communication (section 2), namely, speech act theory, initiated by John L. Austin and John Searle, and the theory of communicative actions as developed by Jürgen Habermas. Finally, in section 3, we examine various aspects of rationality involved in and presupposed by these different approaches.

1. The Individualistic Approach

This chapter is not on communication in general, but only about communicative actions. A general theory of such communicative actions is thus part of a general theory of actions. Its main question is, What is the *differentia specifica* of communicative actions? And regarding this handbook, is the rationality of communicative actions really a form of rationality *sui generis* (as vigorously argued by Habermas)?

Communicative actions are actions aiming at being understood, and understanding an action is just knowledge of its meaning. This is the common ground of all sorts of communication theories. Thus, explicating what is entailed by this common ground is the foremost task of the theory sought after. However, these explications divide over the readings of "action," "action understanding," and "action meaning" they are starting from.

Regarding actions, one has to distinguish between action types (e.g., the action type *singing*) and action tokens (e.g., my singing now). Theories whose basic concepts are primarily defined for action tokens, as performed by their respective agents or subjects at particular

times, follow the so-called research program of *methodological individualism*. Theories following the alternative program of *methodological collectivism* start from concepts basically related to action types. Notice that methodological differences need not entail ontological differences. It is this methodological difference that sets the stage for the following overview.[1]

1.1 Basic Concepts

Communicative actions are actions aiming at being understood. This is enough for them to qualify as instrumental actions, that is, as actions by means of which the agent tries to achieve her respective aim(s).

Instrumental or intentional actions can be defined quite easily by combining the three basic (qualitative) elements of any (individualistic) action theory:

$D(X, f)$:= X *does* or *performs* (at time t) the action f;

$B(X, A)$:= X *believes* (strongly)—i.e., is *convinced*— (at t) that A;

$W(X, A)$:= X *wants* or *desires* (at t) that A.

These are the three basic notions on which everything else will build. First the notion of intentional action:

Definition 1: By doing f, X *intends* to achieve aim A iff (if and only if) X does f, X wants that A, and X believes that A will come about if X does f; in symbols:

$$I(X, f, A) := D(X, f) \wedge W(X, A) \wedge B(X, D(X, f) \leftrightarrow A).$$

Here, $\leftrightarrow$ stands for material equivalence.[2] Here and in the following, it should be strengthened to a subjunctive biconditional. However, we better neglect this complication.

Such an action is *successful* iff X's belief, her expectation of success, is correct, that is, iff $D(X, f) \leftrightarrow A$ indeed holds. Sometimes our aims have a certain hierarchy. If in X's view, her aim A_2 is merely a means to achieve aim A_1, aim A_1 is her *primary aim* regarding A_2. Therefore, *the* primary aim, if it exists, is the aim that is primary regarding all other aims.

What, then, does it mean to understand an action? It is common to say that we understand an action of X iff we know the reasons for which it was done. These reasons should be spelled out more generally with reference to X's beliefs and desires, or preferences and probabilities (see chapter 2.1 by Broome and chapter 8.2 by Peterson, both in this handbook). However, since our focus is on intentions (definition 1), we may conveniently restrict our considerations to the following special case: we understand an action iff we know the intention with which it was performed.

Since intentions are subjective (in the sense that having an intention is being convinced of having it), this agrees with Weber (1913/1988, pp. 429, 432), who says that we understand an action iff we know its subjective meaning. Here, the subjective meaning of an action is nothing but the meaning or function this action has for the agent. And this we know when we know the corresponding intention with which the action was performed.

Let us make our notion of understanding a bit more explicit. For the sake of simplicity, we may equate *knowledge* with true belief (conviction): "X knows (at t) that A" $= K(X, A) := B(X, A) \wedge A$, although we are aware that since Plato, this equation is taken not to be correct. Then our notion comes to

Definition 2: Y (fully) understands X's doing f iff Y knows all the aims which X intends to achieve by doing f; symbolically: $U(Y, D(X, f)) := K(Y, I(X, f, A^*))$, where A^* represents the sum of all aims X intends by doing f.

The upshot is that the understanding of actions and the rationality of actions are two sides of the same coin. Every rational action is understandable, and according to definition 2, *only* rational actions are understandable—trivially so: reasons, and intentions in particular, can be (fully) known to us only if they exist.

1.2 Communicative Actions

Let us apply now these basic notions to communicative actions. Typically, such actions are utterances (of sentences). However, they may also be gestures, facial expressions, flying a flag, and so on. Almost anything may serve communication. We shall focus here, *pars pro toto*, on one general form of communication attempts, namely, so-called *informatives*, where the primary communicative aim of speaker S is to get the hearer H to *believe* some proposition p.

So, S's doing f is an (informative) *communicative attempt* addressed at H with the content p, only if by doing f, S intends to make H believe (at $t' > t$) that p; in symbols: $CA(S, H, f, p) \rightarrow I(S, f, B'(H, p))$.[3]

Communicative aims may also be achieved in a noncommunicative way. The crucial question hence is, What distinguishes a communicative action f from a merely (noncommunicative) instrumental action with the same aim? It is obviously the special communicative way by means of which S expects to achieve her primary communicative aim, namely, through H's very recognition of her communication attempt f. That is, S believes that H comes to believe p (at t') iff H recognizes (knows at t') that S's doing f is such a communicative attempt. Formally: $CA(S, H, f, p) \rightarrow B(S, B'(H, p) \leftrightarrow K'(H, CA(S, H, f, p)))$.

When we take the two conditions on communicative attempts just stated to be jointly sufficient, we arrive at the following adequacy condition:

AC1: $CA(S, H, f, p) \leftrightarrow I(S, f, B'(H, p)) \wedge B(S, B'(H, p) \leftrightarrow K'(H, CA(S, H, f, p)))$.

Note that this is not a definition of CA; as such, it would be circular. Hence, it only serves as a condition that must be derivable from any acceptable definition of CA.

The most important conclusion from AC1 is the following reflexivity condition:

RC: $CA(S, H, f, p) \rightarrow I(S, f, K'(H, CA(S, H, f, p)))$.

In terms of definition 2, RC states the crucial principle that communication always aims at being understood. It follows from AC1, if one accepts the principle that believed means and consequences of intended aims are intended as well.

The next step is to realize that this reflexivity condition in turn spawns further consequences and entails the unlimited transparency or openness of communicative intentions. That is, the communicative attempt has the primary intention $I_0 = I(S, f, B'(H, p))$ that H should believe p. Via RC, this entails that $I_1 = I(S, f, K'(H, I_0))$ also belongs to the intentions of the communicative attempt. Then, so does $I_2 = I(S, f, K'(H, I_1))$, and so on. In general, if I_n is a communicative intention, so is $I_{n+1} = I(S, f, K'(H, I_n))$ for any $n \geq 0$. In this way, the reflexivity condition RC extends to

RC*: $CA(S, H, f, p) \rightarrow I(S, f, K'(H, I^*))$,

where I^* represents the conjunction of the hierarchy just developed; it is, we might say, an unlimitedly transparent communicative intention.

However, we still don't know what an adequate explication of CA looks like. Reviewing the various answers proposed would be a lengthy procedure (sketched in Meggle, 1981). Let us cut the long story short and just look at its beginning and its end. It starts with Grice's basic model (in Grice, 1957, p. 384), which says the following in our present terms:

$$CA(S,H,f,p) := I(S,f,B'(H,p)) \wedge B(S,B'(H,p) \leftrightarrow K'(H,I_0)).$$

And we have just developed the end of the story:

Definition 3: $CA(S,H,f,p) :=$
$\quad I(S,f,B'(H,p)) \wedge B(S,B'(H,p) \leftrightarrow K'(H,I^*)).$

The difference between I_0 and I^* seems small—merely an asterisk. Yet this asterisk harbors what we have called "unlimited transparency" or "openness." Note that definition 3 indeed entails the adequacy condition AC1, since, according to this definition, knowledge of the communicative attempt $CA(S,H,f,p)$ amounts to knowledge of the unlimited intention I^*.

In definition 2, we explicated understanding an action in general. This applies as well to communicative actions and results in

Definition 4: Person Y (who may be the hearer H himself) understands S's doing f as a communicative attempt addressed at H with the content p iff Y knows that S undertakes such a communicative attempt. In symbols: $U(Y,CA(S,H,f,p)) := K(Y,CA(S,H,f,p)).$

The definition takes this simple form because knowing of the communicative attempt $CA(S,H,f,p)$ entails, via the strong reflexivity condition RC*, knowing of the speaker's unlimited intention I^*, with which she performs this attempt.

Let's still attend to the hearer's side. Suppose that you understand S's action as a communicative attempt addressed at you with the content p. When would you actually correspond to S's involved expectation of success, that is, actually believe p? Normally only if you take both conditions,

Sincerity: $CA(S,H,f,p) \rightarrow B(S,p)$, and

Correctness: $B(S,p) \rightarrow p$,

to be satisfied, that is, that S does not tell a lie and does not err. I refer to these conditions as "communicative normality conditions." They may or may not be actually satisfied in a particular given case. On the individual level on which we have moved so far, a theory of communicative actions may be silent on such normality conditions. However, we will see that they play a crucial role on the social level.

1.3 Intersubjective Meaning

If the individualistic approach is to be successful, the social level needs to be construed on its basis. Explaining how this might work in principle is the goal of this section. A first noteworthy point is an ontological one: whereas subjective meanings (i.e., intentions) are those of a communicative action *token* as performed by a subject S (at a time t), intersubjective meanings are attached to action *types* relative to a collective or a population P and, moreover, relative to a certain type of situation Σ. The meaning of a certain gesture or utterance may vary from group to group and even from situation to situation within one group. Again, we better speak of action types generally and not specifically of utterance types.

Now, what might intersubjective meanings be? How to explicate $M(P, \Sigma, F, p)$: that action type F has for P in Σ the *intersubjective meaning* that p? It seems clear that such meanings must be common knowledge in the group P. We follow the usual definition and say that p is *common knowledge* in P—symbolically: $K^*(P, p)$—iff everybody in P knows that p ($K(S, p)$ for all S in P), everybody in P knows that everybody in P knows that p, everybody knows this in turn, and so on (see also chapter 9.2 by Perea on common belief, this handbook). Thus, intersubjective meanings should satisfy the following adequacy condition:

AC2: $M(P,\Sigma,F,p) \rightarrow K^*(P,M(P,\Sigma,F,p)).$

On the basis of our above treatment of the individual level, it is not difficult to arrive at an adequate explication of M. We might start with the proposal of a mere *communicative regularity* (where the index 0 at M_0 signals that there is still a caveat).

Definition 5: $M_0(P,\Sigma,F,p) :=$ for each speaker S in P and each action token f of type F performed in a situation of type Σ, there is a hearer H in P such that doing f is a communicative attempt with the content p: $D(S, f) \rightarrow CA(S,H,f,p).$[4]

We might call M_0 "regular meaning," since it states only a regularity in communication tokens. The caveat is that M_0 does not yet fulfill our adequacy condition AC2. But this is easily filled by requiring that regular meaning be common knowledge:

Definition 6: $M(P,\Sigma,F,p) := K^*(P,M_0(P,\Sigma,F,p)).$

It may be easily shown that this definition satisfies AC2; if the regular meanings are common knowledge, so are the intersubjective meanings (for all details, see Meggle, 2003, pp. 244–251).

Definitions 5 and 6 are the essential steps from (subjective) meanings on the action-*token* level to (intersubjective) meanings on the action-*type* level. Notice that the latter meanings are *defined* in terms of—but *not identified* with—the former ones. (This difference is mostly disregarded or misinterpreted in the criticisms of the individualist approach. This misinterpretation was unfortunately invited by Bennett [1973], one of the champions of the Gricean doctrine. Already the title of this paper, "The Meaning-Nominalist Strategy," suggested that conventional meaning is just a special case of speaker's meaning, CA(S, H, f, p). This is false, of course.)

An essential step in our account is still missing. On the individual level, a speaker S makes a communicative attempt CA(S, H, f, p) with two sorts of expectations, namely, the expectation that the hearer understands her attempt, B($S, D(S, f) \rightarrow K'(H, CA(S, H, f, p))$), and that her attempt is indeed successful, that is, B($S, K'(H, CA(S, H, f, p)) \rightarrow B'(H, p)$). It is not obvious how these expectations transfer to the social level. On which reasons can they be based in order to make them well founded on the action-type level?

Concerning the expectation of understanding, the answer is just M(P, Σ, F, p) itself and its entailed common knowledge; each speaker expects her audience to understand her. However, concerning the expectation of success, there is obviously a rationality gap left open by definition 6. Up to now, there is nothing in M(P, Σ, F, p) on which this expectation could regularly keep relying. But without such a reliable base, the communicative regularity of M(P, Σ, F, p) would be miraculous and could not be stable.

The missing link adding stability to regularity is *conventionality*, which is based not only on common knowledge but also on common interests. For the *locus classicus* of a game-theoretic modeling of how such interests or preferences turn regularities into conventions, see Lewis (1969); see also chapter 9.1 by Albert and Kliemt (this handbook). Lewis's explication of signaling conventions is an explication of conventional meaning as a special case of M(P, Σ, F, p) (see Skyrms, 2010, or the reconstructions in von Kutschera, 1983, and Meggle, 2010).

The way conventionality closes the rationality gap behind M(P, Σ, F, p) is best explained by an idealized story, in which $\Sigma = \Sigma^*$ is a *perfect communication situation*, which means (i) it is a common interest of S and H in Σ^* that B'(H, p) $\leftrightarrow p$; (ii) it is common knowledge among S and H that K(S, p) or K($S, \neg p$), but neither K(H, p) nor K($H, \neg p$); and (iii) it is also common knowledge among S and H that S and H will follow a joint strategy consisting of S's strategy D(S, f) $\leftrightarrow p$ and of H's strategy B'(H, p) $\leftrightarrow$ D(S, f).

Now, suppose that S in P makes the communicative attempt CA(S, H, f, p) toward H in P, in such a perfect situation Σ^*, that f is of type F and that indeed the intersubjective meaning of F in Σ^* is p, that is, M(P, Σ, F, p). Then the normality conditions of sincerity and correctness (section 1.2) do hold. This is entailed by clause (iii) about

common knowledge of the joint strategy, while clauses (i) and (ii) explain why S even makes the effort of a communicative attempt. In this way, the best reasons for the expectation of success at the CA-*token* level (i.e., the conditions of sincerity and correctness) are transformed also into the best reasons at the CA-*type* level.

Clearly, perfect communication situations are very idealized. The actual picture is statistically blurred everywhere. However, the ideal picture is the guideline for all attempts at more realism. Here, the point was only to explain the individualistic strategy for construing intersubjective meanings in terms of subjective communicative actions.

1.4 Toward the Semantics and Pragmatics of Expressions

So far, the individualistic approach has delivered a meaning theory for action performances (e.g., whistling, singing, or flying a flag). However, these performances must be distinguished from their products or results (whistles, songs, or the flags flied). Thus, we get the following matrix comprising the four different levels:

	Action performance	Action product
Type	Action type (II)	Product form (III)
Token	Action token (I)	Product token (IV)

We have dealt with I and II. So, the remaining big question is, How do we get from an action-theoretic semantics for communicative actions to a semantics for action products (signs or expressions)? Let's turn first to level III and then to level IV.

For simple (unstructured) signs (like green or red lights, flags, etc.) the solution is simple, too: their meanings can be identified with the conventional meanings of their performances. So, if F-doing is the way of producing the sign- or utterance-type U, we may equate M(P, Σ, U, p) = M(P, Σ, F, p). However, this trivial identification in the case of unstructured signs is not available when structured signs enter the field. Then the question arises as to how the meaning of a complex sign is a function of the meanings of its parts. And as every linguist will agree, there is no generally accepted answer to this question up to now.

As for structured expressions—linguistic expressions being their most important case—a truly systematic action-theoretic semantics would have to start from an exclusively syntactically characterized language (say, predicate logic amended by all kinds of indexicals and intensional and performative operators) and deliver an interpretation of this language, as paradigmatically designed in modern intensional semantics. And it would have to show how this interpretation can be systematically based on corresponding communicative conventions stated at level II (a task already solved by Lewis [1969], extended by von Kutschera [1983], and embedded into a full account of level I by Meggle [2010]). Eventually, the entire feat would have to be carried over to natural languages. Thereby, however, we would get deeply involved in linguistics.

Let us finally turn to level IV, accounting for the communicative meanings of concrete utterances u of type U. Since a semantics for utterance types U (level III) is already presupposed here, level IV is the field of traditional pragmatics. Thus, the main question on this level is how best to systematize the various literal and nonliteral ways of doing things with signs and words in concrete situations (in a communicative form).

Although pragmatics is the most prolific field in communication theory and has shown already tremendous progress with regards to formalizations in the past decades (see, e.g., Horn & Ward, 2004), a systematic pragmatics based on, and formulated in terms of, a general communication theory is still missing. For quite obvious reasons: conventional sign meanings in the group P are relative to various (idealized) situation-types Σ. Thereby, literal sign-usages in a specific situation σ of type Σ are associated with mutual communicative expectations firmly grounded in the conventions behind $M(P, \Sigma, F, p)$; how this works was indicated at the end of section 1.3 for those perfect situations. However, what the mutual communicative expectations actually are in the situation σ is an entirely open question, because σ has many other features beyond being of type Σ. Each feature is potentially relevant; this might involve everything commonly known from human social life.

Well, relevant for *what* exactly? (See Wilson & Sperber [2012] and chapter 4.3 by Merin, this handbook.) This crucial question is best answered by returning to the beginning of our individualistic account, where we tried to capture communicative understanding.

We started with the question of what communicative understanding means in the case of a totally unspecified action token, and now we have arrived at the same question for concrete utterances of action types with meanings already specified in relation to particular populations, situations, languages, and other structured instruments of human interaction (argumentation rules and other cultural rules and habits in the fields of knowledge, sciences, religion, jurisdiction, arts, etc.). However, all these kinds of factors are relevant only insofar as they contribute to the simple question explicitly left open at the beginning (but required to get a good answer at this final level): how

can speaker S addressing hearer H in this specific situation σ reasonably believe her expectation of understanding and of success to be fulfilled? Well, all these factors are relevant insofar as they are part and parcel of the best rational explanation of S's communicative attempt.

There are, from ancient rhetorics (see chapter 5.6 by Woods, this handbook) over speech act theory, Habermas's universal pragmatics (both to be discussed in a moment), Grice's theory of implicatures (Grice, 1989), and the just-mentioned relevance theory up to modern linguistic pragmatics, many proposals to structure this extremely broad field of possibly relevant factors by various sorts of conversational rules or maxims (see, e.g., Korta & Perry, 2006/2020). However, all of this is beyond the scope of this chapter.

2. Collectivistic Approaches

2.1 Speech Act Theory

Whereas an individualist theory of communicative actions starts with actions, as well as their meaning and understanding, in the subjective sense, speech act theory—which is perhaps the leading collectivist version following the meaning-as-use approach of Wittgenstein (1953)—proceeds from actions already endowed with an intersubjective meaning. In terms of the matrix above, speech act theory starts at level II and not at level I.

Accordingly, the basic question of speech act theory is not how to define communicative actions in general but how to account for the different forms of these acts as specified in the following open list of so-called illocutionary roles: commanding, asking, questioning, thanking, threatening, warning, recommending, dissuading, declaring, excusing, promising, and so on. This has been the central concern of speech act theory as founded by Austin (1962) and developed by Searle (1969) (see also Vanderveken, 1990–1991). Unfortunately, though, no explicit definition has been presented going beyond such an open list of illocutionary roles or acts having these roles. So, on the one hand, speech act theory was very fruitful in many applications in philosophy and linguistics, but on the other hand, it was also in need of conceptual explication from the start.

In fact, it is not quite correct to say that speech act theory starts at level II. Although it focused on illocutionary roles and acts, it presupposed that these acts are utterances of signs already having a conventional linguistic meaning.[5] Thus, it was exclusively interested in linguistic communication. In other words, it started rather at a combination of levels II and III, where the latter was assumed not to be in need of further clarification.

This assumption has initially barred speech act theory from developing a general pragmatic semantics that explicates meaning also for structured (linguistic) action products. The first to notice and try to overcome this bar was Alston (1964; and in more detail in Alston, 2000); he was followed by von Savigny (1983) and in quite different ways by Brandom (1994).

However, let me rather dwell on how speech act theory construes the relation between levels I and II. It is here where its differences with the individualistic approach in section 1 become most conspicuous. A central distinction maintained in all versions of speech act theory is that between illocutionary and perlocutionary acts, the latter ones being acts of actually bringing about something by intentionally doing or uttering something, with or without the help of conventions. And this fits exactly our explication of instrumental actions as presented in definition 1. In short: intentional perlocutionary acts *are* instrumental acts.

Now, the central claim of speech act theory associated with this distinction is the *irreducibility thesis*: illocutionary acts are not definable in terms of perlocutionary acts, and illocutionary forces are not definable in terms of perlocutionary effects; both are *sui generis*. (The locus classicus here is Austin, 1962, chapter 9.)

Is the irreducibility thesis correct? This depends on what kind of illocutionary acts we are talking about. Following earlier suggestions of Strawson (1964) and Schiffer (1972), it is mainly due to Bach and Harnish (1979) that we should distinguish between two kinds of illocutionary acts: they are either communicative acts or institutional acts. Paradigm cases of the latter would be the (procedural) act types of getting married, baptizing, or betting at an auction—the very cases Austin had started from.

Now, by performing such an institutional act, one may indeed try to achieve various things (say, enrichment, in-group acceptance, or eternal redemption). The point, though, is that no special perlocutionary aspect, no special instrumental intention, is conceptually required for an act to count as institutional. Thus, for this kind of illocutionary acts, the irreducibility thesis holds trivially. However, this does not entail that it holds also for the other sort of illocutionary acts, namely, the communicative ones. On the contrary, as shown in section 1, the thesis is simply false at least for the informatives discussed there. There we found that some (communicative) illocutionary intentions *are* a special sort of perlocutionary (instrumentalist) intentions. Hence, at least as far as the informatives, an embracive class of communicative illocutionary acts, are concerned, speech act theory could be individualistically reconstructed on the

basis of the explications given in section 1, as shown by Schiffer (1972) and Meggle (1981).

The most basic issue between an individualist and a collectivist approach to meaning revolves around what kind of explication of $M(P, \Sigma, F, p)$ one is willing to accept. Whereas the individualist would accept a reductionist explication as given in section 1, this very sort of reductionism is denied by the collectivist. He would agree to the adequacy condition AC2 for intersubjective meaning stated in section 1.3. However, he cannot accept the individualistic explication of M as presented in definitions 5 and 6 (section 1.3). He is committed to give M and its corresponding notion of understanding a different reading, one that refers to the relevant rules defining what is or counts as a *correct performance* of the respective action type F in P and Σ. His notion of understanding is simply this: we understand a communicative action iff we know its rules.[6]

2.2 Habermas's Theory of Communicative Actions

Communicative actions, as this term is used in Habermas (1981), are of a very different sort from all those considered so far. Habermas conceives of them as collective actions, performed by a collective P and not by a single individual, as assumed in the individualistic approach (section 1) and in speech act theory (section 2.1). If such individual actions of members of P are to add up to a communicative action in Habermas's sense, they must integrate into some kind of common practice in which the members of P engage with a certain shared attitude. It is this shared attitude that characterizes Habermasian communicative action for a collective P in a situation Σ with respect to some state of affairs p. This attitude is best described as embracing three connected parts: (i) a cognitive, (ii) a voluntative, and (iii) a praxiological part.

Ad (i): it is common knowledge in P_Σ (i.e., the subgroup of P involved in Σ) that it is (maybe only tentatively) an open question for P_Σ whether p or not-p, where p refers to a factual (empirical) or expressive (referring to one's state of mind) or normative state of affairs. Ad (ii): it is a common interest in P_Σ to engage jointly and open-mindedly in trying to reach mutual understanding—in the sense of mutual agreement (consensus) in P_Σ—about whether p or not-p is the case. And ad (iii): the procedures to achieve this common goal are those of giving and taking reasons in consideration of p. They are governed by the essential characteristics of our everyday speech acts, and their rules agree with, and could be explicated in terms of, the argumentation rules of rational theoretical or practical discourse (see also chapter 5.5 by Hahn & Collins and chapter 5.6 by Woods, both in this handbook).

As suggested by part (iii), this characterization of Habermasian communicative actions utilizes, as he intends, some strongly idealized aspects of speech act theory. Following Bühler's (1934) tripartition of the organum model of speech events (with its three poles of sender, receptor, and things and, respectively, its relations of expression, appeal, and representation), Habermas's speech act model distinguishes between three worlds (reflected in part (i) above), with which every speech act is correlated: the objective, the social, and the subjective world, which correspond to the distinction between culture, society, and personality. According to Habermas's theory, these three dimensions show up in three corresponding dimensions of communicative rationality. Communication aims at the discursive redemption of validity claims in ways structurally and universally embedded in our everyday speech act practice. This aim is at the center of Habermas's attempts at a unified concept of rationality. These attempts were strongly assisted by his friend K.-O. Apel. Apel's work on transcendental pragmatics[7] incorporated Peirce's pragmatic conception of truth as consensus reached in the long run by an indefinitely extended community of discourse and thus set the stage for the speech-act-theoretic backings of Habermas's theory.

As should be clear from this very condensed summary, Habermas's project of spelling out the consequences of the above basic characterization of communicative actions in his sense, and applying them to the various possible social and political fields, is much richer than a specific theory such as the one sketched in section 1. It covers a whole research program, which we cannot even sketch here. Moreover, one has to keep in mind how Habermas develops and presents his program, namely, in his typical multiperspective style by means of systematic and historical reconstructions and by combining philosophical accounts with empirical approaches from psychology and the social and political sciences. Commonly, such research programs can only be meant as work in progress from the outset.

Already this very brief outline shows that switching from the notion of communicative actions as developed in section 1 to Habermas's notion implies a radical paradigm shift. The shift may be required, since Habermas intended his theory to be a starting point for a new foundation of social theory, particularly of a theory of modern societies and their transnational extensions. Since this paradigm shift has often been disregarded (maybe even by Habermas himself—see the last paragraph of this section), let us at least notice some of its aspects.

For this purpose, we should keep in mind that we distinguished above between (a) instrumental acts, (b) communicative attempts in general, and (c) conventional and linguistic communicative attempts in particular and, in section 2.1, between (d) communicative and (e) institutional illocutionary acts. Thus, if one would want to systematically reconstruct Habermas's theory, one would have to ask how it relates to these distinctions or, more generally, how Habermas's collective communicative actions are correlated with the various communicative and illocutionary acts on the individual or on the group level. Let me only make a few suggestions in this direction.

For instance, Habermas's theory is committed to the irreducibility thesis (section 2.1). Hence, one would have to know how its speech-act-theoretic foundations look like without this thesis, in order to fairly compare his theory with the individualistic approach at the general level (b).

Next, according to Habermas, communication is essentially a consensus-oriented activity. This famous claim can be subscribed to by a Gricean individualist, too. However, this orientation toward consensus would have to be spelled out quite differently for the two sides. For ease of comparison, let us restrict the group P to S and H. Then, according to the above sketch of Habermas's approach, S and H try to reach common knowledge about the truth of either p or not-p. In the individualistic approach, by contrast, a communicative attempt of the form $CA(S,H,f,p)$ only entails that S intends by doing f that she will bring about common knowledge between her and H that $CA(S,H,f,p)$. So, on both sides, there is necessarily some openness involved. In Habermas's account, however, it is a joint open-mindedness related to some unknown factual or normative state of affairs p of common interest, while in the individualistic approach, this openness is essentially restricted to the communicative state of affairs $CA(S,H,f,p)$ itself.

Moreover, for Habermas, communicative actions are opposed to strategic actions, trivially so, because he defines strategic actions with reference to their missing openness. Therefore, Habermas seems to imply that strategic actions are also the opposite of communicative actions in the individualistic sense. However, there is no such opposition. As explained in section 1, communicative actions in the latter sense are instrumental actions; they need not even be sincere, although they have to be so sufficiently often on the conventional level.

Finally, what about Habermas's central tenet that communicative actions are actions *sui generis*? That is, how do Habermasian communicative actions relate to the various actions types (a)–(e)? There is an easy first answer confirming this tenet: since collective actions in general transcend the contribution of all individual actions, they are by definition *sui generis*. And so are, hence, Habermas's communicative actions. However, nothing follows from this with respect to whether or not even these communicative actions may be ultimately explicated in terms of more basic individualistic concepts.

3. Rationality Assumptions—An Overview

3.1 Rationality in the Individualistic Approach

Let us look back on our route and see what kinds of rationality assumptions we have used at the various stages. And let us first take up the individualistic approach (section 1). There we first came across considerations of rationality when we explained a general concept of understanding an action as knowing its reasons. These reasons lie in the agent's preferences and beliefs, the yardstick by which her action's rationality is measured. This is the concept of *action rationality* (as dealt with in chapter 2.1 by Broome and chapter 2.2 by Wedgwood and the entire section 8 in this handbook).

Our explanation assumed that these reasons in turn correspond to certain standards of rationality that they have to satisfy in order to qualify as reasons in the first place. Which standards? They certainly include the requirement that both, the beliefs and the desires in question, are consistent. One might impose the requirement of deductive closure:

$$B(X,A) \wedge B(X,A \to B) \to B(X,B).$$

One might extend this to the desires. In any case, such principles concern the rationality of our *attitudes*. This is not the place to discuss them in detail, but see, for example, chapter 5.1 by van Ditmarsch, chapter 8.1 by Grüne-Yanoff, and chapter 11.1 by Horty and Roy (all in this handbook). (My own compass has been Lenzen, 1980.) Of course, they would have to include principles stating rationality conditions among the various attitudes, such as the following bridge principle:

$$W(X,A) \wedge B(X,A \to B) \to W(X,B).^8$$

Such principles, which mediate among various attitudes held by a person at one time, need to be complemented by principles that mediate between these attitudes at different times. Again, this is not the place to expand on these issues (see chapter 5.2 by Rott; chapter 5.3 by Kern-Isberner, Skovgaard-Olsen, & Spohn; chapter 5.4 by Gazzo Castañeda & Knauff; and section 5 of chapter 8.1 by Grüne-Yanoff, all in this handbook). Finally, the fact that the agent has sufficient reasons for a given action, and that those reasons in turn satisfy the relevant rationality requirements, does not mean she will

actually do it. She will only do it if she herself *is* rational in the given situation. So, in a nutshell, we must distinguish three aspects of rationality already within general action: action rationality, rationality of an action's reasons, and personal rationality.

Owing to the thoroughly subjective nature of action rationality, the question of whether the agent's attitudes themselves are well founded in turn remains open, especially whether her action (as Weber, 1913/1988, puts it) is rational not only in a subjective means–end sense but also in an objective-correctness sense (by being based on true beliefs). The latter is required only for a successful intentional action.

What do communicative actions add to all this? To wit, nothing! Except certain more special expectations on the part of the speaker. The speaker expects the hearer to understand her communication attempt and thus to assume of her all three forms of rationality. Moreover, as a result of the reflexivity of communication, she expects that the hearer also realizes that the speaker expects him to make these rationality assumptions of her, and so forth. Hence, *communicative rationality* is nothing but a special case of general action rationality plus its openness intended by the speaker.[9] The soundness of the reasons of a communicative action pertains only to its success. However, within the framework of a general theory, one can initially avoid saying anything about whether the expectations of understanding and success have to be sound themselves.

The picture changes as soon as the *types* of communicative actions receive an intersubjective meaning. If *S* and *H* belong to the relevant population, the expectation of understanding is supported by the relevant meaningful situations. Hence, a communicative attempt is not only subjectively rational regarding its aim of being understood but also objectively rational in such a situation. Moreover, the expectation of communicative success must be confirmed with sufficient frequency. Otherwise, the relevant communicative regularity would sooner or later collapse. Communication by means of signs or expressions with a regular, conventional, or linguistic meaning differs from communication without this backup precisely by the regular confirmation of the communicative aims. The structure of this confirmation can be explicated—following the ideas of Lewis (1969)—by applying game theory as a collective version of rational decision theory (see chapter 9.1 by Albert & Kliemt, chapter 9.2 by Perea, chapter 9.3 by Alexander, chapter 10.2 by Schmid, and chapter 10.4 by Raub, all in this handbook). However, note that "collective" is to be understood here only *in sensu diviso*; the agents are, as in decision making in general, individual persons, members of a collective, but not the collective itself.

With this subjective and intersubjective backing of the expectations of understanding, further communicative spaces open up, especially for "communication between the lines," as studied, for example, in Grice's theory of conversational implicatures, which is based on his "cooperative principle." This is the fundamental principle of his theory but a derived principle of cooperative rationality (see Grice, 1989).

3.2 Rationality in Collectivistic Approaches

Let us turn to how communicative rationality may be conceived within speech act theory. If speech act theory were robbed of the irreducibility thesis (section 2.1), then it would coincide with the individualistic theory. Or so we have argued at least with respect to informatives. Hence, the relevant rationality assumptions would coincide as well.

But what about the collectivist version of speech act theory including the irreducibility thesis? In this case, the central concept of understanding, with which our search for rationality assumptions involved in communication started, does not refer to the intention with which the relevant communicative action was performed. It rather relates to the rules with reference to which a correct performance of the relevant action type is defined. Now, just as one might say that rationality is what makes an action correct in a certain sense, one might say that it is the constitutive rules for illocutions that make them correct according to speech act theory. Compare, for example, if you want to play chess (where it would be redundant to add "correctly"), you have to stick to the chess rules. But this would not give us a rational explanation why you would like to play chess right now or why you prefer to make this move rather than that. So, even in speech act theory, one would need to refer to the above notions of action rationality.

Finally, as far as Habermas's theory of communication is concerned, I do not know of any overall analytic reconstruction, although there are some promising beginnings (see Steinhoff [2009] and Heath [2003], the former in a more destructive, the latter in a more constructive way). In particular, I do not know of any serious attempts to connect Habermas's collectivistic account with the field of collective intentionality richly developed in the past 25 years (see the other chapters of section 10 in this handbook). That is regrettable, since communicative actions in Habermas's sense may be expected to play a central role within this field. Without doubt, it would be very promising to develop Habermas's program further within the framework of theories of collective intentionality.

Notes

1. For more details, see Meggle (2010).

2. The point of the equivalence is this: requiring only the weaker belief $B(X, D(X,f) \to A)$ would allow many actions to be believed to be sufficient for the aim A. However, in the theory of communicative actions, we would like to be able to explain why exactly action f and not any other action sufficient for A is performed. And this explanation requires the stronger belief as stated.

3. For directives, the corresponding primary communicative aim would be that H performs some action a. The logical structure of directives and informatives is the same: just substitute the aim $D'(H, a)$ for the aim $B'(H, p)$.

4. This definition, indeed the entire section, is replete with universal quantifiers. Of course, this is only an ideal type. Actually, all these quantifiers should be statistically weakened to high percentages, as has been suggested by Lewis (1975).

5. Cf. Austin's explanation of what he called a "locutionary act" in Austin (1962, pp. 92–93): it is an utterance with a linguistic meaning "in the sense of Frege."

6. For the different sorts of rules constitutive for communicative meanings in language games versus rules of chess, see Meggle (1985) and, in the same spirit, Kemmerling (1992).

7. See Apel (1976, 1981) and Kuhlmann (1992).

8. For instance, Kant in *Groundwork of the Metaphysics of Morals* (1785/1903, B 44–45) states, "He who wants the purpose (given that the reason has a decisive impact on his actions) also wants the means necessary for the purpose." The quote suggests that the material implication within the belief operator needs to be strengthened to a subjunctive conditional or a strict implication.

9. That is, the speaker expects common knowledge of rationality, as studied in epistemic game theory (see chapter 9.2 by Perea, this handbook).

References

Alston, W. (1964). *Philosophy of language*. Englewood Cliffs, NJ: Prentice-Hall.

Alston, W. (2000). *Illocutionary acts and sentence meaning*. Ithaca, NY: Cornell University Press.

Apel, K.-O. (1976). *Transformation der Philosophie* [Transformation of philosophy] (Vol. I). Frankfurt/Main, Germany: Suhrkamp.

Apel, K.-O. (1981). *Transformation der Philosophie* [Transformation of philosophy] (Vol. II). Frankfurt/Main, Germany: Suhrkamp.

Austin, J. L. (1962). *How to do things with words*. Oxford, England: Clarendon Press.

Bach, K., & Harnish, R. M. (1979). *Linguistic communication and speech acts*. Cambridge, MA: MIT Press.

Bennett, J. (1973). The meaning-nominalist strategy. *Foundations of Language, 10*, 141–168.

Brandom, R. B. (1994). *Making it explicit.* Cambridge, MA: Harvard University Press.

Bühler, K. (1934). *Sprachtheorie* [Theory of language]. Jena, Germany: Fischer.

Grice, H. P. (1957). Meaning. *Philosophical Review, 66*, 377–388.

Grice, H. P. (1989). *Studies in the way of words.* Cambridge, MA: Harvard University Press.

Habermas, J. (1981). *Theorie des kommunikativen Handelns* [Theory of communicative action] (2 vols.). Frankfurt/Main, Germany: Suhrkamp.

Heath, J. (2003). *Communicative action and rational choice.* Cambridge, MA: MIT Press.

Horn, L. R., & Ward, G. (Eds.). (2004). *The handbook of pragmatics.* Oxford, England: Blackwell.

Kant, I. (1903). *Grundlegung zur Metaphysik der Sitten* [Groundwork of the metaphysics of morals]. In I. Kant, *Kant's gesammelte Schriften* (Vol. 4, pp. 385–463). Berlin, Germany: Reimer. (Original work published 1785)

Kemmerling, A. (1992). Bedeutung und der Zweck der Sprache [Meaning and purpose of language]. In W. Vossenkuhl (Ed.), *Von Wittgenstein lernen* (pp. 99–120). Berlin, Germany: Akademie-Verlag.

Korta, K., & Perry, J. (2020). Pragmatics. In E. Zalta (Ed.), *The Stanford encyclopedia of philosophy.* Retrieved from https://plato.stanford.edu/archives/spr2020/entries/pragmatics/

Kuhlmann, W. (1992). *Sprachphilosophie – Hermeneutik – Ethik: Studien zur Transzendentalpragmatik* [Philosophy of language – hermeneutics – ethics: Studies in transcendental pragmatics]. Würzburg, Germany: Königshausen und Neumann.

Lenzen, W. (1980). *Glauben, Wissen und Wahrscheinlichkeit* [Belief, knowledge, and probability]. Vienna, Austria: Springer.

Lewis, D. (1969). *Convention: A philosophical study.* Cambridge, MA: Harvard University Press.

Lewis, D. (1975). Languages and language. In K. Gunderson (Ed.), *Minnesota studies in the philosophy of science* (Vol. VII, pp. 3–35), Minneapolis: University of Minnesota Press.

Meggle, G. (1981). *Grundbegriffe der Kommunikation* [Basic concepts of communication]. Berlin, Germany: De Gruyter.

Meggle, G. (1985). Wittgenstein – Ein Instrumentalist? [Wittgenstein—an instrumentalist?] In D. Birnbacher & A. Burkhardt (Eds.), *Sprachspiel und Methode* (pp. 71–88). Berlin, Germany: De Gruyter.

Meggle, G. (2003). Common belief and common knowledge. In M. Sintonen, P. Ylikoski, & K. Miller (Eds.), *Realism in action* (pp. 244–251). Dordrecht, Netherlands: Kluwer.

Meggle, G. (2010). *Handlungstheoretische Semantik* [Action-theoretical semantics]. Berlin, Germany: De Gruyter.

Schiffer, S. R. (1972). *Meaning.* Oxford, England: Clarendon Press.

Searle, J. (1969). *Speech acts: An essay in the philosophy of language.* Cambridge, England: Cambridge University Press.

Skyrms, B. (2010). *Signals: Evolution, learning, and information.* Oxford, England: Oxford University Press.

Steinhoff, U. (2009). *The philosophy of Jürgen Habermas: A critical introduction* (K. Schönner, Trans.). Oxford, England: Oxford University Press.

Strawson, P. F. (1964). Intention and convention in speech acts. *Philosophical Review, 73*, 439–460.

Vanderveken, D. (1990–1991). *Meaning and speech acts* (2 vols.). Cambridge, England: Cambridge University Press.

von Kutschera, F. (1983). Remarks on action-theoretical semantics. *Theoretical Linguistics, 10*(1), 1–11.

von Savigny, E. (1983). *Der Begriff der Sprache* [The concept of language]. Stuttgart, Germany: Reclam.

Weber, M. (1988). Über einige Kategorien der verstehenden Soziologie [On some categories of understanding sociology]. In J. Winckelmann (Ed.), *Gesammelte Aufsätze zur Wissenschaftslehre* (pp. 427–474). Tübingen, Germany: J. C. B. Mohr (Paul Siebeck). (Original work published 1913)

Wilson, D., & Sperber, D. (2012). *Meaning and relevance.* Cambridge, England: Cambridge University Press.

Wittgenstein, L. (1953). *Philosophical investigations.* Oxford, England: Blackwell.

10.4 Rational Choice Theory in the Social Sciences

Werner Raub

Summary

This chapter is on *applications* of rational choice theory in the social sciences rather than on rational choice theory as such, with a focus on social sciences other than economics and, more specifically, on sociology. These applications employ variants of rational choice theory as outlined in the chapters on individual rationality and decision making as well as those on game theory in this handbook (chapters 9.1–9.4). The applications aim at theories in the sense of more or less formalized models yielding explanations and testable predictions. Assumptions on rational choice are important elements of such models but not the only ones. The chapter sketches selected features of applications, including methodological individualism and model building relating micro- and macro-levels of analysis, as well as strategic interdependence and unintended consequences as core ingredients of applications. Some remarks on empirical research and suggestions for further reading conclude the chapter.

1. Applications of Rational Choice Theory in Various Social Science Disciplines

Applications of rational choice theory in the social sciences are not at all restricted to economics. They meanwhile abound, too, in disciplines such as political science and sociology. Economists have applied rational choice theory outside typical domains of economics. Becker's (1976) "economic approach to human behavior" is a prime example. Arrow (1963), Buchanan and Tullock (1962), Downs (1957), and Olson (1971) have been influential in establishing rational choice approaches in political science. An early textbook by political scientists has been Riker and Ordeshook (1973). Ostrom's study on governing the commons (1990 and subsequent contributions) is an example of work by a political scientist who won the Nobel Prize in Economic Sciences. In sociology, rational choice theory and its applications are often associated with Coleman (1990). He built on earlier work (see

Coleman, 1994), including behavioral sociology (Homans, 1961) and social exchange theory (Blau, 1964; Homans, 1958) as well as contributions, starting in the 1970s, by European sociologists such as Boudon, Esser, Lindenberg, Opp, and Wippler, often labeled "structural individualism" or "explanatory sociology" (Wippler & Lindenberg, 1987). Meanwhile, applications of rational choice theory are an established approach in sociology (Wittek, Snijders, & Nee, 2013).

2. RCT, Methodological Individualism, Micro–Macro Models, and an Example

We use "RCT" for, roughly, "theories in the social sciences that include, but are not restricted to, assumptions from rational choice theory" such as standard axioms for rational behavior under certainty, risk, and uncertainty in the sense of Harsanyi (1977) and the corresponding implications on utility maximization, or game-theoretic equilibrium assumptions (see chapter 8.2 by Peterson and chapter 9.1 by Albert & Kliemt, both in this handbook).[1] RCT employs methodological individualism (Udehn, 2001) as an approach to theory construction, accounting not only for the behavior of individual actors but also, and particularly, for phenomena and processes at the level of social systems[2] made up by those actors. Therefore, RCT includes two levels, namely, the level of actors, often referred to as the "micro-level," and the system level, often referred to as the "macro-level." Moreover, and particularly, RCT focuses on how both levels are related.

2.1. Coleman's Diagram

Coleman (1986, 1990) has suggested a popular diagram (figure 10.4.1) summarizing the "logic" of micro–macro models in the social sciences and of RCT explanations (for related earlier work, see Raub & Voss, 2017).

With "macro," Coleman refers to systems, such as a family, a city, a business firm, a school, or a market. "Micro" refers to the actors making up the system. The macro-level

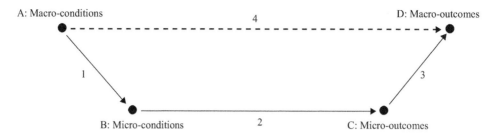

Figure 10.4.1
Coleman's micro–macro diagram.

thus relates to collective phenomena described by concepts referring to properties of social systems, for example, the size of a group. In terms of the number of actors involved, there are large as well as small social systems like dyads, triads, or small groups. The micro-level relates to properties of actors, for example, their preferences, beliefs, and behavior. Typically, the actors making up the system are individuals. However, depending on the application and simplifying assumptions deemed acceptable, the micro-level may include "corporate actors" (Coleman, 1990). For example, in theories of intraorganizational processes, a firm may constitute the system level, with individual employees and managers as actors on the micro-level. Conversely, in theories of interorganizational networks, a set of firms may constitute the system level, with individual firms as corporate actors on the micro-level. Or consider a market model with individual consumers and firms as producers on the micro-level, firms again being conceived as "unitary actors."

In Coleman's diagram, nodes A and D represent propositions describing macro-conditions and macro-outcomes, respectively. Arrow 4 represents propositions about empirical regularities at the macro-level, say, a statistical association between macro-conditions and macro-outcomes. Macro-outcomes (node D) as well as empirical regularities (arrow 4) represent explananda at the macro-level. Node B represents propositions describing micro-conditions. These propositions refer to "independent variables" in assumptions about regularities of actors' behavior, in our case, rational choice theory. Arrow 1 represents "bridge assumptions" (Wippler & Lindenberg, 1987) on how social conditions affect these variables. For example, social conditions such as networks or institutions, but also prices, can be conceived as opportunities or, conversely, constraints affecting the feasible alternatives between which actors can choose. Social conditions likewise shape incentives associated with feasible alternatives and actors' information. Node C represents micro-outcomes and the explanandum on the micro-level, namely, descriptions of actors' behavior. Rational choice assumptions are represented by arrow

2. Thus, arrow 2 represents a micro-theory. Finally, arrow 3 represents assumptions on how actors' behavior generates macro-outcomes. We use "transformation rules" (Wippler & Lindenberg, 1987) as a label for such assumptions on micro-to-macro relations. It is evident from the diagram that the explanandum at the micro-level, descriptions of individual behavior, follows from an explanans comprising assumptions on rational behavior and "initial conditions" on, for example, actors' preferences and beliefs (arrow 2, node B), macro-conditions (node A), and bridge assumptions (arrow 1). The explananda at the macro-level, namely, descriptions of macro-outcomes (node D) and macro-regularities (arrow 4), follow from an explanans comprising rational choice assumptions (node B, arrow 2), macro-conditions (node A), bridge assumptions (arrow 1), and transformation rules (arrow 3).

2.2. An Example: Collective Good Production and the Volunteer's Dilemma

As an illustration, consider the production of collective goods and the empirical regularity at the macro-level that group size is often negatively related to the production of collective goods (Olson, 1971). The core feature of a collective good is that actors who did not contribute to its production cannot be excluded from its consumption. Hence, when costs of individual contribution are high compared to marginal effects of the contribution on individual benefits from the good, actors face incentives not to contribute—the free-rider problem. Assume furthermore that there are no "selective incentives" and, hence, no additional individual benefits that depend on individual contributions to the production of the collective good. Then, Olson argued, collective good production will be negatively related to group size. However, the relationship between group size and collective good production should not be considered a simple macro-law. Rather, the relationship depends on a number of specific conditions such as the absence of selective incentives, the production function for the collective good, and others. Diekmann's (1985) Volunteer's

Dilemma (VOD) is a formal model of a set of conditions that imply the group size effect.

The well-known bystander intervention and diffusion of responsibility problem is a social situation for which VOD is a reasonable model: actors witness an accident or a crime. Everybody would feel relieved if at least one actor would help the victim by calling the police. However, providing help is costly, and each actor might be inclined to abstain from helping, hoping that someone else will help. VOD captures these features in a noncooperative game with N actors. Actors have binary choices. They decide simultaneously and independently whether or not to contribute to the collective good, that is, to provide help. The good is costly and will be provided if at least one actor—a "volunteer"—decides to contribute. Contributions by more than one actor are feasible, and then each actor pays the full costs of providing the good, but contributions of more than one actor do not further improve the utility level of any actor. A core feature of VOD is that the costs K of contributing to the collective good are *smaller* than the gains U from the good. The matrix in table 10.4.1 summarizes the normal form of the game, with rows representing an actor's strategies, namely, to contribute (CONTR) or not to contribute (DON'T), columns indicating the number of other actors who contribute, and cells representing an actor's payoff as a function of his[3] own strategy and the number of other actors who contribute.

In terms of Coleman's diagram (figure 10.4.2), both being a noncooperative game and group size N are macro-conditions (node A). The macro-outcome of interest (node D) is the probability that the collective good will be provided. Arrow 4 now represents the relation between group size and this macro-level probability. Node B represents the micro-conditions (a) that each actor can choose between CONTR and DON'T; (b) actors' information, namely, that actors, when choosing, are not aware of the other actors' choices;[4] and (c) actors' preferences as represented by their payoffs. Note that the

Table 10.4.1

The Volunteer's Dilemma ($U > K > 0$; $N \geq 2$)

	Number of other actors choosing CONTR				
	0	1	2	...	$N-1$
CONTR	$U-K$	$U-K$	$U-K$	...	$U-K$
DON'T	0	U	U	...	U

normal form of the game includes bridge assumptions (arrow 1) on macro–micro transitions. Namely, the normal form specifies how an actor's payoff depends on his own choices as well as those of all other actors—that is, the normal form specifies the structure of the actors' interdependence.

Game-theoretic rationality assumptions such as the assumption of equilibrium behavior are micro-level assumptions (arrow 2). In an equilibrium, each actor's strategy maximizes his own payoffs, given the strategies of the other actors. VOD has N equilibria in pure strategies. These are strategy combinations with exactly one volunteer choosing CONTR with probability 1, while all other actors choose DON'T with probability 1. In each of these equilibria, the collective good is provided with certainty. However, the equilibria involve a bargaining problem: each actor prefers the equilibria with another actor volunteering to the equilibrium where he himself is the volunteer. Moreover, while the game is symmetric, the N equilibria in pure strategies require that actors do not choose the same strategies. It is a rather natural assumption that rational actors, in a symmetric game, play a symmetric equilibrium, that is, choose the same strategies. It can be shown that VOD has a unique symmetric equilibrium in mixed strategies. In this equilibrium, each actor chooses CONTR with probability $p^* = 1 - \left(\dfrac{K}{U}\right)^{\frac{1}{N-1}} < 1$. Under game-theoretic rationality assumptions, the symmetric equilibrium in mixed strategies is a plausible candidate for the "solution" of VOD (for various issues related to

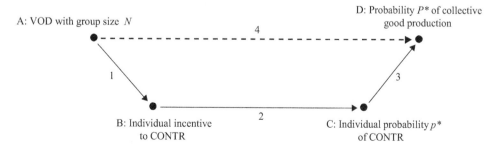

Figure 10.4.2

Micro–macro diagram for the Volunteer's Dilemma.

assuming a mixed strategy equilibrium as the solution of a game, see Camerer, 2003; Osborne & Rubinstein, 1994). Thus, p^* is represented by node C. Note that p^* is a decreasing function of N. Therefore, for VOD and under the assumption of game-theoretic rationality, a testable implication on how group size affects micro-outcomes follows: the individual probability to contribute to the collective good declines with increasing group size.

The individual probability p has to be distinguished from the macro-outcome: the probability P of collective good provision. Note that the normal form of VOD includes a transformation rule (arrow 3). Namely, the normal form specifies how the probability of collective good provision depends on each actor's individual behavior. More specifically, the normal form implies that the collective good is provided if and only if at least one actor chooses CONTR. In VOD, group size affects collective good provision through two different mechanisms. First, there is a positive effect of increasing group size, since it is sufficient that one single actor contributes and since all actors contribute with positive probability in the symmetric mixed equilibrium. After all, with increasing group size, the number of actors increases who may decide to contribute. Second, and conversely, there is a negative effect of increasing group size, since each actor's individual probability p^* to contribute decreases with increasing N. What is the total effect? For the symmetric mixed equilibrium, the probability that there is at least one volunteer and that the collective good will therefore be provided can be shown to be $p^* = 1 - \left(\dfrac{K}{U}\right)^{\frac{N}{N-1}}$. In Coleman's diagram, P^* is represented by node D. Obviously, $P^* < 1$, so that production of the collective good may fail. Moreover, P^* is a decreasing function of N: the negative effect of increasing group size on collective good production outweighs its positive effect. Hence, assuming game-theoretic rationality, a testable implication, represented by arrow 4, follows for how group size affects macro-outcomes in VOD: the probability that the collective good is provided decreases with increasing N.

3. RCT and Model Building

The VOD example illustrates typical features of RCT. First, VOD is a formal theoretical model. RCT has affinity with more or less formalized model building. Coleman's diagram is highly stylized. It looks simple by leaving implicit that full-fledged micro–macro models typically comprise a sizable number of possibly complex assumptions, including assumptions other than those of

rational choice theory as such. The VOD example shows this. The core task in analyzing such models is to identify implications of these assumptions, more specifically implications on how micro-level behavior depends on macro- and micro-conditions as well as implications on how macro-level outcomes and macro-level regularities depend on macro- and micro-level assumptions, including assumptions linking the levels. Deriving such implications is often nontrivial. Therefore, formal model building, or at least the formalization of certain assumptions, can be helpful, if not a prerequisite, for deriving implications. In the VOD example, purely verbal reasoning suffices to see that increasing group size has two opposite effects on the probability that the collective good will be provided. However, formalization seems necessary to derive the total effect. An early version of this argument in sociology is in Coleman's (1964) discussion of "synthetic theories." These theories aim to identify the consequences of a set of assumptions and more precisely to identify consequences on the macro-level for a set of assumptions on the micro-level: "it is characteristic of many of these theories that they begin with postulates on the individual level and end with deductions on the group level" (Coleman, 1964, p. 41). Related arguments include that formalization allows for checking whether implications do follow from assumptions and for identifying assumptions that are implicitly used in merely verbal accounts (for thorough early discussions, see Hummell, 1973; Ziegler, 1972; for a recent outline, see Tirole, 2017). A major advantage of RCT is that it lends itself to formalization of core assumptions and provides analytical power by allowing for the construction of tractable models, specifically models linking macro- and micro-levels.

Second, RCT, like model building in general, typically involves a trade-off: simplifying assumptions preserve tractability and analytical power for the derivation of implications at the cost of (some) assumptions being less realistic empirically, while more complex and realistic assumptions make it more difficult to derive implications. Therefore, RCT typically starts with a model that is as simple as possible, making simplifying assumptions explicit. Subsequently, when simplifying assumptions turn out to be problematic, one introduces more complex assumptions in a stepwise fashion. This is appropriate when implications are highly dependent on such assumptions rather than being robust, or when implications fare badly in the light of empirical evidence. The procedure is known as the "method of decreasing abstraction" (Lindenberg, 1992). For example, the VOD employs the simplifying assumption of a symmetric game. More complex versions of VOD allow for individual heterogeneity with respect to costs of and gains from contributing (Diekmann, 1993). Also, quite a lot of experimental evidence suggests that the individual probability of contributing indeed tends to decline with increasing group size, while a decline is not found for the macrolevel probability that the collective good is provided. Tutić (2014) provides a game-theoretic model accounting for this pattern by employing a solution concept other than the symmetric mixed-strategy equilibrium.

Third, deriving macro-implications from a micro–macro model is a key element of RCT. For example, the core issue in the analysis of VOD is not the individual probability p to contribute but rather the macro-level probability P of collective good production and how that macro-level probability depends on group size as a macro-condition. Tractability requires simplifications, while RCT aims at explanations of macro-outcomes and at incorporating macro-conditions in the explanation rather than at exclusively explaining individual behavior as such. Therefore, it seems advisable to keep micro-assumptions simple and parsimonious. In the VOD example, the analysis becomes feasible through employing micro-assumptions on equilibrium behavior. On the other hand, it seems reasonable to allow, as much as possible, for complexity of macro-assumptions as well as assumptions that provide links between macro- and micro-levels. In the VOD, therefore, the focus is on specifying the structure of interdependence between the actors and on how group size, given this interdependence, affects collective good production. Coleman (1987a, 1993) argues furthermore that careful specification of bridge assumptions and transformation rules is not only a core requirement for RCT but that explanations in the social sciences are often deficient precisely with respect to bridge assumptions and transformation rules. Hence, Coleman assumes that improving bridge assumptions and transformation rules will be more beneficial for theory development than improving micro-assumptions. In the VOD example, it is precisely the explicit specification of the normal form of the game that allows for clearly linking macro- and micro-levels since the normal form comprises the relevant bridge assumptions and transformation rules.

Fourth, RCT is consistent with a modest but realistic view of theory construction in social science, close to Merton's (1968) middle-range theories (Hedström & Udehn, 2009): RCT aims at explanations of macro-outcomes, including macro-regularities, while the use of similar micro-level rational choice assumptions in various models addressing different phenomena at the macro-level allows for a common theoretical core and coherence of the models as well as cumulative growth of knowledge (Diekmann & Voss, 2004). For example, VOD can be seen as one element of a family of models of social dilemmas (Raub, Buskens, & Corten, 2014). Social dilemmas are situations with strategically interdependent actors. Actors can choose between cooperation and defection. Cooperation by all actors on the micro-level implies a Pareto-optimal outcome on the macro-level and is a Pareto-improvement compared to defection by all actors: at least some actors are better off and nobody is worse off when everybody cooperates than when everybody defects. However, cooperation of all actors is typically not an equilibrium, and at least some actors have individual opportunities and incentives to exploit other actors' cooperation by defecting. Conversely, defection by all actors is an equilibrium. The Prisoner's Dilemma is a classic model of a social dilemma. Other well-known models include the Trust Game, the Investment Game, and the Public Goods Game (see also chapter 9.4 by Dhami & al-Nowaihi, chapter 10.2 by Schmid, and chapter 10.5 by Nida-Rümelin, Gutwald, & Zuber, all in this handbook).

For VOD, it follows from properties of equilibria in mixed strategies that the symmetric equilibrium is associated with expected payoffs $U - K$ for each actor. There is another combination of mixed strategies, not being an equilibrium, such that each actor contributes with probability $p^{**} = 1 - \left(\dfrac{K}{NU} \right)^{\frac{1}{N-1}} < 1$. One easily sees that $p^{**} > p^*$.

The strategy combination such that each actor contributes with probability p^{**} is associated with a Pareto-optimal outcome, namely, expected payoffs $U - qK$ for each actor, with $q < 1$ and hence $U - qK > U - K$. Thus, VOD is a social dilemma with the mixed strategy of contributing with probability p^* as defection and the mixed strategy of contributing with probability p^{**} as cooperation.[5]

Models of social dilemmas aim to explain individual cooperation or defection at the micro-level and Pareto-optimal or Pareto-suboptimal macro-outcomes. The models differ by focusing on different macro-conditions and their effects, such as group size effects, the effects of repeated interactions and networks of relations, the effect of various institutions, and so forth. Game-theoretic rationality assumptions, or variants of such assumptions, ensure a common theoretical core and coherence across models of social dilemmas. Moreover, game-theoretic as well as other rational choice assumptions similarly ensure a common theoretical core not only for different models of social dilemmas but also across models for macro-phenomena and macro-regularities in a broad range of other research domains.

4. Variants of RCT

Variants of RCT can be distinguished by considering how they differ from two "benchmarks." One of these comprises commonly used assumptions concerning the micro-level of actors and their behavior. A second one is related to the use of rational choice theory in micro–macro models.

4.1. Variants of RCT: The "Standard Model" and Some Alternatives

Micro-level assumptions of RCT include, but are not restricted to, rationality assumptions. In particular, empirical applications of rational choice theory require additional and substantive assumptions. By way of example, the rational choice assumption of transitive preferences requires that an actor preferring A over B and B over C also prefers A over C, but rational choice assumptions do not presuppose that A is preferred over B and B over C, nor do they comprise assumptions about those properties of A, B, and C due to which A is preferred over B and B over C. More specifically, rational choice theories do not assume at all that actors are exclusively interested in material consequences for themselves and are purely self-regarding in the sense of "utility = own money." Applications in the social sciences do of course require additional and substantive micro-level assumptions about actors' preferences alongside the rationality assumptions (see below). The same applies to substantive assumptions about actors' beliefs concerning the consequences of choosing a certain alternative. However, these assumptions should not be confused with the rationality assumptions themselves. It is true that rational choice assumptions are often combined with assumptions that actors are purely self-regarding and possibly also with assumptions on perfect and complete information—so much so that the combination of these types of assumptions is sometimes referred to as the "standard model" (references in the literature to the "homo oeconomicus" typically concern assumptions closely related to this standard model). This should not distract from the fact that rationality assumptions, on the one hand, and assumptions about self-regarding preferences or about perfect and complete information, on the other, are distinct assumptions of the standard model.

Some variants of RCT focus on the micro-level and offer alternatives to the assumptions of the standard model (Dhami, 2016; see chapter 9.4 by Dhami & al-Nowaihi, this handbook). Some of these variants modify the rational choice assumptions. Experiments like those by Kahneman and Tversky (2000) provided evidence on regularities of behavior ("anomalies") that are hard to reconcile with standard rational choice assumptions, and they induced the development of alternative micro-models of behavior such as prospect theory (Kahneman & Tversky, 1979; see chapter 8.3 by Glöckner, this handbook), models of bounded rationality (e.g., Rubinstein, 1998; see chapter 8.5 by Hertwig & Kozyreva, this handbook), dual-process and two-selves models (Kahneman, 2011), and a variety of models assuming myopic behavior and backward-looking learning (Macy & Flache, 2009). Meanwhile, Esser's model of frame selection (Esser & Kroneberg, 2015; Kroneberg, 2014) and Lindenberg's (2001) theory of social rationality are approaches in sociology that are inspired by such micro-models and aim at systematically developing and employing them in micro–macro models of social phenomena (Bruch & Feinberg, 2017, reviews work in the field of judgment and decision making on micro-models that can be employed in micro–macro models and are based on other than rational choice assumptions).

Coleman's diagram is neutral with respect to the specific assumptions on regularities of individual behavior represented by arrow 2. Hence, alternatives to standard rational choice assumptions such as those mentioned may, in principle, likewise serve as micro-level assumptions. Moreover, regularities of behavior hard to reconcile with standard rational choice assumptions are not only found in experiments. Rather, such regularities have been likewise documented in field research and include micro-level behavior as well as macro-level outcomes in core research fields of economics and other social sciences (Camerer, 2000). This feature has induced attention not only in psychology but also in the social sciences for new theories accounting for such evidence.

Other important variants of RCT that deviate from standard micro-level assumptions retain rational choice assumptions while replacing assumptions such as those on self-regarding behavior and likewise employing assumptions on information imperfections. Influential models combine conventional rationality assumptions with assumptions on inequity aversion or similar motives as elements of an actor's utility function (Cooper & Kagel, 2015; Fehr & Schmidt, 2006). In addition, such models assume heterogeneity between actors with respect to the strength of such motives and incomplete information of actors about other actors' utility functions. Assumptions on other-regarding preferences should be used with care since almost all behavior can be "explained" by assuming the "right" preferences and adjusting utility functions accordingly (Buskens & Raub, 2013). Stigler and Becker (1977) even recommended to abstain from explaining variations in behavior across actors or over time by assuming variation in utility functions. Rather, they propose to assume preferences over fundamental "commodities" as similar across individuals and stable over time, while explaining variations in behavior through variations in opportunities and constraints such as prices, that is, through variations in observable macro-conditions. Lindenberg's theory of "social production functions" is a related approach in sociology (Ormel, Lindenberg, Steverink, & Verbrugge, 1999). One would prefer first of all parsimony with respect to assumptions on other-regarding preferences by adding as few new parameters as possible in micro-level assumptions. Second, assumptions on other-regarding preferences should allow for using the same set of assumptions to explain behavior in a broad range of social situations, thus avoiding "tailor-made" adaptations of assumptions for a specific explanandum that would render independent testability problematic. Third, assumptions on other-regarding preferences should not only account for well-known regularities but should also allow for deriving and testing new predictions. From this perspective, it is reassuring that models combining rational choice assumptions with assumptions on other-regarding preferences, heterogeneity between actors with respect to their preferences, and incomplete information about other actors' preferences do indeed yield implications for micro- as well as macro-outcomes in a broad range of social situations. The models, therefore, preserve tractability and testability, despite being less parsimonious than the standard models with purely self-regarding preferences. The models cover social dilemmas as well as bargaining and distribution problems such as the Ultimatum Game and the Dictator Game, and also include market games (Camerer, 2003). Moreover, these models show that the same set of assumptions implies much cooperative behavior in contexts such as social dilemmas but seemingly selfish behavior in market contexts.

4.2. Variants of RCT: Alternatives to the Perfect Market Model of Neoclassical Economics

A second benchmark for variants of RCT is the perfect market model of neoclassical economics. Coleman (1987b) regarded that model as—in a sense—a paradigmatic example of an RCT micro–macro model. The characteristics of a perfect market (Kreps, 1990) are the relevant macro-conditions. Given rational behavior of actors—producers and consumers—macro-outcomes of individual exchange behavior can be derived, such as the existence of an equilibrium with a set of equilibrium prices for the goods, an equilibrium distribution of goods, and Pareto-optimality

of the equilibrium. The model is exemplary because it is nonobvious that the macro-outcomes are implied by the assumptions. The model highlights that proving theorems, and hence showing that certain consequences follow from assumptions, is a core ingredient of applications of rational choice theory in the social sciences.

Quite a few variants of RCT that focus on the macro-level as well as on how micro- and macro-levels are related can be seen as alternatives to the perfect market model. Often, these variants retain rational choice assumptions and also other assumptions of the standard micro-model sketched above, while adapting assumptions about macro-conditions. Consequently, such variants also focus on bridge assumptions and transformation rules other than in the perfect market model. For example, Granovetter (1985) argued in his programmatic pamphlet that one should replace the assumptions of "atomized" actors on perfect markets, namely, assumptions of the perfect market model due to which strategic interdependence between actors can be neglected. Rather, Granovetter maintains, one should focus on social structures, that is, macro-conditions, that include "embeddedness" of actors. "Embeddedness" refers to ongoing relations between actors as well as networks of actors. Hence, from a rational choice perspective, Granovetter's approach focuses on macro-conditions that do imply strategic interdependence. Likewise, his approach focuses on the effects of strategic interdependence on individual behavior and on macro-effects of that behavior. Granovetter is not usually associated with rational choice approaches in the social sciences. Nevertheless, his approach is strikingly similar to Coleman's heuristic advice to combine robust assumptions on rational choice with more complex assumptions on social structure and carefully designed bridge assumptions and transformation rules. Granovetter opposes "psychological revisionism," characterizing it as "an attempt to reform economic theory by abandoning an absolute assumption of rational decision making" (1985, p. 505). Rather, he suggests maintaining the rationality assumption: "While the assumption of rational action must always be problematic, it is a good working hypothesis that should not easily be abandoned. What looks to the analyst like nonrational behavior may be quite sensible when situational constraints, especially those of embeddedness, are fully appreciated" (1985, p. 506). He argues that investments in tracing the effects of embeddedness are more promising than investments in the modification of the rationality assumption: "My claim is that however naive that psychology [of rational choice] may be, this is not where the main difficulty lies—it is rather in the neglect of social structure" (1985, p. 506).

Variants of RCT in line with Granovetter's approach and Coleman's heuristic advice are often models for social and economic networks. Other variants focus on institutions other than perfect markets, with institutions broadly conceived as constraints for human behavior that result from human behavior itself and that structure the incentives for human behavior (North, 1990). Institutions can be of a formal or informal nature, including legal infrastructure, on the one hand, and informal conventions and codes of behavior, on the other. Still other RCT variants focus on formal and informal organizations. From a sociological perspective, Coleman suggested an analogy between RCT alternatives to the perfect market model and RCT alternatives to assumptions of the standard micro-level model: "Rational choice theory in sociology occupies a position relative to pure neoclassical economics analogous to that of behavioral economics. . . . Analogous to the psychological anomalies of behavioral economics, which show the systematic deviations from rationality that persons exhibit, the social anomalies of sociological rational choice theory show the systemic deviations from the perfect market assumption of neoclassical economies that arise in the linkage between micro and macro levels" (Coleman, 1994, p. 167).

Networks, institutions, and organizations are conceived as exogenous in some variants of RCT, while being endogenous in others. When conceived as exogenous, bridge assumptions can specify how they affect individual behavior. For example, as opportunities and constraints, they may affect actors' feasible choices. They may also affect actors' preferences and information. Networks, institutions, and organizations, assumed to be exogenous, can likewise affect how actors' behaviors "combine" to bring about macro-level outcomes and transformation rules, with transformation rules specifying this micro–macro relation. For example, electoral institutions may not only affect individual voter behavior but also affect the macro-outcomes of individual voter behavior. This includes that macro-outcomes such as the distribution of seats in parliament across political parties and, hence, coalition formation can vary considerably under different electoral institutions, even when controlling for individual voter behavior. Other variants of RCT focus on networks, institutions, and organizations as macro-outcomes themselves. Then, networks, institutions, and organizations are taken as endogenous, and the aim is to explain their emergence and dynamics as a result of actors' rational choice. Often, such models will be dynamic in the sense that they account for

a social process. The burgeoning literature on strategic network formation and the coevolution of networks and behavior offers examples of work in this direction (Buskens, Corten, & Raub, 2014). From the perspective of Coleman's diagram, the macro-outcome (node D) is then simultaneously the "initial" node A of a subsequent micro–macro diagram, and so forth, leading to a sequence of "connected" diagrams as in figure 10.4.1 that account for the development of a social process over periods 1, 2, ...

5. Strategic Interdependence, Game Theory as a Tool, and Unintended Consequences as a Topic

Variants of RCT comprising alternatives to the perfect market model show that applications of rational choice theory in the social sciences often refer to social systems with strategic interdependence between actors. It is therefore not surprising, and illustrated by our VOD example, that game-theoretic tools are often employed. After all, game theory is the branch of rational choice theory that provides concepts, assumptions, and theorems allowing to specify how actors behave in interdependent situations. Game theory assumes that actors behave as if they try to realize their preferences, taking their interdependence as well as rational behavior of the other actors into account (Harsanyi, 1976). Meanwhile, applications in the social sciences employ game theory broadly, including not only "classical games" with complete information in normal form (Harsanyi, 1977) but also, for example, games in extensive form, repeated games, and games with incomplete information (Osborne & Rubinstein, 1994). With respect to RCT in sociology, interdependencies between actors and actors taking their interdependencies into account are likewise the core of Weber's definition of social action: "Sociology . . . is a science which attempts the interpretive understanding of social action in order thereby to arrive at a causal explanation of its course and effects. . . . Action is social in so far as . . . *it takes account of the behaviour of others and is thereby oriented in its course*" (Weber, 1947, p. 88, emphasis added).

Micro- as well as macro-outcomes of an actor's behavior typically depend not only on his own choices and possibly chance events but also on the behavior of other actors. Therefore, outcomes are often unintended consequences of actors' behavior: an actor's intentions need not coincide with the outcomes of his behavior. This applies, too, under rational behavior of actors (Boudon, 1977; Elster, 2007; Hayek, 1967; Merton, 1936). Applications of rational choice theory in the social sciences, certainly applications analyzing situations with strategic interdependence, often focus on outcomes that are unintended consequences. In the VOD example, rational actors maximize their own expected payoffs, given the behavior of the other actors, thus producing a Pareto-suboptimal macro-outcome, but they are not assumed to *intentionally* produce such an outcome. An important distinction is between unintended consequences that are beneficial for the actors and unintended consequences that are unfavorable. The paradigmatic example for beneficial unintended consequences is a Pareto-optimal equilibrium on a perfect market, driven by selfish actors trying to maximizing own benefits, referred to by the "invisible hand" metaphor. The paradigmatic example for unintended consequences that are *unfavorable* for the actors involved are social dilemmas. In these dilemmas, individually rational behavior produces a Pareto-suboptimal macro-outcome. The symmetric equilibrium in mixed strategies in VOD is an illustration.

6.   RCT and Empirical Research

Coleman (1986) argued that applying rational choice theory in the social sciences might contribute to a better integration of theory and empirical research. Nevertheless, RCT has for quite some time been criticized for weak links with empirical research. Focusing on rational choice approaches in political science, Green and Shapiro (1994) argued that much of this work did not induce serious empirical research and that empirical tests have been poorly conducted (the contributions in Friedman, 1996, scrutinize these claims). Meanwhile, the situation has changed. For example, Goldthorpe's (1996) arguments for an alliance involving quantitative analysis of large-scale data sets and rational action theory have been influential in sociology (Blossfeld & Prein, 1998). By now, quite a lot of work by theorists as well as empirical researchers aims at reducing the gap between theoretical models based on rational choice assumptions and empirical research in sociology. Note that empirical research need not and should not be restricted to testing RCT predictions. In addition to testing predictions, and certainly from the perspective of micro–macro models, empirical research is needed to establish macro-level regularities (arrow 4 in Coleman's diagram) that can subsequently serve as RCT explananda (for a similar perspective in demography, see Billari, 2015; Goldthorpe, 1996, 2016).

Today, systematic empirical research based on RCT is common in many fields, including but not restricted to social networks, power and inequality, education and social mobility, deviant behavior and criminology, religion and secularization, migration and immigrant assimilation, families and households, markets and organizations, collective action and collective decision making, ecological behavior, and social dilemmas (Abraham & Voss, 2004; Wittek et al., 2013). While Goldthorpe advocated the use of survey research, other research designs are meanwhile employed as well (Buskens & Raub, 2013; Gächter, 2013). Employing complementary research designs, such as lab and field experiments, surveys and other observational designs, and vignette studies for repeatedly testing the same hypotheses, can yield information on the robustness and replicability of empirical results (Buskens & Raub, 2013; Jackson & Cox, 2013). Other important contributions involve the development of statistical models that facilitate better integration of theoretical models employing rational choice assumptions and empirical research by integrating core assumptions on rational choice into the statistical model itself (McFadden, 1973; Snijders, 2013).

Various issues arise related to empirical research testing predictions based on RCT (Kroneberg & Kalter, 2012). These issues include whether such tests should exclusively focus on predictions themselves or also on empirically assessing assumptions used for deriving predictions, such as assumptions on preferences and beliefs—and on how to do so. Another issue is how to cope with empirical evidence that is hard to reconcile with, or an outright refutation of, predictions based on RCT. Axioms of rational behavior under certainty, risk, and uncertainty à la, for example, Harsanyi (1977) imply that a utility function exists and that behavior in accordance with these axioms can likewise be characterized as maximizing the utility function. In a rather strict sense, an actor behaving according to the axioms behaves *as if* maximizing a utility function: one need not assume that actors consciously calculate (expected) utilities. In this sense, rational choice theory is a theory of behavior rather than of the underlying mental processes. However, behavior may turn out to be more or less systematically inconsistent with predictions based on RCT.

Harsanyi (1976, 1977) argued that although rational choice assumptions may yield empirically problematic conclusions, it is useful to employ them for establishing a "benchmark" so that empirically observed refutations become themselves explananda for more refined models. Other arguments proceed from the idea that macro- rather than micro-level implications of RCT are the core issue. From this perspective, it has been suggested (Goldthorpe, 1996) that errors in predicting individual behavior with RCT will cancel out on the macro-level. Becker (1976) has developed the argument that typical macro-outcomes of micro–macro models are robust to replacing the rational choice micro-model by alternative micro-models. Coleman (1987a, 1990) argues in a similar direction and suggests that replacing simple rational choice assumptions by more complex micro-theories would undermine the tractability of the model because it becomes unfeasible to derive implications for macro-outcomes at all, certainly so when complex bridge assumptions and transformation rules are involved in the model. Both the "errors cancel out" argument and the argument that macro-outcomes are generically robust to varying micro-models of behavior are disputed (Raub, Buskens, & Van Assen, 2011). For example, Thaler (2018) argues that actors' deviations from what rational choice assumptions imply do not exhibit "random error." Rather, Kahneman and Tversky's work, as well as subsequent work building on Kahneman and Tversky's, shows that it is predictable how actors' empirical behavior deviates from implications of rational choice assumptions. Furthermore, Thaler (2018) presents evidence that even macro-outcomes like aggregate market behavior can deviate from what one would expect based on rational choice assumptions. This would contradict the argument of Becker and others that institutions like competitive markets with exchange and division of labor operate so that nonrational behavior of possibly even many actors will be "eliminated" and aggregate outcomes are driven by fully rational behavior of possibly only few actors who thrive on other actors' "errors" (Clement, 2002). Another controversy with respect to RCT and empirical research (Green & Shapiro, 1994; Grofman, 1993) concerns the merits and problems of empirical research that focuses on qualitative predictions of changes "at the margin" using comparative statics rather than on quantitative point predictions.

7. Concluding Remarks and Suggestions for Further Reading

This chapter has been on selected features of contemporary applications of rational choice theory in the social sciences. See Macy and Flache (1995), Hechter and Kanazawa (1997), and Abraham and Voss (2004) for earlier concise reviews of selected applications. Wittek et al. (2013) is a recent handbook on applications in many domains. Ermakoff (2017) is a general discussion of RCT in sociology complementary to this chapter. For reasons of space, we have neglected a "history of ideas" perspective on how contemporary applications relate to, say, British moral and social philosophy of the 17th and 18th centuries (Hobbes, Hume, Smith, Ferguson), the Austrian school of economics (Menger, Schumpeter,

Hayek), and analytical philosophy of science (e.g., Hempel, Nagel) and critical rationalism (Popper, Lakatos). Udehn (2001) is a source with respect to history of ideas. An informative collection of interviews with key scholars in the social sciences who have been influential in promoting and discussing contemporary applications is Swedberg (1990). Wittek (2013) is a bibliography with pointers to additional references on various issues discussed in this chapter. Burgeoning research programs in the social sciences are similar to applications of rational choice theory in that they focus on micro- as well as macro-levels of analysis and try to systematically relate these levels as well as assuming variants of incentive-driven and goal-directed behavior. These research programs include behavioral and experimental game theory (Camerer, 2003), approaches employing models of learning in games and evolutionary game theory (Fudenberg & Levine, 1998; Gintis, 2009a, 2009b; see chapter 9.3 by Alexander, this handbook), analytical sociology (Hedström, 2005; Hedström & Bearman, 2009; Manzo, 2014; for discussion, see Opp, 2013), and applications of agent-based computational modeling (Macy & Flache, 2009).

Acknowledgments

Two reviewers and Wolfgang Spohn provided useful suggestions. Support from NWO (Netherlands Organisation for Scientific Research; grants S 96-168 and PGS 50-370 for the PIONIER-program "The Management of Matches") is gratefully acknowledged. Some parts of the chapter use materials from Raub, Buskens, and Van Assen (2011) and Raub and Voss (2017).

Notes

1. "RCT" is thus *not* used as an abbreviation for "rational choice theory" but refers to social science theories employing rational choice assumptions.

2. We use the label "social system" broadly, including systems of actors typically analyzed in, for example, sociology, political science, and economics.

3. Throughout, we use "he" and "his" to facilitate readability and without intending gender bias.

4. Strictly speaking, we would have to specify the extensive form, including the game tree, rather than only the normal form of VOD, to make its information structure explicit.

5. Note, by the way, that it can also be shown that the macro-level probability that the collective good will be provided increases with increasing group size if each actor contributes with individual probability p^{**} (Diekmann, 1986).

References

Abraham, M., & Voss, T. (2004). Contributions of rational choice theory to modern sociology: An overview. In N. Genov (Ed.), *Advances in sociological knowledge* (pp. 127–150). Wiesbaden, Germany: VS Verlag für Sozialwissenschaften.

Arrow, K. J. (1963). *Social choice and individual values* (2nd ed.). New Haven, CT: Yale University Press.

Becker, G. S. (1976). *The economic approach to human behavior*. Chicago, IL: University of Chicago Press.

Billari, F. C. (2015). Integrating macro- and micro-level approaches in the explanation of population change. *Population Studies, 69*(Suppl 1), S11–S20.

Blau, P. M. (1964). *Exchange and power in social life*. New York, NY: Wiley.

Blossfeld, H.-P., & Prein, G. (Eds.). (1998). *Rational choice theory and large-scale data analysis*. Boulder, CO: Westview.

Boudon, R. (1977). *Effets pervers et ordre social* [Unintended consequences of social action]. Paris, France: PUF.

Bruch, E., & Feinberg, F. (2017). Decision-making processes in social contexts. *Annual Review of Sociology, 43*, 207–227.

Buchanan, J. M., & Tullock, G. (1962). *The calculus of consent*. Ann Arbor: University of Michigan Press.

Buskens, V., Corten, R., & Raub, W. (2014). Social networks. In N. Braun & N. J. Saam (Eds.), *Handbuch Modellbildung und Simulation in den Sozialwissenschaften* [Handbook model-building and simulation in the social sciences] (pp. 663–687). Wiesbaden, Germany: Springer VS.

Buskens, V., & Raub, W. (2013). Rational choice research on social dilemmas. In R. Wittek, T. A. B. Snijders, & V. Nee (Eds.), *Handbook of rational choice social research* (pp. 113–150). Stanford, CA: Stanford University Press.

Camerer, C. F. (2000). Prospect theory in the wild. In D. Kahneman & A. Tversky (Eds.), *Choices, values and frames* (pp. 288–300). Cambridge, England: Cambridge University Press.

Camerer, C. F. (2003). *Behavioral game theory*. New York, NY: Russell Sage.

Clement, D. (2002). Interview with Gary Becker. Retrieved from https://www.minneapolisfed.org/publications/the-region/interview-with-gary-becker

Coleman, J. S. (1964). *Introduction to mathematical sociology*. New York, NY: Free Press.

Coleman, J. S. (1986). Social theory, social research, and a theory of action. *American Journal of Sociology, 91*, 1309–1335.

Coleman, J. S. (1987a). Psychological structure and social structure in economic models. In R. M. Hogarth & M. W. Reder (Eds.), *Rational choice: The contrast between economics and psychology* (pp. 181–185). Chicago, IL: University of Chicago Press.

Coleman, J. S. (1987b). Microfoundations and macrosocial behavior. In J. C. Alexander, B. Giesen, R. Münch, & N. J. Smelser (Eds.), *The micro–macro link* (pp. 153–173). Berkeley: University of California Press.

Coleman, J. S. (1990). *Foundations of social theory*. Cambridge, MA: Harvard University Press.

Coleman, J. S. (1993). Reply to Blau, Tuomela, Diekmann and Baurmann. *Analyse & Kritik, 15*, 62–69.

Coleman, J. S. (1994). A rational choice perspective on economic sociology. In N. J. Smelser & R. Swedberg (Eds.), *The handbook of economic sociology* (pp. 166–180). Princeton, NJ: Princeton University Press.

Cooper, D. J., & Kagel, J. H. (2015). Other-regarding preferences: A selective survey of experimental results. In J. H. Kagel & A. E. Roth (Eds.), *The handbook of experimental economics* (Vol. 2, pp. 217–289). Princeton, NJ: Princeton University Press.

Dhami, S. (2016). *The foundations of behavioral economic analysis*. Oxford, England: Oxford University Press.

Diekmann, A. (1985). Volunteer's dilemma. *Journal of Conflict Resolution, 29*, 605–610.

Diekmann, A. (1986). Volunteer's dilemma: A social trap without a dominant strategy and some empirical results. In A. Diekmann & P. Mitter (Eds.), *Paradoxical effects of social behavior* (pp. 187–197). Heidelberg, Germany: Physica.

Diekmann, A. (1993). Cooperation in an asymmetric volunteer's dilemma game. *International Journal of Game Theory, 22*, 75–85.

Diekmann, A., & Voss, T. (2004). Die Theorie rationalen Handelns: Stand und Perspektiven [The theory of rational action: State of the art and perspectives]. In A. Diekmann & T. Voss (Eds.), *Rational-Choice-Theorie in den Sozialwissenschaften* (pp. 13–29). Munich, Germany: Oldenbourg.

Downs, A. (1957). *An economic theory of democracy*. New York, NY: Harper.

Elster, J. (2007). *Explaining social behavior*. Cambridge, England: Cambridge University Press.

Ermakoff, I. (2017). On the frontiers of rational choice. In C. E. Benzecry, M. Krause, & I. A. Reed (Eds.), *Social theory now* (pp. 162–200). Chicago, IL: University of Chicago Press.

Esser, H., & Kroneberg, C. (2015). An integrative theory of action: The model of frame selection. In E. J. Lawler, S. R. Thye, & J. Yoon (Eds.), *Order on the edge of chaos* (pp. 63–85). Cambridge, England: Cambridge University Press.

Fehr, E., & Schmidt, K. M. (2006). The economics of fairness, reciprocity and altruism—experimental evidence and new theories. In S.-C. Kolm & J. M. Ythier (Eds.), *Handbook of the economics of giving, altruism and reciprocity* (pp. 615–691). Amsterdam, Netherlands: Elsevier.

Friedman, J. (Ed.). (1996). *The rational choice controversy*. New Haven, CT: Yale University Press.

Fudenberg, D., & D. K. Levine (1998). *The theory of learning in games*. Cambridge, MA: MIT Press.

Gächter, S. (2013). Rationality, social preferences, and strategic decision-making from a behavioral economics perspective. In R. Wittek, T. A. B. Snijders, & V. Nee (Eds.), *Handbook of rational choice social research* (pp. 33–71). Stanford, CA: Stanford University Press.

Gintis, H. (2009a). *Game theory evolving* (2nd ed.). Princeton, NJ: Princeton University Press.

Gintis, H. (2009b). *The bounds of reason*. Princeton, NJ: Princeton University Press.

Goldthorpe, J. H. (1996). The quantitative analysis of large-scale data sets and rational action theory: For a sociological alliance. *European Sociological Review, 12*, 109–126.

Goldthorpe, J. H. (2016). *Sociology as a population science*. Cambridge, England: Cambridge University Press.

Granovetter, M. S. (1985). Economic action and social structure: The problem of embeddedness. *American Journal of Sociology, 91*, 481–510.

Green, D. P., & Shapiro, I. (1994). *Pathologies of rational choice theory*. New Haven, CT: Yale University Press.

Grofman, B. (1993). Is turnout the paradox that ate rational choice theory? In B. Grofman (Ed.), *Information, participation, and choice* (pp. 93–103). Ann Arbor: University of Michigan Press.

Harsanyi, J. C. (1976). *Essays on ethics, social behavior, and scientific explanation*. Dordrecht, Netherlands: Reidel.

Harsanyi, J. C. (1977). *Rational behavior and bargaining equilibrium in games and social situations*. Cambridge, England: Cambridge University Press.

Hayek, F. A. (1967). *Studies in philosophy, politics and economics*. Chicago, IL: University of Chicago Press.

Hechter, M., & Kanazawa, S. (1997). Sociological rational choice theory. *Annual Review of Sociology, 23*, 191–214.

Hedström, P. (2005). *Dissecting the social: On the principles of analytical sociology*. Cambridge, England: Cambridge University Press.

Hedström, P., & Bearman, P. (Eds.). (2009). *The Oxford handbook of analytical sociology*. Oxford, England: Oxford University Press.

Hedström, P., & Udehn, L. (2009). Analytical sociology and theories of the middle range. In P. Hedström & P. Bearman (Eds.), *The Oxford handbook of analytical sociology* (pp. 25–47). Oxford, England: Oxford University Press.

Homans, G. C. (1958). Social behavior as exchange. *American Journal of Sociology, 63*, 597–606.

Homans, G. C. (1961). *Social behavior*. New York, NY: Harcourt.

Hummell, H. J. (1973). Methodologischer Individualismus, Struktureffekte und Systemkonsequenzen [Methodological individualism, structural effects, and system-level consequences]. In K.-D. Opp & H. J. Hummell (Eds.), *Probleme der Erklärung sozialer Prozesse II: Soziales Verhalten und soziale Systeme* (pp. 61–134). Frankfurt/Main, Germany: Athenäum.

Jackson, M., & Cox, D. R. (2013). The principles of experimental design and their application in sociology. *Annual Review of Sociology, 39*, 27–49.

Kahneman, D. (2011). *Thinking, fast and slow*. London, England: Allen Lane.

Kahneman, D., & Tversky, A. (1979). Prospect theory. *Econometrica, 47*, 263–291.

Kahneman, D., & Tversky, A. (Eds.). (2000). *Choices, values and frames*. Cambridge, England: Cambridge University Press.

Kreps, D. M. (1990). *A course in microeconomic theory*. Princeton, NJ: Princeton University Press.

Kroneberg, C. (2014). Frames, scripts, and variable rationality: An integrative theory of action. In G. Manzo (Ed.), *Analytical sociology: Actions and networks* (pp. 97–123). Chichester, England: Wiley.

Kroneberg, C., & Kalter, F. (2012). Rational choice theory and empirical research: Methodological and theoretical contributions in Europe. *Annual Review of Sociology, 38*, 73–92.

Lindenberg, S. (1992). The method of decreasing abstraction. In J. S. Coleman & T. J. Fararo (Eds.), *Rational choice theory: Advocacy and critique* (pp. 3–20). Newbury Park, CA: Sage.

Lindenberg, S. (2001). Social rationality versus rational egoism. In J. H. Turner (Ed.), *Handbook of sociological theory* (pp. 635–668). New York, NY: Kluwer.

Macy, M. W., & Flache, A. (1995). Beyond rationality in models of choice. *Annual Review of Sociology, 21*, 73–91.

Macy, M. W., & Flache, A. (2009). Social dynamics from the bottom up: Agent-based models of social interaction. In P. Hedström & P. Bearman (Eds.), *The Oxford handbook of analytical sociology* (pp. 245–268). Oxford, England: Oxford University Press.

Manzo, G. (Ed.). (2014). *Analytical sociology: Actions and networks*. Chichester, England: Wiley.

McFadden, D. (1973). Conditional logit analysis of qualitative choice behavior. In P. Zarembka (Ed.), *Frontiers in econometrics* (pp. 105–142). New York, NY: Academic Press.

Merton, R. K. (1936). The unanticipated consequences of purposive social action. *American Sociological Review, 1*, 894–904.

Merton, R. K. (1968). *Social theory and social structure* (Enlarged ed.). New York, NY: Free Press.

North, D. C. (1990). *Institutions, institutional change and economic performance*. Cambridge, England: Cambridge University Press.

Olson, M. (1971). *The logic of collective action* (2nd ed.). Cambridge, MA: Harvard University Press.

Opp, K.-D. (2013). What is analytical sociology? Strengths and weaknesses of a new sociological research program. *Social Science Information, 52*, 329–360.

Ormel, J., Lindenberg, S., Steverink, N., & Verbrugge, L. M. (1999). Subjective well-being and social production functions. *Social Indicators Research, 46*, 61–90.

Osborne, M. J., & Rubinstein, A. (1994). *A course in game theory*. Cambridge, MA: MIT Press.

Ostrom, E. (1990). *Governing the commons*. Cambridge, England: Cambridge University Press.

Raub, W., Buskens, V., & Corten, R. (2014). Social dilemmas and cooperation. In N. Braun & N. J. Saam (Eds.), *Handbuch Modellbildung und Simulation in den Sozialwissenschaften* (pp. 597–626). Wiesbaden, Germany: Springer VS.

Raub, W., Buskens, V., & Van Assen, M. A. L. M. (2011). Micro–macro links and microfoundations in sociology. *Journal of Mathematical Sociology, 35*, 1–25.

Raub, W., & Voss, T. (2017). Micro–macro models in sociology: Antecedents of Coleman's diagram. In B. Jann & W. Przepiorka (Eds.), *Social dilemmas, institutions, and the evolution of cooperation* (pp. 11–36). Berlin, Germany: De Gruyter.

Riker, W. H., & Ordeshook, P. C. (1973). *An introduction to positive political theory*. Englewood Cliffs, NJ: Prentice-Hall.

Rubinstein, A. (1998). *Modeling bounded rationality*. Cambridge, MA: MIT Press.

Snijders, T. A. B. (2013). Network dynamics. In R. Wittek, T. A. B. Snijders, & V. Nee (Eds.), *Handbook of rational choice social research* (pp. 252–279). Stanford, CA: Stanford University Press.

Stigler, G. J., & Becker, G. S. (1977). De gustibus non est disputandum. *American Economic Review, 67*, 76–90.

Swedberg, R. (1990). *Economics and sociology: Redefining their boundaries*. Princeton, NJ: Princeton University Press.

Thaler, R. H. (2018). From cashews to nudges: The evolution of behavioral economics. *American Economic Review, 108*, 1265–1287.

Tirole, J. (2017). *Economics for the common good*. Princeton, NJ: Princeton University Press.

Tutić, A. (2014). Procedurally rational volunteers. *Journal of Mathematical Sociology, 38*, 219–232.

Udehn, L. (2001). *Methodological individualism*. London, England: Routledge.

Weber, M. (1947). *The theory of social and economic organization*. New York, NY: Free Press.

Wippler, R., & Lindenberg, S. (1987). Collective phenomena and rational choice. In J. C. Alexander, B. Giesen, R. Münch, & N. J. Smelser (Eds.), *The micro–macro link* (pp. 135–152). Berkeley: University of California Press.

Wittek, R. (2013). Rational choice. *Oxford bibliographies online*. DOI: 10.1093/OBO/9780199756384-0070

Wittek, R., Snijders, T. A. B., & Nee, V. (Eds.). (2013). *The Handbook of rational choice social research*. Stanford, CA: Stanford University Press.

Ziegler, R. (1972). *Theorie und Modell* [Theory and model]. Munich, Germany: Oldenbourg.

10.5   Structural Rationality

Julian Nida-Rümelin, Rebecca Gutwald, and Niina Zuber

Summary

This entry focuses on the theory of structural rationality (TSR), which goes beyond the traditional individualist and instrumental approaches to rationality. The starting point of TSR is a shortcoming in traditional rational choice theory (RCT). In RCT, the rationality of an action is evaluated according to its suitability to fulfill the agent's desires at the time of choice, without considering future plans or cooperation. TSR is multidimensional, because it takes at least three dimensions of rational reasoning into account: decisions regarding future aims and plans (diachronic dimension), the plurality of reasons for actions (intrapersonal dimension), and interaction with others (interpersonal dimension). This account of TSR differs from the standard use of "structural rationality" or rather "structural irrationality" in analytic philosophy, since the latter focuses merely on formal relations between different entities such as desires, actions, and intentions. TSR, however, is more radical: it transcends the reductive narrowness of instrumental rationality without denying its practical impact.

1.   Individualistic and Nonindividualistic Approaches to Rationality

There are two major traditions in the theory of practical rationality: the first characterizes practical rationality in terms of the acting individual's properties, capabilities, and states, whereas the second starts from social roles, established practices, and institutions. Within both of these traditions, there are numerous variations but few commonalities. This conflict of paradigms ultimately corresponds to a conflict of disciplines: classical and neoclassical economic theory are committed to the individualistic paradigm, while sociology, cultural studies, and social psychology are committed to the second paradigm (cf. chapter 10.4 by Raub, this handbook). Of course, exceptions do exist: the flourishing of behavioral economics (cf. chapter 9.4 by Dhami & al-Nowaihi, this

handbook) is a reaction to the deficits of the first paradigm, trying to also incorporate consideration of the second paradigm, for instance, by including nonrational behavior in economic explanation or adherence to social norms (Diamond & Vartiainen, 2007; Gigerenzer & Selten, 2001; Kahneman, Slovic, & Tversky, 1982; Kahneman & Tversky, 2000). The theory of structural rationality (TSR) is an account that overcomes this conflict of paradigms. It takes into consideration the embeddedness of individual actions (first paradigm) within the behavioral structures of forms of life (second paradigm). This requires a comprehensive redescription of rational agency.

2.   The Paradigmatic Core of the Theory of Structural Rationality

In his *Nicomachean Ethics*, Aristotle (2000) characterizes *akrasia* (which is usually translated as "weakness of the will," although "lack of self-control" would be more accurate) as a singular action that is not embedded in an intended or desired structure of the agent's course of action. In this context, it is essential to take into consideration that Aristotle and the entire classical antiquity as well as medieval and even early modern practical philosophy do not distinguish between a theory of rationality and a theory of morality: living the righteous life of the *dikaios* consists in realizing *eudaimonia* (usually translated as "happiness"). This characterization of a virtuous life, however, is not individualistic, because it depends on the political community (*polis*) as a whole, in which the lives of individuals unfold. Determining what is good for a community is the precondition for determining what is good for the individual. In other words: aspects of structural rationality can already be found in classical accounts of rationality, ethics, and politics (cf. chapter 12.1 by Fehige & Wessels, this handbook).

The breakup of the unity of morality and rationality that originated in the European era of Enlightenment is most radical in the Kantian and the Utilitarian traditions. Jeremy Bentham (1780/2007) argues that pursuing one's own utility for a maximally advantageous pleasure–pain balance is the decisive motive for individual human action. He deduces the totality of well-being as the guiding moral principle. This tension prevails in contemporary utilitarianism: what is good is determined on the basis of individual well-being or preference fulfillment (Hare, 1963, 1970, 1984; Harsanyi, 1977a, 1977b; Singer, 1993, 2016), while moral obligation does not consist in optimizing individual well-being but in optimizing the *sum* of individual well-being. Rational choice theory (RCT) also focuses on individual preferences: actors do not necessarily optimize their individual well-being but rather strive for the fulfillment of their actual, subjective preferences.

The unity of rationality and morality also disintegrates in the Kantian tradition: rationality amounts to optimizing one's own well-being and is guided by pragmatic imperatives, as Kant (1785) calls them, while moral action is characterized as being exclusively motivated by respect for the Moral Law requiring that the individual maxim of an action is suitable for being transformed into a universal law according to which everybody could act.

In the Kantian tradition, structural embeddedness is realized by adopting only such maxims that are suitable for application as universal rules for action. Kantianism, however, suffers from being underdetermined, because the Moral Law can be met by almost any decision if it is described appropriately. This amounts to the critique of being formalist that Kantianism has been facing since Hegel (1820/2017) and that endures in contemporary communitarianism (Taylor, 1975, 1989). Utilitarianism, in turn, suffers from collectivism, which is incompatible with what John Rawls (1971) called *the separateness of persons*, because it demands maximization of total utility. Individual motives and plans of life are not considered.

It may seem surprising, but ordinary game theory, in tying rational decision making to equilibrium points, also already incorporates an aspect of TSR and is therefore not reducible to decision theory (cf. chapter 9.2 by Perea, this handbook). In Harsanyi's game-theoretic version of rule utilitarianism ("Ethical Bayesianism"; Harsanyi, 1967, 1977c, 1979), the criterion of moral action is the joint optimization of a collective value function through coordinated individual practice. In this understanding, rule utilitarianism is different from act utilitarianism as it requires structurally rational action, that is, to act in such a way that the combination with the actions of others who are also ethically motivated would optimize the common value function (the sum of preference fulfillment), whereas act utilitarianism is about optimizing overall preference fulfillment with every single act.

3. Rational Choice and Structural Rationality

RCT almost never, or at least never directly, includes the embeddedness of pointwise actions in diachronic and interpersonal structures. Other agents are considered part of the environment that influences the fulfillment conditions of personal preferences. However, RCT deals with not only situations of risk and uncertainty but also interaction. Since this branch of RCT emerged historically from the mathematical analysis of games, it is still labeled, misleadingly, as "game theory" (Hargreaves-Heap & Varoufakis, 2004). Rational choice is determined relative to a given value function of consequences and the subjective probability function of the circumstances. In situations of interaction, these circumstances usually consist in the decisions of other agents (possibly plus additional chance events). Hence, the expansion of the rational choice paradigm from decision-theoretic optimization ("playing against nature") to game-theoretic optimization appears to be plausible, at least with regard to interactions in which the decisions of the players are mutually independent from each other (noncooperative games; cf. chapter 9.1 by Albert & Kliemt, this handbook). Problems arise, however, because every agent must assume that the other agents decide rationally, too. The decisions of other players ideally follow the same game-theoretic criteria of rationality. Therefore, game theory renders only those decisions rational that are part of an *equilibrium point*, that is, a combination of decisions in which none of the agents involved has an interest in revoking his decision, as long as all the other agents involved stick to their original decisions (Nida-Rümelin, 1994).

According to game theory, there is only one rational decision for both players, A and B, in the prisoner's dilemma (PD) game described in table 10.5.1: the combination of dominant choices D. A vast evidence (Andreoni & Miller, 1993; Barreda-Tarrazona, Jaramillo-Gutiérrez, Pavan, & Sabater-Grande, 2017; Clark & Sefton, 2001; Cooper, Dejong, Forsythe, & Ross, 1996; Kreps, Milgrom, Roberts, & Wilson, 1982; Pothos, Perry, Corr, Matthew, & Busemeyer, 2001), however, shows that a large number of agents chooses C, even in one-shot prisoner's

Table 10.5.1
Prisoner's dilemma

		Player B	
		C	D
Player A	C	3, 3	1, 4
	D	4, 1	2, 2

dilemmas. The orthodox interpretation takes this as an indicator that irrational behavior prevails, but this is simply not convincing. Decision theory and game theory understand themselves as neutral regarding motivation. Criteria of rationality concern choosing the best means to reach one's goals (i.e., to fulfill one's preferences). Imagine a person who wants to cooperate in a one-shot prisoner's dilemma. Asked about her motivation for choosing C, she might answer, "I choose C because I want that we both choose C and because I expect the other person to choose C, too." A general theory of rationality should allow for a motivation of this cooperative kind (cf. chapter 10.2 by Schmid, this handbook).

Cooperation is a paradigmatic case of structurally rational agency. It is impossible to characterize cooperation adequately if one takes merely outcomes and probabilities into consideration (Nida-Rümelin, 1993, chapter 4; 2019). To render cooperation (i.e., the choice of dominated C) rational in PD, given the cooperative motivations of the players, we cannot dismiss the information of the format itself: the structure of interaction is an essential part of the description of strategies as cooperative or defective. TSR depends on a comprehensive description of strategic options. Cooperative action embeds the individual choice into intended structures of interaction: I do my part, hoping that the other does his part to realize a collective action that secures cooperation (i.e., an outcome that is better for each than the outcome of individual optimization would be).

Assuming a form of "collective" or "shared" intention represents one way of dealing with the PD challenge (Hurley, 2005). Philosophers of collective action and intentionality like Raimo Tuomela (2013), Margaret Gilbert (2013), Michael Bratman (2014), or David Regan (1980; see also Nida-Rümelin, 2014) have therefore introduced a collective perspective on intention, which changes the very outlook on collective-action puzzles like the PD. Collective intention amounts to changing individual utilities based on a group's ethos, as Tuomela has called it.

Empirical findings (Axelrod, 2006; Gintis, Bowles, Boyd, & Richerson, 2005; Rapoport, Chammah, & Orwant, 1956; Seabright, 2004; Sethi & Somanathan, 2005) regarding the behavior of persons in PD situations are best explained by interpreting it as structurally rational. The individual agents prefer C to D in order to realize $\langle C, C \rangle$. This describes why participants in PD games claim to be disappointed if the other person does not choose C and also if they chose C.

4. Meta-Preferences

Structurally rational actions challenge traditional RCT, particularly the revealed-preference concept and the belief–desire model of rationality (Nida-Rümelin, 1991). A prominent approach to remedy this problem is the introduction of meta-preferences, for instance, in the works of Amartya Sen and Harry Frankfurt. Sen (1974) argues that there are three levels of morality: the lowest level, namely, the egoist level, which is captured in PD meta-preferences, that is, meta-preferences that accord with first-order preferences. Second, assurance-game (AG) preferences manifest themselves in favoring cooperation to defection if the other person chooses C, and third, other-regarding meta-preferences (OR) translate as choosing C no matter how the other person decides. Both of these PD variants can be interpreted as disclosing structural rationality: AG players are conditional cooperators, because they do not want to be exploited by egoists, if they cooperate. OR players embed their respective individual choices into the preferred structure of interaction (universal cooperation), possibly with a moral motivation of the Kantian type.

Another prominent account of meta-preferences that is sometimes thought to adequately describe structural aspects of rationality without dismissing the Humean belief–desire model of rationality is presented by Frankfurt (1971). He proposes a concept of a person as an individual who develops meta-preferences, which means that she can distance herself from first-order preferences if she has freedom of the will. Second-order volitions are meta-preferences regarding first-order action-guiding preferences. The idea is that a person is different from a "wanton" insofar as she develops second-order volitions that frame her first-order action-guiding preferences. According to Frankfurt, a "wanton" is not a person in the full sense, since she does not care about her will and only follows her first-order desires.

Both types of meta-preferences display structural aspects of traditional RCT. Without structural reasoning, an individual would be nothing more than a pointwise optimizer and would not be perceived as a person who persists through time and through different moments of interactions (Nida-Rümelin, 2005a, chapter III). We establish and maintain structural deliberation by evaluating our action-guiding preferences. If it turns out that our preferences are structurally incoherent, we are critical about incoherence and wish we had had different first-order action-guiding preferences. This evaluation may influence our first-order action-guiding preferences. If it does not, there is a conflict of the structural dimensions described above and classical rational choice.

5. The Role of Reasons within the Theory of Structural Rationality

The distinction between the theory of rationality and morality collapses if the rationality of an action is characterized in terms of practical reasons (i.e., of different reasons for specific actions). The observation that there are good reasons for actions that are not consequentialist can lead to quite different reactions:

(1) One might assume that these nonconsequentialist reasons are merely *prima facie* reasons that cannot be integrated and thus do not constitute genuine good reasons (this account could be called "consequentialist"). The problem with this account is that some of our most central types of reasons would have to be excluded.

(2) One might confine the range of application of rational choice and exclude moral and other types of good reasons for action (this could be called the "narrow rationality account"). RCT would then not be an all-embracing theory of practical rationality.

(3) One might give up some of the axioms constitutive for standard RCT, as Edward McClennen (1990) demonstrated in his theory of resolute choice (this account could be called "revisionist").

(4) One might redesign the conceptual framework, that is, by reinterpreting the basic concepts of RC such that nonconsequentialist reasons for action can be integrated (this is the strategy of TSR; Nida-Rümelin, 2000, 2005b).

Take the following example: if I have given a promise, I have a reason to keep it even if keeping it has less than optimal consequences for me or the person I gave it to. Speech act theory (Searle, 1969) has analyzed in detail the normative elements that constitute the institution of giving a promise. This analysis reveals the structural character of reasons to act that are connected to the speech act of giving a promise: it maintains the diachronic consistency of one's practice in certain types of interaction. Keeping a promise generates a duty that allows people to relate and forecast the actions of others within a larger context. This institution allows for diachronically and interpersonally structured rationality.

The same applies to communitarian duties one may have because of one's social role, such as parent, teacher, or politician. Communitarian duties constitute these roles and sustain an interpersonally and diachronically consistent social practice. Individual rights can be interpreted structurally, too. They secure the individual's authorship in social contexts. They enable individuals to act within the boundaries of their individual rights. Individual rights thus secure structures of individual agency.

To sum up: the concept of structure is not a kind of external appraisal function that is added to the established life-world practice of giving reasons and accepting reasons. It is a systematization of this practice. We strive for TSR *avant la lettre*. Without a reference to structures, there is no consistent practice, no personhood, and no collective rationality.

6. Is Structural Rationality Instrumental?

The term "structural rationality" can be found in other accounts within the theory of action. The concepts of "structural rationality" in this literature are, however, merely instrumental. Thus, they focus on individual rationality and emphasize the relational and consistent properties of rationality. This instrumental understanding refers to *intra*personal relations between conative attitudes; only some approaches include diachronic dimensions. The analytic philosophical tradition sticks to this account of structural rationality, whereas the TSR also highlights *inter*personal structural features by considering interactions among individuals as a decisive part of rationality. In addition, TSR makes substantial claims about structural rationality.

The planning theory of Michael Bratman (2014) is one of the few accounts that focus on the diachronic and interpersonal aspects of rationality. Bratman thus acknowledges social rationality norms with regard to intentions and policies: friendship, love, dancing together, conversations, getting married, and so on are all practices and forms of action that cannot be explained without reference to a form of "modest sociality" (Bratman, 2007). His concept of modest sociality provides the explanatory basis for our cooperative activities by recourse to conceptual and normative resources that are already in place in our understanding of individual agency. This makes Bratman's explication of cooperation a conceptual extension of his theory of individual agency, since the individual finds himself always already located within social contexts. Bratman argues that collective intentional agency (i.e., cooperative action) is established by shared intentions. A group sharing an intention goes beyond merely intending to do the same thing. They can only act together, or in concert, if they adopt a suitable planning structure that coordinates their actions and deliberations (and if this is common knowledge). If neither of them fulfills the relevant (sub)plan, cooperation fails. Bratman (2007, p. 8), therefore, applies a quite particular notion of "structures": intentions are considered *plan states* that take up a certain position in coordinating plans that, in turn, establish diachronic and interpersonal structures of agency. Intentions have a coordinating role in managing conduct and providing continuity and organization over time—either in collective action or in individual agency.

Thomas Scanlon (2007) explicitly uses the term "structural irrationality" as an identifier for occurrences of irrational behavior. Scanlon is commonly viewed as having coined the term for analytic philosophy of action.[1] In the tradition of analytic philosophy, the term "structural rationality" is used as a formal, relational concept of consistency, for instance, referring to the consistency of the attitudes, intentions, desires, and so on that an individual holds simultaneously at one particular point in time (Wallace, 2018). Structural rationality does *not* make any substantive claim about the desirability of the action's content; it concentrates on the relational requirements that the agent needs to fulfill if she does not want to act irrationally.

To illustrate this relational form of rationality, take the following example, introduced by D. J. Langlois (2014) as one form of irrationality that he calls "intention inconsistency": your intention is to spend the weekend with your family, while you simultaneously have the intention to complete your new manuscript by Sunday. Also, and this is crucial, you believe that you cannot do both. Irrationality arises due to the relational link between the two intentions and the belief about their incompatibility. Structural irrationality comes into existence because it is implausible to state that one considers such an arrangement as consistent. To accept all these propositions simultaneously is not possible since they exclude each other; they are incompatible with each other. This is, however, not due to their content but due to their relations to each other. Neither of them is irrational per se (i.e., can be criticized by its content); they turn out to be irrational because together they cannot form a balanced structure.

Structural rationality is understood in this manner as a nonsubstantial, instrumental, and (at least in some form) normative requirement of rationality. Other forms of structural irrationality are defined as belief inconsistency or *enkrasia*. These types of irrationality have in common that the irrationality arises due to the incompatible relationships between singular entities, such as intentions, attitudes, and beliefs. Structural irrationality is therefore a problem of instrumental rationality since we cannot criticize the contents of our desires. We can only behave irrationally due to not correctly arranging our conative attitudes, including means–end relations.

John Broome (1999), too, describes the concept of structural irrationality as the failure of some people to consistently combine and arrange their attitudes, which prevents them from structuring their attitudes rationally. Thus, irrationality involves not only forms of refraining from action, although one has good reasons to act (*practical irrationality*), or acting without knowledge or on too little information (*theoretical irrationality*). As with Scanlon's account, irrationality affects not only one or several attitudes but also their arrangement. Such an understanding of structural irrationality excludes consideration of reasons and only pertains to the relations of attitudes. Hence, Scanlon claims that there are certain forms of structural requirements that pertain to the relations between an agent's attitudes. This is why Scanlon mainly speaks about forms of structural *irrationality*: irrationality is structural if one fails to consider the adequate relations of attitudes.

Highlighting the consistent internal relations of rationality reduces practical reasoning to instrumental reasoning. Hence, one of the most fundamental questions regarding the conceptualization of rationality pertains to the rules that determine how attitudes should be combined and arranged—without referring to their content or their objectives.[2]

7. Is Structural Rationality Ramsey-Compatible?

Modern RCT is based on the utility theorem that was originally proven by the mathematician and philosopher Frank P. Ramsey (1978). John von Neumann and Oskar Morgenstern (1947) revived this concept of rational choice, which was later implemented in different versions by Marschak (1946), Luce and Raiffa (1957), and others. The utility theorem shows that it is possible to transform the qualitative concept of preference into a quantitative concept of utility if conditions of consistency, such as transitivity of preferences, monotonicity, and continuity of preferences regarding lotteries, are met (cf. chapter 8.1 by Grüne-Yanoff and chapter 8.2 by Peterson, both in this handbook). In economic application, this theory is mainly used to express motivational neutrality of utility functions attributed to individual agents (i.e., it is irrelevant how individuals are motivated). It is viewed as sufficient to observe realized preferences. In fact, utility theory requires persons only to have consistent preferences (in the sense of the axioms); it does not, however, require them to have specific motivations. Furthermore, altruistic preferences can be described by these axioms and therefore incorporated in a cardinal, real-valued utility function.[3] The rationality of expected utility maximization is deduced from the rationality of individual preferences (via the utility theorem), and individual preferences are rational if these preferences are consistent (i.e., meet the axioms of the utility theorem).

TSR is not only instrumental. If expected utility maximization is the criterion of instrumental rationality, then it seems that TSR cannot be Ramsey-compatible. This assumption, however, is fallacious, since it assumes expected utility maximization, as defined by Ramsey, von Neumann, and Morgenstern, to be instrumental. However, preferences can reflect motivations related to the TSR of an individual as well as comply with Rossian *prima facie* duties (Ross, 2002) or with the Kantian categorical imperative or with Aristotelian rules of social practices and still be compatible with the axioms of the utility theorem. Therefore, preferences are compatible with the utility theorem even if they are not instrumental in nature. Hence, expected utility maximization is compatible with TSR. TSR is therefore compatible with rational choice theory. If a structurally rational person makes a decision and the relevant probabilities meet the Kolmogorov axioms and the subjective preferences of the decision maker meet the Ramsey conditions, then this person maximizes expected utility (or, to use a more adequate term: subjective value; cf. chapter 4.1 by Hájek & Staffel, this handbook). Hence, TSR integrates the core idea of RCT: the coherence of two types of subjective attitudes—epistemic (probabilities) and prohairetic (preferences).

8. Structural Rationality—Descriptive or Normative?

There is one salient similarity between traditional RCT and TSR: both are normative and descriptive at the same time. This feature is most conspicuous in economic theory, where economic rationality is used to describe economic behavior and at the same time to advise it (Spohn, 2002).

Looking at the two-sided interpretation of RCT, it comes as no surprise that TSR also qualifies as both normative and descriptive. It starts with our empirically observable behavior and tries to provide an adequate description while simultaneously pointing out and criticizing structurally irrational behavior. It attacks RCT for neglecting structures and therefore disregarding important aspects of any coherent practice. We can see how structures affect behavior in cases of regret or repentance. Structures within theories of rational choice are more successful in describing our life-world reasons for acting. These reasons are, of course, not only empirical but also normative. However, as participants in a specific form of life, we must take these reasons seriously. We can be

convinced by arguments that some of these life-world reasons are inadequate or incompatible with other reasons that we might consider to be more fundamental, but we can never leave the normative frame of the life form we participate in (Nida-Rümelin, 2009). The conceptual frame of TSR is better suited to integrate and systematize life-world practical reasons than conventional RCT.

There are duties that relate to social roles, obligations that are related to personal commitments, and still others that originate from prescriptions or rights. The idea that all of these *prima facie* duties could be reduced to the normative criterion of optimizing states of affairs, as in RCT, has a rationalistic turn of the worst kind. Compared to this, TSR is modest. It avoids philosophical and theoretical hypocrisy; it takes practices and persons seriously but, nevertheless, preserves the mathematical core of decision theory (Nida-Rümelin, 1997a). The agent acts rationally if the series of her singular acts constitutes an intended structure (a pattern of actions) that she can accept as part of her *form of life*.[4] Philosophical or economic theory should not prescribe which practical reasons constitute a form of life.

Notes

The main body of the text was written by Julian Nida-Rümelin, whereas Rebecca Gutwald and Niina Zuber surveyed the literature and completed the text, focusing on alternative accounts of structural rationality in the contemporary debate.

1. In fact, this term was introduced much earlier in German philosophy in the context of the critique of consequentialism in ethics and rationality theory (cf. Nida-Rümelin, 1993; accompanied by various articles in the 1990s such as Nida-Rümelin, 1991, 1994, 1997b, 1997c, 1997d). The core elements of structural rationality were later presented in an essay on structural rationality (Nida-Rümelin, 2001; English version: Nida-Rümelin, 2019, Part I). It has been extended to a general theory of practical reason (Nida-Rümelin, 2020).

2. Benjamin Kiesewetter undertakes a thorough analysis of structural irrationality and structural requirements of rationality in his *The Normativity of Rationality* (Kiesewetter, 2017).

3. In fact, infinitely many of those functions can be transformed into one another by positive linear transformations.

4. See also Korsgaard (1996) for an account of autonomy and rationality.

References

Andreoni, J., & Miller, J. H. (1993). Rational cooperation in the finitely repeated prisoner's dilemma: Experimental evidence. *Economic Journal, 103*(418), 570–585.

Aristotle. (2000). *Nicomachean ethics* (W. D. Ross, Trans.). Adelaide, Australia: University of Adelaide Library.

Axelrod, R. (2006). *The evolution of cooperation*. New York, NY: Basic Books.

Barreda-Tarrazona, I., Jaramillo-Gutiérrez, A., Pavan, M., & Sabater-Grande, G. (2017). Individual characteristics vs. experience: An experimental study on cooperation in prisoner's dilemma. *Frontiers in Psychology, 8*, 1–13.

Bentham, J. (2007). *An introduction to the principles of morals and legislation*. Mineola, NY: Dover. (Original work published 1789)

Bratman, M. E. (2007). *Structures of agency*. Oxford, England: Oxford University Press.

Bratman, M. E. (2014). *Shared agency: A planning theory of acting together*. Oxford, England: Oxford University Press.

Broome, J. (1999). Normative requirements. *Ratio, 12*, 398–419.

Clark, K., & Sefton, M. (2001). The sequential prisoner's dilemma: Evidence on reciprocation. *Economic Journal, 111*(468), 51–68.

Cooper, R., Dejong, D. V., Forsythe, R., & Ross, T. W. (1996). Cooperation without reputation: Experimental evidence from prisoner's dilemma games. *Games and Economic Behavior, 12*, 187–218.

Diamond, P., & Vartiainen, H. (Eds.). 2007. *Behavioral economics and its applications*. Princeton, NJ: Princeton University Press.

Frankfurt, H. G. (1971). Freedom of the will and the concept of a person. *Journal of Philosophy, 68*, 5–20.

Gigerenzer, G., & Selten, R. (Eds.). (2001). *Bounded rationality: The adaptive toolbox* (Dahlem Workshop Reports). Cambridge, MA: MIT Press.

Gilbert, M. (2013). *Joint commitment: How we make the social world*. Oxford, England: Oxford University Press.

Gintis, H., Bowles, S., Boyd, R., & Richerson, P. J. (2005). The evolution of altruistic punishment. In H. Gintis, S. Bowles, R. Boyd, & E. Fehr (Eds.), *Moral sentiments and material interests: The foundations of cooperation in economic life* (pp. 215–227). Cambridge, MA: MIT Press.

Hare, R. M. (1963). *Freedom and reason*. Oxford, England: Oxford University Press.

Hare, R. M. (1970). *The language of morals*. Oxford, England: Oxford University Press.

Hare, R. M. (1984). *Moral thinking: Its levels, method, and point*. Oxford, England: Clarendon Press.

Hargreaves-Heap, S. P., & Varoufakis, Y. (2004). *Game theory: A critical introduction*. London, England: Routledge.

Harsanyi, J. C. (1967). Games with incomplete information played by "Bayesian" players. *Management Science, 14*, 159–182.

Harsanyi, J. C. (1977a). *Morality and the theory of rational behavior*. Oakland: University of California Press.

Harsanyi, J. C. (1977b). *Rational behavior and bargaining equilibrium in games and social situations*. Cambridge, England: Cambridge University Press.

Harsanyi, J. C. (1977c). Rule utilitarianism and decision theory. *Erkenntnis, 11*(1), 25–53.

Harsanyi, J. C. (1979). *Rule utilitarianism, rights, obligations, and the theory of rational behavior*. Berkeley: Center for Research in Management Science, University of California.

Hegel, G. W. F. (2017). *Grundlinien einer Philosophie des Rechts* [Elements of the philosophy of right]. Hamburg, Germany: Meiner. (Original work published 1820)

Hurley, S. (2005). Rational agency, cooperation, and mindreading. In N. Gold (Ed.), *Teamwork: Multi-disciplinary perspectives* (pp. 200–215). New York, NY: Palgrave Macmillan.

Kahneman, D., Slovic, P., & Tversky, A. (1982). *Judgment under uncertainty: Heuristics and biases*. Cambridge, England: Cambridge University Press.

Kahneman, D., & Tversky, A. (2000). *Choices, values, and frames*. Cambridge, England: Cambridge University Press.

Kant, I. (1785). *Grundlegung zur Metaphysik der Sitten* [Groundwork of the metaphysics of morals] (Akademie-Ausgabe, Vol. IV, pp. 385–463). Berlin, Germany: Königlich Preußische Akademie der Wissenschaften.

Kiesewetter, B. (2017). *The normativity of rationality*. Oxford, England: Oxford University Press.

Korsgaard, C. (1996). *The sources of normativity*. Cambridge, England: Cambridge University Press.

Kreps, D. M., Milgrom, P., Roberts, J., & Wilson, R. (1982). Rational cooperation in the finitely repeated prisoners' dilemma. *Journal of Economic Theory, 27*(2), 245–252.

Langlois, D. J. (2014). *The normativity of structural rationality* (Unpublished doctoral dissertation). Harvard University, Cambridge, MA. Retrieved from http://nrs.harvard.edu/urn-3:HUL.InstRepos:13067678

Luce, R. D., & Raiffa, H. (1957). *Games and decision*. New York, NY: Wiley.

Marschak, J. (1946). Neumann's and Morgenstern's new approach to static economics. *Journal of Political Economy, 54*(2), 97–115.

McClennen, E. F. (1990). *Rationality and dynamic choice*. Cambridge, England: Cambridge University Press.

Nida-Rümelin, J. (1991). Practical reason or meta-preferences? An undogmatic defense of Kantian morality. *Theory and Decision, 30*(2), 133–162.

Nida-Rümelin, J. (1993). *Kritik des Konsequentialismus* [Criticizing consequentialism]. Munich, Germany: Oldenbourg.

Nida-Rümelin, J. (1994). Rational choice: Extensions and revisions. *Ratio, 7*(2), 122–144.

Nida-Rümelin, J. (1997a). *Economic rationality and practical reason.* Dordrecht, Netherlands: Kluwer.

Nida-Rümelin, J. (1997b). Structural rationality, democratic citizenship and the new Europe. In P. B. Lehning & A. Weale (Eds.), *Citizenship, democracy and justice in the new Europe* (pp. 34–49). London, England: Routledge.

Nida-Rümelin, J. (1997c). Structural rationality in game theory. In W. Leinfellner & E. Köhler (Eds.), *Game theory, experience, rationality: Foundations of social sciences, economics and ethics: In honor of John C. Harsanyi* (pp. 81–93). Dordrecht, Netherlands: Kluwer.

Nida-Rümelin, J. (1997d). Why consequentialism fails. In G. Holmström-Hintikka & R. Tuomela (Eds.), *Contemporary action theory: Vol. 2. Social action* (pp. 295–308). Dordrecht, Netherlands: Kluwer.

Nida-Rümelin, J. (2000). Rationality: Coherence and structure. In J. Nida-Rümelin & W. Spohn (Eds.), *Rationality, rules, and structures* (pp. 1–16). Dordrecht, Netherlands: Kluwer.

Nida-Rümelin, J. (2001). *Strukturelle Rationalität* [Structural rationality]. Stuttgart, Germany: Reclam.

Nida-Rümelin, J. (2005a). *Über menschliche Freiheit* [On human freedom]. Stuttgart, Germany: Reclam.

Nida-Rümelin, J. (2005b). Why rational deontological action optimizes subjective value. *ProtoSociology, 21*, 182–193.

Nida-Rümelin, J. (2009). *Philosophie und Lebensform* [Philosophy and life-form]. Frankfurt/Main, Germany: Suhrkamp.

Nida-Rümelin, J. (2014). Structural rationality and collective intentionality. In S. R. Chant, F. Hindriks, & G. Preyer (Eds.), *From individual to collective intentionality: New essays* (pp. 207–222). Oxford, England: Oxford University Press.

Nida-Rümelin, J. (2019). *Structural rationality and other essays on practical reason* (Theory and Decision Library A: Rational Choice in Practical Philosophy and Philosophy of Science). Cham, Switzerland: Springer.

Nida-Rümelin, J. (2020). *Eine Theorie praktischer Vernunft* [A theory of practical reason]. Berlin, Germany: De Gruyter.

Pothos, E. M., Perry, G., Corr, P. J., Matthew, M. R., & Busemeyer, J. R. (2001). Understanding cooperation in the prisoner's dilemma game. *Personality and Individual Differences, 51*(3), 210–215.

Ramsey, F. P. (1978). *Foundations: Essays in philosophy, logic, mathematics and economics* (D. H. Mellor, Ed.). London, England: Routledge & Kegan Paul.

Rapoport, A., Chammah, A. M., & Orwant, C. J. (1956). *Prisoner's dilemma: A study in conflict and cooperation.* Ann Arbor: University of Michigan Press.

Rawls, J. (1971). *A theory of justice.* Cambridge, MA: Harvard University Press.

Regan, D. (1980). *Utilitarianism and co-operation.* Oxford, England: Clarendon Press.

Ross, W. D. (2002). *The right and the good.* Oxford, England: Clarendon Press. (Original work published 1930)

Scanlon, T. M. (2007). Structural irrationality. In G. Brennan, R. Goodin, F. Jackson, & M. Smith (Eds.), *Common minds: Themes from the philosophy of Philip Pettit* (pp. 84–103). Oxford, England: Clarendon Press.

Seabright, P. (2004). *The company of strangers: A natural history of economic life.* Princeton, NJ: Princeton University Press.

Searle, J. R. (1969). *Speech acts: An essay in the philosophy of language.* Cambridge, England: Cambridge University Press.

Sen, A. (1974). Choice, orderings and morality. In S. Körner (Ed.), *Practical reason* (pp. 54–67). Oxford, England: Blackwell.

Sethi, R., & Somanathan, E. (2005). Norm compliance and strong reciprocity. In H. Gintis, S. Bowles, R. Boyd, & E. Fehr (Eds.), *Moral sentiments and material interests: The foundations of cooperation in economic life* (pp. 229–250). Cambridge, MA: MIT Press.

Singer, P. (1993). *Practical ethics.* Cambridge, England: Cambridge University Press.

Singer, P. (2016). *The most good you can do: How effective altruism is changing ideas about living ethically.* New Haven, CT: Yale University Press.

Spohn, W. (2002). The many facets of the theory of rationality. *Croatian Journal of Philosophy, 2*(3), 249–264.

Taylor, C. (1975). *Hegel.* Cambridge, England: Cambridge University Press.

Taylor, C. (1989). *Sources of the self: The making of the modern identity.* Cambridge, England: Cambridge University Press.

Tuomela, R. (2013). *Social ontology: Collective intentionality and group agents.* Oxford, England: Oxford University Press.

von Neumann, J., & Morgenstern, O. (1947). *Theory of games and economic behavior.* Princeton, NJ: Princeton University Press.

Wallace, R. J. (2018). Practical reason. In E. N. Zalta (Ed.), *The Stanford encyclopedia of philosophy.* Retrieved from https://plato.stanford.edu/archives/spr2018/entries/practical-reason/

10.6 Adaptationism: A Meta-Normative Theory of Rationality

Leda Cosmides and John Tooby

Summary

Human cognition is often compared—unfavorably—to normative theories of rationality from mathematics, logic, economics, or philosophy. But what justifies *these* normative theories as the proper standard for assessing the rationality of an evolved computational system? The adaptationist program in evolutionary biology provides a meta-normative theory of rationality that is appropriate for evolved organisms. The functional design of our cognitive architecture was built by natural selection, to solve information-processing problems that are strange, exact, and nonintuitive. Theories of adaptive function—task analyses of these problems—provide normative standards of good design for assessing the rationality of human cognition. With five case studies, we show how this approach can reveal sophisticated cognitive mechanisms that would otherwise remain undetected. At the same time, these cases illustrate pitfalls of studying reasoning and choice without reference to the ancestral problems and environments that selected for their design.

1. The Paradox of Human Reasoning

How the mind works can be illuminated by comparing human cognition to standards of good design specified by a normative theory. But which standards are appropriate for an evolved organism? What counts as a rational inference or choice for animals like us, whose minds were designed by natural selection?

The first normative theories used in cognitive psychology were those deemed rational by mathematicians, scientists, economists, and philosophers. Deductive and inductive reasoning were compared to standards of rationality drawn from first-order logic, Bayes' theorem, or statistical principles. The manner in which people evaluate hypotheses was compared to Popper's standard: look for violations, not confirmation. Decision making was compared to Neyman–Pearson decision theory, expected utility theory, or, when choosing among bundles of goods, the axiom of transitivity (e.g., GARP).

In this research tradition, reasoning errors are defined as having occurred whenever human judgment departs from what these normative theories dictate. In their seminal studies of judgment under uncertainty, Kahneman and Tversky (1982) articulated this standard: "The presence of an error in judgment is demonstrated by comparing people's responses either with an established fact (e.g., that the two lines are equal in length) or with an accepted rule of arithmetic, logic, or statistics" (p. 123).

When cognitive performance was assessed by these normative standards, the rationality of human reasoning, judgment, and decision making was found wanting. Psychologists found systematic errors in deductive reasoning and hypothesis testing, violations of probability theory, preference reversals, and "excess" altruism. These "errors" were attributed to a flotilla of heuristics, biases, and emotions—cognitive processes thought to be fast, automatic, unconscious, emotional, associative, and/or stereotypic. This heterogeneous collection of processes was called "System 1." Rational thinking was attributed to "System 2": slow, effortful, conscious deliberations that carry out calculations and logical inferences (Kahneman, 2011).

This research program has produced a formidable paradox. Reviews of human reasoning research are customarily presented as lengthy catalogs of errors, fallacies, biases, heuristics, and emotions. Textbooks organize their discussions around the seemingly endless ways in which human reasoning and decision making depart from the normative ideals of rationality used in science, mathematics, economics, and philosophy. Yet evolved reasoning systems—human and nonhuman minds alike—negotiate the complex natural tasks of their world with a level of operational success far surpassing that of the most sophisticated existing artificial intelligence (AI) systems. It is trivial to equip AI systems with algorithms that implement statistical decision theories, inferences of first-order logic, and other formal methods of rational inference. They can search vast problem spaces in seconds, store massive databases for analysis, and perform lightning calculations. But organisms perform better than these computational systems on virtually every natural inferential problem that has been carefully investigated—the induction of grammar and word meanings, speech perception, vision, color constancy, recognizing objects and making inferences about their interactions, and inferring the beliefs, desires, and behavior of other people, to name a few.

What is the resolution to this paradox? From the perspective of evolutionary biology, the problem is not that our thinking is irrational. The problem is how psychologists have been defining rationality and testing for its presence.

2. A Meta-Normative Theory of Rationality

Let us take one step back and ask a more fundamental question. Why identify rationality with adherence to normative theories from logic, mathematics, economics, and philosophy? What normative theory justifies the choice of these particular normative theories?

In common parlance, a choice, process, or behavior is considered *rational* when it is well designed for achieving a goal, and *irrational* when poorly designed for that function. It would be irrational to eat charcoal instead of fruit to sate hunger (but rational if you had just ingested a poison); it would be irrational to travel to a new location by randomly zigzagging instead of following a straight-line path (but rational if you are evading a predator); it would be irrational to converse out loud while alone (but rational if you are rehearsing a role). There is no goal-independent definition of what counts as rational.

The specialized (and historically recent) tools of scientists, mathematicians, economists, and philosophers were developed for very specific goals: producing knowledge, understanding choice, and guiding behavior in mass societies. From the 1400s to the present, increasing literacy, more reliable data storage technologies, and the decreasing cost of communication and travel have changed the world beyond anything known by our hunter–gatherer ancestors. Because of these developments, hypotheses and data became more widely shared, debated, vetted, and stored, enabling the cultural accumulation of knowledge in ever-larger populations; markets arose in which millions of anonymous individuals cooperated in complex networks; people from locations with different moral norms were brought into contact, in ever-widening circles of interaction. Applying standards of rationality created by scientists,

mathematicians, economists, and philosophers is useful for solving the evolutionarily novel problems created by these conditions, such as producing scientific knowledge, understanding market dynamics, and, perhaps, fostering cooperation in the modern world of mass societies and global contact. But what justifies privileging methods developed for these modern goals as *the* metric against which human rationality is measured?

Nothing. From an evolutionary perspective, these are the wrong goals against which to measure human rationality, because they played no causal role in selecting for the design of the mind. To decide whether a mechanism of human reasoning or decision making is rational—whether it is well engineered for producing a goal or outcome—we need to know what adaptive problem that mechanism was designed (by natural selection) to solve, and what kind of information was available for solving it in the environments that selected for its design.

3. The Adaptationist Program in Evolutionary Biology Provides a Sound Meta-Normative Theory of Rationality

From the molecular machinery of a single cell to the information-processing architecture of the human visual system, the most sophisticated engineering on Earth is found in organisms and was built by natural selection.

Organisms are composed of many machines: systems organized to solve a problem. *These systems evolved to solve problems in ways that improved the ability of the organism's ancestors to produce offspring.* Every piece of organic machinery—the retina, the heart, the lungs—acquired its intricate functional design over deep time, as a downstream consequence of the fact that organisms reproduce themselves.

3.1 Natural Selection Retains Designs That Promoted Their Own Reproduction

When individuals reproduce, replicas of their organic machinery develop in their offspring. But replication is not error-free: chance mutations introduce changes into the design of organic machines, altering their features. These entropic changes usually disrupt the efficiency with which this machinery solves problems, thus interfering with the mutant offspring's ability to produce offspring of its own. Because individuals with the newly modified, but now defective, design produce fewer offspring, on average, than those with the (more efficient) standard design, the defective design eventually disappears from the population—a case of negative feedback.

Occasionally, a mutation will improve a machine's operation in a way that promotes the reproduction of individuals with that mutation—and, therefore, the reproduction of that design. Such improved designs (by definition) cause their own increasing frequency in the population—a case of positive feedback. This increase continues until (usually) the modified design out-reproduces, and thereby replaces, all alternative designs in the population, leading to a new species-standard design. The population- or species-standard design has taken a step "uphill" toward a greater degree of functional organization for reproduction than it had previously.

Over the long run, down chains of descent, this feedback cycle—natural selection—pushes designs through state-space toward increasingly well-engineered functional arrangements. These arrangements are *functional* in a specific sense: the elements are well organized to cause their own reproduction in the environments in which the species evolved.

3.2 Adaptive Problems and Evolved Solutions

Enduring conditions in the world that create reproductive opportunities or obstacles constitute *adaptive problems*. Examples include the presence of predators, the possibility of sharing food to pool foraging risk, or the existence of a cognitive mechanism that could be repurposed for more efficient foraging. Enduring relationships of this kind constitute reproductive opportunities or obstacles in the following sense: *if* the organism had a new property that interacted with these conditions in just the right way, *then* this property would cause individuals who have it to produce more offspring that live to reproductive maturity, relative to those with alternative designs. These reproductive opportunities and obstacles can be thought of as *problems*. A property is a *solution* to such a problem when it allows organisms with this property to take advantage of prevailing conditions, where "advantage" means a reproductive advantage.

Adaptive problems have two defining characteristics. First, they are conditions or cause-and-effect relationships that many or most individual ancestors encountered, reappearing again and again during the evolutionary history of the species. Second, they are that subset of enduring relationships that could, in principle, be exploited by some new property of an organism to increase its reproduction or the reproduction of its relatives (who have a high probability of having inherited the same mutation).

An enduring adaptive problem constantly selects for design features that promote the solution to the problem. Over evolutionary time, more and more design features accumulate that fit together to form an integrated structure or device that is well engineered to solve that adaptive problem. Such a structure or device is called an *adaptation*. An adaptation may have many beneficial effects, but solving the adaptive problem that selected for its design is its *function*. The function of the heart is to pump blood; the function of the liver is to detoxify poisons; the function of the retina is to detect photons and transduce them into neural signals for vision. Adaptive problems that required information processing for their solutions selected for neural circuitry organized to compute these solutions: computational adaptations.

Adaptationism is the name of the research program that explores how natural selection functionally organizes the designs of organisms. Adaptationists start with a careful analysis of an adaptive problem—including information that would have been available in ancestral environments for solving that problem. *A careful task analysis of an adaptive problem serves as a normative theory of good design:* it identifies the problem, allowing one to see what counts as a good solution to it. Because a computational adaptation should be composed of algorithms and representations that are well engineered for solving the problem that selected for its design, a good theory of adaptive function suggests hypotheses about the design of the mind. These can be tested empirically, revealing computational operations that were previously unknown.

The adaptationist perspective provides principled, nonarbitrary criteria for judging rationality. If a choice, process, or behavior is *rational* when it is well designed for achieving a goal, then a computational adaptation that generates inferences or choices is rational when its features are well tailored for solving an adaptive problem faced by the hunter–gatherers from whom we are descended.

The study of reasoning and choice looks very different when viewed from this perspective. Many fallacies disappear, and well-designed reasoning appears, when cognitive performance is compared to theories of adaptive function: normative standards of good design derived from evolutionary biology, behavioral ecology (of hunter–gatherers, nonhuman primates, and other animals), and studies of ancestral environments (from paleoanthropology, archaeology, physics, geology, botany, zoology, and other sources). Using five case studies of reasoning and choice, we illustrate pitfalls that arise from the failure to apply adaptationist thinking. Cases 3–5 tap theories of adaptive function that are normative in evolutionary biology but have no counterpart in traditional normative theories.

4. Case 1: To Discover a Well-Designed Reasoning Mechanism, You Need to Present Stimuli in an Ecologically Valid Format

Behavioral ecologists have sophisticated models of the adaptive problems animals face when foraging. Because efficient foraging requires animals to make judgments under uncertainty, these models incorporate elements of probability theory. Behavioral ecologists typically find that insects, birds, and other animals behave like good Bayesians when they make foraging decisions (e.g., Real, 1991; Stephens & Krebs, 1986). Yet psychologists thought that the human mind was "too limited" in capacity to do the same (Kahneman, Slovic, & Tversky, 1982). Why the difference?

Computational adaptations should be *ecologically rational*: designed to work well in the ecological circumstances that selected for their design. Consider, for example, how well your visual system maintains color constancy given normal variations in light cast by the sun: your green car looks green all day, even at sunset, when it is bathed in "red" (long-wavelength) light. Yet these sophisticated mechanisms fail—your green car looks brown—in the ecologically novel spectrum cast by sodium vapor streetlights. When psychologists first started to study Bayesian reasoning, the stimuli they used were the cognitive equivalent of sodium vapor lights: probabilities of single events (e.g., there is a 4% chance you will find an apple tree in this orchard) and normalized frequencies (36% of the trees are cherry trees). These data formats are the recent cultural product of modern data gathering and statistical techniques.

When asked to compute a conditional probability based on such data (e.g., What is the chance that a tree with red fruit is an apple tree?), people fail spectacularly. But they succeed when given natural frequencies: the *absolute* frequencies of events *as you encounter them* in the world (Gigerenzer & Hoffrage, 1995, 1999). Our minds automatically encode natural frequencies, which were the only kind of probability data available in the ecology of our hominin ancestors (Cosmides & Tooby, 1996). Imagine an orchard with 125 fruit trees—apple, cherry, lemon, and pear. On a stroll, you encounter 50 trees with red fruit: 5 apple trees and 45 cherry trees. Given these natural frequencies, most people realize that 5 out of 50 trees with red fruit are apple trees—the conditional probability that a tree with red fruit is an apple tree (hits / (hits + false alarms)). Note that the low base rate of apple trees (4%) is implicit in the small number of them that you encounter as you walk through the orchard (thereby taking a "natural sample"; Kleiter, 1994). Adaptations for making probability judgments

can safely ignore normalized data about base rates (as people famously do) if they are designed to use natural frequencies derived from natural samples. (On the fit between cognitive mechanisms and their task environment, see chapter 8.5 by Hertwig & Kozyreva, this handbook.)

Our mental mechanisms do produce judgments that match normative standards from probability theory—when they are fed data in the format they were designed to use. The practical relevance for modern environments is clear: a simple change in how data are communicated can help patients and policy makers make rational decisions about risk (Gigerenzer, 2014).

5. Case 2: Judgments That Look Like Normative Violations May Be Very Well Designed for Solving a Recurrent Adaptive Problem—One the Scientist Is Not Considering

In choosing between risky and sure options, many people behave as if "losses loom larger than gains": they are willing to take a bigger risk to avoid *losing* $100 than to *gain* the same amount. To appreciate the puzzle, imagine there is an outbreak of a disease that will kill 600 people if nothing is done. Which of two programs to combat the disease would you favor?

If program A is adopted, *400 people will die.* If program B is adopted, there is 1/3 probability that *nobody will die,* and 2/3 probability that *600 people will die.* (italics added)

Expected utility theory predicts indifference, because both programs have the same expected value. With program A, 400 people die and 200 survive; program B produces the same result, on average. But Tversky and Kahneman (1981) found that people are not indifferent: when the options were framed as lives lost, most people (∼ 80%) chose the risky option, program B.

Even more puzzling: this preference reversed—most people (∼ 70%) chose the sure option, program A—when the same options were framed as gains:

If program A is adopted, *200 people will be saved.* If program B is adopted, there is 1/3 probability that *600 people will be saved,* and 2/3 probability that *no people will be saved.*

Why? The loss and gain frames express logically and mathematically identical situations. Preference reversals are considered irrational: they violate the transitivity axiom of expected utility theory.

Because expected utility theory cannot explain these choices, psychologists and economists proposed that people have a stable taste: an aversion to loss. But a normative theory from behavioral ecology explains these results—and correctly predicts when they will flip.

5.1 Risk-Sensitivity Theory, an Alternative Analysis of Rational Decision Making

Consider a bird who is deciding to forage on one of two patches. Both patches yield about 200 seeds per day of foraging—they have the same expected value—but they differ in variance. The low-variance patch yields close to 200 seeds each day. The yield of the high-variance patch varies wildly from one day to the next, from 50 to 400 seeds.

Now consider the bird's needs. If the bird needs to find 100 seeds today to live to tomorrow—a number below the expected value—the safest bet is to forage on the low-variance patch. But this is a bad bet if the bird needs 300 seeds to live another day: a rational bird will forage on the high-variance patch when its need is above the expected value. According to risk-sensitivity theory, a normatively correct decision takes three variables into account: (1) the expected value of each option, (2) the outcome variance associated with each option, and (3) the decision maker's need level. A rational decision system will be designed to minimize the probability of an outcome that fails to satisfy one's need (Stephens & Krebs, 1986).

To test risk-sensitivity theory in humans, the experimenter must vary the minimum need level before subjects make a decision. When no need level is specified, the subject is free to fill in the blank any way she wishes. For monetary gambles, maintaining the status quo is as good a minimum need level as any; a rational individual who does not want to fall below the status quo will avoid the risky option and choose the sure one. But this is not because she has an aversion to loss—a stable taste—that is independent of context.

Like birds, people choose rationally when their minimum need level is specified (Mishra & Fiddick, 2012; Rode, Cosmides, Hell, & Tooby, 1999). When it is *above* the expected value, people *prefer* the risky option. They choose the sure/low-risk option—thereby appearing to be loss averse—only when they need less than the expected value. Framing affects choice on the disease problem by changing the minimum number of lives people feel they *must* save: saying that 400 people will die leads people to set a higher threshold than saying that 200 will survive (Mishra & Fiddick, 2012). As a bonus, the same theory shows that "ambiguity avoidance" is also a myth (Camerer & Weber, 1992). Ambiguity is usually interpreted as "risky," but when the context implies it is the lower-variance option, people *prefer* the ambiguous option (Rode et al., 1999).

A tangle of results that look irrational when compared to traditional normative theories turn out to be rational responses to a problem that scientists were not considering.

6. Case 3: Cues of a Reasoning System's Proper Domain—the Context for Which It Evolved—May Be Necessary to Activate Its Procedures. To Know Its Proper Domain, One Needs to Correctly Characterize the System's Adaptive Function

An adaptation's *proper domain* is the information that the mechanism was designed by natural selection to process (Sperber, 1994). Ground squirrels, for example, evolved to produce alarm calls when they see a hawk overhead. The sight of a hawk is a cue that the squirrel is in danger from raptors *now*. This cue elicits the correct alarm call from the squirrel's repertoire (snakes and predatory cats elicit different calls). The approach of predatory birds is the adaptation's proper domain.

The raptor call is also activated when zoologists fly a drone with the silhouette of a hawk overhead. The alarm call's *actual* domain is larger than its proper domain: it consists of all cues that activate the call, real hawks (proper domain) and hawk silhouettes alike. In studying a reasoning mechanism in humans, experimenters need to consider (i) the mechanism's evolved function, to ascertain its proper domain, and (ii) whether the experimental stimuli present cues from its actual domain— ones sufficient to activate the mechanism. Mercier and Sperber (2011) have argued that most studies finding poor logical reasoning lack cues to the mechanism's proper domain: devising and evaluating arguments intended to persuade or dissuade another person.

Eliciting poor logical reasoning from people is easy: give them a problem, without context, that requires them to produce a valid inference using modus tollens (MT). MT is an inference rule from first-order logic. Given a conditional rule, such as "If the pipe was fixed, then the bathroom floor will be dry in the morning," and a premise, such as "The bathroom floor is wet this morning," it is logically correct to infer "The pipe was not fixed" (more generally: *If P then Q; not-Q; therefore not-P*). It is trivial to program a computer to produce an MT inference. Yet people fail to do so about 75% of the time (Rips, 1994; Wason & Johnson-Laird, 1972).

But do our minds lack an MT inference rule? Or do indicative conditional rules stripped of context fail to activate it? The mind does seem to implement certain rules of first-order logic (e.g., people usually apply the modus ponens rule, correctly inferring Q from *If P then Q* and *P*). If the human mind has a system designed for reasoning logically, what adaptive problem did it evolve to solve? Mercier and Sperber (2011) have proposed that conscious, deliberative reasoning—including logical reasoning—evolved to solve adaptive problems that arise during communication.

6.1 Biologists Have a Normative Theory of Communication

A seminal paper on the evolution of communication by Krebs and Dawkins (1984) changed how biologists study animal communication. A communication system will not evolve unless it confers a net benefit on both senders and receivers—if a signal benefits only senders, receivers will not evolve mechanisms to decode it. But once a signaling system has evolved, situations will arise in which senders can manipulate receivers, to their own advantage, by sending deceptive signals. If acting on deceptive signals is costly to the fitness of the receiver, selection will favor adaptations in receivers to detect which signals are deceptive. As receivers get better at distinguishing honest from deceptive signals, selection will favor adaptations in senders for producing deceptive signals that are more difficult to detect, and so on. A coevolutionary arms race ensues over generations: receivers get better and better at detecting deceptive signals, and senders get better and better at making deceptive signals resemble honest ones. By contrast, the fitness interests of receivers and honest signalers converge; this favors the evolution of honest signals that are difficult for a deceptive sender to fake (Higham, 2014).

Language presents additional problems, however, because sending a deceptive message is no more costly (in words produced) than sending an honest one. Detecting the veracity of a signal requires logical reasoning, according to Mercier and Sperber, so that receivers can maintain "epistemic vigilance." While the receiver is evaluating the sender's claim, it is stored in a special data format—a meta-representation—which is decoupled from the receiver's semantic memory (her database of knowledge). Logical reasoning is necessary for her to detect whether the claim contradicts other claims made by the sender, claims made by other people, and facts she already knows. This evaluation process is particularly important, they claim, in arguments intended to persuade the receiver to act on beliefs she does not yet hold or dissuade her from actions she wants to take.

6.2 Argumentative Context: A Cue That Activates Logical Reasoning

When the sender and receiver already agree, there is no argument—no attempt to persuade or dissuade. If they disagree and an argument ensues, there are two possibilities: (i) one (or both) of them are misinformed (and they need more accurate information), or (ii) a deceptive message is being sent, intended to manipulate the receiver into doing something against her interests that benefits the sender. Both call for epistemic vigilance by the receiver: the claims must be evaluated for their truth value. The presence of an argument is, therefore, a cue that should activate logical reasoning in the receiver. It should also activate logical reasoning in the sender, who must devise an argument that will persuade the receiver.

This has empirical implications. Divorced from the context of an argument, the MT inference may not be activated. But it can be easily activated in an argumentative context. Let's say I am arguing with the plumber, who claims that he fixed the leaky pipe in my powder room. I doubt him, and deny his claim by saying, "Oh yeah? We'll see! *If the pipe was fixed, then the bathroom floor will be dry in the morning.*" When I discover that the bathroom floor is wet in the morning, we will both conclude that the pipe was not fixed: we easily make the MT inference.

Mercier and Sperber review evidence in favor of their hypothesis that an argumentative context is the proper domain for reasoning, logical and otherwise. Their claim is not that reasoning will be flawlessly logical in all arguments: an argumentative context will activate logic when the reasoner must be epistemically vigilant, but confirmation bias and other infelicities will emerge when sound reasoning would undermine one's attempt to persuade. In their words, "In all these instances traditionally described as failures or flaws, reasoning does exactly what can be expected of an argumentative device: Look for arguments that support a given conclusion, and, ceteris paribus, favor conclusions for which arguments can be found" (Mercier & Sperber, 2011, p. 57).

7. Case 4: A Cognitive Adaptation for Reasoning May Be Specialized for a Specific Domain and, Therefore, Equipped with Procedures That Are Content Rich Rather Than Content Free

Normative theories of rationality from logic, mathematics, and economics typically posit content-free reasoning procedures: ones that operate uniformly on information from every domain. Such procedures—*domain-general* ones—do exist in the human mind (automatic frequency computation is an example), but they cannot solve even routine adaptive problems by themselves (Cosmides & Tooby, 1987; Tooby & Cosmides, 1992). Domain-specialized reasoning systems were also required. These are equipped with content-rich concepts, inference procedures, and decision rules, ones that are superbly engineered for producing fitness-promoting inferences in one ancestral domain but do not apply outside it (e.g., concepts like *belief* and *desire* are useful for predicting the behavior of people, but not rocks). Discovering these systems requires careful analysis of an adaptive problem to see what counts as a functional (i.e., normative) solution.

The evolution of cooperation for mutual benefit—social exchange—poses exacting problems, which have been modeled extensively. In humans, these problems are solved by a functionally specialized reasoning system that deploys content-rich representations and procedures. Its features were revealed by experiments that tested hypotheses derived from evolutionary game theory—hypotheses that were constructed in advance of collecting any data.

7.1 Evolutionary Game Theory Is a Source of Normative Theories

Game theory is a tool for analyzing strategic social behavior—how agents will behave when they are interacting with others who can anticipate and respond to their behavior (chapter 9.3 by Alexander, this handbook). Economists use it to analyze how people respond to incentives present in the immediate situation. Their models typically assume rational actors, who calculate the payoffs of alternative options (anticipating that other players will do likewise) and choose the option most likely to maximize their short-term profits (but see Hoffman, McCabe, & Smith, 1998).

Evolutionary biologists also adopted game theory as an analytic tool, but with a twist (Maynard Smith, 1982). Evolutionary game theory does not assume economically rational agents who can reason about the reasoning of other agents via "backward induction." It can be usefully applied to cooperation among bacteria or fighting in spiders. It is used to model interactions among agents endowed with well-defined decision rules that produce situationally contingent behavior. Although these decision rules are sometimes called "strategies" by evolutionary biologists, no conscious deliberation by bacteria (or humans) is implied (or ruled out) by this term. Sometimes results are derived analytically; in more complex cases, agent-based simulations of natural selection are used.

Whether the decision rules being analyzed are designed to regulate foraging, fighting, or cooperating, the immediate payoffs of these decisions, in food or resources, are translated into the currency of offspring produced by the decision-making agent, and these offspring inherit their parent's decision rule. In evolutionary game theory, a decision rule or strategy that garners higher payoffs leaves more copies of itself in the next generation than alternatives that garner lower payoffs. By analyzing the reproductive consequences of alternative decision rules over generations, evolutionary biologists can determine which strategies natural selection is likely to favor and which are likely to be selected out. This source of normative theories can be tapped for any domain in which organisms make consequential choices.

The evolution of cooperation has been extensively modeled using these tools, with clear results. Strategies that indiscriminately provide benefits to others—including those who do not reciprocate—are eventually eliminated from the population. Because they incur reproductive costs without compensating benefits, they are outcompeted by designs that cheat—that accept benefits without reciprocating them. Selection favors strategies that cooperate conditionally: ones that cooperate with other cooperators and withdraw cooperation from cheaters (e.g., Axelrod, 1984; Trivers, 1971). A *cheater* is an agent endowed with decision rules that accept benefits offered by other agents without satisfying their requirements. Innocent mistakes do not reveal a cheater; a cheater is an agent who violates a social exchange agreement by design (by virtue of its decision rules), not by accident.

This analysis carries many implications about how a system must be designed to implement a strategy for conditional cooperation. The most straightforward: for social exchange to evolve, agents must have mechanisms that make them very good at searching for information that would reveal cheaters. A content-free logic, deontic or otherwise, cannot do the job (e.g., Cosmides & Tooby, 2008).

7.2 Information Search: Looking for Cheaters versus Looking for Logical Violations

The philosopher Karl Popper proposed a normative standard for evaluating hypotheses: look for violations, not confirmation. "All swans are white" is violated by finding a single black swan, no matter how many white swans you have so far observed. (Realizing this requires the MT inference.) Peter Wason developed his four-card selection task to find out if people are natural falsificationists, who look for cases that could violate a hypothesis that is presented as a conditional rule (*If P then Q*). His research suggests that we are not (Wason & Johnson-Laird, 1972).

Imagine, for example, that you are given incomplete information about four birds—two of unknown color (a swan and a parrot) and two of unknown species (one black and one white). Which birds should you investigate further to see if any of them violate the rule "If a bird is a swan, then it is white"? Most people want to investigate the swan, to learn its color. That is logically correct: discovering a black swan would violate the rule. But that implies you should also investigate the black bird, to learn its species: it too might be a black swan. Most people fail to investigate the black bird, and many want to investigate the white one (unnecessary: white birds—swan, dove, parrot—cannot violate the rule). Only cases of *P & not-Q* (here, swans that are not white) can violate *If P then Q*. Most people recognize this when asked, but they do not spontaneously use logic to *search* for violations of indicative rules. Fewer than 25% of people seek information about *P, not-Q*, and no other case.

By contrast, 65% to 80% of people successfully look for violations—cheaters—when the rule involves social exchange. The mind interprets a conditional rule as a *social contract* when it expresses an agreement to cooperate for mutual benefit, for example, "If you borrow my car, then you must fill the tank with gas" (or, more generally, "If you accept *benefit B* from agent *J*, then you must satisfy *J*'s *requirement R*"). It becomes obvious that one needs to investigate the guy who borrowed the car (*P*) and the one who did not fill the tank (*not-Q*).

7.3   A Content-Rich Adaptive Logic

Wason tasks involving social exchange activate a cognitive adaptation that evolved for detecting cheaters. It is part of a computational system that is specialized for reasoning about social exchange (for a review of the evidence, see Cosmides & Tooby, 2015). The social exchange system dissociates, both functionally and neurally, from reasoning systems that are activated by content tapping other domains (including precautionary rules, which are so similar to social contracts that most theories do not distinguish them). It represents social contracts using content-rich proprietary concepts (e.g., *agent_i, benefit to agent_i, requirement of agent_i, obligation, entitlement, cheater*). Its procedures operate on these representations, producing inferences appropriate to social exchange that are not licensed by content-free logics (deontic or otherwise). Its cheater detection mechanism attends to information that would reveal cheaters, whether the resulting answer is logically correct or not. And that mechanism looks for *cheaters*—innocent mistakes do not elicit violation detection. The design features of this content-rich system are normatively correct: they are precisely tailored for their adaptive function.

8.   Case 5: Choices That Deviate from Economic Rationality May Be the Most Adaptive, Fitness-Promoting Strategy—Not a "Bias"

Economists have built models in which rational individuals and firms behave "as if" they were maximizing profits (or, more accurately, utility functions). A utility function is a mathematical device for understanding the dynamics of markets in which millions of anonymous individuals cooperate for mutual benefit: it transforms changes in inputs (e.g., the price of corn) into changes in outputs (e.g., the quantity of tortilla chips produced). These models are successful for their intended purpose. But they fail when they are used to predict the behavior of individuals cooperating in small groups (Smith, 2003). Is this another failure of human rationality?

8.1   A Puzzle from Behavioral Economics

Cooperation can be studied in the laboratory by having people interact in games in which the monetary payoffs for different choices are carefully controlled—dictator games, prisoner's dilemma games, bargaining games (e.g., the ultimatum game), trust/investment games, public goods games, and others. When behavioral economists used these methods to test predictions of game theory, they found that people in small groups do not act as if they are maximizing immediate monetary payoffs (e.g., Hoffman et al., 1998). In a one-shot interaction with anonymous others, *Homo economicus* models predict no generosity, no cooperation, no trust, and no punishment. Yet people give more, cooperate more, trust more, and punish defections more than these models predict, even when the experimenter tells them that the interaction is one-shot and anonymous. But why? Generosity in anonymous, one-shot games is irrational, given the standard theories. According to both economic *and* evolutionary game theory, repeated interactions are necessary for behaviors like this to evolve.

This "excess altruism" is considered irrational on many economic theories, and some social scientists have viewed it as evidence that the psychology of cooperation was shaped by group selection rather than selection operating on individuals (e.g., Bowles & Gintis, 2013). But are these behaviors really *excess* altruism—that is, beyond what can be explained by selection on individuals for direct reciprocity?

8.2   Adaptationist Game Theory: Model the Information-Processing Problem and Let the Psychology Evolve in Response

Selection does not occur in a vacuum: the physical and social ecology of a species shapes the design of its adaptations, and our hunter–gatherer ancestors lived in small, interdependent bands, and had many encounters with individuals from neighboring bands. Adaptations for direct reciprocity evolved to regulate cooperation in an ancestral world in which most interactions were repeated. The high prior probability that any given interaction will be repeated should be reflected in their design. So should the fact that you have interacted at least once with a person: models of this social ecology show that meeting an individual once is itself a good cue that you will meet again (Krasnow, Delton, Tooby, & Cosmides, 2013).

Whether an interaction is one-shot or repeated is a judgment made under uncertainty. What decision rules does selection favor when this judgment-under-uncertainty problem is modeled?

8.3   The Evolution of Generosity under Uncertainty

In most agent-based simulations of the evolution of cooperation, the behavioral strategies are particulate—they do not have internal cognitive components that can evolve—and the simulation environment has either one-shot or repeated interactions, but not both. But what happens if these strategies have a cognitive architecture that can evolve, and the social environment includes both one-shot *and* repeated interactions, as in real life? It turns out that generosity in one-shot interactions evolves easily when natural selection shapes decision systems for regulating two-person reciprocity (exchange) under conditions of uncertainty (Delton, Krasnow, Cosmides, & Tooby, 2011).

In real life, you never know with certainty that you will interact with a person once and only once. Categorizing an interaction as one-shot or repeated is always a judgment made under uncertainty, based on probabilistic cues (e.g., Am I far from home? Did he marry into my band?). In deciding whether to initiate a cooperative relationship, a contingent cooperator must use these cues to make trade-offs between two different kinds of errors: (i) false positives, in which a one-shot interaction is mistakenly categorized as a repeated interaction, and (ii) misses, in which a repeated interaction is mistakenly categorized as one-shot. A miss is a missed opportunity to harvest gains in trade from a long string of mutually beneficial interactions. In a population of contingent cooperators, the cost of a miss is usually much higher than the cost of a false positive.

Using agent-based simulations, Delton et al. (2011) showed that, under a wide range of conditions, individual-level selection favors computational designs that decide to cooperate with new partners, even in a world where most of the interactions are one-shot. Each new partner comes with a number—a cue summary—that serves as a hint to whether an agent's interaction with that partner will be one-shot or repeated. The cue summaries are never perfect predictors: they are drawn from one of two normal distributions (one-shot vs. repeated) that overlap by different amounts.

In one set of simulations, agents evolve a decision threshold determining how strong cues that the interaction is one-shot must be before the agent defects. Selection favored a threshold of evidence so high that most interactions were classified as repeated, triggering cooperation. In other simulations, the agents were perfect Bayesians who developed rational beliefs about whether an interaction is one-shot by using Bayes' rule to integrate (i) cues that the given interaction is one-shot, with (ii) perfect knowledge of the base rate of one-shot interactions in the population. What evolved is a regulatory variable that determines the probability the agent will cooperate *given its rational belief that the interaction is one-shot.*

Selection favored designs with a very high probability (70%–90%) of cooperating given the rational belief that the interaction is one-shot, with modest gains in trade and a modest number of encounters for those interactions that were repeated. This was true even when the base rate of one-shot interactions was unrealistically high (50%–70%).

The simulations with Bayesian agents are particularly apt because most subjects who cooperate in experimental economics games say they believed the experimenter's claim (a cue!) that their interaction would be one-shot. The results show that natural selection can favor a disposition to start out cooperating, even in people who rationally believe an interaction is most likely to be one-shot. No group selection is needed. And this disposition to cooperate in one-shot interactions is not a mistake or an irrational "bias": it is the most adaptive, fitness-promoting decision.

References

Axelrod, R. (1984). *The evolution of cooperation.* New York; NY: Basic Books.

Bowles, S., & Gintis, H. (2013). *A cooperative species: Human reciprocity and its evolution.* Princeton, NJ: Princeton University Press.

Camerer, C., & Weber, M. (1992). Recent developments in modeling preferences: Uncertainty and ambiguity. *Journal of Risk and Uncertainty, 5,* 325–370.

Cosmides, L., & Tooby, J. (1987). From evolution to behavior: Evolutionary psychology as the missing link. In J. Dupré (Ed.), *The latest on the best: Essays on evolution and optimality* (pp. 277–306). Cambridge, MA: MIT Press.

Cosmides, L., & Tooby, J. (1996). Are humans good intuitive statisticians after all? Rethinking some conclusions of the literature on judgment under uncertainty. *Cognition, 58,* 1–73.

Cosmides, L., & Tooby, J. (2008). Can a general deontic logic capture the facts of human moral reasoning? How the mind interprets social exchange rules and detects cheaters. In W. Sinnott-Armstrong (Ed.), *Moral psychology* (pp. 53–119). Cambridge, MA: MIT Press.

Cosmides, L., & Tooby, J. (2015). Adaptations for reasoning about social exchange. In D. M. Buss (Ed.), *The handbook of evolutionary psychology: Vol. 2. Integrations* (2nd ed., pp. 625–668). Hoboken, NJ: Wiley.

Delton, A. W., Krasnow, M. M., Cosmides, L., & Tooby, J. (2011). Evolution of direct reciprocity under uncertainty can explain human generosity in one-shot encounters. *Proceedings of the National Academy of Sciences, 108*, 13335–13340.

Gigerenzer, G. (2014). *Risk savvy: How to make good decisions*. New York, NY: Viking.

Gigerenzer, G., & Hoffrage, U. (1995). How to improve Bayesian reasoning without instruction: Frequency formats. *Psychological Review, 102*, 684–704.

Gigerenzer, G., & Hoffrage, U. (1999). Overcoming difficulties in Bayesian reasoning: A reply to Lewis and Keren (1999) and Mellers and McGraw (1999). *Psychological Review, 106*, 425–430.

Higham, J. P. (2014). How does honest costly signaling work? *Behavioral Ecology, 25*(1), 8–11.

Hoffman, E., McCabe, K., & Smith, V. (1998). Behavioral foundations of reciprocity: Experimental economics and evolutionary psychology. *Economic Inquiry, 36*, 335–352.

Kahneman, D. (2011). *Thinking, fast and slow*. New York, NY: Macmillan.

Kahneman, D., Slovic, P., & Tversky, A. (Eds.). (1982). *Judgment under uncertainty: Heuristics and biases*. Cambridge, England: Cambridge University Press.

Kahneman, D., & Tversky, A. (1982). On the study of statistical intuitions. *Cognition, 11*, 123–141.

Kleiter, G. D. (1994). Natural sampling: Rationality without base rates. In G. H. Fischer & D. Laming (Eds.), *Contributions to mathematical psychology, psychometrics, and methodology* (pp. 375–388). New York, NY: Springer.

Krasnow, M. M., Delton, A. W., Tooby, J., & Cosmides, L. (2013). Meeting now suggests we will meet again: Implications for debates on the evolution of cooperation. *Nature Scientific Reports, 3*, 1747.

Krebs, J. R., & Dawkins, R. (1984). Animal signals: Mind-reading and manipulation. In J. R. Krebs & N. B. Davies (Eds.), *Behavioural ecology: An evolutionary approach* (2nd ed., pp. 380–402). Oxford, England: Blackwell.

Maynard Smith, J. (1982). *Evolution and the theory of games*. Cambridge, England: Cambridge University Press.

Mercier, H., & Sperber, D. (2011). Why do humans reason? Arguments for an argumentative theory. *Behavioral and Brain Sciences, 34*, 57–111.

Mishra, S., & Fiddick, L. (2012). Beyond gains and losses: The effect of need on risky choice in framed decisions. *Journal of Personality and Social Psychology, 102*(6), 1136–1147.

Real, L. A. (1991). Animal choice behavior and the evolution of cognitive architecture. *Science, 253*, 980–986.

Rips, L. (1994). *The psychology of proof*. Cambridge, MA: MIT Press.

Rode, C., Cosmides, L., Hell, W., & Tooby, J. (1999). When and why do people avoid unknown probabilities in decisions under uncertainty? Testing some predictions from optimal foraging theory. *Cognition, 72*, 269–304.

Smith, V. L. (2003). Constructivist and ecological rationality in economics (Nobel Prize Lecture, December 8, 2002). *American Economic Review, 93*(3), 465–508.

Sperber, D. (1994). The modularity of thought and the epidemiology of representations. In L. A. Hirschfeld & S. A. Gelman (Eds.), *Mapping the mind: Domain specificity in cognition and culture* (pp. 39–67). Cambridge, England: Cambridge University Press.

Stephens, D. W., & Krebs, J. (1986). *Foraging theory*. Princeton, NJ: Princeton University Press.

Tooby, J., & Cosmides, L. (1992). The psychological foundations of culture. In J. H. Barkow, L. Cosmides, & J. Tooby (Eds.), *The adapted mind: Evolutionary psychology and the generation of culture* (pp. 19–136). New York, NY: Oxford University Press.

Trivers, R. L. (1971). The evolution of reciprocal altruism. *Quarterly Review of Biology, 46*, 35–57.

Tversky, A., & Kahneman, D. (1981). The framing of decisions and the psychology of choice. *Science, 211*, 453–458.

Wason, P. C., & Johnson-Laird, P. N. (1972). *Psychology of reasoning: Structure and content*. Cambridge, MA: Harvard University Press.

Section 11   Deontic and Legal Reasoning

11.1 Deontic Logic

John Horty and Olivier Roy

Summary

Deontic logic is a general area encompassing many specific logical approaches united by a common concern with normative concepts—concepts, that is, involving norms, such as moral, legal, or rational norms. Based on normative information of this kind, deontic logics, most typically, provide judgments or conclusions about what we ought to do and what we are permitted to do, or about what ought to be the case. In this chapter, we sketch the basic ideas underlying the standard system of deontic logic. Next, just to highlight the interdisciplinary reach of deontic logic, we mention some connections with game theory that we believe may be of particular interest to readers of this volume. Finally, we survey some important variations on standard deontic logic.

1. What Is Deontic Logic?

Deontic logic is a general area encompassing many specific logical approaches united by a common concern with normative concepts—concepts, that is, involving norms, such as moral, legal, or rational norms; the norms of etiquette or aesthetics; or the rules of a game. Based on normative information of this kind, what deontic logics provide, most typically, is judgments or conclusions about what we ought to do and what we are permitted to do, or about what ought to be the case.

Like many areas of logic, deontic logic was studied in a fragmentary way by medieval theorists, but it was not until the 20th century, with the introduction of mathematical techniques into logic, that the subject developed into a systematic discipline. This development can be divided into three stages. In the first, beginning with the work of Mally (1926) early in the 20th century, a number of philosophers and logicians explored ways of extending the axiomatic approach to logic, in particular that of Whitehead and Russell's *Principia Mathematica* (1910–1913), to deontic concepts as well. Although these early axiomatizations were soon shown to be flawed, they

opened up the area for investigation and inspired later researchers. The second stage began near the midpoint of the 20th century with von Wright's (1951) interpretation of deontic logic as a species of modal logic, allowing for the subject to be studied using semantic techniques, and leading, with only slight modifications, to the system of modal logic now known as "standard deontic logic." Although this system stimulated some interesting technical developments, it also served as the primary focus of a philosophical debate concerning the relation between deontic principles and moral concepts that became, over time, increasingly arid and scholastic.

The third stage in the development of deontic logic, which began in the final decade of the 20th century and continues to this day, has been characterized by increasing interdisciplinary connection with many other fields, including artificial intelligence, multiagent systems and computer science more generally, legal theory, linguistics, political science, organizational theory, and economics. These interdisciplinary connections have revitalized the field, leading not only to new areas of application for ideas from deontic logic but also to the incorporation of new techniques into the study of deontic logic itself.

We cannot attempt here either to trace the historical development of deontic logic in any detail or to survey its current range of interdisciplinary connections; an authoritative historical treatment is presented by Hilpinen and McNamara (2014). Our discussion is organized as follows: first, we sketch the basic ideas underlying the standard system of deontic logic. Next, just to highlight the interdisciplinary reach of deontic logic, we mention some connections with game theory that we believe may be of particular interest to readers of this volume. Finally, we survey some important variations on standard deontic logic.

2. Standard Deontic Logic

Standard deontic logic is a species of modal logic, a form of logic in which the truth or falsity of a formula in a particular state of affairs can depend on the truth value of different formulas in different states of affairs. These states of affairs are often referred to, somewhat poetically, as "possible worlds," although the logic itself does not involve any metaphysical commitments, and the possible worlds can be provided with more concrete interpretations, as we shall see.

Let us suppose, then, that W is a set of possible worlds and that v is a valuation function mapping each atomic sentence into the set of possible worlds where it is true, so that, where p is an atomic sentence and w is a possible world, w is in $v(p)$ if p is true in w, and w is not in $v(p)$ if p is false in w. In addition to W and v, the structures against which formulas of standard deontic logic are interpreted contain a further, distinctive component: a function f mapping each possible world w into a nonempty set $f(w)$ of worlds that can be thought of as ideal from the standpoint of w. The basic idea underlying standard deontic logic is that what *ought* to be the case in the world w can be identified with what *is* the case in those worlds that are ideal from the standpoint of w.

To develop the theory formally, we first introduce models of the form $M = \langle W, f, v \rangle$, with W, f, and v defined as above. Next, letting $M, w \vDash A$ indicate that the formula A is true at the world w from the set W of worlds belonging to the model M, and where v is the valuation function from this model, we define truth conditions for atomic sentences through the clause

$M, w \vDash p$ just in case $w \in v(p)$,

according to which the atomic sentence p is true at the world w just in case the valuation function says it is. Once the notion of truth at a world is settled for atomic sentences, it can be lifted to complex sentences formed from the connectives $\neg$, $\wedge$, $\vee$, and $\rightarrow$—representing negation, conjunction, disjunction, and implication, respectively—through the clauses

$M, w \vDash \neg A$ just in case it is not the case that $M, w \vDash A$,
$M, w \vDash A \wedge B$ just in case $M, w \vDash A$ and $M, w \vDash B$,
$M, w \vDash A \vee B$ just in case $M, w \vDash A$ or $M, w \vDash B$,
$M, w \vDash A \rightarrow B$ just in case either not $M, w \vDash A$, or
$\quad M, w \vDash B$.

What these further clauses tell us is simply that the truth value assigned to a complex sentence at a possible world respects the meaning of the connective through which that complex sentence is formed: according to the second clause, for example, a conjunction is true at a world just in case each of its conjuncts is; according to the fourth, an implication is true just in case the consequent of that implication is true whenever its antecedent is.

Finally, we turn to the characteristic deontic connective O, allowing for the construction of sentences of the form OA, to be read "It ought to be the case that A." The clause governing this connective is

$M, w \vDash$ OA just in case
$\quad M, w' \vDash A$ for all w' belonging to $f(w)$,

where of course f is the function from M mapping each world w into the set of worlds that can be considered as ideal from the standpoint of w.

The clause governing the truth value assigned to a deontic statement captures in a formal way the idea that OA holds at a world w just in case A holds at all the worlds w' that are ideal from the standpoint of w. For a concrete illustration, let A represent the statement that there is no war, and suppose we can agree that, from the standpoint of our actual world, any ideal world would have to be one in which there is no war, so that A holds in each of these ideal worlds. Then, what our logic tells us is that, because A holds in each of these ideal worlds, OA is true in the actual world—it ought to be that there is no war.

One natural question to ask at this point is why the worlds considered to be ideal should be relativized to an initial world: why isn't there just a single set of ideal worlds? The answer is that, in general, ought-statements can be contingent, varying from world to world. Consider, for example, the proposition that Jo promises to meet her friend Jack for dinner. Presumably this is a contingent proposition, holding at some worlds but not at others. So imagine that w is a world at which Jo makes this promise and w' is a world at which she does not, and let A stand for the proposition that Jo, in fact, meets Jack for dinner—that she keeps her promise. If we assume that, ideally, promises are to be kept, then it follows that all the worlds ideal from the standpoint of w are worlds in which A holds, so that OA holds at w. However, since there is no particular reason to suppose that A holds at all the worlds ideal from the standpoint of w', where Jo made no promise, it follows that OA fails at that possible world.

Once we have defined truth at a possible world in a model, we can, following the standard pattern in modal logic, define logical implication by stipulating that A logically implies B if B is true at every world from every model at which A is true. To illustrate, we can see that OA and O$(A \rightarrow B)$ implies OB, as follows. Suppose OA and O$(A \rightarrow B)$ are true at w. Then both A and $A \rightarrow B$ must be true at each world from $f(w)$. But it follows from ordinary logic that B is true whenever A and $A \rightarrow B$ are both true, so that B must likewise be true at each world from $f(w)$. From this, we can conclude that OB is true at w.

3. An Interlude: Connections with Game Theory

Although the talk of possible worlds in deontic logic, and in modal logic more generally, can seem hopelessly abstract, it is important to realize that this language can be provided with concrete interpretations in a number of application domains. We cannot describe these various application domains here, of course, but just for the sake of illustration, we briefly sketch how the machinery of deontic logic can be applied in the domain of game theory, where the norms at work are not moral or legal norms but norms of rationality. Our discussion proceeds by way of example; see chapter 9.1 by Albert and Kliemt (this handbook) for formal definitions.

Let us thus start with a simple example of a game in so-called normal or strategic form:

Ann\Bob	L	R
U	1, 1	0, 0
D	0, 1	1, 0

In this game there are two players, Ann and Bob. Ann can choose the upper (U) or the lower (D) row and Bob either the left (L) or the (R) column. Each combination of these choices, for instance, (U,L) or (D,R), corresponds to one possible outcome of the game. Ann and Bob have preferences over these outcomes, which are expressed by the utility numbers in each cell of the matrix, with 1 preferred to 0. The left-hand numbers are Ann's utilities and the right-hand ones Bob's.

What is rational for Ann and for Bob to do in this game? What Ann should rationally do depends on what Bob does. If he plays L, then her best response is to play U. If, however, he plays R, then her best response is D. What is the best response for Bob, however, does *not* depend on what Ann does. If she plays U, he gets 1 by playing L and 0 by playing R, and the same holds if Ann plays D. Game theorists would say that R is strictly dominated by L. Ruling R out as a rational choice for Bob, we conclude that he ought to play L. But then, as we have seen already, if we only consider L as a possible action for Bob, the only rational choice for Ann is to play U, leaving (U,L) as the only rational solution of this game. In other words, Ann ought to play U, and Bob L. This combination of choices also happens to be a Nash equilibrium, meaning that neither Ann nor Bob have an incentive to unilaterally deviate from it. Bob has no incentive whatsoever to play R instead of L. Ann, given that Bob plays L, should not switch to D either.

Let us now look at how deontic logicians would analyze this situation. First, we need to define the set W of

possible worlds. In our example, we can take W to consist of the four possible outcomes of the game, that is, $W = \{(U,L), (U,R), (D,L), (D,R)\}$. Second, we need to define the valuation function v. The atomic sentences that we are interested in describe the strategies that Ann and Bob can choose. Let us use u, d, l, and r for that. The valuation is then defined as the natural one: $v(u) = \{(U,L), (U,R)\}$, $v(l) = \{(U,L), (D,L)\}$, and so on. The final component that we need to define is the function f mapping each possible world w onto its set of ideal worlds. Here, again, there is simple, although of course not a unique, way to define this function f: we can assign the set of rational solutions, $\{(U,L)\}$, as the set of ideal worlds for all the four possible worlds in W. That is, for all $w \in W$, $f(w) = \{(U,L)\}$.

With this in hand, we can see what kind of deontic statements are true in the model that we have just constructed. Recall that u is the statement that Ann plays U. Now, (U,L) is the only ideal world from the standpoint of any world in the model that we constructed above. In that world, u is true, so Ou is true everywhere in the model that we have constructed. Similarly, we have Ol true everywhere, reflecting the fact that in all ideal worlds, Bob plays L.

Now, as we already observed, obligations and permissions might be contingent, varying from world to world. This is *not* the case in the model we just built. By setting $f(w) = \{(U,L)\}$ everywhere, we have identified the set of ideal worlds with the unique outcome that results from first ruling out R as a rational choice for Bob and then identifying Ann's best response *given that Bob does not play R*. As we have seen, however, Ann's best response would be different if Bob would play R after all. To reflect that, we could instead define f as follows: $f(U,L) = \{(U,L)\}$, $f(U,R) = \{(D,R)\}$, $f(D,L) = \{(U,L)\}$, $f(D,R) = \{(D,L)\}$. With this modification, Ou becomes a contingent formula, true only in the possible worlds where Bob plays his only rational strategy L but false in the nonideal case where Bob plays R.

4. Variations

The standard deontic logic sketched earlier is a starting point for the study of deontic logic, but only a starting point—this standard theory has been elaborated along various dimensions, sometimes in response to difficulties or perceived difficulties, and a variety of alternative frameworks have been explored. We cannot hope even to mention all of these developments in this brief chapter but will merely list a few of the most important.

First, there are clear connections between deontic logics and logics of time, or temporal logics. Suppose, as in our earlier example, that Jo has promised to meet Jack for dinner and so ought to have dinner with Jack. But of course, this event must occur at some particular point in the future, relative to the moment of the promise—it will do no good if Jo had dinner with Jack sometime in the past. Further, if Jo's promise is to function as a real constraint on her behavior, we must at least be able to envision alternative futures, or alternative ways in which the world might unfold, in which Jo does *not* meet Jack for dinner. These issues concerning the temporal aspects of ought-statements have been explored in a number of papers beginning with seminal work in the early 1980s by Åqvist and Hoepelman (1981), Thomason (1981), and van Eck (1982); a useful survey can be found in Thomason (1984). Interestingly, similar issues arise in the study of games as well, as illustrated, again, by chapter 9.1 by Albert and Kliemt (this handbook).

Second, deontic logics interact with issues of agency. Consider again our example, leading to the conclusion that Jo ought to dine with Jack. What we naturally mean by this is not simply that the event of Jo dining with Jack should occur but that Jo should bring it about, or see to it, that this event occurs—that this event should occur through Jo's agency. The study of agency in the context of deontic logic began with work by Kanger (1957/1971) and was explored in later work by Pörn (1962/1970) and Lindahl (1977) on the theory of "normative positions," an attempt to survey the various possible normative relations between individuals; this work has now been extended and solidified by Sergot (2014). More recently, Horty (2001) has explored a deontic logic developed within the *stit* analysis of agency introduced by Belnap, Perloff, and Xu (2001). This also has interesting connections with game theory, as illustrated, for instance, in earlier work by Apostel (1960) and van Benthem (1979) as well as more recent work by Tamminga (2013).

Third, although deontic logic typically focuses on the ought-connective introduced above, a connective representing permission can be defined as well, as a dual of the "ought." More precisely, where P represents permission, the statement PA can be defined as $\neg$O$\neg A$—the idea is that A is permissible if it is not required that A not be the case. This standard treatment of permission is sensible and applies naturally in many situations, but there are also cases, known as cases of "free choice permission," in which this standard treatment seems to give the wrong results. Suppose the waiter tells you that, as part of your fixed-price meal, you can have either soup or salad. If we now let A represent the proposition that you have soup and B the proposition that you have salad, then it is reasonable to formalize the waiter's statement as P($A \vee B$)—it is permitted that you have soup or you

have salad. Of course, we would normally conclude from the waiter's statement that P$A \wedge$ PB—you are permitted to have soup and also permitted to have salad. The trouble is that, on the standard definition of permission as the dual of obligation, P($A \vee B$) does not entail P$A \wedge$ PB—we cannot conclude what we want to from what the waiter said. The problem of free-choice permission is the problem of finding another definition of permission that supports the desired inference; a contemporary survey of literature on the problem can be found in Hansson (2014). Again, this problem has also attracted attention from a game theoretic perspective, as illustrated in work by Dong and Roy (2015) and Anglberger, Gratzl, and Roy (2015).

Fourth, there is the problem of contrary-to-duty oughts. To illustrate, return to our example in which, as the result of a promise, Jo ought to have dinner with Jack. Now suppose the time for dinner comes and goes and, for one reason or another, Jo has failed to meet Jack for dinner. What then? It seems sensible to conclude that Jo should try to make amends to Jack for breaking her promise—perhaps she should call up Jack to apologize. The trouble is that since standard deontic logic is concerned only with ideal situations, in which no obligations are violated, it has nothing to say about what Jo ought to do in the various subideal situations in which she has failed to meet some of her obligations. The problem of contrary-to-duty oughts is the problem of designing a deontic logic to deal with these subideal situations, telling us what we ought to do once we have violated our primary oughts. A very readable introduction to the problem, as well as an interesting proposal for a solution, is presented by Prakken and Sergot (1996).

Fifth, and finally, there is the problem of normative conflicts, sometimes known as "moral dilemmas." Let us say that a situation gives rise to a normative conflict if it presents each of two conflicting propositions as obligatory—if, for example, it supports the truth of both OA and OB, where A and B are inconsistent and so cannot hold jointly. We often seem to face conflicts like this in everyday life, and there are a number of vivid examples in philosophy, literature, and law. Perhaps the best known of these is Sartre's (1946) description of a student during World War II who felt for reasons of patriotism that he ought to leave home to join the Free French but who felt also, for reasons of sympathy and personal devotion, that he ought to stay at home to care for his aged mother.

Sartre presents this student's situation in a compelling way that does make it seem as if he had been confronted with conflicting, and perhaps irreconcilable, moral principles. However, if standard deontic logic is correct,

Sartre is mistaken: the student did not face a moral conflict—no one ever does, because according to standard deontic logic, such a conflict is impossible. This is easy to see. For the two statements OA and OB to be supported at a world w, both A and B must hold at each world that is ideal from the standpoint of w. It is a constraint on standard deontic models that the set $f(w)$ of worlds ideal from the standpoint of w must be nonempty, so there must be some world w' belonging to this set. Both A and B, therefore, must hold at w', but this is impossible since, by assumption, A and B cannot hold jointly.

This feature of standard deontic logic—that it rules out normative conflicts—has received extensive discussion in the philosophical literature. There is currently no consensus among moral theorists on the question whether an ideal ethical theory could be structured in such a way that moral dilemmas might arise. The issue has been addressed, for example, by Donagan (1984), Foot (1983), Lemmon (1962), Marcus (1980), and Williams (1965). Still, it can seem like an objectionable feature of standard deontic logic that it rules out this possibility. Because the question is open, and the possibility of moral dilemmas is a matter for substantive ethical discussion, it seems to be inappropriate for a position on this issue to be built into the logic of the subject. And even if it does turn out, ultimately, that research in ethics can exclude the possibility of conflicts in a correct moral theory, it is useful all the same to have a deontic logic that is able to tolerate normative conflicts, since they arise in so many other domains.

The first intuitively adequate account of reasoning in the presence of normative conflicts was presented in van Fraassen (1973). The logic presented there was interpreted in Reiter's (1980) default logic by Horty (1994), establishing the first clear connection between existing deontic and nonmonotonic, or defeasible, logics; this interpretation was later developed from a philosophical perspective in Horty (2012). The connection between deontic and defeasible logics has now been explored in some detail. A useful collection of early papers is found in Nute (1997). Much of the most interesting recent work, summarized by Parent and van der Torre (2014), revolves around the framework of input/output logic, rather than older frameworks for defeasible logic. The most authoritative current treatment of normative conflicts in deontic logic is presented by Goble (2014).

5. Conclusion

The field of deontic logic began early in the 20th century with the introduction of a number of formal systems, many of them flawed, aimed at analyzing normative concepts in philosophy and law. The field was solidified toward the middle of the century with the isolation of the system now known as standard deontic logic, which is sensible and well understood at least from a logical perspective. There followed a period of confusion, and eventual stagnation, centered on the failure of this standard deontic logic to yield what many considered the correct results in a variety of puzzle cases, sometimes known as "deontic paradoxes." As a result of an infusion of new techniques, primarily from computer science, deontic logic is once again active and vibrant, with applications not only in computer science but also in a number of other fields, such as linguistics, political science, organizational theory, and economics.

Our goal here has been to present the central ideas underlying the system of standard deontic logic and to survey some of its more interesting variations. It is impossible, in a brief chapter such as this, to cover the range of current applications for deontic logic, but just to give the reader a sense of these connections, we have sketched a few of the relations between deontic logic and game theory.

References

Anglberger, A. J. J., Gratzl, N., & Roy, O. (2015). Obligation, free choice, and the logic of weakest permissions. *Review of Symbolic Logic, 8*(4), 807–827.

Apostel, L. (1960). Game theory and the interpretation of deontic logic. *Logique et Analyse, 3*(10), 70–90.

Åqvist, L., & Hoepelman, J. (1981). Some theorems about a "tree" system of deontic tense logic. In R. Hilpinen (Ed.), *New studies in deontic logic: Norms, actions, and the foundations of ethics* (pp. 187–221). Dordrecht, Netherlands: Reidel.

Belnap, N., Perloff, M., & Xu, M. (2001). *Facing the future: Agents and choices in our indeterministic world.* Oxford, England: Oxford University Press.

Donagan, A. (1984). Consistency in rationalist moral systems. *Journal of Philosophy, 81,* 291–309.

Dong, H., & Roy, O. (2015). Three deontic logics for rational agency in games. *Studies in Logic, 8*(4), 7–31.

Foot, P. (1983). Moral realism and moral dilemmas. *Journal of Philosophy, 80,* 379–398.

Goble, L. (2014). Prima facie norms, normative conflicts, and dilemmas. In D. M. Gabbay, J. Horty, X. Parent, R. van der Meyden, & L. van der Torre (Eds.), *Handbook of deontic logic and normative systems* (pp. 241–352). London, England: College Publications.

Hansson, S. O. (2014). The varieties of permission. In D. M. Gabbay, J. Horty, X. Parent, R. van der Meyden, & L. van der Torre (Eds.), *Handbook of deontic logic and normative systems* (pp. 195–240). London, England: College Publications.

Hilpinen, R., & McNamara, P. (2014). Deontic logic: A historical survey and introduction. In D. M. Gabbay, J. Horty, X. Parent, R. van der Meyden, & L. van der Torre (Eds.), *Handbook of deontic logic and normative systems* (pp. 3–136). London, England: College Publications.

Horty, J. F. (1994). Moral dilemmas and nonmonotonic logic. *Journal of Philosophical Logic, 23,* 35–65.

Horty, J. F. (2001). *Agency and deontic logic.* Oxford, England: Oxford University Press.

Horty, J. F. (2012). *Reasons as defaults.* Oxford, England: Oxford University Press.

Kanger, S. (1971). New foundations for ethical theory. In R. Hilpinen (Ed.), *Deontic logic: Introductory and systematic readings* (pp. 36–58). Dordrecht, Netherlands: Reidel. (Original work privately distributed in 1957)

Lemmon, E. J. (1962). Moral dilemmas. *Philosophical Review, 70,* 139–158.

Lindahl, L. (1977). *Position and change: A study in law and logic.* Dordrecht, Netherlands: Reidel.

Mally, E. (1926). *Grundgesetze des Sollens: Elemente der Logik des Willens* [Basic laws of ought: Elements of the logic of the will]. Graz, Austria: Leuschner & Lubensky.

Marcus, R. B. (1980). Moral dilemmas and consistency. *Journal of Philosophy, 77,* 121–136.

Nute, D. (Ed.). (1997). *Defeasible deontic logic.* Dordrecht, Netherlands: Kluwer.

Parent, X., & van der Torre, L. (2014). Input/output logic. In D. M. Gabbay, J. Horty, X. Parent, R. van der Meyden, & L. van der Torre (Eds.), *Handbook of deontic logic and normative systems* (pp. 409–544). London, England: College Publications.

Pörn, I. (1970). *The logic of power* (2nd ed.). Oxford, England: Blackwell. (Original work published 1962)

Prakken, H., & Sergot, M. (1996). Contrary-to-duty obligations. *Studia Logica, 57,* 91–115.

Reiter, R. (1980). A logic for default reasoning. *Artificial Intelligence, 13,* 81–132.

Sartre, J.-P. (1946). *L'existentialisme est un humanisme* [Existentialism is a humanism]. Paris, France: Nagel.

Sergot, M. (2014). Normative positions. In D. M. Gabbay, J. Horty, X. Parent, R. van der Meyden, & L. van der Torre (Eds.), *Handbook of deontic logic and normative systems* (pp. 353–406). London, England: College Publications.

Tamminga, A. (2013). Deontic logic for strategic games. *Erkenntnis, 78,* 183–200.

Thomason, R. (1981). Deontic logic as founded on tense logic. In R. Hilpinen (Ed.), *New studies in deontic logic: Norms, actions, and the foundations of ethics* (pp. 165–176). Dordrecht, Netherlands: Reidel.

Thomason, R. (1984). Combinations of tense and modality. In D. M. Gabbay & F. Guenthner (Eds.), *Handbook of philosophical logic: Vol. II. Extensions of classical logic* (pp. 135–165). Dordrecht, Netherlands: Reidel.

van Benthem, J. (1979). Minimal deontic logics. *Bulletin of the Section of Logic, 8*(1), 36–42.

van Eck, J. A. (1982). A system of temporally relative modal and deontic predicate logic and its philosophical applications 2. *Logique et Analyse, 25,* 339–381.

van Fraassen, B. C. (1973). Values and the heart's command. *Journal of Philosophy, 70,* 5–19.

von Wright, G. H. (1951). Deontic logic. *Mind, 60,* 1–15.

Whitehead, A. N., & Russell, B. (1910–1913). *Principia mathematica* (3 vols.). Cambridge, England: Cambridge University Press.

Williams, B. A. O. (1965). Ethical consistency. *Proceedings of the Aristotelian Society, 39*(Suppl.), 103–124.

11.2   Deontic Reasoning in Psychology

Shira Elqayam

Summary

Deontic reasoning is the branch of the psychology of reasoning that studies how we reason with and about normative rules, using logical operators such as "must," "allowed," and "forbidden." Thanks to its dual nature, combining reasoning with social context, deontic reasoning provides a unique, multilevel view on human rationality, bridging between normative and social and pragmatic rationality. This chapter reviews some of the historical as well as contemporary findings regarding deontic thinking, starting with the deontic selection task; considers aspects of the normative, social, and pragmatic rationality of deontic thinking; explores its cognitive variability; and outlines its implications for the great rationality debate. The chapter concludes with some thoughts about the future of research in deontic reasoning.

1.   Reasoning *about* Norms and *with* Norms

Normative rules surround us wherever we turn: as we drive, we conform (hopefully) to traffic laws. We conform to norms when we play games, be it chess, Go, or online role-playing. Indeed, children as young as two years old make spontaneous normative utterances when a player deviates from the rules of a game (Rakoczy, Warneken, & Tomasello, 2008). We have norms in law, in religion, in education, in medical practice. Norms, as Searle (2005) aptly put it, are the glue that holds society together. Normative rules have their own language; we use terms such as *must, ought, allowed,* and *forbidden* to express them: "I *ought* to visit my aunt"; "You *must* support drug recommendations with evidence"; "Smoking is *forbidden* anywhere on this train"; "Students are *allowed* to use a pocket calculator at the exam"; and so on. In logic, this language is called *deontic*, from the Greek *deon*, obligation.[1] Thus, *deontic logic* is the logic of rules, norms, and regulations (McNamara, 2010; and see chapter 11.1 by Horty & Roy, this handbook), and *deontic reasoning* is the branch of the psychology of reasoning that studies how we reason with and about normative rules.

Deontic reasoning provides a unique, multilevel view on human rationality, bridging between normative rationality, on the one hand, and social and pragmatic rationality, on the other: the rationality of conforming to normative standards versus the rationality of achieving one's goals, respectively (Evans & Over, 1996). We can of course ask how well people perform on deontic deductive competence (pretty well), but there is much more to deontic reasoning than straightforward deduction: deontic reasoning is about norms, firmly linking it to social rationality; it also provides the bridge to action, so it is perhaps more directly linked to practical rationality than most other domains of human inference. Moreover, given that deontic reasoning is about norms, it occupies a curious, dual place in the great rationality debate, with both first-order and second-order implications: we can ask whether people reason normatively *about* deontic premises, but we can also explore the type of deontic judgments that scientists make when they theorize about reasoning and decision making; after all, rationality is about *should* and *ought*.

This chapter reviews some of the historical as well as contemporary findings regarding deontic thinking, starting with the deontic selection task; considers aspects of normative, social, and pragmatic rationality of deontic thinking; explores its cognitive variability; and outlines its implications for the great rationality debate. I conclude with some thoughts about the future of research in deontic reasoning.

2.   How It All Started: The Deontic Selection Task

To understand some of the historical context for the appearance of deontic reasoning on the stage of the psychology of reasoning, it helps to keep in mind that for a long time, psychologists of reasoning acknowledged one type of logic only: the two-valued, extensional, first-order classical logic of Introduction to Logic textbooks. Perhaps the most famous (or, if you will, infamous) example of this approach is the Wason selection task (Wason, 1960; for reviews, see Evans & Over, 2004, chapter 5; Ragni, Kola, & Johnson-Laird, 2018). In the abstract version of this fiendishly difficult task, reasoners are presented with (typically) four cards and are told that each card has a letter on one side and a number on the other side. Only one side of each card is visible, for example, A, N, 8, 4. They are also presented with a conditional of the form "If p, then q," for example, "If there is an A on one side of the card, then there is an 8 on the other side." They are told that the conditional is a rule that applies to the cards and may be true or false. Their task is to pick the cards—and only the cards—that need to be turned over in order to test if the conditional is true or false. Wason's normative standard was classical logic and Popper's philosophy of science: with this standard in place, the normatively correct choices are the p cards (A in this case) and the not-q cards (4 in this case) (although cf. Oaksford & Chater, 1994). In this abstract version (technically known as *indicative*), success rates (in this classical normative sense) are extremely low—around 10%, a finding replicated many times since.

Performance on the abstract task was so dismal that psychologists soon started looking for variations of the task that produced better success rates. In the famous "drinking age" problem (Griggs & Cox, 1982), participants are invited to imagine that they are a police officer responsible for ensuring that the following regulation was followed: "If a person is drinking beer, then the person must be over 19." Four cards represented four people sitting at a table and bore the legends "Drinking beer," "Drinking coke," "22 years of age," and "16 years of age" (p, not-p, q, and not-q, respectively). Selections of p and not-q were a staggering 74%—more than seven times as much as in the typical abstract version. This facilitation effect is extremely robust (Ragni, Kola, & Johnson-Laird, 2018).

You might have identified the drinking age rule as deontic: not only is the term "must" mentioned in the conditional itself, but the task also clearly refers to a regulation, identifying the context as deontic. This is typical: deontic versions usually need at least a minimal context to trigger facilitation (Evans & Over, 2004). It was not long before Cheng and Holyoak (1985) identified this version as deontic, thereby introducing the term "deontic" to scientific discourse for the first time. Their theory of *pragmatic reasoning schemas* proposed that people reason by using generalized, yet context-sensitive, knowledge structures (the eponymous *schemas*) defined in terms of classes of goals and relationships to these goals. These schemas

included the deontic rules of permission and obligation as well as the nondeontic schema of causation.

With the theory of pragmatic reasoning schemas criticized for offering little explanation beyond labeling the effect (Cosmides, 1989; Evans & Over, 2004), a new theory emerged: Cosmides and Tooby's evolutionary theory of reasoning (e.g., Cosmides, 1989; Cosmides & Tooby, 1994; see also chapter 10.6 by Cosmides & Tooby, this handbook). Reasoning, according to this theory, consists of *Darwinian algorithms*, specialized information-processing mechanisms that organize behavior meaningfully and adaptively. Deontic reasoning, specifically, was argued to be easier because it triggered the Darwinian algorithm of *social contract*, exchange between individuals for mutual benefit. Cosmides was the first to identify the important role of *utility*, articulated as costs and benefits, in deontic reasoning. In decision theory, "utility" is the technical term for things we want to happen (benefits) or want not to happen (costs). Cosmides showed that participants were quick to identify the p and not-q cards if they referred to cheaters—individuals who violated the social contract by taking a benefit without paying the cost.

Pragmatic reasoning schemas and Darwinian algorithms moved the psychology of deontic reasoning away from classical logic, first by identifying it as deontic and, second, by highlighting the psychological importance of goals and utility. Both theories relied on what came later to be known as "content-specific rules." Manktelow and Over (1991), in contrast, made a direct appeal to a more general, decision-theoretic approach to utility, rather than specialized evolutionary mechanisms. Where utility lies, they argued, depended on perspective (see also Gigerenzer & Hug, 1992)—in this case, social roles. Thus, suppose a mother tells her son:

(a) "If you tidy up, you may go out and play."

The utility for the mother is a tidy room, whereas the utility for the son is the opportunity to play. Participants told to reason from the mother's perspective tended to pick up the not-p & q card (playing without tidying up), whereas participants told to reason from the son's perspective tended to pick up the p & not-q card (that is, not being allowed to play even after tidying up). A more comprehensive decision-theoretic approach was soon to follow: Oaksford and Chater (1994) drew on the constructs of utility and probability to argue that selection of cards in the Wason selection task was driven by the quest for *information gain*. Reasoners picked, they proposed, the cards that provided the most informative answers to the question if q depended on p. These cards differed for the abstract and the deontic version, which explained the differences in performance.

Of course, it is also possible to examine the purely logical relations between deontic operators, detached from considerations of utility. These, too, demonstrate a high degree of conformity to classical logic (Beller, 2008, 2010). Nevertheless, reasoners also deviate occasionally, an effect labeled "illusory inference" in Bucciarelli and Johnson-Laird's (2005) naive deontics theory.

3. The Social and Pragmatic Rationality of Deontic Thinking

Our use of language to draw inferences and make decisions is anchored in social context. Under the heading of "conversational pragmatics," philosophers have highlighted the function of language as a tool of action and communication, in a tradition going back to Austin (1961), Searle (1969), and Grice (1975). Language has a *performative* function: it does things in the world. When we promise a collaborator to finish writing up our share of a manuscript by a particular deadline, we create a new social reality in which we are under an obligation (Beller, Bender, & Kuhnmünch, 2005). A promise is thus a *deontic speech act*: it acts on the world through language, changing it, laying down a new deontic commitment, and people are adept at relating deontic speech acts to situations and actions in the world (Beller, 2008). In psychology of reasoning, the study of conversational pragmatics goes back to Fillenbaum (e.g., 1976), who was the first to identify these functions in his work on *inducement* and *advice conditionals* (promises and threats, and tips and warnings, respectively). Utility is the telltale sign: deontic speech acts create situations in the world where costs and benefits attach to actions and their outcomes (Bonnefon, 2009; see also chapter 6.4 by Bonnefon, this handbook).

Given the rich pragmatic nature of inducement and advice conditionals, a purely deductive approach is unlikely to provide a full psychological understanding of their scope and function: psychologists advocated a multilevel approach (Beller et al., 2005; Ohm & Thompson, 2004; Thompson, 2000) encompassing aspects such as perlocutionary force (the intended effect), emotion, linguistic interpretation, and associated utility. For example, if the mother in (a) violates her promise and does not let the son out to play, the son is likely to feel anger (Beller, 2008; Beller et al., 2005). Furthermore, promises and threats (aka inducements) create stronger obligations than advice conditionals (tips and warnings) because they have stronger perlocutionary force: the

agent has more control over the consequences (Beller et al., 2005; Fillenbaum, 1976).

The same rich pragmatic nature also makes it difficult to disentangle deduction from social pragmatics and normative from practical and social rationality—the latter *leak* into the former (Bonnefon, 2009). For example, the effectiveness and acceptability of deontic speech acts are typically associated both with their truth conditions and their costs and benefits (Evans, Neilens, Handley, & Over, 2008; Evans & Twyman-Musgrove, 1998; Thompson, Evans, & Handley, 2005; for a review, see chapter 6.4 by Bonnefon, this handbook). Deontic speech acts also evoke a complex net of pragmatic implicatures (linguistic social inferences licensed by world knowledge). For example, the promise in sentence (a) invites the implicature that if the son does *not* tidy up, then he will not be allowed to go out and play. Indeed, inducement and advice conditionals, relative to indicative conditionals, tend to inspire a more biconditional interpretation (i.e., "If and only if *p*, then *q*"): participants are more likely to draw all four conditional elimination inferences from threats and promises (Beller, 2005, 2008; Newstead, Ellis, Evans, & Dennis, 1997); but not from the weaker tips and warnings (Ohm & Thompson, 2004). This sometimes takes the form of a *defective* biconditional interpretation. A full biconditional licenses the inference of both the *inverse* (aka *complementary*: "If *not-p*, then *not-q*") and the *converse* (aka *reversed*: "If *q*, then *p*") forms, respectively. However, the inverse inference is more common than the converse (Beller et al., 2005; Newstead et al., 1997), sometimes resulting in what Newstead et al. dubbed as "pattern X," that is, TT and FF cases are judged to be true, TF cases are judged false, and FT cases are judged irrelevant. This does not extend to conditional inference, though: there is no particular preference for Denial of the Antecedent ("If *p* then *q*; not-*p*; therefore not-*q*") as might have been expected given the preference for the inverse implicature.

4. Deontic Reasoning and Cognitive Variability

Deontic thinking might be one of the prime candidates for a cognitive universal, transcending individual and cultural differences. The first line of evidence is cross-cultural comparison: deontic speech acts elicit similar interpretations at least in two WEIRD[2] and two non-WEIRD cultures, Germany and the United Kingdom, as well as China and Tonga, respectively (Song, Bender, & Beller, 2009; Sztencel & Clarke, 2018), although there was also some variation in responses, mainly based on locally accepted behaviors and differing emotional reactions. For example, in a choice task between linguistic formulations, German participants were more likely to choose the explicit threat form in comparison to Tongan participants, who preferred the indirect speech act formulation of promise, while U.K. participants were in between.

An additional line of evidence comes from developmental studies concerning deontic thinking, particularly in early childhood: the normative advantage of deontic rules appears in some studies as early as three to four years of age (Cummins, 1996a, 1996b; Harris & Núñez, 1996; Kanngiesser, Köymen, & Tomasello, 2017). The early development of normativity is of such significance that the *Journal of Experimental Child Psychology* has recently published a special issue on the topic (Paulus & Schmidt, 2018). The precise age is a matter of debate and might be as late as seven years (Dack & Astington, 2011) or as early as two years (Rakoczy et al., 2008); nevertheless, the weight of the evidence is that children do not need very long to develop adult-like sensitivity to deontics. For example, like adults (Kilpatrick, Manktelow, & Over, 2007), children are sensitive to power relations and authority (Chin & Lin, 2018; Dack & Astington, 2011).

In the study of individual differences, Stanovich and West (e.g., 2000) demonstrated a consistent pattern in which able, motivated participants tend to do normatively better on a variety of reasoning and decision-making tasks. These results do not always extend to performance on the deontic selection task. Here the results are more mixed, with some studies finding that cognitive ability had no predictive power (e.g., Stanovich & West, 1998) and some studies establishing some predictive power (Newstead, Handley, Harley, Wright, & Farrelly, 2004).

With these converging lines of evidence, it would seem that deontic thinking has a universal core in understanding and interpreting conditionals, which is in turn tempered by relative sensitivity to cultural and contextual cues.

5. Generating Deontic Norms: Is–Ought Inference

Ben wants to cook an Italian dish to impress his girlfriend. If he uses a certain brand of Italian olive oil, his dish will taste better. Does it then follow that Ben *should* use that brand? Most reasoners agree that he should (Elqayam, Thompson, Wilkinson, Evans, & Over, 2015), and they also think that someone hearing this would be persuaded that he should (Thompson et al., 2005). People readily infer the "ought" from the "is," in effect generating brand-new, bespoke deontic rules, where none existed before. Philosophers sometimes refer to inferring deontic conclusions from indicative premises as the "is–ought fallacy," or the "is–ought problem" (Hudson, 1969; Schurz, 1997).[3] Hume (1739–1740/2000) was the first to identify it—as well as condemn it as invalid: if the premises contain no deontic operators, we cannot validly draw deontic conclusions.

But if we cannot generate new norms, how can we bridge between goals and actions in a fast-changing world? Perhaps more than any other inference, inference from *is* to *ought* (aka deontic introduction) epitomizes the tension between normative rationality, on the one hand, and social and practical rationality, on the other (Evans & Elqayam, 2020). Normatively speaking, deontic introduction is deductively invalid, but pragmatically, we cannot achieve our goals without being able to convert goals to norms. We need to be able to infer "I ought" from the "I want" of goals and desires, and to function in society, we need to be able to convert desires to socially binding rules and norms. Deontic introduction does all this for us. We also need this inference to be flexible, to allow us to act in changing circumstances, and it is. Is–ought inference is defeasible: reasoners withdraw it when the causal link between action and outcome is undermined, or in the presence of conflicting norms or conflicting goals. For example, if we know that the olive oil is manufactured using exploitative agricultural practices, we would no longer think that Ben ought to use it.

Elqayam and Evans (2011) argued that psychologists sometimes draw on various forms of is-to-ought inference to defend their normative stance. For example, consider the following question, which is the appropriate normative solution to the indicative Wason selection task: the classic *p & not-q* selection, or the *p*-only or *p & q* selections? Stanovich (1999; see also Stanovich & West, 1998) argued for the former, based on the positive (albeit moderate) correlation between these selections and cognitive ability and cognitive motivation. Oaksford and Chater (2007), in contrast, argued in favor of the latter, based on the modal responses to the task. These approaches establish the selection among competing normative models—which models we *should* follow—by drawing on empirical evidence, hence arguably drawing an is–ought inference. Normative rationality in psychology is thus arguably difficult to establish, because there are competing normative models, and to support any of them with empirical evidence is self-defeating (for responses to the debate, cf. Oaksford, 2014; Oaksford & Chater, 2011; Stanovich, 2011; and contributions to Over & Elqayam, 2016). Moreover, the inference may be pragmatically infelicitous in the context of normative rationality—recall that people withdraw deontic conclusions in the presence of conflicting norms.

6. Where Do We Go from Here?

The selection task wars are long behind us, but deontic reasoning keeps drawing attention and further research, particularly from the viewpoint of social pragmatics (Hilton, Charalambides, & Hoareau-Blanchet, 2016; Sztencel & Clarke, 2018). Much is now integrated into the useful framework of the theory of utility conditionals (Bonnefon, 2009; chapter 6.4 by Bonnefon, this handbook). There are still some relatively little-explored areas, however, such as cultural differences both in childhood (Miller, Wice, & Goyal, 2018) and adulthood. We are also missing studies that examine directly the link between moral judgment and deontic reasoning, as well as studies that explore further the link to action. Given deontic premises, reasoners are happy to draw inferences about other agents' readiness to act (Beller, 2005, 2008; Sztencel & Clarke, 2018); what we do not know is if endorsement of deontic rules also translates to actual action in the world, or at least to self-reported readiness to act. Similarly, there is some evidence that utilitarian moral judgment draws on deontic conclusions inferred from *indicative* premises (Elqayam, Wilkinson, Thompson, Over, & Evans, 2017), while deontological moral judgment correlates with drawing deontic conclusions from *deontic* premises. We need more studies looking at relations between deontic reasoning and moral judgment.

Another relatively neglected issue is the variety of deontic operators: many studies incorporate just the classic permissions and obligations. Very few studies examine the middle-strength operators such as "ought" and "should not," although these operators are ubiquitous in everyday life.

Lastly, the recent debates on the scientific rigor of psychological science are pretty much deontic debates—that is to say, normative: they are about the *should* and the *ought* (and sometimes downright *must* and *must-not*) just as much as they are about the *is*. Exploring deontic scientific thinking could help inform this important debate.

Notes

1. Not to be confused with deontological morality, the type of moral judgment that relies on preexisting absolute normative rules (see section 12 in this handbook for moral judgment).

2. WEIRD: Western, Educated, Industrialized, Rich, Democratic (Henrich, Heine, & Norenzayan, 2010).

3. A special case of the is–ought fallacy is the naturalistic fallacy, where the "ought" is identified as stemming from natural properties (Ridge, 2019); see the introductory chapter by Knauff and Spohn (this handbook).

References

Austin, J. L. (1961). *How to do things with words*. Cambridge, MA: Harvard University Press.

Beller, S. (2008). Deontic norms, deontic reasoning, and deontic conditionals. *Thinking & Reasoning, 14*, 305–341.

Beller, S. (2010). Deontic reasoning reviewed: Psychological questions, empirical findings, and current theories. *Cognitive Processing, 11*(2), 123–132.

Beller, S., Bender, A., & Kuhnmünch, G. (2005). Understanding conditional promises and threats. *Thinking & Reasoning, 11*, 209–238.

Beller, S., Bender, A., & Song, J. (2009). Conditional promises and threats in Germany, China, and Tonga: Cognition and emotion. *Journal of Cognition and Culture, 9*, 115–139.

Bonnefon, J.-F. (2009). A theory of utility conditionals: Paralogical reasoning from decision-theoretic leakage. *Psychological Review, 116*, 888–907.

Bucciarelli, M., & Johnson-Laird, P. N. (2005). Naïve deontics: A theory of meaning, representation, and reasoning. *Cognitive Psychology, 50*, 159–193.

Cheng, P. W., & Holyoak, K. J. (1985). Pragmatic reasoning schemas. *Cognitive Psychology, 17*, 391–416.

Chin, J.-C., & Lin, M.-H. (2018). Children's understanding of conditional promise contract violations. *Infant and Child Development, 27*.

Cosmides, L. (1989). The logic of social exchange: Has natural selection shaped how humans reason? *Cognition, 31*, 187–276.

Cosmides, L., & Tooby, J. (1994). Beyond intuition and instinct blindness: Toward an evolutionarily rigorous cognitive science. *Cognition, 50*, 41–77.

Cummins, D. D. (1996a). Evidence for the innateness of deontic reasoning. *Mind & Language, 11*, 160–190.

Cummins, D. D. (1996b). Evidence of deontic reasoning in 3- and 4-year-old children. *Memory & Cognition, 24*, 823–829.

Dack, L. A., & Astington, J. W. (2011). Deontic and epistemic reasoning in children. *Journal of Experimental Child Psychology, 110*, 94–114.

Elqayam, S., & Evans, J. St. B. T. (2011). Subtracting "ought" from "is": Descriptivism versus normativism in the study of human thinking. *Behavioral and Brain Sciences, 34*(5), 233–248.

Elqayam, S., Thompson, V. A., Wilkinson, M. R., Evans, J. St. B. T., & Over, D. E. (2015). Deontic introduction: A theory of inference from is to ought. *Journal of Experimental Psychology: Learning, Memory, and Cognition, 41*, 1516–1532.

Elqayam, S., Wilkinson, M. R., Thompson, V. A., Over, D. E., & Evans, J. St. B. T. (2017). Utilitarian moral judgment exclusively coheres with inference from is to ought. *Frontiers in Psychology: Cognitive Science, 8*.

Evans, J. St. B. T., & Elqayam, S. (2020). How and why we reason from is to ought. *Synthese, 197*, 1429–1446.

Evans, J. St. B. T., Neilens, H., Handley, S. J., & Over, D. E. (2008). When can we say 'if'? *Cognition, 108*, 100–116.

Evans, J. St. B. T., & Over, D. E. (1996). *Rationality and reasoning*. Hove, England: Psychology Press.

Evans, J. St. B. T., & Over, D. E. (2004). *If*. Oxford, England: Oxford University Press.

Evans, J. St. B. T., & Twyman-Musgrove, J. (1998). Conditional reasoning with inducements and advice. *Cognition, 69*, B11–B16.

Fillenbaum, S. (1976). Inducements: On phrasing and logic of conditional promises, threats and warnings. *Psychological Research, 38*, 231–250.

Gigerenzer, G., & Hug, K. (1992). Domain-specific reasoning: Social contracts, cheating and perspective change. *Cognition, 43*, 127–171.

Grice, P. (1975). Logic and conversation. In P. Cole & J. L. Morgan (Eds.), *Studies in syntax: Vol. 3. Speech acts* (pp. 41–58). New York, NY: Academic Press.

Griggs, R. A., & Cox, J. R. (1982). The elusive thematic materials effect in the Wason selection task. *British Journal of Psychology, 73*, 407–420.

Harris, P. L., & Núñez, M. (1996). Understanding of permission rules by preschool children. *Child Development, 67*, 1572–1591.

Henrich, J., Heine, S. J., & Norenzayan, A. (2010). The weirdest people in the world? *Behavioral and Brain Sciences, 33*, 61–83.

Hilton, D. J., Charalambides, L., & Hoareau-Blanchet, S. (2016). Reasoning about rights and duties: Mental models, world knowledge and pragmatic interpretation. *Thinking & Reasoning, 22*(2), 150–183.

Hudson, W. D. (Ed.). (1969). *The is–ought question: A collection of papers on the central problem in moral philosophy*. London, England: Macmillan.

Hume, D. (2000). *A treatise on human nature*. Oxford, England: Clarendon Press. (Original work published 1739–1740)

Kanngiesser, P., Köymen, B., & Tomasello, M. (2017). Young children mostly keep, and expect others to keep, their promises. *Journal of Experimental Child Psychology, 159*, 140–158.

Kilpatrick, S. G., Manktelow, K. I., & Over, D. E. (2007). Power of source as a factor in deontic inference. *Thinking & Reasoning, 13*, 295–317.

Manktelow, K. I., & Over, D. E. (1991). Social roles and utilities in reasoning with deontic conditionals. *Cognition, 39*, 85–105.

McNamara, P. (2010). Deontic logic. In E. N. Zalta (Ed.), *Stanford encyclopedia of philosophy*. Retrieved from http://plato.stanford.edu/archives/sum2010/entries/logic-deontic/

Miller, J. G., Wice, M., & Goyal, N. (2018). Contributions and challenges of cultural research on the development of social cognition. *Towards a Cultural Developmental Science, 50*, 65–76.

Newstead, S. E., Ellis, M. C., Evans, J. St. B. T., & Dennis, I. (1997). Conditional reasoning with realistic material. *Thinking & Reasoning, 3*, 49–76.

Newstead, S. E., Handley, S. J., Harley, C., Wright, H., & Farrelly, D. (2004). Individual differences in deductive reasoning. *Quarterly Journal of Experimental Psychology, 57*, 33–60.

Oaksford, M. (2014). Normativity, interpretation, and Bayesian models. *Frontiers in Psychology, 5*.

Oaksford, M., & Chater, N. (1994). A rational analysis of the selection task as optimal data selection. *Psychological Review, 101*, 608–631.

Oaksford, M., & Chater, N. (2007). *Bayesian rationality: The probabilistic approach to human reasoning*. Oxford, England: Oxford University Press.

Oaksford, M., & Chater, N. (2011). The "is–ought fallacy" fallacy. *Behavioral and Brain Sciences, 34*(5), 262–263.

Ohm, E., & Thompson, V. A. (2004). Everyday reasoning with inducements and advice. *Thinking & Reasoning, 10*, 241–272.

Over, D. E., & Elqayam, S. (2016). *From is to ought: The place of normative models in the study of human thought*. Lausanne, Switzerland: Frontiers Media.

Paulus, M., & Schmidt, M. F. H. (2018). The early development of the normative mind. *Journal of Experimental Child Psychology, 165*, 1–6.

Ragni, M., Kola, I., & Johnson-Laird, P. N. (2018). On selecting evidence to test hypotheses: A theory of selection tasks. *Psychological Bulletin, 144*, 779–796.

Rakoczy, H., Warneken, F., & Tomasello, M. (2008). The sources of normativity: Young children's awareness of the normative structure of games. *Developmental Psychology, 44*, 875–881.

Ridge, M. (2019). Moral non-naturalism. In E. N. Zalta (Ed.), *The Stanford encyclopedia of philosophy*. Retrieved from https://plato.stanford.edu/archives/fall2019/entries/moral-non-naturalism/

Schurz, G. (1997). *The is–ought problem: An investigation in philosophical logic*. Dordrecht, Netherlands: Kluwer.

Searle, J. R. (1969). *Speech acts: An essay in the philosophy of language*. Cambridge, England: Cambridge University Press.

Searle, J. R. (2005). What is an institution? *Journal of Institutional Economics, 1*, 1–22.

Stanovich, K. E. (1999). *Who is rational? Studies of individual differences in reasoning.* Mahwah, NJ: Erlbaum.

Stanovich, K. E. (2011). Normative models in psychology are here to stay. *Behavioral and Brain Sciences, 34*(5), 268–269.

Stanovich, K. E., & West, R. F. (1998). Cognitive ability and variation in selection task performance. *Thinking & Reasoning, 4,* 193–230.

Stanovich, K. E., & West, R. F. (2000). Individual differences in reasoning: Implications for the rationality debate. *Behavioral and Brain Sciences, 23,* 645–726.

Sztencel, M., & Clarke, L. (2018). Deontic commitments in conditional promises and threats: Towards an exemplar semantics for conditionals. *Language and Cognition, 10*(3), 435–466.

Thompson, V. A. (2000). The task-specific nature of domain-general reasoning. *Cognition, 76,* 209–268.

Thompson, V. A., Evans, J. St. B. T., & Handley, S. J. (2005). Persuading and dissuading by conditional argument. *Journal of Memory and Language, 53*(2), 238–257.

Wason, P. C. (1960). On the failure to eliminate hypotheses in a conceptual task. *Quarterly Journal of Experimental Psychology, 12,* 12–40.

11.3 Legal Logic

Eric Hilgendorf

Summary

Although the basic method in use for the application of law can be characterized as a form of "legal syllogism," the decision-making procedures raise much more difficult and complex issues, which are debated in legal theory and legal methodology. Formal logic plays virtually no role here; sophisticated types of logic such as deontic logic have so far only found an audience in legal theory. Legal logic should therefore not be understood as the theory of drawing valid logical conclusions but rather as the theory of methods of legal analysis and decision making. These decision-making methods exist along a continuum, with rule-based decision making at one end and the weighing of legal interests at the other end. Uncontrolled "weighing" sometimes threatens to supersede ordinary rule-based decision making.

1. Lawyers and Logic

Nowadays, legal logic is usually understood as a subject in legal theory or legal methodology. It ties in closely with conflict resolution in everyday life, by clarifying and structuring the forms of argument that can be found there.[1] Without this close connection between legal logic and our understanding of everyday life, the law would not be able to regulate and resolve social conflicts. Application of the law is always an exercise of power; this may explain why the persuasiveness of legal logic is ever being tested. Its practical orientation constitutes a central difference between legal logic and the (often highly formalized and thus otherworldly) logic that is practiced and taught mostly at philosophical faculties.[2]

"Legal logic" can be defined as the sum of all forms of argument and rules that lawyers orient themselves toward, or at least ought to orient themselves toward, in their legal activities (e.g., legal analyses or judicial decision making). This proposed definition opens up a wide range of applications for legal logic. It is not surprising that terms like "judicial logic" (Klug, 1982; Neumann, 2016), "legal logic"

(Hage, 2005; Joerden, 2010; Weinberger, 1970/1989), "normative logic" (Kalinowski, 1972), or "deontic logic" (Gabbay, Horty, & Parent, 2013; Joerden, 2017; Navarro & Rodríguez, 2014; von Wright, 1977)[3] are used in a variety of different ways today, sometimes employing irreconcilable definitions. For a very long time, the extreme positions between which debates on legal logic were held have been, on the one hand, the notion of strictly formal logic-based reasoning and legal decision making using the "legal syllogism" and, on the other hand, "free judicial determination" (*freie Rechtsfindung*), which exempts decision makers from any obligation to comply with rules or requirements of law previously conceived.

The way lawyers think is considered particularly "logical"; legal texts are often extolled for their "logic." Nevertheless, (deductive) logic, understood as a philosophical discipline, hardly plays any role in legal science and law. Even "deontic logic" was received only in legal theory and only to a limited degree. Together with general propositional logic and predicate logic, it has been made the subject of some interesting research during the renaissance in legal theory (Hilgendorf, 2005) in the 1960s and 1970s (Horovitz, 1972; Kalinowski, 1972; Koch & Rüßmann, 1982; Rödig, 1980; Tammelo, 1969; Wagner & Haag, 1970; Weinberger, 1974, 1979), without this work, however, being able to exert any relevant influence on legal practice. It is particularly noteworthy how lawyers, including almost all lawyers at university law faculties, keep their distance from any form of symbolic or mathematical logic. The same seems to be true for Anglo-Saxon countries. The rule *judex non calculat* still applies. Even a theory as committed to logical structures as the *Pure Theory of Law* by Hans Kelsen can dispense with formal logic (Kelsen, 1934/1960; cf. also Kelsen & Klug, 1981).

However, there are some indications that such reservation with respect to formal approaches to logic will no longer be sustainable in the future. Particularly in view of the progress made in the development of artificial (machine) intelligence, the question arises as to whether, and to what extent, legal norms can be represented in such a way that they can be processed by computer programs (Ashley, 2017). Such legal algorithms could be of great importance in practice. This would be true, for example, for automated driving: in the context of driving using automated systems, it is being increasingly discussed how traffic rules could be transformed so that they become "computer-readable." Cars could then be programmed with the applicable traffic law, rather than relying on the vehicles themselves to detect traffic instructions to drivers (e.g., street signs) via optical sensors, as has been the case up to now. An "algorithmization of the law" would presuppose, however, that legal rules could be structured and represented using formal logic to a far greater degree than has previously been done, so that computers can process them.[4]

The task of the law is to regulate conflicts and create legal certainty and predictability. Social conflicts should be resolved according to rules in an intersubjectively understandable manner, and there seems to be a fundamental human need for equal constellations of facts to be treated equally. For that reason, legal decision making right from the start has something inherently conservative about it. The search for consistent solutions to social conflicts requires that decision makers have a methodological tool at their disposal to help analyze conflicts, identify key issues, and develop solutions that correspond to the degree of consistency our sense of justice requires. This methodological instrument is "legal logic."[5]

The special reputation lawyers have of being able to argue logically is due to several factors: in order to analyze and solve a legal case in a systematic, rule-based manner, one has to ignore many peculiarities of the specific situation and instead focus on the abstract (and legally relevant) features of the case. This process of abstraction distinguishes the methodology of lawyers from unreflected conflict resolution, which seeks to intuitively settle individual cases. Legal reasoning consists in concentrating on what is legally relevant for coming to a decision, ignoring "purely subjective" issues. The lawyer, according to a widely held view of good legal practice, should "withdraw behind the law."

Rules of law used for dealing with social conflicts consist of a general abstract description of a factual situation (e.g., person X killed a human being) and a legal consequence (e.g., person X should be punished with life imprisonment).[6] The two elements, "factual situation" and "legal consequence," are linked in an "if–then relationship": "If situation X exists, then legal consequence Y should occur." Within the abstractly formulated situational description, several individual elements can be distinguished, for example, a description of the actor

(a person), a description of the target (object) that was affected by the actor's behavior (a human being), and a description of the action in question itself (here: killing). For legal assessment, only the elements of the offense(s) at issue are relevant; all other elements in the description of the facts are generally ignored as immaterial and irrelevant. Not giving consideration to nonmaterial facts can be interpreted as the special "objectivity" of legal factual analysis.

The judicial determination of the existence of individual elements of criminal offenses ought to be made in accordance with a system of evidence, which is described in detail in the rules of procedure, or, in some legal systems, in special rules governing the proceedings of the court. Violation of the rules of logic, which, in contrast to modern formal logic, are mostly understood as the "laws of sound thinking,"[7] makes a court decision *per incuriam* ("through lack of care") and is considered a ground for appeal. This expresses the special appreciation of logic in lawyers' thinking. However, it is not absolutely clear what "logic" is supposed to mean here. What is clear is that it includes deductive logic; errors of this kind of logic make a judicial decision *per incuriam*. But the conception of "legal logic" proposed in this chapter will be wider than the concept of logic employed by appellate courts.

Astonishingly, the concept of truth, that is, the idea of what it means to be "true," is little discussed in law. Lawyers follow, as a matter of course, the correspondence theory of truth (see, e.g., David, 2002/2016), which shows again the close connection between law and the concepts and methods of everyday life. More recent theories of truth, such as the "consensus theory of truth," have at best found an audience in legal philosophy and legal theory.[8]

Other topics from logic and the theory of science, which are also usually discussed in connection with theories of argumentation and reasoning, have so far received little attention in legal methodology. This applies, for example, to the analysis of counterfactual reasoning (Evans, 2017, pp. 46ff.),[9] to the use of heuristic methods,[10] and above all to the study of typical errors of reasoning, which include, for example, confounding causation and correlation (Evans, 2017, pp. 44ff.),[11] the distortion of rationality standards in argument by the inadvertent influence of stereotypes and cognitive biases (Evans, 2017, pp. 38ff., 54ff.),[12] and the notoriously error-prone use of statistics and probability (Evans, 2017, pp. 60ff.).

Any analysis of "legal logic" faces the difficulty that, unlike the patterns of argument of natural scientists or even mathematicians, patterns of argument of lawyers differ markedly according to the jurisdiction where they were trained. At least two major legal traditions, each with characteristic patterns of argument, can be identified around the world, namely, the civil law legal tradition (Germany, France, Italy, Spain, in a broader sense continental Europe, Russia, Central and East Asia including Japan, South Korea, and China, as well as Latin America), on the one hand, and the common-law legal tradition (England and many of its former colonies, including most of Canada, India, and the United States), on the other hand.

While positive law predominates in countries following the civil law legal tradition (i.e., statutes, codes, regulations according to our usual understanding), common-law countries rely much more on case law (precedents), also referred to as "court jurisprudence."[13] This chapter has been written by a lawyer trained in the civil law tradition. Learned colleagues trained in common-law jurisdictions will probably miss something or other and may also see things a bit differently. It needs to be made clear, however, that the differences between civil law and common-law legal reasoning are in many ways far smaller than is sometimes thought.

In the next section, the essential forms and structures of legal argumentation and reasoning,[14] the observance of which is regarded as a necessary condition of legal rationality in civil law legal systems, are presented. The German legal system, as one of the core legal systems of the civil law legal tradition, has been used for the purposes of illustration and comparison. The discussion is supplemented by references to the common-law legal model as well as to some subject areas that, in the author's view, have so far often been neglected in the literature on legal logic, such as research on "thinking and reasoning,"[15] the history of legal thinking, and the social and political significance of legal reasoning and argumentation.

2. The Legal Syllogism

The rules according to which lawyers deal with social conflicts today basically have the structure of "if–then" sentences: if X is the case, Y should happen. The first part of the sentence is the description of the relevant facts, and the second part refers to the legal consequences. In civil law jurisdictions, both the relevant facts and the legal consequences are fixed in positive law. Since the Enlightenment, lawyers and judges have been required to follow the written law; this was intended to restrict the judicial arbitrariness that was widespread in the early modern period, often accompanied by great cruelty, especially in the criminal law (Shaffern, 2009, pp. 203ff., 209).

The obligation to follow the law is to be ensured via a logically derivable relationship: a legal decision must be represented in the form of a syllogism with the law as its major premise, a description of the facts as its minor premise, and a concrete legal decision as its conclusion.

This concept of application of the law spread in Europe as early as the 16th century[16]; Thomas Hobbes (1588–1679) explicitly referred to the principle of the application of a general rule to the individual case (Hobbes, 1651/1841, pp. 193, 245). In Germany, the model of deductive legal analysis (*deduktive Rechtsdarstellung*) was discussed in the context of Christian Wolff's (1679–1754) rationalism. The logical structure of the process of application of the law was described in detail by Heinrich Köhler (1685–1737), a student of Wolff. Since then, it seems to have become the common intellectual property of legal philosophy, legal methodology, and practical philosophy.

The classic German-language treatise on legal logic, as it is understood to this day, is by Karl Engisch (1899–1990). His *Logische Studien zur Gesetzesanwendung* (*Logical Studies on the Application of Law*, 1963), first published in 1943, is still a standard work on the application of the law. According to Engisch, the goal of every application of the law is "to obtain concrete judgments on compliance with the law, linked with consequences for failure to comply" (Engisch, 1943/1963, p. 3). Engisch understood these judgments as being conditional statements of fact that have a truth value, that is, they can be verified as true or false (*Sollensurteile*). Thus, they are not commands (imperatives), which cannot be judged to be "true" but are merely "just" or "purposeful" statements.

Engisch's reconstruction attempt operates completely in the context of propositional logic. The logical structure of the application of the law can be represented as a syllogism:

If someone kills a human being, he should be punished.
Person A killed a human being.

∴ Person A should be punished.

Engisch's logical model of the application of law has been examined and restated several times; among these works, the treatise *Juristische Logik* (*Legal Logic*) by Ulrich Klug (1982) has attained special prominence.[17] In the 1980s, Hans-Joachim Koch and Helmut Rüßmann sought a more precise explanatory model using formal language (Koch & Rüßmann, 1982, pp. 31ff.). Their explanations, however, were hardly taken notice of outside legal theory. The same applies to prominent non-German legal theorists of the time, such as Aulis Aarnio, Carlos

Alchourrón, Eugenio Bulygin, Neil MacCormick, and Jerzy Wróblewski.

3. Legal Interpretation

Within the description of the relevant facts of a given legal norm, it is possible to further differentiate individual elements, for example (in criminal law), the perpetrator, the victim of the action, and the manner in which the action (killing, assault, insult, etc.) was performed. Where law is in written form, the elements are sometimes extremely complex. For example, the offense of theft in German criminal law is defined as follows: "Whosoever takes chattels (i.e., movable property, E.H.) belonging to another away from another with the intention of unlawfully appropriating them for himself or a third person" (§ 242 German Criminal Code [StGB]). This legal rule includes the following elements: "property," "belonging to another," "movable," and "takes away." To these objective elements are added the subjective elements "intention to appropriate" and, as an unwritten element, intention (*Vorsatz*) to steal (i.e., to "fulfill" all the objective elements).

By contrast, in the common-law legal tradition, the relevant rules are to a large extent not fixed in writing by the legislature but must be sought out by the lawyer through the analysis of previous case decisions made by judges with the help of "their own sense of fairness, reasonableness, custom, and good policy" (Schauer, 2012, p. 104). Special emphasis is placed on not deviating from the fundamental rules established in earlier cases (the principle of *stare decisis et non quieta movere*). Structurally, therefore, the difference between legal systems based on written law and those based on case law is far smaller than it might seem at first sight.

After determining what the individual elements of the applicable rule of law are, the next question is whether in the present case all elements of the offense are present, that is, whether person A killed person B or whether person A took from person B a movable object that was not his own property with the intention of claiming it for himself. Most difficulties in conducting the legal analysis of a practical case are to be found in this second step, that is, in the determination of whether the elements of the offense are present, rather than in the (ultimately trivial) logical structure of the application of the law.

In civil law systems, the interpretation of the law—or, better, the interpretation of the individual linguistic elements of the law—generally takes place using four methods, typically done in succession, which are usually called "canons of interpretation" (Larenz, 1991,

pp. 320ff.). The first step is to interpret the words in the legal norm in question according to their normal meaning. The second method is an analysis of the legislative history of a statute. The third method, usually referred to as "systematic interpretation," is used when the lawyer tries to gain information from the interpretation of other legal norms in order to interpret the statute at hand. The last method is the interpretation of the law according to its "sense and purpose," also known as "teleological interpretation." The wording of legal texts can usually be interpreted in a number of different ways. If the range of possible interpretations is not restricted by the historical or systematic methods of interpretation, then the lawyer is faced with the decision of choosing a particular interpretation. One rational way to choose would be to look at the expected consequences of the decision, to evaluate those consequences, and to make a decision in the light of that assessment. Legal interpretation thus presents itself as a kind of "social technology" (Albert, 1968/1991, pp. 207ff.).[18]

By means of the methods of legal interpretation, the application of law can be structured. These various steps and methods of interpretation have heterogeneous methodological sources. For example, while the interpretation of the wording (including its interpretation according to legal usage) can be seen as an essentially hermeneutical activity, historical interpretation is a historical-empirical procedure, namely, the examination of source material for information on a particular topic. This, in turn, needs to be differentiated from the systematic-interpretation approach, which requires a linguistic analysis of the relevant law and its associated interpretations in other legal statutes, court jurisprudence, and legal doctrine. The teleological-interpretation approach contains, on the one hand, a significant element of subjective interpretation; on the other hand, this subjective interpretation can be structured rationally, for example, on the basis of prognoses about the expected effects of certain interpretative approaches. In light of these prognoses, it is possible to make rational decisions on which interpretive approaches are preferred.

The process by which law is applied, as it has developed in the most important jurisdictions following the civil law legal tradition, does not allow for the law to be applied in a purely deductive way. The same is true for the Anglo-Saxon legal systems. It certainly does not seem possible to algorithmize the process by which the law is applied on the basis of current legal rationality and argumentation standards. All the same, the instruments used for the application of the law permit individual elements of the application process to be differentiated, the overall process to be structured, and the individual steps in decision making to be made transparent and thus criticizable and controllable. Therein lie the significant advantages of rational (i.e., systematic) legal decision making over chance, "gut feelings," or spontaneous emotions.[19]

4. Weighing of Interests

"Weighing" plays a special role in legal analysis and legal argument today, and its relevance even seems to be growing (Klatt, 2013). The concept of "weighing" is nearly omnipresent in law and legal science today. At the same time, the concept lacks a clear methodological structure; what the lawyer does when he "weighs" is therefore often unclear (also and especially for the lawyer himself). Therefore, from a "rule of law and not of men" perspective, a logical analysis of "weighing" is very important in order to better understand, predict, and control current practice.[20]

The focus of such a weighing process is the decision between two conflicting interests. In many cases, the "weighing" concerns rights. "Rights" may be defined as "legally protected interests": Should this chattel belong to person *A* or to person *B*? When a decision is being made on planning permission for a new construction project, does the public interest or do private interests prevail? Whose interests, those of patient *A* or those of patient *B*, get priority in a decision on the allocation of medical resources? Such questions can be decided by a weighing of interests.

In some cases, weighing of interests is required by law. A well-known example of this can be found in § 34 German Criminal Code (StGB), which states that an action is not unlawful in an emergency situation (i.e., based on necessity), "if upon weighing the conflicting legal interests, in particular the affected legal goods (*Rechtsgüter*) and the degree of danger facing them, the protected legal interest substantially outweighs the one interfered with." So, if a person takes the walking stick from a pedestrian in order to defend a child against a dog attacking the child, that person does not act unlawfully because the protected interests (the physical integrity and possibly even the life of the child) are more important ("weigh more") than the impaired interest (the interest of the pedestrian in having the use of his walking stick at this moment).

When a weighing of interests is conducted, however, what is compared are not simply the legal interests in themselves but also the probabilities with which a violation of those legal interests might take place. The parallels to classical decision theory are obvious.[21] Thus, a situation could exist in which one legal interest would

almost certainly be violated, while the infringement of a different legal interest appears possible but unlikely. An example would be a situation in which a motorist swerves to avoid hitting an injured person lying on the street, thereby opening up the possibility of hitting a pedestrian, injury to whom, however, seems unlikely. In such a situation, and taking into account the low risk of hitting the pedestrian, there is much that speaks in favor of choosing to swerve in order to save the injured person on the ground from an (almost) certain death. The result of the weighing of interests would therefore be that the swerving maneuver should be carried out in preference over killing the injured person.

A particularly difficult category of cases are situations in which life must be weighed against life. The debate about what to do in such situations can be traced all the way back to antiquity. The most famous example showing this weighing of interests is called the "Plank of Carneades" (Aichele, 2003). After their ship has sunk, two shipwrecked seamen are adrift on a wooden plank floating in the Mediterranean. As the plank becomes sodden with seawater, they can foresee that, at some point, it will not be able to carry both men any longer. May person *A* push person *B* into the water to save himself? Or is it person *B* who is allowed to push person *A* into the water? Is there any relevant difference between *A* and *B*? Maybe it is better that both drown? And what is the logic behind such decisions from a more theoretical perspective?

There are more recent variations of the problem of weighing life against life: the cases of cannibalism on the high seas in the 19th century (for example, the much-discussed English case referred to after the name of the ship as the "Mignonette Case"[22]); the hypothetical "Switchman's Problem," formulated by the German criminal lawyer Hans Welzel (which the British philosopher Philippa Foot transformed into the "Trolley Problem," generating through her writings a wide response in the academic literature in English[23]); and the decision of the German Federal Constitutional Court on the Aviation Security Act (2006),[24] where the question was whether the police could shoot down a passenger plane that has been hijacked by terrorists, causing the deaths of many innocent passengers, in order to save the lives of a significantly larger number of innocent people on the ground.

Such "life-against-life" constellations could also be resolved in principle on the basis of formulated rules of law (e.g., "If in a life-against-life situation only one person's life can be saved, then the person with characteristic *X* should be saved"). Apparently, it was already difficult for ancient ethicists and lawyers to agree on what characteristic *X* should be. In contrast, a weighing of interests is much more flexible and allows consideration of a wide variety of factors, such as the specifics of the individual case but also the specific preferences of the individual who is doing the weighing.

Solving such weighing-of-interests problems is a controversial issue. In most civil law countries, a justification for the killing of innocent people has been rejected and instead attention has been drawn to the possibility of lack of culpability (*Entschuldigung*) on the part of the perpetrators. Such inculpability presupposes, however, that the perpetrators themselves were in a desperate predicament (i.e., under duress), so that they cannot be held personally responsible despite having committed an unlawful act. In countries with less developed logical (or "structural") models of legal analysis (lawyers trained in continental law refer to this as "legal dogmatics" or "legal doctrine"), the distinction between justification and culpability is not (yet) made,[25] and the actor is simply held not to have incurred criminal liability, which, however, leaves unanswered the question as to whether the action as such is considered to be in conformity with the law, that is, "lawful," or whether the judicial system merely refrains from holding the actor to be personally blameworthy, that is, "culpable" of committing the offense.

Like hardly any other instrument in the lawyer's toolkit, the concept of the weighing of interests is based both on historically developed practice and on a more or less commonly accepted foundation of values and general societal preferences. It is obvious that with the pluralization of society, this common ground may become increasingly unstable. Weighing of interests is a competitive alternative to rule-based decision making in conflict situations. This is particularly problematic because there is no consensus about the basic rules and criteria for weighing interests. The sheer amount of individual subjective evaluation is very large. Therefore, the results of balancing-of-interests exercises are often extremely difficult to predict. In consequence, the often quite unreflected weighing of interests is regarded by many—and not without reason—as a severe threat to rule-based, reviewable, and thus also criticizable, legal decision making (Rückert, 2011).[26]

Along a hypothetical continuum with rule-based legal decision making at one end and legal decision making by means of the weighing of interests at the other, there is a wide range of specific forms of legal argumentation. This includes *argumentum e contrario* (argument from the contrary), *argumentum a fortiori* (argument from the stronger), and *argumentum ad absurdum* (argument to absurdity). None of these forms of legal argument is compelling from the perspective of logic.[27] Rather, their value

lies in the fact that certain patterns of argumentation that are accepted and used not only in law but also in day-to-day life can be structured more clearly and thus made reviewable.

5. Analogy

Analogy, meaning "comparison based on similarity," plays a special role as a legal argumentation method. Common lawyers will be reminded here of *persuasive precedent*.[28] Analogy may well be the most important method of legal reasoning. Assume there is a rule of law according to which the legal consequence R should occur if elements of the offense T_1 to T_n are present. In a second case constellation, the existence of the elements of the offense T_2 to T_n is established, but instead of T_1, T_1' is present. In such a situation, the question arises as to whether the legal consequence R should also occur. If one wishes to argue by analogy, this depends on whether T_1 and T_1' are sufficiently similar to each other. This way of argumentation seems to be acceptable not only in everyday practice but also in most fields of the law, even though from a theoretical perspective, the concept of "similarity" is extremely vague.[29]

In criminal law, however, under most legal systems, the argument by means of analogy to the detriment of the accused is prohibited. In order for criminal liability to be incurred, the exact elements of the offense required under the (written) criminal legal provisions must be met. This close connection between criminal liability and the precise letter of the criminal law dates back to the Enlightenment and is intended to impose limits on the arbitrary application of the law by judges.[30]

Looking at the details, the key elements of comparisons based on similarity are as much in need of clarification as are the structural elements of decisions made by balancing: the logic involved in decisions made by weighing and by analogy remains obscure.[31] One might even start with the assumption that the relevant factors influencing decisions vary from person to person. Thus, the psychology of decision making by weighing of interests or by analogy represents a promising future research question. A rationalization effect probably does occur, however; due to similar patterns of socialization, one would expect similar decisions based on the use of weighing and analogy. In addition, the key elements of legal decision making based on either method must be made explicit and thus become reviewable. In this way, the goal of subjecting legal decision making to rational criticism and control can be achieved.

6. Outlook

What is remarkable about legal logic is its close connection to the kinds of reasoning used in everyday life. One could regard legal logic as an abstracted structuring, in the form of legal rules, of real-life conflict-resolution strategies. Without this strong link, the law and its champions (i.e., lawyers) would not be accepted by society. The special character of legal logic is thus determined by its function. Nevertheless, legal logic does not seem arbitrary, nor is it merely an expression of specific cultures. Its basic elements—the syllogism by which the law is applied to the facts and the requirement that the facts be correctly established—seem to be valid in all cultures, even if there are enormous differences in the ways by which fact is ascertained. These structural similarities suggest that the different types of legal logic have a common root but that its further exploration falls within the domain, not of formal logic, but rather of empirical anthropology and psychology.

Notes

1. For a long time, research into everyday as well as scientific reasoning fell within the domain of rhetoric. In the 1970s and 1980s, a (general) theory of reasoning established itself on this basis, from which specifically legal theories of reasoning soon developed (cf. Bongiovanni et al., 2018; Feteris, 1999/2017; van Eemeren et al., 2014). What is common to almost all of these different notions is that deductive logic is considered to play a very small role in factual reasoning.

2. An excellent overview is found in Prakken and Sartor (2015; cf. also Hage, 2005). For a historical overview, see Krimphove (2017).

3. For more details, see chapter 11.1 by Horty and Roy (this handbook).

4. Preliminary work on this took place in the early years of legal informatics (e.g., Rödig, 1980). An excellent example for recent work is Prakken (2017).

5. On the meaning of "legal logic" as part of the doctrine of legal argumentation, see Perelman (1979), Feteris (1999/2017), and Bongiovanni et al. (2018).

6. This is, so to speak, the "standard" for the drafting of legal rules (cf. Larenz, 1991, pp. 250ff.; Zippelius, 2012, § 5).

7. The legal concept of "laws of sound thinking" seems extremely naive, considering the now very sophisticated literature on "thinking and reasoning." For a good introduction, see Evans (2017).

8. For a critical evaluation of the "consensus theory of truth," see Rescher (1995, pp. 44ff.).

9. For more detail, see chapter 6.1 by Starr (this handbook).

10. Strictly speaking, the flowcharts for the analysis of cases so popular in legal education and legal science could also be called "heuristic methods." For more details on heuristics and the law, see Gigerenzer and Engel (2006).

11. Compare also with section 7 in this handbook.

12. Compare also with chapter 2.4 by Fiedler, Prager, and McCaughey (this handbook).

13. It should be noted that precedent plays an important role both in common and civil law systems (cf. Glendon, Carozza, & Picker, 2015, chapter 11/2; Harris, 2016, pp. 190ff., 202ff.).

14. Compare with details in chapter 11.4 by Prakken, chapter 5.5 by Hahn and Collins, and chapter 5.6 by Woods (all in this handbook).

15. For a good summary, see Evans (2017); for legal perspectives, see Spellman and Schauer (2012) and Warner (2005).

16. See citations in Hilgendorf (1991, pp. 26ff.).

17. For a more modern and detailed analysis, see chapter 11.4 by Prakken (this handbook).

18. For the English translation, see Albert (2014, chapter 28).

19. Such forms of judicial decision making are often referred to as "qadi jurisprudence." (A "qadi" [or "kadi"] is a judge of a traditional sharia court.)

20. For a detailed analysis of "weighing," see Möllers (2017/2019, pp. 331–364). For an Anglo-Saxon perspective, see Broome (2006) and Lord and Maguire (2016).

21. Compare with chapter 8.2 by Peterson (this handbook).

22. R. v. Dudley and Stevens (1884) 14 QBD 273 DC.

23. See Hilgendorf (2018) for more details and references.

24. *Entscheidungen des Bundesverfassungsgerichts*, vol. 115, pp. 118–166 (judgment of 15.2.2006, Az. 1 BvR 357/05).

25. In countries belonging to the common-law legal tradition, the distinction between "justification defenses" and "excuse defenses" has still not been fully accepted in criminal legal theory, although it has been dealt with in more sophisticated treatments for decades (cf. Robinson, 1997, pp. 379ff., 477ff.).

26. Compare also with Ladeur (2004). For the Anglo-Saxon perspective on weighing, see Broome (2006) and Lord and Maguire (2016).

27. Klug (1982, pp. 137ff.).

28. Under the doctrine of binding judicial precedent, a decision by a superior court will bind lower courts in the same jurisdiction in later cases where the material facts are exactly the same. Sometimes, however, the material facts are slightly different. Then the superior court's decision is said to be a *persuasive* precedent

rather than a binding precedent, meaning that the lower court should consider it but is under no obligation to follow it.

29. For a thorough analysis, see Goldstone and Son (2012).

30. It is significant that one of the first legislative measures undertaken by the National Socialists after their seizure of power in Germany was the lifting of the ban on the use of analogy in criminal law. The rule "No punishment without law" was replaced with "No offense without punishment."

31. For an ambitious attempt to rationalize the concepts of "weighing" and "reasoning by analogy" from a logical perspective, see Hage (1997).

References

Aichele, A. (2003). Was ist und wozu taugt das Brett des Karneades? Wesen und ursprünglicher Zweck des Paradigmas der europäischen Notrechtslehre [What is the board of Carneades and what is it good for? Essence and original purpose of the paradigm of the European doctrine of the law of necessity]. In B. S. Byrd, J. Hruschka, & J. C. Joerden (Eds.), *Jahrbuch für Recht und Ethik 11* (pp. 246–268). Berlin, Germany: Duncker & Humblot.

Albert, H. (1991). *Traktat über kritische Vernunft* [Treatise on critical reason] (5th ed.). Stuttgart, Germany: UTB. (Original work published 1968)

Albert, H. (2014). *Treatise on critical reason*. Princeton, NJ: Princeton University Press.

Ashley, K. D. (2017). *Artificial intelligence and legal analytics: New tools for legal practice in the digital age*. Cambridge, England: Cambridge University Press.

Bongiovanni, G., Postema, G., Rotolo, A., Sartor, G., Valentini, C., & Walton, D. (Eds.). (2018). *Handbook of legal reasoning and argumentation*. Dordrecht, Netherlands: Springer.

Broome, J. (2006). *Weighing lives*. Oxford, England: Oxford University Press.

David, M. (2016). The correspondence theory of truth. In E. N. Zalta (Ed.), *The Stanford encyclopedia of philosophy*. Retrieved from https://plato.stanford.edu/archives/fall2016/entries/truth-correspondence/

Engisch, K. (1963). *Logische Studien zur Gesetzesanwendung* [Logical studies on the application of law] (3rd ed.). Heidelberg, Germany: Winter. (Original work published 1943)

Evans, J. St. B. T. (2017). *Thinking and reasoning: A very short introduction*. Oxford, England: Oxford University Press.

Feteris, E. T. (2017). *Fundamentals of legal argumentation: A survey of theories on justification of judicial decisions* (2nd ed.). Dordrecht, Netherlands: Springer. (Original work published 1999)

Gabbay, D. M., Horty, J., & Parent, X. (Eds.). (2013). *Handbook of deontic logic and normative systems*. Milton Keynes, England: College Publications.

Gigerenzer, G., & Engel, C. (Eds.). (2006). *Heuristics and the law*. Cambridge, MA: MIT Press.

Glendon, M. A., Carozza, P., & Picker, C. (Eds.). (2015). *Comparative legal traditions in a nutshell* (4th ed.). St. Paul, MN: West Academic.

Goldstone, R. L., & Son, J. Y. (2012). Similarity. In K. J. Holyoak & R. G. Morrison (Eds.), *The Oxford handbook of thinking and reasoning* (pp. 155–176). Oxford, England: Oxford University Press.

Hage, J. C. (1997). *Reasoning with rules*. Dordrecht, Netherlands: Springer.

Hage, J. (2005). *Studies in legal logic*. Dordrecht, Netherlands: Springer.

Harris, P. (2016). *An introduction to law* (8th ed.). Cambridge, England: Cambridge University Press.

Hilgendorf, E. (1991). *Argumentation in der Jurisprudenz: Zur Rezeption von analytischer Philosophie und kritischer Theorie in der Grundlagenforschung der Jurisprudenz* [Argumentation in jurisprudence: On the reception of analytic philosophy and critical theory in basic research in jurisprudence]. Berlin, Germany: Duncker & Humblot.

Hilgendorf, E. (2005). *Die Renaissance der Rechtstheorie zwischen 1965 und 1985* [The renaissance of legal theory between 1965 and 1985]. Würzburg, Germany: Ergon.

Hilgendorf, E. (2018). Dilemma-Probleme beim automatisierten Fahren: Ein Beitrag zum Problem des Verrechnungsverbots im Zeitalter der Digitalisierung [Dilemma problems in automated driving: A contribution to the problem of balancing in the age of digitization]. *Zeitschrift für die gesamte Strafrechtswissenschaft, 130*, 674–703.

Hobbes, T. (1841). *Philosophical rudiments concerning government and society*. In W. B. Molesworth (Ed.), *The English Works of Thomas Hobbes of Malmesbury* (Vol. 2). London, England: Longman, Brown, Green, and Longmans. (Original work published 1651)

Horovitz, J. (1972). *Law and logic: A critical account of legal argument*. Vienna, Austria: Springer.

Joerden, J. C. (2010). *Logik im Recht: Grundlagen und Anwendungsbeispiele* [Logic in law: Fundamentals and examples of application] (2nd ed.). Berlin, Germany: Springer. (Original work published 2005)

Joerden, J. C. (2017). Deontische Logik [Deontic logic]. In E. Hilgendorf & J. C. Joerden (Eds.), *Handbuch Rechtsphilosophie* [Handbook of legal philosophy] (pp. 242–245). Stuttgart, Germany: Metzler.

Kalinowski, G. (1972). *Einführung in die Normenlogik* [Introduction to the logic of norms]. Frankfurt/Main, Germany: Athenäum.

Kelsen, H. (1960). *Reine Rechtslehre* [Pure theory of law] (2nd ed.). Vienna, Austria: Mohr Siebeck. (Original work published 1934)

Kelsen, H., & Klug, U. (1981). *Rechtsnormen und logische Analyse: Ein Briefwechsel 1959–1965* [Legal norms and logical analysis: A correspondence 1959–1965]. Vienna, Austria: Deuticke.

Klatt, M. (Ed.). (2013). *Prinzipientheorie und Theorie der Abwägung* [Principle theory and theory of balancing]. Tübingen, Germany: Mohr Siebeck.

Klug, U. (1982). *Juristische Logik* [Legal logic] (4th ed.). Berlin, Germany: Springer.

Koch, H.-J., & Rüßmann, H. (1982). *Juristische Begründungslehre* [Theory of legal reasoning]. Munich, Germany: Beck.

Krimphove, D. (2017). A historical overview of the development of legal logic. In D. Krimphove & G. M. Lentner (Eds.), *Law and logic: Contemporary issues* (pp. 12–52). Berlin, Germany: Duncker & Humblot.

Ladeur, K.-H. (2004). *Kritik der Abwägung in der Grundrechtsdogmatik: Plädoyer für eine Erneuerung der liberalen Grundrechtstheorie* [Critique of weighing in the dogmatics of fundamental rights: A plea for a renewal of the liberal theory of fundamental rights]. Tübingen, Germany: Mohr Siebeck.

Larenz, K. (1991). *Methodenlehre der Rechtswissenschaft* [Methodology of law]. Berlin, Germany: Springer.

Lord, E., & Maguire, B. (Eds.). (2016). *Weighing reasons*. Oxford, England: Oxford University Press.

Möllers, T. M. J. (2019). *Juristische Methodenlehre* [Legal methodology] (2nd ed.). Munich, Germany: Beck. (Original work published 2017)

Navarro, P. E., & Rodríguez, J. L. (2014). *Deontic logic and legal systems*. New York, NY: Cambridge University Press.

Neumann, U. (2016). Juristische Logik [Judicial logic]. In A. Kaufmann, W. Hassemer, & U. Neumann (Eds.), *Einführung in die Rechtsphilosophie und Rechtstheorie der Gegenwart* (9th ed., pp. 272–290). Heidelberg, Germany: Müller.

Perelman, C. (1979). *Juristische Logik als Argumentationslehre* [Judicial logic as argumentation theory]. Freiburg, Germany: Alber.

Prakken, H. (2017). On making autonomous vehicles respect traffic law: A case study for Dutch law. In *Proceedings of the 16th International Conference on Artificial Intelligence and Law, London (UK) 2017* (pp. 241–244). New York, NY: ACM Press.

Prakken, H., & Sartor, G. (2015). Law and logic: A review from an argumentation perspective. *Artificial Intelligence, 227*, 214–245.

Rescher, N. (1995). *Pluralism: Against the demand for consensus*. Oxford, England: Oxford University Press.

Robinson, P. H. (1997). *Criminal law*. New York, NY: Aspen Law and Business.

Rödig, J. (1980). *Schriften zur juristischen Logik* [Studies on legal logic]. Berlin, Germany: Springer.

Rückert, J. (2011). Abwägung—die juristische Karriere eines unjuristischen Begriffs oder: Normenstrenge und Abwägung im Funktionswandel [Balancing—the legal career of an unlegal term or: Strictness of norms and balancing in the change of function]. *Juristenzeitung, 66*(19), 913–923.

Schauer, F. (2012). *Thinking like a lawyer: A new introduction to legal reasoning*. Cambridge, MA: Harvard University Press.

Shaffern, R. W. (2009). *Law and justice from antiquity to enlightenment*. Lanham, MD: Rowman & Littlefield.

Spellman, B. A., & Schauer, F. (2012). Legal reasoning. In K. J. Holyoak & R. G. Morrison (Eds.), *The Oxford handbook of thinking and reasoning* (pp. 719–735). Oxford, England: Oxford University Press.

Tammelo, I. (1969). *Outlines of modern legal logic*. Wiesbaden, Germany: Steiner.

van Eemeren, F. H., Garssen, B., Krabbe, E. C. W., Snoeck Henkemans, A. F., Verheij, B., & Wagemans, J. H. M. (2014). *Handbook of argumentation theory*. Dordrecht, Netherlands: Springer.

von Wright, G. H. (1977). *Handlung, Norm und Intention: Untersuchungen zur deontischen Logik* [Action, norm, and intention: Studies in deontic logic] (H. Poser, Ed.). Berlin, Germany: De Gruyter.

Wagner, H., & Haag, K. (1970). *Die moderne Logik in der Rechtswissenschaft* [Modern logic in law]. Bad Homburg, Germany: Gehlen.

Warner, R. (2005). Adjudication and legal reasoning. In M. P. Golding & W. A. Edmundson (Eds.), *The Blackwell guide to the philosophy of law and legal theory* (pp. 259–270). Oxford, England: Blackwell.

Weinberger, O. (1974). *Studien zur Normenlogik und Rechtsinformatik* [Studies in the logic of norms and legal informatics]. Berlin, Germany: Schweitzer.

Weinberger, O. (1979). *Logische Analyse in der Jurisprudenz* [Logical analysis in law]. Berlin, Germany: Duncker & Humblot.

Weinberger, O. (1989). *Rechtslogik: Versuch einer Anwendung moderner Logik auf das juristische Denken* [Legal logic: An attempt to apply modern logic to legal reasoning] (2nd ed.). Berlin, Germany: Springer. (Original work published 1970)

Zippelius, R. (2012). *Juristische Methodenlehre* [Legal methodology] (11th ed.). Munich, Germany: Beck. (Original work published 1971 as *Einführung in die juristische Methodenlehre*)

11.4 Logical Models of Legal Argumentation

Henry Prakken

Summary

This article reviews logical models of legal argumentation as they are being developed in the field of artificial intelligence and law. One aim is to show that several aspects of legal reasoning that have often been claimed to escape formalization can very well be formalized. First the limitations of logical deduction as a model of legal reasoning are discussed. Then models of defeasible legal reasoning and legal argumentation, which focus on the generation and comparison of reasons or arguments for and against legal claims, are discussed. Models of rule-based defeasible reasoning address exceptions and rule conflicts. Models of legal interpretation emphasize the interplay between rules and cases and the role of principles, purposes, and values. Models of legal proof account for the uncertainty in legal proof in three alternative ways: with Bayesian probability theory, with stories, and with argumentation. Finally, dialogue models of legal argument address issues of procedural rationality and strategic choice.

1. Logic, Argumentation, and Law

Models of rationality in the law must be diverse. In a legal case, various types of decisions must be made, including determining the facts, classifying the facts under legal concepts, and deriving legal consequences from the classified facts, and each has its own modes of reasoning. Therefore, in the law, theoretical and practical reasoning must be combined. When determining the facts, the modes of reasoning are often probabilistic and may involve reasoning about causation and about mental attitudes such as intent. Classifying the facts under legal concepts involves interpretation. Here the prevailing modes of reasoning are analogy, appeals to precedent, and the balancing of interests, purposes, and values. Finally, when deriving legal consequences from the classified facts, the main modes of reasoning are deductive but leave room for exceptions to rules and for the resolution of rule conflicts.

While legal reasoning is diverse, still a unified account can be given in terms of argumentation. A simple but naive model of legal reasoning sees it as logically deducing legal consequences from a precisely stated body of facts and legal rules. On this account, once a legal text and a body of facts have been clearly represented in a logical language, the valid inferences are determined by the meaning of the representations, and so techniques of automated deduction apply. However, this mechanical approach leaves out much of what is important in legal reasoning. To start with, the law is not just a conceptual or axiomatic system but has social objectives and effects, which must be taken into account when applying the law. For example, in the well-known *Riggs v. Palmer* case in American inheritance law (discussed by Dworkin, 1977), a grandson had killed his grandfather and then claimed his share in the inheritance. The court made an exception to inheritance law based on the principle that one cannot profit from their own wrongdoing. Moreover, legislators cannot fully predict in which circumstances the law will have to be applied, so legislation has to be formulated in general and abstract terms, such as "duty of care" or "misuse of trade secrets," and qualified with general exception categories, such as "self-defense" or "unreasonable." Such concepts and exceptions must be interpreted in concrete cases, which creates room for disagreement. This is the more so since legal cases often involve conflicting interests of opposing parties. The prosecution in a criminal case wants the accused convicted while the accused wants to be acquitted, and the plaintiff in a civil lawsuit wants to be awarded compensation for damages while the defendant wants to avoid having to pay. These three aspects of the law—the possibility that legal rules have unjust outcomes, the tension between the general terms of the law and the particulars of a case, and the adversarial nature of legal procedures—cause legal reasoning to go beyond the meaning of the legal rules. It involves appeals to precedent, principle, policy, and purpose, as well as the consideration of reasons for and against drawing certain conclusions. Therefore, law relies not just on deduction but also on argument.[1] This chapter aims to show that these aspects of legal reasoning, which have often been claimed to escape formalization (see, e.g., chapter 11.3 by Hilgendorf, this handbook) can very well be formalized.

Accordingly, this chapter[2] reviews logical models of legal argumentation, which have mostly been developed in the field of artificial intelligence (AI) and law, as formal underpinnings of implemented AI systems for supporting legal argumentation and decision making. In reviewing the various approaches, we will focus more on their underlying ideas and concepts than on their technical particulars. First applications of nonmonotonic logics for dealing with rule exceptions and rule conflicts are discussed. Then, argumentation-based models of legal interpretation that emphasize the interplay between rules and cases, as well as the role of principles, purposes, and values, are reviewed. As regards fact finding, argumentation-based, story-based, and probabilistic models of legal proof are briefly compared. Finally, dialogical models of legal argument and their use in modeling legal-procedural notions are discussed.

2. A First Challenge for Deductive Models: Exceptions and Rule Conflicts

Any attempt to logically model reasoning with legal rules faces the challenge that legal rules can have exceptions and that several conflicting legal rules can apply to a case. One problem is that the law often makes legal effects dependent on the nonavailability of evidence for exceptions. Consider a legal rule "An offer and an acceptance create a binding contract" with an exception in case the offeree was insane when accepting the offer: as long as there is no evidence that the offeree was insane, the general rule can be applied to defeasibly conclude that there is a binding contract. If later evidence is provided that the offeree was insane, this conclusion must be retracted. The same holds for exceptions that come from sources other than statutes, as in the abovementioned *Riggs v. Palmer* case, where a statutory rule was overridden by an unwritten principle. The making and retracting of such "defeasible" inferences goes beyond standard deductive models of reasoning. Instead, nonmonotonic logics[3] should be used. Such logics allow the making of inferences subject to evidence to the contrary and the retraction of such inferences when evidence to the contrary comes in. They thus formalize one influential idea introduced by Toulmin (1958), namely, that "good" arguments can be defeasible in that it is sometimes rational to accept all their premises but not their conclusion.

For representing legal rules and exceptions, two different logical techniques have been used. The first technique adds an additional condition "unless there is an exception" to every rule and combines it with a nonprovability operator, such as logic programming's negation-as-failure (Sergot et al., 1986). Thus, for example, the above legal contract rule and its exception can be represented as follows (where "$\sim$" stands for nonprovability):

r_1:   Offer & Acceptance & $\sim$Exception(r_1) $\Rightarrow$ Binding Contract

r_2:   Insane $\Rightarrow$ Exception(r_1)

r_3:   Insane & $\sim$Exception(r_3) $\Rightarrow$ Not BindingContract

The second technique states a priority between a general rule and an exception:

r_1:   Offer & Acceptance $\Rightarrow$ BindingContract

r_2:   Insane $\Rightarrow$ Not BindingContract

$r_1 < r_2$

Several nonmonotonic logics can also model reasoning about the resolution of rule conflicts. Legal systems have general principles for resolving rule conflicts, such as preferring the more specific or more recent rule, or preferring the rule that is hierarchically superior (e.g., "The constitution is superior to statutes"). Moreover, specific statutes can have specific conflict resolution rules such as "Rules concerning labor contracts override rules from general contract law." If no clear conflict resolution principles apply or if these principles themselves conflict, then some approaches allow reasoning about which rule preference holds (e.g., Gordon, 1994; Prakken & Sartor, 1997; Verheij, Hage, & van den Herik, 1998).

Although rule-based nonmonotonic techniques technically deviate from deductive logic, their spirit is still the same, namely, that of deriving consequences from clear and unambiguous representations of legal rules and facts. The logical consequences of a representation are still clearly defined. While technically, most nonmonotonic logics allow for alternative conclusion sets (see further section 4 below), in legal practice, statutory rule–exception structures and legislative hierarchies still usually yield unambiguous outcomes. More often, conflicts arise not from competing norms but from the variety of ways in which they can be interpreted. A real challenge for deductive accounts of legal reasoning is the gap between the general legal language and the particulars of a case. Because of this gap, disagreement can arise, and it will arise from the conflicts of interest between the parties.

3. A Second Challenge for Deductive Models: Bridging the Gap between Legal Language and the World

One might think that disagreements about interpretation are resolved in concrete cases by courts, so that additional rules can be found in case law. If different courts disagree on an interpretation, such disagreements could be represented with nonmonotonic techniques for handling conflicting rules. This was indeed the approach taken by Gardner (1987), who designed a program for so-called issue spotting in law school contract-law exam problems. Given an input case, the task of the program was to determine which legal questions involved were easy and which were hard, and to solve the easy ones. If all the questions were found easy, the program reported the case as clear, otherwise as hard. The system contained domain knowledge of three different types: legal rules, commonsense rules (e.g., on the interpretation of utterances like "Will you supply . . . ?"), and rules extracted from cases. The program considered a question hard if either "the rules run out" or different rules or cases point at different solutions, unless it had reasons to prefer one solution over the other. For example, it preferred case-law rules over conflicting legal or commonsense rules.

A problem with such approaches is that rules derived from precedents are often specific to the case, so a new case will rarely exactly match the precedent, and techniques for handling conflicting rules fall short. Instead, reasoning forms are called for in which case-law rules can be refined and modified. Here, factors and reasons play an important role, and analogies between cases are drawn and criticized. Such observations led to a shift in focus away from "traditional" nonmonotonic logics and toward argumentation-based approaches, in part drawing on general AI models of argumentation such as Dung (1995), Pollock (1995), and later Modgil and Prakken (2013). Another approach is reason-based logic (Hage, 1996; Verheij et al., 1998), a nonmonotonic logic that models defeasible reasoning as the weighing of reasons for and against a conclusion, where rules are just one source of reasons, along with, for example, precedents or principles.

4. Intermezzo: Logical Models of Argumentation

Logical models of argumentation as developed in AI view defeasible inference as an inherently dialectical matter: an argument warrants its conclusion if it is acceptable, and an argument is acceptable if, first, it is properly constructed and, second, it can be defended against counterarguments. They thus formalize the second influential idea introduced by Toulmin (1958), namely, that outside mathematics, the validity of arguments does not depend on their syntactic form but on whether they can be defended in a rational dispute. Argumentation logics define three things: how arguments can be constructed, how they can be attacked by counterarguments, and how they can be defended against such attacks. Argumentation logics are a form of nonmonotonic logic, since new information may give rise to new counterarguments against arguments that were originally acceptable.

The basis of much logical work reviewed below is Dung's (1995) famous abstract account of argumentation, which just assumes a set of *arguments* (whatever they look like) with a binary relation of *attack* (on whatever grounds). Dung's theory characterizes various sets S of arguments that are *admissible* in that

1. S is *conflict free* (i.e., no argument in S attacks an argument in S), and

2. S *defends* all its members (i.e., all attackers of a member of S are attacked by a member of S).

Each admissible set is a coherent point of view. Reasoners are usually interested in subset-maximal points of view, called *extensions*. Sometimes a theory has a unique extension and sometimes multiple, mutually incompatible ones. Consider the two example graphs in figure 11.4.1, in which the nodes are arguments and the links are attack relations.

In figure 11.4.1a, there is a unique maximal admissible set, namely, {A,C} ({C} and the empty set are also admissible but they are included in {A,C}). Note that {A,B} is not admissible since it is not conflict free, while {B} is not admissible since it does not defend B against C. So in figure 11.4.1a, both A and C are warranted while B is not warranted.

In figure 11.4.1b, there are two maximal admissible sets, namely, {A} and {B}. In the first, A defends itself against B while in the second, B defends itself against A. Note that {A,B} is not admissible since it is not conflict free. So in figure 11.4.1b, neither argument is warranted, but both A and B are "defensible" in that they belong to alternative coherent points of view.

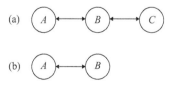

Figure 11.4.1
Two abstract argumentation frameworks.

As for the structure of arguments and the nature of attack relations, various approaches exist. An early example is Prakken and Sartor's (1997) model of reasoning with legal rules mentioned above in section 2. More recent approaches are inspired by the concept of argument, or "argumentation," schemes from argumentation theory[4] (cf. Walton, 1996). Argument schemes model stereotypical forms of defeasible, or "presumptive," reasoning, such as reasoning with defeasible legal rules, reasoning with defeasible generalizations; reasoning from expert or witness testimony; various forms of analogical, causal, or temporal reasoning; and schemes for practical reasoning (in terms of goals fulfilled or violated by decision options). Each scheme comes with a set of "critical questions," which point at possible exceptions to the scheme (e.g., Does applying this legal rule violate its purpose? Was the witness biased? Are there relevant differences between these analogous cases? Are there better ways to fulfill the same goal?). In the ASPIC+ framework (Modgil & Prakken, 2013), argument schemes can be modeled as deductive or defeasible inference rules, and applications of such rules can be chained into tree-style arguments. All arguments can be attacked on their premises while arguments using defeasible rules can also be attacked on the use of these rules. Then a preference relation on arguments can be defined to see which attacks succeed as defeats, after which Dung's theory of abstract argumentation can be used with attacks replaced by defeats. There are similar approaches that are not within ASPIC+ or Dung's theory, such as Carneades (Gordon & Walton, 2009).

5.   Interpreting Legal Concepts with Cases

Ways of reasoning "when the rules run out" can be found in their most explicit form in common-law jurisdictions, which traditionally were not statute- but precedent-based. This led to reasoning forms where legal rules are formulated by courts in the context of particular cases and are constantly refined and modified to fit new circumstances that were not originally taken into account. However, these reasoning forms can also be found in civil law jurisdictions, since the concepts in statutory rules also have to be interpreted.

5.1.   System-Oriented Work on Legal Case-Based Reasoning

A recent attempt to formalize traditional canons of statutory interpretation (see chapter 11.3 by Hilgendorf, this handbook) with argument schemes is Walton, Sartor, and Macagno (2016). A limitation of this work is that interpretation canons are themselves often vague and conflicting. Much AI & law work on the interpretation of legal concepts instead centers on the notions of a dimension, introduced in the HYPO system of Ashley (1990), and the related notion of a factor, originating from the CATO system of Aleven (2003) (for a detailed recent review, see Bench-Capon, 2017). *Factors* are abstractions of fact patterns that favor ("pro-factors") or oppose ("con-factors") a conclusion. Factors are thus in an intermediate position between the specific facts of a case and the legal predicates to which such facts may be relevant. For example, in CATO, which like HYPO argues about misuse of trade secrets, some factors pro misuse are that a nondisclosure agreement was signed, that the plaintiff had made efforts to maintain secrecy, and that the copied product was unique, and some factors con misuse are that disclosures were made by the plaintiff in negotiations and that the information was reverse-engineerable. While in CATO, factors are a basic notion, in HYPO, they are defined in terms of *dimensions*, which can have a range of values. Examples of dimensions are the number of people to whom a trade secret has been disclosed and the severity of the security measures. Different dimension values can favor different sides and to varying degrees, depending on their place in the range of values. Factors are then dimension–value pairs that favor a particular side to some degree. Generally no clear rules can be given on how dimensions and factors relate to legal predicates, since each case can have different constellations of factors with different degrees of favoring a side. Hence the importance of past cases.

HYPO and CATO generate disputes between a plaintiff and a defendant of a legal claim concerning misuse of a trade secret. Each move conforms to certain rules for analogizing and distinguishing precedents given that cases contain sets of factors (in HYPO dimension–value pairs) for and against a decision, plus the decision that resolves the conflict between the competing factors. A precedent is *citable* for a side if it has the decision wished by that side and shares with the current case at least one factor that favors that decision. Thus, citable precedents do not have to exactly match the current case, which is a way of coping with the case-specific nature of case-law decisions. A citation can be countered by a *counterexample*, that is, by producing a citable precedent that has the opposite outcome. A citation may also be countered by *distinguishing*, that is, by indicating a factor in the current case that is absent in the cited precedent and that supports the opposite outcome, or a factor in the precedent that is missing in the current case and that supports the outcome of the cited case. Dimensions allow an additional way to distinguish a precedent, namely, on a shared pro-decision factor that more strongly favors the decision in the precedent than in the current case.

CATO (Aleven, 2003) also has a "factor hierarchy," which expresses expert knowledge about the relations between the various factors: more concrete factors are reasons for or against the more abstract factors to which they are linked. Thus, the factor hierarchy can be used to explain why a certain decision was taken, which in turn facilitates debate on the relevance of differences between cases. For instance, the hierarchy positively links the factor *security measures taken* to the more abstract concept *efforts to maintain secrecy*. Now if a precedent contains the first factor but the current case lacks it, then the precedent can be distinguished on the absence of *security measures taken*, and this distinction can be emphasized by saying that thus no efforts were made to maintain secrecy. However, if the current case also contains a factor *agreed not to disclose information*, then the factor hierarchy enables downplaying this distinction, since it also positively links this factor to *efforts to maintain secrecy*: the party that cited the precedent can say that in the current case, just as in the precedent, efforts were made to maintain secrecy.

5.2. Logical Accounts of Legal Case-Based Reasoning

Various logical accounts of factor-based reasoning with cases have been given. A key idea here is that case decisions give rise to conflicting rules (or conflicting sets of reasons) plus a preference rule expressing how the court resolved this conflict. In the notation of Prakken and Sartor (1998), inspired by Hage (1996) and Loui and Norman (1995):

r_1: *Pro-factors* $\Rightarrow$ *Decision*

r_2: *Con-factors* $\Rightarrow$ *Not Decision*

p: $\ldots \Rightarrow r_1 > r_2$

The priority expresses the court's decision that the pro-factors in the body of rule r_1 together outweigh the con-factors in the body of rule r_2. Since the preference is expressed as a rule, arguments can be modeled about why one set of factors outweighs another set. This approach also allows for "a fortiori" reasoning in that adding factors to a pro-decision rule or removing factors from a con-decision rule does not affect the rule priority. It allows for analogical uses of a pro-decision rule by deleting pro factors, which is called "broadening a rule." A broadened rule does not inherit the priority relations of the rule from which it is obtained. So if rule r_1 in our schematic example is broadened to rule r_1' by deleting one of the pro-factors in rule r_1, then one cannot

conclude from the priority $r_1 > r_2$ in the precedent that $r_1' > r_2$, since the deleted factor might have been essential in preferring r_1 over r_2.

Later work has formalized these ideas in an argument-scheme approach. For example, Prakken, Wyner, Bench-Capon, and Atkinson (2015) model argument schemes for deriving preferences between factor sets from cases, for citing cases, for distinguishing case citations, and for downplaying distinctions within the ASPIC+ framework. Uses of these schemes yield a Dung-style argumentation framework that can be used for evaluating the arguments.

Horty (2011) addresses the issue of *precedential constraint*, that is, under what conditions is a decision in a new case allowed or forced by a body of precedents? Consider the *Keeble* case from the common law of property, part of a well-known series of cases on ownership of wild animals that are being chased. In *Keeble*, a pond owner placed a duck decoy in his pond with the intention to sell the caught ducks for a living. The defendant used a gun to scare away the ducks, for no other reason than to damage the plaintiff's business. Here the court held for the plaintiff. The pro-plaintiff factors were that the plaintiff was hunting for a living (*PlLiving*) and was hunting on his own land (*OwnLand*). The single pro-defendant factor was that the animals were not yet caught (*Not Caught*). The issue is whether the plaintiff became the owner of the ducks (*Owner*). In the notation of Prakken and Sartor (1998), we have:

Keeble

k_1: *PlLiving, OwnLand* $\Rightarrow$ *Owner*

k_2: *Not Caught* $\Rightarrow$ *Not Owner*

$\Rightarrow$ $k_1 > k_2$

In another precedent, *Young*, both plaintiff and defendant were fishermen fishing in the open sea. Just before the plaintiff closed his net, the defendant came in and caught the fishes with his own net. Here, not only the plaintiff but also the defendant were hunting for a living (*DefLiving*). Then we have:

Young

y_1: *PlLiving* $\Rightarrow$ *Owner*

y_2: *Not Caught, DefLiving* $\Rightarrow$ *Not Owner*

To decide *Young* for the plaintiff, the required priority $y_1 > y_2$ cannot be based on the precedent, since y_1 lacks one antecedent of k_1 and also since y_2 adds a con-factor to k_2. However, deciding *Young* in accordance with *Keeble* is still allowed by *Keeble*, since that leaves the decision in the precedent unaffected. Horty calls this *following* the precedent.

The situation is different if the case base also contains a second precedent that is almost like *Keeble* except that the defendant also hunted for a living and in which the defendant won:

Precedent 2

p_1: *PlLiving, OwnLand ⇒ Owner*

p_2: *Not Caught, DefLiving ⇒ Not Owner*

⇒ $p_2 > p_1$

The priority $p_2 > p_1$ then implies $y_2 > y_1$, since y_2 has the same con-decision factors as p_2 while y_1 lacks one pro-decision factor of p_1. But $y_2 > y_1$ is inconsistent with $y_1 > y_2$, so deciding *Young* as *Keeble* is not allowed by the extended case base, since it would amount to *overruling* the second precedent.

In Horty's approach, not all deviations from a precedent are overrulings. Suppose that precedent 2 is not in the case base but is a new case, so its decision is not yet known. Then, $p_2 > p_1$ is consistent with $k_1 > k_2$, so deciding the new case differently than in the precedent is allowed by the case base, since it leaves all decisions in precedents unaffected. Horty here says that deciding con the original decision in the new case *distinguishes* the precedent. Horty has thus given precise logical formalizations of the important common-law notions of following, distinguishing, and overruling a precedent.

The work described so far does not address why factors are pro or con a decision. Berman and Hafner (1993) argued that often a factor favors a decision by virtue of the purposes served or values promoted by taking that decision because of the factor. Cases are thus not compared on the factors they contain but on the values they promote or demote through these factors. Bench-Capon and Sartor (2003) computationally modeled this approach and illustrated it with the series of cases from American property law on ownership of wild animals that are being chased. For example, the plaintiff in *Keeble* could argue that people should be protected when pursuing their livelihood, since society benefits from their activities. He could also argue that he was hunting on his own land, so that the value of protection of property is another reason why he should win. The defendant in *Keeble* could argue that since the plaintiff had not yet caught the ducks, he had no right to the ducks, since if such rights depended on who first saw the animals, there would be no clear criterion and the courts would be flooded with cases. Thus, the defendant argues that deciding for him promotes the value of legal certainty. Since the plaintiff won in *Keeble*, we can on this interpretation of the case say that the court found that the combination of the values of protecting property and protecting the pursuit of livelihood outweighs the single value of legal certainty. This value preference can then be cited in new cases where the same values are at stake.

The attention paid to the role of value and purpose led to accounts of legal interpretation as a decision problem, that is, as a choice between alternative interpretations on the basis of the values promoted and demoted by these interpretations.[5] The same account can be given of the choice whether to follow a legal rule or to distinguish it by formulating a new exception. Philosophers call this "practical reasoning." In AI & law, this line of research has especially been pursued by Bench-Capon and Atkinson and colleagues (e.g., in Atkinson & Bench-Capon, 2005). For example, deciding *Young* for the defendant can be reconstructed as follows. First, practical-reasoning arguments for following or distinguishing *Keeble* are stated. Then the argument for distinguishing *Keeble* is preferred over the argument for following *Keeble*, on the ground that following *Keeble* only promotes the value of protecting livelihood, while distinguishing *Keeble* in addition promotes the value of legal certainty. This account can be refined in terms of the relative preferences between goals and values, as well as in terms of the extent to which they are promoted or demoted.

6. Establishing the Facts of a Case

While legal education and scholarship mostly focus on reasoning with and about the law, in practice, most cases are decided on the facts, so insight as to how facts can be proven is crucial for legal practice. AI & law research has addressed two main questions: which model of rational proof can best be applied to the law, and what is the logical nature of important legal evidential constructs like burdens of proof and presumptions? We address these issues in turn.

6.1 Models of Legal Proof

Theoretical models of rational legal proof are generally of three kinds: probabilistic, story based, or argument based. (In Prakken, Bex, & Mackor [2020], these three approaches and combinations of them are illustrated in several reconstructions of an actual Dutch murder case.) All three approaches acknowledge that evidence cannot provide watertight support for a factual claim but always leaves room for doubt and uncertainty, but they account for this in different ways. *Probabilistic* approaches express uncertainty in terms of numerical probabilities attached to hypotheses given the evidence. Often a

Bayesian approach is taken, nowadays more and more with Bayesian networks (Fenton & Neil, 2011).[6] Probabilistic approaches are by no means uncontroversial. One objection is that in legal cases, the required numbers are usually not available, either because there are no reliable statistics or because experts or judges are unable or reluctant to provide estimates of probabilities. Another objection is that probability theory imposes a standard of rationality that cannot be attained in practice, so that its application would lead to more instead of fewer errors. To overcome these limitations of probabilistic models, argumentation-based and story-based models have been proposed.

Story-based approaches go back to the work of the psychologists Bennett and Feldman (1981), who observed that the way judges and prosecutors tend to make factual judgments is not by probabilistic or logical reasoning but by constructing and comparing alternative plausible stories about what might have happened. In these approaches, the story that best explains the evidence must, if it does so to a sufficient degree, be adopted as true. These approaches thus model forms of inference to the best explanation (Lipton, 1991).

Both Bayesian and story-based approaches reason from hypotheses to the evidence in that, to assess alternative hypotheses, they model how likely the evidence is under the various hypotheses. In contrast, *argumentation-based* approaches reason from the evidence to the hypothesis, by stepwise building evidential arguments from the available evidence to the hypotheses. For example, Bex, Prakken, Reed, and Walton (2003), inspired by Pollock's (1995) theory of defeasible reasons, model evidential reasoning as the application of various argument schemes, such as schemes for perception, memory, induction, applying generalizations, reasoning with testimonies, and temporal persistence.

6.2 Burdens of Proof and Presumptions

To deal with uncertainty and defeasibility, legal systems use such notions as presumptions and burdens of proof. Research has been done on modeling these notions with techniques from nonmonotonic logic and argumentation (Gordon & Walton, 2009; Governatori & Sartor, 2010; Prakken & Sartor, 2009).

Legal *presumptions* obligate a fact-finder to draw a particular inference from a proved fact. Typical examples are a presumption that the one who possesses an object in good faith is the owner of the object, or a presumption that when a pedestrian or cyclist is injured in a collision with a car, the accident was the driver's fault.

Some presumptions are rebuttable while others are irrebuttable. Prakken and Sartor (2009) argue that rebuttable presumptions can be logically interpreted as defeasible rules in a nonmonotonic logic. They also argue that to fully understand notions of *burden of proof*, logical and dialogical models of argumentation must be combined. For example, they say that a so-called burden of persuasion for a claim is fulfilled if at the end of a proceeding, the claim is acceptable according to the argumentation logic applied to the then available evidence. This brings us to the dialogical aspects of legal argument.

7. Legal Reasoning and Legal Procedure

Legal reasoning usually takes place in the context of a dispute between adversaries, bound by legal procedures. This makes the setting inherently dynamic and multiparty. For example, the facts and opinions are not given at the start of a case, but the adversaries provide their evidence and arguments at various stages, and the adjudicator can allocate burdens of proof before deciding the dispute. Thus, the quality of a legal decision depends not only on its grounds but also on how it was reached, which raises issues of procedural rationality (cf. Alexy, 1978; Toulmin, 1958).

AI & law researchers have studied these issues by combining defeasible-reasoning models and dialogue models of argumentation in formal models of legal-procedural notions. Thus, properties of legal procedures can be formalized and verified. For example, relevance of dialogue moves can be formally characterized, and it can be verified whether a procedure always allows the adducing of relevant information possessed by a dialogue participant. Insights thus obtained about formalized procedures can then be used in analyzing or even (re)designing actual procedures.

An influential computational model of a legal procedure was Gordon's (1994) Pleadings Game, which, inspired by Alexy (1978), formalized a normative model of pleading founded on first principles. It was meant to identify the issues to be decided at trial, given what the parties had claimed, conceded, challenged, and denied in the pleadings phase and what (defeasibly) follows from it. Other games define the outcome in terms of whether the adversaries in the end agree on the main issue or in terms of a decision by an adjudicator.

The dynamic and multiparty setting of legal procedures also raises issues of strategic choice. They are addressed in a legal context by Riveret, Prakken, Rotolo, and Sartor (2008), who apply a combination of game theory and argumentation logic to the problem of determining optimal strategies in debates with an adjudicator. In such debates, the opposing parties must estimate the probability that the premises of their arguments will be accepted by the adjudicator. Moreover, they may have preferences over the outcome of a debate, so that optimal strategies are determined by two factors: the probability of acceptance of their arguments' premises by the adjudicator and the costs/benefits of such arguments.

8. Conclusion

Logical models of legal argument aim to respect that the central notion in legal reasoning is not deduction but argument. Deduction has its place in legal reasoning, but only as part of a larger model of constructing, attacking, and comparing arguments. Moreover, logical models of legal argument reflect that rules are not the only source of legal knowledge: the roles of cases, principles, purposes, and values should not be ignored, and these models stress the importance of dynamics, procedure, and multiparty interaction. While most work has addressed legal interpretation and normative determinations, some work addresses legal proof. Here, too, deduction is just one of the tools, as part of probabilistic, argumentation-based, or story-based approaches. A main concern here is developing models of legal proof that are rationally well founded but respect the practical and cognitive constraints faced by the participants in court proceedings. Finally, dialogical models of legal argument can be used to address issues of procedural rationality and strategic choice.

Notes

1. See chapter 5.5 by Hahn and Collins and chapter 5.6 by Woods (both in this handbook) for introductions to argumentation theory and Feteris (2017) for an introduction to the theory of legal argumentation.

2. Several parts of this chapter are taken or adapted from Bench-Capon, Prakken, and Sartor (2009); Prakken (2015); and Prakken and Sartor (2015). Trevor Bench-Capon and Giovanni Sartor gave useful feedback on a version of this chapter.

3. Some ideas concerning nonmonotonic logics are reviewed in chapter 5.2 by Rott; chapter 5.3 by Kern-Isberner, Skovgaard-Olsen, and Spohn; chapter 5.4 by Gazzo Castañeda and Knauff; and chapter 6.1 by Starr (all in this handbook).

4. See chapter 5.5 by Hahn and Collins and chapter 5.6 by Woods (both in this handbook).

5. For decision theory, see section 8 of this volume.

6. Bayesian networks are discussed in chapter 4.2 by Hartmann (this handbook).

References

Aleven, V. (2003). Using background knowledge in case-based legal reasoning: A computational model and an intelligent learning environment. *Artificial Intelligence, 150,* 183–237.

Alexy, R. (1978). *Theorie der juristischen Argumentation: Die Theorie des rationalen Diskurses als eine Theorie der juristischen Begründung* [Theory of legal argumentation: The theory of rational discourse as a theory of legal justification]. Frankfurt/Main, Germany: Suhrkamp.

Ashley, K. D. (1990). *Modeling legal argument: Reasoning with cases and hypotheticals.* Cambridge, MA: MIT Press.

Atkinson, K. D., & Bench-Capon, T. J. M. (2005). Legal case-based reasoning as practical reasoning. *Artificial Intelligence and Law, 13,* 93–131.

Bench-Capon, T. J. M. (2017). HYPO's legacy: Introduction to the virtual special issue. *Artificial Intelligence and Law, 25,* 205–250.

Bench-Capon, T. J. M., Prakken, H., & Sartor, G. (2009). Argumentation in legal reasoning. In I. Rahwan & G. Simari (Eds.), *Argumentation in artificial intelligence* (pp. 362–382). Berlin, Germany: Springer.

Bench-Capon, T. J. M., & Sartor, G. (2003). A model of legal reasoning with cases incorporating theories and values. *Artificial Intelligence, 150,* 97–143.

Bennett, W. L., & Feldman, M. S. (1981). *Reconstructing reality in the courtroom: Justice and judgment in American culture.* London, England: Methuen-Tavistock.

Berman, D. H., & Hafner, C. D. (1993). Representing teleological structure in case-based legal reasoning: The missing link. In *Proceedings of the Fourth International Conference on Artificial Intelligence and Law* (pp. 50–59). New York, NY: ACM Press.

Bex, F. J., Prakken, H., Reed, C. A., & Walton, D. N. (2003). Towards a formal account of reasoning about evidence: Argumentation schemes and generalisations. *Artificial Intelligence and Law, 12,* 125–165.

Dung, P. M. (1995). On the acceptability of arguments and its fundamental role in nonmonotonic reasoning, logic programming and *n*-person games. *Artificial Intelligence, 77,* 321–357.

Dworkin, R. M. (1977). *Taking rights seriously.* Cambridge, MA: Harvard University Press.

Fenton, N., & Neil, M. (2011). Avoiding legal fallacies in practice using Bayesian networks. *Australian Journal of Legal Philosophy, 36,* 114–151.

Feteris, E. T. (2017). *Fundamentals of legal argumentation* (2nd ed.). Dordrecht, Netherlands: Springer.

Gardner, A. v. d. L. (1987). *An artificial intelligence approach to legal reasoning.* Cambridge, MA: MIT Press.

Gordon, T. F. (1994). The pleadings game: An exercise in computational dialectics. *Artificial Intelligence and Law, 2*, 239–292.

Gordon, T. F., & Walton, D. N. (2009). Proof burdens and standards. In I. Rahwan & G. Simari (Eds.), *Argumentation in artificial intelligence* (pp. 239–258). Berlin, Germany: Springer.

Governatori, G., & Sartor, G. (2010). Burdens of proof in monological argumentation. In R. F. Winkels (Ed.), *Legal knowledge and information systems: JURIX 2010: The twenty-third annual conference* (pp. 57–66). Amsterdam, Netherlands: IOS Press.

Hage, J. C. (1996). A theory of legal reasoning and a logic to match. *Artificial Intelligence and Law, 4*, 199–273.

Horty, J. (2011). Rules and reasons in the theory of precedent. *Legal Theory, 17*, 1–33.

Lipton, P. (1991). *Inference to the best explanation.* London, England: Routledge.

Loui, R. P., & Norman, J. (1995). Rationales and argument moves. *Artificial intelligence and Law, 3*, 159–189.

Modgil, S., & Prakken, H. (2013). A general account of argumentation with preferences. *Artificial Intelligence, 195*, 361–397.

Pollock, J. L. (1995). *Cognitive carpentry: A blueprint for how to build a person.* Cambridge, MA: MIT Press.

Prakken, H. (2015). Legal reasoning: Computational models. In J. D. Wright (Ed.), *International encyclopedia of the social & behavioral sciences* (2nd ed., pp. 784–791). Oxford, England: Elsevier.

Prakken, H., Bex, F. J., & Mackor, A. R. (Eds.). (2020). Models of rational proof in criminal law [Special issue]. *Topics in Cognitive Science, 12*.

Prakken, H., & Sartor, G. (1997). Argument-based logic programming with defeasible priorities. *Journal of Applied Nonclassical Logics, 7*, 25–75.

Prakken, H., & Sartor, G. (1998). Modelling reasoning with precedents in a formal dialogue game. *Artificial Intelligence and Law, 6*, 231–287.

Prakken, H., & Sartor, G. (2009). A logical analysis of burdens of proof. In H. Kaptein, H. Prakken, & B. Verheij (Eds.), *Legal evidence and proof: Statistics, stories, logic* (pp. 223–253). Farnham, England: Ashgate.

Prakken, H., & Sartor, G. (2015). Law and logic: A review from an argumentation perspective. *Artificial Intelligence, 227*, 214–245.

Prakken, H., Wyner, A. Z., Bench-Capon, T. J. M., & Atkinson, K. D. (2015). A formalisation of argumentation schemes for legal case-based reasoning in ASPIC+. *Journal of Logic and Computation, 25*, 1141–1166.

Riveret, R., Prakken, H., Rotolo, A., & Sartor, G. (2008). Heuristics in argumentation: A game-theoretical investigation. In P. Besnard, S. Doutre, & A. Hunter (Eds.), *Computational models of argument: Proceedings of COMMA 2008* (pp. 324–335). Amsterdam, Netherlands: IOS Press.

Sergot, M. J., Sadri, F., Kowalski, R. A., Kriwaczek, F., Hammond, P., & Cory, H. T. (1986). The British Nationality Act as a logic program. *Communications of the ACM, 29*, 370–386.

Toulmin, S. E. (1958). *The uses of argument.* Cambridge, England: Cambridge University Press.

Verheij, B., Hage, J. C., & van den Herik, H. J. (1998). An integrated view on rules and principles. *Artificial Intelligence and Law, 6*, 3–26.

Walton, D. N. (1996). *Argumentation schemes for presumptive reasoning.* Mahwah, NJ: Erlbaum.

Walton, D. N., Sartor, G., & Macagno, F. (2016). An argumentation framework for contested cases of statutory interpretation. *Artificial Intelligence and Law, 24*, 51–91.

Section 12   Moral Thinking and Rationality

12.1 Rationality and Morality

Christoph Fehige and Ulla Wessels

Summary

Is practical rationality on the side of morality? Is it even the benchmark or the ground of morality? Why do what morality requires you to do? The diversity of answers that are still in the running and of considerations for and against them is astonishing. Our aim is to delineate the structure of the debate and to locate and clarify some major questions, options, and moves. We organize the presentation around a pair of prominent sample views, linking the rationality of an action to the agent's desires but its morality to the general welfare. We expound how different the matter looks for other views of rationality or of morality. On the whole, thoughts about each of the two normative domains and about conflicting norms in general suggest that, even regarding well-informed agents, rationality and morality cannot be fully harmonized. To some extent, convergence will remain gappy and contingent.

1. The Question

How do the judgments of practical rationality relate to those of morality? Let us call that question the RM question, with "R" for practical rationality and "M" for morality. The RM question is complex because each of the two relata is controversial in its own right—what *is* the rational thing to do, and what *is* the morally right thing to do?—and so is the pecking order among them: does an action have to be morally right in order to qualify as rational, or vice versa, or is there no such connection?

Judgments of the two kinds appear at variance in many cases that have existential weight. It may well happen, for example, that morality appears to require an affluent person to donate eighty percent of her income to a charity that saves lives efficiently, whereas practical rationality appears to require her to use the same money for completing her beloved collection of abstract paintings. We would expect a satisfactory answer to the RM question to get some kind of grip on such constellations. The answer should either show that R and M are in concord after all or tell us, insofar as they are not, what follows from the discord for theories of normativity and for thoughtful agents.

2. Actions and Requirements

As usual, it makes sense to restrict the inquiry. The plan is to look at rationality and morality in relation to actions. There are other items that beckon for our attention—think of the rationality and morality of beliefs, decisions, desires, emotions, intentions, maxims, or ways of life—but we will not extend the discussion to them. One consequence of the focus on actions is that the rationality that pertains is *practical* rationality by definition, which enables us to omit the adjective "practical" most of the time.

We will restrict the scope further by looking at only one kind of assessment by R and M, that of actions as (rationally or morally) required, permitted, optional, or forbidden. These are known as the "deontic" assessments, and it is the "all things considered" versions of them that we will be concerned with. We present the logical relations among the deontic terms in figure 12.1.1. There is one couple of terms, not in the diagram, that we will reserve for use in the domain R: the terms "rational" and "irrational" for actions that are rationally permitted and rationally forbidden, respectively. There is another couple, in the diagram,

permitted, right		forbidden, wrong
required, obligatory ought to be done	optional	

Figure 12.1.1
The main deontic terms and the logical relations among them. Terms written in the same field are roughly synonymous. Each term can occur in discourse about R and in discourse about M ("rationally permitted" vs. "morally permitted," etc.).

that today we will reserve for use in the domain M: in this chapter, the terms "right" and "wrong" will be used for moral assessments.

3. Two Sample Doctrines: Instrumentalism and Utilitarianism

It will help to look at the relations between one common criterion of R and one common criterion of M and to widen the view from there. As to the rationality of actions, many criteria that have been proposed are variations of one simple tenet: that it is rational for a person to try to get what she wants. There are competing ways of refining that outlook, and here is the version that will serve as our sample of a criterion: *An action that the agent can perform is rational if and only if the agent believes that no other action that she can perform brings about more fulfillment of her intrinsic desires.*

The performing, believing, and desiring in the criterion should all be understood as happening or obtaining at the same moment, but the believing and desiring don't have to occur in the agent's consciousness; they may be purely implicit. An intrinsic desire should be understood as a desire of something for its own sake, not of it as a means to something else; henceforth in this chapter, whenever we write "desire," we will mean "intrinsic desire." The quantitative notion of fulfillment takes into account both the number and the strengths of desires.

We will treat the criterion as the defining feature of "instrumentalism."[1] The label is apt because the criterion codifies a view of actions as tools, assessing them with respect to their putative efficiency in achieving the agent's ends. Instrumentalism is a close relative of that theory of rational decision-making that puts the maximization of expected utility (MEU) center stage and plays a large role in the behavioral sciences. The MEU theorist's probabilities and amounts of utility correspond, by and large, to the instrumentalist's beliefs and amounts of desire fulfillment.

In the moral domain, our sample criterion will be utilitarian: *An action is morally right if and only if no other action that the agent can perform brings about more welfare, worldwide.* The utilitarian message is that welfare counts, no matter whose welfare it is, and that nothing else does. What is that all-important stuff called "welfare"? One widely held view, and one that we will assume here, is that a person's welfare is the fulfillment of her desires and that in consequence, due to conceptual connections between pleasure and desire fulfillment, pleasure is an important part of welfare.[2]

4. Convergence

To what extent do our sample criteria of R and M move in sync? One source of hitches can be an agent's beliefs, which play a role in one criterion but not in the other. For example, even the most fervent utilitarian can have erroneous beliefs about the impact of her actions on the amount of general welfare, and those beliefs can make it rational for her to do what is morally wrong.

If we leave aside the threat posed by deficient beliefs, we reach more significant notions of convergence and divergence. Let us call an agent well-informed if her beliefs are in such a good state that the rational thing for her to do would not change if we improved them further (that is, if we corrected false beliefs or added true ones). Using well-informedness as a stepping-stone, let us understand "convergence" and "divergence" as follows: R and M converge insofar as the actions of well-informed agents that are morally required are also rational; R and M diverge insofar as the opposite is the case.

4.1 Convergence through Moral Desires

If instrumentalism is on the right track, the principal forces of convergence will have to be desires that point in the right direction. Less metaphorically speaking, they will have to be desires that, provided the agent is well-informed, have a propensity to make it rational for her to perform an action that is morally right. Let us call them RM desires. Life abounds with such desires, and one challenge is to produce a helpful classification.

Some RM desires aim directly at something that would be morally positive in itself. Figure 12.1.2 presents them as the "moral desires." Some of those attitudes even "bring up" the topic of morality. Examples are desires that the world be a better place, to do the right thing, or to be a virtuous person. Other moral desires do not invoke morality as such but exhibit the relevant directness all the same. Depending on what is morally positive in itself, examples might be desires that some specific people be happy, that there be a lot of happiness, that everybody be treated equally—or desires to keep promises and to refrain from telling lies. The propensity that defines RM desires is present in either case. Applied to our sample moral doctrine, utilitarianism: both if Mary desires to do the right thing and if Mary desires to maximize welfare worldwide, it holds true that, provided she is well-informed, the desire will tend to make it rational for her to do the right thing.

Sympathy and moral sentiment, widespread as they are, serve as high-yield sources of those moral desires that do not invoke morality.[3] If Mary sympathizes with others,

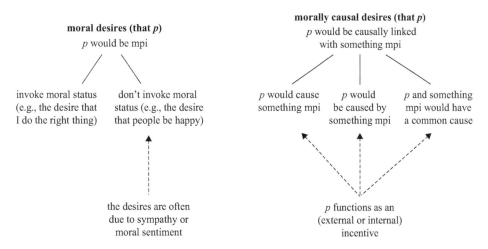

Figure 12.1.2
Some important kinds of RM desires. The abbreviation "mpi" stands for "morally positive in itself." One and the same desire can belong to more than one kind, even on the same level.

her sympathy is likely to give rise to or even to constitute a desire that others fare well. And if Mary has moral sentiments (for example, sentiments of moral admiration, indignation, or satisfaction) with regard to acts of a certain kind, the sentiments are likely to give rise to or even to constitute desires for or against performing acts of that kind. The two sources are so rich that there have been proposals to feed morality from them alone, giving us an "ethics of sympathy" or a "sentimentalist ethics."[4]

Are all moral desires contingent? Is it just as possible for a person to have them as to lack them? The rationality of morality would be a more robust affair if there were numerous strong moral desires that we cannot fail to have. An argument has been devised that purports to establish the existence of those resources. Everybody, the argument aims to establish, necessarily desires that other people have pleasure, with the strength of those desires proportional to that of the pleasure at issue. The idea is that, if you fully represented to yourself that another person experiences a specific pleasure, you would (this being entailed by full representing) experience that very pleasure yourself—and would thus yourself be pleased while representing. And since a disposition to be pleased when fully representing a state of affairs to oneself is a desire that the state of affairs obtain, you desire that the other experience the pleasure.[5] If the argument works, those desires—which on conceptual grounds everybody holds regarding everybody else all the time—will do a sizable part of the work that we are anxious to see done.

4.2 Convergence and Rationality without Egoism
Our glance at other-regarding moral desires, no matter whether they are necessary or contingent, is a good occasion for setting aside egoism. One might think on the following grounds that the relations between rationality and morality are particularly strained: (i) they would be strained if rationality required us to act egoistically, and (ii) according to instrumentalism rationality requires exactly that, because the word "egoistic" stands for the property of acting in the light of one's own desires.

The train of thought is misguided. The notion of "egoism" that is advanced in claim (ii) is both unusual and apt to weaken claim (i). The usual understanding is that acting on one's own desires is not a sufficient condition for being egoistic but that it also matters what those desires are. The usual understanding is that, if a good-hearted person strongly desires that other people fare well and acts on *such* desires, she is not an egoist but an altruist, while an egoist is a person who lacks or fails to put into action *such* desires. To be sure, somebody might go along with that understanding itself but still link instrumentalism to egoism by adding the claim that such desires do not exist. However, we know of no sound argument for the additional claim.[6]

The misclassification of rationality as egoistic can also result from sloppy thinking about the *combination* of instrumentalism and the theory of welfare as desire fulfillment. It is true that the combination entails some connection between rational action and the agent's welfare: in acting rationally at point of time t, the well-informed person-stage Mary-at-point-of-time-t maximizes the fulfillment of its desires and cannot help thereby maximizing its own welfare. We should beware of letting that connection blur the picture. In the first place, the rational person (or person-stage) at issue need not be concerned with her (or its) own welfare and can act, due to the nature of her (or its) desires, highly altruistically. Secondly, the person-stage that acts can still fail to maximize the welfare of the entire person, who is extended over time and may have different desires later. Thirdly, even the limited connection we are looking at requires as one ingredient a certain view of welfare; if we adopt a hedonistic view instead (seeing the welfare of an entity as the pleasure that the entity has and not as the fulfillment of her desires) and keep instrumentalism, the connection vanishes.

According to most conceptions of rationality, instrumentalism included, rationality does not require us to act egoistically. We need to distinguish the question how rationality relates to morality from the questions how egoism, how prudence, and how an agent's self-interest, self-love, or welfare relate to morality.[7] Many well-known discussions from the history of philosophy are largely about questions of the second kind. Plato, for example, assigns a key role in the *Republic* to the case of Gyges, who is ruthless in using for his own advantage his power to become invisible; Aristotle writes about virtue as a constituent of an agent's flourishing in the *Nicomachean Ethics* and Henry Sidgwick about individual versus universal happiness in *The Methods of Ethics*. Those discussions apply to our question at best partly or indirectly.

4.3   Convergence through Morally Causal Desires
Some RM desires, also indicated in figure 12.1.2, are related to the good or the right in a different way. If a person desires p and p would be causally linked to something that is morally positive in itself, we will speak of a "morally causal desire." We distinguish subgroups of such desires by distinguishing three kinds of the linkedness.

A desire that p is in the first subgroup if p would cause something morally positive. For example, a person desires to cultivate a garden, to keep the kitchen clean, or to mend broken bones, and her doing these things would cause pleasure in the beholders, eaters, and patients and would thus cause something that is morally positive in itself. A desire that p is in the second subgroup if, conversely, something that is morally positive in itself would cause p. For example, Mary desires that her parents treat her to cake, which the parents do only when, and in that case because, they are happy. The treat is desired and is the effect of the happiness, with the happiness being morally positive in itself. A desire that p is in the third subgroup if p and something that is morally positive in itself have a common cause.

Consider, for example, heartless Mary, who does not care about the victims of malaria but who desires to be praised at a charity ball for donating to the fight against malaria. If Mary donates, neither does the praise cause the thing that is morally positive in itself (the praise does not causally affect the saving of the lives) nor vice versa, but the two have a common cause: Mary's donation. In such cases, too, the one comes with the other, and that matters for RM purposes.

The key feature, wherever morally causal desires are in play, is the indirectness. A desire can fail to be directed at anything that would be morally positive in itself and still tend to make actions rational that are (or that cause things that are) morally positive in themselves. Links of the relevant kind are legion: you do the right thing and some "other" desire of yours is fulfilled. In that sense, large chunks of morality are connected to rationality through carrots and sticks.

The incentives can be external—think of fellow-citizens who in response to your morally positive behavior honor and help you and refrain from ignoring, deriding, jailing, or lynching you. The incentives can also be internal—think of the joys of believing to have done what was good or right and of the absence of pangs and remorse. Even when the incentives are internal, the constellation differs from the one that characterizes the first main group of RM desires, the moral desires. Two distinctions apply. In the first place, when you desire *your joy* of doing good and when you desire *to do good*, those are two different desires.[8] Secondly, we should in some cases distinguish even for one and the same desire between the reason to put it into one group and the reason to put it into another. Consider, for example, your desiring your own joy of having done good. The reason to count that desire as a morally causal one (your joy is caused by something morally positive) is different from the reason to count it as a moral one (your joy, too, is something morally positive in itself). One desire can be both.

Incentives are studied by game theorists in particular. We understand quite well by now, with regard to several moral standards, how even agents who have no moral desires and are at the same time without ifs and buts instrumentally rational find themselves drawn into actions and outcomes that are morally positive. Under various conditions, repeated encounters in the same group of such agents become, not least because rewards and punishments can emerge, a breeding ground for various amounts of cooperation, equality, justice, solidarity, trust, and public good. While significant general results, most prominently a family known as the "folk theorems" of game theory, have been established by mathematical methods of the more traditional kind, there is also evidence from agent-based computer simulations, sometimes involving entire artificial societies: the rational thing to do for the virtual agents, it turns out, is often to spare or even to assist each other.[9]

So much for the many devices in human minds and societies that make it rational for people to do the right thing a lot of the time. On the other hand, given criteria of R or M resembling our two samples, there is not much hope of full convergence. In the example from our opening paragraphs, the agent might favor abstract paintings over human lives, which in the absence of competing considerations would make it the rational thing for her to violate her moral obligation. When incentives and the agent's moral desires do not add up, rationality will take a stand against morality.

5. Resisting Divergence by Aligning Rationality with Morality

The dominant impetus among philosophers is to keep the divergence of R and M in check. A proof that there is no divergence is the Holy Grail of practical philosophy, yearned for but hard to attain. Some philosophers go as far as to let their thinking on R itself or on M itself be governed by the premise of "moral rationalism"—the claim that every action that is morally required is rational.[10] With that sweeping premise or without, answers to questions like "Why be moral?" or "Why *act* morally?" are sought after, and the risk that, even with regard to well-informed persons, sometimes no good answer emerges is often perceived as an invitation to rethink R or M. Those who want to rethink in order to reduce or even eliminate divergence have essentially two options: they assimilate rationality to morality or vice versa. They could also combine the two moves, moralizing rationality *and* rationalizing morality, but for the bigger picture it will suffice if we treat each of the two in turn.

5.1 Smaller Departures from Instrumentalism
As to alternative views of practical rationality, modest deviations from instrumentalism make some difference regarding divergences. One example is the temporal extension of the conative basis. We could modify the instrumentalist criterion so that it covers not just the desires that the agent has at the time of acting but all those that she had, has, or will have. The modification would prevent the rational agent from behaving ruthlessly toward her past self or her future self. Morally speaking, it would be a step in the right direction—but

only a small, intrapersonal one. Anchoring in a rational agent regard for the welfare of all her own person-stages is a far cry from anchoring in that agent regard for everybody's welfare.

A second example originates in the observation that, if instrumentalism is correct, rational agents can get caught in traps of practical rationality. Such traps are situations in which, if every agent acts rationally, each of them receives, predictably, less fulfillment of her desires than she would if everyone acted irrationally. General compliance with the requirements of rationality thus leads to the opposite of the moral goal, which is general welfare.

Many real-life situations appear to be traps of practical rationality. Consider societies that become much nastier because people carry weapons. Each individual reasons that to carry a weapon is likely to be more conducive to the fulfillment of her desires either way—that is, both if people whom she encounters carry a weapon and if they don't. Because of the individual decisions for weapons, the considerable advantages of a weaponless society and of avoiding an arms race remain out of reach for all. Even if a weaponless society could still be achieved by changing the individual decisions through the imposing of sanctions, the sanctioning itself would gobble resources and thus welfare. The ubiquity of such traps in human interaction and the unequivocalness of the moral setbacks—less welfare for one and all—have incited a vast complex of research in ethics, economics, and psychology.[11]

Can we devise a criterion of rationality that preserves the spirit of instrumentalism but spares us the traps? David Gauthier is one of the theoreticians who have tried.[12] Gauthier suggests, controversially, that a rational agent can choose, "on utility-maximizing grounds, not to make further choices on those grounds," but to adopt a certain stance that will determine her behavior. She will then be a "constrained maximizer." The stance that Gauthier has in mind is roughly this: I will do, provided that so will the others who are involved, my part in making possible an outcome that is better for everybody than the outcome that unconstrained maximizers could achieve. Since an agent who has recognizably adopted such a stance will sometimes produce and reap fruits of cooperation that are not available to one who hasn't, it is rational to become such an agent. Individual instrumental rationality is declared to be a tad more collective than we thought.

If Gauthier's claims are sound and help us make headway with the RM question, they do so within limits. Gauthier's ambition is restricted to establishing rational support for a morality of *mutual* benefiting. He has no ambition to provide support for moral obligations toward beings who are in need but have nothing to offer.

5.2 Larger Departures from Instrumentalism

We find more radical consequences for the shape and extent of divergence when we turn our attention to more radically different pictures of practical rationality. The main move is the dethroning of desire.

Suppose that we break with instrumentalism by replacing the appeal to desire fulfillment with an appeal to conformity to reasons. Let us agree that facts can be reasons and that constellations of facts (for example, regarding pains and cures) can make it the case that there is, altogether, for a person more reason to perform one action (say, to see her cardiologist) than another (say, to see her homeopath). The new criterion of rationality could then read as follows: *An action that the agent can perform is rational if and only if she believes that there is not more reason for her to perform another action instead.*[13] Suppose further that we take many reasons to be "worldly" in that they do not involve the agent's desires. Perhaps, for example, *that Peter's education would be finished if he received a donation* is a reason to donate, even if the agent desires no such finishing or donating and desires nothing that comes with it; or perhaps *that serenading the moon would increase the glory of the moon* is a reason to serenade the moon.[14]

The impact on the RM nexus would be momentous. The task of showing that it is rational for a well-informed agent to do what morally she ought to do no longer has as its central component the task to track down a fitting constellation of desires of hers but the task to identify a fitting constellation of reasons for her to do things. In order for the entire scheme to be successful, worldly reasons for actions need to exist *and* to have a considerable affinity to morality *and* to be connected to rationality as stated by the new criterion. Whether all or at least some of the three conditions are fulfilled is controversial.

Which morality you take to be connectible to rationality-in-the-new-spirit will depend on your views, possibly your intuitions, about the realm of reasons. If morality is concerned with promise-keeping, loyalty, or the increase of human knowledge, and so are reasons to do things and in the same proportion, then every well-informed rational agent will do the right thing: keep her promises and so on. Our sample moral doctrine, utilitarianism, is no exception. If there is always most reason to maximize the amount of the fulfillment of everybody's desires, then every well-informed rational agent will do the right thing and maximize that amount.

Following a markedly different path, Immanuel Kant, too, ends up advocating some variety of rational benevolence, maximization included. We read that a rational being would try, "as far as he can, to advance the ends of others" and would come to the conclusion: "the ends of a subject that is an end in itself must, as much as possible, also be *my* ends."[15] In which way according to Kant rationality secures so much morality is extraordinarily contentious, even among his followers. He draws on the claim that a rational being is autonomous in the literal sense of giving herself a law. To Kant, that feature appears imbued with moral significance. Being autonomous, the agent's rational will is not pushed around by anything, not even by the agent's own inclinations; it finds itself with nothing left to be constituted by than the respect for rationality itself and for people who have it and for their ends; and, such being the character of laws, that will is general, not concerned with one person or group in particular. Sound statements of those connections remain a desideratum.

As expected, the picture of the relation between R and M changes when that of R changes. Our brief encounters with the "worldly reasons, not desires" approach and the "autonomy, not desires" approach have illustrated tectonic movements and the hopes of convergence that can be associated with them.

6. Resisting Divergence by Aligning Morality with Rationality

If divergence is to be avoided, what about getting to work at the other end? We could truncate morality. The fewer actions are morally obligatory in the first place, the smaller the risk that an action is morally obligatory but irrational.

The demands of some moralities are so removed from most agents' conative constitution that cutting down on the demanding looks particularly promising. According to utilitarianism, for example, an agent ought to give the same weight to her own welfare and to that of her nearest and dearest as to everybody else's. Since hardly an agent's desires manifest such impartiality, utilitarian obligations that it is irrational for the obligated agent to comply with are thick on the ground.

Various moral prerogatives for agents have been suggested. We could permit the agent, for some factor $k > 1$, to attach up to k times as much weight in her decisions and actions to her own welfare as to the welfare of everybody else; or permit her, for some $l > 0$, never to give up more than l units of her own welfare; or permit her, for some threshold value m, never to make her own welfare fall below m.[16] We could think of other permissions, too, concerning her projects rather than her welfare or concerning the welfare of those she is close to rather than just her own. Every such prerogative is a loosening of utilitarianism; it would allow the agent to be partial in the sense of granting in her decisions extra force to this or that, even if the amount of worldwide welfare suffers.

Prerogatives are apt to narrow the gap between R and M but will not close it. A morality that involves a prerogative will still demand *something* (for instance, that the interests of others be respected to *some* extent), and there is bound to be some well-informed agent who in some situation has desires that make it irrational for her to comply even with those moderate demands.

Structurally, contract theories in the Hobbesian tradition inhabit the same middle ground. Those theories are reciprocitarian. They say, with various qualifications, that what a person morally ought to do in relation to another person is to play by rules that satisfy the following condition: the fulfillment of the desires of each of the persons would increase if each of them, rather than none, played by those rules. For example, if both refrain from insulting each other, that will save both of them some distress.

Once again, we find the curtailing of moral obligation. If contractarians are right, there are fewer moral obligations than we thought and thus fewer that it might be irrational to meet. Most notably, about all persons (and other beings) who are not in a position to increase the fulfillment of her desires, the contractarian will say that she owes them nothing. And once again we also find that the curtailing does not suffice to provide full alignment with rationality. Sometimes a well-informed agent will have and see the possibility of maximizing the fulfillment of her desires by breaking even the few rules of that "minimal morality" and getting away with it. And no earthly regime of sanctions will eliminate such possibilities, since all such regimes are gappy.[17]

Some thinkers have suggested that we go one step further and prevent all divergence by fully rationalizing morality. The proposal is to combine moral rationalism with the claim that instrumentalism is by and large on the right track. Gilbert Harman endorses the combination. He writes: "If S says that (morally) A ought to do D, S implies that A has reasons to do D." And he continues: "I assume that the possession of rationality is not sufficient to provide a source for relevant reasons, that certain desires, goals, or intentions are also necessary." Since agents might lack the relevant attitudes, Harman infers "that there might be no reasons at all for a being from outer space to avoid harm to us" and "that, for Hitler, there might have been no reason at all not to order the extermination of the Jews."[18]

The moral consequences are remarkable. If the moral "ought" requires reasons, and reasons require desires, but the desires aren't in place and thus neither are the reasons, then neither is the moral "ought." While Harman invites us to say other things about Hitler (for example, that Hitler is evil or our enemy), he claims that the moral "ought" is out of place. It is "odd to say," Harman asserts, that "it was wrong of Hitler to have acted as he did" or that "Hitler ought morally not to have ordered the extermination of the Jews."

And so the rationalization of the moral "ought" would be completed. The approach is Procrustean. There is no mismatch between the moral "ought" and the practical rationality of its addressees because every instance of a moral "ought" that would not conform is discarded.

7. Accepting Divergence

Should we accept that R and M diverge? We should if it seems to us that they do and we see no reason to resist the claim. The two parts of that condition deserve separate treatments.

7.1 Finding Some Divergence in the First Place

Not many of us enter the inquiry holding fully convergent views of R and of M, each of which they deem plausible in its own right, independently of any pressure to see the two in line. It is true that there is full convergence according to some views of R and of M that we have encountered, but the independent plausibility of those views is the crux.

On the rational side, how plausible is the claim that something other than the agent's aims and projects rules the roost? The subjective picture exerts a considerable pull: if you've set your heart on something, it is rational for you to go after it, and if you haven't, it is not. On the moral side, the questions are inverse. How plausible is the claim that, provided you are indifferent to other people's welfare and have to fear no backlash from wrecking it, you are morally permitted to wreck it? On either side, the claims that would secure full convergence do not have the ring of truth.

The problem does not just arise when, at least regarding well-informed agents, one of the two domains is claimed to fully look the way we always took the other one to look, with the implausibility due to the fact that *one* view does all the reaching out. If a view of R and a view of M met halfway, the implausibility would be distributed evenly but not lessened. Although we pointed out that some independently plausible components with a conciliatory effect may well be missing from our two sample doctrines (which we didn't call sample doctrines for nothing), no such components are in sight that would happen to dovetail, resulting in a perfect fit.

7.2   Not Theorizing Away the Divergence We Find

Matters would look different if our thinking were driven by the will to rule out divergence. Maybe there are general considerations that speak against divergence and that should be taken into account when we form our views of R and M? It is surprisingly difficult to find distinct statements of such considerations in the literature, but here is a brief attempt to articulate and assess some candidates.

The argument from the *negativity of non-compliance* says: "Since people by and large act rationally, every morally required act that is irrational is a morally required act that is unlikely to be performed—which is a bad thing." We respond that any such badness would be a sad truth and that good theorizing should acknowledge truths, sad ones included. The badness at issue does not provide a right kind of reason for changing our views of R or M.

A follow-up argument adduces the *pointlessness of moral judgment due to the negativity of non-compliance*: "The point of engaging in moral judgment is to avoid the said negativity. Why bother if the project does not boost the right and the good through compliance?" Part of the answer is that not only are there many different functions that moral thinking, judging, and speaking have, but also many different paths, including indirect ones, on which the function of boosting the good and the right can be fulfilled. Some of the moral point of considering or making or uttering the moral judgment that, say, Mary ought to spend half of her spare time fighting for a certain cause may well be independent of the factual question whether Mary ends up conforming to the judgment. Sorting out and signaling our moral view of the matter can help shaping decisions, education, outlooks, politics, relationships, and sanctions in a myriad of ways, many of which do not even relate to Mary in particular.

Next, there is the threat of the *unjustifiedness of moral judgments*: "To 'provide morality with a foundation' or to 'justify moral judgments' is or includes showing that it would be rational to act, if the opportunity arose, in line with the moral judgments at issue. Moral judgments that are subject to divergences are therefore unfounded and unjustified." The complaint is worth pondering. Still, you justify something *to somebody*, and so the premiss

of the complaint licenses at best the conclusion that, in cases of divergence, the "ought" judgment has not been justified to *all* persons who, according to the judgment, ought to do a certain thing under certain circumstances. But maybe to some of them. Secondly, the premiss of the complaint is controversial in that there are other conceptions of what it is to justify a moral judgment. Thirdly, let's not forget that generally speaking, since justification stops somewhere, the use of unjustified items might be respectable.

A final argument asserts the *inacceptability of normative impasses*: "When an agent grasps the fact of the situation and the morally required action is irrational, what is there left for us to tell her? We can tell her that two kinds of norms that apply to her impending action point in opposite directions, irreconcilably so, and that we have no third kind of norm that would adjudicate between them. We can wish her good luck at the normative crossroads and move on. None of that is of any help to her, and theorizing about normativity should do better than that. It should avoid divergences."

The most general reply to the objection is that theorizing about normativity should "do better" only if there is independent evidence that it got the lay of the land wrong. That evidence would need to be produced—and will hardly consist in the fact that something is or feels awkward for a well-informed agent. Moreover, the objection misses its target, divergence, because divergence does not entail the alleged source of awkwardness, the absence of normative guidance. Divergence allows for the possibility that there is a boss: one of the divergers or some adjudicator. The possibility is very much alive in the literature on normative pluralism, where the view is not rare that divergences between kinds of norms coexist with unequivocal overall norms.[19]

When it comes to R and M, the observation that divergence and guidance can coexist gains in stature, since one of the two divergers is practical reason itself. There is a fairly straightforward sense in which practical reason is always in charge. Practical reason deals with practical reasons—with all of them. The fascinating question whether all of them involve agents' desires makes no difference for the following consideration, which is quite general. Neither does it make a difference whether some kind of incommensurability threatens to hold *among* an agent's practical reasons. We may ignore that possibility here because we're grappling with the specter of lacks of guidance that originate *between* R and M. Our question is whether the risk of such lacks exists and justifies the axiomatic excluding of a divergence of R and M. Surely a risk of such lacks could not justify such an excluding if there is lack of guidance even *within* R. Thus, what remains to be looked at is only the other case, in which all is well within R: there is some balance of all practical reasons that an agent has and thus something that the agent has most reason to do.

And now to the question where the balance leaves *moral* practical reasons. They can relate to the "most reason" verdict in two different ways but can escape neither the verdict nor the guidance it gives. We can understand moral practical reasons either as being practical reasons of a certain kind (think of yellow bicycles, which are bicycles), in which case the balance of all practical reasons will have taken them into account—or as *not* being practical reasons (think of "root beer," which is not beer, and of "toy money," which is not money), in which case the balance of all practical reasons will not have taken them into account. There is guidance by the balance either way. It is guidance on the level "most practical reason," a level on which all practical reasons—that is, all reasons to do things—have been taken into account. That seems guidance enough.

From the list of objections against divergence, not much is left. It seems that we should bemoan, but not deny, the existence of divergence.

8. Conclusion

We have explored in which sense and why it is in many cases rational for people to do what morally they ought to do but also why with respect to many cases, even of well-informed agents, the diagnosis is controversial. We have sided with the common-sense view that performing actions that are stupid (to use the laity's term for "irrational") and performing actions that are morally wrong are two very different kinds of shortcomings. The action that is stupid can be altruistic, benevolent, beneficent, and morally right, and the action that is not stupid can be egoistic, malevolent, maleficent, and morally wrong. Given the duality, theoreticians of the rational and the moral will keep or turn their spotlights on the kinds, mechanics, and extents of convergence and divergence, including the metanormative challenges posed by competing norms. In practice, both desires and ways of fulfilling them ought to be shaped—and that "ought" is a moral one—so that divergence is reduced.

Acknowledgments

More people have given us a hand regarding some aspect or other of this chapter than we can list here, and we are grateful to every one of them. Special thanks go to

Kevin Baum, Inga Lassen, Susanne Mantel, Helge Rückert, Rudolf Schüßler, Stephan Schweitzer, and Christian Wendelborn.

Notes

1. More on doctrines like instrumentalism and on their rivals in Hutcheson (1728, section 1 of treatise 2), Millgram (2001), Schroeder (2007), and Schmidt (2016). For further references, see Fehige (2001, note 1).

2. For the equating of welfare with desire fulfillment, see von Wright (1963, esp. sections 5.9 and 5.11) and Carson (2000, chapter 3); see also Wessels (2011), with numerous sources. For the claim that pleasure is one kind of desire fulfillment, see Fehige (2004, esp. pp. 143–145) and Heathwood (2007), both of which give further references. As to normative ethics: utilitarianism and its rivals are sketched in Vaughn (2010/2013, chapter 2) and treated more thoroughly in Copp (2006, part 2); a helpful introduction to utilitarianism is Bykvist (2010).

3. Hutcheson (1728) covers in some detail the desires associated with the "publick sense," which is the disposition "to be pleased with the *Happiness* of others, and to be uneasy at their *Misery*" (art. 1.1 of treatise 1), and with the "moral sense," which is the disposition to have moral sentiments. More on the connections between morality, desire, sentiment, and sympathy in Bricke (1996, esp. chapter 6), Fehige (2004), and Fehige and Frank (2010).

4. Single-source approaches to morality have been championed, for example, by Arthur Schopenhauer (1841, esp. section 16), who counts on sympathy, and by Francis Hutcheson (1725, preface and treatise 2) and David Hume (1751, appendix 1), who count on the moral sentiments.

5. The argument is put forth in Fehige (2004); the book includes a discussion, in chapter 6, of the limits of even that connection between R and M.

6. Lucid treatments include Hutcheson (1728, esp. art. 1.3 of treatise 1 and the beginning of treatise 2) and Sharp (1923, section 2).

7. Gregory Kavka captures the difference with maximum clarity when he distinguishes the "Wider Reconciliation Project" (1985, section 5) from a narrower one. More on handling egoism and its relation to morality can be found in Cholbi (2011).

8. More on the important distinction, for example, in Hutcheson (1728, art. 1.4 of treatise 1), Rashdall (1907, esp. pp. 17–18, 28–32), and Schlick (1930, sections 2.6–2.8).

9. For the first kind of approach, see Fudenberg and Tirole (1991, chapter 5, esp. section 5.1.2) and Maschler, Solan, and Zamir (2008/2013, chapter 13, esp. sections 13.5 and 13.6). The agent-based simulation approach is surveyed by Gotts, Polhill, and Law (2003); a telling example is Hegselmann (1998).

10. Moral rationalism is also known as the "claim of over-ridingness" and is highly controversial; contributions to the dispute include Scheffler (1992a, chapter 4), Cholbi (2011, esp. section 1), Portmore (2011), and Dorsey (2012). An illuminating critical study of ways of engineering convergence is Brink (1992).

11. Peterson (2015) can serve as a gateway to the area; see also Dawes (1980) and Binmore (1994).

12. A good starting point is Gauthier (1986, esp. chapter 6, which has the upcoming quotation on p. 158). Appeals to "resolute choice" (McClennen, 1985) bear some resemblance to Gauthier's "constrained maximization." Gauthier later espouses a conception of rationality that he acknowledges is morally charged to begin with (2013, p. 624); for a similar step, see McClennen (2012).

Other devices, too complex to sketch here, have been invoked against the traps: the claim that people who are in similar circumstances are bound to act similarly (see Davis, 1985, and, on "mirror strategies," J. V. Howard, 1988), the proposal to shift our attention from Nash equilibria to "dependency equilibria" (Spohn, 2009, foreshadowed by Aumann, 1987), and, although with an emphasis on the explanatory rather than the normative dimension, the conceptualization of a relevant decision situation as a "metagame" that involves partial commitments (N. Howard, 1971, esp. sections 2.5, 3.1, 3.2) or of the decision-makers as members of a group (Bacharach, 2006, esp. section 4 of the "Conclusion," and Butler, 2012). For critical thoughts on some of the approaches, see, e.g., Binmore (1994, chapter 3).

13. The view that rationality is the corresponding of actions to beliefs about reasons can be spelt out in quite different ways; see, for example, Scanlon (1998, sections 1.1.3–1.1.5) and Parfit (2011, sections 1 and 17).

14. The claim that there are worldly normative reasons for actions has acquired quite a few supporters. Among them are Thomas Nagel (1970, esp. chapter 10, and 1986, esp. sections 8.4, 8.5, 9.2, 9.3), Thomas Scanlon (1998, sections 1.9–1.11), Jonathan Dancy (2000, chapters 2 and 5), Philippa Foot (2001, chapter 4), Frederick Stoutland (2001, sections 3.2 and 3.3), and Derek Parfit (2011, sections 1–15).

15. The quotations are from Kant (1785/1903, p. 430); for the moral impact of autonomy, see, for example, pp. 405, 428–434, 444, 447–452. Kant later invokes the moral law as a "fact of reason" (1788/1908, pp. 31–32, 42–43); whether in doing so he renounces, summarizes, or supplements his justificatory efforts is controversial. After Kant, attempts to ground morality in the respect for rationality itself have been numerous and varied; Smith (2011) is one case in point.

16. The three kinds of prerogatives are considered, one each, in Scheffler (1992b, section 1), Mulgan (2001, section 5.5), and Nagel (1986, p. 202). For further reflections on possible kinds and shapes of prerogatives, see Wessels (2002) and Stroud

(2010). Some prerogativists make it clear that anti-divergence is where they come from. It is for the sake of keeping M in the orbit of R that they conceive of M as partial and as in that respect non-utilitarian (see Nagel, 1986, pp. 202–203, and Portmore, 2011).

17. A particularly acute analysis of the connections between R and M in the contractarian project is provided by Rainer Trapp (1998), who also explains (pp. 339, 356–359) that it will not always be rational for a person to do what by contractarian lights she morally ought to do. Contractarians who acknowledge the divergence include Peter Stemmer (2017, pp. 646–648, esp. note 34) and Gregory Kavka (1985, pp. 305–308 and, most clearly in terms of rationality, section 5). That a contractarian like Gauthier might avoid the divergence by operating with a modified conception of rationality (Gauthier, 1986, chapter 6) is a different matter.

18. The preceding quotations are from Harman (1975, p. 9); the subsequent ones are from Harman (1977, p. 107). Stemmer has retracted the crucial claims (2017, note 34) but used to travel a very similar path. For a while, he saw norms geared so radically to the addressee's wanting that even an accidental hole in the sanctioning was considered to constitute a hole in the norm itself. Insofar as an individual action would not be followed by a sanction that the agent herself wants to avoid, so the retracted claims run, the norm not to perform that action "does not exist," and the action "is not really forbidden" (Stemmer, 2008, p. 181).

19. That domain-specific "ought" judgments relate to overarching ones, which have the last word, is argued by McLeod (2001) and Woods (2018, section 10.2.2). In a similar spirit, Case (2016) provides a powerful argument for the conditional claim that, if you accept "source pluralism" and "conflict," you are committed to accepting "authoritative adjudication." However, support for the claim that "ought" judgments from different domains diverge is wider and comes also from authors who deny that there is an adjudicative level—see Baker (2018) and the sources given there.

References

Aumann, R. J. (1987). Correlated equilibrium as an expression of Bayesian rationality. *Econometrica, 55*, 1–18.

Bacharach, M. (2006). *Beyond individual choice*. Princeton, NJ: Princeton University Press.

Baker, D. (2018). Skepticism about ought *simpliciter*. *Oxford Studies in Metaethics, 13*, 230–252.

Binmore, K. (1994). *Playing fair*. Cambridge, MA: MIT Press.

Bricke, J. (1996). *Mind and morality*. Oxford, England: Oxford University Press.

Brink, D. O. (1992). A puzzle about the rational authority of morality. *Philosophical Perspectives, 6*, 1–26.

Butler, D. J. (2012). A choice for 'me' or for 'us'? Using we-reasoning to predict cooperation and coordination in games. *Theory and Decision, 73*, 53–76.

Bykvist, K. (2010). *Utilitarianism*. London, England: Continuum International.

Carson, T. L. (2000). *Value and the good life*. Notre Dame, IN: University of Notre Dame Press.

Case, S. (2016). Normative pluralism worthy of the name is false. *Journal of Ethics and Social Philosophy, 11*, 1–19.

Cholbi, M. (2011). The moral conversion of rational egoists. *Social Theory and Practice, 37*, 533–556.

Copp, D. (Ed.). (2006). *The Oxford handbook of ethical theory*. Oxford, England: Oxford University Press.

Dancy, J. (2000). *Practical reality*. Oxford, England: Oxford University Press.

Davis, L. H. (1985). Is the symmetry argument valid? In R. Campbell & L. Sowden (Eds.), *Paradoxes of rationality and cooperation* (pp. 255–263). Vancouver, Canada: University of British Columbia Press.

Dawes, R. M. (1980). Social dilemmas. *Annual Review of Psychology, 31*, 169–193.

Dorsey, D. (2012). Weak anti-rationalism and the demands of morality. *Noûs, 46*, 1–23.

Fehige, C. (2001). Instrumentalism. In E. Millgram (Ed.), *Varieties of practical reasoning* (pp. 49–76). Cambridge, MA: MIT Press.

Fehige, C. (2004). *Soll ich?* [Should I?]. Stuttgart, Germany: Reclam.

Fehige, C., & Frank, R. H. (2010). Feeling our way to the common good. *The Monist, 9*, 141–165.

Foot, P. (2001). *Natural goodness*. Oxford, England: Oxford University Press.

Fudenberg, D., & Tirole, J. (1991). *Game theory*. Cambridge, MA: MIT Press.

Gauthier, D. (1986). *Morals by agreement*. Oxford, England: Oxford University Press.

Gauthier, D. (2013). Twenty-five on. *Ethics, 123*, 601–624.

Gotts, N. M., Polhill, J. G., & Law, A. N. R. (2003). Agent-based simulation in the study of social dilemmas. *Artificial Intelligence Review, 19*, 3–92.

Harman, G. (1975). Moral relativism defended. *Philosophical Review, 84*, 3–22.

Harman, G. (1977). *The nature of morality*. New York, NY: Oxford University Press.

Heathwood, C. (2007). The reduction of sensory pleasure to desire. *Philosophical Studies, 133*, 23–44.

Hegselmann, R. (1998). Experimental ethics. In C. Fehige & U. Wessels (Eds.), *Preferences* (pp. 298–320). Berlin, Germany: De Gruyter.

Howard, J. V. (1988). Cooperation in the prisoner's dilemma. *Theory and Decision, 24*, 203–213.

Howard, N. (1971). *Paradoxes of rationality*. Cambridge, MA: MIT Press.

Hume, D. (1751). *An enquiry concerning the principles of morals*. London, England: Millar.

Hutcheson, F. (1725). *An inquiry into the original of our ideas of beauty and virtue*. London, England: Darby.

Hutcheson, F. (1728). *An essay on the nature and conduct of the passions and affections: With illustrations on the moral sense*. London, England: Darby & Browne.

Kant, I. (1903). *Grundlegung zur Metaphysik der Sitten* [Groundwork of the metaphysics of morals]. In *Kant's gesammelte Schriften* (Vol. 4, pp. 385–463). Berlin, Germany: Reimer. (Original work published 1785. English quotations from *Groundwork of the Metaphysics of Morals*, trans. M. Gregor & J. Timmermann, 2011, Cambridge, England: Cambridge University Press.)

Kant, I. (1908). *Kritik der praktischen Vernunft* [Critique of practical reason]. In *Kant's gesammelte Schriften* (Vol. 5, pp. 1–163). Berlin, Germany: Reimer. (Original work published 1788)

Kavka, G. S. (1985). The reconciliation project. In D. Copp & D. Zimmerman (Eds.), *Morality, reason and truth* (pp. 297–319). Totowa, NJ: Rowman & Allanheld.

Maschler, M., Solan, E., & Zamir, S. (2013). *Game theory* (Z. Hellman, Trans.). Cambridge, England: Cambridge University Press. (Original work published in Hebrew 2008)

McClennen, E. F. (1985). Prisoner's dilemma and resolute choice. In R. Campbell & L. Sowden (Eds.), *Paradoxes of rationality and cooperation* (pp. 94–104). Vancouver, Canada: University of British Columbia Press.

McClennen, E. F. (2012). Rational cooperation. *Synthese, 187*, 65–93.

McLeod, O. (2001). Just plain "ought." *Journal of Ethics, 5*, 269–291.

Millgram, E. (Ed.). (2001). *Varieties of practical reasoning*. Cambridge, MA: MIT Press.

Mulgan, T. (2001). *The demands of consequentialism*. Oxford, England: Oxford University Press.

Nagel, T. (1970). *The possibility of altruism*. Oxford, England: Oxford University Press.

Nagel, T. (1986). *The view from nowhere*. Oxford, England: Oxford University Press.

Parfit, D. (2011). *On what matters*. Oxford, England: Oxford University Press.

Peterson, M. (Ed.). (2015). *The prisoner's dilemma*. Cambridge, England: Cambridge University Press.

Portmore, D. W. (2011). Consequentialism and moral rationalism. *Oxford Studies in Normative Ethics*, *1*, 120–142.

Rashdall, H. (1907). *The theory of good and evil* (Vol. 1). Oxford, England: Clarendon Press.

Scanlon, T. M. (1998). *What we owe to each other*. Cambridge, MA: Harvard University Press.

Scheffler, S. (1992a). *Human morality*. Oxford, England: Oxford University Press.

Scheffler, S. (1992b). Prerogatives without restrictions. *Philosophical Perspectives*, *6*, 377–397.

Schlick, M. (1930). *Fragen der Ethik* [Problems of ethics]. Vienna, Austria: Springer.

Schmidt, T. (2016). Instrumentalism about practical reason: Not by default. *Philosophical Explorations*, *19*, 17–27.

Schopenhauer, A. (1841). Preisschrift über die Grundlage der Moral [Prize essay on the basis of morals]. In *Die beiden Grundprobleme der Ethik* (pp. 101–280). Frankfurt/Main, Germany: Joh. Christ. Herrmannsche Buchhandlung.

Schroeder, M. (2007). *Slaves of the passions*. Oxford, England: Oxford University Press.

Sharp, F. C. (1923). Some problems in the psychology of egoism and altruism. *Journal of Philosophy*, *20*, 85–104.

Smith, M. (2011). Deontological moral obligations and non-welfarist agent-relative values. *Ratio*, *24*, 351–363.

Spohn, W. (2009). Wider Nash-Gleichgewichte [Against Nash equilibria]. In C. Fehige, C. Lumer, & U. Wessels (Eds.), *Handeln mit Bedeutung und Handeln mit Gewalt* (pp. 131–149). Paderborn, Germany: Mentis.

Stemmer, P. (2008). *Normativität* [Normativity]. Berlin, Germany: De Gruyter.

Stemmer, P. (2017). Moral, moralisches Müssen und Sanktionen [Morality, the moral "must," and sanctions]. *Deutsche Zeitschrift für Philosophie*, *65*, 621–656.

Stoutland, F. (2001). Responsive action and the belief–desire model. *Grazer Philosophische Studien*, *61*, 83–106.

Stroud, S. (2010). Permissible partiality, projects, and plural agency. In B. Feltham & J. Cottingham (Eds.), *Partiality and impartiality* (pp. 131–149). Oxford, England: Oxford University Press.

Trapp, R. W. (1998). The potentialities and limits of a rational justification of ethical norms. In C. Fehige & U. Wessels (Eds.), *Preferences* (pp. 327–360). Berlin, Germany: De Gruyter.

Vaughn, L. (Ed.). (2013). *Contemporary moral arguments* (2nd ed.). Oxford, England: Oxford University Press. (Original work published 2010)

von Wright, G. H. (1963). *The varieties of goodness*. London, England: Routledge.

Wessels, U. (2002). *Die gute Samariterin* [The good Samaritan]. Berlin, Germany: De Gruyter.

Wessels, U. (2011). *Das Gute* [The good]. Frankfurt/Main, Germany: Klostermann.

Woods, J. (2018). The authority of formality. *Oxford Studies in Metaethics*, *13*, 207–229.

12.2 Moral Reasons

Michael Smith

Summary

How are moral reasons to be integrated into empirical psychology? The answer depends on our views about both morality and reasons for action. As regards morality, there is disagreement about whether we should be Rationalists or Antirationalists, and hence about whether one mark of the moral is that moral requirements entail reasons for action. As regards reasons for action, although it is widely agreed that we have a reason to do what we would be motivated to do if we deliberated well, there is disagreement about what it is to deliberate well. Should we be Humeans about deliberating well or Reasons Primitivists, or Constitutivists? The main difference between these views lies in their accounts of the rational evaluation of intrinsic desires. Psychologists who talk about moral reasons and the philosophers who collaborate with them therefore need to be clear about which of these assumptions they are making in their research.

1. Different Views about Morality, Reasons for Action, and Their Relationship

Moral reasons must in some way be integrated into empirical psychology, but the manner of their integration will depend in large part on which of the competing philosophical views about morality and reasons for action we should accept. Our views about morality will depend on what we think moral knowledge is knowledge of, and hence on whether we think that basic moral facts are knowable a priori. Our views about reasons for action will depend on our understanding of the constitutive norms governing belief and desire, and hence on our understanding of the functional roles that these states play, where their playing these roles is what makes them the psychological states that they are. Unfortunately, few views about these matters are uncontroversial, so how should we proceed?

At a minimum, psychologists should be explicit about which of the controversial views concerning these matters their theories and hypotheses assume, and this in turn suggests that they should engage with philosophers who can help them navigate the conceptual minefields. To make clear the nature of these conceptual minefields, this chapter will describe two views about morality—*Rationalism* and *Antirationalism*—and three views about reasons for action—*Humeanism* and two anti-Humean views: *Reasons Primitivism* and *Constitutivism*—and consider how these views are to be combined with each other. Some criticisms will be offered, but since these criticisms are themselves controversial, readers will be left to judge the overall merits of the views for themselves.

2. Morality

Let's begin by getting clear about the nature of morality. What is the mark of moral principles?

Moral principles are practical principles that tell us, *inter alia*, what we ought to do. Although there is disagreement about their substance, it is more or less universally acknowledged that the mark of moral principles—that is, what distinguishes them from practical principles of other kinds like aesthetic principles, principles of etiquette, and rules of games—is, *inter alia*, that they are principles of *impartial* conduct (Gauthier, 1986; Hare, 1981; Scanlon, 1982). Put slightly differently, the moral viewpoint is the viewpoint we adopt when we think of ourselves as just one among many, each of whom is to be accorded the same consideration as everyone else. It is in this sense that moral principles tell us that we ought to treat each other, ourselves included, impartially, under some interpretation of what it is to treat each other impartially.

Disagreement about the substance of morality, so understood, is disagreement about either the scope of principles of impartial treatment or the conception of impartiality. Perhaps we're required to treat impartially *all sentient beings*, or *all rational agents*, or *all those to whom we can justify our conduct*, or all of the members of some other group of which we are members. Answering this question about the scope of principles of impartial treatment requires us to give an account of which of the many kinds of which we are members is the kind in virtue of which we are owed impartial treatment. The idea that moral principles are by nature impartial leaves this open.

Similarly, the idea that moral principles are by nature impartial leaves it open whether treating each other impartially requires us to give the similar interests of each the same weight in decisions about how we act so as to promote, as best we can, the welfare of all; to act in ways that leave people free to choose how to live on condition that they leave others similarly free; or to be able to justify our conduct to each of the parties affected by our conduct. The idea that moral principles are by nature impartial is thus consistent with utilitarianism, deontology, and contractualism and no doubt with many other candidate moral views too.

If the mark of moral principles is, *inter alia*, that they are principles telling us to act impartially, then what is ruled out? What is ruled out is the idea that moral principles might tell us, at the most fundamental level, that certain individuals count more than others simply in virtue of being the individuals that they are. It is thus a conceptual confusion to suppose that *our* welfare, *our* freedom, or *our* recognition by others counts for more morally than that of others simply because it is ours. Egoism in all its forms is thus a nonmoral view par excellence. It should therefore come as no surprise that the paradigmatic immoral person is someone who is utterly selfish.

In this respect, moral principles stand in stark contrast to principles of epistemic and practical rationality—these are the constitutive norms governing belief and desire mentioned earlier—at least on the conception of such principles we have inherited from Hume (1739–1740/1969). This is important because, as we will see shortly, it is widely agreed that there is a tight connection between reasons for action and how we would be motivated to act if our motivations conformed to principles of epistemic and practical rationality. The stark contrast between moral principles and Humean conceptions of epistemic and practical *rationality* thus suggests a stark contrast between moral principles and Humean conceptions of *reasons for action*.

3. Humeanism about Reasons for Action

According to Humeans, principles of epistemic rationality specify how we ought to form our beliefs in response to the evidence available *to us*, and practical principles specify how we ought to be motivated given *our* intrinsic desires for different outcomes, which may vary in their strength, and our beliefs about which of *our* options satisfy these intrinsic desires—let's call these *instrumental beliefs*—where these may vary in their associated confidence levels. People are thus epistemically and practically rational, according to Humeans, to the extent that they have and exercise the capacity to form beliefs and be motivated in ways that conform to these principles, principles that are themselves thoroughly self-referential.

Note that Humean principles of practical rationality provide us with a way of assessing the rationality of agents' *motivations* in the light of whatever intrinsic desires and instrumental beliefs they happen to have and that Humean principles of epistemic rationality provide us with a way of assessing the rationality of agents' *beliefs* in the light of the evidence available to them. Insofar as attempts have been made to codify these Humean principles of epistemic and practical rationality, the codification has come in the form of the probability calculus as a way of modeling rational belief revision and expected utility as a way of modeling rational decision making.

Importantly, this means that the Humean view that there are no principles in terms of which we can assess the rationality of an agent's intrinsic desires has itself leaked all the way into theoretical psychology. This is not surprising given the very different understandings of intrinsic desires, on the one hand, and beliefs, on the other, we have inherited from Hume—understandings that have become part of common sense. Although intrinsic desires and beliefs both have representational content, Hume thought that intrinsic desires are unlike beliefs in not representing anything to be the case. Instead, they represent a way the world could be associated with positive affect, or a way that we would be disposed to make the world if we believed we had the option to make it that way. Moreover, Hume thought it followed from this that intrinsic desires are unlike beliefs in another respect as well, namely, in not being sensitive to evidence of truth and falsehood, and so in not being assessable for their reasonableness. This is why Humeans deny that there are any principles in terms of which we can assess the rationality of intrinsic desires.

According to Humeans, intrinsic desires thus stand in stark contrast not just to beliefs but also to instrumental desires, which they take to be amalgams of intrinsic desires and instrumental beliefs, amalgams that are apt to cause action. Since the instrumental beliefs that partially constitute instrumental desires are sensitive to evidence of truth and falsehood about the nature of the possible worlds in which we pursue our options, Humeans think that the instrumental desires thus partially constituted are also sensitive to evidence of truth and falsehood, and hence assessable for their reasonableness. This is a big difference between instrumental desires and intrinsic desires, but it is a difference explained by the fact that Humeans think that the former are, whereas the latter are not, amalgams partially constituted by beliefs.

We have focused on Humean conceptions of epistemic and practical rationality because, as signaled earlier, on many views of reasons for action, both Humean and anti-Humean, there is an intimate connection between the reasons for action agents have and the principles of epistemic and practical rationality to which they are subject. This is because of a platitude about reasons for action. Reasons for action are those considerations that would motivate us to act if we were to deliberate well, where deliberating well is a matter of our meeting three conditions: first, our beliefs must conform to all principles of epistemic rationality; second, our motivations must conform to all principles of practical rationality; and third, our beliefs and motivations must conform to these principles in circumstances in which the world is maximally and reliably revealed to us, that is, in circumstances in which we can be certain about everything that's relevant to the formation of our beliefs and resultant motivations (Korsgaard, 1986; Williams, 1981).

Note that this platitude about reasons for action helps us understand the often-made distinction between *subjective* and *objective* reasons for action (Sepielli, 2018). Objective reasons for action are those considerations that meet all three conditions just mentioned, whereas subjective reasons for action are those considerations that meet just the first two conditions. Subjective reasons for action are thus relative to the available evidence, where that evidence may be misleading. Objective reasons for action, by contrast, are relative to the facts. The platitude also helps us to understand the commonsense distinction between what there is *some* reason to do and what there is *most* reason to do. Since agents may have multiple motivations, some weaker and others stronger, it follows that they have *some* reason to do whatever they would have some motivation to do if they deliberated well, where that motivation may be weak or strong, and that they have *most* reason to do whatever they would be most strongly motivated to do if they deliberated well.

We are now in a position to see that what gets codified in expected utility conceptions of rational decision making is the Humean's conception of what we have *most subjective* reason to do. In what follows, we will mostly ignore such reasons and focus on what we have *some objective* reason to do, as this focus will make it easier for us to see the contrast between Humean conceptions of reasons for action and moral requirements. Unless flagged otherwise, all talk of reasons for action in the remainder should therefore be understood accordingly. How do Humeans think about such reasons? They think we have (some objective) reason to do each of the things that we would have some motivation to do if we were to deliberate well on the basis of the facts about what would satisfy our various intrinsic desires (Schroeder, 2007; Williams, 1981).

One immediate consequence of this Humean view of reasons for action is that, in certain circumstances, there are some people who have no reason at all to act in accordance with moral principles. Imagine someone utterly selfish whose only intrinsic desire is to promote his own welfare. In circumstances in which the evidence reveals to him that his own welfare won't be promoted by his acting impartially, under any conception of impartiality, he will violate no principle of epistemic or practical rationality by having no motivation to so act. According to Humeans, he therefore has no reason to act impartially. Whether people have a reason to act impartially will therefore depend entirely on what intrinsic desires they happen to have and whether their acting impartially will lead to the satisfaction of one of these desires.

4. Rationalism versus Antirationalism about Moral Reasons for Action

Is this an objection to the Humeans' view of reasons for action, and hence to their conception of principles of practical rationality? Not necessarily.

I said at the outset that we will consider two views of morality. One of these views is Rationalism, the view that if there are any moral requirements, then the basic requirements are knowable a priori and entail corresponding reasons for action (Kant, 1785/1948). The other view is Antirationalism, which is just the denial of Rationalism (Brink, 1989; Railton, 1986). If Rationalism is true, and if those who have no intrinsic desires that would be served by their acting impartially are nonetheless subject to moral requirements, then this is a decisive objection to the Humean view of reasons for action. But if Antirationalism is true, it is no objection at all.

Moreover, at first sight, there seems to be nothing implausible about Antirationalism. Consider norms of other kinds. People can be subject to rules of games—"Thoroughly shuffle the deck before dealing the cards in poker"—and social rules like requirements of etiquette—"Reply in the third person to an invitation sent in the third person"—without having any reason at all to conform to these rules or requirements (Foot, 1978). A magician who has been secretly hired to entertain the other players with their card tricks need have no reason at all to shuffle the cards when playing poker, and a maverick who receives an invitation written in the third person need have no reason at all to reply in the third person. Since the basic requirements of poker and etiquette can change over time, knowledge of such requirements also seems to be a posteriori knowledge par excellence, not a priori knowledge. Antirationalists are therefore within their rights to ask why we should suppose that requirements of morality are any different.

How will Humeans who are Antirationalists think about moral reasons? They will think that people have moral reasons only contingently, and they will think that they have them in virtue of having some intrinsic desire that would be served by their acting impartially in the sense of "impartiality" picked out by the correct conception of morality. Humean Antirationalists may disagree among themselves about what the empirical facts are that make a conception of morality the correct conception. However, they will all agree that in circumstances in which people have no such intrinsic desires, they have no moral reasons.

Note that Humean Antirationalists needn't deny that some people will have moral reasons more robustly than others. For example, those who happen to have an intrinsic desire to act in accordance with moral requirements will have moral reasons even when their acting in accordance with such requirements doesn't serve any other intrinsic desire they happen to have, whereas those who lack such a morally loaded intrinsic desire will only have reasons to act impartially when they do have some other such intrinsic desire. But even those who have morally loaded intrinsic desires, and so have moral reasons robustly, will still only have such reasons contingently, as they wouldn't have had such reasons if they hadn't had an intrinsic desire to act in accordance with moral requirements.

5. The Problem with Humean Antirationalism about Moral Reasons

How plausible is Humean Antirationalism about moral reasons? The main attraction of this view is its modesty.

As regards psychology, Humean Antirationalism about moral reasons assumes only easily understandable conceptions of belief and desire, ideas that have been codified in the probability calculus and expected utility conceptions of rational decision making, and as regards norms, it assumes that moral requirements are no different from

other, more familiar norms like rules of games and social rules like requirements of etiquette. If we are to reject Humean Antirationalism, then moral requirements must be very different from these other norms, and belief and desire must be rather different from the way they're ordinarily taken to be.

Rationalists think that moral requirements are indeed very different. The problem, as they see things, is that Humean Antirationalism cannot be squared with the conditions under which we hold people morally responsible. Let's begin with people's responsibility for having knowledge of basic moral facts. We assume that such knowledge is within the grasp of anyone with normal powers of reasoning independently of where or when they grew up. In this respect, it is more like knowledge of basic mathematical facts. Although this doesn't entail that basic moral facts are, like mathematic facts, knowable a priori, it does rule out many of the obvious empirical alternatives. For example, if we assume that intelligent extraterrestrials would have knowledge of basic moral facts, as it seems we do, then that rules out moral facts being empirical facts that are peculiar to human beings or to life on Earth.

Now consider people's responsibility to act in accordance with moral requirements. When people fail to act in accordance with a moral requirement, we take them to be fit candidates for blame if they don't have an exemption or an excuse. Exemptions include the fact that they lack the normal powers of reasoning required to have knowledge of basic moral requirements—perhaps they are infants or insane—and excuses include their being nonculpably ignorant of the fact that the circumstances that they find themselves in are circumstances in which the moral requirement applies to them.

Importantly, however, someone's lacking any intrinsic desire that would be served by their acting in accordance with a moral requirement is neither an exemption nor an excuse. Imagine an immensely callous person who fails to help someone in need when their helping them is morally required. The fact that they don't care is neither an excuse nor an exemption for their failure to help. Those who fail to act in accordance with a moral requirement simply because they lack any desire that would be served by their doing so are therefore blameworthy. The question is whether this can be squared with the Humean's conception of moral reasons.

There is a difficulty here, as it seems that someone couldn't be blameworthy for failing to help if they had no reason to help (Darwall, 2006; Portmore, 2011). They couldn't be blameworthy because it would be completely unreasonable to expect anyone to do something that they have no reason to do. Someone's being a fit candidate for blame thus seems to presuppose that there was at least some reason for them to do what they're blameworthy for failing to do. Indeed, it seems to presuppose something much stronger, namely, that there was a decisive reason for them to do what it would be fit to blame them for failing to do. Someone's failing to act in accordance with a moral requirement is in this respect very different from their failing to act in accordance with the rules of a game or with requirements of etiquette.

Failures to abide by rules of games or requirements of etiquette do not in general imply that violators are fit candidates for blame, as the examples given earlier amply illustrate. Of course, blame is appropriate in such cases if the violation of the rule of the game or the requirement of etiquette is, in the circumstances, also the violation of a moral requirement. But that just reinforces the point that there is an important difference between moral requirements and other norms like rules of games and requirements of etiquette, and that the difference lies in the fact that moral requirements do, whereas rules of games and requirements of etiquette do not, entail reasons for action.

Here, then, lies the problem with the Humean conception of moral reasons. According to Humeans, those who lack intrinsic desires that would be served by their acting in accordance with a moral requirement have no reason at all to so act, but this cannot be squared with the conditions under which we hold people morally responsible. This is because, on the one hand, we don't blame people for acting in ways that they have no reason to act, while on the other, we do blame them for failing to act in accordance with moral requirements, absent an excuse or an exemption, where lacking an intrinsic desire that would be served by their so acting is neither. Something has to give, and it appears to be the Humeans' view of reasons for action. But what is the alternative to the Humeans' view?

Let's return to the platitude about reasons for action. It tells us that people have reason to do what they would be motivated to do if they deliberated well. Where Humeans seem to go wrong is in their conception of what it is to deliberate well, particularly in their view that there are no principles of practical rationality in terms of which we can assess an agent's intrinsic desires. But Humeans are surely right that intrinsic desires don't themselves represent anything to be the case but instead represent things as being either a way things could be that is associated with positive affect or a way that we would be disposed to make them if we believed that we

had the option of doing so. If they aren't truth-assessable, then how can there be principles of practical rationality in terms of which we can assess their rationality? We will consider two answers Anti-Humeans have given to this question, a radical answer and a not-so-radical answer. Only the radical answer requires us to reject the Humeans' conception of intrinsic desire.

6. Reasons Primitivism about Reasons for Action

According to Reasons Primitivists, Humeans go wrong at the very beginning (Parfit, 2011; Scanlon, 2013). At the most fundamental level, the considerations that support the truth of our beliefs do so because they *count in favor* of believing, where counting in favor is a *primitive* relation that considerations stand in to states of believing, a relation that is not further analyzable in terms of being truth-supporting. If we ask what it is for considerations to count in favor of believing, Reasons Primitivists tell us that all we can say is that it is for them to provide reasons, and if we ask what it is for considerations to provide reasons, they tell us that it is for them to count in favor.

Of course, Reasons Primitivists don't deny that the considerations that count in favor of believing support the truth of what's believed. What they deny is just that we could analyze the reason-relation in terms of truth-supportingness. In their view, it is a substantive truth about a certain class of reasons, reasons for belief, that they are truth-supporting. We cannot turn this into an analysis of the reason-relation, they say, because not all of the considerations that count in favor of psychological states for which reasons can be given—call these *judgment-sensitive attitudes*—are truth-supporting considerations. The judgment-sensitive attitudes include not just believing but also intrinsically desiring, intending, trusting, admiring, fearing, and so on. Although the considerations that count in favor of believing support the truth of what's believed, the considerations that count in favor of intrinsically desiring, admiring, and trusting don't count in favor of the truth of what's intrinsically desired, what's admired, and what's trusted. Reasons for beliefs are thus outliers in this group.

Reasons Primitivists thus recommend radical alternatives to the Humeans' principles of epistemic and practical rationality. In their view, ideally rational people are those whose beliefs and intrinsic desires—and their intentions, attitudes of trust, feelings of admiration, and so forth—come and go to the extent that they take there to be considerations that count sufficiently in favor of them. Someone who is ideally rational thus acquires intrinsic desires, just like they acquire beliefs, when they take there to be sufficient reasons for them, and they lose their intrinsic desires, just like they lose their beliefs, when they don't take there to be sufficient reasons for them or when they take there to be insufficient reasons for them. Where Humeans see a striking difference between beliefs and intrinsic desires, Reasons Primitivists see a striking similarity.

This, in turn, suggests a very different Reasons Primitivist account of what it is for there to be (some objective) reason for action. Remember again the platitude about reasons for action. It tells us that what we have a reason to do is what we would be motivated to do if we were to deliberate well, where deliberating well is a matter of our beliefs conforming to all principles of epistemic rationality, our motivations confirming to all principles of practical rationality, and our beliefs and motivations conforming to these principles in circumstances in which the world is maximally revealed to us.

Reasons Primitivists think that the first and second conditions are met when agents are ideally rational—that is, when they have beliefs and intrinsic desires, and hence resultant motivations, that they take there to be sufficient reasons for—and they think that the third condition is met when the considerations that agents take to be sufficient reasons for believing and intrinsically desiring are sufficient reasons. In other words, a crucial aspect of the world that Reasons Primitivists think needs to be maximally and robustly revealed to us if we are to deliberate well is the normative part, the part that concerns which considerations are sufficient reasons for believing and intrinsically desiring, and hence for being motivated.

The difference between the Humean conception of reasons for action and the Reasons Primitivist conception can thus be summed up as follows. Whereas Humeans think that there are reasons for agents to do whatever will satisfy *their intrinsic desires*, Reasons Primitivists think that there are reasons for agents to do whatever will satisfy *the intrinsic desires that they have sufficient reason to have*. In their view, the concept of a reason is therefore polysemous. The primitive reason-relation holds between considerations and judgment-sensitive attitudes. Since actions are not themselves such attitudes, they do not figure as a relatum of the primitive reason-relation. The concept of *a reason for action* is defined in terms of the concepts of *a reason for believing* and *a reason for intrinsically desiring*, where the latter is the primitive reason-relation, the relation of counting in favor. This relation is primitive because no general account can be given of why certain considerations count in favor of

judgment-sensitive attitudes rather than others. They are just the ones that count in favor, in the sense of providing reasons for them.

What do moral reasons look like, according to Reasons Primitivists? Agents have moral reasons for action, according to Reasons Primitivists, if and only if they would be motivated to act impartially, given the correct conception of impartiality, if they deliberated well, where this requires that there are considerations that count sufficiently in favor of their being motivated to so act. These moral reasons will be noninstrumental if the intrinsic desires that they have sufficient reasons to have, and that are satisfied by their acting impartially, themselves have impartial contents; otherwise, they will be instrumental. This, in turn, means that agents will have moral reasons robustly—which is to say independently of there being sufficient reasons for them to have intrinsic desires that just so happen to be served by their acting impartially—just in case they are noninstrumental.

Note that this Reasons Primitivist account of what it is for agents to have noninstrumental moral reasons combines readily with Rationalism. What is it for an action of a certain kind to be morally required in certain circumstances? The most plausible answer for Rationalists to give is that an action of that kind is morally required in certain circumstances just in case there is a *decisive* noninstrumental moral reason to perform an action of that kind in those circumstances. According to the Rationalist who is also a Reasons Primitivist, this in turn requires that there be a sufficient reason for an agent to have an intrinsic desire with the right kind of impartial content, that acting in that way satisfies that intrinsic desire, and that that sufficient reason is itself strong enough, given the other reasons in play, for the agent to be most strongly motivated to act in that way in those circumstances.

The Rationalist Reasons Primitivist thus has no problem at all making sense of the conditions under which we hold agents responsible. Imagine again a callous person who is morally required to help someone but who fails to help not because he has an excuse or an exemption but just because he doesn't care. We blame him for his failure. According to the Rationalist Reasons Primitivist, we blame him because being morally required to help entails that he has a decisive moral reason to help, and he has this reason because there is a sufficient reason for him to have an intrinsic desire with impartial content that will be satisfied by his helping, an intrinsic desire sufficiently strong to make him most motivated to do so. The fact that the callous man lacks any desire that will be satisfied by his helping is thus neither here nor there as regards his having such a reason.

We have spent some time outlining the Reasons Primitivists' alternative to the Humeans' account of what it is to deliberate well. The main advantage of Reasons Primitivism is that it combines so well with Rationalism, as this allows it to square with the conditions under which we hold people morally responsible. The main disadvantage is that it does so at the cost of postulating a primitive normative relation of a consideration's *counting in favor* of an attitude. This isn't just metaphysically profligate but also psychologically demanding, as Reasons Primitivists have to think that people have attitudes toward this primitive relation—the attitude of taking considerations to count in favor of their believing and desiring—and that they are epistemically and practically rational to the extent that their beliefs and desires are sensitive to these attitudes. It is thus worth noting that one upshot of Reasons Primitivism is that those who deny the existence of such a primitive relation are apparently irrational by default.

This forces us to ask an obvious question: is a less radical alternative available that combines just as well with Rationalism? If so, parsimony and plausibility will tell in favor of preferring that view to Reasons Primitivism.

7.   Constitutivism about Reasons for Action

Anti-Humean Constitutivists—hereafter "Constitutivists," for short—think that such an alternative is available (Korsgaard, 1986; Smith, 2013). In their view, the Humeans' principles of epistemic and practical rationality are fine as far as they go. The problem isn't that we should reject these principles, as Reasons Primitivists do, but rather that we need to add more principles. But what are these extra principles, and why should we add them?

Constitutivists begin by reminding us that the complete set of rational principles governing beliefs and desires are those principles, whatever they are, that fix the roles these states play when we deliberate well. They then point out that it follows from this that if having certain intrinsic desires is itself necessary in order to deliberate well, then these intrinsic desires will be rationally required. In their view, Humeans therefore go wrong in moving from the premise that intrinsic desires cannot be assessed for their truth to the conclusion that they cannot be assessed for their reasonableness. They point out that they would still be so assessable if satisfying them were partially constitutive of what it is to deliberate well.

For Rationalism to be true, intrinsic desires with impartial contents would be required for us to deliberate well. Is this plausible? Constitutivists think it is. They point out that deliberating well is a robust activity, not something that we manage to do by chance or happenstance. This is why one of the conditions on correct deliberation is that the world *reliably* manifests itself to us. But this means that other wordly conditions are necessary too. After all, we will not deliberate well if we are in the company of others who deceive us, or if we deceive ourselves, or if others coerce us, or we coerce ourselves, or if others stand idly by while we succumb to some preventable deterioration in our deliberative capacities, or if we stand idly by ourselves. Nor will we deliberate well if we or others aren't *reliably* disposed not to deceive, coerce, or stand idly by while we succumb to some preventable deterioration in our deliberative capacities.

What this suggests to Constitutivists is that deliberating well must take place in a social context in which others deliberate well too, a social context in which each deliberator has a strong intrinsic desire that they do not interfere with anyone's exercise of their deliberative capacities—neither their own nor anyone else's—and a strong intrinsic desire that they do what they can to ensure that everyone has deliberative capacities to exercise. Let's call these "the desires to help and not interfere." The role of these intrinsic desires to help and not interfere, to repeat, is to make sure that when we act, we not only don't undermine but also sustain the possession and exercise of the deliberative capacities that make it possible for us to deliberate well.

Note that the intrinsic desires to help and not interfere have impartial contents. If having such intrinsic desires is constitutive of deliberating well, then it follows that all agents, independently of having reasons to satisfy whatever other intrinsic desires they might happen to have, will have noninstrumental moral reasons to help and not interfere. Since all agents have such moral reasons, Constitutivism entails Rationalism, with moral requirements understood in terms of reasons in the way suggested earlier. Constitutivism therefore squares well with the conditions of moral responsibility.

Constitutivism also entails that, to the extent that people's satisfying whatever intrinsic desires they happen to have doesn't undermine anyone's helping and not interfering, they will also have reasons to satisfy whatever intrinsic desires they happen to have. This is why Constitutivism is a less radical alternative to the Humeans' account of what it is to deliberate well. It tells us that it is rationally permissible to satisfy whatever intrinsic desires we happen to have, so long as doing so is consistent with our acting on our moral reasons to help and not interfere when these reasons are decisive.

We saw earlier that the mark of moral principles is that they are impartial, where this leaves it open what impartiality consists in. Note that Constitutivism takes a stand on this issue. The idea that we all have reasons to satisfy whatever intrinsic desires we happen to have, so long as our doing so is consistent with our acting on our moral reasons to help and not interfere when these reasons are decisive, sits happily alongside the familiar deontological conception of moral principles as protecting the freedom of each person to live a life of their own choosing, so long as their doing so is consistent with everyone else's doing the same thing. According to this view, welfare is not of intrinsic moral significance. Welfare matters morally only to the extent that it affects freedom.

8. Conclusion

The question we began with is how moral reasons are to be integrated into empirical psychology. We have seen that the answer we give to this question will depend on our views about both morality and reasons for action.

As regards morality, although it is widely agreed that the mark of moral requirements is their impartiality, there is disagreement about whether we should be Rationalists or Antirationalists and hence whether a further mark of moral requirements is that they entail reasons for action, as Rationalists think they do. As regards reasons for action, although it is widely agreed that we have a reason to do what we would be motivated to do if we deliberated well, there is a great deal of disagreement about what it is to deliberate well. Should we be Humeans about deliberating well, Reasons Primitivists, or Constitutivists? The main difference between these views lies in their very different accounts of the rational evaluation of intrinsic desires.

Given the disagreement, psychologists who talk about moral reasons need to make clear the assumptions they make about morality and reasons for action. Do they assume Rationalism or Antirationalism about morality, and do they assume Humeanism, Reasons Primitivism, or Constitutivism about reasons for action? If they make these assumptions clear, then, with the passage of time, the answer to a further important question will hopefully emerge: do certain of these assumptions rather than others make for a more productive research program? The answer to this question will be of great significance to psychologists and philosophers alike.

References

Brink, D. O. (1989). *Moral realism and the foundations of ethics*. Cambridge, England: Cambridge University Press.

Darwall, S. (2006). *The second-person standpoint: Morality, respect, and accountability*. Cambridge, MA: Harvard University Press.

Foot, P. (1978). Morality as a system of hypothetical imperatives. In *Virtues and vices* (pp. 157–173). Berkeley: University of California Press.

Gauthier, D. (1986). *Morals by agreement*. Oxford, England: Clarendon Press.

Hare, R. M. (1981). *Moral thinking*. Oxford, England: Oxford University Press.

Hume, D. (1969). *A treatise of human nature*. Oxford, England: Clarendon Press. (Original work published 1739–1740)

Kant, I. (1948). *Groundwork of the metaphysics of morals*. London, England: Hutchinson. (Original work published 1785)

Korsgaard, C. (1986). Skepticism about practical reason. *Journal of Philosophy, 83*, 5–25.

Parfit, D. (2011). *On what matters*. Oxford, England: Oxford University Press.

Portmore, D. W. (2011). *Commonsense consequentialism: Wherein morality meets rationality*. Oxford, England: Oxford University Press.

Railton, P. (1986). Moral realism. *Philosophical Review, 95*, 163–207.

Scanlon, T. (1982). Contractualism and utilitarianism. In A. Sen & B. Williams (Eds.), *Utilitarianism and beyond* (pp. 103–128). Cambridge, England: Cambridge University Press.

Scanlon, T. (2013). *Being realistic about reasons*. Oxford, England: Oxford University Press.

Schroeder, M. (2007). *Slaves of the passions*. Oxford, England: Oxford University Press.

Sepielli, A. (2018). Subjective and objective reasons. In D. Star (Ed.), *The Oxford handbook of reasons and normativity* (pp. 784–799). Oxford, England: Oxford University Press.

Smith, M. (2013). A constitutivist theory of reasons: Its promise and parts. *LEAP: Law, Ethics, and Philosophy, 1*, 9–30.

Williams, B. (1981). Internal and external reasons. In *Moral luck* (pp. 101–113). Cambridge, England: Cambridge University Press.

12.3 The Psychology and Rationality of Moral Judgment

Alex Wiegmann and Hanno Sauer

Summary

The field of moral psychology has become increasingly popular in recent years. This chapter focuses on two interrelated questions: first, how do people make moral judgments? It will review the most prominent theories in moral psychology that aim to characterize, explain, and predict people's moral judgments. Second, how should people's moral judgments be evaluated in terms of their rationality? This question is approached by reviewing the debate on the rationality of moral judgments and moral intuitions, which has been strongly influenced by findings in moral psychology but also by recent advances in learning theory.

1. Rationality and Morality

The rationality of moral judgment is one of the most contested issues in moral psychology. In this chapter, we will outline the cutting edge of contemporary work in moral psychology, convey a sense of the most promising recent developments, and explain where this research is currently heading. We will also address some of the philosophical implications these theories are supposed to yield: Are our moral intuitions robust and trustworthy? Or are they frail and unreliable? Ultimately, these *normative* questions regarding the rationality of moral judgment are what most philosophers and probably also many psychologists are interested in.

In the first part, we will summarize theories that make specific predictions about which moral judgments people are likely to make in response to which types of cases and what the psychological processes are that explain the patterns so discovered. In particular, we will explain how a theory based on the distinction between *model-free* and *model-based* information processing and decision making has come out as the main contender among such predictive theories. In doing so, we will integrate both insights from computational approaches

to reinforcement learning as well as recent developments in dual-process models of cognition. This concludes the first, *descriptive* part of our overview.

The second, *normative* part addresses the issue of the rationality of moral judgments head-on.[1] The recent debate suggests that this question might not be best addressed in terms of the psychological foundations of moral cognition: Are moral judgments intuitive or controlled? Are they based on emotion or reason? Does reasoning produce, or merely rationalize, moral judgments? Rather, addressing the question of whether the processes generating moral beliefs are capable of incorporating and updating on morally salient information, that is, capable of *learning*, seems the most promising line of research. We will explain why some authors argue that the sophistication of processes of moral learning provides reasons for optimism regarding the rationality of moral judgments and why some authors continue to strike a more pessimistic note.

2. Predictive Theories of Moral Judgment

In this descriptive part of the chapter, we will review three prominent predictive theories of moral judgments, namely, Greene's dual-process model, Mikhail's universal moral grammar theory, and Cushman's and Crockett's model-free versus model-based approach, with a focus on the latter since it is the most recent and probably the most promising theory. It should be noted that the distinction between "descriptive" and "normative" accounts of moral cognition is at least somewhat artificial, as some of the models discussed in the former also have bearing on the latter issue. What we mean to emphasize here is that some theories in moral psychology make specific, testable predictions regarding how subjects will actually respond to certain issues and why, while the latter focus on deriving possible normative conclusions from an overall assessment of the state of the art in the psychology of moral judgment.

2.1   Dual-Process Models

The contemporary debate on the rationality of moral judgment is dominated by *dual-process* models of moral cognition (Greene, 2013). According to such models, cognitive operations are carried out by two fundamentally different types of processing, one quick, intuitive, automatic, evolutionarily old, and frequently unconscious (System 1), the other slow, effortful, analytic, evolutionarily young, and consciously controlled (System 2; Evans & Stanovich, 2013; Kahneman, 2011).

Some of these theories hold that moral judgment is System 1 all the way down (see section 3.1). Others have argued that the two types of processing can be mapped onto two different subsets of moral judgment: deontological and consequentialist moral judgments (Greene, 2008, 2013, 2014). This distinction is imported from philosophical moral theory and refers, roughly, to the views that the outcome of an action alone determines its moral status (consequentialism) or that other factors—such as a person's intentions or the way an outcome was brought about (e.g., as a means or a foreseen side effect)—matter as well (May, 2014b). This alternative is typically operationalized by classifying subjects' responses to certain well-known moral problems ("trolley cases"; for an overview, see Waldmann, Nagel, & Wiegmann, 2012), in which participants have to imagine a train about to run over five people and are asked whether it would be OK to perform a certain action in order to save the lives of the five people. The two most prominent variants of such trolley cases are usually referred to as *Switch* and *Push*, respectively. In Switch, one can, by flipping a switch, divert the train onto a different track, where it would run over and kill a single person. In Push, the train can be stopped by pushing a large man, who is standing on a bridge crossing the tracks, in front of it. While interventions in these cases (flipping the switch and pushing the man, respectively) are classified as *consequentialist* responses, not performing the action necessary to save the five people is usually considered the *deontological* option.[2]

Some want dual-process accounts to *debunk* deontological theories and *vindicate* consequentialist ones. The former are purportedly based on crude, alarm-like emotions (see also the moral-heuristics approach by Gigerenzer, 2008, or Sunstein, 2005) and the latter on flexible, sophisticated cost–benefit analyses (Greene, 2008). This pattern seemed to be reflected by patterns of brain activation, differences under cognitive load, and greater response times for consequentialist judgments, indicating controlled intuition override (Greene, Morelli, Lowenberg, Nystrom, & Cohen, 2008; Greene, Nystrom, Engell, Darley, & Cohen, 2004;

Greene, Sommerville, Nystrom, Darley, & Cohen, 2001). Moreover, the etiology of deontological intuitions seemed to show that they pick up on morally irrelevant differences such as how "up close and personal" a person is harmed (Greene et al., 2001).

Criticism of the dual-process model has focused on four issues: first, that the empirical data from neuroimaging (Machery, 2014; Poldrack, 2006; Sauer, 2012b) or response times (McGuire, Langdon, Coltheart, & Mackenzie, 2009) are more ambiguous than originally thought. Second, some have argued that the empirical data from neuroscience or certain evolutionary considerations play no actual role in the argument against deontology (Berker, 2009; cf. Greene's [2010] response). Third, it's not obvious why the emotional or intuitive basis of deontological judgments should pose any epistemic problem for them at all (Railton, 2014). Fourth, and most important, many have argued that even *if* the empirical data are sound, and even *if* they do the moral heavy lifting, and even *if* they do identify a rationally defective process, it remains unclear whether they really target deontological theories. The reason for this doubt is that the link between consequentialist moral judgment and controlled cognition can be reversed when the consequentialist option is the intuitive one and the deontological option requires override (Kahane et al., 2011). Moreover, in order for people's moral judgments to count as genuinely consequentialist, they would have to stem from an impartial concern for the greater good. This seems not to be the case (Kahane, Everett, Earp, Farias, & Savulescu, 2015). In general, the Trolleyological paradigm in moral psychology tends to misconstrue consequentialism (Kahane et al., 2018; but see also Conway, Goldstein-Greenwood, Polacek, & Greene, 2018; see also endnote 2).

2.2   Universal Moral Grammar

The dual-process model assumes that moral judgments result from domain-general processes that are not specific to morality. A further possibility is that people are endowed with an innate moral faculty that allows them to cognitively structure actions in terms of moral categories and to assign deontic statuses (right or wrong, permissible or impermissible) to those actions depending on this structure. Drawing on ideas pioneered in a linguistic context, various authors have tried to show that moral cognition is based on such a *universal moral grammar* (Hauser, Young, & Cushman, 2008; Mikhail, 2007, 2011; cf. Prinz, 2008).

People start to appreciate certain moral distinctions at a very early age, many of those distinctions seem to be culture-invariant, and people have a hard time articulating the rules structuring their moral thinking (Hauser, Cushman, Young, Jin, & Mikhail, 2007). Also, some patterns in people's moral judgment are difficult to account for from within a dual-process framework, such as the distinction between harming as a means and harming as a foreseen side effect (e.g., Switch versus TrapDoor; see next section). Primarily, however, the nativist case for a universal grammar is supported by so-called poverty-of-the-stimulus arguments, according to which the moral rules that implicitly inform people's moral judgments are too subtle to be read off the scarce available evidence; subjects *couldn't* learn these rules empirically. We will return to this question after a short detour.

2.3 Model-Free versus Model-Based Moral Judgment

Crockett (2013) and Cushman (2013) independently developed a dual-system account of moral psychology that has already had an impact on the field of moral psychology and will likely continue to do so (e.g., Greene, 2017; Railton, 2017). According to this approach, two distinct systems evaluate actions: while the model-based system selects actions based on inferences about its (expected) outcomes in a specific situation, the model-free system assigns value to actions intrinsically, based on past experience (Crockett, 2013; Cushman, 2013).

Such dual-system frameworks have been argued to be of great value for studying and understanding judgment and decision making in general (Evans & Stanovich, 2013; Kahnemann, 2011; see Kruglanski & Gigerenzer, 2011, for a critical assessment), and Greene's emotional-versus-cognitive framework has been very influential in the field of moral psychology (e.g., Greene et al., 2001). However, it has been argued that this particular distinction of processes (i.e., emotional versus cognitive) might not be fully adequate (Cushman, 2013; Pessoa, 2008) and computationally too crude (Crockett, 2013; Mikhail, 2007; see Crockett, 2016, for the advantages of formal models) to characterize and explain moral judgments. Crockett and Cushman's distinction of action evaluation versus outcome evaluation is supposed to avoid the shortcomings of Greene's characterization of the two systems while keeping the benefits of a dual-system framework (Greene [2017] seems to agree). Findings from studies dissociating the aversiveness of actions from outcomes provide some initial evidence for separate action- and outcome-based (model-free versus model-based) value representations (Cushman, Gray, Gaffey, & Mendes, 2012; Miller & Cushman, 2013; Miller, Hanni-kainen, & Cushman, 2014).

The distinction between action- and outcome-based value representations resembles a distinction from computational approaches to reinforcement learning (cf. Cushman, 2013), and recent advances in neuroscience have identified corresponding neural signatures (Dayan & Niv, 2008; Dolan & Dayan, 2013; Gläscher, Daw, Dayan, & O'Doherty, 2010; Wunderlich, Dayan, & Dolan, 2012). Reinforcement learning investigates how agents (humans, rats, algorithms) can learn to predict the consequences of and optimize their actions in interaction with the environment (Dayan & Niv, 2008). Learning algorithms can be classified broadly as model free or model based (Sutton, 1988; Sutton & Barto, 1999, 2018), depending on how they learn and select actions in a given situation (for an example in human behavior, see Dayan & Niv, 2008). Put simply, a model-based algorithm builds a causal representation of its environment, uses this model to consider the whole range of effects that different actions would have, and chooses the action with the highest expected value. In contrast, a model-free algorithm does not rely on a model of the environment and thus cannot make far-sighted choices. In a given situation, it simply selects the action with the best track record, which can be acquired by rather simple means such as prediction error (Gläscher et al., 2010) and temporal difference learning (Sutton, 1988; cf. Cushman, 2013). Although a model-free algorithm is computationally cheap, it can often lead to successful behavior. However, if the environment changes, it cannot adapt to these changes as readily as the computationally rather expensive model-based algorithms since it needs to gather experience in the new environment to update the track record accordingly (Crockett, 2013; Cushman, 2013; Greene, 2017).

Can the model-free-versus-model-based (or action-versus-outcome) framework explain prominent findings in moral psychology? Let us start with applying the framework to three variants of the well-known trolley dilemma: Switch, Push, and TrapDoor (the latter being similar to Push, but rather than the heavy person being pushed from the bridge, the flip of a switch opens a trap door through which the large man falls onto the tracks). The relation among the approval rates for the respective actions in these dilemmas is Switch > TrapDoor > Push (Mikhail, 2007, 2011). The model-free-versus-model-based framework can provide a straightforward explanation for the fact that Push is usually evaluated worse than the other two cases: although the model-based evaluation of the actions does not differ (since they lead to the same outcome), pushing a person usually leads to negative experiences (punishment, distress cues), whereas flipping a switch is not associated with a negative action value. In other words, the model-free evaluation prefers flipping a switch to pushing a person (Crockett, 2013; Cushman, 2013). Things get more complicated when comparing the Switch and the TrapDoor variant. Here, not only the outcome (saving five lives) but also the required action (flipping a switch) is the same. What differentiates these two cases is the causal role of the victim. In Switch, the one person getting hit by the train is not causally necessary for saving the five lives (imagine the one person jumping off the track before getting hit), whereas in TrapDoor, the five people could not be saved without making use of the heavy person as a trolley stopper (Mikhail, 2007; Wiegmann & Waldmann, 2014). The model-free-versus-model-based framework attempts to capture this difference by postulating that the model-based system represents harming a person as a subgoal in TrapDoor but not in Switch. Subgoals, in turn, can be treated by the model-free system as if they were actions—and if the subgoal is aversive, as it is in the case of a person getting hit by a train, the model-free system assigns a negative value to it.[3] It is this negative value assignment of the model-free system that differentiates side effect cases from means cases and can explain why the latter are often judged as more aversive. A further popular finding in moral psychology is that usually people consider harmful actions worse than omissions with identical effects (Baron & Ritov, 2004). This asymmetry can be explained by pointing out that model-free values are only assigned to actions but not to the always available option of *not* performing an action.

To sum up, the model-free-versus-model-based account by Crockett (2013) and Cushman (2013) offers a promising domain-general theory of moral judgment.

3.   The Rationality of Moral Judgment

In the first part of this chapter, we described prominent theories, findings, and developments in moral psychology. Not surprisingly, such insights came along with discussions about how to evaluate people's moral judgments and moral intuitions. In this second part of the chapter, we will address this normative question by first retracing the debate on the rationality of moral judgments from the beginning of moral psychology to the current state of the field on a rather abstract level. We then narrow down the considered time frame and content by focusing on the rationality of moral intuitions and how this debate has evolved in recent years due to advances in learning theory.

3.1   Classic Rationalism, Social Intuitionism, and Sentimentalism

Classical rationalism   According to what has been the dominant paradigm in moral psychology for the second half of the 20th century, the rationality of moral judgment—or lack thereof—has to be sought in how the capacity for moral judgment and reasoning *develops*. Moral reasoning becomes increasingly sophisticated with ontogenetic stages. While infants may start out with a *preconventional* moral code in terms of external norms and sanctions, older children construe morality as a *conventional* matter of playing one's role and sticking to the rules; with adolescence, some people come to see morality as a postconventional affair regulated by universal principles (Kohlberg, 1969).

On the other hand, one of the hallmarks of moral judgment—such as the distinction between moral and conventional norms—is already appreciated at a very early age (Smetana & Braeges, 1990; Turiel, 1983), and the Kohlbergian account implicitly operates within a male perspective, thereby dismissing moral outlooks that are focused on maintaining a complex web of social interactions between concrete individuals, rather than on preserving abstract rights, principles of justice, or procedures of deliberation, as somehow undeveloped and immature (Gilligan, 1982).

The antirationalist turn   With the beginning of the 21st century, rationalist dominance began to erode. The social intuitionist model, for instance, holds that moral judgment is a largely intuitive and automatic enterprise; reasons are typically produced post hoc, if at all; and when they are, it is not in the unmotivated pursuit of truth but for the purpose of social persuasion (Haidt, 2001; Haidt & Björklund, 2008). When people's moral judgments are debunked, they defend rather than question them. When this turns out to be unsuccessful, they don't abandon their beliefs but enter a state of being *morally dumbfounded*. Moral judgment, it now seemed, is not just largely intuitive, but which intuitions people have is due to differences in their *moral foundations*, with important political consequences (Graham, Haidt, & Nosek, 2009; Iyer, Koleva, Graham, Ditto, & Haidt, 2012; cf. Curry, 2016; Sauer, 2015).

Sentimentalism   Others are less concerned with the proximal processes that yield particular moral judgments than with the psychological foundations of valuing itself. According to *sentimentalism* about moral judgment, values are distally grounded in (dispositions for) emotional reactions (Nichols, 2004; Nichols, Kumar, Lopez, Ayars, & Chan, 2016; Prinz, 2006, 2007, 2016). The empirical evidence suggests that moral judgments are constituted by emotions, because emotions are both necessary and sufficient for moral judgment. Conditions such as psychopathy suggest that without the appropriate affective reactions, moral judgment is impaired (Blair, 1995). A battery of studies suggests that changing someone's emotions affects their moral beliefs. Dispassionate moralizing seems to be nonexistent. On the other hand, both the interpretation of this evidence (Sauer, 2012a) and its robustness have been called into question. The effect of (incidental) affect on moral judgment seems to be meagre (Landy & Goodwin, 2015; May, 2014a). Psychopaths do master the moral–conventional distinction after all, despite their affective impairments.

Rationalist replies Partly due to these developments, rationalists about moral judgment have started to speak up again. For some, the issue is partly conceptual: moral judgments must in some way be responsive to reasons; otherwise, they don't qualify as genuinely moral (Kennett & Fine, 2009). Sentimentalists are also accused of misidentifying their subject matter. What they have studied are time-slice responses to far-fetched toy problems (Gerrans & Kennett, 2010) rather than people's temporally extended moral agency. Others emphasize that rational processes can "migrate" into intuitive ones (Sauer, 2017). Still others have offered hybrid theories according to which moral judgments are compound cognitive–motivational states in which belief and emotion homeostatically cluster together (Kumar, 2016).

A third interesting development is due to research on how moral judgment affects various *other*, seemingly nonmoral, cognitive domains. Many studies conducted in the so-called experimental philosophy (Kauppinen, 2007) paradigm, for instance, focus on how causal or social cognition lies downstream from moral cognition: the moral, evaluative, or otherwise normative beliefs people hold influence their judgments about what causes what (Knobe & Fraser, 2008), who did what intentionally or not (Knobe, 2003), or what someone's personal identity is (Newman, De Freitas, & Knobe, 2015). This debate, too, has led to interesting discussions regarding the legitimacy of such an influence (Sauer & Bates, 2013) and whether it makes sense for us to be "moralizers through and through" (Knobe, 2010).

3.2 Moral Intuitions and Rational Learning

Recently, the debate on the rationality of moral judgments has focused on moral intuitions. Do our moral intuitions track morally relevant features in a reasonable way, suggesting that we should grant them high evidentiary status and be optimistic about their ability to

successfully navigate us through, and solve problems in, the social world? Or does the relevant evidence advise us to paint a darker picture? In this section, we trace how these discussions have evolved in recent times due to advances in understanding the origins of our intuitions in general and of our moral intuitions in particular. For what follows, we closely follow Haidt's (2001, p. 818) characterization of the subject matter, according to which moral intuition is the sudden appearance in consciousness of a moral assessment, including an affective valence (good–bad, like–dislike), without any conscious awareness of having gone through steps of searching, weighing evidence, or inferring a conclusion—and these moral intuitions are assumed to play a crucial role for moral judgments and decisions, although they can be overridden by other considerations.

Especially with the publication of the seminal articles by Haidt (2001) and Greene and colleagues (2001) at the beginning of the 21st century, the hitherto dominant rationalist view of moral judgments (Kohlberg, 1969; Piaget, 1932; Turiel, 1983) was seriously called into question by highlighting the role of moral intuitions, which were characterized as a rather bad guide for moral judgments and decision making. Moreover, this view seemed to be well supported by decades of psychological research indicating that intuitions in general are often misleading (Kahneman, 2003, 2011).

Relatedly but separately, there are interesting recent developments in philosophical moral epistemology that can both learn from and inform current psychological research. In particular, philosophers are keen to explore the issue of whether we can trust our moral intuitions and under what conditions and whether doing so requires a robust notion of mind-independent moral truth. Many philosophers argue that, despite what we know about the distal and proximal origins of our moral beliefs, we need not abandon most, or at least only some, of them (Bengson, Cuneo, & Shafer-Landau, 2019; Enoch, 2011; Huemer, 2008).

Recent empirical accounts of intuitive cognition also seem to somewhat vindicate this philosophical optimism regarding the rationality of moral judgment. Due to advances in research on the computational underpinnings of learning, especially in reinforcement learning (e.g., Mnih et al., 2015) and Bayesian learning (e.g., Tenenbaum, Kemp, Griffiths, & Goodman, 2011), it does not seem unreasonable to assume that our intuitions are the output of highly sophisticated learning systems that are rather flexible and domain general (cf. Railton, 2017). The findings suggest that our intuitions can be finely attuned to even subtle contours of the decision landscape and often align with normative models of judgment and decision making such as, for example, Bayesian updating, although we often do not have deliberative access to these learning processes.

These advances in understanding the origins of our intuitions have been used by several authors to argue that the rather pessimistic assessment of moral intuitions may have to be revised as well. Perhaps most prominently, Railton (2014, 2017) argues that some popular findings in moral psychology that have been used to paint people's moral intuition in an irrational light should be reinterpreted. For instance, it was argued that people enter a state of moral dumbfounding (see above) in response to scenarios featuring taboo cases (such as sibling incest) even when most plausible reasons against the taboo action have been ruled out by the details of the vignette (such as consensual sex between siblings with birth control and in secret). However, rather than characterizing our aversive responses to such taboo cases as misguided, they could be considered well attuned to the fact that the two siblings played Russian roulette with their emotional health—the fact that everything turned out fine does not justify judging their risky action as "okay" (cf. Jacobson, 2012; for criticisms of the moral dumbfounding effect, see Guglielmo, 2018; Royzman, Kim, & Leeman, 2015). In a similar vein, Allman and Woodward (2008; see also Woodward & Allman, 2007) argue that our affective intuitions often provide us with important information in moral settings, such as signaling the intentions of others and their likely responses to our actions, and they can thereby lead to better moral decision making. Their view is underscored by findings demonstrating that people with damages in areas crucial to emotional processing exhibit moral behavior and decision making that would not be considered appropriate by any reasonable standard (e.g., Damasio, 1994).

The view that moral reasoning can "migrate" into people's moral intuitions is now gaining momentum (Sauer, 2017). To give a concrete example of how a rational learning process can lead to moral intuitions, let us consider a study by Nichols and colleagues (2016), who have argued that the acquisition of deontological rules may be based on a Bayesian principle called the "size principle." To illustrate this principle, imagine that there are four fair dice, each with a different denomination: 4, 6, 8, and 10. Out of your sight, one dice is picked at random and rolled 10 times, and you are told the outcomes: 2, 2, 1, 4, 3, 1, 2, 2, 3, and 4. Your task is to figure out which dice has been rolled. The formula of the size principle provides exact probabilities for the four

hypotheses (4, 6, 8, 10) and favors the narrowest one (4), which corresponds to our intuitive assessment that we would expect some outcomes to be greater than 4 if a dice with a denomination greater than 4 was rolled. The size principle is able to explain children's fast learning of word categories: a few positive examples of different dogs that are called "dogs" are sufficient to infer that the term refers to dogs and not to a broader category of animals, and some examples of Dalmatians are sufficient to let them infer that the term "Dalmatian" refers to this breed of dogs and not to dogs in general (again favoring the narrower hypothesis). Nichols and colleagues provide experimental evidence indicating that deontological principles, such as the act–allow and the intended–foreseen distinction, can be learned by the same mechanism. Moreover, given the learning input children receive (as indicated by a corpus analysis), they would naturally acquire such deontological rules by approximating Bayesian inference—it might even be statistically irrational for them to instead infer utilitarian rules. Hence, Nichols and colleagues provide a rational learning mechanism for the acquisition of deontological principles that can undercut attempts to characterize deontological rules and principles as the product of irrationality. Moreover, it also offers an empiricist explanation of how rational learners can retrieve sophisticated moral rules from seemingly limited information in a way that escapes nativist arguments from the poverty of moral stimuli.

Let us assume, for the sake of the argument, that intuitions are the output of rational learning mechanisms. Does this fact mean that we should stop worrying that moral intuitions might be misleading and grant them moral authority? Some authors believe that such an optimistic view would not be justified. For instance, Gjesdal (2018) points out that since the learning mechanisms underlying our moral intuitions are domain general, they are potentially sensitive to morally relevant features—but they might also pick up morally irrelevant or even immoral information.[4] We should not consider them as morally authoritative on their own, Gjesdal argues, as long as we cannot rule out the possibility that moral intuitions are affected by morally nonrelevant features. This skeptical stance gains traction by the plausible assumption that our learning systems are well attuned and sensitive to information that is important for an agent's welfare. The reason is the—unfortunate—fact that what promotes an agent's welfare and what morality requires from her often come into conflict.[5] Hence, it cannot be ruled out that our moral intuitions are tainted by self-interested considerations. One might address this worry by pointing out that in an appropriately structured environment, our learning mechanisms and moral intuitions could become properly attuned to morally relevant features. For instance, we might learn that a certain degree of impartiality is necessary for us to successfully navigate through the social world, and prejudices might be reduced by coming into contact with the respective people. But making the case for moral intuitions in this way would let it rest on shaky ground, since it is far from obvious that our environment was, is, or will become structured in a way that would guarantee that our moral intuitions are (or will become) adapted in a morally appropriate way. Hence, Gjesdal concludes, the finding that our intuitions are rationally attuned to reality does not warrant that they are morally appropriate. Greene (2017) seems to agree with this assessment and also points out that no matter how sophisticated learning mechanisms might be, they crucially depend on the input they get and the feedback they receive—as the saying goes in contexts that deal with data: garbage in, garbage out. He thus maintains his rather pessimistic view and concludes that moral intuitions might offer reasonable advice in our everyday social life, such as coordinating social interactions within a group ("Me versus Us," as Greene puts it), but are likely to fail when it comes to solving problems between groups with competing interests ("Us versus Them"), the moral issues surrounding novel or unfamiliar problems that originate in distinctively modern institutions, practices or technologies, or making progress in moral philosophy.

What can we conclude from these findings and reflections? On the one hand, the worry that people's moral judgments are not rational *because* they are based on intuitions seems misguided. Intuitions can be the output of highly sophisticated learning systems, and they probably have the potential to reflect moral truths and to successfully navigate us through, and solve problems in, the social world. On the other hand, there is no moral safeguard in these systems ensuring that only morally relevant information is picked up and that only morally adequate outputs are produced. Hence, the prospects of the rational-learning strategy for vindicating people's moral intuitions seem to depend on the extent to which the sophisticated internal learning mechanisms shaping our moral intuitions can avail themselves of morally relevant information in the external social environment (Gjesdal, 2018; Kumar, 2016). Investigating the conditions under which this endeavor can succeed will probably (and hopefully) be a major focus of future moral psychology.

Notes

1. To avoid misunderstandings, we do not mean to imply that the studies presented in the normative part were designed with the purpose of answering normative questions (although some may have been). Rather, most of them can be characterized as descriptive work that has more less direct implications for normative issues.

2. Unfortunately, the labels "consequentialist" and "deontological" are often used in a crude and loose way in moral psychology (e.g., intervening in Switch is actually consistent with most deontological moral theories). See Sinnott-Armstrong (2019) and Alexander and Moore (2016) for a comprehensive review of consequentialist and deontological ethics, respectively.

3. Crockett (2013) explains such means–versus–side effect cases by postulating an interaction of the model-based system and a third, Pavlovian system, which responds reflexively to aversive and rewarding states.

4. Railton (2017) reviews research specifically on moral learning. See also a recent special issue of *Cognition* on moral learning (Cushman, Kumar, & Railton, 2017).

5. Fehige and Wessels (see chapter 12.1 in this handbook) discuss at greater length the relationship between morality and an instrumental view of rationality and conclude that they cannot be brought into full harmony.

References

Alexander, L., & Moore, M. (2016). Deontological ethics. In E. N. Zalta (Ed.), *The Stanford encyclopedia of philosophy*. Retrieved from https://plato.stanford.edu/archives/win2016/entries/ethics-deontological/

Allman, J., & Woodward, J. (2008). What are moral intuitions and why should we care about them? A neurobiological perspective. *Philosophical Issues, 18*, 164–185.

Baron, J., & Ritov, I. (2004). Omission bias, individual differences, and normality. *Organizational Behavior and Human Decision Processes, 94*, 74–85.

Bengson, J., Cuneo, T., & Shafer-Landau, R. (2019). Trusting moral intuitions. *Noûs, 54*(4).

Berker, S. (2009). The normative insignificance of neuroscience. *Philosophy & Public Affairs, 37*(4), 293–329.

Blair, R. J. R. (1995). A cognitive developmental approach to morality: Investigating the psychopath. *Cognition, 57*(1), 1–29.

Conway, P., Goldstein-Greenwood, J., Polacek, D., & Greene, J. D. (2018). Sacrificial utilitarian judgments do reflect concern for the greater good: Clarification via process dissociation and the judgments of philosophers. *Cognition, 179*, 241–265.

Crockett, M. J. (2013). Models of morality. *Trends in Cognitive Sciences, 17*(8), 363–366.

Crockett, M. J. (2016). How formal models can illuminate mechanisms of moral judgment and decision making. *Current Directions in Psychological Science, 25*(2), 85–90.

Curry, O. S. (2016). Morality as cooperation: A problem-centred approach. In T. K. Shackelford & R. D. Hansen (Eds.), *The evolution of morality* (pp. 27–51). Cham, Switzerland: Springer.

Cushman, F. (2013). Action, outcome, and value: A dual-system framework for morality. *Personality and Social Psychology Review, 17*(3), 273–292.

Cushman, F., Gray, K., Gaffey, A., & Mendes, W. B. (2012). Simulating murder: The aversion to harmful action. *Emotion, 12*(1), 2–7.

Cushman, F., Kumar, V., & Railton, P. (2017). Moral learning: Psychological and philosophical perspectives. *Cognition, 167*, 1–10.

Damasio, A. (1994). *Descartes' error: Emotion, reason, and the human brain*. New York, NY: Putnam.

Dayan, P., & Niv, Y. (2008). Reinforcement learning: The good, the bad and the ugly. *Current Opinion in Neurobiology, 18*, 185–196.

Dolan, R. J., & Dayan, P. (2013). Goals and habits in the brain. *Neuron, 80*(2), 312–325.

Enoch, D. (2011). *Taking morality seriously: A defense of robust realism*. Oxford, England: Oxford University Press.

Evans, J. St. B. T., & Stanovich, K. E. (2013). Dual-process theories of higher cognition: Advancing the debate. *Perspectives on Psychological Science, 8*(3), 223–241.

Gerrans, P., & Kennett, J. (2010). Neurosentimentalism and moral agency. *Mind, 119*(475), 585–614.

Gigerenzer, G. (2008). Moral intuition=fast and frugal heuristics? In W. Sinnott-Armstrong (Ed.), *Moral psychology: Vol. 2. The cognitive science of morality: Intuition and diversity* (pp. 1–26). Cambridge, MA: MIT Press.

Gilligan, C. (1982). *In a different voice*. Cambridge, MA: Harvard University Press.

Gjesdal, A. (2018). Moral learning, rationality, and the unreliability of affect. *Australasian Journal of Philosophy, 96*, 460–473.

Gläscher, J., Daw, N., Dayan, P., & O'Doherty, J. P. (2010). States versus rewards: Dissociable neural prediction error signals underlying model-based and model-free reinforcement learning. *Neuron, 66*(4), 585–595.

Graham, J., Haidt, J., & Nosek, B. A. (2009). Liberals and conservatives rely on different sets of moral foundations. *Journal of Personality and Social Psychology, 96*, 1029–1046.

Greene, J. D. (2008). The secret joke of Kant's soul. In W. Sinnott-Armstrong (Ed.), *Moral psychology: Vol. 3. The neuroscience of morality: Emotion, brain disorders, and development* (pp. 35–79). Cambridge, MA: MIT Press.

Greene, J. D. (2010). Notes on "The normative insignificance of neuroscience" by Selim Berker. Retrieved from https://static1.squarespace.com/static/54763f79e4b0c4e55ffb000c/t/54cb945ae4b001aedee69e81/1422627930781/notes-on-berker.pdf

Greene, J. D. (2013). *Moral tribes: Emotion, reason, and the gap between us and them*. New York, NY: Penguin.

Greene, J. D. (2014). Beyond point-and-shoot morality: Why cognitive (neuro)science matters for ethics. *Ethics, 124*(4), 695–726.

Greene, J. D. (2017). The rat-a-gorical imperative: Moral intuition and the limits of affective learning. *Cognition, 167*, 66–77.

Greene, J. D., Morelli, S. A., Lowenberg, K., Nystrom, L. E., & Cohen, J. D. (2008). Cognitive load selectively interferes with utilitarian moral judgment. *Cognition, 107*, 1144–1154.

Greene, J. D., Nystrom, L. E., Engell, A., Darley, J. M., & Cohen, J. D. (2004). The neural basis of cognitive conflict and control in moral judgment. *Neuron, 44*, 389–400.

Greene, J. D., Sommerville, R. B., Nystrom, L. E., Darley, J. M., & Cohen, J. D. (2001). An fMRI study of emotional engagement in moral judgment. *Science, 293*, 2105–2108.

Guglielmo, S. (2018). Unfounded dumbfounding: How harm and purity undermine evidence for moral dumbfounding. *Cognition, 170*, 334–337.

Haidt, J. (2001). The emotional dog and its rational tail: A social intuitionist approach to moral judgment. *Psychological Review, 108*, 814–834.

Haidt, J., & Björklund, F. (2008). Social intuitionists answer six questions about moral psychology. In W. Sinnott-Armstrong (Ed.), *Moral psychology: Vol. 2. The cognitive science of morality: Intuition and diversity* (pp. 181–217). Cambridge, MA: MIT Press.

Hauser, M., Cushman, F., Young, L., Jin, R. K.-X., & Mikhail, J. (2007). A dissociation between moral judgments and justifications. *Mind & Language, 22*(1), 1–21.

Hauser, M. D., Young, L., & Cushman, F. (2008). Reviving Rawls's linguistic analogy: Operative principles and the causal structure of moral actions. In W. Sinnott-Armstrong (Ed.), *Moral psychology: Vol. 2. The cognitive science of morality: Intuition and diversity* (pp. 107–144). Cambridge, MA: MIT Press.

Huemer, M. (2008). Revisionary intuitionism. *Social Philosophy and Policy, 25*(1), 368–392.

Iyer, R., Koleva, S., Graham, J., Ditto, P., & Haidt, J. (2012). Understanding libertarian morality: The psychological dispositions of self-identified libertarians. *PLoS One, 7*(8), e42366.

Jacobson, D. (2012). Moral dumbfounding and moral stupefaction. *Oxford Studies in Normative Ethics, 2*, 289–316.

Kahane, G., Everett, J. A. C., Earp, B. D., Caviola, L., Faber, N. S., Crockett, M. J., & Savulescu, J. (2018). Beyond sacrificial harm:

A two-dimensional model of utilitarian psychology. *Psychological Review*, *125*(2), 131–164.

Kahane, G., Everett, J. A. C., Earp, B. D., Farias, M., & Savulescu, J. (2015). "Utilitarian" judgments in sacrificial moral dilemmas do not reflect impartial concern for the greater good. *Cognition*, *134*, 193–209.

Kahane, G., Wiech, K., Shackel, N., Farias, M., Savulescu, J., & Tracey, I. (2011). The neural basis of intuitive and counterintuitive moral judgment. *Social Cognition and Affective Neuroscience*, *7*(4), 393–402.

Kahneman, D. (2003). A perspective on judgment and choice: Mapping bounded rationality. *American Psychologist*, *58*, 697–720.

Kahneman, D. (2011). *Thinking, fast and slow*. New York, NY: Farrar, Straus & Giroux.

Kauppinen, A. (2007). The rise and fall of experimental philosophy. *Philosophical Explorations*, *10*(2), 95–118.

Kennett, J., & Fine, C. (2009). Will the real moral judgment please stand up? *Ethical Theory and Moral Practice*, *12*(1), 77–96.

Knobe, J. (2003). Intentional action and side effects in ordinary language. *Analysis*, *63*(3), 190–194.

Knobe, J. (2010). Person as scientist, person as moralist. *Behavioral and Brain Sciences*, *33*(4), 315–329.

Knobe, J., & Fraser, B. (2008). Causal judgment and moral judgment: Two experiments. In W. Sinnott-Armstrong (Ed.), *Moral psychology: Vol. 2. The cognitive science of morality: Intuition and diversity* (pp. 441–447). Cambridge, MA: MIT Press.

Kohlberg, L. (1969). Stage and sequence: The cognitive-developmental approach to socialization. In D. A. Goslin (Ed.), *Handbook of socialization theory and research* (pp. 347–480). Chicago, IL: Rand McNally.

Kruglanski, A. W., & Gigerenzer, G. (2011). Intuitive and deliberate judgments are based on common principles. *Psychological Review*, *118*(1), 97–109.

Kumar, V. (2016). The empirical identity of moral judgment. *Philosophical Quarterly*, *66*(265), 783–804.

Landy, J. F., & Goodwin, G. P. (2015). Does incidental disgust amplify moral judgment? A meta-analytic review of experimental evidence. *Perspectives on Psychological Science*, *10*(4), 518–536.

Machery, E. (2014). In defense of reverse inference. *British Journal for the Philosophy of Science*, *65*(2), 251–267.

May, J. (2014a). Does disgust influence moral judgment? *Australasian Journal of Philosophy*, *92*(1), 125–141.

May, J. (2014b). Moral judgment and deontology: Empirical developments. *Philosophy Compass*, *9*(11), 745–755.

McGuire, J., Langdon, R., Coltheart, M., & Mackenzie, C. (2009). A reanalysis of the personal/impersonal distinction in moral psychology research. *Journal of Experimental Social Psychology*, *45*, 577–580.

Mikhail, J. (2007). Universal moral grammar: Theory, evidence and the future. *Trends in Cognitive Sciences*, *11*, 143–152.

Mikhail, J. (2011). *Elements of moral cognition: Rawls' linguistic analogy and the cognitive science of moral and legal judgment*. Cambridge, England: Cambridge University Press.

Miller, R., & Cushman, F. (2013). Aversive for me, wrong for you: First-person behavioral aversions underlie the moral condemnation of harm. *Social and Personality Psychology Compass*, *7*(10), 707–718.

Miller, R. M., Hannikainen, I. A., & Cushman, F. A. (2014). Bad actions or bad outcomes? Differentiating affective contributions to the moral condemnation of harm. *Emotion*, *14*(3), 573–587.

Mnih, V., Kavukcuoglu, K., Silver, D., Rusu, A. A., Veness, J., Bellemare, M. G., . . . Hassabis, D. (2015). Human-level control through deep reinforcement learning. *Nature*, *518*, 529–533.

Newman, G. E., De Freitas, J., & Knobe, J. (2015). Beliefs about the true self explain asymmetries based on moral judgment. *Cognitive Science*, *39*(1), 96–125.

Nichols, S. (2004). *Sentimental rules: On the natural foundations of moral judgment*. Oxford, England: Oxford University Press.

Nichols, S., Kumar, S., Lopez, T., Ayars, A., & Chan, H. Y. (2016). Rational learners and moral rules. *Mind & Language*, *31*, 530–554.

Pessoa, L. (2008). On the relationship between emotion and cognition. *Nature Reviews Neuroscience*, *9*, 148–158.

Piaget, J. (1932). *The moral judgment of the child*. Oxford, England: Harcourt, Brace.

Poldrack, R. A. (2006). Can cognitive processes be inferred from neuroimaging data? *Trends in Cognitive Sciences*, *10*(2), 59–63.

Prinz, J. (2006). The emotional basis of moral judgments. *Philosophical Explorations*, *9*(1), 29–43.

Prinz, J. (2007). *The emotional construction of morals*. Oxford, England: Oxford University Press.

Prinz, J. (2008). Resisting the linguistic analogy: A commentary on Hauser, Young, and Cushman. In W. Sinnott-Armstrong (Ed.), *Moral psychology: Vol. 2. The cognitive science of morality: Intuition and diversity* (pp. 157–170). Cambridge, MA: MIT Press.

Prinz, J. (2016). Sentimentalism and the moral brain. In S. M. Liao (Ed.), *Moral brains: The neuroscience of morality* (pp. 45–74). Oxford, England: Oxford University Press.

Railton, P. (2014). The affective dog and its rational tale: Intuition and attunement. *Ethics*, *124*, 813–859.

Railton, P. (2017). Moral learning: Conceptual foundations and normative relevance. *Cognition*, *167*, 172–190.

Royzman, E. B., Kim, K., & Leeman, R. F. (2015). The curious tale of Julie and Mark: Unraveling the moral dumbfounding effect. *Judgment & Decision Making*, *10*(4), 296–313.

Sauer, H. (2012a). Educated intuitions: Automaticity and rationality in moral judgment. *Philosophical Explorations*, *15*(3), 255–275.

Sauer, H. (2012b). Morally irrelevant factors: What's left of the dual-process model of moral cognition? *Philosophical Psychology*, *25*(6), 783–811.

Sauer, H. (2012c). Psychopaths and filthy desks: Are emotions necessary and sufficient for moral judgment? *Ethical Theory and Moral Practice*, *15*(1), 95–115.

Sauer, H. (2015). Can't we all disagree more constructively? Moral foundations, moral reasoning, and political disagreement. *Neuroethics*, *8*(2), 153–169.

Sauer, H. (2017). *Moral judgments as educated intuitions*. Cambridge, MA: MIT Press.

Sauer, H., & Bates, T. (2013). Chairmen, cocaine, and car crashes: The Knobe effect as an attribution error. *Journal of Ethics*, *17*(4), 305–330.

Sinnott-Armstrong, W. (2019). Consequentialism. In E. N. Zalta (Ed.), *The Stanford encyclopedia of philosophy*. Retrieved from https://plato.stanford.edu/archives/sum2019/entries/conseq uentialism/

Smetana, J. G., & Braeges, J. L. (1990). The development of toddlers' moral and conventional judgments. *Merrill Palmer Quarterly*, *36*(3), 329–346.

Sunstein, C. R. (2005). Moral heuristics. *Behavioral and Brain Sciences*, *28*(4), 531–541.

Sutton, R. S. (1988). Learning to predict by the methods of temporal differences. *Machine Learning*, *3*(1), 9–44.

Sutton, R. S., & Barto, A. (1999). Reinforcement learning. *Journal of Cognitive Neuroscience*, *11*, 126–134.

Sutton, R. S., & Barto, A. G. (2018). *Reinforcement learning: An introduction* (2nd ed.). Cambridge, MA: MIT Press.

Tenenbaum, J. B., Kemp, C., Griffiths, T. L., & Goodman, N. D. (2011). How to grow a mind: Statistics, structure, and abstraction. *Science*, *331*, 1279–1285.

Turiel, E. (1983). *The development of social knowledge: Morality and convention*. Cambridge, England: Cambridge University Press.

Waldmann, M. R., Nagel, J., & Wiegmann, A. (2012). Moral judgment. In K. J. Holyoak & R. G. Morrison (Eds.), *The Oxford handbook of thinking and reasoning* (pp. 364–389). New York, NY: Oxford University Press.

Wiegmann, A., & Waldmann, M. R. (2014). Transfer effects between moral dilemmas: A causal model theory. *Cognition*, *131*, 28–43.

Woodward, J., & Allman, J. (2007). Moral intuition: Its neural substrates and normative significance. *Journal of Physiology*, *101*(4), 179–202.

Wunderlich, K., Dayan, P., & Dolan, R. J. (2012). Mapping value based planning and extensively trained choice in the human brain. *Nature Neuroscience*, *15*(5), 786–791.

Part IV Facets of Rationality

Section 13   Visual and Spatial Reasoning

13.1 Logical Reasoning with Diagrams

Mateja Jamnik

Summary

This chapter examines the role of diagrams in reasoning. It exposes the tension between formal and informal use of diagrams. Diagrams are a complementary representation to symbolic formulae. Since choosing the right representation may be essential to solving a problem, we study diagrams as one such suitable choice. We analyze algorithmic and computational approaches to formalizing the logic of diagrams and present some modern implementations of diagrammatic reasoning systems.

1. Diagrams through Time

Humans have used diagrams to convey reasoning since the ancient times of Aristotle and Euclid. In fact, until the end of the 19th century, diagrams were considered a legitimate formal tool for explaining the rationale of the solution to a problem. They were a solution in themselves. But while Euler (1768), Venn (1881), and Peirce (1933) were devising logics for their diagrams, Hilbert, Gödel, Russell, and Wittgenstein formalized symbolic logic. This formal movement cemented the notion that only sentences that can be deduced from core axioms through a process of inference are considered a proof of a theorem:

> A theorem is only proved when the proof is completely independent of the diagram. The proof must call step by step on the preceding axioms. The making of figures is [equivalent to] the experimentation of the physicist, and experimental geometry is already over with [laying down of the] axioms. (Hilbert, 1894/2004)

Diagrams, in any form, have until the rise of symbolic logic not been formalized in this Hilbertian way: there were no axiomatic diagrams,[1] there were no formalized inference steps of diagrammatic manipulations, and moreover, mathematics was full of erroneous solutions due to faulty or deceptive diagrams.[2] Yet mathematicians continued to use diagrams to aid their reasoning and to convey the intuition of a solution.

The prejudice against the formal status of diagrams and visual methods of reasoning with them was finally overturned by Shin's seminal work on the formalization of Venn diagrams (Shin, 1994). Shin devised a formal syntax and semantics of particular forms of Venn diagrams, and proved her logic sound and complete. This legitimized and spurred on renewed interest in studying and devising systems for reasoning with diagrams, some of which will be surveyed here.

This chapter examines the role that diagrams play in human reasoning from the *computational, algorithmic*, and *implementational* points of view.[3] We argue that diagrams can encode the part of human reasoning that carries the intuition of the solution to a problem and can, moreover, convey the proof without needing to be supplemented by symbolic logic. Human rationality and reasoning are intimately related, as the reasoning process provides the rationale for a human understanding of, and belief in, the correctness of a solution. Our interest does not lie in machine-oriented approaches to reasoning, for which symbolic logic is typically used (with perhaps a small exception in the case of the motivation for natural deduction). Instead, we study diagrammatic reasoning and systems as one plausible way to model human-oriented approaches to rationales and therefore reasoning. Machine-oriented approaches to reasoning are typically motivated by the search for a categorical answer to the question of whether a conjecture is a theorem, by the speed of finding a proof, or by the number of proofs that the system is able to find. By contrast, human-oriented approaches to reasoning seek solution explanations that are understandable to humans. This human-oriented approach to modeling reasoning is motivated by the goal of general artificial intelligence (AI) to make machines more human-like in their reasoning and in the solutions that they find.[4] The hope is that diagrammatic reasoning systems will give us insights into human reasoning, which can then be modeled on machines to make them more human-like. As AI systems become ever more ubiquitous in our interactions with the world, it is important to build systems that humans find amenable and understandable. The new legal requirement for explainability makes the combination of sound reasoning and accessibility very timely. In part, this means capturing the human intuition and rationale of a solution in some way—the diagram presents not only a solution but also an explanation of why the solution is a correct one. We aim to demonstrate that this human-like reasoning with diagrams can be computationally modeled in logically formal ways on machines.

2. Representations

The importance of problem representation has been acknowledged by many researchers (Amarel, 1968; Simon, 1996). Pólya wisely pointed out in his books *How to Solve It* (1957) and *Mathematical Discovery* (1965) that reasoning about a problem depends on how one represents the problem in the first place. J. R. Anderson's (1978) mimicry argument adds the inseparability of the format of a representation and the processes that operate on this representation. This is even more immediately apparent when modeling reasoning with diagrams: representing the problem in the right way may lead to a trivial solution that captures the intuition and truthfulness in an obvious and accessible way. Representing the problem inadequately may prevent us from ever finding a solution. In contrast to machines, which typically use a single fixed representation for a problem, most humans are good at choosing the representation that is just right to enable them to solve a new problem. Moreover, given the right representation, human problem solving is dramatically improved, and humans recognize the benefits of such changes (Cheng, Lowe, & Scaife, 2001). Take, for example, the famous problem of the mutilated checkerboard in figure 13.1.1. It is called "mutilated" because two diagonal corners have been removed. The question is, Can you cover this mutilated checkerboard with dominoes?

One could try many different ways to tile the board to find out if it can be covered. However, if we change the representation and color the board black and white like a chessboard (figure 13.1.2), then it is immediately obvious that the mutilated checkerboard cannot be covered with dominoes, since by mutilating it, we have removed two squares of the same color. But we need the same number of black and white squares since each domino covers one black and one white square. This is obvious for humans, but how would a machine go about finding this helpful representation?

It is difficult to see how to use Pólya's advice about representations in computational reasoning systems.

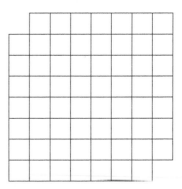

Figure 13.1.1
Mutilated checkerboard. Can it be covered with dominoes?

Figure 13.1.2
No, this mutilated checkerboard cannot be covered with dominoes, because it has more black than white squares, but there must be the same number of black and white squares since a domino covers one of each.

Unlike humans, machines' representations are usually fixed by their programmer, and while they can be from a wide range, they are typically a single symbolic representation. What machines don't currently do is reason about what system of representation to adopt. They deal in local parameters of whatever representation they are programmed in. Enabling a reasoning system to choose the right representation in problem solving is a big open challenge that has only recently started to be addressed (Raggi et al., 2020). Its importance lies in the relation between the problem representation and the computational complexity of finding its solution: imagine symbolically describing and solving the mutilated checkerboard example, and contrast this with the diagrammatic solution in figure 13.1.2. Larkin and Simon (1987) extensively discussed the importance of diagrams in many domains of reasoning: they claim that "a diagram is (sometimes) worth ten thousand words." Many so-called diagrammatic proofs without words can be found in Nelsen's books *Proofs without Words: Exercises in*

Visual Thinking (1993) and *Proofs without Words II: Exercises in Visual Thinking* (2001).

There is strong evidence from cognitive psychology that diagrams have many properties that often enable humans to process them with immediate comprehension. Johnson-Laird (1983) and Hegarty and Just (1993) argue that humans, at least in some cases, use diagrams in their mental models of a situation. It is important to recognize that the wrong level of granularity or too much visual clutter can impede cognitive processes (Alqadah, Stapleton, Howse, & Chapman, 2014; Knauff, 2013), but if diagrams capture the logical structure of the problem, they can be helpful in reasoning and problem solving (see chapter 13.3 by Knauff, this handbook). Diagrams concisely store information, explicitly represent the relations among the elements of the diagram, and support a lot of perceptual inferences that are very easy for humans. Shimojima (1996), for example, characterized a property of diagrams called "free rides": expressing some chosen information in a diagram often results in expressing other pieces of information that are consequences of the chosen information; informally, these other pieces of information are free rides. Figure 13.1.3b gives an example of how a Euler diagram representation of expressions in symbolic logic in figure 13.1.3a gives rise to a free ride about the fact that R and Q are disjoint: this can directly be read off from the diagrams in figure 13.1.3b, whereas it must be inferred from the symbolic sentences in figure 13.1.3a. Notions such as free rides give one possible explanation of why humans find diagrams so accessible.

Modeling this inherently human ability to choose or change appropriate representations in computational systems is an open research question. To address it, we need to find out which cognitive processes humans use to select representations, what criteria they use to choose them, and how we can model this ability on machines. The choice of representation, whether symbolic or diagrammatic (or even among a number of different symbolic or diagrammatic representations), will depend on whether the target audience is a human (and on their level of expertise) or a machine. Such a computational system could then readily assess the human and capitalize on the explanatory power of diagrammatic representations.

3. Logical Systems of Diagrams

Despite the persistent mistrust in the logical status of diagrams, the end of the 20th century started to see a redressing of this issue (M. Anderson, Meyer, & Olivier, 2001; Chandrasekaran, Glasgow, & Narayanan, 1995). Examples include formalized logical systems of diagrams (Hammer, 1995; Howse, Stapleton, & Taylor, 2005; Shin, 1994). This directly abolished the widely held Hilbertian theoretical objections to diagrams being used in proofs. Some of this seminal work is presented here.

While Peirce (1933) began to formalize a variant of Venn diagrams and diagrammatic inference rules on them, it was Shin (1994) who was the first to show that two particular systems of Venn diagrams can have a formal syntax, semantics, and model theory. A *Venn diagram* (Venn, 1881) expresses all possible logical relations between a fixed finite number of sets. It does so by depicting sets as regions inside closed curves, which overlap in all possible ways (see figure 13.1.4). Venn diagrams use shading to represent empty sets.

Shin (1994) formalized a form of Venn diagrams, called "Venn-I," where she used shading to express set-emptiness (as in the original diagrams by Venn), x-sequences to express that a set is *not* empty, and linked x-sequences to express disjunction (similar to Peirce[5]). Shin's further Venn-II system introduced connecting lines between diagrams to express disjunctive information. Figure 13.1.5 shows statements in both Venn-I and Venn-II systems.

Importantly, Shin equipped both systems with formal syntax, semantics, as well as sound and complete inference rules and showed that Venn-II is expressively equivalent to monadic first-order logic. This was the first

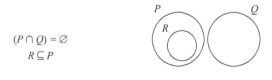

$$(P \cap Q) = \varnothing$$
$$R \subseteq P$$

(a) Set-theoretic symbolic formulae.   (b) Euler diagram.

Figure 13.1.3

Free ride: in Euler diagrams, the fact that R is disjoint from Q can be read off, whereas this has to be inferred from the symbolic formulae.

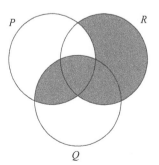

Figure 13.1.4

Venn diagram representation of the information in figure 13.1.3.

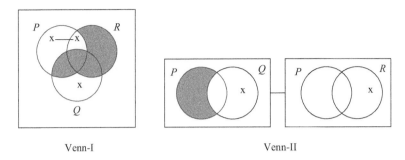

Venn-I   Venn-II

Figure 13.1.5

The Venn-I diagram expresses the information in figure 13.1.4 and also that there is at least one Q that is neither P nor R and there is at least one P that is not Q. The Venn-II diagram expresses that either P is a subset of Q and Q\P is not empty, or R\P is not empty.

time that a diagrammatic system for reasoning received a Hilbertian treatment. Shin showed that diagrams in a purely diagrammatic reasoning system can be used as formal tools.

Hammer (1995) extended Shin's work to *Euler diagrams*: he defined their formal syntax and semantics, and proved that a set of diagrammatic inferences were sound and complete. Euler diagrams (Euler, 1768) use closed curves that can overlap, be totally enclosed by other curves, or do not overlap at all, that is, exclude each other, to represent intersection, subset, and disjoint sets, respectively (see figure 13.1.3b).

Spider diagrams adapt Hammer's Euler and Shin's Venn-II diagrams. They are based on Euler diagrams, but instead of using x-elements to indicate nonempty regions, they use "spiders," that is, dots connected by lines, to represent the existence of distinct elements. Spider diagrams use shading: in a shaded region, all elements are denoted by spiders. So, a shaded region with no spiders represents the empty set (see figure 13.1.6).

The syntax and semantics of spider diagrams were formally defined in Howse et al. (2005). The language of spider diagrams is also accompanied by sound and complete inference rules, which result in the logic of spider diagrams. This logic is expressively equivalent to monadic first-order logic with equality (Stapleton, Howse, Taylor, & Thompson, 2004). A set of sound inference rules for spider diagrams was defined and proved complete in Urbas, Jamnik, and Stapleton (2015). Thus, spider diagrams are more expressive than Venn-II diagrams.

The revived interest in diagrams has seen other domains formalizing a rigorous syntax and semantics of diagrams for reasoning. Some are equipped with proved formal properties of soundness (which is a minimal requirement for using them in proofs) and some even with completeness. Examples include analogical reasoning in Grover (Barker-Plummer & Bailin, 1997), diagrammatic reasoning about arithmetic in Diamond (Jamnik,

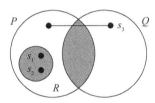

Figure 13.1.6

Spider diagram: R is a subset of P and has exactly two elements, s_1 and s_2; Q is disjoint from P (and R); element s_3 is either in P or in Q, but not in both.

2001), ontology reasoning with concept diagrams in iCon (Shams, Sato, Jamnik, & Stapleton, 2018), and computer science specification with UML diagrams based on higraphs (Rumbaugh, Jacobson, & Booch, 1999). Similar to the formalization of Venn, Euler, and spider diagrams, all of these systems provide evidence that diagrams can be used in formal proofs and thus show that diagrams can be made Hilbertian.

4. Implementations of Diagrammatic Reasoning Systems

In parallel with theoretical interest in the logic of diagrams revived at the end of the 20th century, several diagrammatic mechanized reasoning systems have been implemented. The motivations for their implementations differ: from cognitive ones about improved accessibility of reasoning systems to proof of concept that diagrams can be formally used for mechanized machine reasoning. They provide one model of reasoning that is more human-like than traditional symbolic machine-oriented approaches: they are both formal and explainable.

In implementing diagrammatic reasoning systems, there are many design choices that need to be carefully considered. For example, how do we devise algorithms for laying out diagrams in the best and clearest possible way, is the set of inference rules sound and complete, what level of abstraction of inference rules is most accessible to humans, what is the notion of a diagrammatic proof/solution, and should diagrams be general or should we exploit their concreteness? Here we demonstrate how some of these choices were made for three particular implementations of diagrammatic reasoning systems: Hyperproof, Diamond, and Speedith.

4.1 Hyperproof

Hyperproof (Barker-Plummer, Barwise, & Etchemendy, 2017; Barwise & Etchemendy, 1991) was the first system that implemented the use of both diagrammatic and symbolic representations to derive conclusions. The diagram is a blocks-world situation depicted in a graphical display: the upper part of Hyperproof's screen in figure 13.1.7.[6] In addition, some symbolic formulae of first-order logic might be given (in the lower part): $Dodec(c) \rightarrow Dodec(d)$ and $Small(c)$. The aim is to show that some conjecture about the given information, usually expressed symbolically, is a consequence or a nonconsequence of the given information. In the example in figure 13.1.7, we want to determine whether block c and block d are of the same shape: $SameShape(c,d)$. Hyperproof combines inspecting the diagram and applying logical rules in order to derive the conclusion.

The proofs in Hyperproof are guaranteed to be correct if every step, diagrammatic or symbolic, is evaluated to be true according to the evaluation schema in Kleene logic. They are heterogeneous since they combine symbolic and diagrammatic information. This notion of a heterogeneous proof is novel and stands in contrast to the Hilbertian view of formal proof. Using the blocks-world situation exploits not only diagrammatic representations but also their concreteness property—concrete proof objects can be named and their shape and size inspected. Hyperproof has been successfully deployed in an educational setting to teach first-order logic precisely for these properties: they make the reasoning more accessible to most users.[7]

4.2 Diamond

Diamond is the first system that implemented the construction of purely diagrammatic proofs (Jamnik, 2001). Its domain is natural number arithmetic for inductive theorems. Uniquely, Diamond exploits the concreteness property of diagrams to express a particular concrete instance of a theorem and its proof, rather than the general, universally quantified case. For example, figure 13.1.8 gives a diagrammatic proof for a theorem about the sum of odd natural numbers for a concrete instance $n = 6$. The proof cuts a square of size 6 into a sequence of so-called $ells$,[8] each representing a subsequent odd natural number since it consists of two edges, $2n$, but the joining vertex is counted twice, so $2n - 1$. In

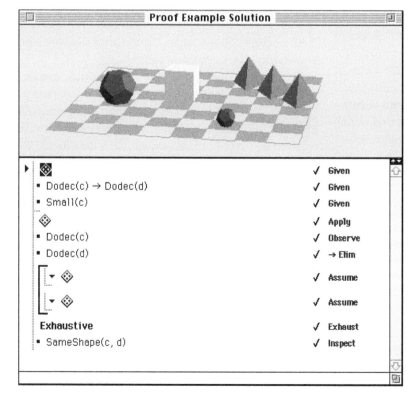

Figure 13.1.7
Hyperproof's proof.

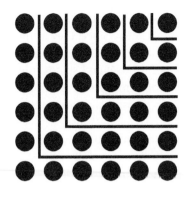

$$1 + 3 + 5 + \cdots + (2n - 1) = n^2.$$

Figure 13.1.8
Diagrammatic proof of the theorem about the sum of odd natural numbers.

the concrete case of $n = 6$, the inference *lcut* was carried out five times.

Diamond's approach to capturing the generality of the proof while using concrete instances is theoretically justified by the constructive ω-rule (Baker, Ireland, & Smaill, 1992). This rule states that if there is a uniform effective procedure generating every instance of a proof, then we can conclude the universally quantified theorem. One such effective procedure is a recursive program. Therefore, Diamond's general proof is encoded in the recursive program that, upon an input for every value of the universal quantifier, produces a proof for that instance of the theorem.

The work on automation of diagrammatic proofs in Diamond provides important information on proof procedure construction. It demonstrates the importance of representing diagrammatic expressions so that general reasoning techniques can be applied to them. Furthermore, it provides insight into how diagrams and purely diagrammatic inferences can be used in formal proofs.

4.3 Speedith

Speedith (Urbas et al., 2015) is an interactive diagrammatic theorem prover for spider diagrams. It is the first implemented system with the same notion of proof as in a typical symbolic theorem prover: apply inference rules until tautology or contradiction is reached. Instead of symbolic formulae, the theorem is expressed using diagrams. The inference rules are diagrammatic (and some symbolic) and form a sound and complete set—this guarantees that diagrammatic proofs are correct.

Figure 13.1.9 shows an example of a diagrammatic proof as constructed in Speedith. Here, the antecedent of the *Initial proof goal* is a spider diagram that conveys

information about the relationships between two elements and two sets, and it proves that the consequent of the *Initial proof goal* follows logically. In particular, the proof establishes that given sets A and B, if there are two elements, one of which is in both A and B and the other is either only in A or only in B, then we can deduce that one element is in A and the other is in B.

5. Discussion

The rationality of human thought has been studied from a number of different angles. Cognitive, psychological, and philosophical perspectives include investigations into mental models, heuristics, deduction, argumentation, beliefs, social epistemology, morality, and learning, among others. This chapter presented an angle stemming from the perhaps less mainstream, yet foundational, artificial intelligence view of one aspect of human rationality, namely, that of human-oriented visual mathematical reasoning. We approached rationality from the algorithmic and implementational view to computationally model reasoning with diagrams. The invention of Hilbertian reasoning at the start of the 20th century firmly rooted the formality of human thought in symbolic logic. This was despite centuries of mathematicians using visual tools such as diagrams to convey theorems and their proofs. The human ability to "see" the problem and its solution in a diagram is one of the fundamental components of the human mathematical cognitive repertoire. Nevertheless, there is a clear tension between symbolic and diagrammatic representations when computationally modeling human intuition in problem solving.

Unlike sentential, symbolic, or linguistic representations, which represent relations between objects in a single dimension by concatenation, diagrammatic representations use multidimensional relations between objects. This has both advantages and drawbacks. On the plus side, once the relations between objects are correctly represented, their properties or new relations can be just read off from the diagram. Indeed, diagrams may give you new information for free that was not originally declared and represented—referred to as "free rides." On the negative side, some relations cannot be expressed in diagrammatic representations by utilizing spatial properties, and if relations are not represented completely, this may lead to unintended and erroneous conclusions. Stenning and Oberlander (1995) argue that diagrams are necessarily concrete, with just a limited capacity for abstraction. Knauff discusses this tension between concreteness and abstraction, and the level of granularity in diagrams, in chapter 13.3 of this handbook. The

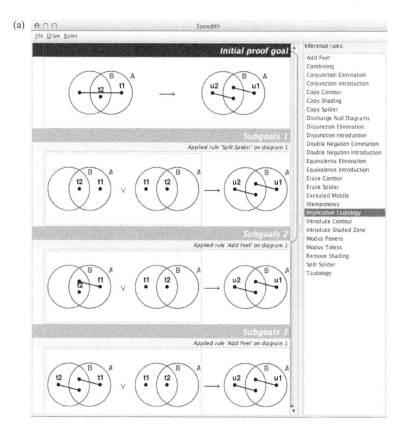

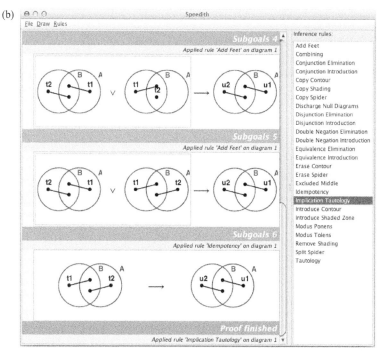

Figure 13.1.9

Screenshot of a proof of a spider-diagrammatic statement as constructed in Speedith.

consequence is that the user has to distinguish intended from unintended concreteness (e.g., the use of $n = 6$ above). Getting this wrong often causes erroneous conclusions. There are numerous examples in the history of mathematics of faulty solutions—Lakatos (1976) and Maxwell (1959) illustrate some better-known ones.

When are diagrammatic representations not expressive enough? The systems we mentioned here are all limited in expressivity in different ways. For example, Venn-I cannot express disjunctive information, spider diagrams can only express monadic first-order logic with equality, and Diamond can only tackle arithmetic theorems expressed as three-dimensional shapes. Diagrams are concrete objects and are thus subject to spatial constraints that sentential representations do not have. This is clearly limiting, and there are sentential logics, and their computational implementations in theorem provers, that are more expressive and indeed very impressive (e.g., Coq's proof of the famous four-color theorem; Gonthier, 2008).

However, purely sentential logics are limited in representational scope and often unilluminating with regard to solution explanations. The diagrammatic approaches to reasoning presented in this chapter provide a vehicle for widening the scope and expanding the explanatory power of computational systems for reasoning. They work with users in accessible and explanatory ways and can be formal in the Hilbertian sense. They enable us to explore the importance of representational choice on the diversity of inference methods and the ability to actually solve problems in mathematics. Ultimately, the goal is to understand human rationality and, from the AI point of view, to build computational systems that reflect it—this is going to make AI machines more human-like in internal workings and in external interactions with humans.

Notes

1. The distinction between symbolic and diagrammatic methods is not straightforward. While Newton's famous *Philosophiæ Naturalis Principia Mathematica* (*Mathematical Principles of Natural Philosophy*, 1726/2020) is resolutely based on geometric techniques, its contributions are little understood for their diagrammatic beauty—they are usually demonstrated and analyzed using a symbolic calculus into which they were translated very early on.

2. One of the best-known examples of erroneous proofs is Cauchy's visual proof of Euler's theorem, which is well documented in Lakatos's (1976) seminal work on the analysis of mathematical discovery. This "proof" stood for over 50 years before a Hilbertian symbolic formalization was devised that could readily be checked for errors.

3. The computational, algorithmic, and implementational views of rationality are defined in the Introduction to this volume by Knauff and Spohn. Here, our implementational view does not refer to the hardware that embodies rationality but rather to the software that executes reasoning tasks.

4. Current mainstream AI is mostly interested in machine-oriented statistical learning. Nevertheless, modeling human-like reasoning on machines remains one of the foundational motivations of the AI field.

5. Peirce did not use shading for set emptiness, but he introduced x (for nonemptiness) and o (for emptiness) elements that can be linked with lines to express disjunctive information in the diagrams. These diagrams, however, lost somewhat the spatial appeal and ease of reading that Venn diagrams have.

6. The diamond in the lower part of the display denotes the blocks-world situation—clicking the diamond displays the situation graphically in the upper part.

7. Stenning and Oberlander (1995) showed that there are individual differences in users' performance: users who are more visual reasoners do better with visual representations, and sentential reasoners do better with sentential representations.

8. Two perpendicularly joined sides of a square look like a letter L, so they are referred to as *ells*. Splitting an *ell* from a square is thus referred to as an *lcut*.

References

Alqadah, M., Stapleton, G., Howse, J., & Chapman, P. (2014). Evaluating the impact of clutter in Euler diagrams. In T. Dwyer, H. Purchase, & A. Delaney (Eds.), *Diagrams. LNCS 8578* (pp. 108–122). Berlin, Germany: Springer.

Amarel, S. (1968). On representations of problems of reasoning about actions. In D. Michie (Ed.), *Machine intelligence 3* (pp. 131–171). Edinburgh, Scotland: Edinburgh University Press.

Anderson, J. R. (1978). Arguments concerning representations for mental imagery. *Psychological Review, 85,* 249–277.

Anderson, M., Meyer, B., & Olivier, P. (2001). *Diagrammatic representation and reasoning.* London, England: Springer.

Baker, S., Ireland, A., & Smaill, A. (1992). On the use of the constructive omega rule within automated deduction. In A. Voronkov (Ed.), *International Conference on Logic Programming and Automated Reasoning (LPAR). LNAI 624* (pp. 214–225). Berlin, Germany: Springer.

Barker-Plummer, D., & Bailin, S. (1997). The role of diagrams in mathematical proofs. *Machine Graphics and Vision, 6*(1), 25–56.

Barker-Plummer, D., Barwise, J., & Etchemendy, J. (2017). *Logical reasoning with diagrams & sentences: Using Hyperproof.* Stanford, CA: CSLI.

Barwise, J., & Etchemendy, J. (1991). Visual information and valid reasoning. In W. Zimmerman & S. Cunningham (Eds.), *Visualization in teaching and learning mathematics* (pp. 9–24). Washington, DC: Mathematical Association of America.

Chandrasekaran, B., Glasgow, J., & Narayanan, N. H. (1995). *Diagrammatic reasoning: Cognitive and computational perspectives.* Cambridge, MA: AAAI Press/MIT Press.

Cheng, P. C.-H., Lowe, R., & Scaife, M. (2001). Cognitive science approaches to understanding diagrammatic representations. *Artificial Intelligence Review, 15*, 79–94.

Euler, L. (1768). *Lettres à une Princesse d'Allemagne* [Letters to a German princess]. St. Petersburg, Russia: l'Academie Imperiale des Sciences.

Gonthier, G. (2008). Formal proof—the four-color theorem. *Notices of the American Mathematical Society, 11*, 1382–1393.

Hammer, E. M. (1995). *Logic and visual information.* Stanford, CA: CSLI Press.

Hegarty, M., & Just, M. (1993). Constructing mental models of amchines from text and diagrams. *Journal of Memory and Language, 32*, 717–742.

Hilbert, D. (2004). *Grundlagen der Geometrie.* In M. Hallet & U. Majer (Eds.), *David Hilbert's lectures on the foundations of geometry, 1891–1902* (pp. 65–144). Berlin, Germany: Springer. (Original work published 1894)

Howse, J., Stapleton, G., & Taylor, J. (2005). Spider diagrams. *LMS Journal of Computation and Mathematics, 8*, 145–194.

Jamnik, M. (2001). *Mathematical reasoning with diagrams: From intuition to automation.* Stanford, CA: CSLI.

Johnson-Laird, P. (1983). *Mental models: Towards a cognitive science of language, inference and consciousness.* Cambridge, England: Cambridge University Press.

Knauff, M. (2013). *Space to reason—a spatial theory of human thought.* Cambridge, MA: MIT Press.

Lakatos, I. (1976). *Proofs and refutations: The logic of mathematical discovery.* Cambridge, England: Cambridge University Press.

Larkin, J. H., & Simon, H. A. (1987). Why a diagram is (sometimes) worth ten thousand words. *Cognitive Science, 11*(1), 65–99.

Maxwell, E. A. (1959). *Fallacies in mathematics.* Cambridge, England: Cambridge University Press.

Nelsen, R. B. (1993). *Proofs without words: Exercises in visual thinking.* Washington, DC: Mathematical Association of America.

Nelsen, R. B. (2001). *Proofs without words II: Exercises in visual thinking.* Washington, DC: Mathematical Association of America.

Newton, I. (2021). *The mathematical principles of natural philosophy* (C. Leedham-Green, Ed., Trans.) (3rd ed.). Cambridge, England: Cambridge University Press. (Original work published 1726)

Peirce, C. S. (1933). *Collected papers.* Cambridge, MA: Harvard University Press.

Pólya, G. (1957). *How to solve it: A new aspect of mathematical method.* Princeton, NJ: Princeton University Press.

Pólya, G. (1965). *Mathematical discovery.* New York, NY: Wiley.

Raggi, D., Stapleton, G., Stockdill, A., Jamnik, M., Garcia Garcia, G., & Cheng, P. C.-H. (2020). How to (re)represent it? In *32nd IEEE international conference on tools with artificial intelligence (ICTAI)* (pp. 1224–1232). Baltimore, MD: IEEE.

Rumbaugh, J., Jacobson, I., & Booch, G. (1999). *The Unified Modeling Language reference manual.* Reading, MA: Addison-Wesley.

Shams, Z., Sato, Y., Jamnik, M., & Stapleton, G. (2018). Accessible reasoning with diagrams: From cognition to automation. In P. Chapman, G. Stapleton, A. Mokfeti, S. Perez-Kriz, & F. Bellucci (Eds.), *Diagrammatic representation and inference (Diagrams). LNCS 10383* (pp. 247–263). Cham, Switzerland: Springer.

Shimojima, A. (1996). Operational constraints in diagrammatic reasoning. In G. Allwein & J. Barwise (Eds.), *Logical reasoning with diagrams* (pp. 27–48). New York, NY: Oxford University Press.

Shin, S.-J. (1994). *The logical status of diagrams.* Cambridge, England: Cambridge University Press.

Simon, H. A. (1996). *The sciences of the artificial.* Cambridge, MA: MIT Press.

Stapleton, G., Howse, J., Taylor, J., & Thompson, S. (2004). The expressiveness of spider diagrams. *Journal of Logic and Computation, 14*(6), 857–880.

Stenning, K., & Oberlander, J. (1995). A cognitive theory of graphical and linguistic reasoning: Logic and implementation. *Cognitive Science, 19*, 97–140.

Urbas, M., Jamnik, M., & Stapleton, G. (2015). Speedith: A reasoner for spider diagrams. *Journal of Logic, Language and Information, 24*(4), 487–540.

Venn, J. (1881). *Symbolic logic.* London, England: Macmillan.

13.2 Rational Reasoning about Spatial and Temporal Relations

Marco Ragni

Summary

Reasoning about spatial and temporal relations is ubiquitous in everyday life. It is important that such reasoning processes proceed rationally; otherwise, we will arrive at the wrong place or appear there at the wrong time. To demonstrate the specifics of the underlying cognitive processes, human spatial and temporal reasoning is compared to formal approaches from artificial intelligence (AI). The underlying cognitive processes have particular characteristics: First, they are *qualitative*, that is, metrical information is taken into account only when necessary. Second, they realize *resource-bounded rationality*, that is, due to limited resources, only parsimonious mental representations are built: often a preferred mental model is constructed and, based on it, putative conclusions are drawn. Third, they are *sensitive to the mental effort* problems require, that is, often mental models are constructed that require a small number of operations only. This often leads to rational spatial and temporal inferences but can also lead to systematic errors.

1. Space and Time Are Fundamental in Everyday Life

Humans are always within the realm of space and time. But we are not always consciously aware of both dimensions of our existence. Kant (1781/1787/1995, A23/B38) claimed that space and time are not empirical (in the sense that "abstract" space or time is not perceivable by our senses) but rather "a priori representations." We also cannot perceive abstract space and time directly but become aware of the reality of space and time whenever we navigate, communicate, or reason about our environment or about events in time. In the following, we start by considering a simple spatial reasoning problem concerning four entities and consisting of two pieces of relational information. Please read the following information and try to answer the question afterward.

You are now at the train station.

Your bicycle is in front of the north entrance of the church.

The church is north of the train station.

What, if anything, can be inferred from the given information?

An immediate conclusion, that is, an inference that can easily be drawn, is that you have to head north to find your bicycle. Analogously, let us assume that you know that

The complete decoding of the human genome took place after the first moon landing.

The invention of the computer took place before the first moon landing.

It is easy for you to infer from these two statements that the computer was invented before the complete decoding of the human genome took place. With some background knowledge, such as that computers are nowadays often involved in scientific progress, you can even form a putative conclusion that the invention of the computer was very likely important in the complete decoding of the human genome.

These two examples are similar to a multitude of other problems we encounter in everyday life. Let us consider some characteristics of such problems: first, the relations are qualitative, that is, no numerical information is given, and you do not need any numerical information or may have even thought about it. Second, all information is relevant to solving the problem. Third, in our example, the four terms in the spatial problem and the three events in the temporal problem were relatively simple, but it is not hard to imagine that additional information may make such problems more difficult to solve.

In solving such problems, you may have had the impression that you formed a vivid mental picture or that you used an abstract relational representation of the described spatial situation. Or, instead, you may have applied logical inference rules (e.g., a transitivity rule) to the two relations "in front of (the north entrance of)" and "north of" (despite being nonidentical, they do not differ in their direction), without forming an analogous model-like representation, to infer that you have to head north. Or you may have applied a heuristic to generate a conclusion. An example of such a heuristic might be to use the only cardinal direction relation ("north") mentioned in the given information, giving you the general direction in your conclusion. While for this sample problem, each type of inference process can be applied, all leading to the same answer, it is very unlikely that a human reasoner employs all of these processes. It is much more plausible that there are specific processes that most reasoners tend to use. Probably an important observation is that humans often use qualitative spatial and temporal relations, which do not represent exact, metrical information in a continuous way. Rather, they only represent those relations that are necessary for the inference. This allows humans to represent spatial and temporal information even if exact information is not available or would require too much effort to acquire. An advantage of qualitative relations is that they do not require any artificial external metrical scale (Freksa, 1992). Therefore, qualitative spatial and temporal relations are more practical in an environment where exact information is missing. A qualitative relation can be considered an "abstraction summarizing similar quantitative states into one qualitative characterization" (Moratz & Ragni, 2008, p. 76; see also Rauh et al., 2005). In this sense, qualitative descriptions are compressed representations and are much more useful in a complex environment for an agent that has a bounded rationality only.

An underlying assumption of this chapter is that the way in which human reasoning proceeds depends on the need to be applicable and useful in a complex world, given that cognitive resources, such as working memory, are limited. We will consider the underlying processes of reasoning about spatial and temporal beliefs both from a *normative* perspective (i.e., looking at logical formalisms capturing idealized knowledge representation and reasoning) and from a *descriptive* perspective (i.e., how humans actually represent knowledge and reason about it). By contrasting what *ought to be* and what *is*, it is possible to identify specific differences between formal and human reasoning. The chapter addresses this question on a *formal* level, by introducing artificial intelligence (AI) systems based on qualitative relations; on an *empirical* level, to gain access to human reasoning; and on an *algorithmic* level, by explaining the cognitive processes via cognitive modeling.

2. Formal Models for Reasoning about Spatial and Temporal Relations

Since it is natural for humans to draw inferences such as the ones above, we often consider these problems as simple. From a formal perspective, however, they are certainly not. In the past years, cognitively inspired calculi (e.g., the double-cross calculus; Zimmermann & Freksa, 1996) have been introduced that help us to reconstruct how the example problems can be solved. Two famous calculi are the interval calculus introduced by Allen (1983) and the region connection calculus introduced by Randell, Cui, and Cohn (1992).

The interval calculus was initially developed for reasoning with temporal relations but was soon transferred to spatial reasoning (Guesgen, 1989). In the spatial domain, the calculus works with solid objects. The calculus can also be extended into more complex formalisms (Ligozat, Mitra, & Condotta, 2004). An event can be represented by its start time and its end time (see table 13.2.1). Accordingly, two events can be qualitatively related by the relations between the start and end times of the two events, for example, that the first event ends before the second starts, or that the first event overlaps with the second, or that both events start at the same time, or that both finish at the same time, or that the first event takes place during the second event. If the qualitative relations between start and end points are known, then 13 different relations can be identified (see table 13.2.1). Imprecise knowledge about the start and end times of the events is represented by disjunctions of these 13 relations (e.g., an event finishes before the other starts *or* they overlap).

The region connection calculus (RCC) has been developed in the spatial domain and is similar to the interval calculus, but instead of orientational information, it just uses topological information, that is, the relations between the boundaries of objects (e.g., two regions can be disjoint, or they can overlap, or the first region can be a proper part of the second, or vice versa, or they can be identical). The calculus using these five basic topological relations is called RCC-5. If the boundaries of the regions are relevant too (e.g., when two countries share a common border), then three more relations can be formed: a region touches another from the outside or they are disjoint (and not even their borders overlap), or the first region is a proper part of the second and touches the latter's border or not, or this relation holds for the second region. With the resulting eight relations, we have the calculus RCC-8 (see figure 13.2.1). Based on these base relations, one can again express indeterminate

Table 13.2.1

The 13 qualitative interval relations of the interval calculus, their names for temporal reasoning, natural language expressions for the spatial domain, and an interval representation

Temporal	Spatial	Relation	Inverse	Pictorial Representation
before	*I* lies to the left of *J*	*I* b *J*	*J* bi *I*	——— *I* ——— *J*
meets	*I* touches *J* at the left	*I* m *J*	*J* mi *I*	——— *I* ——— *J*
overlaps	*I* overlaps *J* from the left	*I* o *J*	*J* oi *I*	——— *I* ——— *J*
during	*I* is completely in *J*	*I* d *J*	*J* di *I*	— *I* — *J*
starts	*I* lies left-justified in *J*	*I* s *J*	*J* si *I*	— *I* — *J*
finishes	*I* lies right-justified in *J*	*I* f *J*	*J* fi *I*	— *I* — *J*
equals	*I* equals *J*	*I* e *J*	*J* e *I*	— *I* — *J*

Note: Adapted from Knauff (1999).

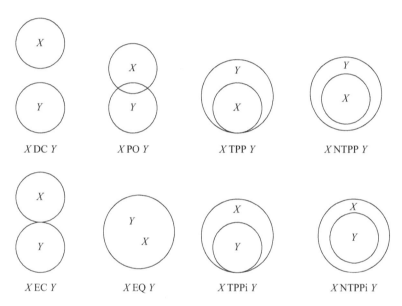

X DC Y   X PO Y   X TPP Y   X NTPP Y
X EC Y   X EQ Y   X TPPi Y   X NTPPi Y

Figure 13.2.1

The region connection calculus with eight base relations (RCC-8). *X* and *Y* are regions. The eight base relations are DC = disconnected, PO = partially overlaps, (N)TPP = (non)tangential proper part, EC = externally connected, EQ = equal, and the two inverse relations NTTPi and TPPi.

descriptions reflecting imprecise knowledge (e.g., that a region contains the other or that they are disjoint).

These are just two of many approaches that have been suggested in AI for reasoning in a rational way with spatial and temporal relations. These approaches can serve as normative accounts of how correct reasoning with qualitative spatial and temporal relations should proceed. They have many interesting formal characteristics and have been used in many domains of application, for example, in robotics (Moratz & Ragni, 2008), plan generation (e.g., Westphal, Dornhege, Wölfl, Gissler, & Nebel, 2011), sea navigation (e.g., Wolter et al., 2008), and language processing (e.g., Mast, Wolter, Klippel, Wallgrün, & Tenbrink, 2014). These calculi are of interest in this chapter also because many cognitive psychologists have used them to compare human reasoning with reasoning in technical systems (Knauff, Rauh, Schlieder, & Strube, 1998; Rauh et al.,

2005; Schlieder & Berendt, 1998) and also to determine the correct solutions to inference problems. The results of these empirical studies show that the formal approaches capture some aspects of human reasoning but that people also quite often reason with spatial and temporal relations differently, and there are cross-cultural similarities as well as differences (e.g., Knauff & Ragni, 2011).

The problem with many formal calculi is that they can be computationally expensive (for an overview, see Renz & Nebel, 2007), that is, the number of simple computations and/or the amount of memory space needed to solve the problem rise exponentially with the size of the problem description. This makes "large" problems practically intractable. A central empirical finding, however, is that, in contrast to formal approaches, humans do not consider all possible models but rather try to find a useful specific model, especially when the verbal description is indeterminate (i.e., allows for multiple models). In the following, we describe this and related findings from empirical research and computational modeling.

3. Cognitive Findings on Spatial and Temporal Reasoning

Let us first look at the behavioral level by studying which conclusions people draw from a set of given premises. Such questions have been in the center of reasoning research for several decades, and many effects are so reliable that they have been replicated across many studies. Here are a few of these robust findings:

(1) The *indeterminacy effect*: if a problem description is indeterminate, that is, if it allows for multiple scenarios, the problems are on average more difficult to reason through. Indeterminate descriptions can appear quite often in everyday life (e.g., in cases where information is missing). This effect has been observed repeatedly and has been explained via both mental models and mental logic approaches (e.g., Johnson-Laird & Byrne, 1991; Kießner & Ragni, 2017; Van der Henst, 2002).

(2) The *relational complexity effect*: if a problem description contains relations of a higher arity (e.g., "between," which is a ternary relation), or the relational information needs to be decomposed into facts involving simpler relations, then the reasoning difficulty increases. Hence, relational complexity influences the mental representation that is constructed (e.g., Goodwin & Johnson-Laird, 2005; Halford, Wilson, & Phillips, 1998; Jahn, Knauff, & Johnson-Laird, 2007).

(3) The *visual impedance effect*: if a problem description uses relations that are easier to visualize than more abstract relations, then this visualizability can impede reasoning (e.g., Knauff, 2013). This indicates that the mental representation most human reasoners use is rather abstract and relational and less a visual image (e.g., Pylyshyn, 2002).

(4) The *continuity effect*. Consider the following two descriptions:

(*continuous*): A is left of B; B is left of C; C is left of D.

(*discontinuous*): A is left of B; C is left of D; B is left of C.

In the first description (*continuous*), each new premise connects directly to the previous one, such that the new information can be easily mentally integrated into a new representation. In the second description (*discontinuous*), the information cannot

be directly related. The first type of description is easier for reasoners to process than the second (e.g., Knauff et al., 1998; Nejasmic, Bucher, & Knauff, 2015). Since the information given is identical, an explanation for the difference is that for the continuous problem, there is an immediate successive integration of the premise information into a mental model, while in the discontinuous case, the processing requires to first store two independent pieces of information.

(5) The *relevance of information effect*: if reasoners receive information that is partially irrelevant for the conclusion (e.g., because the conclusion is already given), then such problems are easier than problems with the same amount of information where all of it is relevant (e.g., Schaeken, Girotto, & Johnson-Laird, 1998; Van der Henst, 1999). This indicates that spatial and temporal reasoning difficulty depends on the amount of relevant information, not on the total amount of information, and that the model contains only the relevant information.

These five effects (for an overview, see, e.g., Ragni, 2008) have been repeatedly reported and support the hypothesis that spatial beliefs are represented by mental models based on the principle of minimal operations (see below). We will later see that an explanation for how humans draw inferences is that often, instead of constructing all possible models, they construct just one specific mental model, the "preferred" model. This does not only reduce the memory load incurred by *storing* models but also reduces the mental effort that comes with *constructing* models. This resource awareness is a realization of *bounded rationality*. Bounded rationality is an adaptation of our mind to these limitations, for example, of the available information and of our cognitive resources (see Gigerenzer & Selten, 2002; Simon, 1955). Hence, another robust result is the following:

(6) The *preference effect*. Consider the following example from Ragni and Knauff (2013):

The Ferrari is parked to the left of the Porsche.

The Beetle is parked to the right of the Porsche.

The Porsche is parked to the left of the Hummer.

The Hummer is parked to the left of the Dodge.

The description allows for the following three models:

Model 1: Ferrari Porsche Beetle Hummer Dodge

Model 2: Ferrari Porsche Hummer Beetle Dodge

Model 3: Ferrari Porsche Hummer Dodge Beetle

While logically, all three arrangements are models of the situation described in the premises, experimental studies showed that human reasoners accept model 1 (which has been called the preferred mental model) more often and faster than model 2, which they in turn accept more often and faster than model 3 (which are called alternative mental models). To make logically correct inferences, the construction of *all* alternative models is relevant, because an alternative to the preferred model can be a counterexample, which would make the inference based on the preferred model wrong. For example, the previous premises might be presented to reasoners, and then they must answer the question, "Is the Beetle (necessarily) parked to the left of the Dodge?" In this case, most people do construct the first model but not the other two models. Hence, they give a wrong answer.

The preference effect has been reported for the first time in Knauff, Rauh, and Schlieder (1995). These authors also suggested the term "preferred mental model" to explain the cognitive foundations of the preference effect. Since then, the preference effect was repeatedly found in many experiments and has been replicated several times (e.g., Knauff, 2013; Knauff et al., 1998; Ragni & Knauff, 2013; Rauh et al., 2005; Schultheis, Bertel, & Barkowsky, 2014). We return to this in the next section.

4.   Neural Findings and Their Impact for a Theory of Spatial and Temporal Reasoning

The way our brain processes spatial and temporal information can explain how and why we deviate from formal approaches to reasoning in AI. For instance, we know that, in contrast to formal models of computation, like Turing machines, the human mind is not monolithic, that is, there is no single storage space; rather, for different tasks, different regions are relevant. Core regions (cf. chapter 1.4 by Goel, this handbook) identified for spatial and temporal reasoning are the prefrontal cortex (especially the dorsolateral and the left prefrontal cortex), mainly implementing control functions (e.g., the integration of different premises; Fangmeier, Knauff, Ruff, & Sloutsky, 2006), and the parietal cortex, mainly storing the mental model (Crone et al., 2009; Knauff, 2006; Ragni, Franzmeier, Wenczel, & Maier, 2014). A recent meta-analysis (Wertheim & Ragni, 2018) of functional magnetic resonance imaging studies was able to identify the involvement of two frontal areas (Brodmann areas 6 and 45, with a role in complex planning and semantic understanding, respectively) and one parietal area (Brodmann area 7). For spatial reasoning, especially the superior parietal lobe (SPL) is relevant.

The theory of preferred mental models (Ragni & Knauff, 2013) claims that different phases in spatial reasoning exist: first constructing the preferred mental model and then searching for counterexamples by constructing alternative mental models. Now, a study shows that disturbing the information processing in SPL by transcranial magnetic stimulation (TMS) leads to a decrease in people's ability to construct the preferred mental model, but the stimulation of the SPL has no impact on the generation of alternative models, that is, models 2 and 3 above (Ragni, Franzmeier, Maier, & Knauff, 2016). This supports that these processes are different, because they are realized differently on the neural level. Additionally, the study supports the claim that bounded rationality is a relevant factor. Reasoners with a smaller working memory capacity made more errors in accepting the preferred and alternative models, which are correct models. Reasoners with a smaller working memory capacity are more susceptible to disturbance effects by TMS and subsequently accept the preferred model less often. In fact, the disturbance effect by application of TMS was inversely correlated with working memory capacity: it was lower for participants with greater working memory capacity and higher for those with smaller working memory capacity.

The results show that reasoning about spatial and temporal relations depends on the preferred mental models that are formed in the parietal cortex and on control processes that are relevant for integrating new information and are located in the frontal regions. Finally, limited working memory capacity can explain why not all mental models are constructed.

5.   Computational Models and the Relevance of Complexity for Rationality

We now turn to cognitive computational systems that aim to model, reconstruct, and explain the processes underlying human reasoning. Some of these cognitive computational systems for spatial reasoning use Cartesian coordinate systems (Johnson-Laird & Byrne, 1991; Schultheis & Barkowsky, 2011), whereas others use discrete representational structures such as lists or grids (e.g., Krumnack, Bucher, Nejasmic, Nebel, & Knauff, 2011; PRISM, Ragni & Knauff, 2013). Some systems assume the construction of mental models while others rely on visual images (e.g., Casimir; Schultheis & Barkowsky, 2011). Yet others use abstract relational information (e.g., PRISM), or are inspired by the linguistic task of encoding and reencoding spatial relations (Krumnack et al., 2011).

Those computational models that successfully solve problems like humans do shed some light on the underlying cognitive processes. For instance, most cognitive models assume that human reasoners construct mental models and make transitive inferences based on these models. For the visual impedance effect, current cognitive models assume that people initially create a visual mental image but then have to extract the relevant spatial information from this image, which can be time-consuming and error-prone (Knauff, 2013). Others try to explain the effect by assuming that the visual relations activate additional features in declarative memory (Albrecht, Schultheis, & Fu, 2015).

There exist only a few systems for *temporal* relational reasoning, and these are based on mental models. The cognitive processes are similar to those in processing spatial relations: humans construct preferred mental models and then vary them to test putative conclusions (Johnson-Laird & Byrne, 1991; Juhos, Quelhas, & Johnson-Laird, 2012). Kelly and Khemlani (2019) have demonstrated that many features that have been identified in spatial reasoning also transfer to durative temporal relations, for instance, one event happening during another event. Examples of features that transfer from the spatial to the temporal domain are that the underlying mental models are iconic, that they rely both on intuitive and on deliberative processes, and that problems eliciting more models have a higher reasoning difficulty. Moreover, the consideration of cognitive costs can explain why some reasoners are able to mentally simulate alternative models that falsify the conclusion suggested by the preferred model.

We already mentioned that the findings from research on spatial and temporal reasoning can be explained within the conception of *bounded rationality*. In particular, in cases where several models are possible, humans systematically prefer one model while ignoring others (Knauff et al., 1998; Ragni & Knauff, 2013; Rauh et al., 2005). The explanation is that such preferred mental models are constructed according to principles of *operational parsimony*: the human mind builds the mental model that needs the smallest number of cognitive operations.

An important point that these findings demonstrate is that human reasoners do not generate all possible mental models of given information with equal ease and equal likelihood. And this in turn shows that logical formalisms such as formal qualitative calculi (see section 2 above), which do not make any difference between mental models, are descriptively inadequate.

The theory of preferred mental models, in contrast, assumes that reasoners successively insert each new object from the premises into their mental model, filling a slot that is not already occupied by another object, such that the resulting model still matches the premises. This process has been algorithmically realized in a computational model called PRISM (Ragni & Knauff, 2013), which indeed implements bounded rationality and preference effects, based on the idea of simple spatial operations such as inserting an object into a mental model or changing the position of a mental focus. The number of such operations needed for solving a problem forms a measure of its cognitive complexity. By considering the number of operations in a spatial working memory representation, this measure can explain the preference effect and which alternative model is constructed when (Ragni & Knauff, 2013). The number of such operations explains which mental model is constructed and inspected, as well as which putative conclusion is considered in the reasoning process. If the number of operations is high, often only a single model is constructed (Ragni & Knauff, 2013). Which model is constructed depends on the specific way the information is mentally processed (e.g., whether the information is presented in a continuous way, as in the continuity effect), on background knowledge (e.g., whether we have a preferred arrangement of objects, say, of the silverware on a table), and on the underlying spatial domain. For example, it is easier to mentally insert objects in large-scale scenarios (e.g., cardinal directions; Ragni & Becker, 2010) than in small-scale ones (e.g., left–right relations among fruits on a table; Ragni & Knauff, 2013).

The preference effect is ubiquitous and has been replicated for small-scale spaces (Ragni, Fangmeier, Webber, & Knauff, 2006; Rauh et al., 2005), for large-scale spaces (with respect to the cardinal directions; Ragni & Becker, 2010; Schultheis & Barkowsky, 2011), for topological relations (even cross-culturally; Knauff & Ragni, 2011), and also for other reasoning domains (e.g., Johnson-Laird, 2006). That the preference effect is so prevalent indicates that in the domain of spatial and temporal relational reasoning, rationality is bounded not only by working memory limitations but also by limitations on the number of mental operations one can perform. Of course, the two are related, but we will not discuss this here (see Knauff, 2013). In the following, some further heuristics of human spatial and temporal reasoning are described.

6. Further Characteristics of Human Spatial and Temporal Reasoning

Although many kinds of spatial and temporal relations exist, a particularly important kind of such relations are *transitive inferences*. In such inferences, a conclusion "*A* is north of *C*" is drawn from the premises "*A* is north of *B*" and "*B* is north of *C*." When humans try to solve such inferences, they often fall prey to illusions even for nonspatial relations such as blood-relatedness: many reasoners infer from "*A* is blood-related to *B*" and "*B* is blood-related to *C*" that *A* is blood-related to *C*. In fact, there is a type of counterexample known to almost all reasoners: *A* is the father, *B* is the daughter, and *C* is the mother. Still, the internal mental representation and the cognitive inference system prompt many human reasoners to make this inference (Goodwin & Johnson-Laird, 2008). Such illusions demonstrate that our internal inference mechanism is overeager to apply transitivity. And they demonstrate that we internally represent even nonspatial information in a way that allows drawing transitive inferences based on spatially organized models. Many of the following results are from experiments with transitive relations.

6.1 Human Reasoning about Space and Time Employs Mental Models and Manifests Preference Effects

There is broad agreement among current theories and computational models that spatial and temporal beliefs are represented by mental models (Krumnack et al., 2011; Ragni & Knauff, 2013; Schultheis & Barkowsky, 2011; Johnson-Laird & Byrne, 1991), which are analogical representations of states of affairs. But there is also some disagreement about how much visual information is present in mental models and in what form it is represented (e.g., Pylyshyn, 2002). Experimental findings indicate that our internal representation is rather abstract-relational and less visual. In fact, if additional visual features must be encoded, an impedance effect on reasoning can occur (Knauff, 2013). The underlying representation is again an abstraction comprising only the information relevant for the specific task. This principle of minimizing cognitive effort holds also for cases where multiple models are possible: usually only one particular model is constructed, which comprises relevant beliefs. Given limited cognitive resources and time pressure, this parsimony is rational. In the following, results are reported that support this claim.

6.2 Human Reasoning Is Influenced by Problem Content and Embodiment

Experimental studies have shown that problem content and the reasoner's background information about the given domain can have strong effects on the construction of the preferred mental model (e.g., Rauh, 2000). Accordingly, the mechanisms for the construction of spatial representations even extend to other domains (e.g.,

mathematical cognition). In the spatial–numerical association of response codes (SNARC) effect, small numbers are responded to faster with the left hand, and large numbers show a similar proclivity with regard to right-sided responses and presentations (Hubbard, Piazza, Pinel, & Dehaene, 2005). This indicates that people internally represent numbers in a spatial ordering from left to right by ascending size. For relational reasoning, the same effect has been identified for the relational information "*A* is left of *B*" and "*B* is left of *C*" in responding with the left hand for items further left and with the right hand for items further right (Prado, Van der Henst, & Noveck, 2008). Thus, even for arbitrary relational information, our internal mental representation is not independent from our body; hence, our "embodiment" matters.

6.3 Human Rationality Is Bounded

Empirical findings demonstrate that human reasoners often do not generate all possible models but only a specific one—the preferred mental model. Additionally, not all alternative models are constructed, only small modifications of the preferred model. Findings from neuroscience demonstrate that mental processes are not monolithic: alternative mental models are generated in other brain regions than the preferred mental model. Working memory capacity in turn is a limiting factor for the construction of mental models. An analysis of core empirical effects demonstrates that problems that are more difficult (e.g., the discontinuous case or multiple-model problems) require more mental operations than easier problem descriptions—supporting an operational parsimony. Together, working memory limitations and operational parsimony are core aspects constituting *bounded rationality* in humans.

7. Conclusion

We can draw several conclusions from this chapter. Probably the most important are that human reasoning about spatial and temporal relations deviates from classical conceptions of rationality and that it is multifaceted: we do not simply apply a few general principles. Beyond the limited information capacity of our senses, human rationality is bounded in a twofold way: first, our working memory capacity is limited, forcing our brains to construct and manipulate specific mental models; second, our computing capacity is limited, whence our brains cannot exhaustively survey all possibilities—this operational parsimony explains many, if not all, of the cited effects. For estimating operational effort, there is a cognitive complexity measure that is based on the number of simple operations and explains especially why some models are constructed and others are neglected.

While humans have sometimes been considered irrational, they are actually not: they are quite sensible about how much mental effort—given their limited resources—they do invest. Often a first mental model is sufficient to understand the information, and very often finding an optimal path or time plan takes too much effort. Optimality is not necessary in our everyday lives; avoiding mental depletion is—it can be dangerous. From a strictly normative perspective, humans may not always find the correct inference—but they often find one that suffices.

References

Albrecht, R., Schultheis, H., & Fu, W. T. (2015). Visuo-spatial memory processing and the visual impedance effect. In D. C. Noelle, R. Dale, A. S. Warlaumont, J. Yoshimi, T. Matlock, C. D. Jennings, & P. P. Maglio (Eds.), *Proceedings of the 37th Annual Meeting of the Cognitive Science Society* (pp. 72–77). Austin, TX: Cognitive Science Society.

Allen, J. F. (1983). Maintaining knowledge about temporal intervals. *Communications of the ACM, 26*(11), 832–843.

Crone, E. A., Wendelken, C., Van Leijenhorst, L., Honomichl, R. D., Christoff, K., & Bunge, S. A. (2009). Neurocognitive development of relational reasoning. *Developmental Science, 12*(1), 55–66.

Fangmeier, T., Knauff, M., Ruff, C. C., & Sloutsky, V. (2006). fMRI evidence for a three-stage model of deductive reasoning. *Journal of Cognitive Neuroscience, 18*(3), 320–334.

Freksa, C. (1992). Temporal reasoning based on semi-intervals. *Artificial Intelligence, 54*(1–2), 199–227.

Gigerenzer, G., & Selten, R. (Eds.). (2002). *Bounded rationality: The adaptive toolbox.* Cambridge, MA: MIT Press.

Goodwin, G. P., & Johnson-Laird, P. N. (2005). Reasoning about relations. *Psychological Review, 112*(2), 468–475.

Goodwin, G. P., & Johnson-Laird, P. N. (2008). Transitive and pseudo-transitive inferences. *Cognition, 108*(2), 320–352.

Guesgen, H. W. (1989). *Spatial reasoning based on Allen's temporal logic.* Berkeley, CA: International Computer Science Institute.

Halford, G. S., Wilson, W. H., & Phillips, S. (1998). Processing capacity defined by relational complexity: Implications for comparative, developmental and cognitive psychology. *Behavioral and Brain Sciences, 21*(6), 803–831.

Hubbard, E. M., Piazza, M., Pinel, P., & Dehaene, S. (2005). Interactions between number and space in parietal cortex. *Nature Reviews Neuroscience, 6*(6), 435–448.

Jahn, G., Knauff, M., & Johnson-Laird, P. N. (2007). Preferred mental models in reasoning about spatial relations. *Memory & Cognition, 35*(8), 2075–2087.

Johnson-Laird, P. N. (2006). *How we reason.* Oxford, England: Oxford University Press.

Johnson-Laird, P. N., & Byrne, R. M. J. (1991). *Deduction.* Mahwah, NJ: Erlbaum.

Juhos, C., Quelhas, A. C., & Johnson-Laird, P. N. (2012). Temporal and spatial relations in sentential reasoning. *Cognition, 122*(3), 393–404.

Kant, I. (1995). *Kritik der reinen Vernunft* [Critique of pure reason] (W. Weischedel, Ed.) (9th ed., Vol. 1). Frankfurt/Main, Germany: Suhrkamp. (Original work published 1781/1787)

Kelly, L., & Khemlani, S. (2019). The consistency of durative relations. In A. Goel, C. Seifert, & C. Freksa (Eds.), *Proceedings of the 41st Annual Conference of the Cognitive Science Society* (pp. 1998–2003). Austin, TX: Cognitive Science Society.

Kießner, A.-K., & Ragni, M. (2017). Processing spatial relations: A meta-analysis. In G. Gunzelmann, A. Howes, T. Tenbrink, & E. J. Davelaar (Eds.), *Proceedings of the 39th Annual Conference of the Cognitive Science Society* (pp. 2401–2406). Austin, TX: Cognitive Science Society.

Knauff, M. (1999). The cognitive adequacy of Allen's interval calculus for qualitative spatial representation and reasoning. *Spatial Cognition and Computation, 1*(3), 261–290.

Knauff, M. (2006). Deduktion, logisches Denken [Deduction, logical thinking]. In J. Funke (Ed.), *Denken und Problemlösen* [Thinking and problem solving] (Enzyklopädie der Psychologie, Themenbereich C: Theorie und Forschung, Serie II: Kognition, Vol. 8, pp. 2–97). Göttingen, Germany: Hogrefe.

Knauff, M. (2013). *Space to reason: A spatial theory of human thought.* Cambridge, MA: MIT Press.

Knauff, M., & Ragni, M. (2011). Cross-cultural preferences in spatial reasoning. *Journal of Cognition and Culture, 11*(1), 1–21.

Knauff, M., Rauh, R., & Schlieder, C. (1995). Preferred mental models in qualitative spatial reasoning: A cognitive assessment of Allen's calculus. In D. Kirsh (Ed.), *Proceedings of the Seventeenth Annual Conference of the Cognitive Science Society* (pp. 200–205). Mahwah, NJ: Erlbaum.

Knauff, M., Rauh, R., Schlieder, C., & Strube, G. (1998). Mental models in spatial reasoning. In C. Freksa, C. Habel, & K. F. Wender (Eds.), *Spatial cognition* (pp. 267–291). Berlin, Germany: Springer.

Krumnack, A., Bucher, L., Nejasmic, J., Nebel, B., & Knauff, M. (2011). A model for relational reasoning as verbal reasoning. *Cognitive Systems Research, 12*(3–4), 377–392.

Ligozat, G., Mitra, D., & Condotta, J. F. (2004). Spatial and temporal reasoning: Beyond Allen's calculus. *AI Communications, 17*(4), 223–233.

Mast, V., Wolter, D., Klippel, A., Wallgrün, J. O., & Tenbrink, T. (2014). Boundaries and prototypes in categorizing direction. In C. Freksa, B. Nebel, M. Hegarty, & T. Barkowsky (Eds.), *International conference on spatial cognition* (pp. 92–107). Cham, Switzerland: Springer.

Moratz, R., & Ragni, M. (2008). Qualitative spatial reasoning about relative point position. *Journal of Visual Languages & Computing, 19*(1), 75–98.

Nejasmic, J., Bucher, L., & Knauff, M. (2015). The construction of spatial mental models—A new view on the continuity effect. *Quarterly Journal of Experimental Psychology, 68*(9), 1794–1812.

Prado, J., Van der Henst, J.-B., & Noveck, I. A. (2008). Spatial associations in relational reasoning: Evidence for a SNARC-like effect. *Quarterly Journal of Experimental Psychology, 61*(8), 1143–1150.

Pylyshyn, Z. (2002). Mental imagery: In search of a theory. *Behavioral and Brain Sciences, 25*(2), 157–237.

Ragni, M. (2008). *Räumliche Repräsentation, Komplexität und Deduktion: Eine kognitive Komplexitätstheorie* [Spatial representation, complexity, and deduction: A cognitive complexity theory]. Mannheim, Germany: Akademische Verlagsgesellschaft.

Ragni, M., & Becker, B. (2010). Preferences in cardinal direction. In S. Ohlsson & R. Catrambone (Eds.), *Proceedings of the 32nd Annual Conference of the Cognitive Science Society* (pp. 660–666). Austin, TX: Cognitive Science Society.

Ragni, M., Fangmeier, T., Webber, L., & Knauff, M. (2006). Preferred mental models: How and why they are so important in human reasoning with spatial relations. In T. Barkowsky, M. Knauff, G. Ligozat, & D. R. Montello (Eds.), *Spatial cognition V: Reasoning, action, interaction: International conference Spatial Cognition 2006, Bremen, Germany, September 24–28, 2006* (pp. 175–190). Berlin, Germany: Springer.

Ragni, M., Franzmeier, I., Maier, S., & Knauff, M. (2016). Uncertain relational reasoning in the parietal cortex. *Brain and Cognition, 104*, 72–81.

Ragni, M., Franzmeier, I., Wenczel, F., & Maier, S. (2014). The role of the posterior parietal cortex in relational reasoning. *Cognitive Processing, 15*(1), 129–131.

Ragni, M., & Knauff, M. (2013). A theory and a computational model of spatial reasoning with preferred mental models. *Psychological Review, 120*(3), 561–588.

Randell, D. A., Cui, Z., & Cohn, A. G. (1992). A spatial logic based on regions and connection. In B. Nebel, C. Rich, & W. Swartout (Eds.), *Principles of Knowledge Representation and Reasoning: Proceedings of the 3rd International Conference* (pp. 165–176). San Mateo, CA: Morgan Kaufmann.

Rauh, R. (2000). Strategies of constructing preferred mental models in spatial relational inference. In W. Schaeken, G. De Vooght, A. Vandierendonck, & G. d'Ydewalle (Eds.), *Deductive reasoning and strategies* (pp. 177–190). Mahwah, NJ: Erlbaum.

Rauh, R., Hagen, C., Knauff, M., Kuss, T., Schlieder, C., & Strube, G. (2005). Preferred and alternative mental models in spatial reasoning. *Spatial Cognition & Computation, 5*(2–3), 239–269.

Renz, J., & Nebel, B. (2007). Qualitative spatial reasoning using constraint calculi. In M. Aiello, I. Pratt-Hartmann, & J. van Benthem (Eds.), *Handbook of spatial logics* (pp. 161–215). Dordrecht, Netherlands: Springer.

Schaeken, W., Girotto, V., & Johnson-Laird, P. N. (1998). The effect of an irrelevant premise on temporal and spatial reasoning. *Kognitionswissenschaft, 7*(1), 27–32.

Schlieder, C., & Berendt, B. (1998). Mental model construction in spatial reasoning: A comparison of two computational theories. In U. Schmid, J. F. Krems, & F. Wysotzki (Eds.), *Mind modelling: A cognitive science approach to reasoning, learning and discovery* (pp. 133–162). Lengerich, Germany: Pabst Scientific Publishers.

Schultheis, H., & Barkowsky, T. (2011). Casimir: An architecture for mental spatial knowledge processing. *Topics in Cognitive Science, 3*(4), 778–795.

Schultheis, H., Bertel, S., & Barkowsky, T. (2014). Modeling mental spatial reasoning about cardinal directions. *Cognitive Science, 38*(8), 1521–1561.

Simon, H. A. (1955). A behavioral model of rational choice. *Quarterly Journal of Economics, 69*, 99–118.

Van der Henst, J.-B. (1999). The mental model theory of spatial reasoning re-examined: The role of relevance in premise order. *British Journal of Psychology, 90*(1), 73–84.

Van der Henst, J.-B. (2002). Mental model theory versus the inference rule approach in relational reasoning. *Thinking & Reasoning, 8*(3), 193–203.

Wertheim, J., & Ragni, M. (2018). The neural correlates of relational reasoning: A meta-analysis of 47 functional magnetic resonance studies. *Journal of Cognitive Neuroscience, 30*(11), 1734–1748.

Westphal, M., Dornhege, C., Wölfl, S., Gissler, M., & Nebel, B. (2011). Guiding the generation of manipulation plans by qualitative spatial reasoning. *Spatial Cognition & Computation, 11*(1), 75–102.

Wolter, D., Dylla, F., Wölfl, S., Wallgrün, J. O., Frommberger, L., Nebel, B., & Freksa, C. (2008). SailAway: Spatial cognition in sea navigation. *KI, 22*(1), 28–30.

Zimmermann, K., & Freksa, C. (1996). Qualitative spatial reasoning using orientation, distance, and path knowledge. *Applied Intelligence, 6*(1), 49–58.

13.3 Visualization and Rationality

Markus Knauff

Summary

The epistemic function of visualization was always controversial. This chapter is organized around the important, but often blurred, distinction between external and internal visualization. *External* visualizations are visual aids such as pictures, graphics, or diagrams, which are designed to help people to reason accurately. The question is whether such external visualizations are useful or harmful when people solve epistemic problems. *Internal* visualizations are picture-like mental images, a special form of mental representation, characterized as analogous to representations that arise in the mind as a result of actual visual perception. The question is whether such visual mental images are useful or harmful in rational reasoning. The chapter presents the empirical findings from cognitive psychology, cognitive brain research, and computational modeling.

1. Two Types of Visualization

Visualization is an important concept in the cognitive sciences. Many experiments show that mental visualization can help remembering objects or events, or when objects must be mentally inspected or manipulated (Kosslyn, 1980). Such visualizations are correlated with activity in early visual cortices, which might explain the experienced similarity between mental visualization and visual perception (Kosslyn, 1994). However, there is also evidence that visualization can disrupt cognitive processes and impede thinking. For instance, individuals on the autism spectrum often posit that they use visualization for tasks that typically developing individuals perform verbally (Kunda & Goel, 2011). In her famous autobiography *Thinking in Pictures*, Temple Grandin (1995/2006) describes how her visual thinking style supported her practice but also resulted in problems in dealing with more abstract thoughts. Certain psychotic drugs that suppress visual mental images (e.g., benzodiazepine) lead to better reasoning performance if the problems would otherwise elicit visual images (Pompéia, Manzano, Pradella-Hallinan, & Bueno, 2007). And the role of visualization in mathematics is still controversial (Arana, 2016; Mancosu, 2005). Some scholars assert that it helps. For instance, the influential mathematician Felix Klein (1979) claimed that "mathematics is not merely a matter of understanding but quite essentially a matter of imagination" (p. 207). Other eminent mathematicians argued that visualization is a nuisance in getting mathematical insight or proofs. The most prominent advocate of this view was Hilbert, who stated that deduction should be independent of figures in order to be rigorous (Hilbert, 1899).

This chapter focuses on the relationship between epistemic rationality and visualization. The first concept, *epistemic rationality*, is concerned with belief acquisition, formation, and revision. A central component of these processes is reasoning, a cognitive process that leads from given premises to a rationally justified conclusion, that is, a new belief or, at least, a belief that is made explicit in the process of inference. To be epistemically rational, the conclusion should conform to certain normative ideals, which often come from logic or other formal systems for rational reasoning. In chapter 5.4 by Gazzo Castañeda and Knauff (this handbook), it is shown that classical logic might not always be the best normative standard for everyday human reasoning. Yet, understanding how humans reason logically is still important for a comprehensive cognitive theory of human rationality.

The second concept, *visualization*, is often used in a very broad sense. Here I use it with two specific meanings. *External* visualizations are visual aids such as pictures, graphics, or diagrams, which are designed to help people to reason accurately. The question is whether such external visualizations are useful or harmful when people have problems to solve or make inferences. *Internal* visualizations are picture-like mental images, a special form of mental representation, characterized as analogous to representations that arise in the mind as a result of visual perception. The question is whether such visual mental images are epistemically useful or harmful in human logical reasoning. Does visualization help people in drawing rationally justified inferences, in reasoning accurately and without logical errors? In other words, what are the epistemic benefits and limitations of mental imagery in human reasoning? To answers these questions, the chapter reports empirical evidence from cognitive psychology, cognitive brain research, and computational modeling. It closes with some general corollaries on the epistemic role of visual thinking in human rationality.

2.   Epistemology of Pictures, Graphics, and Diagrams

Philosophers, logicians, and mathematicians have always had an ambiguous relationship with external visualization, that is, graphics or diagrams (Krämer, 2016). Take, for example, the following problem: imagine a square with another square inside it, where the corners of the inner square are exactly at the midpoints of the sides of the outer square. What is the size of the inner square in comparison to the outer square? It will be hard for you to solve this problem in your head alone. Now the picture in figure 13.3.1 is presented to you. You will notice that you can now solve the problem more easily—the size of the outer square is exactly twice as large as that of the inner one, because the four triangles in the corners of the outer square together cover exactly the inner square. You can easily see this if you imagine folding over the corner triangles along the sides of the inner square.

Pedagogues and teachers often use such visualizations in their classes, because they believe that they help students to understand complex mathematical issues (Kadunz & Yerushalmy, 2015). Yet, in geometry, diagrams often have been distrusted and considered inadequate and inappropriate for constructing proofs (Giaquinto, 2016). Imagine, for instance, that I have been just a bit

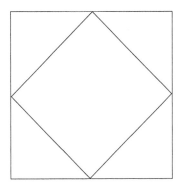

Figure 13.3.1
What is the size of the inner square in relation to the outer square?

sloppy with the drawing—then it could be that the four triangles do not exactly fit into the inner square, and you may not understand the general mathematical principle. Apparently, the formal laws of geometry allow you to say something about all squares and rectangular triangles, independently from the concrete visualization—your insights are a priori and hence independent from perceptual experience, as Kant would say. The long and interesting history of the controversies about the epistemology of visualization in mathematics is described in Arana (2016), Giaquinto (2016), and Mancosu (2005).

Since logical reasoning is a particularly important foundation of epistemic rationality, it is not surprising that philosophers and logicians have paid particular attention to the relation between logic and visualization. On the one hand, up to the turn of the 20th century, mathematicians and logicians were concerned with the formal and normative dimension of visualization in logical reasoning. Euler and Venn developed logical systems for reasoning with diagrams. Most notably, Peirce introduced the notion of *iconicity* in his nonverbal logic. An *icon*, in Peirce's semiotics, is a sign that is perceptibly similar to the object it represents. More recently, Shin (1994) has devised a formal syntax and semantics of particular forms of Venn diagrams that has also inspired much research in artificial intelligence. A good overview of this work is given in chapter 13.1 by Jamnik (this handbook).

On the other hand, the majority of scientists and philosophers questioned the importance of visualizations for logical reasoning. Driven by Ryle (1949) and notably the later Wittgenstein (1953/2001), this formal movement resulted in picture-free logical systems that are inextricably linked to language. Since then, the dogma for most logicians and philosophers has been that rational reasoning relies on the sentences of a language from which further valid sentences of this language are derived. The language need not be a natural language but rather can be any language that has a formal grammar, such as propositional or predicate logic. Thus, external visualizations (e.g., diagrams) are often accepted as aids to foster logical intuitions but not as an independent tool for conducting logical proofs.

The empirical evidence from cognitive research is similarly ambiguous. On the one hand, some experiments show that diagrams can help in logical thinking. Bauer and Johnson-Laird (1993) presented their participants conditional reasoning problems as sentences and additionally clarified the logical connections using different types of diagrams. The diagrams were relatively abstract line-drawings that represented the logical relations between different sets of entities in a spatial way. In another experimental condition, there were no diagrams; the tasks were presented only as sentences. The participants solved the tasks better with diagrams than without. On the other hand, Stenning (2002) has explored in detail the logical role of visual presentation formats in Hyperproof, a tutorial system developed by the logicians Barwise and Etchemendy (1994) to teach students formal logic through language and visualization.

Stenning (2002) found that students with poorer logical abilities use visualizations, whereas students with better logic skills use more nonvisual reasoning strategies. In fact, strong "visualizers" could benefit the most from learning to use abstract strategies instead of visualization. Similar results were reported by Sato, Sugimoto, and Ueda (2017), who explored the usefulness of different grades of concreteness in conditional reasoning. They either presented photos of real objects to their participants, or the objects were presented more abstractly in an annotated virtual environment. Reasoning performance was better with the more abstract objects in the annotated virtual environment. The visual details of the photos seem to distract people from what is actually relevant for performing the inference. Similar results were found in spatial cognition research, where people have to make inferences to understand verbal route descriptions or navigate through photorealistic virtual environments (Meilinger, Knauff, & Bülthoff, 2008).

Overall, the cognitive advantages and disadvantages of external visual aids seem to depend on the iconicity, concreteness, and abstractness of the form of representation. Too many visual details seem to hinder the process of reasoning, whereas more abstract diagrams that aid in understanding the underlying logical structure of the problem can help people to reason more accurately and to grasp the intuition of the proof, that is, to understand why the theorem and even the proof are correct (see chapter 13.1 by Jamnik, this handbook). In any case, the common intuition that it is always helpful for understanding and reasoning to make abstract issues more concrete in pictures or diagrams is certainly wrong. Images, photos, and other highly visual presentation formats can actually lead to the deterioration of epistemic rationality. A possible explanation is that people learn better when pictures are excluded from a learning system rather than included. The reason is that visual overload may result in redundancy, for instance, if texts and images are presented simultaneously, which can hinder learning and reasoning effectiveness (Mayer, 2009; Schnotz & Bannert, 2003). This also agrees with

the *seductive details effect*: textbooks are often enriched with nice-looking pictures or other eye-catching details that are hardly relevant for meaningful learning. But empirical findings show that students display poorer learning outcomes when they learned with this kind of seductive information (Eitel, Bender, & Renkl, 2019; Harp & Mayer, 1998).

3. Epistemology of Visual Mental Imagery

We have seen that the epistemic role of external visualization for human reasoning is complex. But how helpful are *internal* visualizations for logical reasoning? This is even more difficult, as we will now see. The idea of thinking in the mind's eye already existed long before it received attention in psychology. In philosophy, the concept already appears in the oldest writings about cognition—the works of Plato and Aristotle (Nigel, 2018). Later, philosophers like Descartes, Locke, and Hume also developed different thoughts about the structure and function of mental imagery. Although their conceptions were very different, all had the idea that human thought relies on pictorial mental images (Thomas, 1997/2019; Tye, 1991). This idea was also supported by the introspective reports of scientists, for example, the distinguished Austrian physicist Ludwig Boltzmann, who wrote, "I am of the opinion that the task of theory consists in constructing an image of the external world that exists purely internally and must be our guiding star in thought and experiment" (Boltzmann, 1890, p. 33). Similarly, the chemist August Kekulé reported that visualization had suggested to him that the benzene molecule might have a cyclic structure (Kekulé, 1865). Today, some contemporary scholars even allege that vivid imagination was crucial in solving the key problems of chemistry in the 19th century (Rocke, 2010).

In psychology, several academic debates revolved around the role of visual images in human cognition. On the one hand, in 1910, studies by Cheves Perky revealed that people often merge mental images and what is actually seen. Obviously, visualizations can be so similar to real perceptions that they can be mistaken for the latter (Perky, 1910). On the other hand, in particular the "Würzburg School" argued that thinking is possible without imagination. These researchers conducted several experiments in which they asked the participants what had happened in their mind when solving different kinds of reasoning problems. Not one of them reported experiencing visual images. Karl Bühler, one of the main figures of this school, concluded that thinking is possible without seeing in the mind's eye. Human thinking, he argued, often is "pure thinking," which is independent from concrete imagination or visualization (Bühler, 1909). Other authors have questioned this approach and had no doubt that thinking calls for visual imagination (e.g., Titchener, 1909).

Later, psychologists avoided the concept of visual mental imagery, given the hostile criticism it had received from behaviorists (Watson, 1913). In his famous "manifesto," Watson (1913) argued that the controversy about visual mental images was a prime example of the maladies of psychology for which behaviorism would be the therapy. For Watson, the speculations about mental images in the mind (as all other mentalistic concepts) were only "old wives' tales" and nothing more than the sentimental "dramatizing" of verbally mediated memories (Watson, 1928).

This radically changed with the rise of cognitive psychology, which put mental representations and processes center stage. One of the most striking arguments for the role of visual mental images in human cognition was the seminal experiment by Kosslyn, Ball, and Reiser (1978), who asked people to memorize and visualize the map of a fictitious island. In a later memory test, the time to mentally move from one object to another was proportional to the distance between the objects on the map. Such results were taken as evidence that visual mental images are a distinct kind of mental representation, which is similar to representations resulting from visual perception (e.g., Pearson, Naselaris, Holmes, & Kosslyn, 2015).

Other researchers are, on the whole, dismissive of the role of visualization in human thought. Largely in parallel to the arguments of logicians against external diagrams for logical reasoning, these scholars argue that mental reasoning is solely based on propositional representations as a language of thought. Visual imagery is a mere epiphenomenon that does not play a causal role in mental processes (e.g., Adler & Rips, 2008; Pylyshyn, 2002; Rips, 1994). Under the label "mental imagery debate," these controversies addressed some of the most fundamental questions at the frontiers of the cognitive sciences. They are still not resolved. In the following, the empirical results from the area of logical reasoning are reported.

3.1 Results from Cognitive Psychology

A pioneering study on the role of visual images in logical reasoning was carried out by De Soto, London, and Handel (1965), who argued that reasoners visualize the content of reasoning problems in a mental image and then "read off" the conclusion by inspecting this visual image. Following this idea, several authors assumed that if reasoning relies on visual mental images, then reasoning with problems that are easy to visualize should be easier and accompanied by better performance than reasoning with problems that are hard to visualize. Yet, until recently, the evidence was equivocal. Some researchers found that the ease of visualization helps people to reason accurately, whereas others did not find such an effect. A detailed overview of these ambiguous findings is given in Knauff (2013).

Research from our own laboratory helps to resolve the inconsistency in the previous results and draws an opposite picture to what the mental imagery account suggests: visual mental images can actually impede reasoning and hinder the process of inference. If the content of a reasoning process is easy to imagine visually, people need more time and make more errors than with less visual problems (Knauff, 2013). In Knauff and Johnson-Laird (2002), we empirically identified four sorts of reasoning problems: (1) *visuospatial problems*, which are easy to envisage visually and spatially; (2) *visual problems*, which are easy to envisage visually but hard to envisage spatially; (3) *spatial problems*, which are hard to envisage visually but easy to envisage spatially; and (4) *abstract problems*, which are hard to envisage either visually or spatially. We then asked our participants to solve reasoning problems of these four sorts and measured the error rates and how much time they needed to solve the problem. A theory based on visual mental images would predict that visual (and probably visuospatial) problems are easier to solve. Our prediction, however, was that problems that elicit visual images containing details that are irrelevant to an inference should impede the process of reasoning. Our findings supported this prediction: in several experiments, we found that the ease of visualization impaired reasoning: participants were significantly slower with these problems than with the other sorts of problems. In other experiments, we also found that visualization results in lower reasoning accuracy: people make more errors with visual problems than with the other sorts of problems (Knauff, 2009; Knauff & May, 2006). This is called the *visual impedance effect* (Knauff, 2013; Knauff & Johnson-Laird, 2002).

The visual impedance effect was corroborated in research from several other groups. For instance, studies have shown that one of the main causes for dyslexia (a reading deficit) is the strong tendency of dyslexics to represent the information from a text in a visual, rather than a verbal, way. In a series of experiments, Bacon and coworkers could show that dyslexics are also handicapped in logical reasoning with highly visual materials. The authors explain this with the visual impedance effect (Bacon & Handley, 2010; Bacon, Handley, & McDonald, 2007). Panagiotidou, Serrano, and Moreno-Rios (2018) asked adult participants with reading deficits to solve reasoning problems, which were represented either visually or verbally. They also found a strong visual impedance effect and argue that the visual materials help to study the reasoning abilities of people with and without reading deficits. Tse, Ragni, and Lösch (2017) found that the forming of excessive visual images induced by the premises can impede reasoning and that this interacts with the number of relations used in the problem. Gazzo Castañeda and Knauff (2013) used the well-known Verbalizer–Visualizer Questionnaire (VVQ; Richardson, 1977) to identify two groups of individuals with different cognitive styles. One group consisted of "visualizers," with a strong tendency toward reporting visualization during thinking, and the second group consisted of "verbalizers," with a strong bias toward reporting verbalization during thinking. We found that visualizers showed a strong visual impedance effect with the visual problems but not with nonvisual problems. This effect was stronger than in the nonvisualizers. In Mathias, Vogel, and Knauff (2019), we found similar effects of visual thinking styles in the learning of minimally invasive surgery.

3.2   Results from Cognitive Neuroscience

Several studies showed that areas of the primary and secondary visual cortex are active during visual mental imagery. This supports the assumption that visual perception and visual imagination also overlap on the cortical level (Kosslyn, Ganis, & Thompson; 2001; Kosslyn & Thompson, 2003). Other studies explored the neural correlates of human reasoning. A core result is that reasoning problems activate a complex and widespread network of cortical areas, covering regions in the frontal, temporal, and parieto-occipital lobes (Goel, 2007; Knauff, 2009; Prado, Chadha, & Booth, 2011).

Our group was the first that studied the connection between reasoning, visual mental imagery, and the visual cortices. In one functional magnetic resonance imaging (fMRI) study, we used the same materials as Knauff and Johnson-Laird (2002). While lying in the scanner, participants saw the visual, the visuospatial, the spatial, and the abstract problems and had to evaluate conclusions by mouse clicks (Knauff, Fangmeier, Ruff, & Johnson-Laird, 2003). Two results were important: first, during the solving of all problems, we found activity in areas of the parietal cortex. This indicates that people solved all kinds of problems by constructing and inspecting

spatially organized mental representations, which are more abstract than visual images (Knauff, 2013). However, the second result was that only the visual problems resulted in additional activity in the early visual cortex. These areas are related to visual mental imagery in the literal sense, because they are the only regions of the brain that are retinotopically organized, that is, a stimulation of the retina is (almost) analogically represented in these brain areas. Obviously, these areas can be activated not only by visual perception but also by the process of thought. Our interpretation is that, first, the visual problems activate visual mental images in these visual areas but, second, as we found before, these mental images can impede the process of reasoning.

In another study, we used transcranial magnetic stimulation (TMS; Hamburger et al., 2018). TMS uses strong magnetic fields to hinder the neural processing in a well-defined cerebral area and to study how this suppression influences participants' performance in cognitive tasks. In the experiment, we applied the magnetic field to the primary visual cortex. Our assumption was that this should hinder participants' ability to create visual mental images in the visual cortex, and this in turn should help them to reason better than with the visual images they would spontaneously construct, if the activity in these areas were not suppressed. That is exactly what we found. When people could not visualize the content of the reason problem, they reasoned better—which is exactly the opposite of what the visualization theory would assume. In another TMS study, we applied the TMS signal to the parietal cortex and found that this hinders people's reasoning performance. They reasoned less accurately when they could not construct spatial mental models in the parietal cortex (Ragni, Franzmeier, Maier, & Knauff, 2016).

We also used patient studies to explore the visual impedance effect. In one of the studies, we asked congenitally totally blind people to solve our reasoning problems. People who are blind from birth do not experience visual mental images. However, they are as good as sighted people in constructing and utilizing spatial representations. Again, the participants solved visual, visuospatial, spatial, and abstract problems. Interestingly, the visual impedance effect disappeared in those participants. Obviously, congenitally totally blind people are immune to the visual impedance effect, as they do not construct impeding visual mental images (Knauff & May, 2006).

To more precisely study the relation between visualization and reasoning processes, we conducted a further fMRI study (Fangmeier & Knauff, 2009; Fangmeier, Knauff, Ruff, & Sloutsky, 2006). This was the first experiment that was conducted in a way that allowed us to distinguish between different phases of a reasoning process. We distinguished between three different phases: first, people have to understand the premises. Second, they have to integrate the information from the premises into one unified mental representation. Third, they are confronted with a conclusion and have to say whether or not this conclusion logically follows from the given premises. As shown in figure 13.3.2, we obtained different patterns of neural activity for all three phases. In the first phase, we found activity in the primary visual cortex, although all activities related to the purely visual presentation of premises were excluded. In the second phase, this activity was still visible, together with some other activities that were related to executive control, and so forth. In the third phase, which is the actual core of reasoning, something interesting happened: now the activity in the visual cortex disappeared, but another activity in a large cluster in the parietal cortex arose. We argued that this pattern of neural activation explains why people experience visual mental images during reasoning. During the reading and processing of the premises, such visual mental images are present in the visual cortex. This might then help people to maintain the information active in working memory. Hence, people have the feeling that they can see the reasoning problem in their inner eye. However, during the actual reasoning, this activity was not present anymore. Instead, cortical areas were active that are related to spatial processing from different modalities. Typically, people do not become aware of such spatial representations, because they do not have conscious access to

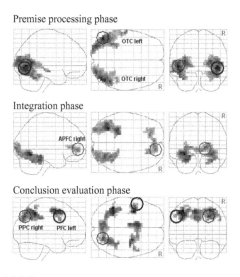

Figure 13.3.2
Neural activation in the three phases of reasoning (from Fangmeier et al., 2006).

representations and processes in the parietal cortex (for details, see Knauff, 2013).

3.3 Results from Computational Cognitive Modeling
The idea that reasoning is linked to visualization also aroused researchers' interest in developing computer programs of imagery-based inference. A classic example is the WHISPER system, which could observe and manipulate diagrams to solve problems in the blocks-world environment (Funt, 1980). Kosslyn and Shwartz (1977) programmed a system in which the locations of objects were represented metrically in a matrix of filled and unfilled cells, simulating the use of visual images. Today, many different imagery-based approaches in artificial intelligence exist, notably in the area of diagrammatic reasoning, where researchers seek to use visual images to solve complex problems. A good overview is given in chapter 13.1 by Jamnik (this handbook).

Some computer scientists have developed systems for logical reasoning with visual mental images and with more abstract spatial representations. These computer simulations are helpful in exploring whether visual mental images or spatial representations are more useful on the algorithmic level of human reasoning. One such system was developed by Berendt (1996), who tried to simulate results from our research group. In these experiments, we showed that in the case of ambiguous sets of premises, people prefer one model but ignore other interpretations. The *preferred model* is the one that is easier to construct and manipulate in working memory (Knauff, 1999; Knauff & Ragni, 2011; Knauff, Rauh, & Schlieder, 1995; Rauh et al., 2005). Berendt could show that a system that uses visual mental images is able to reconstruct some of these preferences (Berendt, 1996). However, Schlieder (1999) developed another computational model that just used the order of starting points and endpoints of the object's edges. This means that this representation is spatial in nature—it just needs ordering information—and it is more parsimonious than Berendt's visual system, which uses excessive visual details to perform the inference. Nevertheless, the abstract spatial model was better able to reconstruct the empirical data than the visual model (Schlieder & Berendt, 1998).

In Ragni and Knauff (2013), we developed a computational theory and implemented computer program for reasoning with spatial mental models. The PRISM system can reconstruct many empirical findings from the literature on human reasoning. The basic idea is that working memory is conceptualized as a spatial array of filled and unfilled cells. A spatial scanning mechanism inspects this array to find new information, which is not explicitly given in the premises. In this way, the system can draw logically valid conclusions without representing visual details. It can also reconstruct why people sometimes err and draw inferences that are invalid from a logical point of view. Based on this work, two approaches tried to simulate the visual impedance effect in computational systems (Albrecht, Schultheis, & Fu, 2015; Boeddinghaus, Ragni, Knauff, & Nebel, 2006). The approaches are an interesting starting point, although still much work has to be done to understand the algorithmic details of the visual impedance effect.

4. Corollaries and Future Work

The goal of this chapter was to develop a better empirical understanding of the epistemic role of visualization in human rationality. This is important, as the field is full of speculations from folk psychology and misleading subjective experiences. Sometimes these may even bias scientists toward certain theories. So, what can we really learn from this chapter?

The first corollary is that introspection can be fatally misguiding. This is by no means new. It has long been known that people do not have conscious access to most of their cognitive processes.

The second corollary is that the visual impedance effect provides a new view on the ongoing debates on the epistemic role of visualization in logical reasoning and rationality. Most of these controversies have taken place in philosophy, mathematics, and other areas addressing the *normative function* of visualization. Instead, the present chapter is concerned with *actual* human reasoning. As such, it is part of the goal to develop a comprehensive *descriptive* theory of human thinking. Such a theory also accounts for errors, that is, deviations from what we normatively would call rational.

The third corollary is that visual mental imagery in reasoning is not a mere epiphenomenon: it has a causal power. But this effect is not in the direction that the orthodox idea of visual imagery suggests. Visualization does not help people to reason; actually, it can hinder human reasoning. The reason is that, in the "foreground" of conscious experience, people seem indeed to envisage the content of a reasoning problem in a visual image, at least if the situation is easily visualizable. Since the visual cortex is linked to visual experience, these mental images might be pushed into consciousness. However, in the "background" of conscious experience, the actual logical work is based on more abstract mental representations and processes. Hence, premises that

almost automatically evoke vivid visual images initiate a "detour" via the visual cortex, which requires additional time but is useless for the actual reasoning process that works on these abstract representations. This detour may cause the visual impedance effect and may also cause the low utility of external visualizations in the form of pictures or other highly visual illustrations (Knauff, 2013). This largely resembles the normative view of most logicians and philosophers that reasoning should be independent of visualization in order to be rational.

The fourth corollary concerns the impact of the current results for education in logical reasoning. This chapter does not underestimate the power of mental visualization in training and education. We know from many different areas that mental imagery can accelerate learning and leads to better performance in different tasks (Nigel, 2018). The point in this chapter is, however, that the power of visualization depends on many different factors. In particular, very concrete, vivid mental imagery must be distinguished from more abstract spatial representations. Although many people are confident that the former is important for human cognition, it might actually be the case that the facilitating effects of imagination are mostly caused by more abstract—albeit not consciously experienced—spatial representations and processes.

The fifth corollary is that the relation between external and internal visualizations is complex. Sometimes the former can trigger the latter. And this means that it is important how much visual detail is given in external visualizations such as diagrams, figures, pictures, and so forth. Too many seductive visual details in external visualizations may trigger the construction of visual mental images, which can impede cognition. But if external visualizations really focus on the information relevant for the task, they can support thinking and learning (see chapter 13.1 by Jamnik, this handbook). Teachers and pedagogues should carefully consider this important trade-off between eye-catching, nice-looking pictures and meaningful learning.

The sixth corollary from this chapter is that still much work has to be done in the field. On the one hand, logical reasoning is a very particular domain of human cognition. Hence, we probably cannot simply transfer the corollaries from this chapter to all kinds of problems. On the other hand, however, the reported results challenge the positive image that visualization has in science and folk psychology. I hope that this result will trigger further research that goes beyond the area of logical reasoning and human rationality.

References

Adler, J. E., & Rips, L. J. (Eds.). (2008). *Reasoning: Studies of human inference and its foundations.* Cambridge, England: Cambridge University Press.

Albrecht, R., Schultheis, H., & Fu, W. T. (2015). Visuo-spatial memory processing and the visual impedance effect. In D. C. Noelle, R. Dale, A. S. Warlaumont, J. Yoshimi, T. Matlock, C. D. Jennings, & P. P. Maglio (Eds.), *Proceedings of the 37th Annual Conference of the Cognitive Science Society* (pp. 72–77). Austin, TX: Cognitive Science Society.

Arana, A. (2016). Imagination in mathematics. In A. Kind (Ed.), *Routledge handbook on the philosophy of imagination* (pp. 463–477). London, England: Routledge.

Bacon, A. M., & Handley, S. J. (2010). Dyslexia and reasoning: The importance of visual processes. *British Journal of Psychology, 101,* 433–452.

Bacon, A. M., Handley, S. J., & McDonald, E. L. (2007). Reasoning and dyslexia: A spatial strategy may impede reasoning with visually rich information. *British Journal of Psychology, 98*(1), 79–92.

Barwise, J., & Etchemendy, J. (1994). *Hyperproof.* Cambridge, England: Cambridge University Press.

Bauer, M. I., & Johnson-Laird, P. N. (1993). How diagrams can improve reasoning. *Psychological Science, 4*(6), 372–378.

Berendt, B. (1996). Explaining preferred mental models in Allen inferences with a metrical model of imagery. In G. W. Cottrell (Ed.), *Proceedings of the Eighteenth Annual Conference of the Cognitive Science Society* (pp. 489–494). Mahwah, NJ: Erlbaum.

Boeddinghaus, J., Ragni, M., Knauff, M., & Nebel, B. (2006). Simulating spatial reasoning using ACT-R. In D. Fum, F. Del Missier, & A. Stocco (Eds.), *Proceedings of the Seventh International Conference on Cognitive Modeling* (pp. 62–67). Mahwah, NJ: Erlbaum.

Boltzmann, L. (1890). Über die Bedeutung von Theorien [On the significance of theories]. In L. Boltzmann, *Populäre Schriften* [Popular writings] (pp. 76–82). Leipzig, Germany: Barth.

Bühler, K. (1909). Zur Kritik der Denkexperimente [A critique of reasoning experiments]. *Zeitschrift für Psychologie, 51,* 108–118.

De Soto, C. B., London, M., & Handel, S. (1965). Social reasoning and spatial paralogic. *Journal of Personality and Social Psychology, 2,* 513–521.

Eitel, A., Bender, L., & Renkl, A. (2019). Are seductive details seductive only when you think they are relevant? An experimental test of the moderating role of perceived relevance. *Applied Cognitive Psychology, 33,* 20–30.

Fangmeier, T., & Knauff, M. (2009). Neural correlates of acoustic reasoning. *Brain Research, 1249,* 181–190.

Fangmeier, T., Knauff, M., Ruff, C., & Sloutsky, V. (2006). fMRI evidence for a three-stage model of deductive reasoning. *Journal of Cognitive Neuroscience, 18*(3), 320–334.

Funt, B. V. (1980). Problem-solving with diagrammatic representations. *Artificial Intelligence, 13,* 201–230.

Gazzo Castañeda, L. E., & Knauff, M. (2013). Individual differences, imagery and the visual impedance effect. In M. Knauff, M. Pauen, N. Sebanz, & I. Wachsmuth (Eds.), *Proceedings of the 35th Annual Conference of the Cognitive Science Society* (pp. 2374–2379). Austin, TX: Cognitive Science Society.

Giaquinto, M. (2016). The epistemology of visual thinking in mathematics. In E. N. Zalta (Ed.), *The Stanford encyclopedia of philosophy.* Retrieved from https://plato.stanford.edu/archives/win2016/entries/epistemology-visual-thinking/

Goel, V. (2007). Anatomy of deductive reasoning. *Trends in Cognitive Science, 11,* 435–441.

Grandin, T. (2006). *Thinking in pictures, and other reports from my life with autism* (Expanded ed.). New York, NY: Vintage Books. (Original work published 1995)

Hamburger, K., Ragni, M., Karimpur, H., Franzmeier, I., Wedell, F., & Knauff, M. (2018). TMS applied to V1 can facilitate reasoning. *Experimental Brain Research, 236,* 2277–2286.

Harp, S. F., & Mayer, R. E. (1998). How seductive details do their damage: A theory of cognitive interest in science learning. *Journal of Educational Psychology, 90*(3), 414–434.

Hilbert, D. (1899). *Grundlagen der Geometrie* [Foundations of geometry]. Leipzig, Germany: Teubner.

Kadunz, G., & Yerushalmy, M. (2015). Visualization in the teaching and learning of mathematics. In S. Cho (Ed.), *Proceedings of the 12th International Congress on Mathematical Education* (pp. 463–467). Berlin, Germany: Springer.

Kekulé, A. (1865). Sur la constitution des substances aromatiques [On the constitution of aromatic substances]. *Bulletin de la Société Chimique de Paris, 3*(2), 98–110.

Klein, F. (1979). *Development of mathematics in the 19th century.* Brookline, MA: Mathematical Science Press.

Knauff, M. (1999). The cognitive adequacy of Allen's interval calculus for qualitative spatial representation and reasoning. *Spatial Cognition and Computation, 1,* 261–290.

Knauff, M. (2009). A neuro-cognitive theory of deductive relational reasoning with mental models and visual images. *Spatial Cognition and Computation, 9*(2), 109–137.

Knauff, M. (2013). *Space to reason: A spatial theory of human thought.* Cambridge, MA: MIT Press.

Knauff, M., Fangmeier, T., Ruff, C. C., & Johnson-Laird, P. N. (2003). Reasoning, models, and images: Behavioral measures and cortical activity. *Journal of Cognitive Neuroscience, 15,* 559–573.

Knauff, M., & Johnson-Laird, P. N. (2002). Visual imagery can impede reasoning. *Memory & Cognition, 30,* 363–371.

Knauff, M., & May, E. (2006). Mental imagery, reasoning, and blindness. *Quarterly Journal of Experimental Psychology, 59*(1), 161–177.

Knauff, M., & Ragni, M. (2011). Cross-cultural preferences in spatial reasoning. *Journal of Cognition and Culture, 11,* 1–21.

Knauff, M., Rauh, R., & Schlieder, C. (1995). Preferred mental models in qualitative spatial reasoning: A cognitive assessment of Allen's calculus. In J. D. Moore & J. F. Lehman (Eds.), *Proceedings of the Seventeenth Annual Conference of the Cognitive Science Society* (pp. 200–205). Mahwah, NJ: Erlbaum.

Kosslyn, S. M. (1980). *Image and mind.* Cambridge, MA: Harvard University Press.

Kosslyn, S. M. (1994). *Image and brain.* Cambridge, MA: MIT Press.

Kosslyn, S. M., Ball, T. M., & Reiser, B. J. (1978). Visual images preserve metric spatial information: Evidence from studies of image scanning. *Journal of Experimental Psychology: Human Perception and Performance, 4,* 47–60.

Kosslyn, S. M., Ganis, G., & Thompson, W. L. (2001). Neural foundations of imagery. *Nature Reviews Neuroscience, 2,* 635–642.

Kosslyn, S. M., & Shwartz, S. P. (1977). A simulation of visual imagery. *Cognitive Science, 1,* 265–295.

Kosslyn, S. M., & Thompson, W. L. (2003). When is early visual cortex activated during visual mental imagery? *Psychological Bulletin, 129*(5), 723–746.

Krämer, S. (2016). *Figuration, Anschauung, Erkenntnis: Grundlinien einer Diagrammatologie* [Figuration, imagination, cognition: Foundations of diagrammatology]. Berlin, Germany: Suhrkamp.

Kunda, M., & Goel, A. K. (2011). Thinking in pictures as a cognitive account of autism. *Journal of Autism and Developmental Disorders, 41,* 1157–1177.

Mancosu, P. (2005). Visualization in logic and mathematics. In P. Mancosu, K. F. Jørgensen, & S. A. Pedersen (Eds.), *Visualization, explanation and reasoning styles in mathematics* (pp. 13–30). Berlin, Germany: Springer.

Mathias, A. P., Vogel, P., & Knauff, M. (2019). Different cognitive styles can affect the performance in minimally invasive surgery skill training. *Surgical Endoscopy, 34*(11), 4866–4873.

Mayer, R. E. (2009). *Multimedia learning* (2nd ed.). New York, NY: Cambridge University Press.

Meilinger, T., Knauff, M., & Bülthoff, H. H. (2008). Working memory in wayfinding—A dual task experiment in a virtual city. *Cognitive Science, 32,* 755–770.

Panagiotidou, E., Serrano, F., & Moreno-Rios, S. (2018). Reasoning and reading in adults: A new reasoning task for detecting the visual impendance [*sic*] effect. *Advances in Cognitive Psychology, 14*(4), 150–159.

Pearson, J., Naselaris, T., Holmes, E. A., & Kosslyn, S. M. (2015). Mental imagery: Functional mechanisms and clinical applications. *Trends in Cognitive Sciences, 19*(10), 590–602.

Perky, C. W. (1910). An experimental study of imagination. *Journal of Psychology, 21,* 422–452.

Pompéia, S., Manzano, G. M., Pradella-Hallinan, M., & Bueno, O. F. A. (2007). Effects of lorazepam on deductive reasoning. *Psychopharmacology, 194*(4), 527–536.

Prado, J., Chadha, A., & Booth, J. R. (2011). The brain network for deductive reasoning: A quantitative meta-analysis of 28 neuroimaging studies. *Journal of Cognitive Neuroscience, 11,* 3483–3497.

Pylyshyn, Z. W. (2002). Mental imagery: In search of a theory. *Behavioral and Brain Sciences, 25,* 157–238.

Ragni, M., Franzmeier, I., Maier, S., & Knauff, M. (2016). Uncertain relational reasoning in the parietal cortex. *Brain and Cognition, 104,* 72–81.

Ragni, M., & Knauff, M. (2013). A theory and a computational model of spatial reasoning with preferred mental models. *Psychological Review, 120,* 561–588.

Rauh, R., Hagen, C., Knauff, M., Kuss, T., Schlieder, C., & Strube, G. (2005). From preferred to alternative mental models in spatial reasoning. *Spatial Cognition and Computation, 5,* 239–269.

Richardson, A. (1977). Verbalizer–visualizer: A cognitive style dimension. *Journal of Mental Imagery, 1,* 109–125.

Rips, L. J. (1994). *The psychology of proof: Deductive reasoning in human thinking.* Cambridge, MA: MIT Press.

Rocke, A. J. (2010). *Image and reality: Kekulé, Kopp, and the scientific imagination.* Chicago, IL: University of Chicago Press.

Ryle, G. (1949). Meaning and necessity. *Philosophy, 24*(88), 69–76.

Sato, Y., Sugimoto, Y., & Ueda, K. (2017). Real objects can impede conditional reasoning but augmented objects do not. *Cognitive Science, 42*(2), 691–707.

Schlieder, C. (1999). The construction of preferred mental models in reasoning with interval relations. In G. Rickheit & C. Habel (Eds.), *Mental models in discourse processing and reasoning* (pp. 333–357). Amsterdam, Netherlands: Elsevier.

Schlieder, C., & Berendt, B. (1998). Mental model construction in spatial reasoning: A comparison of two computational theories. In U. Schmid, J. F. Krems, & F. Wysotzki (Eds.), *Mind modelling: A cognitive science approach to reasoning, learning, and discovery* (pp. 133–162). Lengerich, Germany: Pabst Science Publishers.

Shin, S.-J. (1994). *The logical status of diagrams.* Cambridge, England: Cambridge University Press.

Schnotz, W., & Bannert, M. (2003). Construction and interference in learning from multiple representation. *Learning and Instruction, 13*(2), 141–156.

Stenning, K. (2002). *Seeing reason: Image and language in learning to think.* New York, NY: Oxford University Press.

Thomas, N. J. T. (2019). Mental imagery. In E. N. Zalta (Ed.), *The Stanford encyclopedia of philosophy.* Retrieved from http://plato.stanford.edu/archives/sum2019/entries/mental-imagery/

Titchener, E. B. (1909). *Lectures on the experimental psychology of the thought processes.* New York, NY: Macmillan.

Tse, A. P. P., Ragni, M., & Lösch, J. (2017). Beyond the visual impedance effect. In M. K. van Vugt, A. P. Banks, & W. G. Kennedy (Eds.), *Proceedings of the 15th International Conference on Cognitive Modeling* (pp. 115–120). Coventry, England: University of Warwick.

Tye, M. (1991). *The imagery debate.* Cambridge, MA: MIT Press.

Watson, J. B. (1913). Psychology as the behaviorist views it. *Psychological Review, 20,* 158–177.

Watson, J. B. (1928). *The ways of behaviorism.* New York, NY: Harper & Brothers.

Wittgenstein, L. (2001). *Philosophical investigations/Philosophische Untersuchungen* (3rd ed.). Oxford, England: Blackwell. (Original work published 1953)

Section 14   Scientific Rationality

14.1 Scientific Rationality and Objectivity

Line Edslev Andersen and Hanne Andersen

Summary

This chapter provides an overview of the accounts of scientific rationality and objectivity that have been offered by philosophers of science during the 20th century. We begin by presenting different accounts of how *individual* scientists should act to be rational and objective. In the latter half of the 20th century, some philosophers argued that the rationality and objectivity of science are obtained at the level of communities of scientists rather than at the level of individual scientists. Hence, we also present accounts of how *communities* of scientists can be rational and objective. Finally, we illustrate how the philosophical discussions about scientific objectivity may be relevant for scientific practice.

1. Models of Individual Rationality and Objectivity

This chapter provides a brief overview of the various accounts of scientific rationality and objectivity that have been offered by philosophers of science during the 20th century. We start with models of individual rationality and objectivity, which dominated philosophy of science until the mid-20th century. We then move to models of group rationality, which became popular during the second half of the 20th century. We also examine recent models of group objectivity, focusing on feminist approaches to objectivity in science. Finally, we illustrate how the philosophical discussions about objectivity may be relevant for scientific practice. In line with most English-language accounts of rationality and objectivity in science, "science" is used to refer to the natural sciences.

Science is often considered the epitome of rationality and objectivity. The view that science is *objective* can be understood in two different ways. It can be understood either in the way that scientific *theories* faithfully describe facts about the world or in the way that the *methods* employed in science are independent of social and personal values and biases (see Reiss &

Sprenger, 2014/2017). In this section, we focus on models of rationality and objectivity from the first half of the 20th century, and at that time, science was generally seen as objective in both these ways. The methods employed by scientists were usually seen to be objective and to lead to objective theories. In fact, insofar as scientists acted *rationally*, they were usually considered to adhere to objective methods that led to objective theories. Thus, the central question was what it means for scientists to act rationally.

In the first half of the 20th century, many philosophers of science linked questions about scientific rationality to the question of how individual scientists validate their scientific beliefs. This approach was based on the assumption that there is a clear distinction between discovery and justification in science.[1] On this view, how scientists *form* new scientific hypotheses is seen as an empirical question to be relegated to disciplines such as psychology, sociology, and history of science, while in contrast, the question of how scientists should *evaluate* scientific hypotheses and make choices between competing theories is seen as a normative question, to be answered by philosophy. The standard view was that science could be characterized by its special method that specified how scientific theories could be assessed. On this view, scientists act rationally by adhering to this scientific method. Consequently, an important philosophical question was how to define this method.

Some of the dominant accounts of the unique scientific method in the 20th century have been Hempel's (1966) hypothetico-deductive model, Popper's (1963/1972, 1934/2014) falsificationist model, and Bayesianism (we will not examine Bayesianism in the present chapter but refer the reader to chapter 4.1 by Hájek & Staffel, this handbook).

Hempel's hypothetico-deductive, or H-D, model presents a logic of confirmation. On this model, scientists assess a hypothesis by first deducing from it the observable events that would take place if the hypothesis were true and then conducting experiments to observe whether the predicted events occur. If the predicted events occur, this provides supporting evidence for the hypothesis, where the degree of support depends on the quantity, variety, and precision of the supporting evidence. In contrast, if the predicted events do not occur, scientists should reject the hypothesis in question.

In contrast to the H-D model's logic of confirmation, Popper's falsificationist model presents a logic of refutation. On the falsificationist model, like on the H-D model, scientists start by deducing from the hypothesis under test the observable events that would occur if the hypothesis were true. However, Popper stressed that, regardless of the amount of supporting evidence for the hypothesis, the hypothesis can never be confirmed with certainty. By contrast, it follows logically that the hypothesis is false if it has false implications: if events can be deduced from it that do not in fact occur. This led Popper to suggest that, for scientists to act rationally, they must proceed by trying to *falsify* hypotheses. Thus, Popper focused on the potential falsifiers of a hypothesis and applauded scientists' sincere attempts at falsification. On this view, the more a hypothesis claims, the better it is, in the sense that it is more open to falsification. If a hypothesis has been subjected to numerous tests and still has not been falsified, this does not justify scientists in accepting the hypothesis. It only means that the hypothesis should be refined to be more open to falsification and be made subject to even more stringent tests. On this view, the scientific method can be described as a method of conjectures and refutations.

Both the H-D model and falsificationism have had a huge influence on the perception of science and scientific rationality. In general, the view that there is a unique scientific method, to which all rationally behaving scientists adhere, has been strong not only in philosophy but also in other fields. For example, in science education, the scientific method is often presented as a fixed procedure from the description of a phenomenon, over the formulation of a hypothesis that explains the phenomenon, to the conduct of an experiment that tests the hypothesis (Blachowicz, 2009; see also Bauer, 1992, for a historical overview).[2] The literature on scientific method is vast, but overviews can be found, for example, in H. Andersen and Hepburn (2015/2016), Laudan (1968), and Losee (2004, 2005).

We end this section by discussing some challenges to the presented models of individual rationality.

First, it has been argued that the rational acceptance or rejection of a hypothesis is not solely based on experiments but is to be seen as a value-based decision between different courses of action, where the values may be epistemic or nonepistemic, namely, social, ethical, and political values. An aspect of this debate is therefore also whether to understand rationality in *epistemic* terms as a question of which theory or hypothesis to accept, given the evidence, or if rationality can also be understood in *instrumental* terms as a question of which actions to perform, given specific beliefs and ends. Among philosophers of science, Churchman (1948) and Rudner (1953) have argued that an analysis of the methods of scientific inference includes both an assessment of the strength of the evidence in favor of the hypothesis and an assessment of the consequences of making a mistake in accepting the hypothesis. A similar debate has been seen within statistics. Here, Fisher (1955) has argued that a theory of inductive inference should provide a numerical expression of disbelief in a hypothesis, while Neyman (1956) and Pearson (1955) have argued that the consequence of error also needs to be included (for details on this debate, see, e.g., Howie, 2002; Lenhard, 2006; Marks, 1997). In recent decades, these issues have been pursued by discussing whether there is a distinction between rational and irrational influences on theory choice and, more specifically, between the roles played by epistemic and nonepistemic values (see, e.g., Douglas, 2009; Machamer & Wolters, 2004). For further reading on this topic, the reader is referred to chapter 14.2 by Bueter (this handbook) on the value-freedom of science.

Second, it has been argued that there is no single rational method for scientists to follow, because scientific norms and aims differ over time and across disciplines. Hence, it has been argued, it is not possible to provide a universal and ahistorical account of the rational method of scientific inquiry. During the 1960s and 1970s, historically inclined philosophers of science such as Feyerabend, Hanson, Kuhn, Laudan, Shapere, and Toulmin argued from studies of the history of science that scientific norms and aims may differ across disciplines or over time.[3] The most radical position was advanced by Feyerabend, who in a monograph with the provocative title *Against Method* (1975/1993) argued for a methodological anarchism on which any methodological rule can be violated fruitfully in some contexts.

Finally, some, such as Hanson (1958), argued that observations are often theory-laden (i.e., they depend on theoretical assumptions). The relation between observation and theory is therefore more complex than assumed both by the accounts of the rationality of science examined above and by the view that science is objective in the sense that scientific theories faithfully describe facts about the world. Like the previous paragraph, this provides a reason for believing that it is not possible to provide a universal and ahistorical account of the rationality of science.

2. Models of Group Rationality

In this section, we consider models of rationality that can accommodate the challenges to the traditional models of rationality just presented. As opposed to the traditional models, these are models of group rationality. In the next section, we consider models of group objectivity. During the second half of the 20th century, philosophers of science started questioning whether the rationality of science should be addressed from the level of the individual scientist and not rather from the level of the scientific community.

Especially the work of Thomas S. Kuhn was foundational for this discussion. Kuhn developed an account of the scientific enterprise on which individual scientists work on a set of received beliefs that they do not question. An individual scientist will sometimes replace this set of beliefs with another set of beliefs, and the time at which she does so is partially determined by personal preferences. Given the traditional conception of scientific rationality and objectivity introduced in the previous section, this makes science seem irrational and subjective. However, on Kuhn's model, science *is* a rational and objective enterprise, but it is the scientific community that serves as the agent responsible for the rationality and objectivity of science.

Kuhn is well known for the phase model of the development of science that he advanced in his monograph *The Structure of Scientific Revolutions* (1962/2012). On this model, science develops through successive periods of cumulative "normal science" separated by noncumulative "revolutions." Normal science is dependent on some set of received beliefs, a "paradigm," which marks out what the acceptable research problems are and what acceptable approaches and solutions to these problems must look like. Yet, some of the scientific problems defined by a paradigm may turn out recalcitrant. If such anomalies cannot be resolved, they may cause a crisis in the scientific community, and this crisis may eventually lead to a revolution, after which the community will have adopted a new paradigm. Such a revolution is not a cumulative process: the new paradigm is not just an extension of the old; on the contrary, the previous paradigm may have precluded solutions of the sort provided by the new. Thus, for Kuhn, competing paradigms were incommensurable and could not be compared in any straightforward manner.

This phase model of the development of science has often by itself been taken to imply that science is irrational. To understand why it does not, one must look at the "essential tension," as Kuhn called it, between the tradition-*preserving* activity of normal science and the tradition-*shattering* activity of scientific revolutions. On the one hand, the tradition-preserving phases of normal science are very effective for solving the problems that the existing paradigm defines. When scientists within a community adhere to the same paradigm, science can move faster and penetrate deeper than if they continuously questioned the paradigm. In this respect, normal science increases the efficiency with which scientific problems are solved.

On the other hand, science would not be able to produce fundamental innovations if normal science were the only mode of doing science. A tradition-shattering complement is needed as well. Kuhn introduced a mechanism that could ensure that the change from one paradigm to another happens neither too quickly nor too late. Thus, if there is a distribution of conservative and innovative dispositions among the members of a scientific community, individual scientists may react differently to encountered anomalies, such that only some start considering alternatives to the reigning paradigm while others continue working within the established paradigm.[4]

To explain both how individual scientists can come to divergent decisions and how the scientific community can, in the end, come to a common decision on which theory to prefer, Kuhn suggested that theory choice was guided by a set of epistemic values that the members of the scientific community share, such as accuracy, consistency, scope, simplicity, and fruitfulness. While these values provide the basis for theory choice, they do not determine it: individuals may legitimately interpret and weigh the values differently. The fact that individuals may weigh the values differently matters to theory choice because the values may conflict. For example, the simpler of two competing theories is not necessarily also the most accurate. Further, Kuhn even claimed that not only epistemic values such as accuracy, consistency, scope, simplicity, and fruitfulness may play a role in theory choice; also, personal preferences about, for example, originality, risk taking, aesthetics, or philosophical convictions may influence theory choice (see Hoyningen-Huene, 1993, chapter 4.3.c). As a result, although individual members of the scientific community may employ the same set of shared epistemic values, they can nevertheless still reach different conclusions as to which of the competing theories to pursue.

Hence, Kuhn's model of the development of science explicitly "*requires* a decision process which permits rational men to disagree" (Kuhn, 1977, p. 332). This stands in stark contrast to the accounts of philosophers like Hempel, who, as described above, linked the rationality of science to individual scientists' adherence to a unique method that determines the rational theory choice. On this view, two rational scientists will always choose the same theory. On Kuhn's view, the fact that theory choice is guided by values that differ from scientist to scientist, as well as by shared values that may be interpreted differently by different scientists, contributes to the development of science. This does not mean that science is irrational; rationality just has to be seen from a social rather than from an individual perspective. As a new, alternative theory is developed, it may prove capable of solving encountered anomalies in such a way that it is gradually accepted by more and more scientists until, eventually, a consensus on the new theory is established among members of the scientific community. Dissent continues until all members of the community, based on their individual interpretation and weighing of the shared values, prefer the same paradigm. Hence, "it is the community of specialists rather than its individual members that makes the effective decision" (Kuhn, 1962/2012, p. 199). A new consensus in the scientific community reflects that arguments in favor of the new paradigm have proliferated.

It is not clear where this leaves us in terms of how exactly to conceive of scientific objectivity. Kuhn does not go much further than to suggest that an account of scientific objectivity should be informed by the fact that the criteria of accuracy, consistency, scope, simplicity, and fruitfulness constrain, but do not determine, theory choice:

> Objectivity ought to be analyzable in terms of criteria like accuracy and consistency. If these criteria do not supply all the guidance that we have customarily expected of them, then it may be the *meaning* rather than the limits of objectivity that my argument shows. (Kuhn, 1977, p. 338, emphasis added)

Hence, Kuhn claims to have shown that science is objective in a different sense than traditionally thought, rather than having shown that science is subjective. But while Kuhn develops an account of scientific rationality on which the rationality of science is obtained at the level of scientific communities, he at most hints at an account on which the objectivity of science is likewise obtained at the level of scientific communities. As will be described in the next section, others have developed such accounts of scientific objectivity.

When Kuhn introduced his model, many critics found that it made theory choice seem irrational (see, e.g., Lakatos, 1970; Scheffler, 1967/1982; or Shapere, 1966).

This perception was further amplified when sociologists of science, partially inspired by Kuhn's account, started arguing that scientists' beliefs should be explained in terms of the social conditions that bring them about. This "strong" program of sociology of science claimed to be impartial with respect to rationality and irrationality, meaning that it set out to explain successful as well as unsuccessful knowledge claims in science in the same type of way, with reference to the same set of social causes (Bloor, 1976/1991).

However, other philosophers of science, such as David Hull (1988) and Philip Kitcher (1993), have followed Kuhn in arguing that the social structure of a scientific community contributes to the effective realization of the epistemic goals of science (for a detailed analysis, see also Wray, 2011). Hull (1988) argued that scientists are motivated by the credit they get for contributing true theories. If they would get credit for contributing fraudulent theories, they would cheat. But the structure of reward and punishment in science ensures that scientists will not cheat, because the punishment if they are caught is very severe. On Hull's account, science is so successful because the social structure of science is such that it is in the self-interest of scientists to pursue true theories.

Kitcher (1993) has similarly argued that nonepistemic pressures on individuals, such as desire for fame, can help a scientific community to effectively realize its epistemic goals. It may be tempting for all the scientists working on a given problem to use the same method to pursue the problem: the available method whose intrinsic qualities are highest. But for a scientific community to effectively realize its epistemic goals, it is important that scientists choose *different* methods to pursue a problem. For it may be an unorthodox method that can provide the solution. The nonepistemic pressures can help prevent that all scientists working on the problem use the same method. For example, if a scientist has a desire for fame, it may be better to pick an unorthodox method since there *is* a chance that it will provide the solution, and there are fewer direct competitors in the sense of other scientists pursuing the same strategy. In this way, nonepistemic considerations at the level of individual scientists can contribute to the attainment of the epistemic goals of the scientific community.

On the accounts considered in this section, the agent that is responsible for the rationality of science is the scientific community as a whole, and consequently, the rationality of scientific beliefs should be examined using the tools of social epistemology. For the topic of social epistemology, the reader is referred to chapter 10.1 by Dietrich and Spiekermann (this handbook).

Similarly, philosophers of science such as D'Agostino (2010), De Langhe (2010, 2012, 2014; see also De Langhe & Rubbens, 2015), Weisberg (2013; see also Weisberg & Muldoon, 2009), or Zollman (2007, 2010, 2018; see also Mayo-Wilson, Zollman, & Danks, 2011) have drawn on game theory, economics, and mathematical modeling of social behavior in examining how various forms of heterogeneity in the scientific community can be beneficial or detrimental for the progress of science.

3. Models of Group Objectivity

In recent decades, feminist philosophers of science have examined different ways in which traditional models of the objectivity of science have disadvantaged women and other minority groups (for a review, see Anderson, 2000/2020). In their view, scientific knowledge has a situated and perspectival nature. What is known reflects the perspective of the knower; there is no view from nowhere. This has led some to embrace subjective ways of knowing nature (Irigaray, 1987). Others have argued, similarly to Kuhn, that the objectivity of science should be assessed at the level of scientific communities (Haraway, 1988; Longino, 1989, 1990; Solomon, 2001).

In Helen Longino's view, observational data are evidence for a theory only relative to a context of background assumptions. This is so because the theory says more than the statements describing the data. The theory is, in other words, underdetermined by the data (see Longino, 1990, pp. 38–61, for her defense of a contextualist account of epistemic justification). On Longino's account, the background assumptions are often biased.[5] So she raises the question of how scientific inquiry can be objective if the background assumptions of individual scientists are biased. She argues that the production of scientific knowledge can be said to be objective only if scientific results are subjected to evaluation by a plurality of individuals and formed by the ensuing discussions. Only then can we see how the influence of subjective assumptions and preferences can be blocked or mitigated (Longino, 1989, p. 266).[6]

Like Kuhn, Longino believes that scientific inquiry as practiced *is* generally objective. Longino has pointed out that scientific knowledge is not produced by just collecting the products of individuals but through a process in which those products are critically examined and modified by other scientists. Experiments thus "get repeated with variations by individuals other than their originators, hypotheses and theories are critically examined, restated, and reformulated before becoming an accepted part of the scientific canon" (Longino, 1989, p. 265).

Hence, scientific inquiry is objective because what is ultimately accepted as a piece of knowledge is the result of a process formed by multiple individuals with different perspectives.

At the same time, Longino makes the case that scientific communities should strive for a *higher* degree of objectivity. Objectivity comes in degrees, and the more perspectives represented in the critical discussions that form scientific results, the more objective the process. This has important implications for the organization of science. The degree to which scientific inquiry is objective depends on the extent to which the scientific community is organized in such a way that a multiplicity of perspectives on a theory is elicited. Longino tentatively suggests a list of criteria that a scientific community must meet in order to support effective criticism. Included on the list are appropriate venues for criticism and that the community responds to criticism (Longino, 2006, pp. 172–173).

Other scholars have argued for a conception of objectivity comparable to Longino's (e.g., Haraway, 1988; Harding, 1992; Solomon, 1992, 2001; Wylie, 2013). For example, Donna Haraway (1988) speaks of "feminist objectivity" and states, "Feminist objectivity means quite simply *situated knowledges*" (p. 581). Haraway insists on "the embodied nature of all vision" and rejects that there is such a thing as "a conquering gaze from nowhere" (p. 581). It is the recognition of the contingency of knowledge claims that enables objectivity in science. Sandra Harding's (1992) "strong objectivity" requires not just the recognition of the situatedness of knowledge claims but also the recognition that certain marginalized perspectives are epistemically privileged.

If we accept a Longino-style account of scientific objectivity, peer review and post-publication criticism are important parts of the processes that secure the objectivity of science. Peer review serves not only to check that data seem right and that arguments are sound but also to introduce alternative viewpoints and thereby to minimize subjective biases. Similarly, post-publication reception in the form of replications, modifications, integration with other results, and so on also contributes to refining new pieces of knowledge and shaping them into what is ultimately accepted (Longino, 1990, p. 69). The process of peer review is examined in more detail in the next section.

4. Objectivity in Philosophy of Science in Practice

In this section, we provide an example of how our conception of scientific objectivity matters to current discussions of how science should be organized. We focus on the question of how peer review should be organized, a question that has attracted a lot of attention in the past four decades.

There are three common types of peer review: single-blind, double-blind, and open. The most commonly used is single-blind peer review, where the identity of the reviewer is concealed from the author, but the reviewer knows the identity of the author (Lee, Sugimoto, Zhang, & Cronin, 2013, p. 10; Ware, 2008, pp. 6–7). The main argument made for concealing the reviewer's identity to the author is to allow the reviewer to provide honest commentary without fear of repercussions. Double-blind peer review is also common, especially in the humanities and social sciences (Ware, 2008, pp. 6–7).[7] The main argument made for concealing the identity of the author to the reviewer is to protect the author against social bias on the part of the reviewer. When peer review is open, finally, authors and reviewers know each other's identities, and reviews may even be put online. In most such systems, the name of the reviewer is made public (Lee et al., 2013, p. 10).

Peer review can now be considered from two perspectives: (a) from the point of view that the objectivity of science is a function solely of the methods employed by the individual scientists and (b) from the point of view that the objectivity of science is also a function of how values and biases are distributed in the scientific community and brought to bear on the process of knowledge creation.

From the former perspective, concealing the identity of the author and/or reviewer contributes to the objectivity of science. This is so because in practice, scientists are not as objective as legend has claimed. Scientists sometimes fail to follow the scientific methods that ensure the objectivity of science on the former conception of objectivity. If reviewers always assessed papers with no regard to who the authors were, and if reviewers had no reason to fear retaliation from authors, there would be no need for concealing the identities of either author or reviewer. In sum, blinding the peer-review process contributes to maintaining the objectivity of science because scientists do not always follow scientific methods.

From the latter perspective, the objectivity of science is a function of whether the perspectives of multiple individuals with different values and biases are brought to bear on the process of knowledge creation. Hence, concealing the identities of authors and/or reviewers may mask the distribution of values and biases and may therefore *not* support the effort to maintain the objectivity of science. Not knowing who the author is will

make it harder for the reviewer to uncover problematic background assumptions of the author. Likewise, not knowing who the reviewer is will make it harder for the author to uncover problematic background assumptions of the reviewer. Hence, on a Longino-style account of objectivity, disclosing author and reviewer identities can be an advantage. On the other hand, disclosing reviewer identities may work against the honest articulation of criticism because the reviewer fears retaliation from the author. Weighing such considerations is a complex task, and we shall not do so here. We merely want to emphasize that different conceptions of objectivity call for different approaches to the question of how peer review should be organized.

The last example we will consider involves open peer review of a certain kind. Some journals with open peer review make reviewer names and reviews *publicly* available. For example, the prominent medical journal *BMJ* publishes everything online: the reviewer comments and identities, the authors' responses, and the different versions of the articles (Groves & Loder, 2014; Smith, 1999, 2006). The aim of publishing everything online is to promote transparency and accountability. This transparency about reviewer comments and identities allows members of the community to assess the review while taking into account the theoretical and cultural perspective it represents, that is, while having a way of recognizing the background assumptions of the reviewer. In this way, reviews can be perspectives in the critical discussion of research on a par with other perspectives that are presented publicly (see also Lee et al., 2013, p. 11). These perspectives may even be especially helpful voices in the discussion since public peer review provides an incentive for reviewers to be both thorough and fair. Hence, if one conceives of objectivity as obtained at the level of scientific communities and as a function of the different perspectives that are brought to bear on the process of knowledge creation, this extreme kind of open peer review may help promote objectivity.

How we conceive of scientific objectivity not only matters to the question of how peer review should be organized but also to the question of how to *measure* the reliability of peer review of the different kinds. In particular, whether the level of agreement among reviewers is a good measure of the reliability of peer review depends on how we conceive of objectivity. Much theorizing about peer review is based on the assumption that agreement among reviewers is something desirable (cf. Lee, 2012, pp. 861–862). However, Kuhn's and Longino's accounts of science indicate that this is not necessarily the case. With reference to Kuhn, Lee (2012, pp. 863–867) argues that reviewers can disagree in normatively appropriate ways about how to interpret and apply evaluative criteria such as soundness, significance, and novelty. To use Kuhn's (1977, p. 325) own example, some scientists value originality more and are accordingly more prepared to take epistemic risks. They may reasonably have a lower evidential threshold for accepting a hypothesis. Furthermore, if Longino is right about the role of biases in the choices of individual scientists, a high level of inter-reviewer agreement could just as well be an indication that the referees have common biases. By the same token, low inter-reviewer agreement may be an indication that the scientific community in question supports effective criticism. Consequently, the fact that inter-reviewer agreement has been found to be low in empirical studies of peer review is not necessarily a cause for concern.

5. Conclusion

In this chapter, we have reviewed a number of accounts of scientific rationality and objectivity. Some philosophers consider scientific rationality and objectivity to be obtained at the level of individual scientists when the individual scientists follow a particular method for doing science. The chapter has presented divergent views on what that method consists in. Others argue that scientific rationality and objectivity are obtained at the level of communities of scientists. These accounts emphasize that individual scientists will be guided by different values, but that this in fact contributes to the rationality and objectivity of science at the level of scientific communities. We have presented different views on what this process looks like.

Notes

1. See Hoyningen-Huene (1987) for a detailed account of different interpretations of the distinction between the context of discovery and the context of justification.

2. For an example of a textbook aimed at university students that avoids conveying the idea that there is one scientific method, see Potochnik, Colombo, and Wright (2019).

3. For detailed accounts of the various positions advanced during these decades, see, for example, Nickles (2017/2020), Losee (2004, 2005), or Newton-Smith (1981).

4. For further details, see H. Andersen (2013).

5. Longino examines numerous cases of research in which she claims biases have played an important role. Among other things, she studies cases of research in human evolution and

behavioral neuroendocrinology (see Longino, 1990, pp. 103–161). For a review of empirical studies of the role of bias in the peer review process in particular, see Lee, Sugimoto, Zhang, and Cronin (2013).

6. Popper (1934/2014) similarly claimed that "the objectivity of scientific statements lies in the fact that they can be *intersubjectively tested*" (p. 44). There is also an interesting similarity between Longino's account of scientific objectivity and Jürgen Habermas's account of how truth claims are justified. On Habermas's (1973) account, truth claims are justified in a critical discussion among free and uncoerced participants. The reasons for the claim should in principle be acceptable to any reasonable person (see also Bohman & Rehg, 2007/2017).

7. It is worth noting that double-blind peer review is sometimes not possible: the content of a manuscript will sometimes give away the author, which is presumably particularly true in highly specialized disciplines (L. E. Andersen, 2017, p. 178; Brown, 2007, p. 133; for further references, see Lee et al., 2013, p. 10).

References

Andersen, H. (2013). The second essential tension. *Topoi, 32*(1), 3–8.

Andersen, H., & Hepburn, B. (2016). Scientific method. In E. N. Zalta (Ed.), *The Stanford encyclopedia of philosophy*. Retrieved from https://plato.stanford.edu/archives/sum2016/entries/scientific -method/

Andersen, L. E. (2017). On the nature and role of peer review in mathematics. *Accountability in Research, 24*(3), 177–192.

Anderson, E. (2020). Feminist epistemology and philosophy of science. In E. N. Zalta (Ed.), *The Stanford encyclopedia of philosophy*. Retrieved from https://plato.stanford.edu/archives /spr2020/entries/feminism-epistemology/

Bauer, H. H. (1992). *Scientific literacy and the myth of the scientific method*. Urbana: University of Illinois Press.

Blachowicz, J. (2009). How science textbooks treat scientific method: A philosopher's perspective. *British Journal for the Philosophy of Science, 60*(2), 303–344.

Bloor, D. (1991). *Knowledge and social imagery* (2nd ed.). Chicago, IL: University of Chicago Press. (Original work published 1976)

Bohman, J., & Rehg, W. (2017). Jürgen Habermas. In E. N. Zalta (Ed.), *The Stanford encyclopedia of philosophy*. Retrieved from https://plato.stanford.edu/archives/fall2017/entries/habermas/

Brown, R. J. C. (2007). Double anonymity in peer review within the chemistry periodicals community. *Learned Publishing, 20*(2), 131–137.

Churchman, C. W. (1948). Statistics, pragmatics, induction. *Philosophy of Science, 15*(3), 249–268.

D'Agostino, F. (2010). *Naturalizing epistemology: Thomas Kuhn and the 'Essential tension'*. New York, NY: Palgrave MacMillan.

De Langhe, R. (2010). The division of labour in science: The tradeoff between specialisation and diversity. *Journal of Economic Methodology, 17*(1), 37–51.

De Langhe, R. (2012). The problem of Kuhnian rationality. *Philosophica, 86*(3), 11–31.

De Langhe, R. (2014). To specialize or to innovate? An internalist account of pluralistic ignorance in economics. *Synthese, 191*(11), 2499–2511.

De Langhe, R., & Rubbens, P. (2015). From theory choice to theory search: The essential tension between exploration and exploitation in science. In W. J. Devlin & A. Bokulich (Eds.), *Kuhn's Structure of scientific revolutions—50 years on* (pp. 105–114). Cham, Switzerland: Springer.

Douglas, H. E. (2009). *Science, policy, and the value-free ideal*. Pittsburgh, PA: University of Pittsburgh Press.

Feyerabend, P. K. (1993). *Against method* (3rd ed.). London, England: Verso. (Original work published 1975)

Fisher, R. A. (1955). Statistical methods and scientific induction. *Journal of the Royal Statistical Society, Series B, 17*(1), 69–78.

Groves, T., & Loder, E. (2014). Prepublication histories and open peer review at The BMJ. *BMJ, 349*, g5394.

Habermas, J. (1973). Wahrheitstheorien [Theories of truth]. In H. Fahrenbach (Ed.), *Wirklichkeit und Reflexion* (pp. 211–265). Pfüllingen, Germany: Neske.

Hanson, N. R. (1958). *Patterns of discovery*. Cambridge, England: Cambridge University Press.

Haraway, D. (1988). Situated knowledges: The science question in feminism and the privilege of partial perspective. *Feminist Studies, 14*(3), 575–599.

Harding, S. (1992). Rethinking standpoint epistemology: What is "strong objectivity?" *Centennial Review, 36*(3), 437–470.

Hempel, C. G. (1966). *Philosophy of natural science*. Englewood Cliffs, NJ: Prentice Hall.

Howie, D. (2002). *Interpreting probability: Controversies and developments in the early twentieth century*. Cambridge, England: Cambridge University Press.

Hoyningen-Huene, P. (1987). Context of discovery and context of justification. *Studies in History and Philosophy of Science, 18*(4), 501–515.

Hoyningen-Huene, P. (1993). *Reconstructing scientific revolutions: Thomas S. Kuhn's philosophy of science* (A. T. Levine, Trans.). Chicago, IL: University of Chicago Press.

Hull, D. L. (1988). *Science as a process: An evolutionary account of the social and conceptual development of science*. Chicago, IL: University of Chicago Press.

Irigaray, L. (1987). Is the subject of science sexed? *Hypatia, 2*(3), 65–88.

Kitcher, P. (1993). *The advancement of science*. Oxford, England: Oxford University Press.

Kuhn, T. S. (2012). *The structure of scientific revolutions* (4th ed.). Chicago, IL: University of Chicago Press. (Original work published 1962)

Kuhn, T. S. (1977). Objectivity, value judgment, and theory choice. In *The essential tension: Selected studies in scientific tradition and change* (pp. 320–339). Chicago, IL: University of Chicago Press.

Lakatos, I. (1970). Falsification and the methodology of scientific research programmes. In I. Lakatos & A. Musgrave (Eds.), *Criticism and the growth of knowledge* (pp. 91–196). Cambridge, England: Cambridge University Press.

Laudan, L. (1968). Theories of scientific method from Plato to Mach. *History of Science, 7*(1), 1–63.

Lee, C. J. (2012). A Kuhnian critique of psychometric research on peer review. *Philosophy of Science, 79*(5), 859–870.

Lee, C. J., Sugimoto, C. R., Zhang, G., & Cronin, B. (2013). Bias in peer review. *Journal of the American Society for Information Science and Technology, 64*(1), 2–17.

Lenhard, J. (2006). Models and statistical inference: The controversy between Fisher and Neyman–Pearson. *British Journal for the Philosophy of Science, 57*(1), 69–91.

Longino, H. E. (1989). Feminist critiques of rationality: Critiques of science or philosophy of science? *Women's Studies International Forum, 12*(3), 261–269.

Longino, H. E. (1990). *Science as social knowledge: Values and objectivity in scientific inquiry*. Princeton, NJ: Princeton University Press.

Longino, H. E. (2006). Philosophy of science after the social turn. In M. C. Galavotti (Ed.), *Cambridge and Vienna: Frank P. Ramsey and the Vienna Circle* (pp. 167–177). Dordrecht, Netherlands: Springer.

Losee, J. (2004). *Theories of scientific progress*. London, England: Routledge.

Losee, J. (2005). *Theories on the scrap heap: Scientists and philosophers on the falsification, rejection, and replacement of theories*. Pittsburgh, PA: University of Pittsburgh Press.

Machamer, P. K., & Wolters, G. (Eds.). (2004). *Science, values, and objectivity*. Pittsburgh, PA: Pittsburgh University Press.

Marks, H. M. (1997). *The progress of experiment: Science and therapeutic reform in the United States, 1900–1990*. Cambridge, England: Cambridge University Press.

Mayo-Wilson, C., Zollman, K. J. S., & Danks, D. (2011). The independence thesis: When individual and social epistemology diverge. *Philosophy of Science, 78*(4), 653–677.

Newton-Smith, W. H. (1981). *The rationality of science*. London, England: Routledge and Kegan Paul.

Neyman, J. (1956). Note on an article by Sir Ronald Fisher. *Journal of the Royal Statistical Society, Series B, 18*(2), 288–294.

Nickles, T. (2020). Historicist theories of scientific rationality. In E. N. Zalta (Ed.), *The Stanford encyclopedia of philosophy*. Retrieved from https://plato.stanford.edu/archives/spr2020/entries/rationality-historicist/

Pearson, E. S. (1955). Statistical concepts in their relation to reality. *Journal of the Royal Statistical Society, Series B, 17*(2), 204–207.

Popper, K. R. (1972). *Conjectures and refutations: The growth of scientific knowledge* (4th ed.). London, England: Routledge. (Original work published 1963)

Popper, K. R. (2014). *The logic of scientific discovery* (reprint of the 1st English ed.). Mansfield Centre, CT: Martino Publishing. (Original work published in German 1934, English ed. published 1959)

Potochnik, A., Colombo, M., & Wright, C. (2019). *Recipes for science: An introduction to scientific methods and reasoning*. New York, NY: Routledge.

Reiss, J., & Sprenger, J. (2017). Scientific objectivity. In E. N. Zalta (Ed.), *The Stanford encyclopedia of philosophy*. Retrieved from https://plato.stanford.edu/archives/win2017/entries/scientific-objectivity/

Rudner, R. (1953). The scientist *qua* scientist makes value judgments. *Philosophy of Science, 20*(1), 1–6.

Scheffler, I. (1982). *Science and subjectivity* (2nd ed.). Indianapolis, IN: Hackett. (Original work published 1967)

Shapere, D. (1966). Meaning and scientific change. In R. Colodny (Ed.), *Mind and cosmos* (pp. 41–85). Pittsburgh, PA: Pittsburgh University Press.

Smith, R. (1999). Opening up *BMJ* peer review: A beginning that should lead to complete transparency. *BMJ, 318*, 4–5.

Smith, R. (2006). Peer review: A flawed process at the heart of science and journals. *Journal of the Royal Society of Medicine, 99*, 178–182.

Solomon, M. (1992). Scientific rationality and human reasoning. *Philosophy of Science, 59*(3), 439–455.

Solomon, M. (2001). *Social empiricism*. Cambridge, MA: MIT Press.

Ware, M. (2008). *Peer review: Benefits, perceptions and alternatives*. London, England: Publishing Research Consortium. Retrieved from http://citeseerx.ist.psu.edu/viewdoc/download?doi=10.1.1.214.9676&rep=rep1&type=pdf

Weisberg, M. (2013). Modeling herding behavior and its risks. *Journal of Economic Methodology, 20*(1), 6–18.

Weisberg, M., & Muldoon, R. (2009). Epistemic landscapes and the division of cognitive labor. *Philosophy of Science, 76*(2), 225–252.

Wray, K. B. (2011). *Kuhn's evolutionary social epistemology*. New York, NY: Cambridge University Press.

Wylie, A. (2013). Why standpoint matters. In R. Figueroa & S. Harding (Eds.), *Science and other cultures: Issues in philosophies of science and technology* (pp. 26–48). New York, NY: Routledge.

Zollman, K. J. S. (2007). The communication structure of epistemic communities. *Philosophy of Science, 74*(5), 574–587.

Zollman, K. J. S. (2010). The epistemic benefit of transient diversity. *Erkenntnis, 72*, 17–35.

Zollman, K. J. S. (2018). The credit economy and the economic rationality of science. *Journal of Philosophy, 115*(1), 5–33.

14.2 Rationality and the Value-Freedom of Science

Anke Bueter

Summary

The value-freedom of science has traditionally been regarded as a presupposition of scientific rationality. However, during the past decades, in addition to numerous empirical counterexamples of value-laden science, systematic arguments have put the adequacy of value-freedom as an ideal into doubt. This chapter presents the most important debates on the ideal of value-freedom, which concern the epistemic impact of values in the discovery and justification of theories, the distinction between epistemic and nonepistemic values, and the argument from inductive risk. Taken together, these arguments call for new normative models of how to deal with values in science, which no longer equate value-laden science with bad science or irrationality. Rather, they suggest that scientific rationality is highly complex, since epistemic issues are interwoven with practical, sociopolitical, institutional, and ethical ones.

1. Values and the Rationality of Science

Science is often seen as an exemplarily rational endeavor, ideally generating knowledge untainted by extra-scientific interests, prejudices, or wishful thinking. As such, it can support rational beliefs about the world and serve as the basis for rational decision making in policy matters (e.g., concerning the mitigation of climate change). As Newton-Smith (1981) puts it, "The image that the scientific community likes to project of itself, and indeed the image that most of us accept of that community, is that of rationality par excellence" (p. 1).

This rationality of science is secured by the methodical production and testing of scientific knowledge that minimize random errors as well as systematic distortions—for instance, by value-laden perspectives. The value-freedom of science is therefore traditionally considered a precondition of its rationality.[1] Such a view is particularly plausible if one takes value judgments to be mere expressions of subjective preferences, incapable of rational justification. It is, however, equally compatible with all positions that consider value judgments to allow for rational discussion, yet not as the kind of judgment that can be supported by scientific methods. Essential to the ideal of value-free science is a separation of the realms of *is* and *ought*: empirical conclusions do not follow from normative premises, and neither can empirical judgments legitimize normative conclusions.

1.1 From Is to Ought

Important predecessors of the current debate on science and values have often focused on this latter question of whether science can provide us with normative guidance in the ethical or political realm. David Hume is famous for criticizing deductions from *is* to *ought* (Hume, 1739/1975, p. 469). In the context of the developing social sciences in the 19th century, Max Weber has argued for the importance of keeping science and politics separate. Expressing normative, political judgments in the role of (social) scientist, he says, is misguiding and only serves to immunize the respective judgments from criticism by other citizens (Weber, 1917/1973). The moral and rational thing to do would be to keep the roles of scientist and citizen distinct and to clearly separate descriptive and normative judgments (on the relation of normative and descriptive theories, see also section 4 of the Introduction by Knauff & Spohn, this handbook). What science can do, according to Weber, is to study values and value systems. Their relations and coherency can be analyzed, and science can clarify whether and how certain political goals are empirically realizable. For example, economics might inform us about the most effective ways to increase production, but it cannot tell us whether or not we should prioritize productivity over fair working conditions. Science may give us the means to realize certain practical ends, but it cannot justify these ends (cf. Weber, 1904). In other words: science should be politically neutral.

This aim of neutrality has also been debated in the positivism dispute between proponents of critical rationalism (Popper, Albert) and the Frankfurt school in sociology (e.g.,

Adorno, Habermas) during the 1960s. The latter argued that a critique of the societal status quo is an important task of the social sciences, and the scientific justification of political goals is possible through historical materialism. Critical rationalists criticized this approach for purporting normative assertions to be scientifically justified even though they are not testable by scientific methods. The Frankfurt school, they argued, borrowed the authority of science for their political positions but without submitting them to the rigorous critical process that gives science its epistemic authority. Adorno and others replied that a political neutrality of science serves and perpetuates given power structures by failing to offer a critique and an alternative (cf. Adorno et al., 1969; see also Bueter, 2012).

1.2   From Ought to Is

Since the 1980s, the value-freedom of science has yet again been subjected to intense debate. The current debate mostly focuses on the impact of value judgments on scientific theories and processes, asking whether such influences are avoidable—and, if not, how to deal with them. Here, the terms "value" and "value judgment" are often used in a very broad sense, in which "a value is something that is desirable or worthy of pursuit" (Elliott, 2017, p. 11). This may include religious or social values but also economic interests, political goals, or individual preferences. Often, "value judgment" is characterized in a negative manner as comprising all judgments that cannot be justified by logic and empirical evidence alone. Prima facie empirical judgments can be "value-laden"; for instance, if they rely on an interpretation of the evidence from a certain interested perspective or result from biased methodological standards.

Much research in philosophy and history of science as well as in science studies has brought forward examples of science that is value-laden in this sense. In particular, feminist scholars have exposed a multitude of cases where social prejudices related to gender have found their way into accepted scientific theories (cf., e.g., Keller, 1985; Kourany, 2010; Schiebinger, 1999). While such *empirical* cases of value-ladenness were acknowledged, they have often been taken to underline, not undermine, the importance of value-freedom from a *normative* point of view. In other words, the *ideal* of value-freedom is often held on to: "Although complete value-freedom cannot be achieved, it ought to be a goal or ideal of science" (Shrader-Frechette, 1991, p. 44).

This contemporary ideal of value-freedom[2] is usually based (implicitly or explicitly) on a distinction between the contexts of discovery, justification, and application.[3]

There should be no compromise when it comes to the proposal that ideological factors be invited into the Context of Justification. . . . We should make every attempt to keep politics and religion out of the laboratory. We may not always be successful, but that simply means that we should try harder, not that we should give up the attempt. (Koertge, 2000, p. 53)

"Context of discovery" usually refers to the generation of hypotheses. Whether these hypotheses are the result of a long thinking process, misogynism, or a nightmare about snakes is considered irrelevant for their epistemic status, which has to be assessed via rigorous and systematic testing in the "context of justification." Finally, empirically supported hypotheses and theories can be implemented in practice ("context of application"). Yet again, such an implementation is not considered a scientific decision. Instead, it is delegated to policy-makers and democratic deliberation, as it often requires the negotiation of different values and practical goals.

The current ideal of value-freedom thus does not require science to be completely independent of any values or interests; rather, these may impact the direction and application of research as well as decisions on the ethical admissibility of methods. However, the epistemic assessment (i.e., justification) of theories is supposed to be value-free.

In the past decades, the feasibility of this normative ideal has been thoroughly questioned by systematic arguments that concern both the value-freedom of justification and its independency from values at play in discovery and application. In the following, I will present the most prominent debates on value-freedom: from the relevance of (values in) the context of discovery (section 2) and the distinction between epistemic and nonepistemic values (section 3) to the problem of inductive risk (section 4). The respective criticisms of the value-free ideal call for a refined understanding of scientific rationality that does not presuppose value-freedom. I will conclude with a brief outlook on resulting questions and emerging topics.

2.   Discovery, Pursuitworthiness, and Significance

Restricting value-freedom to the testing and evaluation of theories presumes that this stage of the research process is epistemically independent from discovery and application. To put it another way, it assumes that there is a rational, scientific core of research, on the one hand, and more external stages not amenable to scientific rationality, on the other. Accordingly, the context of justification has often been treated as the sole proper subject of philosophy of science.

Against this, it has been argued that while there may not be an unambiguous *logic* of discovery, the generation of hypotheses and the selection of research questions are also not totally random or even irrational. Rather, they depend on ascriptions of significance that can be influenced by epistemic as well as sociopolitical, economic, or ethical considerations.

Philip Kitcher argues that any assumption of a super-ordinate goal of science (such as truth, successful predictions, etc.) is incapable of efficiently directing research. Scientists are not equally interested in anything that may be true; rather, they search for *significant* truths (or significant predictions, explanations of significant phenomena, etc.). Such ascriptions of significance depend on more specific epistemic goals (e.g., making successful predictions about climate change or explaining the existence of our solar system). These specific epistemic goals are often informed by values. As Kitcher argues, even what seems epistemically significant (rather than practically important) today does so as result of a specific scientific history, which may have been influenced by value-laden concerns in earlier phases (Kitcher, 2001, chapters 6 and 7).

Based on this, Kitcher develops the ideal of a *well-ordered science*, which broadens the focus of philosophy of science to include questions traditionally relegated to the context of discovery. Well-ordered science calls for political responsibility and accountability in the direction of research, making it subject to a process of democratic deliberation by representatives of different social interests (to be informed by impartial scientific experts; Kitcher, 2001, chapter 10; cf. also Kitcher, 2011).

An important point to make here is that ascriptions of significance do not stop at the choice of a research question. To get from a specific question to a testable theory is a process that involves more than just the logical deduction of empirical consequences from a certain hypothesis. It may involve questions of how exactly to frame the hypothesis, which variables to focus on, what concepts to employ,[4] and which empirical consequences of a hypothesis to test and how. Developing a testable theory, moreover, also depends on sufficient funding as well as the time and willingness to persevere in the face of setbacks.

This has two important consequences: first, the distinction between discovery and justification tends to make all of these important steps invisible, even though many of them are capable of rational evaluation. We could thus add a *context of theory pursuit*, where we judge theories not in terms of their eventual justification but their (epistemic and/or practical) prospects and fertility as well as initial plausibility (cf., e.g., Kordig, 1978; Nickles, 1980; Šešelja & Straßer, 2014; Whitt, 1992).

[A] careful examination of scientific practice reveals that there are generally *two* quite different contexts within which theories are evaluated. . . . Even if we had an adequate account of theory choice within the context of acceptance . . . we would still be very far from possessing a full account of rational appraisal. (Laudan, 1977, pp. 108–109)

Second, several authors have argued that values in discovery and/or pursuit have impacts on the evaluation of theories, putting the epistemic independence of the context of justification in doubt. The general idea here is that ascriptions of significance impact which theories are developed and tested (and which not), as well as which data are generated and what is considered noise. This, in turn, impacts the assessment of existing theories. For example, Okruhlik (1994) argues that theory assessment is usually comparative: what is the best theory depends on the range of available theories. In a thoroughly sexist culture, though, "[n]on-sexist rivals will never even be generated. Hence, the theory which is selected by the canons of scientific appraisal will simply be the best of the sexist rivals; and the very *content* of science will be sexist" (Okruhlik, 1994, pp. 201–202).

The choice between competing theories may thus be based on rational criteria and yet unable to screen out earlier social impacts—a point that Ohkruhlik takes to underline a need for more enhanced models of scientific rationality, which include the contexts of theory generation and attend to the social embeddedness of science. A similar point has been made by Elliott and McKaughan (2009). Based on case studies from toxicology, they show how social values and interests impact the availability of theories and data—and, consequentially, theory assessment:

The degree of evidential support for a theory clearly depends both on the *array of available theories* and on the *set of data at hand*. Therefore, to the extent that the nonepistemic values associated with discovery and pursuit influence the available theory and data, they affect theory appraisal. (Elliott & McKaughan, 2009 p. 600)

It is necessary to make choices about significance in scientific research. However, if all of these choices are guided by the *same* value-laden background assumptions, this can lead to bias in our results, even if these are tested methodically and meticulously. As Anderson (1995) explains, a partial truth can be thoroughly misleading. Bueter (2015) argues that values in discovery and pursuit can create blind spots such that alternative theories or data, which would undermine currently accepted theories if they existed, are never even generated. For instance, this can happen due to widely shared sexist or racist prejudices, or due to a domination of research fields by agents with particular interests, for example, when medical research is driven by the pharmaceutical industry that aims to make profits and therefore does not explore non-pharmaceutical treatment alternatives (cf. Jukola, 2019).

3.   Values in Justification: The Social, the Epistemic, and the Nonepistemic

Section 2 summarized arguments against a restriction of scientific rationality and value-freedom to the context of justification. For one, decisions concerning agenda-setting and pursuitworthiness are said to allow for rational discussion; for another, value-impacts on these decisions have been argued to affect theory assessment via the (non)existence of alternatives. However, a large part of research on science and values has focused on questions relating to justification. Prominently, it has been debated whether epistemic values can be distinguished from non-epistemic ones and, if so, whether the latter can or should be eliminated from theory assessment.

The modern ideal of value-freedom is often taken to be essentially this: the evaluation of scientific contents is supposed to rely exclusively on epistemic criteria. Kuhn (1977) has been the first to call such criteria *values*, thereby starting a still-ongoing debate on whether one can distinguish between epistemic (or cognitive)[5] and nonepistemic (noncognitive) values. Kuhn's original list contained five such criteria that serve as indicators of good theories: *accuracy* (or *empirical adequacy*), requiring that a theory's deducible consequences are in line with empirical observations; *consistency* (*internal*, i.e., logical, as well as *external*, i.e., in relation to other currently accepted theories); *scope*; *simplicity*; and *fruitfulness*.

Kuhn calls these criteria "values" because they do not provide algorithmic rules for theory choice, only guidance. They need to be interpreted and balanced against each other in specific research contexts. Moreover, their interpretation and weighing can be influenced by other factors such as the personality of a researcher (for a more detailed account of Kuhn's position, see chapter 14.1 by Andersen & Andersen, this handbook).

McMullin (1982) calls Kuhn's values *epistemic*, arguing that they "promote the truth-like character of science" (p. 18), and distinguishes them from nonepistemic (e.g., religious or sociopolitical) values. Against this, Rooney (1992) points out the existence of different lists of such epistemic values, taking this variety to indicate a lack of clarity in the distinction between epistemic and non-epistemic. She further argues that this lack of clarity can

sometimes turn sociopolitical (e.g., androcentric) values into part of what it means to be epistemically successful.

This resonates with Longino (1996), who argues that the function of epistemic values is to promote the goal(s) of science.[6] Such goals can be value-laden, for instance, when feminists aim to understand mechanisms of gender oppression or to produce nonsexist theories about women's health. Thus, we could have research goals that are both feminist and epistemic—a status that translates accordingly to the values guiding theory evaluation. Starting from case studies on research with feminist goals, Longino creates an alternative list of such (feminist) epistemic values. She agrees that *internal consistency* and *empirical adequacy* are necessary features of a good theory.[7] In contrast to Kuhn, however, she adds *novelty*, *ontological heterogeneity*, *complexity of relation*, *applicability to human needs*, and *diffusion of power*. Furthermore, she argues that the Kuhnian values can function as promoters of androcentric theories in particular research contexts (e.g., when simplicity serves as a justification of considering only men's perspectives and concerns).[8]

Others have defended the distinction between epistemic and nonepistemic values. For instance, Ruphy (2006) proposes an empirical evaluation of epistemic values in terms of their contribution to empirically successful theories. Douglas (2013) argues that a more fine-grained classification of kinds of epistemic values can help both to reduce tensions between them and to support their epistemic status. She does so by distinguishing epistemic values on two axes: (1) ideal desiderata versus minimal criteria and (2) values that apply to theories per se versus to theories in relation to evidence. Some, but not all, of the resulting four groups of epistemic values are genuinely truth-assuring; others can be considered strategic or pragmatic epistemic values in that they make it more likely (e.g., via testability) that flaws will be discovered (Douglas, 2013).

Hugh Lacey argues that, when considering the role of epistemic and nonepistemic values in justification, it is necessary to distinguish between different kinds of attitudes with regard to the assessment of theories, which tend to be blurred under the common term of "accepting" a theory. According to him, a theory may be "adopted" (i.e., used to frame and direct future research), it can be "endorsed" (i.e., considered well enough supported to put it into practical use), or it may be "impartially held of a set of phenomena" (i.e., considered so well established that it does not need further testing). In this third case, only epistemic values have legitimate roles to play (Lacey, 2017; cf. also Lacey, 1999). Rooney (2017) replies that Lacey's argument is question-begging, because it presupposes, rather than grounds, the distinction between epistemic and nonepistemic values.

Steel (2010) distinguishes between "intrinsic" and "extrinsic" epistemic values. *Intrinsic* epistemic values, such as empirical adequacy, are those that are necessary for the attainment of truth. *Extrinsic* epistemic values promote such an attainment only indirectly; examples here are testability or the ability to obtain funding. For Steel, nonepistemic values are values that are either irrelevant or detrimental to the attainment of truth.

Even if we assume the possibility of distinguishing epistemic from nonepistemic values though, further problems remain. In recent years, a controversy has emerged about the appropriate normative role of nonepistemic values in theory assessment, which focuses on the priority of epistemic over nonepistemic values. For instance, Steel's account represents an "epistemic constraints" approach: nonepistemic values are illegitimate if they interfere with or override the standards of adequate science expressed in extrinsic and intrinsic epistemic values. On the other hand, they are allowed to play a role as "tie-breakers" if epistemic values are insufficient to guide theory choice in particular cases.

Steel (2017) contrasts such epistemic-constraints approaches with aim-oriented ones, which do not ascribe unambiguous priority to epistemic values. For instance, Elliott and McKaughan (2014) argue that the goals of inquiry are underdetermined if characterized as attainment of truth in general—rather, research aims for truths that are considered significant. The relevant goals here are therefore particular research goals that integrate practical aims and concerns, which can be promoted by both epistemic and nonepistemic values. Nonepistemic values may therefore have legitimate roles to play in theory assessment, as long as scientists are maximally transparent about their goals and values. Relatedly, Brown (2013) argues that it is unnecessary to assign priority to epistemic values in order to avoid wishful thinking, as long as the value judgments involved are subjected to rational discussion and are considered in principle revisable (cf. also Brown, 2017). A question for future research that arises here is how to determine the legitimacy of such epistemic-social goals (and thus values) and how exactly to avoid a corruption of research by political or economic interests.

4. Application, the Consequences of Error, and Inductive Risk

Another important strand in the debate on value-freedom focuses on the problem of inductive risk. The central question here is whether all responsibility for decisions on applying research results can be externalized or whether scientists themselves need to take into account potential practical consequences of their work—at least in the case of uncertainty. The argument from inductive risk says that practical consequences of mistaken scientific results are relevant to the setting of evidentiary standards, thereby rejecting the epistemic independence of justification from the context of application.

Rudner (1953) famously argued that in deciding to accept or reject any hypothesis, scientists always face two different kinds of potential errors: false-negative results (rejecting a true hypothesis) versus false-positive results (accepting a false hypothesis). This gives rise to the question of how much evidence is enough to decide upon the acceptance or rejection of individual hypotheses in the light of error risks. Consequently, Rudner argues, scientists need to refer to value judgments concerning the potential consequences of different kinds of errors. For instance, if the acceptance of a hypothesis about the safety of a new drug will likely lead to its approval and use in a large population, a false positive may lead to many patients experiencing severe side effects, while a false negative would mean the loss of a potential treatment to them and of financial profits to the pharmaceutical company. Which kind of error is considered worse in a specific situation depends on nonepistemic value judgments and impacts the epistemic standards for acceptance (e.g., if it is most important to us to avoid a false-positive result, these standards will be set very high).

Rudner (1953) concludes that this means that the scientist *qua* scientist necessarily makes value judgments. Against this, Levi (1962) has argued that Rudner conflates accepting a hypothesis in terms of its epistemic validity and thus *believing* in its truth, with *acting* based on such a belief. Relatedly, Jeffrey (1956) has objected that scientists often cannot foresee potential practical applications of their work and thus consequences of error. Therefore, they can hardly decide upon the acceptance/rejection of a hypothesis based on values referring to these consequences. Instead, Jeffrey proposes that scientists should not make such decisions but confine themselves to ascribing probabilities to hypotheses—leaving it to policy-makers to decide how high a probability is high enough in a particular situation. Pointing out the problems with leaving such value judgments to scientists rather than democratic decision making, Betz (2013) similarly argues that scientists should not accept or reject plain hypotheses but rather "hedged hypotheses," which articulate the amount and kind of uncertainty involved in making a certain claim (for a response, see John, 2015a).

In one of the most influential articles in the debate, Douglas (2000) has both resuscitated Rudner's argument and strengthened it. First, she confines the argument to policy-relevant research, where practical consequences of scientists' assertions often are predictable. Second, she shows how trade-offs between the probabilities of false positives versus false negatives occur not only at the point of a hypothesis's final evaluation but at several methodological stages before. For instance, such trade-offs relate to the setting of a significance level, the gathering and interpretation of data, or the interpretation of results. This makes it more difficult to externalize all the decisions facing inductive risk and implies a thorough value-ladenness of science. For Douglas, this amounts to an ethical obligation of scientists to consider the potential practical consequences of errors. For her, the respective ethical considerations *should* have an impact on evidential standards in science. She distinguishes such legitimate *indirect* value influences from illegitimate ones, where nonepistemic values *directly* impact the rejection/acceptance of a hypothesis:

> Two clear roles for values in reasoning appear here, one legitimate and one not. The values can act as reasons in themselves to accept a claim, providing direct motivation for the adoption of a theory. Or, the values can act to weigh the importance of uncertainty about the claim, helping to decide what should count as sufficient evidence for the claim. (Douglas, 2009, p. 96)

Douglas's work has been the beginning of an ongoing and very lively debate about inductive risk in science. For one, the argument of inductive risk has been applied in a variety of case studies from medicine over climate science to theoretical physics (cf., e.g., Elliott & Richards, 2017). For another, its soundness, its scope, and its impact on the ideal of value-free science continue to be discussed (cf., e.g., Biddle, 2013; de Melo-Martín & Intemann, 2016; Hicks, 2018). Biddle (2016) as well as Biddle and Kukla (2017) have recently called for, and started on, a more fine-grained differentiation of kinds of risk involved in inductive risk.

Douglas's normative answer—distinguishing between indirect value influences on evidential standards as legitimate and direct influences on acceptance decisions as illegitimate—has also been called into question, in terms of both its conceptual clarity and its normative sufficiency (e.g., Elliott, 2011). In particular, indirect influences can also sometimes be problematic, for example, if they overly tip the balance between error probabilities in

favor of sponsor interests (cf. Wilholt, 2009, for an example). Wilholt, moreover, provides an alternative proposal for identifying biased research, which relies on the role of conventions in the scientific community. Such conventions serve as a means to standardize methodological decisions that can indirectly affect acceptance decisions. Deviations from such commonly employed standards can thus make value influences visible. In addition, Wilholt argues that the distribution of error probabilities is not only a problem in cases of applied research with foreseeable social consequences. It is rather an in-principle epistemological issue, since false-positive or false-negative results differ in relevance and importance even if we consider only epistemic utilities, such as the overall importance of a certain hypothesis for our belief systems and theories (Wilholt, 2009, 2013).

5. Conclusion

To sum up, the value-freedom of science has been the subject of intense debate in the past decades—both in terms of numerous empirical counterexamples and regarding its adequacy as a normative ideal. At the same time, most philosophers do not equate value-ladenness with bad science (anymore). Instead, they aim for more sophisticated approaches, supposed to be able to distinguish between illegitimate and legitimate value influences; for example, in terms of kinds of values (i.e., epistemic vs. nonepistemic), indirect or direct roles of values, or stages of scientific research in which they may or may not play a role. While this debate on normative alternatives is still ongoing, the emerging consensus seems to be both that the traditional value-free ideal is doomed—and that this does not imply that science lacks rationality. Rather, it indicates that the kind of rationality involved is highly complex, since epistemic issues are thoroughly interwoven with practical, sociopolitical, institutional, and ethical concerns.

Since one of the purported functions of the ideal of value-free science was to provide the public with factual, untainted information to be used in democratic deliberations on practical concerns, an emerging topic now is how this new image of science relates to issues of public trust (see also chapter 14.3 by Bromme & Gierth, this handbook). In particular, if science is shot through with ethical and social values, the question arises whose values these are, how they are to be legitimated, and how this can be compatible with rational trust in the results of such science. Furthermore, this leads to the question of how to communicate to the public the role of values in scientific research. While, for instance, McKaughan and

Elliott (2013) have pleaded for transparency here, John (2015b) warns against a further erosion of public trust caused by such transparency. To conclude, the question at this point in time no longer seems to be whether science is value-free or whether it must be so in order to be rational—but rather how to deal with value-laden science in a sensible manner, not least on the institutional level of organizing scientific research and science–public relations.

Notes

1. It has also usually been treated as a necessary, yet not sufficient, condition for objectivity. For a more detailed look at objectivity, see chapter 14.1 by Andersen and Andersen (this handbook).

2. Regarding different elements and historical versions of the value-free ideal, see, for example, Proctor (1991).

3. For a closer analysis and discussion of the context distinction, see Schickore and Steinle (2006). I am referring to it here not in order to support it but because it has had a considerable impact on the ideal of value-freedom and constitutes a useful means to organize different discussions in the science-and-values debate.

4. Another discussion I will not go into here for reasons of space concerns the semantic value-ladenness of scientific concepts, classifications, and terminologies (cf., e.g., Dupré, 2007).

5. The terminology differs here. For the sake of readability, I will use "epistemic" throughout in the following.

6. Helen Longino's work has played a very decisive role in debates on values in science; this holds in particular for her account of *social objectivity*, which does not presume value-freedom. Rather, social objectivity requires scientific communities to be organized in a way that enhances the efficiency of critical discourse (Longino, 1990; for more details, see chapter 14.1 by Andersen & Andersen, this handbook).

7. Similar to Kuhn, Longino takes consistency and empirical adequacy to be insufficient to determine theory choice. In both cases, this assumption is based on moderate versions of the underdetermination thesis. Kuhn argues that rival theories are often not empirically equivalent, yet both more or less empirically adequate, differing in what areas of evidence they match better (Kuhn, 1977, p. 323). Longino proceeds on the basis of an underdetermination thesis regarding evidential relations. As she argues, the relation between theoretical hypotheses and (significant) empirical consequences is mediated by substantial and methodological background assumptions, which are possibly value-laden (Longino, 1990, pp. 40–48).

8. Such a diversity in values guiding theory assessment might be taken to undermine the possibility of rational discourse

in the case of dissent. Carrier (2013) argues that pluralism extending to the epistemic values used in theory assessment is unproblematic if the scientific community shares a fundamental "epistemic attitude" toward the goals of research, which enables a joint striving for consensus and rational discussion. This epistemic attitude is not implemented at the level of standards for theory assessment but at the level of procedural rules for debating knowledge claims.

References

Adorno, T. W., Dahrendorf, R., Pilot, H., Albert, H., Habermas, J., & Popper, K. R. (1969). *Der Positivismusstreit in der deutschen Soziologie* [The positivism dispute in German sociology]. Neuwied, Germany: Luchterhand.

Anderson, E. (1995). Knowledge, human interests, and objectivity in feminist epistemology. *Philosophical Topics, 23*(2), 27–58.

Betz, G. (2013). In defence of the value free ideal. *European Journal for Philosophy of Science, 3*(2), 207–220.

Biddle, J. (2013). State of the field: Transient underdetermination and values in science. *Studies in History and Philosophy of Science, Part A, 44*(1), 124–133.

Biddle, J. (2016). Inductive risk, epistemic risk, and overdiagnosis of disease. *Perspectives on Science, 24*(2), 192–205.

Biddle, J., & Kukla, R. (2017). The geography of epistemic risk. In K. C. Elliott & T. Richards (Eds.), *Exploring inductive risk: Case studies of values in science* (pp. 215–237). New York, NY: Oxford University Press.

Brown, M. J. (2013). Values in science beyond underdetermination and inductive risk. *Philosophy of Science, 80*(5), 829–839.

Brown, M. J. (2017). Values in science: Against epistemic priority. In K. C. Elliott & D. Steel (Eds.), *Current controversies in values and science* (pp. 64–78). New York, NY: Routledge.

Bueter, A. (2012). *Das Wertfreiheitsideal in der Sozialen Erkenntnistheorie: Objektivität, Pluralismus und das Beispiel Frauengesundheitsforschung* [The value-free ideal in social epistemology: Objectivity, pluralism, and the example of women's health research]. Frankfurt/Main, Germany: Ontos.

Bueter, A. (2015). The irreducibility of value-freedom to theory assessment. *Studies in History and Philosophy of Science, Part A, 49*, 18–26.

Carrier, M. (2013). Values and objectivity in science: Value-ladenness, pluralism and the epistemic attitude. *Science & Education, 22*(10), 2547–2568.

de Melo-Martín, I., & Intemann, K. (2016). The risk of using inductive risk to challenge the value-free ideal. *Philosophy of Science, 83*(4), 500–520.

Douglas, H. (2000). Inductive risk and values in science. *Philosophy of Science, 67*(4), 559–579.

Douglas, H. (2009). *Science, policy, and the value-free ideal*. Pittsburgh, PA: University of Pittsburgh Press.

Douglas, H. (2013). The value of cognitive values. *Philosophy of Science, 80*(5), 796–806.

Dupré, J. (2007). Fact and value. In J. Dupré, H. Kincaid, & A. Wyli (Eds.), *Value-free science? Ideals and illusions* (pp. 27–41). Oxford, England: Oxford University Press.

Elliott, K. C. (2011). Direct and indirect roles for values in science. *Philosophy of Science, 78*(2), 303–324.

Elliott, K. C. (2017). *A tapestry of values: An introduction to values in science*. New York, NY: Oxford University Press.

Elliott, K. C., & McKaughan, D. J. (2009). How values in scientific discovery and pursuit alter theory appraisal. *Philosophy of Science, 76*(5), 598–611.

Elliott, K. C., & McKaughan, D. J. (2014). Nonepistemic values and the multiple goals of science. *Philosophy of Science, 81*(1), 1–21.

Elliott, K. C., & Richards, T. (Eds.). (2017). *Exploring inductive risk: Case studies of values in science*. New York, NY: Oxford University Press.

Hicks, D. J. (2018). Inductive risk and regulatory toxicology: A comment on de Melo-Martín and Intemann. *Philosophy of Science, 85*(1), 164–174.

Hume, D. (1975). *A treatise of human nature* (L. A. Selby-Bigge, Ed.). Oxford, England: Clarendon Press. (Original work published 1739)

Jeffrey, R. C. (1956). Valuation and acceptance of scientific hypotheses. *Philosophy of Science, 23*(3), 237–246.

John, S. (2015a). The example of the IPCC does not vindicate the Value Free Ideal: A reply to Gregor Betz. *European Journal for Philosophy of Science, 5*(1), 1–13.

John, S. (2015b). Inductive risk and the contexts of communication. *Synthese, 192*(1), 79–96.

Jukola, S. (2019). Commercial interests, agenda setting, and the epistemic trustworthiness of nutrition science. *Synthese*. Online publication.

Keller, E. F. (1985). *Reflections on gender and science*. New Haven, CT: Yale University Press.

Kitcher, P. (2001). *Science, truth, and democracy*. New York, NY: Oxford University Press.

Kitcher, P. (2011). *Science in a democratic society*. Amherst, NY: Prometheus Books.

Koertge, N. (2000). Science, values, and the value of science. *Philosophy of Science, 67*, 45–57.

Kordig, C. R. (1978). Discovery and justification. *Philosophy of Science, 45*(1), 110–117.

Kourany, J. A. (2010). *Philosophy of science after feminism*. Oxford, England: Oxford University Press.

Kuhn, T. S. (1977). Objectivity, value judgment, and theory choice. In *The essential tension: Selected studies in scientific tradition and change* (pp. 320–339). Chicago, IL: University of Chicago Press.

Lacey, H. (1999). *Is science value free? Values and scientific understanding*. London, England: Routledge.

Lacey, H. (2017). Distinguishing between cognitive and social values. In K. C. Elliott & D. Steel (Eds.), *Current controversies in values and science* (pp. 15–30). New York, NY: Routledge.

Laudan, L. (1977). *Progress and its problems: Towards a theory of scientific growth*. London, England: Routledge & Kegan Paul.

Levi, I. (1962). On the seriousness of mistakes. *Philosophy of Science, 29*(1), 47–65.

Longino, H. E. (1990). *Science as social knowledge: Values and objectivity in scientific inquiry*. Princeton, NJ: Princeton University Press.

Longino, H. E. (1996). Cognitive and non-cognitive values in science: Rethinking the dichotomy. In L. H. Nelson & J. Nelson (Eds.), *Feminism, science, and the philosophy of science* (pp. 39–58). Dordrecht, Netherlands: Springer.

McKaughan, D. J., & Elliott, K. C. (2013). Backtracking and the ethics of framing: Lessons from voles and vasopressin. *Accountability in Research, 20*(3), 206–226.

McMullin, E. (1982). Values in science. In P. D. Asquith & T. Nickles (Eds.), *PSA: Proceedings of the Biennial Meeting of the Philosophy of Science Association, 1982* (Vol. 2, pp. 3–28). Chicago, IL: University of Chicago Press.

Newton-Smith, W. H. (1981). *The rationality of science*. Boston, MA: Routledge & Kegan Paul.

Nickles, T. (1980). Introductory essay: Scientific discovery and the future of philosophy of science. In T. Nickles (Ed.), *Scientific discovery, logic, and rationality* (pp. 1–59). Dordrecht, Netherlands: Springer.

Okruhlik, K. (1994). Gender and the biological sciences. *Canadian Journal of Philosophy, 24*(Suppl. 1), 21–42.

Proctor, R. N. (1991). *Value-free science? Purity and power in modern knowledge*. Cambridge, MA: Harvard University Press.

Rooney, P. (1992). On values in science: Is the epistemic/non-epistemic distinction useful? In D. Hull, M. Forbes, & K. Okruhlik (Eds.), *PSA: Proceedings of the Biennial Meeting of the Philosophy of Science Association, 1992* (Vol. 1, pp. 13–22). Chicago, IL: University of Chicago Press.

Rooney, P. (2017). The borderlands between epistemic and non-epistemic values. In K. C. Elliott & D. Steel (Eds.), *Current controversies in values and science* (pp. 31–45). New York, NY: Routledge.

Rudner, R. (1953). The scientist *qua* scientist makes value judgments. *Philosophy of Science, 20*(1), 1–6.

Ruphy, S. (2006). "Empiricism all the way down": A defense of the value-neutrality of science in response to Helen Longino's contextual empiricism. *Perspectives on Science, 14*(2), 189–214.

Schickore, J., & Steinle, F. (Eds.). (2006). *Revisiting discovery and justification: Historical and philosophical perspectives on the context distinction*. Dordrecht, Netherlands: Springer.

Schiebinger, L. (1999). *Has feminism changed science?* Cambridge, MA: Harvard University Press.

Šešelja, D., & Straßer, C. (2014). Epistemic justification in the context of pursuit: A coherentist approach. *Synthese, 191*(13), 3111–3141.

Shrader-Frechette, K. S. (1991). *Risk and rationality: Philosophical foundations for populist reforms*. Berkeley: University of California Press.

Steel, D. (2010). Epistemic values and the argument from inductive risk. *Philosophy of Science, 77*(1), 14–34.

Steel, D. (2017). Qualified epistemic priority: Comparing two approaches to values in science. In K. C. Elliott & D. Steel (Eds.), *Current controversies in values and science* (pp. 49–63). New York, NY: Routledge.

Weber, M. (1904). Die „Objektivität" sozialwissenschaftlicher und sozialpolitischer Erkenntnis [The "objectivity" of knowledge in social science and social policy]. *Archiv für Sozialwissenschaft und Sozialpolitik, 19*(1), 22–87.

Weber, M. (1973). Der Sinn der „Wertfreiheit" der soziologischen und ökonomischen Wissenschaften [The meaning of "value-freedom" in the sociological and economic sciences]. In *Gesammelte Aufsätze zur Wissenschaftslehre* (4th ed., pp. 146–214). Tübingen, Germany: Mohr. (Original work published 1917)

Whitt, L. A. (1992). Indices of theory promise. *Philosophy of Science, 59*(4), 612–634.

Wilholt, T. (2009). Bias and values in scientific research. *Studies in History and Philosophy of Science, Part A, 40*(1), 92–101.

Wilholt, T. (2013). Epistemic trust in science. *British Journal for the Philosophy of Science, 64*(2), 233–253.

14.3 Rationality and the Public Understanding of Science

Rainer Bromme and Lukas Gierth

Summary

Science[1] and rationality are strongly linked. Therefore, the *public understanding of science*[2] is crucial for rationality as an everyday mode of reasoning. However, the knowledge provided by science is distinct from everyday thinking to a large degree, and at least nonscientists' understanding of it is limited. This chapter will provide an overview of the tension between the significance of science as a source for rational reasoning and the boundedness of nonscientists' understanding of science. It will be argued that *trust* is a key concept for analyzing how people cope with this boundedness. It will be elaborated why asking *Whom to believe?* is a rational way for citizens to reason about the truth of scientific knowledge claims. Then, *motivated reasoning* and *knowledge* as psychological conditions for judging what to believe and whom to trust will be discussed. Finally, the question about rationality within the public's understanding of science will be placed into the context of controversies between normative and descriptive accounts of rationality.

1. The Nexus between Science and Rationality

Rational inferences as well as rational beliefs are deemed valid because their validity has been established by scientific methods. Why is it justified to evaluate a Bayesian inference about the probability of an event as a rational inference? Why is it justified to evaluate the belief that objects move at a constant speed as a rational belief? In both cases, these justifications have been established by the scientific community as *scientific* knowledge (Bayes' theorem, Newton's first law). Rational decisions about the acceptance or rejection of such inferences and beliefs are therefore embedded in the acceptance of scientific background knowledge. In this regard, science is a source for establishing normatively what is "rational." Conversely, the production and justification of knowledge is accepted as *scientific* only if it can be described as rational.

The relationship between science and rationality is strong, but they are not equivalent concepts: it is possible to make rational inferences without reference to scientific knowledge. Conversely, rationality is necessary for science but not a sufficient condition for science.[3] This can be illustrated by the problem of demarcating between science and pseudoscience (e.g., creationism, climate change denialism, homeopathy, flat-earth theories, chem-trail theories). Pseudoscientists claim to adhere to scientific criteria of truth, they provide supposedly supporting evidence, and they mimic scientific discourse, especially when attacking real science. In this regard, pseudoscience is different from conspiracy theories and mere superstition, although the boundaries between these kinds of belief frameworks are fuzzy, and belief in one is correlated with belief in the other. There is no set of precise necessary and sufficient criteria for the demarcation between pseudoscience and science, but it is possible to tell them apart because pseudosciences typically ignore some of the criteria of science, for example, the need for coherence among its theoretical propositions (Law, 2020; Mahner, 2013). The validity of scientific inferences is not guaranteed solely by their adherence to normative rules of rational reasoning. All rational reasoning refers to some background knowledge about the world (Chater & Oaksford, 2012). The problem of demarcating between pseudoscience and science illustrates that "rationality" is not just an issue of adhering to norms of reasoning but also an issue of adopting shared assumptions about the world. For example, the inference that the dilution of a pharmaceutical agent increases its potency (a basic tenet of homeopathy) could be deemed rational as long as it is consistent with background beliefs about a world including the capacities of substances to store and to transmit health-related information irrespective of molecular concentrations. Hence, there could be a *local* rationality within the worldview of pseudoscientists. Nevertheless, in a more global sense it is not rational, because these background beliefs can be justified only by selectively ignoring contradicting evidence and arguments.

The validity of this background knowledge about the world does not *have to* be established as scientific knowledge, but with regard to many domains of our lives, it *can*[4] only be established as scientific knowledge. The strong nexus between science and rationality follows from the fact that the understanding of the actual structure of our environment (the natural and the social world) is formed by science. Modern societies rely on all kinds of artifacts (technological as well as social) that have been developed by means of scientific knowledge. Moving successfully in this world requires at least a certain degree of understanding and acceptance of this scientific knowledge, although it is possible (for some people within some contexts) to live without such acceptance, as the case of pseudoscience believers shows. In this regard, rationality also rests on the (public) acceptance of what can be taken as "true" (in the sense of scientifically justified) worldviews, not only on the local adherence to certain rules of rational reasoning. This does not mean that such a public acceptance can be taken for granted; it has to be maintained continually, for example, via schooling and science communication. In other words, a "public understanding of science" (PUS) (in a sense that will be detailed below in this chapter) is crucial for rationality.

Because of this nexus, analyzing the public understanding of science could contribute to understanding rationality and its limitations among the general public. There are roughly three strands of research providing empirical insight into conditions and processes of citizens' understanding of science: (a) research on the emergence and fostering of *scientific literacy* (SL), typically focused on children and young adults in formal learning contexts; (b) studies about the processing and acceptance of scientific beliefs, typically focused on adults encountering scientific claims via the media; and (c) studies about the personal stances toward science held by different segments of the general public, typically based on representative surveys.

2. What Is Meant by *Public Understanding of Science*? In Which Sense Is This Understanding *Bounded*?

Scientific literacy (SL) and the public's bounded[5] understanding of science are related but distinct concepts. According to a recent definition, covering widely shared notions of SL, a scientifically literate person has the competencies to "explain phenomena scientifically—recognise, offer and evaluate explanations for *a range of natural and technological phenomena*. Evaluate and design scientific enquiry—describe and appraise scientific investigations and propose ways of addressing questions scientifically. Interpret data and evidence scientifically—analyse and evaluate data, claims and arguments in a variety of representations and draw appropriate scientific conclusions" (OECD, 2016, p. 13, emphasis added). Based on this definition, international studies within the PISA project revealed a large variance in the degree of SL, between countries as well as between individuals. In this regard, it can be assumed that—at least for large parts of the general public—the individual understanding of science is limited.

However, there are further limitations. SL typically refers to knowledge items and to the kinds of scientific reasoning that are taught in schools (DeBoer, 2000). These knowledge items are in principle *understandable* for the general public without specialized science training. Furthermore, they contain knowledge that has been established as *certain*, at least within the pertinent scientific community. Understandability and certainty are conditions for including a piece of knowledge in the school curriculum. Therefore, only a narrow selection of all the scientific knowledge and all the kinds of scientific reasoning (Kind & Osborne, 2017) can be included. In contrast, everyday encounters with natural and technological phenomena are confined neither to *understandable* nor to *certain* scientific knowledge. Parents reason about the actual or supposed scientific evidence put forward in favor of certain educational practices, patients reason about contradictions between their doctors' arguments and what they read on the Internet, and citizens reason about contradicting arguments concerning the feasibility of carbon storage as a means of reducing global warming. In this regard, public understanding of science refers to a much broader spectrum of citizens' encounters with science than the concept of SL. Furthermore, which topics emerge within the public discourse is not limited to topics that are understandable or certain. Which topics are receiving attention by the general public reflects practical problems on a personal (Feinstein, 2011) as well as on a societal level and—in a mutual relationship—the media coverage of certain topics (National Academies of Sciences, Engineering, and Medicine, 2017). Thus, neither understandability nor certainty of knowledge is a limiting factor for which topics emerge within the public discourse about science.

There are further epistemological as well as psychological reasons to conceive the boundaries of understanding science as a constraint on the public's encounters with science. One epistemological reason is the complexity of scientific knowledge. This refers to the possibly unlimited depth of scientific explanations (Keil, 2010).

Additionally, there are the epistemic, social, and technological complexities of the methods and procedures necessary to establish a scientific knowledge claim as a true belief about the world. Nowadays, the production of scientific evidence is based on an elaborate *division of cognitive labor*, which is mirrored in the disciplinary structure of science as well as in the volatility of this structure.

A further epistemological reason refers to the qualitative differences between the objects, models, and methods of science, which refer to a certain piece of the natural or the social world and the everyday experiences with this piece. The emergence of science in modern history has also been a history of separating a scientific understanding from an everyday understanding of the world (Wolpert, 1992). Many entities of science could not be experienced with the unaided senses. Even concepts that are well established in our everyday reasoning, like "bacterium," "virus," or "gene," refer to entities that we cannot experience without technology. Some processes can only be modeled mathematically, for example, in nuclear physics. Others extend the number and kinds of dimensions that make up our everyday experience of the world. Such models can be linked to everyday reasoning only by way of analogies (Boudry, Vlerick, & Edis, 2020).

Research on the psychological differences between scientific and everyday knowledge has found systematic misconceptions (from a normative, i.e., scientific, point of view) in several domains of knowledge (Shtulman, 2017). Such misconceptions are robust, and even after science instruction, they still coexist with elements of scientifically justified knowledge. Two prominent features of everyday thinking illustrate that these differences are not due to a void within individual knowledge structures (and should thus not be explained in terms of a deficit model) but due to the stability and local coherence of the already existing knowledge. Both features refer to ontological differences between the kinds of categories used in scientific modeling and in everyday thinking: everyday reasoning prefers explanations in terms of substances and their features, while scientific thinking requires an understanding of relationships. For example, concepts like mass, energy, and heat are—from a physics point of view—concepts about relationships, while in everyday reasoning, they are dealt with as features of objects (Slotta & Chi, 2006). Everyday reasoning accepts "essences" as sufficient causes for explaining phenomena. In this vein, genetic explanations are popular in everyday reasoning about developmental and behavioral patterns of humans because they causally link the essence-like genes with these outcomes (Heine, Dar-Nimrod, Cheung, & Proulx, 2017).

This boundedness of science understanding is not fixed, neither historically nor individually. Specialist knowledge can become part of the school curriculum, as has been the case with concepts from nuclear physics or molecular genetics. These are examples for historic shifts of the boundary between specialist knowledge and public knowledge, at least for well-educated citizens. Appropriate instructional environments (e.g., science education in schools, lab experiences in science museums) can increase individual scientific literacy. Well-designed information about health issues, nuclear physics, or climate change is available. It is possible to increase scientific knowledge about topics of public interest. This is what strategic public-understanding-of-science campaigns are aiming for.

Nevertheless, this does not nullify the constraining effect of the previously outlined boundaries of understanding science for nonscientists. Such boundaries matter especially when it comes to *conflicting* scientific knowledge claims. When citizens look up health-related information on the Internet (e.g., about the relationship between nutrition, body weight, and illnesses), they will come across claims that do not align with each other or with claims made by their medical specialist. Even when citizens look up evidence for the existence of climate change, they will come across contradictory voices. Hence, there are not only conflicts between claims put forward by different scientists but also—and practically more important—conflicts between the claims of scientists and those of actors who would not be conceived as members of the scientific community, for example, deliberate science deniers, quacks, preachers of esoteric belief systems, conspiracy theorists, and "merchants of doubt" (Oreskes & Conway, 2010).

Often, the claims themselves and the reasoning by which they follow from the evidence are understandable. When they are presented well, they can be understood by nonexperts, even with a low level of SL. However, when reasoning about which of the conflicting knowledge claims is true and which is false, the abovementioned boundaries of public understanding matter. In a British survey (Castell et al., 2014), 70% of respondents agreed that there is so much conflicting information that it is hard to know what to believe.

3. Whom to Believe? The Role of Trust in Reasoning about Scientific Knowledge Claims

When coming across conflicting claims, one could evaluate these claims as more or less plausible and then accept one as a subjective belief. But with regard to many topics, most members of the general public are not able to decide by virtue of their own understanding of the relevant topic *which* scientific knowledge claim should be adopted as a true belief (in the sense of being justified scientifically).

Another way of coping with the encounter of competing knowledge claims could be a shift from the question of the plausibility of the content ("What to believe?") toward the question of the trustworthiness of its respective sources ("Whom to believe?"). The latter question is both logically and empirically quite different from the former. *Trust* is a kind of willingness to depend on others. Whenever people are dependent on other agents (persons or organizations), and whenever they are willing to accept the risks that come along with this dependency, they (the "trustors"; Blöbaum, 2016) put trust into these agents (the "trustees"). In the context of science, the good that the trustee provides to the trustor is "knowledge," and the risk to the trustor is his or her vulnerability to a lack of truth or validity of that knowledge (Hendriks, Kienhues, & Bromme, 2016). This is called *epistemic trust* (Sperber et al., 2010). Judging the validity of a claim on the basis of an ascription of trustworthiness to the source has been called the "argument from authority" or "argumentum ad verecundiam." Its logical form is "According to person *A*, who is an expert on the issue of *Y*, *Y* is true. Therefore, *Y* is true." This is, from a classical logical perspective, a fallacious inference (Cummings, 2014), and it seems to be at odds with scientific reasoning and argumentation. A core principle of modern science is to *know* the truth instead of just trusting what you are told. "Historically, this individualistic stance could be seen as a reaction against the pervasive role in Scholasticism of arguments from authority" (Sperber et al., 2010, p. 361). Only a minority of philosophers and historians of science have emphasized the role of *trust* in science, and the ones who do focus mostly on how it relates to doing science and only to a lesser degree on science communication with the public (Hardwig, 1991; Irzik & Kurtulmus, 2019; Wilholt, 2013). However, the main argument for trust within science, the division of cognitive labor,[6] is the same reason why trust is essential in the general public's encounter with science (Bromme & Goldman, 2014; Levy, 2019).

Within informal reasoning, the argument from authority is a legitimate inference (Cummings, 2014; Keren, 2018). It is a shift of the *topic* of rational reasoning (from "What to believe?" to "Whom to believe?"), not the *denial* of rationality. Reasoning about "Whom to believe?" requires thinking about science, not as a cognitive structure, but as a social system.

Research has revealed that judgments about trust in scientists involve three dimensions (although the terminology is not consistent), which are in line with trust research in other domains: expertise, integrity, and benevolence (Cummings, 2015; Hendriks, Kienhues, & Bromme, 2015; Peters, Covello, & McCallum, 1997). *Expertise* refers to someone's amount of knowledge and skill. The dimension of expertise also encompasses the aspect of pertinence (Bromme & Thomm, 2016): the person must have the *relevant* expertise. *Integrity* refers to adherence to the rules of the profession and—in the context of science—to the intention to produce "true" (unbiased) knowledge. *Benevolence* refers to taking into account the interests of the trustor or (more generally) the goods of the scientist's work for society. Importantly, people recognize that a scientist can have his or her own interests in mind and still be benevolent (Critchley, 2008).

When do people shift from a "What to believe?" toward a "Whom to believe?" strategy? Recent research on the conditions under which people pay attention to source information to assess the quality of knowledge claims (Bråten, Stadtler, & Salmerón, 2017; Bromme, Stadtler, & Scharrer, 2018) has corroborated two main results: first, there is no *spontaneous* attention to sources, and thus perceived content quality is unaffected by source information. Second, people only actively process source information when encountering some kind of *discrepancy*. Such discrepancies can be perceived when claims are inconsistent, either *within the same* piece of information or *across two or more different* pieces of information or *between the piece of information and prior beliefs* held by the recipient. When such discrepancies occur and people turn to the source information to determine which claim is correct, both types of judgment inform each other: on the one hand, people make inferences about the trustworthiness of sources based on their willingness to accept the messages provided by these sources. On the other hand, assumptions about a source affect judgments about the plausibility of their claims (Hahn, Harris, & Corner, 2009; Lombardi, Seyranian, & Sinatra, 2014; Thomm & Bromme, 2016).

There is a preference for immediate judgments about the plausibility or believability of claims; people seem to be *epistemic individualists* (Levy, 2019). There are different approaches to explaining this default preference. It could be explained as a case of a general *overconfidence regarding one's own knowledge* (Atir, Rosenzweig, & Dunning, 2015), as a case of a more specific *illusion of explanatory depth*—confined to the assumed understanding of causal mechanisms (Mills & Keil, 2004)—or as a kind of fluency effect, called the *easiness effect*. Scharrer, Stadtler, and Bromme (2014) asked laypersons to read short texts containing medicine-related knowledge claims. In all conditions, judging the validity of the medical claims was clearly beyond laypersons' medical or biological knowledge, but the texts varied regarding their lexical simplicity (easiness). Subjects who read the easier (but still conceptually very difficult) texts reported a lesser need for further expert advice to judge the credibility of the claim. However, this *easiness effect* was mitigated when subjects were explicitly told about the epistemic complexity of the medical topic. Awareness of the epistemic complexity matters for people's beliefs about the explanatory power of science. It seems to be a condition for *deference to science as an epistemic authority*.

4. Conditions for Judging What to Believe and Whom to Trust: Motivated Reasoning and Knowledge

Recent research on the public understanding of science—especially within the abovementioned strands (b) (studies about the processing of scientific beliefs) and (c) (studies about the personal stances toward science held by different segments of the general public)—has put a strong emphasis on the public's personal *acceptance* of specific scientific claims, on their *general appreciation* of science, and on the public's *trust* in science and scientists (Hendriks et al., 2016; Rutjens, Heine, Sutton, & van Harreveld, 2017).

This research focuses more on *attitudes* about science rather than on the *understanding* of science in the sense of the abovementioned research strand (a), regarding scientific literacy. Nevertheless, since attitudes refer to beliefs about the world, they matter for reasoning about the world and hence for rationality. This shift in focus has been fueled by evidence about the weak and sometimes inverse relationship between knowledge *about* science and *acceptance* of scientific knowledge as valid beliefs. Additionally, it has been fueled by findings pointing to the impact of attitudes, religious and political beliefs, group norms, personal identity, self-concepts, and personal goals on the processing of science information. Recent research on public understanding of science conceptualizes such processing as *motivated reasoning* (Kunda, 1990), subsuming these personal goals and beliefs as "motivations" for (biased) reasoning in favor of results that align with these motivations. Both the impact of knowledge and that of such motivations on reasoning are crucial for rationality and will hence be discussed in the following.

Motivations matter for the *selection* of information as well as for its subsequent processing. Selective exposure

and selective avoidance, respectively, can lead to so-called echo chambers, in which people only hear what they already believe. Regarding the relative importance of these processes, people's preference for belief-confirming information seems to be stronger than their avoidance of belief-disconfirming information (Garrett, 2009). Motivated reasoning[7] in general, and so-called defensive motivations in particular, could lead people to discount belief-disconfirming evidence by devaluating science in general or by doubting its ability to provide true knowledge about critical issues (Munro, 2010). For example, computer gamers disparage research on negative consequences of playing violent video games due to anger about the goal of this research and the stigmatization of gamers that they perceive to go along with it. This devaluation of the critical research findings generalizes to the entire research about video games (Nauroth, Gollwitzer, Bender, & Rothmund, 2014). Indeed, being a member of a certain social group, be it gamers, parents, or supporters of a certain political party, often provides motivations that affect reasoning. In the United States, Republicans and Democrats differed in their opinions on 16 socioscientific issues, from mandatory childhood vaccinations to climate change and the teaching of evolutionary theory (Blank & Shaw, 2015). In fact, that Democrats are more likely to accept the existence of anthropogenic global warming is the most extensively studied case of how political opinions affect science attitudes and trust.

Typically, the main focus of such controversies is rather political or cultural than scientific (Who will have to pay, in any sense, for human-caused climate change?), but citizens' beliefs about the validity of the implied scientific knowledge claims (e.g., about effects of carbon emissions on the mean temperature) are affected by these debates and by the political and ideological landscape of a society (Gauchat, 2012). For example, Lewandowsky and Oberauer (2016) have shown that beliefs in the importance of free markets are negatively correlated with the acceptance of human agency as the cause of climate change. Some researchers on PUS argue that the affiliation with a certain *culture* (a set of related political, ideological, and religious stances) is the main predictor for the appreciation of science and for the acceptance of scientific knowledge claims as valid (Kahan, Jenkins-Smith, & Braman, 2011). It is questionable if these patterns of beliefs and affiliations to social groups within a society are properly described as *cultures* (van der Linden, 2016), but there is enough evidence that such belief systems impact directly on the acceptance of scientific claims as true beliefs and on trust in science. Examples for the impact of belief systems are *framing effects* in reasoning. In a study in the United States, by simply substituting the term "climate change" for "global warming" on a survey questionnaire, Republicans were rendered more willing to indicate agreement with its existence (Schuldt, Roh, & Schwarz, 2015).

These belief systems and reasoning directed by them also modify the role of knowledge for the public's trust in science as well as for the acceptance of specific scientific claims. Early strategic endeavors for an improvement of the public understanding of science followed a coarse enlightenment heuristic: "If the public *understands* science, it will *believe* in science and support scientists." However, it rapidly became evident that this heuristic is flawed. There is a positive but small correlation between general knowledge about science and general trust in science (Allum, Sturgis, Tabourazi, & Brunton-Smith, 2008; Shi, Visschers, Siegrist, & Arvai, 2016). In contrast, when it comes to topics that are linked to political, religious, or cultural debates, the relationship between specific knowledge about the critical topic and trust in scientific evidence about this topic is mediated by the relevant belief systems: Kahan et al. (2012) found higher levels of science knowledge being related to increased polarization between Republicans and Democrats in regard to global warming belief. Similarly, accuracy of vaccination beliefs was most polarized among well-educated Americans (Joslyn & Sylvester, 2019). Broader science knowledge could enable people holding extreme beliefs (e.g., climate change deniers) to employ analytic reasoning strategies in order to reject belief-disconfirming evidence (Lewandowsky & Oberauer, 2016). Nevertheless, it is important to emphasize that these findings apply only to those who hold extreme beliefs about publicly debated topics and cannot be taken as a complete rebuttal of the abovementioned enlightenment heuristic. Carefully constructed and compelling counterarguments, such as the mechanistic explanation of global warming (Ranney & Clark, 2016), could lead to attitudinal change, as they are difficult to refute (Lewandowsky & Oberauer, 2016).

5.   Public Understanding of Science as a Play and a Stage for the Tension between Normative and Descriptive Accounts of Rationality

The relationship between actual reasoning and normative models of reasoning is a core issue within controversies about the nature of rationality (see also the Introduction by Knauff & Spohn, chapter 1.1 by Sturm, and chapter 1.2 by Evans, all in this handbook), even if science communication and citizens' understanding of

science are mostly not explicitly discussed in controversies about the nature of rationality. The boundedness of nonexperts' understanding of science points to a tension between the way in which science (as an ideal epistemic agent) reasons about the world and the way nonscientists do so in everyday contexts. In a sense, the public understanding of science is a case of this tension (merely a further play), just as reasoning within other domains of knowledge and of life are other plays. In another sense, the public understanding of science is also the overarching societal context (the *stage* for all these plays) for rationality, because—as was argued at the beginning of this chapter—there is a strong nexus between rationality and science. If it is true that science provides the means for discerning between different beliefs about the world on a rational basis, then the public's capability for rational reasoning is inevitably entrenched in its capability to understand science and its models of the world (i.e., scientific knowledge). When descriptive accounts of rationality aim to scrutinize the variance of rational reasoning in the general public, they do so on the stage of public understanding of science. Therefore, assessments of this variance (Stanovich, 2016) include subscales measuring scientific beliefs about the world.

Within the controversies about the nature of rationality, there have been several accounts arguing that the boundedness of reasoning (from a normative point of view) is no limitation for other kinds of rationality (see chapter 8.5 by Hertwig & Kozyreva and chapter 10.6 by Cosmides & Tooby, both in this handbook). Like others within research on the public understanding of science, we, too, argue in this vein. Judging the trustworthiness of researchers and of science in general is a rational strategy for overcoming the boundedness of a direct understanding of science. The topic of this reasoning about trustworthiness is science as a social system, not science as a cognitive structure with its unlimited depth of possible causal relations.

Citizens' capability to make such judgments is less bounded than their capability to judge which claims from a set of competing claims are true.[8] Science is a social subsystem of modern societies. Judgments about the trustworthiness of its actors require knowledge about this subsystem. Knowledge about the scientific consensus regarding a certain topic belongs to this knowledge. Knowledge about science institutions, about their affiliation with different (vested) interests, about reasons for dissent among scientists (Thomm, Hentschke, & Bromme, 2015), and about limitations and failures of science experts (Burgman, 2016) are further examples. The very fact that especially deniers of climate change underestimate the consensus about this topic (van der Linden, Leiserowitz, Feinberg, & Maibach, 2015), even if they have some understanding of its mechanisms, points to the role of this kind of knowledge for trust judgments. Of course, such knowledge must be acquired, and its acquisition is related to a wide range of educational experiences.[9] Nevertheless, this knowledge is more general than knowledge about specific science topics, because the criteria for judging the trustworthiness of different sources are less domain specific than the criteria for judging the validity of a specific knowledge claim. Furthermore, the social subsystem of science has common features with other subsystems within a society (education, health, industry) in which citizens have personal experience. Hence, knowledge about criteria for making rational judgments about the trustworthiness of science could be linked to these experiences. For example, a possible detrimental effect of vested interests on a source's epistemic trustworthiness could be experienced also in everyday contexts. Another example: consensus matters also for establishing what is true within other domains of rationality (Mercier & Sperber, 2017). This is not to argue that such reasoning is infallible and immune to the previously described effects of motivated reasoning. However, reasoning about science as a trustworthy source of knowledge contributes to maintaining a broad stage for rationality within society.

Acknowledgments

Many thanks to the Center for Advanced Studies (CAS) of the Ludwig Maximilian University Munich, which provided R.B. a perfect environment for working on this chapter.

Notes

1. "Science" is used here in the comprehensive sense of the German notion *Wissenschaft*, including the social sciences as well as the humanities.

2. Here we use the *analytic* concept "public understanding of science," which embraces all kinds of citizens' encounters with science, while the *strategic* notion of "Public Understanding of Science" typically refers to educational attempts (e.g., campaigns, books, events) to improve the public appreciation of science or to deliberate marketing attempts for increasing the public's acceptance of science (Bromme & Goldman, 2014).

3. Peels (2020), in a critique of scientism, provides convincing arguments why it is not justifiable to claim that *only* scientific propositions could be deemed rational. Gottlieb and Lombrozo (2020) provide examples of, as well as reasons for, the *limits* of

scientific explanations as well as people's intuitions about such limits.

4. The concept of *epistemic rationality* is even more strongly linked with science; it refers to "how well beliefs map onto the actual structure of the world" (Stanovich, 2012, p. 434), and beliefs about this *actual* structure could not be as arbitrarily established as beliefs about aliens.

5. The emphasis on the boundaries of the public's understanding of science should not be mistaken for the *deficit model* of science communication. This model assumes that it is just a lack of appropriate information that predicts people's rejection of science. Recent reports on PUS start with a stark rejection of this model, motivating a focus on other predictors like values and attitudes (e.g., National Academies of Sciences, Engineering, and Medicine, 2017; but see Suldovsky, 2016). Such predictors will be discussed below, but we would argue that the interplay between attitudes, values, and knowledge within reasoning could not be understood if the very fact that there are boundaries of understanding is downplayed.

6. Keren (2018) provides arguments (based on a social epistemology approach) why reliance on the experts' consensus is a rational way of reasoning for nonexperts, while it would be inappropriate for science as an epistemic community.

7. Druckman and McGrath (2019) argue that *all* reasoning is motivated and that these cases of denying scientific evidence are not due to a deliberate rejection of evidence but to different criteria for what counts as evidence.

8. Interestingly, representative democracy is based on a similar assumption. All citizens—as voters—are deemed to be able to make informed choices between different politicians (Goldman, 1999; critically Baurmann & Brennan, 2009). It is assumed that they have sufficient capability to judge rationally whom to trust for solving political problems, even when they could not solve these problems by virtue of their own capacities.

9. Recent educational approaches for analyzing and fostering citizens' capacity to deal with information found in the Internet put a strong emphasis on such trust judgments (Barzilai & Chinn, 2019; Stadtler, Bromme, & Rouet, 2018; Tabak, 2015).

References

Allum, N., Sturgis, P., Tabourazi, D., & Brunton-Smith, I. (2008). Science knowledge and attitudes across cultures: A meta-analysis. *Public Understanding of Science*, 17(1), 35–54.

Atir, S., Rosenzweig, E., & Dunning, D. (2015). When knowledge knows no bounds: Self-perceived expertise predicts claims of impossible knowledge. *Psychological Science*, 26(8), 1295–1303.

Barzilai, S., & Chinn, C. A. (2019). Epistemic thinking in a networked society: Contemporary challenges and educational responses. In Y. Kali, A. Baram-Tsabari, & A. Schejter (Eds.),

Learning in a networked society (Vol. 17, pp. 57–77). Cham, Switzerland: Springer.

Baurmann, M., & Brennan, G. (2009). What should the voter know? Epistemic trust in democracy. *Grazer Philosophische Studien, 79*(1), 157–186.

Blank, J. M., & Shaw, D. (2015). Does partisanship shape attitudes toward science and public policy? The case for ideology and religion. *Annals of the American Academy of Political and Social Science, 658*(1), 18–35.

Blöbaum, B. (Ed.). (2016). *Trust and communication in a digitized world: Models and concepts of trust research.* Berlin, Germany: Springer.

Boudry, M., Vlerick, M., & Edis, T. (2020). The end of science? On human cognitive limitations and how to overcome them. *Biology & Philosophy, 35*, 18.

Bråten, I., Stadtler, M., & Salmerón, L. (2017). The role of sourcing in discourse comprehension. In M. Schober, D. N. Rapp, & M. A. Britt (Eds.), *Handbook of discourse processes* (pp. 141–168). New York, NY: Taylor & Francis.

Bromme, R., & Goldman, S. R. (2014). The public's bounded understanding of science. *Educational Psychologist, 49*(2), 59–69.

Bromme, R., Stadtler, M., & Scharrer, L. (2018). The provenance of certainty: Multiple source use and the public engagement with science. In J. L. G. Braasch, I. Bråten, & M. T. McCrudden (Eds.), *Handbook of multiple source use* (pp. 269–284). New York, NY: Routledge.

Bromme, R., & Thomm, E. (2016). Knowing who knows: Laypersons' capabilities to judge experts' pertinence for science topics. *Cognitive Science, 40*(1), 241–252.

Burgman, M. A. (2016). *Trusting judgements: How to get the best out of experts.* Cambridge, England: Cambridge University Press.

Castell, S., Charlton, A., Clemence, M., Pettigrew, N., Pope, S., Quigley, A., . . . Silman, T. (2014). *Public attitudes to science 2014.* London, England: Ipsos Mori Social Research Institute.

Chater, N., & Oaksford, M. (2012). Normative systems: Logic, probability, and rational choice. In K. Holyoak & R. Morrison (Eds.), *The Oxford handbook of thinking and reasoning* (pp. 11–21). New York, NY: Oxford University Press.

Critchley, C. R. (2008). Public opinion and trust in scientists: The role of the research context, and the perceived motivation of stem cell researchers. *Public Understanding of Science, 17*(3), 309–327.

Cummings, L. (2014). The "trust" heuristic: Arguments from authority in public health. *Health Communication, 29*, 1043–1056.

Cummings, L. (2015). *Reasoning and public health: New ways of coping with uncertainty.* Heidelberg, Germany: Springer.

DeBoer, G. E. (2000). Scientific literacy: Another look at its historical and contemporary meanings and its relationship to science education reform. *Journal of Research in Science Teaching, 37*(6), 582–601.

Druckman, J. N., & McGrath, M. C. (2019). The evidence for motivated reasoning in climate change preference formation. *Nature Climate Change, 9*(2), 111–119.

Feinstein, N. (2011). Salvaging science literacy. *Science Education, 95*(1), 168–185.

Garrett, R. K. (2009). Echo chambers online? Politically motivated selective exposure among Internet news users. *Journal of Computer-Mediated Communication, 14*(2), 265–285.

Gauchat, G. (2012). Politicization of science in the public sphere: A study of public trust in the United States, 1974 to 2010. *American Sociological Review, 77*(2), 167–187.

Goldman, A. (1999). *Knowledge in a social world.* Oxford, England: Oxford University Press.

Gottlieb, S., & Lombrozo, T. (2020). What are the limits of scientific explanations? In K. McCain & K. Kampourakis (Eds.), *What is scientific knowledge? An introduction to contemporary epistemology of science.* (pp. 260–273). New York, NY: Routledge.

Hahn, U., Harris, A. J. L., & Corner, A. (2009). Argument content and argument source: An exploration. *Informal Logic, 29*(4), 337–367.

Hardwig, J. (1991). The role of trust in knowledge. *Journal of Philosophy, 88*(12), 693–708.

Heine, S. J., Dar-Nimrod, I., Cheung, B. Y., & Proulx, T. (2017). Essentially biased: Why people are fatalistic about genes. In J. Olson (Ed.), *Advances in experimental social psychology* (Vol. 55, pp. 137–192). Cambridge, MA: Academic Press.

Hendriks, F., Kienhues, D., & Bromme, R. (2015). Measuring laypeople's trust in experts in a digital age: The Muenster Epistemic Trustworthiness Inventory (METI). *PLoS ONE, 10*(10), e0139309.

Hendriks, F., Kienhues, D., & Bromme, R. (2016). Trust in science and the science of trust. In B. Blöbaum (Ed.), *Trust and communication in a digitized world: Models and concepts of trust research* (pp. 143–159). Berlin, Germany: Springer.

Irzik, G., & Kurtulmus, F. (2019). What is epistemic public trust in science? *British Journal for the Philosophy of Science, 70*(4), 1145–1166.

Joslyn, M. R., & Sylvester, S. M. (2019). The determinants and consequences of accurate beliefs about childhood vaccinations. *American Politics Research, 47*(3), 628–649.

Kahan, D. M., Jenkins-Smith, H., & Braman, D. (2011). Cultural cognition of scientific consensus. *Journal of Risk Research, 14*(2), 147–174.

Kahan, D. M., Peters, E., Wittlin, M., Slovic, P., Ouellette, L. L., Braman, D., & Mandel, G. (2012). The polarizing impact of science literacy and numeracy on perceived climate change risks. *Nature Climate Change*, *2*(10), 732–735.

Keil, F. C. (2010). The feasibility of folk science. *Cognitive Science*, *34*(5), 826–862.

Keren, A. (2018). The public understanding of what? Laypersons' epistemic needs, the division of cognitive labor, and the demarcation of science. *Philosophy of Science*, *85*(5), 781–792.

Kind, P., & Osborne, J. (2017). Styles of scientific reasoning: A cultural rationale for science education? *Science Education*, *101*(1), 8–31.

Kunda, Z. (1990). The case for motivated reasoning. *Psychological Bulletin*, *108*(3), 480–498.

Law, S. (2020). How can we tell science from pseudoscience? In K. McCain & K. Kampourakis (Eds.), *What is scientific knowledge? An introduction to contemporary epistemology of science* (pp. 100–116). New York, NY: Routledge.

Levy, N. (2019). Due deference to denialism: Explaining ordinary people's rejection of established scientific findings. *Synthese*, *196*(1), 313–327.

Lewandowsky, S., & Oberauer, K. (2016). Motivated rejection of science. *Current Directions in Psychological Science*, *25*(4), 217–222.

Lombardi, D., Seyranian, V., & Sinatra, G. M. (2014). Source effects and plausibility judgments when reading about climate change. *Discourse Processes*, *51*, 75–92.

Mahner, M. (2013). Science and pseudoscience: How to demarcate after the (alleged) demise of the demarcation problem. In M. Pigliucci & M. Boudry (Eds.), *Philosophy of pseudoscience: Reconsidering the demarcation problem* (pp. 29–43). Chicago, IL: University of Chicago Press.

Mercier, H., & Sperber, D. (2017). *The enigma of reason: A new theory of human understanding*. Cambridge, MA: Harvard University Press.

Mills, C. M., & Keil, F. C. (2004). Knowing the limits of one's understanding: The development of an awareness of an illusion of explanatory depth. *Journal of Experimental Child Psychology*, *87*, 1–32.

Munro, G. D. (2010). The scientific impotence excuse: Discounting belief-threatening scientific abstracts. *Journal of Applied Social Psychology*, *40*(3), 579–600.

National Academies of Sciences, Engineering, and Medicine. (2017). *Communicating science effectively: A research agenda*. Washington, DC: Academies Press.

Nauroth, P., Gollwitzer, M., Bender, J., & Rothmund, T. (2014). Gamers against science: The case of the violent video games debate. *European Journal of Social Psychology*, *44*(2), 104–116.

OECD. (2016). *PISA 2015 assessment and analytical framework: Science, reading, mathematic, financial literacy and collaborative problem solving*. Paris, France: OECD Publishing.

Oreskes, N., & Conway, E. M. (2010). *Merchants of doubt: How a handful of scientists obscured the truth on issues from tobacco smoke to global warming*. London, England: Bloomsbury.

Peels, R. (2020). Should we accept scientism? The argument from self-referential coherence. In K. McCain & K. Kampourakis (Eds.), *What is scientific knowledge? An introduction to contemporary epistemology of science* (pp. 274–287). New York, NY: Routledge.

Peters, R. G., Covello, V. T., & McCallum, D. B. (1997). The determinants of trust and credibility in environmental risk communication: An empirical study. *Risk Analysis*, *17*(1), 43–54.

Ranney, M. A., & Clark, D. (2016). Climate change conceptual change: Scientific information can transform attitudes. *Topics in Cognitive Science*, *8*(1), 49–75.

Rutjens, B. T., Heine, S. J., Sutton, R. M., & van Harreveld, F. (2017). Attitudes towards science. *Advances in Experimental Social Psychology*, *57*, 125–165.

Scharrer, L., Stadtler, M., & Bromme, R. (2014). You'd better ask an expert: Mitigating the comprehensibility effect on laypeople's decisions about science-based knowledge claims. *Applied Cognitive Psychology*, *28*(4), 465–471.

Schuldt, J. P., Roh, S., & Schwarz, N. (2015). Questionnaire design effects in climate change surveys: Implications for the partisan divide. *Annals of the American Academy of Political and Social Science*, *658*(1), 67–85.

Shi, J., Visschers, V. H. M., Siegrist, M., & Arvai, J. (2016). Knowledge as a driver of public perceptions about climate change reassessed. *Nature Climate Change*, *6*, 759–762.

Shtulman, A. (2017). *Scienceblind: Why our intuitive theories about the world are so often wrong*. New York, NY: Basic Books.

Slotta, J. D., & Chi, M. T. H. (2006). The impact of ontology training on conceptual change: Helping students understand the challenging topics in science. *Cognition and Instruction*, *24*, 261–289.

Sperber, D., Clément, F., Heintz, C., Mascaro, O., Mercier, H., Origgi, G., & Wilson, D. (2010). Epistemic vigilance. *Mind & Language*, *25*(4), 359–393.

Stadtler, M., Bromme, R., & Rouet, J.-F. (2018). Learning from multiple documents: How can we foster multiple document literacy skills in a sustainable way? In E. Manalo, Y. Uesaka, & C. A. Chinn (Eds.), *Promoting spontaneous use of learning and reasoning strategies: Theory, research, and practice for effective transfer* (pp. 46–61). London, England: Routledge.

Stanovich, K. E. (2012). On the distinction between rationality and intelligence: Implications for understanding individual differences in reasoning. In K. Holyoak & R. Morrison (Eds.), *The Oxford handbook of thinking and reasoning* (pp. 343–365). New York, NY: Oxford University Press.

Stanovich, K. E. (2016). The comprehensive assessment of rational thinking. *Educational Psychologist, 51*(1), 23–34.

Suldovsky, B. (2016). In science communication, why does the idea of the public deficit always return? Exploring key influences. *Public Understanding of Science, 25*(4), 415–426.

Tabak, I. (2015). Functional scientific literacy: Seeing the science within the words and across the web. In L. Corno & E. M. Anderman (Eds.), *Handbook of educational psychology* (3rd ed., pp. 269–280). London, England: Routledge.

Thomm, E., & Bromme, R. (2016). How source information shapes lay interpretations of science conflicts: Interplay between sourcing, conflict explanation, source evaluation, and claim evaluation. *Reading and Writing, 29*(8), 1629–1652.

Thomm, E., Hentschke, J., & Bromme, R. (2015). The Explaining Conflicting Scientific Claims (ECSC) questionnaire: Measuring laypersons' explanations for conflicts in science. *Learning and Individual Differences, 37*, 139–152.

van der Linden, S. (2016). A conceptual critique of the cultural cognition thesis. *Science Communication, 38*(1), 128–138.

van der Linden, S. L., Leiserowitz, A. A., Feinberg, G. D., & Maibach, E. W. (2015). The scientific consensus on climate change as a gateway belief: Experimental evidence. *PLoS ONE, 10*(2), e0118489.

Wilholt, T. (2013). Epistemic trust in science. *British Journal for the Philosophy of Science, 64*(2), 233–253.

Wolpert, L. (1992). *The unnatural nature of science*. Cambridge, MA: Harvard University Press.

15.1 The Development of Basic Human Rationality

Henry Markovits

Summary

Understanding how rationality develops through child-hood to adulthood is important for both theoretical and practical reasons. However, existing theories provide very different and seemingly conflicting descriptions, with some postulating the existence of very early forms of biologically based rationality, while others describe a long developmental process leading to explicit principles guiding rational thought. In this chapter, these different approaches will be described. An initial distinction is made between explicit, metacognitive forms of rational thinking and the implicit structure of inferences. Within each, theories and data that describe the developmental course of reasoning are analyzed. While it is not possible to point to a clear synthesis, the continual interaction between empirical knowledge and logical reasoning is a consistent component in all of these approaches.

1. Understanding the Development of Rationality

Development is generally characterized by progression from simple to more complex forms of cognitive functioning, with or without the aid of biological underpinnings. By analogy, one might expect children to become increasingly more rational. But this is far from universally supported. As with the adult literature, there is little consensus about the nature of rationality either in a broad or in a more restricted sense. The different developmental approaches that will be examined here, while seemingly divergent, indicate the complexity of the underlying developmental trajectory. As will be seen, the existence of irrational behavior in adults is coupled with evidence of forms of logical reasoning in very young children. Minimally, attempting to trace the development of rational thinking must require incorporating different systems, which might well have very different developmental paths.

Understanding the development of rationality is further complicated by the difficulty of clearly defining what rationality is in adults and even more in children. This is in fact the aim of many of the chapters in this handbook, and it would be impossible to repeat these arguments in a developmental context. However, most developmental studies that have looked at rationality specifically examine the ability of children to make inferences or judgments that correspond to standard logical norms, and for the purposes of this discussion, we will mostly use this general definition (for a broader perspective, see chapter 2.4 by Fiedler, Prager, & McCaughey, this handbook). However, even with this relatively straightforward definition, there is little consensus as to how, or even whether, the ability to make rational judgments develops. This inconsistency is probably most clearly shown by the existence both of theories that project clear developmental increases in children's rationality and of theories that suppose that even adults are not rational. In addition, developmental studies have examined very different phenomena, which are often difficult to compare. However, we can make one useful distinction that reflects different ways of examining this question. Specifically, we can distinguish between explicit and implicit rationality. The former refers to the study of the reasons that people give either to justify their own inferences or to explicitly account for differences in the level of adequacy of different forms of inference. The latter refers to the logical adequacy of judgments or inferences that people actually make.

2. The Development of Explicit Rationality

The study of the development of explicit rationality has two major poles, metalogical and argumentation.

2.1 Metalogical Development

"Metalogic" refers to the ability to think about one's own reasoning processes and can be seen as a subset of more general metacognitive processes (Flavell, 1979). The most direct theory describing the development of reasoning abilities is Moshman's account of reasoning

(Moshman, 1990, 2004, 2013; Moshman & Franks, 1986). Moshman makes a distinction between the implicit processes that are used to make specific inferences (which will be discussed later) and rationality, which can be defined as the ability to explicitly represent higher-level metalogical components of reasoning, such as concepts of inferential validity and logical necessity (Christoforides, Spanoudis, & Demetriou, 2016; Demetriou, Spanoudis, & Shayer, 2014; Moshman, 2011; Moshman & Franks, 1986). Metalogical understanding is the result of self-reflective processes that allow the abstraction of general principles from more specific forms of inference. In other words, rationality is developed through thinking about implicit inferences, by the reasoner trying to understand why these inferences are good or bad. The process is conceived of as a continuous one, leading to complex forms of epistemological understanding whose development goes well through adulthood (Hallett, Chandler, & Krettenauer, 2002) but with no necessary endpoint. More specifically, Moshman (2014) defines reasoning as involving inferential processes within the more general constraints of a metacognitive understanding of the goals and aims that animate a reasoner. Such forms of reasoning are inherently contextual, with the criteria defining "good" reasoning varying according to the domain in question. In other words, inferential processes are cognitive tools in the service of broader metacognitive goals and understanding, that is, rationality comprises the broader principles that guide specific inferences. Both the cited empirical studies and theoretical underpinnings suggest that with this definition, rationality undergoes an ongoing developmental process, such that the ability to be rational increases with experience and age. The clear emphasis on reasoning as inferences under metacognitive control thus suggests that rationality is late-developing, since it requires fairly complex forms of metacognitive understanding. Evidence of early forms of logical inference are simply building blocks in a complex developmental progression. For example, young children can consistently go from premises of the form "If P then Q; P is true" to correctly infer "Q is true" but do not necessarily understand why this inference is necessary. The basic metalogical understanding of the notion of logical necessity (i.e., to explicitly distinguish between empirical truth and validity) undergoes a clear developmental increase and is not typically shown before 10 to 12 years of age (Moshman & Franks, 1986).

There is one important question that is left unanswered by this theory, which is the relationship between rationality and inferential reasoning. Moshman's approach suggests that inferential processes are guided by metacognitive analyses. Indeed, one would expect a stronger relationship, so that the increased metacognitive awareness of the characteristics of reasoning that underpins the development of rationality should have a clear effect on the nature of the inferential processes that are used. There are almost no empirical data in a developmental context that addresses this question. It is worth citing recent work by Thompson (Newman, Gibb, & Thompson, 2017; Thompson, Prowse Turner, & Pennycook, 2011) that has examined the relationship between explicit degree of confidence in the rightness of judgments, which is one evaluation of metacognitive processes, and their logicality in adults. In fact, this work has found very little relationship between the two and certainly suggests that explicit metacognitive judgments do not necessarily correlate with the extent to which actual reasoning is logical (although see Markovits, Thompson, & Brisson, 2015). Thus, although it is very clear that the ability to explicitly analyze reasoning in terms of logical components develops in a consistent manner, it is less clear that rationality defined in this manner has a specific impact on actual judgments.

2.2 Argumentation

Another important form of explicit understanding that shows clear developmental patterns is derived from the notion of argumentation (see also chapter 5.5 by Hahn & Collins, this handbook). "Argumentation" refers to the way that specific inferences and accompanying justifications can be marshaled in order to either support a point of view or dispute an alternative viewpoint (Perelman, 1971). Ideally, argumentation skills allow a conscious weighting of the relative logical strength of different forms of inference, which can be seen as a building block for rationality. More important, argumentation requires an interaction between people having different points of view. Although it is possible to conceive of argumentation as an essentially one-way process (Mercier & Sperber, 2011), or a Vygotskian interaction between an adult and a child (Berk & Winsler, 1995), a stronger developmental viewpoint would suggest that the continual confrontation between opposing forms of inference would generate some form of metacognitive resolution of the resulting conflict. In fact, this form of interaction places the role of argumentation within the general context of social constructivism (Doise & Mugny, 1981), which suggests that the social confrontation of different points of view may be a critical component of cognitive development in general. There are in fact several studies that have shown the beneficial effect of interactions between children of different cognitive levels on basic Piagetian concepts (Ames & Murray, 1982; Murray, 1972). Such interactions have been shown to improve individual levels of logical reasoning (Moshman & Geil, 1998). More specifically, Kuhn and colleagues have shown that argumentation skills have a clear developmental trajectory (Kuhn, 1991, 1993; Kuhn & Udell, 2003). Promoting argumentation within the context of a collaborative social context produces a clear improvement in the ability of individuals to use argument strategies of increasing sophistication (Crowell & Kuhn, 2014; Kuhn & Crowell, 2011; Kuhn, Zillmer, Crowell, & Zavala, 2013). In other words, this general approach considers that rationality is a natural end state of a process by which different forms of arguments are directly confronted. This in turn supposes, consistent with the general tenets of social constructivism, that such a confrontation allows explicit acknowledgment of arguments that are logically superior and promotes metacognitive development leading to a more explicit understanding of what *makes* arguments logically superior.

As with the metalogical conception of rationality, this conception of argumentation sees it as a developmental process that acts on explicit representations, one that produces a consistent evolution toward greater rationality. However, it is also worth noting that a recent approach to argumentation proposed by Mercier and Sperber (2011) suggests a more unilateral conception with a variable relationship to rationality. They suggest that argumentation has evolutionary underpinnings for which the main aim is to achieve a form of dominance, by convincing others of the strength of an individual's point of view. In fact, this approach suggests that logical reasoning has evolved in order to support argumentation and that logic and rationality are rhetorical devices intended to construct an individual point of view strong enough to overcome other arguments (Mercier, 2011a, 2011b). Within this general perspective, rationality is seen as a skill that can be deployed or not, depending on the specific aims and context. There is some evidence that supports this point of view. For example, Klaczynski (2001) asked adolescents to evaluate arguments in terms of their logical adequacy. Older adolescents were quite good at picking out logical inconsistencies in relatively neutral arguments. However, where the same inconsistencies were used to support conclusions for which these adolescents had emotional beliefs, they tended very strongly to consider these arguments to be logically appropriate.

Both metalogical and argumentative approaches consider that there is a clear developmental progression in

the ability of children and adolescents to explicitly consider the logical status of inferences and arguments. Both see this progression to be long and relatively difficult, suggesting that rationality requires a long developmental sequence. One major difference between these two points of view is the locus of this development. Moshman's approach suggests that metacognitive control and understanding is essentially an individual process, mediated by thinking about the structure of inferences. The notion of argumentation, by contrast, suggests that the development of an explicit rationality is the product of an ongoing process of social interaction. However, in both cases, there are reasons to believe that the kinds of inferences and judgments that people make are not fully accounted for by their explicit understanding.

In fact, there is a great deal of evidence that the inferences people actually make are highly variable. Both of the approaches that we have examined clearly suggest that there is an ongoing developmental progression in the abilities of people to consciously understand the basic principles underlying explicit rationality. If this was accompanied by a corresponding control of the actual inferences and judgments that people make, then we would expect a clear developmental progression in the nature of the inferences that were made, with adults showing a clearly consistent pattern of rational judgments. However, there is a great deal of empirical evidence that adults are much less rational in their judgments than one would expect from such a point of view. Both the biases-and-heuristics program of Tversky and Kahneman (2004) and more current dual-process theories (Evans & Stanovich, 2013; Sloman, 1996) have clearly shown that people's judgments and inferences are far from logically appropriate. In fact, much of the research on the development of reasoning has examined what kinds of inferences people make irrespective of their explicit reasons, with the idea that the structure of these inferences gives an implicit picture of the underlying level of rationality.

3. The Development of Implicit Rationality

Before examining the different developmental theories that have attempted to explain how inferential reasoning develops in children and adolescents, it is useful to present the general interplay between logic and knowledge, which has been one of the key debates in this field. As mentioned previously, research on reasoning in adults has shown a remarkable level of what can be construed as irrationality. Although there are many factors underlying these effects, one of the more important of these is the strong influence of people's knowledge and beliefs on the inferences that they make. This has led some current theorists to suggest that classical "logic" may not be the best model to understand the adequacy of people's reasoning but that some form of Bayesian probability calculus might be better adapted to the reality of human reasoning (e.g., Evans, Over, & Handley, 2005; Oaksford & Chater, 2007; see also chapter 4.5 by Chater & Oaksford, this handbook). In other words, rationality might be construed as the ability to reason in a way that is consistent with one's knowledge. Developmental theories have, in contrast, tended to examine changes in the ability to generate inferences that correspond to logical norms, in a way that is relatively independent of empirical knowledge. Nonetheless, the interplay between logic and knowledge is critical to understanding how reasoning develops, and this is one of the major themes of the developmental theories that will be described.

There are, however, different mechanisms that could explain interactions between contextual knowledge and reasoning. On the one hand, it has been claimed that biology can account for at least some of the ways that people reason, since evolutionary mechanisms could create ways of thinking that are rational in the sense that they allow inferences and judgments that directly reflect important characteristics of evolved performance and social structures. Other explanations involve the development of specific cognitive processes that follow a developmental trajectory and results in developmental changes in reasoning abilities. We will examine different approaches to development of reasoning within this general distinction. We start by examining Piaget's theory, since this combines both biological and developmental perspectives, and it was the impetus for many subsequent developmental theories.

3.1 Piaget's Theory

Piaget's theory of cognitive development proposes that development occurs through the constant interaction between cognitive assimilation and accommodation, which are biological processes the use of which leads to a slow but progressive increase in the power of reasoning processes (Piaget, 1972). Without going into too much detail, one of the key factors characterizing the increasing ability to make logical inferences is the existence of three major levels of abstraction, with the transition being mediated by a process referred to as reflexive abstraction (Piaget, 2001). This is the ability to think about lower-level reasoning and represent the commonalities in more abstract form. The initial sensorimotor level uses action and perception-based

schemas. A second level, concrete operations, uses an initial form of abstraction that is linked to the general characteristics of physical objects. The third level, formal operations, uses a higher level of abstraction that allows the manipulation of general concepts that are no longer tied to physical objects. Progression through each of these three general levels leads to an increasing ability to make inferences and judgments that, at least in theory, reflect the logical structure appropriate to a given level. The key to understanding this progression is the idea that at any given level, children are "locally rational." That is, they make inferences that are consistent with the structural information available at any given level. Thus, sensorimotor inferences are consistent with the structure of physical action schemas, and so on. It is not until the formal operational level that inferences become rational in a larger sense, since the structure of formal operational schemas functions at a very high level of abstraction. Interestingly, empirical examination of this model has led to two very different criticisms. On the one hand, there is research that suggests that even very young children are capable of being logical in a way that should only be accessible to formal operational adolescents and adults (Hawkins, Pea, Glick, & Scribner, 1984). On the other hand, as we have already seen, much research suggests that even educated adults are not consistently logical.

One theory that has attempted to account for the latter results within the Piagetian framework is Overton's competence performance model (Overton & Ricco, 2011). This theory makes a distinction between the theoretical competence to reason logically, which follows the basic trajectory described by Piaget, and use of this competence, which is subject to a variety of performance-related factors. Consistent with this approach, several studies have found that manipulations designed to improve reasoning have effects that are clearly age related (O'Brien & Overton, 1980, 1982).

However, other approaches have focused on empirical results that indicate that very young children can sometimes reason logically. These have sometimes been interpreted within a biological perspective.

3.2 Evolution and Reasoning

Natural-logic theory A very different approach is provided by Braine's natural-logic theory (Braine, 1978; see also chapter 3.2 by O'Brien, this handbook). This theory has an empirical basis in studies that show that under certain circumstances, and at least with a limited range of inferences, even quite young children are able to make logical deductions (e.g., Hawkins et al., 1984). There are two major components to the theory. The first is the idea that people have specific algorithmic inference rules embedded into the cognitive system. These rules are essentially syntactic in form, and they allow analysis of the syntax of a given inference in order to generate a conclusion that is theoretically independent of context. The second component is a biological underpinning. The basic argument is that in certain circumstances, it is absolutely critical for individual survival to be able to make very rapid and accurate inferences. Thus, biology equips people with selected inference rules that allow consistent production of logical inferences in situations where these inferences are critical. One such example is the modus ponens inference, which is a basic inference from conditional premises of the form "If P then Q; P is true" to "Q is true."

Although this model of reasoning suggests that people are equipped with rules of inference that always give the logically correct response, existing empirical data show much more variability than would be suggested by the model. In order to account for such variability, the theory adds an interpretational module to the basic mechanisms underlying reasoning. This in turn creates an interaction between empirical knowledge, especially knowledge of the pragmatics of a given situation, and reasoning.

Cheater detection modules Another strong biological approach to reasoning was suggested by Cosmides (1989; see also chapter 10.6 by Cosmides & Tooby, this handbook). This theory had its origins in studies of the Wason selection task (Wason, 1968), which was a task first conceived as a way of showing that even very well-educated adults did not always make formally correct inferences. People were shown four cards, each of which had a number on one side and a letter on the other. A visible top of each card showed, respectively, a vowel, a consonant, an even number, and an odd number. Subjects were presented with a conditional rule of the form, "If a card has a vowel on one side, then it has an even number on the other." They were asked to choose which cards they would turn over in order to either confirm or disconfirm the rule. The logically correct answer is, "The card with a vowel and the card with an odd number." Both the original study, which was done with philosophy students, and several replications consistently found remarkably low levels of correct responding. Subsequent studies examined conditions under which the rate of correct responses was higher. One of the more intriguing results was that use of certain kinds of social rules led to very high levels of correct responding, for example, "If a person wants to drink at a bar, then they need to be over

21" (Cheng & Holyoak, 1985). With such a rule, people have no problems choosing cards that represent a person drinking and a person who is less than 21. In order to explain these results, Cosmides proposed that these social rules lead to good logical performance because they tap into a specific cheater detection module that has biological origins. The general proposition was based on the idea that social exchange and explicit or implicit contracts are a vital part of human social structure and have been so for a very long time. However, cheating is also a basic part of social interactions. Thus, it is argued that, in order to maintain a social structure based on exchange, people have developed a specific, specialized way of identifying cheaters. Consistent with this point of view, studies have shown that even very young children are able to produce logically correct inferences on some forms of the Wason selection task (Cummins, 1996; Girotto, Light, & Colbourn, 1988). Critically, such a module is biologically designed to allow people to interact in social situations in a way that maximizes the probability of successfully concluding a social contract but is not designed to reason logically. The fact that use of this module leads to the logically correct response in specific situations is thus basically incidental. In other words, this approach suggests that biology may have equipped people with inference mechanisms that are what we might call contextually rational. This means a kind of reasoning that accurately reflects the specific dimensions of a given social situation, which can be seen as rational, in the sense that it accurately reflects the specific parameters of the situation in question, but is not necessarily logical in a more formal sense.

Biologically based social reasoning It is interesting to examine some possible extensions of this notion of social reasoning, which shows the extent to which implicit inferences are affected by social schemas. One good example of this mechanism is given by attachment theory (Bowlby, 1969). The attachment system is biologically based and is designed to ensure the proximity of offspring to their principal caregiver during the period in which the offspring are most physically vulnerable. By itself, the system is a behavioral one that functions through an intricate mechanism of signals and internal emotions (Ainsworth, Blehar, Waters, & Wall, 1978). Depending on the reactions of the caregiver in situations of stress and/or anxiety, children create differing expectations, which results in very different forms of attachment behavior. For example, if a child's caregiver is physically affectionate and comforting in situations of stress, they will have a secure attachment. If the caregiver is cold and distant, this will result in an avoidant form of attachment. Critically, this behavioral cycle is internalized in the form of an internal model of a child's attachment figure (Bretherton & Munholland, 2008). This internal model is then used to interpret the behavior of potential romantic partners (Hazan & Shaver, 1987). In other words, the internal model generates inferences about how important people in a person's life will react, which in turn reflects their early life experiences (Baldwin, 1992).

As we can see from these different approaches, biology has been claimed to underlie both a universal system of strictly logical deduction and specialized systems of contextual inferences that accurately mirror the constraints of specific social situations. These theories are designed to explain empirical data that show that certain types of inferences (which can be considered rational in the sense that they correspond to either logical or social norms) are consistently made by even very young children. However, there is also a very strong body of empirical evidence that shows important developmental changes in children's and adolescent reasoning. What we might call information-processing theories attempt to specify the underlying mechanisms that could explain these developmental patterns.

3.3 Information-Processing Approaches to the Development of Reasoning

As we have previously indicated, these approaches can be seen as the result of two tendencies. On the one hand, there is clear evidence that the efficiency of cognitive processing increases over time. On the other hand, it is also the case that people's knowledge about the empirical world increases concurrently. Both of these factors are conceivably linked to the development of reasoning abilities. The following theories put different weights on these factors.

Barrouillet's mental model theory   One of the key factors to understanding the efficiency of cognitive processes is the capacity of working memory. This is a critical factor in the developmental approach to reasoning proposed by Barrouillet and colleagues (Barrouillet, Grosset, & Lecas, 2000; Barrouillet & Lecas, 1999; Gauffroy & Barrouillet, 2011). Their theory is based on mental model theory (Johnson-Laird & Byrne, 2002, which is a semantics-based approach to understanding reasoning (see chapter 2.3 by Johnson-Laird, this handbook). This attempts to model the processes that are used when people make deductions on the basis of a given set of major and minor premises. Since one of the principal forms of reasoning that have been examined by this

theory, and other developmental theories, is conditional (if–then) reasoning, we will use this as an example.

Conditional reasoning examines the inferences that are made on the basis of a major premise of the form "If P then Q." There are four inference forms that characterize such reasoning, constructed by minor premises having true or false forms of the antecedent (P) or consequent (Q) terms. Two of these lead to logically necessary conclusions. These are modus ponens (MP), "If P then Q; P is true," which leads to the necessary conclusion "Q is true," and modus tollens (MT), "If P then Q; Q is false," which leads to the necessary conclusion "P is false." The other two forms have what can be referred to as invited conclusions, which are however not logically necessary. These are affirmation of the consequent (AC), "If P then Q; Q is true," which leads to the invited conclusion "P is true," and denial of the antecedent (DA), "If P then Q; P is false," which leads to the invited conclusion "Q is false." However, for both the AC and the DA inferences, the logically correct response is to deny the invited conclusions.

Mental model theory proposes that people have a theoretical representation of conditionals that include combinations of antecedent and consequent terms that correspond to the true rows of the logical truth-table representation. However, the actual process people use is complicated by the fact that constructing such internal representations requires cognitive resources. Thus, a two-stage process is proposed that allows maximal use of memory in particular. When reasoning with a conditional statement "If P then Q," people start with an initial representation (model) of the statement, which simply represents the conjunction of the antecedent and the consequent terms, in the following way:

$$P \quad Q$$

for each of the four different inferences, then the model is scanned for cases involving the minor premise, particularly for cases in which the minor premise is associated with two different possible states. With this initial representation, people will make the MP inference and will accept the invited conclusion to the AC inference.

This initial representation can in some circumstances be fleshed out by the addition of other models that are also consistent with the conditional relation. The simplest of these leads to the following representation:

$$P \qquad Q$$
$$\text{not-}P \qquad \text{not-}Q$$

These models correspond to a biconditional interpretation, that is, "P is true if and only if Q is true." With this representation, people will make the correct inferences for the MP and the MT forms, and they will accept the invited conclusions for the AC and the DA forms. Such an interpretation corresponds to a pattern of inferences very frequently observed in children.

Finally, the full interpretation of conditionals requires the addition of a further model, leading to the following representation:

P Q
not-P Q
not-P not-Q

With such a representation, people will make the correct inferences for the MP and the MT forms, and they will reject the invited conclusions for the AC and the DA forms. Mental model theory supposes that the interpretation given to conditional statements is a basically semantic process based on understanding of language, modulated by pragmatic considerations (Johnson-Laird & Byrne, 2001). In other words, people will generate models of full conditionals unless cognitive and/or contextual constraints limit their ability to do so.

Barrouillet's theory makes the very strong claim that working-memory constraints are the primary developmental factor determining how children and adults reason (Barrouillet & Lecas, 1999). The key explanatory factor is the idea that working-memory capacity constrains the number of models that people can actively manipulate. Young children have very limited working-memory capacity and are limited to a single model. This leads to a conjunctive interpretation of conditionals. Older children have increasing working-memory abilities and are able to manipulate two models, which leads to the generation of a biconditional interpretation. Finally, much older adolescents and adults have sufficient working-memory capacity to generate the three models required for the full interpretation of conditionals.

Empirical evidence for this model has used tasks that attempt to directly examine the interpretation that is made of conditional statements. One such task is the truth-table task, in which people are given conditional statements and are asked whether each of the four combinations composed of true or false antecedent and consequent terms makes the conditional true, false, or neither true nor false. Several studies have found a clear developmental progression from a conjunctive interpretation, which is predominant until 8 or 9 years of age, to a biconditional interpretation, predominant until 11 to 12 years of age, to a conditional interpretation, which appears in mid-adolescence, with an additional relationship to measures of working memory as hypothesized

(Barrouillet & Lecas, 1999; although see Markovits, Brisson, & de Chantal, 2016).

This theory suggests that the ability to make logical inferences resides primarily in the increasing capacity of the cognitive system. Interestingly, it supposes that the underlying logic is determined by the way that statements are interpreted and that such interpretations are theoretically accessible even to young children. It should be noted that this theory does acknowledge the existence of interpretations reflecting contextual, pragmatic factors, which lead to inferences that are not logically adequate but pragmatically appropriate (Barrouillet, 2011). Nonetheless, it does incorporate one of the more controversial aspects of mental model theory, which is the idea that in ideal circumstances, reasoning will conform to the norms of standard logic. Thus, rationality in a larger sense can be seen as the result of a more adequate cognitive apparatus that can increasingly profit from the basic capacity for logical reasoning.

Semantic redescription theory Another approach that is based on mental model theory has been developed by Markovits and colleagues (Markovits, 2004; Markovits & Barrouillet, 2002). This theory postulates that the basic interpretation of conditionals is used as a retrieval cue for the activation of information present in semantic memory, which is then incorporated into a mental model. Two major classes of information have been shown to impact the way that people reason. The first of these are alternative antecedents. These are propositions A that differ from P and for which "If A then Q" is true. For example, consider the conditional rule "If a rock is thrown at a window, then the window will break," for which examples of alternative antecedents are "if a chair is thrown at a window," "if any hard object is thrown at a window," and so on. A great many empirical studies have shown that the tendency to accept or to reject the invited conclusions to the two invalid inferences (AC, DA) is directly related to the accessibility of alternative antecedents in memory (Cummins, 1995; Cummins, Lubart, Alksnis, & Rist, 1991; Markovits, Fleury, Quinn, & Venet, 1998; Markovits & Vachon, 1990; Thompson, 1995). Another form of information that is cued by conditional rules are disabling conditions (Cummins et al., 1991). These are conditions that allow P to be true without Q being true. For example, a disabling condition for the conditional rule presented above would be "The window is made of Plexiglas." Studies have shown that the tendency to accept the logical conclusions to the MP and MT inferences is related to the relative accessibility of disabling conditions in memory (Cummins, 1995; Cummins et al., 1991; De Neys, Schaeken, & d'Ydewalle,

2003). One of the first implications of this theory is that the ability to reason logically is associated with the efficiency of people's retrieval processes (Markovits & Quinn, 2002). However, the retrieval processes used during reasoning are not targeted but tend to result from activation of all the information that is associated with the premises (Janveau-Brennan & Markovits, 1999; Markovits & Potvin, 2001). Thus, the tendency to reject the AC and DA inferences is also associated with the tendency to reject the MP and MT inferences. This in turn suggests the importance of processes that are used to inhibit the retrieval of information that is inconsistent with the basic premises (Handley, Capon, Beveridge, Dennis, & Evans, 2004; Simoneau & Markovits, 2003). On a general level, this theory suggests that the ability to reason logically depends on the amount of information stored in semantic memory, the ability to retrieve this information, and the ability to inhibit retrieval of inappropriate information (i.e., information that contradicts the major premise). Since all of these factors increase developmentally, this results in a general improvement in the ability of children and adolescents to reason logically (Janveau-Brennan & Markovits, 1999).

However, this theory also postulates the existence of a redescriptive process as suggested by Karmiloff-Smith (1995), which produces descriptions of alternative antecedents of an increasingly abstract nature, resulting in a developmental increase in the ability to be logical with premises that are more abstract. Empirical studies that have examined the ability of children and adolescents to make consistently logical inferences have shown a clear age-related progression related to the content of the conditional rules, which in turn is related to the extent to which these rules allow accessibility of alternative antecedents and/or disabling conditions. These can be summarized by the following sequence:

- category-based conditional rules (if an animal is a dog, then it has legs),
- causal conditional rules (if a rock is thrown at a window, then the window will break),
- contrary-to-fact conditional rules (if a feather is thrown at a window, then the window will break), and
- abstract conditional rules (if XY, then it will blrp).

Studies have shown that children as young as seven or eight years of age are able to reason logically with category-based conditionals (Markovits, 2000; Markovits & Thompson, 2008). However, reasoning logically with causal conditionals is more difficult; it is not observed

before 11 or 12 years of age (Janveau-Brennan & Markovits, 1999; Markovits, 2017). Reasoning with contrary-to-fact conditionals and with abstract conditionals is much more complex, with the former showing high rates of logical reasoning only at late adolescence (Markovits, 2014; Markovits & Vachon, 1989), while abstract reasoning is not found at a high rate even with educated adults (Markovits & Lortie-Forgues, 2011; Markovits & Vachon, 1990). The redescriptive theory assumes that people think about the kinds of reasoning that they can do well and can use this to construct a more abstract representation of the processes involved, which then allows a higher level of reasoning.

This further supposes that the four categories of premise described previously correspond to two major levels of abstraction. The first two, comprising reasoning with category-based and causal premises, require retrieval of information that is directly or indirectly available in semantic memory. The second two, reasoning with contrary-to-fact and abstract premises, require the active construction of relevant information. Evidence for this comes from studies that have shown that it is possible to improve reasoning by asking reasoners to simply generate explicit alternatives for premises that are at a higher level of abstraction than those used for reasoning (Markovits, 2014; Markovits & Lortie-Forgues, 2011), while generating alternatives for highly familiar premises has no effect. In addition, these studies have found clear suppression effects showing that levels of logical reasoning with concrete and familiar premises are lowered when reasoners are first asked to reason with abstract premises (Markovits & Vachon, 1990).

3.4 Divergent Thinking and Reasoning

This theory also suggests an interesting relationship between logical reasoning and creativity. One of the more important factors that allow people to make logical inferences based on conditional rules is the ability to maintain the conditional in memory and at the same time to activate alternative antecedents. This process is made more difficult when premises contain terms that are already strongly associated (Quinn & Markovits, 1998), so that overall rates of logical reasoning are related to the ability of reasoners to retrieve information that is outside the usual range suggested by the premises (Markovits et al., 1998; Markovits & Quinn, 2002). This ability is very similar to that involved in divergent thinking, which is a key component of creativity (McCrae, 1987). There is in fact some evidence that differences in divergent thinking drive the very early ability of children to reason logically (de Chantal,

Gagnon-St-Pierre, & Markovits, 2020; de Chantal & Markovits, 2017; Markovits & Brunet, 2012). For example, very young children have great difficulty in accepting the uncertainty of even a very simple category-based inference, such as "If an animal is a dog, then it has four legs. An animal has four legs. Is it a dog?" However, if they are asked to do a divergent-thinking task (without examples) such as "We can make noise with many things. I'd like to know some ways of making noise," their ability to accept uncertainty increases greatly. The general idea that rationality in a broader sense depends upon the growing ability of reasoners to envisage possibilities (which is important for divergent thinking) and then to account for these possibilities when making inferences, is very similar to the interplay between possibility and necessity that was identified by Piaget in his later works (Gauffroy & Barrouillet, 2011; Piaget, 1987).

4. Conclusion

While there is no real consensus about the way that rationality develops, the variety of approaches does suggest a multidimensional trajectory that might usefully be extrapolated from these theories and the associated empirical data. One of the themes is the interaction between context-dependent forms of reasoning and more general forms of reasoning, of which the strongest version is metacognitive and abstract. The simplest version of this would suggest that the earliest form of rational thinking is contextual, either due to biological underpinnings or early experience, with rationality implying reasoning that reflects specific situational constraints. Subsequent development would involve a gradual replacement of this form of reasoning with more abstract and general forms.

However, one interpretation of the many studies that have examined the development of reasoning is that inferences and judgments can be seen as interactions between some form of rationality and a growing understanding of the structure of the empirical world, with the latter anchored in a biological substrate. Actual inferences can then be seen as the product of an ongoing exchange between children's logical abilities and their increasing understanding of the empirical world. Each of these components is subject to developmental variation underpinned by clear increases in the computational power of the cognitive system. Thus, development allows children to become both more logical and more conversant with the way that the empirical and social worlds are constructed. The paradox of development is that real-world knowledge is necessary for the construction of an increasingly powerful ability to make rational judgments, and it is at the same time a driver of judgments that reflect the often irrational structure of the social and possibly the physical world (Artman, Cahan, & Avni-Babad, 2006). Accounting for this complex interaction is one of the major challenges of understanding the development of rational thought.

References

Ainsworth, M. D. S., Blehar, M. C., Waters, E., & Wall, S. N. (1978). *Patterns of attachment: A psychological study of the strange situation*. New York, NY: Psychology Press.

Ames, G. J., & Murray, F. B. (1982). When two wrongs make a right: Promoting cognitive change by social conflict. *Developmental Psychology*, *18*(6), 894–897.

Artman, L., Cahan, S., & Avni-Babad, D. (2006). Age, schooling and conditional reasoning. *Cognitive Development*, *21*(2), 131–145.

Baldwin, M. W. (1992). Relational schemas and the processing of social information. *Psychological Bulletin*, *112*(3), 461–484.

Barrouillet, P. (2011). Dual-process theories of reasoning: The test of development. *Developmental Review*, *31*(2), 151–179.

Barrouillet, P., Grosset, N., & Lecas, J.-F. (2000). Conditional reasoning by mental models: Chronometric and developmental evidence. *Cognition*, *75*(3), 237–266.

Barrouillet, P., & Lecas, J.-F. (1999). Mental models in conditional reasoning and working memory. *Thinking & Reasoning*, *5*(4), 289–302.

Berk, L. E., & Winsler, A. (1995). *Scaffolding children's learning: Vygotsky and early childhood education* (NAEYC Research into Practice Series, Vol. 7). Washington, DC: Education Resources Information Center (ERIC).

Bowlby, J. (1969). *Attachment and loss: Vol. 1. Attachment*. London, England: The Hogarth Press and the Institute of Psycho-Analysis.

Braine, M. D. (1978). On the relation between the natural logic of reasoning and standard logic. *Psychological Review*, *85*(1), 1–21.

Bretherton, I., & Munholland, K. A. (2008). Internal working models in attachment relationships: Elaborating a central construct in attachment theory. In J. Cassidy & P. R. Shaver (Eds.), *Handbook of attachment: Theory, research, and clinical applications* (pp. 102–127). New York, NY: Guilford Press.

Cheng, P. W., Holyoak, K. J. (1985). Pragmatic reasoning schemas. *Cognitive Psychology*, *17*, 391–416.

Christoforides, M., Spanoudis, G., & Demetriou, A. (2016). Coping with logical fallacies: A developmental training program for learning to reason. *Child Development*, *87*(6), 1856–1876.

Cosmides, L. (1989). The logic of social exchange: Has natural selection shaped how humans reason? Studies with the Wason selection task. *Cognition, 31*(3), 187–276.

Crowell, A., & Kuhn, D. (2014). Developing dialogic argumentation skills: A 3-year intervention study. *Journal of Cognition and Development, 15*(2), 363–381.

Cummins, D. D. (1995). Naive theories and causal deduction. *Memory & Cognition, 23*(5), 646–658.

Cummins, D. D. (1996). Evidence for the innateness of deontic reasoning. *Mind & Language, 11*(2), 160–190.

Cummins, D. D., Lubart, T., Alksnis, O., & Rist, R. (1991). Conditional reasoning and causation. *Memory & Cognition, 19*(3), 274–282.

de Chantal, P.-L., Gagnon-St-Pierre, É., & Markovits, H. (2020). Divergent thinking promotes logical reasoning in very young children. *Child Development, 91*(4), 1081–1097.

de Chantal, P.-L., & Markovits, H. (2017). The capacity to generate alternative ideas is more important than inhibition for logical reasoning in preschool-age children. *Memory & Cognition, 45*(2), 208–220.

Demetriou, A., Spanoudis, G., & Shayer, M. (2014). Inference, reconceptualization, insight, and efficiency along intellectual growth: A general theory. *Enfance, 2014*(3), 365–396.

De Neys, W., Schaeken, W., & d'Ydewalle, G. (2003). Causal conditional reasoning and strength of association: The disabling condition case. *European Journal of Cognitive Psychology, 15*(2), 161–176.

Doise, W., & Mugny, G. (1981). *Le développement social de l'intelligence* [The social development of intelligence] (Vol. 1). Paris, France: InterEditions.

Evans, J. St. B. T., Over, D. E., & Handley, S. J. (2005). Suppositionals, extensionality, and conditionals: A critique of the mental model theory of Johnson-Laird and Byrne (2002). *Psychological Review, 112*, 1040–1052.

Evans, J. St. B. T., & Stanovich, K. E. (2013). Dual-process theories of higher cognition advancing the debate. *Perspectives on Psychological Science, 8*(3), 223–241.

Flavell, J. H. (1979). Metacognition and cognitive monitoring: A new area of cognitive–developmental inquiry. *American Psychologist, 34*(10), 906–911.

Gauffroy, C., & Barrouillet, P. (2011). The primacy of thinking about possibilities in the development of reasoning. *Developmental Psychology, 47*(4), 1000–1011.

Girotto, V., Light, P., & Colbourn, C. (1988). Pragmatic schemas and conditional reasoning in children. *Quarterly Journal of Experimental Psychology, 40*(3), 469–482.

Hallett, D., Chandler, M. J., & Krettenauer, T. (2002). Disentangling the course of epistemic development: Parsing knowledge by epistemic content. *New Ideas in Psychology, 20*(2), 285–307.

Handley, S. J., Capon, A., Beveridge, M., Dennis, I., & Evans, J. St. B. T. (2004). Working memory, inhibitory control and the development of children's reasoning. *Thinking & Reasoning, 10*(2), 175–195.

Hawkins, J., Pea, R. D., Glick, J., & Scribner, S. (1984). "Merds that laugh don't like mushrooms": Evidence for deductive reasoning by preschoolers. *Developmental Psychology, 20*(4), 584–594.

Hazan, C., & Shaver, P. (1987). Romantic love conceptualized as an attachment process. *Journal of Personality and Social Psychology, 52*(3), 511–524.

Janveau-Brennan, G. M., & Markovits, H. (1999). The development of reasoning with causal conditionals. *Developmental Psychology, 35*(4), 904–911.

Johnson-Laird, P. N., & Byrne, R. M. J. (2002). Conditionals: A theory of meaning, pragmatics, and inference. *Psychological Review, 109*(4), 646–678.

Karmiloff-Smith, A. (1995). *Beyond modularity: A developmental perspective on cognitive science.* Cambridge, MA: MIT Press.

Klaczynski, P. A. (2001). Analytic and heuristic processing influences on adolescent reasoning and decision-making. *Child Development, 72*(3), 844–861.

Kuhn, D. (1991). *The skills of argument.* Cambridge, England: Cambridge University Press.

Kuhn, D. (1993). Science as argument: Implications for teaching and learning scientific thinking. *Science Education, 77*(3), 319–337.

Kuhn, D., & Crowell, A. (2011). Dialogic argumentation as a vehicle for developing young adolescents' thinking. *Psychological Science, 22*(4), 545–552.

Kuhn, D., & Udell, W. (2003). The development of argument skills. *Child Development, 74*(5), 1245–1260.

Kuhn, D., Zillmer, N., Crowell, A., & Zavala, J. (2013). Developing norms of argumentation: Metacognitive, epistemological, and social dimensions of developing argumentive competence. *Cognition and Instruction, 31*(4), 456–496.

Markovits, H. (2000). A mental model analysis of young children's conditional reasoning with meaningful premises. *Thinking & Reasoning, 6*(4), 335–347.

Markovits, H. (2004). The development of deductive reasoning. In J. P. Leighton & R. J. Sternberg (Eds.), *The nature of reasoning* (pp. 313–338). New York, NY: Cambridge University Press.

Markovits, H. (2014). On the road toward formal reasoning: Reasoning with factual causal and contrary-to-fact causal premises during early adolescence. *Journal of Experimental Child Psychology, 128*, 37–51.

Markovits, H. (2017). In the beginning stages: Conditional reasoning with category based and causal premises in 8- to 10-year olds. *Cognitive Development*, *41*, 1–9.

Markovits, H., Barrouillet, P. (2002). The development of conditional reasoning: A mental model account. *Developmental Review*, *22*(1), 5–36.

Markovits, H., Brisson, J., & de Chantal, P.-L. (2016). How do pre-adolescent children interpret conditionals? *Psychonomic Bulletin & Review*, *23*(6), 1907–1912.

Markovits, H., & Brunet, M.-L. (2012). Priming divergent thinking promotes logical reasoning in 6- to 8-year olds: But more for high than low SES students. *Journal of Cognitive Psychology*, *24*(8), 991–1001.

Markovits, H., Fleury, M.-L., Quinn, S., & Venet, M. (1998). The development of conditional reasoning and the structure of semantic memory. *Child Development*, *69*(3), 742–755.

Markovits, H., & Lortie-Forgues, H. (2011). Conditional reasoning with false premises facilitates the transition between familiar and abstract reasoning. *Child Development*, *82*(2), 646–660.

Markovits, H., & Potvin, F. (2001). Suppression of valid inferences and knowledge structures: The curious effect of producing alternative antecedents on reasoning with causal conditionals. *Memory & Cognition*, *29*(5), 736–744.

Markovits, H., & Quinn, S. (2002). Efficiency of retrieval correlates with "logical" reasoning from causal conditional premises. *Memory & Cognition*, *30*(5), 696–706.

Markovits, H., & Thompson, V. (2008). Different developmental patterns of simple deductive and probabilistic inferential reasoning. *Memory & Cognition*, *36*(6), 1066–1078.

Markovits, H., Thompson, V. A., & Brisson, J. (2015). Metacognition and abstract reasoning. *Memory & Cognition*, *43*(4), 681–693.

Markovits, H., & Vachon, R. (1989). Reasoning with contrary-to-fact propositions. *Journal of Experimental Child Psychology*, *47*(3), 398–412.

Markovits, H., & Vachon, R. (1990). Conditional reasoning, representation, and level of abstraction. *Developmental Psychology*, *26*(6), 942–951.

McCrae, R. R. (1987). Creativity, divergent thinking, and openness to experience. *Journal of Personality and Social Psychology*, *52*(6), 1258.

Mercier, H. (2011a). On the universality of argumentative reasoning. *Journal of Cognition and Culture*, *11*(1–2), 1–2.

Mercier, H. (2011b). Reasoning serves argumentation in children. *Cognitive Development*, *26*(3), 177–191.

Mercier, H., & Sperber, D. (2011). Why do humans reason? Arguments for an argumentative theory. *Behavioral and Brain Sciences*, *34*(2), 57–74.

Moshman, D. (1990). The development of metalogical understanding. In W. F. Overton (Ed.), *Reasoning, necessity, and logic: Developmental perspectives* (pp. 205–225). Hillsdale, NJ: Erlbaum.

Moshman, D. (2004). From inference to reasoning: The construction of rationality. *Thinking & Reasoning*, *10*(2), 221–239.

Moshman, D. (2011). Evolution and development of reasoning and argumentation: Commentary on Mercier (2011). *Cognitive Development*, *26*(3), 192–195.

Moshman, D. (2014). Epistemic cognition and development. In P. Barrouillet & C. Gauffroy (Eds.), *The development of thinking and reasoning* (pp. 13–33). London, England: Psychology Press.

Moshman, D., & Franks, B. A. (1986). Development of the concept of inferential validity. *Child Development*, *57*(1), 153–165.

Moshman, D., & Geil, M. (1998). Collaborative reasoning: Evidence for collective rationality. *Thinking & Reasoning*, *4*(3), 231–248.

Murray, F. B. (1972). Acquisition of conservation through social interaction. *Developmental Psychology*, *6*(1), 1–6.

Newman, I. R., Gibb, M., & Thompson, V. A. (2017). Rule-based reasoning is fast and belief-based reasoning can be slow: Challenging current explanations of belief-bias and base-rate neglect. *Journal of Experimental Psychology: Learning, Memory, and Cognition*, *43*(7), 1154–1170.

Oaksford, M., & Chater, N. (2007). *Bayesian rationality*. Oxford, England: Oxford University Press.

O'Brien, D. P., & Overton, W. F. (1980). Conditional reasoning following contradictory evidence: A developmental analysis. *Journal of Experimental Child Psychology*, *30*(1), 44–61.

O'Brien, D. P., & Overton, W. F. (1982). Conditional reasoning and the competence–performance issue: A developmental analysis of a training task. *Journal of Experimental Child Psychology*, *34*(2), 274–290.

Overton, W. F., & Ricco, R. B. (2011). Dual-systems and the development of reasoning: Competence–procedural systems. *Wiley Interdisciplinary Reviews: Cognitive Science*, *2*(2), 231–237.

Perelman, C. (1971). The new rhetoric. In Y. Bar-Hillel (Ed.), *Pragmatics of natural languages* (pp. 145–149). Dordrecht, Netherlands: Reidel.

Piaget, J. (1972). Intellectual evolution from adolescence to adulthood. *Human Development*, *15*, 1–12.

Piaget, J. (1987). *Possibility and necessity: Vol. 1. The role of possibility in cognitive development* (H. Feider, Trans.). Minneapolis: University of Minnesota Press.

Piaget, J. (2001). *Studies in reflecting abstraction*. Hove, England: Psychology Press.

Quinn, S., & Markovits, H. (1998). Conditional reasoning, causality, and the structure of semantic memory: Strength of association as a predictive factor for content effects. *Cognition, 68*, B93–B101.

Simoneau, M., & Markovits, H. (2003). Reasoning with premises that are not empirically true: Evidence for the role of inhibition and retrieval. *Developmental Psychology, 39*(6), 964–975.

Sloman, S. A. (1996). The empirical case for two systems of reasoning. *Psychological Bulletin, 119*(1), 3–22.

Thompson, V. A. (1995). Conditional reasoning: The necessary and sufficient conditions. *Canadian Journal of Experimental Psychology/Revue canadienne de psychologie expérimentale, 49*(1), 1–60.

Thompson, V. A., Prowse Turner, J. A., & Pennycook, G. (2011). Intuition, reason, and metacognition. *Cognitive Psychology, 63*, 107–140.

Tversky, A., & Kahneman, D. (2004). *Judgment under uncertainty: Heuristics and biases*. New York, NY: Psychology Press.

Wason, P. (1968). Reasoning about a rule. *Quarterly Journal of Experimental Psychology, 20*, 273–281.

15.2 Rationality and Intelligence

Keith E. Stanovich, Maggie E. Toplak, and Richard F. West

Summary

There are individual differences in rational thinking that are less than perfectly correlated with individual differences in intelligence because intelligence and rationality occupy different conceptual locations in models of cognition. A tripartite extension of currently popular dual-process theories is presented in this chapter that illustrates how intelligence and rationality are theoretically separate concepts. The chapter concludes by showing how this tripartite model of mind, taken in the context of studies of individual differences, can help to resolve the Great Rationality Debate in cognitive science—the debate about how much irrationality to attribute to human cognition.

1. What Intelligence Is—and Why It Is Not the Same as Rationality

Because both intelligence and rationality have a plethora of definitions that differ across the various disciplines, it is not surprising that our chapter will begin with some definitional clarifications. We take our definitions from cognitive science, so those defaulting to other fields and disciplines may be confused if they insist on stipulating definitions we are not using. Also, if rationality and intelligence are to be compared and contrasted, each must be defined at a similar grain size, which we do in this chapter. That is, rationality defined in some ways would be a category error if directly compared with intelligence. Because intelligence is an individual difference concept to a psychologist, rationality must be too. That rules out certain definitions that are popular in philosophy and in lay discourse.

We get surprised when someone whom we consider to be smart acts stupidly. When someone we consider to be not so smart acts stupidly, we tend not to be so surprised. But why should we be so surprised in the first case? A typical dictionary definition of the adjectival form of the word "smart" is "characterized by sharp quick thought;

bright" or "having or showing quick intelligence or ready mental capacity." Thus, being smart seems a lot like being intelligent, according to the dictionary. Dictionaries also tell us that a stupid person is "slow to learn or understand; lacking or marked by lack of intelligence." Thus, if a smart person is intelligent and "stupid" means a lack of intelligence, then the "smart person being stupid" phrase seems to make no sense.

However, a secondary definition of the word "stupid" is "tending to make poor decisions or careless mistakes"—a phrase that attenuates the sense of contradiction. Thus, the phrase "smart but acting dumb"—intelligent people taking injudicious actions or holding unjustified beliefs—means that folk psychology is picking out two different traits: mental brightness (intelligence) and making judicious decisions (rational thinking). If we were clear about the fact that the two traits are different, the sense of paradox or surprise at the "smart but acting foolish" phenomenon would vanish. What perpetuates the surprise is that we tend to think of the two traits as one, or at least that they should be strongly associated. The confusion is fostered because psychology has a measurement device for the first (the intelligence test) but not the second. Psychology has a long and storied history (over one hundred years old) of measuring the intelligence trait. Although there has been psychological work on rational thinking, this research started much later and was not focused on individual differences.

Many treatments of the intelligence concept could be characterized as permissive rather than grounded conceptualizations. Permissive theories include in their definitions of intelligence aspects of functioning that are captured by the vernacular term "intelligence" (adaptation to the environment, showing wisdom, creativity, etc.) whether or not these aspects are actually measured by existing tests of intelligence. Grounded theories, in contrast, confine the concept of intelligence to the set of mental abilities actually tested on IQ tests. Adopting permissive definitions of the concept of intelligence serves to obscure what is absent from extant IQ tests. Instead, in order to highlight the missing elements in IQ tests, we adopt a thoroughly grounded notion of the intelligence concept in this chapter—one that anchors the concept in what actual IQ tests measure. Likewise, we ground the concept of rationality in operationalizations from current cognitive science.

2. Intelligence and Rationality in Cognitive Science

The closest thing to a consensus, grounded theory of intelligence in psychology is the Cattell–Horn–Carroll (CHC) theory of intelligence (Carroll, 1993; Cattell, 1963, 1998; Horn & Cattell, 1967). It yields a scientific concept of general intelligence, usually symbolized by g, and a small number of broad factors, of which two are dominant. Fluid intelligence reflects reasoning abilities operating across a variety of domains—including novel ones. It is measured by tests of abstract thinking such as figural analogies, Raven Matrices, and series completion. Crystallized intelligence reflects declarative knowledge acquired from acculturated learning experiences. It is measured by vocabulary tasks, verbal comprehension, and general knowledge measures.

"Rationality" is a tortuous and tortured term in intellectual discourse (Stanovich, West, & Toplak, 2016). Many philosophical notions of rationality are crafted so as to equate all humans—thus, by fiat, defining away the very individual differences that a psychologist wishes to study. In contrast, rationality in the sense employed in cognitive science—and in this book—is a normative notion. Rationality thus comes in degrees defined by the distance of the thought or behavior from the optimum defined by a normative model (Etzioni, 2014). Thus, when a cognitive scientist terms a behavior less than rational, he or she means that the behavior departs from the optimum prescribed by a particular normative model.

We follow many cognitive science theorists in recognizing two types of rationality, instrumental and epistemic (Manktelow, 2004; Over, 2004). The simplest definition of instrumental rationality is the following: behaving in the world so that you get exactly what you most want, given the resources (physical and mental) available to you. Epistemic rationality concerns how well beliefs map onto the actual structure of the world.

More formally, economists and cognitive scientists define instrumental rationality as the maximization of expected utility. Expected utility is calculated by taking the utility of each outcome and multiplying it by the probability of that outcome occurring and then summing those products over all of the possible outcomes. In practice, assessing rationality in this computational

manner can be difficult because eliciting personal probabilities can be tricky. Also, getting measurements of the utilities of various consequences can be experimentally difficult. Fortunately, there is another useful way to measure the rationality of decisions and deviations from rationality. It has been proven through several formal analyses that if people's preferences follow certain consistent patterns (the so-called axioms of choice), then they are behaving as if they are maximizing utility (Dawes, 1998; Edwards, 1954; Jeffrey, 1983; Luce & Raiffa, 1957; Savage, 1954; von Neumann & Morgenstern, 1944). These analyses have led to what has been termed the axiomatic approach to whether people are maximizing utility. It is what makes people's degrees of rationality more easily measurable by the experimental methods of cognitive science. The deviation from the optimal choice pattern according to the axioms is an (inverse) measure of the degree of rationality.

A substantial research literature—one comprising literally hundreds of empirical studies conducted over several decades—has firmly established that people's responses sometimes deviate from the performance considered normative on many reasoning tasks. For example, people assess probabilities incorrectly, they test hypotheses inefficiently, they violate the axioms of utility theory, they do not properly calibrate degrees of belief, their choices are affected by irrelevant context, they ignore the alternative hypothesis when evaluating data, and they display numerous other information-processing biases (Baron, 2008, 2014; Evans, 2014; Kahneman, 2011; Stanovich, 1999, 2011; Stanovich et al., 2016). Much of the operationalization of rational thinking in cognitive science comes from the heuristics and biases tradition, inaugurated by Kahneman and Tversky in the early 1970s (Kahneman & Tversky, 1972, 1973; Tversky & Kahneman, 1974). Thus, as measures of rationality, the tasks in the heuristics and biases literature, while tapping intelligence in part, actually encompass more cognitive processes and knowledge than are assessed by IQ tests. In the next section, we will outline the functional cognitive theory that we will use to interpret the rational thinking tasks in this literature and show how they relate to intelligence. We will show that rationality is actually a more encompassing mental construct than is intelligence.

3. Dual-Process Theory: The First Step toward a Model of Cognitive Architecture

There is a wide variety of evidence that has converged on the conclusion that some type of dual-process model of the mind is needed in a diverse set of specialty areas not limited to cognitive psychology, economics, social psychology, naturalistic philosophy, decision theory, and clinical psychology (Chein & Schneider, 2012; De Neys, 2018; Evans, 2008, 2010, 2014; Evans & Stanovich, 2013; Sherman, Gawronski, & Trope, 2014; Stanovich, 1999, 2004). Because there is now a plethora of dual-process theories (see Stanovich, 2011, 2012, for a list of the numerous versions of such theories), there is currently much variation in the terms for the two processes. For the purposes of this chapter, we will most often adopt the Type 1/Type 2 terminology discussed by Evans and Stanovich (2013) and occasionally use the similar System 1/System 2 terminology of Stanovich (1999) and Kahneman (2011). The defining feature of Type 1 processing is its autonomy—the execution of Type 1 processes is mandatory when their triggering stimuli are encountered, and they are not dependent on input from high-level control systems. Autonomous processes have other correlated features—their execution tends to be rapid, they do not put a heavy load on central processing capacity, and they tend to be associative—but these other correlated features are not defining (Stanovich & Toplak, 2012).

In contrast with Type 1 processing, Type 2 processing is nonautonomous. It is relatively slow and computationally expensive. Many Type 1 processes can operate at once in parallel, but Type 2 processing is largely serial. One of the most critical functions of Type 2 processing is to override Type 1 processing. This is because Type 1 processing heuristics depend on benign environments providing obvious cues that elicit adaptive behaviors.

In hostile environments, reliance on heuristics can be costly (see Hilton, 2003; Over, 2000; Stanovich, 2004).

Once detection of the conflict between the normative response and the response triggered by System 1 has taken place (De Neys & Pennycook, 2019; Stanovich, 2018), Type 2 processing must display at least two related capabilities in order to override Type 1 processing. One is the capability of interrupting Type 1 processing. The second is to enable processes of hypothetical reasoning and cognitive simulation that are a unique aspect of Type 2 processing (Evans, 2007, 2010; Evans & Stanovich, 2013). In order to reason hypothetically, we must, however, have one critical cognitive capability—we must be able to prevent our representations of the real world from becoming confused with representations of imaginary situations. The so-called cognitive decoupling operations (Stanovich, 2011; Stanovich & Toplak, 2012) are the central feature of Type 2 processing that make this possible, and they have implications for how we conceptualize both intelligence and rationality, as we shall see. The important issue for our purposes is that decoupling secondary representations from the world and then maintaining the decoupling while simulation is carried out is a Type 2 processing operation. It is computationally taxing and greatly restricts the ability to conduct any other Type 2 operation simultaneously.

A preliminary dual-process model of mind, based on what we have outlined thus far, is presented in figure 15.2.1. The figure shows the Type 2 override function we have been discussing, as well as the Type 2 process

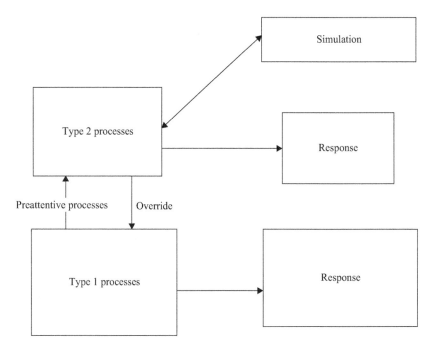

Figure 15.2.1
A preliminary dual-process model.

of simulation. Also rendered in the figure is an arrow indicating that Type 2 processes receive inputs from Type 1 computations. These so-called preattentive processes (Evans, 2008) establish the content of most Type 2 processing.

4.   Differentiating Type 2 Processes: The Reflective and Algorithmic Minds

In this section, we will explain why rational thinking stresses a level in the hierarchical control system of the brain that is only partly tapped by IQ tests. This is because the override mechanism depicted in figure 15.2.1 needs to be conceptualized in terms of two levels of processing— what are sometimes termed the algorithmic and reflective levels of processing (Stanovich, 2011).

Our attempt to differentiate the two levels of control involved in Type 2 processing is displayed in figure 15.2.2. The psychological literature provides much converging evidence and theory to support such a structure. First, psychometricians have long distinguished typical performance situations from optimal (sometimes termed "maximal") performance situations (Ackerman & Kanfer, 2004). Typical performance measures implicate, at least in part, the reflective mind—they assess goal prioritization and epistemic regulation. In contrast, optimal performance situations are those where the task interpretation is determined externally. The person performing the task is told the rules that maximize performance. Thus, optimal performance tasks assess questions of the efficiency of goal pursuit—they capture the processing efficiency of the algorithmic mind. All tests of intelligence or cognitive aptitude are optimal performance assessments, whereas

measures of critical or rational thinking are often assessed under typical performance conditions.

The difference between the algorithmic mind and the reflective mind is captured in another well-established distinction in the measurement of individual differences— the distinction between cognitive ability (intelligence) and thinking dispositions. The former are, as just mentioned, measures of the efficiency of the algorithmic mind. The latter travel under a variety of names in psychology—thinking dispositions or cognitive styles being the two most popular. Examples of some thinking dispositions relevant to rationality that have been investigated by psychologists are actively open-minded thinking, need for cognition, consideration of future consequences, need for closure, and dogmatism (see Stanovich et al., 2016).

In short, measures of individual differences in thinking dispositions are assessing variation in people's goal management, epistemic values, and epistemic self-regulation— differences in the operation of the reflective mind. People have indeed come up with definitions of intelligence that encompass the reflective level of processing, but nevertheless, the actual measures of intelligence in use assess only algorithmic-level cognitive capacity.

Figure 15.2.2 represents the classification of individual differences in the tripartite view. The broken horizontal line represents the location of the key distinction in older, dual-process views. Figure 15.2.2 identifies variation in fluid intelligence with individual differences in the efficiency of processing of the algorithmic mind. To a substantial extent, fluid intelligence measures the ability to cognitively decouple—to suppress Type 1 activity and to enable hypothetical thinking. The raw ability to sustain such simulations while keeping the

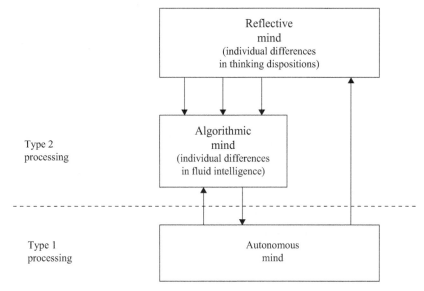

Figure 15.2.2
The tripartite structure and the locus of individual differences.

relevant representations decoupled is one key aspect of the brain's computational power that is being assessed by measures of fluid intelligence. This is becoming clear from converging work on executive function and working memory, which both display correlations with fluid intelligence that are quite high (Duncan et al., 2008; Jastrzębski, Ciechanowska, & Chuderski, 2018; Stanovich et al., 2016). This is because most measures of executive function, such as working memory, are direct or indirect indicators of a person's ability to sustain decoupling operations. Thus, Type 2 processes are strongly associated with fluid intelligence. Finally, the reflective mind is identified with individual differences in thinking dispositions related to beliefs and goals.

5.   Why Rationality Is a More Encompassing Construct than Intelligence

Figure 15.2.2 highlights an important sense in which rationality is a more encompassing construct than intelligence: as previously discussed, to be rational, a person must have well-calibrated beliefs and must act appropriately on those beliefs to achieve goals—both of these depend on the thinking dispositions of the reflective mind. The types of cognitive propensities that these thinking disposition measures reflect are the tendency to collect information before making up one's mind, the tendency to seek various points of view before coming to a conclusion, the disposition to think extensively about a problem before responding, the tendency to calibrate the strength of one's opinion to the degree of evidence available, the tendency to think about future consequences before taking action, the tendency to explicitly weigh pluses and minuses of situations before making a decision, and the tendency to seek nuance and avoid absolutism.

In order to achieve both epistemic and instrumental rationality, an individual must also, of course, have the algorithmic-level machinery that enables him or her to carry out the actions and to process the environment in a way that enables the correct beliefs to be fixed and the correct actions to be taken. Thus, individual differences in rational thought and action can arise because of individual differences in fluid intelligence (the algorithmic mind) or because of individual differences in thinking dispositions (the reflective mind), or from a combination of both.

To put it simply, the concept of rationality encompasses thinking dispositions and algorithmic-level capacity, whereas the concept of intelligence (at least as it is commonly operationalized) is largely confined to algorithmic-level capacity. Intelligence tests do not attempt to measure aspects of epistemic or instrumental rationality, nor do they examine any thinking dispositions that relate to rationality. Thus, as long as variation in thinking dispositions is not perfectly correlated with variation in fluid intelligence, there is the statistical possibility of rationality and intelligence explaining at least partially separable variance.

In fact, substantial empirical evidence indicates that individual differences in thinking dispositions and intelligence are far from perfectly correlated. Studies (e.g., Ackerman & Heggestad, 1997; Cacioppo, Petty, Feinstein, & Jarvis, 1996) have indicated that measures of intelligence display only moderate-to-weak correlations with some thinking dispositions (e.g., actively open-minded thinking, need for cognition) and near-zero correlations with others (e.g., conscientiousness, curiosity, diligence). Other important evidence supports the conceptual distinction made here between algorithmic cognitive capacity and thinking dispositions. For example, across a variety of tasks from the heuristics and biases literature, it has consistently been found that rational thinking dispositions will predict variance after the effects of general intelligence have been controlled (for a discussion and citations, see Stanovich et al., 2016).

6.   The Fleshed-Out Model

The functions of the different levels of control are illustrated more completely in figure 15.2.3. There, it is clear that the override capacity itself is a property of the algorithmic mind, and it is indicated by the arrow labeled A. However, previous dual-process theories have tended to ignore the higher-level cognitive operation that initiates the override function in the first place. This is a dispositional property of the reflective mind that is related to rationality. In the model in figure 15.2.3, it corresponds to arrow B, which represents the instruction to the algorithmic mind to override the Type 1 response by taking it offline. This is a different mental function than the override function itself (arrow A), and the evidence cited above indicates that the two functions are indexed by different types of individual differences.

The override function has loomed so large in dual-process theory that it has somewhat overshadowed the simulation process that computes the alternative response that makes the override worthwhile. Thus, figure 15.2.3 explicitly represents the simulation function as well as the fact that the instruction to initiate simulation originates in the reflective mind. The decoupling operation itself (indicated by arrow C) is carried out by the algorithmic mind. The instruction to initiate

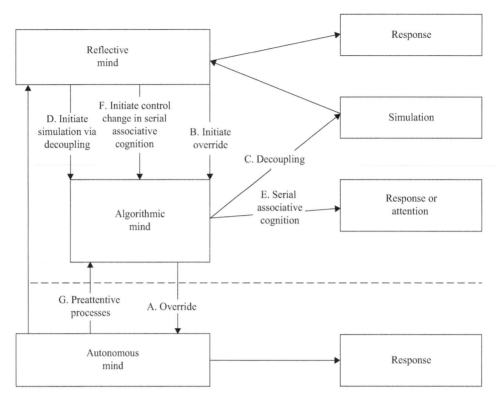

Figure 15.2.3
A more complete model of the tripartite structure.

simulation (indicated by arrow D) is carried out by the reflective mind. Again, two different types of individual differences are associated with the initiation call and the decoupling operator—specifically, thinking dispositions with the former and fluid intelligence with the latter. Also represented is the fact that the higher levels of control receive inputs from the autonomous mind (arrow G) via so-called preattentive processes.

The arrows labeled E and F reflect the decoupling and higher-level control of a kind of Type 2 processing (serial associative cognition) that does not involve fully explicit cognitive simulation, but we will not review its function here (see Stanovich, 2011). In figure 15.2.3, we can now identify a third function of the reflective mind: initiating an interrupt of serial associative cognition (arrow F). This interrupt signal alters the next step in a serial associative sequence that would otherwise direct thought. This interrupt signal might stop serial associative cognition altogether in order to initiate a comprehensive simulation (arrow C) or start a new serial associative chain (arrow E) from a different starting point.

Although taking the Type 1 response priming offline might itself be procedural, the process of synthesizing an alternative response often utilizes stored knowledge of various types. During the simulation process, declarative

knowledge and strategic rules (linguistically coded strategies) are used to transform a decoupled representation. The knowledge, rules, procedures, and strategies that can be retrieved and used to transform decoupled representations have been referred to as "mindware," a term coined by David Perkins in a 1995 book. The mindware available for use during cognitive simulation is in part the product of past learning experiences.

Because the Cattell–Horn–Carroll (CHC) theory of intelligence is one of the most comprehensively validated theories of intelligence available, it is important to see how two of its major components miss critical aspects of rational thought. Fluid intelligence will, of course, have some relation to rationality because it indexes the computational power of the algorithmic mind to sustain decoupling. Because override and simulation are important operations for rational thought, fluid intelligence will definitely facilitate rational action in some situations. Nevertheless, the tendency to initiate override (arrow B in figure 15.2.3) and to initiate simulation activities (arrow D in figure 15.2.3) are both aspects of the reflective mind not assessed by intelligence tests, so the tests will miss these components of rationality. Such propensities are instead indexed by measures of typical performance (cognitive styles and thinking dispositions) as opposed to measures of maximal performance such as IQ tests.

The situation with respect to crystallized intelligence is a little different. Rational thought depends critically on the acquisition of certain types of knowledge. That knowledge would, in the abstract, be classified as crystallized intelligence. But is it the kind of crystallized knowledge that is assessed on actual tests of intelligence? The answer is "no." The knowledge structures that support rational thought are specialized. They cluster in the domains of probabilistic reasoning, causal reasoning, and scientific reasoning. In contrast, the crystallized knowledge assessed on IQ tests is deliberately designed to be nonspecialized. The designers of the tests, in order to make sure the sampling of vocabulary and knowledge is fair and unbiased, explicitly attempt to broadly sample vocabulary, verbal comprehension domains, and general knowledge. In short, crystallized intelligence, as traditionally measured, does not assess individual differences in rationality.

Finally, there is one particular way the autonomous mind supports rationality that we would like to emphasize. It is that the autonomous mind contains rational rules and normative strategies that have been tightly compiled and are automatically activated due to overlearning and practice. This means that, for some people, in some instances, the normative response emanates directly from the autonomous mind rather than from the more costly Type 2 process of simulation.

Figure 15.2.4 illustrates more clearly the point we wish to make here. This figure has been simplified by the removal of all the arrow labels and the removal of the boxes representing serial associative cognition, as well as the response boxes. In the upper right is represented the accessing of mindware that is most discussed in the literature. In the case represented there, a nonnormative response from the autonomous mind has been interrupted, and the computationally taxing process of simulating an alternative response is under way. That simulation involves the computationally expensive process of accessing mindware for the simulation.

In contrast to this type of normative mindware access, indicated in the lower left of the figure, is a qualitatively different way that mindware can determine the normative response. The figure indicates the point we have stressed earlier: that within the autonomous mind can reside normative rules and rational strategies that have been practiced to automaticity and can automatically compete with (and often immediately defeat) any alternative nonnormative response that is also stored in the autonomous mind (De Neys & Pennycook, 2019; Stanovich, 2018).

So it should be clear from figure 15.2.4 that it does not follow from the output of a normative response that System 2 was necessarily the genesis of the rational responding. According to the model just presented, rationality requires three different classes of mental characteristics:

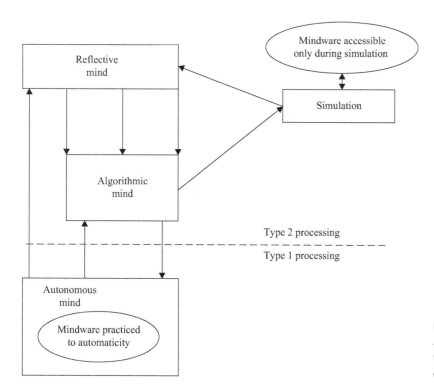

Figure 15.2.4
A simplified model showing both automatized mindware and mindware accessible during simulation.

first, algorithmic-level cognitive capacity (fluid intelligence) is needed for override and sustained simulation activities. Second, the reflective mind must be characterized by the tendency to initiate the override of suboptimal responses generated by the autonomous mind and to initiate simulation activities that will result in a better response. Finally, the mindware that allows the computation of rational responses needs to be available and accessible during simulation activities or be accessible from the autonomous mind because it has been highly practiced (see figure 15.2.4).

7. Our Tripartite Cognitive Architecture Reconciles the Opposing Positions in the Great Rationality Debate

Researchers working in the heuristics and biases tradition tend to be so-called Meliorists (Stanovich, 1999, 2004, 2010). They assume that human reasoning is not as good as it could be and that thinking could be improved. Thus, a Meliorist is one who feels that education and the provision of information could help make people more rational. This optimistic part of the Meliorist message derives from the fact that Meliorists see a large gap between normative models of rational responding and descriptive models of what people actually do. Over the past several decades, an alternative interpretation of the findings from the heuristics and biases research program has been championed. Contributing to this alternative interpretation have been philosophers, evolutionary psychologists, adaptationist modelers, and ecological theorists (Cohen, 1981; Cosmides & Tooby, 1996; Gigerenzer, 2007; Oaksford & Chater, 2007, 2012; Todd & Gigerenzer, 2000). They have reinterpreted the modal response in most of the classic heuristics and biases experiments as indicating an optimal information-processing adaptation on the part of the subjects. This group of theorists—who argue that an assumption of maximal human rationality is the proper default position to take—have been termed the Panglossians. The Panglossian theorists often argue either that the normative model being applied is not the appropriate one because the subject's interpretation of the task is different from what the researcher assumes it is or that the modal response in the task makes perfect sense from an evolutionary perspective. The contrasting positions of the Panglossians and Meliorists define the differing poles in what has been termed the Great Rationality Debate in cognitive science—the debate about how much irrationality to attribute to human cognition (Gigerenzer, 1996; Kahneman & Tversky, 1996; Kelman, 2011; Lee, 2006; Polonioli, 2015; Samuels & Stich, 2004; Stanovich, 1999, 2004; Stanovich & West, 2000; Stein, 1996).

A reconciliation of the views of the Panglossians and Meliorists is possible, however, if we take two scientific steps. First, we must consider data patterns long ignored in the heuristics and biases literature: individual differences on rational thinking tasks. Second, we must understand the empirical patterns obtained through the lens of the modified and updated dual-process theory we outlined in this chapter.

We have argued (Stanovich & West, 2000) that the statistical distributions of the types of goals being pursued by Type 1 and Type 2 processing are different. Specifically, there is a difference between the goals at the level of the gene and the goals at the level of the individual, and important consequences for the pursuit of rationality follow from this fact. The greater evolutionary age of some of the mechanisms underlying Type 1 processing accounts for why it more closely tracks ancient evolutionary goals (i.e., the genes' goals), whereas Type 2 processing instantiates a more flexible goal hierarchy that is oriented toward maximizing overall goal satisfaction at the level of the whole organism. Type 2 processing (especially at the reflective level) is more attuned to the person's needs as a coherent organism than is Type 1 processing. As a result, in the minority of cases where the outputs of the two systems conflict, people will often be better off if they can accomplish a system override of the Type 1–triggered output (the full argument is contained in Stanovich, 2004).

Instances when there is a conflict between the responses primed by Type 1 and Type 2 processing are thus interpreted as reflecting conflicts between two different types of optimization: fitness maximization at the subpersonal genetic level and utility maximization at the personal level. A failure to differentiate these interests is at the heart of the disputes between researchers working in the heuristics and biases tradition and their critics in the evolutionary psychology camp. First, it certainly must be said that the evolutionary psychologists are on to something with respect to the tasks they have analyzed, because in each case, the adaptive response is the modal response in the task—the one most subjects give. Nevertheless, this must be reconciled with a triangulating data pattern relevant to this discussion—an analysis of patterns of covariation and individual differences across these tasks. Specifically, we have found that cognitive ability often (but not always) dissociates from the response deemed adaptive from an evolutionary analysis (Stanovich & West, 1998, 1999, 2000).

The evolutionary psychologists are probably correct that most Type 1 processing is evolutionarily adaptive in the ancestral environment. Nevertheless, their evolutionary interpretations do not impeach the position of the heuristics and biases researchers that the alternative response given by the minority of subjects is rational at the level of the individual. Subjects of higher analytic intelligence are simply more prone to override Type 1 processing in order to produce responses that are epistemically and instrumentally rational. This rapprochement between the two camps was introduced by Stanovich (1999), and subsequent research has only reinforced it (see Kahneman & Frederick, 2002; Kelman, 2011; Samuels & Stich, 2004; Stanovich, 2004, 2011).

References

Ackerman, P. L., & Heggestad, E. D. (1997). Intelligence, personality, and interests: Evidence for overlapping traits. *Psychological Bulletin, 121*, 219–245.

Ackerman, P. L., & Kanfer, R. (2004). Cognitive, affective, and conative aspects of adult intellect within a typical and maximal performance framework. In D. Y. Dai & R. J. Sternberg (Eds.), *Motivation, emotion, and cognition: Integrative perspectives on intellectual functioning and development* (pp. 119–141). Mahwah, NJ: Erlbaum.

Baron, J. (2008). *Thinking and deciding* (4th ed.). Cambridge, MA: Cambridge University Press.

Baron, J. (2014). Heuristics and biases. In E. Zamir & D. Teichman (Eds.), *The Oxford handbook of behavioral economics and the law* (pp. 3–27). Oxford, England: Oxford University Press.

Cacioppo, J. T., Petty, R. E., Feinstein, J. A., & Jarvis, W. B. G. (1996). Dispositional differences in cognitive motivation: The life and times of individuals varying in need for cognition. *Psychological Bulletin, 119*, 197–253.

Carroll, J. B. (1993). *Human cognitive abilities: A survey of factor-analytic studies*. Cambridge, England: Cambridge University Press.

Cattell, R. B. (1963). Theory for fluid and crystallized intelligence: A critical experiment. *Journal of Educational Psychology, 54*, 1–22.

Cattell, R. B. (1998). Where is intelligence? Some answers from the triadic theory. In J. J. McArdle & R. W. Woodcock (Eds.), *Human cognitive abilities in theory and practice* (pp. 29–38). Mahwah, NJ: Erlbaum.

Chein, J. M., & Schneider, W. (2012). The brain's learning and control architecture. *Current Directions in Psychological Science, 21*, 78–84.

Cohen, L. J. (1981). Can human irrationality be experimentally demonstrated? *Behavioral and Brain Sciences, 4*, 317–370.

Cosmides, L., & Tooby, J. (1996). Are humans good intuitive statisticians after all? Rethinking some conclusions from the literature on judgment under uncertainty. *Cognition, 58*, 1–73.

Dawes, R. M. (1998). Behavioral decision making and judgment. In D. T. Gilbert, S. T. Fiske, & G. Lindzey (Eds.), *The handbook of social psychology* (Vol. 1, pp. 497–548). Boston, MA: McGraw-Hill.

De Neys, W. (Ed.). (2018). *Dual process theory 2.0*. London, England: Routledge.

De Neys, W., & Pennycook, G. (2019). Logic, fast and slow: Advances in dual-process theorizing. *Current Directions in Psychological Science, 28*, 503–509.

Duncan, J., Parr, A., Woolgar, A., Thompson, R., Bright, P., Cox, S., Bishop, S., & Nimmo-Smith, I. (2008). Goal neglect and Spearman's *g*: Competing parts of a complex task. *Journal of Experimental Psychology: General, 137*, 131–148.

Edwards, W. (1954). The theory of decision making. *Psychological Bulletin, 51*, 380–417.

Etzioni, A. (2014). Treating rationality as a continuous variable. *Society, 51*, 393–400.

Evans, J. St. B. T. (2007). *Hypothetical thinking: Dual processes in reasoning and judgment*. New York, NY: Psychology Press.

Evans, J. St. B. T. (2008). Dual-processing accounts of reasoning, judgment and social cognition. *Annual Review of Psychology, 59*, 255–278.

Evans, J. St. B. T. (2010). *Thinking twice: Two minds in one brain*. Oxford, England: Oxford University Press.

Evans, J. St. B. T. (2014). *Reasoning, rationality and dual processes*. London, England: Psychology Press.

Evans, J. St. B. T., & Stanovich, K. E. (2013). Dual-process theories of higher cognition: Advancing the debate. *Perspectives on Psychological Science, 8*, 223–241.

Gigerenzer, G. (1996). On narrow norms and vague heuristics: A reply to Kahneman and Tversky (1996). *Psychological Review, 103*, 592–596.

Gigerenzer, G. (2007). *Gut feelings: The intelligence of the unconscious*. New York, NY: Viking Penguin.

Hilton, D. J. (2003). Psychology and the financial markets: Applications to understanding and remedying irrational decision-making. In I. Brocas & J. D. Carrillo (Eds.), *The psychology of economic decisions: Vol. 1. Rationality and well-being* (pp. 273–297). Oxford, England: Oxford University Press.

Horn, J. L., & Cattell, R. B. (1967). Age differences in fluid and crystallized intelligence. *Acta Psychologica, 26*, 1–23.

Jastrzębski, J., Ciechanowska, I., & Chuderski, A. (2018). The strong link between fluid intelligence and working memory cannot be explained away by strategy use. *Intelligence, 66*, 44–53.

Jeffrey, R. C. (1983). *The logic of decision* (2nd ed.). Chicago, IL: University of Chicago Press.

Kahneman, D. (2011). *Thinking, fast and slow*. New York, NY: Farrar, Straus & Giroux.

Kahneman, D., & Frederick, S. (2002). Representativeness revisited: Attribute substitution in intuitive judgment. In T. Gilovich,

D. Griffin, & D. Kahneman (Eds.), *Heuristics and biases: The psychology of intuitive judgment* (pp. 49–81). New York, NY: Cambridge University Press.

Kahneman, D., & Tversky, A. (1972). Subjective probability: A judgment of representativeness. *Cognitive Psychology, 3,* 430–454.

Kahneman, D., & Tversky, A. (1973). On the psychology of prediction. *Psychological Review, 80,* 237–251.

Kahneman, D., & Tversky, A. (1996). On the reality of cognitive illusions. *Psychological Review, 103,* 582–591.

Kelman, M. (2011). *The heuristics debate.* New York, NY: Oxford University Press.

Lee, C. J. (2006). Gricean charity: The Gricean turn in psychology. *Philosophy of the Social Sciences, 36,* 193–218.

Luce, R. D., & Raiffa, H. (1957). *Games and decisions.* New York, NY: Wiley.

Manktelow, K. I. (2004). Reasoning and rationality: The pure and the practical. In K. I. Manktelow & M. C. Chung (Eds.), *Psychology of reasoning: Theoretical and historical perspectives* (pp. 157–177). Hove, England: Psychology Press.

Oaksford, M., & Chater, N. (2007). *Bayesian rationality: The probabilistic approach to human reasoning.* Oxford, England: Oxford University Press.

Oaksford, M., & Chater, N. (2012). Dual processes, probabilities, and cognitive architecture. *Mind & Society, 11,* 15–26.

Over, D. E. (2000). Ecological rationality and its heuristics. *Thinking & Reasoning, 6,* 182–192.

Over, D. E. (2004). Rationality and the normative/descriptive distinction. In D. J. Koehler & N. Harvey (Eds.), *Blackwell handbook of judgment and decision making* (pp. 3–18). Malden, MA: Blackwell.

Perkins, D. N. (1995). *Outsmarting IQ: The emerging science of learnable intelligence.* New York, NY: Free Press.

Polonioli, A. (2015). Stanovich's arguments against the "adaptive rationality" project: An assessment. *Studies in History and Philosophy of Biological and Biomedical Sciences, 40,* 55–62.

Samuels, R., & Stich, S. P. (2004). Rationality and psychology. In A. R. Mele & P. Rawling (Eds.), *The Oxford handbook of rationality* (pp. 279–300). Oxford, England: Oxford University Press.

Savage, L. J. (1954). *The foundations of statistics.* New York, NY: Wiley.

Sherman, J. W., Gawronski, B., & Trope, Y. (2014). *Dual-process theories of the social mind.* New York, NY: Guilford.

Stanovich, K. E. (1999). *Who is rational? Studies of individual differences in reasoning.* Mahwah, NJ: Erlbaum.

Stanovich, K. E. (2004). *The robot's rebellion: Finding meaning in the age of Darwin.* Chicago, IL: University of Chicago Press.

Stanovich, K. E. (2010). *Decision making and rationality in the modern world.* New York, NY: Oxford University Press.

Stanovich, K. E. (2011). *Rationality and the reflective mind.* New York, NY: Oxford University Press.

Stanovich, K. E. (2012). On the distinction between rationality and intelligence: Implications for understanding individual differences in reasoning. In K. Holyoak & R. Morrison (Eds.), *The Oxford handbook of thinking and reasoning* (pp. 343–365). New York, NY: Oxford University Press.

Stanovich, K. E. (2018). Miserliness in human cognition: The interaction of detection, override, and mindware. *Thinking & Reasoning, 24,* 423–444.

Stanovich, K. E., & Toplak, M. E. (2012). Defining features versus incidental correlates of Type 1 and Type 2 processing. *Mind & Society, 11,* 3–13.

Stanovich, K. E., & West, R. F. (1998). Individual differences in rational thought. *Journal of Experimental Psychology: General, 127,* 161–188.

Stanovich, K. E., & West, R. F. (1999). Discrepancies between normative and descriptive models of decision making and the understanding/acceptance principle. *Cognitive Psychology, 38,* 349–385.

Stanovich, K. E., & West, R. F. (2000). Individual differences in reasoning: Implications for the rationality debate? *Behavioral and Brain Sciences, 23,* 645–726.

Stanovich, K. E., West, R. F., Toplak, M. E. (2016). *The Rationality Quotient: Toward a test of rational thinking.* Cambridge, MA: MIT Press.

Stein, E. (1996). *Without good reason: The rationality debate in philosophy and cognitive science.* Oxford, England: Oxford University Press.

Todd, P. M., & Gigerenzer, G. (2000). Précis of Simple heuristics that make us smart. *Behavioral and Brain Sciences, 23,* 727–780.

Tversky, A., & Kahneman, D. (1974). Judgment under uncertainty: Heuristics and biases. *Science, 185,* 1124–1131.

von Neumann, J., & Morgenstern, O. (1944). *Theory of games and economic behavior.* Princeton, NJ: Princeton University Press.

15.3   How to Improve Rational Thinking?

Stephanie de Oliveira and Richard Nisbett

Summary

Current perspectives on what constitutes "good thinking" suggest that it should include proper process (e.g., be internally coherent/rational) and some pragmatic element of usefulness (e.g., it should produce actions and results that are accurate and/or beneficial to the thinker). Although people's reasoning in economic and social domains has various shortcomings on both counts, it can be improved through training. Such training does not need to be intensive or time-consuming in order to produce measurable improvements in reasoning. However, it typically must go beyond simply telling people to be more rational, to "try harder," or to avoid particular biases. The most effective training seems to be the type that uses diverse examples and offers practice and feedback opportunities. Although some research demonstrates correlational benefits of "good thinking" for various life outcomes, future work should examine whether training has causal effects on improving outcomes for people over time.

1.   What Is Good Thinking?

Consider the following scenario:

> Imagine you spent $100 on a ticket for a ski trip to Michigan. Several weeks later you buy a $50 ticket for a ski trip to Wisconsin. You think you will enjoy the Wisconsin ski trip more than the Michigan ski trip. As you are storing your Wisconsin ski trip ticket in your wallet, you notice that both trips are for the same weekend. It's too late to sell either ticket, and you cannot return either one. You must use one ticket and not the other. Which trip will you take? (adapted from Arkes & Blumer, 1985, p. 126).

Which is the better decision? Why?
  Now, consider the following statements:

> If it's raining, then the streets must be wet.
> It's not raining.
> Therefore, the streets must not be wet. (Nisbett, 2015, p. 218)

Is it better to agree or disagree with the conclusion? Why?

Such exercises help us grapple with questions about what constitutes normative thinking. Are there better and worse ways of thinking? And if there are better ways of thinking, how can people improve? We begin this chapter with some definitions of good reasoning that have been popular across different cultures and disciplines. We then summarize research on human judgment in the social sciences, focusing on diagnosis ("How good is human judgment?") and potential solutions ("Can it be improved?").

Greek philosophers played a major role in defining "good thinking" in Western societies. Their rules of logic are used to this day. The second opening example of this chapter shows the commonly made "inverse error"— reasoning that violates the rules of propositional logic. The first argument takes the form "If P then Q." This means that if we observe P (rainy weather), we can conclude Q (wet streets). The second statement says that we observe "*not P*." Firm conclusions cannot in fact be drawn because P did not occur.

Conclusions can be logically derived yet wrong, and they can be illogically derived yet right. The internal validity of an argument and its truth are two entirely separate things. Thus, modern psychologists' definitions of good reasoning typically take both factors into account. One prominent perspective states that good reasoning should include "correspondence" and "coherence" (Hammond, 1996); its conclusions should generally *correspond* with reality, and it should also use *coherent* processes to produce such conclusions. As another prominent psychologist put it, "Good judges should both 'get it right' and 'think the right way'" (Tetlock, 2005, p. 7).

Nevertheless, people in different societies have very different ideas about what constitutes proper, or normative, reasoning. For example, dialectical systems of thought tend to emphasize the shifting nature of reality, acknowledging that reality is full of contradictions, changes, and coexisting yet opposing forces (DaMatta, 1995; Nisbett, Peng, Choi, & Norenzayan, 2001; Peng & Nisbett, 1999). Correspondence and coherence may be hard or even impossible to achieve, since what is true at one moment may be false a moment later. The dialectic system of thought is more heavily adopted in some East Asian and Latin American societies (de Oliveira & Nisbett, 2017a; Nisbett et al., 2001), and the difference may be partially traceable to differences in how philosophy developed. For example, in contrast to Greek philosophy, which was developed in the service of guiding debate, East Asian philosophy was developed to serve society. Thus, good thinking was not necessarily about being right in some logical abstract sense but about managing the vicissitudes of life in a pragmatically useful and socially harmonious way.

Some Westerners have also argued that thinking should not be assessed merely by its adherence to abstract principles (Keys & Schwartz, 2007). Instead, one should consider how decision outcomes are experienced by the thinker and "how the decision-making process fits into the decision maker's life as a whole" (p. 165). From an economic perspective, the correct answer to this chapter's opening question about the ski tickets is to go on the trip that one would typically enjoy more since the cost of the $100 ticket cannot be recovered. Going on the less enjoyable, more expensive trip amounts to wasting time in addition to wasting money since one has thereby forgone a more enjoyable way to spend one's vacation. Yet most untrained people who answer that question choose the more expensive and less enjoyable trip to Michigan. Although that decision does not adhere to economic principles, choosing Wisconsin may make people feel like they are violating another cherished rule, "Waste not, want not" (Arkes & Ayton, 1999). The unpleasantness of violating that rule may carry over from the decision process into the decision experience, making Wisconsin the less appealing choice at the moment even if it is the more appealing choice when detached from the decision dilemma (Keys & Schwartz, 2007).

To summarize, there are diverse perspectives on what constitutes good thinking across cultures and within societies. Most psychological research on judgment and decision making has studied Western thinking and evaluated it by Western standards of coherence and correspondence. Thus, our chapter adopts a similar Western focus, but we have reviewed important cultural variation in reasoning elsewhere (de Oliveira & Nisbett, 2017b; Yates & de Oliveira, 2016).

2. How Good Is Human Judgment?

Psychologists began to empirically study reasoning relatively recently. Initial studies from the 1960s through the 1980s suggested that although human judgment has been adequate for survival, it is characterized by errors in logic and nonadherence to normative economic principles. In social judgment studies, people paid too much attention to the actor and not enough attention to the actor's context (Ross, 1977). They based social judgments and frequency estimates not on base rates or valid predictors but on whether examples of similar cases easily came to mind (Tversky & Kahneman, 1973). In economic studies, people were loss-averse, leading them to make contradictory decisions when identical options were framed either as gains or as losses (Kahneman & Tversky, 1982). People were also poor logicians. For example, they were terrible at testing conditional propositions. Wason's card selection task tested how well people could falsify *if P then Q* propositions (Wason, 1960). In the task, people received a rule (e.g., "If the front of the card has a vowel, then the back has an even number"). Participants would then examine four cards displaying letters and numbers (e.g., an A, a B, a 4, and a 7). Their task was to turn over only as many cards as necessary to test whether the rule was violated. Few people would correctly turn over the A card and the 7 card.

Compendiums of reasoning errors were published in books like *Human Inference* (Nisbett & Ross, 1980). Such works were not optimistic about human reasoning: "We insist that the errors demonstrated in the laboratory and chronicled in this book are the ingredients of individual and collective human tragedy" (p. 251). But after several years of experiments demonstrating reasoning errors, Nisbett and his colleagues failed to replicate one of Kahneman and Tversky's findings. Specifically, for a problem in which people typically misapplied the law of large numbers, they found that a large share of their participants responded accurately (Nisbett, 1993, p. 4). Participants who had taken a statistics course did particularly well, whereas those who had not taken statistics performed poorly, like Kahneman and Tversky's original participants.

Taking a different approach, another scholar found that although people were bad at the standard Wason card selection task, they were better at testing conditional propositions when the task mapped onto certain real-world scenarios. For example, when asked to turn over as many sales receipts as necessary to establish that, if the receipt is for more than $20, it has a signature on the back, people largely understood that they should check receipts with large amounts and unsigned backs (D'Andrade, 1982, as cited in Lehman, Lempert, & Nisbett, 1988).

These findings indicated that mental tools necessary to engage in sound reasoning were not altogether *absent* from the human toolbox. Rather, they were neglected for certain problems, and absent concepts could perhaps be taught via formal training. Scholars hoping to improve cognition would need to be able to teach new strategies but also to find ways of encouraging the use of already existing strategies and rules across domains.

In summary, there is ample research documenting reasoning "failures" in humans, but there are diverse views on just how much of a problem this is and on whether observations of such failures are partly due to how the questions are asked. Taking abstract logical tasks like Wason's card selection task and couching them in more practical terms can improve people's performance. Nevertheless, people—even experts—still make many errors in common everyday judgments. Judgment errors can materialize in social domains, in health care (e.g., Gigerenzer, Gaissmaier, Kurz-Milcke, Schwartz, & Woloshin, 2007), in economic domains (e.g., Arkes & Blumer, 1985), and in political domains (Tetlock, 2005). Although we are less gloomy about the state of reasoning than psychologists were in the 1980s, there is substantial room for improvement.

3. Can Reasoning Be Improved?

There have been many proposed approaches to improving reasoning (for a general review, see Larrick, 2004). Approaches include changing the structure of the question or the task (e.g., Thaler, Sunstein, & Balz, 2012), using groups rather than individuals (Sunstein & Hastie, 2015; Surowiecki, 2005), or even jettisoning the human altogether in favor of mathematical models (Dawes, 1979). In this chapter, we focus on training.

Consider the following problem (adapted from Gigerenzer et al., 2007):

> Assume you conduct breast cancer screening in a certain region. You know the following information about this region: (1) The probability that a woman has breast cancer is 1%; (2) if a woman has breast cancer, the probability that she tests positive is 90%; and (3) if a woman does not have breast cancer, the probability that she nevertheless tests positive is 9%. Now imagine that a woman tests positive. What are the chances that she has breast cancer?

Laypeople and professionals struggle with this type of problem. In a survey of 160 gynecologists, only 21% responded correctly when given four answer choices (Gigerenzer et al., 2007). (She has a 10% chance; out of 10 women with a positive mammogram, about 1 has cancer.) Those doctors received a training session in which they were taught to convert percentages into frequencies

(e.g., think of 10% as 1 in 10, or 10 out of 100). Afterward, 87% were able to give the correct answer.

Research on training has generally found that reasoning can be improved through training and that this training does not need to be terribly extensive. We review some representative studies below that also address additional questions: Are some types of training more effective than others? How long do effects last? Can training on one rule in one domain transfer such that those newly learned skills are more broadly applied to other problems? We begin with some findings from statistical reasoning and then report some results on how the environment and feedback can improve rational reasoning.

3.1 Learning Statistics

Although early psychologists were skeptical about the effectiveness of training, research on training has produced encouraging results. For example, one study assessed people's reasoning by asking them to explain why a traveling saleswoman's second visit to a restaurant typically leads to disappointment when her first meal had been outstanding (Fong, Krantz, & Nisbett, 1986). On such questions, respondents with no statistical background gave exclusively nonstatistical answers (e.g., suggesting that the chefs change a lot or that her expectations were too high). Respondents who had taken one statistics course gave statistical answers about 20% of the time (e.g., suggesting she was lucky the first time, noting that restaurant meal quality varies). New graduate students in psychology gave statistical answers about 40% of the time, and doctoral-level scientists at a research institution gave such answers about 80% of the time. Giving a statistical answer to questions such as these reflects better thinking because single extreme observations are typically followed by less extreme observations; extreme events "regress" toward the mean value more often than not.

Further research suggested that the effects of formal training on reasoning depended on how much the program linked abstract principles, including statistical ones, to everyday life problems. Skills taught in a domain-restricted manner may show less generalizability than skills taught with examples of diverse applications. Lehman and colleagues (1988) found that getting a graduate degree in medicine or psychology was associated with substantial increases in conditional reasoning and statistical/methodological reasoning, but getting a graduate chemistry degree was not. Chemistry students typically study very little statistics or scientific methodology with direct relevance to everyday life. Psychology and medical students study a great deal of statistics and scientific methodology having relevance to everyday-life domains.

This idea—that people can generalize learned reasoning rules across domains if they are shown how and get practice examples with it—was tested in a more formal experiment (Fong et al., 1986). In their first study, the researchers tested people's reasoning through problems that varied from seeming relatively objective (e.g., judging the characteristics of a lottery) to relatively subjective (e.g., choosing which college to attend). Participants were randomly assigned to one of five conditions. In one condition, people read a packet explaining statistical principles such as the law of large numbers. (The law of large numbers in statistics states that smaller samples are more likely than larger samples to vary from the population distribution.) After reading the packet, they watched the researcher demonstrating the principle by drawing balls from an urn. In another condition, participants worked through example problems in which those abstract principles were applied to everyday problems. For example, they read about hypothetical ballet auditions and learned to think of the observed dance moves as a sample of a larger "population" of dance moves. They were reminded that smaller samples—like dance moves observed at an audition—are unreliable estimates of population distributions—such as all of a dancer's moves. They then considered how, based on this observation, it would make sense that some of the best dancers at an audition would turn out to be average performers after all; their good performance at the audition was a statistical fluke due to small sample size. In a third condition, participants read the informational packet *and* worked through the example problems. There were two control conditions; in one, participants were given no training. In another, they were simply given the definitions of each of the statistical principles.

Training influenced the frequency and quality of statistical responses. In both control conditions, fewer than half of the responses invoked statistical principles, and the quality of the statistical answers they gave was low. In the full training condition, about 64% of responses invoked statistical principles, and the quality of the answers was relatively high. The examples-only training and the packet-only training conditions fell in between.

A similar study tested whether training in one domain generalized to other domains and whether effects persisted over time (Fong & Nisbett, 1991). Participants were trained on the law of large numbers through example problems. They were tested immediately or after a two-week delay, either in the same domain as the examples or in a different domain. The training was equally effective immediately whether participants were tested in the same or a different domain. Participants tested later in the same domain showed no decline in their performance. Participants tested later in a different domain showed some decline, but they still performed better than participants who had received no training. Similar work examining financial training found domain-independent gains as well: people trained in normative economic reasoning with financial (vs. nonfinancial) examples did better on both financial and nonfinancial measures of reasoning (Larrick, Morgan, & Nisbett, 1990), and training gains were still observed four to six weeks post-training. It's notable that both statistical training and cost–benefit analysis training improved answers to questions in non-laboratory, nonacademic settings such as alleged telephone surveys about politics or consumer preferences.

3.2 The Learning Environment and Feedback

Competence is best developed in contexts where the learners can get relatively fast, frequent, and complete feedback on their performance—this allows people to know whether there is a problem with their reasoning in the first place (Hogarth, 2001; Larrick, 2004). For example, if one wants to know how good one is at spotting academic talent in the graduate admissions process, one would need to track outcomes of one's decisions—both the accepted and rejected students—over many admissions cycles. Some real-world tasks provide people with feedback-heavy learning environments (e.g., weather forecasting, sports training), but many tasks provide slow and incomplete feedback (e.g., admissions and hiring decisions).

Training programs can artificially structure the learning environment so that it provides fast and complete feedback on tasks that otherwise are learned in feedback-poor real-world contexts. Morewedge and colleagues (2015) compared the effectiveness of video training and video *game* training for improving reasoning. The video game allowed participants to not only learn about the biases but also engage in problem-solving tasks that elicited those biases. After responding to the task, they received feedback on their performance as well as suggestions for how to improve. Participants trained with videos merely received information about those biases and did not receive performance feedback. The interventions targeted bias blind spot (i.e., thinking you are less biased than others), confirmation bias (i.e., overweighting confirming vs. disconfirming evidence), the fundamental attribution error (i.e., neglecting situational influences when attributing events to a person), anchoring (i.e., overweighting initial information when making a subsequent judgment), the representativeness heuristic (i.e., relying on judgments of similarity to make inferences about probability), and social projection (i.e., assuming others' emotions and values are like one's own).

In both studies, comparisons of pre- and post-test reasoning revealed that both the video and the video game reduced bias. Follow-up assessments two to three months later showed that the improvements were sustained for both methods. However, the video game debiasing effects were much larger than the video effects at both post-intervention assessment times. These results suggest that personal feedback and practice greatly enhance the effects of training.

One research team recently demonstrated increases in geopolitical forecasting competence both by training people with instructional modules and also by placing workers in more interactive, feedback-rich team environments (Mellers et al., 2014). The training participants received either 45-minute "scenario training" or "probability training" of similar length. The "scenario training" module taught forecasters to generate new futures, actively entertain more possibilities, use decision trees, and avoid certain biases. The "probability training" module taught forecasters to average multiple estimates from models, polls, and expert panels; extrapolate over time when variables were continuous; and avoid errors like base-rate neglect. Forecasters working in teams (vs. individually) had ample experiential learning opportunities; they could make forecasts as often as they wished, interact with other forecasters, share information via a web-based platform, get feedback on their thoughts and on their forecasts, and learn from that feedback. Training improved performance in both the first year and the second year of the tournament. Teaming also helped: forecasters working in teams did better than those working individually. Remarkably, although the authors did not offer formal statistical analyses on the matter, the benefit of teaming seemed to be larger than that of training.

4. Conclusion

Training can improve reasoning, at least by Western definitions, and the type of training matters. Training with examples and practice helps people generalize specific skills to a variety of domains, and feedback-rich environments likely boost training effects and help maintain gains over time. Programs do not need to be intensive or long in order to produce lasting effects on reasoning,

but they do need to go beyond merely telling people about common judgment errors or instructing them in abstract principles.

There is good reason to believe training effects would carry over to everyday life problems that do not closely resemble problems used for training. Research has linked higher normative reasoning ability to lower risky behavior and juvenile delinquency (Parker, Bruine de Bruin, Fischhoff, & Weller, 2018) as well as higher salaries for a sample of professors and higher GPA for a sample of students (Larrick, Nisbett, & Morgan, 1993).

References

Arkes, H. R., & Ayton, P. (1999). The sunk cost and Concorde effects: Are humans less rational than lower animals? *Psychological Bulletin, 125*(5), 591–600.

Arkes, H. R., & Blumer, C. (1985). The psychology of sunk cost. *Organizational Behavior and Human Decision Processes, 35*, 124–140.

DaMatta, R. A. (1995). For an anthropology of the Brazilian tradition, or "A virtude está no meio." In D. J. Hess & R. A. DaMatta (Eds.), *The Brazilian puzzle: Culture on the borderlands of the Western world* (pp. 270–292). New York, NY: Columbia University Press.

Dawes, R. M. (1979). The robust beauty of improper linear models in decision making. *American Psychologist, 34*(7), 571–582.

de Oliveira, S., & Nisbett, R. E. (2017a). Beyond East and West: Cognitive style in Latin America. *Journal of Cross-Cultural Psychology, 48*(10), 1554–1577.

de Oliveira, S., & Nisbett, R. E. (2017b). Culture changes how we think about thinking: From "Human Inference" to "Geography of Thought." *Perspectives on Psychological Science, 12*(5), 782–790.

Fong, G. T., Krantz, D. H., & Nisbett, R. E. (1986). The effects of statistical training on thinking about everyday problems. *Cognitive Psychology, 18*(3), 253–292.

Fong, G. T., & Nisbett, R. E. (1991). Immediate and delayed transfer of training effects in statistical reasoning. *Journal of Experimental Psychology: General, 120*(1), 34–45.

Gigerenzer, G., Gaissmaier, W., Kurz-Milcke, E., Schwartz, L. M., & Woloshin, S. (2007). Helping doctors and patients make sense of health statistics: Toward an evidence-based society. *Psychological Science in the Public Interest, 8*(2), 53–96.

Hammond, K. R. (1996). *Human judgment and social policy: Irreducible uncertainty, inevitable error, unavoidable injustice.* New York, NY: Oxford University Press.

Hogarth, R. M. (2001). *Educating intuition.* Chicago, IL: University of Chicago Press.

Kahneman, D., & Tversky, A. (1982). The psychology of preferences. *Scientific American, 246*(1), 160–173.

Keys, D. J., & Schwartz, B. (2007). "Leaky" rationality: Normative standards of rationality. *Perspectives on Psychological Science, 2*(2), 162–180.

Larrick, R. P. (2004). Debiasing. In D. J. Koehler & N. Harvey (Eds.), *Blackwell handbook of judgment and decision making* (pp. 316–337). Maiden, MA: Blackwell.

Larrick, R. P., Morgan, J. N., & Nisbett, R. E. (1990). Teaching the use of cost–benefit reasoning in everyday life. *Psychological Science, 1*(6), 362–370.

Larrick, R. P., Nisbett, R. E., & Morgan, J. N. (1993). Who uses the cost–benefit rules of choice? Implications for the normative status of microeconomic theory. *Organizational Behavior and Human Decision Processes, 56*, 331–347.

Lehman, D. R., Lempert, R. O., & Nisbett, R. E. (1988). The effects of graduate training on reasoning: Formal discipline and thinking about everyday life events. *American Psychologist, 43*, 431–443.

Mellers, B. A., Ungar, L. H., Baron, J., Ramos, J., Gurcay, B., Fincher, K., . . . Tetlock, P. E. (2014). Psychological strategies for winning a geopolitical forecasting tournament. *Psychological Science, 25*(5), 1106–1115.

Morewedge, C. K., Yoon, H., Scopelliti, I., Symborski, C. W., Korris, J. H., & Kassam, K. S. (2015). Debiasing decisions: Improved decision making with a single training intervention. *Policy Insights from the Behavioral and Brain Sciences, 2*(1), 129–140.

Nisbett, R. E. (1993). Reasoning, abstraction, and the prejudices of 20th-century psychology. In R. E. Nisbett (Ed.), *Rules for reasoning* (pp. 1–14). Hillsdale, NJ: Erlbaum.

Nisbett, R. E. (2015). *Mindware: Tools for smart thinking.* New York, NY: Farrar, Straus and Giroux.

Nisbett, R. E., Peng, K., Choi, I., & Norenzayan, A. (2001). Culture and systems of thought: Holistic versus analytic cognition. *Psychological Review, 108*(2), 291–310.

Nisbett, R. E., & Ross, L. (1980). *Human inference: Strategies and shortcomings of social judgment.* Englewood Cliffs, NJ: Prentice-Hall.

Parker, A. M., Bruine de Bruin, W., Fischhoff, B., & Weller, J. (2018). Robustness of decision-making competence: Evidence from two measures and an 11-year longitudinal study. *Journal of Behavioral Decision Making, 31*, 380–391.

Peng, K., & Nisbett, R. E. (1999). Culture, dialectics, and reasoning about contradiction. *American Psychologist, 54*(9), 741–754.

Ross, L. (1977). The intuitive psychologist and his shortcomings: Distortions in the attribution process. In L. Berkowitz (Ed.),

Advances in experimental social psychology (Vol. 10, pp. 173–220). New York, NY: Academic Press.

Sunstein, C. R., & Hastie, R. (2015). *Wiser: Getting beyond groupthink to make groups smarter.* Boston, MA: Harvard Business Review Press.

Surowiecki, J. (2005). *The wisdom of crowds.* New York, NY: Anchor Books.

Tetlock, P. E. (2005). *Expert political judgment.* Princeton, NJ: Princeton University Press.

Thaler, R. H., Sunstein, C. R., & Balz, J. P. (2012). Choice architecture. In E. Shafir (Ed.), *The behavioral foundations of public policy* (pp. 428–439). Princeton, NJ: Princeton University Press.

Tversky, A., & Kahneman, D. (1973). Availability: A heuristic for judging frequency and probability. *Cognitive Psychology, 5*(2), 207–232.

Wason, P. C. (1960). On the failure to eliminate hypotheses in a conceptual task. *Quarterly Journal of Experimental Psychology, 12*, 129–140.

Yates, J. F., & de Oliveira, S. (2016). Culture and decision making. *Organizational Behavior and Human Decision Processes, 136*, 106–118.

Contributors

Rakefet Ackerman, Faculty of Industrial Engineering and Management, Technion—Israel Institute of Technology, Technion City, Haifa, Israel

Max Albert, Department of Economics, Justus Liebig University Gießen, Germany

J. McKenzie Alexander, Department of Philosophy, Logic and Scientific Method, London School of Economics, London, UK

Ali al-Nowaihi, Division of Economics, School of Business, University of Leicester, UK

Hanne Andersen, Department of Science Education, University of Copenhagen, Denmark

Line Edslev Andersen, Department of Mathematics—Science Studies, Aarhus University, Denmark

Johan van Benthem, Department of Philosophy, Stanford University, Stanford, USA; Department of Philosophy, Tsinghua University, Beijing, China; Institute for Logic, Language and Computation, University of Amsterdam, the Netherlands

Jean-François Bonnefon, Toulouse School of Economics & Centre National de Recherche Scientifique, France

Rainer Bromme, Institute of Psychology, University of Münster, Germany

John Broome, Faculty of Philosophy, University of Oxford, UK; School of Philosophy, Australian National University, Canberra, Australia

Anke Bueter, Department of Philosophy and History of Ideas, Aarhus University, Denmark

Ruth M. J. Byrne, School of Psychology and Institute of Neuroscience, Trinity College Dublin, University of Dublin, Ireland

Nick Chater, Behavioural Science Group, Warwick Business School, University of Warwick, UK

Peter Collins, School of Human Sciences, University of Greenwich, UK; Munich Center for Mathematical Philosophy, Ludwig Maximilian University Munich, Germany

Leda Cosmides, Department of Psychological & Brain Sciences, Center for Evolutionary Psychology, University of California, Santa Barbara, USA

Nicole Cruz, Department of Psychological Sciences, Birkbeck College, University of London, UK

Stephanie de Oliveira, Decision Lab, Department of Psychology, University of Michigan at Ann Arbor, USA

Sanjit Dhami, Division of Economics, School of Business, University of Leicester, UK

Franz Dietrich, Paris School of Economics & Centre National de Recherche Scientifique, France

Hans van Ditmarsch, Centre National de Recherche Scientifique, Laboratoire Lorrain de Recherche en Informatique et ses Applications, University of Lorraine, France

Didier Dubois, Emeritus Research Director, Centre National de Recherche Scientifique and Université Paul Sabatier, Institut de Recherche en Informatique de Toulouse, France

Shira Elqayam, School of Applied Social Sciences, Faculty of Health and Life Sciences, De Montfort University, Leicester, UK

Orlando Espino, Department of Cognitive, Social and Organizational Psychology, University of La Laguna, Tenerife, Spain

Jonathan St. B. T. Evans, Emeritus Professor of Psychology, University of Plymouth, UK

Christoph Fehige, Department of Philosophy, Saarland University, Saarbrücken, Germany

Klaus Fiedler, Department of Psychology, Heidelberg University, Germany

Lupita Estefania Gazzo Castañeda, Department of Psychology, Justus Liebig University Gießen, Germany

Lukas Gierth, Institute of Psychology, University of Münster, Germany

Andreas Glöckner, Department of Psychology, University of Cologne, Germany; Max Planck Institute for Research on Collective Goods, Bonn, Germany

Vinod Goel, Department of Psychology, York University, Toronto, Canada; Department of Psychology, Capital Normal University, Beijing, China

Till Grüne-Yanoff, Division of Philosophy, Royal Institute of Technology, Stockholm, Sweden

Rebecca Gutwald, Munich School of Philosophy, Munich, Germany

Ulrike Hahn, Birkbeck College, University of London, UK, Munich Center for Mathematical Philosophy, Ludwig Maximilian University Munich, Germany

Alan Hájek, School of Philosophy, Australian National University, Canberra, Australia

Stephan Hartmann, Munich Center for Mathematical Philosophy, Faculty of Philosophy, Philosophy of Science, and the Study of Religion, Ludwig Maximilian University Munich, Germany

Ralph Hertwig, Center for Adaptive Rationality, Max Planck Institute for Human Development, Berlin, Germany

Eric Hilgendorf, Department of Jurisprudence, Julius Maximilian University Würzburg, Germany

Brian Hill, Department of Economics and Decision Sciences, HEC Paris & Centre National de Recherche Scientifique, France

John Horty, Department of Philosophy, University of Maryland, College Park, USA

Mateja Jamnik, Department of Computer Science and Technology (Computer Laboratory), University of Cambridge, UK

P. N. Johnson-Laird, Department of Psychology, Princeton University, Princeton, NJ; Department of Psychology, New York University, USA

Gabriele Kern-Isberner, Department of Computer Science, Technical University Dortmund, Germany

Sangeet Khemlani, Navy Center for Applied Research in Artificial Intelligence, Naval Research Laboratory, Washington, DC, USA

Karl Christoph Klauer, Institute of Psychology, Albert Ludwig University Freiburg, Germany

Hartmut Kliemt, Department of Economics, Justus Liebig University Gießen, Germany

Markus Knauff, Department of Psychology, Justus Liebig University Gießen, Germany

Anastasia Kozyreva, Center for Adaptive Rationality, Max Planck Institute for Human Development, Berlin, Germany

Fenrong Liu, Department of Philosophy, Tsinghua University, Beijing, China; Institute for Logic, Language and Computation, University of Amsterdam, the Netherlands

Henry Markovits, Department of Psychology, University of Québec at Montréal, Canada

Ralf Mayrhofer, Department of Psychology, University of Göttingen, Germany

Linda McCaughey, Department of Psychology, Heidelberg University, Germany

Björn Meder, Health and Medical University Potsdam, Germany; iSearch and Center for Adaptive Rationality, Max Planck Institute for Human Development, Berlin, Germany

Georg Meggle, Institute for Philosophy, University of Leipzig, Germany

Arthur Merin †, Department of Philosophy, Philosophy of Science, and the Study of Religion, Munich Center for Mathematical Philosophy, Ludwig Maximilian University Munich, Germany

Julian Nida-Rümelin, Department of Philosophy, Philosophy of Science, and the Study of Religion, Ludwig Maximilian University Munich, Germany

Richard Nisbett, Institute for Social Research, Department of Psychology, University of Michigan at Ann Arbor, USA

Mike Oaksford, Department of Psychological Sciences, Birkbeck College, University of London, UK

Klaus Oberauer, Department of Psychology, University of Zurich, Switzerland

David P. O'Brien, Department of Psychology, Baruch College & the Graduate Center of the City University of New York, USA

David E. Over, Department of Psychology, Durham University, UK

Judea Pearl, Computer Science Department, University of California, Los Angeles, USA

Andrés Perea, Department of Quantitative Economics, Maastricht University, the Netherlands

Danielle Pessach, Institute for Psychology, Albert Ludwig University Freiburg, Germany

Martin Peterson, Department of Philosophy, Texas A&M University, College Station, USA

Niki Pfeifer, Department of Philosophy, University of Regensburg, Germany

Henri Prade, Emeritus Research Director, Centre National de Recherche Scientifique and Institut de Recherche en Informatique de Toulouse, Université Paul Sabatier, France

Johannes Prager, Department of Psychology, Heidelberg University, Germany

Henry Prakken, Department of Information and Computing Science, Utrecht University; Law Faculty, University of Groningen, the Netherlands

Marco Ragni, Department of Computer Science, Albert Ludwig University Freiburg, Germany; Department of Psychology, Justus Liebig University Gießen, Germany

Werner Raub, Department of Sociology/ICS, Utrecht University, the Netherlands

Hans Rott, Department of Philosophy, University of Regensburg, Germany

Olivier Roy, Faculty of Humanities and Social Sciences, University of Bayreuth, Germany; Munich Center for Mathematical Philosophy, Germany

Hanno Sauer, Department of Philosophy, Utrecht University, the Netherlands

Hans Bernhard Schmid, Department of Philosophy, University of Vienna, Austria

Gerhard Schurz, Düsseldorf Center for Logic and Philosophy of Science, Heinrich Heine University Düsseldorf, Germany

Niels Skovgaard-Olsen, Department of Psychology, University of Göttingen, Germany

Sonja Smets, Institute for Logic, Language and Computation, University of Amsterdam, the Netherlands; Department of Information Science and Media Studies, University of Bergen, Norway

Michael Smith, Department of Philosophy, Princeton University, NJ, USA

Kai Spiekermann, Department of Government, London School of Economics, UK

Wolfgang Spohn, Department of Philosophy, University of Konstanz, Germany; Eberhard Karl University, Tübingen, Germany

Julia Staffel, Department of Philosophy, University of Colorado at Boulder, USA

Keith E. Stanovich, Department of Applied Psychology and Human Development, University of Toronto, Canada

William B. Starr, Sage School of Philosophy, Cornell University, Ithaca, NY, USA

Florian Steinberger, Department of Philosophy, Birkbeck College, University of London, UK

Thomas Sturm, Catalan Institute for Research and Advanced Studies (ICREA), and Department of Philosophy, Autonomous University of Barcelona, Spain; Kantian Rationality Lab, Immanuel Kant Baltic Federal University, Kaliningrad, Russia

Valerie A. Thompson, College of Arts and Science, University of Saskatchewan, Canada

John Tooby, Department of Anthropology, Center for Evolutionary Psychology, University of California, Santa Barbara, USA

Maggie E. Toplak, Department of Psychology, Faculty of Health, LaMarsh Centre for Child and Youth Research, York University, Canada

Michael R. Waldmann, Department of Psychology, University of Göttingen, Germany

Ralph Wedgwood, School of Philosophy, University of Southern California, Los Angeles, USA

Ulla Wessels, Department of Philosophy, Saarland University, Saarbrücken, Germany

Richard F. West, Department of Graduate Psychology, James Madison University, Harrisonburg, VA, USA

Alex Wiegmann, Department of Psychology, University of Göttingen, Germany; Institute for Philosophy II, Ruhr University Bochum, Germany

John Woods, Department of Philosophy, University of British Columbia, Vancouver, Canada

Niina Zuber, Bavarian Research Institute for Digital Transformation, Munich, Germany

Name Index

Italic numbers indicate that the author is denoted by "et al." or a similar phrase.

Subject Index

Italic page numbers refer to definitions or explanations.